Advanced Composites X

Proceedings of the 10th Annual ASM/ESD
Advanced Composites Conference
Dearborn, Michigan, USA
7-10 November 1994

Sponsored by

With the cooperation of
American Society of Composites
Automotive Composites Consortium
Edison Materials Technology
Federal Laboratory Consortium
GreatLakes Composites Consortium, Inc.
Michigan Materials Processing Institute
National Institute of Standards and Technology
National Research Council of Canada
Suppliers of Advanced Composite Materials
Society for the Advancement of Material and Process Engineering
Society of Manufacturing Engineers

Published by
ASM International
Materials Park, Ohio 44073-0002

ORGANIZING COMMITTEE

D. Hagen
Ford Motor Company
Dearborn, MI
Executive Chairman

K. Ashida
University of Detroit
Detroit, MI

R. Bauerle
GM-Delco Chassis
Dayton, OH
Chassis Track Chair

G. Bretz
Ford Motor Co.
Dearborn, MI

N.M. Chavka
Vetrotex Certain Teed Corp.
Maumee, OH

P. Dantzer
Davidson-Textron
Walled Lake, MI

C. Di Natale
General Motors Corporation
Warren, MI
Leader, Best Paper Subcommittee

C. Grant
GM Powertrain Headquarters
Pontiac, MI
Powertrain/Propulsion Systems
Applications Track Chair

C. Johnson
Ford Motor Company
Dearborn, MI

T. Kosakowski
Miles, Inc.
Troy, MI

R. Jeryan
Ford Motor Company
Dearborn, MI
General Chairman

E.O. Ayorinde
Wayne State University
Detroit, MI
Infrastructure Track Chair

W. Berrached
ESD-The Engineering Society
Ann Arbor, MI
ESD Staff Representative

N.G. Chavka
Ford Motor Co.
Dearborn, MI

H. Couch
GM-Delco Chassis
Dayton, OH
Leader, 1994 ACCE Proceedings

D. Denton
Chrysler Corp.
Auburn Hills, MI

A.P. Divecha
Naval Surface Warfare Center
Silver Spring, MD

D. Houston
Ford Motor Company
Dearborn, MI

D. Kenerson
Menasha Corporation
Bloomfield, MI
Manufacturing Processes Track Chair

J. Lees
DuPont Composites
Newark, DE
Material Sciences Track Chair

M.J. Martin
Michigan State University
East Lansing, MI
Leader, 1994 ACCE Tutorials

B. Mellian
Owens-Corning Fiberglas
Oak Park, MI
Exhibits Chairman

I. Poston
General Motors Corp.
Warren, MI

K.C. Rusch
The Budd Company
Madison Heights, MI
Body/Fuselage Applications Track Chair

M. Weir
ASM International
Materials Park, OH
ASM Staff Representative

M. Mehta
Environmental Research Institute of Michigan
Ann Arbor, MI

D. Peterson
Chrysler Corp.
Highland Park, MI

M.J. Rokosz
Ford Research Laboratory
Dearborn, MI
Leader, 1994 ACCE Poster Session

J. Scrivo
Menasha Corporation
Watertown, MI

Composites in Transportation: Ten Years of Progress

In the past ten years, high volume, high performance applications of advanced composites in transportation have grown substantially. Migrating from the extremes of exotic aerospace applications and low volume marine uses, these materials now provide commercial users performance and durability improvements, weight reduction, part integration and investment cost benefits. Over ten years of research in materials, processing, engineering mechanics and design have yielded growing applications in the automotive and commercial transportation, aerospace, defense, marine and recreational industries. The 10th annual ASM/ESD Advanced Composites Conference and Exposition presents the latest developments in composite applications and technologies in a series of technical tracks, tutorials, exhibits and plenary sessions.

The conference is organized in tracks covering body, chassis, powertrain and infrastructure applications, material sciences, manufacturing processes and recycling. Polymer composite and metal matrix composite technologies are included throughout. Body sessions feature adhesive bonding, analysis and test methods and crash energy absorption. The Chassis sessions showcase polymer and metal composite applications. The Powertrain/Propulsion track includes emerging materials as well as design and processing case studies. The Materials Science track features papers on new materials, their performance and theoretical treatment. Manufacturing Processes sessions cover process, modelling, fiber preforming and emerging manufacturing methods. The Infrastructure and Recycling track includes a panel discussion of infrastructure applications and technical papers on the recycling of polymer composites and nondestructive testing. The Conference also offers introductory and advanced tutorials on polymer composite component design and a tutorial on aluminum metal matrix composites.

The ACCE Exposition runs concurrently with the technical sessions, giving attendees the opportunity to view a wide range of materials, equipment and services. Along with the exposition, there is a Product Showcase displaying exceptional engineering efforts in new and innovative uses applications in structural composites.

Four outstanding plenary speakers from aerospace, automotive and government sectors, including Dr. Mary Good from the US Department of Commerce, highlight the developments in composite technology over the last decade and present a vision of the opportunities and benefits in crticical transportation applications.

In its 10th year, the ACCE continues to function as a forum to support the exchange of advanced composites technology.

We wish to personally thank the members of the planning committee, staff members from ASM and ESD, and the track and session chairs for their enthusiastic support and their commitment to the preparation and execution of this conference.

David Hagen
Executive Chairman

Richard Jeryan
General Chairman

Table of Contents

Infrastructural Applications and Recycling

Chassis

Powertrain/Propulsion Systems Applications

Manufacturing Processes

Materials Sciences

Body/Fuselage Applications

FRP Structures for Infrastructural Applications

I.E. Harik, B.N. Robson, T. Hopwood II
University of Kentucky
Lexington, Kentucky

Abstract

The U.S. Department of Defense (DOD) is funding research into the utilization of high performance fiber-reinforced composite materials, primarily developed for military applications, in infrastructure projects. Through the DOD's Advanced Research Projects Administration (ARPA) and Technology Reinvestment Project (TRP) a consortium of industries and universities will demonstrate the applicability of composite materials in traditional civil engineering environments. The University of Kentucky (UK) will be introducing the analysis, design and construction techniques of composite components to the local, national and international engineering community.

For UK's contract, the College of Engineering formed a partnership with the Kentucky Transportation Cabinet, the Daniel Boone National Forest, composite manufacturers, coating manufacturers, prestressing manufacturers, and consulting engineers. The partners are providing a major part of the In-Kind matching funds in personnel, materials, facilities and development. UK's project involves several firsts for fiber reinforced plastics (FRP): 1. Construction of the U.S.'s longest nonsuspended FRP pedestrian/vehicular bridge 2. The world's first FRP highway signpost 3. The world's first prestressed concrete box-beam bridge using FRP prestressing cables and 4. The world's longest composite bridge. Additionally, UK will conduct component testing and instrument the structures for continuous monitoring of performance.

THE USE OF COMPOSITE MATERIALS in traditional civil engineering projects is relatively new. Bridges, in particular, are one area of civil projects where composite materials (other than steel reinforced concrete) are uncommon. The University of Kentucky under a contract with the Department of Defense will demonstrate the feasibility of using composite materials in three different bridge applications. Further, a cantilevered highway sign support will be designed and constructed. These demonstration projects will provide the engineering community with impetus to consider composite materials in novel applications.

Much of the material used throughout the project will be commercially available fiber reinforced plastics (FRP's). FRP's have the advantage of being high strength, lightweight and highly corrosion resistant. These properties allow a structure to be much lighter than a conventional steel or concrete structure. Also, the perpetual problem of protecting the structural components from corrosion for exterior exposure is eliminated.

On the other hand, the designer must also consider an FRP's low stiffness and property degradation at moderately elevated temperature. Both the modulus of elasticity (E) and the shear modulus (G) are substantially lower than a steel with equivalent ultimate strength. Because shearing deformations are not considered, classical Euler-Bernoulli bending theory may not be appropriate for FRP beams. Deflections or buckling can govern the design of an FRP structure whereas strength criteria could govern a similar steel or concrete structure.

Environments where the temperature exceeds 32 degrees C (90 deg. Fahrenheit) can have a marked effect on the properties of an FRP. Design of exposed structures can, therefore, be controlled by reduced strength or stiffness effects due to thermal degradation.

Bridge Applications

FRP Girder Bridge. The longest FRP girder bridge in the U.S. will be constructed in the Daniel Boone National

Forest in eastern Kentucky. Figure 1 shows the configuration of this common bridge type with uncommon materials. It will span 18.3 meters (60') and will be designed for both pedestrian traffic and light vehicular traffic.

Prestressed Concrete Box Beams. Precast, prestressed concrete beams are used in many short and medium span highway bridges. High-strength steel prestressing cables are susceptible to corrosion when beams are exposed to saline solution from the salting of roadways during winter snows. Using FRP prestressing cables can eliminate the possibility of a corrosion induced repair or replacement problem.

One particularly cost effective bridge type for short span, low traffic roads is the side-by-side box beam bridge. The beams serve as the structural members as well as the riding surface of the bridge. Joints between the box beams often allow saltwater to attack the steel the beams. The world's first bridge using aramid fiber reinforced plastic (AFRP) prestressing cables will demonstrate the viability of replacing steel with composites.

World's Longest Composite Bridge. A recent newspaper article described a swinging (suspension) pedestrian bridge in Johnson County, Kentucky that has fallen into a state of disrepair. The county maintains the bridge but does not have the economic resources to replace it. A replacement has been designed that utilizes composite materials wherever possible.

This 128 meter (420') suspension bridge (Figures 3 & 4) will be the longest composite bridge in the world. Currently, the world's longest composite bridge is a 120 meter (395') cable-stayed pedestrian bridge at Aberfeldy, Scotland.

Additional Applications and Testing

Highway Sign Support. The world's first composite highway sign support will be designed and erected during this project. This structure will consist of a space truss or beam and column which will be cantilevered from the side of the highway. It will validate the use of FRP's in a transportation application where steel has been the material of choice.

Monitoring of the FRP structures over the life of the structures will provide invaluable information. Instrumentation will be affixed to the structures at each site to enable continual monitoring of performance with time. Critical stresses and deflections will be measured to quantify creep rates and/or stress relaxation of actual in-service structures.

Conclusion

Fiber reinforced composite materials (e.g. plastics) will be the construction material of the 21st century. The University of Kentucky will research innovative uses of composites for the nation's infrastructure through several demonstration projects. These efforts will expand the realm of composite materials to include many transportation components. Transfer of composite material technology to the engineering community will be accomplished through a partnership with industry, government and research facilities. UK's research will establish a solid foundation to promote composite materials use in infrastructural applications.

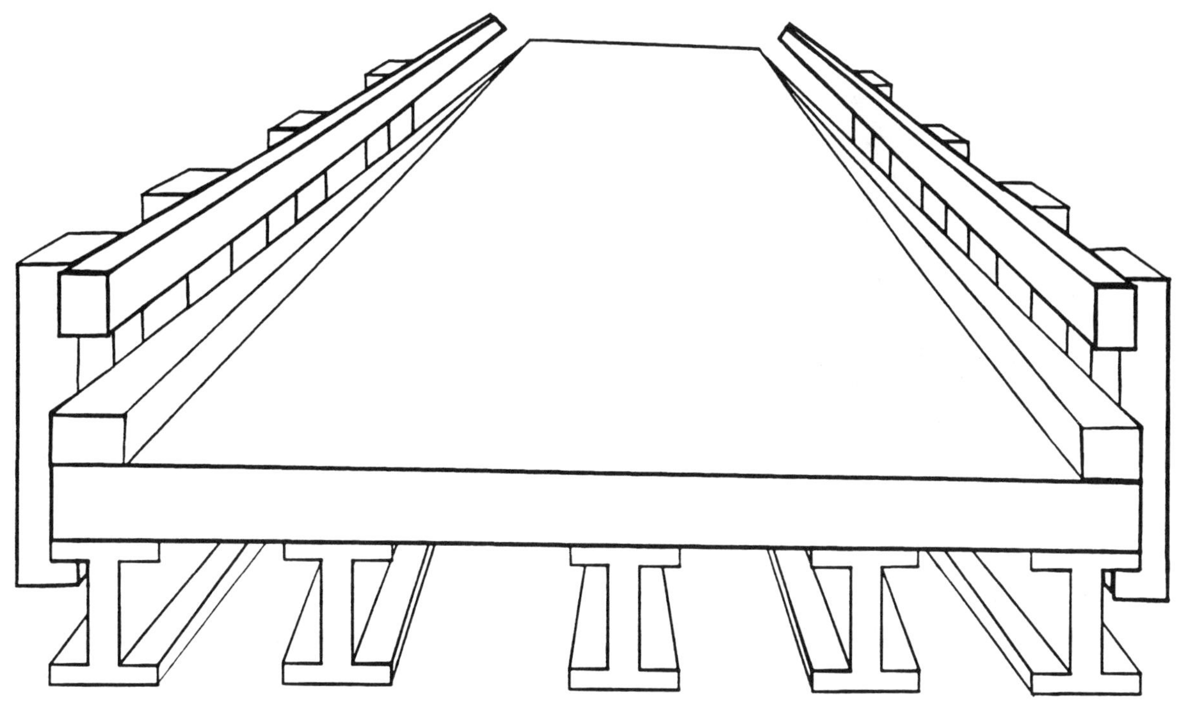

Figure 1: Fiber Reinforced Plastic Girder Bridge

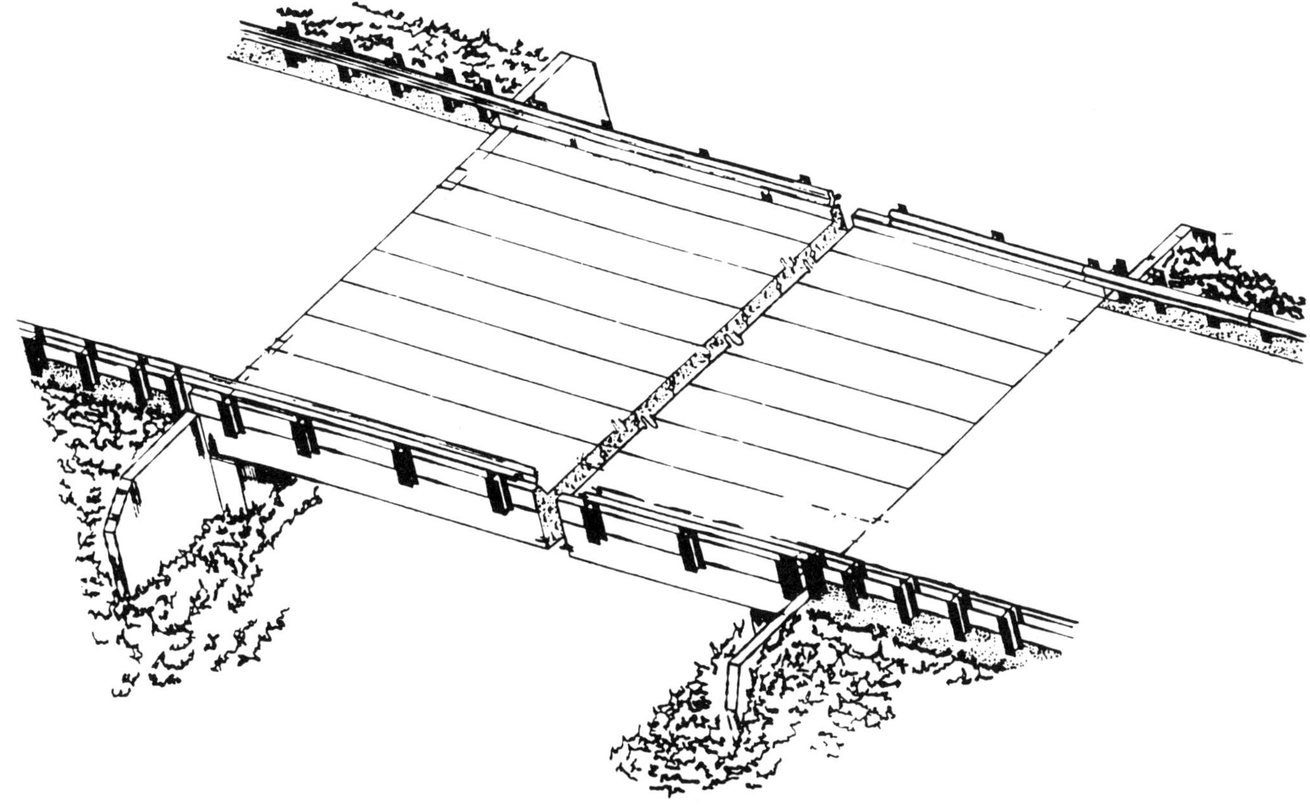

Figure 2: Side-By-Side Concrete Box Beam Bridge Using FRP Prestressing Cables

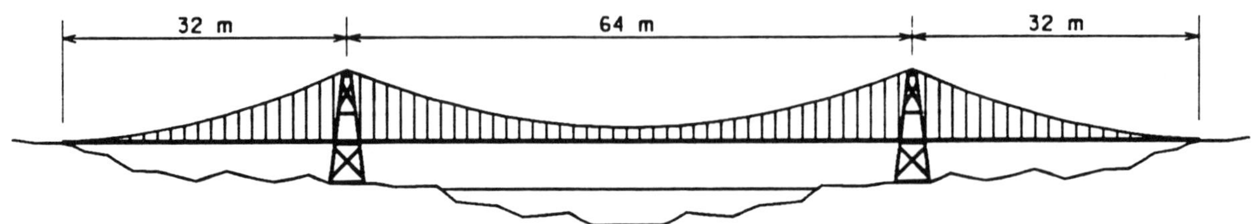

Figure 3: Pedestrian Suspension Bridge Using Fiber Reinforced Composites

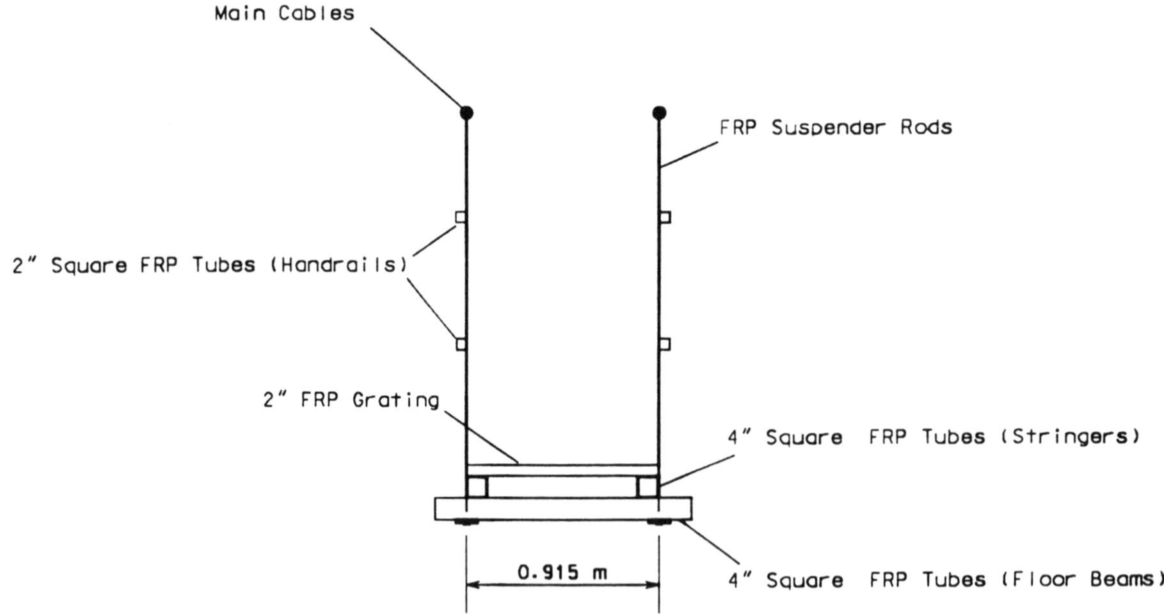

Figure 4: Cross Section of Superstructure

Interface Strength of Epoxy Overlays in Concrete

B. Deming, H. Aktan, M. Usmen
Wayne State University
Detroit, Michigan

Abstract

Chemical and physical attack on reinforcing steel and concrete deteriorate existing bridge decks and pavements, thereby shortening useful life. The high cost of total deck replacement requires finding cost-effective repair alternatives. Polymer concrete(PC) is a fast setting and durable repair material for portland cement concrete(PCC), which can be such an alternative.

Any repair or overlay material must develop a very good interface tensile and shear strength to be successful. The design of the PC overlay must be based on rational values of interface tensile and shear strength. The purpose of this project is to develop a laboratory direct test procedure for evaluating the interface strength of polymer concrete to portland cement concrete. The objective is to develop interface tensile strength that can be used in design of PC overlays. The advantage of a direct test is that the results provide the strength experimentally, without relying on transformations based upon assumptions of stress distributions and material elasticity or plasticity.

REINFORCED CONCRETE(RC) deteriorates when exposed to harsh environmental conditions. RC transportation infrastructure showing signs of extensive deterioration are a common sight, especially in urban areas. This infrastructure includes RC pavements, bridge decks, piers, abutments, etc. The deterioration initiates from or near the surface often due to corrosion of reinforcing steel. Useful life of the bridges and pavements can be increased by placing a carefully engineered overlay or fill replacing the deteriorated portion(1-3). The selection of material for the overlay is often dictated by its curing time, because of the need to put the pavement back in service on heavily travelled highways.

Polymer concrete(PC) is marketed as a useful material for overlay applications on RC(1,2). The material appears to be very fast setting and non-shrinking. However, there is significant discrepancy between the thermal expansion coefficients and elasticity moduli of polymer concrete and portland cement concrete(PCC)(2). The differences in thermal expansion coefficient will generate a significant demand on interface shear strength between polymer concrete and existing concrete during temperature variations. The difference between the elasticity modulus will also generate interface shear strength demand between the overlay and parent concrete in cases where the overlay is relatively thin.

The objective of this study is to develop a testing procedure to investigating the polymer concrete and parent concrete interface. The immediate objective here is to evaluate all the testing procedures developed for this purpose and modify and/or propose a new testing procedure that best can determine the interface shear strength.

Background

Polymer concrete refers to a group of composite materials that use a polymer monomer as the binder instead of portland cement as in traditional concrete. The reaction to cure the polymer is started with chemical initiators or hardeners. Graded aggregates serve as the filler material(4).

Polymer Concrete Materials. The types of polymer concrete are named by the monomer of the composite. There are four main types of monomers used for bridge decks(4). They are polyesters, epoxies, methacrylates, and polyurethanes. Aggregate is either added to the mix before placement or it is broadcast over the surface after the polymer.

Table 1. Properties of PC and PCC(2)

Property	PCC	Polyester	Epoxy	Methacrylate	Polyurethane
Modulus of elasticity, compression 10^3 MPa	9.7 - 20.7	7 - 35	0.7 - 40	7- 25	0.7 - 40
Compressive strength MPa (cure time)	13.8 - 41.4 28 day	20 - 35 24 hr	55 - 80 24 hr	55 - 65 24 hr	3 - 35 24 hr
Thermal Coefficient $10^{-6}/°C$	9 - 12.6	32 - 54	27 - 54	13 -23	54 - 126

The curing reaction is started by the addition of an initiator such as benzoyl, lauroyl, and methyl ethyl ketone peroxides. These materials can be formulated with different viscosities depending on the application. Promoters in the form of tertiary amines are added to insure complete mixing. The different monomers and aggregate used to formulate PC are described below(1,2):

Polyester. This polymer is used in thick overlays and large repairs, because of its relatively low cost. It is sometimes used in conjunction with a methacrylate. The methacrylate acts as a primer to seal the surface before the polyester PC is applied. One future source of monomer of this type that is being researched is using recycled plastics.

Epoxy. This is a well known bonding material in all construction. PC made with epoxy is a two component system of epoxy and a hardening agent. It is important to have the correct mix ratio for epoxy to harden correctly, requiring close inspection in preparation. An advantage of epoxy PC is higher tolerance to moisture on the surface of the substrate.

Methacrylate. These polymers typically have a very low viscosity that seals bridge deck well and forms excellent bond to substrate. It is also very volatile, requiring ventilators for workers. One of the major advantages of this polymer is that it will harden in temperatures ranging from 15 to 100 degrees Fahrenheit. This allows permanent repairs to be made year round.

Polyurethane. This PC is most often used as a water barrier on bridge and parking decks. It is applied in multiple layers with aggregate imbedded in the surface for skid resistance.

Aggregate. Aggregate used for PC must be clean and dry. Any moisture will weaken polymer bond to aggregate and deck surface. The void ratio should be minimized to reduce quantity of binder required. Gradation is dependent on application. Silica or basalt aggregate is considered suitable for polymer concrete.

Properties of PCC and PC. The portland cement concrete which is a major material in civil infrastructure has a range of properties. The two that are most significant are the compressive strength(f_c) and elasticity modulus(E_c). Representative values are: f_c'=13.8-41.4 MPa (2000-6000 psi) and E_c=9.7-20.7 GPa (1400-3000 ksi). Another significant property of portland cement concrete which is becoming very important when overlaid by a dissimilar material (in this case PC) is the coefficient of thermal expansion which ranges from 9 to $12.6/°C$ (5-7E-6/°F).

The properties of PCC and PC are summarized in Table 1 for comparison between the two materials.

Existing Test Procedures for Interface Bond Strength

There are a few tests currently used for determining the bond interface strength. The American Concrete Institute's <u>Manual of Practice(5)</u> specifies a field test for surface soundness and adhesion. This test measures the strength required to pull a 3.7 cm (1.5 in) diameter machined pipe cap that has been bonded to a core of the same diameter, off the surface of the overlay. The minimum strength specified is 1.7 MPa (250 psi). The purpose of this test is primarily quality control and the results should not be used for design purposes. Other uncertainties affecting the interface bond strength measured by this test are the adhesive and heat curing method used in attaching the pull plate to the specimen. Since the PC overlay is quite sensitive to heat, any adhesive requiring heat curing should not be used for this purpose.

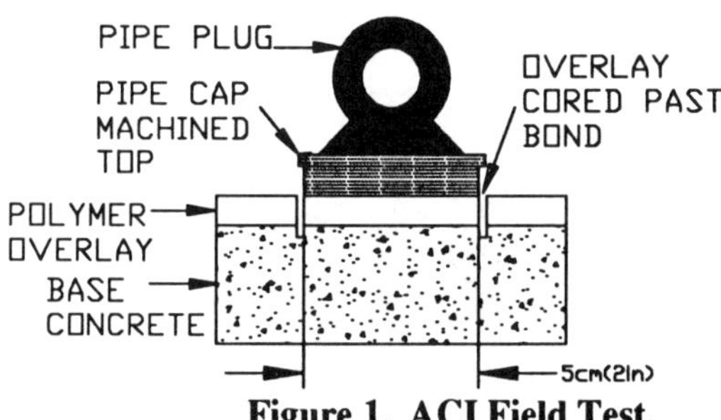

Figure 1. ACI Field Test.

ASTM specifies a test for determining bond strength of epoxy resin systems by slant shear(6). Although this a test for resin only it was modified to test PC according to a NBS report(7). The minimum bond strength specified for this test by ASTM is 6.9-10.3 MPa (1000-1500 psi)(8). The test correlates shear strength to tensile strength. There is an empirical relationship for PCC developed between its tensile and shear strength. However, the correlation between shear and tensile strength at the interface of two dissimilar materials especially with significantly different elasticity moduli, may not be very good. This underscores the importance of determining interface tensile strength by a direct test.

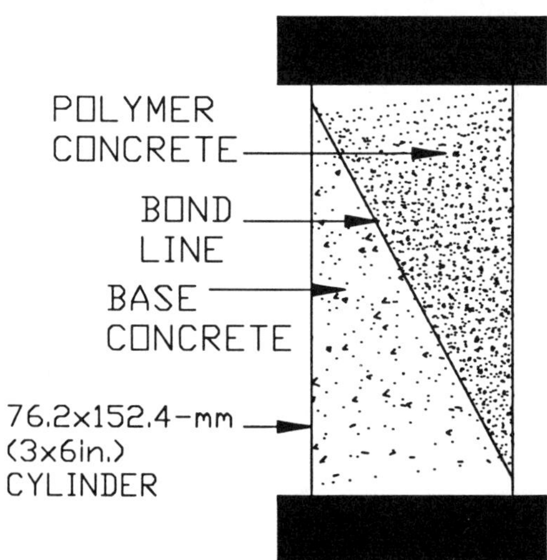

Figure 2. ASTM Slant Shear Test.

Kuhlmann suggests a direct bond interface test for latex-modified concrete and mortar(9). This test procedure bonds two pipe nipples to the circumference of a specimen one to each of the two materials that are bonded together. Pipe caps attached to universal joints connect the ends of the specimen to the testing machine. A variation of this test using friction grips instead of pipe nipples was tested as described in reference 7. The specimen of the direct test should be able to develop a uniform tensile stress at the interface. However, casing the specimen, and applying tensile loads through the casing concentrates the stresses

near the edges of the interface. This distribution will force the delamination of the interface near the specimen circumference which will lead to tearing of the interface. The strength obtained from these tests will be too low to be used for economical design of PC overlays.

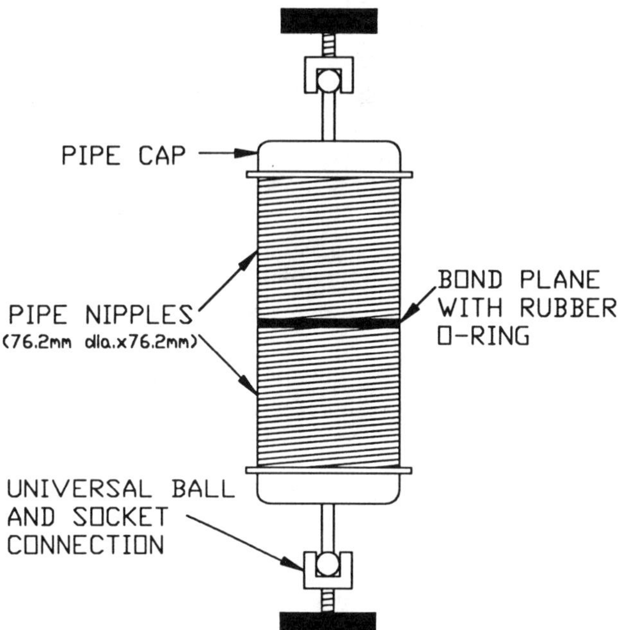

Figure 3. Pipe Nipple Test Apparatus.

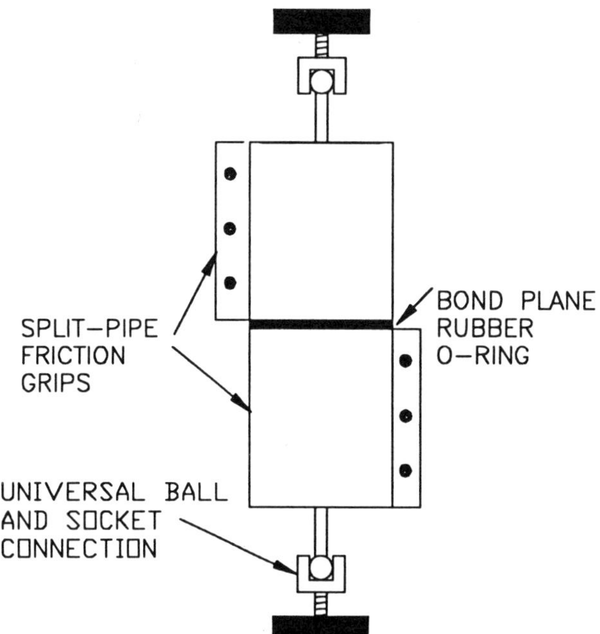

Figure 4. Split-Pipe Friction Grip Test Apparatus

Proposed Test Procedure

The basic testing apparatus consists of steel pull plates uniformly attached to the top and bottom of the specimen with structural adhesive. The interface between PC and PCC is transverse to the loading direction. This apparatus is then connected to the testing machine with clevices at top and bottom to prevent any overturning moment on the interface during load application. The test is conducted under strain control due to the fact that the delamination is very rapid and will be hard to detect under traditional displacement or force control modes.

Specimen Preparation. The specimen is prepared by applying a 12.7 mm (0.5 in) thick PC (epoxy mortar) overlay to a 152.4 mm diameter portland cement concrete cylinder. The PCC substrate used is 152.4 x 304.8 mm (6 x 12 in) cylinders sawn in half. Both surfaces are blade sawn, allowing easily repeatable surface preparation. The epoxy used is Transpo's T45, a high strength, low modulus material. Properties of the material as listed by Transpo are given in Table 2(10).

The epoxy is prepared as a slurry according to manufacturer's instructions. The two part epoxy is mixed in a 2:1 ratio in a large mixer. A uniformly graded basalt aggregate (maximum size pases #8 sieve) is added in a volume ratio of 2.2:1 to the mixed epoxy volume. It should be noted that this product is not recommended for use on an nonroughened surface. Our objective however using a blade sawn surface in order to obtain results that are repeatable and near the lower bound of expected interface tensile strength. A roughened surface will enhance bonding properties considerably by improving the mechanical bond along the interface.

The substrate is normal PCC with no additives. This concrete was prepared with a dense-graded aggregate, consisting of crushed limestone (maximum size 19 mm), pea gravel and river sand. A 101.6 mm (4 in) slump was achieved using a w/c ratio of 0.52. Twenty-five specimens were cast and moist cured for 28 days. The first nine were tested for compressive strength immediatly after curing to determine f'_c. The average 28 day compressive strength was 37.5 MPa (5450 psi). The rest of the specimens were stored in the laboratory until further testing.

Table 2 Properties of the Epoxy Binder Tested(10).

Properties at 25 °C	Transpo T-45	Test Method
Tensile Strength, 7 days	17.2 + 3.4 MPa	ASTM D638
Tensile Elongation, 7 days	35 + 5 %	ASTM D638
Tensile Modulus, 7 days	550 - 620 MPa	ASTM D638
Compressive Strength, 7 days	41.4 MPa (min.)	ASTM C109

Procedure Development. After compressive testing the next four specimens (#10-13) were tested without an overlay, only PCC. The testing rate and data collection parameters were obtained at this time. During these tests the specimen lost its stability after failure. It is essential to retain specimen stability during strain control since the strain sensor is connected between the pull plates. This problem was resolved by attaching four springs evenly spaced around the specimen attached to extensions on the pull plates. The spring load observed after failure is then subtracted from the maximum load.

The specimen preparation protocol was changed after the next set of three tests (#14-16). First specimens were made with the PC overlay bonded over the entire 182.39 cm^2 (28.27 in^2) interface area of the PCC specimen. When tested, these specimens all failed locally in the PCC near the surface of the pull plate. The problem is addressed in a paper by Hughes and Chapman(11) on direct tensile testing of PCC . It is reported that the failure occurs near the pull plate surface because the different modulus of elasticity and Poisson's ratio of the two materials causes a stress concentration near the bonded surface during the direct tensile testing of concrete using adhesives to attach specimens.

The problem is overcome by reducing the bonded surface area between the PCC and PC while retaining the same area bonded to the plates. This forces failure to occur at or near the bond line of the two materials, giving a more representative strength of the bond interface. The bonded area of our specimen was reduced by 75%, to 45.61 cm^2 (7.07 in^2), by applying a donut shaped tar paper to the surface of the PCC before overlaying to prevent polymer from bonding elsewhere.

After modifying the specimen preparation procedure, the testing method and procedure ran quite well but the strength was low. This was determined to be caused by the method for the attachment of the specimen to the pull plates. We were using an epoxy (3M #2216), which requires 7 days to achieve full strength at room temperature. In order to acieve full cure the specimen was heated to 93°C (200⁰F) to speed up the curing reaction. The heating of the specimen and/or handling the heated specimen apparently weakens the bond causing poor results. Therefore, the results of test numbers 17-20 are not presented. This problem was resolved by using a different 3M epoxy #3501 which has similar strength characteristics but achieves full cure at room temperature in 24 hours.

Test Apparatus and Procedure. The testing proce dure begins by attaching the specimen to the pull plates with a two part epoxy adhesive. The objective in pull plate attachment is to assure the tensile stress distribution to be uniform along the PC to PCC interface. The problems associated with an uneven stress distribution will not produce representative tensile strength results,as addressed addressed in the preceding section.

In this investigation the pull plates are attached using 3M gray epoxy, number 3501. After the adhesive has cured a set of four springs are attached between the plates at equal distances. This insures the specimen stability after failure. A rod is attached from each plate parallel to the loading direction for the extensometer to be mounted on to the specimen. Figures 5 and 6 illustrate the specimen and apparatus setup.

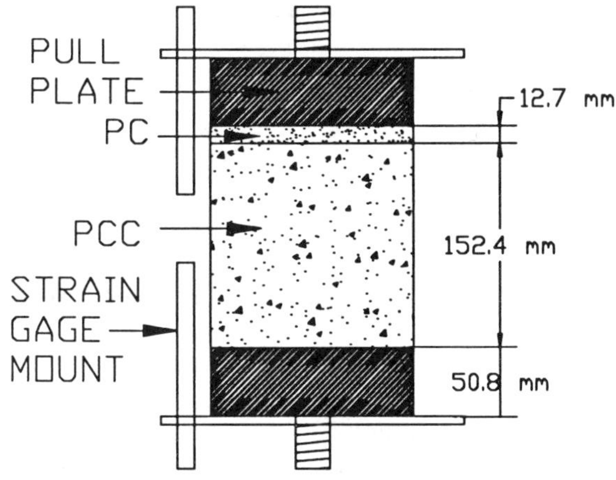

Figure 5 Specimen Dimensions

Data Acquisition and Test Control. The extensometer used for this test is MTS model 658.20 with a gage length of 50.8 mm (2 in). A Sensotec load cell model 41 with a capacity of 22680 kg (50000 lb) load is used. The apparatus is assembled into a MTS uniaxial test frame attached by clevices. The MTS hydraulics is controlled by the 458.20 microconsole which has three built in transducers to control testing by displacement, force or strain. The test is conducted in strain control using the extensometer connected to the specimen as the feed back sensor. The loading rate used for the test is 5.291E-4 mm/sec (2.08E-5 in/sec). The reason for testing in strain

control is to see the stress-strain relationship and to obtain the ultimate strength of the material at failure.

Figure 6 Test Apparatus

Data Analysis and Test Results. Table 3 shows the preliminary results of our testing, including strain and strength at failure, and where the failure occured. The strength was obtained by subtracting the spring load from the maximum load and dividing by the interface cross-sectional area of 45.61 cm^2 (7.07 in^2). Figure 7 shows a representative failure in the PCC where the pieces of concrete are pulled out on the overlay. The expected failure loctions are in the PC, at the Interface or in the PCC. Failure in the PCC is most desirable because it is an indicator that the repair is stronger than the subsrate. It is important that the strength of the substrate (PCC) that is tested is equal to or stronger than what is to be expected in the field.

Figure 7 Typical Failure.

The data presented in Table 3 indicates that the test produces relatively consistent results. Because of a limited number of tests (with reliable results) we were able to perform, the results obtained may not be totally representative of the true interface strength. These values, however, show that appreciable tensile strength is developed.

The fact that a majority of the failures have occured in the PCC rather than at the interface indicates that the interface tensile strength exceeds the tensile strength of the PCC. One limiting factor on the testing procedure is that the bonding adhesive requires 24 hous to cure.

Table 3 Test Results

Test #	Strain at Failure mm/mm	Strength at Failure MPa (psi)	Failure Location
21	0.105	3.10 (509)	in PCC
22	0.140	3.51 (450)	in PCC
23	0.130	3.03 (440)	in PCC
24	0.240	3.47 (503)	in PCC
25	0.010	2.47 (358)	interface

Summary

This study has fulfilled its primary objective of developing a direct tensile testing procedure for PCC to PC interface. The results appear consistent and repeatable. Relatively high tensile strengths are measured at the interface. Further research should confirm the repeatability of the testing procedure and generate more representative values of tensile strength.

Acknowledgments

This paper is part of ongoing research in the area of testing advanced civil engineering materials. The authors would like to thank Transpo Industries and Frank Constantino for supplying the materials and information on mix design for the polymer concrete which was tested. We would also like to thank 3M, Inc. and Arlettta Harrell for supplying samples of the epoxy adhesive used for the investigation. The help of graduate students Jaiminkumar Pandya and Jin Lee, and technician Radu Iacoban in this investigation was very much appreciated.

References

1   Sprinkel, Michael M., "Proceedings of International Congress Polymers in Concrete," ACI, Detroit, MI, 67-85 (1991)

2   Krauss, Paul D., "Proceedings of International Congress on Polymers in Concrete," ACI, Detroit, MI, 160-181 (1991)

3   Fontana, Jack J. and John Bartholomew, "Applications of Polymer Concrete," ACI, Detroit, MI, 23-31 (1981)

4   ACI Committee 548, ACI Materials Journal, 90(5), 499-522 (1993)

5   American Concrete Institute , "Manual of Concrete Practice, Part 5," Detroit, MI, 503R-30 to 31, (1991)

6   ASTM, "Standard Test Method for Bond Strength of Epoxy-Resin Systems Used with Concrete," ASTM   C 882-91 (1992)

7   Knab, L.I. and Spring, C.B., "Evaluation of Test Methods for Measuring the Bond Strength of Portland-Cement Based Repair Materials to Concrete," National Bureau of Standards Report, NBSIR 88-3746 (1988)

8   ASTM, "Standard Specification for Epoxy-Resin-Base Bonding Systems for Concrete," ASTM C 881-90, (1992)

9   Kuhlmann, Louis A., ACI Materials Journal, 87(4), 387-394, (1990)

10   Transpo Product Data, "Transpo T-45 Epoxy Overlay and Mortar Binder," Transpo Industries Inc.

11   Hughes, B.P. and Chapman, G.P., "Direct Tensile Test for Concrete Using Modern Adhesives," Bulletin Rilem, No. 26, 77-80, (March 1965)

An Integrated System for Evaluating Defects in Composite Structures

R.S. Frankle
Failure Analysis Associates, Inc.
Menlo Park, California

Abstract

Nondestructive test (NDT) techniques, such as radiography, ultrasonics, and eddy current, have traditionally been used to detect defects in materials, including composite materials. In composite structures, more than one NDT inspection is frequently required to detect all of the defects of interest. For example, radiography can be used to detect variations in density, ultrasound to detect delaminations, and eddy current to detect broken fibers. One of the advantages of digital NDT techniques, such as X-ray computed tomography (CT) and digital ultrasound, is that computers can be used to process the NDT data, which aids in understanding the state of the material and assessing the integrity of the structure. The quantity and complexity of digital NDT data, especially from more than one type of inspection, make it difficult to process the data in a practical, engineering environment. To address this problem, an Integrated Part Evaluation System (IPES), was developed. IPES is a software-based system for visualizing and analyzing data from different NDT techniques which provides capabilities to visualize NDT data, automatically detect defects, and predict the response of structures containing defects. Within IPES, these capabilities can be used separately or in combination. This paper describes the IPES capabilities and how they can be used to help evaluate structural reliability.

ONE CHALLENGE OF APPLYING DIGITAL NDT TECHNIQUES, such as X-Ray computed tomography (CT), digital ultrasonics (UT), and digital eddy current (ET), is effectively using digital NDT data once you have acquired it. The amount of data produced by digital techniques can make effective use of the data intimidating and impractical. CT inspection can easily produce 100 megabytes of data. Even 10 megabytes of UT data can overwhelm conventional data processing techniques. Evaluation of these large data sets requires specialized software and powerful computer hardware that are typically not part of the NDT equipment.

Fortunately, one of the advantages of digital NDT techniques is that the resulting digital data can be evaluated using computers. Current computer graphics workstations have the performance and storage capabilities required to process the typically large NDT data sets. Software is available to better visualize and analyze the NDT data. Data visualization tools provide a convenient mechanism to apply the knowledge and experience of engineers and scientists, who may know a great deal about defects in materials but very little about NDT and computer graphics. On the other hand, software for automated NDT data analysis, such as for defect detection, improves productivity by taking the human element out of the loop. By inputting material characteristics derived from NDT data into engineering analysis, the predicted margin of safety can be used as a quantitative measure of structural integrity and part acceptance. These are examples of ways in which NDT data can be beneficial, if the needed tools are available to conveniently and efficiently use the data.

Software for visualization and analysis of digital NDT data was developed and incorporated into a workstation-based Integrated Part Evaluation System (IPES) (1,2). This system provides a practical means to visualize data from various NDT techniques, automatically identify anomalies based on the data, and perform engineering analyses of parts containing anomalies.

Integrated Part Evaluation System

IPES consists of separate software packages used in an integrated system, as shown in Figure 1. These software packages provide the following capabilities:

- **Visualization** of NDT data and the inputs and outputs of the other software packages.

- **Automated defect detection** using the CLASSIFY code to automatically identify features, such as anomalies, by analyzing NDT data. Accept/reject decisions can be made by applying an acceptance criteria to the results of the data analysis.

- **NDT-analysis interface** for utilizing NDT data in engineering analysis. NDT data is assigned to a finite element model and used to determine part specific characteristics, such as material properties, anomalies, and geometry, which can then be accounted for in the analysis.

- **Engineering analysis**, including finite element and fracture mechanics analysis, of parts utilizing part-specific characteristics derived from NDT data.

The software packages can be used either independently or in combination. If used in combination, the output of one package is automatically accessed as input to a subsequent package.

A basic input to IPES is digital NDT data. Current experience includes use of data from X-ray computed tomography, digital ultrasonics and digital eddy current inspections. A finite element model is also required input to perform a finite element analysis. The user interacts with the IPES system through a UNIX-based engineering workstation. The various software packages that comprise the system can all be resident on the workstation. Alternatively, various software packages can be operating on other computers and accessed with the workstation via a network.

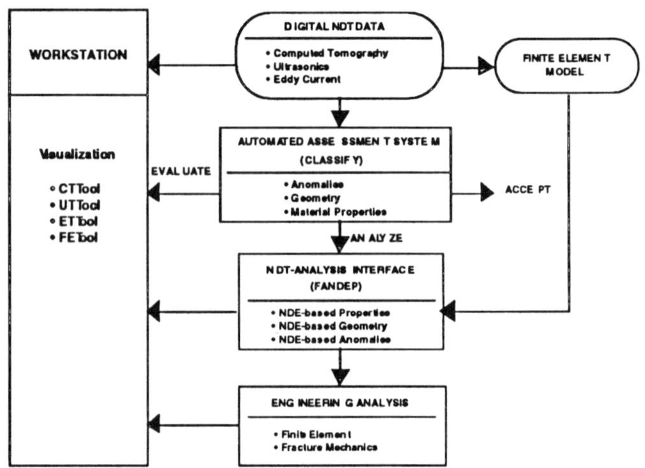

Figure 1. Integrated Part Evaluation System

Data Visualization

IPES has software tools for visualizing CT, ultrasonic, eddy current, and finite element analysis data, which are referred to as CTTool, UTTool, ETTool, and FETool, respectively. Each of the software tools (CTTool, UTTool, ETTool, and FETool) is tailored to the individual characteristics and requirements of the different NDT techniques. For CT data, the parameter of interest is either X-ray intensity or material density, which can be correlated with intensity. UT data may include a number of different parameters that are derived from the ultrasonic waveforms. Ultrasonic attenuation and time flight are typically used to detect defects in composites. Eddy current response is sensitive to reinforcement in composite materials. Therefore, broken fibers can be detected and fiber orientation can be derived from ET data and visualized. FETool provides capabilities to visualize finite element models and finite element analysis results.

In addition, all of the data visualization tools contain comprehensive visualization capabilities. These include data classification, image processing, image animation, and 2-D and 3-D image display operations. In particular, 3-D volumetric images can be created and manipulated to present user-defined views, planes and surfaces of interest. Convenient interfaces are provided for inputting data from specific NDT equipment. However, inputs from other sources can easily be accommodated by appropriate translation of the input data format.

The IPES visualization capabilities are based on the commercial AVS (Application Visualization

System) scientific visualization software from AVS, Inc. IPES includes new visualization capabilities, such as measurement and region of interest operations, which are not available within AVS. A turnkey-like system was created to provide the user with easy access to IPES operations. IPES provides capabilities to display NDT data in ways that make it easier to visualize and comprehend. Due to limitations inherent in the format of this paper, such as inability to reproduce color figures, it is not possible to adequately illustrate the available visualization capabilities. The following examples are presented in gray scale images.

Volumetric Visualization of Ultrasonic Data. IPES has capabilities for visualizing UT data as a three-dimensional volume. Ultrasonic time-of-flight data are used to locate anomalies through the thickness of a part. By combining ultrasonic time-of-flight and attenuation data, a realistic image of the inspected part can be created. For example, Figure 2 is a volumetric image of a laminated panel containing four Teflon shims between the composite plies. In Figure 2, different gray levels correspond to variations in ultrasonic attenuation. In this image, the part geometry and the shim locations are clearly visible.

Figure 2. Volumetric ultrasonic image of shims in a laminated panel.

Multi-modal Data Visualization. IPES can be used to superimpose multi-modal NDT data in a single image. For example, Figure 3 presents CT and UT data from a composite cylinder containing two wrinkles approximately 180° apart. This image of multi-modal NDT data delineates nominal material (dark gray) and anomalies that were detected with CT inspection (light gray) or UT inspection (white). Note that the wrinkles appear as anomalies in the UT data but not in the CT data. This approach for visualizing multi-modal NDT data improves the information content that can be derived from the individual inspections alone.

Figure 3. CT and UT data of wrinkled cylinder.

Automated Defect Detection

The IPES automated defect detection and assessment capabilities are based on the CLASSIFY code (2). CLASSIFY is a feature recognition code that can be used within the IPES system to identify variations in NDT responses which may be caused by different materials, variations in a material, or the presence of anomalies. CLASSIFY accepts as input a field of digital NDT data, analyzes the NDT data, and locates regions of different NDT response according to a user-defined criteria. Output is available as either a table of results or images highlighting identified regions of interest.

Inputs to CLASSIFY are digital NDT data and a user-defined criteria for anomalies. Anomalies are first defined in terms of their NDT response. CLASSIFY combines into regions all contiguous pixels within the same range of specified NDT response. The acceptability of these regions can then be determined by consideration of additional factors, such as the size, shape, and location of a region. CLASSIFY automatically applies the criteria to the NDT data and identifies all the regions corresponding to anomalies. The results are presented in a table, which lists the size and location of each anomaly. The IPES visualization tools can be used to produce images of the part with highlighted anomalies identified by CLASSIFY.

Figure 4 is an example of automated defect detection using CLASSIFY. A composite part was ultrasonically inspected following impact loading, which produced no visible damage. The UT data were analyzed with CLASSIFY to characterize varying levels of internal damage. The different regions shown as different gray levels in Figure 4 correspond to different ranges of UT attenuation and corresponding levels of damage.

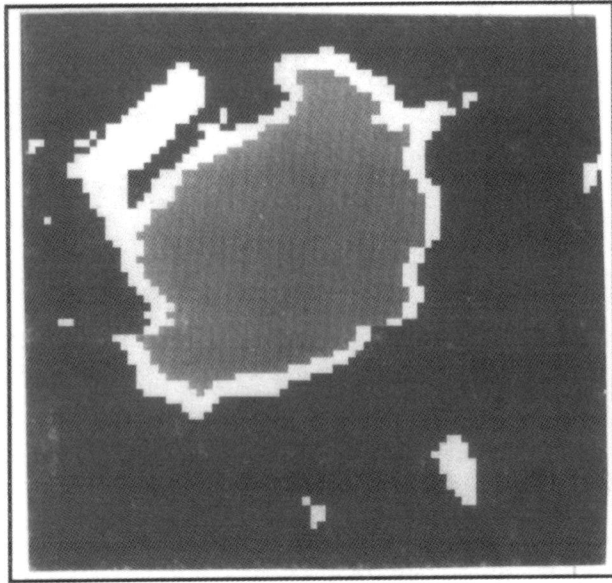

Figure 4. Impact damage detected using CLASSIFY.

Part-Specific Integrity Analysis

The concept of part-specific integrity analysis is based on the ability to derive characteristics of an individual part from NDT data and include these characteristics in an engineering analysis of the part's performance. These characteristics can include material properties, anomalies, and part geometry derived from various nondestructive test techniques. The result of the engineering analysis can be expressed in terms of predicted reduction in margin of safety, which is a quantitative value that can be considered in making a part acceptance decision.

Analysis of Structures Containing Defects. The implementation of part-specific integrity analysis in IPES is illustrated in Figure 5, which shows how NDT data are input to and used in engineering analysis. The software constituents of this integrated analysis system (shown in Figure 5) include:

- A code to assign NDT data to a finite element model. Within IPES, the FANDEP code performs this function.

- A finite element code. The finite element code is not included within IPES, but an existing code can be interfaced to IPES through the PATRAN® pre- and post-processing code.

- A fracture mechanics code. IPES does not include a fracture mechanics code, but an existing code can be interfaced to IPES.

- Tools to visualize the various analysis inputs and outputs. In IPES, CTTool, UTTool, ETTool, and FETool provide the necessary data visualization capabilities.

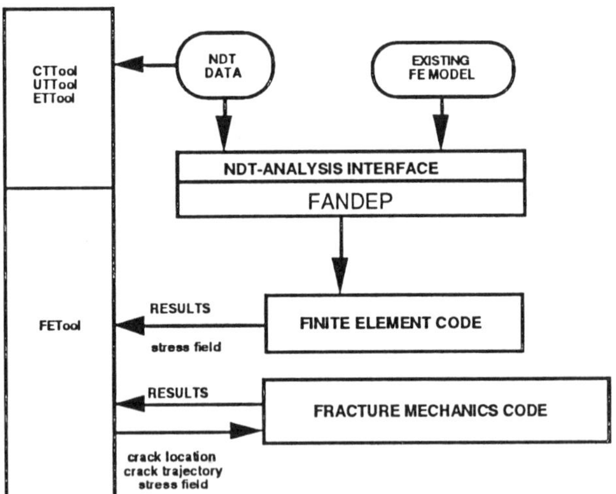

Figure 5. IPES approach to engineering analysis of defective structures.

The general approach for using IPES to analyze the integrity of parts containing defects is as follows :

1. Use CTTool, UTTool, and/or ETTool to detect, visualize, and identify anomalies in a part.

CLASSIFY can also be used within IPES for automated defect detection.

2. Use FANDEP to assign the NDT data, especially anomalies, to a finite element model of the part. Display the assigned data with FETool.

3. Perform a finite element stress analysis of the part, using the NDT data to account for the presence of the anomalies by modifying the material properties and/or the model geometry. Use FETool to display and evaluate the stress analysis results.

4. Use FETool to input stresses from the finite element analysis into a fracture mechanics analysis of the part. For metals, fracture mechanics analysis is a mature technology for predicting structural integrity. For composite materials, the IPES system provides the codes and interfaces required to develop needed analytical models and techniques to predict the integrity of parts containing defects.

With IPES, the user has a comprehensive capability for analysis of parts containing defects. Not only can defects be detected, but their potential for adversely affecting a part's performance and integrity can be directly evaluated using state-of-the-art structural analysis tools. The current capabilities of IPES are intended to represent the state of the art and to be easily upgradable as needed improvements become available.

Use of NDT Data in Finite Element Analysis. Within IPES, the FANDEP code assigns digital NDT data to a finite element model, which permits changes to be made in the model that correspond to changes in the NDT data (3,4). For example, material properties may be modified to reflect local material anomalies. Figure 6 illustrates the FANDEP approach for utilizing NDT data in analysis, including FANDEP's two primary inputs:

1. Parameters derived from digital NDT inspection. FANDEP currently accepts as input CT, UT and ET data. Data is easily input to FANDEP using CTTool, UTTool, and ETTool, which have capabilities to translate the data into FANDEP input format.

2. An existing finite element model, which may have been previously created for design analysis. In order to account for three-dimensional variations in a part, FANDEP operates on a fully three-dimensional finite element model. It currently supports an 8-noded brick element, which is commonly used in finite element codes. FANDEP functions on axisymmetric shapes or shapes of closed simple curvature.

For each integration point in the finite element model, FANDEP locates the NDT pixel at the same physical location and assigns the NDT parameters associated with that pixel to the integration point. If necessary, the NDT parameters are then interpolated to the nodal locations. A set of material properties required for the analysis can then be assigned to the integration points or nodes based on specified relationships between the properties and the NDT parameters. With this approach, spatial variations detected in the part by NDT can be reflected in corresponding spatial variations in the material properties used in the analysis.

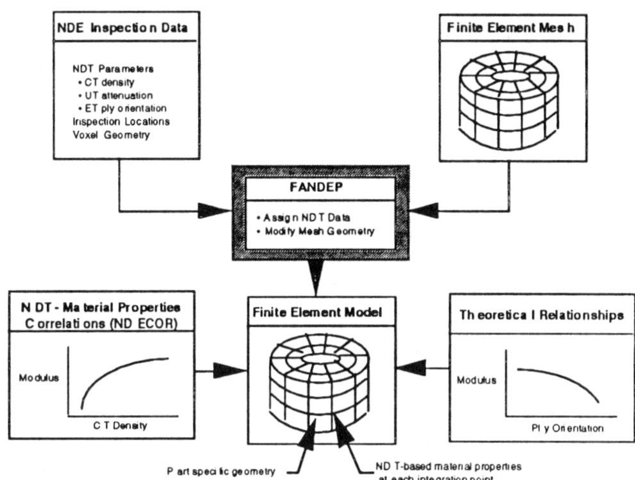

Figure 6. The FANDEP NDE preprocessor.

Prior to performing the NDT data assignment, FANDEP first modifies the geometry of the finite element model to account for differences between the model and the actual part. Most finite element models are generated from design drawings. However, most parts are not exactly the same as the drawing geometry. FANDEP overcomes this discrepancy by automatically modifying the finite element model to match the part geometry derived from CT data.

Once NDT data has been assigned to a finite element model, the challenge is to use the NDT data in the analysis to model the response of the part. Unfortunately, finite element analysis does not accept parameters resulting from NDT inspections, such as CT density or UT attenuation, as input. However, NDT data can be used to obtain estimates of material properties and to identify the presence of local anomalies. In this way, nonuniform material properties and local material anomalies can be accounted for in finite element analysis.

Figure 6 shows two approaches for modifying material properties based on NDT data values assigned by FANDEP. One approach is to use empirical correlations between NDT data and the material properties input to the finite element analysis. For example, it may be possible to correlate CT density with elastic modulus by measuring the elastic modulus of materials with varying CT density. The second approach is to modify material properties based on theoretical models. For example, composite material theory relates material properties with fiber orientation through well-known transformation equations. By correlating eddy current response with fiber orientation, these equations can be used to transform the material properties to the local conditions.

**An IPES Example:
Fiber Misalignment in Thin Tubes**

As an example of the way in which IPES can be used to evaluate composite structures containing defects, consider the problem of misaligned fibers in thin carbon-carbon tubes. The challenge is to measure the fiber orientation nondestructively and determine the effect of the misalignment on the structural integrity of the tubes. To address this problem, an eddy current inspection method was developed for measuring fiber orientation in the tubes (5, 6, 7). Measured fiber orientations were assigned to a finite element model of one of the tubes with misaligned fibers. This model was then used to predict the response of the tube and evaluate the effect of fiber misalignment.

Test Samples. The thin carbon-carbon tubes, manufactured by Kaiser Aerotech, all have the same material system and nominal dimensions of 5.2 cm (6-inch) axial length, 3.8 cm (1.5-inch) inside diameter, and 0.07 cm (0.028-inch) wall thickness. Each tube consists of one fabric ply wrapped three times circumferentially around a male mandrel. The Karbon 656 fabric has T-300 fibers in a 5-harness satin weave with 30 warp and 15 fill end-counts per inch. There are 3000 filaments in the warp bundles and 1000 filaments in the fill bundles. The matrix is a K641 phenolic resin with Kaiser Code 88A liquid pitch. The densification process consisted of four carbonization and three impregnation cycles to a target density of 1.65 g/cc.

In the baseline tube, the warp fibers were aligned with the longitudinal direction of the tube and the fill fibers were aligned with the circumferential direction. Tubes 5W5F-1, 5W5F-2, 2W2F, and 0W5F were intentionally manufactured with misaligned warp and fill fibers. Warp and fill are perpendicular in tubes 2W2F, 5W5F-1, 5W5F-2, but each is skewed either 2 degrees (2W2F) or 5 degrees (5W5F-1 and 5W5F-2) from the axial and circumferential directions. In tube 0W5F, warp is aligned with the axial direction, but fill is skewed 5 degrees from the circumferential direction. The manufacture of tube 0W5F involved stretching the prepreg to obtain non-orthogonal warp and fill fibers.

Measurement of Fiber Misalignment. Warp and fill fiber orientations were derived from eddy current inspection of the tubes. To inspect the tubes for fiber misalignment, a 1.9 cm (0.75-inch) diameter shielded horseshoe eddy current probe was placed in contact with the part, and data were collected for approximately one second from that location. (All inspections were performed using the smartEDDY 3.0 digital eddy current instrument.) A circular table supporting the tube was then rotated under the probe to change the angle of the probe (and the primary eddy current) to the tube. The resistive, reactive, magnitude, and phase angle components of the eddy current response were measured at each probe angle. This procedure was repeated at three distinct axial and circumferential locations on tubes 0W0F, 2W2F, 5W5F-1, 5W5F-2, and 0W5F.

The eddy current data were statistically analyzed to derive fiber orientation. We computed an eigenvector of the four eddy current components that accounted for most of the variability in the data. Linear regression models were developed for predicting the nominal warp and fill fiber directions from the variation of this eigenvector with the angle of the eddy current probe to the tube. The predictions of these models are illustrated graphically in Figure 7, which shows good agreement between predicted and design fiber orientations. Both models explained at least 98% of the variability in nominal fiber orientation values. However, the model for predicting nominal fill direction appeared less stable in a statistical sense. The accuracy of the predicted fiber orientations could not be fully evaluated because no independent orientation measurements were made at each eddy current inspection site.

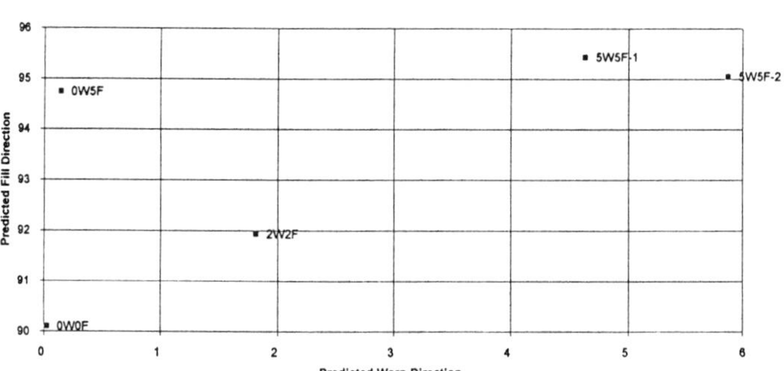

Figure 7. Predicted vs. design fiber directions.

Finite Element Analysis. FANDEP was used to assign eddy current-derived warp and fill fiber orientations for carbon-carbon tube 5W5F-2 to a finite element model (5).   In this tube, the warp and fill fibers are orthogonal but they are each misaligned by 5 degrees. Eddy current measurements were made at twelve different locations on the outer surface of the tube. The local warp and fill fiber angles were then computed from the eddy current data using a relationship derived from statistical analysis of the data. As shown in Table 1, the average measured warp and fill fiber orientations were within 10% of the design values. FANDEP was used to assign the twelve values of warp and fill fiber orientation to a finite element model of the tube. An FETool plot of the warp fiber orientations is shown in Figure 8. Notice that FANDEP interpolated the local measurements to compute values at other locations on the tube.

Finite element analyses were performed of the baseline tube and tube 5W5F-2 under uniform tension. The predicted response of the misaligned tube (5W5F-2) was compared to the response of the baseline tube, assumed to have no fiber misalignment. The comparison revealed that the misalignment significantly affected the stresses in the tube. As expected, the greatest effect was observed in the in-plane shear stress, which increased due to the fiber misalignment.

Table 1. Predicted Fiber Orientations for Tube 5W5F-2

Site	Warp, deg	Fill, deg
1b	6.0	95.5
1m	4.2	94.2
1t	4.4	92.7
2b	5.4	96.0
2m	6.2	94.4
2t	6.3	94.5
3b	5.2	94.5
3m	5.7	95.9
3t	5.4	95.2
4b	4.8	95.7
4m	5.1	94.6
4t	5.5	94.0
Average	5.4	94.8
Range	4.2-6.3	92.7-96.0
Design	5	95

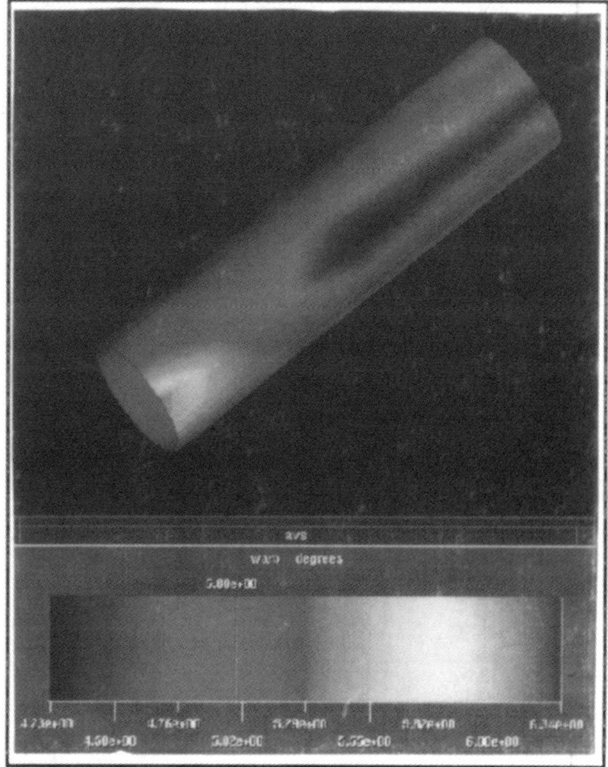

Figure 8. Warp fiber orientations in carbon-carbon tube.

Conclusions

The IPES system is designed to provide state-of-the art capabilities for visualizing NDT data, for detecting defects by analyzing NDT data, and for predicting the response of parts containing defects. Of particular interest are the following features:

- The CLASSIFY code to automatically detect features of interest, such as material anomalies, by analysis of NDT data.

- The FANDEP code to assign NDT data to a finite element model. This provides the capability to modify a three-dimensional finite element model so that it more accurately represents the geometry and properties of a specific part.

- An integrated capability to perform finite element and fracture mechanics analyses. Inputs to the fracture mechanics analysis, such as the crack trajectory and stress field, are conveniently available to the fracture mechanics code.

The IPES software is designed to be easily tailored for specific applications and implemented on engineering workstations. This technology is ready to be applied to the solution of real world problems.

Acknowledgments

IPES was developed, in part, under the NDE Data Feature Extraction project, which was a Small Business Innovative Research (SBIR) project conducted for the Air Force Phillips Laboratory. The authors would also like to acknowledge the support of Olin Corporation and FaAA, who have sponsored the development of the technology described in this paper.

The work on eddy current detection of fiber misalignment in thin carbon-carbon tubes was performed for Materials Sciences Corporation under an SBIR project for the Phillips Laboratory entitled Development of Acceptance Criteria in Thin 2D Carbon-Carbon Materials for Space Structures. FANDEP was originally developed and used to analyze wrinkled carbon-carbon rings in the NDE Data Application project for the Phillips Laboratory.

REFERENCES

1. R. Frankle, S. Warmbrodt, and P. Woytowitz, "IPES - An Integrated Part Evaluation System," Paper presented at the JANNAF Nondestructive Evaluation Subcommittee Meeting, Boeing Aerospace Co, Kent, WA (July 15, 1992).

2. R. Frankle, "NDE Data Feature Extraction," Failure Analysis Associates, Inc., Final Report to the Air Force Phillips Laboratory, Contract F04611-90C-0010, PL-TR-93-3021 (May 1993).

3. R. Frankle, "Application of NDE Data to Finite Element Analysis of Parts Containing Defects," ASTM International Symposium On Damage Detection and Quality Assurance in Composite Materials, ASTM STP 1128, John E. Masters, Editor, 1992, pp. 85-100.

4. R. Frankle, D. Jones, B. Roberts, and L. Shusto, "Analysis of Composite Material Containing Defects," 10th ASTM Symposium on Composite Materials: Testing and Design, ASTM STP 1120, Glenn C. Grimes, Editor, 1992, pp. 320-329.

5. R. Menich, R. Frankle, and Y. Krampfner, "Eddy Current Detection of Defects in Carbon-Carbon Materials", Paper presented at the JANNAF Rocket Nozzle Technology Subcommittee Meeting, Lockheed Missiles and Space Company, Sunnyvale, CA (December 1992).

6. R. Frankle, "Application of Eddy Current Inspection to Thin 2D Carbon-Carbon Materials for Space Structures," Failure Analysis Associates, Inc., Final Report to Materials Sciences Corporation, (1993).

7. R. Frankle, R. Menich, "Eddy Current Inspection of Composite Materials," 39th International SAMPE Symposium and Exhibition, Anaheim, CA (April 14, 1994).

Improved Characterization of Thick Composite Panels

E.O. Ayorinde
Wayne State University
Detroit, Michigan

Abstract

A method of obtaining the three dimensional elastic constants of a completely free orthotropic plate from experimental plate vibration data, using an optimized three-mode Rayleigh formulation that incorporates through-the-thickness shear and rotatory inertia, is described in this paper. The author had earlier utilized the classical lamination theory (which does not include these two effects) with both three and six optimized modes. The forward problem of computing plate vibration frequencies from given elastic constants by the Rayleigh method is solved for three thickness ratios (.075, .1, .125) for an isotropic (aluminum) and an orthotropic (graphite/epoxy) material. The inverse method for extracting the elastic constants from experimental vibration data is then applied to two thick plates fabricated from these materials. The results suggest that the proposed method is potentially useful for rapid, inexpensive extraction of in-situ elastic constants of composite plates.

THERE IS A NEED for easy and reliable methods of obtaining the properties (including the elastic properties) of composite materials. Many static methods for obtaining the elastic properties of such materials exist[1,2], but it is better if the elastic properties can be obtained more rapidly without significantly sacrificing accuracy. For example, a major advantage would be the possibility of using such a method for in-process monitoring of composite material properties as the material or structure was being manufactured.

Different dynamic measurement methods have been proposed [3] for the elastic constants. These include a method [3] based on the classical lamination theory and a potential-energy-optimized three-mode representation of the transverse displacement of a completely free plate, using Rayleigh's method. A version of this method for isotropic materials [4], and an extension the plate vibration model to six optimized modes [5] have also been given. In all these works, transverse shear and rotatory inertia were disregarded, and thus, the models could only be expected to be good for thin plates. For such thin orthotropic plates, four independent elastic constants, such as the longitudinal and transverse Young's moduli, the in-plane shear modulus, and the major Poisson's ratio are sufficient to uniquely characterize the elasticity of the material.

It becomes necessary to include transverse shear and rotatory inertia in the vibration model because practical structural construction sometimes requires thick plates. For such thick orthotropic plates, nine independent elastic constants - the Young's modulus in each of the three orthogonal directions, and the shear modulus and major Poisson's ratio defined for each of the three direction pairs, are required for complete elastic definition. If transverse isotropy exists, only five independent elastic constants, such as the longitudinal and transverse Young's moduli, the shear moduli in two orthogonal planes, and the in-plane major Poisson's ratio, are now required to fully characterize the material's elasticity. The method developed in this paper is capable of obtaining up to nine independent elastic constants for the orthotropic material, provided that enough experimental frequency values are available. The major constraint therefore arises from the experimental difficulty of obtaining so many resonance frequencies with reasonable accuracy.

The frequency equation for the proposed shear-inclusive model is first derived. This is then applied to an aluminum (isotropic) and a graphite/epoxy (highly orthotropic) plate, for various thickness-to-width ratios.

Analysis

Forward Problem. A Bernoulli-Euler model is no longer adequate for beam normal mode vibration analysis, and a Timoshenko model is used here. This constitutes the major difference between this and previous work [3-5], the transverse vibration of the completely free plate lying in the x-y plane is modelled with the combined motion of free-free beams lying in the x- and y-directions respectively.

The transverse deflection, W, is represented by a three-mode expression that is optimized for minimum potential energy [6]. For the (i,j)th mode, this may be written as

$$W_{i,j}(x,y) = A\,\theta_i(x)\,\varphi_j(y) - c\,\theta_i(x)\,\varphi_n(y) - d\,\theta_m(x)\,\varphi_j(y) \tag{1}$$

In this expression, $\theta(x)$ represents the transverse deflection of a beam lying along the x-direction, and $\phi(y)$ that of a beam lying along the y-direction. The indices i and j are the modal indices that identify the mode under reference. They may represent the number of half-waves along their respective directions, or the number of nodal lines, or, as in this paper, the number-of-nodal-lines-plus-one encountered along their particular directions, depending on the particular usage. If the support conditions are similar at opposite edges, then m = (i+2), on account of symmetry (else m = i+1). Similarly, n = (i+2) here.

The motion of a thick beam (i.e. where transverse shear and rotatory inertia are significant) lying along the x-axis can be modelled by the two standard Timoshenko equations [7] that follow, in contrast to the single equation when using the Bernoulli-Euler theory for thin beams.

$$EI\frac{\partial^2 \psi}{\partial x^2} + \kappa^2 AG\left(\frac{\partial w}{\partial x} - I\varrho\frac{\partial^2 \psi}{\partial t^2}\right) = 0 \tag{2}$$

$$\varrho A\frac{\partial^2 w}{\partial t^2} - \kappa^2 AG\left(\frac{\partial^2 w}{\partial x^2} - \frac{\partial \psi}{\partial x}\right) = 0 \tag{3}$$

where E = modulus of elasticity, G = modulus of rigidity, I = area moment of inertia of cross-section, A = cross-sectional area, ρ = material mass density, and κ^2 = shear coefficient, ψ = bending slope of the beam. For the orthotropic case, E, G take on the appropriate values such as E_x, G_{xy}. A similar equation may be written in y for a beam along the y-axis.

The frequency equation is obtained as [8],

$$2 - 2\cosh b\alpha \cos b\beta + b\frac{[b^2 r^2(r^2 - s^2)^2 + (3r^2 - s^2)]}{[1 - b^2 r^2 s^2]^{1/2}} \bullet$$

$$\sinh b\alpha \sin b\beta = 0 \tag{4}$$

where,

$$b^2 = \omega^2 \rho\, AL^4 / EI$$

$$r^2 = I / A L^2$$

$$s^2 = EI / \kappa^2 AGL^2$$

$$\binom{\alpha}{\beta} = \frac{1}{\sqrt{2}}[\mp(r^2 + s^2) + [(r^2 - s^2)^2 + 4/ b^2]^{\frac{1}{2}}]^{\frac{1}{2}}$$

$$k_1 = b (\beta^2 - s^2) / \beta L$$

$$k_2 = b (\alpha^2 + s^2) / \alpha L \tag{5}$$

For a particular plate, its beam frequencies are computed when it is treated as a beam, first along one direction, and then along the other. These frequencies are used to obtain the beam modes, which are utilized in obtaining the plate deflection and slope equations, which are in turn used to obtain the potential and kinetic energy expressions of the Rayleigh method.

The potential energy of a vibrating rectangular plate lying in the x-y plane, when thickness shear is included, may be written as [9]

$$P.E._{max} = \frac{1}{2} \int_0^a \int_0^b [D_{11}(\frac{\partial \psi_x}{\partial x})^2 + D_{22}(\frac{\partial \psi_y}{\partial y})^2 +$$
$$2 D_{12} \frac{\partial \psi_x}{\partial x}\frac{\partial \psi_y}{\partial y} + D_{66}(\frac{\partial \psi_x}{\partial y} + \frac{\partial \psi_y}{\partial x})$$
$$+ 2D_{16}\frac{\partial \psi_x}{\partial x}(\frac{\partial \psi_x}{\partial y} + \frac{\partial \psi_y}{\partial x}) + 2 D_{26}\frac{\partial \psi_y}{\partial y}(\frac{\partial \psi_x}{\partial y} + \frac{\partial \psi_y}{\partial x})$$
$$+ A_{44}(\psi_y + \frac{\partial W}{\partial y})^2 + A_{55}(\psi_x + \frac{\partial W}{\partial x})^2$$
$$2 A_{45}(\psi_x \psi_y + \psi_x \frac{\partial W}{\partial y} + \psi_y \frac{\partial W}{\partial x} + \frac{\partial W}{\partial x}\frac{\partial W}{\partial y})]\ dydx$$

$$(6)$$

where,

$$D_{ij} = \int_{-h/2}^{h/2} Q_{ij} z^2\ dz = \frac{1}{3} \sum_{m=1}^{NL} (Q_{ij})_m (h_m^3 - h_{m-1}^3)$$

$$(7)$$

$$A_{ij} = k_i k_j \int_{-h/2}^{h/2} Q_{ij} dz = k_i k_j \sum_{m=1}^{NL} (Q_{ij})_m (h_m - h_{m-1})$$

$$(8)$$

and [10]

$$Q_{ij} = C_{ij} - (C_{i3} C_{3j})/C_{33}\ ,\ \text{ for } i,j = 1,2,6$$

$$Q_{ij} = C_{ij}\ ,\ \text{ for } i,j, = 4,5 \qquad (9)$$

where k_i, k_j are shear correction factors. For an orthotropic plate, $D_{16} = D_{26} = A_{45} = 0$.

When the deflection and slope expressions are substituted into the potential energy equation, the differential of the energy expression with respect to each of the nine constant A, c, d, B, G, P, Z, Q, V may be set to zero, in order to optimize the potential energy.

If a typical equivalence of the pattern

$$\int \theta_r(x)^l \phi_s(y)^m\ d\xi = \theta_{rs}^{lm} \qquad (10)$$

is made, where r and s are modal indices and l and m indicate powers of differentiation with respect to ξ, then the potential energy expression may then be written as

$$P.E.=(1/2)L_x L_y[E_{ij}A^2 + E_{in}c^2 + E_{mj}d^2 + F_{ij}B^2 +$$
$$F_{in}G^2 + F_{mj}P^2 + G_{ij}Z^2 + G_{in}Q^2 + G_{mj}V^2$$

$$- 2H_{in}Ac - 2H_{mj}Ad + 2R_{ij}AB - 2M_{in}AG -$$
$$2M_{mj}AP + 2T_{ij}AZ - 2N_{in}AQ - 2N_{mj}AV$$
$$+ 2H_{mn}cd - 2M_{in}cB + 2R_{in}cG + 2M_{mn}cP -$$
$$2S_{in}cZ + 2T_{in}cQ + 2S_{mn}cV - 2Q_{mj}dB$$
$$+ 2Q_{mn}dG + 2R_{mj}dP - 2N_{mj}dZ + 2N_{mn}dQ +$$
$$2T_{mj}dV - 2J_{in}BG - 2J_{mj}BP + 2W_{ij}BZ$$
$$- 2P_{in}BQ - 2X_{mj}BV + 2J_{mn}GP - 2U_{in}GZ +$$
$$2W_{in}GQ + 2V_{mn}GV - 2P_{mj}PZ + 2P_{mn}PQ$$
$$+ 2W_{mj}PV - 2K_{in}ZQ - 2K_{mj}ZV + 2K_{mn}QV](11)$$

where L_x and L_y represent the x- and y-direction plate dimensions, and, for example,

$$E_{ij} = (A_{44}/(L_y)^2)\theta_{ii}^{00}\phi_{jj}^{11} + (A_{55}/(L_x)^2)\theta_{ii}^{11}\phi_{jj}^{00}$$

$$F_{in} = (D_{11}/(L_x)^2)\alpha_{ii}^{11}\phi_{nn}^{00} + (D_{66}$$
$$/(L_y)^2)\alpha_{ii}^{00}\phi_{nn}^{11} + A_{55}\alpha_{ii}^{00}\phi_{nn}^{00}$$

$$G_{mj} = (D_{22}/(L_y)^2)\theta_{mm}^{00}\gamma_{jj}^{11} +$$
$$(D_{66}/(L_x)^2)\theta_{mm}^{11}\gamma_{jj}^{00} + A_{44}\theta_{mm}^{00}\gamma_{jj}^{00}$$

$$H_{in} = (A_{44}/(L_y)^2)\theta_{ii}^{00}\phi_{jn}^{11} + (A_{55}/(L_x)^2)\theta_{ii}^{11}\phi_{jn}^{00}$$

$$J_{mj} = (D_{11}/(L_x)^2)\alpha_{im}^{11}\phi_{jj}^{00} +$$
$$(D_{66}/(L_y)^2)\alpha_{im}^{00}\phi_{jj}^{11} + A_{55}\alpha_{im}^{00}\phi_{jj}^{00}$$

$$K_{mn} = (D_{22}/(L_y)^2)\theta_{im}^{00}\gamma_{jn}^{11} +$$
$$(D_{66}/(L_x)^2)\theta_{im}^{11}\gamma_{jn}^{00} + A_{44}\theta_{im}^{00}\gamma_{jn}^{00}$$

$$R_{ij} = (A_{55}/L_x)(\theta\alpha)_{ii}^{10}\phi_{jj}^{00}$$

$$M_{in} = (A_{55}/L_x)(\theta\alpha)_{ii}^{10}\phi_{jn}^{00}$$

$$N_{mj} = (A_{44}/L_y)\theta_{im}^{00}(\phi\gamma)_{jj}^{10}$$

$$T_{ij} = (A_{44}/L_y)\theta_{ii}^{00}(\phi\gamma)_{jj}^{10}$$

$$S_{in} = (A_{44}/L_y)\theta_{ii}^{00}(\gamma\phi)_{jn}^{01}$$

$$Q_{mn} = (A_{55}/L_x)(\alpha\theta)_{im}^{01}\phi_{jn}^{00}$$

$$W_{ij} = (D_{12}/L_x L_y)(\theta\alpha)_{ii}^{01}(\phi\gamma)_{jj}^{01} +$$

$$(D_{66}/L_xL_y)(\theta\alpha)_{ii}{}^{10}(\phi\gamma)_{jj}{}^{10}$$

$$X_{mj} = (D_{12}/L_xL_y)(\alpha\theta)_{im}{}^{10}(\phi\gamma)_{jj}{}^{01} +$$

$$(D_{66}/L_xL_y)(\alpha\theta)_{im}{}^{01}(\phi\gamma)_{jj}{}^{10}$$

$$U_{in} = (D_{12}/L_xL_y)(\theta\alpha)_{ii}{}^{01}(\gamma\phi)_{jn}{}^{10} +$$

$$(D_{66}/L_xL_y)(\theta\alpha)_{ii}{}^{10}(\gamma\phi)_{jn}{}^{01}$$

$$P_{mn} = (D_{12}/L_xL_y)(\theta\alpha)_{im}{}^{01}(\phi\gamma)_{jn}{}^{01} +$$

$$(D_{66}/L_xL_y)(\theta\alpha)_{im}{}^{10}(\phi\gamma)_{jn}{}^{10}$$

$$V_{mn} = (D_{12}/L_xL_y)(\alpha\theta)_{im}{}^{10}(\gamma\phi)_{jn}{}^{10} +$$

$$(D_{66}/L_xL_y)(\alpha\theta)_{im}{}^{01}(\gamma\phi)_{jn}{}^{01} \qquad (12)$$

Since the mode shapes are only relative in magnitude, we assign an arbitrary value to one of them (here A is taken to be unity), thus leaving only eight independent constants.

The total kinetic energy is the sum of the transverse translational, and the rotatory inertia components, and may be written as [9]

$$K.E. = \frac{1}{2}\int_0^a\int_0^b[m_1\,(\frac{\partial w}{\partial t})^2 + m_2\,\{(\frac{\partial \psi_x}{\partial t})^2 + (\frac{\partial \psi_y}{\partial t})^2\}]\,dy\,dx$$

$$(13)$$

In this equation, m_1 and m_2 are the equivalent vibratory and rotatory masses, given by

$$m_1 = \int_{-\frac{h}{2}}^{\frac{h}{2}}\rho\,dz = \sum_{l=1}^{NL}\rho_l\,(h_l - h_{l-1})$$

and

$$m_2 = \int_{-\frac{h}{2}}^{\frac{h}{2}}\rho z^2\,dz = \frac{1}{3}\sum_{l=1}^{NL}\rho_l(h_l^3 - h_{l-1}^3)$$

$$(14)$$

where h_l is the thickness of the lth layer, and NL is the total number of layers.

Finally, for each mode, the frequency is obtained according to the Rayleigh formula, as

$$\omega^2 = P.E._{max}/K.E._{max} \qquad (15)$$

Inverse Problem. The inverse problem is to obtain estimates of the elastic constants from an input of the plate geometry and the experimentally-determined frequencies and mode shape indices. The methodology of the inverse problem has been described in detail [3]. The focus of the method is on how well the frequency is satisfied by the experimentally obtained frequencies and trial values of elastic constants. The objective function is the residual on the satisfaction of the frequency equation. The trial values of the elastic constants are varied until the minimum average residual is obtained.

Results

The frequencies predicted by the theory developed here, for plates of various thicknesses are shown in Tables 1 and 2 for the lowest six modes for the aluminum and graphite/epoxy plates respectively. The plates have a square 254mm x 254mm planform. The thicknesses of the three sizes of plates investigated are 19.05mm, 25.4mm and 31.75mm respectively, while the densities of the aluminum and graphite/epoxy plates are 2770 kg/m^3 and 1584 kg/m^3 respectively. The values of elastic constants for these two materials in the forward program are the static values used in the earlier thin-plate works of Ayorinde and Gibson [3-5,12]. The longitudinal and transverse Young's moduli, the longitudinal-transverse and transverse-transverse shear moduli, and the in-plane major Poisson's ratio are 127.9 GPa, 10.27 GPa, 7.3 GPa, 7.3 GPa and 0.22 respectively for the graphite/epoxy. These constants are 72.4 GPa, 72.4 Gpa, 28 GPa, 28 GPa and 0.33 respectively for the aluminum. It is observed that the influence of transverse shear and rotatory inertia becomes more pronounced as the plate becomes thicker, and as the mode number increases. The frequency values have been contrasted with those obtained from the application of the optimized three-mode formulation based on the Bernoulli - Euler theory (which does not include shear deformation or rotatory inertia) [3].

From the results shown in Tables 1 and 2, it is clear that the inclusion of shear deformation and rotatory inertia in the vibration model leads to a decrease in frequency. The experimentally obtained frequencies would be used to solve the inverse problem for the elastic constants, as explained in the previous section and the quoted reference [3]. The only difference here is that we are solving for five, rather than four constants. Since the inverse problem solution method is already established, the most important requirement for its execution is the establishment of the frequency equation.

In order to investigate the full problem (forward and inverse), two thick plates were fabricated - a graphite/epoxy plate, approximately 5 inches square (exact dimensions 119.53 mm square x 11.81mm thick) from 85 Fiberite Hy-E1034C Prepreg layers and cured in an autoclave-style press, and a 5 inch square x 0.5 inch thick aluminum plate machined from stock.

Tables 3 and 4 show the frequencies obtained for the two test plates. Results from a 400-element Nastran FEM analysis, the 3-mode classical lamination theory, the present work, and the vibration experiment are given. As in the previous results in tables 1 and 2, it is seen that the shear-and-rotary-inertia model yields lower frequencies. The experimental frequencies obtained for the two plates were utilized in the solution of the inverse problem. The elastic constants E, G and ν for the aluminum plate were found to be 65.5.5MPa, 25.3MPa, and 0.30 respectively. The corresponding static values are 72.4MPa, 28.0MPa, and 0.33 respectively. For the graphite/epoxy plate, the elastic constants E_x, E_y, G_{xy}, G_{yz} and ν_{xy} were found to be 134.5MPa, 10.5MPa, 6.57MPa, 0.45MPa and 0.20 respectively. The corresponding static values are 127.9MPa, 10.3MPa, 7.3MPa, 0.5MPa and 0.22 respectively. These results appear to be good approximations to the static values. Better agreement could be expected if the experimental values were more accurately obtained, and also if a more accurate representation for the transverse vibration shape (e.g. six optimized modes instead of three) is used.

Conclusion

A method for obtaining the transverse vibration frequencies of a thick orthotropic plate from a modified Rayleigh method utilizing three optimized modes and including thickness shear and rotatory inertia, has been presented in this paper. The use of this method in the solution of the inverse problem of predicting elastic constants from experimentally determined transverse vibration frequencies of a thick plate has also been demonstrated by its application to two samples - one isotropic and one orthotropic. The results suggest that the proposed method could be useful as a rapid and inexpensive method of extracting elastic constants from the vibration test data of thick plates.

Acknowledgement

The author acknowledges with gratitude grants from the Ford Motor company and the Wayne State University Research and Sponsored Programs Office that supported this work.

Table 1. Natural frequencies of thick aluminum plates

Mode	Classical lamination theory	Present (shear +R.I.) theory
	Frequencies	(Hertz)
2,2	990.27^{+}	969.96
	1642.65^{*}	1274.13
	$1650.45^{\#}$	1563.37
3,2	2598.42	2483.97
	4310.24	3210.13
	4330.70	3808.91
2,3	2598.42	2483.97
	4310.24	3210.13
	4330.70	3868.91
1,4	4526.98	4235.44
	7509.32	5400.58
	7544.97	6416.30
4,1	4526.98	4235.44
	7509.32	5400.58
	7544.97	6416.30
3,3	4745.56	4388.72
	7871.91	5573.35
	7909.27	6594.32

[+] thickness ratio, $h/l=0.075$
[*] thickness ratio, $h/l=0.1$
[#] thickness ratio, $h/l=0.125$

Mode	Classical lamination theory	Present (shear +R.I.) theory
	Frequencies (Hertz)	
2,2	669.39[+]	649.42
	892.53[*]	847.68
	1115.66[#]	1033.46
1,3	774.60	759.68
	1032.80	997.88
	1291.00	1222.72
2,3	1573.84	1465.33
	2098.46	1864.11
	2623.07	2218.09
1,4	2134.51	2085.76
	2846.02	2677.32
	3557.52	3165.80
3,1	2734.07	2518.09
	3645.35	3179.70
	4556.69	3737.98
2,4	2945.60	2689.97
	3927.47	3378.67
	4909.34	3941.96

[+] thickness ratio, h/l=0.075
[*] thickness ratio, h/l=0.1
[#] thickness ratio, h/l=0.125

Table 3. Natural Frequencies of the aluminum test plate

Mode	Natural	Frequencies	(Hertz)	
	Nastran FEM	CLT (3-mode)	SH+RI (3-mode)	Expt.
2,2	2481.8	2630.3	2550.7	2511
2,3	6302.6	6901.8	6458.6	6391
3,2	6302.3	6901.8	6458.6	6391
1,4	11155.3	12024.4	10899.8	11069
4,1	11155.3	12024.4	10899.9	11069
3,3	11309.9	12605.0	11281.5	11141

Table 4. Natural frequencies of the graphite/epoxy test plate.

Mode	Natural	Frequencies	(Hertz)	
	Nastran FEM	CLT (3-mode)	SH+RI (3-mode)	Expt.
2,2	1533.4	1957.2	1673.2	1490
1,3	2010.6	2264.9	1946.7	1940
2,3	3108.5	4601.8	3554.4	3465
1,4	4398.2	6241.1	4148.1	4000
3,1	6592.0	7994.0	6992.2	6901
3,2	7323.9	8908.0	7407.2	6989

References

1. American Society for Testing and Materials , *ASTM Standards and Literature Reference for Composite Materials*, ASTM, Philadelphia, PA., (1987).

2. Automotive Composite Consortium . *Test Procedures for Automotive Structural Composite Materials*, A.C.C., Troy, MI, (1990).

3. Ayorinde, E.O., and Gibson, R.F., "Elastic Constants of Orthotropic Composite Materials Using Plate Resonance Frequencies, Classical Lamination Theory and an Optimized Three-mode Rayleigh Formulation", *Composites Engineering*, Vol. 3, No. 5, 395-407, (1993).

4. Ayorinde, E.O., Gibson, R.F., and Y.F. Wen, "Elastic Constants of Isotropic and Orthotropic Composite Materials from Plate Vibration Test Data", Composite Materials: Testing and Design (Eleventh Volume), *American Society for Testing and Materials (ASTM) Special Technical Publication (STP)* 1206, Philadelphia, (1993).

5. Ayorinde, E.O., and Gibson, R.F., "Optimized Six Mode Rayleigh Formulation for Determination of Elastic Constants of Orthotropic Composite Materials Using Plate Resonance Data", *Vibroacoustic Characterization of Materials and Structures*, NCA Vol. 14, 167-175, ASME, (1992).

6. Kim, C.S. and Dickinson, S.M., "Improved Approximate Expressions for the Natural Frequencies of Isotropic and Orthotropic Rectangular Plates", *Journal of Sound and Vibration*, **103**(1), 142-149, (1985).

7. Weaver, W., Timoshenko, S.P., and Young, D.H., "*Vibration Problems in Engineering*", John Wiley, (1990).

8. Huang, T.C., "The Effect of Rotatory Inertia and of Shear Deformation on the Frequency and Normal Mode Equations of Uniform Beams With Simple End Conditions", *Journal of Applied Mechanics*, Trans. ASME, December, 579-584, (1961).

9. Craig, T.J., and Dawe, D.J., "Flexural Vibration of Symmetrically Laminated Composite Rectangular Plates Including Transverse Shear Effects", *International Journal of Solids and Structures*, Vol. 22, No. 2, 155-169, (1986).

10. Jones, D., *Mechanics of Composite Materials*, Hemisphere, New York, (1975).

11. Dawe, D.J., and Roufaeil, O.L. (1980), "Rayleigh-Ritz Vibration Analysis of Mindlin Plates", *Journal of Sound and Vibration*, **69**(3), 345-359

12. Ayorinde, E.O. and Gibson, R.F.(1993), "Improved Method for In-Situ Elastic Constants of Isotropic and Orthotropic Materials Using Plate Modal Data with Trimodal and Hexamodal Rayleigh Formulations", submitted to *Journal of Vibration and Acoustics, Transactions of the ASME*.

Fail-Safe Liquid Containment Materials

D.F. Watt, R. Pan
University of Windsor
Ontario, Canada

Abstract

The transportation of hazardous liquids poses a problem if the vehicle hauling the liquid is involved an accident. The current work is attempting to develop an inexpensive tough flexible sparsely-reinforced elastomer sheet to incorporate inside the containers, to contain leakage should the outer shell fail. The thermoplastic polyurethane elastomer is chosen as an ideal candidate for at least one side of a two-sheet matrix material because of its unique physical properties. This matrix is reinforced with webs of continuous fiber of various types and patterns. The reinforcement becomes a load carrying structure element designed to resist both crack initiation and propagation, in the accident event. In this paper, we focus on studying the deformation behaviour of the sparsely reinforced elastomer sheet by means of tensile tests, tear tests, snag tests, slow puncture tests, large missile impact tests and hydraulic bulge tests.

Scope of the Problem

Accidents in the transportation of hazardous goods take their toll on people as well as the environment. It is desirable to find ways to lesson their effect, and to that end we have made a preliminary collection of some statistics to estimate the extent of the need for action. The data gathered were for marine, road and rail transport, but only the latter two will be discussed in this paper. The Canadian Government categorizes hazardous goods into the 9 classes listed in Table I. The most common of these in terms of tonnage shipped by road or rail is Class 3; flammable liquids. As an example of the scale of the problem, in Canada in 1986 51.3 million tonnes of dangerous goods were transported, of which 63% was Class 3. A comparable figure for the transport of hazardous goods by road in the U.S. was 930 million tonnes in 1982. This does not include automotive gas tanks, and we do not see elastomer liners as necessary in automobile or small

Table I. Classifications of Dangerous Goods

Classifications	Definition
Class 1	Explosive, including explosives within the meaning of the Explosives Act;
Class 2	Gases: compressed, deeply refrigerated, liquified or dissolved under pressure;
Class 3	Flammable and combustible liquids;
Class 4	Flammable solids; substances liable to spontaneous combustion; substances that on contact with water emit flammable gases;
Class 5	Oxidizing substances; organic peroxides;
Class 6	Poisonous (toxic) and infectious substances;
Class 7	Radioactive materials and prescribed substances within the meaning of the Atomic Energy Control Act;
Class 8	Corrosives; and
Class 9	Miscellaneous products, substances or organisms considered by the Governor in Council to be dangerous to life, health, property or the environment when handled, offered for transport or transported and prescribed to be included in this class.

truck gas tanks. In 1979, Transport Canada, an agency of the Canadian Government, established "CANUTEC", the Canadian Transport Emergency Centre to provide advice to shippers 24 hours a day. They also compiled accident statistics. The sub-division of these by transportation mode and Class of Hazard are given in Table II, for a typical fiscal year.

Table II. Accidents handled by CANUTEC 1986-1987

Mode	Number of Accidents	Class	Number of Accidents
Road	250	1	7
Rail	39	2	60
Marine	8	3	93
Air	14	4	14
Facilities	151	5	21
		6	57
		7	12
		8	119
		9	58
		NR	71

We envisage that inexpensive, tough, lightweight liners could ameliorate the effect of these accidents for all but Class 1, Class 7 and some parts of Class 9 types of materials. While the total fraction of vehicles and rail cars releasing hazardous chemicals remains a very small percentage, the effects of a single release can be very serious, and expensive in terms of damage at the time of the accident, and in terms of later insurance coverage. A rail car accident in Mississauga, Ontario caused a city of 200,000 to be evacuated. Luckily no one was hurt, but one should strive to keep the luck factor to a minimum.

A different aspect of the scope of the problem is the actual design and fabrication of a liner for a tanker-trailer or a railcar. It must be inexpensive and easy to install, and cause no interference with filling or draining the tank. The mating of the liner to these connector fittings will be a challenging problem for someone else. But it seems to us that someone should start somewhere, and so we have chosen to develop a material that would be suitable for such a liner.

Types of Materials Tested To Date

The materials tested to date include an unreinforced polyester based urethane with a Shore A hardness of 85, both in single thickness (0.3 mm) and double thickness (0.6 mm). This urethane material has also been reinforced with either an elastomer (flexible) reinforcement, namely 1 mm diameter thermoplastic polyurethane (TPU) rod, or carbon (stiff) fibres with either 1000 or 3000 fibers per strand. These sheets have been produced by the authors using a thermoforming technique to imbed the reinforcements (1).

For comparative purposes, a commercially available polyester scrim material coated on both sides with urethane was also tested. In fairness to the producers of this material, its properties have been developed for other applications, and certainly do not represent the optimum capability of this type of material in this type of application. However it is a type commonly used for related purposes, and is used for comparison because design engineers may be familiar with it.

Among the major differences between our material and the commercial scrim is the size and spacing of the reinforcement. The commercial material has a fine screen of reinforcement about 0.12-0.13 mm in diameter spaced in a square grid of wavelength 0.5 x 0.5 mm. The locally developed material has reinforcements in the order of 1 mm in diameter with a reinforcement spacing of about 1 cm for both the TPU and the C fiber reinforcements, in an variety of patterns.

Comparative Test Results

(1) Tensile Tests. Tensile tests have been conducted using die-cut samples whose gage length is 102 mm and gage width is 45 mm. The unreinforced urethane samples, as well as the samples reinforced with TPU rod stretched out to the limit (350 %) of the tensile tester without failure. The main results, shown in Table III and in Figure 1 are that the commercial polyester scrim material has a much higher initial strength (at a cross-head displacement of 5 cm) than the locally produced materials, but at that point the scrim strength drops significantly as the first fibers tear. In contrast, the coarse carbon fiber reinforced materials, after suffering an early drop in strength, later recover and slowly exceed their initial strength as the urethane stretches out to the limit of the tensile test machine crosshead travel. The commercial scrim material has by far the poorest elongation to complete failure. The best strengths with the local materials is achieved by the TPU rod reinforced sheets with the TPU in an equilateral triangular pattern.

(2) Tear Tests. Tear tests were conducted using a modified ASTM test D624-89, Die C. The results are shown in Table IV and Figure 2. The poorest tear strength was found in the unreinforced TPU sheet. Adding a stiff reinforcement, whether it be polyester scrim, or 1K or 3K C fibers toughened the material to about the same extent. The best tear resistance and elongation to failure was shown by the TPU rod reinforced TPU. The initiator crack present at the start of the test was blocked at the first reinforcement until very high elongations were applied.

The tear strength of the commercial scrim was higher than the unreinforced TPU, and was about equal to the carbon fiber reinforced TPU, and these were all significantly lower than TPU rod reinforced sheet. The latter had the highest strength and the best elongation.

Table III. Tensile test results:

Patterns	Fibers	Thickness (mm)	Force (N)		Displacement (mm)	Broken
			1st Peak	at Stop		
DSH	None	0.558	/	433.0	535.7	NO
Square	TPU-rod	0.448	/	450.5	514.0	NO
Triangle		0.442	/	490.5	517.0	YES
Square	1K Carbon Fiber	0.485	81.0	419.0	538.5	NO
Triangle		0.454	189.0	369.0	538.2	NO
Square	3K Carbon Fiber	0.489	320.0	341.0	536.7	NO
Triangle		0.447	239.0	325.5	536.3	NO
Square	PEst Scrim	0.409	648.0	376.5	121.4	YES

Note: DSH is double sheet; Thickness is measured away from reinforcement; Displacement is cross-head displacement.

Table IV. Tear test results: (ASTM D624-89, Die C)

Samples	Reinforcement	Thickness (mm)	Tear Resistance (N/mm)	Elongation (%)
SHT	None	0.321	183.0	140.0
SHL		0.302	198.7	108.5
DSH		0.547	223.0	133.1
Square	1K Carbon Fiber	0.561	253.1	233.3
Triangle		0.540	253.7	137.0
Square	3K Carbon Fiber	0.537	249.5	170.0
Triangle		0.421	310.0	133.6
Square	TPU Rod	0.498	467.9	286.8
Triangle		0.450	644.2	304.1
Square	PEst Scrim	0.419	278.1	133.1

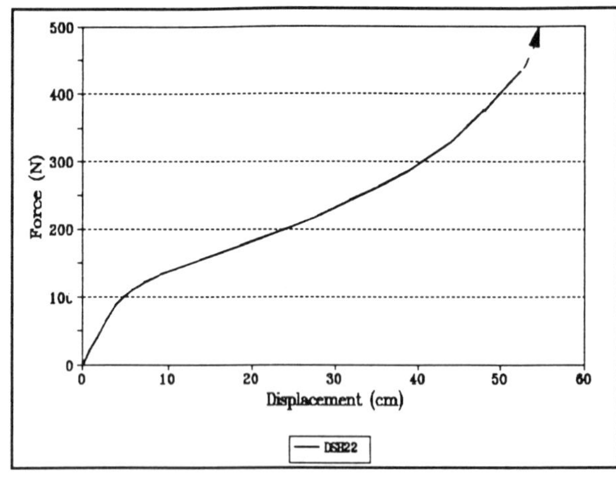

(a) Thermoformed double-sheet

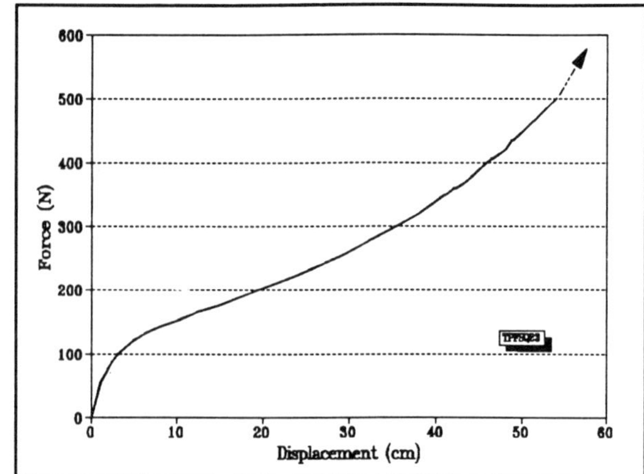

(b) Square, TPU-rod

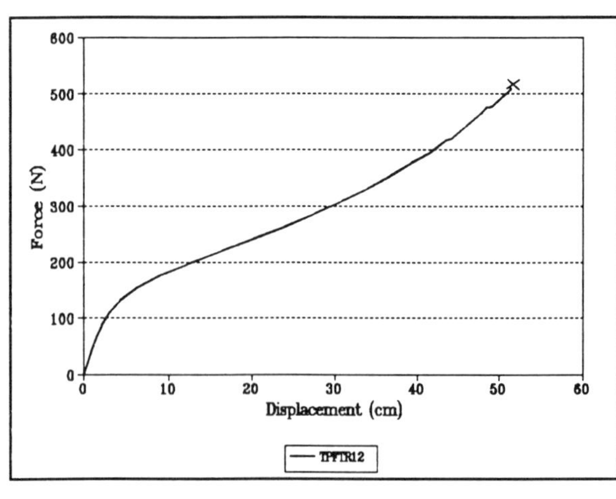

(c) Triangle, TPU-rod

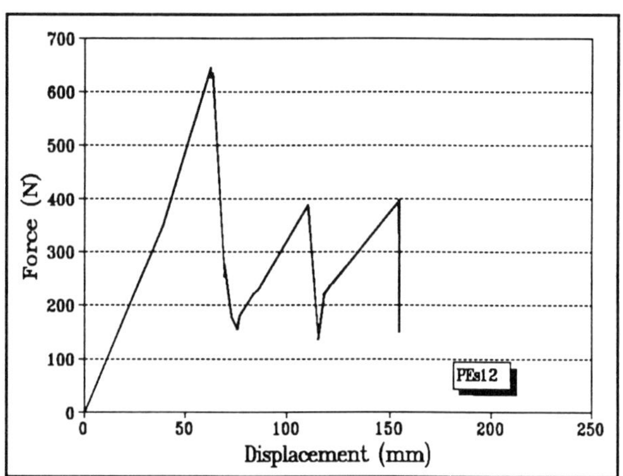

(d) Polyester scrim

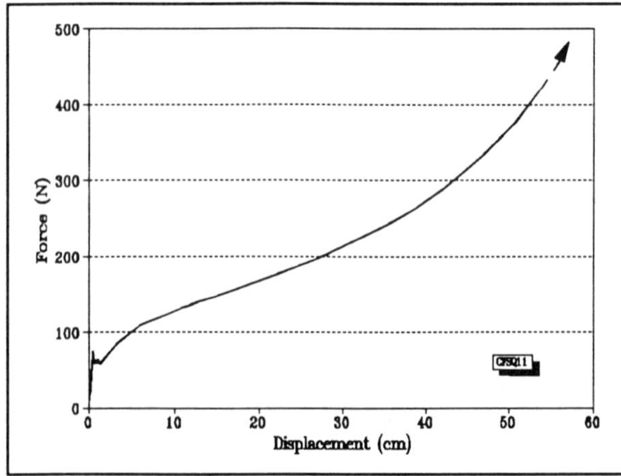

(e) Square, 1K CF

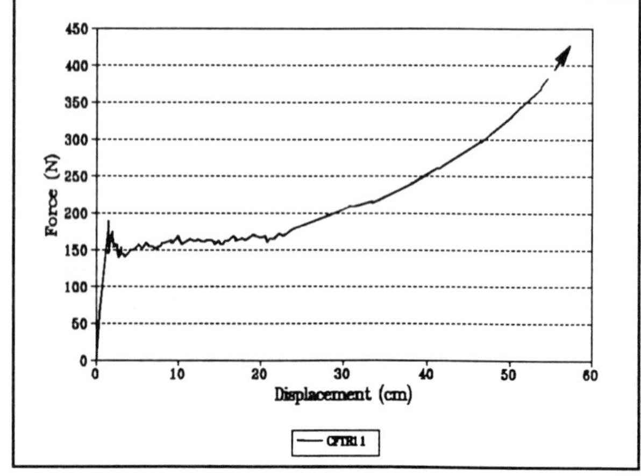

(e) Triangle, 1K CF

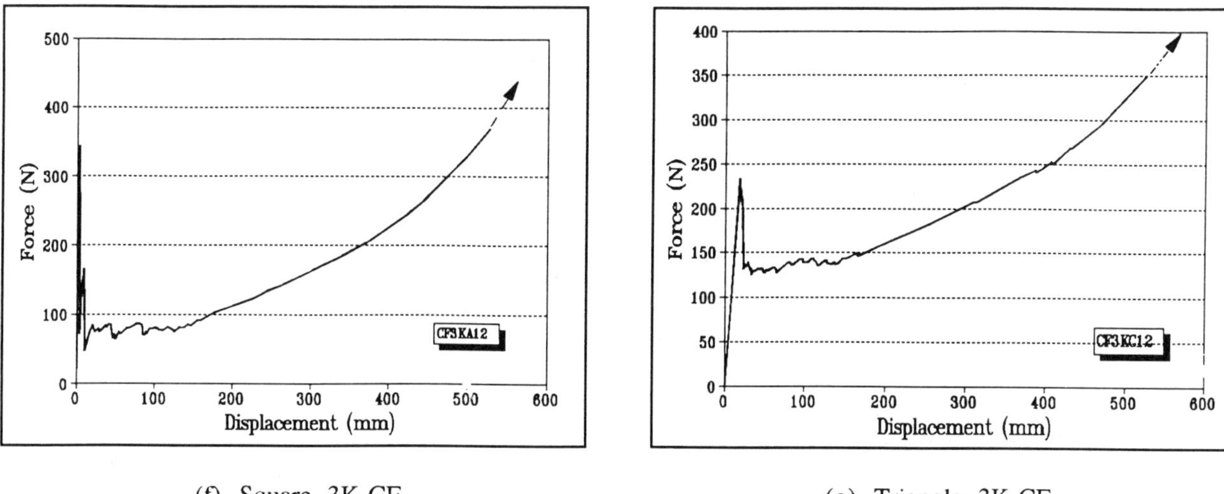

(f) Square, 3K CF

(g) Triangle, 3K CF

Fig. 1 - Tensile test results for various types of materials. The arrow at the end of the curve indicates no failure. Square or triangle (equilateral) reinforcement patterns used.

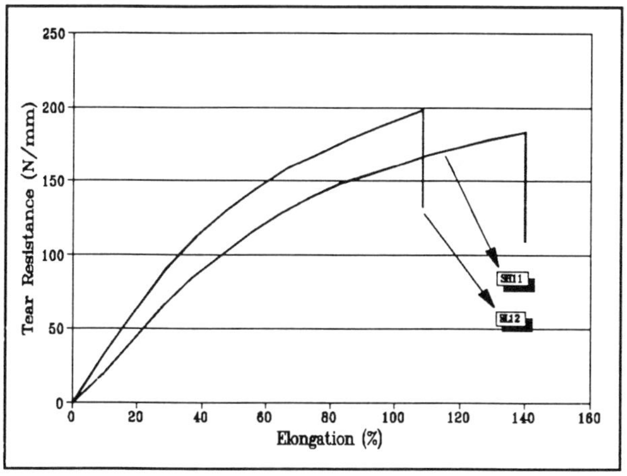

(a) As-received

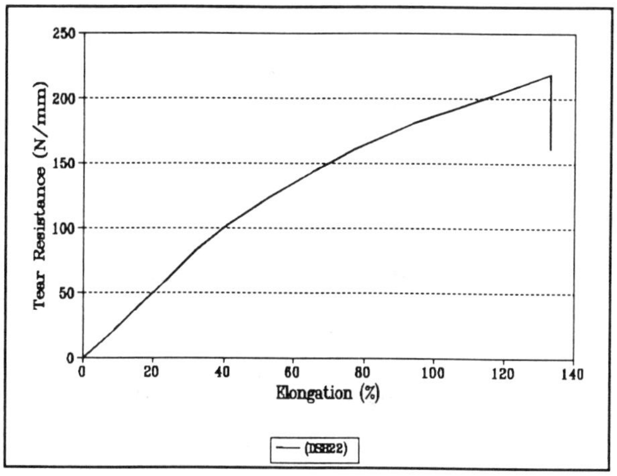

(b) Thermoformed double sheet

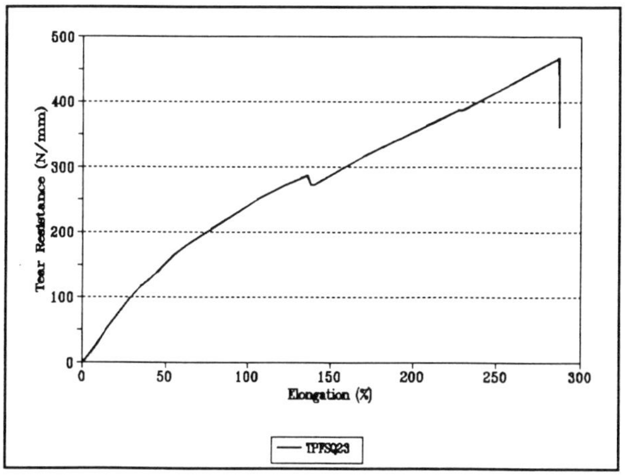

(c) Square, TPU-rod

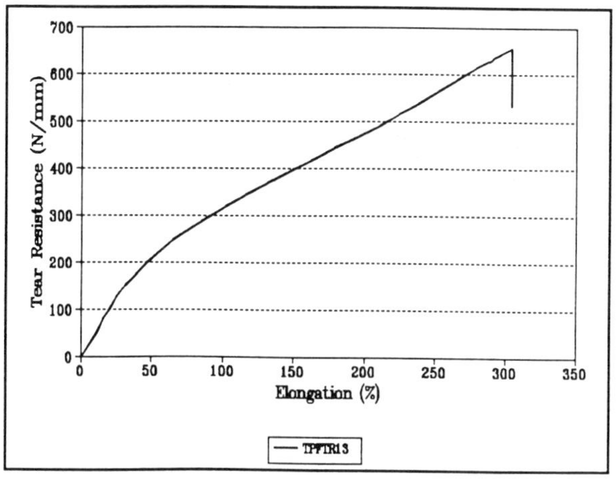

(d) Triangle, TPU-rod

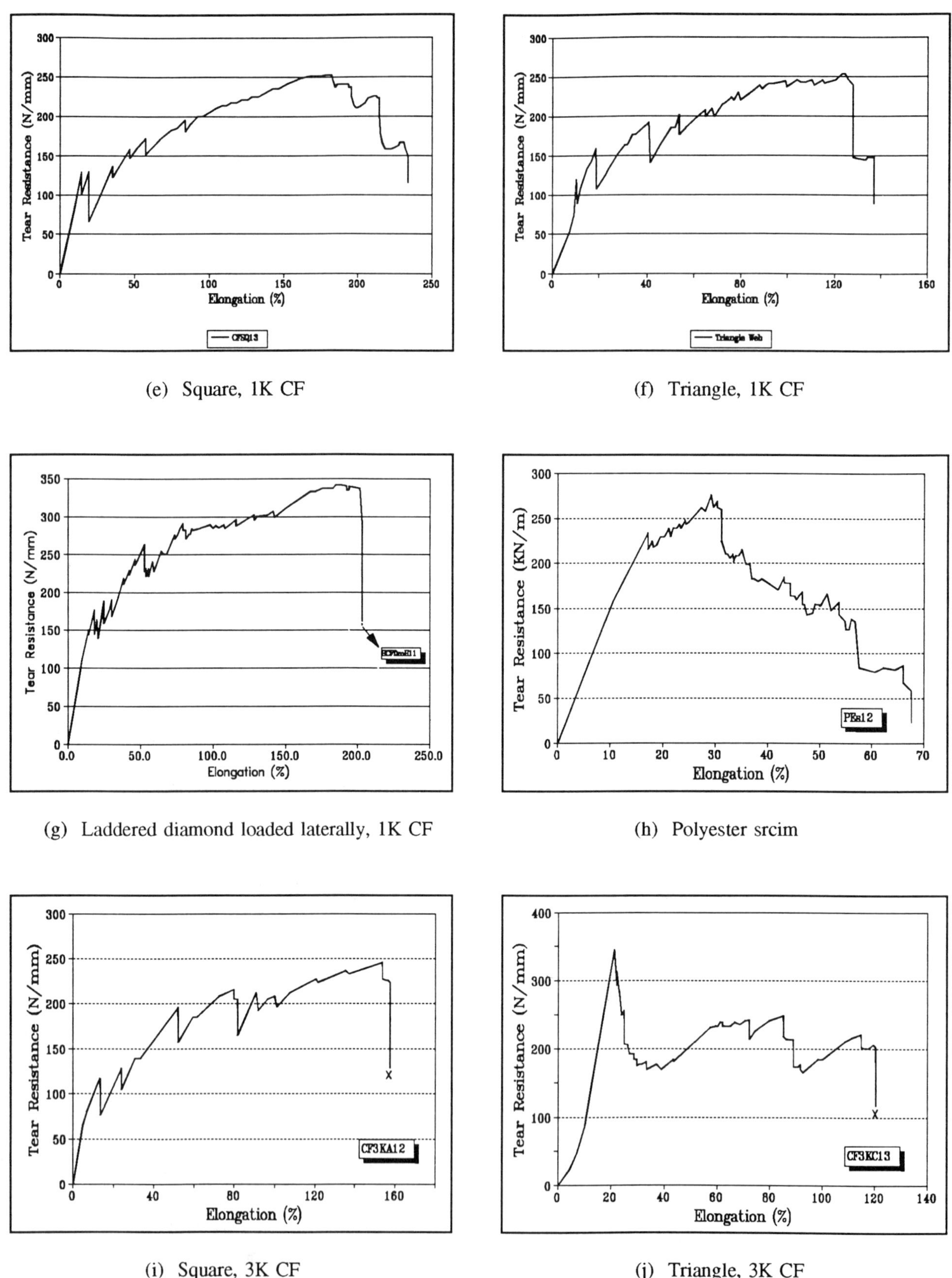

(e) Square, 1K CF

(f) Triangle, 1K CF

(g) Laddered diamond loaded laterally, 1K CF

(h) Polyester srcim

(i) Square, 3K CF

(j) Triangle, 3K CF

Fig. 2 - (a)-(j) Tear test results for different types of materials.

36

(3) Slow Penetration Tests. In the ASTM F1306-90 slow rate penetration resistance test, a relatively sharp pointed indentor, Figure 3 is pushed into a membrane tightly clamped in a ring-shaped holder. The resistance force as a function of penetration distance is recorded. The energy stored in the sheet at failure can also be found by integrating the force-distance curve. In this test, Table V shows that the TPU unreinforced double-sheet had the best penetration resistance as judged by a combined resistance force and energy criterion. The commercial polyester scrim had the poorest properties. The locally produced TPU material reinforced with carbon fiber or TPU rod all gave good results though not quite as high as the unreinforced material.

A modified test in which a screwdriver-like indentor (Figure 4) was pushed into the materials using the same clamping device gave slightly different results (Table VI). The TPU rod reinforced material now gave both the highest resistance force and the largest stored energy. Adding carbon fibres to unreinforced TPU generally slightly decreased the peak force attained, and decreased the penetration depth, but had little effect on the stored energy because the resistance force built up more quickly as a function of penetration depth when the carbon fibers were added. The commercial polyester scrim material had a better resistance force than the

Table V. Sharp point slow rate test results:

Samples	Thickness [mm]	Force [N]	Preload [mm]	Penetration Depth [mm]	Energy [J]
SH1	0.302	69.5	3.0	30.5	0.978
DSH1	0.532	122.3	2.2	30.2	1.728
TPU-SQ	0.539	111.7	3.0	26.3	1.304
TPU-TR	0.495	108.5	2.1	25.3	1.257
CF1K-SQ	0.440	106.8	3.5	28.8	1.418
CF1K-TR	0.465	98.0	2.8	24.1	1.044
CF3K-SQ	0.475	108.5	0.9	21.5	1.120
CF3K-TR	0.497	122.0	0.6	18.9	1.124
PEst SCRIM	0.406	57.6	2.1	9.2	0.203

Depth is the depth of penetration at failure.

Table VI. Dull edge slow rate test results:

Sample	Thickness [mm]	Force [N]	Preload [mm]	Penetration Depth [mm]	Energy [J]	Comments [Compared with DSH]
DSH	0.504	187.0	2.4	42.3	3.70	
CF3K-SQ	0.468	174.5	0.0	37.8	3.71	Force : lower
CF3K-TR	0.511	165.8	0.0	27.8	3.48	Penetrating Depth: lower Energy: lower
CF1K-SQ	0.494	193.3	0.8	40.7	4.11	Force: 2-higher, 1-lower
CF1K-TR	0.447	164.8	1.1	38.3	3.36	Penetrating Depth: approx. same Energy: 2-higher, 1-lower
TPU-SQ	0.474	233.0	2.6	44.0	5.08	Force: higher
TPU-TR	0.481	223.5	2.0	44.3	5.31	Penetrating Depth: higher Energy: higher
PEst SCRIM	0.409	206.0	1.0	15.9	1.43	Force: higher, Penetrating Depth: low, Energy: low.

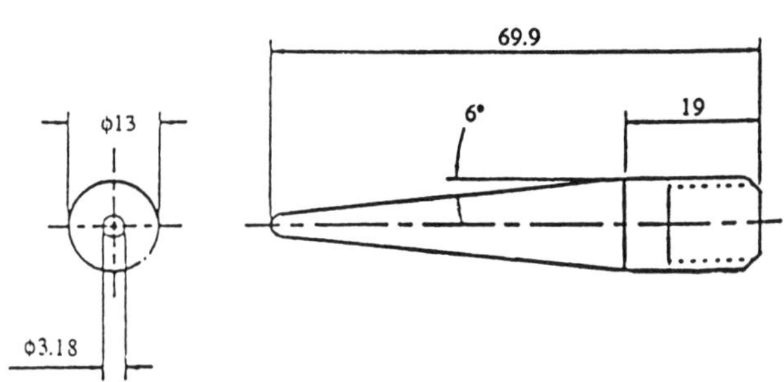

Fig. 3 - (a) The shape of the indentor. Dimemsions are in mm;

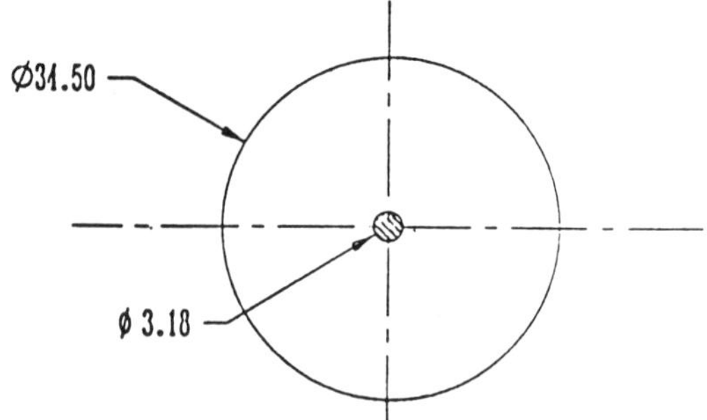

(b) The outer circle is the inner edge of the clamping ring, the inner circle is the area of the tip of the indentor at the end of the cone.

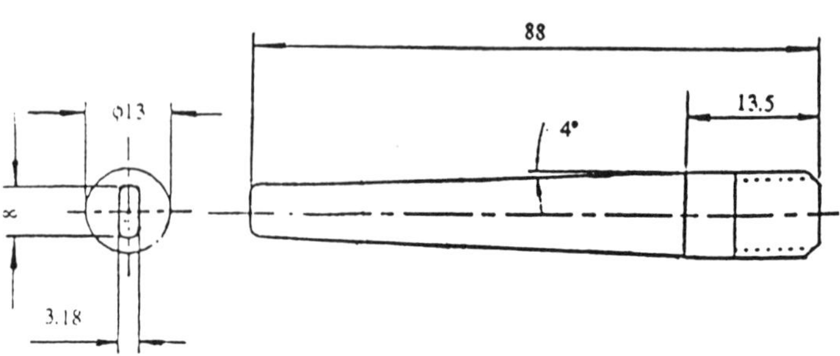

Fig. 4 - (a) The shape of the the dull-edge indentor;

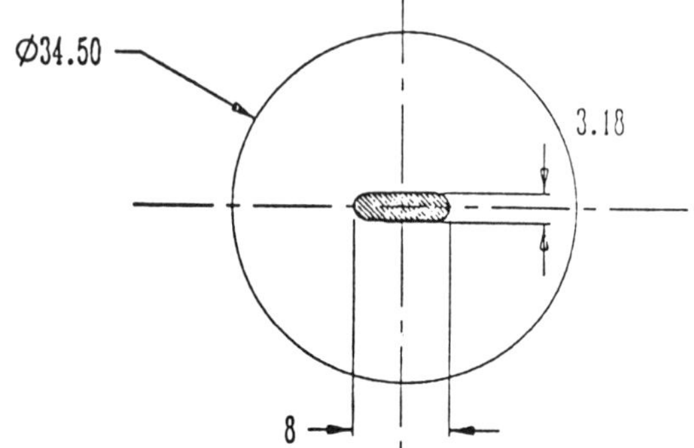

(b) The outer circle is the inner egde of the clamping ring, the inner area is the tip of the indentor at the end of the dull-edge.

unreinforced TPU sheet, but failed after a low penetration distance so that it had the lowest energy to failure.

(4) Snag Tests. A snag test apparatus conforming to ASTM D2582-90 has been built and some preliminary test results have been obtained.

In this test a trolley with a sharp horizontal needle is dropped vertically along guides to contact sheet held taut over an arc of sheet metal. The sheet metal support has a vertical slot to accommodate the needle. The trolley can be dropped, using a controlled release mechanism, from different heights and can be loaded with different weights to accommodate materials with widely different snag resistance.

For the results to date, the trolley has been dropped from one height, but with 3 different weight levels aboard. In Table VII, W is the mass of the trolley and added weights (kg), H is the distance the trolley falls before making contact with the sheet (mm), L is the vertical distance over which the sheet is torn (mm), and F is the mean tear resistance of the sheet given by F=9.8065W(1+[H/L]) in N.

The desirable result is a high F and a low L value. The best performance was from TPU rod strengthened material with a triangular reinforcement pattern. The poorest material was the commercial scrim with the reinforcements aligned in the horizontal and vertical directions. This produced a low energy tear separating along the cloth thread direction. When the scrim sheet was rotated 45°, the failure changed to a V-type failure with significantly improved tear resistance. No tests have been done as yet on carbon reinforced TPU, though these are scheduled to be done shortly.

(5) Large Missile Impact Test. In this test a piece of "2 by 4" lumber with a mass varying from 1.35 kg to 4.05 kg is accelerated to a velocity of 15 m per second end-first into a tightly clamped membrane panel. The membrane is pierced by the missile or it rebounds the board. Only a few tests have been carried out, but the to date the only material to survive the heaviest missile is 3.5 mm thick unreinforced TPU. The lightest missile easily penetrated the commercial scrim material, but it was stopped by the TPU rod reinforced TPU sheet. Tests using combinations of these materials will be carried out in the near future.

(6) Hydraulic Bulge Test. The membranes are clamped in a ring 28 cm in diameter and then are subjected to a uniform pressure causing a biaxial strain bulge in the material. The pressure vs the displacement of the middle of the membrane are measured to failure defined as leakage through the membrane. To date only the unreinforced TPU sheet and the TPU rod reinforced sheet have been tested. Both of these bulge out into a large bubble without failure at relatively low pressures. The temperature sensitivity of bulge resistance will be examined by changing the water temperature.

be examined by changing the water temperature.

Table VII. Snag test results:

Samples	Thickness (mm)	Height (mm)	Weight (Kg)	Tear Length L(mm)	Tear Resis. F(N)
Single Sheet	0.28	508.0	0.1134	7.0	81.81
			0.4536	16.0	145.68
			0.9072	28.9	165.82
Double Sheet	0.60	508.0	0.1134	4.0	142.34
			0.4536	13.0	178.27
			0.9072	20.0	234.87
TPU Triangle	0.60	508.0	0.1134	2.0	283.57
			0.4536	5.0	456.39
			0.9072	12.0	385.51
PEst Scrim			0.1134	8.0	71.73
			0.4536	27.0	88.14
			0.9072	58.3	86.42
PEst Scrim (45°)	0.41	508.0	0.1134	3.0	189.42
			0.4536	10.0	230.42
			0.9072	21.0	224.10

Anticipated Relevance of These Tests to Real Accidents

At this time we do not have detailed information on the modes in which the leaks have occurred from tanks involved in accidents. We envisage the following possibilities:

(a) The Tank Splits Open Cleanly. The dynamic forces developed in the tank and the fluids during the accident cause the tank to tear relatively cleanly. This means that tensile stresses or bending moments, perhaps acting at welds, cause something close to a simple tensile failure such that an opened crack would allow the tank to leak fluid.

The tests most relevant here would be the bulge test for a large gap in the crack faces, and the tensile ductility for the case where the internal membrane sticks to the inner tank wall, perhaps for inertial reasons. Penetration resistance, tear initiation and propagation resistance would only be an issue if the crack opened out onto a sharp object.

(b) The Tank is Penetrated by a Sharp-edged Object. In the crash, the tank is raked along its side by some strong object which folds the sharp broken edges of the torn tank inward onto the membrane. Alternatively the corner of something like an I-beam is driven into the tank wall.

The slow penetration test results are a guide to resistance here, as is the snag test. The former with the screwdriver indentor allows the reinforcements (TPU rod or C fiber) to help support the load, as they would if the edge were not more ragged than a carpenter saw blade. The snag test is a dynamic test against the penetration by a sharp snagging point on a rough edge. It becomes relevant when the tank continues to move past a sharp edged intruding object. The tear resistance test is also relevant if the membrane is torn locally by the sharp edge, as this could determine the rate of fluid loss. Sheet polymers generally have good tear initiation resistance but relatively poor tear propagation resistance. Another type of potentially useful test we are now considering would be one where a piece of sheet metal is put over the membrane material which has water on its inner side. This combination could then be hit by an standard axe swinging from a predetermined height.

(c) The Tank is Penetrated by A Sharp-pointed Object. In the accident, the tank falls onto an exposed rod or spike which pierces the tank wall.

Again the snag test and the slow penetration tests are the most relevant of the current tests. Another new test comprised of a sharp pick falling into sheet metal backed by the membrane full of fluid would be more representative for this type of failure mode.

(d) The Tank Filling or Drain Connections are Torn Off. There is no doubt that making suitable connections to the tank filling and drain connection points will be a big challenge in the detailed design of bladder inserts for liquid hauling vehicles. Provision to accommodate sensors and monitors are also required. Because the fill and drain points usually extend outside the general tank body, they appear to be vulnerable to being sheared off in an accident. If these connections are directly connected to an elastomer fitting, it is conceivable that they may be able to withstand larger displacements without loss of liquid than the welded connections now found on tanks. The details of such fill point to membrane connections remain to be worked out, but crash worthiness will have to be taken into account.

Conclusions

The use of tanker trucks and railcars to haul hazardous liquids poses a danger to the environment. Ideally, a flexible tear resistant liner could lessen the loss of liquids when these vehicles are involved in accidents. This paper makes a modest start toward the development of such liners by testing models of candidate materials. Our conclusion from this early data is that flexible reinforcements, represented here by TPU rods, significantly raise the tear propagation resistance and general toughness of relatively lightweight sheet, and appear to represent the best choice of the types of materials examined. The issues of how to produce this material on a large scale, and how to fabricate it into a suitable liner remain to be addressed.

References

1.   D. F. Watt and R. Pan, ANTEC 94, San Francisco, U.S.A., May 1-5, 1994, p. 860-864.

Flexural and Toughening Behavior of Cementitious Composites Reinforced by Waste Prepreg and Recycling Issues Thereof

B.P. Lennon, V.M. Karbhari, H.E. Allen
University of Delaware
Newark, Delaware

Abstract

Recycling has become an increasingly important issue in the area of composites. Prepreg scrap is one type of composite waste that presents considerable difficulty in recycling. One strategy that shows promise is the use of the prepreg scrap as a reinforcement in concrete. The scrap has the potential to resolve the problem of poor tensile and flexural performance of concrete. Test results on scale size beams show that the scrap can increase the flexural capacity of the concrete. The good chemical resistance of fiber reinforced plastic makes this method a good candidate for replacement of steel reinforcement in adverse environments. The environmental impact of using this type of waste in concrete also will be assessed.

FIBER REINFORCED PLASTICS (FRP's) continue to find uses in new and innovative designs. Sales of advanced composites components was exceeded 15,000 metric tons in 1991, but, FRP users/suppliers must demonstrate their recyclability to maintain this market acceptance[1]. Consumer preferences as well as legislation mandating recycle levels threaten to replace any material deemed unrecyclable with a more 'green' material. This paper will discuss the options available for FRP recycling as well as one method which is currently under investigation. The method under investigation is the use of scrap prepreg as a reinforcement in concrete.

FRP Recycling Options

When a material reaches the end of its design life it may get disposed of in any of three ways; it can be placed in a landfill, incinerated or reused/recycled. Reuse and recycling is the option that is receiving the most attention at the current time. For example, in 1992 the state legislature of Rhode Island passed a law banning the construction of waste incinerators in the state and required the state to sign contracts for recycling 70 percent of its municipal solid waste.[2] In the case of recycling FRP's, there are many different methods to consider but they all can be classified into three main categories: 1) thermal recycling, 2) chemical recycling, and 3) mechanical recycling. Thermal recycling recovers the energy contained in the material, chemical recycling recovers the basic chemical constituents for use as feed stocks in new processes, and material recycling physically recovers the material for reuse in one form or another.

Thermal recycling (energy recovery) is a disposal method applicable to a wide variety of wastes. One of its great advantages is that it can accommodate a mixed stream of waste. In addition to recovering energy it greatly reduces the volume of material that needs to be disposed of in the landfills. The energy content of plastics is comparable to that of fuel oil, however the presence of inorganic fibers could reduce the energy content significantly depending on the fiber fraction. But the presence of fibers is not always considered bad. When combined with other wastes for energy recovery, the melted glass fibers can assist in fixing toxic metals in an inert slag.[3]

Chemical recycling differs from thermal recycling in the fact that it requires a well sorted waste stream of compatible resins. The idea is to break down the resins into the constituent molecules which the can be fed back into refineries or petrochemical processes as feed stocks. Several methods show potential. including cracking, gasification hydrogenation, pyrolysis, and hydrolysis. Cracking operates at 400-600° C and pressures slightly higher then atmospheric and produces gases to heat the process and oligomeric liquids which can be further broken down. Japan's Fuji Recycle has been operating a cracking plant since 1992 that converts polyolefin and polystyrene into gasoline kerosene and gas oil.[4] Gasification incinerates the plastics at 900-1400°C in a pure oxygen atmosphere to produce mostly carbon monoxide and hydrogen. The pure oxygen atmosphere used makes it easier to recover the gases than if they were diluted with nitrogen from the air. Thermoselect of Switzerland has a process which can treat 4.2 tons/hr of municipal waste in Italy.[4] In gasification fiber reinforcement is either burned or melted to form a glass slag that helps fix toxic substances. Hydrogenation uses a hydrogen atmosphere at 300-500°C to produce hydrocarbon oil gases and solid residues. The process, which is still under trials, yields good output but is susceptible to fluctuations in the crude oil market. Pyrolysis is a process that runs at temperatures of 500-900°C in an oxygen free atmosphere and produces pyrolysis oil, gases and some solid residues (soot). The SMC-Automotive alliance (in the US)has been running a test plant for SMC and GRP waste and reports positive results.[5] Hydrolysis uses high temperatures and water to dissociate polyurethanes, polyesters, and polyamides to their basic monomers. These can be re polymerized into new products. If dihydric alcohols are used instead of water the process is called glycolsis. In the case of gasification fibers may be beneficial, however in the other processes fibers can lead to down time for the cleaning and recovery. Often chemical recycling processes have little to distinguish them from energy recovery because the many of the oils recovered are used as fuel oils or compete directly with crude (90% of which is used for heating and fuel).[6]

Mechanical material recycling must be divided into categories by resin type: thermoplastics which can be remelted and thermosets which cannot. Typically, thermoplastics are chopped into granules, remelted and reintroduced into a process that makes the same or similar parts. The problem is that the chopping of the scrap shortens the fiber length and the remelting degrades the resin as well as the fiber/matrix interface. Recycling thermoplastic composites in this manner requires a single sort waste stream and usually the reclaimed FRP must be mixed with virgin matrix to maintain sufficient properties. The primary method thermoset composites are recycled is by regrinding them and reusing them as a filler. One of the biggest advantages (over thermoplastics) is the fact that these thermoset composites appear to have the ability to be reground and reused any number of times without any loss in property of the filler.[7] Unlike thermoplastics, reground thermosets don't require a single sort, pure waste stream when they are used as fillers. The use of reground FRP in place of other traditional fillers, like calcium carbonate, can result in a reduced density as well as increased tensile strength.[8] Several systems are already in place that use this technology. ERCOM in Europe has a plant that accepts scrap SMC from suppliers/collectors, grinds it into a powder and returns it to the supplier for use as a filler in new SMC and BMC parts.[5] In Farmington Hills, MI GenCorp Automotive uses recycled SMC to produce a body panel for the (1993) Corvette.[9] Another widely considered option for the use of regrind in construction materials, concrete being the most obvious. Studies have shown that concrete made with FRP powder as a replacement for sand has a lower density than normal concrete, however the mechanical strength decreases as the percent replacement increases.[8]

When looking for disposal or recycle options for FRPs there are many factors that must be balanced. Menges outlined the requirements he deemed most important for the disposal of problematic wastes:[3]

1. only minimal costs to be incurred in preparation
2. no sorting required
3. highest possible volume reduction
4. no residuals materials to be dumped
5. minimal generation of pollutants
6. recovery of valuable components
7. suitability for small volumes

Each method consists of trade-offs between one requirement and the next. A method which can reintroduce the scrap into the same process from where it came (closed loop recycling) may be a tradeoff in terms of preparation and sorting but recovers most or all of the valuable components.

Motivation

The purpose of this paper is to evaluate one option for the management of FRP waste; the use of scrap prepreg as a reinforcement in concrete. Although a closed loop recycling scheme may fulfill the requirements best, scrap prepreg would be difficult, if not impossible, to recycle in such a manner. Grinding the scrap prepreg to a powder, or thermal, or chemical recycling fails to recover the valuable combination of properties that prepregs give, namely the high stiffness to weight ratio and excellent tensile and flexural strength. We propose that these properties can be recovered and used to solve the problems of poor performance of concrete. Concrete performs well in compression, but behaves poorly in tension and flexure. The combination of both materials can solve both the problems of waste disposal and material performance. The scrap would require minimal preparation and sorting, the method would reduce the volume that gets landfilled, little or no residuals would be dumped, and it would generate a minimal amount of air pollutants as compared to incineration or chemical recovery. Furthermore, it would not require large capital investment in equipment, thus making it suitable for small volumes. The emphasis of this investigation is to demonstrate that prepreg can improve the performance of concrete and assess the environmental impact (if any) of disposing of the scrap in this way.

Most manufactures make prepreg parts primarily by hand, and as a result they generate a large portion of scrap. The scrap can be a problematic waste to dispose of because uncured prepregs are often classified as hazardous wastes, and have to specially handled. The manufacturers can render the scrap non hazardous by curing before disposal, but under current law this classifies as a treatment technology and would subject the manufactures to the permit requirements of a treatment facility.[8] The reuse of prepreg scrap in concrete presents a way around this problem.

Concrete is widely used in construction applications and performs well in compression, but it has poor tensile and flexural properties and fails in a brittle manner. For this reason it is often reinforced with steel, but steel has its own disadvantages as well. Steel is vulnerable to corrosion, especially in high chloride environments. Corrosion of steel reinforcement is considered to be a major cause of the deterioration of many bridges in the US. FRP reinforcement would offer a corrosion resistant alternative to steel. The first objective of this study will demonstrate that the prepreg scrap offers viable solutions to these problems.

The second part of this study will assess the potential environmental impact of using scrap prepreg in concrete. Many uncured prepregs can be classified as hazardous wastes. That being the case, is there a danger that these components may leach out of the concrete? Very few studies have been done on the leaching potential of composites but those that have been done show that further investigation is merited. In water immersion studies with aged and cured E-glass epoxy prepregs Roylance et al. noted a 1% loss of weight due to leaching.[10] In boiling water experiments Birger et al. identified partially reacted components of the dicyanadiamide system as having leached out.[11] Koga et al. studied the leaching of components of epoxy resins in aqueous solutions and proposed models to predict it. They noted the leaching of a reactive diluent (*n*-butyl glycidyl ether) out of the raw resin system and p-diamino-diphenyl methane out of the hardener system used.[12]

Materials and Test Specimen Fabrication

The basic materials used for the study were scrap prepreg and concrete mortar. The mortar was composed of water, Portland cement and sand (.46: 1: 2 ratio) which was used to make small beams 50.8 mm x 25.4mm x 330.2mm (2" x 1"x 13") in size. The scrap prepreg used was Hercules IM7-8551 unidirectional graphite epoxy tape. The strips of reinforcement were cut from a 16 ply 0° unidirectional lay-up and were approximately 2.5mm x .6.4mm x 304.8mm (.1" x .25" x 12") in size. Both cured and uncured scrap reinforcement was used in order to simulate the extremes of scrap that might be encountered. The mortar for the beams was mixed by hand and the reinforced and unreinforced (control) specimens were molded from the same batch. After mixing, a 6.4mm (.25") layer of mortar was placed on the bottom of the mold and the reinforcement was placed by hand (see figure 1 for orientation). Then, the molds were filled and vibrated to remove any large voids. After one day the beams were removed from the molds and then cured for a minimum of 28 days in a water saturated air environment. Care was taken to ensure that the control and the reinforced specimens were subjected to the same conditions throughout the experiment.

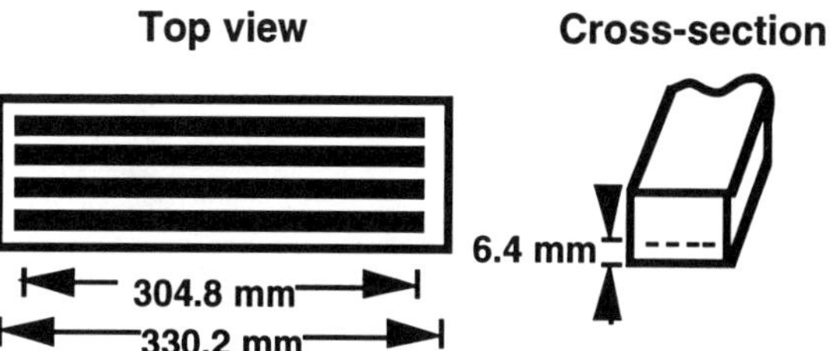

Figure 1: Placement of scrap prepreg in beams.

Results

After curing the beams were exposed to either of three environmental conditions, atmospheric, distilled water, or salt water. The beams were removed for periods of testing after periods of 2, 4 and 8 weeks. The distilled water and salt water soaking solutions were saved for further quantitative analysis with attenuated reflectance FTIR and the beams tested in flexure (3 point bending). The flexure tests were conducted on an Instron 1125 screw driven tension/compression machine. A span length of 11" was used, with a crosshead speed of .5 mm per min. (.02 in/min). The midpoint deflection was measured using an LVDT and a computer data acquisition system.

Only the results of the uncured scrap reinforced beams specimens are the available at the time of submission. As hoped, the specimens with prepreg scrap reinforcement far outperformed the unreinforced controls They held higher loads then the controls even after large deflections. Nearly all reinforced specimens held 5-7 times the maximum load of their unreinforced counterparts. Table 1 has data tabulating the ratio of the maximum loads held by the reinforced vs. the load held by the control specimens.

All of the reinforced specimen demonstrated a mode of failure fitting the description of a shear tension failure as described in reinforced concrete design texts. Due to the scatter in the data no trends could be discerned when comparing the reinforced beams under the different environmental conditions. Figures 2, 3 and 4 show load deflection curves for beams after exposure to each of the environmental conditions for 8 weeks.

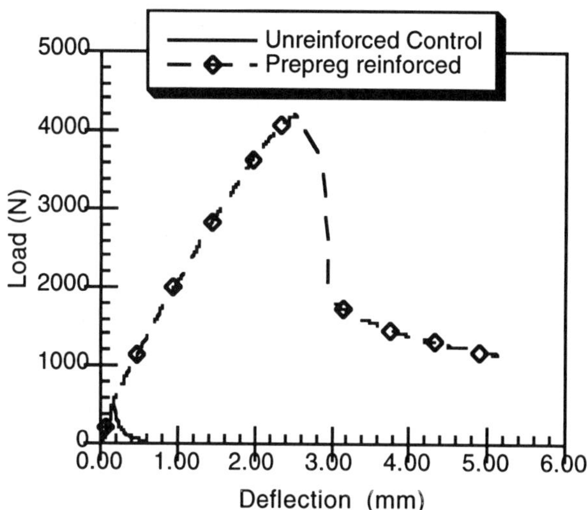

Figure 2: Load deflection curve after 8 weeks exposure to atmosphere

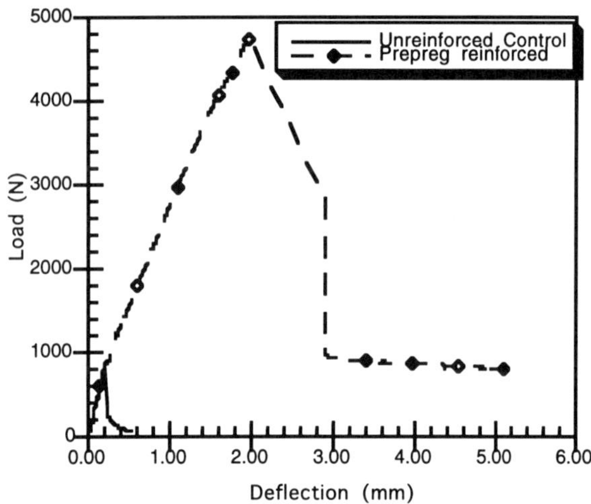

Figure 3: Load deflection curve after 8 weeks exposure to distilled water.

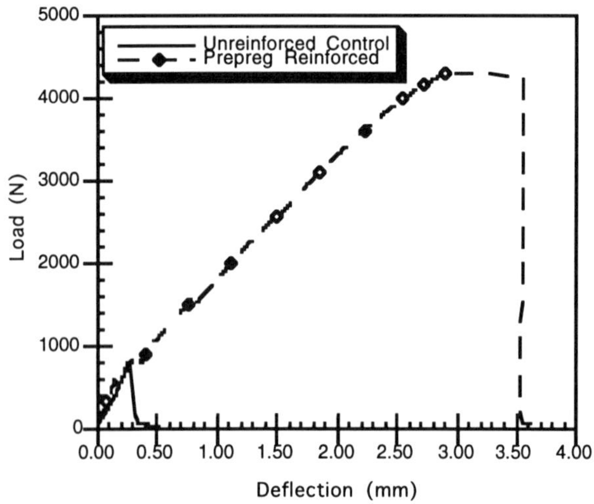

Figure 4: Load deflection curve after 8 weeks exposure to salt water.

Table 1: Comparison of load carrying capacity

Exposure Conditions	Time (weeks)	Max load ratios prepreg/ctrl
Atm	2	6.51
	4	5.73
	8	7.50
Dist Water	2	6.41
	4	8.38
	8	5.44
Salt Water	2	5.40
	4	9.63
	8	5.12

Quantitative analysis of the soaking solutions using the attenuated reflectance FTIR proved inconclusive. However, gravametric analysis on strips of uncured prepreg soaked in water for 1 week consistently showed about a 5% weight loss with a corresponding amount of residue left from the solution after evaporation. Gravametric analysis of the soaking solutions is not viable due to the large amount of salts in solution from the mortar and the salt water. Attempts will be made to use a Total Organic Carbon (TOC) analysis of the solutions to quantify the amount leached from the beams (if any).

Discussion

As can be seen from the data scrap reinforcement shows very positive results and has the potential for use in construction applications. Potentially, it has many of the same advantages as FRP rebar namely an ultimate tensile strength stronger then steel, a low specific gravity (25% that of steel), electromagnetic insulating properties, and high resistance to chemical attack. Potential disadvantages of FRP rebar include the low modulus (compared to steel) and the fact that it does not exhibit yielding or ductility at failure.[13] However, it should be noted that scrap composite reinforcement cannot be designed as traditional concrete and steel structures. Design for use would have to follow guidelines like those developed by the Constructed Facilities Center (CFC) at West Virginia University for concrete reinforced with FRP bars. These include charts tables and design equations for ultimate moment, crack width and deflection analysis.[14] The main challenge facing the acceptance of FRP bars for use is to make them available in high volume and at competitive costs to conventional materials. Bedard identified four categories of applications that showed potential for FRP bars to replace steel bars. They include: 1) reinforced concrete exposed to deicing salts such as parking structures, bridge decks, and at grade road slabs, 2) structures built in or close to sea water such as bridges, canals and roads, 3) applications subject to corrosive agents like waste water treatment plants, 4) Applications where low electric conductivity or electromagnetic neutrality is required like manholes for electric and telephone equipment.[13] Scrap prepreg could present a low cost alternative to steel in many of these applications.

The unique challenge presented in this reuse strategy is to assess the environmental impact. One way would be to model the diffusion of any hazardous constituents out of the prepreg and out of the concrete into the environment. Such a model would have to include or account for several components such as the diffusion of the constituents to the surface of the polymer composite, the partitioning of those components between the polymer surface and the pore water of the concrete, and then the diffusion of those components to the surface of the concrete. The first part would incorporate the diffusion to the polymer surface and partitioning to the pore water. Studies on the leaching behavior of polymer composites are difficult to find, but one can locate information in alternative sources such as moisture absorption studies and the leaching of additives from food packaging. Both describe the leaching process using Fick's second law:

for a plane sheet

$$\frac{\partial C}{\partial t} = D\frac{\partial^2 C}{\partial x}$$

or for solid cylinders

$$\frac{\partial C}{\partial t} = \frac{1}{r}\cdot\frac{\partial}{\partial r}\left(rD\cdot\frac{\partial C}{\partial r}\right)$$

The solutions to these equations and the mass flux across the polymer composite surface is dependent upon the diffusion coefficient, partition coefficient, mass transfer (convective) resistance, possible chemical reactions and their rate, as well as boundary conditions (such as initial concentrations, and the size and mixing characteristics of the solvent). Once the constituents diffuse across the polymer surface their movement would then be controlled by diffusion processes in the concrete. Concrete, with its high pH, acts as an effective medium for solidifying and stabilizing hazardous metals, however, it is not often used in the case of organics. Available data on volatile and semivolatile organics has yielded unclear results on the effectiveness of cement based solidification processes. Application of the EPA's Toxicity Characteristic Leaching Procedure (TCLP) often proves inconclusive in evaluating the leaching potential because low solubility organics quickly saturate the TCLP liquid.[15] Sequential batch extraction tests appear to give a better measure of the total leachability then do single exposure tests. Even so, Bart et al. take a conservative approach to stabilization of organics in cement and recommend the assumption that the all organics will be released from the cured concrete or volatilized during mixing. A program which combones existing concrete and polymer diffusion models as well as extraction tests on the scrap and the reinforced concrete would predict the long term release and the release under worst case scenario.

Conclusions

While there remains much work to be done, the potential to develop a new reuse strategy for prepreg scrap is significant. Recent estimates place the bill for US. highway bridge repair at $50 billion and for Canadian parking structures alone at $4 to 6 billion. Much of the deterioration is attributed to corrosion of reinforcement.[13] Scrap prepreg reinforcement could be one way to help avoid such corrosion problems in the future and find a reuse alternative to a problematic waste. These preliminary results already show its potential, and what remains is to be done is to investigate more thoroughly potential uses, design options, and fully assess the environmental impact.

Acknowledgments

The authors would like to gratefully acknowledge the support of the University of Delaware Center for Composite Materials University/Industry Research Consortium

References

1   McDermott, Joe. "The structure of the advanced composites industry." Advanced Composites: 1993 Bluebook, pp 6-15.

2   Alexander, Judd H. In Defense of Garbage. Praeger, Westport Connecticut, 1993, p. 189

3   Menges, G. "New developments in Chemical Recycling as Sink for Problematical Plastics Waste like from Shredded Cars and Waste from Fiber Reinforced Plastics". Presented at the 1993 CCM Symposium, CCM University of Delaware, Sept 1993

4   Mapletston, Peter. "Chemical recycling may be an option to meet mandated reclaim levels." Modern Plastics, November 1993, pp 58-61.

5   Menges, G. "The recycling of Fibre Composite Plastics" Presented at the 1993 CCM Symposium, CCM University of Delaware, Sept 1993

6   Weber, A. "Aspects on Recycling Plastics Engineering Parts." Developments in the Science and Technology of Composite Materials, Fourth European Conference on Composite Materials. September 1990, pp 15-35.

7   Grahm, David W., David L Shipp, Ralph B. Jutte and M Hapi Cummons. "Recyclability of Glass Reinforced Composites." Advanced Composites Technologies, Proceedings of the 9th Annual ASM/ESD Advanced Composites Conference, Dearborne, Michigan, November, 1993, pp 753-765.

8   Oyanagi, Y. "Crashing and Recycling of FRP waste." International Workshop on Environmentally Compatible Materials and Recycling Technology. Japan, November 1993, pp 119-125.

9   Rogers, Jack K. "Environmental issues are show stoppers at SAE." Modern Plastics, April 1993, pp 69-71.

10   Rolance, Margret E. E.R. Pattie, and L.L. Ghiorse. "The Effects of Environmental Exposure on an E-Glass/Epoxy Prepreg." 22nd International Sampe Technical Conference. November 6-8, 1990, 693-708.

11   Birger, S., A. Moshnov, and S. Kenig. "The effects of thermal and hygrothermal ageing on the failure mechanisms of graphite-fabric epoxy composites

subjected to flexural loading." <u>Composites</u>. Volume 20, Num 4, July 1989, pp 341-348.

12 Koga, Mitsuo, S. Terasaki, M. Emori, and S. Satoh. "Migration of Organic Soluble Subsatnces from Epoxy Resin into Surrounding Water Accompanied by a Chemical Reaction." <u>Journal of Chemical Engineering of Japan</u>. Vol 23, No 1, 1990, pp 103-105.

13 Bedard, Claude, "Composite Reinforcing Bars: Assesing Their Use in Construction." <u>Concrete International</u>. V14 n1 Jan. 1992, pp 55-59.

14 Faza, Salem S. and Hota V.S. GangaRao, "Composite Materials as Concrete Reinforcement in the Next Decade." <u>Plastics Composites Composites for 21st Century Construction</u>., Proceedings of a Session Sponsored by Materials Engineering Divisio of ASCE in conjunction with the ASCE Convention Dallas,Texas Oct 1993. pp 15-22

15 Wiles, Carlton C. and Edwin Barth, "Solidification/Stabilization: Is It Always Appropiate?" <u>Stabilization and Solidification of Hazardous, Radioactive and Mixed Wastes, 2nd Volume</u>, ASTM STP 1123, T.M. Gilliam and C. C. Wiles, Eds. American Society for Testing and Materials, Philadelphia, 1992, pp. 18-32.

Low Temperature Pyrolysis of SMC Scrap

S.K. Soh, D.K. Lee, Q. Cho, Q. Rag
University of Detroit Mercy
Detroit, Michigan

Abstract

It is shown by cost analysis that glass fiber used as reinforcing filler in sheet molding compound (SMC) scrap possesses most of recoverable economic value of the scrap. Therefore recycling/reuse of SMC scraps must concentrate on recovering glass fiber in good condition. A process consisting of steam pyrolysis of SMC scraps followed by acid digestion, separation and cleaning is presented. This proposed process recovers glass fiber and produces calcium chloride dihydrate flakes from calcium carbonate filler. Technical challenges of this process are studied: 1) Thermal embrittlement of glass fiber during pyrolysis, which reduces economic value of recovered glass fiber : The need for reusable glass fiber establishes an operating window of pyrolysis temperature and time. 2) The kinetics of pyrolysis process under nitrogen atmosphere and steam atmosphere is studied. The latter produces pyrolysis ash (PSMC) which cleans easier, but no significant difference in pyrolysis kinetics was observed by gravimetric analysis. The improved cleanabilty enables recovery of glass fiber from partially pyrolyzed PSMC while preventing severe degradation of glass fiber properties by thermal embrittlement.

Introduction

Pyrolysis of SMC scrap has been studied in order to reduce landfill volume with possible energy recovery (1,2,3). The use of PSMC in various applications has also been investigated (4-8). The present paper proposes a recycling process which can recover glass fiber in marketable form. Preliminary economic analysis will be presented to show that the proposed process may be profitable and self-sustaining, in addition to solving waste disposal problem for SMC scrap.

For success of this proposed process, the glass fiber recovered must have acceptable physical properties. However, previous pyrolysis studies employed high pyrolysis temperature (more than 600 C). At these temperatures, the irreversible thermal embrittlement phenomenon may render the recovered glass fiber useless for reinforcing applications. The present study advocates using low pyrolysis temperature (about 400 C) and short exposure time (5-10 minutes) in order to preserve reinforcing properties of recovered glass fiber. On the other hand, acid digestion and cleaning requires sufficient organic phase removal by pyrolysis, which prefers higher temperature and longer exposure time. We have found that conducting pyrolysis under steam atmosphere instead of the conventional nitrogen atmosphere may help overcome this dilemma by improving the cleanability of PSMC. Comparative pyrolysis data with nitrogen atmosphere and steam atmosphere are presented.

Glass Fiber Recovery Process

Fig. 1 shows a simplified process flow sheet of the proposed process: SMC scrap is fed into a steam pyrolysis furnace where steam pyrolysis reaction volatilizes organic resin component. Pyrolysis gas and oil produced supplies thermal energy needed to run the process. Surplus pyrolysis oil may be sold as fuel oil substitute.

The pyrolysis ash (PSMC) is from the furnace is fed into the acid digestion reactor, where the calcium carbonate filler reacts with hydrochloric acid and dissolves into the aqueous phase as calcium chloride. The solid residue that contain glass fiber and PSMC is filtered and further separation and cleaning produces clean glass fiber. The filtrate is concentrated to dihydrate composition by evaporation.

Feeding the hot solution into cold roller produces calcium dihydrate flakes commonly used for deicing applications.

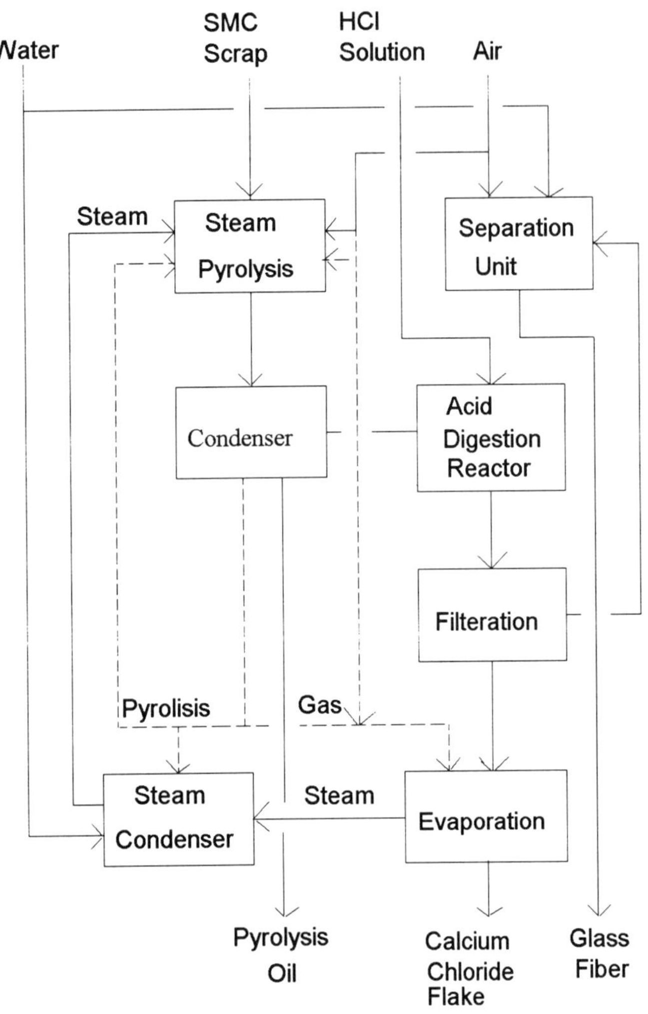

Fig. 1. Proposed Glass Fiber Recovery Process

Economics of Recovery process

Filler (mostly calcium carbonate), glass fiber, and resin are the three major ingredients of typical SMC formulation. Typical material cost of these components are listed in the second column of Table 1. This brings the average material cost of SMC at about 60 cents/lb. As usual for scrap materials, the cost of processing and forming is not recoverable but only a fraction of the original material cost may be. In the fifth column of Table 1, estimated market value of ingredients are listed if they are separated and cleaned. Even if allowance is made for variation of these estimates, it is obvious that glass fiber possesses nearly 90 % of the salvage value of SMC scrap. Therefore recovering glass fiber will be economically far superior to grinding SMC scraps and using them as filler substitute as currently practiced (9).

Table 1. SMC Value Analysis

Comp.	wt. %	cost c/lb	% cost	salvage value (c/lb)	% salvage value
Filler	40	5	3.2	3.0	5.8
Glass	30	100	48.4	60.0	87.0
Organic	30	100	48.4	5.0	7.2
Total	100	62	100.0	20.7	100.0

Table 2. shows a simplified material balance of the proposed recovery process per 1 M/T SMC scrap and their estimated market value. After subtracting the cost of acid used, net margin of $448.70 per ton of SMC is calculated, which should be sufficient to cover capital and operating cost and produce profit.

Table 2. Typical Material Balance

comp.		Quantity (kg)	Unit Price (@ $/kg)	Value ($)
Input	SMC scrap	1,000	0	0
	Hydrochloric acid[1]	1,030	0.07	72.10
Output	Pyrolysis Oil[2]	150	0.20	30.00
	Glass Fiber	300	1.30	390.00
	Calcium Chloride[3]	560	0.18	100.80
	subtotal			520.80

1. 35.5 % aq. solution.
2. The ratio of pyrolysis gas to oil increases with increasing pyrolysis temperature. At 400 C, more than pyrolysis oil yield of 90% or more can be achieved. After satisfying energy need of the process, such as heating pyrolysis reactor and generating process steam, the excess oil may be sold as fuel oil substitute. 50 % of pyrolysis fluid generated is listed here as marketable output.
3. Dihydrate flake

Pyrolysis Condition and Thermal Embrittlement of Glass Fiber

When the pyrolysis temperature is higher than 300 C, the glass fiber in SMC scrap undergoes an irreversible embrittlement process by which it gradually loses mechanical properties. After some trial and error it was found that % SFER, the percentage of specific fracture energy retained after heating is a suitable measure of thermal embrittlement process (10). The fracture energy of virgin and heat treated fibers were measured by the area under the stress-strain curve as sample glass fiber strand is pulled at 0.508 cm/min (0.2 in/min) until all fiber strands are broken (11).

Other measures which may be used to quantify thermal embrittlement include elongation at break and tensile strength from same type of experiments. Reduction of % SFER by thermal embrittlement is approximately the same as the reduction in elongation at break of glass fiber strands. The tensile strength does not deteriorate as much as % SFER or elongation at break.

Glass fiber used had silane type sizing as supplied by manufacturer. Bare glass fibers (or water sized) become so brittle after heating no usable fibers can be recovered. Even though the exact formulation of sizing agents is considered proprietary, most manufactuers employ similar formulations for SMC grade glass fibers.

Fig. 2 shows % SFER data for 38 mm (1.5 inch) glass fibers, 15 μm. fibers from Owens-Corning (yield 222 m/kg) and 12 μm. Certainteed fibers (yield 437 m/kg).

Whether glass strands are imbedded in resin matrix or exposed to air during heat treatment does not seem to produce any significant difference in the % SFER data at 500 C and 550 C. The resin matrix used was Derakane 470-36 from the Dow Chemical Company, which is epoxy novolac based vinyl ester blended with styrene. However, during pyrolysis of thick SMC samples, the endothermic heat of pyrolysis may have cooling effect on glass fiber and it may appear that glass fiber in SMC resist thermal embrittlement better than bare fiber strands. For example, a temperature difference of about 30 C between SMC surface and center was observed for pyrolysis of 3.2 mm (1/8 in.) thick SMC parts at pyrolysis temperature of 400 C.

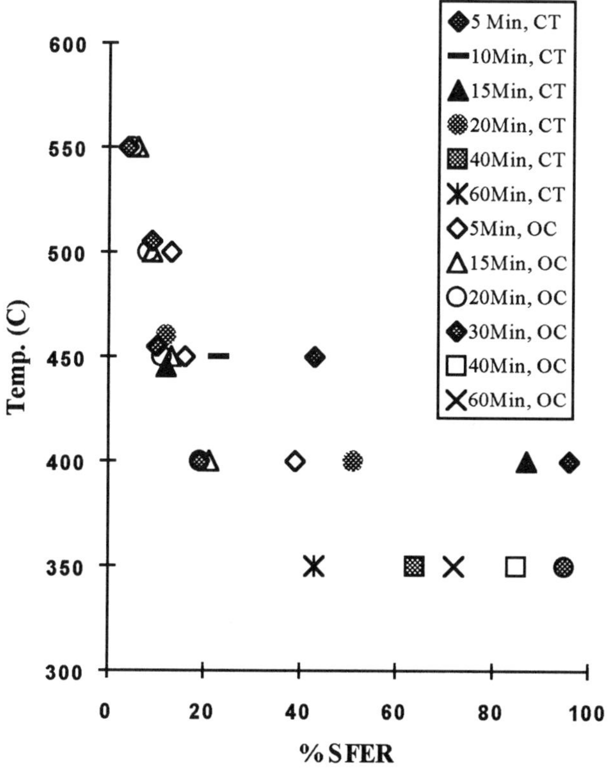

Fig. 2. Thermal Embrittlement of Glass Fiber

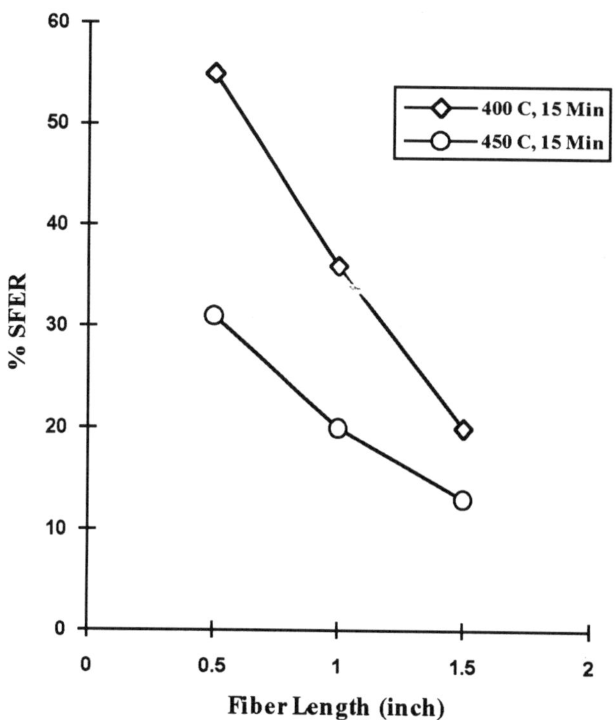

Fig. 3. Effect of Fiber Length on Thermal Embrittlement

In Fig. 2, considerable sample to sample variations are observed, which is understandable as there exist similar variations among virgin samples before heat treatment. Each datum point in the Fig. 2 is an average of at least seven measurements from randomly selected samples. In addition, it was found that glass fiber with prior thermal history (initiation of thermal embrittlement) embrittles faster than virgin fibers. It is presently unknown to the authors how to model or predict the progress of thermal embrittlement under unsteady and non uniform temperature distribution expected during industrial pyrolysis condition.

The effect of fiber length is studied at 400 C and 450 C for Owens-Corning glass fiber samples and the results are shown in Fig. 3. As expected, shorter fibers show higher retention of % SFER. As most of chopped glass fibers used in SMC are about 1 inch long, the data of Fig. 2 based on 1.5 inch samples will be conservative estimates. In order to retain about 80 % SFER at 400 C, which appears to be the lowest temperature to produce cleanable PSMC, the pyrolysis time must be limited to less than about 10 minutes.

Kinetics of Pyrolysis Reaction

The kinetics of SMC pyrolysis are determined by chemical composition and pyrolysis condition. The SMC scraps used in this study was supplied by Rockwell International, and had the approximate composition of 43.2 % organic, mostly unsaturated polyester crosslinked with styrene monomer, 17.0 % E-glass fiber, and 39.8 % inorganic filler, as determined in the authors' laboratory. The rate of pyrolysis at a given temperature depends primarily on composition of the organic phase, especially the ratio of aromatic and aliphatic component. Therefore the kinetics of each batch of SMC scraps must be determined individually. Intrinsic rate that is free from non isothermal and mass transfer effects could be measured with thermogravimeter (TG), with heated oven or with reactor setup (Fig. 4) as long as sample thickness is less than 2 mm, as determined by preliminary tests. The apparatus depicted in Fig. 4 was used for pyrolysis studies under controlled atmosphre and for determining nonisothermal effects.

Pyrolysis experiments in a heated oven with sample thickness of 3.175 mm (1/8 inch) and 6.35 mm (1/4 inch) under nitrogen atmosphere in an oven indicated that the kinetics of 1/8 inch samples do not differ from intrinsic rate data from TGA, while data from 1/4 inch thick samples show substantial deviation. this seems reasonable considering calculated Biot numbers for the thicknesses. (12)

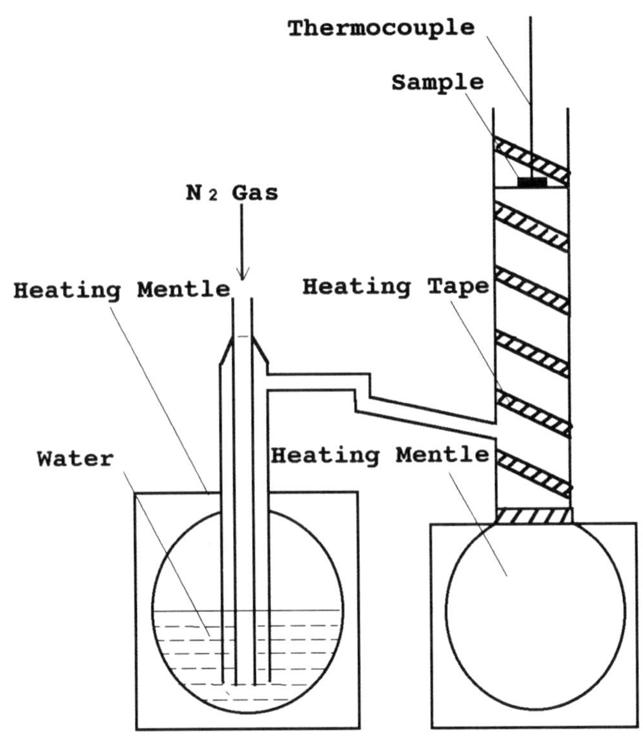

Fig. 4. Reactor Setup for Steam Pyrolysis

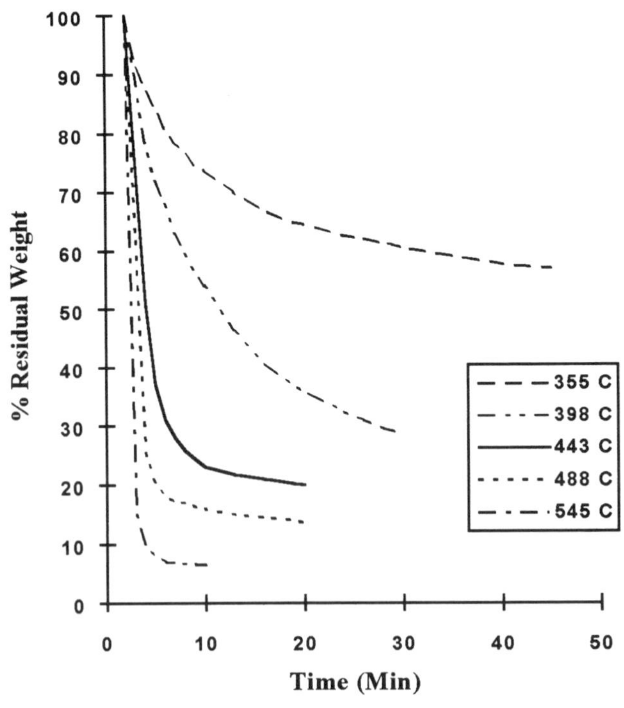

Fig. 5. Thermogravemetric Analysis of SMC

Fig. 5 shows a typical TGA weight loss curves which show the progress of pyrolysis reaction at different temperatures. % residual weight is calculated with respect to pyrolyzable organic weight.

To investigate the reaction order of pyrolysis reaction, natural logarithm of reciprocal fractional residual weight, - ln (1-x), is plotted against time as shown in Fig. 6. First order kinetics would produce straight line in such a plot. Fig. 6 seem to fit first order kinetics in the earlier stage of pyrolysis, but the reaction order appear to change after pyrolysis has progressed further. Similar variable order kinetics are observed for crosslinked polymethyl siloxane (13), melamine-formaldehyde and urea--formaldehyde resins (14).

The initial rate constants obtained from Fig. 6 fits Arhenius equation very well. The determined Arhenius equation constants are:

$$k = 2.5 \times 10^6 \exp(-11,400/T) \qquad [1/min]$$

where T is in K.

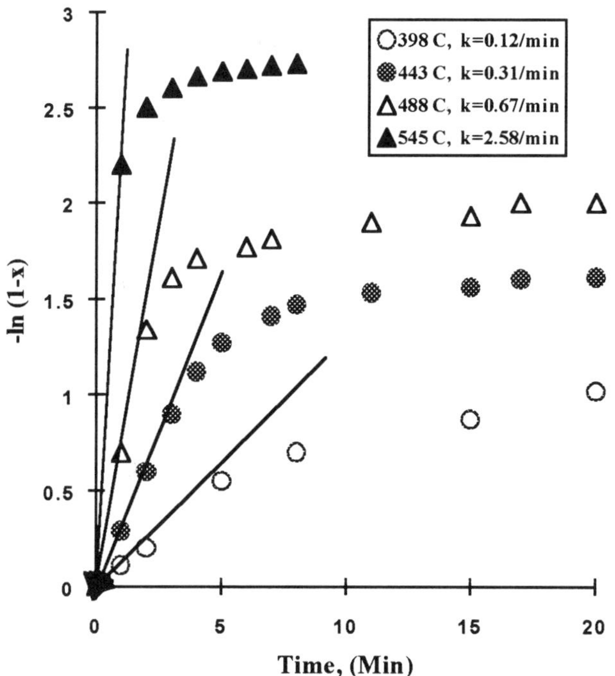

Fig. 6. Determination of Pyrolysis Rate Constants

Carbon residue in PSMC contains fixed carbon and partially pyrolyzed or unpyrolyzed resin. For a given resin, carbon residue content decreases with increasing pyrolysis temperature. The effect of resin composition may be estimated by char forming tendency (CFT) calculations using group contribution method (15). For example, maleic anhydride-polypropylene glycol-styrene type polyester has close to zero CFT value, while replacing some maleic anhydride with phthalic anhydride may increase CFT value to about 5%, and vinyl ester based polyesters may typically have near 10% CFT value. Any exposure to oxygen partial pressure may reduce carbon residue considerably, so that air leak of pyrolysis chamber may distort carbon residue data considerably.

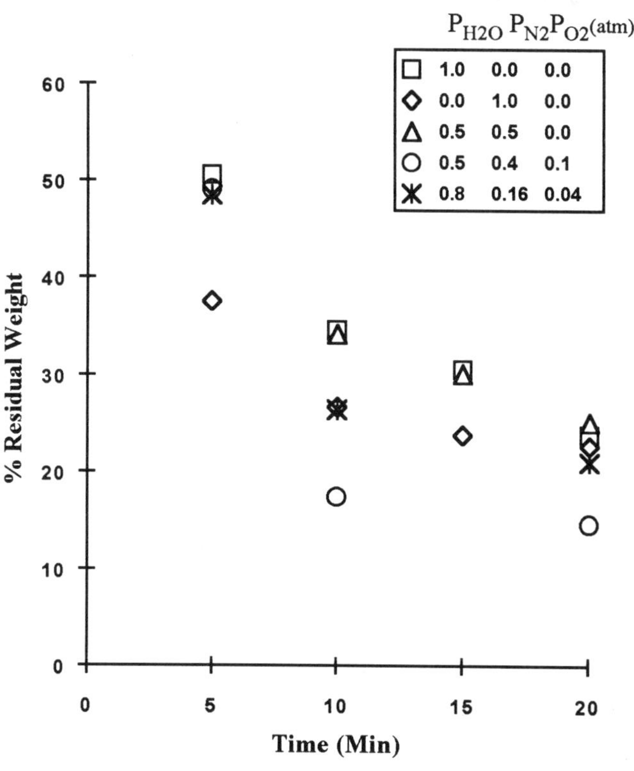

Fig. 7. Effect of Pyrolysis Atmosphere

No significant differences in pyrolysis rate are found when the tests were conducted under helium atmosphere, nitrogen atmosphere, and steam atmosphere. Fig. 7 shows comparison among different atmospheres. As pyrolysis reaction is endothermic and the temperature of thick SMC parts is lowered considerably, controlled introduction of small amount of oxygen during pyrolysis may help maintain more uniform temperature distribution and produce more uniformly pyrolyzed product. Oxidation also reduces char formation. In addition, preliminary experiments indicate that pyrolysis under steam atmosphere may produce SMC pyrolysis residue with looser structures, which is easier for digesting acid to penetrate into, dissolving the filler component, calcium chloride. As this inorganic filler occupy up to 50% of SMC weight, dissolution of calcium chloride results in complete disintegration of SMC structure and freeing of glass fibers from the matrix. More investigation into these matters may help improve pyrolysis process for glass fiber recovery.

Conclusions

Based on the presented results, it may be concluded that with further improvements in pyrolysis conditions, possibly by introducing controlled side reactions such as steam hydrolysis and partial oxidation, it may become possible to recover clean glass fiber with acceptable level of thermal embrittlement. Such fibers would be readily marketable at profitable price. Such development will provide a closed-loop physical/chemical recycling which is economically self-sustaining and without generating landfill or pollution.

Acknowledgments

The authors would like to greatly acknowledge the support of this research by the U.S. Environmental Protection Agency, assistance I.D. #CR820478-01-0 and Michigan Materials Processing Institute.

The authors would also like to acknowledge donating material, samples, and providing valuable discussions by Dr. W. D. Graham of Owens-Corning Company, Mr. Bruce Greve of the Budd Company, Mr. Daniel Dowdall of Rockwell International and Mr. Charles Biliu of Ticom Corporation, a subsidiary of Northrop Corporation.

References

1. Irvin Poston, "Can Pyrolysis Be Used to Recycle SMC?", SPI 45th Annual Conference (1990).

2. Donald R. Norris, "Development of the Pyrolysis Process for Recycling of SMC", SPI 45th Annual Conference (1990).

3. SMCAA, "Pyrolysis as a Means of Recycling Thermoset Composites", SPI 46th Annual Conference (1991).

4. Kurt Butler, "Recycling of Molded SMC and BMC Materials", SPI 46th Annual Conference (1991).

5. D. F. Watt et al., "On the Use of PSMC as a Concrete Additive", ASM/ESD Advanced Composite Conference proceedings, vol. 7, p615 (1991).

6. P. P. Hudec, "Reduction of Alkali Reactivity and De-icing Salt Damage in Concrete by Addition of PSMC", ibid., vol. 7, p611 (1991).

7. R. B. Jutte and .W D. Graham, "Recycling of SMC", SPI 46th Annual Conference (1991).

8. R. B. Jutte and W. D. Graham, "Recycling of SMC Scrap as a reinforcement", Plastics Engineering, vol. 47, p13 (1991).

9. Roger Rewand, "GM Using Recycled SMC in '93 Corvette", Plastics News, p3, October 12 (1992).

10. W. Zurek et al., "Mechanical Properties of Multifilaments in Different Measurement Conditions", J. Applied Polymer Science, vol. 43, p1343 (1991).

11. ASTM D-2101.

12. J. P. Holman, "Heat Transfer", McGraw Hill (1981).

13. D. Li and S. Hwang, "Pyrolysis Kinetics of Highly Crosslinked Polymethylsiloxane by TGA", J. Applied Polymer Science, vol. 44, p1979 (1992).

14. T. Hirata et al., "Pyrolysis of Melamine-Formaldehyde and Urea-Formaldehyde Resins", ibid., vol. 42, p3147 (1991).

15. D. W. V. Krevelen, "Some Basic Aspects of Flame resistance of Polymeric Materials", Polymer, vol. 16, p615 (1975).

Proceedings of the 10th Annual ASM/ESD Advanced Composites Conference, Dearborn, Michigan, USA, 7-10 November 1994

Thermoplastic Stampable Sheet with Post Consumer Recycle Resin

G.P. Weeks, A.E. Rabe, J.G. Lertola, J.M. Fisher, G.W. Fenrick, N.J. Croft, Sr., E.C. Roman
DuPont Company
Wilmington, Delaware

Abstract

The multiple ways in which thermoplastics can be recycled make them increasingly valuable in an environmentally intensive world. DuPont XTC™ Thermoplastic Composite Sheet is an excellent illustration of the flexible use and re-use of thermoplastics. XTC™ 025R, a glass reinforced thermoplastic stampable sheet, is produced using post- consumer-recycle resin. The flow compression molding of XTC™ yields excellent mechanical properties and is inherently low waste. Molded parts from XTC™ can be recycled by either regrind/remelt, or through methanolysis back to the original monomers. XTC™ can be combined with other thermoplastic polyester products such as DuPont polyester clusters and upholstery fabrics to create fully recyclable sub-assemblies, for example automotive seating.

IN 1992 DUPONT introduced an new compression molded glass mat thermoplastic sheet product to the automotive market designated DuPont XTC™. (1, 2) Aimed at Class-A as well as non-Class-A panel applications, this new material system offers excellent surface quality, high temperature stability, reduced density, reduced emissions, and fast mold cycle time. Because XTC™ meets the tough cost and property hurdles of the automotive industry, interest in this flexible new thermoplastic material offering has continued to increase.

A further reason for this success is that XTC™ is based on the most recycled resin available in the North American plastics market: thermoplastic polyester. Studies show that thermoplastics show a favorable life time energy balance over metals and require less energy to recycle. (3, 4) Taking advantage of this, XTC™ with consistent properties can be fabricated from post-consumer resin sources, and can itself be recycled for continued use into primary and secondary end uses, or the resin returned to the original monomer state.

In this paper we review XTC™-025R technology emphasizing its fabrication from post-consumer resin sources, its usefulness for compression molding a wide variety of automotive parts, and show technical data supporting the viability of post-use recycling via regrind and chemical monomer recovery routes.

XTC™ Thermoplatic Composite Sheet

XTC™ polyester thermoplastic composite is sold in mill rolls of a flexible, porous sheet with a basis weight of 450 grams per square meter (Figure 1). The XTC™-025R sheet consists of 25% by weight glass fiber reinforcement and 75% by weight polyester matrix resin. The internal structure of the XTC™ preform mat consists of discrete glass filaments dispersed primarily in the plane of the mat, each coated with tiny globules of thermoplastic resin (Figure 2). DuPont XTC™ does not require special storage facilities or predrying as is often the case with polyester injection molding pellets. This is made possible by the unique, patented sheet morphology, and the use of a forced air preheat oven system prior to molding.

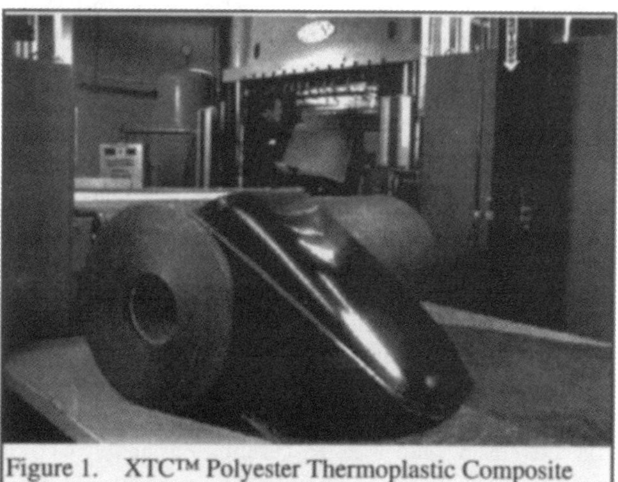

Figure 1. XTC™ Polyester Thermoplastic Composite

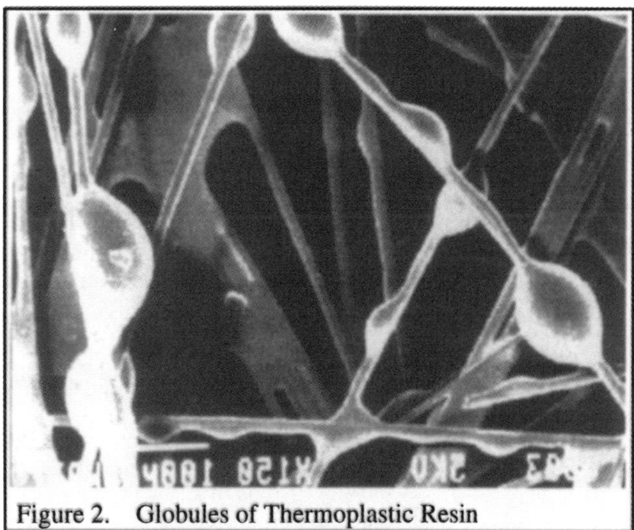
Figure 2.   Globules of Thermoplastic Resin

Properties Using Post-Consumer Polyester Raw Materials

XTC™-025R sheet is created using a process in which chopped glass fibers and resin fiber are co-slurried, formed into sheet, and continuously melt impregnated. The original XTC™ formulation was based on virgin polyester resin. However, the process is flexible and consistent post-consumer sources of polyester resin are now used routinely (Figure 3). Because the resin for XTC™ is first extruded as fiber, the quality systems inherent in the preparation for resin fiber production also help insure the quality of the XTC™-025R composite sheet. Research over the last two years has shown that post-consumer recycled polyester resin results in XTC™-025R parts with very consistent properties.

Figure 4 shows the physical and mechanical properties of XTC™-025R parts. Polyester has a high melting point and the resulting heat distortion temperature (251 C) is much higher than polypropylene and higher than automotive-grade SMC. As a result XTC™-025R is a viable material choice for panels which will be painted and baked. The impact strength, tensile, and stiffness are all well within the range of automotive panel applications.

Figures 5 compares the flexural modulus properties of molded composites made using XTC sheet processing and molding technology. Each item is based on 25% glass fiber reinforcement and 75% polyester matrix resin comparing three resin sources: (1) virgin polyester having Mn of 10,300 and Mw of 40,500, (2) recycle bottle resin having Mn of 11,700 and Mw of 39,100, and (3) recycled post-consumer film resin having an Mn of 9810 and Mw of 38,500. In each case the polyester resin has incorporated in it an antioxidant to enhance performance. These properties are indicative of viability of post-consumer polyester raw materials sourcing and the robust character of the resulting XTC™ mechanical and physical properties. The higher DEG content of the post-consumer resin sources makes a useful contribution to com-

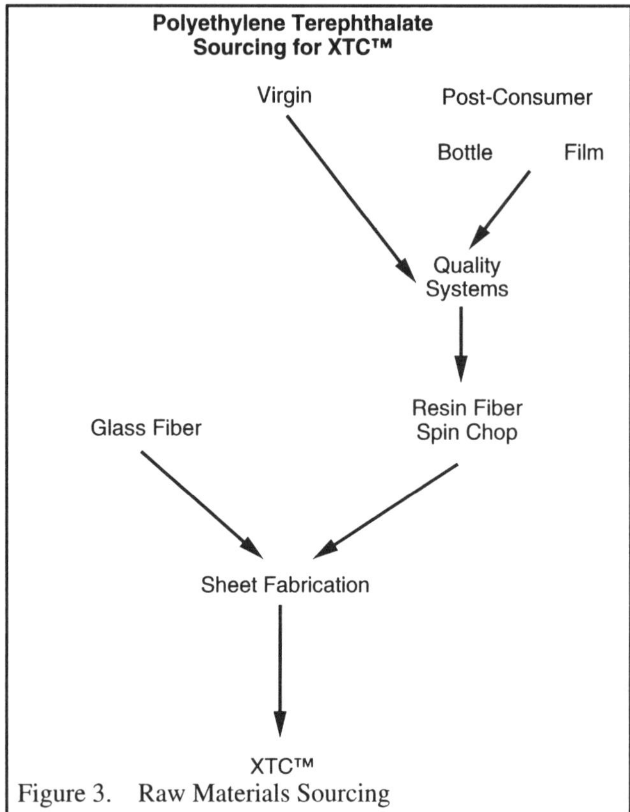

Figure 3.   Raw Materials Sourcing

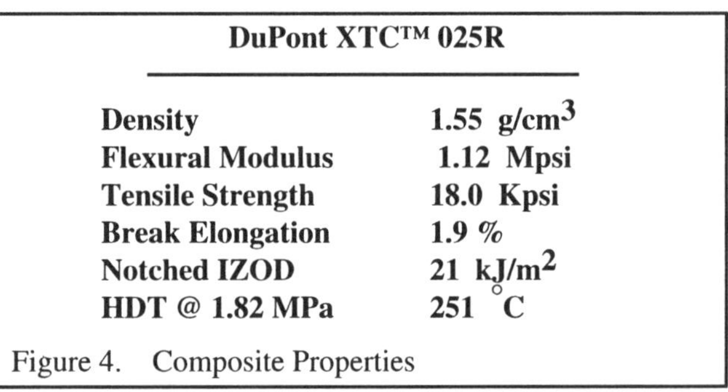

DuPont XTC™ 025R

Density	1.55 g/cm^3
Flexural Modulus	1.12 Mpsi
Tensile Strength	18.0 Kpsi
Break Elongation	1.9 %
Notched IZOD	21 kJ/m^2
HDT @ 1.82 MPa	251 °C

Figure 4.   Composite Properties

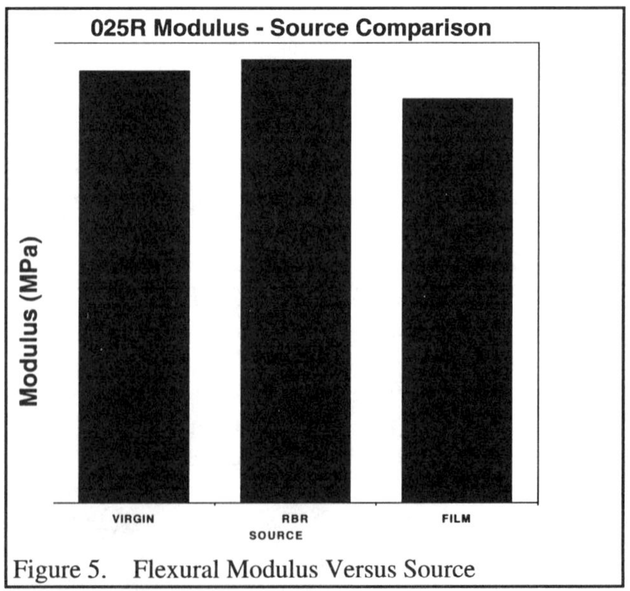

Figure 5.   Flexural Modulus Versus Source

posite toughness as seen in the relative improvement in notched and un-notched Izod properties as shown in Figure 6.

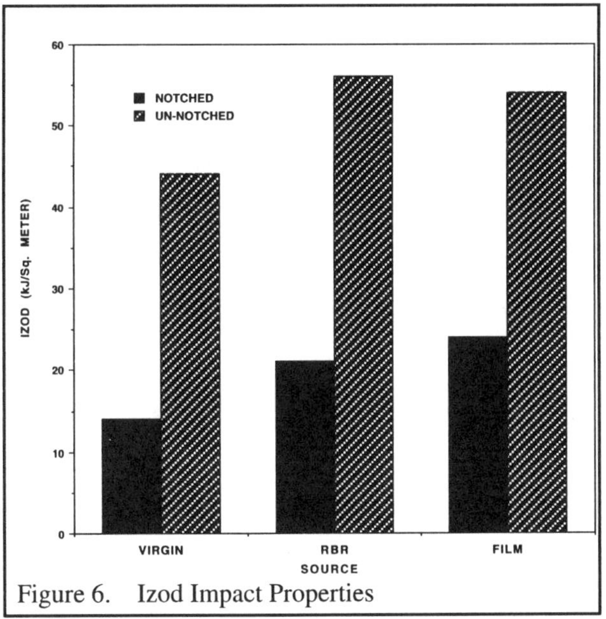

Figure 6.   Izod Impact Properties

Figure 7.   Automated Molding Scheme

XTC™ Molding Process

DuPont XTC™ compositions are specifically designed for rapid cycle compression molding (Figure 7). To mold XTC™, mill rolls are first slit and sheeted. A stack of rectangular sheets comprises a preform which can be heated rapidly, then manually or robotically placed in a matched compression die tool, and flow molded under pressure to form a complex part. The number of preform layers is selected according to the desired weight of the final part. For example, 18 layers might be used to produce a part 2 mm thick with a cavity coverage of about 75%.

Just prior to molding, the charge is heated above the melt point of the resin using forced air convection heating. The molten charge is then transferred to the compression mold. Stacks of preform sheets are generally molded at between 1000 and 2000 psi for best results, depending on the characteristics of a particular part. A high speed, programmable press with close parallelism is recommended for most applications. The molded part is removed after a typical cycle time of about 60 seconds. For class A surface panels, an in-mold coating process is used. Cooling fixtures are used to retain precise dimensions during post-mold cool down.

XTC™ Flow Molded Parts

Since its introduction in 1992 DuPont engineers have successfully demonstrated XTC™ in a wide variety of useful applications and are continuing to expand the horizons of extremely flexible product offering. The following examples are offered to highlight some of the unique features of the XTC™ technology.

Figure 8 shows a GM Impact electric car roof panel molded from XTC™. Electric vehicles are being touted as one solution to the environmental crisis in large urban areas. This application requires stiff, lightweight, recyclable panel components which can be produced cost efficiently in low and moderate volumes. Other requirement include good dimensional stability during painting and baking and, of course, excellent painted surface finish. Molded in XTC™-025R, the part is approximately 1.14 x 1.15 meters with an average thickness of 3.28 mm. It was molded using a preform with 90% tooling coverage and 15 preform layers. Dimensional tolerances in the plane of the part of +- 2.4 mm were demonstrated.

Other exterior panel applications for XTC™ including vertical panels such as a door panel, and horizontal hood panels, have also been demonstrated.

Figure 8. Electric Vehicle Exterior Roof Panel

Demonstrating the ability to flow mold complex parts from XTC™ preforms, Figure 9 shows a fuel tank heat shield molded from XTC™. This part, intended to protect a polyethylene gas tank from heat and road wear, required 12 inches of vertical wall mold flow with a 2.5 mm average wall thickness. Such parts have been molded using robotic charge placement in a continuous molding cycle of less than 80 seconds with a tool coverage of only 60%. The part weighs 15% less than the equivalent SMC part and passes key tests for the end use.

Combining the new structural composite forms of reinforced thermoplastic polyester with other polyester-based fiber materials can lead to completely recyclable sub-assembles (5). As an example, a complete car seat can now be manufactured of DuPont thermoplastic polyester. DuPont XTC™ and Rynite™ can be used for seat backs and frames, Dacron™ fiber for upholstery, and polyester fiber clusters for cushions. This allows the complete seat to be unbolted and chemically recycled without the need to sort polymers.

Figure 9. Fuel Tank Heat Shield

Recycle of XTC™ Parts

Figure 10 shows a hierarchy of recycling routes generally discussed in literature on recyclable materials. Condensation polymers such as thermoplastic polyester have a wide range of recycle options including direct reuse, re-use after modification, chemical monomer recovery, thermal cracking, or use as a filler. Thermosets have fewer options, generally thermal cracking or use as filler.

For XTC™ we have explored the technical feasibility of three options (Figure 11) at the high end of the recycle hierarchy: (1) regrind as a recycled feedstock for injection molded polyester resin products, (2) regrind and recycle back into XTC™ sheet as a partial ingredient, (3) chemical monomer recovery.

Recycling Hierarchy

	Condensation Polymers (Polyesters, Nylons)	Polyolefins (HDPE, PP, TPO)	Polystyrene PVC ABS	Acrylics	Thermosets	Cured Rubbers
Direct Re-Use	×	×				
Re-Use After Modification	×	×	×	×		
Chemical Monomer Recovery	×					
Thermo Monomer Recovery				×		
Thermal Cracking	×	×	×	×	×	×
Use as Filler	×	×	×	×	×	×

Figure 10. Recycling Hierarchy

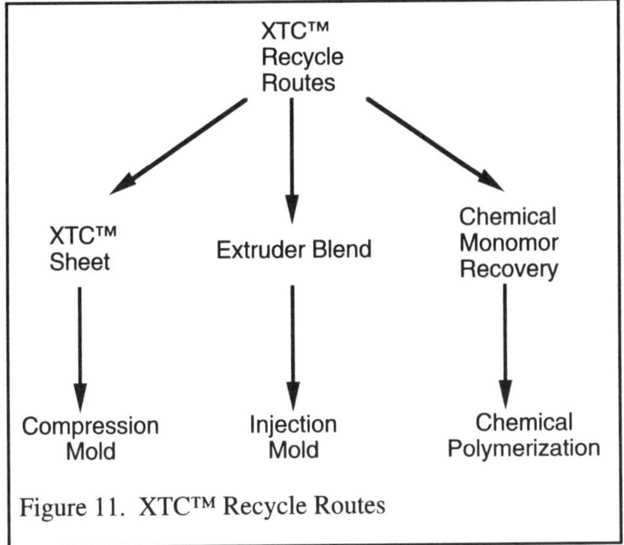

Figure 11. XTC™ Recycle Routes

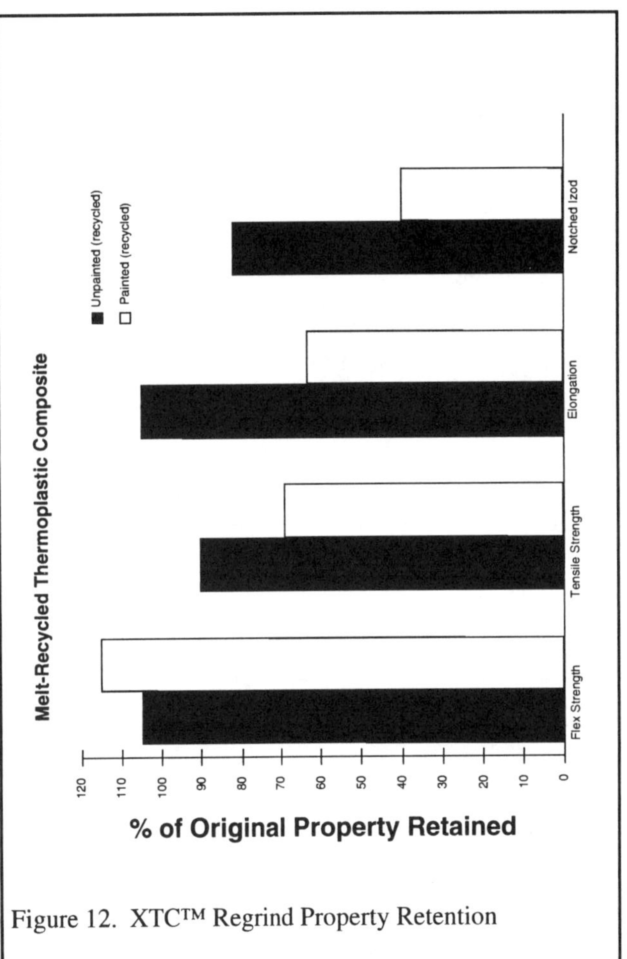

Figure 12. XTC™ Regrind Property Retention

Regrind/Remelt Recycling

Molders currently regrind millions of pounds of scrap parts made from injection molded thermoplastic, along with mold runners and sprues, and reprocess them for the same application. Similar options are available for XTC™. Several different types of commercial grinders can be used to reduce XTC™ parts to feedstock material. Such materials can be combined with other thermoplastic polyester resin materials, extruder blended, and repelletized or injection molded. Since the resin component of XTC™ is nearly 100% polyester thermoplastic, it makes an ideal source of resin for these products.

One issue is the retention of properties when recycling painted parts. Removing the paint chemically or via abrasive techniques prior to regrinding is one option. However, this adds cost. Alternatively, one can leave the paint in the blend to act as a filler in the recycled component. Figure 12 shows the retention of properties for 100% XTC™ regrind injection molded test bars, comparing unpainted XTC™, and XTC™ painted with a standard automotive paint system. Good retention of flex modulus is indicated, while tensile and impact properties deteriorate somewhat for reground painted parts indicating some reactive degradation due to paint content.

Alternatively, we have also demonstrated that it is feasible to regrind XTC™ parts in such a way that they can be reslurried in the XTC™ sheet manufacturing operation as a partial constituent. In this way reground materials can be returned to the original source and re-used.

Methanolysis of XTC™

DuPont has developed propriety chemical polymer recovery technologies for polyester based on methanolysis and is expanding its polyester recycling capabilities. Methanolysis depolymerizes polyester to dimethyl terephthalate and ethylene glycol monomers which can be separated from glass and other contaminants as vapors, and then further purified. The recovered monomers retain their original properties and can be repolymerized into the highest quality thermoplastic with no cost penalty. These new polymers can be returned to any first quality application. Current investment is in place for chemical monomer recovery from relatively pure feed streams, and new technology has been developed specifically for recovery of polyester from glass reinforced materials such as XTC™. Small prototype methanolysis reactor units charged with XTC™ regrind solids have demonstrated efficient recovery to 100% of theoretically calculated yields.

XTC™ Life Cycle

DuPont XTC™ is an good example of how a new material offering can be designed from its inception with its complete life cycle in mind. Figure 13 shows the complete value added chain for XTC™. Post-consumer bottle and film and virgin sources of polyester thermoplastic resin are combined with a simple additive package and glass fiber to create the XTC™ preform mat. Edge trim from mat fabrication can be recycled back into the XTC™ sheet, or can be injection molded. Similarly, molded part waste can be recycled. Parts molded from XTC™ work their way from the automotive OEM, to the consumer, to the automotive post-consumer recycling infra-structure. These post-consumer thermoplastic parts can be recycled via remelt into injection molding or via grinding into compression molded sheet applications. Finally, methanolysis can be used to return the polymer to its original monomers. These monomers can be used to again create virgin polyester for any application.

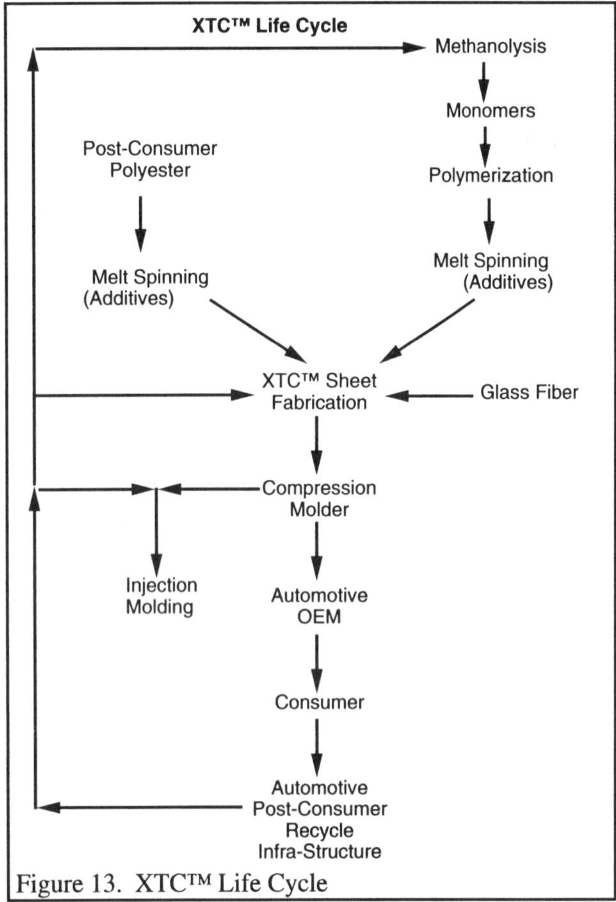

Figure 13. XTC™ Life Cycle

Conclusions

DuPont XTC™ is an excellent illustration of a product designed to take advantage of the flexible use and re-use of thermoplastics. XTC™ 025R, a glass reinforced thermoplastic stampable sheet, is produced using post-consumer-recycle resin. The flow compression molding of XTC™ 025R yields excellent mechanical properties and is inherently low waste. Molded parts from XTC™ can be recycled by either regrind/remelt, or through methanolysis back to the original monomers with excellent conversion efficiency. XTC™ can be combined with other thermoplastic polyester products such as DuPont polyester clusters and upholstery fabrics to create fully recyclable sub-assemblies, such as automotive seating.

References

1. SAE Detroit 1992 - Publication 920377, "Thermoplastic Compression Molded Horizontal Automotive Panels," M.D. Drigotas, G.P. Weeks, J.M. Fisher.
2. Proceedings of the 8th Advanced Composites Conference, Chicago, Illinois, USA, 2-5 November 1992, "Advances in Thermoplastic Compression Molding Material and Technology," G.P. Weeks and A.H. White.
3. DuPont Technology Brochure H48048.
4. Plastics World/October 1992, "Life-cycle Savings Make Strong Case for Plastics," J.R. Dieffenbach.
5. Automotive and Transportation Interiors/March 1994, "Recycling Pressures Affect Seating Materials," R. Eller.

Comparing C-Shaped and I-Type Bumpers Molded in Thermoplastic Composites

C.L. Clark, P. Bejin
GE Plastics
Southfield, Michigan

Abstract

The C-shaped bumper design is regarded as a reasonable geometry for front bumper impacts, especially when used with glass-filled, sheet-stampable, thermoplastic composite materials. C-section bumpers are well proven and accepted in the automotive industry, performing well in a variety of crashes. A new, more complicated I-type cross-section bumper (with multiple ribbing) has recently been proposed for glass-filled thermoplastic composites. These bumper cross-sections can be useful in certain specialized cases; however, they may not perform adequately in all crash situations. In addition, it is important to consider the manufacturing limitations and the realities of material performance in the complex I-type bumper design. This paper presents data comparing the use of ribbed bumper designs and the traditional C-section beam. Analysis of this data, along with consideration of the processing difficulties inherent with the I-type design, suggest that the traditional C-section bumper should remain the logical choice for mass-production vehicle use.

THE FRONT AND REAR OF MANY VEHICLES produced in the 1920s and 1930s incorporated leaf-spring type bumpers. In a collision, deformation of the spring absorbed impact energy, reducing the stress that was transferred to the car and its occupants.

The primary function of automotive bumpers remained impact attenuation until after World War II. In the 1950s and 1960s, the emphasis changed from impact attenuation to styling enhancement.

Recently, bumpers have again become more functional. Federal law requires that the sheet metal and safety-related components of a vehicle remain undamaged after a 1.1-m/sec (2.5-mi/hr) impact. Most American passenger vehicles are designed to withstand a higher 2.2-m/sec (5-mi/hr) collision without damage to safety-related components or sheet metal. This performance is achieved through bumper engineering.

One primary requirement of a bumper reinforcement beam is stiffness (spring rate). Stiffness is important because vehicle design considerations limit the space for the bumper to deform under load.

A second requirement of a bumper reinforcement beam is energy absorption. Energy absorption is important because the bumper must limit the amount of impact force that it transmits to the mounting rails, which are attached to the vehicle frame.

To determine the stiffness and energy absorption of a bumper, a graph can be used to plot the load imposed on the bumper versus the bumper's deformation or deflection in response to that load. The stiffness of the bumper is the slope of this load/deflection curve; the total energy absorbed by the bumper is the area under the curve [1,3].

Most automotive bumpers today are formed in a C-shaped cross-section. This design has evolved over many years of engineering effort and is relatively simple to manufacture. These bumpers have performed satisfactorily in a variety of real-world collisions.

Overview

A previous technical paper [2] proposed that the I-type bumper fabricated of thermoplastic composite materials offered weight-saving advantages over the C-section bumper. This paper addresses issues raised by two statements in that document:

- "If loads are not central to the section, however, capacity is substantially reduced." The loss of capacity that might be expected under these non-ideal impacts is investigated in this paper.
- "The I-section drawback of eccentric load limitations were (sic) addressed by the addition of a rib pattern."

The effects of adding ribs to enhance structural performance is also discussed.

It is the contention of this paper that the C-section beam offers superior all-around performance in real-life impact situations. The C-section beam also offers processing advantages that ensure the as-molded performance of the beam is much closer to theoretical calculations, particularly when using thermoplastic composite materials.

Two main sections make up this paper. The first section (Defining Bumper Performance) consists of a general discussion of the performance of the geometry of C-section and I-type bumper beams in three different impact events without consideration for a particular material's properties. The material properties used for the computer-aided structural analyses in this section were those of a homogeneous, isotropic material. Even though the specific results shown are dependent on material properties, the ranking of the results should remain the same unless highly nonisotropic materials with nontypical Poisson's ratios are used in the analysis.

The second section (Rib Effectiveness Study) investigates the effects of these beam designs relative to the known performance of commercially available grades of sheet-stampable thermoplastics. The practical aspects of material and processing techniques are factored into the discussion of the comparative performance of the two beam designs.

Defining Bumper Performance

A study of comparable beams was conducted to compare the performance of the two bumper geometries. Beams of the same weight, rail span, and package space (height, length, and depth) were defined and impact-location considerations were specified. Both beams were designed to provide overall performance while still passing Federal Motor Vehicle Safety Standards (FMVSS). Neither beam was designed solely to pass a specific FMVSS test. In addition, neither beam was optimized for any particular impact (e.g., through local thinning or thickening of wall sections, changes in rib pattern density, or changes in gross geometry), and both beams were designed to be as similar as possible.

After the equivalent beams were defined, performance of the two geometries could be evaluated using either computer-based, finite-element analysis (FEA) or actual part testing. Currently, no commercial vehicles use the I-type bumper design. Therefore, this study was restricted to the use of computer analyses and results obtained from molding studies using four types of thermoplastic composite materials.

DETERMINING BUMPER GEOMETRY. The evaluation of impact location on the C-section and I-type bumpers began by establishing dimensions for the two beams. Both bumpers weighed 3.3 kg (7.2 lb). Wall thickness for the C-section beam was nominally 6 mm (0.24 in.). Overall dimensions for the beams were 152 mm (6 in.) tall and 76 mm (3 in.) deep. The C-section beam, from rail center to rail center, was 1,016 mm

(40 in.) long, with an additional 178 mm (7 in.) overhang beyond the rail centers. Beam flanges were out-turned 25 mm (1 in.).

For an I-type bumper that fit into the same package space, wall thicknesses varied from 3 mm (0.1 in.) for the ribs to 5 mm (0.2 in.) for the webs to 4 mm (0.16 in.) for the flanges. Figure 1A-C indicates the cross-sections and dimensions of the two beams.

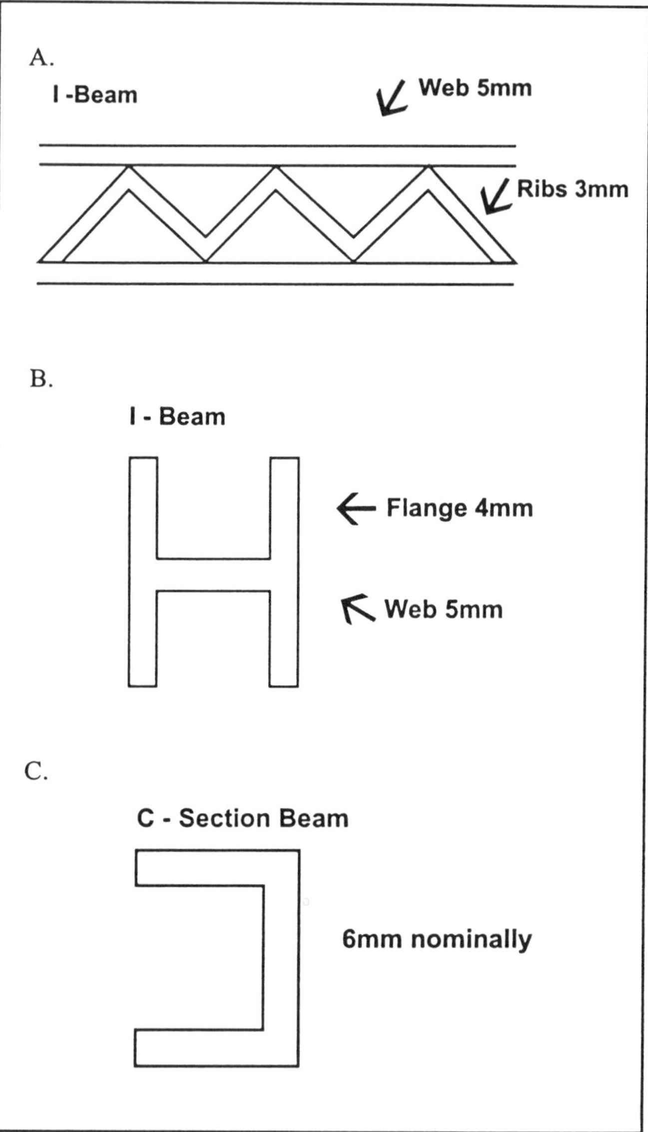

Figure 1: C-Section and I-Beam Cross-Sections and Geometry

DEFINING BUMPER IMPACT LOCATIONS. The next step in the study was to select impact locations for which performance of the two beams could be evaluated:

- Center-Center Impact - At the horizontal centerline of the bumper and the center of the span;
- High-Center Impact - At the center of the span, 76 mm (3 in.) above the center of the horizontal centerline;

- <u>High-Outboard Impact</u> - At 300 mm (11.8 in.) from the center of the span and 76 mm (3 in.) above the horizontal centerline.

Figure 2B indicates a center-center hit, Figure 2C a high-center hit, and Figure 2D a high-outboard hit.

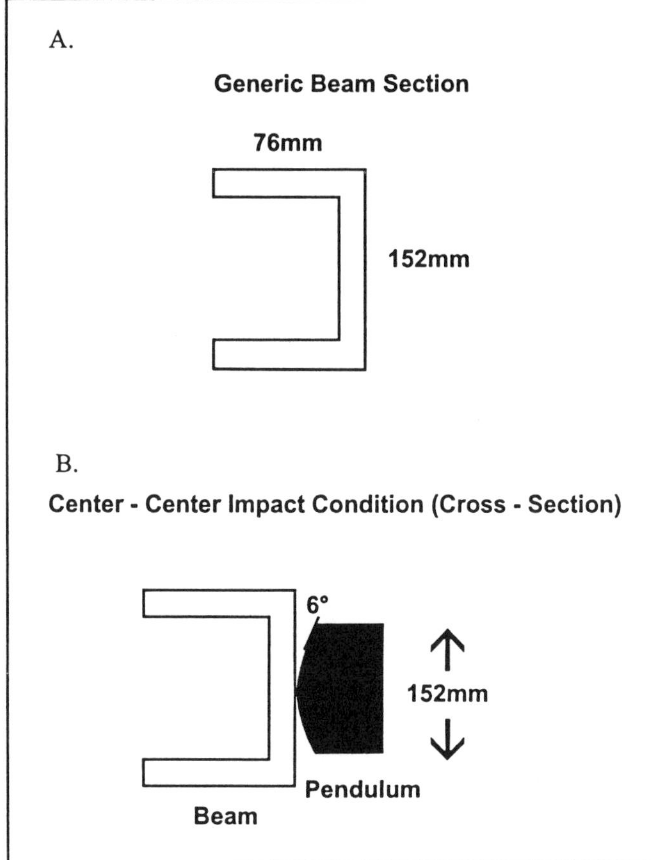

Figure 2: Impact-Location Considerations

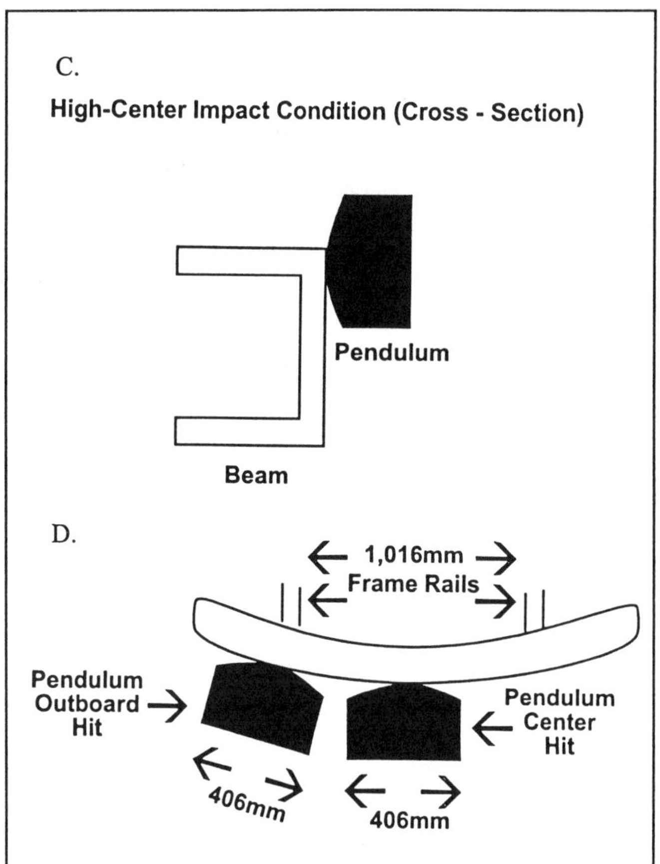

Figure 2: Impact-Location Considerations (cont'd)

ANALYZING BUMPER PERFORMANCE. Analysts next created computer models of the two bumpers. Rectangular thin-shell elements were used in meshing the models because of the relatively thin walls of these parts in proportion to their overall lengths and thicknesses. Due to its internal ribbed structure, the I-type geometry was about twice as element-intensive as the C-section beam.

Since the I-type beam defined in Reference 2 was symmetrical, only half of the beam was meshed and analyzed for the center-center and high-center analyses, which reduced model complexity and CPU time. The half-section that was used extended from one outboard end of the bumper to the beam's vertical centerline. Since the high-outboard impact was not symmetrical around the bumper's frame rails, a whole I-type beam was modelled for that analysis. With the C-section bumper, the entire beam was modelled because it was simple to mesh and contained about the same number of elements as the I-type half-model.

The models were then imported into the ADINA (R) structural analysis package [3]. This software contains both a dynamic and a quasistatic code. Prior evaluations of bumper impacts had revealed a resemblance to quasistatic events [4] because the time scale of the impact event is long in comparison to the time it takes a stress wave to propagate through the structure [4]. Since stress-wave propagation within the part does not make significant contributions to the load that is causing failure — rather, gross deformation does — the problem can be evaluated using a quasistatic code. Non-linear dynamic analyses can be CPU-intensive and costly to run.

A pendulum impact was also modelled (based on FMVSS 215 bumper test setup). The pendulum measured 114 mm (4.5 in.) high and 610 mm (24 in.) long — 406 mm (16 in.) of that was a flat region and 102 mm (4.08 in) was a radiused corner on each end. The pendulum was assumed to be infinitely rigid. To introduce displacement, the applied displacement technique was used in the analysis. This technique begins by displacing nodes on the pendulum, pushing it into the finite-element model of the bumper. (A more detailed explanation is given in References [4] and [18].) As the pendulum intrudes further and more elements are affected, they are also progressively displaced. In the author's experience, the applied displacement technique provides the best correlation of results between computer simulations and values measured from actual large-part testing.

The two bumpers were then analyzed at the impact locations. Outputs from the finite-element evaluations are shown graphically as deformed geometries and load displacement curves. Results are grouped by impact location.

Center-Center Impacts. Figures 3 through 6 relate to the simulated center-center impact on the two bumper geometries:

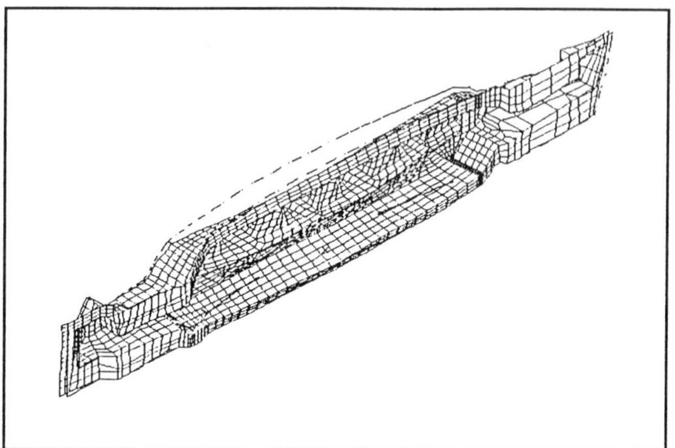

Figure 3: Deformed Geometry for C-Section Beam After Center-Center Impact

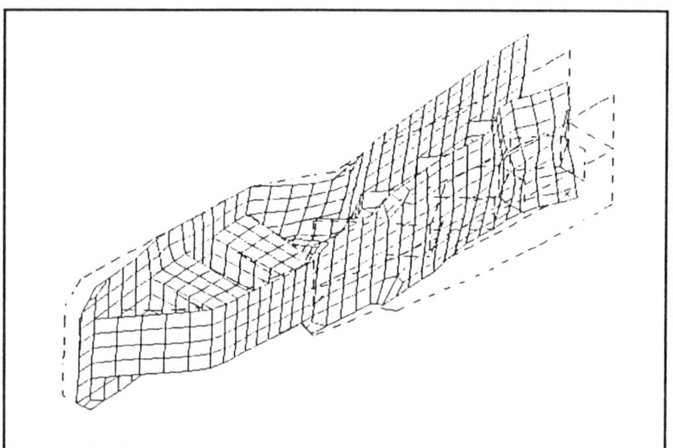

Figure 4: Deformed Geometry for I-Beam After Center-Center Impact

- Figure 3 depicts deformed geometry (with initial geometry provided as a baseline) for the C-section beam after impact; Figure 4 shows similar information for the I-type. The flanges of the C-section beam spread during failure, but the bumper continues to absorb energy. The I-type fails in web collapse.
- Figure 5 depicts a load-vs.-displacement curve for the C-section beam; Figure 6 shows a similar diagram for the I-type beam. The I-type is stiffer (the slope of the curve is steeper and the energy absorbed is greater at lower pendulum loads). On the other hand, the C-section beam continues to deform and absorb energy at 60 mm (2.4 in.) displacement, while the I-type has failed at 25 mm (1.0 in.) displacement.
- The strain at final displacement for the I-type beam in this impact is 23.1%. For the C-section beam, it is

21.4%. Both beams provide roughly equivalent performance in this impact situation. In other words, they will fail at about the same time or displacement level.

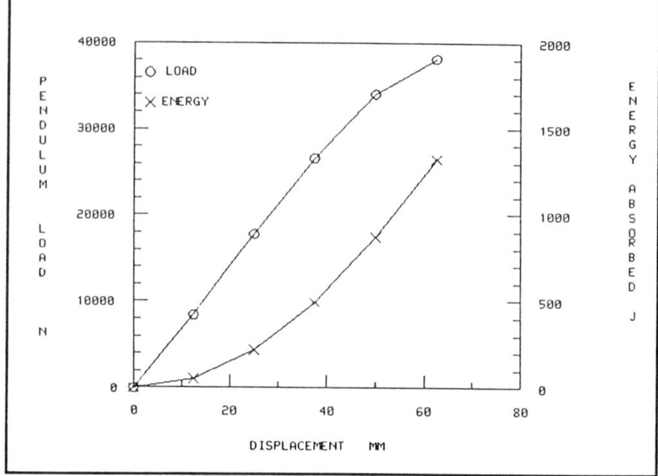

Figure 5: Load-vs.-Displacement Curve for C-Section Beam Under Center-Center Impact

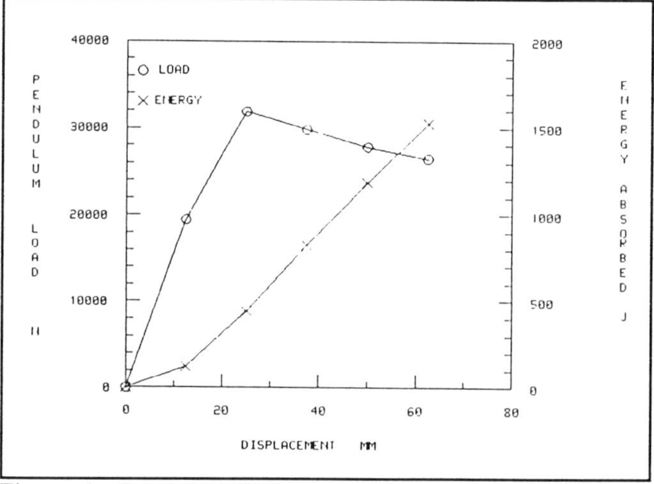

Figure 6: Load-vs.-Displacement Curve for I-Beam Under Center-Center Impact

High-Center Impacts. Figures 7 through 10 relate to the high-center impact for the two bumper geometries:

- Figure 7 depicts deformed geometry (with initial geometry provided as a baseline) for the C-section beam after impact; Figure 8 shows similar information for the I-type beam. In both figures, since the pendulum is no longer striking on the centroidal axis of the part, the section being impacted is rotating. In the case of the I-type beam (Figure 8), the impact is driving the flange directly onto the webs, crushing the ribs. With the C-section beam (Figure 7), the empty space inside the "C" provides more room for flange deformation to take place. This enables stress to be spread across the entire cross-section instead of being restricted to a localized area where strains will accumulate rapidly. The C-beam experiences global or gross deformation rather than local deformation, as does the I-type. Therefore, strains for the C-section beam will be lower for a given displacement, and failure should occur later or at a greater displacement level.

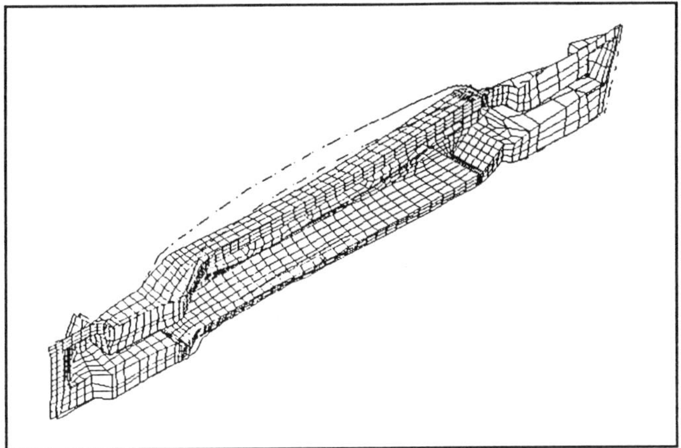

Figure 7: Deformed Geometry for C-Section Beam After High-Center Impact

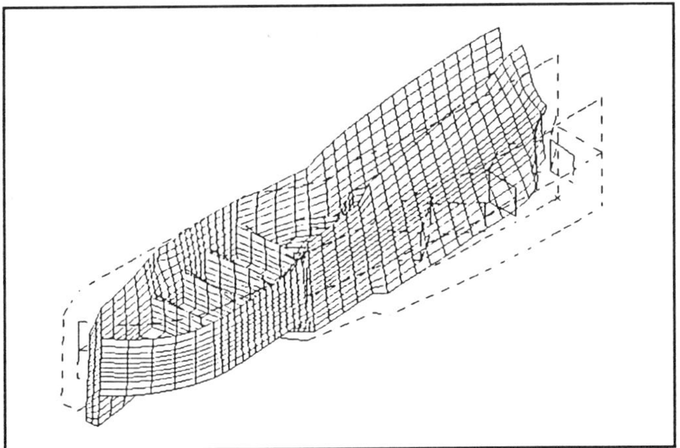

Figure 8: Deformed Geometry for I-Beam After High-Center Impact

- Figure 9 is a diagram of the load-vs.-displacement results for the C-section beam; Figure 10 is a similar diagram for the I-type. The linear response of the C-section shows a maximum displacement of 60 mm (2.4 in.), where it has withstood a pendulum impact of 35,000 N (7,870 ft-lb). The I-type begin to fail at 24,000 N (5,400 ft-lb) and 40 mm (1.6 in.) displacement, and has withstood only 26,000 N (5,845 ft-lb) at 60 mm (2.4 in.) displacement. Also note the change in slope on the load curve for the I-beam in this impact situation. Where the slope of the curve decreases and then begins to increase again (indicating the beam is getting stiffer), two walls are striking each other. This situation would not likely occur in real life because local strains would exceed the material's

yield strain and the part would break before the walls met. The analysis code cannot distinguish this type of situation by itself. Therefore, it is important to have a trained analyst interpret the results.

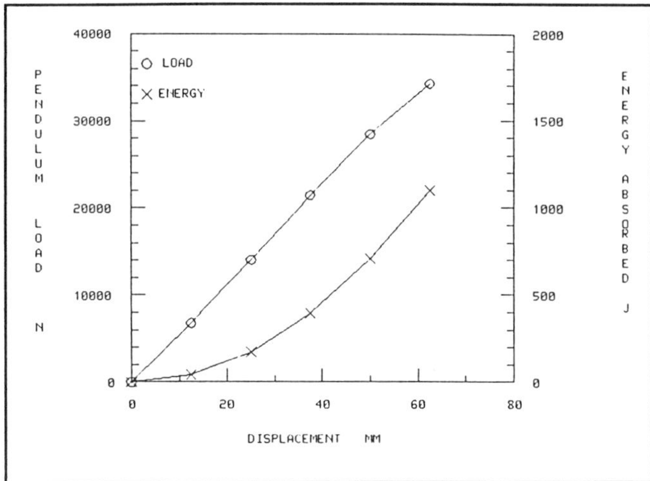

Figure 9: Load-vs.-Displacement Curve for C-Section Beam Under High-Center Impact

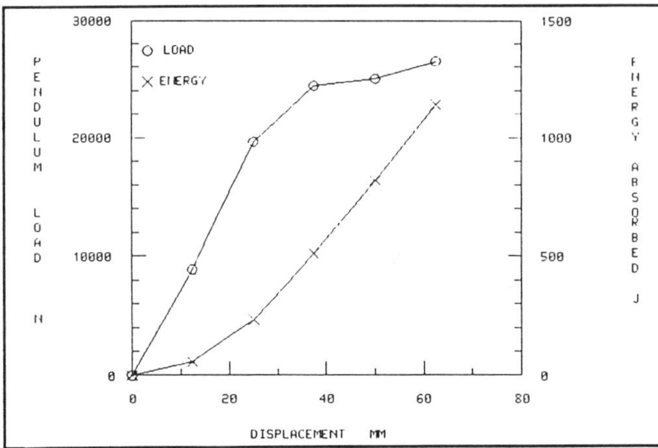

Figure 10: Load-vs.-Displacement Curve for I-Beam Under High-Center Impact

- The strain at final displacement for the I-type beam in this displacement is 56.06% versus 17.86% for the C-section beam. Performance is not equivalent in this impact situation. Maximum strain for the I-type is almost three times higher than for the C-section beam. Therefore, the I-type geometry will fail at an earlier time or displacement level.

High-Outboard Impacts. Figures 11 through 14 relate to the high-outboard tests on the two bumper types:

- Figure 11 depicts deformed geometry (with initial geometry provided as a baseline) before and after impact for the C-section bumper; Figure 12 shows similar information for the I-type. The pendulum impact causes the I-type's flanges to be driven into its ribs. These ribs prevent stresses from being dispersed quickly across the full length of the part, leading to rapid strain accumulation over a small area. As strains increase, plastics stiffen; since both types of materials are already high-modulus polymers, the increased local strains cause failure of the I-type at an earlier time and lower deflection than the C-section beam.

- Figure 13 shows a load-vs.-displacement diagram for the C-section beam; Figure 14 is a similar diagram for the I-type. Load curves for both beams at this impact show a sharp increase in slope: between 50-60 mm (2.0-2.4 in.) for the C-section and 40-50 mm (1.6-2.0 in.) for the I-type beam. Again, the code thinks two walls are striking each other, reinforcing the beam. In real life, failure would have occurred before this situation became apparent.

- In this impact, the strain at final displacement for the I-type is 47.67%; it is 16.21% for the C-section beam. Performance of the two beams is not equivalent, and maximum strain is about three times higher for the I-type than the C-section beam. Again, the I-type can be expected to fail at an earlier time or at a lower displacement level.

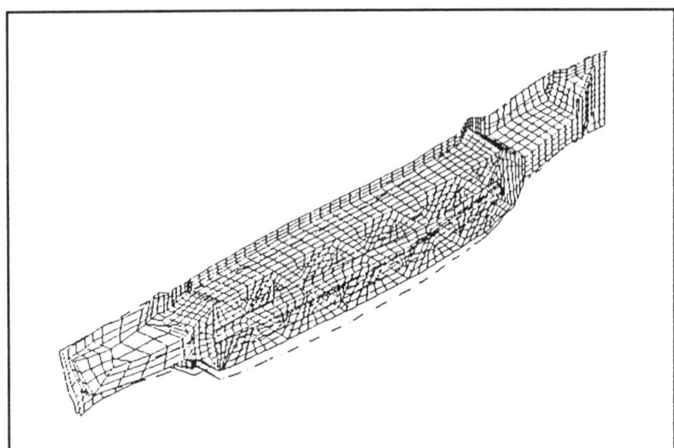

Figure 11: Deformed Geometry for C-Section Beam After High-Outboard Impact

The experience of the author and his GE colleagues indicates that, once a fracture has been initiated in a continuous-glass-mat thermoplastic composite part, by the time the predicted strain level reaches 5%, failure has begun to occur. The author recognizes that some unfilled, injection molded materials can go to higher strain levels before failure.

<u>Impact-Location Summary</u>. The impact-location study produced the following:

- Under center-center impacts, performance of the two beams is generally equivalent. The I-type absorbs more energy, and it also incurs higher strains.

- In high-center impacts, the C-section beam outperforms the I-type. At 60 mm (2.4 in.) of

displacement, the C-section beam has absorbed 35% more pendulum energy and continues to absorb impact energy. The I-type begins to fail at 40 mm (1.6 in.) displacement.

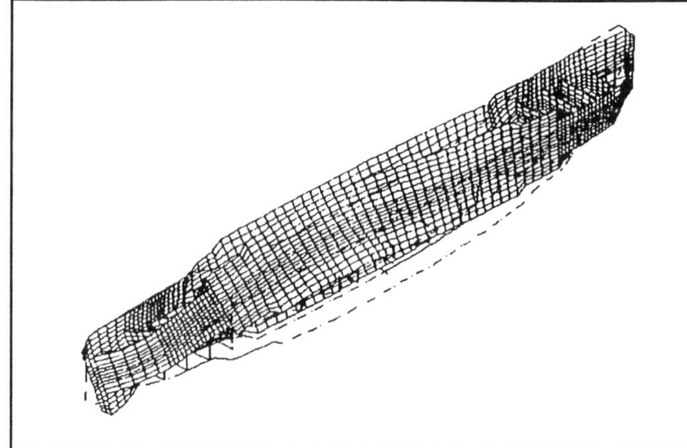

Figure 12: Deformed Geometry for I-Beam After High-Outboard Impact

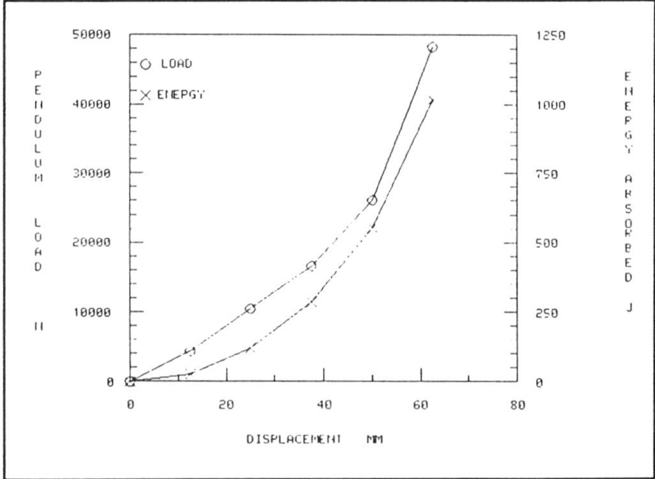

Figure 13: Load-vs.-Displacement Curve for C-Section Beam Under High-Outboard Impact

- In all cases, maximum strain at failure for the I-type is higher than for the C-section (Table I). In high-center and high-outboard impact conditions, strain is about three times higher for the I-type than for the C-section. The differences in geometry between the two beams most likely accounts for the higher strain in the I-type beam. While the C-section beam allows stress and strain to dissipate across the full beam, the I-type tends to concentrate strain in a localized area, causing earlier material failure. In addition, the empty space inside the "C" of the C-section beam appears useful for high-center and high-outboard impacts because it provides more room for deformation to occur, preventing local strain and failure (crushing) for a longer period of time or a deeper intrusion.

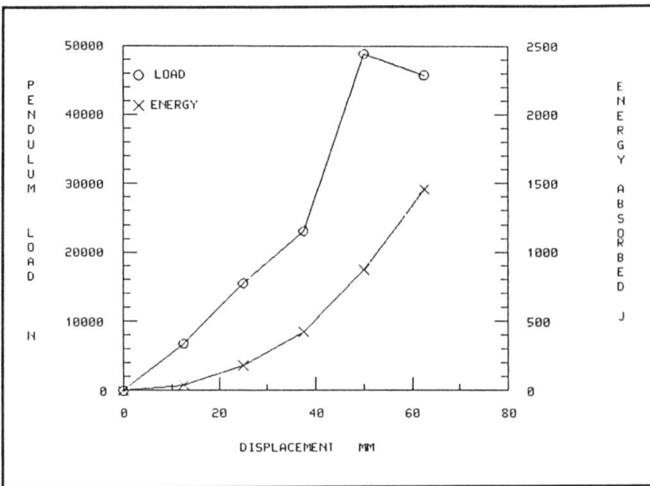

Figure 14: Load-vs.-Displacement Curve for I-Beam Under High-Outboard Impact

Table I: Maximum Strain at Final Displacement for the I-Type and C-Section Beam at Three Impact Locations

Geometry	Maximum Strain		
	Center-Center	High-Center	High-Outboard
I-Type	23.1%	56.06%	47.67%
C-Section	21.4%	17.86%	16.21%

- Because the flanges of the I-type beam have rigid spacers front-to-back, it is much more difficult for this beam to change geometry. The beam cannot move out of its original geometry because it is being constrained and held in place. Since the geometry cannot easily change, the material in the beam must go to a higher stress or strain level, eventually leading to failure of the high stress regions. With the C-section beam, however, there are no equivalent constraints holding the beam in place, so it is free to move out of its original moment of inertia, leading to a rotation of the flanges outward. This is why there is less strain in the C-section beam for an equivalent displacement.

Physical Testing

Ideally, computer-aided analyses will be conducted at the beginning of a product-development cycle to optimize a part's final design. Physical testing of actual molded parts should also be used to correlate theoretical performance from analyses with actual performance in production parts. Flow and orientation effects and molded-in stresses can influence the physical properties of a material and therefore a molded part.

For this study, it would have been desirable to have been able to perform physical testing on actual C- and I-type beams using pendulum impact testers. Unfortunately, there are no production vehicles that incorporate the I-type geometry. Future impact testing in this area should concentrate on correlating theoretical and actual performance of molded beams.

Since it was not possible to evaluate the effects of processing on actual molded beams, a test was conducted on molded plaques with various ribbing patterns to evaluate the effects of processing four types of compression-moldable, glass-filled thermoplastic composite materials [5]. The study was conducted to determine if flow effects limited the amount of glass that penetrated into a given rib geometry. Based on previous studies of these types of materials [6-15], it was known that, if glass penetration was reduced from the ideal assumed during analyses, the strength and stiffness of a molded part would be lower than predicted.

Rib Effectiveness Study

STRUCTURAL - Reference [2] indicates that the structural deficiencies of I-type beams can be addressed by adding rib stiffeners. The effect of using ribs in thermoplastic composites to enhance structural performance will be analyzed relevant to claims made in Reference 2.

Theoretical Rib Performance. Theoretical performance of a rib-stiffened part can be predicted using several techniques:

- Three-dimensional, finite-element analysis techniques.
- Automated two-dimensional solutions, such as the "Rib Stiff" program found in the Engineering Design Database. (Published by GE, this database predicts properties of plastic materials at conditions other than those specified as standard under ASTM procedures [16].).
- The closed-form linear equation:

$$Y = PL^3/48\ EI \qquad (1)$$

In Equation 1, Y = deflection recorded during testing; L = span between supports; E = published material modulus; I = centroidal moment of inertia; and P = load.

Equation 1 was solved using P as the unknown; in other words, it was used to calculate the load that would produce a given deflection. The resulting theoretical values were then compared with values determined through actual physical testing of molded plaques.

Test Procedure. Four different thermoplastics were used to mold the test plaques:

- Material A - a thermoplastic composite of 30% (by weight) randomly oriented, continuous glass-fiber mat in a polypropylene matrix
- Material B - a thermoplastic composite of 40% (by weight) randomly oriented, continuous glass-fiber mat in a polypropylene matrix
- Material C - same as Material B with the change that the needling process for the glass mat was modified

to break up the strands to a higher degree.

- Material D - a thermoplastic composite of 40% (by weight) randomly oriented, chopped glass fiber in a polypropylene matrix.

Table II depicts certain physical properties for the four materials.

TableII: Select Physical Properties of Sheet-Stampable Thermoplastic Composite Materials [17]

Property	Material A	Material B	Material C	Material D
Tensile Strength, break				
(MPa)	85	95	93	77
(Psi)	12,000	14,000	13,400	11,200
Tensile Modulus				
(MPa)	4,600	5,850	5,796	5,800
(Psi)	670,000	850,000	840,000	845,000
Flexural Modulus				
(MPa)	4,300	5,500	6,279	6,700
(Psi)	620,000	800,000	910,000	845,000
Flexural Strength, break				
(MPa)	140	160	170	150
(Psi)	20,000	24,000	24,600	22,000
Notched Izod @ 23C/73F				
(J/m)	651	748	678	502
(Ft-Lb/In)	12.2	14	12.7	9.4
Instrumented Impact @ 23C/ 73F				
(J)	20.2	21.7	25.3	23.4
(Ft-Lb)	14.9	16	18.7	17

These materials were processed using compression molding. They represent the full range of sheet-stampable thermoplastic composites available commercially. Among some of the materials, glass content varies while glass geometry is held constant (e.g., Materials A and B). With other grades, the percentage of glass content is constant (40%) and the glass geometry varies (Materials C and D).

The typical cross-section of the 480 plaques fabricated for structural testing is shown in Figure 15. Each plaque measured 610 mm (24 in.) long and 152 mm (6.0 in.) wide. The rib heights and rib and plate thicknesses were typical of the dimensions found in production parts. The ribs and plate thicknesses included 3.15 mm (0.125 in.), 5.53 mm (0.218 in.), and 7.93 mm (0.312 in.). The height of the ribs varied as

a percentage of rib width. Following are the rib heights tested: 1.5 x rib width, 2.5 x rib width, 3.5 x rib width, and 4.5 x rib width.

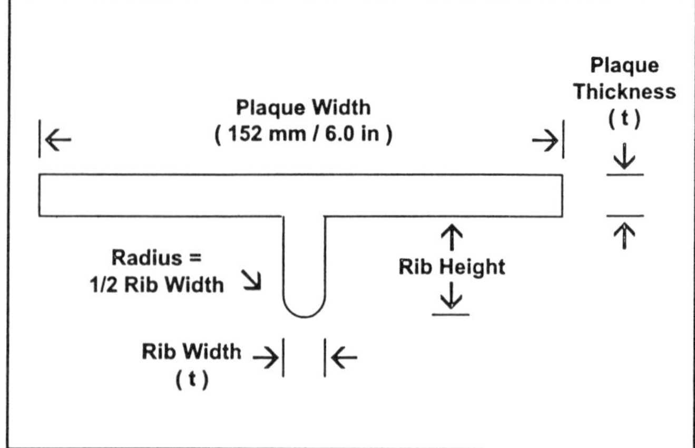

Figure 15: Cross-Section of Plaque Used in Rib Study

Since the molds were drape-charged, little material flowed during densification (compression). Material blanks were loaded into the tool so that the ribs were filled from the top. Ribs were aligned parallel to the sheet's machine direction in one set of samples and transversely in another set. (Subsequent work should focus on the effects of flowing into ribs from different directions and distances from the rib.) Testing included five samples of each material, rib orientation, and rib geometry.

Loading the plaques in a three-point bending fixture required the use of an Instron test frame (Model 1123) with a 454 kg (1,000 lb) load cell calibrated with a shunt resistor. The test span was set at 19 x rib height, in accordance with ASTM D790, to ensure that shear was minimized and the primary loading condition was bending. Load-vs.-displacement data were preserved on a chart recorder.

The data were transferred to a spreadsheet, which averaged the five test results for each configuration and compared them to the calculated theoretical loads. Comparisons were made at six points between initial load and the proportional limit, so all points fell on the linear region of the load/deflection curve where the modulus is calculated. Data were also recorded at maximum load and at failure.

Test Results. The following graphs demonstrate that, due to real-life processing considerations, the use of ribs in glass-mat thermoplastic composite materials actually provides far less structural benefit than theoretical calculations would indicate. This paper includes representative samples of these graphs:

- Figures 16-18 show measured stiffness as a percent of theoretical stiffness at various rib aspect ratios for plaques cut in the machine direction. The fit lines are of an exponential form that follow the general equation:

$$y = a * b^x \qquad (2)$$

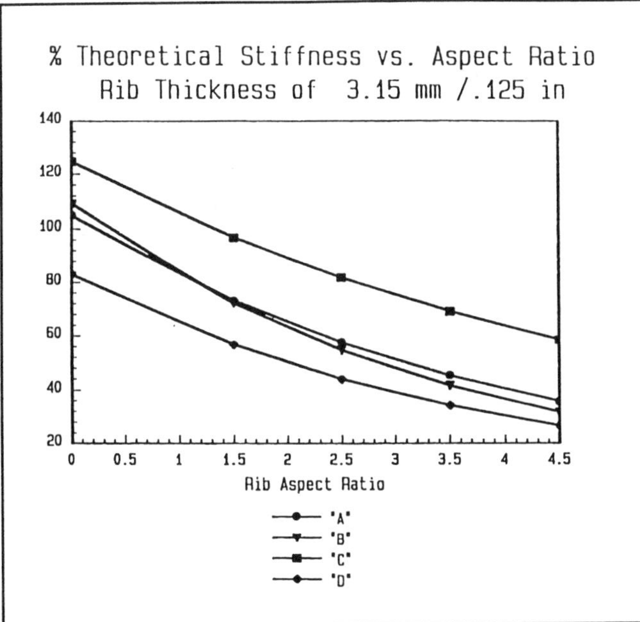

Figure 16: Measured Stiffness vs. Percent of Theoretical Stiffness for 3.15 mm (0.125 in.) Ribs

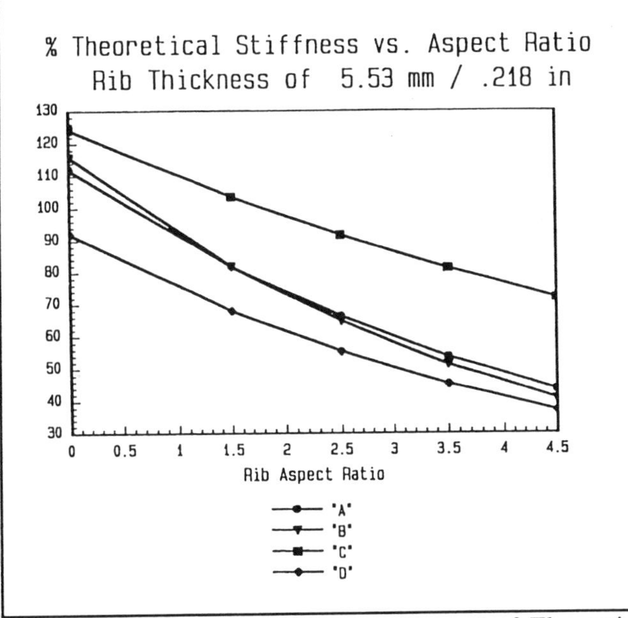

Figure 17: Measured Stiffness vs. Percent of Theoretical Stiffness for 5.53 mm (0.218 in.) Ribs

STRUCTURAL SUMMARY. The study of measured rib effectiveness versus theoretical rib effectiveness produced the following conclusions:

- Actual performance is far less than theoretical performance in terms of stiffness or spring rate, and effective material modulus at a rib aspect ratio (height to thickness) of 2.5-to-1 (Table III).
- Rib aspect ratios greater than 2.5 are not recommended because of the decrease in the spring rate.

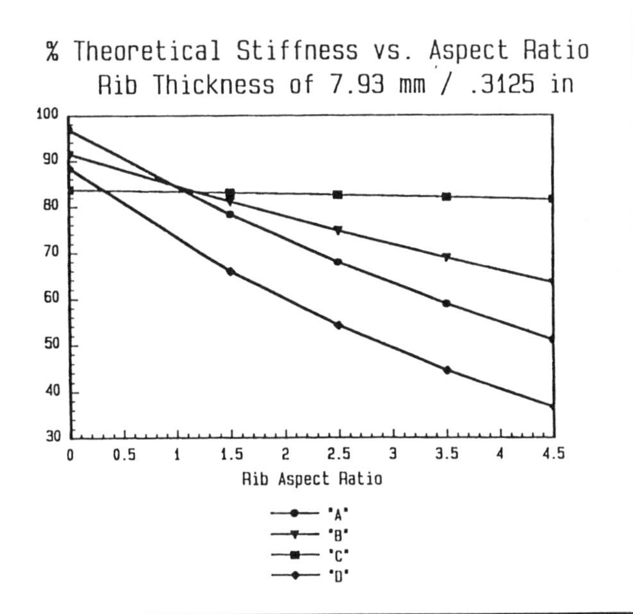

Figure 18: Measured Stiffness vs. Percent of Theoretical Stiffness for 7.93 mm (0.312 in.) Ribs

Table III. Actual Structural Performance as a Percent of Calculated Performance for Various Rib Thicknesses

Rib Thickness		Calculated Performance
(mm)	(in.)	
3.15	0.125	40%
5.53	0.218	50%
7.93	0.312	60%

- Rib strength is proportional to the penetration of glass into the ribs. However, increasing rib thickness from 5.5 to 8.0 mm (0.2 to 0.3 in.) does not increase glass penetration. It appears unlikely that any available material or practical molding technique could result in uniform glass percentages throughout ribbed parts. This is the reason that measured performance is inferior to theoretical performance, which assumes uniform glass content throughout the ribbed component.
- With three materials, rib performance is not related to transverse or machine-direction alignment due to the random orientation of the glass mat and due to the fact that the fibers can span or bridge the rib gap and block further glass flow into the structure. In Material C, however, the modified mat provides better filling because its glass strands are more broken up.
- Increasing rib thickness from 5.5 to 8.0 mm (0.2 to 0.3 in.) has little effect on rib efficiency, but increasing both rib and plate thicknesses has a linear effect on part stiffness.

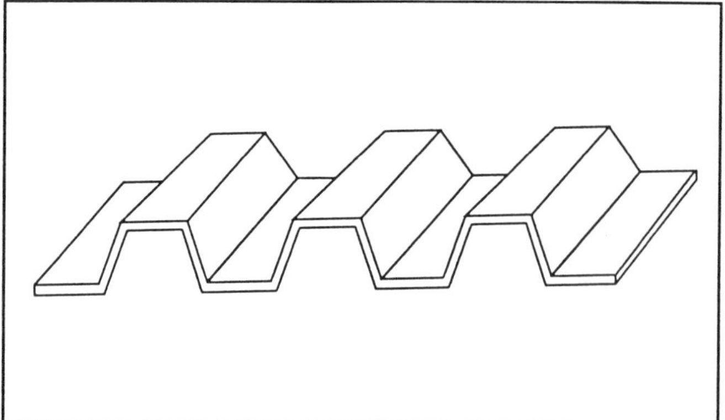

Figure 19: Typical Corrugated Geometry Recommended for Stiffening of Compression Molded, Glass-Filled Thermoplastic Composites

Many previous documents [1,5-18] have addressed the fact that the properties of molded ribs are rarely equivalent to the properties of the parent plate in compression-molded thermoplastic composite materials due to nonhomogeneous glass distribution. In cases where a plate of compression-molded thermoplastic composite material requires stiffening by some technique, GE has found that geometries approximating corrugations have proved the most successful (Figure 19). When sufficient package space and design freedom are available in an application, this stiffening method is preferred for flat plates molded from glass-filled thermoplastic composites — whether the glass is short, chopped fiber or a continuous mat. Note that each rib in this design is a mini C-section connected in sequence.

Other Processing Issues

Other processing issues may need to be considered with the I-type bumper design. As the rib study clearly showed, strength of ribs in thermoplastic composite materials could be a concern if glass penetration does not reach the full rib depth. The study also suggests that incomplete glass penetration frequently occurs.

In addition, intricate compression-molded shapes are often susceptible to dieseling. Dieseling is gas entrapment and automatic ignition of volatiles in the gas. This effect can occur between the plastic flow front and the mold. As occurs in the cylinder of a diesel engine, compression pressures alone cause the gases to ignite, which can lead to burning of nearby plastic. This condition can also prevent proper filling of a section of the mold [5,18,19].

Finally, the intricacy of the I-type design could also make the mold susceptible to non-uniform cooling, which can lead to longer molding and ejection cycles and ultimately to part warpage. Achieving proper cooling of a mold like this would be more involved than usual because it would be difficult to get cooling lines into the ribbed sections to cool the plastic (and cooling lines cannot be run through the ribs) [5,18,19].

Conclusions

Computer-aided analysis of C-section and I-type bumpers fabricated in glass-reinforced thermoplastic composites shows that C-section bumpers outperform I-type bumpers by a wide margin except when the impact is ideally located on the bumper's centerline and at the center of the span between bumper rails. In this center-center impact, the I-type beam absorbs slightly more energy per millimeter of pendulum penetration and incurs higher strains.

In addition, the I-type beam requires rib stiffeners to perform adequately. However, testing shows that ribs in compression-molded thermoplastic composite materials do not necessarily provide the increased performance a designer might expect from theoretical calculations. If complete glass penetration into the ribs is not achieved, the stiffness and strength of those ribs may be lower than desired.

Finally, the I-type design may be susceptible to the processing issues of dieseling and non-uniform cooling. Careful attention needs to be paid to these issues when designing the tool and molding parts.

Under these conditions, the minimal weight savings that might be obtained by abandoning the proven C-section bumper for the newer I-type design appears insufficient to justify a change at this time.

# # # # #

References

1. Bals, C., C. Clark, and M. Layson, "Effects of Fiber and Property Orientation on "C" Shaped Cross Sections," SAE, Paper 910049, Warrendale, Penn, (1991)

2. Crandall, J. and D. Bhutani, "Design of a New Bumper Beam Using Structural Thermoplastic Composite," SAE, Paper 930542, Warrendale, Penn, (1993)

3. *ADINA (R) Theory and Modelling Guide*, Report AE-84-4, ADINA Engineering, Inc., Watertown, Mass. (1984)

4. Rawson, J., "Comparison of Analysis Results to Physical Testing for the Performance of Engineering Thermoplastic Bumper Beams," SAE, Technical Paper Series, Warrendale, Penn, (1994)

5. Clark, C. "AZDEL Rib Plaque Study, Final Report, June 1993," GE Plastics, Southfield, Mich, (1993), (unpublished company white paper)

6. Stokes, V. K., "Random Glass Mat Reinforced Thermoplastic Composites—Part I: Phenomenology of Tensile Modulus Variations," *Pol. Comp*, 11 (1): 32-44, (1990)

7. Stokes, V. K., "Random Glass Mat Reinforced Thermoplastic Composites—Part II: Analysis of Model Materials," *Pol. Comp.*, 11 (1): 45-55, (1990)

8. Stokes, V. K., "Random Glass Mat Reinforced Thermoplastic Composites—Part III: Characterization of the Tensile Modulus," *Pol. Comp.*, 11 (6): 342-353, (1990)

9. Stokes, V. K., "Random Glass Mat Reinforced Thermoplastic Composites— Part IV: Characterization of the Tensile Strength," *Pol. Comp.*, 11: 354-367, (1990)

10. Bushko, W. C. and V. K. Stokes, "Statistical Characterization of the Tensile Moduli of Random Glass Mat Reinforced Thermoplastic Composites," *1991 SPE ANTEC*, Brookfield Center, Conn, 1991, pp. 2097-2101

11. Tomkinson-Walles, G. D., "Performance of Random Glass Mat Reinforced Thermoplastics," *J. Therm. Comp. Mat.*, 1: 94, (1988)

12. Stokes, V. K. and H. F. Nied, "Solid Phase Sheet Forming of Thermoplastics—Part I: Mechanical Behavior of Thermoplastics to Yield," *J. Eng. Mat. & Tech.*, Trans. ASME, 108 (2): 107-112, (1986)

13. Buskko, W. C. and V. K Stokes, "Strength of Glass Mat Reinforced Thermoplastic Composites—A Statistical Approach," *1992 SPE ANTEC*, Brookfield Center, Conn, 1992, pp. 783-787

14. Kolberg, R. F.,"Structural Designs in Thermoplastic Composite Structures," *1987 SPE RETEC*, Brookfield Center, Conn, 1987, pp. 295-298

15. Kolberg, R. F., "Investigation Into Finite-Element Analyses of Designs with Randomly Oriented, Continuous Fiber Reinforced Thermoplastic Composite Sheet," Masters Program, Rensselaer Polytechnic Institute, Troy, NY (1987)

16. Engineers may obtain a complimentary copy of information on the Engineering Design Database and an application for access by calling GE Plastics at 1-800-845-0600

17. *Thermoplastic Composites Design & Processing Handbook*, AZDEL, Inc. Shelby, NC, TPS-351, (1993)

18. Gilliard, B., "AZDEL (R) Bumper Design Guide," AZDEL, Inc., Southfield, Mich. (unpublished company white paper)

LITECAST[TM]: A Novel Solution for Attachment Problems with Structural Composites

B.P. Graham, J.R. Gentry
Delco Chassis Division
Dayton, Ohio

ABSTRACT

Delco Chassis has patented a process that involves casting attachment fittings directly onto a polymeric composite in a manner that is not only economical, but also provides distinct advantages to design and function. In this process, a molding alloy with a relatively low temperature melt point (aluminum, magnesium, etc.) is brought into contact with a highly filled polymer matrix composite utilizing a rapid casting technique. Although the melt temperature of the metal alloy is significantly higher than the degradation temperature of the polymer matrix, decomposition is limited by 1) rapid cooling of the metal and thus a limited extreme temperature excursion, and 2) the insulating effect of the fiber that limits any degradation beyond the composite surface.

INTRODUCTION

The automotive manufacturer must offer the consumer more value each year or face the prospect of losing market share. Value is a balance among performance, quality and cost. Typically, one or two of these value measurements must be optimized against the other(s).

The choice of materials and processes used to manufacture automotive components has a major impact on the overall value of the automobile. Traditional materials and manufacturing techniques are being challenged by those once considered to be exotic. The objective is to increase vehicle performance through mass reduction. The challenge is to provide these components without sacrificing cost or quality.

Composite materials offer significant mass reduction potential:

$$\rho_{STEEL} = 7.8 \ g/cc$$
$$\rho_{ALUMINUM} = 2.7 \ g/cc$$
$$\rho_{FIBERGLASS/EPOXY} = 1.8 \ g/cc$$
$$\rho_{GRAPHITE/EPOXY} = 1.6 \ g/cc$$

Many automotive companies are using or considering the use of aluminum as a replacement for steel. The use of composite materials has been limited by the cost of the material, processing difficulties and the unknown quality factor.

BACKGROUND

Delco Chassis efforts in the area of advanced composites have been focused primarily upon suspension components. Having successfully developed the LITEFLEX[TM] fiberglass leaf spring for mass production[1,2], the next focus was other structural suspension members, specifically suspension links and control arms.

The functional viability of a composite suspension link was proven in 1986[3]. The struggle over the next 10 years was to devise an economical way of manufacturing the end attachments. With conventional link materials, this is typically done by pressing a rubber bushing into holes which have been directly pierced or drilled through the link ends. Taking a similar approach with composite materials would significantly weaken the link.

Work in the composites industry over the last decade has tried to solve similar attachment problems. A study of these developments shows 5 different families of composite link design. Examples of family #1 are shown in Fig. 1. Preimpregnated fiberglass is filament wound over a core. The viability of this design depends upon the cost of the core. The core is required to have sufficient hoop strength for the subsequent operation of bushing installation. This requires the utilization of core manufacturing techniques which are uneconomical.

Family #2 is a maturation of family #1. The core is eliminated. A filament wound composite is compression molded with a knob on each end. This knob is used to attach a metal bushing housing. Although economical, the bushing attachment technique is considered to be nonconventional.

An example of family #3 is shown in Fig.2. This uses a core onto which fiberglass is braided. The braided pattern around the eyelet area provides sufficient hoop strength for bushing installation, thereby reducing the structural requirement for the core. After braiding, the fiberglass preform is then impregnated with resin through a process called Structural Reaction Injection Molding (S-RIM). This link can meet structural requirements which far exceed those for a suspension link. However, the cost and complexity of braiding and S-RIM make this option unrealistic. Although a method of wet braiding has been developed, thereby eliminating the S-RIM (see Fig. 3), the cost is still prohibitive.

Family #4 tries to solve the problems of family #3 by utilizing a less costly method of manufacturing the preform. Typically, this involves a preform made simply of random chopped glass fiber. This may then be impregnated using S-RIM. Holes may then be drilled in the ends for bushing installation, or metallic inserts may be used. Material properties utilizing this manufacturing process are significantly lower than

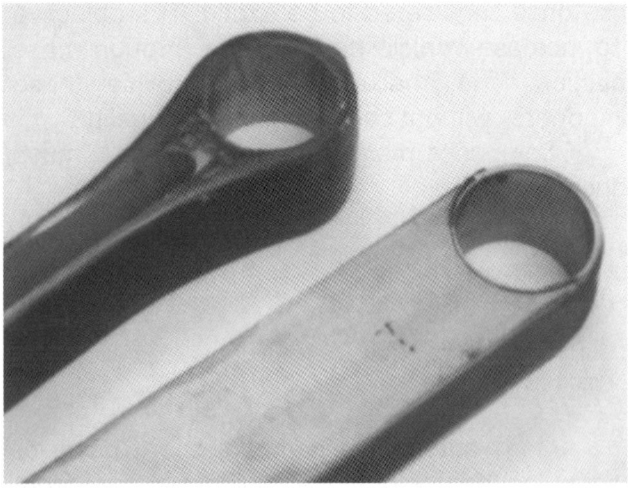

Fig. 1 Two examples of unidirectional reinforcement filament wound around a preformed core.

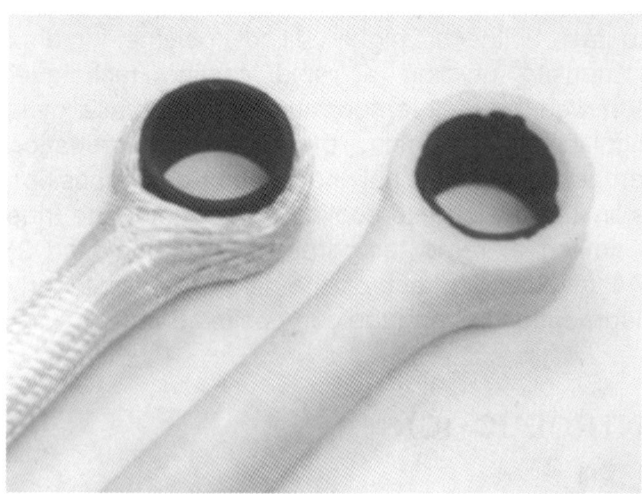

Fig. 2 A braided fiberglass preform (left) and the composite structure after S-RIM (right).

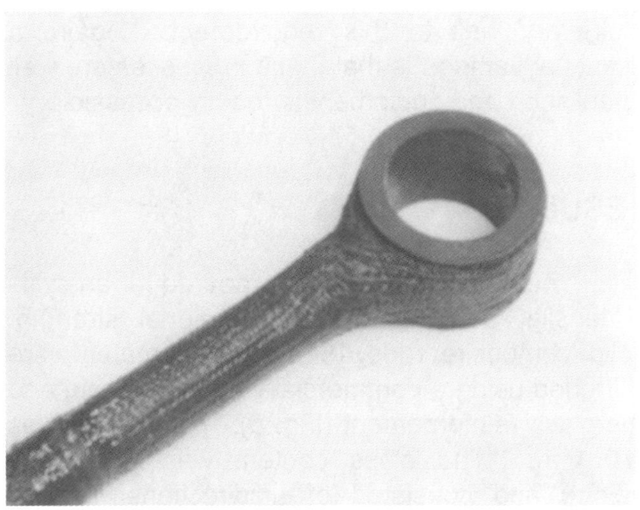

Fig. 3 A wet-braided structure similar to that of Fig. 2. Wet-braiding eliminates the extra step if S-RIM impregnation.

LITECAST TECHNOLOGY

Having already established the disadvantages of conventional composite fabrication techniques in the previous discussion, a method was needed to enable the lowest cost technologies of composite fabrication and fitting attachment to be utilized. The process patented by Delco Chassis involves casting attachment fittings directly onto a polymeric composite in a manner that is not only economical, but also provides distinct advantages to design and function.

In this process, a molding alloy with a relatively low temperature melt point (aluminum, magnesium, etc.) is brought into contact with a highly filled polymer matrix composite utilizing a rapid casting technique. Although the melt temperature of the metal alloy is significantly higher than the degradation temperature of the polymer matrix, decomposition is limited by 1) rapid cooling of the metal and thus a limited extreme temperature excursion, and 2) the insulating effect of the fiber that limits any degradation to the composite surface. Figs. 4 & 5 show the surface region of a fiberglass reinforced composite before and after encapsulation by metal. Note that the thin surface layer of the composite is removed by the process, but there is little evidence of degradation beyond that. This leaves the bulk composite properties relatively unaffected.

the first three families. Some even less demanding applications use thermoplastic random fiber injection moldings.

Family #5 involves the manufacturing of a metallic end fitting which is then attached to the composite material through bonding, pressure or a combination of the two. This eliminates the need for complex composite structures, taking full advantage of a unidirectional composite material. This technique carries with it the labor disadvantage generally associated with bonding operations.

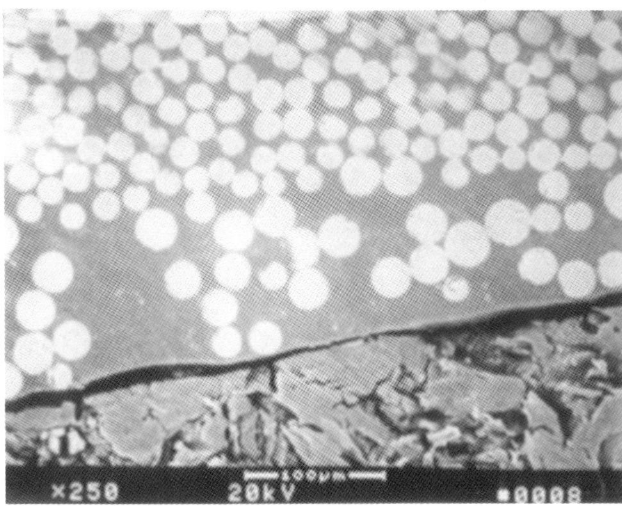

Fig. 4   Cross section of composite (in phenolic mount compound) prior to encapsulation. Note the resin rich outer region towards the bottom of the photo.

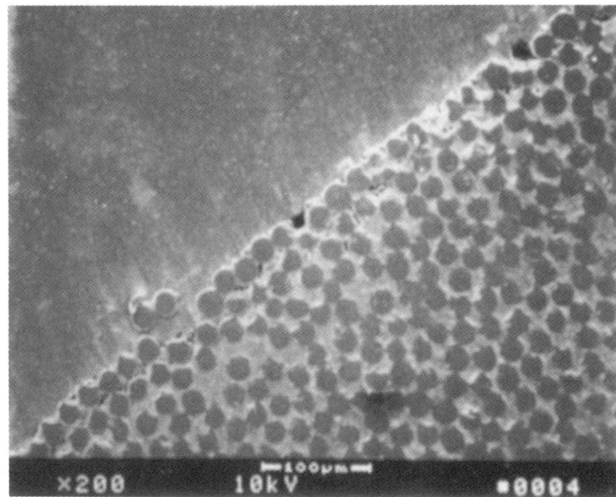

Fig. 5   Cross section of composite (lower right) after encapsulation by metal (upper left). Note the absence of the resin rich layer and the intimate contact between the outer fibers and the metal.

<u>Composite Materials</u>

The types of applications of greatest interest to Delco Chassis are structural, where fiber contents are typically in excess of 50% (by volume). As mentioned above, highly filled composites are desired in this process to resist thermal degradation of the polymer and allow conventional low cost materials to be employed. Resins successfully evaluated to date have been primarily thermosets: polyesters, vinyl esters and epoxies with moderate to high glass transition temperatures. There are a number of thermoplastic systems that would also be good candidates, with cost being negatively impacted due to the necessity of utilizing materials with higher temperature resistance and the difficulty of processing highly filled thermoplastic composites.

Fiberglass is the fiber of choice from the standpoint of cost, thermal conductivity and thermal stability. The thermal conductivity of glass can be 8-10 times less that of graphite, thus providing a more effective barrier to thermal penetration into the composite[4,5].

<u>Composite Processing</u>

In general, virtually any composite processing technique can be utilized to produce the composite element of the structure. Processes that lend themselves to producing structural composites such as hand lay-up, filament winding, SRIM and pultrusion can be utilized with little or no modification in respect to use of the structure in the casting process. Thus, established process technology and equipment can be selected on the basis of design requirements and economics.

<u>Casting Material</u>

Low melting point alloys of greatest interest to date have been aluminum and magnesium, materials that are enjoying increased usage in light-weight applications. Melting point for these materials is in the 550-650 C range, and although several times greater than the Tg of most resin systems evaluated thus far, forms an excellent structure when the temperature excursion is limited as described earlier.

<u>Casting Processes</u>

Any number of commercial casting technologies can be considered, with the primary goal being to maintain a rapid solidification time so as to limit polymer degradation. Die casting in its various forms of hot-chamber, cold-chamber, permanent mold , squeeze casting, etc., lends itself very well to this requirement. Again, a distinct advantage is that these processes are well established and equipment is readily accessible.

RESULTS

Test specimens were prepared for analysis of tensile, compression and torsional strength. Solid composite rods 19 mm in diameter were pultruded using a commercial vinyl ester resin and fiberglass reinforcement (Fig. 6). Rod length was 270 mm. The glass content was ~75% (by weight) and consisted of unidirectional rovings oriented along the axial direction of the rod. The rods were processed using temperatures ranging from 150-220 C with no additional postcure. Prior to die casting, the rods were notched in such a manner as to provide a mechanical interlock between the rod and casting and thus increase load transfer efficiency.

End fittings were then die cast onto the rods using a conventional permanent mold die cast technology. A commercial aluminum alloy was selected having a melt temperature ~650 C. Process cycle times and die temperatures were established such that the time of exposure of the composite rod to high temperature liquid alloy was limited to less than two seconds.

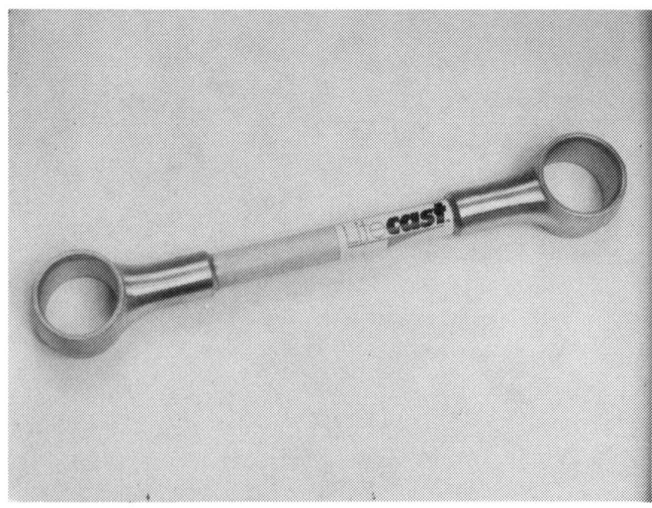

Fig. 6 LITECAST[TM] test specimen. Composite rod is unidirectional fiberglass reinforced vinyl ester. Encapsulating metal is aluminum.

The results are tabulated in Fig. 7. The compression values are primarily determined by the buckling characteristics of the composite, whereas the tensile strength is a function of the aluminum end attachment design. The torsional response depends on the composite construction, and since a unidirectional composite is being tested in this case, the values are relatively low. It is evident that there is a wide range of properties obtainable by tailoring materials, design and construction to fit specific product needs.

Fig. 7 Ultimate strength ranges for 19 mm unidirectional composite rods with die cast aluminum ends.

Compression	48-52 KN
Tension	25-30 KN
Torsion	1085-1250 Nm

CONCLUSIONS

A unique processing technology has been developed that allows casting attachment fittings directly onto polymeric composites to produce high strength structures. Although developed with automotive chassis use in mind, the process and ensuing structural properties are versatile enough to apply to a wide variety of end uses. This technology could very easily be applied to linkage applications in aerospace, marine and recreational areas, to name only a few.

ACKNOWLEDGMENTS

The authors would like to recognize the efforts of the members of the LITECAST[TM] Product Development Team, who have been responsible for advancing this technology to its present state. As in any endeavor, it is the collective effort of dedicated individuals that leads to success. Appreciation is also extended to the Delco Chassis division of the Automotive Components Group for allowing this work to be published.

REFERENCES

1   Kirkham, B.E., L.S.Sullivan and R.E. Bauerle; "Development of the LITEFLEX TM Suspension Leaf Spring"; Proceedings of the SAE International Congress and Expo; Detroit, Mich; Feb. 22-26, 1982; Paper No. 820160.

2   Mutzner, J.E., and D.S. Richard, "Development and Testing of Composite Truck Trailer Springs"; Proceedings of the SAE International Congress and Expo; Detroit, Mich.; Feb. 27-Mar.3, 1989; Paper No. 890545.

3   Hurtubise, D.W., Design of a Pultruded Core with Isolation Bushings for a Composite Suspension Link, Inland Division, GMC Thesis for General Motors Institute, Flint, Mich., May 1986.

4   Nomura, S. and Chou, T.W., "Bounds of Effective Thermal Conductivity of Short Fiber Composites", Environmental Effects on Composite Materials - Vol. 2, (1984), pp 271-279.

5   American Society for Metals, 1987, Engineers' Guide to Composite Materials, pp 5-34.

Design and Manufacturing Pilot of an SRIM Composite Crossmember

R.K. Wlosinski
Ford Motor Company
Dearborn, Michigan

T.W. Kniveton
Rockwell International
Newton, North Carolina

Abstract

This paper describes the design, fabrication and performance of a composite crossmember for a full frame vehicle. The transmission crossmember support demonstrates the design, weight, and performance benefits of composite materials in a structural application. Program drivers, (cost, weight save, and vehicle performance) led to the selection of the manufacturing processes and materials. Structural reaction injection molding (SRIM), foam cores, braided preforming, and molded in heat shields are used in construction of the crossmember. This paper also describes the high volume manufacturing pilot and experimental test verification program.

STRUCTURAL COMPOSITE applications in automotive vehicle structures have been of interest for several years. Ford in combination with the Automotive Composites Consortium and it's suppliers has been pursuing generic research of composite applications to develop an understanding of material properties, manufacturing processes, material and fabrication cost, and energy management. The benefits of using composites in a structural application are:

- Fully accounted cost reduction at lower volumes
- Improved durability and corrosion resistance.
- Improved NVH (noise/vibration/harshness)
- Cost-effective weight reductions for CAFE improvements.

Objective

The purpose of the composite crossmember pilot is to demonstrate the manufacturing feasibility and the functional performance of the structural crossmember, understand the cost of manufacture and establish assembly plant feasibility. The program was executed by a crossfunctional team including all the affected areas within Ford; the molder, Rockwell International; and the supporting Tier 2 suppliers, Owens Corning and ICI.

Vehicle System Design Requirements

The #3 crossmember, (Figure #1) in the full frame vehicle, was selected for trial because of it's major structural and functional requirements and has the advantage of being bolted onto the frame rail brackets. A center transmission mount isolates the powertrain from the chassis, and reduces the noise and vibrations transmitted to the passenger compartment. Package limitations and the proximity to the exhaust presented a realistic challenge for composite materials.

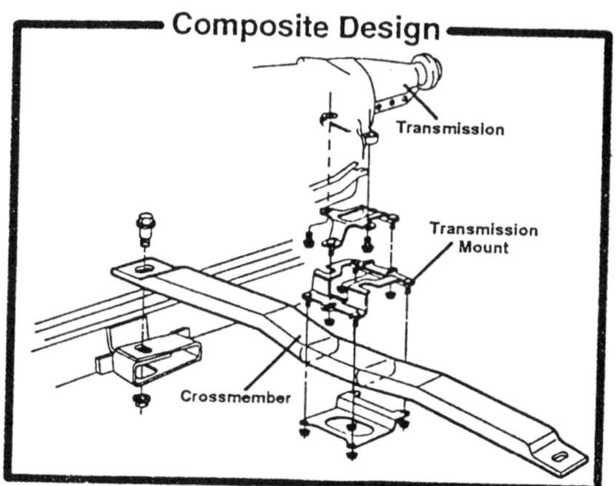

Fig. 1 Composite crossmember design
The transmission mounting point and clearances to the floorpan and the exhaust system established

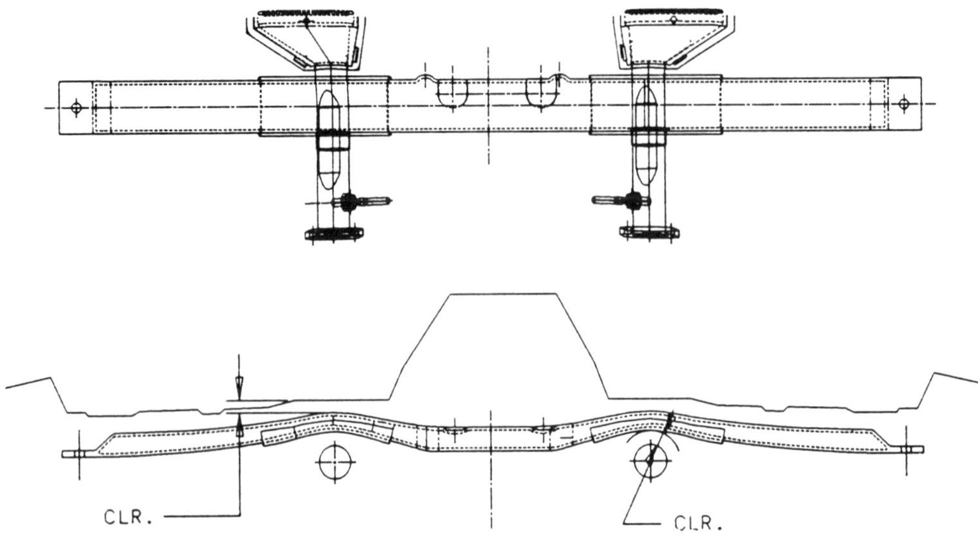

Fig. 2 Clearance Envelope

the package envelope for the #3 crossmember. Standard design clearances from the floorpan to the crossmember and from the exhaust system to the crossmember were maintained. (Figure #2)

Structural Design Requirements

The structural requirements for the #3 crossmember included stiffness, strength, durability, spring rate, and NVH (noise, vibration, and harshness) performance. The basis for the structural loading in the composite crossmember design analysis is to be equal to the production vehicle performance. The stiffness of the composite system matched the stiffness of the steel system in the fore/aft and vertical planes.

Strength requirements included the peak loads, durability, and 35 mph frontal crash. The peak operating loads are the maximum loads that the crossmember is subjected to without degradation. Design values from previously run 5 mph collisions, pot hole and rail simulator tests with production steel crossmembers established the maximum load levels.

The crossmember has two 35 mph frontal crash requirements: 1) maintain alignment of the driveshaft and rear axle during crash, and 2) support the static weight of the transmission after crash (Figure #3). During crash, the crossmember is allowed structural degradation. Crash strategy allows the driveline to move rearward during frontal impact; part of the displacement by the crossmember deformation and the remaining by deformation of the frame rail bracket. The frame bracket is a roll strap allowing the driveline system to displace without the composite crossmember reaching its elastic limit.

Operating / Durability / Crash Design

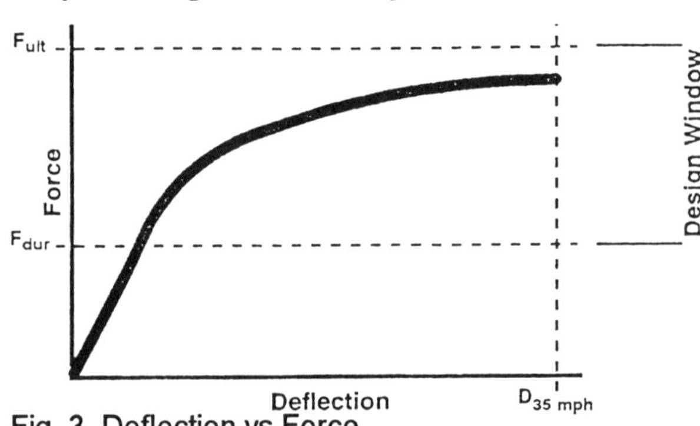

Fig. 3 Deflection vs Force

Durability loads are the fatigue forces that the crossmember system experiences during the high mileage durability test cycles. The crossmember must withstand the maximum normal operating force without experiencing damage.

Body and Assembly Operations established vehicle assembly requirements for the crossmember in the assembly plant. Assembly tolerances and build sequences could not be changed from the current steel crossmember vehicle assembly process.

Vehicle serviceability is also cited as an issue in the design of the crossmember. Replacement of the transmission mount must be accomplished without removing the transmission.

Extreme heat generated by the close proximity of the catalytic convertors made it necessary to package heat shields. Chassis Engineering Office required that the heat shield be an integral part of the crossmember to avoid the potential removal by the customer.

Composite Crossmember Design

The pilot program was set up to prove out and achieve realistic cycle times, part to part variability, cost, process capability and process requirements. Evaluation of a variety of cores, glass preforming, cost, and liquid molding techniques led to the process designed for the crossmember. The processing issues were divided into six groups.

1) Liquid Molding Process
2) Resin Selection
3) Core Type
4) Fiber Preform
5) Heat Shield
6) Mold Design (Core and Part)

Molding Process

SRIM (Structural Reaction Injection Molding) The high production volumes for the composite #3 crossmember and the capital investment costs steered the team to select SRIM as the liquid molding process. The SRIM process allows for rapid cure rates and the faster cycle times reduces the number of molds needed to meet the volumes. The two-piece steel mold provides even temperature distribution across the face of the mold and allows for geometry and gate changes. Mold heat accelerates the reaction although it is not required for initiation. Molding of the crossmember has been optimized to minimize the amount of material used for the crossmember. Flow analysis was used to predict the flow pattern for the SRIM process. Pressure transducers and thermocouples were added to the mold to record the in-mold pressures and exotherm temperatures during the process. Real-time process monitoring ensures repeatable manufacturing quality.

Resin Selection. The SRIM process and vehicle temperature requirements narrowed the resin selection to a manageable number of candidates. Key issues for the selection criteria included high temperature performance, cost, and processing. Other characteristics taken into consideration were creep behavior, fatigue strength, chemical resistance, and glass wetability. Vehicle performance requirements for high temperature, made resin temperature capability and cost the highest priority criterion for making the resin selection decision. An elevated temperature isocyanurate urethane resin system was selected as the resin of choice for the pilot and possible high volume production.

Foam Core. The foam core is a mandrel for winding the fiber reinforcement preform. The core controls the thickness of the molded laminate and provides a small degree of the crossmember stiffness. A two-component polyurethane foam system is directly injected into aluminum molds to produce the core. The aluminum molds are heated and maintain a uniform temperature across their face. Proper mixing of the foam constituents and control of the mold temperature reduce the chance for large voids and produce the proper core skin thickness. The elimination of the resin intrusion into the foam core allows better SRIM process control and consistent mechanical properties of the crossmember. A foam core density of 14 pounds per cubic foot was required to withstand the in-mold pressures during the SRIM process. The weight of the foam core was held to ±3 percent during the manufacturing pilot.

Preforming. Preforming is the shaping of dry glass fibers into the net shape of the molded part. Common forms of preforming are hand lay-up, thermoformable stamped mats, braiding, spray-up, or a slurry process. Braiding (Figure #4) was selected to preform the composite #3 crossmember because it allows minor changes in the fiber architecture to be made without affecting the preforming system. Braiding also eliminates joints that are prevalent in other preforming processes. Waste is minimized by close tolerance trimming of the preform length. Biased and axial fiberglass rovings are braided directly onto the foam cores. Changes to the roving sizes allow the strength and stiffness to be tailored to the vehicle needs. The glass volume percentage can also be tailored by adjusting the angle of the braid, thus controlling the SRIM flow process by selectively varying permeability. Several braid architectures were tested and results correlated with predicative capabilities of the computer simulation at North Carolina State University.

Fig. 4 Braider

Heat Shields. An earlier version of the composite #3 crossmember was tested for heat resistance in the wind tunnel using a 4.6L engine with a 1994 exhaust system. (Figure #5) The #3 crossmember used was constructed of a RTM composite material with a open hat section design. The test was run to measure temperatures on the composite material with a drivetrain and exhaust system programmed for the 1994 Crown Victoria. Thermocouples were used to map the temperature generated by the powertrain and catalytic converters during severe operating conditions. Operating conditions simulated in the test included normal operation, trailer towing, and engine malfunction. Thermocouples between the heat shield and the laminate demonstrated adequate temperature reduction. Results of this test, and subsequent field tests, determined the operating temperatures of the #3 crossmember environment in the vehicle.

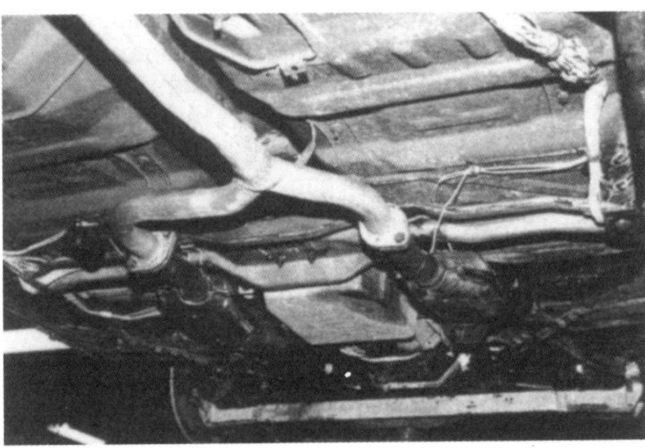

Fig. 5 Under body view of heat management testing

Laboratory experiments were conducted to determine the concentration of the heat generated by the exhaust at the maximum surface temperature. It was determined from these experiments that 100 mm each side of the centerline of the exhaust pipe and the forward edge of the crossmember required protection.

Heat shields must remain intact on the crossmember during all foreseeable operating conditions. The integrity of the attachment of the heat shield to the crossmember was verified by subjecting it to thermal cycling, cyclic bending, road durability, and the gravelometer test. No delamination or detachment was found. In addition to the vehicle requirements, compatibility with the SRIM process was considered in the selection.

Manufacturing Pilot Production

A 5000 piece pilot production run is being conducted in Rockwell Plastic Products plant in Newton, North Carolina. The run is conducted in three phases. First phase consists of 500 pieces, second phase 2000 pieces and the third phase 2500 pieces to prove out the manufacturing sequence and ergonomics of the pilot. Every foam core, preform, and final part is being weighed. One out of 10 cores and SRIM parts are being measured and tested for significant characteristic measurements and part to part consistency. Studies of molding cycle times, molding pressures and operator ergonomics are being conducted. The results of the pilot will be used to improve the manufacturing processes and make recommendations for production implementation.

Vehicle DVP&R (Design Verification Plan and Report): (figure #6)

NVH Testing. NVH testing was conducted on a 1994 Lincoln Town Car. Four composite crossmembers, with varying spring rates, were tested at the Ford Evaluation Test Track in Naples, Florida. Subjective and objective measurements were made. The subjective evaluations rated the vehicle for shake and ride characteristics. For the objective measurements, triaxial accelerations were measured at the seat track and specifically reflected vehicle shake. The composite system provided a level of overall shake lower than the production Town Car (measured subjectively and objectively), with slightly increased harshness (measured subjectively). It is evident the recommended crossmember and transmission mount provide equal acceleration levels at the seat track in the fore-aft and vertical directions, while decreasing lateral acceleration levels. The recommended crossmember and mount provide a higher rating for mid car shake, jiggle, after shake, front end sheet metal, plushness and rolling feel.

Barrier Crash Testing. Three vehicles were crashed with the composite #3 crossmember system. The system held the driveline in position and remained connected to the frame. Deflection of the crossmember was controlled by the design features in the frame rail bracket. Measurements of the crossmember deflection and deformation to the frame rail brackets were within the predicted ranges.

Further laboratory and vehicle testing, including high mileage fleet testing, will be completed within one year. The completion of the design verification plan and report (DVP&R) will meet the objectives for first proof of the structural composite technology in a frame crossmember.

DESIGN VERIFICATION PLAN AND REPORT
COMPOSITE #3 CROSSMEMBER; TRANSMISSION
MOUNT AND ATTACHMENT SYSTEM

I. GENERAL DESIGN CRITERIA
    1. FMEA
    2. FEA
    3. DRAFTING CLEARANCE CHECK
    4. DESIGN AID INSTALLATION /
        CLEARANCE CHECK
    5. ASSEMBLY
    6. POWERTRAIN DECKING
    7. SERVICEABILITY
    8. ROAD LOAD DATA ACQUISITION
    9. QUALITY CONTROL (CONTROL
        PLAN)
    10. MODAL ANALYSIS

II. ENVIRONMENTAL
    1. HEAT PROTECTION TESTING
    2. LABORATORY CORROSION
        TESTING

III. IN-VEHICLE TESTING
    1. HIGH MILEAGE FLEET TESTING
    2. DURABILITY
    3. 5 MPH BUMPER IMPACT TEST
    4. BARRIER CRASH TEST
    5. TRANSMISSION MOUNT /
        CROSSMEMBER CLEARANCE TO
        FLOOR PAN,
        EXHAUST AND DRIVE SHAFT
        MOVEMENT TEST
    6. BROKEN C/M; SEVERED
        ELASTOMER TEST
    7. NVH PERFORMANCE

IV. LABORATORY TESTING
    1. DURABILITY SIMULATION
          LOW TEMPERATURE
          NORMAL TEMPERATURE
          HIGH TEMPERATURE
    2. CREEP TESTING
          CHEMICAL, MOISTURE
          ENVIRONMENT
          HIGH TEMPERATURE
    3. FASTENER TORQUE / TENSION
        TEST
    4. C/M END PULL TEST
    5. COMPONENTS TESTING TO
        DESTRUCTION

V. ENGINEERING SPECIFICATION (ES)
    TESTING
    1. STATIC RATE TEST PROCEDURE
    2. DIMENSIONAL CHECKS
    3. MATERIAL SPECIFICATIONS:

          FOAM CORE
          RESIN SYSTEM
          BRAIDING
    4. FATIGUE TEST PROCEDURE
          NORMAL TEMPERATURE
          HIGH TEMPERATURE
    5. END PULL TEST PROCEDURE
    6. NON DESTRUCTIVE TEST
        PROCEDURE

Fig. 6 Design Verification Plan & Report

Weight and Cost Status

The composite crossmember has undergone several design changes since the beginning of the program. Material changes and process modifications led to a lighter than expected crossmember. Laminate thickness decreased by 20% from the original estimate. The foam core weight is at the original estimate with the opportunity to reduce the weight with further testing and process development. Total weight of the composite crossmember was reduced by 5% from the estimated weight.

The cost of the composite crossmember is influenced by raw material cost. Approximately 40% of the total piece price is the material cost. Processing and careful design considerations reduced the amount of flash and waste in SRIM and foam core molding. The current preform method reduced the glass waste and cost by using less expensive rovings and net part preform trimming. The total cost of the crossmember is 5.5% lower than the original estimate. Comparison of the composite system to the steel system shows a net reduction in weight of 33% and a piece cost increase of 85%.

Conclusions

A structural composite crossmember has been designed built, piloted, and tested in a joint effort between Ford and Rockwell International. The objectives of the program were met or exceeded. Functional and durability issues of composite materials in a realistic environment were demonstrated. Thermal related issues were identified and solved. Attachment issues concerning pullout strength during normal operating conditions, crash issues, safety issues, and other performance related issues were addressed and solved. A complete FMEA (failure mode effects analysis) for design, manufacturing, and assembly was completed.

The 5000 piece manufacturing pilot is being completed to demonstrate full production capability. A control plan for the manufacturing process was developed to insure consistent parts through each of the manufacturing stations. Significant characteristics of the crossmember were established and measured during the pilot to satisfy the quality requirements.

Acknowledgements

The authors would like to thank the team members from Ford and Rockwell International, Rockwell Plastic Products Newton, NC, Owens Corning and ICI Polyurethanes. A special thanks to the Ford management and Rockwell management team who continued to support the program financially and with encouragement. And to Bill Mellian (Owens Corning) for breaking-in his lap-top computer by taking minutes at the program meetings.

Proceedings of the 10th Annual ASM/ESD Advanced Composites Conference, Dearborn, Michigan, USA, 7-10 November 1994

Embedded Acoustic Waveguides for Monitoring
Composite Material Processing and NDE

R.T. Harrold, R. Brynsvold, D.J. Hanko
Westinghouse Science & Technology Center
Pittsburgh, Pennsylvania

Abstract

Embedded acoustic waveguides (AWG) of 10 mil diameter Nichrome for example, can be used to guide acoustic waves through composite parts during processing and the wave attenuation data can be interpreted in terms of real time viscosity, gelation and modulus information. Consequently, embedded AWG offer a means of feedback and control of composite molding processes. Following the manufacture of a part, measurements of acoustic wave velocity can yield approximate measurements of internal residual strain and material modulus and help in quality assessment. During the material lifetime the embedded AWG are available for internal strain sensing (via attenuation) and modulus measurement (via wave velocity) which should help in assessing damage and remaining life.

In a program funded by the National Automotive Center[*] it is planned to refine and further develop this AWG technology for production line use during the manufacture of both military and civilian composite vehicles. Specifically, this program addresses the control of resin transfer molding processes with emphasis on monitoring complete cure, void

formation, complete mold filling, bondline integrity and N.D.E. In the work reported here it is shown that the embedded AWG can monitor the cure of parts with right-angled bends, yield credible values for resin viscosity and modulus, and have the potential to determine the relative shear strengths of quasi kissing-type of bonds between resin parts.

ACOUSTIC WAVEGUIDES (AWG) EMBEDDED WITHIN MATERIALS allow the monitoring of material properties during manufacturing processes and also throughout the material lifetime (1 to 4). In addition, they can be used for non-destructive evaluation (N.D.E.), for example, for measuring internal strain and sensing and locating damage. The AWG may be of many sizes and materials, provided the AWG and host material have similar thermal expansion characteristics, and the acoustic waves may have different values of frequency. In the work reported here, 10 mil diameter Nichrome AWG were embedded within resin specimens and the transmitted acoustic signals had frequencies ranging from ~60 to 80 kHz. Although AWG theory can be very complex, their function within materials can be understood by considering the acoustic impedance (density (ρ) x acoustic wave velocity (c)) of both the host material and AWG. If the acoustic impedance of the host material matches that of the AWG, then only a small signal will be transmitted within the AWG as most

---

*.   Contract: DAAE07-93-C-R121
    National Automotive Center/U.S. Army
    (TACOM)/Westinghouse STC
    C.O.T.R.: Capt. Richard Brynsvold*

of the signal is attenuated. However, if the acoustic impedance of the host material is different to that of the AWG, then due to reflections at the interface, a larger signal level will be transmitted within the AWG and there is less attenuation. Consequently, during the processing of a material, such as a resin, as this changes state from a liquid to a highly viscous solution; to gelation (transition from a liquid state to that of a rubbery gel); to a more rigid material; to a high modulus state; the attenuation of signals within the AWG tracks these different states.

Several studies (1 to 4) relating to the use of embedded acoustic waveguides for monitoring material processes and properties have been reported on previously and most of these are related to flat panels of graphite-epoxy composite materials and straight lengths of AWG sensors.

In the new work reported on here the AWG sensor was used to monitor the cure of a complex shaped resin part, i.e., a part with two right-angled bonds; a more detailed analysis was made of AWG measurements of the resin viscosity and modulus; and the AWG were used to interrogate bondlines between resin parts with different degrees of bonding, or quasi-kissing bonds. A kissing bond, for example between two clear resin parts, would have the appearance of complete bonding, but in reality, a considerable area at the bondline may not actually be bonded, and consequently, the bondline would have a poor mechanical shear strength.

AWG Cure Monitoring of Resin Parts with Two Right-Angled Bends:

During this study, in order to reduce time and cost, and also to obtain practical data from a large number of experiments, a simple mold was designed and constructed for making small-size (10 cm wide with a 1 cm cross section) resin specimens with two right-angled bends and a single embedded AWG, Figure 1.

In Figure 1 it can be seen that the AWG exits at each end of the mold and the AWG terminations are bonded to acoustic transducers, one acting as a transmitter and one as a receiver. At the transmitter end, pulsed ~70 kHz wavetrains from a signal generator with a

repetition rate of ~100 pulses per second are used to activate the transmitter.

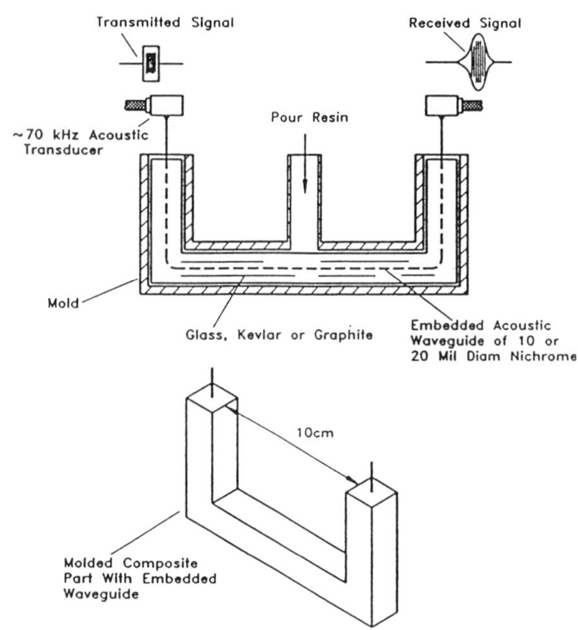

Figure 1. Arrangement for acoustic waveguide cure monitoring of molded composite part.

The acoustic transducer converts these electrical signals to acoustical pulses which are transmitted through the AWG to the receiving transducer at the opposite end of the AWG. Here, the acoustic signals are converted back to electrical signals and measured on an oscilloscope. The important measurement parameters are, the peak height of the received signal and the acoustic wave transit time from one end of the embedded AWG to the other. During resin processing, the changes in value of the received signal yield attenuation data which are used for monitoring the curing cycle. From the wave transit time data, the acoustic wave velocity within the embedded AWG can be obtained and these velocity data can also be used for monitoring cure. Usually, however, this is not as accurate a method as using the AWG signal attenuation, which is the preferred method for AWG cure monitoring.

Typical AWG cure curves of AWG signal levels and AWG wave velocity values for a Shell 815 resin (a thermosetting modified bisphenol-A epoxy resin) are shown in Figure 2. Note that instead of attenuation values, just the changes in the AWG transmitted signal levels with time are plotted on a logarithmic scale. Attenuation (α) at any time (t) into the curing process is readily obtained as it is simply the logarithm to the base e of the ratio of the signal at time (t) to the signal of time zero and is expressed in nepers. From the Figure 2 curves it can be seen that in the first part of the curing cycle, as the resin viscosity increases, the AWG transmitted signal level falls several orders of magnitude in value to reach a minimum at gelation (the transition from a liquid state to that of a rubbery gel). Following gelation, as the resin becomes more rigid and hardens the AWG transmitted signal level increases by two orders of magnitude.

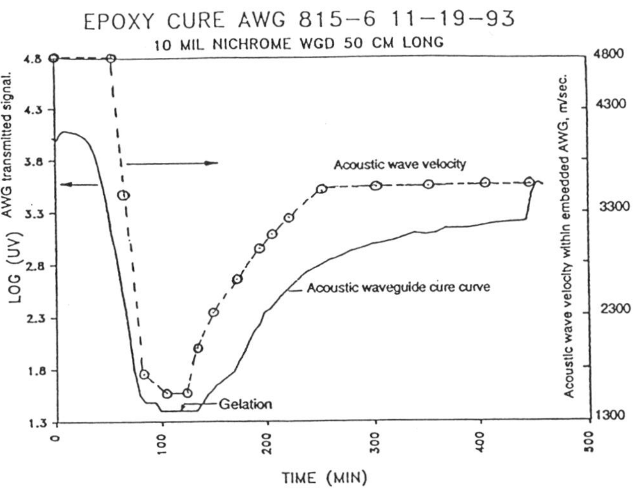

Figure 2. Acoustic waveguide cure curve for part with two right angle bends molded using Shell 815 resin with accelerator and heated via infra-red.

As explained earlier, the AWG transmitted signal changes during the cure cycle because the acoustic impedance of the resin is changing and restricting the transmission of signals within the acoustic waveguide. Actually, during the curing cycle the acoustic wave, which was initially travelling at a velocity of ~5000 m/sec within the AWG; travels at a slower velocity in the resin at the interface with the AWG surface, Figure 2.

From Figure 2 it can be seen that during the cure cycle the initial acoustic wave velocity is ~4800 m/sec and this surprisingly falls to ~1300 m/sec at gelation, and near the end of cure at ~400 minutes, the velocity is near 3300 m/sec.

The important aspect of this study is that although the resin part has two right-angle bends which the embedded AWG must pass through, the sensitivity of the AWG is not diminished and the cure curve is similar to those found in previous studies[4]. Also, as the resin is clear and transparent, residual optical strain patterns can be viewed on a Polaroscope. In Figure 3, which is a picture of the residual strain patterns in the right-angle bends of resin specimens both with and without an embedded AWG, it does not appear that the AWG causes any strain problems. In addition, it has also been demonstrated that measurements of acoustic wave velocity also can be used to monitor the resin curing cycle. The experiments described and discussed were repeated using resin reinforced with ~27% by weight of "S" glass fibers, and a similar looking signal level cure curve obtained, although the acoustic wave velocity data were of different values. This would be expected due to a higher acoustic wave velocity in the composite material compared with resin only.

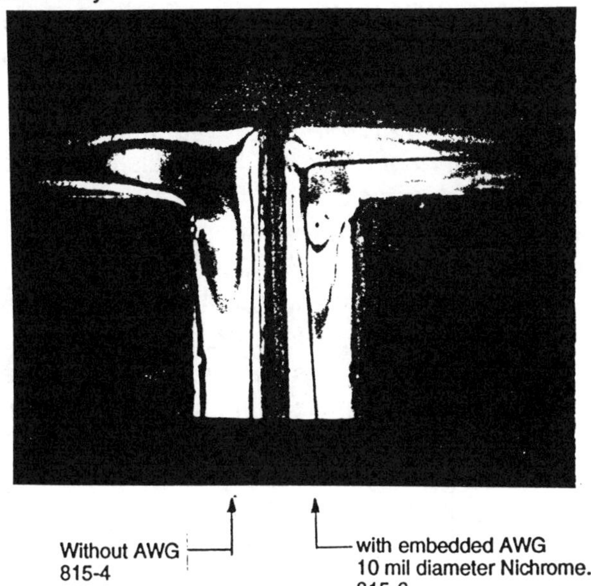

Figure 3. Residual strain patterns within molded right-angle specimens of clear Shell 815 resin both with and without embedded AWG.

Acoustic Waveguide Measurements of Resin Viscosity and Modulus

Resin Viscosity. It is a characteristic of curing resins that prior to gelation, they proceed through a period of rapidly increasing viscosity values extending over several orders of magnitude. As can be seen in Figure 2, the embedded AWG sensor readily provides a measure of these changes. However, in order to obtain calibration estimates for the actual resin viscosity values prior to gelation, a 10 cm length of 10 mil diameter Nichrome AWG was calibrated by measuring the reduction (attenuation in nepers) in value of the AWG transmitted acoustic signal as the castor oil was frozen to -49°C. Castor oil was chosen as its viscosity values, at least to -20°C, can be obtained from the literature.

    The castor oil attenuation data were plotted against the logarithm of the absolute temperature, not shown here, and a linear relationship was evident, at least from +5°C to -20°C. This type of relationship is typical for most viscous fluids. Also, from prior studies[4] it is known that AWG signal attenuation (α) in nepers is related to resin viscosity prior to gelation and resin modulus after gelation. Consequently, for estimating resin viscosity values prior to gelation, it is reasonable to plot the castor oil values of AWG attenuation in nepers versus the logarithm of castor oil viscosity (P), and this resulted in the straight line quasi-calibration curve of Figure 4. From this curve the viscosity (P) equation $P = 10^{\alpha+1.323}$ Poise, where α is the AWG attenuation in nepers, can be derived.

    Based on the measured attenuation values (α) from different materials and using the derived viscosity equation, the estimates, Table I, of the viscosity values at gelation (the transition from a liquid state to that of a rubbery gel) were made.

Resin Modulus. The AWG cure curves, Figure 2, illustrate the AWG response to increasing resin viscosity prior to gelation and to increasing resin modulus after gelation. A simple means of estimating the actual resin modulus values is to use the equation for the acoustic wave velocity;

$$c = \sqrt{Y/\rho} \,,$$

Table I. Estimated Viscosity Values at Gelation

Material	Viscosity in Poise:
Jello	134
Polyurethane	2×10^5
Shell 815 Resin	10^6 to 3×10^7
Concrete	3×10^9
Polymer Concrete	10^{11}

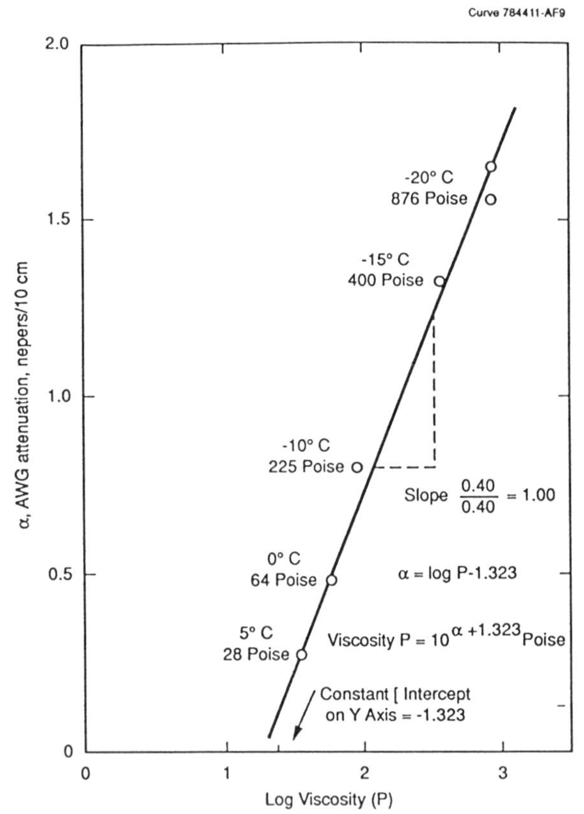

Figure 4. Acoustic waveguide attenuation (nepers/10 cm) versus the logarithm of Castor Oil viscosity over the temperature range of 5° to -20°C.

where c is the longitudinal wave velocity; ρ is density and Y is Young's Modulus. Actually, a term for Poisson's ratio is missing from the denominator of this equation, but that does not make a great deal of difference. The acoustic wave velocity was measured within a resin part reinforced with 33% by weight of "S" glass fibers and was found to be ~5050 m/sec. Its density value was ~1.28 gm/cm^3 and from the velocity equation it was calculated that the reinforced resin part had a modulus value of ~32 GPa. Similar measurements on different materials yielded the modulus values, listed in Table II:

Table II. Estimated Modulus Values at the End of "Curing"

Material:	Modulus, GPa:
Shell 815 Resin	~ 8
Shell 815 Resin with 26.6% by weight of "S: glass fibers	~ 21
Shell 815 Resin with 33% by weight of "S" glass fibers	~ 32

AWG Sensing of Bondline Integrity:

A technical challenge in the science of materials is the bonding together of parts, especially where complex shapes and cylindrical structures are involved. One problem is that parts which have the appearance of being well bonded, or were processed to be well bonded, may have very little mechanical strength. These types of false bonds are called "kissing bonds". In applying the embedded AWG to the kissing bond problem it was thought that acoustic wave transmission between two AWG, one each side of a bondline could be used to assess the degree of bonding. Also, if the AWG were positioned parallel to, and close to (within one or two cm, or a wavelength) the bondline, then the signal transmission level along a single AWG may also respond to the bondline strength or degree of bonding.

In order to test this hypothesis, resin parts were molded with bondlines of varying bonded areas equispaced between two AWG embedded parallel to the bondline. This was achieved by first molding resin parts 1 x 2 x 10 cm with a 10 mil diameter Nichrome AWG passing through the center in the long direction, and then partially joining these parts with different areas coated with resin to form varying bondline adhesion areas, Figure 5. This configuration allowed acoustic interrogation of the bondline area, not only across the bondline from AWG to AWG, but also by transmission along each AWG located parallel to the bondline. Many interrogation paths are available, for example, using the AWG terminations marked 1, 2, 3, and 4 in Figure 5, acoustic transmission through the bondline can be from 1 to 4, 4 to 1, 3 to 2, 2 to 3, 2 to 4, 4 to 2 and from 1 to 3 and 3 to 1. Interrogation parallel to the bondline can be from 1 to 2, 2 to 1 and 3 to 4, 4 to 3. These types of measurements were made on twelve pairs of specimens bonded together with different degrees of bonding. The bonded areas ranged from zero bond (surfaces pressed together with an acoustic coupling agent, called Aquagel), to several different bonding areas with between 15% to 95% of the total area between parts bonded. The bonding was not precise as it was carried out by applying small drops of resin equispaced along the resin surfaces which were then pressed together and allowed to cure at room temperature. The areas covered by the resin spread as the parts were pressed together and after cure, as the resin is transparent the bonded areas were visually estimated. Next, acoustic interrogation measurements of the various bondlines of different bonding areas were carried out.

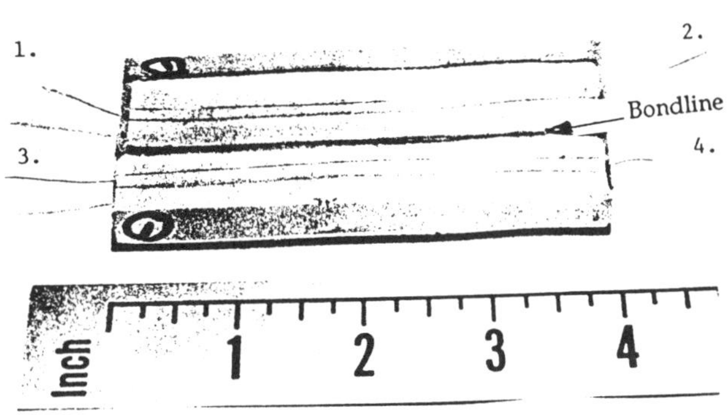

Two resin parts bonded together

Figure 5. Two resin parts, each with an embedded AWG, bonded together.

The maximum, average and minimum values of the acoustic data for transmission across the bondlines from AWG to AWG, are plotted against, the percentage of bonded area in Figure 6. Following the acoustic measurement, mechanical shear strength tests using the apparatus illustrated in Figure 7 were performed on all the bonded specimens and these data are also plotted in Figure 6. The most striking result which is apparent in Figure 6 is that resin parts with ~50% to ~70% of the area between them bonded have maximum shear strength, while parts with greater or lower percent bonding areas have less shear strength. Furthermore, most of the acoustic bondline measurement data correlates with the shear strength data.

Although these are encouraging results for demonstrating that AWG acoustic interrogation of bondlines can yield credible information in regard to the actual shear strengths of "kissing-type" of bonds, much more experimental confirmatory measurements are required. In addition, the fact that resin specimens with bondlines ~50% to ~70% bonded yielded maximum shear strengths, raises fundamental questions in regard to the bonding of composite materials. Preliminary analysis of the fractured specimens suggests tensile failure of the bonds and not shear failures. It is thought that this was due to the brittle or glassy nature of the resin used for bonding the parts together. Consequently, it is planned to repeat the tests using a less glassy bond in order to obtain true shear failures.

The data taken for signal transmission along AWG parallel to the bondline are not shown here, but were similar to those plotted in Figure 6 for transmission from AWG to AWG. These data are important because a single AWG embedded within a composite material would not only have the potential to monitor cure and lifetime NDE, but also be able to measure bondline integrity. For example, for inspection of a composite cylinder with a reinforcing ring bonded inside, an AWG could be embedded circumferentially either within the cylinder or the ring.

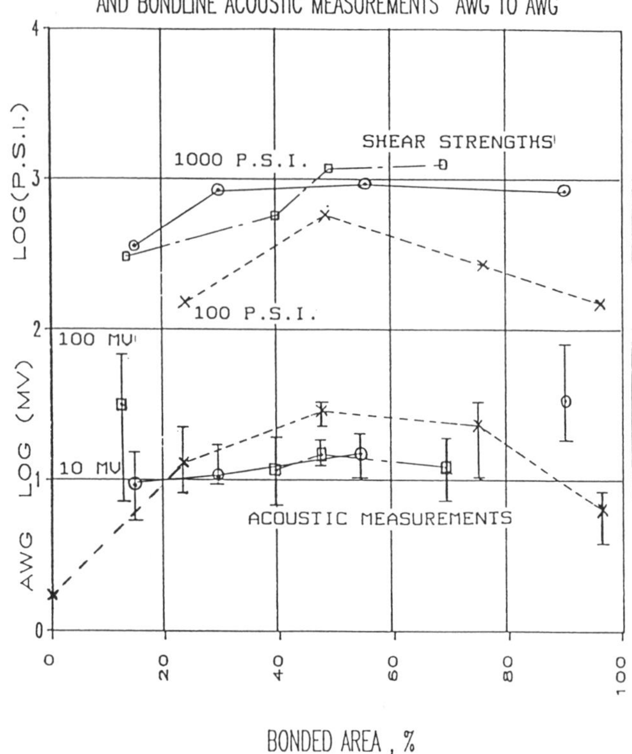

Figure 6. Comparison of bondline shear strength and measurements of acoustic wave transmission through bondline, AWG to AWG, versus percent bonded area for pairs of resin parts bonded together.

Conclusions and Discussion:

It has been demonstrated that the acoustic wave signal changes or attenuation within an embedded acoustic waveguide can be used for monitoring the cure cycle of resin parts with two right-angled bends. Even though the embedded AWG had to pass through the right-angles, its sensitivity was not reduced, and it responded to the orders of magnitude changes in resin viscosity prior to gelation and similar changes in resin modulus after gelation. In addition, it was found that the values of acoustic wave velocity within the embedded AWG also gave a measure of the cure cycle.

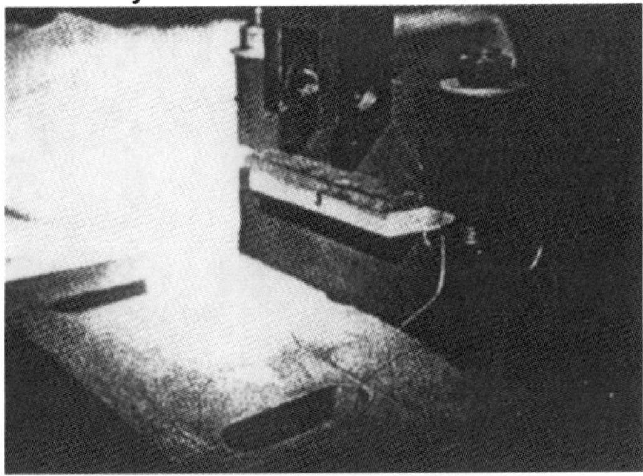

Steel plate is placed on specimen surface prior to shear strength test.

Resin specimen clamped in position for testing.

Figure 7. Bonded resin specimen clamped in position for shear strength test.

Based on AWG measurements of castor oil of known viscosities and taken to -20°C, an equation was derived for estimating the viscosity values of resins and other materials at gelation. This yielded credible estimates for several materials. Modulus values of resins and other materials after gelation and at the end of cure were estimated using measured acoustic wave velocities and density values and the equation for longitudinal wave velocities in materials, and credible values were obtained for several materials.

The most important part of this study may be the AWG interrogation of bondline integrity in an attempt to measure and locate "kissing-type" or false bonds with little mechanical strength.

Initial experiments for both bondline interrogation across a bondline from one embedded AWG to another, and for interrogation by examining signal transmission along an embedded AWG close to and parallel to a bondline, generally have a good correlation with the bondline mechanical shear strengths. The most striking result is that parts with approximately 50% to 70% of the area between them bonded have the highest shear strengths, while parts with greater or less percent bonding area have less strength. The correlation between acoustic measurements and bondline shear strength (or tensile strength) is encouraging, but much more experimental confirmation measurements are required before it can be claimed that AWG sensing systems can identify kissing bonds. In addition, as noted earlier, the tests need to be repeated with a less glassy bonding material in order to obtain true shear failure.

ACKNOWLEDGMENT:

Ed Diaz of (W) STC for mechanical test analysis and facilities.

REFERENCES:

1.   Harrold and Sanjana, J. Polym. Sci. 26.5 (1986).

2.   Harrold, U.S. Pat. 4,590,803 (1986).

3.   Harrold and Sanjana, Rev. Prog. Quan. NDE, 10B, ed., D.O. Thompson (1991).

4.   Harrold and Sanjana, ADPA/AIAA/ ASME/SPIE Conf., Session 26. Institute of Physics Publishing Ltd. (1992).

Evaluation of Manufacturing Variations in XMC Composites Using Modal Parameters

P.K. Mallick
University of Michigan
Dearborn, Michigan

Abstract

This study presents experimental data on the vibration characteristics of XMC composite plates containing both x-patterned continuous fibers and short random fibers. Some of the composite plates contain large manufacturing variations intentionally introduced during compression molding of these plates. These variations are thickness variation, fiber misorientation, improper plying and overlaps. The modal parameters investigated are resonant frequencies, damping factors, mode shapes and vibration amplitude. Based on these observations, the use of modal testing for non-destructive evaluation of manufacturing defects is explored.

VIBRATION CHARACTERISTICS OF COMPOSITE MATERIALS have been studied, both theoretically and experimentally, by many researchers for a number of years. Reviews of these studies have been published by Adams (1) and Gibson (2). In a series of articles, Gibson and his co-workers have successfully explored the possibility of using vibration measurements for characterization of composite materials, including determination of their elastic constants, damping factor, creep behavior and interlaminar fracture toughness (3).

Recently, the availability of modal testing and associated softwares has made the vibration characterization of materials much simpler and less time consuming than before. Several investigators have taken advantage of modal testing to investigate the possibility of using modal parameters, such as natural frequencies, damping factors and mode shapes, for nondestructive evaluation of damage in composite materials. For example, Sanders et al. (4) performed modal tests on $[0/90_3]_S$ laminated composite beams in the undamaged state as well as with various damaged conditions. The damage in their specimens was in the form of transverse cracks in the 90° layers with varying degrees of crack density. Their experiments show a steady decrease in the first seven natural frequencies measured and corresponding increase in damping ratios with increasing amount of damage. Tracy and Pardoen (5) also used experimental modal analysis to study the effect of prescribed delamination on the natural frequencies of $[90/\pm45/0]_{2S}$ beams. Various lengths of delaminations were introduced by placing a 0.013 mm fluorinated ethylene propylene tape at the laminate midplane during the molding of these laminates. Their experiments also show a decrease in natural frequencies, particularly at higher modes, with increasing length of delamination. Similar observations were made by Terek et al. (6) for $[90/0/90]_S$ beams. A reduction in natural frequencies proportional to the delamination size and location occurred at high frequencies, whereas relatively large delamination size had no significant effect on the lower natural frequencies.

This paper presents experimental modal analysis data of compression molded XMC composites containing both continuous and discontinuous fibers. XMC is a structural sheet molding compound first introduced by PPG, Inc. in the mid-70's. Modal tests were performed on "apparently" defect-free XMC plates as well as on XMC plates with intentionally introduced variations, such as thickness variation, fiber misorientation, improper plying and overlaps. These can be considered manufacturing defects in compression molded XMC parts, some of which may be created due to the way XMC layers are stacked prior to placing them in the mold.

Material

The material used in this study is an XMC sheet molding compound, containing 42 weight percent of x-patterned continuous E-glass fibers and 30 weight percent of randomly oriented 25 mm long chopped E-glass fibers. The matrix is a vinyl ester resin. The angle between the continuous fibers in the uncured XMC sheet is 9-10 degrees (Fig. 1). The XMC sheet contains red tracer fibers which help in measuring the included angle between the fibers.

The XMC material is manufactured by filament winding continuous strand rovings on a large rotating drum. The chopped fibers are sprayed on the continuous fiber layers as they are formed. After the desired thickness is obtained, the wound-up material is cut by a knife along a longitudinal slit in the drum to form a flat XMC sheet. As with other sheet molding compounds, XMC sheets are placed in a maturation room to allow it to "thicken" before they are compression molded into the cured shape.

In the present work, a 305 mm x 305 mm flat steel mold was used to compression mold several cured XMC plates. The mold temperature was 144°C and the molding pressure was 9.65 MPa (1,400 psi). The molding time was 2.5 minutes.

The XMC plates used for experimental modal analysis are listed in Table 1. Plates 14, 16, 17 and 29 were molded with a single 279 mm x 279 mm layer and are considered here to have no major defects. Plates 28 and 39 were molded with two 279 mm x 279 mm layers with (0/0) orientation between the layers. Plates 42 also had two 279 mm x 279 mm layers, but with (0/90) orientation between the layers. Plates 30, 31 and 32 were molded with single layers, but lower width of these layers caused various degrees of resin flow in the transverse direction. The resin flow was accompanied by buckling of continuous fibers as well as a change in angle between the continuous fibers in the x-pattern (Fig. 2). Plates 24, 25 and 26 were molded with various lengths of overlap between the two layers placed side by side in the mold (Fig. 3). The molded dimensions of all plates are 305 mm x 305 mm.

Modal Tests

Modal tests were performed on compression molded XMC plate specimens using a Hewlett Packard 3562A dynamic signal analyzer, a Bruel & Kjaer (B&K) type 8202 impact hammer (capped with a metal tip), a B & K miniature accelerometer type 4393 and an IBM PC equipped with the Star modal software package.

Square grid lines were drawn on one surface of each plate (Fig. 1). A total of 49 grid points were used in the measurement. The grid dimensions were 49 mm x 49 mm. The plates were placed on a 381 mm x 280 mm x 76 mm (thickness) semi-rigid foam. The hammer was used to impact the top face of the plate at grid point 25. Parameters were set on the signal analyzer to measure the frequency response functions up to a frequency of 1500 Hz. Five FRF's were averaged at each point and accepted into the Star modal system, provided the coherence was over 90 percent. After the measurements were accepted from all 49 points, the Star modal system estimated the natural frequencies and damping for each mode and constructed an animated shape for each mode.

Results

Table 2 lists the first four fundamental frequencies and damping factors for the plates molded with 279 mm x 279 mm layers. The weight and average thickness of each plate are listed Table 1. Bending is the principal mode of deformation at all the frequencies (Fig. 4).

Plates 14, 16, 17 and 29 contain single layers. The slight difference between their frequencies is due to the difference in their thickness. Plates 28 and 39 contain two layers and are 1.54 times thicker than the previous plates. The natural frequencies of these two plates are also greater than the previous plates, principally due to their increased thickness. When the frequency values are divided by the respective plate thickness (Table 2), the first two frequencies are within the same range as those for the single layer plates; however, the second two frequencies are considerably lower.

Plate 42 contains two layers that are 90 degrees to each other. The thickness of this cross-plied plate is the same as that of plate 28 which contains two layers that are 0 degrees to each other. The first and third natural frequencies of plate 42 is 25 percent and 12 percent higher than the corresponding natural frequencies of plate 28. The second and fourth natural frequencies are relatively closer.

Table 3 lists the natural frequencies and damping ratios for plates 30, 31, and 32 which have fiber misorientation due to transverse flow. These plates were molded with single XMC layers that were 18, 36 and 54.5 percent less in width than the previous plates. The single layer was centrally placed in the mold and forced to spread over the entire mold width by the application of pressure. The flow in the transverse direction caused buckling of continuous fibers as well as an increase in the angle between x-patterned fibers (Table 3). It was shown in Ref. 7 that the misorientation of continuous fibers causes a significant decrease in tensile strength as well as a greater degree of tensile strength variation across the width of the plate. The data in Table 3 show that all four natural frequencies increase with decreasing layer width, i.e., with increasing misorientation of continuous fibers.

Table 4 presents the natural frequencies and damping factors of the three plates that were molded with 0, 25 and 76 mm overlap at the center. The first and second natural frequencies of these plates are nearly the same as those with no overlaps, such as plate 14. The third and fourth natural frequencies are also not affected for plate 26 with zero overlap, although a straight knit line is clearly visible across the length of this plate. The third natural frequency of plate 25 with a 25 mm overlap is in the same range as that observed with a plate with no overlap, such as plate 14; but its fourth natural frequency is clearly diminished. Both third and fourth natural frequencies are reduced significantly for plate 24 which had a 76 mm overlap. Both plates 24 and 25 had curved knit lines and severe misorientation of continuous fibers.

Unlike the natural frequency, the damping factor reported in Tables 2, 3, and 4 does not show any trend. Plates 14, 16, 17 and 29 are apparently similar plates. But their damping factors show a wide variation at all four modes. Their natural frequencies at each mode are reasonably close, particularly when the difference in their thickness is taken into account.

The mode shapes of plates 14, 28, 42, 31 and 24 are shown in Fig. 4. There are minor differences in the mode shapes of plates 14 and 28. Even though plate 28 is 1.54 times thicker, both plates had 279 mm x 279 mm layer size and basically contain x-patterned fibers in one direction. Plate 42 has a 0/90 orientation of layers. Although its deflection pattern is the same as that of plate 28, the axes of bending are not the same. Plate 31 had a single 279 mm x 178 mm layer and contains fiber misorientation due to transverse flow. Its mode shapes are much more complex than those of plate 14 which a single 279 mm x 279 mm layer. Plate 24 had a 76 mm overlap. The first two mode shapes of plate 24 are similar to those of plate 14; however, there are differences in the third modal shapes.

In order to examine modal shapes more closely, the deflections (δ) of plates 14, 24, 25 and 26 in mode 1 at fourteen grid points across the horizontal and vertical center lines are considered. The average thicknesses (t) of these plates are 3.61, 4.55, 3.73 and 3.25 mm, respectively. Since there are differences in thicknesses, δt^3 of these plates are compared in Fig. 5 instead of δ. Even though all these plates have similar modal shapes in mode 1, there are clearly some differences in their deflection values. Whether these differences are of significance from the standpoint of evaluating these plates should be further examined.

Conclusions

This exploratory study has shown that natural frequencies are sensitive to cross-plying, fiber misorientation and ply overlap. Mode shapes and deflections are also sensitive to these parameters, but less conspicuous than the natural frequencies.

Damping ratio, on the other hand, does not indicate any trend.

The manufacturing variations considered here are not like the internal defects seen in compression molded parts, e.g., delaminations and voids. A separate study is required to examine how these defects may influence the modal parameters of compression molded composites. From the studies on laminated composites mentioned in the Introduction section, it is anticipated that delamination will have discernible effects on some of the modal parameters.

There is a great need for developing a fast, relatively inexpensive nonrestrictive technique for evaluating the manufacturing defects in compression molded composites. To be successful, such a method must be easy to implement in a quality control environment where an operator will be able differentiate quickly between good and bad parts. Although this study is very preliminary, it shows some promises in that direction. More research is needed to prove the validity of modal testing as a quality control tool.

Acknowledgment

The author wishes to acknowledge the help of Ms. Michelle Jacobs who performed the modal experiments as part of an undergraduate guided study under the author's direction.

References

1. Adams, R.D., Damping 1986 Proc., Vol. 2, AFWAL-TR86-3059, May (1986).
2. Gibson, R.F., "Principles of Composite Material Mechanics", P. 306, McGraw Hill, Inc., N.Y. (1994).
3. Gibson, R.F., Proc. 9th Annual ASM/ESD Advanced Composites Conference, 321-331 (1993).
4. Sanders, D.R., Y.I. Kim and N. Stubbs, Experimental Mechanics, 32, 3, 240-251 (1992).
5. Tracy, J.J. and G.C. Pardoen, J. Composite Matls., 23, 1200-1215 (1989).
6. Terek, L.H., E.G. Henneke II and M.D. Gunzburger, Composite Structures, 23, 253-262 (1993).
7. Mallick, P.K., Polymer Composites, 7, 1, 14-18 (1986).

Table 1. Compression molded XMC plates* used in modal tests

Plate No.	Layer Size	Number of Layers	Percent Mold Width Covered	Avg. Thickness (mm)	Weight (lb)	Comments
14	279 x 279	1	91.5	3.61	1.24	
16	279 x 279	1	91.5	3.24	1.20	
17	279 x 279	1	91.5	3.32	1.17	
29	279 x 279	1	91.5	3.19	1.18	
28	279 x 279	2	91.5	5.06	1.92	0/0 orientation
39	279 x 279	2	91.5	5.21	1.88	0/0 orientaion
42	279 x 279	2	91.5	5.03	1.82	0/90 orientation
30	279 x 229	1	75	2.59	0.96	
31	279 x 178	1	58.4	2.07	0.74	
32	279 x 127	1	41.6	1.39	0.52	
26	279 x 139.7	2	91.5	3.25	1.18	zero overlap
25	279 x 152.4	2	91.5	3.73	1.28	25 mm overlap
24	279 x 178	2	91.5	4.55	1.60	76 mm overlap

*Plate dimensions after molding = 305 x 305 mm for all plates

Table 2. Natural frequency and damping factor of XMC plates with 279 mm x 279 mm layer size.

Plate No. (Thickness)	Natural Frequency (Hz)				Natural Frequency Thickness				Damping Factor (%)			
	Mode 1	Mode 2	Mode 3	Mode 4	Mode 1	Mode 2	Mode 3	Mode 4	Mode 1	Mode 2	Mode 3	Mode 4
1. Single Layer												
14 (3.61 mm)	103.45	186.18	352.94	576.47	28.66	51.57	97.77	159.69	8.144	8.061	7.265	8.860
16 (3.24 mm)	99.15	178.89	331.34	550.66	30.60	55.52	102.26	169.96	8.216	4.666	2.706	1.933
17 (3.32 mm)	97.25	177.02	334.53	544.88	29.29	53.32	100.76	164.12	5.635	2.832	2.212	1.745
29 (3.19 mm)	98.29	179.23	326.65	545.38	31.01	56.18	102.40	170.97	7.237	3.843	2.224	1.624
2. Two Layers (0/0)												
28 (5.06 mm)	165.81	276.38	344.55	549.22	32.77	54.62	68.09	108.54	3.238	3.843	2.224	1.624
39 (5.21 mm)	162.49	273.44	344.01	460.73	31.19	52.48	66.03	88.43	4.258	2.653	1.997	1.871
3. Two Layers (0/90)												
42 (5.03 mm)	207.27	253.63	302.03	522.45	41.21	50.42	60.04	103.87	2.683	2.111	2.633	1.210

Table 3. Natural frequency and damping factor of XMC plates with various layer width.

Plate No.	Natural Frequency (Hz)				Natural Frequency Thickness				Damping Factor (%)			
	Mode 1	Mode 2	Mode 3	Mode 4	Mode 1	Mode 2	Mode 3	Mode 4	Mode 1	Mode 2	Mode 3	Mode 4
14 (279 mm x 279 mm) $\theta=9.5^{o}$	103.45	186.18	352.94	576.47	28.66	51.57	97.77	159.69	8.144	8.061	7.265	8.860
30 (279 mm x 229 mm) $\theta=14.5^{o}$	84.45	151.66	267.32	435.04	32.61	58.56	103.21	167.97	10.340	5.989	3.720	2.390
31 (279 mm x 178 mm) $\theta=15.5^{o}$	116.12	223.48	337.22	401.96	56.09	107.96	162.91	194.18	7.178	3.121	2.385	1.888
32 (279 mm x 127 mm) $\theta=23.5^{o}$	94.17	179.03	259.85	386.03	67.75	128.80	186.94	277.72	8.983	4.781	3.229	2.704

Table 4. Natural frequency and damping factor of XMC plates with various overlaps.

Plate No.	Natural Frequency (Hz)				Natural Frequency Thickness				Damping Factor (%)			
	Mode 1	Mode 2	Mode 3	Mode 4	Mode 1	Mode 2	Mode 3	Mode 4	Mode 1	Mode 2	Mode 3	Mode 4
14 (No overlap)	103.45	186.18	352.94	576.47	28.66	51.57	97.77	159.69	8.144	8.061	7.265	8.860
26 (zero overlap)	99.03	162.66	327.24	544.39	30.47	50.05	100.69	167.50	1.851	0.877	0.364	1.078
25 (25 mm overlap)	104.18	174.27	360.39	487.73	27.93	46.72	96.62	130.76	8.170	4.501	2.567	1.784
24 (76 mm overlap)	136.60	208.44	284.74	402.77	30.02	45.81	62.58	88.52	3.886	2.404	1.731	1.558

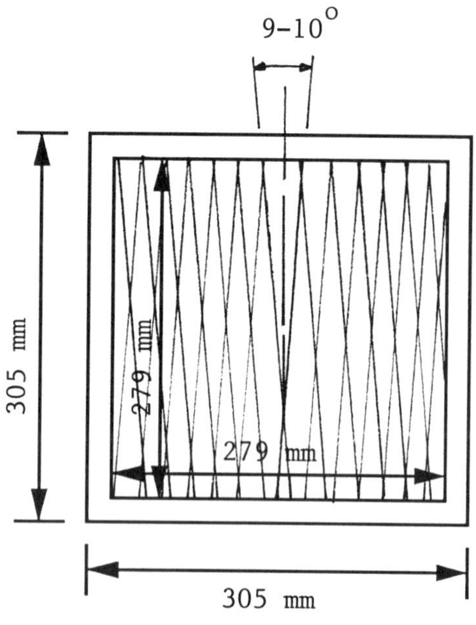
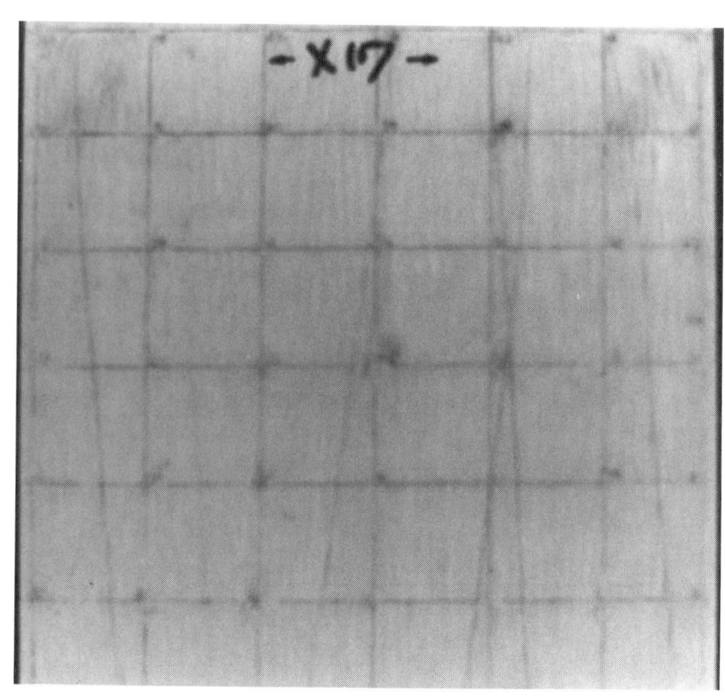

Fig. 1: XMC sheet with 91.5 percent mold width coverage (left) and
photograph of a molded plate. X-patterned tracer fibers as
well as horizontal and vertical grid lines can be seen in
the photograph.

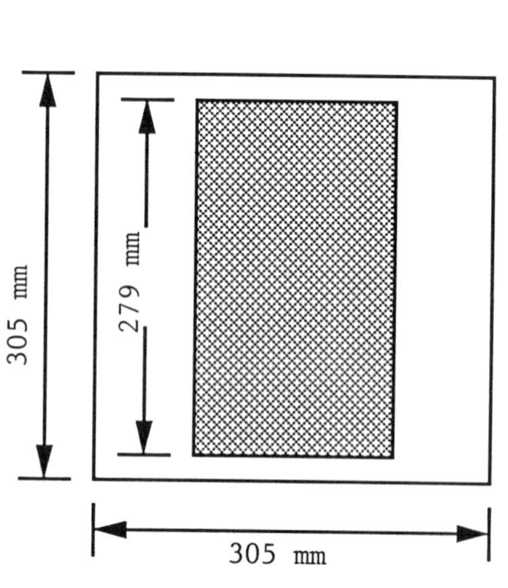
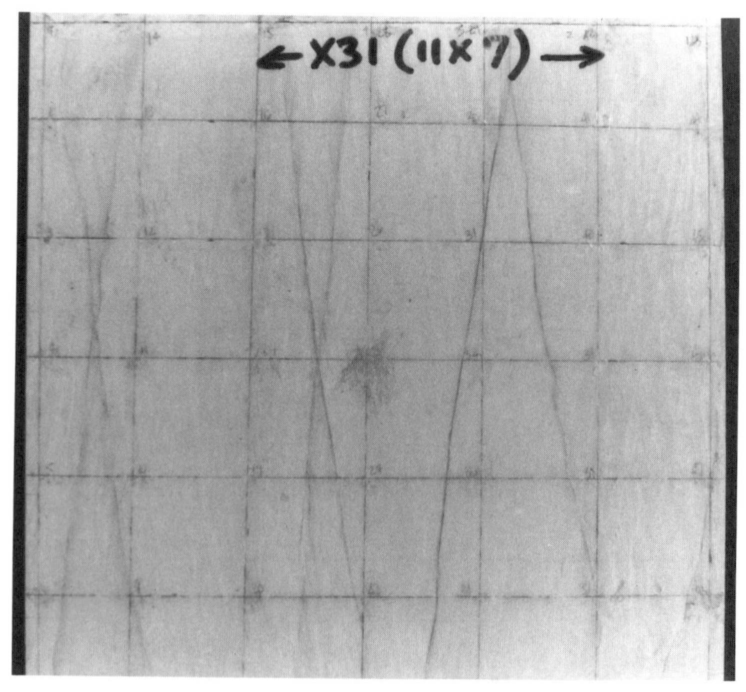

Fig. 2: XMC sheet with less than 91.5 percent mold width coverage
as in plates 30, 31 and 32. Photograph shows the fiber orientation in plate 31 (75 percent mold width coverage) after
molding.

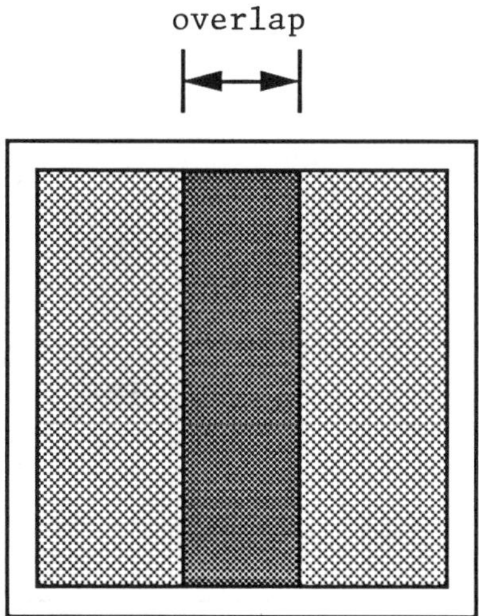

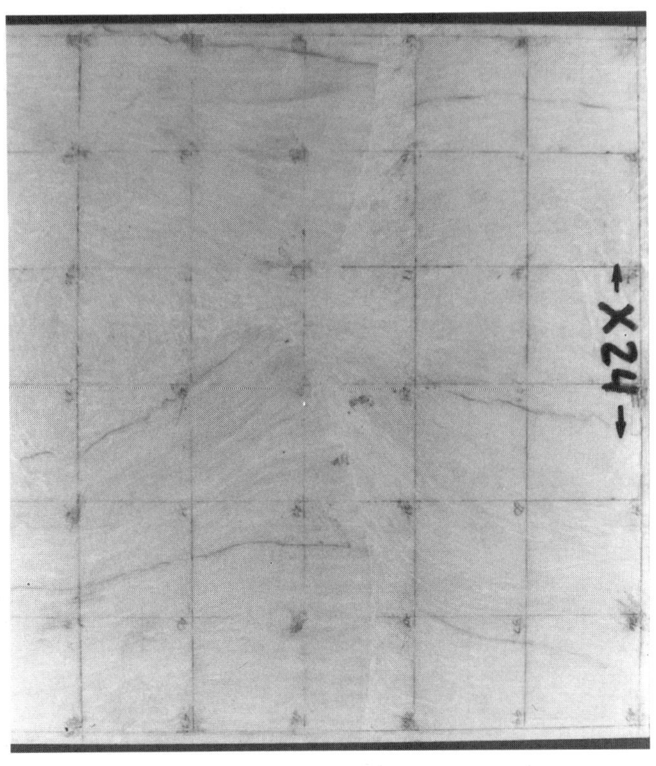

Fig. 3: XMC sheets with overlap at the center. Plates 24, 25 and 26 were molded with overlaps. Photograph shows the fiber orientation in plate 24 after molding.

Plate No.	Mode 1	Mode 2	Mode 3

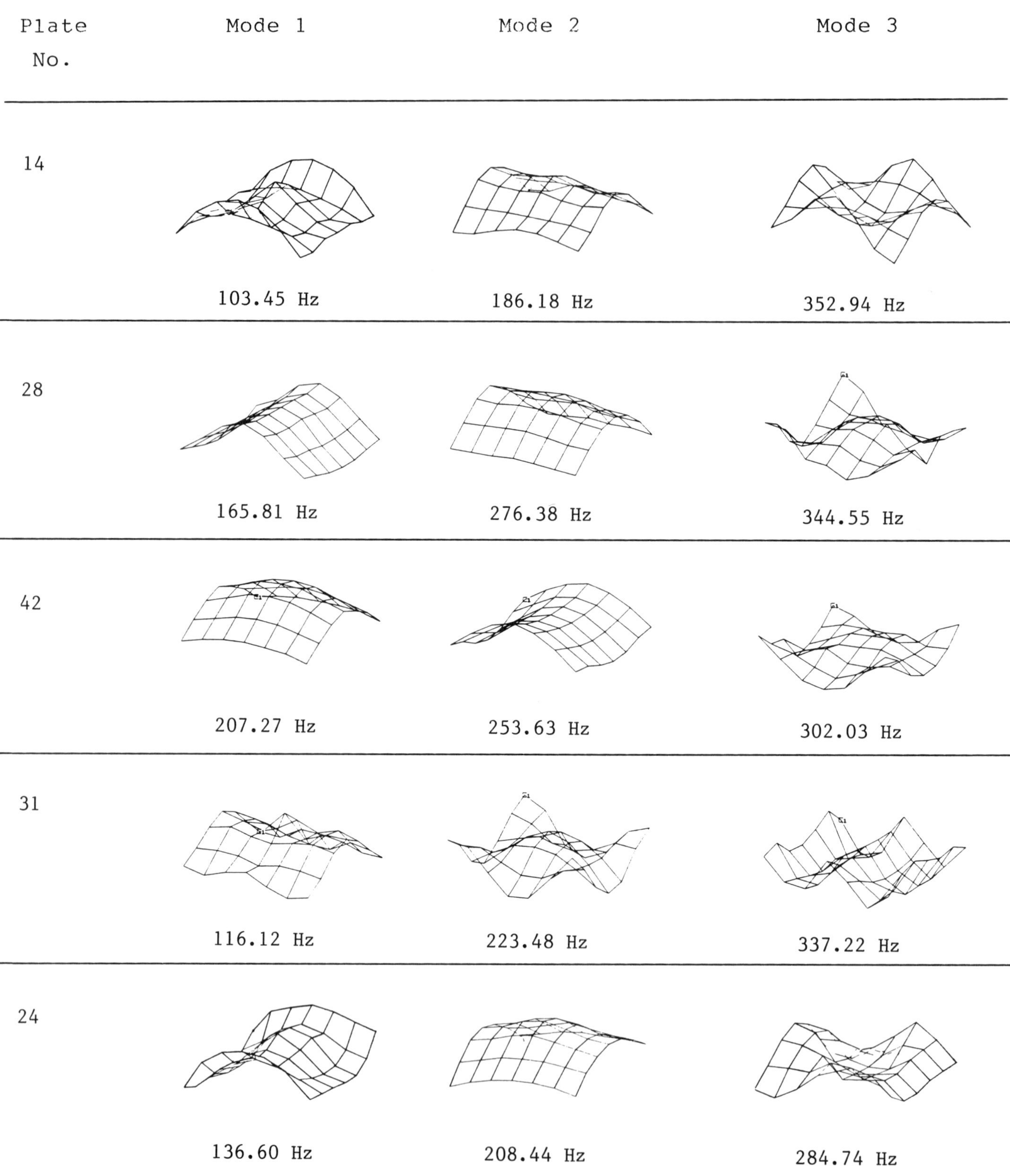

Fig. 4: Mode shapes of various XMC plates (14:single layer,279mmx279mm, 28:two layers,279mmx279mm,0/0, 42:twolayers,279mmx279mm,0/90, 31:single layer,279mmx178mm, 24:single layers,279mmx178mm, 76mm overlap)

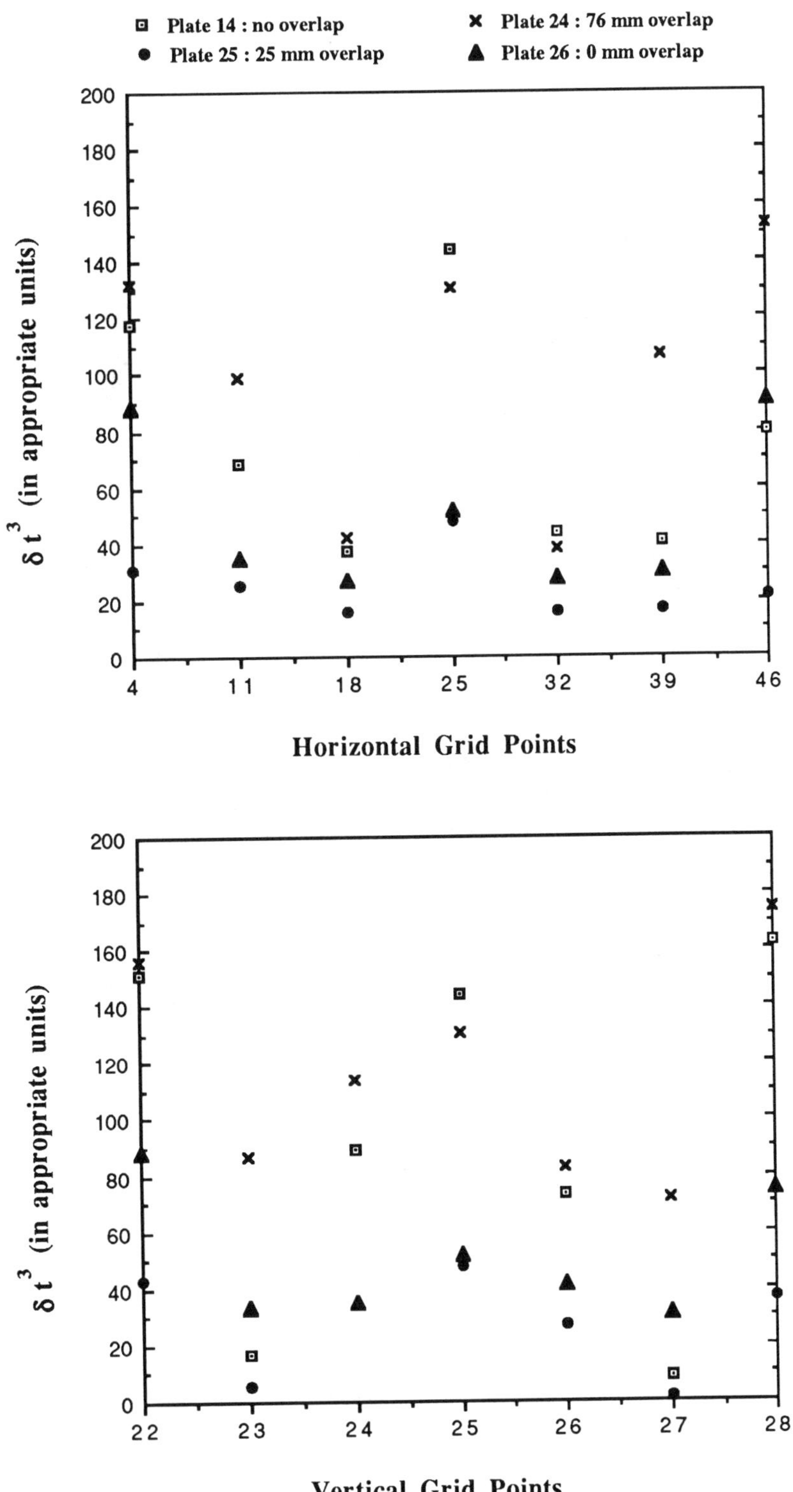

Fig. 5: Comparison of vibration amplitudes at various grid points along the horizontal and vertical center lines of XMC plates with different overlaps.

Property Distribution Determination for Nonuniform Composite Beams from Spectral Data and Galerkin's Method

W.-H. Chen, R.F. Gibson
Wayne State University
Detroit, Michigan

Abstract

This paper extends the authors' previous work regarding the determination of the property distributions in nonuniform composite beams by the use of the impulse-frequency response technique and Galerkin's method. In our previous work, the required frequency spectral data was generated by shifting the clamping location of the cantilever beam specimens. Attention is now focused on the generation of frequency data by the use of two types of clamped-mass beam configurations. One configuration has different masses attached at the free end of a cantilever beam and the other has the same mass located at different positions along the cantilever beam. The elastic constants and the density are all assumed to be functions of fiber volume fraction, while the spatial distribution of the fiber volume fraction is assumed to be given by a polynomial function. The concept of an effective density is employed to obtain the coefficients of the polynomial as the solutions in the inverse problem. Results show that the fundamental mode gives rise to better predictions than the higher modes do. Also, the error in the solution increases as the weight of the added mass increases for a fixed mode. Results are also compared with those obtained from the earlier model which was developed on the basis of shifting of the clamping location.

CHARACTERIZATION OF THE ELASTIC PROPERTIES of a machine or structural part by employing fast, inexpensive dynamic experiments has been one of the most challenging nondestructive testing problems in recent years. This kind of engineering problem is classified as an "inverse problem", and could be of great practical importance because of its potential applications in quality control, damage detection and optimal design. The impulse-frequency response technique is the most widely used experimental method for generating the required frequency data, and has been adopted for the determination of globally averaged properties of orthotropic composite panels in conjunction with certain theoretical models like Galerkin's method (1), the Rayleigh-Ritz method (2), the Rayleigh method (3,4), and the finite element method (5).

The inverse problem for the vibration of a nonuniform beam appears to have been first investigated by Barcilon (6,7). Gladwell (8,9) extended this work and established the necessary and sufficient conditions for a unique solution. According to these analyses, three sets of distinct frequency spectra corresponding to three different end conditions were required to reconstruct a nonuniform, isotropic and homogeneous beam, and these spectra must satisfy certain inequalities involving the interlacing of eigenvalues.

The present work is concerned with the determination of property distributions in heterogeneous composite beams. In order to obtain distinct frequency spectra, Barcilon and Gladwell changed the end conditions at the free end of a cantilever beam. The other two configurations in addition to clamped-free were clamped-pinned and clamped-sliding. This procedure was not only time consuming but was also difficult to conduct experimentally. In our earlier work (10,11), we developed the "shifting method" by retaining the clamped-free configuration and shifting the clamping location to obtain the required distinct frequency spectra. The generated eigenvalues satisfied Gladwell's necessary and sufficient conditions and the results were quite encouraging. However, shifting the clamping locations may not be easily carried out when the structure geometry is complicated. In the present work, we investigated two different configurations under the general category of the "added-mass method" : (1) the clamped-mass configuration, which involves clamping one end of the beam and adding different masses at the free end; and (2) the clamped-mass-free configuration, which involves clamping one end of the beam, retaining the other end free and adding the same mass at different positions along the beam. The added masses are small and light in comparison with the beam itself in order to avoid the rotatory inertia effect, and are cylindrical for the purpose of simulating distributed load exerted across the beam widths.

As with our most recent previous work (11)], analytical models are again derived from the Bernoulli-Euler beam theory and Galerkin's method. Although we had initially used an "effective modulus" approach (10), we had concluded that the Galerkin's approach was better. Composite modulus and density are calculated from the fiber volume fractions based on the micromechanics of composite materials. The distribution of fiber volume fraction is assumed to be given by a polynomial function. The coefficients are obtained by solving a set of simultaneous equations which could be linear or nonlinear. The appropriate solution is then found by employing the concept of an "effective density".

Analysis

Based on the Bernoulli-Euler beam theory, the equation of motion for the flexural vibration of a composite beam with constant cross-sectional area is

$$\frac{\partial^2}{\partial x^2}\left[E_c(x)I\frac{\partial^2 w(x,t)}{\partial x^2}\right] + \rho_c(x)A\frac{\partial^2 w(x,t)}{\partial t^2} = 0 \qquad (1)$$

where $E_c(x)$ = local composite modulus

$\rho_c(x)$ = local composite density

$w(x,t)$ = transverse displacement of beam

x = distance along beam (Fig.1)

t = time

I = area moment of inertia of beam cross section about neutral axis

A = cross sectional area of beam

The local longitudinal modulus and density of a unidirectional fiber reinforced laminated beam with nonuniformly distributed fiber volume fraction along the length can be expressed by the rule of mixtures as

$$E_c(x) = E_f v_f(x) + E_m v_m(x) = v_f(x)(E_f - E_m) + E_m \qquad (2(a))$$

$$\rho_c(x) = \rho_f v_f(x) + \rho_m v_m(x) = v_f(x)(\rho_f - \rho_m) + \rho_m \qquad (2(b))$$

where $v_f(x)$ = local fiber volume fraction, and the subscripts c, f and m refer to composite, fiber and matrix, respectively. It is assumed that fiber and matrix materials are isotropic, and that $v_f(x) + v_m(x) = 1$. The spatial distribution of fiber volume fraction in this work is assumed to be of polynomial type as

$$v_f(x) = \sum_{n=0}^{K} v_n x^n \qquad (3)$$

where the v_n are the coefficients to be determined and K refers to the order of the polynomial function. For free harmonic vibration at natural frequency, ω, the transverse displacement is separable in time and space as

$$w(x,t) = W(x)e^{i\omega t} \qquad (4)$$

where

$$W(x) = \sum_{i=1}^{N} C_i \phi_i(x) \qquad (5)$$

and

$\phi_i(x)$ = the ith mode shape function

C_i = the ith magnitude which indicates the contribution of the ith mode

Substitution of Equation (4) into Equation (1) gives the characteristic equation

$$\frac{d^2}{dx^2}\left[E_c(x)I\frac{d^2 W(x)}{dx^2}\right] - \rho_c(x)A\omega^2 W(x) = 0 \qquad (6)$$

and from Galerkin's method, Equation (6) becomes

$$\int_0^L \frac{d^2}{dx^2}\left[E_c(x)I\frac{d^2 W(x)}{dx^2}\right]\phi(x)dx - A\omega^2 \int_0^L \rho_c(x)W(x)\phi(x)dx = 0 \qquad (7)$$

For a cantilever beam with a mass M added at the free end (Fig.1), the boundary conditions are

$$\phi_i(0) = \frac{d\phi_i(0)}{dx} = 0$$

$$\frac{d^2\phi_i(L)}{dx^2} = 0 \qquad (8)$$

$$E_c(L)I\frac{d^3\phi_i(L)}{dx^3} = -M\omega^2\phi_i(L)$$

Integrating by parts over the domain $0 < x < L$ and employing the boundary conditions, Equation (7) becomes

$$\int_0^L E_c(x)IW''(x)\phi''(x)dx$$
$$- \omega^2\left[\int_0^L \rho_c(x)AW(x)\phi(x)dx + MW(L)\phi(L)\right] = 0 \qquad (9)$$

where the superscript (") implies d^2/dx^2. Substituting Equation (5) into Equation (9), we have

$$\sum_{i=1}^{N} C_i\left\{\int_0^L E_c(x)I\phi_i''(x)\phi_j''(x)dx\right.$$
$$\left. - \omega^2\left[\int_0^L \rho_c(x)A\phi_i(x)\phi_j(x)dx + M\phi_i(L)\phi_j(L)\right]\right\} = 0 \qquad (10)$$

where $j = 1,2,\cdots,N$
or alternatively,

$$[k_{ij} - \omega^2\rho_{ij}]\{C_i\} = 0 \qquad (11)$$

where

$$k_{ij} = \int_0^L E_c(x) I \phi_i''(x) \phi_j''(x) dx \qquad (12(a))$$

and

$$\rho_{ij} = \int_0^L \rho_c(x) A \phi_i(x) \phi_j(x) dx + M \phi_i(L) \phi_j(L) \qquad (12(b))$$

are the components of the $N \times N$ symmetric stiffness and density matrices, respectively. Equation (11) represents the eigenvalue problem for an N degree-of-freedom discrete system. The determinant of the coefficient matrix should be equal to zero for nontrivial solution of eigenvectors $\{C_i\}$, and by substituting Equations (2) and (3) into Equation (11), we have

$$\left| k_{ij} - \omega^2 \rho_{ij} \right| = 0 \qquad (13)$$

Accordingly, a set of $K+1$ homogeneous algebraic equations in terms of coefficients $\{v_n\}$ will be generated. For the forward problem, the natural frequencies, ω, can be calculated from the known properties k_{ij} and ρ_{ij}. For the inverse problem, these properties are determined from the measured natural frequencies, ω, and from Equations (12), (3) and (2), the coefficients $\{v_n\}$ can be found.

To be more realistic, this paper employs the mode shape functions and eigenvalues for clamped-mass beam configuration base on the work of Laura, et al. (12). These exact functions were established for different mass ratios (added mass / beam mass) and for beams with uniformly distributed properties. The physical eigenvalues are calculated by Newton-Gregory backward interpolation of order 4 from Laura's eigenvalue table. The nonuniform property distributions are assumed to have little or no effect on the mode shapes at this point. As shown later, the experimental measurements verify the validity of this assumption. For the ith mode, the eigenvalue, λ_i, is calculated by the following transcendental equation

$$\frac{1}{\lambda_i} \frac{1 + \cos \lambda_i \cosh \lambda_i}{\sin \lambda_i \cosh \lambda_i - \cos \lambda_i \sinh \lambda_i} = \frac{M}{M_b} \qquad (14)$$

For different mass ratios (M/M_b), the eigenvalues are different. The corresponding mode shape function for mode i is

$$\phi_i(x) = \cosh \frac{\lambda_i x}{L} - \cos \frac{\lambda_i x}{L} - \left(\frac{\cosh \lambda_i x + \cos \lambda_i x}{\sinh \lambda_i x + \sin \lambda_i x} \right) \left(\sinh \frac{\lambda_i x}{L} - \sin \frac{\lambda_i x}{L} \right) \qquad (15)$$

where L is the free length of the beam. When the mass ratio is zero, Equation (14) becomes $1 + \cos \lambda_i \cosh \lambda_i = 0$, which is the well known frequency equation for a cantilever beam. In this work, we use the eigenfunctions corresponding to the clamped-free configuration for a uniform beam (13) in the analysis of clamped-mass-free case. Substitution of Equation (15) into Equation (13) yields

$$\left| \frac{\lambda_i^2 \lambda_j^2 I}{L^4} \int_0^L \left[\sum_{n=0}^K v_n x^n \left(E_f - E_m \right) + E_m \right] \phi_i''(x) \phi_j''(x) dx \right.$$
$$\left. - \omega^2 \left\{ A \int_0^L \left[\sum_{n=0}^K v_n x^n \left(\rho_f - \rho_m \right) + \rho_m \right] \phi_i(x) \phi_j(x) dx + M \phi_i(L) \phi_j(L) \right\} \right| = 0 \qquad (16)$$

Experiments

The two types of composite beam arrangements that were investigated in this research are shown schematically in Fig.2. Case 1 stands for the beams having fiber volume fraction distribution decreasing along the length in a stepwise linear fashion, while Case 2 represents the beams with fiber volume fraction increasing along the length in a similar fashion. There were two specimens in each case. These laminated beams were made of 10 plies of 3M Scotchply 1003 unidirectional glass/epoxy prepreg tape by using an autoclave-style press cure. The micromechanical properties for use in Equations (2(a)) and (2(b)) are assumed to be as follows :

$$E_f = 10.5 \times 10^6 \; psi \quad (72.4 \; GPa)$$
$$\rho_f = 0.092 \; lb/in^3 \quad (2540 \; kg/m^3)$$
$$E_m = 0.55 \times 10^6 \; psi \quad (3.79 \; GPa)$$
$$\rho_m = 0.045 \; lb/in^3 \quad (1230 \; kg/m^3)$$

Specimen dimensions, for both cases, were

free length $L = 9 \; in \quad (228.6 \; mm)$

width $= 0.6 \pm 0.005 \; in \quad (15.24 \pm 0.127 \; mm)$

thickness$= 0.082 \pm 0.0015 \; in \quad (2.0828 \pm 0.0381 \; mm)$

The natural frequencies and mode shapes were measured by using the impulse-frequency response technique. The flexural vibration apparatus for cantilever beam specimens is shown in Fig.3. The specimen is excited by an electromagnetic hammer with a force transducer attached at its tip. The response is detected by a noncontacting eddy current proximity transducer, and both input force and displacement response signals are fed into the Fast Fourier Transform (FFT) analyzer which displays the transfer function on the screen in real time. The peak frequencies, or the natural frequencies of the corresponding modes are thus determined. The test fixture was calibrated by testing a uniform isotropic aluminum beam specimen in the clamped-free configuration. The error between the estimated and physical moduli for the calibration specimen was within 3.5%.

The transfer function $H(\omega)$ is defined as the ratio of the frequency spectrum of the response, $X(\omega)$, to the frequency spectrum of the input excitation, $F(\omega)$, and may be expressed in complex form,

$$H(\omega) = \frac{X(\omega)}{F(\omega)} = |H(\omega)|e^{i\theta} = H_R(\omega) + iH_I(\omega) \tag{17}$$

where

$$H_R(\omega) = |H(\omega)|\cos\theta = \text{real part of transfer function}$$
$$H_I(\omega) = |H(\omega)|\sin\theta = \text{imaginary part of transfer}$$
$$\text{function}$$
$$\omega = \text{frequency}$$
$$\theta = \text{phase angle}$$

From the Nyquist plot (Fig.4), the imaginary part of the transfer function is the only non-zero term at the resonant state, $\omega = \omega_n$, where ω_n is the natural frequency for the nth mode. By taking multiple transfer function measurements along the specimens, the mode shapes can be generated for each mode. Recall that in the discussion leading to Equation (14), it was assumed that the mode shapes were not affected significantly by nonuniform property distributions. Figures 5 and 6 show the mode shapes for the first three flexural modes of a nonuniform composite beam in Case 1 under both the clamped-mass and clamped-free configurations, respectively. It is seen that the nonuniform property distributions do have negligible influence on the mode shapes.

The true physical fiber volume fractions were obtained by using a burn-out test. In this test, the matrix resin from small samples taken from different positions along the beam specimens is burned away in a furnace at a temperature of 1050°F (565°C) for two hours. From the weight differences of these small pieces before and after the test, the in situ fiber volume fractions are computed. The scatter of these measured data was within 1.3 % of the mean value in all cases. The physical fiber volume fractions of the present beam specimens were higher than those reported in our previous work (10,11) because we used a higher laminating pressure during the fabricating process in this work.

Discussion

Since the specimens were fabricated with fiber content varying as shown in Fig.2, we selected the order of the polynomial function (Equation (3)) as 1, so that

$$v_f = v_0 + v_1 x \tag{18}$$

Before working on the inverse problem, it is worthwhile to carefully study the forward problem because the result for the forward problem is an indication of the validity of the theory and the experiment and is a necessary prerequisite to the solution of the inverse problem. For the forward problem, the coefficients v_0 and v_1 were found by curve-fitting Equation (18) to the burn-out test data. The predicted natural frequencies were computed based on Equation (11). The resulting differences between the predicted and the measured

natural frequencies for different modes and for different mass ratios in Case 1 are shown in Fig.7. These differences were judged to be within the acceptable range. The difference increases with increasing mass ratio and increasing mode number.

For the inverse problem, in order to compute the coefficients of the polynomial function, the number of the measured natural frequencies corresponding to different structures (different masses for clamped-mass or different positions for clamped-mass-free conditions) must be greater than the order of the polynomial function. Generally, $n+1$ frequencies are required for the nth order function assumption. Therefore, two natural frequencies, one from each of two distinct structures are needed to compute the coefficients v_0 and v_1. Each combination of two different added masses will give rise to a solution. The concept of an effective density is then employed for the selection of the coefficients. The effective density of an equivalent homogeneous beam with a constant cross-sectional area is defined as

$$\rho_{c,eff} = \frac{\int_0^L \rho_c(x)dx}{\int_0^L dx} \tag{19}$$

where $\rho_c(x)$ is the local composite density defined in Equation (2(b)). The effective specimen density is easily measured. The best solution for v_0 and v_1 is therefore the one which yields the least difference between the calculated and the measured densities.

Figure 8 shows the differences between the measured and predicted fiber volume fractions and effective densities with respect to different mass ratios based on the first mode in the case of the clamped-mass configuration. The mass ratio here is defined as the summation of the individual mass ratio which generates a specific structure. The difference in fiber volume fraction is defined as the average of the absolute values of the differences between the measured, v_{fm}, and predicted, v_{fp}, fiber volume fractions at 8 different locations ($x/L = 0, 0.17, 0.31, 0.44, 0.58, 0.72, 0.83, 1$) along the beam specimen, and can be expressed as

$$\frac{1}{8}\sum_{j=1}^{8} \frac{\left|v_{fp}^{(j)} - v_{fm}^{(j)}\right|}{v_{fm}^{(j)}} \times 100 \tag{20}$$

where j refers to the jth location. The difference in effective density is defined as

$$\left|\frac{\rho_p - \rho_m}{\rho_m}\right| \times 100 \tag{21}$$

where ρ_p is the predicted density and ρ_m is the measured density. The differences increase slightly with increasing mass ratio, but not in a very strict sense. This may be because the added mass would dominate the free vibration as the mass increases and the rotatory inertia effect should be taken into account in the theoretical model. Figures 9 and 10 show the results for the measured and predicted fiber volume fraction distributions in Case 1 and Case 2, respectively. These results correspond to the second points from the left in Figure (8), and are deduced from two distinct structures. In Case 1, structures were built with mass ratios of 0 and 0.1234, while in Case 2, structures were generated with mass ratios of 0 and 0.1205. Results obtained from the shifting method are also sketched together for comparison. The shifting value C represents the distance between the new and the original clamping locations. The averaged differences between the measured and predicted fiber volume fractions, as defined in Equation (20), for different modes are listed in Table 1, which shows that the fundamental mode gives rise to better prediction of fiber volume fraction distribution than the higher modes do. This can be explained from the following two aspects.

(a) *"Ill-condition" analysis*. The systems of simultaneous equations resulting from inverse problems are often ill-conditioned (14). For a system of the form

$$\cdot [D]\{X\} = \{B\} \qquad (22(a))$$

or

$$\{X\} = [D]^{-1}\{B\} \qquad (22(b))$$

the nonsingular matrix $[D]$ is said to be ill-conditioned for inversion when the solution vector $\{X\}$ is overly sensitive to small errors in the vector $\{B\}$. The condition number, $Q(D)$, of the coefficient matrix $[D]$ is used to quantify the degree of ill-conditioning (14),

$$Q(D) = \|D\| \, \|D^{-1}\| \qquad (23)$$

where

$$\|D\| = \underset{i=1,\cdots,N}{Max} \sum_{j=1}^{N} |D_{ij}| \qquad (24)$$

is defined as the norm of an $N \times N$ matrix $[D]$, and $\|D^{-1}\|$ is the norm of the inverse matrix $[D^{-1}]$. If the condition number is less than or equal to one, the system is said to be well-conditioned. If the condition number is greater than one, the system is ill-conditioned. The greater the condition number, the worse the ill-conditioning. Thus, the condition number is an indication of the potential for output error due to input error.

Notice that this ill-condition analysis is applicable only to the linear equations in Equations (22(a)) and (22(b)). In this work, Equation (13) is linear in coefficients v_0 and v_1 when only first mode is to be considered, and is nonlinear when higher modes are to be investigated even when the order of polynomial $K = 1$. It is necessary to linearize the nonlinear equations in order to conduct the test for ill-conditioning. We linearize Equation (13) by expanding it in a Taylor series about an equilibrium solution and looking at the first order terms. The following is a demonstration of the ill-condition analysis for the first mode. Based on the first order polynomial assumption for the fiber volume fraction distribution (Equation (18)), Equation (16) becomes

$$\left| \frac{\lambda_i^2 \lambda_j^2 I}{L^4} \int_0^L \left[(v_0 + v_1 x)(E_f - E_m) + E_m \right] \phi_i''(x) \phi_j''(x) dx \right.$$

$$\left. - \omega^2 \left\{ A \int_0^L \left[(v_0 + v_1 x)(\rho_f - \rho_m) + \rho_m \right] \phi_i(x)\phi_j(x) dx \right.\right.$$

$$\left.\left. + M\phi_i(L)\phi_j(L) \right\} \right| = 0 \qquad (25)$$

Under the first mode, Equation (25) becomes

$$\frac{\lambda_1^4 I}{L^4} \int_0^L \left[(v_0 + v_1 x)(E_f - E_m) + E_m \right] {\phi_1''}^2 (x) dx$$

$$- \omega^2 \left\{ A \int_0^L \left[(v_0 + v_1 x)(\rho_f - \rho_m) + \rho_m \right] \phi_1^2(x) dx + M\phi_1^2(L) \right\} = 0 \qquad (26)$$

or, in the form as Equation (22(a)) with

$$[D] = \left\{ \frac{\lambda_1^4 I}{L^4}(E_f - E_m)P_{11}^{(0)} - \omega^2 A(\rho_f - \rho_m)R_{11}^{(0)}, \right.$$

$$\left. \frac{\lambda_1^4 I}{L^4}(E_f - E_m)P_{11}^{(1)} - \omega^2 A(\rho_f - \rho_m)R_{11}^{(1)} \right\}$$

$$\{X\} = \{v_0, \ v_1\}^T$$

$$\{B\} = \left\{ -\frac{\lambda_1^4 I}{L^4}E_m P_{11}^{(0)} + \omega^2 \left[\rho_m A R_{11}^{(0)} + M\phi_1^2(L) \right] \right\}$$

where

$$P_{11}^{(k)} = \int_0^L x^k {\phi_1''}^2 (x) dx$$

and $\qquad R_{11}^{(k)} = \int_0^L x^k \phi_1^2(x) dx \qquad k = 0, \ 1$

It can be proved that matrix $[D]$ is positive definite in this physical problem. In this case, Equation (26) is linear in v_0 and v_1, and the ill-condition analysis can be easily conducted. In the case of higher modes, each element in matrix $[D]$ would have the similar form as the left hand side of Equation (26). The determinant of $[D]$ is thus nonlinear in v_0 and v_1, and the linearization routine is needed before the ill-condition analysis. The condition numbers for different modes and for different orders of the polynomial function assumptions are listed in Table 2. It is shown in Table 2(a) that higher modes give rise to worse ill-conditioning than the fundamental mode does. It was also found that for $K = 1$, the condition numbers for different mass ratios are very small (near zero) for the first mode. For K higher than one, as shown in Table 2(b), the condition numbers are very large, and a small amount of error in the measured frequencies would cause large errors in the solution. In fact, the results we obtained in this case are very unrealistic. Due to the limitation of our test fixture, we were not able to obtain precise enough frequency measurements for a satisfactory solution when the order of the polynomial function was higher than one.

(b) *Limitation of Bernoulli-Euler beam theory.* It is known that the transverse shear deformation and rotatory inertia effects are more pronounced in higher modes. These effects are not included in the Bernoulli-Euler beam theory. This is one reason why the fundamental mode is better than the higher modes in predicting the physical properties. This has been verified by the forward problem results which were shown in Figure 7.

Good results were also obtained for the clamped-mass-free configuration, as shown in Fig.11, where the mass was located at $x/L = 0.33$ and 0.78. When using the clamped-mass-free configuration, attention must be paid to the position of added mass along the beam specimen. For mode number higher than one, there are nodes along the beam axis when the beam is in free vibration. If the mass is added at or near one of these nodes, the necessary and sufficient condition of interlacing of eigenvalues would break down, and a unique solution would not be assured. The higher modes could then no longer be used to solve the inverse problem. This is just the case in this work. The best solution occurs when the masses are added near the nodal points for both the second and third modes, so only the first mode can be used to back out the coefficients of the polynomial function.

Conclusion

Analytical models based on measured frequency spectra and Galerkin's method have been developed for both the clamped-mass and clamped-mass-free beam configurations to determine the fiber volume fraction distributions in composite beams. The ill-condition analysis and the limitation of slender beam theory lead to the conclusion that the fundamental mode is better than the higher modes in the solution to the inverse problem. When the order of the polynomial for the distribution of fiber volume fractions is higher than 1, the ill-condition analysis verifies the infeasibility of the present methodology. This means that one would likely obtain an incorrect solution to the inverse problem by solving a set of simultaneous linear equations directly. Due to the imperfection of the theoretical model and the experimental apparatus, the errors resulting from these sources should be reduced as much as possible. An optimization scheme to minimize the discrepancy between the measured and predicted responses would be a possible solution to these problems. Discretizing the model on the basis of Galerkin's method and employing an optimization

procedure appears to be a feasible approach, and we will report on this method in future papers.

Acknowledgements

The authors gratefully acknowledge a graduate assistantship funded by the Institute for Manufacturing Research at Wayne State University. Additional support from a grant from Ford Motor Company is also greatly appreciated.

References

1. DeWilde, W. P., Narmon, B., Sol, H. and Roovers, M., Proceedings of the 2nd International Modal Analysis Conference, Orlando, FL, I, 44-49 (1984).
2. Deobald, L. R. and Gibson, R. F., Journal of Sound and Vibration, 124(2), 269-283 (1988).
3. Ayorinde, E. O. and Gibson, R. F., Vibro-Acoustic Characterization of Materials and Structures, P. K. Raju, Editor, NCA-Vol.14, 167-175, ASME (1992).
4. Ayorinde, E. O. and Gibson, R. F., Composites Engineering, 3(5), 395-407 (1993).
5. Nelson, M. F. and Wolf, J. A., Vibro-Acoustic Characterization of Materials and Structures, P. K. Raju, Editor, NCA-Vol.14, 227-233, ASME (1992).
6. Barcilon, V., Journal of Applied Mathematics and Physics (ZAMP), Vol.27, 347-357 (1976).
7. Barcilon, V., Royal Society of London, Philosophical Transactions, Ser.A, 304, 211-251 (1982).
8. Gladwell, G. M. L., "Inverse Problems in Vibration", Martinus Nijhoff Publishers (1986).
9. Gladwell, G. M. L., Journal of Sound and Vibration, 119(1), 81-94 (1987).
10. Gibson, R. F., Hwang, S. J. and Chen, W. H., Proceedings of the 8th ASM/ESD Advanced Composites Conference, Chicago, IL, 2-5 Nov., 317-326 (1992).
11. Gibson, R. F., Chen, W. H. and Hwang, S. J., Dynamic Characterization of Advanced Materials, P. K. Raju and R. F. Gibson, Editor, NCA-Vol.16/AMD-Vol.172, 33-40, ASME (1993).
12. Laura, P. A. A., Pombo, J. L. and Susemihl, E. A., Journal of Sound and Vibration, 37(2), 161-168 (1974).
13. Blevins, R. D., "Formulas for Natural Frequency and Mode Shape", Van Nostrand Reinhold (1979).
14. Hensel, E., "Inverse Theory and Applications for Engineers", Prentice Hall, Englewood Cliffs, NJ (1991).

Table 1 - Averaged (%) differences between the measured and predicted fiber volume fraction for different modes.

	First mode	Second mode	Third mode
Case 1	4.97	11.88	13.17
Case 2	6.06	9.79	10.53

Table 2 - Condition numbers for (a) different modes and $K = 1$, and (b) first modes and $K = 1,2$.

Mode number	Case 1	Case 2
1	4.82 ± 0.12	3.29 ± 0.02
2	78.67 ± 1.34	93.06 ± 2.70
3	262.42 ± 0.34	218.83 ± 41.33

(a)

K	Case 1	Case 2
1	4.82 ± 0.12	3.29 ± 0.02
2	18906 ± 1058	10704 ± 203

(b)

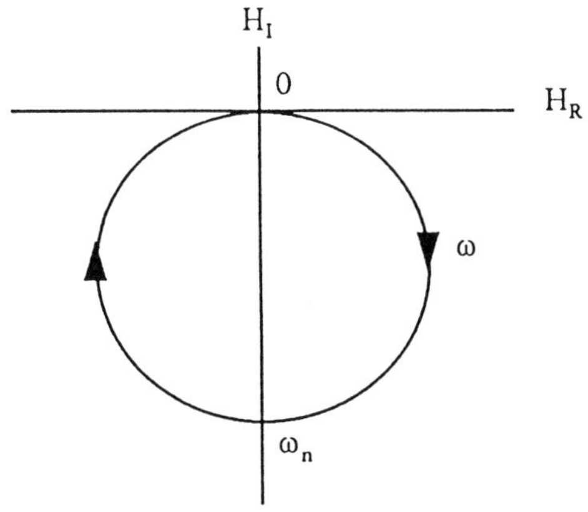

Fig.4 - Nyquist plot

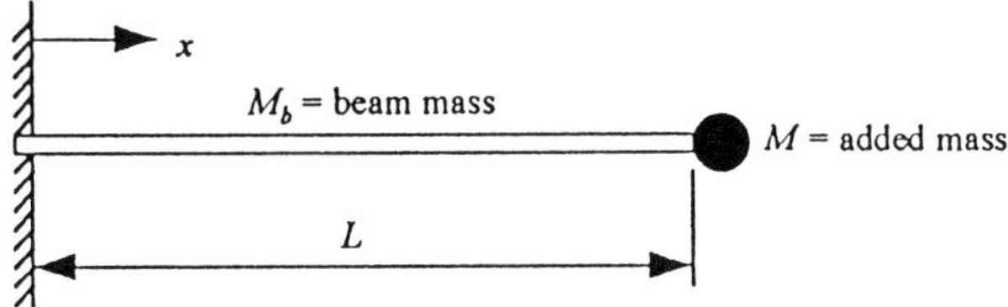

Fig.1 - Clamped-mass beam configuration.

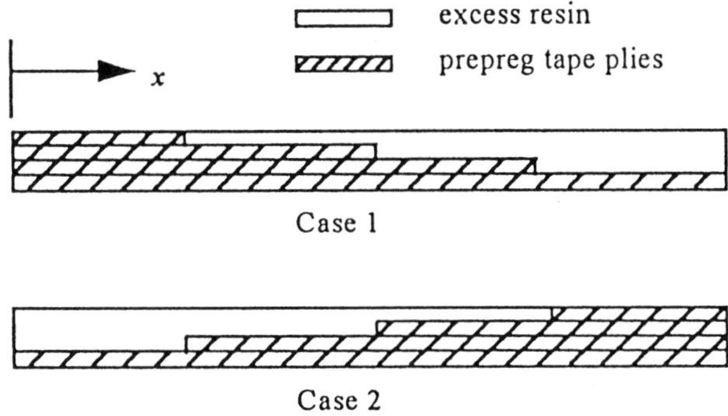

Fig.2 - Fabrication of beam specimens.

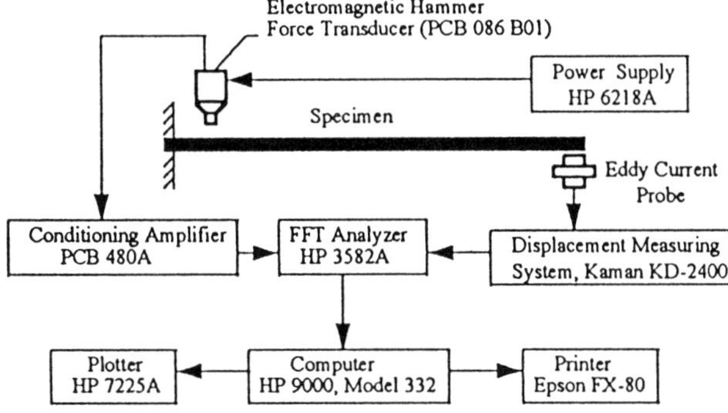

Fig.3 - Flexural vibration apparatus for beam specimens.

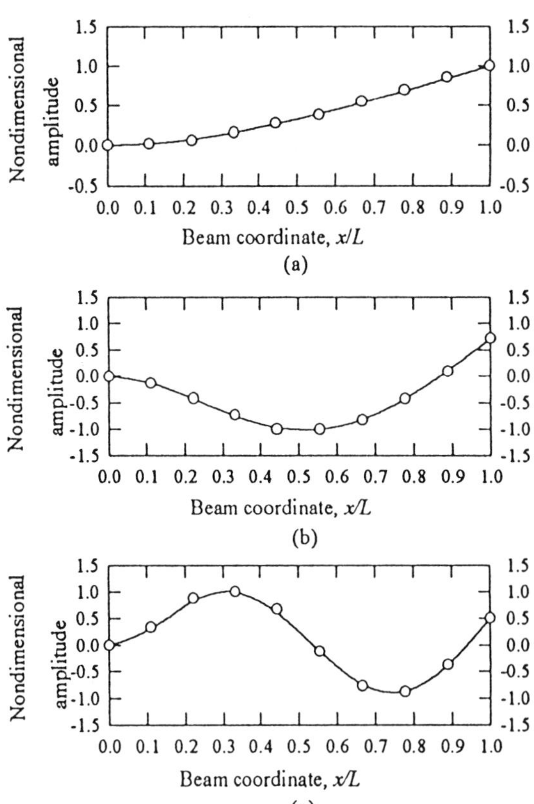

Fig.5 - Mode shapes for clamped-mass beam configuration with mass ratio $M/M_b = 0.1712$: (a) first mode, (b) second mode, (c) third mode.

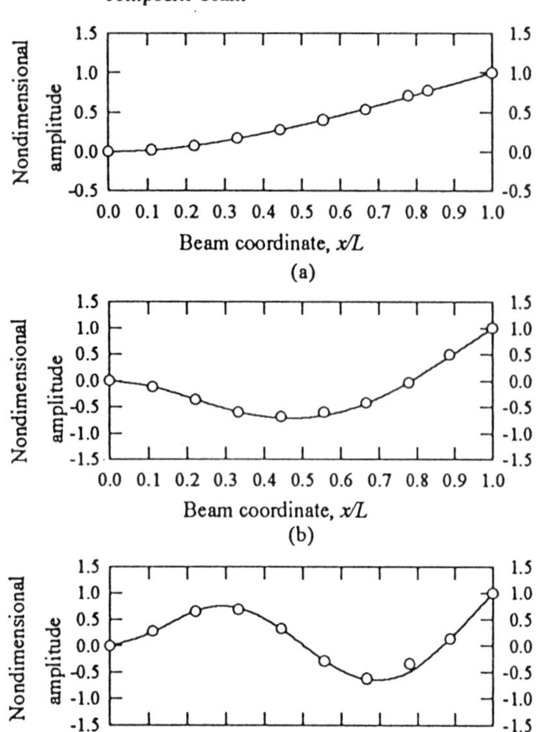

Fig.6 - Mode shapes for clamped-free beam configuration : (a) first mode, (b) second mode, (c) third mode.

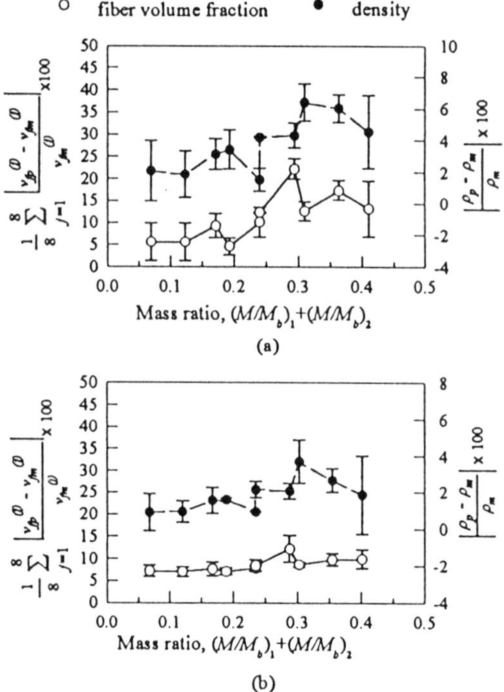

Fig.8 - Difference between measured and predicted fiber volume fractions and densities : (a) Case 1; (b) Case 2.

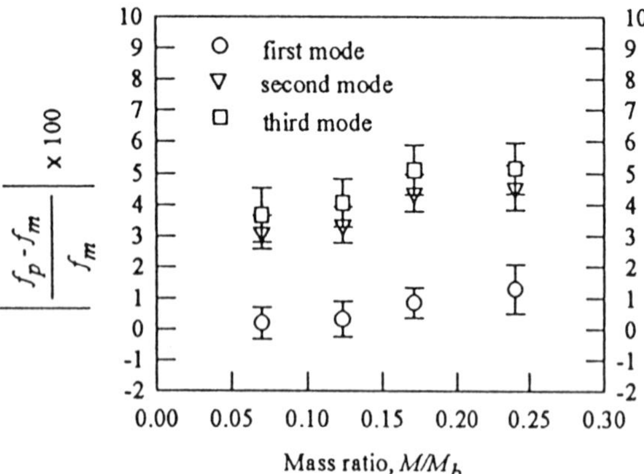

Fig.7 - Difference between measured and predicted natural frequencies for different modes with respect to mass ratio. (f_p - predicted frequency, f_m - measured frequency)

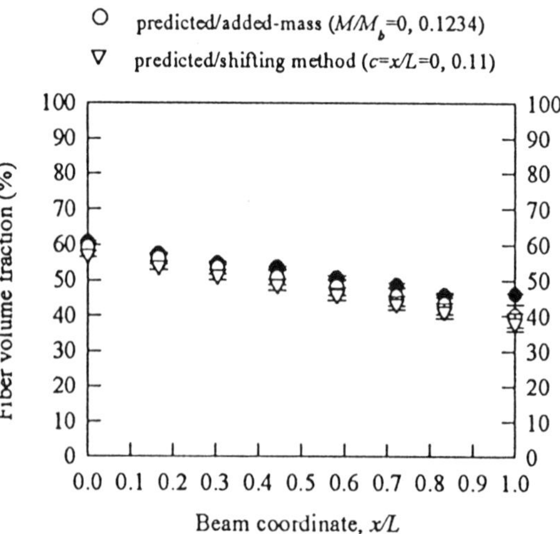

Fig.9 - Comparison between the measured and predicted fiber volume fraction distributions in Case 1.

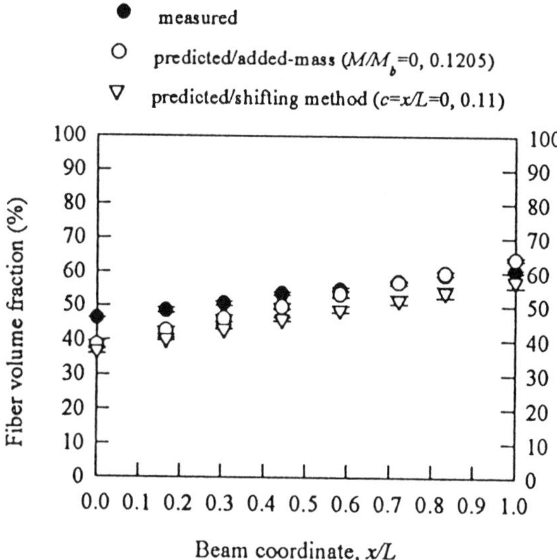

Fig.10 - Comparison of the measured and predicted
fiber volume fraction distributions in Case 2.

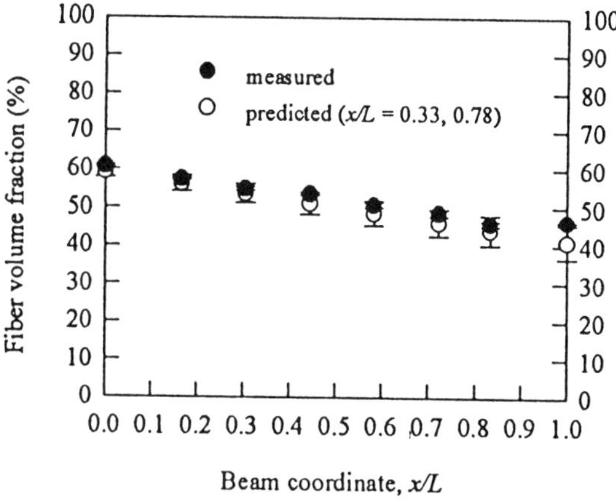

Fig.11 - The predicted fiber volume fraction distribution
from clamped-mass-free beam configuration in Case 1.

Damping Composite Tubes and Plates
Through Segmented Shear Coupling

K.C. Womack, T.H. Fronk, K. Shah, D.P. Widauf
Utah State University
Logan, Utah

Abstract

Damping through shear-coupling utilizes a viscoelastic layer sandwiched between composite laminates, the uni-directional fibers on either side of the viscoelastic having an opposite angular orientation. Under load the composite laminates orient themselves with the applied load, resulting in a scissoring motion that shears the viscoelastic layer. This deformation of the viscoelastic serves as a passive damping mechanism.

One modification to shear-coupling is to create chevroned laminate segments. The viscoelastic is sandwiched between two layers of the laminated chevrons, each chevron having the opposite angle orientation to the layer on the other side of the viscoelastic. These chevron segments increase the shear deformation within the viscoelastic, improving the damping.

Several composite tubes, of designs varying the number and thickness of viscoelastic layers, were tested for damping of free vibrations in the axial mode. Damping levels of close to 5 percent of critical damping were achieved in damped tubes, compared to 1.5 percent of critical damping in undamped tubes.

Additionally, two plates were tested for FMC Corporation as part of the Composite Armored Vehicle program. One plate was damped using segmented shear-coupling, the other undamped. The damped plate had four times the damping of the undamped plate for a transverse excitation.

Introduction

The idea of passive damping in composite structures is not new. However, in the past the main approach has been to use the constrained layer approach or a modification of that technique. One example in the use of constrained layer damping is the work performed by Quezon et al. [1]. Constrained layer damping consists of placing a viscoelastic layer between two composite laminates of a typical composite lay-up. Under a transverse dynamic load the bending of the composite structural element creates strains in the two composite layers that differ with their distance from the neutral axis. This difference in strain deforms the viscoelastic layer; energy dissipated in deforming the viscoelastic layer results in a damped structural response. For transverse dynamic loads this is a practical approach to passive damping. However, constrained layer damping is not functional for axial dynamic loads due to the fact that there is no difference in strain between the composite laminates.

A newer form of damping composite structural elements, one that can be used to dampen response to both transverse and axial excitations, is known as shear-coupling and comes from Barrett's work [2,3,4,5]. The results of the experimental work discussed in this paper focuses on the damping performance of composite tubes that utilize segmented shear-coupling as a passive damping mechanism.

Shear-Coupling

In this approach a viscoelastic is placed between two composite laminates, with the difference between shear-coupling and constrained layer damping being in the design of the laminates themselves. Constrained layer damping uses laminates made of laminae that have varying angles of orientation for uni-directional fibers; all laminae contained in a laminate used for shear-coupling have the same fiber orientation. However, the laminates on either side of the viscoelastic layer have an opposite angular orientation for the fibers (see Figure 1). When placed under an axial tensile load, for example, the fibers in each laminate will move to align themselves with the applied load--the fibers in one laminate rotating in a direction opposite to that of the fibers in the other laminate. Under a dynamic axial load this results in a scissoring movement of the two laminates that constantly shears the viscoelastic, providing a significant damping mechanism.

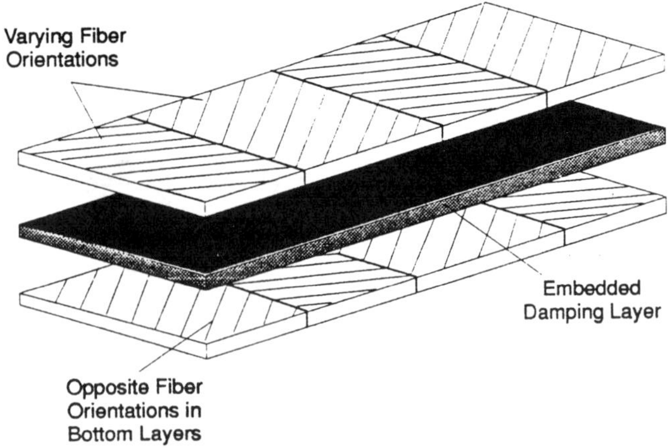

Figure 2. Segmented Shear Coupling, Varied Orientations of Composite Fibers

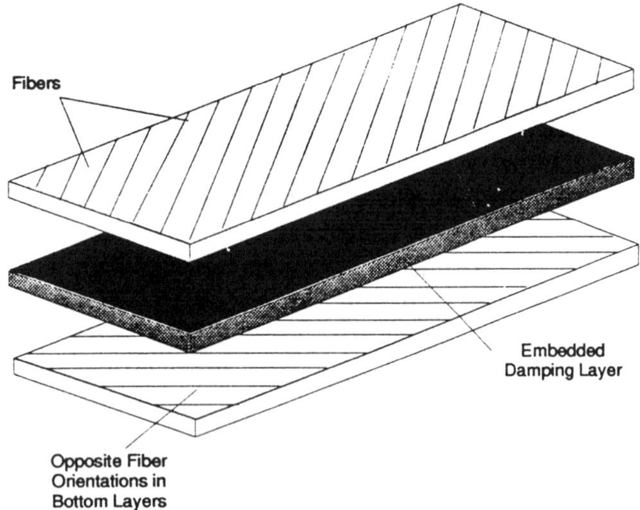

Figure 1. Damping by Shear Coupling

For transverse dynamic loads the damping mechanism for shear-coupling is the same as for constrained layer damping--the varying shears due to the difference in distance from the neutral axis for each composite layer. What is not known at this point is if the shear-coupling design will have improved damping characteristics over the constrained layer approach for transverse loads.

Segmented Shear-Coupling. One significant modification to the shear-coupling approach is the concept of using segmented laminates within each composite layer. This idea was first espoused by Dolgin [6] in his work at the Jet Propulsion Laboratory. The concept is to segment the laminates surrounding the viscoelastic, each segment having its composite fibers oriented at an angle negative to that of the two neighboring segments; the result being either a sinusoidal or chevron pattern within the laminate. The two laminates on either side of the viscoelastic maintain an opposite fiber orientation from each other (see Figure 2). This chevron pattern, with its opposite fiber orientations, causes an increase in the scissoring action with the result being more shear deformation and, thereby, increased damping.

Two of the most critical factors that influence the amount of damping gained from shear-coupling are the number of viscoelastic layers within a composite section and the thickness of these layers. Previous work by Finlinson et al. [7] indicates that there is an optimum thickness for these viscoelastic layers beyond which the damping either does not increase or actually decreases. This work looks to corroborate Finlinson's work and also look at the effect the number of viscoelastic layers has on the damping of composite tubes.

Given that damping is best determined through physical testing, an experimental program was developed to examine the effect that the number and thickness of viscoelastic layers has on the damping of composite tubes and plates. A total of 12 composite tubes and two composite plates were tested in free-vibration to determine their damping characteristics.

Test Specimen Design

Composite Tube Design. Six different composite tube designs were utilized, two tubes tested for each design. All of the tubes had an inner diameter of 1.25 inches and were laid up at a length of 27 inches, then cut down to 24 inches after curing. The composite used was a uni-directional Hercules IM6 graphite/epoxy prepreg, 0.170 mm (6.7 mils) thick. The viscoelastic layers were made of an Avery self-adhesive viscoelastic tape, 0.127 mm (5 mils) thick, that had a compatible cure temperature with the composite prepreg. All of the tubes used a total of 12 laminae to produce the graphite composite laminates.

The composite fibers within the tubes were oriented at an angle of $\pm 22.5°$, the angle alternating from plus to minus with each chevron. The angle of orientation for the composite fibers that will result in maximum damping is a function of the coefficient of mutual influence (cross-coupling coefficient). For the greatest damping the orientation of the fibers should be such that the difference in shear between the composite laminates is a maximum. Research conducted at Brigham Young University [8] has shown that for the Hercules IM6 angles of orientation between $\pm 20.0°$ and $\pm 25.0°$ for the composite fibers will result in near maximum differences between the shears in each composite laminate.

The number and thickness of viscoelastic layers were the design parameters altered for each design. One design was a control, having no viscoelastic layers. Three of the designs contained a single viscoelastic damping layer, these layers having thicknesses of 0.127 mm (5 mils), 0.254 mm (10 mils), or 0.381 mm (15 mils). The last two designs varied the number of viscoelastic layers, having either two or three layers of viscoelastic material, each layer being 0.127 mm (5 mils) thick. The overall thickness of the tubes and each composite laminate depends on the number of viscoelastic layers used. With one viscoelastic layer the two laminates were 1.02 mm (40.2 mils) thick, for two viscoelastic layers, the three composite laminates were 0.681 mm (26.8 mils) thick and for three viscoelastic layers the four composite laminates were 0.511 mm (20.1 mils) thick.

Composite Plate Design. The two composite plates used in this study are intended to serve as armor attachment stiffeners within an experimental armor system for the Composite Armored Vehicle Program, FMC Corporation being the contractor. These channel shaped plates have overall dimensions of 0.61 m × 0.91 m × 0.076 m (24 inches × 36 inches × 3 inches), see Figure 3. The thickness of each plate is approximately 6.35 mm (0.25 inches), the damped plate being somewhat thicker due to the viscoelastic layers. These plates are made of the same Hercules IM6 and Avery viscoelastic materials as the tubes discussed previously. After the plates were cured and before the experiments run, each of the plates was epoxied to a 0.91 m × 0.91 m × 6.35 mm (36 inch × 36 inch × 0.25 inch) graphite composite back-up plate that served as the composite vehicle skin.

The two channel shaped armor stiffener plates used the same symmetric lay-up, the differences between the undamped and damped plates being the use of the viscoelastic layers and segmented laminates

Figure 3. Armor Attachment Stiffeners

to create the chevrons in the damped plate. This symmetric lay-up consisted of (from the center-line of the plate to the outer skin): 3 laminae of composite prepreg at 0°, two laminae of composite prepreg at 90°, a 0.254 mm (10 mils) thick layer of viscoelastic for the damped plate, four laminae of composite prepreg at $\pm 25°$, a 0.254 mm (10 mils) thick layer of viscoelastic for the damped plate, four laminae of composite prepreg at $\pm 25°$, a 0.254 mm (10 mils) thick layer of viscoelastic for the damped plate, four laminae of composite prepreg at $\pm 25°$, and lastly a single laminae of composite prepreg at 90°.

Testing Procedure

The objective of the tests conducted was to determine the damping characteristics of each of the twelve composite tubes and two composite plates. This was done by monitoring the response of each tube and plate during a free-vibration response to a short duration impulse load.

The amount of damping characteristic to a structural element may be expressed in several ways: quality factor, reverberation time, damping loss factor, damping ratio (fraction of critical damping), etc. The damping of the tubes and plates tested in this experimental program was determined from frequency response curves using the half-power bandwidth method [9], this method calculates the damping loss factor, η. Within this paper the damping of a tube or plate will be expressed as the damping ratio, ζ, this is the factor most commonly used in structural applications and is multiplied by 100 to express the percent of critical damping. The damping ratio is related to the damping loss factor by a factor of 2; the damping loss factor being twice the damping ratio [10].

Test Equipment and Set-Up. The Andriulli test set-up was used to conduct the free vibration tests of the composite tubes (see Figure 4). Cap plates were placed on each end of the hollow tubes to provide mounting and striking surfaces. The tubes were hung from elastic supports and an accelerometer (PCB Model 309A) placed on one end. An impact hammer

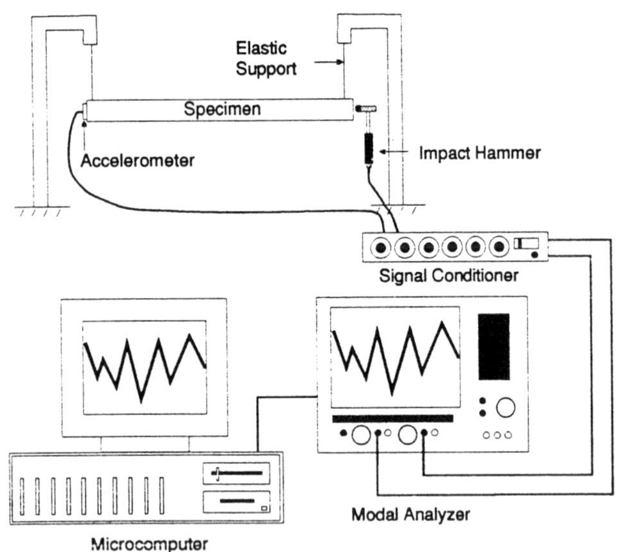

Figure 4. Andriulli Test Configuration

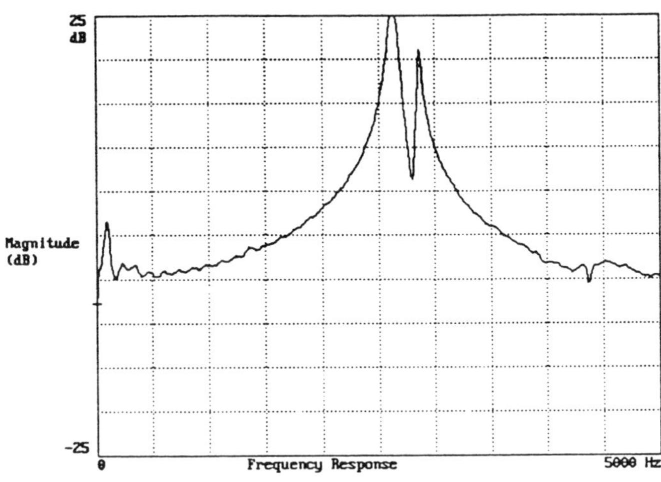

Figure 5. Frequency Response for Undamped Tube

(PCB Model 086C03) was used to provide the dynamic excitation of the tube. Both the impact hammer and accelerometer were connected to a modal analyzer (Hewlett-Packard Model 3560A), this analyzer collected the data provided by both the hammer and the accelerometer, the accelerometer data being the most critical. Each tube was struck 10 times and the data recorded, the modal analyzer averaged the 10 responses for each tube to determine a frequency response. Striking the tubes many times helped to reduce the variability of the response that comes from striking the specimen by hand. The data recorded for each tube was then downloaded to a computer and frequency response curves plotted. Two examples of these curves are shown in Figures 5 and 6. The response of an undamped tube is shown in Figure 5 and Figure 6 shows the response of a tube with two 0.127 mm (5 mils) thick layers of viscoelastic material; the wider spikes on the diagrams indicate a higher damping capacity.

For the two plates tested a modification of the Andriulli set-up was used (see Figure 7). The plates were suspended and an accelerometer attached to the back side of the plate. A small steel disk was used as a strike plate for the impact hammer on the front side of the composite plate. Other than these small deviations no changes were made in the test procedures from the tubes to the plates. The frequency response curves for each plate are shown in Figures 8 (undamped plate) and 9 (damped plate).

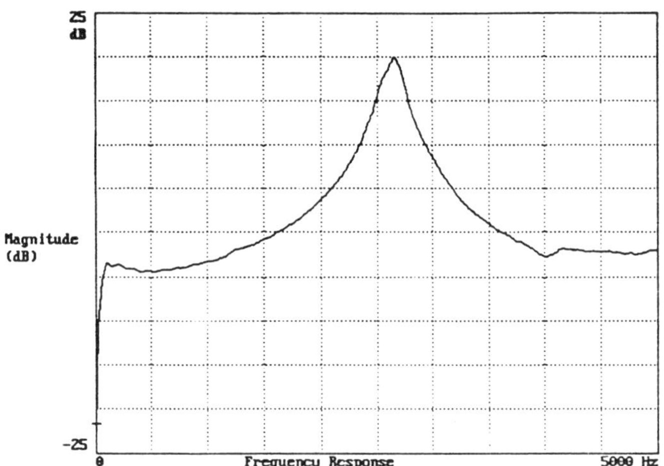

Figure 6. Frequency Response for Damped Tube
(Two Viscoelastic Layers, 0.127 mm Each)

Test Results

The results of the free-vibration tests performed on the 12 tubes, in terms of percent critical damping, are shown graphically in Figure 10. The damping for each pair of tubes is fairly consistent except for the pair of tubes containing the single, 0.381 mm (15 mils) thick viscoelastic layer. The extremely high damping for the one tube in this pair does not conform to the trend indicated by the data and is most likely due to a delamination within this one particular tube. The outcome of these tests does indicate that there is an optimal design for the damping of composite tubes, any alteration of which (either in the amount or arrangement of the viscoelastic layers) decreases the

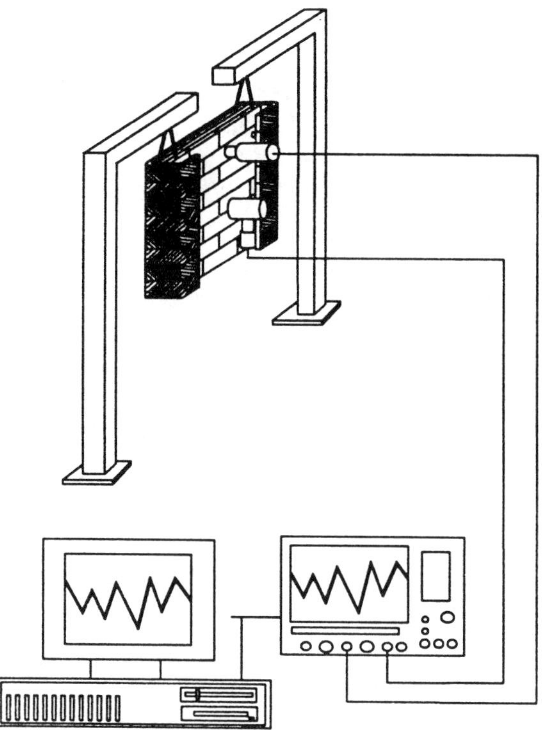

Figure 7. Modified Andriulli Test Set-Up

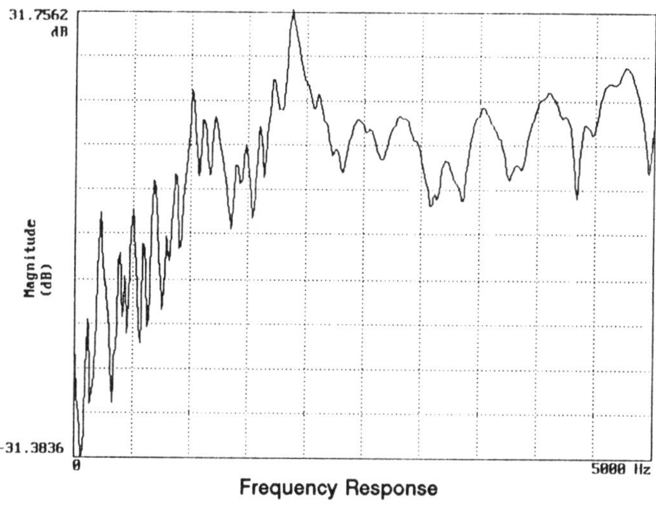

Figure 8. Frequency Response for Undamped Armor
Attachment Stiffener Plate

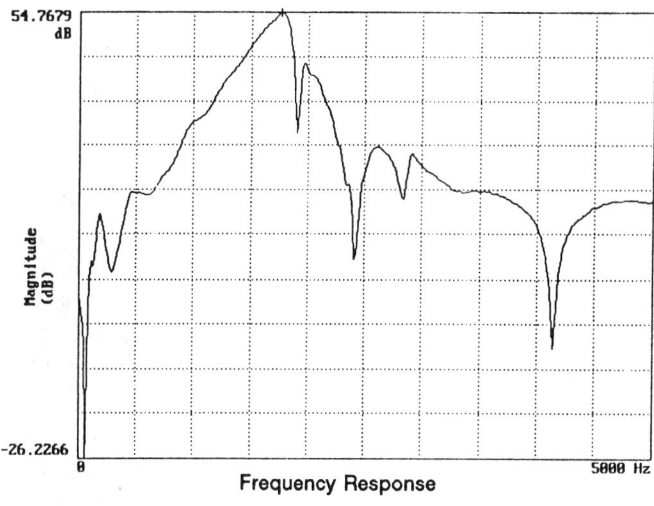

Figure 9. Frequency Response for Damped Armor
Attachment Stiffener Plate

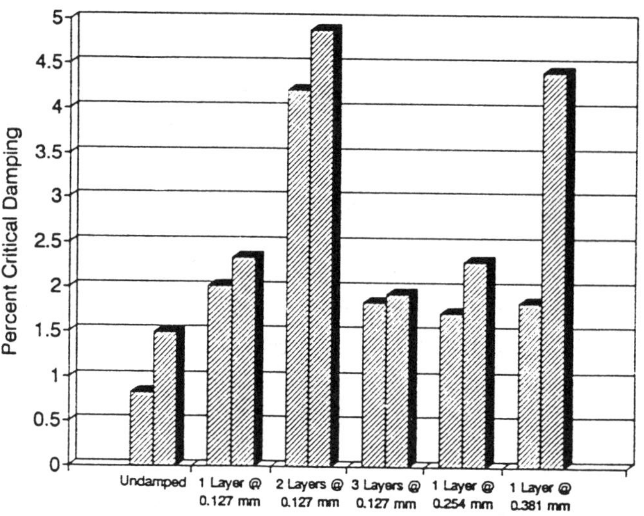

Figure 10. Comparison of Damping Levels
In Composite Tubes

damping capability of the tube. In the case of these tubes the optimal design was one that used two viscoelastic layers, each layer 0.127 mm (5 mils) thick, the viscoelastic being 10.5 percent of the total material volume. The fact that the damped tubes with the same amount of viscoelastic material, 10.5 percent of the material volume, placed in a single layer had a damping ratio nearly half of the two layer design indicates that not only is the amount of viscoelastic material critical to the damping performance, but the manner in which it is placed within the composite is critical as well. This concept of an optimal damping design is supported by Finlinson et al. [7].

The outcome of the damping tests on the plates is equally impressive to that of the tubes. The damped plate had a damping ratio of 0.0625 (6.25% of critical damping) compared to the undamped plate with a damping ratio of 0.015 (1.5% of critical damping), a four-fold increase in damping.

Conclusions

The tests conducted show without a doubt that segmented shear-coupling is an effective way of providing passive damping for axial and transverse vibrations of composite structural elements. The fact that this approach dampens axial vibrations is significant in that previous to this point in time little has been done to provide axial damping for composites.

The results also show, in conjunction with previous research [7], that the maximum effect of damping through shear-coupling can be attained by an optimal design. Simply providing more damping material (more or thicker viscoelastic layers) does not mean an improvement in damping characteristics. However, research up to this point does not provide enough information to be able to positively determine what the optimum damping design would be for a specific composite structural element. This should be the major area of future research, determining design criteria and methodologies for developing optimal designs for the damping of composite structural elements.

Acknowledgements

The authors would like to express their appreciation for the partial funding of this research provided by the FMC Corporation.

References

1. Quezon, A.J., D.C. Warwick and M.J. Martin, Naval Surface Warfare Center Report CDNSWD-SD- 92/60 (1992).

2. Barrett, D.J., Journal of Sound and Vibration 160(1), 187-191 (1993).

3. Barrett, D.J. and C.A. Rotz, Proceedings Second International Congress of Recent Developments in Air- and Structure-Borne Sound and Vibration, Auburn University, March 1992, pp. 257-263.

4. Barrett, D.J., U.S. Patent 5,087,491.

5. Barrett, D.J., Journal of Sound and Vibration 154(3), 453-465 (1992).

6. Dolgin, B.P., U.S. Patent Application.

7. Finlinson, L.W., Masters Thesis, Utah State University, 1994.

8. Olcott, D.D., Doctoral Dissertation, Brigham Young University, 1992.

9. Suarez, S.A., R.F. Gibson and L.R. Debold, Experimental Techniques, 19-24, October, 1994.

10. Gade, S. and H. Herlufen, Sound and Vibration 24(3), 24-32 (1990).

A Description of Impact Induced Damage Mechanisms in RTM Composites

R.W. Rydin, V.M. Karbhari
University of Delaware
Newark, Delaware

Abstract

Investigations of the impact response in many different glass reinforced vinyl-ester RTM composites have revealed several previously unreported damage types. Damage in classical laminates consists of matrix cracking, delamination and fiber breakage. These three damage types are observed in impacted RTM plates, however, the extent of delamination is less and there is evidence of interbundle, intrabundle and void pocket cracking. Understanding the sequencing of such failures and their interactions is critical to successful modelling of damage propagation during impact. Untrabundle cracking may be linked to residual processing induced stresses and may not be unique to glass reinforcement.

EXPERIMENTAL OBSERVATIONS OF DAMAGE induced by dynamic loading such as low velocity impact began in the early 1970's [1, 2] and has become a focal point of many investigators during the last two decades [3-9]. The composite systems studied most extensively were unidirectional and crossply laminates, typically consisting of carbon or glass reinforced plastics. Three classic damage mechanisms ubiquitous to laminated systems including delamination, matrix cracking and fiber breakage are discussed. For a more detailed account of these mechanisms we direct the interested reader to a review article by Abrate [10]. Following this introduction several additional failure modes are presented that we have observed in resin transfer molded (RTM) composites consisting of biaxial woven and knit glass reinforcement.

Before reviewing dynamic failure modes in laminated composites and comparing them with RTM behavior, several of the physical features distinguishing these two processing methods should be discussed. Laminates are simply stacks of prepreg material consisting of a planar fibrous reinforcement encased within a thermoplastic or thermoset resin. During a controlled pressure and temperature cure cycle in an autoclave, individual lamina are diffusion bonded into a continuous composite structure illustrated in Figure 1. Reinforcement in adjacent layers does not necessarily make intimate contact, hence there is a distinct layering of fiber and matrix.

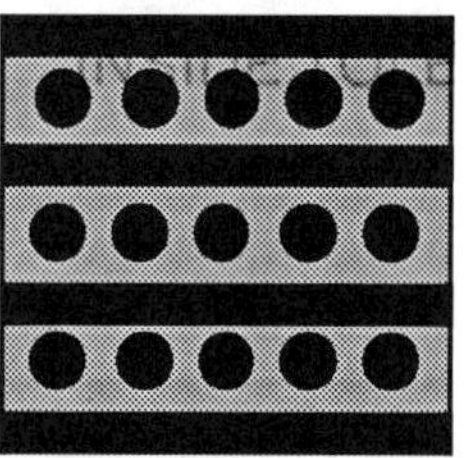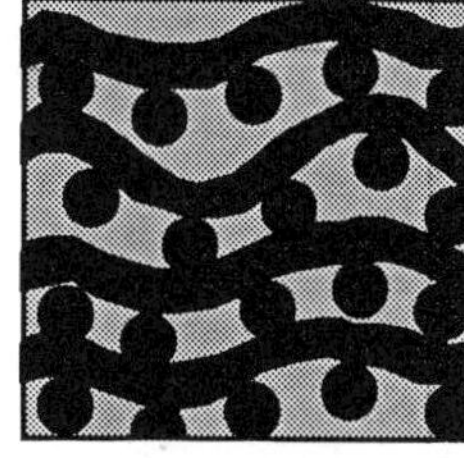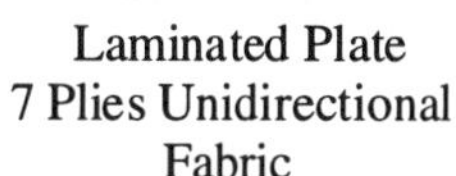

Laminated Plate	RTM Plate
7 Plies Unidirectional	4 Layers
Fabric	Biaxial Knit

Fig. 1 - Comparison of laminated and RTM microstructure

The RTM process is fundamentally different from autoclaving. Dry fiber preforms are compacted in a mold cavity which brings the fabric layers into varying degrees of intimate contact before any resin is injected. This step permits a certain amount of mechanical interlocking between layers, consequently, there are no distinct macroscale regions of matrix or reinforcement within this interconnected network. There is no guarantee that one resin interlayer will be as thick as another, or that the entire resin network is even contiguous. Establishing a homogenous void free part requires complete wetout between fiber bundles and between the fibrils within bundles. Small defects such as voids are present even under the best processing conditions, however, such defects are certainly not unique to RTM.

Classical Laminate Failures Modes

In the literature delamination is defined as the separation of adjacent reinforcing layers, however it seems to have become an all encompassing description of damage which behaves in a numerically manageable fashion and is casually associated with projected damage areas. Two different delamination initiation mechanisms are depicted in Figure 2, both originate from matrix cracks which propagate to an interface between reinforcing layers [11]. Transverse shear force resultants and transverse normal components are held responsible for matrix cracks which propagate rapidly along laminate intersections at speeds approaching 300 m/s [12, 13]. Several investigators originally suggested that reduced bending stiffness mismatch between adjacent plies causes delamination [14] and now it is almost universally accepted that reinforcement orientation differences between adjacent plies controls such growth [1, 10]. Delamination maps further indicate that the major delamination axis lies parallel to the fiber axis of the lower reinforcing layer [12].

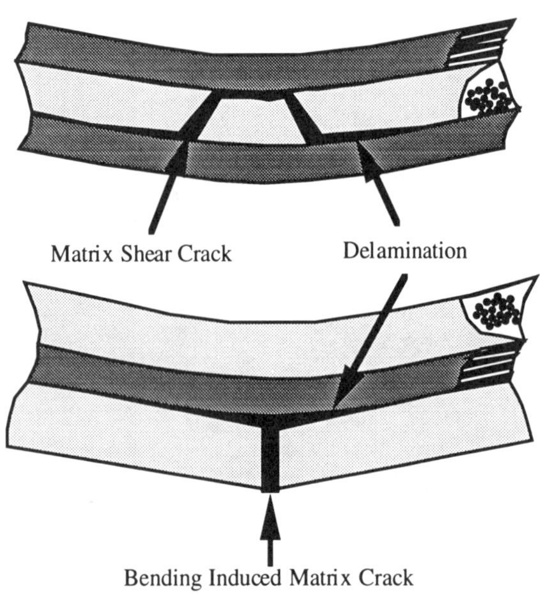

Fig. 2 - Delamination mechanisms in laminated composites

Recent work on woven systems [15] suggests that delaminations may also initiate from microcracks originating at localized regions of debonded fiber. The driving force for debonding is excessive shear stress developed during load transfer, unfortunately, predicting such behavior becomes a complicated issue for interlocked bundles in many fabric architectures. The authors believe that non-uniform sizing thickness and a distribution of surface flaws determines the location of such failures. A basic tenet of fracture mechanics is that crack growth is preceded by crack initiation. In some cases critical flaws are inherent in the material, otherwise a local magnification of stress is required to induce crack initiation. Understanding what represents an inherent critical flaw in composites is a key step in describing impact behavior.

It is documented that even small delaminations can drastically reduce interlaminar shear strength [16] and in plane compression strength [17] which explains much of the effort extended to model impact induced delamination growth. We present this paper in hopes that attention might also be focused on several other extensively observed failure modes that have received limited attention. Of the specific types discussed, ply separation is most similar in nature to delamination while interbundle, intrabundle and void pocket cracking share some characteristics of matrix cracking.

Matrix cracking is a complex fracture process that depends on local geometry and the externally applied stresses. Sih [18] discusses linear elastic fracture mechanics solutions to numerous fiber/matrix orientations and external loading scenarios. It is generally accepted that matrix splitting parallel to fibers is a low energy fracture path which causes crack blunting [16]. Internal matrix cracks inclined at 45 degrees are typically compressive shear bands while transverse cracks have been attributed to tensile stresses developed by large membrane reaction displacements [19]. Distinctions are also made for cracks which occur on either surface of the plate. According to Liu and Malvern [20], cracks in non impacted surface layers are due to tensile stress waves generated by bending during the first quarter of flexural wave propagation and cracks on the impact face are a result of transient tensile waves. In isotropic plates this transient tensile pulse produces circular crack rings.

The principal reason behind such extensive study of dynamic failure has traditionally been to develop an experimentally verified series of dynamic failure criteria which can then be more confidently applied in numerical simulations. Such criteria must account for strain rate effects, damage propagation speeds, interactions, etc. The first step for any new system is to adequately characterize the types of failures which occur. This paper describes a number of failure types observed in plain weave and biaxial knit RTM composites which are distinct from those damage modes reported in laminated systems.

Experimental Method

Since the objective of this study was to identify generic failure mechanisms in RTM composite plates a certain degree of freedom in selecting variables such as plate thickness and fiber volume fraction is allowed, however, the fabric architectures selected were restricted to commercially available E-glass biaxial knits and plain weaves. A Dow Derakane 411-C50 thermosetting vinyl-ester resin was incorporated because of its low viscosity which facilitates injection into a dense preform.

Resin was injected at pressures between 2-3 atm (30 -- 55 psi) into a 24.5 cm square mold cavity containing the compacted reinforcement. Following resin infusion the tool was heated to 98° Celsius for 30 minutes to achieve green strength and final cure was completed in an oven at 120° Celsius for 3 hours. The plates were then cut into 8.3 cm square specimens and impacted with a drop weight impact tower at energies up through complete puncture. The types of damage reported occur at various energy levels, however, no relationships to such are drawn in this paper.

Impact damage was observed using a combination of evaluation techniques. The projected damage area was obtained first using through transmission of light and NDE pulse echo apparatus. Destructive evaluation of internal damage included the use of dye penetrant wells to trace crack paths through matrix and fiber bundles and highlight delamination areas. Closer inspection required sectioning the plates for both macroscope and SEM characterization. Fiber breakage was investigated by burning off the resin at 600° Celsius and deplying the fabric.

Results and Discussion

Intrabundle Cracking

The distinguishing feature of intrabundle cracks is that they appear to be contained within a given bundle and propagate distances approaching several cm along the bundle as it undulates within its fabric layer. These parallel cracks are uniformly spaced at about 1-2 mm. We estimate the crack widths to be less than 0.1 mm which may be responsible for their transparency to normal incidence acousto-ultrasonic waves used in our NDE efforts. Once these intrabundle cracks propagate to the bundle surface they may act as initiation sites for interbundle cracking discussed later.

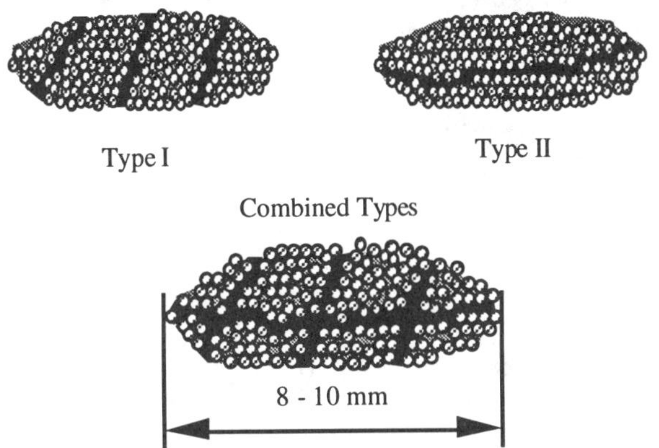

Fig. 3 - Types of intra-bundle cracks

We have identified two types of intrabundle cracking depicted in Figure 3. Type I cracks are oriented at an angle to the bundle major axis approaching 45° while the less

common Type II cracks run more or less parallel to the bundle axis. Type II cracks were seldom observed alone, rather they exist in a combined state with Type I. It appears from the micrograph in Figure 4 that the fibrils debond rather than fracture.

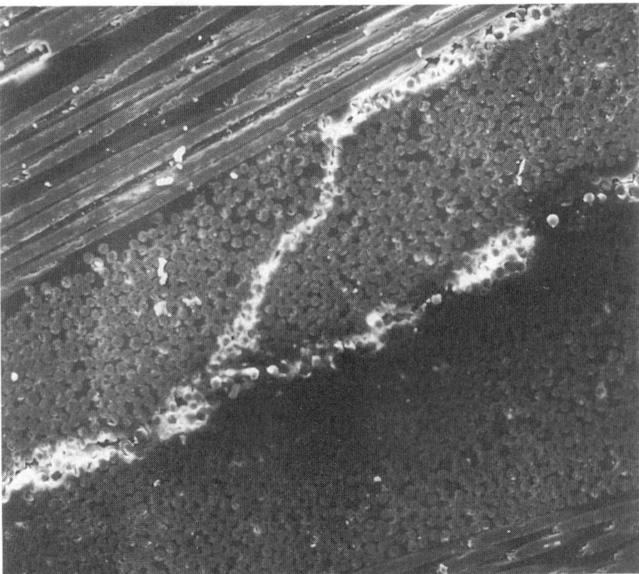

Fig. 4 - SEM micrograph of intra-bundle cracks at 80x

Two mechanisms are proposed for intrabundle cracking and both relate to deformations of the fiber bundle which is basically a microcomposite composed of glass fibrils, sizing and matrix. This microcomposite is somewhat unique because the fibril volume fraction may locally exceed the hexagonal close packing limit of 73% for spheres, which is much higher than typical composite fiber volume fractions of 30-60%. We believe that the commensurately small volume fraction of sizing and matrix facilitates the development of critical strains when the bundle is deformed. Two specific mechanisms are put forth to explain the origin of bundle deformation.

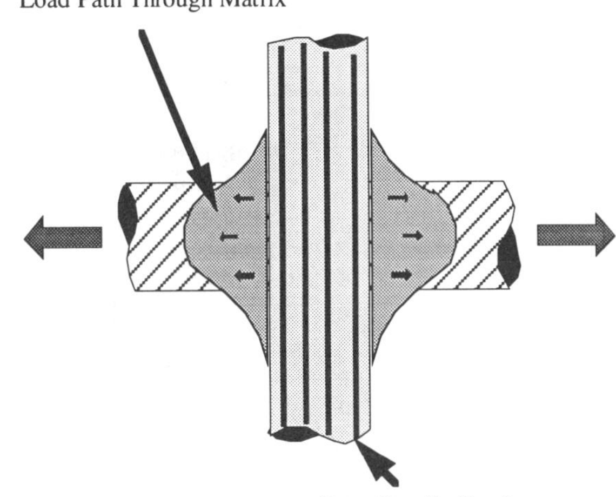

Fig. 5 - Proposed mechanism I for intra-bundle cracking

The first bundle deformation mechanism illustrated in Figure 5 is based on transverse tensile strains that result from load transfer through adjoining matrix at bundle crossovers. An orthotropic biaxial plain weave preferentially transfers load in the warp and weft directions. Since the bundles are mechanically interlocked, there is almost continuous load transfer between adjacent bundles and the resolved stresses are normal to the adjacent bundle fiber axes. This may explain why the intrabundle cracks are all parallel.

The second proposed bundle deformation mechanism shown in Figure 6 is based on the redistribution of fibrils to accommodate a change in bundle aspect ratio as a compressive load is applied. The degree of bundle spreading is exaggerated in the figure to illustrate the deformation mechanism which basically relies on the assumption that individual fibrils are only elastically strained while the small volume of connective phase must undergo significant plastic deformation. The 45° incline of these cracks suggests a shear type failure which could result from compression, tension or a combined biaxial loading if both of the proposed mechanisms operate simultaneously.

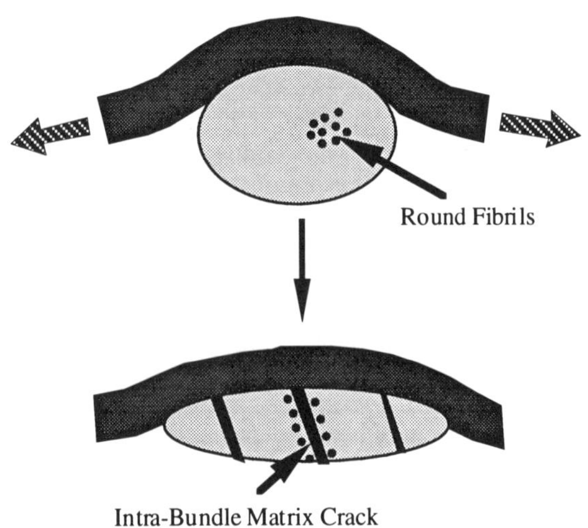

Fig. 6 - Proposed mechanism II for intra-bundle cracking

We have observed similar intra-bundle cracking immediately after processing in both biaxial and triaxial woven fabrics but to a lesser extent than following impact. Intrabundle cracks may be the result of residual stresses within individual fiber bundles that are induced by a combination of cure shrinkage and thermal strain mismatch between the glass and vinyl-ester. The strains associated with these two processing elements could develop sufficient residual stresses that intrabundle fracture commences at low impact loads and propagates to exterior bundle interfaces where catastrophic cracks may develop prematurely.

It was fortuitous that vinyl-ester is transparent and we were able to see the dye penetrant wick along intrabundle cracks. Similar damage may exist in graphite tows and result in catastrophic failure of composites made from them as well. Unfortunately, these cracks are practically invisible to NDE and their full extent may have been predominantly hidden during casual observations of cross-sectioned damage areas.

Interbundle Cracking

The second class of observed fracture labeled interbundle cracking comprises those cracks which run along a bundle interface either with the surrounding matrix or another bundle in intimate contact. Figure 7 illustrates three distinct types of interbundle cracking, two of which share many features of classical debonding and delamination. These generic cracks likely result from a combination of Mode I and Mode II type loading with matrix or interphase shear strength being the governing material parameter.

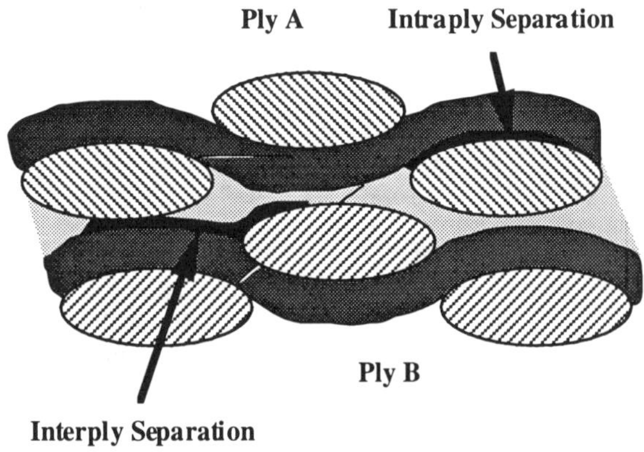

Fig. 7 - Types of inter-bundle cracks

The first type of interbundle fracture is observed principally at intersections of bundles that share a common reinforcing layer. Recalling the inherent reinforcement compaction during RTM, it is expected that many bundle intersections present a configuration that does not leave much extra space for matrix. If complete wetout is assumed then a finite interfacial matrix layer must exist, however, the inherent thinness of this layer makes it susceptible to fracture if the bundles shift or strain relative to each other. If wetout is incomplete then some fiber bundles in intimate contact may remain unbonded and thus form a crack at their interface. The presence of intrabundle cracks that have propagated to a bundle surface also provide a crack initiation source.

The actual fracture path is rough as shown in Figure 8. It appears that individual fibrils are debonded from the thin resin interlayer which is of the same order of magnitude thickness as fibrils themselves. This would appear to be a high energy fracture with a Mode I type behavior since the mating resin interlayer is not sheared off. There is a kind of nested similitude in the partial debonding of individual fibrils and the macroscale partial debonding of the bundle itself. Since the bundles are mechanically interwoven there is no clear driving force for complete debonding within a layer, rather, isolated sites of partial debonding are observed.

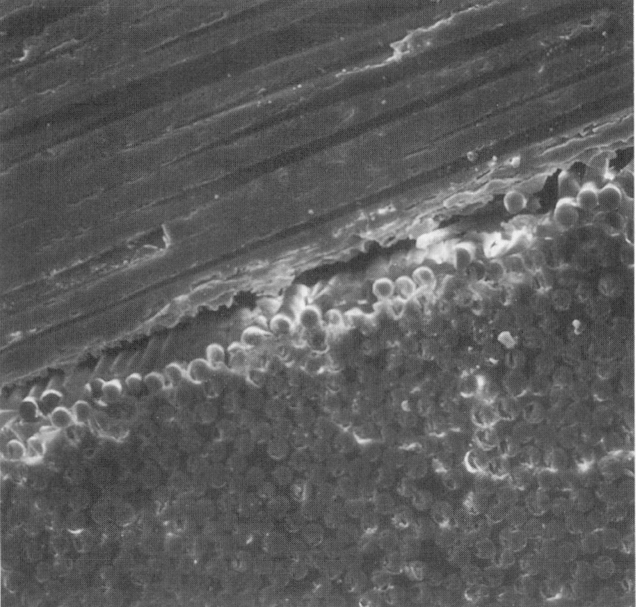

Fig. 8 - SEM micrograph of inter-bundle cracks at 160x

The second type of interbundle cracking describes cracks occurring between bundles that are resident within separate fabric layers. A general label would be ply-separation or delamination. Recalling the inherent differences between laminates and RTM composites one would expect to see a much higher energy fracture surface for ply-separation than delamination and Figures 9 and 10 portray this. The close up shows a hackled matrix that has been sheared after extensive plastic deformation.

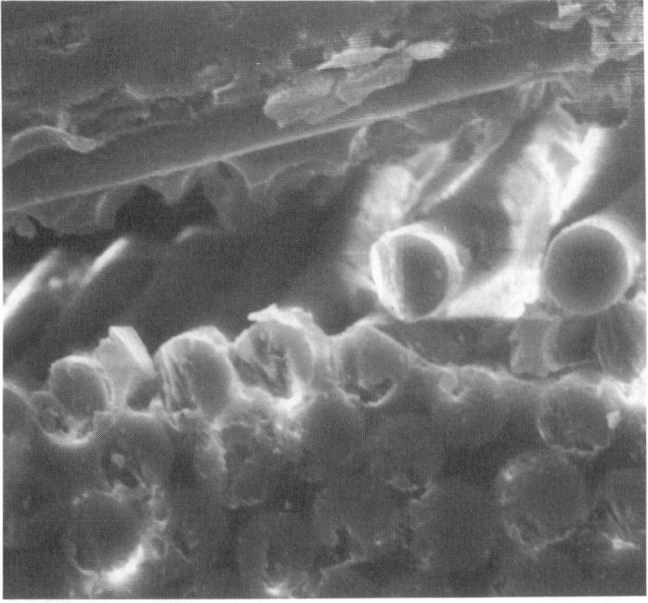

Fig. 9 - SEM micrograph of inter-bundle crack at 640x

The presence of fiber bundles debonded to varying depths also indicates the complex nature of this fracture process. Observations of dye penetration suggest a limited extent of delamination. Rather than one continuous planar delamination we observe separated islands with areas of unbroken composite in between. One can look through a damage area, for example, and see many areas where the dye has not penetrated. In general there appears to be more dye concentrated on the back half of the plate.

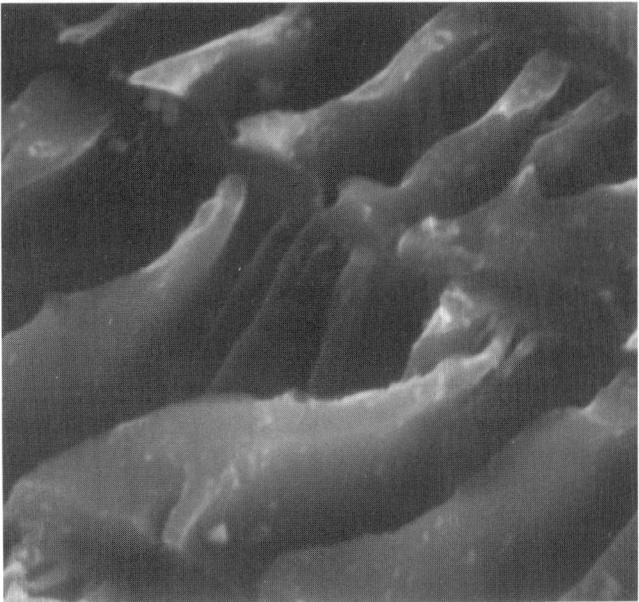

Fig. 10 - SEM micrograph of matrix fracture surface at 1250x

Void Pockets

Voids within the matrix of various dimensions and shapes were observed scattered throughout a typical cross-section. The largest of these ellipsoidal defects that we observed had a major axis length of approximately 0.2 mm shown in Figure 11. Although voids are found predominantly in resin rich pockets formed adjacent the intersection of bundles as shown in Figure 12, we cannot rule out the possibility that smaller scale voids might have been responsible for some interlayer separation. Similar defects have also been observed in the vicinity of thermoplastic threads that hold biaxial knits together as shown in Figure 13.

Since the primary objective is to introduce and describe the nature of several unique failures within RTM composites, we evade the question of how to eliminate voids in the first place. Suffice to say that void formation during thermoset processing is a complicated issue involving resin purity, thermal gradients, volatile production, cure shrinkage, degree of wetout and many other considerations. A well consolidated part should be void free while in fact less that 1% is considered ideal under laboratory conditions. The reality is that voids are an integral part of commercial composites so let us strive to understand their effect on material behavior.

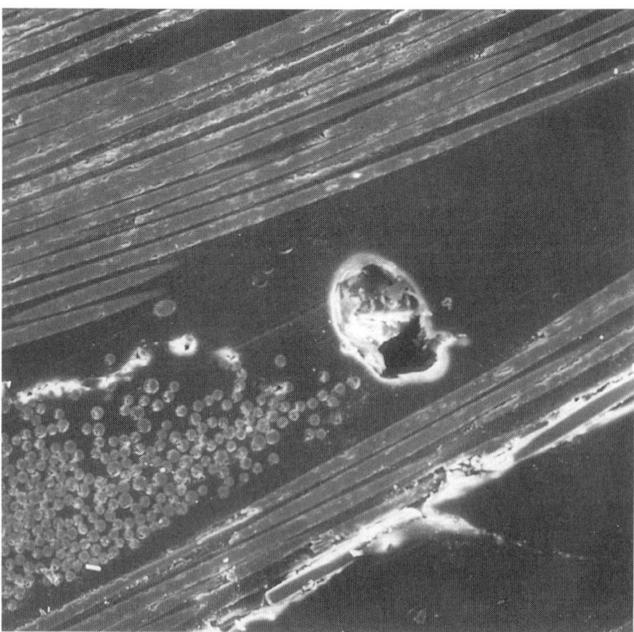

Fig. 11 - SEM micrograph of partially fractured void Pocket at 80x

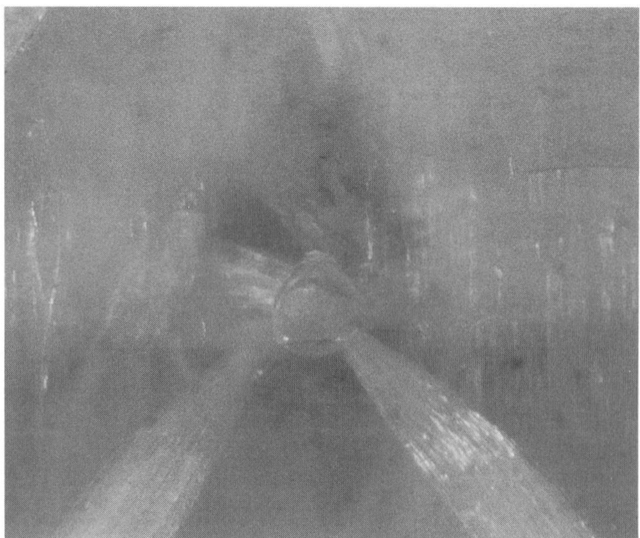

Fig. 12 - Depiction of void pocket in matrix rich region

Fig. 13 - Macroscope micrograph of void adjacent knitting thread at 50x

The presence of voids in itself is not necessarily debilitating to macro properties of a composite. Problems arise, however, when the unsupported regions of matrix act as stress concentrators and become sites of crack initiation. Figure 12 shows a fractured zone of matrix surrounding a void; the multitude of cracks are aligned in all directions which facilitates propagation under practically any external loading. The degree of stress concentration is linked to the size and shape of the defect, or more generally the sharpness which depends on the void location. Voids in contact with two intersecting fiber bundles could be extremely sharp compared to one in bulk matrix.

Matrix and Fiber Failure

The final two failures types are described only briefly because they have been documented quite extensively in the literature. These are matrix cracking and fiber breakage. Matrix cracking seems to be a catchword to describe those types of damage which are neither delamination or fiber breakage. This is understandable in light of maintaining simplicity in order to make numerical modeling tractable, however, there are several distinctions that need to be made for RTM composites. The first is related to the small volume fraction of non-contiguous matrix phase relative to laminates where the matrix is homogeneously interconnected. Matrix cracks do form in RTM plates but it appears that the extent is mitigated somewhat by the many other fracture types described. Matrix cracking is observed within matrix rich regions, but certainly not all such regions are fractured. In the limit one might say that interbundle and intrabundle cracking are subsets of matrix cracking but these are distinct from the matrix cracks observed in laminates.

A second issue relating to matrix damage is crazing, which takes on a characteristic whitish haze as the molecular chains are deformed enough to get an interference in the diffraction of light. This damage is not necessarily debilitating and certainly does not absorb a significant fraction of energy, however, it does provide a useful marker of the extent of internal strain. We have performed post impact static compression tests on several samples and even observed a shrinking and disappearance of the haze region following static puncture in some instances. The implication is that projected damage area while a meaningful indicator of damage extent should be understood as representing many interacting mechanisms. It does not provide a-priori a means of quantitatively measuring the extent of any one type of damage.

Fiber breakage is ubiquitous to composite plate penetration. The number of fibers broken in the path of an indentor is primarily a function of fabric architecture, tup geometry and size, stacking sequence and bundle diameter [21, 22]. Knit fabrics differ from wovens in that the weak knitting threads may be easily broken and allow the bundles to be pushed aside while woven fabric interlocking prevents this. A result of this is that woven fabric penetration is a sequential process of bundle fracture occuring on each fabric layer as depicted in Figure 14. The load path along principle reinforcing fibers determines which neighboring bundles will fail and we observe very little fracture outside a couple of bundle diameters from the impact locus.

We do observe some instances of fibril tearing as shown in Figure 15, which may be one mechanism of bundle fracture. The tear commences at one edge of the bundle where there are fewer fibrils. Fiber bundles are also observed to commence fracture from the middle which may result from thier being bent to excessive curvature strain by the advancing tup.

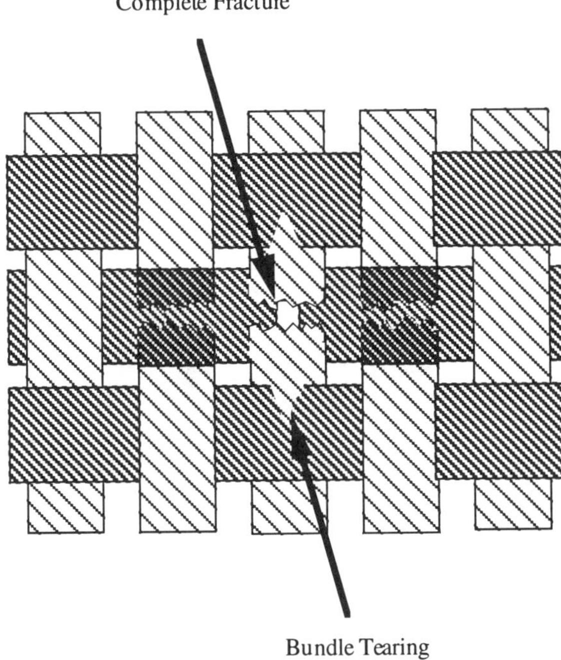

Fig. 14 - Depiction of sequential bundle fracture in plain weave fabric

Conclusions

This work represents a collection of insights into the damage mechanisms that occur during impact of RTM polymer reinforced composites. The traditional damage modes of matrix cracking, delamination and fiber breakage were observed in addition to several previously unreported behaviors. These include five identified types of intrabundle and interbundle cracking. It is important to adequately understand the interaction of these damage types in order to model the impact response of these composites systems. RTM is a regaining acceptance as a commercially viable composite processing technique so it makes sense to evaluate material behavior that might be specific to RTM consolidated parts. While interbundle cracking shares some similarities with classical laminate failure modes, intrabundle cracking has not been reported for laminates. Intrabundle cracking may be unique to RTM, however,

there is a possibility that tows used for autoclave processing are also subject to such cracking and investigators have been unable to observe this visually or by using NDE and sectioning techniques.

Fig. 15 - Depiction of bundle tearing

Acknowledgment

The authors acknowledge partial support from the ARO/URI program at the Center for Composite Materials.

References

1. Cristescu, N., L.E. Malvern, and R.I. Sierakowski, *Failure mechanisms in composite plates impacted by blunt-ended penetrators,* in *ASTM STP 568.* 1975, p. 159-172.

2. Sierakowski, R.L., L.E. Malvern, and C.A. Ross, *Dynamic failure modes in impacted composite plates,* in *Failure Modes in Composites III,* T.T. Chiao, Editor. 1976, AIME: New York. p. 73-88.

3. Malvern, L.E., *et al., Impact failure mechanisms in fiber reinforced composite plates,* in *High velocity Deformation of solids.* 1978, Springer-Vorlag: Berlin. p. 120-130.

4. Ross, C.A., *et al., Finite element analysis of interlaminar shear stress due to local impact,* in *ASTM STP 864,* ASTM, Editor. 1985, p. 355-367.

5. Cantwell, W.J. and J. Morton. *Low velocity impact damage in carbon fibre reinforced plastic laminates.* in *Vth Int Congr Exp Mech.* 1984. Montreal:

6. Cantwell, W.J. and J. Morton, *Detection of impact damage in CFRP laminates.* Composite Structures, 1985. **3**: p. 241-257.

7. Greszczuk, L.B., ed. *Damage in Composite Materials due to Low Velocity Impact.* Impact Dynamics, ed. L.A. Zukas. 1982, John Wiley & Sons: New York. 55-95.

8. Sjoblom, P., J. Hartness, and T. Cordell, *On low-velocity impact testing of composite materials.* J. Composite Materials, 1988. **22**(1): p. 30-52.

9. Winkel, J.D. and D.F. Adams, *Instrumented drop weight impact testing of cross-ply and fabric composites.* Composites, 1985. **16**(4): p. 268-278.

10. Abrate, *Laminated Composite Materials.* Applied Mech Rev, 1991. **44**(4): p. 155-190.

11. Choi, H.Y. and F.-K. Chang, *A Model for Predicting Damage in Graphite/Epoxy Laminated Composites Resulting from low-Velocity Impact.* Journal of Composite Materials, 1992. **26**(14): p. 2134-2169.

12. Joshi, S.P. and C.T. Sun, *Impact Induced Fracture in a Laminated Composite.* J. of Composite Materials, 1985. **6**: p. 51-66.

13. Takeda, N., R.L. Sierakowski, and L.E. Malvern, *Delamination crack propagation studies of ballistically impacted composite laminates.* Exp Mechanics, 1982. **22**(1): p. 19-25.

14. Liu, D., *Impact induced delamination - a view of bending stiffnes mismatching.* J. Composite Materials, 1988. **22**: p. 674-692.

15. Dvorak, G.J. and N. Laws, *Anaylsis of Progressive Matrix Cracking In Composite Laminates II. First Ply Failure.* Journal of Composite Materials, 1987. **21**(April): p. 309-329.

16. Dorey, G. *Impact and Crashworthiness of Composite Structures.* in *Int. Conference on Structural Impact and Crashworthiness.* 1984. Imperial College, London: Elsevier Applied Science Publishers LTD.

17. Wyrick, D.A. and D.F. Adams, *Residual Strength of a Carbon/Epoxy Composite Material Subjected to Repeated Impact.* Journal of Composite Materials, 1988. **22**(August): p. 749-765.

18. Sih, G.C., *Dynamics of Composites with Cracks,* in *Handbook of Composites,* G.C. Sih and A.M. Skudra, Editor. 1985, Elsevier Science Publishers: p. 127-176.

19. Takeda, N., R.L. Sierakowski, and L.E. Malvern, *Studies of impacted glass fiber reinforced laminates.* SAMPE Quart, 1981. **12**(2): p. 9-17.

20. Liu, D. and L.E. Malvern, *Matrix cracking in impacted glass/epoxy plates.* J. Composite Materials, 1987. **21**: p. 594-609.

21. Elber, W., *Failure Mechanics in Low-Velocity Impacts on Thin Composite Plates.* 1983, NASA:

22. Rydin, R.W., A. Locorcio, and V.M. Karbhari, *The influence of reinforcement layer orientation on impact resistance of RTM woven glass reinforced vinyl-ester composite plates.* J. Reinforced Plastics, (To Be Published in 1994).

Structural Ceramics and Their Potential for Noise and Vibration Control in Engine Components

S. Yang, R.F. Gibson
Wayne State University
Detroit, Michigan

G.M. Crosble, R.L. Allor
Ford Motor Company
Dearborn, Michigan

Abstract

This paper discusses the potential application of structural ceramics in the control of noise and vibration. High temperature damping experiments have been conducted at temperatures up to 1,100°C on cantilever beam samples of silicon nitride (Si3N4) ceramics and ceramic composites reinforced with silicon carbide (SiC) whiskers, while room temperature tests have been carried out on silicon nitride, silicon carbide, steel, titanium and titanium aluminide engine intake and exhaust valves. Experimental results show that ceramic structural materials (both monolithic and composites) have great potential and can play important roles in increasing quietness and durability in view of their unique combinations of large high temperature damping factors and high ratio of stiffness to density, which may result in inaudible (above audio frequency range) vibrations for some structural components. Since these ceramic materials are strong candidates for many structural applications in advanced heat engines which operate under high temperature, vibratory conditions, these characteristics are very significant.

ADVANCED CERAMICS AND CERAMIC MATRIX COMPOSITES have seen rapid development and increasing applications as structural materials in recent decades. One such application is their use in advanced combustion engines and other energy generating equipment in which the operating conditions of high temperatures and vibratory loading have almost driven the development of metals and super-alloys to their limits. But most of the gains from using advanced ceramics and ceramic matrix composites have so far been attributed to their high temperature stability, high strength and hardness, good wear and corrosion resistance and low density, and studies and improvements of these properties have been the subjects of numerous investigations and publications. However, few reports have been found regarding the vibration characteristics of these structural ceramics, even though this behavior plays an important role in reducing noise and vibration levels and hence increasing performance as well as durability and reliability for extreme environments such as in heat engines. This paper explores the potential of these materials for noise and vibration control by summarizing two discoveries from our recent research: (1) some natural frequencies of ceramic engine valves lie well above the audio frequency range due to the high stiffness to density ratio and hence the noises emitted from such vibrations may be inaudible; (2) damping values of ceramic materials at high temperatures are much greater than room temperature values.

Frequency, Damping and Vibration

Frequency is one of the basic characteristics of an oscillating object. Since the audible frequency range of the human ear is below 20,000 Hz, any noises resulting from frequency beyond that range behave "quietly" just because they cannot be heard. Damping is another important vibration characteristic. From the mechanics points of view, the damping properties of a material are closely related to its noise and vibration attenuation behavior. Figures 1 and 2 show illustrative examples regarding the roles of material damping in attenuating amplitudes of free and forced vibrations. In Figures 1 and 2, the parameter η is the loss factor, a measure of damping which is defined in the next section. It is seen in Figure 1 that a damped free oscillation decays rapidly with time but an undamped oscillation does not. Figure 2 shows that a resonant amplitude is smaller with heavier damping. Thus, the existence of damping is significant not only for the quietness but also for the durability and reliability of any vibrating components.

Calculations and Measurements

Frequency and damping of a vibrating component can be described by many different models. For example, the damping loss factor η, which is the damping parameter used in this paper, can be either expressed by the half power band width method [1] as

$$\eta = \frac{\Delta f}{f_n} \qquad (1)$$

or by the logarithmic decrement method [2] as

$$\eta = \frac{1}{n\pi} \ln \frac{x_0}{x_n} \qquad (2)$$

where f_n is the resonance frequency of the nth vibration mode, Δf is the half power band width of the nth mode resonant peak and n is the mode number (Figure 3); x_0 is the initially measured amplitude, and x_n is the amplitude measured after n cycles have elapsed from x_0 (Figure 1). The measurements in this research were realized by the impulse-response frequency technique [3], which was used to excite specimens, collect vibration signals, and measure f_n and Δf for the half power band width method or x_0 and x_n for the logarithmic decrement method. A block diagram for the impulse-frequency response measuring system built up at WSU is shown in Figure 4, with details being described in the authors' previous publications [4,5].

Description of Samples

Samples are divided into two groups, i.e. engine intake and exhaust valve samples and rectangular beam samples. Tables I and II list the identifications of the valve samples and the beam samples, respectively. All the samples were obtained from Ford Motor Company. It should be pointed out that valves made from materials other than ceramic materials are investigated here because engine tests seem to indicate that ceramic valves "run quieter" than metal valves. The detailed descriptions of the materials from which the samples are made can also be found in References [4,5].

Table I. Identification of valve samples

No.	Function	Material
1	Intake	SAE 1547 Steel
2	Exhaust	Head : Cr, Mn, Ni, and Fe Alloy; Stem: SAE 8645 Steel
3	Intake	Ti-6Al-4V
4	Exhaust	Ti-6Al-2Sn-4Zr-2Mo-.08-.12Si
5	Intake	Si_3N_4 Ceredyne ceralloy 147-31-A
6	Exhaust	Si_3N_4 Ceredyne ceralloy 147-31-A
7	Exhaust	Gamma-TiAl (48a/oAl, 1a/oV, bal Ti)
8	Exhaust	Sintered SiC

According to the composition of the materials, the samples in Table I can be sorted into metallic or alloy valves (nos. 1-4, 7) and ceramic valves (nos. 5-6, 8); and the samples in Table II are ceramics (without SiC whiskers) and ceramic composites (with SiC whiskers), respectively.

Table II. Identification of rectangular beam samples

Name	Material	Whisker	Additive	Processing
KNw/SiC		SiC (5)	Y_2O_3	
KNw/oSiC	Si_3N_4	--	(6)	Hot-pressed
UBEw/SiC		SiC (5)	Al_2O_3	
UBEw/oSiC		--	(3)	

In Table II, KNw/SiC is KemaNord low purity silicon nitride powder matrix with silicon carbide whiskers, KNw/oSiC is KemaNord low purity silicon nitride powder matrix without silicon carbide whiskers; UBEw/SiC is high purity silicon powder matrix with silicon carbide whiskers, UBEw/oSiC is high purity silicon nitride powder matrix without silicon whiskers; and the numbers in parentheses represent volume percentage or weight percentage for whisker and additive, respectively.

Set up of Experiments

Vibration characteristics of the valve samples were measured at room temperature under free-free beam extensional vibration test and free-free beam flexural vibration test. Figures 5 and 6 show the corresponding experiment set ups, respectively. Since the major source of excitation of these valves (combustion in the cylinder) is expected to be along the longitudinal axis of the valves, the extensional testing mode should be practically important.

Similar measurements were conducted on the beam samples but at the temperatures from room temperature to 1,100°C and under cantilever beam flexural vibration tests (Figure 7). Cantilever beam tests were found to be more practical than extensional tests for high temperatures. Even though the samples are prepared specimens instead of real components, as in the cases of valves, the results are still considered to be of practical significance due to the fact that, for example, a blade of a turbine engine is always modeled as a cantilever beam [6].

Results of Experiments

All the experiments were done in the first mode of vibration and the corresponding frequencies and loss factors were recorded. The half power band width method was used for most of the measurements except for the three ceramic valves under the free-free beam extensional vibration because their first mode frequencies were found to be too high for accurate analysis of the frequency response function with an audio frequency spectrum analyzer. So for those samples, the logarithmic decrement method was used with a digitizing oscilloscope.

The experimental results for the vibration characteristics of the two groups of the samples are shown graphically in Figures 8 to 10 and numerically in Tables III to V. Figure 8 shows the measured damping loss factors versus frequency for all the valves under free-free beam extensional vibration. Figure 9 shows the corresponding results for the valves under the free-free beam flexural vibration. The corresponding numerical values of the measurements are listed in Table III. Similarly, Figure 10 shows the damping loss factors for the ceramic beam samples made from four materials measured as a function of temperature up to 1,100°C. The corresponding numerical statistics for room temperature measurements and the 900°C peak value measurements are listed in Table IV and V, respectively.

Table III. Numerical values* of frequency
and damping for valve samples

Sample	Extensional vibration		Flexural vibration	
No.	f_n (Hz)	η (10^{-3})	f_n (Hz)	η (10^{-3})
1	13,620	0.266	1,498	0.211
2	14,360	0.348	1,594	0.233
3	12,161	1.460	1,321	2.256
4	13,034	1.736	1,414	2.755
5	23,370	1.176	2,710	0.647
6	25,750	1.096	2,925	0.448
7	17,520	2.140	1,956	3.335
8	28,450	3.404	3,410	1.493

* Tabulated values represent the average of eight tests on each samples.

Table IV. Damping of ceramic beam
samples at room temperature

Material	No. of Tests	Frequency (Hz) Mean	Loss Factor			
			Max (10^{-3})	Min (10^{-3})	Mean (10^{-3})	Std. Dev (10^{-3})
KNw/SiC	32	119.69	0.67	0.34	0.49	0.11
KNw/oSiC	32	116.22	0.57	0.42	0.51	0.06
UBEw/SiC	32	118.21	0.58	0.36	0.49	0.06
UBEw/oSiC	32	117.56	0.63	0.40	0.52	0.08

Table V. Peak damping of ceramic beam
samples at 900°C

Material	No. of Tests	Frequency Mean (Hz)	Loss Factor			
			Max (10^{-3})	Min (10^{-3})	Mean (10^{-3})	Std. Dev. (10^{-3})
KNw/SiC	16	108.02	21.95	20.38	21.00	0.545
KNw/oSiC	16	104.67	20.35	17.42	18.44	0.723
UBEw/SiC	16	107.79	24.01	22.72	23.51	0.448
UBEw/oSiC	16	104.04	21.97	20.33	20.93	0.500

Discussion

From the experimental results of valve samples presented in Figures 8-9 and Table III, it is seen that the room temperature damping of all the samples is relatively small, with damping loss factors falling in the range of 10^{-4} to 10^{-3}, under both vibration modes. Since the damping for the ceramic valves in extensional vibration was measured from the digitizing oscilloscope due to their higher frequencies, the loss factor values are believed to be a little bit higher than those measured from the audio frequency spectrum analyzer [4]. Otherwise, the damping of ceramic valves is not that different from that of the others in magnitude, just as shown in the case of flexural vibration. The most interesting result comes from the comparison of the frequencies of the metal valves and ceramic valves in the extensional vibration. It is found that such frequencies of all of the metal valves (steel, titanium, titanium aluminide and numbered as 1-4 and 7, respectively) lie in the audio frequency range (<20,000 Hz), while the corresponding frequencies for all of the ceramic valves (silicon nitride, silicon carbide and numbered as 5-6 and 8, respectively) lie above the audio frequency range. As previously mentioned, the major source of excitation of these valves (combustion in the cylinder) is along the longitudinal axis of the valves, the extensional modes should be very important. Thus, any noises emitted due to extensional vibration of ceramic valves should be inaudible to the human ear, regardless of the damping behavior. The extensional vibration frequencies are directly proportional to the square root of the E/ρ ratio, where E is the elastic modulus and ρ is the density. Silicon nitride and silicon nitride, from which the ceramic valves are made, all have higher E/ρ ratios than metallic materials the other valves are made from.

The experimental results for the ceramic beam samples are shown in Figure 10 and Tables IV and V. The room temperature results are presented in Table IV, where it is seen that the damping is very small, with room temperature loss factors are all in the range of 10^{-4} to 10^{-3}, which is consistent with the results obtained from ceramic valves. The curves of damping loss factors as a function of temperature up to 1,100 °C are shown in Figure 10, and the numerical results for the peak values are listed in Table V, for all the four materials. It is seen that the damping behavior is quite stable up to about 700°C, then the damping rises and reaches its peak value near 900°C. Peak loss factors for the four materials are all near 0.02, which is much greater than the room temperature values. More importantly, this large damping is comparable with the damping of polymer and polymer matrix composites which have been considered to be good damping materials but usually fail at such high temperatures. The high temperature damping peak is also expected to occur for extensional vibration, although some frequency dependence of the peak temperature is supposed. Even though the laboratory samples have been tested here, the practical significance of the results obtained is obvious due to the fact that (1) the samples are made from silicon nitride based ceramics and ceramic matrix composites which have been considered to be the most practical high temperature materials for advanced heat engine uses and (2) the cantilever flexural vibration modes are more important than others for simulating the vibration of the blades of turbine engines.

The reasons why ceramic valves seem to "run quieter" than metallic valves may be explained, in part by the key findings of this study, that is: (1) most of the noises emitted from extensional vibrations of ceramic valves are inaudible because their frequencies are beyond the audible range; and (2) the damping of flexural (and extensional) vibrations at engine working temperatures (600-800°C for most automotive engines) is large, so the vibration amplitude and noise level are reduced.

Concluding Remarks

Due to their unique combination of high stiffness to density ratio and large internal damping at high temperatures, structural ceramics have great potential for noise and vibration control in components of combustion engines. This behavior is of special importance for advanced heat engines and their components which are subject to increasing demands for improving performance, reliability and durability. More effort should be devoted to integrating this unique behavior with other desirable properties of advanced ceramics and ceramic composites in practical applications.

Acknowledgments

The authors gratefully acknowledge the support of the Ford Motor Company through an unrestricted grant. Assistance in sourcing the valves for test is due to L. R. Swank, T. J. Whalen, and W.E. Dowling; Whisker dispersion and hot pressing for the KNw/SiC and UBEw/SiC were carried out at Cercom, Inc., Vista, California, under the direction of A. Ezis.

References

1. Gibson, R. F., "Damping Characteristics of Composite Materials and Structures," Journal of Materials Engineering and Performance," 1(1), pp. 11-20, 1992.
2. Thomson, W. T., "Theory of Vibration with Applications," Third Edition, Prentice Hall, 1988.
3. Suarez, S. A. and Gibson, R. F., "Improved Impulse-Frequency Response Techniques for Measurement of Dynamic Mechanical Properties of Composite Materials," Journal of Testing and Evaluation, 15(2), pp. 114-121, 1987.
4. Gibson, R. F., Yang, S. and Crosbie, G. M., "Vibration Characteristics of Intake and Exhaust Valves Made from Different Materials," Proceedings of the 1993 SEM "50th Anniversary" Spring Conference on Experimental Mechanics, pp. 8-18, 1993.
5. Yang, S., Gibson, R. F., Crosbie, G. M. and Allor, R. L., "Internal Damping of Silicon Nitride and Silicon Nitride Composites with Silicon Carbide Whiskers to 1100°C," Presented at the American Ceramic Society 96th Annual Meeting and Exposition, Indianapolis, IN, April 24-28, 1994, to appear in Proceedings of ACers Symposium on Ceramic Composites, 1994.
6. Afolabi, D., "Natural Frequencies of Cantilever Blades with Resilient Roots," Journal of Sound and Vibrations, Vol. 110 (3), pp. 429-441, 1986.

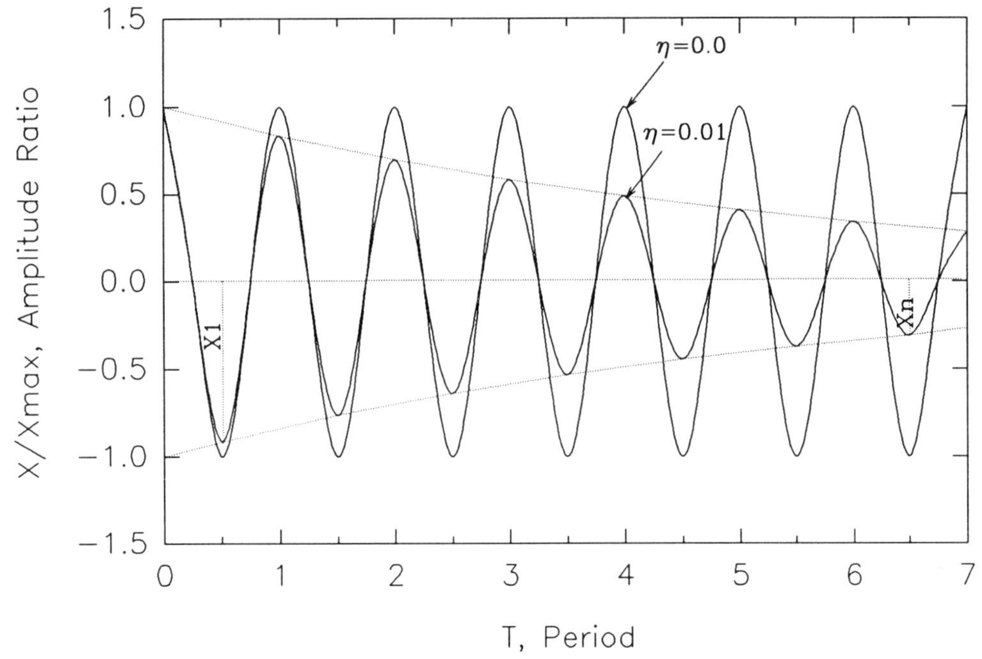

Figure 1. Free vibrations decay faster with damping; n=7 for Eq. 2.

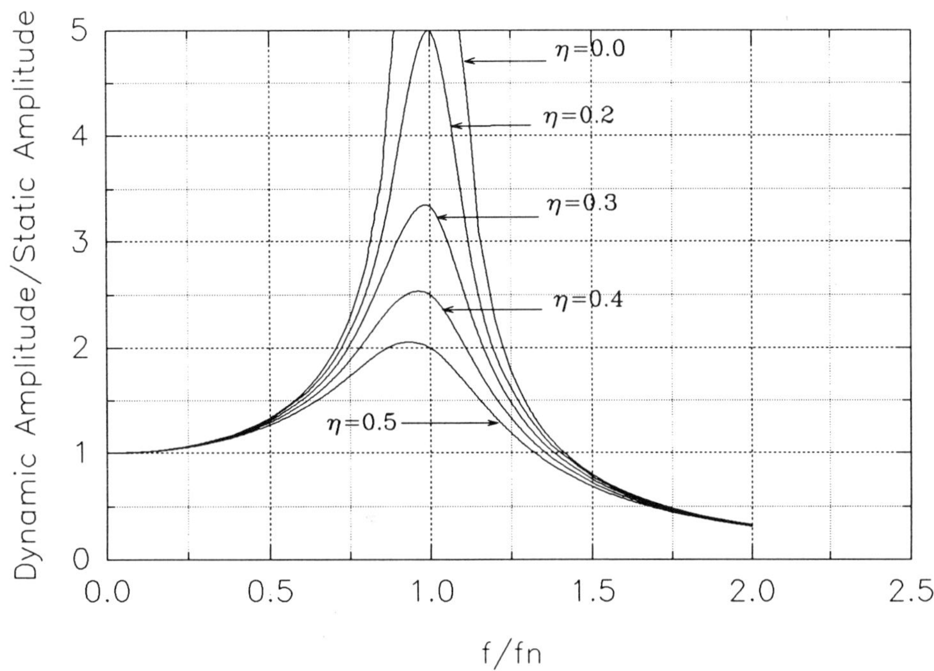

Figure 2 Resonant amplitude is reduced with increased damping

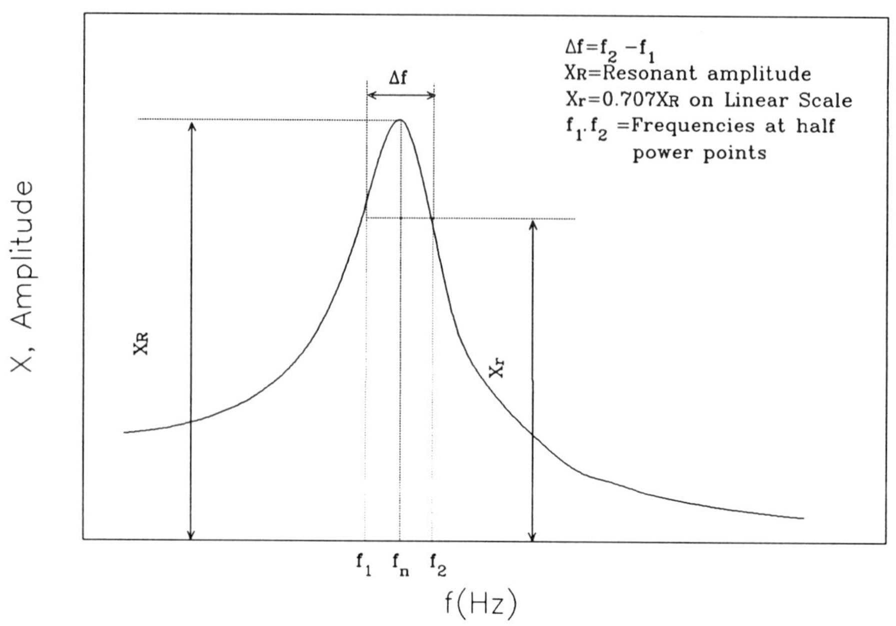

Figure 3. Definition of Eq. 1 from frequency response spectrum

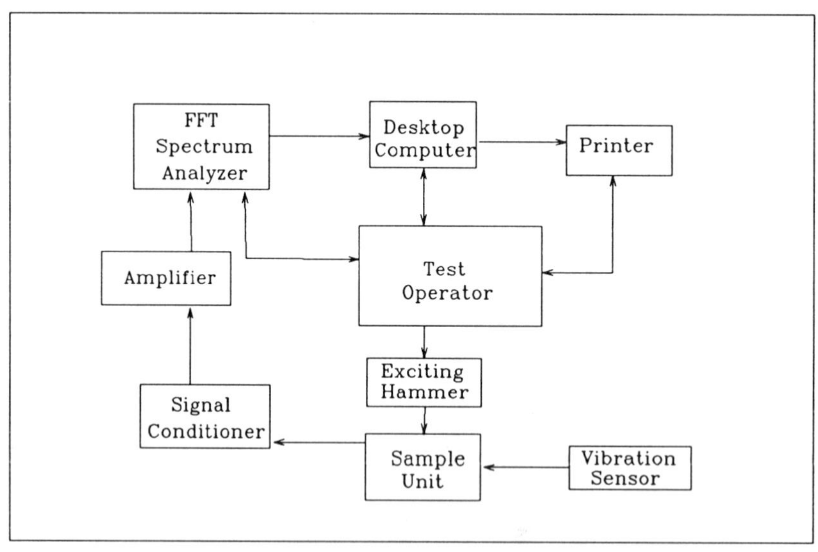

Figure 4. Block diagram of impluse−frequency response
experimental system

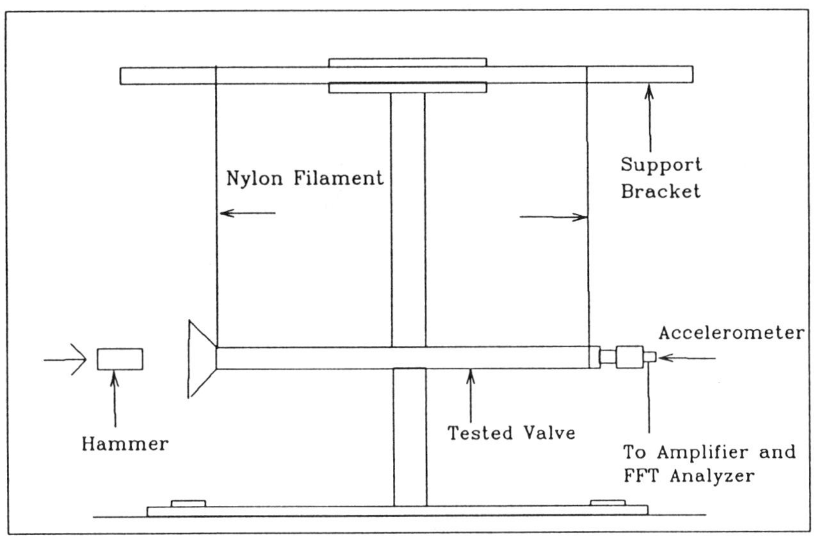

Figure 5. Apparatus for free-free beam extensional vibration test

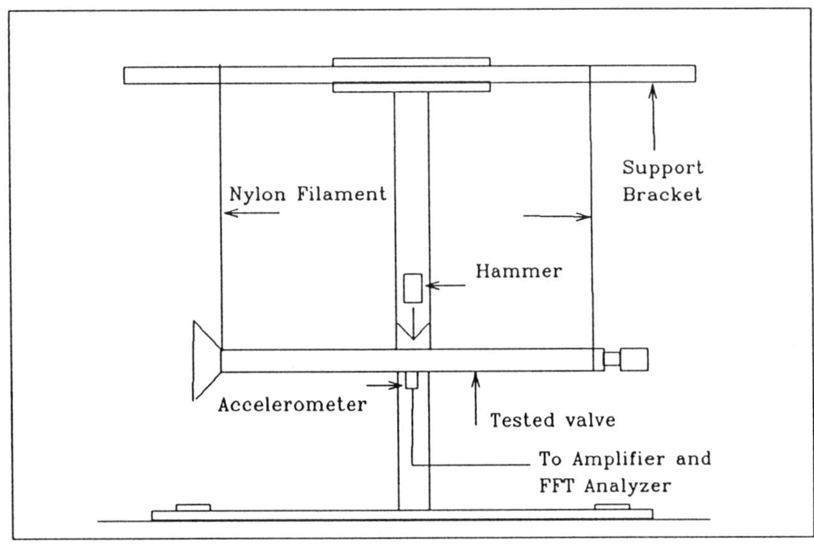

Figure 6. Set up for free-free beam flexural vibration test

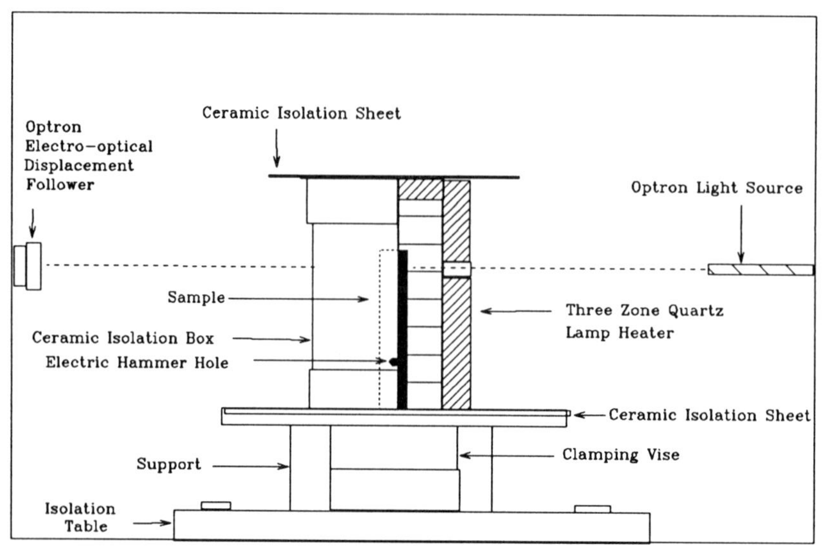

Figure 7. Cutaway drawing of set up of cantilever beam
test at elevated temperatures

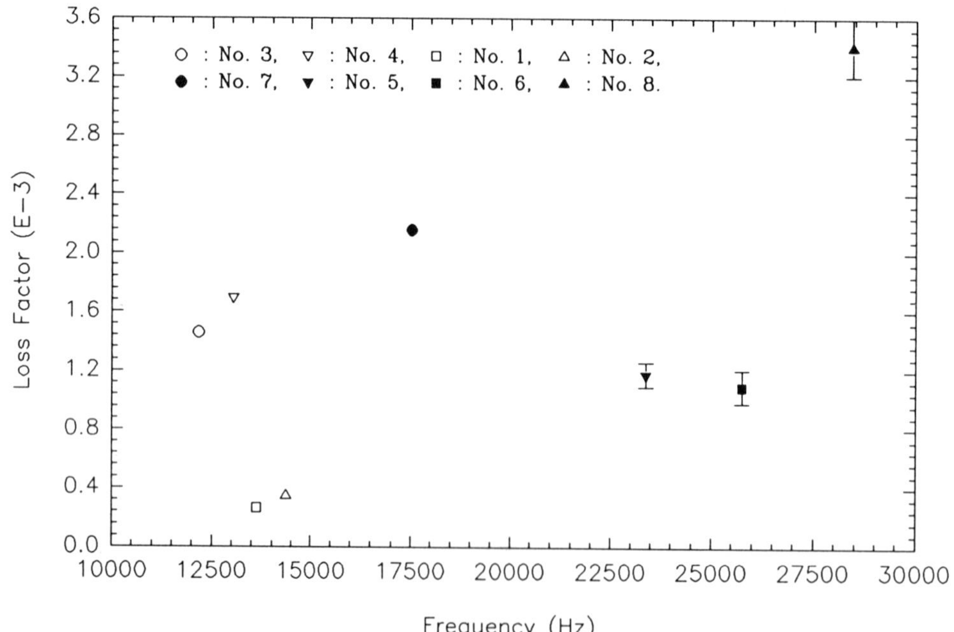

Figure 8. Loss factor of valves under free-free beam
extensional vibration

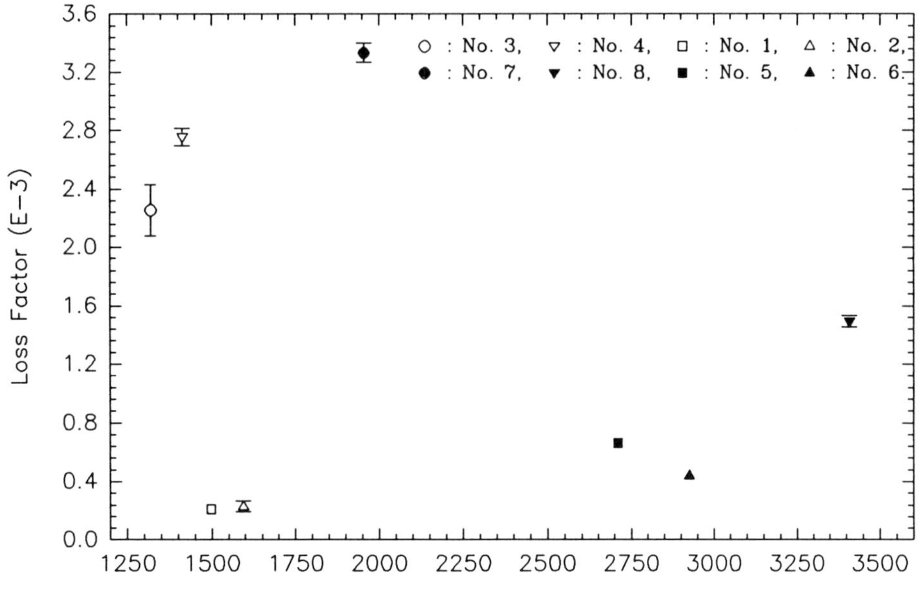

Figure 9. Loss factors of valves under free–free beam
flexural vibration

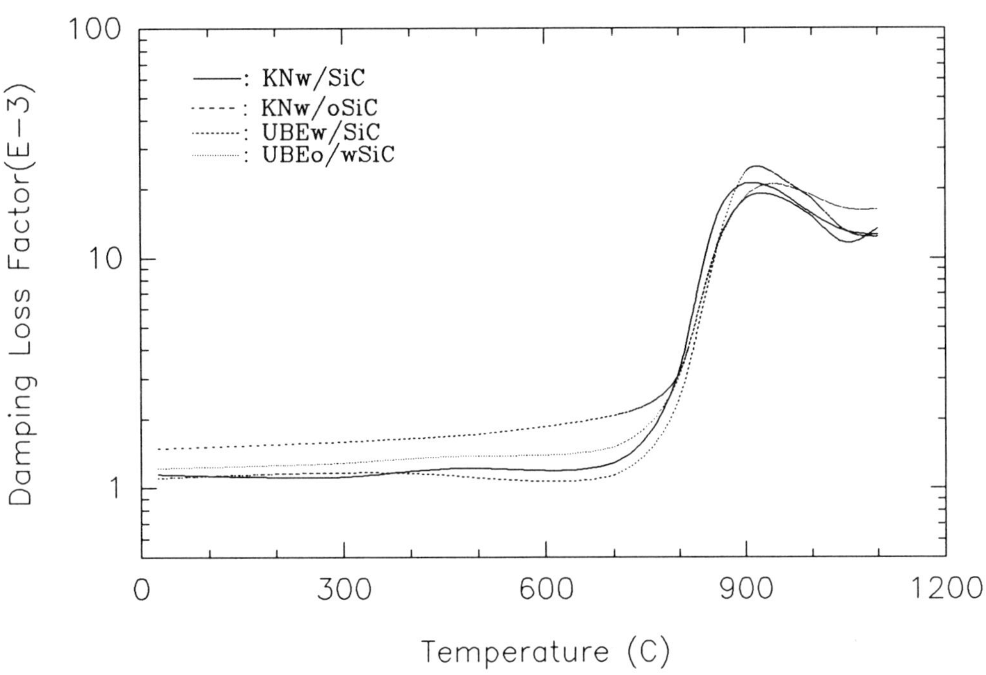

Figure 10. Damping as a function of temperature for four
ceramic materials

The Application of Composites in Diesel Engines for Naval Use

G.D. Duvall, D.P. Guimond
Naval Surface Warfare Center
Annapolis, Maryland

Abstract

The high speed diesel engine is the preferred primemover for a wide variety of military Naval and land applications. Its high fuel efficiency, low cost and high reliability make it an attractive choice. Diesel engines are, however, relatively heavy and require large foundations for both support and to reduce structureborne noise. There are military applications, such as high speed patrol craft, minesweepers, and amphibious vehicles, where reductions in engine weight, magnetic signature and noise would greatly enhance its effectiveness as a primemover. To facilitate these reductions, the Navy is examining the application of emerging composite materials and fabrication methods to diesel engines. The scope of the effort includes vibration modeling, materials evaluation, and full scale engine demonstrations. Material candidates include metal and ceramic matrices and polymerics, for engine structural and reciprocating components, along with various non-structural covers and housings. The cornerstone of this effort will be close working partnerships with other services and with industry, where economic benefits are significant.

MARINE DIESEL ENGINES offer high fuel economy, but are heavy and require large foundations and mounts to reduce vibration and noise. For smaller craft, the engine becomes a large part of the total craft weight, and in mine warfare, it becomes a significant magnetic source. The Navy seeks to exploit the rapid development of new composite materials and their potential for mitigating the traditional disadvantages associated with diesel engines. Engine developments would be targeted for retrofit into existing applications and for future designs and applications.

Diesel Engine Applications

There are a wide variety of Naval uses of diesel engines. The inherent fuel efficiency, durability, response and low cost make diesels attractive primemovers. Sizes range from a few hundred horsepower to several thousand horsepower and include naturally aspirated to turbo intercooled/after cooled engines. In Table 1 the various applications are listed along with pertinent ship and engine parameters.

Table I. Naval diesel engine applications

DESIGNATION		ENGINE	POWER - KW
Ships			
Frigate	FFG 7	DDA 16V149TI	1342
Amphibious	LSD 41	C-P 16PC25	7755
Fleet Tug	ATF	GM EMD V20	2685
Cutter (CG)	WMEC	16V-251	2416
Light Forces			
Patrol	Mark III	DDC 8V92	246
	PC 1	Paxman-Valenta	2498
	WPB (CG)	CAT 3412	2535
Mine Warfare	MCM	I-F ID36SS6V	447
	MCH	I-F ID36SS6V	447
Small Boats			
7m Rigid Infl	RIB	Cumins Series B	134
10m Utility	UB	Cumins Series B	86
Harbor Craft			
Yard Utility	YFU	GM 12V-71	250

For the ship applications, except for the FFG 7, the engines are used as main propulsion. With these larger engine blocks, it would be difficult, due to the manufacturing processes involved, to incorporate composite materials. Also the weight and volume fractions for these vessels is not as critical. The physical characteristics of the diesel engine do become increasingly important in the light force and small boat applications. The patrol craft (Figure 1), both Navy and Coast Guard, place great emphasis on high speed, >30 knots. To accomplish this, the power to weight ratio is high, 30.8 kW per metric ton (42 hp per long ton) of displacement. As the power demand increases for upgrades and future designs, engine weight fraction becomes critical and it becomes more difficult to maintain a proper balance. Currently available, conventional design engines in the 1200 to 1800 RPM category have a weight of 3-6 kg/kW (5-10 lb/hp). Higher speed (>2600 RPM) engines are available at 2.4 kg/kW (4 lb/hp), but they are expensive, specialty engines and their cost effectiveness for these applications would be questionable.

A recent internal study was conducted to quantify the impact of lighter weight composite engines for a possible Naval combatant application. A 2032 metric ton displacement corvette type ship was chosen. The impact was evaluated for a Combined Diesel and Diesel (CODAD) propulsion plant configuration. Conventional standard engines at 6 kg/kW (10 lb/hp) were compared with conventional design composite engines which were 40% lighter. Results showed a savings of 5% ($4M) in total ship costs, and 7% (142 MT) in total displacement. Advanced design composite engines with significantly lower weight would show even greater savings. It was clearly indicated that domestic diesels could become very competitive with their foreign specialty engine rivals by using composite technology.

Fig. 1. Mark III patrol boat.

Fig. 2. Mine countermeasures (MCM) vessel.

The AVENGER (MCM 1) class mine countermeasures vessels, (Figure 2), and the OSPREY (MHC 51) minehunter coastal vessels are the newest minewarfare vessels in the US fleet. The hulls are wood and GRP respectively, and much attention has been given to machinery. The current minewarfare engines are manufactured by Isotta-Francini of Italy. They are based on an existing conventional design, and achieve a lower magnetic signature with an austenitic cast iron block and cylinder head. The weight, however, is relatively high at 4.9 kg/kW (8 lb/hp). A lighter weight, non-ferrous composite engine would achieve two goals simultaneously.

The small boats represent a very large and varied population. They are carried on larger vessels and perform many different functions. On these boats, the engine weight can be as high as 16% of the total craft weight. This is a significant factor in speed and carrying capacity. Since these boats are usually secured on deck and lowered into the water when needed, overall weight is especially critical to the deck machinery and support structure. There is currently little or no margin in these systems. Lighter engines for backfitting into existing craft and also for future craft, would be a distinct advantage.

Current Approaches

For today's Naval vessels avoiding detection and targeting from enemy weapons systems, is key to survival in the battlespace. Ship signatures can be made up of various elements, including acoustic, magnetic, infrared and radar cross-section. The propulsion and ship service plants are potential contributors to the first three elements. Advanced structural modifications to diesel engines can have a significant impact on the overall acoustic and magnetic ship signature.

Acoustic Signature. Machinery vibrations by virtue of their location and attachment to the ship structure, contribute to the acoustic or noise signature emitted into the water. Standard procedures for mitigating vibration effects involve engine mounts, base structure and supporting mounts as shown in Figure 3. Depending on the application, engine support systems can become quite elaborate. This is especially true in mine warfare ships, as illustrated by Figure 4. All internal equipment, even major items such as the propulsion and ship service engines, are underhung from the main deck or cradled between the bulkheads. This arrangement minimizes the "hard points" and better absorbs the effect of any underwater explosion. It also reduces the transmission of noise and vibration to the surrounding water, thus decreasing ship vulnerability (1). The needed ship structure, however, is driven by the weight and inherent vibration characteristics of the engines.

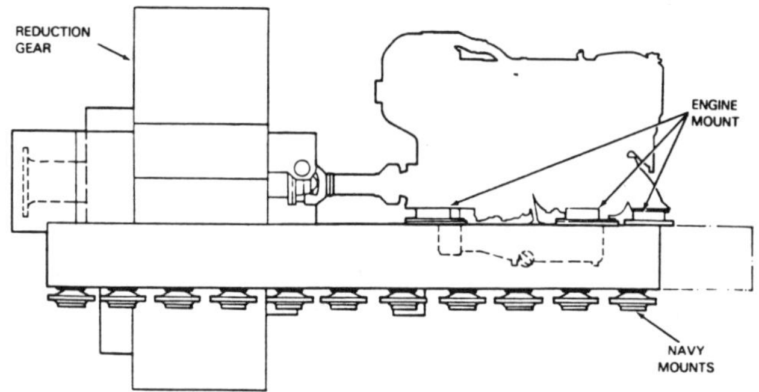

Fig. 3. Conventional diesel mounting scheme.

The US Coast Guard conducted an extensive program for vibration noise isolation for diesel engines used on the 95 ft Patrol Boats. The program was initiated under the premise that "Since the diesels were already selected, it was not possible to reduce source vibration levels through the selection of low-vibration-level diesels"(2). To reduce structural excitation from the diesel to acceptable levels, specially designed mounting systems and additional foundation stiffening (Figure 5) were evaluated. The mounts were effective but again they represent an approach to mitigate an inherent disadvantage as apposed to addressing the root cause, high engine weight and vibration levels.

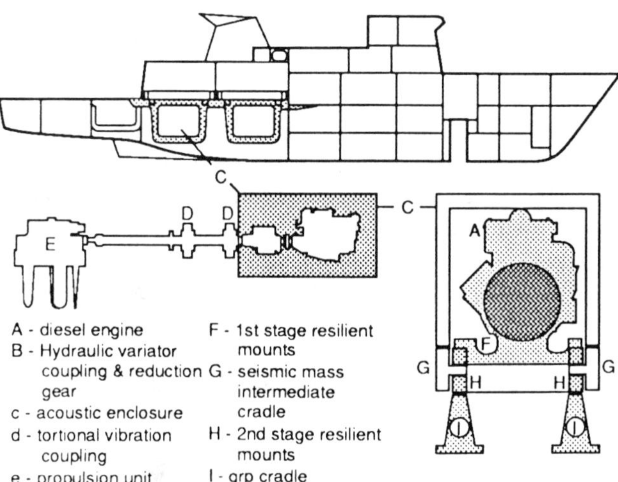

Fig. 4. Mine countermeasures ship machinery support arrangement.

Magnetic Signature. A minehunting vessel's primary mission involves exposure to deployed mines, and survival in that environment in order to neutralize the threat. The ships magnetic signature must be reduced as low as possible to avoid mine detonation and possible damage to the ship and crew (Figure 6). Minesweeper hulls are constructed of non-ferrous materials and as far as possible all machinery onboard must be non-magnetic. The diesel propulsion engines are the major machinery items. Their signature can be lowered by protecting an off-the-shelf engine with electrical compensation loops, called degaussing coils, or by making the engine inherently non-magnetic by maximizing the use of non-ferrous materials (3).

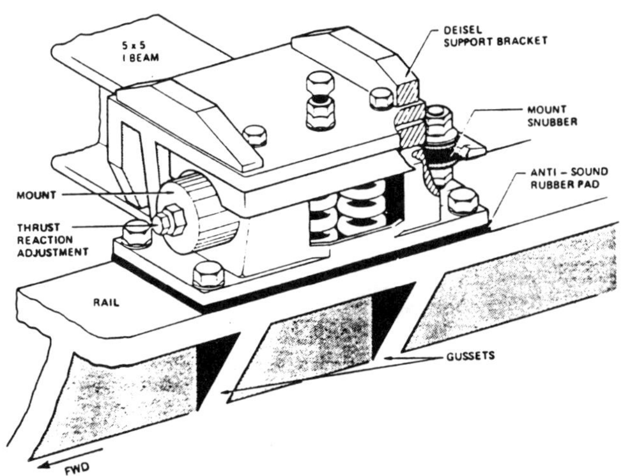

Fig. 5. Diesel vibration mount.

The various "artificial" means of reducing magnetic signatures have several drawbacks. Degaussing coils must be adjusted with respect to the earth's field, a sensitive and complicated procedure (4). These can prevent access to the engine for maintenance, and present a real danger to the mission of the ship if the system were to fail in a critical situation. Reducing the permanent magnetization of each individual magnetic part, is called deperming. This is a time-consuming and complicated procedure done in a specially designed facility.

In order to minimize the magnetic contribution of the engines, it has often been the practice to fabricate engine housings and other parts, if feasible, of nonferromagnetic materials such as aluminum and austenitic steel. Using these materials sometimes compromises the reliability and economic operation of the engine because tradeoffs must be made between magnetic effects and material strength, durability, initial cost, and engine efficiency. Moreover, parts selected for their magnetic properties alone are often unique to low population military applications, and engine manufacturers are less inclined to mass produce them. The cost of such parts is, therefore, frequently very high and the availability very low (5).

Advanced Material Solution

The incorporation of advanced materials into the design and construction of diesel engines has a direct and favorable impact on the weight and the signature, both vibration and magnetic. Reduced weight reduces mount requirements and accompanying base and ship structure as well as the driving force into the structure. Vibration reduction lowers the engine contribution to the ship signature and improves habitability in the machinery spaces, by lowering airborne noise. The reduced magnetic signature is now stable, non-directional and as reliable as the primemover itself.

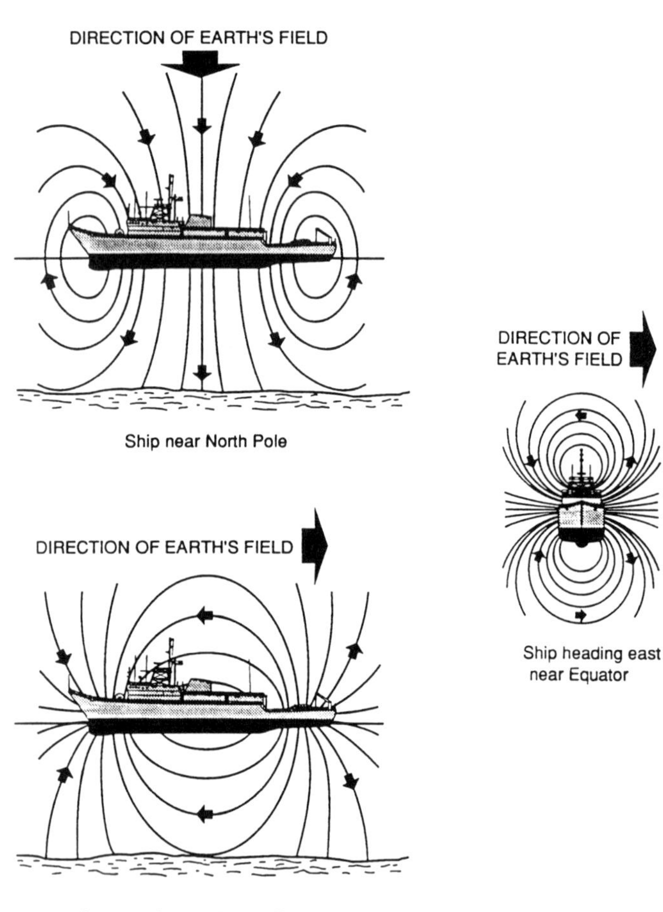

Fig. 6. Distortion of the earth's field due to the presence of a mine countermeasure vessel.

Traditionally, heavy-duty diesel engines have been made from predominantly ferrous materials because of the severe performance and durability demands of the applications in which they are used. This means that the engine head and block are made from cast iron, the crankshaft, camshaft and connecting rods are made of steel. Due to their large size, the block and head have been identified as the primary candidates for large weight and magnetic signature reduction by material substitution. For example, substituting high strength aluminum alloys for the cast iron could provide a 60% weight savings for these two components, producing an overall weight reduction of 25%. In recent years, numerous formulations of metal matrix composites (MMC) have emerged as candidates for engine components, primarily in the automotive industry. Their high specific modulus and fatigue strength, wear resistance and tailorable properties make them highly attractive for future automobiles (6). Ceramic matrix composites (CMC) have already appeared in engines designs, where high wear resistance is needed, such as injector links and timing plungers. Ceramics have also been considered as thermal barriers to improve engine efficiency. Polymeric matrix composites (PMC) offer light weight and durability in non-structural, low temperature areas such as valve covers and pans, and are beginning to appear in engine designs (7).

Metal and ceramic matrix composites can be used in a variety of locations, offering significant advantages in achieving the low magnetic, low vibration and light weight engine goals for Naval applications. There exists limited experience, however, with these materials in diesel engines produced by US manufacturers. Foreign companies, specifically the Japanese and Germans, are currently using advanced materials to produce lighter weight, reliable engines at a reasonable cost. Producing diesel components at a competitive cost is a major barrier to US manufacturers. This is of particular concern to the military, which has a strong impetus to reduce acquisition cost while increasing overall performance to counter emerging threats. It appears that mutual needs exist for the military and private sectors in the arena of advanced materials in diesel engines. Several consortia have been established to address various issues associated with composite materials in engines. Under the auspice of the US Council for Automotive Research (USCAR), a composites consortium was established to conduct joint research programs on structural polymers. The Great Lakes Composite Consortium (GLCC) was established under the Navy MANTECH program to promote composite manufacturing processes. These are two of many examples of cooperation and collaboration underway to bring about significant changes in diesel engine material selections.

Navy Approach

The primary goal of the Navy effort is to develop the technology to produce a non-magnetic, lightweight, reduced vibration/noise signature diesel engine. This will be accomplished by integrating advanced composite materials and nonferrous alloys into commercial, high production base engines, with selected Navy specific components developed and substituted when necessary. Teaming with The Department of Energy and vehicle/engine manufacturers will allow utilization of commercial/dual-use equipment and technology. This will conserve R&D resources, focus industry on Navy requirements and provide more cost effective options for future ship designs. The thrust of this expected three year effort will be on maturing the technology at an accelerated pace by conducting several demonstrations leading to engine prototyping. The major technical challenge is to integrate advanced ceramic and metal matrix composites along with more conventional nonferrous metal alloys into a coherent, durable engine structure while containing cost and vibration (Figure 7).

This concept for producing lightweight, high strength, structural engine components, poses many technical challenges in areas such as mechanical design and casting technology. Basic science and technology issues are also involved such as elevated temperature material characterization and the development of techniques to obtain metallurgical bonds between the high strength inserts and the aluminum block material.

A low magnetic signature engine demonstration will build upon commercially available engines provided through partnership agreements with DOE and major engine/component manufacturers. After the first generation engine blocks and cylinder heads are fabricated, they will be

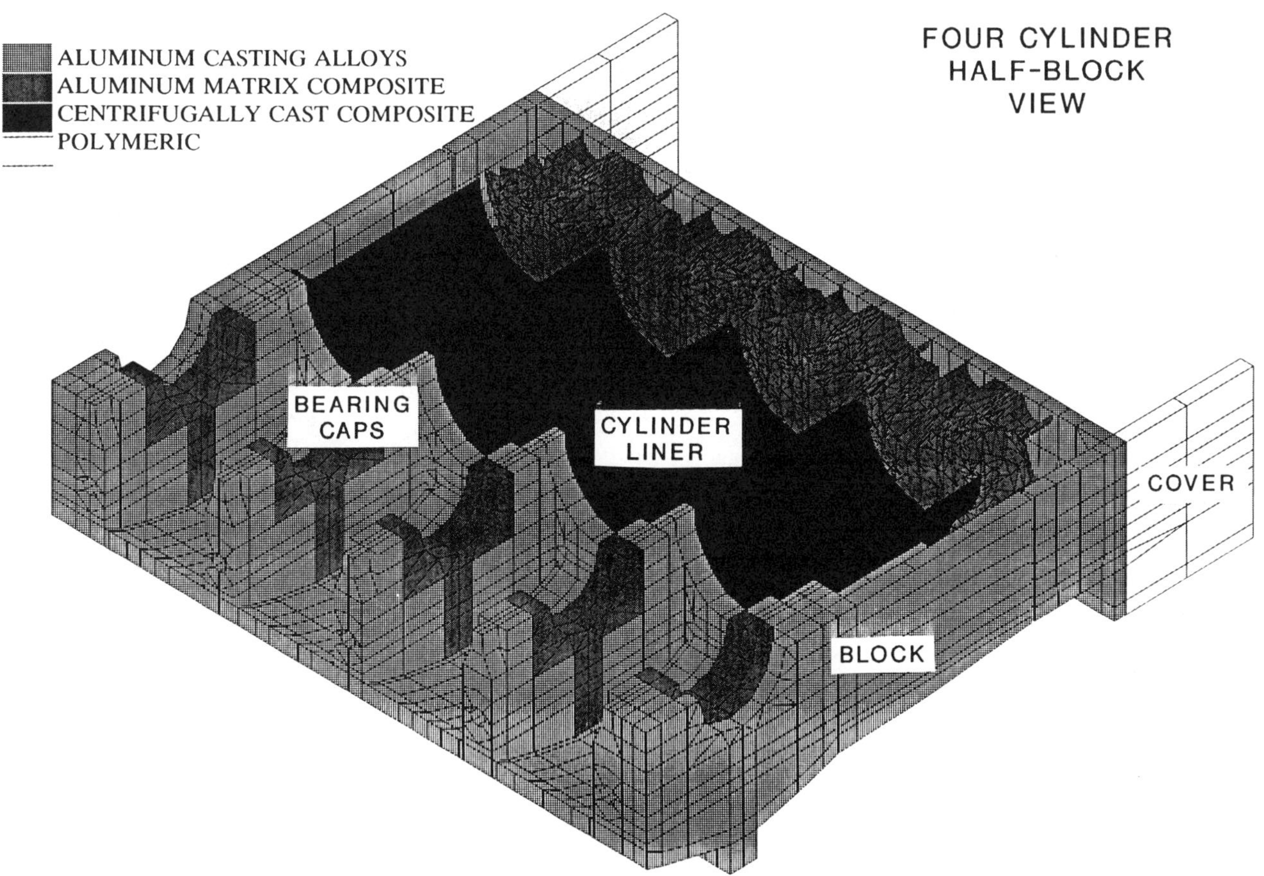

Fig. 7. Candidate composite engine components.

The starting point will be to develop the technology necessary to produce an engine block and cylinder head that are lightweight and nonmagnetic without compromising strength and durability. This will be accomplished by utilizing aluminum alloy castings reinforced by inserts made of a high strength material which are placed in areas subjected to high stress or temperatures. Initially, these inserts may be made of conventional cast iron, but they will be eventually replaced with non-magnetic metal matrix composite inserts.

assembled along with components from conventional baseline engines to produce operational demonstration engines. When second generation blocks and heads utilizing metal matrix inserts are available, they will be used in the demonstration engines. In addition to the structural components, non-magnetic, lightweight, rotating and reciprocating components will also be developed. These components such as connecting rods, valvetrain components and possibly the crankshaft will be retrofitted on the demonstration engines as they become available. The demonstration engines will be evaluated by the Navy and industry to gather experimental data such as stresses and temperatures in the newly developed components under expected operating conditions. The machinery silencing and magnetic field test facilities of the Navy will determine the signature benefits compared to conventional baseline diesel technology at various stages of the demonstration engine's evolution.

In addition to the above mentioned technology demonstrations, the Navy is also pursuing the development of a software based analytical simulation of a running diesel engine capable of predicting noise and vibration signatures called the Engine Signature Simulation Model (ESSM). The major elements of the ESSM include experimentally correlated dynamic finite element models of each engine component, strategies for assembling structural components which maintain dynamic properties, strategies for simulating crankshaft bearing loads, piston/cylinder contact forces and valvetrain loads under normal operating conditions and acoustic prediction using the boundary element technique. When completed the ESSM will be a valuable tool for tailoring component design and material selection to achieve reductions in noise and vibration levels along with meeting the necessary strength and durability requirements.

Concluding Remarks

Composite materials are on the cutting edge of technology. In the past, the use of composite materials was primarily limited to the aerospace industry where the high performance characteristics of the materials overshadowed the high cost. Presently, new materials and processing methods are being developed which will make composite materials more cost competitive. Also, new composite fabrication and production technologies are unfolding which enhance the feasibility of producing engine and automotive components in large quantities at a reasonable cost. For diesel engines, the potential for reduced weight, vibration and magnetic signature is dramatic. The program described is an innovative approach to Navy applications. It allows the military to achieve performance improvements by leveraging a technology that is already being embraced by industry and that will proceed rapidly with many opportunities for collaboration and dual-use.

References

1   Maritime Defense, pgs 116-119, May 1992
2   Fischer, W. SNAME Transactions, Vol 92, pgs. 223-239, 1984
3   Kamal, U. SAE 861270, Sept 1986
4   Wingo, R.A., J.J. Holmes, M.H. Lackey, Nav Eng J., May 1992
5   White, LCDR James W., Nav Eng J., May 1984
6   Allison, J.E. and G. S. Cole, J. Metals, Jan 1993
7   Plastics World 46 (5), May 1988

Composite Materials in Reciprocating Engines: A State-of-the-Art Review

M. Mehta
Environmental Research Institute of Michigan
Ann Arbor, Michigan

Abstract

An historic perspective and state-of-the-art review of the use of advanced composite materials in gasoline and diesel engines is presented, including application information on polymeric, metal matrix and ceramic composite materials. The review considers some of the main technology drivers that have contributed to the application of composites in automotive on-engine applications. The paper will also discuss some of the key challenges facing the commercial proliferation of composite components in reciprocating engines, such as (1) lack of design experience with composite engine components, (2) absence of relevant material and processing databases, and (3) lack of cost-effective manufacturing methods and automation.

Introduction

ADVANCED COMPOSITE MATERIALS may be engineered to not only reduce the weight of engine structures, but also possess the potential to improve fuel efficiency and engine life. For many mission critical military applications, there is interest in applying them to reduce signature and improve stealth characteristics. The US Navy recently sponsored a technology survey and review of applications of composite materials to reciprocating internal combustion engines, in order to establish feasibility of incorporating similar advanced materials on a generic diesel engine, illustrated in Figure 1, for demonstration of optimized noise-vibration-harshness (NVH) performance, weight reduction and reduced magnetic signature, at reasonable cost.

This state-of-the-art review covers the general composite material applications in the automotive engine arena, including significant case studies and advances in both, gasoline and diesel fuelled reciprocating engines.

Methodology and Scope

The literature review was conducted primarily through search and access of published information documenting the application of advanced composite components in reciprocating internal combustion engines, and further substantiated through interviews with selected material suppliers, in order to assess the state-of-the-art. This review covers polymeric, metal matrix and ceramic composites and coating technologies for engine components.

Study Findings

General Industry Trends. The automotive industry is extremely cost and risk conscious in the application of advanced materials, cost in order to be competitive, and risk through direct sales to the public consumer. Regardless of the composite materials application potential in reciprocating engines, three major areas of concern are common for component manufacturers, OEMs and end users. They are:
- predictability of material performance,
- the cost of composite components, and
- long-term reliability.

Whereas considerable resources are being expended by the OEMs in the development and optimization of materials, processing and production technologies for composite components in the passenger car and truck industries, innovative on-engine applications are frequently demonstrated earlier in the race car engine and specialty vehicle markets where costs and risks are more acceptable.

Traditionally, the requirement for high power density has been the single most important factor differentiating combat vehicle engine developments from their commercial counterparts (both, diesel and gasoline) [1]. In fact, many performance characteristics of combat vehicle diesels approach those of race car engines. In recent years, the 'race' for cleaner, light weight, more fuel efficient commercial and passenger vehicles, driven by both competition and regulation, has resulted in an increasingly converging trend in composite applications to engine components for civilian and military dual-use, even though the goals are divergent, often more mission specific in case of military land vehicles and marine vessels.

A Delphi study of materials applications in the US automotive industry [2] forecasts a 70% increase in the use of aluminum for engine cylinder heads and a 20% increase for blocks for weight reduction. It also predicts dramatically increased use of powder metals for connecting rods, valve seats and valve guides, as well as coatings for reducing emissions. Another trend from the study concerns the increased use of magnesium as an automotive powertrain material. The current limitations to its use include the relatively high cost per pound, inadequate design and production experience, as well as commercially viable supplies of magnesium for automotive applications. The study also forecasts 10-15% increases in metal matrix composite uses for engine pistons and connecting rods. By 2000 AD, nearly 60% of all gasoline engine valve covers, 10% of oil pans, and 30% of water pump housings are expected to be fabricated from plastics, according to the study. It is predicted that only 5% of valve components will be made of ceramics and 4% of titanium, compared to 90% valves made of steel; and that about 10% of all camshafts will be made by powder metallurgy using steel-composite systems.

Many of these application trends were recently paralleled in a recent Mitsubishi Motors Corp. undertaking [3] of one of the most extensive experimental studies of engine performance through use of a variety of alternate lightweight materials substituted for conventional engine materials. The results were demonstrated on a 4-cylinder in-line gasoline engine. The study included connecting rods, valvetrain components and air intake ports, demonstrating the overall benefits of weight, noise and friction reduction through use of combinations of

conventional steel, aluminum and cast iron as well as metal matrix, ceramics and polymer matrix composites.

Significant case studies accessed in each composite materials category are discussed in the following sections.

Polymer Matrix Composites (PMCs).

<u>PMC Performance Requirements</u>. The general performance requirements of composite materials for structural engine parts such as the oil pan, timing chain, valve covers, manifolds and flywheel housings include the following:

- Low thermal expansion coefficient
- Low density
- High chemical resistance to automotive fluids
- High corrosion resistance to oxidizing environments
- High temperature resistance (upto 200C)
- Medium to high Young's modulus, and
- High shock-absorption/damping.

Advantages of plastic composites include light weight, creep resistance, chemical resistance, sound and vibration damping, parts consolidation, design flexibility and lower finishing costs. The same chemistry that makes them stable at high temperatures also makes them difficult to synthesize and process, resulting in increased costs over conventional engineering plastics such as nylon 6/6, polycarbonate, polyphenylene oxides and thermoplastic polyesters. Other performance concerns include the higher coefficients of thermal expansion (CTE) that 'neat resins' have over metals. While fiber reinforcements reduce the CTE of the composite, thermal mismatches often induce high internal stresses. Moisture absorption, while it improves the material loss factor, also reduces the T_g (glass transition temperature) rapidly, resulting in structural degradation.

<u>State-of-the-Art in PMC On-Engine Applications.</u> Early attempts in under-the-hood applications of high temperature polymeric composites were driven in the 1970s and 1980s by the need to reduce automobile weight and improve fuel efficiency in response to the 'oil crisis'. In the US, these efforts have resulted in the production of chopped glass reinforced, mineral filled polyester (Cyglas[1] 685 from American Cyanamid) valve covers for 17 GM passenger car models [4]. The covers are made by compression injection molding. In case of diesel

---

[1]   The names of products mentioned herein are used for identification purposes only, and may be trademarks or registered trademarks of their respective supplier companies.

engines, due to the higher heat and stability requirements, Cyglas 695 vinyl ester appears to be a promising candidate. Similarly, Ford and Chrysler are also evaluating Cyglas for valve covers. Cyglas vinyl ester is based on the Dow Chemical Derakane 790 family of thermoset resin systems. A high heat resistant experimental version of Derakane 790 with 32% glass content has already been successfully demonstrated in sheet molding compound (SMC) timing chain covers for a Ford 4.6L gasoline engine, and a prototype Cummins L10 diesel engine valve cover made by resin transfer molding of a production SMC epoxy vinyl ester formulation [5]. Detroit Diesel Corp. has introduced Derakane 790 vinyl ester valve covers and oil pans made by SMC at the Budd Company since 1987, though the designs have not been optimized to address noise and vibration concerns [6,7].

In the Japanese and European car markets, a Japanese polymer product of polyarylamide (trade name Solvay IXEF 2032) has become a popular choice for automotive OEMs such as Citroen, Honda, BMW and Peugeot. The Solvay IXEF 2032 material appears to have superior vibration damping properties vs. temperature over the vinyl esters, though it commands a higher per pound price. Examples of structural components proliferate in automotive on-engine applications of polymeric composites throughout the literature. Some of these findings are presented in summary form in Table 1.

A large commercial and military market is seen for plastic composite reciprocating engine intake manifolds of higher temperature resins. In case of fighting vehicles, General Dynamics Land Systems has prototyped a composite air intake plenum and manifold for the M1 Abrams tank, which used Tactix epoxy-carbon-E-glass material system, fabricated by resin transfer molding [8]. The benefit of composite manifolds for gasoline engines, is increased engine efficiency (including maximum power, torque and improved transient knocking at lower temperatures primarily due to the improved surface finish of the inner walls of the composite plastic manifold over the die cast metal manifolds conventionally used [3]. General Motors has introduced a 33% glass filled Zytel Nylon 6/6 injection molded composite manifold for its 3.8L V6 engines [9].

Polyphenylene Sulfide (PPS) has significant promise for automotive underhood application due to several key strengths. Firstly, it is the most reasonably priced of all high temperature resins (about $1.50/lb for 35% glass/35% filler grades, which is comparable to the Cyglas family). Higher glass content (40%) PPS is priced at under $3.00/lb, and may be used at up to 240C with a heat deflection temperature of 260C. Secondly, PPS is available either as linear or branched molecular structure. The highly crystalline nature of the linear PPS inherently resists flames, chemical attack and solvents at temperatures below 204C, while the material itself is dimensionally stable, an excellent insulator, and can be easily injection molded into long, thin-wall section components for valve and timing chain covers. The US suppliers of PPS include Hoechst-Celanese, and Quadrax Corp.

In the early 1980s, Polimotor Research demonstrated the Lola race car engine, made primarily of plastics [10]. The Lola engine is a 2L, 4-cylinder power plant that produces 315 HP at 9500 rpm; it weighs 160 lbs- about half the weight of its metal predecessor, with plastics. Most of the plastics demonstrated are from the Torlon PAI (Poly-amide-imide) thermoplastic family from Amoco Chemicals Corp. Torlon materials have also been used for the engine's timing gears, valve spring retainers, tappets, piston pins, intake valve stems, piston skirts and connecting rods, all produced by injection molding. The engine block is made of graphite fiber reinforced phenolic, which keep their strength up to 500F. The main attributes of phenolic composite aplication include its high elastic stiffness and creep resistance which enable the design of bolted engine components without requiring costly and heavy metal mounting boss inserts [11]. On an experimental scale, Easwaran [12] has also used injection-molded Torlon 7130/chopped graphite powertrain components to demonstrate the weight savings and survivability of polymer composites in extended engine tests on a 5 HP two-cycle, air-cooled Jacobson engine. The components demonstrated included a connecting rod, piston (coated with steel and nylon powder for thermal protection) and crankshaft (pultruded vinyl ester with 70% glass and Torlon counterweights). Recently, Rogers Corp. has developed a glass-reinforced phenolic composite material for use in the fuel rails of Chrysler's Concorde, Dodge Intrepid and Eagle Vision gasoline engines [13]. The fuel rails are multicavity, injection molded to net shape and fastened to the engine head. Reduced engine operating noise, improved fuel temperature management, and reduction in labor cost and fabrication steps are reported advantages of using the phenolic composite.

Polysulfones are a resin family that have high molecular weights, are amorphous, clear, rigid and tough for impact resistance. Also, they resist oil and gasoline, flame spread and smoking at elevated temperatures. Candidate diesel engine applications of polysulfones exist for engine turbochargers and supercharger parts and bearing cages. Key suppliers include Amoco (Radel A and R, Mindel), BASF (PES polyethersulfone), ICI (PES), and Sumitomo-Mitsui.

Bank et al. [14] describe initial tests results with a new, semi-crystalline thermoplastic polymer material system known as syndiotactic polystyrene (SPS) for combined high temperature (up to 230C), moisture absorption (virtually none) and chemical reaction resistance in under-hood applications. Applications are projected for several high and low temperature electrical/electronic components and fluid contact underhood devices. Damping properties of SPS do not appear to have been investigated.

GM's 1993 gasoline engine models for passenger cars are the first production engines to feature valve lifter guides of structural thermoplastic. The guides are molded in Nylatron GS-51, a glass reinforced, molybdenum disulfide-filled nylon 66 [15]. The function of the molybdenum in the filler is for added lubrication at high temperatures. The one-piece molded guides replace six metal guides and a couple of retainer stampings, with component costs upto 60% lower accompanied by a net weight savings as well as elimination of assembly steps for the metal predecessors [16].

Metal Matrix Composites (MMCs). Discontinuously reinforced aluminum MMCs offer to date the greatest commercial viability in reciprocating engine components in the automotive propulsion markets, provided per pound prices of MMC components can be reduced to under $2.00 by use of high volume-amenable manufacturing processes such as squeeze or vacuum casting. The application of MMCs has been driven primarily by weight reduction motivations. However, several studies have demonstrated that the use of stiffer, low mass MMC moving parts also creates fewer vibrations and reduces friction, so that engine parts may be downsized, and support components and counterweights may be eliminated. The types of weight reduction methods investigated to date have included one or more of the following:

- direct substitution of light-weight materials for conventional materials,
- decrease in specific gravities of conventional materials by alternate processing technologies,
- parts re-design/integration/consolidation, and
- downsizing the volumes of critical components.

The aerospace and aeroengine industry has traditionally led research and development in MMCs. Relatively little information is available in the literature on the fatigue and damping behavior of MMCs, because most users regard high temperature strength of MMCs more important - this may be acceptable in aerospace applications of MMCs. However, in case of automotive applications and other such high volume uses, the fatigue and damping behavior of MMCs need to be investigated and documented in considerably more detail to enhance the confidence levels of designers.

MMC Applications in Engine Components. Table 2 provides examples of MMC engine components researched by the OEMs; details of selected component programs are discussed in this section.

Piston: The chief requirements of MMC components for diesel engine piston applications include:

- Low thermal expansion coefficient, CTE
- Low density, weight
- High stiffness/Young's modulus
- High chemical resistance to engine fuels and fluids
- High corrosion resistance to oxidizing environments
- High temperature resistance and thermal stability
- High wear resistance
- High fatigue cracking resistance

Satisfaction of these requirements would result in greater design freedom, better wear characteristics, less vibration and ultimately, greater fuel economy of reciprocating engines. These requirements have also been corroborated by Mingledorff [1] in assessing materials needs for new diesel engine programs of the US Army.

Considerable research has been conducted during the last decade in Japan, Europe and the US in MMC and metal-ceramic piston development. The use of MMC pistons, promises more efficient combustion bowl shapes and the ability to run the engine hotter so that less fuel would be consumed over the life of the engine. Walzer et al. [17] have tested aluminum titanate as a piston crown insert material in high speed diesel engines. However, they report that the low CTE of aluminum titanate makes it difficult to fix the ceramic insert to the piston body. Baker et al. [18] have reported that at AE Developments, United Kingdom, aluminum pistons with zirconia crown inserts have run successfully in a single cylinder diesel engine.

Toyota Motors has pioneered the use of discontinuously reinforced aluminum in production parts for its light truck diesel engines since the mid-1980s for Japanese markets. Toyota's piston crowns are fabricated from Kaowool or Saffil alumina fiber preforms for its high speed engines. The trend in the US automotive and off-highway diesel industry is towards larger diesel engine models (8L to 12L) which have lower rotation speeds than Japanese or European diesels, and hence, less of a wear problem.

Squeeze casting is the most common fabrication method for piston components from MMCs, though other methods such as vacuum casting and gravity casting are used for other engine components. It involves the application of high pressures to the liquid metal during solidification over a reinforcing fibrous preform inside a mold [25, 26]. Toaz and Smalc [20] have shown a significant weight reduction by substitution of a Ni-Resist cast iron insert (heavy) with an alumina fiber Al_3Ni particulate reinforced aluminum MMC (light) which reduced piston weight by 10%. This is an advantage for high speed diesel engines, where inertial loads are critical to performance. Toyota has mass-produced these selectively reinforced MMC pistons at a cost claimed to be 15% less than conventional pistons [21, 22]. In the US, Zollner Pistons, Inc., Fort Wayne, IN, specializes in squeeze cast MMC pistons for heavy duty diesel engines, which utilize ceramic alumina fiber preforms [23]. Several studies have reported that ceramic fiber reinforced pistons can operate at higher temperatures, allowing the engine to burn fuel more cleanly and efficiently. One study, by Chi et al. [24], used F332 aluminum alloy reinforced with 15% (by volume) of discontinuously reinforced Fiberfax[R] aluminosilicate ceramic fibers, concluding that long term thermal exposure at 400C indicated no difference in fatigue strength between 100 hr and 1,000 hr aged samples. These results provide further strong incentive for piston manufacturers to use MMCs in piston applications.

Connecting Rod: MMC applications to connecting rods are numerous throughout Europe (UK, Germany, France) and the US, though early innovations originated in Japan. The application is driven by the possibility of reduced noise and vibrations in the various engine reciprocating parts, lower inertia and mass resulting in higher rpm and improved efficiency of engines.

Toyota has demonstrated reinforced Al and Mg connecting rods, though at a significant cost premium (approx. $25/rod, compared to $4 for conventional steel rods). Both, Duralcan and DWA Composite Specialties have demonstrated MMC connecting rods that are fabricated to net shape by extrusion and machining. Ricardo Consulting Engineers [25] have demonstrated a hybrid composite connecting rod application in their experimental 1.6L Pegasus gasoline engine, at costs which are reported to be suitable for low cost, high volume production. The Pegasus program requirements were primarily to meet fuel efficiency and emissions requirements. The high stiffness engine block and cylinder head are of aluminum-magnesium alloy castings. Other relevant features include damped covers and oil pan, a dual mode crankshaft damper, a 2-ring piston and a hybrid MMC connecting rod which is comprised of hand laid-up bolt bosses, and filament-wound rod ends.

A recent Office of Naval Research (ONR) Report [26] describes current research at Nissan Motor Co. in SiC fiber whisker reinforced aluminum alloy connecting rods for reciprocating engines. Nissan's experience has resulted in forged Al MMC connecting rods which are 45% lighter than steel rods. In piston research, Nissan has demonstrated a lightweight aluminum oxide short-fiber reinforced magnesium alloy piston (10% by volume fibers) for improved engine noise and efficiency. The material is reported to be five times more abrasion-resistant than conventional aluminum alloy pistons, and also 40% lighter.

Tanaka et al. [3] compared the performance of several alternate light weight material substitutes for a 4-cylinder in-line Mitsubishi gasoline engine, including Steel (baseline), Free Cutting Titanium Alloy (Ti-3Al-2V), rapidly solidified (RS) Aluminum Alloy (Al-8Cr-2Fe-1Ti), SiC Whisker/Al 6061 MMC and Free-Cutting Steel (S48CL2). In the study, both, Titanium alloy and the MMC connecting rods were found to significantly improve engine performance characteristics, as well as noise and vibration. The titanium and MMC connecting rods were also found to have the highest specific fatigue strengths from all the candidate materials tested.

Engine Block: Honda Motor Co. has already been mass-producing pressure cast aluminum MMC engine blocks for their Honda Prelude models, featuring alumina (Kaowool and Saffil) and carbon short fiber reinforced cylinder liner preforms, selectively reinforcing a cast matrix of ADC12 aluminum (JIS). By adopting the partial reinforcement method, which strengthens only the cylinder bore of the engine block, an inexpensive, light, compact and more producible engine block has resulted. Considerable progress has been reported by Honda from supporting studies on engine seizure limits, sliding wear and optimal fiber reinforcement configurations, using several 2.0L engines on which extensive operating and field tests were run. The use of MMC engine block materials, instead of cast iron liners, allows a shorter distance between the centers of the cylinder bores due to improved cooling capacity, resulting in a more compact engine having the same power density as one with cast iron liners [27, 28, 29]. Using powder metallurgy techniques, Honda is also mass-producing RS Al alloy/alumina-zirconia-silica particulate reinforced retainer rings for valve springs, saving 40% by weight over steel, as well as pressure casting stainless steel filament reinforced aluminum connecting rods. Significant research is underway in prototyping Al and Mg MMC pistons.

Other MMC Applications: Other candidate MMC components in diesel applications include the cylinder head, block and valve train components. No significant publications or descriptions were found on MMC cylinder head or engine block programs for diesel engines, though the potential for weight, signature (in case of military engines) and noise-vibration reduction is high. For both of these applications, highly advanced, cost effective processing technology is necessary to reduce the possibility of introducing porosities in MMC castings of such complex load bearing structures. Peugeot of France has begun the use of an ICI-produced fiber preform insert at each of the critical inter-exhaust valve bridges on its 16-valve, high performance cylinder heads [30]. MMC valve seat inserts, for wear resistance and thermal expansion that is better matched to the cylinder head, may also have significant effects on noise and vibration reduction in future engines. The trend to a complete MMC aluminum cylinder head will sooner or later render the MMC valve seat insert redundant. Intermetallic compounds of TiAL and $TiSi_2$ are being actively studied for automotive valve applications at Nissan due to their high temperature resistance and high specific strengths [26]. A similar processing research and cost modeling thrust is underway at Ford Motor Co. in titanium alloy valve train components.

Chrysler Motors has recently reported studies in modified sand casting of MMC components for several production engine components [31]. The process design changes were accomplished primarily through new gating and riser configurations, and validated by demonstration components that included camshaft thrust plates and cylinder liners, both of Aluminum MMC -T61 matrix with 10% and 20% SiC whisker reinforcements. In case of the camshaft thrust plate, tests have shown that the 20% SiC reinforced Aluminum MMC underwent little wear when tested for 800 hours, compared with the 10% SiC MMC, though the high cost prevented its release for production use. Research is currently in progress at Chrysler on combinations of the cast MMC cylinder liner with piston coatings and piston ring pack variations. The Ohio State University [32] reports a newly developed process which allows the near-net shape fabrication of alumina/aluminum composites by immersion of a sacrificial oxide preform into a molten aluminum alloy bath. The process appears to offer better economy in the manufacture of engine cylinder liners made of co-continuous ceramic composites without sacrificing material properties and performance.

Advanced Ceramics, Composites and Treatments. Advanced monolithic ceramics and ceramic composites for diesel engine applications include alumina (Al_2O_3), silicon nitride (Si_3N_4), silicon carbide (SiC), partially stabilized zirconia (PSZ), sialon (silicon aluminum nitride) etc. [18]. Compared to conventional automotive metals, ceramics are lighter, harder, and more resistant to wear and heat, and also excellent insulators to both heat and electricity. Their use in land and marine engines would permit operation at higher temperatures, so that fuel efficiency and power are improved while particulate emissions (a serious problem in diesels) are significantly reduced [33]. Designers could substantially reduce engine size and weight by elimination of cooling systems from ceramic engines.

Ceramics Implementation Issues. The chief disadvantages include their high susceptibility to mechanical and thermal stress, catastrophic failure modes, and the difficulty in making reliable reproducible components by casting and finish machining to net shape. Because of their high hardness, ceramics are extremely difficult to machine. Other design for manufacturability issues concern the lack of room temperature ductility, so that parts that do not fit perfectly in an assembly may become highly stressed, and rupture catastrophically. Complicated shapes, as often encountered in engine design, are difficult to make economically by hot pressing. Sintering or sintering followed by hot isostatic pressing are the alternate processing routes for non-glassy ceramics, but often add to final cost. Machining may consume upto 90% of the total cost of ceramic parts. While worldwide efforts to exploit the promise of ceramic and ceramic matrix composites in engines have proliferated over the years, few ceramic parts have reached production in diesel (or gasoline) engines to date. Despite these drawbacks, the application of ceramic and ceramic composite components in reciprocating engines is expected to become more economically viable in the near-term, because it is anticipated that many end-users would be willing to trade the higher up-front costs for fuel and operating savings over the life of the vehicle.

Ceramic and Composite Applications.
Table 3 presents a summary of recent ceramic applications in reciprocating engines. In the US,

Caterpillar, Cummins, Detroit Diesel, Ford, GM, Lanxide and Norton Co., all have strong ceramic composite research programs for engine applications. Considerable research investments have been made by the Japanese OEMs (Toyota, Isuzu, Mazda) and their suppliers (Kyocera, NGK and NTK among others) in ceramic gasoline engine components manufacture for passenger cars.

In general, most automotive research in ceramics has focused on monolithic ceramics rather than ceramic composites which are considerably more expensive to use. Zirconia-based materials have been evaluated on several advanced diesel engine development programs by Cummins Engine Co [34, 35]. The low thermal conductivity of zirconia is combined with a relatively high CTE which eases the difficulties of attaching the ceramic component to metal. In the Cummins programs, zirconia cylinder liners and cylinder head face inserts have operated successfully in an engine which has run without the water cooling system [18]. Cummins currently also has a silicon nitride fuel injector link for use in a commercial diesel engine, for which field use results indicates that the ceramic links provide improved dimensional tolerances and reduced wear [36]. Detroit Diesel Corp. has recently introduced ceramic valve seats in a limited commercial engine family (the Series 60 engines), though none have been introduced in military engines [6].

Other market projections on ceramic engine technology come from a 1989 study by Argonne National Laboratory, which estimated that ceramic intensive heat engines will account for 1% of light and heavy duty diesel engine sales in 1995. The study also projected that adiabatic diesel engines will be available in 2000 AD. Currently, from all accounts, Japan leads in ceramic engine design, component processing and fabrication technology. Early successes that support this assessment include their introduction of diesel glow plugs and precombustion and vortex (swirl) chambers (demonstrated by Mazda in 1987 on a light 1200cc diesel), as well as turbocharger rotors for reciprocating engines (Mitsubishi and NGK). One of the few success stories with larger ceramic components comes from Isuzu Motors, Japan, which has developed a demonstration turbocompound engine that operates on gasoline, methanol and pulverized coal.

Valve train components (such as valve guides, valve seats, etc.) appear to be receiving the greatest research focus due to the need to withstand high stresses and wear rates. Silicon nitride (also known as Nitrasil) valves have nearly half the weight of their metal components, and are known to require lower spring forces or permit higher engine speeds at lower noise and vibration levels. Cam roller followers, rocker arm tips of silicon nitride or silicon carbide represent additional opportunities in engines. Ceramtec Division of Hoechst AG, Germany have demonstrated silicon nitride ceramic intake valves in Diamler Benz cars [37]. The prototypes are 60% lighter than those of conventional steel, permitting reductions in both, the size of the valve spring and stem length, so that the cylinder head too may be made more compact and lighter in design. Test vehicles modified with ceramic valves have been shown to run smoother (30% less inner friction), with much lower noise levels (a decrease of 18dB at 3000 rpm), as well as 30% lower CO and HC emissions, accompanied by 3-4% less fuel consumption.

Few published studies are available on noise-vibration performance of ceramic or ceramic composite automotive components, despite the worldwide proliferation of R&D in development of ceramic reciprocating engine components. However, one study by Gibson et al. and Ford Motor Co. [38] reports that the main vibration frequencies of ceramic intake and exhaust valves lie well above the audio frequency range so that the noises emitted should be inaudible to the human ear. The study compared results of flexural and extensional vibration tests to determine natural frequencies and damping properties of engine intake and exhaust valves of several metal (steel and titanium aluminide alloys) and ceramic (silicon nitride and silicon carbide) valves. In flexural vibration, the titanium aluminide valves had the greatest damping, while the Silicon carbide valve had the greatest damping in extensional vibration. The natural frequencies of all metal and intermetallic valves tested were found to lie in the audio frequency range ($<$20,000Hz).

Tanaka et al. [3] at Mitsubishi Motors have studied effects of direct silicon nitride ceramic substitution versus several alternative metals and intermetallics (Ti-Al, Ti-6Al-4V, and Heat-resisting Steels) for a 4-cylinder passenger car gasoline engine intake and exhaust valve application. They report superior heat resistance of the SiN valves at exhaust temperatures of over 800C, as well as superior wear resistance on the weight-reduced engine.

In case of alternately fuelled engines, the use of monolithic ceramics such as Si_3N_4 appears to be undesirable due to chemical reactions between the fuel and component materials. Caterpillar has considered SiC/Alumina glow plugs for alcohol engines because the high sodium content of ethanol fuel degrades Si_3N_4. In the marine propulsion environment, there is an important benefit to be gained from greater use of ceramics which resist high-temperature corrosion that metallic engine components of both, gas turbines and diesels are so vulnerable to. The use of ceramics and ceramic composites in reciprocating engines for marine propulsion also represents an opportunity in reduction of engine rebuilds.

Advanced Engine Coatings and Treatments. Significant research in high performance engines has been undertaken in the development and use of ceramic coatings on exposed surfaces of engine components made of conventional materials such as steel and cast iron. Most frequently, the lack of maturity of composite materials and of economical fabrication techniques were found to be important drivers for pursuit of advanced coatings for enhanced strength, thermal-oxidation resistance and wear characteristics of engine components.

The low coolant, heat rejection diesel engine has been under investigation for many years by many investigators worldwide. The early strategy of using monolithic ceramics for these insulated engines soon gave way to thermally sprayed ceramic coatings due to lower cost, greater reliability and ease of design of engine components. Piston coating applications are indicated in Table 3. Typical coating materials include slurries of zirconia, chromium oxide or cobalt oxide, etc., applied to a component by either dipping, spraying, painting or drain casting, followed by curing to obtain densification and finish machining. Potential for property improvements is presently being investigated in the application of newer beam (laser, ion and electron) technologies that use chemical vapor deposition, direct metal deposition, dispersion hardening, as well as functionally gradient material casting for developing heat and wear resistance on both, metal and ceramic matrices and substrates.

Few studies have reported on the noise-vibration response characteristics of advanced ceramic coatings and treatments applied on engine components such as piston crowns, cylinder liners and cylinder heads. Peripheral benefits of coatings that are cited include maintenance of dimensional stability and oil leak prevention in combustion chambers [39].

Conclusions

Several composite materials have been shown to be technically acceptable for application to on-engine parts. Cost studies have narrowed these down significantly. Nearly all material alternatives experienced early development problems. In the US, few composite materials have reached volume production, due to factors which include vehicle lead times, individual designer's risk profiles, capital investment needs, and absence of proven production viable processes. The survey findings indicate that the Japanese automakers have the clear lead in developing and deploying metal matrix and ceramic composites technology into engine components. Application successes with polymeric matrix composite components to engines are shared equally by US, European and Japanese manufacturers, though the specific materials and processing technologies are largely pioneered by multi-national materials and molding suppliers.

The scale-up to production applications of advanced composites is expensive, requiring collaboration between supplier and user, and the timing is critical. To ensure US domestic industry competitiveness, it is essential that dedicated industry consortia and recent industry-government collaborations such as the Partnership for a New Generation of Vehicles address these issues in their totality.

Acknowledgements

The author gratefully acknowledges the support of this study provided by Roush Anatrol Division of Roush Industries and the Naval Surface Warfare Center under contract # 1998/N61533-92-R-0062.

References

1.　Mingledorff, MS, US Army Materials Technology Lab Report No. MTL-TR-87-1, Watertown, MA, (1987)

2.　Cole, D., D.J.Andrea and R.L.Doyle, "Delphi VI Forecast and Analysis of the US Automotive Industry Through the Year 2000", Vol. 3: Materials, The University of Michigan Transportation Research Institute, Ann Arbor, MI, (1992)

3　Tanaka, I., T.Shimamoto, T.Yamaguchi and J.Noguchi, SAE Paper 922090, Dearborn, MI, March 1992

4.　Krigbaum, R.S., SAE Paper 930089, Detroit, MI, March 1993

5.　Personal Communications, Dr. James Patten, Director of Materials and Manufacturing Research, Cummins Engines Co., Columbus, IN, June 1993

6.　Personal Communications, Dr. Tony Kaushal, Materials and Manufacturing R&D, Detroit Diesel Corp., Detroit, MI, March 1993

7.　Personal Communications, Dr. Marie Winkler,

Technical Specialist, Dow Chemical Co., Freeport, TX, July 1993

8. Rock, D.K. and W.S.Mosset, in Proceedings, Advanced Composites Conf. and Exhibition, Detroit, MI, September 1990, p.83-96

9. Pawl, TE, D.Dalo and D.Tres, SAE Paper 930087, Detroit, MI, March 1993

10. Anonymous, Modern Plastics, December 1984, p.32

11. Arimond, J., SAE Paper No. 931027, Detroit, MI, March 1993

12. Easwaran, J., in Proceedings, Advanced Composites Conf. and Exhibition, Detroit, MI, 1990, p.523-527

13. Anonymous, "Composite Fuel Rails", Automotive Engineering, June 1993, pp.84

14. Bank, D.H., T.E.Wessel and J.J.Kolb, SAE Paper No.930088, Detroit, MI, March 1993

15. Brooke, L., Automotive Industries, p.30, July 1993

16. Birch, S., J.Yamaguchi, A.Demmler and K.Jost, Automotive Engineering, p.43, December 1992

17. Walzer, P., H.Heinrich and M.Langer, SAE Paper 850567, Detroit, MI, March 1985

18. Baker, A.R., D.J.Dawson and D.C.Evans, Materials and Design, 8(6), p.315-323, 1987

19. Donomoto, T., N.Miura, K.Funatani and N.Miyake, SAE Paper No.83052, Detroit, MI 1983

20. Toaz, M.W. and M.D.Smalc, Diesel Progress in North America, June 1985.

21. Suganuma,T., and A.Tanaka, J. of Iron and Steel Institute of Japan, Vol.75, p.376-383, 1989

22. Kubo, M., A.Tanaka and T.Kato, Japan Society of Automotive Engineers Review, Vol.9, p.56-61, 1988

23. Personal communication, Jeffrey Kasselman, Structures Analyst, Zollner Pistons, Inc., Fort Wayne, IN, November 1993

24. Chi, F.K., SAE Paper 930183, Detroit, MI, 1993.

25. Ricardo Marketing Information, Ricardo News, Issue 45, March 1993

26. Vedula, K., Scientific Research Bulletin, Office of Naval Research Asian Office, NAVSO P-3580 18(2), April-June 1993

27. Ebisawa, M., T.Hara, T.Hayashi and H.Ushio, SAE Paper 1991

28. Hayashi, T., H.Ushio and M.Ebisawa, SAE Paper 1989

29. Ushio, H., T.Hayashi, K.Shibata, Y.Fujisawa and T.Hata, Japan Patent 4,817,578 (1989)

30. Lovell, B., Racecar Engineering, 2(6), p.37-41, 1993

31. Chang, N.S., G.P.Faubert, K.A.Goulait, J.C.Grebetz and J.R.Huth, SAE Paper 930179, Detroit, MI, 1993

32. Breslin M.C., Liang Xu, G.S.Daehn and H.L.Fraser, SAE Paper No. 930184, Detroit, MI, 1993

33. Myers, P.S., Applied Mechanics Review, 42(3) ASME, p.53-69, 1989

34. Kamo, R., in Proceedings of International Symposium on Ceramic Components for Engines, Hakone, Japan, 1984, Editors Sowmiya, S., E.Kanai and K.Ando, D.Ridel Publishing Co. (1984)

35. Woods, M.E., and T.Schofield, in Proceedings of International Symposium on Ceramic Components for Engines", Hakone, Japan, 1984, Editors Sowmiya, S., E.Kanai and K.Ando, D.Ridel Publishing Co. (1984)

36. Nestlerode, S., Ceramic Industry, p.40-44, July 1992

37. Anonymous, Automotive Engineering, p.14, August 1993

38. Gibson, R.F., S.Yang and G.M.Crosbie, in Proceedings of Spring Conference on Experimental Mechanics, SEM, Dearborn, MI, June 7-9 1993

39. Bryzik W., E.Schwarz, R.Kamo and M.Woods, SAE Paper 931021, Detroit, MI, 1993

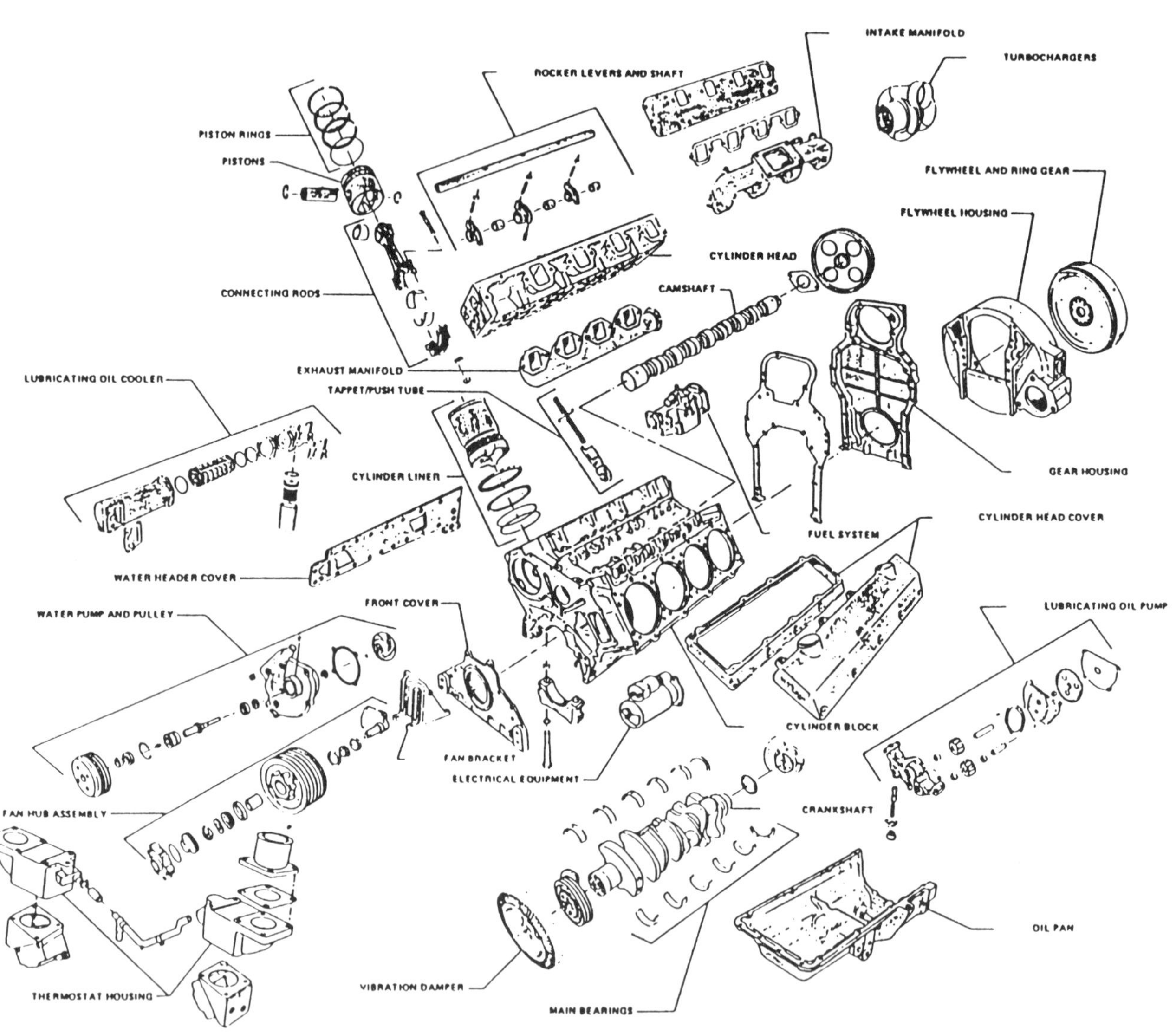

Figure 1. Components of a Generic V-8 Diesel Engine.
[source: Naval Surface Warfare Center]

Table 1. Examples of Recent Polymeric Composite Applications in Reciprocating Engines

Component Application	OEM/ Supplier	Material System	Manufacturing Process	Diesel/ Gasoline Engine	Application Status (Prototype or Production)
VALVE COVER	Cummins/ Sewards	Derakane vinyl ester/ glass	RTM	Diesel	Prototype
	GM/ Cytec	Cyglas 685 polyester/ glass	Compr. injection molded	Gasoline	Production
	Hyundai/ Du Pont	Minlon 2C/ glass	Injection molded	Gasoline	Production
	Citroen/ Solvay	IXEF 2030/ glass	Injection molded	Diesel	Production
	Detroit Diesel/ Dow	Derakane 790/ glass	SMC	Diesel	Production
INTAKE MANIFOLD	Handy & Harman/ Hoechst Celanese	Fortron PPS H40L4/ glass	Injection molded	Gasoline	Production
	Laval/ Solvay	IXEF 1022/ glass	Fusible core molded	Gasoline	Prototype
	General Motors/ Du Pont	Zytel nylon 66/ glass	Injection molded	Gasoline 3800 V6	Production
	Mitsubishi	Phenolic/ glass	Lost core molded	Gasoline	Prototype
	Mitsubishi	Polyamide 66/glass	Lost core molded	Gasoline	Prototype
OIL PAN	Detroit Diesel/ Dow	Derakane 790/glass	SMC	Diesel	Production

Table 2.   Examples of Recent Metal Matrix Composite Applications in Reciprocating Engines.

Component Application	OEM/ Supplier	Material System	Manufacturing Process	Diesel/ Gasoline Engine	Application Status (Prototype or Production)
PISTON	Daimler-Benz	Al/Al Titanate crown insert	Squeeze cast	Gasoline	Production
	AE Dev., UK	Al/Zirconia crown insert	Unreported	Diesel	Prototype
	Toyota	Al/Saffil-alumina	Squeeze cast	Diesel	Production
	Zollner	Al/SiC-alumina	Squeeze cast	Diesel	Production
	Nissan	Mg/Alumina		Gasoline	Prototype
CONNECTING ROD	Toyota	Al/SiC Mg/SiC	Extrusion + machining	Gasoline	Production
	Nissan	Al/SiC		Gasoline	Prototype
	Ricardo	unpublished	Filament winding/ Layup	Gasoline Pegasus 1.6L	Prototype
	Mitsubishi	Al 6061/SiC	Forged + machined	Gasoline	Prototype
	Mitsubishi	Titanium	Forged + machined	Gasoline	Prototype
ENGINE BLOCK	Honda	Al ADC12/ Fibers: Kaowool, Saffil, Carbon	Injection Die Casting	Gasoline, 2.0L	Prototype
VALVETRAIN	Ford	Titanium TiAl	Powder Metallurgy	Gasoline	Prototype
CAMSHAFT THRUST PLATE	Chrysler	Al/SiC	Sand Casting	Gasoline, V-6	Prototype
CYLINDER LINERS	Chrysler	Al/SiC	Sand Casting	Unknown	Prototype
	Honda	Al/Alumina + carbon	Pressure Casting	Gasoline	Production

Table 3. Examples of Recent Ceramic and Ceramic Composite Applications in Reciprocating Engines.

Organization	Con-rod	Cylr liner	Exh. port liner	Glow plugs	Piston pin	Pre-combus. chamber	Rocker arm pads	Turbo-charger rotor	Valves	Valve guides/ seats	Piston coating
Detroit Diesel										X	X
Caterpillar				X							
Cummins Engines		X									
Daimler-Benz			X		X			X			X
Hoechst-Ceramtec			X					X		X	
Isuzu				X	X	X	X	X	X	X	
Kyocera				X		X	X				
NGK-Sparkplug	X					X	X	X	X	X	
Peugeot						X	X				
NTK						X	X	X	X	X	X
Toyota						X					

Cost Effective Discontinuously Reinforced Aluminum Via Near-Net Shape P/M Processes

W. Hunt Jr., T.J. Rodjom, M.E. Hyland
Alcoa Technical Center
Alcoa Center, Pennsylvania

Abstract

The P/M approach has traditionally offered the highest quality Discontinuously Reinforced Aluminum materials, which can be cost-effective for high performance applications. Recent breakthroughs in net shape processing have enabled P/M DRA materials which can address lower cost requirements yet still meet the technical needs of demanding applications in the automotive engine.

OVER THE COURSE OF THE PAST TWENTY or so years of development of aluminum metal matrix composite materials, a subclass, commonly known as Discontinuously Reinforced Aluminum (DRA), has emerged as a leading group of materials being considered for commercial application. This is a result of a combination of desirable technical attributes, available materials at a commercial scale, and promise of sufficient cost-effectiveness for a range of applications. A key element in the evolution of DRA materials has been the development and refinement of the primary processing methods for composite manufacture. While blending of aluminum powders and ceramic reinforcements was one of the earliest methods considered, a myriad of other processing routes, including liquid metal stirring (Duralcan), spray forming (Osprey, CoSpray), and *in situ* reactions (XD, LSM, Sutek) have been developed and in some cases carried to the pilot or commercial scales. In parallel with the development of these processes, the powder metallurgy (P/M) route has continued to be studied and refined.

The objective of this paper is to highlight recent developments in the P/M DRA process, emphasizing the flexible capabilities of powder metallurgy-based processes for the production of DRA materials to meet automotive application needs. The first part of this paper will provide a summary of the conventional processing route for P/M DRA materials, with the specific example of an automotive engine application highlighted. This will be followed by a discussion of recent work directed towards developing more net shape processing routes for P/M DRA materials in order to reduce final part cost. Finally, some future directions for the development of the P/M DRA process and products will be suggested.

The Conventional Powder Metallurgy DRA Process

A schematic outline of the conventional P/M DRA processing route for high performance structural components is shown in Figure 1. A description of the key elements will be given to illustrate the critical aspects of this processing route.

Component Materials. High performance P/M DRA materials primarily involve the following two starting components:

<u>P/M Aluminum Alloy Matrix</u>. The matrices are in the form of prealloyed, atomized aluminum powders typically in a size range of >90% -325 mesh. An oxide layer of 5-15 nm in thickness consisting of hydrated alumina as well as physically adsorbed water and oxygen is found on air atomized powders.[1,2] In addition, MgO is observed in alloys containing Mg. These oxides are an important consideration in the processing and performance of powder metallurgy aluminum materials and will be discussed further in a subsequent section. In addition, through the rapid solidification of the powders, a refined microstructure with homogeneity of the scale of the powder particles is achieved. This microstructure provides for improved properties and consistency. A recent review discusses many aspects of importance for aluminum powder.[3]

Research on DRA materials has shown that the matrix characteristics play a dominant role in determining many properties of the composite. Al-Cu-Mg (2XXX) matrix

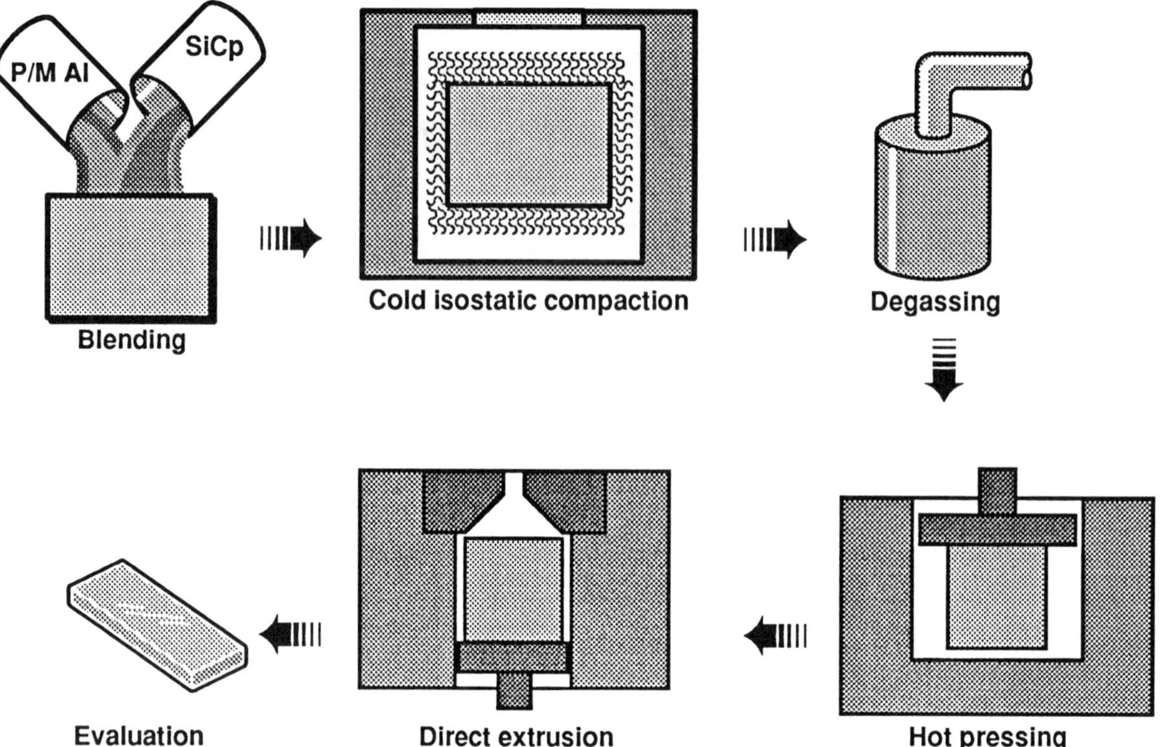

Fig. 1 – Conventional P/M DRA Process

systems provide excellent combinations of strength and damage tolerance. Al-Zn-Mg-Cu (7XXX) matrix systems offer higher strength potential. Al-Mg-Si-Cu (6XXX) systems can provide improved corrosion resistance for severe environments as well as improved product fabricability. Al-Fe-X (8XXX) systems provide the opportunity for higher temperature performance. The P/M DRA processing route is essentially alloy insensitive, allowing incorporation of the full range of aluminum alloy matrix alloys to meet a broad spectrum of performance needs.

In initial DRA materials, standard "garden variety" alloys such as 6061 and 2024 were used as matrices. While these alloys had been developed to have a good balance of properties in the unreinforced condition, there were no changes to accommodate the addition of the reinforcement particles. Recently, more composite-specific alloy development work has led to alloy matrices which better account for the metallurgical changes which occur with the addition of reinforcement. Recent alloys registered with the Aluminum Association specifically for use in composites include X2080, 2009, 6113, 6091, and 6092.

<u>Ceramic Reinforcement Particles.</u> Selection of the reinforcement, including composition, morphology, and volume loading, is primarily determined by the balance of technical and economic characteristics desired. Typically, the properties most directly influenced by reinforcement selection are elastic modulus, strength, fracture toughness, and fatigue resistance, along with physical properties such as coefficient of thermal expansion. Selection of the reinforcement type and geometry is critical in obtaining

the best combination of properties and cost. Continuous fibers provide high levels of properties, but are anisotropic, have high costs, and preclude further fabrication of the composite. Whisker reinforcements also offer the potential for enhanced properties, however, they are expensive and can lose much of their advantage due to breakage during secondary composite fabrication. Additionally, the processing of whisker reinforcement often have significant health issues. Particle reinforcements provide isotropic stiffness improvements with generally lesser degradation in fracture properties than whiskers. In addition, because there is a large established commercial base for the production of ceramic particle materials in the abrasives industry, this form of reinforcement can be quite cost effective. Of the available particle reinforcement candidates, including SiC, Al_2O_3, TiC, TiB_2, and B_4C, SiC is currently the most widely used material due to its favorable combination of mechanical properties, density, availability, and cost.

Standard ceramic reinforcement particles used in P/M DRA materials, such as SiC and Al_2O_3, are produced through large-scale processes and are typically sized to a range roughly similar to that of the aluminum matrix powder. Due to the grinding process used in their manufacture, standard SiC reinforcement particles typically exhibit sharp corners and faceted faces. Substantial effort has been made by the suppliers of reinforcement materials to tailor them specifically for use in DRA materials. This includes controls on surface chemistry, size distribution, and, in some cases, particle shape. Understanding the importance of these features of the reinforcement on importance of these features of the reinforcement on microstructure and properties is important in providing the desired product characteristics consistently.

Blending. Blending of the metallic and ceramic powder components is carried out with the objective of distributing the ceramic particles uniformly in the aluminum powder matrix. Typically, structural composites are made with ceramic volume loadings of 10%-40% by volume. The details of the blending process are traditionally proprietary, and only recently have scientific studies of the blending process been published.[4]

Assessment of the effectiveness of the blending process can be made through measurements on both the macroscopic and microscopic scales. Macroscopically, measurements of the volume fraction of ceramic phase from region to region in the blend can be made by obtaining blended powder samples, dissolving the aluminum matrix powder, and determining the amount of remaining ceramic particle amount. Calculation of the standard deviation of the ceramic particles loadings from typically 10-20 tests provides a measure of the uniformity of macroscopic distribution of the ceramic phase. As a result of process development and a fundamental understanding of the blending dynamics, the standard deviation values have decreased to ±0.2%, indicating the progress toward more macroscopically uniform blended material.

It is also important for structural properties that the dispersion of the ceramic phase be as uniform as possible on a microscopic level as well. Even with good uniformity of dispersion on a macroscopic level, the potential for reinforcement particle clustering on a microscopic level exists. It is possible to have similar macroscopic levels of reinforcement distribution while on a microscopic scale the distribution is substantially different, as discussed in other papers.[5,6] It is at the microscopic level that the understanding and control of the blending process becomes critical for the achievement of high performance properties on a consistent basis.

Billet Consolidation. After blending it is necessary to consolidate the powder into a fully dense product. For materials involving prealloyed aluminum powders, these consolidation processes specifically need to deal with the presence of a hydrated aluminum oxide film on the powder particles. Therefore, steps must be included in this process to fracture and disperse this film in order to ensure particle-to-particle bonding during consolidation.

The conventional methods for production of P/M DRA billet products involve a three step process of cold isostatic pressing of the powder to an approximately 70% dense compact, canning and degassing the compact at elevated temperature under vacuum to remove adsorbed gases and moisture, then consolidating it under high pressure to full density. In the elevated temperature degassing step, the aluminum oxide is converted from a hydrous to an anhydrous form, which makes it more brittle, such that upon subsequent consolidation and

deformation processing, the oxides can be dispersed in fine form. Prior work on unreinforced aluminum alloy P/M products has shown that the degassing step is important in these materials in order to achieve the desired level of structural characteristics, especially fracture-related properties.[7] Insufficient degassing leaves a more continuous oxide film on the powder surfaces which does not produce the desired level of metal-to-metal bonding and can result in premature fracture along prior particle boundaries.

Secondary Processing. The consolidated billet materials can be subsequently fabricated by conventional wrought processes, with adaptation for the different deformation characteristics of the ceramic particle reinforced materials. Prior work on aluminum powder metallurgy alloys without ceramic particle reinforcement demonstrated the improvement in structural properties which resulted from additional hot working of the materials through forging or extrusion processing.[1] The hypothesis is that even though the billet was 100% dense after consolidation, the additional deformation was required in order to further break up and distribute the oxide particles. For unreinforced aluminum P/M materials, "critical" values of post-consolidation deformation have been established to produce optimum properties. It has also been shown that the details of the deformation processing process are important in determining the properties of the P/M DRA materials as well.[6]

Advantages of the P/M Process. Through intelligent control of the P/M DRA process, repeatable, high quality products ranging from powder blends to fully-dense billets to final fabricated extrusions and forgings have been demonstrated.[8] Production of DRA materials via this route offers a number of advantages over competitive routes, including:

• <u>A wider range of reinforcement levels</u>. Stir casting and spray casting tend to be limited to lower reinforcement levels (typically 15%-20% by volume) due to inherent process restrictions. P/M DRA materials have been produced to volume fractions as high as 55% by volume with structural properties achieved repeatability at levels of 40% by volume. This extended range of reinforcement volume loadings is important to address customized applications requiring the higher levels of stiffness and lower levels of coefficient of thermal expansion which can be obtained with increased reinforcement levels.

• <u>A wider range of aluminum matrix alloys</u>. Since the processes used in the production of DRA materials by the powder metallurgy route are primarily carried out in the solid state, issues related to reactivity between the ceramic reinforcement particles and the matrix are minimized. In addition, the rapid solidification resulting from the atomization process can produce unique aluminum microstructures,[3] which enable improved properties both at ambient and moderately elevated temperatures.

• Improved uniformity of the reinforcement spatial distribution. A homogeneous distribution of the reinforcement phase throughout the matrix is paramount to the achievement of enhanced, repeatable property combinations. A well-controlled powder blending process will produce a more random distribution of the ceramic particles in the aluminum matrix than can be achieved by liquid metal processes, due to the well-documented segregation which occurs in the latter due to particle pushing phenomena. As the scale of the billet increases, this effect becomes more pronounced while the powder blending process is scale-independent. This uniformity is important not only to achieve optimized structural properties but also for obtaining repeatable property levels. In segregated materials, properties especially important in structural applications such as toughness and fatigue resistance can vary significantly from test to test.

Application Example: Connecting Rod.

The P/M DRA connecting rod shown in Figure 2, designed to replace a powder steel part, has been highlighted as a potential application for P/M DRA materials because of the potential of reduced secondary shaking forces and, therefore, improved noise, vibration, and harshness (NVH) characteristics through lightweight drive train components.[9]

Table I shows a comparison of key properties of X2080/SiC, one P/M DRA material especially well suited to demanding automotive engine applications, with steel as well as other candidate materials for this application.

Table I. Specific Properties of Selected Materials (taken from Ref. 9)

Material	Specific Modulus $(x10^7$ mm$)$	Specific Fatigue Strength $(x10^6$ mm$)$
2080/SiC/15p	3.5	7.6
6061/Al$_2$O$_3$/20p	3.5	3.8
6061	2.6	4.0
1040 Steel	2.6	3.0
Ti-6A1-4V	2.5	7.0
Ductile Cast Iron	2.2	7.0

Note: Fatigue strength is strongly influenced by metallurgical factors and testing methods. Values given for specific fatigue strength are typical values for comparison.

The primary factor making X2080/SiC of interest for this application is the high cycle fatigue performance of the P/M DRA material, which is greatly improved compared to conventional aluminum alloys and suitably high to substitute for steel in selected applications. The S-N fatigue performance of X2080/SiC/15p-T6 extrusions in comparison to conventional 2XXX aluminum alloys at room temperature is shown in Figure 3a and at the temperature of interest for connecting rod applications (150°C) in Figure 3b. In both cases, the high cycle fatigue resistance for the DRA material is superior to that of the unreinforced alloy. These levels of fatigue performance are not achievable in DRA materials produced by casting routes,[10] hence making P/M DRA the material of choice. Bench and engine tests have confirmed the performance of the X2080/SiC connecting rods.

The X2080/SiC connecting rod discussed above was produced using the conventional P/M process route. The hot consolidated billets were extruded to round forging stock and then forged to rough configuration through a series of steps, illustrated in Figure 4. The resulting forging was trimmed and machined to the connecting rod component which was subsequently tested and proved to have satisfactory performance. While producing the desired part performance, the multi-step conventional forging process results in low material recovery, which adversely impacts the final part cost. Thus, it is desirable to be able to achieve the improved performance of this material with higher material recovery.

The Direct Powder Metallurgy DRA Process

Driven mainly by the need to reduce the cost of P/M DRA materials for the automotive industry, experiments were undertaken to examine the effect of more direct processing routes on the microstructures and properties of P/M DRA.

As mentioned earlier, conventional wisdom has dictated the use of an elevated temperature vacuum degassing step in the billet consolidation process to modify the oxide on the aluminum powder followed by an amount of subsequent deformation processing through extrusion, forging, rolling, etc., in order to develop microstructures which achieve the desired levels of damage tolerance. Most of this technology was primarily based on experience with unreinforced P/M aluminum alloys. The situation appears to be somewhat different when ceramic particle reinforcements are added to the aluminum alloy powder matrix. Tests on consolidated billets of X2080/SiC/15p DRA material, without any subsequent deformation processing, had demonstrated that the billet properties in the DRA were only slightly inferior to those in the fabricated extrusion product.[6]

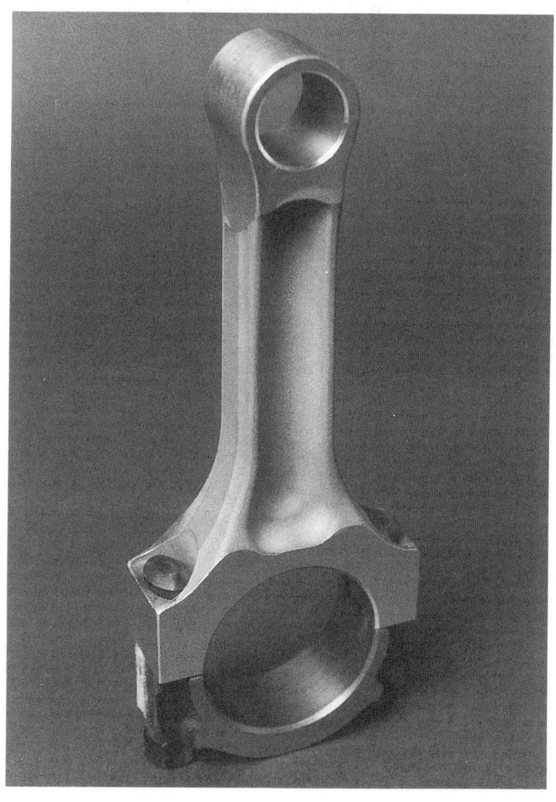

Fig. 2 – Prototype X2080/SiC DRA Connecting Rod

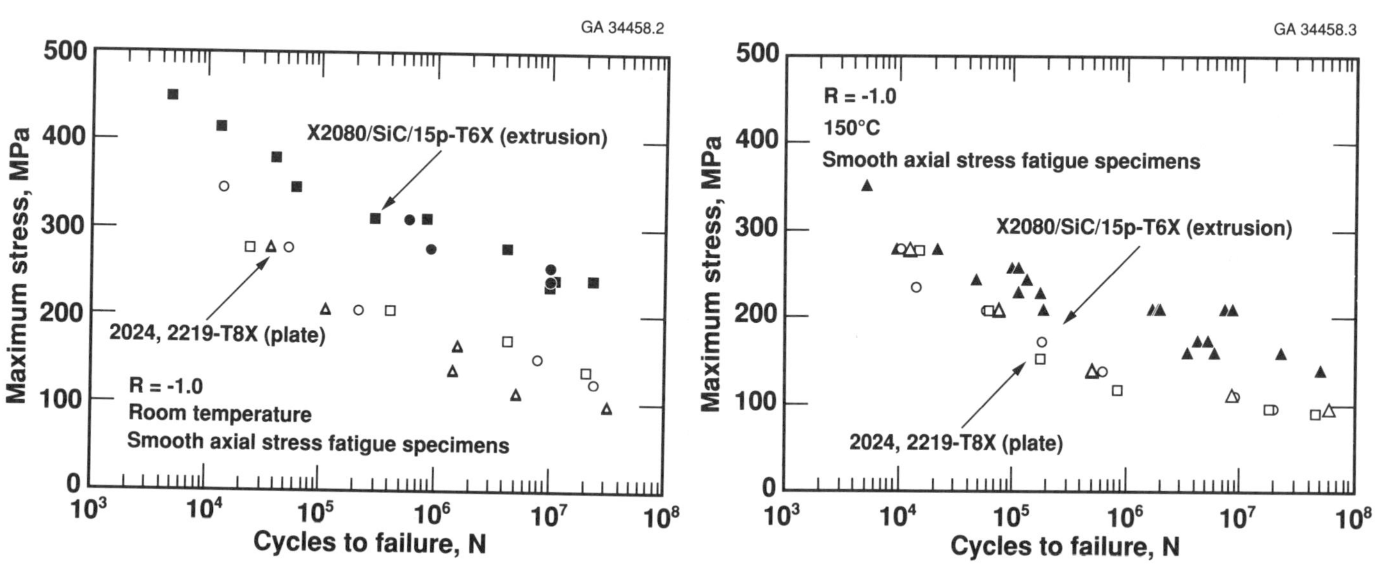

Fig. 3 – S-N Fatigue Performance of P/M X2080/SiC DRA at a) Room Temperature and b) 150°C

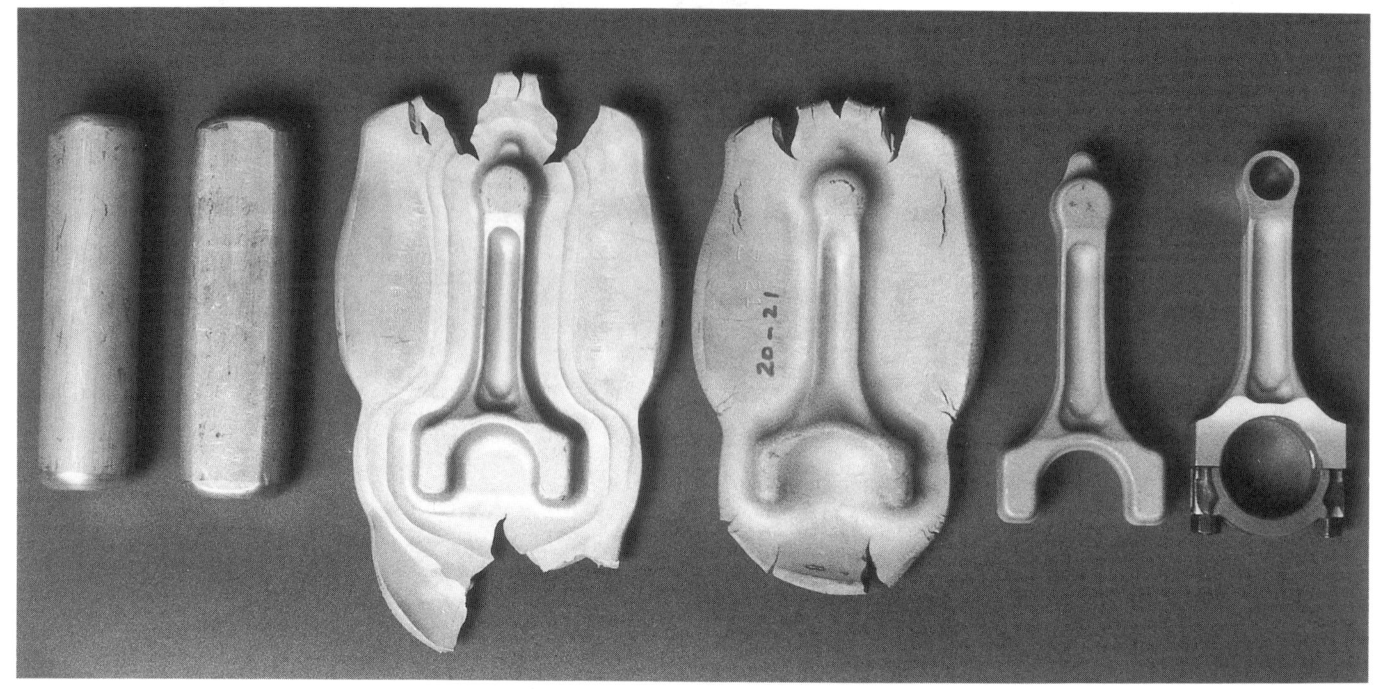

Fig. 4 – Fabrication Sequence of Conventionally Processed X2080/SiC DRA Connecting Rod Illustrating Number of Process Steps and Material Recovery

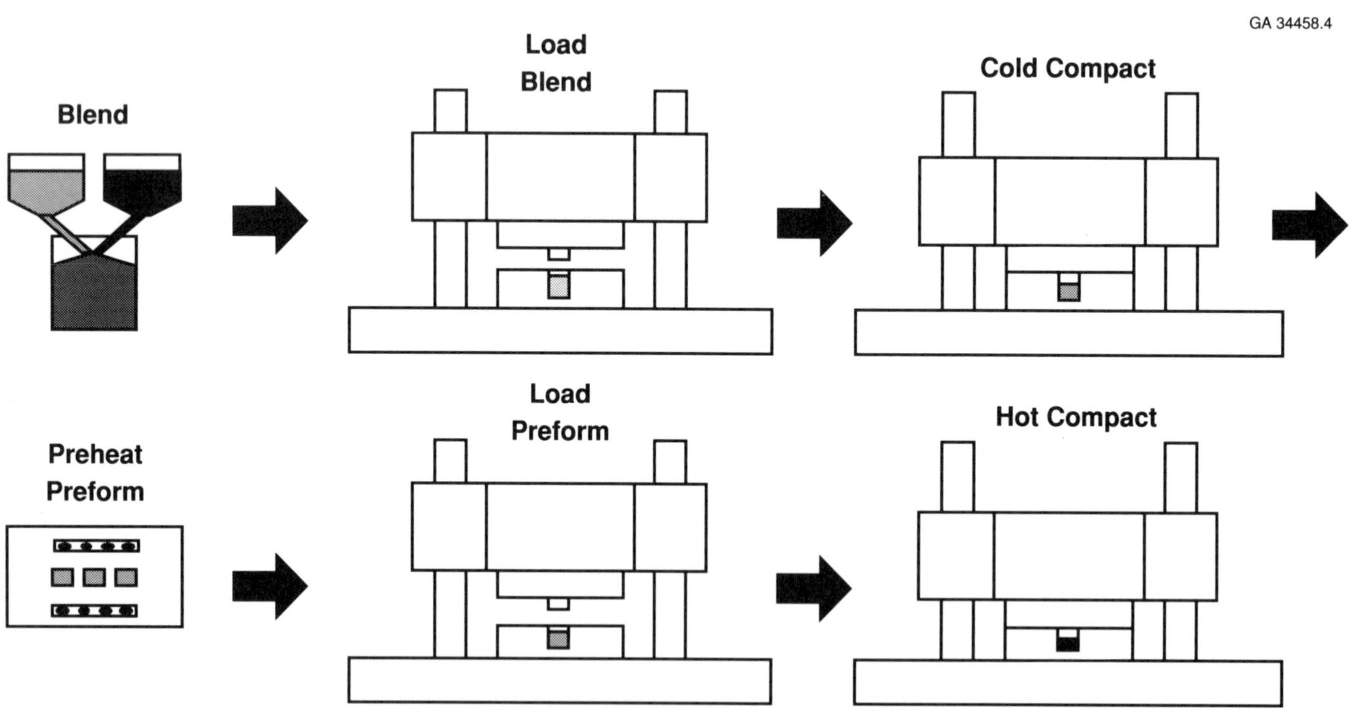

Fig. 5 – Direct Compaction P/M DRA Process

Subsequent experiments have examined the properties of X2080/SiC/20p billets which have been consolidated without the vacuum degassing step by a process called direct compaction shown schematically in Figure 5. Table II shows the strength and ductility properties of the direct compacted billet are similar to a conventionally degassed and hot pressed billet. For the automotive engine connecting rod application, the critical property is elevated temperature fatigue performance.

Table II. Tensile Properties of X2080/SiC/20p-T6 Billet

Process	Tensile Strength, MPa (ksi)	Yield Strength, MPa (ksi)	Elongation %
Direct*	514 (74.6)	432 (62.7)	1.8
Conventional+	522 (75.7)	415 (60.3)	2.0

* Average of four tests
+ Average of two tests

Figure 6 indicates that the fatigue performance of the direct compacted billet material is on par with that of a conventionally produced X2080/SiC/15p-T6 extrusion.

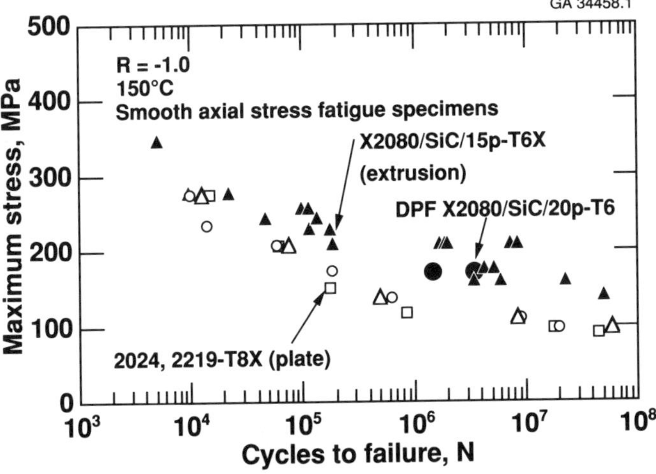

Fig. 6 – S-N Fatigue Performance of Direct Compacted P/M X2080/SiC DRA in Comparison with Conventional Process P/M DRA at 150°C

The direct compaction process can also be extended to a processing route in which billets are direct compacted then extruded in a single operation, shown schematically in Figure 7. This process route offers a number of advantages over conventional processing. In addition to the obvious reduction of processing steps and improved flow times, it eliminates the aluminum can required for the vacuum degassing step in the conventional process. Not only is the cost of the can and its removal saved, but, more importantly, the recovery of the billet improves by up to 30% due to the elimination of the need to remove the can (which due to the consolidation process has collapsed and is mingled with the outer portion of the DRA billet) by machining. From an economics viewpoint, a direct compaction approach, through the elimination of process steps as well as improved material recovery and throughput, is expected to result in production cost reduction on the order of 40% relative to billets produced by the conventional P/M route.

In particular applications, the need for low cost drives the process toward consolidation of net shape components. This further streamlined process maximizes material utilization while reducing processing steps. Recent reports of direct formed products from P/M aluminum and P/M DRA materials have indicated that there is potential for this route.[11,12]

One possible approach to direct forged parts is shown schematically in Figure 8. Research and development for direct forging in the case of P/M steel components has demonstrated that an important technological aspect for success is the intelligent design of a cold pressed preform shape such that subsequent forging achieves adequate consolidation and shape control. Thus, one element critical to success for direct forged DRA parts will be a fundamental understanding of the consolidation process in these materials.

Direct forging of parts offers further potential for considerable cost reduction through the elimination of process scrap from the forging process. For the connecting rod application discussed earlier, estimates suggest that a 60% reduction in unmachined can rod cost could be achievable through a direct forging route, presuming that fatigue property levels can be maintained. While the potential benefits of the direct forming approaches are apparent, substantial development work will be required to characterize and commercialize this process.

Concluding Remarks

Through the continued development of the P/M process for DRA materials, application opportunities are becoming more realistic. The improved fatigue properties of the P/M DRA materials in particular have opened up the potential application in the automotive engine such as the connecting rod discussed. In this area, cost effectiveness is predicated on the replacement of a low cost steel component, and thus, as with nearly all automotive applications, minimum cost at acceptable performance is paramount. To address this need, more direct processing routes have been investigated for P/M DRA materials, taking advantage of an apparent benefit of the ceramic particles in assisting the bonding of the metal matrix powder particles. This exciting development promises to open applications not only in the automotive area but also in other applications such as recreational and industrial components, previously thought to be the province only

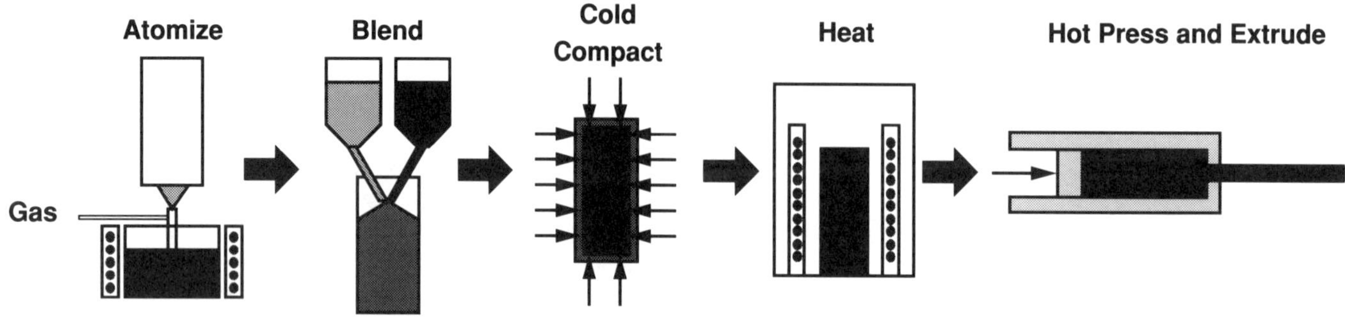

Fig. 7 – Direct Extrusion P/M DRA Process

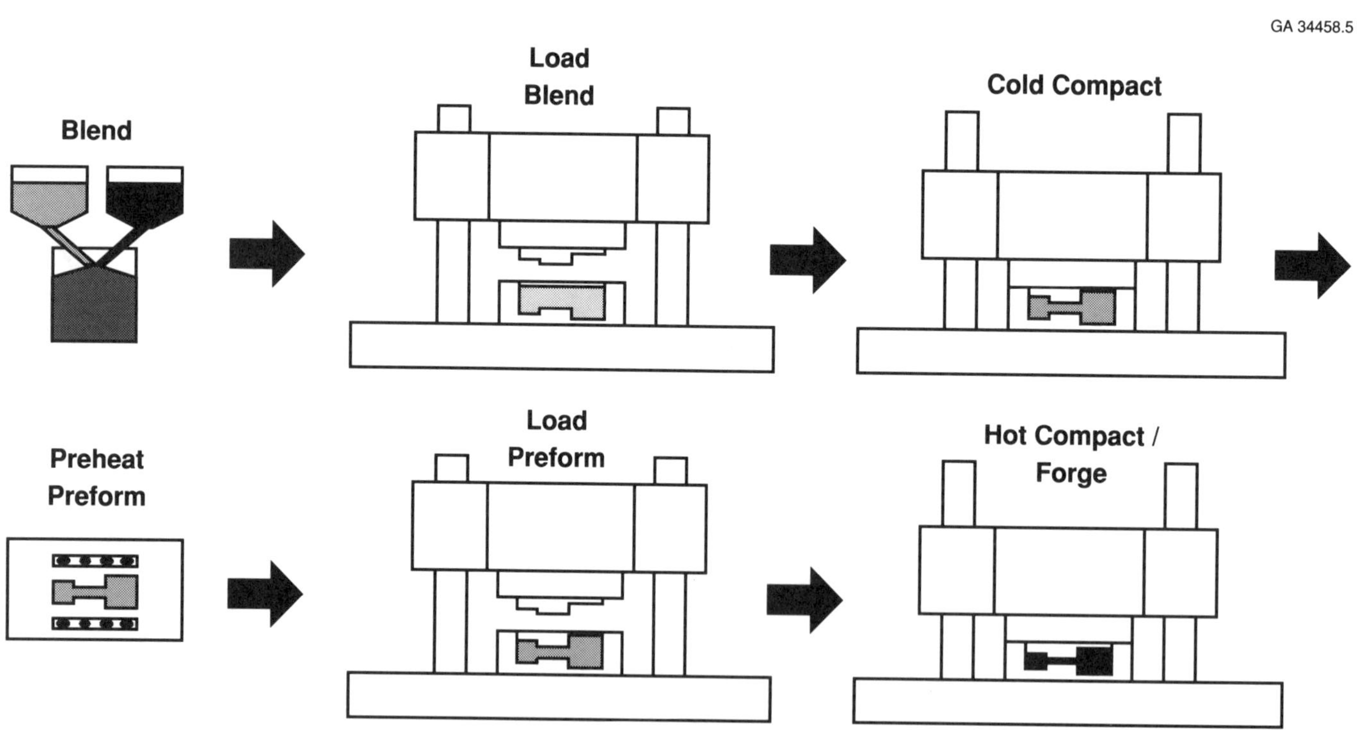

Fig. 8 – Direct Forged P/M DRA Process

of cast DRA products. In fact, if one considers the impact of direct P/M DRA processing routes on cost, the typical picture of cost and performance trade-offs is changed, as illustrated in Figure 9. This suggests that direct processed P/M DRA materials could be cost-competitive with cast DRA materials while meeting higher performance objectives.

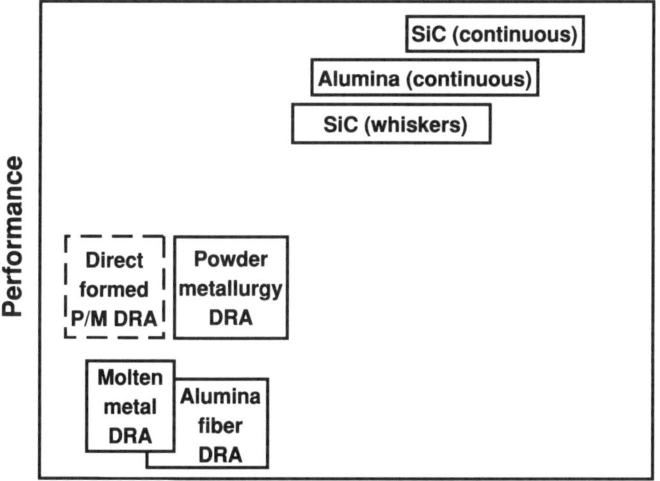

Fig. 9 – Cost-Performance Space for Metal Matrix
Composites, Illustrating the Promise for Direct
Formed P/M DRA Materials

Continued progress in all of these areas will require the perseverance and partnership of both materials producers and users. The recent departure of some key players in the DRA area due to lack of commercial market is of concern. In addition, there are still many hurdles to clear on the path to commercial implementation, not the least of which are the secondary processing methods required to convert DRA materials into components. These issues can be addressed and overcome through the coordinated efforts of both producers and potential users. The assistance of outside funding sources such as the U.S. Government as well as industry-government consortia are necessary to maintain the momentum in this important area.

References

1.   Kim, Y.-W., Griffith, W. M. and Froes, F. H., *Journal of Metals,* August 1985, p. 27.

2.   Estrada, J., Duszczyk, J. and Korevaar, B., *J. Materials Science,* 1991, vol. 26, p. 1431.

3.   Lavernia, E. J., Ayers, J. D. and Srivatsan, T. S., *Int. Mat. Rev.,* 1992, vol. 37, p. 1.

4.   Parent, J. O. G., Iyengar, J. and Henein, H., *Int. J. of Powder Metallurgy,* 1993, vol. 29, pp. 353-365.

5.   Lewandowski, J. J., Liu, C. and Hunt, W. H., Jr., *Processing and Properties of Powder Metallurgy Composites*, TMS, Denver, CO, 1988, p. 117.

6.   Osman, T. M., Lewandowski, J. J., and Hunt, W. H., Jr., *Fabrication of Particulates Reinforced Metal Composites*, ASM, Montreal, Quebec, Canada, 1990, p. 209.

7.   Estrada, J. and Duszczyk, J., *J. Materials Science,* 1991, vol. 26, p. 4203.

8.   Hunt, W. H., Jr. and Rodjom, T. J., *Particulate Materials and Processes*, Metal Powder Industries Federation, 1992, p. 21-32.

9.   Allison, J. E. and Cole, G. S., *JOM,* 1993, vol. 45, pp. 19-24.

10. Allison, J. E. and Jones, J. W. in *Fundamentals of Discontinuously Reinforced MMCs*, S. Suresh, A. Mortensen and A. Needleman, Ed., Butterworth-Heinemann, Guildford, U.K., 1993.

11. Morimoto, H., Iwamura, H. and Jyono, K., *ICCM-VIII*, Honolulu, HI, 1991, p. 17-L-1.

12. Otsuki, M., Kakehashi, S. and Kohno, T., *Metal Powder Reports*, 1991, p. 30.

Proceedings of the 10th Annual ASM/ESD Advanced Composites Conference, Dearborn, Michigan, USA, 7-10 November 1994

The Effect of Cold Drawing on the Elastic Modulus
of Aluminum Composite Driveshaft Tube

T.F. Klimowicz, W. Dixon, S. Inouye
Duralcan USA
San Diego, California

Abstract

The effect of cold draw reduction on the elastic modulus of extruded aluminum composite driveshaft tube was examined. The driveshaft tube, which consisted of 6061 aluminum alloy reinforced with nominally 20 v/o Al_2O_3 particulate ("*DURALCAN* W6D.20A"), was subjected to a number of different cold draw schedules with varying severity of individual pass reduction, as well as varying amounts of total reduction. The elastic moduli of the drawn tubes were then measured using both mechanical and dynamic methods. It was found that even moderate levels of reduction (i.e., less than 10%), produced a decrease in the modulus. In addition, there was little difference in the modulus loss between material which was given more numerous, light draw passes (with interpass anneals after each draw pass) and material which was given fewer, but heavier, draw passes- provided that the total reductions were equivalent. Metallographic examination showed that the modulus loss was caused by fracture of the Al_2O_3 particulate. The results indicate that, in order to maximize the modulus of the finished driveshaft, cold drawing must be minimized, and consequently, the driveshaft tube should be extruded to near-final dimensions.

A CRITICAL PARAMETER in the design of a vehicular driveshaft is its critical speed. This is the speed at which the rotating shaft becomes dynamically unstable due to transverse vibrations which can result in catastrophic failure. While the exact vibratory characteristics of a driveshaft are dependent on the end support conditions, for rigid, "pinned" connections, the critical speed is given by (1):

$$N_c = 30 \; \pi \; \sqrt{\frac{EIg}{WL^3}} \tag{1}$$

where
$$
\begin{array}{rcl}
N_c & = & \text{Revolutions per minute, rpm} \\
E & = & \text{Modulus of elasticity} \\
I & = & \text{Moment of inertia} \\
g & = & \text{Acceleration due to gravity} \\
W & = & \text{Total weight of shaft} \\
L & = & \text{Shaft length between supports}
\end{array}
$$

In the case of tubular driveshafts, this equation becomes:

$$N_c = \frac{15\,\pi}{L^2} \; \sqrt{\left(\frac{E}{\rho}\right)(R_0^2 + R_i^2)} \tag{2}$$

where R_0 and R_i are the outer and inner radii of the tube and ρ is the material density. As can be seen, the only material property which affects the critical speed is the specific stiffness, E/ρ. Obviously, for any given geometry, in order to increase the critical speed, it is necessary to increase the specific stiffness of the material from which the tube is made.

Although steel and aluminum have vastly different elastic moduli and densities, they, somewhat surprisingly, have the same specific stiffness- 2.56 X 10^{10} m^2/sec^2. *DURALCAN* W6D.20A ($6061/Al_2O_3/20p$) however, has a substantially higher specific stiffness- 3.31 x 10^{10} m^2/sec^2- which translates to a 14% increase in critical speed, as compared with either aluminum or steel. Because of this higher critical speed capability, *DURALCAN* W6D.20A is currently undergoing intensive evaluation for driveshafts in several passenger cars and light trucks. (2)

The established fabrication method for close tolerance aluminum driveshafts involves the extrusion of an oversize tubular "bloom", followed by cold drawing to final dimensions. Because of economic considerations relating to extrusion productivity, the bloom is sometimes extruded substantially larger than the final tube, requiring correspondingly large amounts of cold (i.e., room temperature) drawing. In the case of aluminum composites, this cold drawing operation is of particular concern because it has been well established (3,4,5) that cold deformation can lead to cracking of the ceramic particulate and degradation of the elastic modulus.

Because of this concern, the present investigation was undertaken in order to examine the effect of cold drawing on the elastic modulus of driveshaft tube. In addition to varying the total cold reduction given to the tubes, the severity of the individual draw passes was also varied. Because of the strong dependence of elastic modulus on the amount of ceramic reinforcement, tests were performed on material containing both 18 and 21 v/o Al_2O_3. (The producer specifications for the material allow a normal variation of +/- 2 v/o from the nominal value.)

Materials and Experimental Procedure

Tube Fabrication. Direct chill-cast billets of material containing both 18 and 21 v/o Al_2O_3 were extruded to tubular bloom of dimensions that resulted in a finished drawn tube of 102 mm nominal outside diameter X 2.1 mm nominal wall thickness after cold draw reductions of approximately 38%.

The extrusion and drawing processes and heat treatment of the tubes were carried out at Alcoa, Lafayette, Indiana. The seamless extruded blooms were produced by direct extrusion utilizing a press with an integral piercing mandrel (Figure 1). Cold drawing was performed at different levels of reduction, varying from nominally 7% to 21% per pass and to either 2 or 5 passes in total. The tubes were annealed prior to each draw operation. The required cumulative reduction of 38% was therefore achieved by 5 relatively gentle pass reductions or by 2 more severe reductions. After drawing, the tubes were subjected to solution heat treatment, straightened, and finally aged. For the purposes of this study samples were taken from a specifically selected tube after each pass, so eliminating any influence of variability across tubes in any one group.

Tensile tests were performed on longitudinally oriented coupons cut from the drawn tubes in order to assess the impact of the cold drawing on the tensile behavior of the material. The gage section of the samples used was 13 mm wide X 50 mm long. The inner and outer surfaces of the tube were left unmachined, and testing was conducted in accordance with ASTM E8.

Modulus Measurement. The elastic moduli of the extruded bloom and drawn tubes were measured both mechanically (using strain gages) and dynamically (using impulse excitation/resonant frequency). The measurements were performed on machined, flat rectangular strips (12 mm wide X 150 mm long X approximately 2 mm thick) oriented in the longitudinal direction. Dynamic modulus measurements were made using a commercial apparatus- the "Grindosonic" instrument (from J.W. Lemmens, Inc.). The machined strip specimens were "rung" by tapping the ends with a 6 mm steel ball attached to a flexible plastic stem. The acoustic signal was picked up by a microphone positioned beneath the strip and the determination of the fundamental resonance frequency was performed automatically by the Grindosonic instrument. The calculation of the modulus from the resonant frequency was made using the methods outlined in references (6) and (7).

Once the dynamic modulus measurements had been completed, holes for loading pins were drilled into the strips, and uniaxial strain gages were bonded onto each side. The strips were then cycled three times between 0 and 1500 microstrain (.15 %) in a servohydraulic test frame while monitoring load, and the Young's modulus was determined by performing a linear regression between 500 and 1500 microstrain. These strain levels correspond to stress levels of approximately 50 and 150 MPa, respectively. 500 microstrain was chosen as the lower limit in order to avoid the non-linearities sometimes seen at the very early stages of the tests. 1500 microstrain was chosen as the upper limit, in order to avoid straining the material beyond its proportional limit of 180 MPa. In all cases, the reported moduli are based on the average of the outputs from the two strain gages on each strip.

In addition to modulus measurements on machined strips, dynamic modulus measurements were also performed on curved strips on which the as-drawn tube surfaces were left intact. The purpose of these measurements was to assess the impact of the surface tearing, when present, on the modulus of the tubes by comparing the moduli of the flat, machined strips with those of the curved, unmachined strips.

Metallography and S.E.M. Optical metallography was conducted on longitudinal samples of the drawn tube in order to assess particle cracking and surface tearing. The tubes which were to be metallographically sectioned and examined for surface tearing were plated with electroless nickel prior to mounting and polishing in order to minimize edge rounding and loss of detail at the tear surfaces. In addition, the electroless nickel provided some contrast which was beneficial for imaging the very fine tears. The degree of particle cracking was quantified by measuring the area fraction of cracked particles and dividing by the area fraction of all particles, to give the fraction of cracked particles. SEM examination of the tube surfaces was conducted using a Cambridge S360 scanning electron microscope operated at 15 kv.

Volume fraction measurements were made via gravimetric determination. Drillings from the test coupons were chemically digested in a sodium hydroxide/hydrogen peroxide solution, and the Al_2O_3 powder was then filtered, dried, and weighed. Weight percents were converted to volume percents using the densities of the constituents.

Results and Discussion

Dependence of Modulus on V/O. A comparison of the mechanical and dynamic moduli for both the 18 and 21 v/o extruded bloom (prior to any draw reduction) is shown in Figure 2. As can be seen, the dynamic modulus is consistently greater than the mechanical modulus- on average, 1.7 GPa greater. This systematic difference is probably related to the fact that the strain levels which the material experiences during dynamic modulus testing are exceedingly small- certainly far smaller than those which the material experiences during the mechanical measurement. Because of the microscopic curvature of the stress-strain curve, even in the "linear" portion, the dynamically measured modulus would be expected to be somewhat greater than the mechanical value.

For the mechanically determined moduli, a least squares fit of the data yields the following dependence of the modulus on volume fraction:

$$E = 1.82 \ (v/o) + 61.07 \qquad (3)$$

where E is in GPa and v/o is in percent. For 20.0 v/o, this equation predicts a modulus of 97.5 GPa, which is in good agreement with the manufacturer's published value of 97.2 GPa (8). Of course, for large changes in v/o, the dependence of modulus on v/o may be non-linear, and this equation should not be used to predict the modulus of material with substantially different levels of Al_2O_3 reinforcement.

Modulus Loss Due to Deformation. Figures 3 (a) and (b) show the dynamic and mechanical moduli of the drawn tubes as a function of total cumulative reduction. As is readily apparent, both measurement techniques show a steady loss of modulus with increasing amounts of cumulative reduction. There does not appear to be any significant difference in the rate of modulus degradation between the 18 and 21 v/o materials. Overall, the average modulus loss for both materials- measured with both techniques- is .15 MPa for each percent of reduction. Furthermore, there does not appear to be a large difference in the rate of modulus loss between the 2-pass and the 5-pass material. This is particularly significant with regard to manufacturing because it indicates that, even with relatively modest draw reductions (8%), it is not possible to avoid modulus loss. In order to maximize the modulus of the drawn tube, total deformation must simply be minimized.

The loss of modulus, as has been shown in previous studies (3,4,5), is related to the cracking of the ceramic particulate, examples of which are shown in Figures 4 a and b. An interesting aspect of the cracking is that, because of the repetitive draw/anneal cycles and the hydrostatic stresses present during drawing, there is often significant intrusion of the matrix aluminum alloy between the broken halves of the particle (Fig. 4 b).

Figure 5 shows the amount of cracked particulate as a function of cumulative reduction for the 21 v/o material which was put through the two heavy draw passes. The extruded bloom starts out with virtually no particle cracking (0.4 area pct.). After the first 20% draw pass, approximately 5.4 percent of the particulate is cracked. And finally, after the second draw pass (for a total of 37% reduction), 13.1% of the particulate is cracked.

Obviously, in order to minimize cold drawing, a bloom must be extruded to near-final dimensions, and close to the tolerances required by the automotive industry in terms of roundness and, particularly, concentricity. This presents extruders with a challenge - direct seamless extrusion has limitations in tolerance control, typically corrected, in the case of close tolerance tubes, by drawing. In addition, near-net shape extrusions invariably result in higher extrusion costs, only partially offset by reduced draw operations. An opportunity may exist if indirect seamless extrusion is employed. Indirect extrusion has been proven to produce product to closer tolerances and at higher extrusion ratios, so producing nearer net-size blooms requiring less draw reduction and tolerance correction. Indirect extrusion and the associated benefits may need to be further examined if composite tube processing is to be optimized to produce maximum available modulus with good tolerance control.

Tensile Properties of the Drawn Tube. Despite the significant amount of particulate cracking and modulus loss seen in the heavily drawn tubes, there is only a relatively small change in the tensile properties, as shown below:

Table I

Tensile Properties of Extruded Bloom
and Drawn Tube

Al_2O_3(v/o)	18 v/o		21 v/o	
Draw Reduct. (%)	0	37	0	37
Yield Str. (MPa)	297	276	313	272
Ult. Str. (MPa)	348	338	348	324
Elongation (%)	8	6	5	6

Surface Tearing. Some of the tube, particularly the heavily drawn, 21 v/o material, had fine surface tearing (Figure 6d). In order to measure the depth of tearing, samples were plated with electroless nickel, and then polished in cross section (Figure 7). In the most severe case (the 21 v/o material with two heavy draw passes), the tears extended up to 0.13 mm into the wall thickness. Given that the wall thickness of this tube was 2.1 mm, a .13 mm tear represents a significant portion of the total wall thickness (6%).

Interestingly, the tears seem to have originated principally during the extrusion of the bloom, and not during the cold drawing. During cold drawing, however, the tears do tend to grow somewhat. Figure 6 compares the surface of the extruded bloom and drawn tube for the two materials, and clearly the bloom with most surface tearing produces the drawn tube with the most surface tearing.

There are two principal causes of tearing of composite during extrusion. One is attributable to hot-shortness of the matrix alloy and occurs under combined conditions of high extrusion temperature and strain rate, initiated due to heat build-up during deformation but aggravated by local friction at the die bearing. Excessive hot shortness may result in incipient melting of matrix solute phases. The second cause, peculiar to low ductility materials such as composites, is known commonly as "stick-slip" tearing. Cracking results when tensile surface forces due to friction at the die exceed the capacity of the composite. Lower extrusion temperatures and speed result in stick-slip tearing, whereas higher temperatures and speed cause hot shortness cracking. The challenge to any extruder of composite is to operate consistently under the safe conditions between these two defect windows.

In an attempt to quantify the impact of these surface tears on the modulus of the tube, curved strips were cut from both the extruded bloom and the drawn tube, and tested dynamically. (The inner and outer diameter tube surfaces were not machined.) In the limited number of measurements made, however, the modulus loss (between bloom and drawn tube) measured on the curved strips was not significantly different than the modulus loss measured on the flat, machined strips. Consequently, based on the experimental data, no modulus loss could be ascribed to the presence of these surface tears. Nonetheless, the presence of surface tears must certainly decrease the modulus somewhat, and our inability to measure it is probably due to inadequate experimental precision.

Conclusions

The degradation of elastic modulus in cold drawn aluminum composite driveshaft tube has been found to be caused by particle cracking which takes place during drawing. The modulus loss depends on the total cold reduction given to the material. More numerous, light draw reductions seem to be about as damaging as fewer, but heavier draw reductions, provided that the total cumulative reductions are equivalent. In addition, relatively small changes in particle content (from 18 to 21 v/o) seem to have no effect on the rate of modulus degradation.

To maximize the modulus of a driveshaft tube, then, the amount of cold drawing should be minimized, and consequently, the bloom should be extruded to near-final dimensions. Because of this requirement for better dimensional control in the extruded bloom, indirect extrusion offers potential advantages as compared with conventional, direct extrusion, and may need to be further explored as a manufacturing route.

Acknowledgements

The authors would like to thank John Dickson and Dan Collins (Alcoa) for their help in extruding and drawing the tubes, and Rob Dickinson (Duralcan U.S.A.) for performing the strain gage modulus testing.

References

1. Society of Automotive Engineers, Inc., "Universal Joint and Driveshaft Design Manual", p. 267, Society of Automotive Engineers, Inc., Warrendale, Pa (1991).

2. Duggan, J. A., Dickson, J. A., and Hoover, W., Automotive Engineering, 102 (2), 87-90 (1994).

3. Mochida, T., Taya, M., and Obata, M., JSME International Journal, 34 (10), 187-93 (1991).

4. Mochida, T., Taya, M., and Lloyd, D., Materials Transactions, JIM, 32 (2), 931-42 (1991).

5. Hunt, W.H., Brockenbrough, J.R., and Magnusen, P.E., Scripta Met., 25, 15-20 (1991).

6. Lemmens, J.W., in Dynamic Elastic Modulus Measurements in Materials, ASTM STP 1045, Philadelphia, PA, 1990, p. 90-99.

7. Spinner, S., and Tefft, W.E., ASTM Proceedings, 61, 1221-1238 (1961).

8. "*DURALCAN* Composites for Wrought Products-Property Data", Duralcan USA, San Diego, CA (1993).

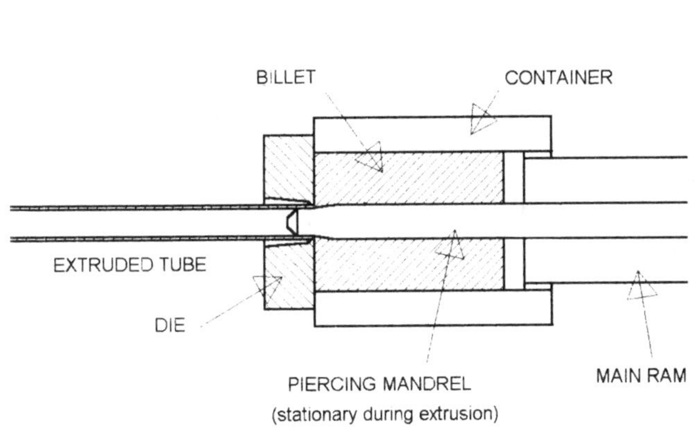

Figure 1. Schematic diagram of the direct extrusion process for production of tubular bloom.

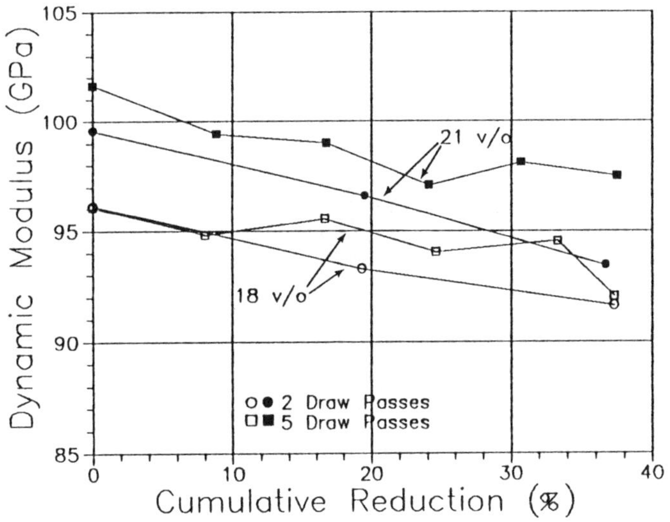

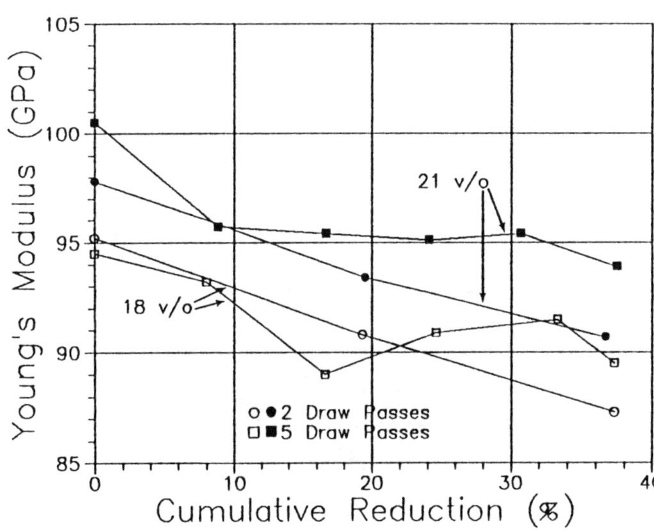

Figure 3a. Dynamic modulus as a function of cumulative reduction.

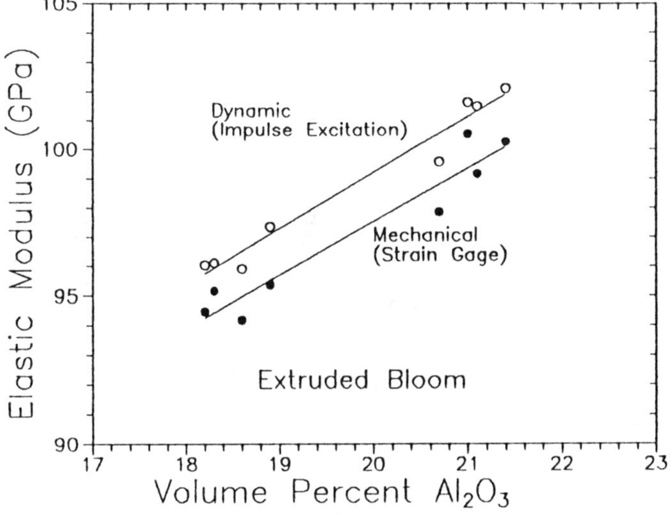

Figure 2. Comparison of the dynamic and mechanical moduli of extruded bloom with various levels of Al$_2$O$_3$ particulate.

Figure 3b. Mechanical (Young's) modulus as a function of cumulative reduction.

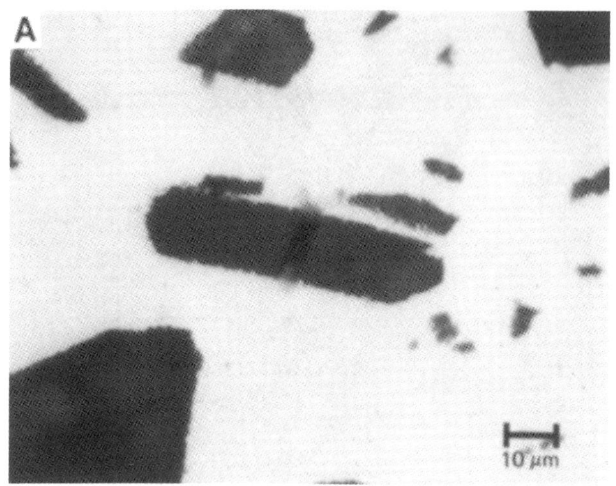

Figure 4a. A cracked Al$_2$O$_3$ particle.

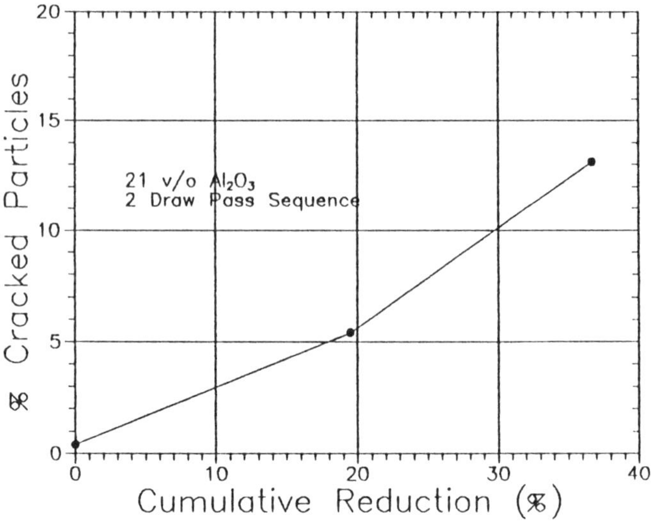

Figure 5. The amount of cracked particulate as a function of cumulative reduction, for the 21 v/o material in the 2 draw pass sequence.

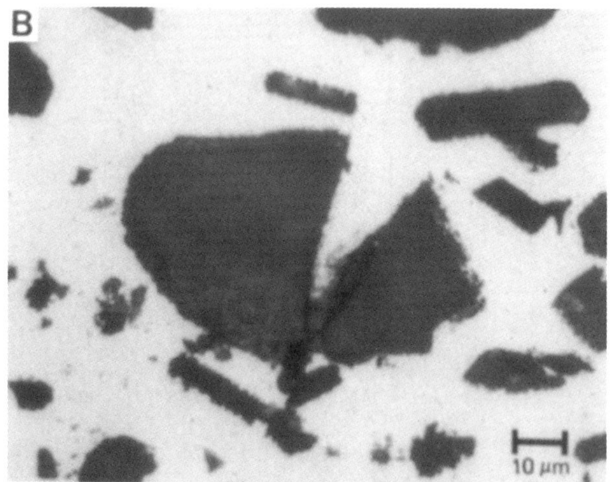

Figure 4b. A cracked Al$_2$O$_3$ particle, with matrix intrusion.

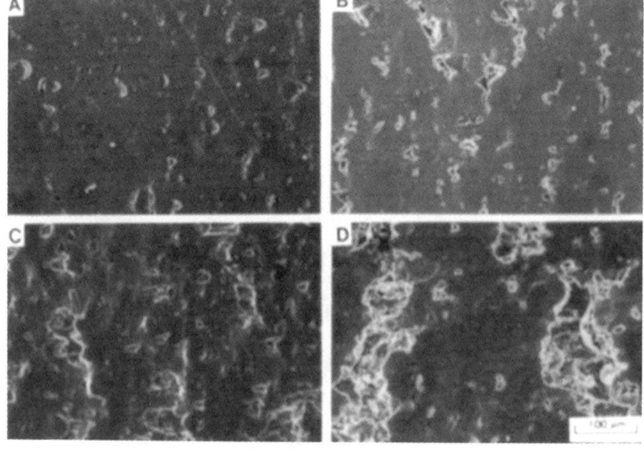

Figure 6. SEM micrographs of tube surfaces, (a) 18 v/o bloom, (b) 21 v/o bloom, (c) 18 v/o, 37% cold drawn tube, and (d) 21 v/o, 37% cold drawn tube.

170

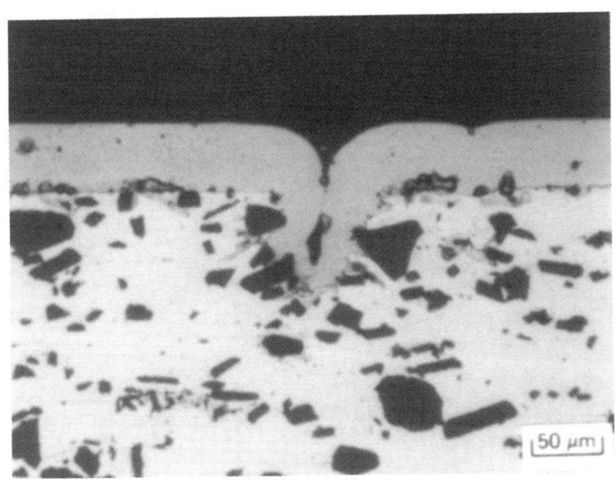

Figure 7. Metallographic cross section of a surface tear on 21 v/o, 37% cold drawn tube. A 50 μm layer of electroless nickel has been plated onto the surface to preserve detail.

Nondestructive Evaluation of Discontinuous Reinforced Aluminum Metal Matrix Composites Requirements for New Technologies

E.Y. Chen, L. Lawson, M. Mesu
Northwestern University
Evanston, Illinois

Abstract

Discontinuous reinforced aluminum (DRA) metal matrix composites represent a promising class of advanced composites with growing use in transportation. The employment of DRA, however, presents new challenges for nondestructive evaluation (NDE). Fatigue failure in DRA is governed by the initiation and growth of many microcracks which lead to widespread coalescence and ultimately fatal crack formation. The interval from which the fatal crack grows as a long crack and can be detected by NDE is relatively short. As a result, present techniques of NDE inspection and scheduling are not always reliable. Quality assurance inspections, nevertheless, can be improved if the microcrack length distributions at various stages in the fatigue are known. Such information pinpoints the timing for NDE inspection and prescribes the sensitivity of crack detection depending on needed reliability. This study reports experimental data for microcrack distributions in smooth specimens of 2xxx-type Al-SiC composites accompanied with simulation results used to estimate the initiation likelihood of cracks of particular lengths. A microcrack-based NDE inspection philosophy encompassing not only DRA, but other advanced materials and common metals as well, is presented.

THE CONSIDERABLE INTEREST IN DISCONTINUOUS REINFORCED ALUMINUM (DRA) metal matrix composites in recent years arises from their vast potential for transportation applications requiring high stiffness, low weight, and low density. When matched against monolithic metals, DRA also offer other tailorable physical properties including thermal expansivity, electrical and thermo-conductivities, in addition to better wear, corrosion, and high temperature resistance [1,2]. DRA generally exhibit superior stress level fatigue properties [3-6] with higher volumes of reinforcement typically leading to longer fatigue lives [7]. Compared with other types of advanced composites, particularly long, continuous fiber composites, the advantages of DRA include lower processing costs, higher machinability and workability, and can be manufactured with existing techniques [2].

The use of DRA composites for transportation purposes is expanding. In the aerospace industry, DRA have been utilized for aircraft structures such as the vertical tail stabilizer of the U. S. Air Force F-22 fighter [8,9] and the escape hatch for C-141 transports [10]. In the automotive industry, DRA have been mass produced by Toyota and Honda for components in diesel engines and engine cylinder blocks, respectively [1,2,11]. DRA are also being considered for rotor brake parts, drive shafts, connecting rods, pistons, and piston pins [2,12,13]. General Motors and its subsidiary, Hughes Aircraft, have funded a program for assessing DRA for dual use commercial and U. S. Army automotive needs [14]. In addition, DRA are becoming present in an ever-growing number of specialty products such as bicycle frames [15] and outboard engine pistons [1].

As DRA composites are placed in applications in which catastrophic failure may result in the loss of human lives, an understanding of how these materials behave in fatigue becomes vital. Studies by the authors on a 2124 Al alloy reinforced with SiC whiskers [16-18] have indicated that the fatigue resistance of DRA is largely influenced by the nucleation and growth of a very large number of microcracks* for most of the fatigue life (Figure 1). This high density of microcracks, in which most microcracks do not exceed a crack length of 100 m, leads to widespread coalesence and ultimately the fatal crack formation. The

---

* Crack sizes in this work will be distinguished by a terminology similar to [19]: microcracks refer to mechanically- and microstructurally-small cracks which are < 100 m in length and depth, and of sizes comparable to their own crack tip plastic zones; short cracks are physically-small cracks which are 100-500 m in either directions; long cracks have lengths and depths which are greater than 500 m and have growths which can be described by the Paris-Erdogan equation [20].

fatal crack is primarily the result of linkages between the largest microcracks present as well as many smaller microcracks in their paths. Long crack propagation (i.e. fatal crack growth) does not come into play until as late as 95% of the fatigue life (see also [4,5]). In some instances, the effective critical crack size as measured from the fracture surface is just 200 µm. The corresponding critical crack size as obtained from the linear elastic fracture mechanics (LEFM) fracture toughness value (12.7 MPa$\sqrt{m}$ for the 2124 Al-SiC$_W$ composite [21]) is expected to be 1.2 mm. This large discrepancy between experimental and predicted values of the critical crack size is most likely due to the fact that LEFM are based on long crack growth behavior. Knowing that the smallest crack which can be detected by present-day nondestructive evaluation (NDE) methods under optimum conditions is about 200 µm (e.g. eddy current), it would not be possible to detect the fatal crack until just prior to instability. As a result, this renders most current NDE inspection techniques useless for reliability inspections of DRA composite components.

The purpose of this work is to examine the precise requirements for NDE and unified life cycle engineering (ULCE) methods for advanced materials being applied in transportation based on the fatigue characteristics of DRA composites. More accurate, cost effective quality assurance inspections are always desired as evident in that available predictive and NDE detection methods are still not fully capable of treating crack growth processes that occur in service despite the vast amount of data collected and the identification of the basic mechanisms for crack growth [22]. In fact, the cost of fatigue in the United States has been estimated to be as high as $119 billion (in 1982 dollars), or about 4% of the gross national product [23]. With the introduction of DRA and other advanced materials into ULCE applications, the challenges faced by NDE have become even more rigorous. To set the pace for future developments in the field, this study presents an experimental-theoretical philosophy for NDE and ULCE which includes the initiation, growth, and distribution of fatigue microcracks. Microcracks are significant to NDE

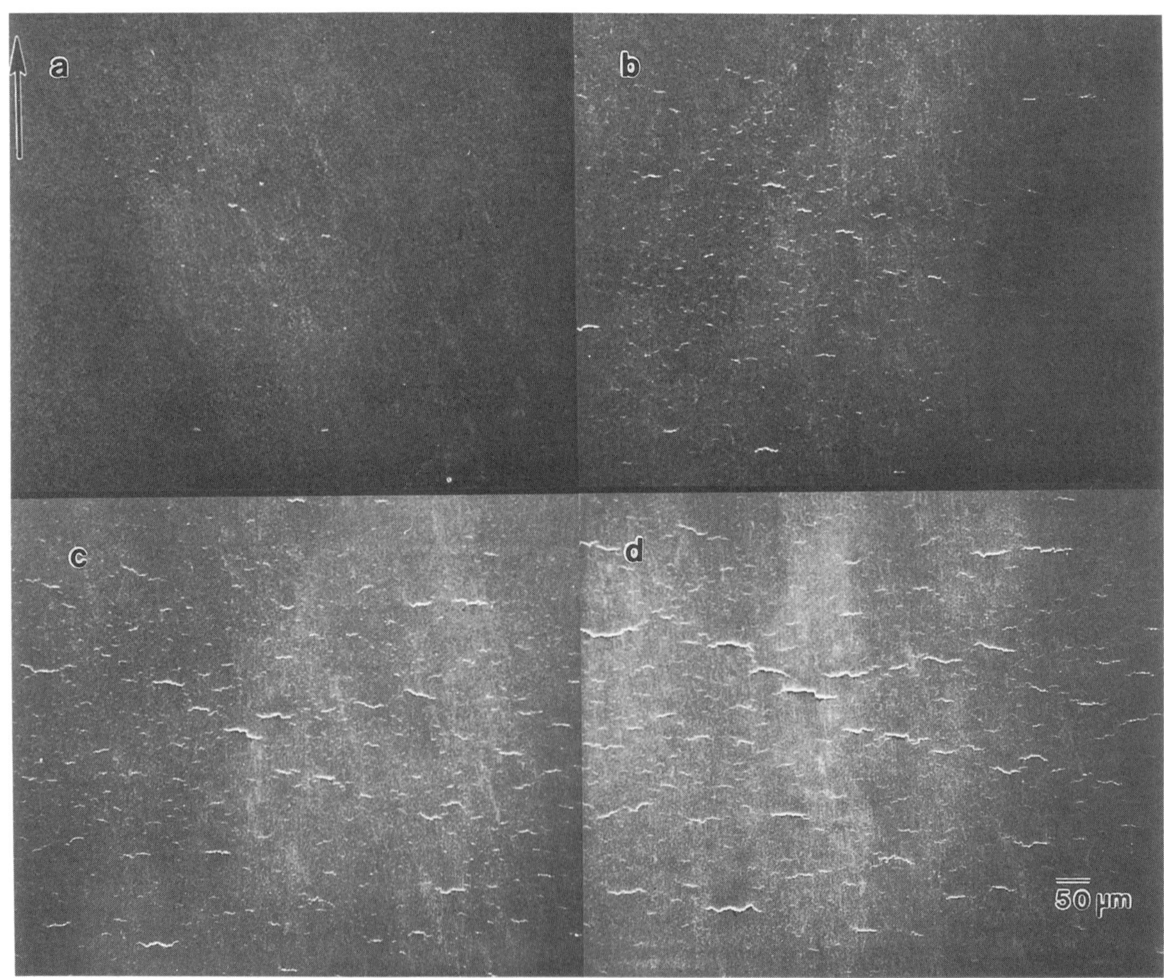

Fig. 1 - Initiation, growth, and distribution of surface fatigue microcracks in the 2124 Al-SiC$_W$ composite at a) N/N$_f$ = 0.33, b) N/N$_f$ = 0.58, c) N/N$_f$ = 0.80, and b) N/N$_f$ = 0.95. Arrow indicates loading direction. R = -1, σ_{max} = 500 MPa, N$_f$ = 13,718 cycles.

as all detectable macrocracks are the result of their formation and growth. To illustrate this, experimental data on fatigue microcracks as obtained from smooth specimens of a 2124 Al-20 vol.% SiC_W composite are presented and provide a foundation for a microcrack-based reliability inspection philosophy being developed as a part of an actual ULCE project [24]. The methodology devised enables the timing for NDE inspection to be determined and prescribes the sensitivity of crack detection depending on needed reliability. It is applicable not only for DRA, but for similar advanced materials and even commercial metals and corrosion/high temperature fatigue as well. A Monte Carlo simulation was constructed to model the experimental data. The simulation allows the probability of cracks of particular lengths initiating in a given part of a component to be estimated as a function of fatigue loading and number of cycles.

Significance of Microcracks in Reliability

In the search for more economical and reliable ULCE methods, the use of fracture mechanics, notably LEFM, have been expanded. Fracture mechanics are vital for making fatigue crack growth rate (da/dN) predictions which determine the remaining life and inspection scheduling. The long crack Paris-Erdogan equation [20] is most commonly used for such predictions. It adequately describes long crack growth in the linear portion of the sigmoidal log-log scaled da/dN versus stress intensity factor (ΔK) relation for monolithic metals as well as for some composites (Figure 2). In this linear region, long crack da/dN are empirically fitted to a power law behavior. The Paris equation does, however, overestimate da/dN for lower values of ΔK approaching the long crack growth threshold (ΔK_{th}) where long cracks cannot grow since the local crack tip stress intensities are too small [25]. Nevertheless, it is still often extended into and beyond the "thresholded" regime for short crack (and microcrack) da/dN predictions (dashed line in Figure 2).

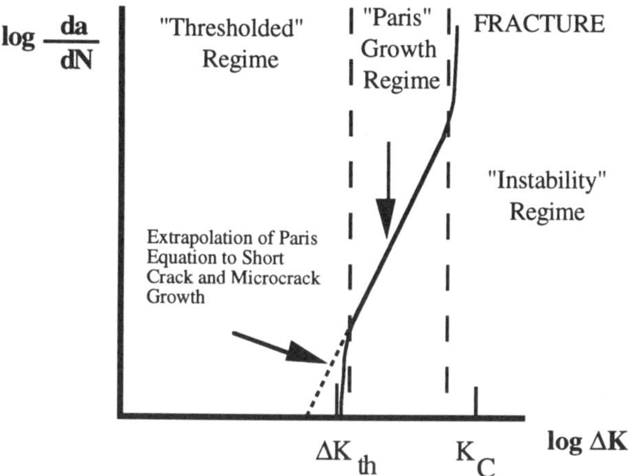

Fig. 2 - Schematic of long crack growth behavior in metals and some composites.

The validity of extending the Paris equation to microcracks to estimate their growth rates is controversial. It is well known that short cracks can behave much differently from long cracks and deviate in da/dN from those predicted by the Paris equation. With regard to reliability, this would be particularly dangerous as catastrophic failure could occur while within the expected service life of a component. Short cracks can grow below ΔK_{th} and at rates faster than long cracks at the same ΔK [19,26-30]. Attempts to correlate differences in short and long crack growth rates with the Paris equation using modified versions of ΔK or ΔJ (the elastic-plastic method) have yielded mixed results [28,31-33]. For example, when ΔK is modified as an effective stress intensity factor (ΔK_{eff}) to account for crack closure effects, some studies found the short and long cracks da/dN to be the same [31,32] while others still observed differences [33]. This lack of agreement may be due to the use of ΔK itself since similitude between short and long cracks does not exist (thus violates the basic crack tip requirements of LEFM) [34] or the semiempirical nature of its modified versions (e.g. ΔK_{eff}). It may also come from the fact that while deterministic expressions of the crack driving force such as ΔK can treat mechanically-induced microcrack growth effects, they are inadequate in their present forms to treat the microstructurally-induced factors which greatly influence microcrack growth [35].

The Paris equation considers cracks to be growing consistently, individually, and isolated. Microcrack growth is highly intermittent in reality with large da/dN scatters due to the extensive influence of microstructure (e.g. arrest) and the other microcracks present (e.g. coalescence). As a result, there are statistical variations in microcrack growth not accounted for with the use of the Paris equation. The growth behavior of one microcrack is not necessarily carried over to different microcracks. A probabilistic approach in characterizing microcrack growth, therefore, may be more appropriate than a deterministic one [19]. In materials similar to DRA composites where a dominant flaw does not exist for most of the fatigue life and failure takes place from a distribution of microcracks, a probabilistic approach may be the only alternative [22]. Some microcracks ultimately become long cracks and therefore have growth rates which can be described by the Paris equation. The exact instance in which this occurs, however, is still unknown.

Fatigue of DRA Metal Matrix Composites

Studies by the authors [16-18, 36, 37] have shown significant effects of microcracks on the fatigue behavior of DRA composites. These works have primarily tracked microcrack growth throughout the fatigue life from scanning electron microscopy (SEM) micrographs of the specimen surfaces. Figure 1 shows an example of a set of replica micrographs at different stages of fatigue for a powder metallurgy 2124-T6 aluminum alloy reinforced with 20 volume percent silicon carbide whiskers. Four major processes are seen to occur in this composite as a result of fatigue: microcrack initiation, growth, arrest, and coalescence. A large number of newly-initiated microcracks continually emerge over time (Figure 3). These microcracks initiate at fractured SiC whiskers, debonded matrix-reinforcement interfaces, whisker clusters, and inclusions. After initiation, microcracks can grow on their own and/or by linking with other microcracks nearby. The high density of microcracks in DRA makes coalescence a much larger factor in growth than in the case of commercial metals. Most microcracks remain rather small and many arrest soon after initiating. Others grow for some period and then arrest. Near the end of life, some microcracks grow so large that linkages with other microcracks in their paths occur to form the fatal crack. The effective critical crack lengths as measured from the fracture surface range from just 68 to 250 μm while those of the critical crack depths vary from 20 to 140 μm, both for an applied stress amplitude range of 600 to 400 MPa [18]. The largest crack observed near the end of life ($N/N_f = 0.95$) in [17] was just 60 μm long. Knowing that the critical crack length in this case was about 200 μm, this corresponds to the linkage of 2-4 of the largest microcracks. The extremely short period from the formation of the fatal crack until failure indicates that instability was sudden and rapid with little time spent in stable fatal crack growth. With regard to current NDE detection capability, there would be little warning time from initial detection (if any detection at all) of the fatal crack prior to instability, thus the possibility of catastrophic failure.

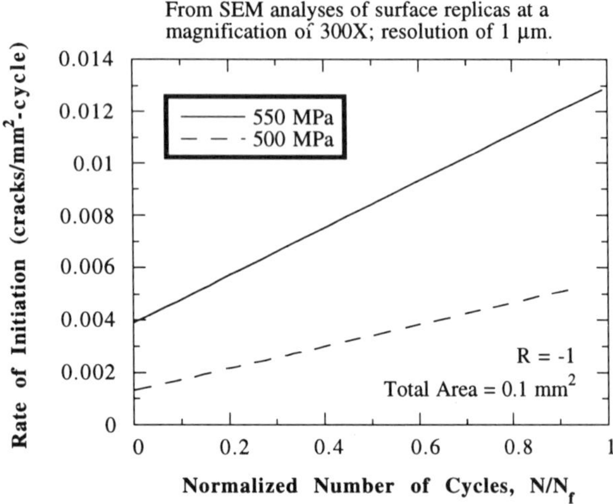

Fig. 3 - Rate of initiation of microcracks in the 2124 Al-SiC$_W$ composite.

By measuring the lengths of all microcracks in a designated area on the replicas, the microcrack length distributions (MLD) at different instances in the fatigue life can be obtained. The MLD at one stress level were previously reported for up to 420 microcracks in an area of 0.1 mm² (i.e. density of up to 4,000 cracks/mm²) [17]. In attempt to fit the MLD with various commonly-used

distribution functions for prediction purposes, the MLD were found best fitted with a lognormal distribution. The lognormal fit, however, did not always satisfy statistical χ^2 goodness of fit tests at a confidence level of 95% with the fits being rather poor late in the fatigue life. An explanation for this was linked to the exponential nature of the lognormal distribution which allows it to closely fit the distribution of smallest crack lengths, but not the longest cracks at the tail [36]. Since the tail represents the cracks of lengths bearing most significance to failure, the poor overall fits of common distribution functions suggest that using them to represent a population of cracks, although convenient, can be inaccurate.

The growth behavior of individual microcracks was followed by backtracking from the last to the first replica in sequences of replicas. From the micrographs, the change in crack length (b) over time (in number of fatigue cycles, N) can be determined. The growth rates along the surface crack length, db/dN, are then found by curve-fitting the b vs. N data and differentiating the best-fitting curve. The Paris equation is applied by matching db/dN with ΔK for semi-elliptical cracks [38]. Examples of the results are shown in Figure 4. There are large fluctuations in the db/dN of individual microcracks not seen for long crack growth. Coalescence and arrest occur frequently in the growth of some microcracks, leading to periods of accelerated or little/no growth, respectively. The entire growth spectrum cannot be fitted empirically to a single Paris equation if each data point was followed precisely. Some microcracks did exhibit a linear growth curve on the log db/dN vs. log ΔK plots, but they were very few in number. The growth rates of most individual microcracks cannot be accurately predicted with the deterministic Paris equation as with long cracks.

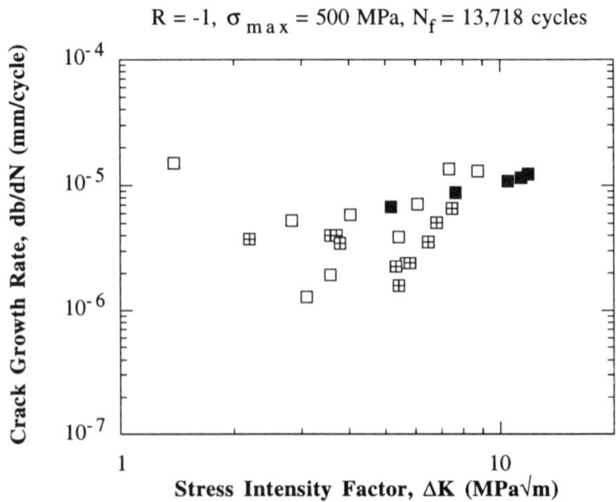

Fig. 4 - Individual microcrack growth behavior in the 2124 Al-SiC$_W$ composite.

Figure 5 shows the microcrack growth rates obtained in the authors' studies as compared with those found in the literature for long cracks in similar composites and an unreinforced 2124 matrix using compact tension (CT) specimens [39,40]. Microcrack growth rates were higher than those of long cracks at the same ΔK at low values of ΔK approaching ΔK_{th} for long cracks in composites. The opposite is seen at larger values of ΔK. Large scatters in the growth rates extending to both ends of ΔK were seen at all stress levels with some small microcracks growing extremely fast. No growth thresholds were observed for microcracks. The smaller slopes seen for the clustered microcrack db/dN vs. ΔK data at each different stress level reflect the ability of DRA to resist microcrack growth better than long cracks. This is most likely due to the crack trapping capability of the reinforcement for microcracks [18]. When cracks are long, the crack tip stress intensity becomes so large that cracks can propagate directly through whiskers by fracturing them. It is noted that long crack data from CT specimens are commonly employed for crack growth rate measurements even though cracks of such lengths (on the order of several millimeters) may never appear, as seen for the DRA composites. Basing reliability predictions on the growth rates of long cracks which may never emerge, therefore, can lead to poor predictions and ineffective inspection scheduling.

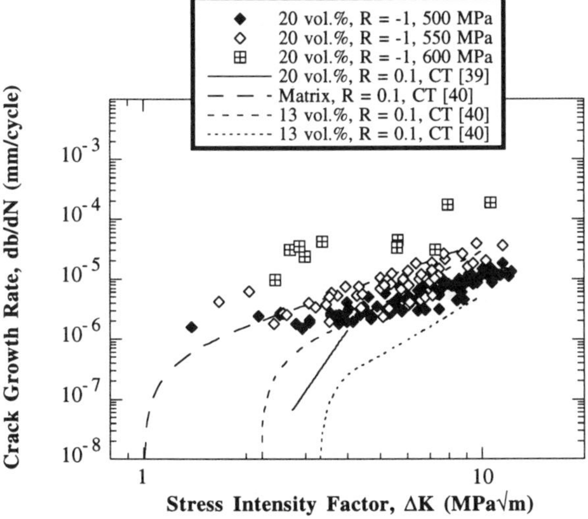

Fig. 5 - Microcrack growth behavior in the 2124 Al-SiC$_W$ composite as compared with long crack growth data in similar composites and matrices obtained from compact tension (CT) specimens.

Microcrack Distributions and Reliability

With the importance of microcracks in the fatigue of DRA composites (and the difficulties of detecting these microcracks by NDE) emphasized, a reliability philosophy centered on their growth can be devised to allow these materials to be used in ULCE. The basis for this philosophy is that the probability of an NDE-detectable crack appearing depends on the population of undetectable microcracks. By knowing how and at what rate these microcracks grow and interact (e.g. coalescence) in fatigue,

the rate of emergence of detectable cracks can be estimated to properly schedule the inspections. It is noted that microcracking occurs in the fatigue, whether at room or high temperature in air or corrosive environments, of almost all structural materials prior to the formation of a fatal crack. Consequently, the reliability model constructed in this work can be extended beyond the DRA composites to other materials, such as a 304 stainless steel [16]. Information on microcrack growth behavior was primarily gathered from the studied DRA composite since the high density of microcracks present allows statistical adequate numbers of microcracks to be more easily counted and measured.

To devise the foundation of the reliability model, some clarifications in current reliability literature are necessary with regard to the initiation and growth of cracks and the relationship of their initiation to the hazard function. A recent paper on aircraft reliability describes the stochastic approach to durability prediction for aging aircraft being developed by the U. S. Air Force [41]. This approach is based on identifying an Equivalent Initial Flaw Size Distribution (EIFSD) for a structural element by cycling the element and measuring the time it takes for cracks to reach a reference size. These data are then used to "grow the distribution of cracks backward" to obtain the EIFSD by means of a Paris-Erdogan-type rule containing random variables (Figure 6). Growing the EIFSD forward by the same rule then is used to calculate the crack size distribution at any point in service life. This approach conflicts with the experimental results obtained for MLD in DRA mentioned in the previous section, as well as studies by other authors with a similar material [4], and a 304 stainless steel [16]. These works indicate that measurable cracks derive from smaller microcracks, not measurable by conventional NDE methods, viz. less than 200 μm, which themselves are created continuously at an increasing rate throughout the life. As discussed in the previous section, microcracks in these MLD obey a Paris equation only some of the time when they are growing. Otherwise microstructural arrest and coalescence dominate in the microcrack regime leading to intermittent crack growth. These facts suggest that the only way an EIFSD would be realistic is if it was obtained at the end of life. Since predictions are based on a fixed initial distribution and Paris-type growth, at times before the end of life this durability model would predict the presence of cracks of particular lengths that in reality would not exist due to the effects of arrest and coalescence in retarding or accelerating crack growth, respectively.

The reason for concern over cracks which do not exist is that they cannot be culled out by inspection. A repair action, using the terminology of Berens et al. [41], may include replacement or merely assurance by inspection that no cracks below a certain size exist apart from those which are overlooked. Consequently, unless the repair action results in the actual replacement of a member, the evolution of the microcrack distribution goes unchecked. Since, both the rate of new cracks appearing in the microcrack distribution and the rate of detectable cracks emerging out of the microcrack distribution are increasing functions of time, it is reasonable to assume that the hazard function for an old but crack-free structural member is not going to be the same as that for a new one. Yet, in common with many textbooks and papers, Berens et al. [41] show graphs indicating that after inspection the hazard function behaves exactly the same as it does for a new part. The consequence of this is that the inspection interval tends to be equispaced for a constant risk. What follows here is a discussion indicating that the characteristics of actual MLD lead to conditions requiring inspections to be closer together as the structure ages in order to keep the risk constant. The terminology and notation used will be similar to those found in [42] (see also Appendix A) and the basic assumptions of the theory is briefly described in Appendix B.

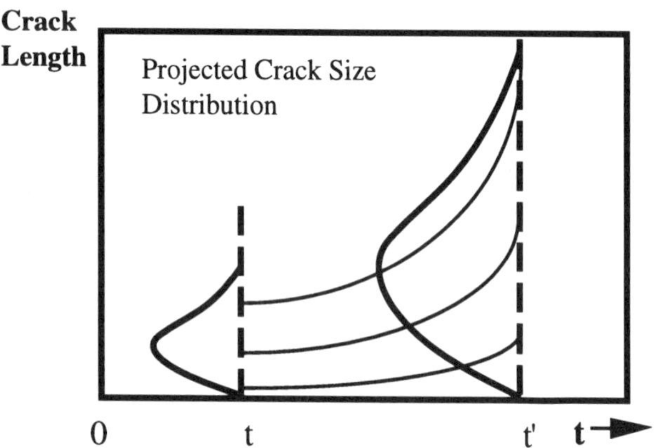

Fig. 6 - Schematic of projected crack size distribution as obtained from growing EIFSD by the Paris equation with random variables.

To begin, some discussions about MLD based on the experimental data presented earlier will be introduced. Microcracks do not experience the same thresholds as larger cracks do. Long cracks extend over many microstructural domains and average the local thresholds. Local thresholds vary widely and naturally-occurring microcracks only grow where conditions favor them with very low thresholds. The pattern of periods of arrest has stronger effects on microcrack growth than does their ΔK. As cracks reach an NDE-detectable size, they begin to obey a Paris growth relationship. This relationship, however, is subject to some variability and needs a stochastic description when considered for reliability. Figure 7 schematically illustrates the variability of typical growth paths for a crack of a threshold size (e.g. 200 μm) under constant cyclic or equivalent spectrum loading. When the crack reaches a critical length, it grows at an uncontrolled rate (i.e. instability) and fracture ensues. A failure function, similar to those in current reliability theory (e.g. [42]), for a single, isolated crack in a member may be defined as:

$$F(t) \equiv P\{t \le t\} = H(t)$$
$$\equiv \{l \ge l_c | l = l_o, t = t_o\} \tag{1}$$

The actual critical length, l_c, in this function is variable in spectrum loading, thus does not need to be specified explicitly. H is then the probability of such a crack causing failure under some continuous specified loading process.

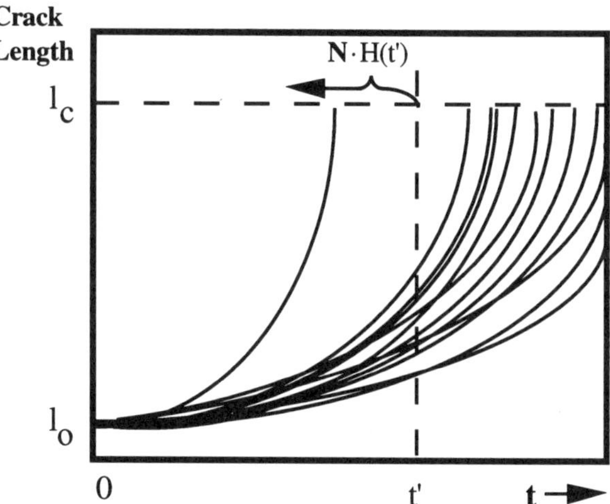

Fig. 7 - Growth curves and H, the probability of a given crack of length l_c causing failure at time t' under some continuous specified loading process. N is the total number of cracks studied.

We may next consider the rate of initiation of just-detectable cracks having a length of l_o. The rate of initiation of cracks into microcrack distributions has been found to increase with time [16]. The rate of emergence of cracks of size l_o also increases with time and has a positive second derivative. The expected rate of appearance or "initiation" of cracks of size l_o is:

$$I(t')$$
$$\equiv \lim_{\Delta t \to 0} \frac{1}{\Delta t} n\left(\left\{\begin{matrix} l_o - \varepsilon < l < l_o \\ t = t' \end{matrix}\right\} \cap \left\{\begin{matrix} l_o \le l < l + \varepsilon \\ t = t' + \Delta t \end{matrix}\right\}\right) \tag{2}$$

where n is the number of cracks in the specified set. This simply expresses "the number of cracks growing through size l_o per unit time". If a small time interval, Δt, is considered, we will have:

$$m = I\left(\frac{\Delta t}{2}\right) \cdot \Delta t \tag{3}$$

cracks introduced. At some later time, t, these cracks will have grown and could cause failure. We may write a general expression for the probability of failure at time, t, as a consequence of these cracks created near t = 0, by noting that, were the cracks independent, failure would follow a binomial distribution based on H, the probability of failure by each crack independently. Because cracks are not independent, this binomial distribution is modified by an interaction term which lowers the probability of failure for each crack so as to insure that multiple-crack failure is as rare in the model as it is in reality.

$$F(t)\Delta t$$

$$= \sum_{k=1}^{m} \binom{m}{k} (H(t) - J(m,H))^k \qquad (4)$$

$$(1 - H(t) + J(m,H))^{m-k}$$

where J is the interaction term. From the remarks above J is approximately H at the end of life. Both H and J increase with cycling and are always very small compared to unity in any regime which could be even roughly described as operationally safe. An upper bound on F may be obtained by simplification of the above expression using these conditions and the fact that multiple-crack failures are rare (not to be confused with "multi-site damage" in aircraft):

$$F(t)\Delta t \underset{\sim}{<} I\!\left(\frac{\Delta t}{2}\right) H(t) \cdot \Delta t \qquad (5)$$

The above expression may now be used to generate a Rieman sum to obtain F for the real case under the conditions of time-dependent crack initiation. Noting that both H and J are zero for negative arguments, this expression is a convolution of I and H.

$$F(t) \underset{\sim}{<} \int_0^t I(t')H(t-t')dt' = I \otimes H \qquad (6)$$

Differentiation of F gives the hazard function,

$$\lambda(t) \underset{\sim}{<} \frac{I \otimes H'}{1 - I \otimes H} \qquad (7)$$

(the "prime" indicates differentiation with respect to time) and the reliability follows,

$$R(t) \underset{\sim}{\geq} 1 - I \otimes H. \qquad (8)$$

We wish to inquire after the effect of inspection on the hazard function without knowing the details of I and H. We note the function H by definition and because of the uncertainty of the Paris equation when applied to microcracks, only operates on cracks larger than l_o. After inspection, there should be none of these cracks. Some large cracks, of course, may be overlooked, but including these only intensifies the argument that equispaced inspection is a poor idea and intensifies the math as well. Leaving these cracks out will not change the nature of the result. Since they are not there, the hazard function goes essentially to zero after inspection. The time reference for H in the hazard function is reset to zero since all cracks in its domain are yet to be, but the time reference in I is not reset. From the negative inspection result, probability may seem to have hindered the appearance of larger cracks. That, however, says nothing about the evolution of the

microcrack distribution. One such large crack is the most that is likely to ever be found on inspection. Its appearance is therefore chancy. Microcracks exist often in numbers of thousands per square millimeter. The laws of large numbers can be expected to hold in the microcrack regime.

To further examine the hazard function, an expression for its time derivative is needed:

$$\lambda'(t)$$

$$\underset{\sim}{<} \frac{[1 - 1 \otimes H][I \otimes H'' + I(t)H'(0)] + [I \otimes H']^2}{[1 - I \otimes H]^2} \qquad (9)$$

We would like to know what this derivative is for a new part and for an old inspected part right after inspection in order to compare the two. The derivative for a new part is, from above, simply:

$$\lambda'(0) \underset{\sim}{<} I(0)H'(0). \qquad (10)$$

When inspection takes place at time, t*,

$$\lambda'_{t*}(t - t* = 0) = I(t*)H'(0). \qquad (11)$$

Remember that I increases with time, often substantially due to its positive second derivative [16]. The slope of the hazard function after inspection therefore increases with the time of inspection in proportion to the increase in the function, I. For constant risk, the spacing of the inspections, the inspection period, ought to be inversely proportional to the slope of the hazard function.

The rate of appearance of NDE-detectable cracks from evolving microcrack distributions is not consistent with the notion of an EIFSD -- at least unless that distribution first evolves by microcrack processes before anything like a Paris equation is applied to it. Microcracks evolve at first slowly due to frequent arrests and then rapidly due to coalescence. In the microcrack regime, the effect of length and hence ΔK on growth velocity is too weak to allow a Paris growth description. Furthermore, new smallest-discernable microcracks appear over time with a nearly fractal character as the magnification is increased. The net result is a tendency for NDE-detectable cracks to appear ever-increasing rates with additional fatigue. The apparent initial growth speed of these late-appearing cracks may be quite high since they often result from coalescence. Therefore, to keep up with these late-appearing cracks, inspection intervals need to be shortened as the structure ages.

Simulation of Microcrack Distributions

To show how MLD as discussed in the previous section may be used for reliability predictions, a Monte Carlo simulation was employed [16,36] and the flow chart is shown in Figure 8. A Monte Carlo simulation was chosen for this purpose since it represents a step toward describing the MLD and microcrack growth as Markov

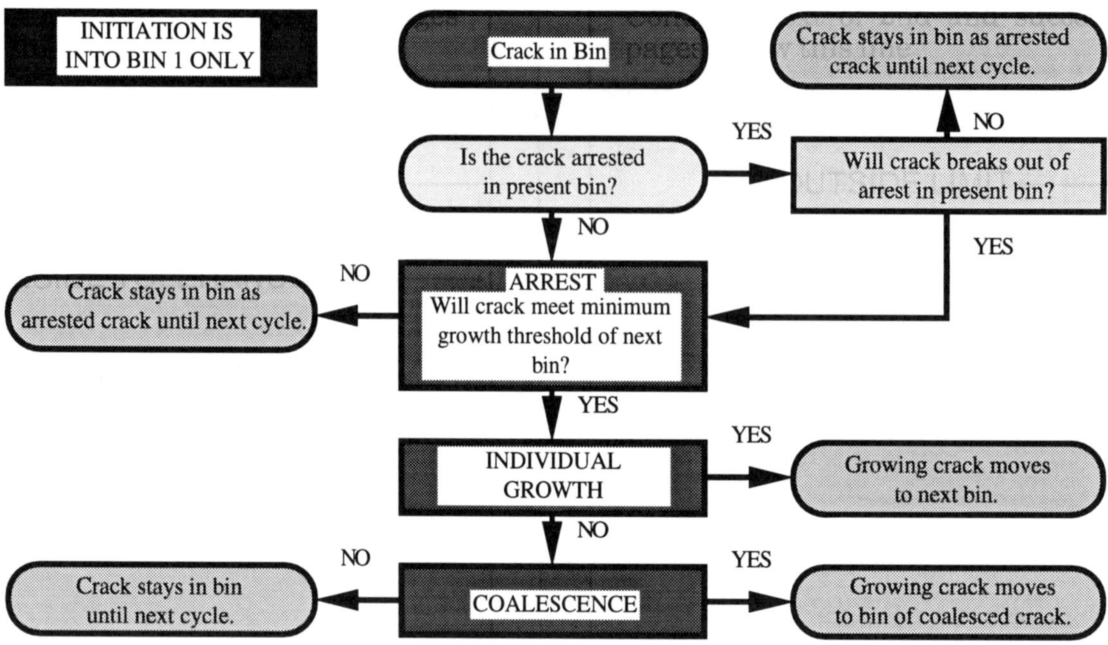

Fig. 8 - Flow chart for the Monte Carlo simulation used in the reliability modeling [36].

processes. Markov processes are important in the modeling for finding simple closed-form expressions which can describe the entire crack distributions (e.g. calculate the coefficients in a lognormal approximation). The simulation considers the four major processes highlighted in the Fatigue of DRA Composites section (i.e. initiation, growth, arrest, coalescence). The crack growth law is taken as the Paris equation in the present calculation *only* when the cracks do grow. In other words, cracks grow by the Paris equation when they are not arrested or growing by coalescence.

The simulation results as compared with the lognormal and observed distributions at different stages of fatigue are shown in Figure 9. The lognormal distribution was selected for comparison since it was the best-fitting commonly-used distribution function. Both the simulation and lognormal distributions agree reasonably well with the experimental data although they did not always satisfy χ^2 tests at a confidence level of 95% (Table I). Considering the simple assumptions used, however, the simulation fits are actually quite good and even improve the lognormal distribution fits most of the time, particularly late in life. The simulation can predict cracks at both ends of the MLD curve. It can closely fit both the exponential nature of the smallest microcracks, to which the lognormal distribution is limited, and the lower tail of long microcracks. With further refinement, the simulation should be capable of predicting MLD in both the DRA and commercial metals closely. From the simulation, the fraction of cracks of given lengths can be estimated as a function of fatigue loading and number of cycles, thus provide needed information for more accurate reliability scheduling.

Concluding Remarks

Present-day reliability methods rely on NDE to provide input on the size of flaws at a given point in life. However, as shown in this paper, the unusually small sizes of most microcracks and the lack of a detectable long, stable-growing fatal crack for much of the fatigue life severely limits the use of NDE in ULCE assessments of DRA composites. Without the capability to detect microcracks accurately in tens of microns, it will not be possible to utilize NDE for detecting individual cracks in these and similar materials. One possible approach being examined is to detect a population of microcracks rather than individual ones. Future studies by authors will involve NDE techniques such as eddy currents, ultrasonics, and acoustic microscopy in attempt to assess how the presence of many microcracks affect materials properties (e.g. elastic modulus, electrical conductivity) of the composites through variations in the signal responses of the NDE method.

This work also presented a new philosophy in designing future ULCE methods based on the growth behavior of microcracks. This philosophy shows that it is more feasible to have quality assurance inspections spaced in shorter intervals late in life rather than equispaced. The main reasons for this are that new cracks are continually being initiated as a result of additional cyclic loading and that late in life coalescences between microcracks leading to higher crack growth rates becomes important. Thus, to avoid missing a significant crack, it is necessary to schedule more frequent, shorter spaced inspections later in life. By knowing the distribution of microcrack lengths present and the rate cracks of a particular size appear, it will be possible to schedule the inspections more effectively. The simulation presented above shows that

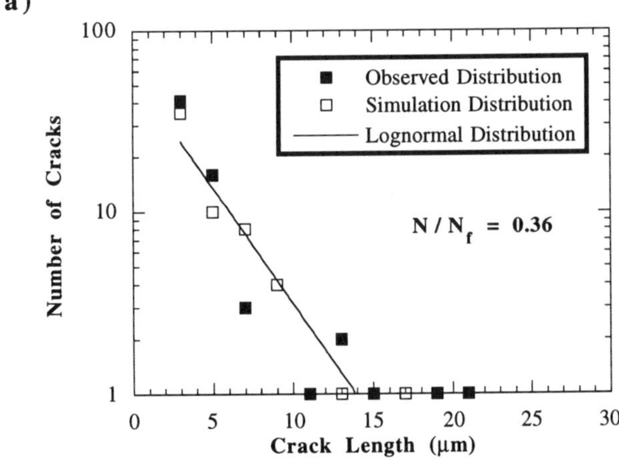

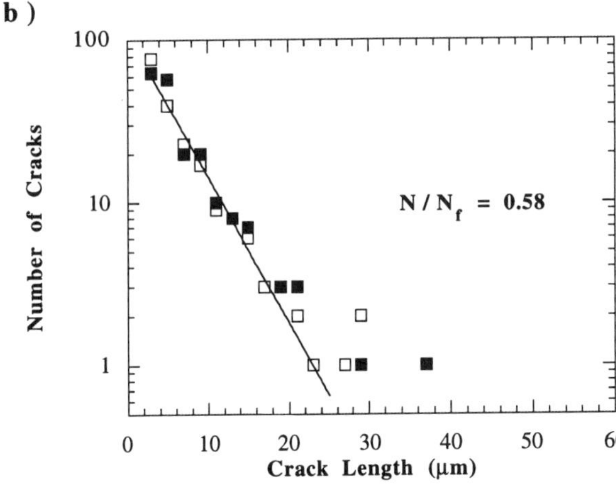

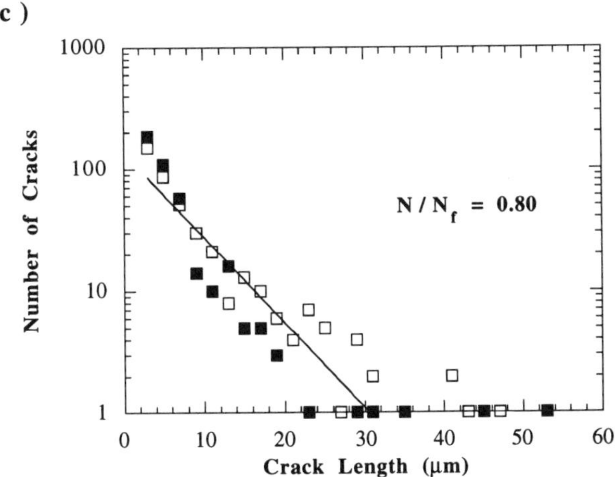

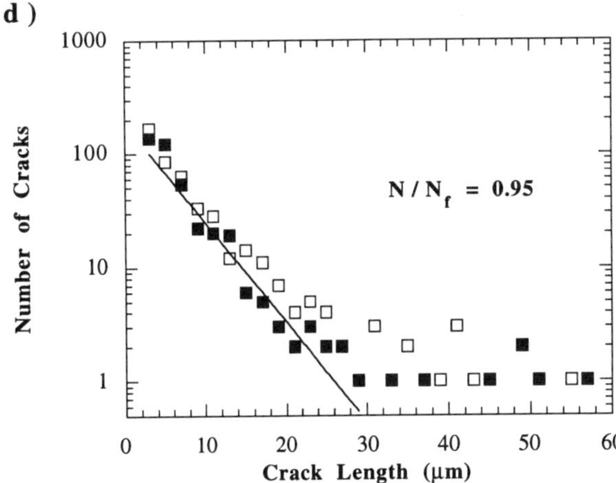

Fig. 9 - Simulation of microcrack length distributions as compared with experimentally observed and lognormal distributions at a) $N/N_f = 0.36$, b) $N/N_f = 0.58$, $N/N_f = 0.80$, and d) $N/N_f = 0.95$. R = -1, $\sigma_{max} = 500$ MPa, $N_f = 13,718$ cycles.

this MLD approach to inspection scheduling can be modeled closely and matched with actual experimental data, thus allowing for improved reliability.

Acknowledgments

This work is a part of the Program for Integrated Design, NDE and Manufacturing Sciences supported by the National Institute of Standards and Technology under Contract No. 70NNANB9H0916. Thanks are due M. Sasaki for performing many of the fatigue tests, and to M. Seniw, R. Kraemer, and A. Bilyk for excellent technical assistance. The generosity of Advanced Composite Materials Corporation in donating the material studied is deeply appreciated. This work made use of MRL Central Facilities supported by the National Science Foundation, at the Materials Research Center of Northwestern University, under Award No. DMR-9120521.

References

1. Zweben, C., Adv. Mater. Processes, 145, 1, 28-30 (1994).

2. Allison, J. E. and G. S. Cole, J. Met., 45, 1, 19-24 (1993).

3. Nair, S. V., J. K. Tien and R. C. Bates, Int. Metals Rev., 30, 6, 275-90 (1985).

4. Williams, D. R. and M. E. Fine, in Int. Conf. Compos. Mater., ICCM-5 Conf. Proc., 5th, W. G. Harrigan, Jr., J. Strife, and A. K. Dhingra, eds., TMS-AIME, Warrendale, PA, 639-70 (1985).

5. Sasaki, M., M.S. Thesis, Northwestern University (1992).

6. Bonnen, J. J., J. E. Allison and J. W. Jones, Metall. Trans. A, 22A, 1007-19 (1991).

7. Masuda, C. and Y. Tanaka, J. Mater. Sci., 27, 413-22 (1992).

8. Lavoie, F. J., Mater. Eng., 9, 27-33 (1986).

9. Froes, F. H., Mater. Des., 10, 3, 110-20 (1989).

Table I - Goodness of Fit of the Monte Carlo Simulation Distributions with the Observed and Lognormal Microcrack Length Distributions by χ^2 Tests

Distribution \ N/N$_f$	0.36 (6)	0.58 (11)	0.80 (12)	0.95 (18)
χ^2 95% Probability Point Table Value	18.31	28.87	38.88	41.34
Simulation	16.75	15.18	60.05	50.34
Lognormal	20.89	29.19	59.86	85.18

( ) = Degrees of Freedom

10. Lavitt, M. O., Aviat. Week Space Technol., 140, 3, 13 (1994).
11. Noguchi, M. and K. Fukizawa, Adv. Mater. Processes, 143, 6, 20-1 (1993).
12. Adv. Mater. Processes, 143, 6, 20 (1993).
13. "Aluminum and Aluminum Alloys", J. R. Davis, ed., ASM International, Materials Park, OH, 160-79 (1993).
14. Ceram. Ind., 142, 2, 27 (1994).
15. Hunt, M., Mater. Eng., 1, 37-40 (1989).
16. Chen, E. Y., L. Lawson and M. Meshii, presented at the Fatigue Crack Initiation Symposium, TMS Fall Meeting, Pittsburgh, 1993 (Manuscript in preparation).
17. Chen, E. Y., L. Lawson and M. Meshii, Scr. Metall. Mater., 30, 6, 737-42 (1994).
18. Sasaki, M., L. Lawson and M. Meshii, Metall. Mater. Trans. A (In press).
19. Ritchie, R. O. and J. Lankford, in "Small Fatigue Cracks", R. O. Ritchie and J. Lankford, eds., TMS-AIME, Warrendale, PA, 1-5 (1986).
20. Paris, P. C. and F. Erdogan, J. Basic Eng., 85D, 528-534 (1963).
21. Lawson, L., E. Y. Chen, M. Sasaki, and M. Meshii, in "Mechanisms and Mechanics of Composites Fracture", R. B. Bhagat, S. G. Fishman, and R. J. Arsenault, eds., ASM International, Materials Park, OH, 65-75 (1993).
22. Kanninen, M. F. and C. H. Popelar, "Advanced Fracture Mechanics", Oxford University Press, NY, 60 (1985).
23. Duga, J. J., W. H. Fisher, R. W. Buxbaum, A. R. Rosenfield, A. R. Burh, E. J. Honton, S. C. McMillan, "The Economic Effects of Fracture in the United States", NBS Special Publication 647-2, March 1983, U. S. Department of Commerce, Washington, D.C. (1983).
24. Schmerr, L. W. and D. O. Thompson, in "Review of Progress in Quantitative Nondestructive Evaluation", Vol. 12, D. O. Thompson and D. E. Chimenti, eds., Plenum Press, NY, 2325-31 (1993).
25. Suresh, S. and R. O. Ritchie, Int. Met. Rev., Vol. 29, 6, 445-76 (1984).
26. Pearson, S., Eng. Fract. Mech., 7, 235-47 (1975).
27. Heubaum, F. and M. E. Fine, Scr. Metall. Mater., 18, 1235-40 (1984).
28. Klesnil, M., J. Polak and P. Liskutin, Scr. Metall. Mater., 18, 1231-34 (1984).
29. Lankford, J. and D. L. Davidson, in "Small Fatigue Cracks", 51-71.
30. Lankford, J., Fatigue Fract. Eng. Mater. Struct., 5, 3, 233-48 (1982).
31. Davidson, D. L., Acta Metall. Mater., 36, 2275-82 (1988).
32. Shang, J. K., J. L. Tzou, and R. O. Ritchie, Metall. Trans. A, 18A, 1613-1627 (1987).
33. Ebi, G. and P. Neumann, Mater. Technol., 10, 498-503 (1990).
34. Obrtlik, K. and J. Polak, Fatigue Fract. Eng. Mater. Struct., 8, 1, 23-31 (1985).
35. Hudak, S. J. and K. S. Chan, in "Small Fatigue Cracks", 379-405.
36. Chen, E. Y., L. Lawson and M. Meshii, Metall. Mater. Trans. A, 1994 (Submitted).
37. Chen, E. Y., L. Lawson and M. Meshii, in "High Performance Composites: Commonality of Phenomena", K. K. Chawla, ed., TMS-AIME, Warrendale, PA (In press).
38. Murakami, Y. and M. Isida, Trans. Japan Soc. Mech. Engrs., 51, 464, Ser. A, 1050-56 (1985).
39. Logsdon, W. A. and P. K. Liaw, Eng. Fract. Mech., 24, 5, 737-51 (1986).
40. Christman, T. and S. Suresh, Mater. Sci. Eng., A102, 211-16 (1988).

41. Berens, A. P., J. G. Burns and J. L. Rudd, in
     "Structural Integrity of Aging Aircraft", S. N. Althuri,
     S. G. Sampath, and P. Tong, eds., Springer, NY, 37-
     51 (1991).
42. Lewis, E. E., "Introduction to Reliability
     Engineering", John Wiley & Sons, New York, NY
     (1987).

Appendix

Appendix A - Terminology and Notation

$Al-SiC_W$ = aluminum alloy reinforced with silicon carbide whiskers composite
CT = compact tension specimen
DRA = discontinuous reinforced aluminum
EIFSD = equivalent initial flaw size distribution
LEFM = linear elastic fracture mechanics
MLD = microcrack length distribution.
NDE = nondestructive evaluation
SEM = scanning electron microscope
ULCE = unified life cycle engineering

a = crack depth, or crack length into surface direction

b = crack length along surface direction

da/dN = crack growth rate of a crack into the depth

db/dN = crack growth rate of a crack along the surface length

ΔJ = cyclic J-integral (for elastic plastic fracture mechanics)

ΔK = stress intensity factor

ΔK_{eff} = effective stress intensity factor (for crack closure effects)

ΔK_{th} = long crack growth threshold

Δt = time interval

$F(t) \equiv P\{t \leq t\}$ = probability that failure takes place at a time less than or equal to t

H = probability of a critical crack causing failure under some continuous specified loading process.

I = rate of initiation of cracks of size l_0

J = interaction term which reduces the probability of failure by any one crack in the presence of other

   cracks

K_C = critical stress intensity

l = crack length

l_c = critical crack length

l_0 = NDE crack detection threshold (i.e. 200 μm)

$\lambda(t)$ = failure rate

m = number of cracks in time interval Δt

n = number of cracks in a specified set

N = number of fatigue loading cycles

N = total number of cracks studied

N/N_f = normalized number of fatigue loading cycles

N_f = number of fatigue loading cycles to failure

R = stress ratio

$R(t) = P\{\mathbf{t} > t\}$ = probability that a system operates without failure for a length of time t, or reliability

σ_{max} = maximum applied stress

t = time in the service life of a component

$\mathbf{t}$ = time to system failure

t_o = time of initial inspection

Appendix B - Theory on the Role of Time-dependent Initiation on Reliability

1.   A homogeneous component having uniformly distributed stresses undergoing cyclic stresses is considered. Such a component may be, in real life, a part of some larger structure over which loading is sufficiently uniform. Such a component may also be an ensemble of components where the time-dependency of initiation is sufficiently uniform in the sum. This is to allow that inspection may reduce or eliminate initiated cracks (when infrequent) without altering the initiation process constants themselves.

2.   It is taken that the smallest detectable crack by current NDE methods is sufficiently large to be described by the Paris-Erdogan equation (i.e. 200 μm). This length is called l_0.

3.   The component is initially inspected. There are no detectable cracks with length, $l \geq l_0$.

4.   A function H(t) is defined for a single crack,

$$H(t) \equiv P(l \geq l_c | l = l_0, t = t_o) = \overline{G(t)} \tag{A1}$$

where l_c is the critical length.

5.   I(t) is defined as the number of cracks per unit time growing in the component to or beyond l_0 from a size less than l_0. Between $t = 0$ and $t = \Delta t$,

$$m = I(\frac{\Delta t}{2}) \cdot \Delta t \tag{A2}$$

"new" cracks are initiated at size l_0.

6.   At time $t > \Delta t$, these new cracks, m, will contribute to the probability of failure,

$$F = \sum_{k=1}^{m} \binom{m}{k}(H(t) - J(m,H))^k (1 - H + J)^{m-k} \tag{A3}$$

where J is an interaction term which reduces the probability of failure by any one crack in the presence of other cracks.

NIST-Industry Collaborations with Emphasis on the Automotive Sector

D. Hunston
National Institute of Standards and Technology
Gaithersburg, Maryland

Abstract

NIST has a wide range of collaborations with industry, and since the automotive sector has been an important participant in these interactions, it serves as an excellent example of what is possible. The interactions fall into two groups: those which involve NIST's extramural activities and those which collaborate with our intermural programs. The two most important extramural activities are the Advanced Technology Program (ATP), and the Manufacturing Extension Partnership (MEP). ATP provides cost sharing support to industry or industry led teams for the development of high-risk, high-payoff, civilian technologies. MEP provides funding to facilitate technology transfer to small and mid-size companies. The second category of interactions at NIST involves cooperation with our internal research programs. These interactions range from informal information exchange to organized programs with specific tasks and exchange of staff. In the area of composites, NIST has had a major internal research program since 1988. The focus is commercial applications, and the automotive industry has been involved with the planning of the program since its inception. As a result, there are major interactions with this industrial sector. Although both the internal and external programs focus on industry, the emphasis on research and development means that universities and government laboratories often have an important role to play as well.

NIST HAS A LONG HISTORY OF INTERACTIONS AND COLLABORATIONS WITH INDUSTRY. Such collaborations have always been an important part of NIST mission. In 1988, however, the mission statement was reprioritized to make assisting U.S. industry maintain and improve its competitiveness a high priority. Associated with the revised mission was a change in name from the National Bureau of Standards (NBS) to the National Institute of Standards and Technology (NIST) and the development of a plan for a strong extramural program with industry. In addition to the extramural program, interactions with industry have also expanded through collaborations with our intramural research effort. The purpose of this paper is to discuss NIST's experience with government-industry interactions and illustrate such collaborations with examples involving polymer composites and the automotive industry.

Polymer composites is a particularly appropriate area to illustrate such interactions because 1988 also marked the initiation of a major expansion for NIST's internal program on polymer composites. One of the major ways that NIST tries to promote competitiveness is to focus in-house research on the critical technologies that hold the keys to future growth. There was unanimous agreement expressed in a wide range of industry and government reports that advanced composites represented such a technology (1-3). The government had a wide variety of in-house programs to assist the development of these materials for defense and aerospace applications but almost no work focusing on the great potential such materials offered for civilian applications like transportation, infrastructure, construction, and electronics. The goal in NIST's program was to address the technical barriers that hinder the introduction of composites in such industries. Although many civilian applications offer tremendous market potential long term, it was felt that the automotive industry had the greatest short term potential, and thus this industry was the initial priority in NIST's program.

Extramural Programs

Interactions with industry through NIST's extramural programs will be discussed first. The two most important extramural activities are the Manufacturing Extension Partnership (MEP) and the Advanced Technology

Program (ATP). Details on these programs can be obtained in a variety of NIST publications (4-6). The brief summary given below is largely taken from these sources.

Advanced Technology Program: The ATP provides multi-year cost-sharing support for high-risk, high-payoff, civilian technology developed by individual companies or industry-led joint ventures. Because the projects involve high-risk research and development, it is not unusual for the participation by universities and government laboratories. It is essential, however, that industry take the lead.

The program began in 1990 but has grown rapidly since then. Moreover, the President's plan as outlined in "a Vision of Change for America," proposes continued expansion (1) as illustrated in Figure 1.

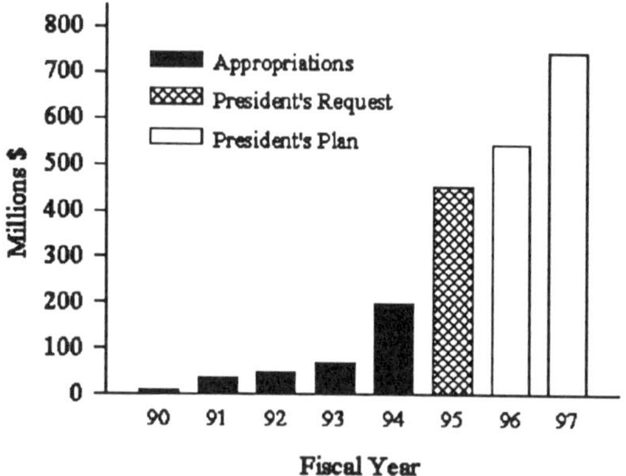

Figure 1: ATP Budget Profile.

Initial ATP Competitions. In the first four ATP competitions, project proposals from any area of technology were accepted and considered in open competitions. The automotive sector has done well in this format with two of the largest ATP projects involving automakers and suppliers. The first which is entitled Cyclic Thermoplastic Liquid Composite Molding for Automotive Structures involves Ford Motor Company and General Electric. This project is developing the needed technologies and materials data for the use of cyclic thermoplastic composites in automobile structural components. The second ATP project which comes from General Motors and General Electric is entitled Engineering Design with Injection-Molded Thermoplastics. The objective is to develop a scientific understanding of the relationship between processing, part geometry, microstructure, and part performance for fiber-reinforced molded thermoplastic parts and embody this knowledge in an integrated thermoplastic engineering design methodology.

Focus Programs. In the initial years when the ATP was a relatively small pilot program the idea of general completions open to all areas of technology was appropriate. As the program expands to full-scale, however, a strategy was needed to retain the emphasis on leadership by the private sector while realizing the benefits to be gained by focusing major resources on areas with strong potential for economic benefit.

To accomplish this, NIST requested industry and other organizations to provide suggestions for focus areas through white papers and workshops. In the initial effort, NIST received over 550 individual proposals, mostly from industry. Many of the suggestions overlapped, and from this input, five programs were developed by combining elements of about 150 different suggestions. These programs were further refined through a series of public workshops and consulting with experts from industry, academia, and other federal agencies. One of the five topics chosen for a focus program is "Manufacturing Composite Structures."

This program is directly relevant to the discussion here and will therefore be discussed in more detail. It should be noted, however, that additional focus programs will be organized in the future based on the information already obtained as well as new suggestions that can be provided at any time. Moreover, the general competition will also continue to be a strong part of the ATP effort.

The vision in the focus program on Manufacturing of Composite Structures is to create a first and second tier manufacturing structure for high performance commercial composite parts in the United States early in the 21st Century. As shown in Table I, the effort will focus on three business sectors and three enabling technologies. The competition will seek projects that address one or more of the items in this matrix.

The focus program anticipates a five year effort with three solicitations. The first competition closed in July of 1994. The government funding for this first solicitation is $25M while the government anticipates committing $160M for the full 5 years. It is expected that industry will approximately match this amount with cost sharing so the total effort will be in excess of $300M.

By the end of the five year ATP program, the technical objectives are to demonstrate commercial benchmarks in the three business sectors. The scope include high-risk, generic research and development projects that address topics which are specifically and directly linked to the commercial business opportunities associated with the use of composite materials in these business sectors. Since surface transportation is a major part of this program, it is hoped that the automotive industries will participate in this program in a major way.

Table I: Program on Manufacturing Composite Structures

	Surface Transportation	Infrastructure	Marine Structures
Low Cost Manufacturing			
Integrated Design			
Risk Reduction Sensors			

Manufacturing Extension Partnership: The second major extramural Program is the Manufacturing Extension partnership (MEP). The MEP program provides funding for regionally based extension centers that help small and medium sized businesses adopt modern technologies. MEP also provides matching grants for state-based technology extension efforts and develops linkage between federal, state, local, and other technology extension efforts. The partnerships include four major elements:

● Regionally based Manufacturing Technology Centers (MTCs) providing hands-on technical assistance to small and mid-sized manufacturers.

● Smaller, satellite operations called Manufacturing Outreach Centers, some affiliated with an MTC.

● The State Technology Extension Program (STEP) providing grants to help states build the infrastructure needed for technology transfer efforts.

● The Links Program to pull together--both electronically and otherwise--not only the NIST affiliated offices but also all other federal, state, local, and university technology entities into one national network.

The philosophy of the MEP is to take maximum advantage of programs already in place. It avoids duplication of effort among existing technology assistance organizations and concentrates on matching company needs to available help regardless of the source.

Like the ATP, the MEP is young but growing rapidly. To date, the impact on composites in the automotive industry have been quite limited, but as the program grows, there should be significant opportunities for interaction with the small and mid-sized companies that make up an important part of this industry segment.

Intramural Program

The other opportunity for NIST-industry interaction is through cooperation with our in-house research programs. Polymer composites has been a major research effort at NIST since the work was expanded in 1988. The emphasis has been commercial, mass-market applications. The goal is to facilitate the introduction of these light-weight, strong, corrosion resistant materials in civilian applications that include infrastructure, electronics, medical devices, robotics structures, etc., but the initial focus was automotive applications.

The two thrusts of NIST's program are based on the two most important barriers to expanded use of composites: the lack of rapid, reliable, cost-effective processing, and the inability to understand and predict long term performance. The emphasis in the 1988 expansion was processing science, and most of the new work addressed that area. A smaller expansion occurred in 1992, and this permitted the addition of work on durability.

The program's planning and direction relies heavily on industry input. For example, the first step in organizing the 1988 program expansion was to host and industry workshop to seek their advice on priorities and to assure that the work would be relevant to industrial needs. The workshop involved representatives from a wide range of suppliers, fabricators, and users of composites. They were ask to identify the most promising processing methods for the future and the scientific and technical barriers that limit the use of these methods today. The top choice (7) was liquid composite molding (LCM), and as a result, this became the center for NIST's research program. LCM includes resin transfer molding and structural reaction injection molding. Representative of the automotive industry were active participants in the workshop, and LCM was also their choice as the method with the most potential for fabrication of structural composites parts. Since 1988, three additional industry workshops (8,9) have been held to be sure the program stays directed at the most important problems. The results indicate that LCM continues to be the most promising processing method for many applications.

Liquid Composite Molding: NIST is studying LCM with a three task program (10). The first task is materials characterization, and it is developing and utilizing measurement methods to determine the properties that are important for manufacturing. The prime example is permeability of preform materials. Characterization of the microstructure for preform materials is also important since a long term goal in the program is to develop the capability to predict the permeability of a preform material from a knowledge of its microstructure. The second task is developing process simulation models for LCM. Two types of models are under development: macro-models and micro-models. The first is used to describe macroscopic events like mold filling. These models incorporates all the details of the reinforcement microstructure by using an empirically determined permeability, and as a result, such models do well for predicting mold filling but not describe local events like fiber wetting and micro-void formation. The micro-models, on the other hand, include the microstructure details so they are much more complex, and the size of the system they can analyze is limited. Nevertheless, they have the potential to do things like predict permeability from a knowledge of reinforcement microstructure. The third task is investigating process monitoring with two objective. First, the results provide feedback that test the predictions of the process simulations so the models can be refined and improved. Second, the monitoring capability provides a key element for the development of an on-line process control technology. This task benefits from NIST's past work on study process monitoring technology for composites (11).

Because the automotive companies have had a great role in the planning of NIST's program through participation in the workshops and a variety of direct interactions, there has been a strong collaboration with this industry sector since the program expansion in 1988. This collaboration has evolved, and in 1990, a Cooperative Research and Development Agreement (CRADA) was signed between NIST and the Automotive Composites Consortium (ACC) to work together on LCM. The areas of interaction are shown in Figure 2. The objective is to use NIST's process simulation models to study and optimize the fabrication of specific parts being made by the ACC as part of their program on structural composite components. The ACC

provides feedback to NIST on how well the simulations describe the actual fabrication process so the models can be refined and improved. As part of this effort, NIST is measuring permeability data on materials of interest to the ACC in our materials characterization task. The ACC is providing the materials to be characterized and details on the part design for input to the simulation models.

The first part examined was the ACC's focal project I, i.e. the front end structure of the Ford Escort. The results (12) from one process simulation run are shown in Figure 3. The finite element mesh for this part is displayed in Figure 3A while the fill pattern for resin injection at the indicated location is shown in Figure 3B.

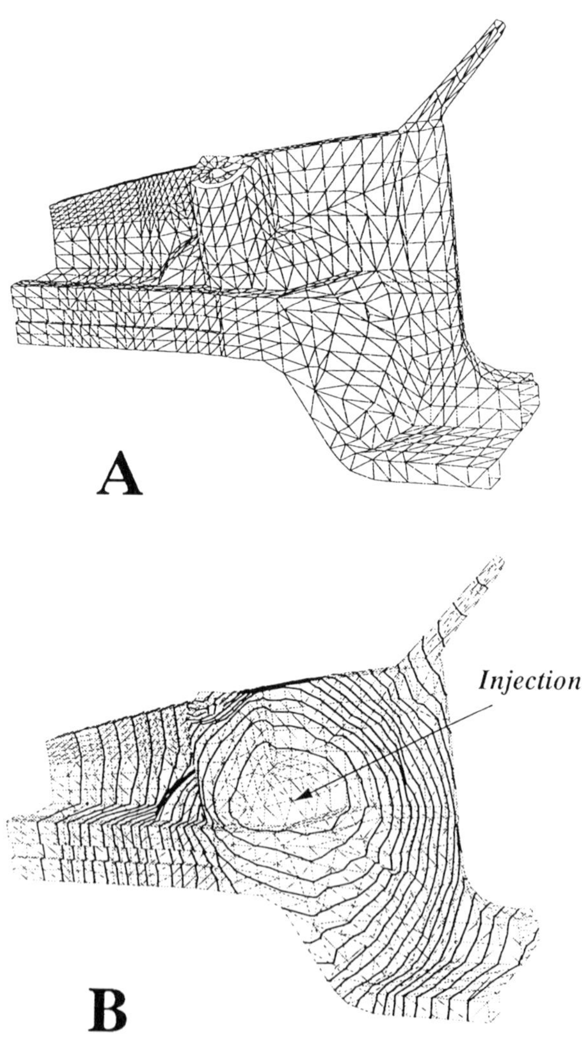

Figure 3: The finite element mesh is shown in A while B illustrates the fill pattern--the solid lines are the flow front position at various times.

By conducting simulations such as that shown in Figure 3, the ideal locations for injection ports and vents can

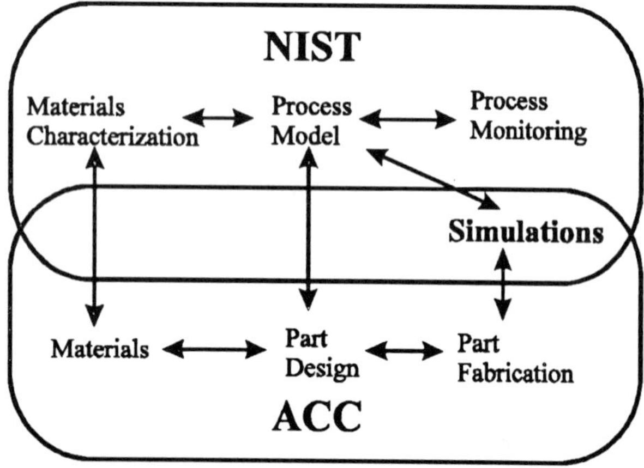

Figure 2: NIST program on processing science of LCM and the collaboration with the ACC in this area.

be found rapidly and inexpensively. Flow patterns that lead to incomplete filling or the capture of large air bubbles (or voids) can be avoided. Since the simulations give the pressures involved, the results can be used to estimate the required mold stiffness and wall thicknesses. The fill time is also obtained so the trade-off among resin viscosity, flow rate, and pressure can be examined. Increased flow rate leads to reduced fill times but also increased pressure. This can mean larger molds and presses. The most important output of the cooperation on the Focal Project, however, is that it provides a concrete example to show how simulations can be used in the future to rapidly and inexpensively optimize a process. In this sense, it helps the automotive industry and their suppliers prepare for the next generation of fabrication technology where simulations will play a critical role.

A second major collaborative project with the ACC on LCM is now underway. It is examining the fabrication of a cross-member for a van. The mold for this part is being instrumented so that data on the changes that occur during processing can be monitored. This will provide critical information for improving the process and refining the simulation models. Several preliminary mold filling simulations were performed to help identify the ideal points for placement of pressure sensors. This information was combined with the fabrication expertise in the ACC to select the positions which provided the most information and yet were easiest to implement in the mold. Manufacturing trials have been planned and should be conducted within the next six months.

Durability: The second major thrust in the NIST composites program is durability. Work in this area was added in the 1992 program expansion. Based on the results of NIST's third industry workshop on polymer composites, the initial focus in this program is the attack of moisture on composites. The approach is to seek information not just on the changes in mechanical properties that occur on long term exposure to moisture but also on the mechanisms of degradation that are responsible for these changes. Thus both chemical and mechanical events are of interest as well as any interactions that may occur between them. The goals are to develop test methods for material selection, models for service life prediction, and accelerated testing procedures.

To accomplish these goals, a three task program has been organized (13). The first task prepares samples with systematic variations in materials, fiber surface treatments, and processing conditions. The second task measures the properties of the composites made in task one with macro-tests (test on full composites) as a function of exposure to moisture. To help understand this behavior, a third task is conducting micro-tests (tests on simplified specimens) that try to isolate the influence of moisture on the constituents: the fiber, the resin, and the fiber-matrix interface. Experiments for the first two constituents are straight

forward, but to examine the interface, a single fiber fragmentation test is being used (Figure 4).

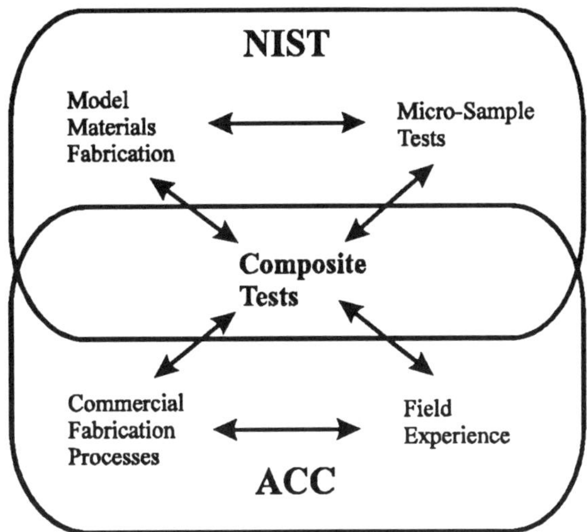

Figure 4: NIST program to study moisture attack on polymer composites and its interaction with the ACC.

In the initial experiments, the samples were made with a model epoxy and glass fibers. This system has the advantage that much data is already available. The first goal was to develop an experimental protocol, while the second is to investigate the connection between the micro- and macro-tests. This connection has been established for tests on fresh samples but not for experiments related to durability and moisture attack. Such a connection is important since without it, the relevance of the micro-tests to real world performance is in question.

To date, the program has produced a number of important results (13). First, the experimental protocol for such experiments has been established. Second, the results have shown that moisture attacks both the fiber and the interface so experiments must be able to monitor both changes. The usual analysis for the single fiber fragmentation tests assumes that the fiber strength is measured independently since this value is required to calculate the interface strength. As part of the NIST program, an analysis technique was developed to estimate both the fiber and interface strengths from the single fiber test (14).

In 1993, NIST and the ACC developed a cooperative program on durability. In this collaboration, materials and test conditions of interest to the automotive companies are being included in NIST's test program for durability (Figure 4). To compliment this work, the ACC is providing samples fabricated by standard commercial methods and will help conduct tests on the composite samples. In addition, the ACC is interested in how the results compare to what is seen in the field since ultimately, that is where the results must be relevant. The NIST program and the interaction with the ACC is illustrated in Figure 4.

Although the epoxy system used in the NIST program is ideal for some applications, the automotive companies are much more interested in vinyl esters and urathanes. As a result, NIST is now trying to develop a model urethane. The results of this effort will be discussed in another paper at this meeting. If this material system can be developed, it will be incorporated into the on-going effort on durability.

Closing Comments

The cooperative programs described above represent only a small fraction of the industry interactions with NIST's composites program. Other organizations involved include companies in the automotive sector, but since much of the work is very generic a wide range of industries participate. These range from suppliers to users and from aerospace to electronics. A number of universities are also collaborating with NIST either by direct interactions or by participation as part of a team involving industry. With the growing importance of composites in commercial applications, these interactions will continue to grow.

References

1. Department of Commerce, "Emerging Technologies - A Survey of Technical and Economic Opportunities," 1990.

2. Aerospace Industries Association, "Key Technologies for the 1990's," 1988.

3. Chemical and Engineering News, "Key Technologies for the 1990's Targeted," p. 24, April 3, 1989.

4. "Guide to NIST," U.S. Department of Commerce, Washington, DC, 1994.

5. "Advanced Technology Program, Proposal Preparation Kit," Department of Commerce, Technology Administration, NIST, Feb. 1994.

6. Commerce Business Daily, "Announcement of Advanced Technology Program: Manufacturing Composite Structures," May 6, 1994.

7. Beardmore, P. and Hunston, D., "An Industry Workshop on Polymer Composite Processing," Oct. 7, 1987, NBSIR 87-3686, NBS, Department of Commerce, Feb., 1988.

8. Johnson, C., Chang, S. S., and Hunston, D. L., "2nd Industry Workshop on Polymer Composite Processing," May, 1990, NISTIR 4461, NIST, Department of Commerce, Dec., 1990.

9. Parnas, R. S., Salem, A. J., Kendall, K. N., and Bruschke, M. V., "Report on the Workshop on Manufacturing Polymer Composites by Liquid Molding," Sept., 1993, NISTIR 5373, NIST, Department of Commerce, Feb., 1994.

10. Hunston, D. L., Phelan, F., Parnas, R., "Flow Behavior in Liquid Molding," Proc. FiberTex, 1991, & p. 23-42, NASA Conf. Pub. 3176, 1991.

11. Hunston, D. L., McDonough, W. G., Fanconi, B. M., Mopsik, F. I., Wang, F. W., Phelan, F. R., and Chiang, M. Y., "Assessment of the State-of-the-Art for Process Monitoring Sensors for Polymer Composites," NISTIR-4514, NIST, Department of Commerce, June, 1991.

12. Phelan, F. R., "Flow Simulation of the Resin Transfer Molding Process," Proc. Amer. Soc. of Composites, 7th Annual Technical Conference, 1992.

13. Schutte, C. L., McDonough, W., Shioya, M., Hunston, D. L., "The Use of a Single-fiber Fragmentation Test to Study Environmental Durability of Interfaces-Interphases between Epoxy and a Glass Fiber," Mat. Res. Soc., Symp. Proc., Vol. 304, 43-48, 1993.

14. Shioya, M., McDonough, W. G., Schutte, C. L., and Hunston, D. L., "Test Procedure for Durability Studies of the Fiber-Matrix Interface," Proc. Adhesion Society Meeting, Feb., 1994.

Mechanical Properties of Micron-Sized Pultruded Composite Profiles

K.C. Kennedy, T. Chen, R.P. Kusy
University of North Carolina
Chapel Hill, North Carolina

Abstract

Continuous, unidirectional, fiber-reinforced-plastic composites were formed via a proprietary pultrusion process to a finished round profile of 0.5mm in diameter. Commercially available quartz, S2-glass, and E-glass yarns were used as reinforcement in a 3-D network-polymer matrix that is commonly used for dental restorations. Materials were produced with various amounts of reinforcement, and evaluated in tension and in three-point bending. Interfacial bond quality was evaluated indirectly by calculating an efficiency as the ratio of experimental to theoretical property values. Cross-sectional morphology was documented via reflected light microscopy.

Fiber content reached as high as 79% (by volume). Mechanical properties varied linearly with fiber content; however, flexure properties dropped at reinforcement levels above 70%. The elastic modulus approached theoretical rule of mixture predictions in both tension and in flexure, and efficiency in strength ranged from about 0.2 to 0.9. The S2-glass reinforced materials exhibited superior properties, with an elastic modulus in flexure as high as 59.7 GPa, and tensile and flexure strengths of 2.68 GPa and 1.66 GPa respectively. At low to moderate fiber content, cross-sectional morphology was typically inhomogeneous, with reinforcement located around the perimeter of the profile. The finished material had the ability to be bonded to other materials with a common quick-setting epoxy, illustrating their potential utility as building-blocks in the assembly of larger composites.

COMPOSITE MATERIALS CAN be found in a wide and varied array of scientific disciplines, and many different types have been developed to meet a broad range of material needs. As the sciences become more interdisciplinary, many materials and processing techniques are finding applications in other areas. New ideas and innovative technologies are continually emerging, and one of the important and ongoing challenges is to integrate these technologies in novel ways, and to investigate the utility of the outcome.

Modern composites were being developed for use in the field of restorative dentistry as early as 1958.[1] Since then, the focus has primarily been centered around particulate-type composites of silica-based reinforcement in an acrylic-based matrix. Over the last few decades, improvements in the polymer-resin and interfacial-bonding chemistries have led to materials with outstanding strength and durability.[2-9] Though the fundamental stress mode for dental composites is compression, a similar composite might be expected to perform well in tension or in flexure if the particulate reinforcement were replaced with fibers.

Pultrusion has been used since the 1950's to form composites of unidirectional-fiber-reinforced plastic (UFRP).[10,11] While polyesters and epoxies have been the most widely used matrix materials, polycarbonate, polyethylene-terepthalate-glycol, phenolics, methacrylates, and many other polymers have also been used.[12,13] Profiles have typically been no smaller than 3mm in diameter.

A small UFRP profile with a high aspect ratio has potential to be used in the lay-up of larger composites with more complex shapes. Though UFRP profiles on the order of 0.6mm have been produced by both direct pultrusion and by slitting thin pultruded tapes, fiber content has been limited to about 60% by volume (hereafter denoted as %v, or %wt to indicate volume or weight percentages), and the resulting properties have typically been lower than what has been obtained for larger profiles.[14-17]

An experimental process has been developed, whereby pultrusion of very small profiles is possible. The process has been used to produce glass- and quartz-based UFRP profiles measuring 0.5mm in diameter. A common dental-restorative polymer was used as the matrix phase.

Preliminary mechanical testing was conducted to determine the most basic limitations of the materials.

This investigation highlights the high stiffness and strength of the composites produced by the novel process, as well as the efficiencies of the interfacial bonds. The mechanical properties are highly predictable, and the process produces a unique cross-sectional morphology. The use of a simple gripping method for tensile testing reveals the ability of the material to form strong bonds with common epoxy adhesives. This result carries strong implications concerning the potential utility of these materials as "pre-composites" to be used in the formation of larger structural composites.

Materials and Methods

Reinforcement. Reinforcement consisted of either quartz, S2-glass, or E-glass fibers in the form of commercially available yarns. All yarns were received pre-sized with epoxy-compatible-organo-silane binders. Yarns were re-wound for processing, but were otherwise used as-received. Specifications for each of the yarn types are shown in Table I, and mechanical properties of the bare filaments are shown in Table II.

Table I. Reinforcement yarn specifications.

Material	Manufacturer and Product Designation	Filament Size (μm)	Filament Count	Twist $\left(\frac{\text{turns}}{\text{inch}}\right)$
Quartz	Quartz Products Co. Quartzel™ 300 1/0 0.5Z QS13	9	120	0.5
S2-glass	Owens Corning S2 CG150 1/0 1.0Z 493	9	204	1.0
E-glass	Owens Corning E CG150 1/0 0.7Z 603-O	9	204	0.7

Table II. Reinforcement filament mechanical properties as specified by the manufacturers.

Material	Elastic Modulus (GPa/Msi)	Tensile Strength (GPa/Msi)	Maximum Elongation (%)
Quartz	78/11	6.0/0.87	—
S2-glass	85.5/12.4	4.59/0.665	5.7
E-glass	72.4/10.5	3.45/0.500	4.8

Matrix. The matrix material for these composites was a glassy network copolymer consisting of the following components: 61%wt BIS-GMA, and 39%wt triethylene-glycol-dimethacrylate (TEGDMA). Initiators, accelerators, and all other additives accounted for less than 1%wt of the mixture, and are proprietary at the time of this writing. The chemical structures of the two primary components are shown in Figure 1, and the mechanical properties of the cured copolymer are shown in Table III.[18,19]

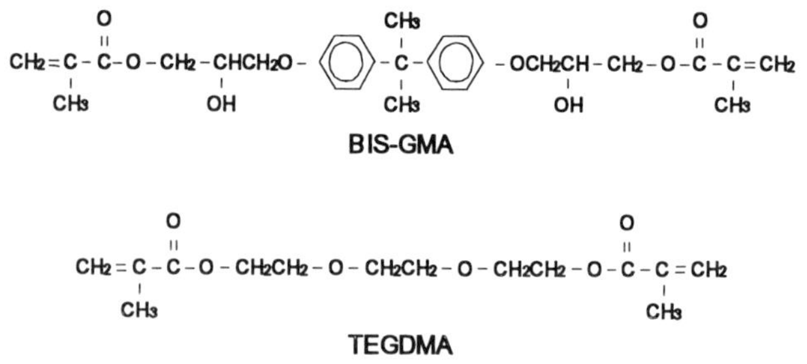

Fig. 1 - Chemical structures of the primary matrix monomers.

Table III. Mechanical properties of the cured polymer.

Elastic Modulus (GPa/Msi)	Strength (MPa/ksi)		
	Tensile	Compressive	Shear
2.42/0.351	21/3.0	83/12	42/6.1

Processing. Composite material was formed continuously via a novel pultrusion process to a finished round profile measuring 0.5mm in diameter. Further details of the process are forthcoming, pending receipt of a U.S. patent filing date. Materials were made with a range of fiber content varying between 33.0%v and 79.3%v for each of the three reinforcement types.

Mechanical Testing. Composites were tested in tension and in flexure using an Instron™ Universal Testing Machine. All tests were conducted at room temperature (approximately 25˚C). The time between material production and testing varied from one to several days. Prior to testing, materials were stored at room temperature and humidity in brown paper envelopes.

Tensile Tests. For the quartz-reinforced materials, nine levels of reinforcement were chosen for tensile testing; five were chosen for each of the S2-glass- and E-glass-reinforced materials. Three specimens were prepared for each of the 19 combinations for a total of 57 specimens.

Tensile specimens were cut from the bulk material in 35.5cm lengths. An epoxy adhesive was used to glue each specimen end into a 4.5cm long, loose-fitting brass tube (1.59mm outside diameter) to form a machine grip. A mechanical lock was formed between the tube and the epoxy-coated composite by crimping the tube every 6-7mm along its length before the epoxy set. Crimping depth was kept shallow so that only light, non-binding contact was made between the wall of the tube and the composite.

Machine grips were positioned 25.4cm apart. A 1.26cm, 50% extensometer was placed near the center of the mounted specimen, which was then elongated at a rate of 1cm/min. Each specimen was subjected to at least two loading cycles up to about 120N, and then loaded a final time until failure. The elastic modulus in tension, Et, was computed from the average slope of the loading cycles (including the portion of the failure cycle below 120N). The ultimate tensile strength, UTS, was computed from the maximum load divided by a nominal cross-sectional area, which was based on a perfectly round profile of 0.5mm in diameter.

Flexure Tests. Composite flexure properties were examined in three-point bending, with a span length of 8.89mm. Details of the testing apparatus are reported elsewhere.[20] For each of the three reinforcement types, five different levels of fiber content were selected for testing. Four specimens were tested for each of the 15 combinations for a total of 60 specimens.

Flexure specimens were cut to approximately 5cm in length, and the central 8.89mm were marked for identification. The diameter was measured within the central section at four different locations using a Sony μ-mate™ Digital Micrometer. Care was taken to make each measurement at a different radial orientation. The four measurements were averaged to obtain the mean diameter for that specimen.

A bending load was applied at a deflection rate of 0.1cm/min until the onset of failure, which was indicated by a drop in the load. The slope was extrapolated from the linear portion of the load-deflection curve. Elastic modulus in flexure, E_f, and flexure strength, FS, were calculated from Eqns. 1 and 2, respectively, according to simple elastic-beam theory.[21]

$$E_f = \frac{PL^3}{48I\delta} \tag{1}$$

$$FS = \frac{Mc}{I} \tag{2}$$

where
P = applied load,
L = span length,
I = moment of inertia,
δ = beam deflection,
M = maximum bending moment, and
c = sectional radius.

Morphological Study. Ordinary shears were used to cut short samples approximately 2.5cm in length. Specimens were mounted into an epoxy plug, with interspersed glass beads to keep the surface flat during polishing so that the specimen edges would be retained. In order to remove any damage caused by the crushing action of the shears, at least 4.5mm of material was removed from the end of each plug by wet-grinding with 240-grit silicon-carbide paper in water. Successively finer papers were used, followed by 1.0μm, 0.3μm, and 0.05μm dispersed alumina particles. The polished sections were examined with a Zeiss™ Universal Light Microscope. A 35mm

Nikon™ Nikkormat camera was mounted on the microscope to record the images.

Results

Mechanical Testing. Figure 2 shows representative tensile and flexure load-deflection curves as recorded by the Instron™ Universal Testing Machine. The curves shown are for two different E-glass specimens, but the overall shape and characteristics of the curves are typical of all specimens tested. In Figure 3 the results of the mechanical testing are summarized in four plots, where each data point represents a mean value (the marker legend of Fig. 3a applies to all parts of the figure). Actual mean values and standard deviations are summarized in Tables IV and V. Each table entry consists of a mean followed by the standard deviation, s.d., and in parentheses the number of samples tested.

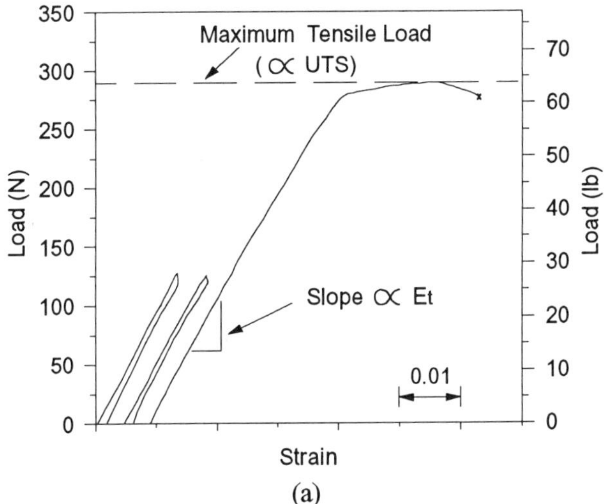

(a)

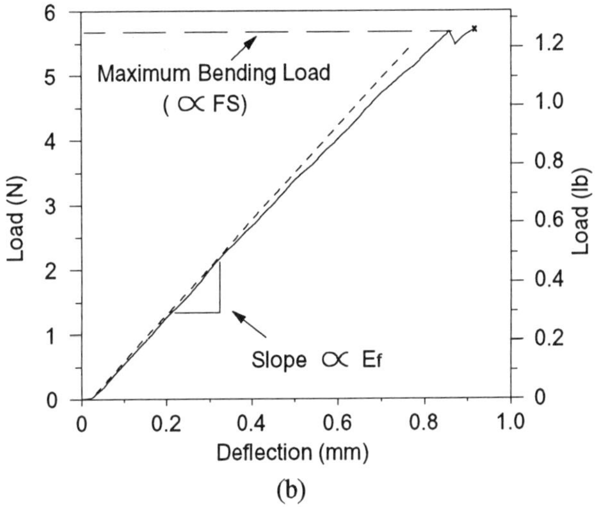

(b)

Fig. 2 - Load deflection curves for E-glass specimens in: (a) tension, 66%v; and (b) flexure, 53%v.

Tensile Properties. Figures 3a and 3b show the respective Et and UTS of the different composite materials as a function of fiber content. In both cases the S2-glass reinforced materials exhibited superior properties, followed by quartz, and then E-glass. The Et for the S2-glass materials ranged from 28.3 GPa to 62.5 GPa, and for the quartz and E-glass materials the respective ranges were 31.7 GPa to 51.7 GPa and 24.0 GPa to 50.4 GPa. The UTS for the S2-glass materials ranged from 1.42 GPa to 2.68 GPa, and the respective ranges for the quartz and E-glass materials were 1.43 GPa to 2.11 GPa and 0.77 GPa to 1.36 GPa.

The variability among samples at a particular volume fraction was low, with the s.d. typically falling below 6% of the mean value, and falling below 10% of the mean value in all cases. All three materials showed a strong correlation between fiber content and Et, and between fiber content and UTS. In each case the probability of the regression, p, was less than 0.01.

Flexure Properties. Figures 3c and 3d show the respective Ef and FS of the three materials. Again, the S2-glass reinforced materials were both stiffer and stronger than the quartz and E-glass materials, with an Ef ranging from 20.1 GPa to 59.7 GPa, and a FS between 1.20 GPa and 1.66 GPa. The values of Ef for the quartz and E-glass materials were between 20.9 GPa and 48.1 GPa, and 16.9 GPa and 45.6 GPa, respectively; the FS ranged from 0.84 GPa to 0.97 GPa and 0.79 GPa to 1.29 Gpa, respectively.

The variability among the samples in flexure was greater than it was in tension, though more notably so for the FS than for the Ef. Values of the s.d. for Ef were typically between 5% and 15% of the mean value, with the exception of one case where the s.d. was 29% of the mean. The s.d. of the FS was typically between 10% and 20% of the mean.

For all three materials, Ef was strongly correlated to fiber content at levels below 70%v. The values of Ef for the quartz and E-glass materials were similar, and for clarity the data were combined into one regression. The data for S2-glass was regressed seperately. In both cases the value of p was less than 0.01.

In the case of FS for the E-glass material, a good correlation was found with fiber content at levels below 67%v ($p < 0.01$). The flatness of the regression line for the FS of the quartz material reveals that the FS is nearly independent of fiber content; thus little statistical significance can be drawn from the correlation ($p > 0.1$). The correlation for the S2-glass materials was only slightly significant ($p < 0.1$), owing to the fact that only three mean values were regressed.

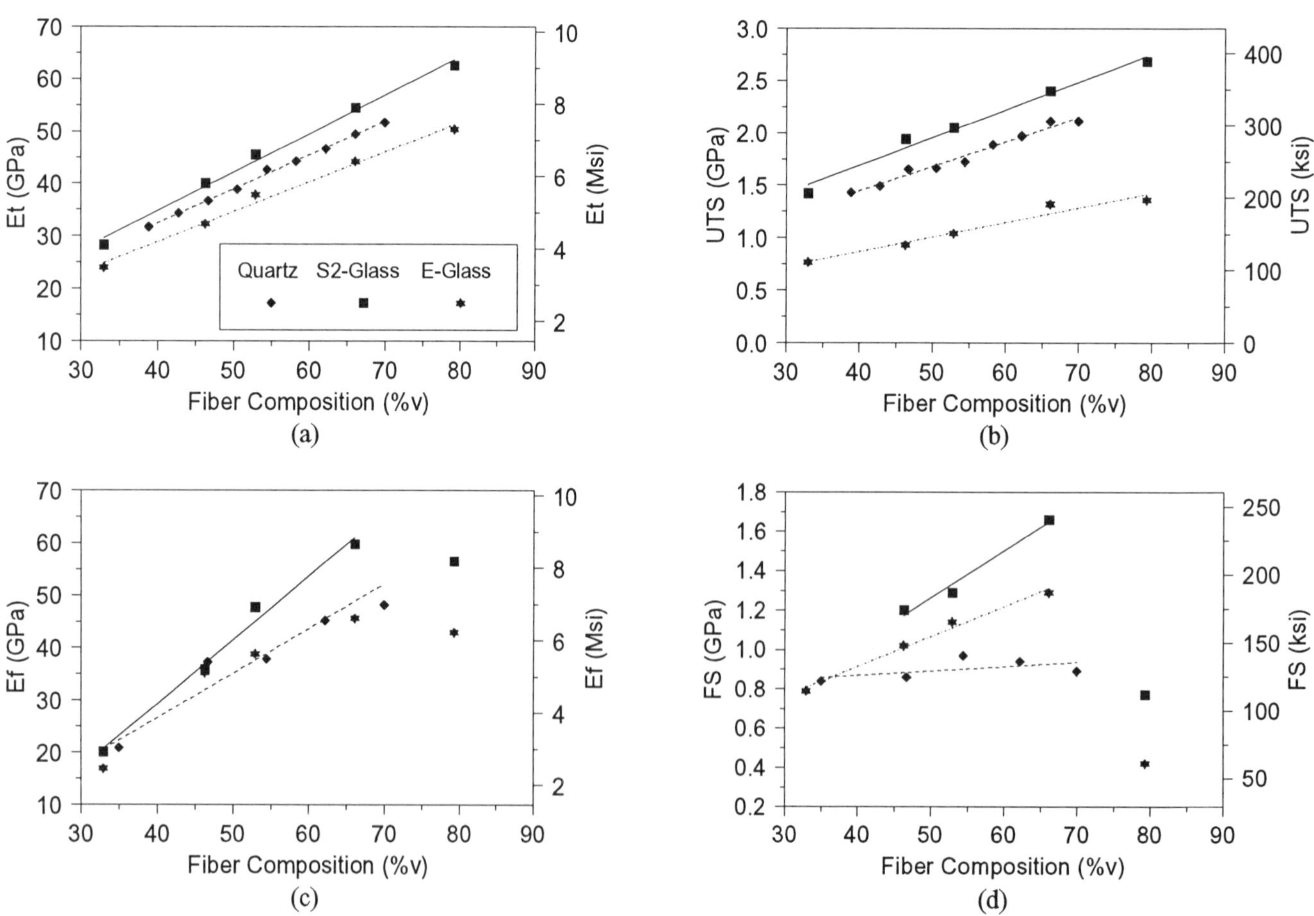

Fig. 3 - Mean values for experimentally determined mechanical properties: (a) Et; (b) UTS; (c) Ef; (d) FS. Marker legend in (a) applies to all.

Morphology. Figure 4 is a photograph of several tensile-test specimens. At the right of the figure an untested specimen is shown for comparison. Tensile failures were generally explosive in nature, with the material failing in a global fashion rather than locally. Ultimate separation usually occurred near the center of the specimen where the strain gauge was attached.

Table IV. Summary of UFRP tensile properties.

Reinforcement Material	Fiber Content (%v)	Elastic Modulus (GPa)	Tensile Strength (GPa)
Quartz	38.9	31.7±1.3 (3)	1.43±0.07 (3)
	42.8	34.3±1.0 (3)	1.49±0.11 (3)
	46.7	36.6±1.0 (3)	1.65±0.05 (3)
	50.5	38.8±1.5 (3)	1.66±0.04 (2)
	54.4	42.6±0.7 (3)	1.72±0.17 (3)
	58.3	44.2±0.8 (3)	1.89±0.08 (3)
	62.2	46.6±0.8 (3)	1.97±0.11 (3)
	66.1	49.5±0.1 (3)	2.11±0.05 (3)
	70.0	51.7±1.0 (3)	2.11±0.03 (3)
S2-glass	33.0	28.3±1.2 (3)	1.42±0.02 (2)
	46.3	40.0±0.4 (3)	1.94±0.05 (3)
	52.9	45.5±1.0 (3)	2.05±0.05 (3)
	66.1	54.5±3.6 (3)	2.40 (1)
	79.3	62.5±0.6 (3)	2.68±0.03 (3)
E-glass	33.0	24.0±0.3 (3)	0.77±0.05 (3)
	46.3	32.3±0.3 (3)	0.93±0.06 (3)
	52.9	37.8±0.4 (3)	1.04±0.07 (3)
	66.1	44.2±2.6 (3)	1.32±0.08 (3)
	79.3	50.4±1.3 (3)	1.36±0.13 (3)

Table V. Summary of UFRP flexure properties.

Reinforcement Material	Fiber Content (%v)	Elastic Modulus (GPa)	Flexure Strength (GPa)
Quartz	35.0	20.9±0.8 (4)	0.84±0.14 (4)
	46.7	37.3±6.5 (4)	0.86±0.16 (4)
	54.4	37.9±11.0 (4)	0.97±0.10 (4)
	62.2	45.2±2.7 (4)	0.94±0.15 (4)
	70.0	48.1±3.1 (4)	0.89±0.18 (4)
S2-glass	33.0	20.1±2.6 (4)	—
	46.3	35.9±2.4 (4)	1.20±0.08 (4)
	52.9	47.7±6.9 (4)	1.29±0.10 (4)
	66.1	59.7±4.4 (4)	1.66±0.24 (4)
	79.3	56.4±6.3 (3)	0.77±0.17 (3)
E-glass	33.0	16.9±2.5 (4)	0.79±0.14 (2)
	46.3	35.2±7.0 (4)	1.02±0.11 (4)
	52.9	38.8±2.4 (4)	1.14±0.13 (4)
	66.1	45.6±2.4 (4)	1.29±0.16 (4)
	79.3	42.9±3.5 (4)	0.42±0.05 (4)

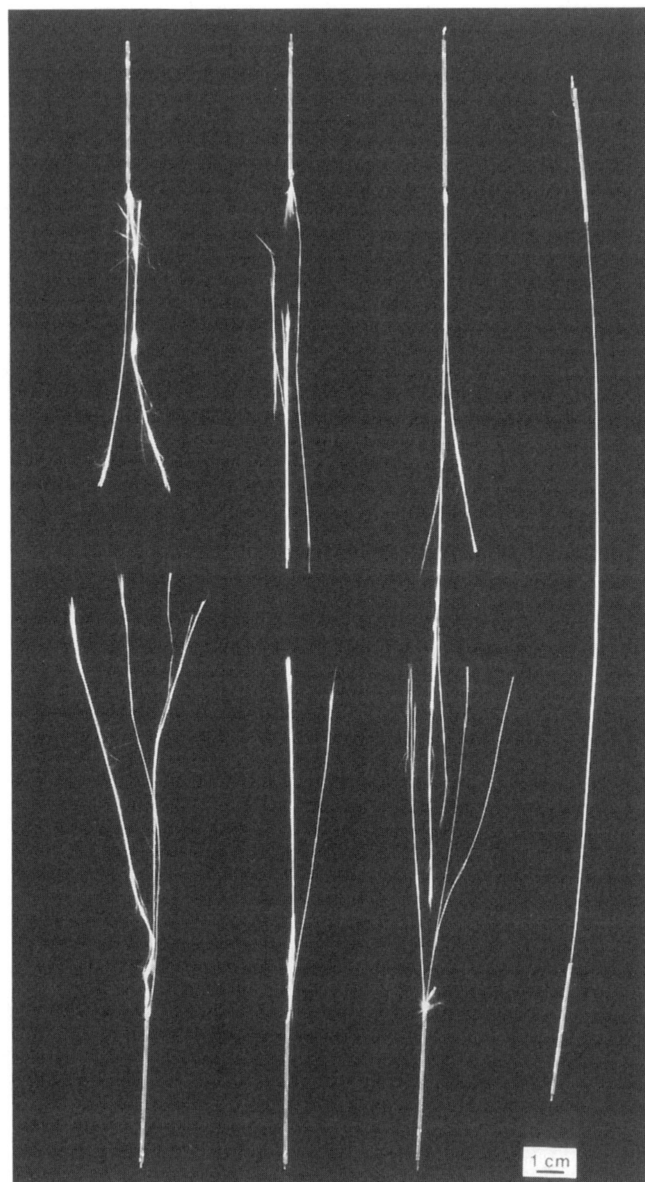

Fig. 4 Tensile-test specimens. From left to right: Quartz, S2-glass, E-glass, and untested specimen.

Figure 5 is a series of reflected-light photomicrographs showing the cross-sections of several quartz-reinforced composites at various fiber volume fractions. The number at the top of each frame indicates the number of reinforcement yarns included. The characteristics of these sections are typical of all three types of material.

At lower reinforcement levels, filaments tended to pack together tightly near the perimeter of the profile, leaving a polymer-rich zone in the center. Here the individual yarns are discernible. In some cases one or more yarns were displaced towards the center of the section, and the circular profile was filled out with resin alone (see Fig. 5, X=9). The polymer-rich zone became

195

smaller as more filaments were included. Overall the shapes of the sections are less than ideal, but in the most highly reinforced materials the shape appears nearly circular.

Discussion

Mechanical Properties. As well as considering the experimentally determined mechanical properties of these materials, an efficiency, φ, has been defined as the ratio of an experimental value to a theoretical value. Theoretical values were predicted by the rule of mixtures for fiber reinforced composites according to Eqn. 3.[22]

$$X_c^n = v_1 X_1^n + v_2 X_2^n \qquad (3)$$

where X_c = a mechanical property of the composite,
$\quad X_i$ = a property of constituent material i (i = 1 or 2),
$\quad v_i$ = volume fraction of constituent material i, and
$\quad n$ = exponent for fiber orientation ($-1 \leq n \leq 1$).

More specifically, the isostrain form of the general equation was used ($n=1$), which was to assume a negligible influence of any off-axis fiber orientation due to yarn twist. Efficiencies were computed using the mean values of Tables IV and V. Figures 6a through 6d are plots of φ for each of the mechanical properties examined (marker legend in Fig. 6a applies to all parts of the figure). The horizontal dashed line in each frame indicates the ideal situation, where $\varphi = 1$.

Tensile Properties. Figure 6a shows the efficiency of E_t for the three types of material. The value of φ ranges from about 0.85 to 1.0. All three of the materials approximated ideal behavior. The high efficiency substantiates the efficacy of the gripping motif as well as highlighting the ability of the material to behave cohesively and to effectively utilize the inherent stiffness of the reinforcing fibers. Even at higher levels of reinforcement, the stiffness of the materials was relatively close to the theoretically ideal values. Though measuring the elastic modulus in axial loading is probably the least

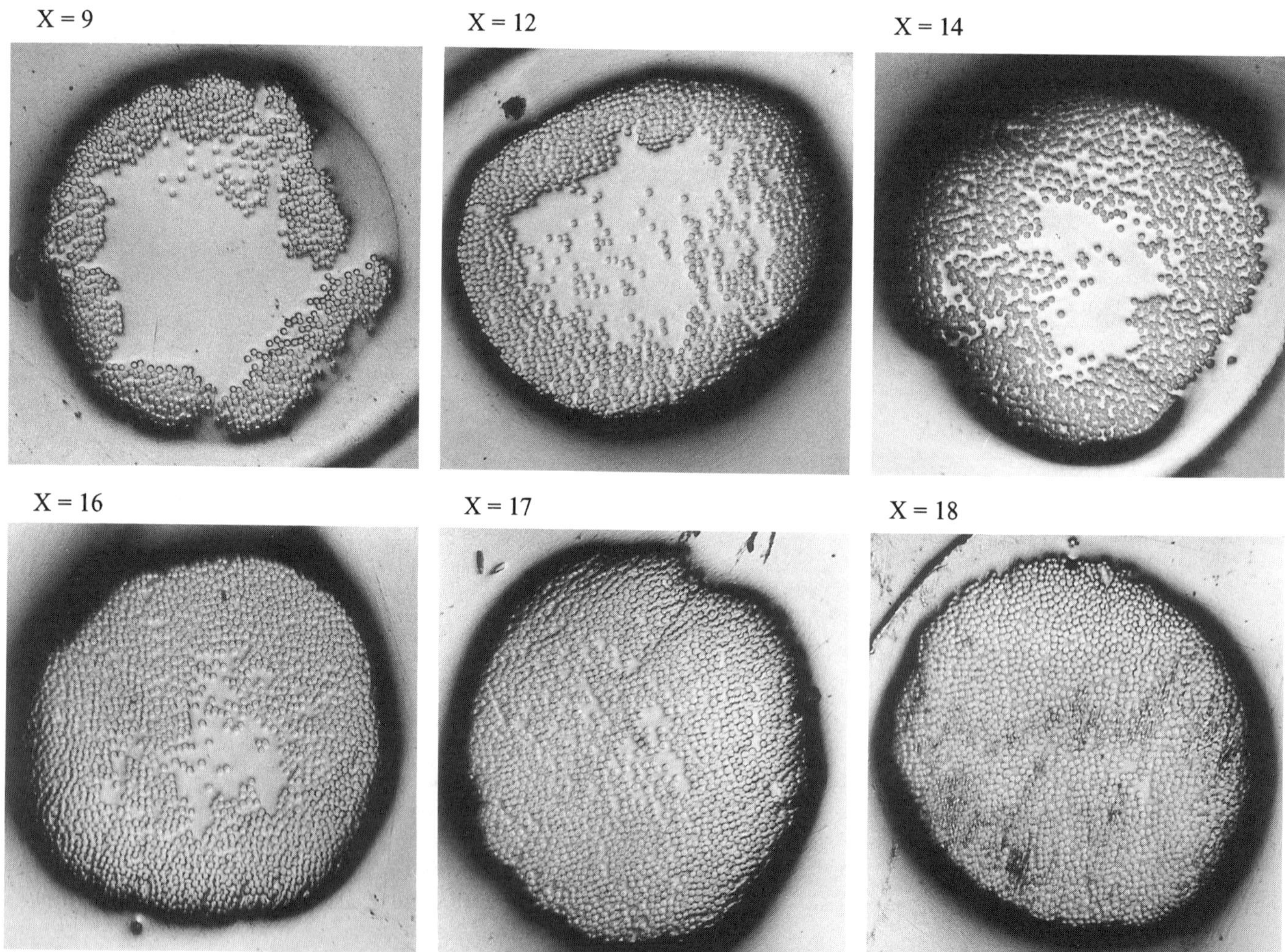

Fig. 5 - Cross-sectional morphology of quartz-reinforced materials at various levels of reinforcement. The number in the upper left-hand corner of each frame indicates the number of reinforcement yarns included in that specimen. Nominal section diameter is 0.5mm.

rigorous test for a UFRP composite, the matrix material and the interfacial bonds must still be capable of transferring load from the perimeter of the profile (where the grips are applied) to the rest of the reinforcing material throughout the center of the mass. Thus, the test is indicative of bond strength.

The tendency for φ to increase at low reinforcement levels is probably related to the morphological characteristics of Figure 5. Because the reinforcement materials tend to remain tightly packed around the perimeter of the profile, the average radial distance between adjacent fibers remains relatively small and constant rather than varying with the inverse of fiber content (as would be the case for a reinforcement phase that was more homogeneously distributed). The matrix material filling those spaces must transfer the load between adjacent fibers. Keeping the average distance small and constant assists in achieving the isostrain situation, which was assumed for the theoretically ideal properties used in calculating φ.

The lines of Figure 3a illustrate that the stiffness of these composites in tension is highly predictable. This is further supported by the low s.d. of the means at each level of reinforcement, and the high probabilities obtained from the regression analyses.

As with E_t, the UTS of these composites varies linearly with fiber composition (Fig. 3b). Again, the low variability among samples together with the high probability of the regression lines mean that the materials will behave in a predictable manner with respect to UTS. The superior strength of the S2-glass materials was a surprise, considering that the purported strength of the quartz filaments is over 30% higher than that for the S2-glass filaments.

In addition to being stronger in absolute terms, the efficiency of the UTS for the S2-glass composites was also superior (Fig. 6b), with φ ranging between 0.73 and 0.93. A high efficiency in tension is indicative of a material with relatively few flaws. Taken together with the fact that no special handling of the fiber yarns was attempted, the high value of φ suggests that the novel processing technique is relatively harmless to the vulnerable fibers. Efficiencies for the other two materials were similar to one another, but only about 2/3 as high as for the S2-glass. Again, the sense of the slope of these lines is probably due to the morphology of the fiber packing, and the associated variation in the radial distances over which load must be transferred.

Unlike metals, which are limited to service below their yield strength, the almost purely elastic nature of

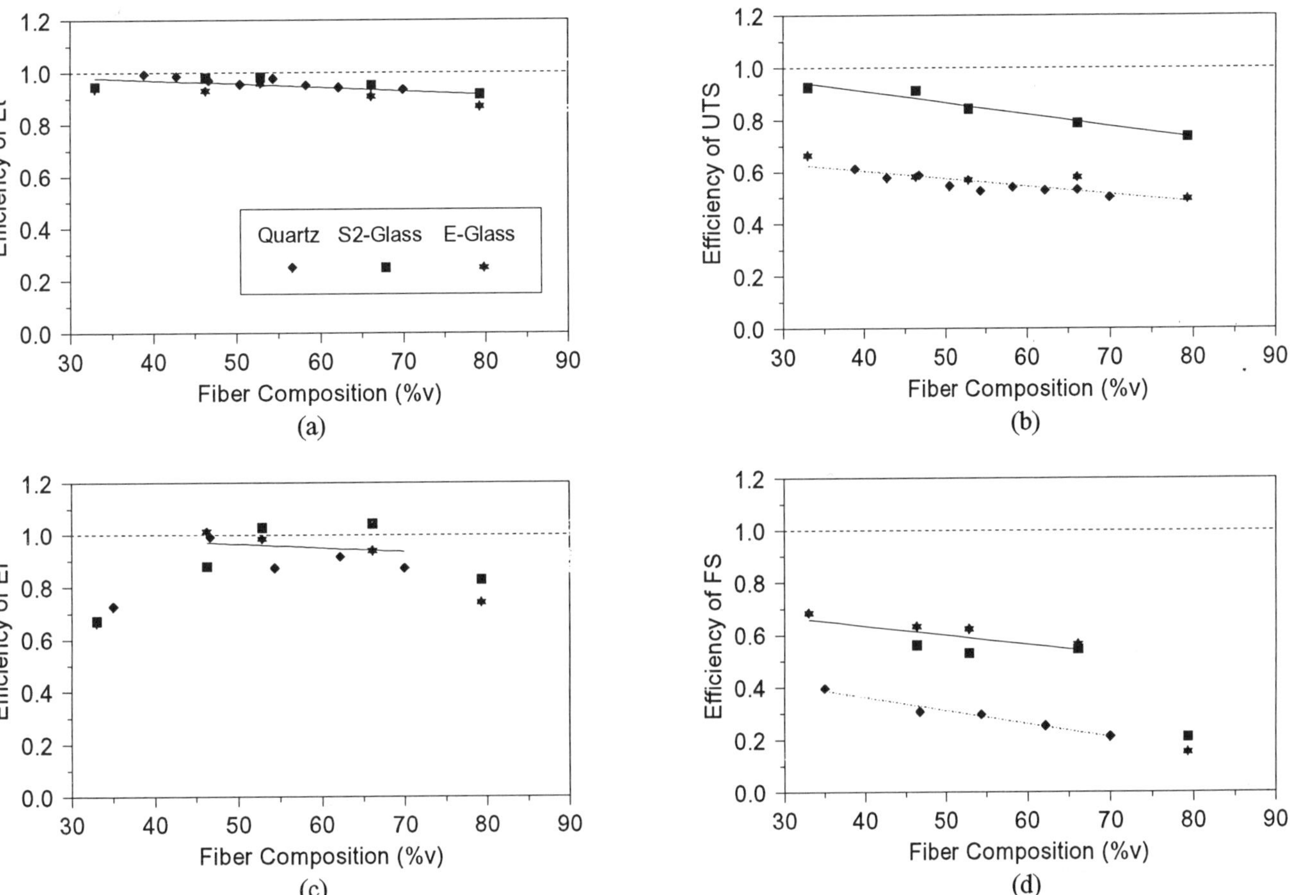

Fig. 6 - Mean values for efficiencies of mechanical properties: (a) Et; (b) UTS; (c) Ef; (d) FS. Marker legend in (a) applies to all.

these materials allows them to be utilized up to the point of failure. This means that a high UTS translates into a high load carrying capacity under normal working conditions. The utility of this feature is further emphasized by the fact that these high strengths could be realized with such a simple gripping technique.

Flexure Properties. Up to a reinforcement level of about 70%v, the flexure behavior of these materials was proportional to fiber content (Figs. 3c & 3d). The most highly reinforced materials at 79%v (glass only) failed to maintain the proportionality in both E_f and FS. This is also reflected in the efficiencies of Figures 6c and 6d. Microscopic investigation of the polished cross-sections of the 79%v reinforced materials revealed the presence of tiny voids dispersed throughout the matrix. These voids can be seen as black features in the photomicrograph of Figure 7. The voids are believed to be the result of poor wetting. This situation is currently being addressed through proposed improvements in the processing equipment. Based on the data regression lines of Figures 3c and 3d, the E_f and the FS may potentially be increased above 68 GPa and 2.0 GPa, respectively.

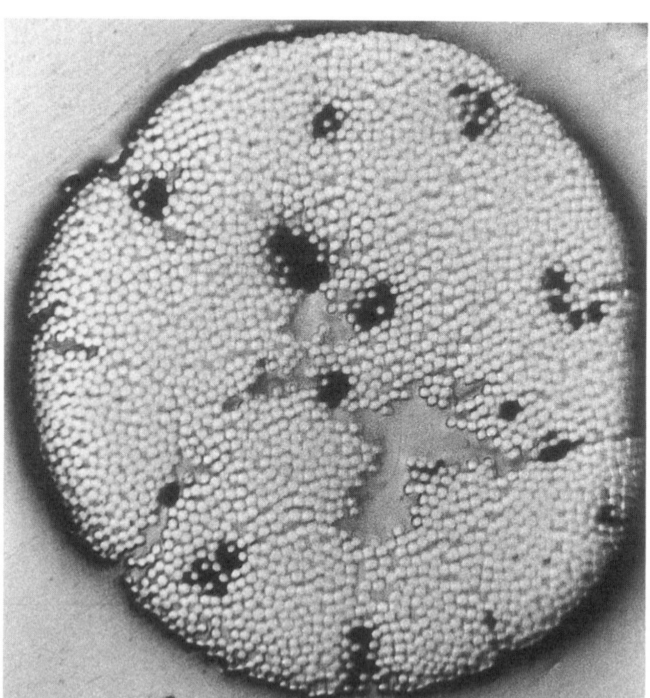

Fig. 7 - Cross-section of a composite reinforced with 79%v S2-glass. Note the black voids in the matrix. Nominal section diameter is 0.5mm.

Figure 6c also reflects a drop in efficiency for the E_f of materials with fiber levels below about 45%v. One possible cause of this is measurement errors resulting from imperfections in the shapes of the less-reinforced profiles. With fewer yarns packed-in, the profiles are more susceptible to distortion. The moment of inertia term that was used to compute E_f and FS (Eqns. 1&2) is proportional to the fourth power of the diameter. By comparison, the cross-sectional area that was used to compute the tensile properties is proportional to just the square of the diameter, which was assumed to be nominal. As a result, any errors in the tensile data associated with the assumption of a nominal diameter are much smaller than the errors in the flexure data, because of the exponential amplification. Another potential cause of the drop in efficiency is the occasional asymmetric distribution of the reinforcement yarns (see Fig. 5, X=9). Like the imperfections in shape, the flawed distribution is more prevalent in the profiles with less reinforcement. Most probably, these two factors are also responsible for the overall increase in the s.d. that was observed for the flexure data compared to the tensile data.

Fiber push-out tests and transverse flexure strengths are currently being used to objectively evaluate interfacial bond strengths for UFRPs of continuous aligned fibers.[23,24] Nevertheless, a simple longitudinal bending test like the one used here can reveal useful information about the bonds between the phases. A high efficiency in E_f indicates a positive transfer of load between adjacent filaments of reinforcement. The E_f of Figure 3c and the φ of Figure 6c clearly illustrate that interfacial bonds of high strength can be obtained with these materials.

Three-point bending is not pure bending– it also includes transverse and longitudinal shear stresses, and localized compressive-stress concentrations at the points of load application.[25] The testing apparatus was originally designed to measure only the stiffness of metallic specimens not the flexure strengths.[20] The striker used to impose deflection was a small cylinder measuring 0.75mm in diameter. Due to the small contact area of this cylinder, the localized compressive loads translate into high compressive stresses. All flexure specimens ultimately failed in a compressive buckling mode at the point where the bending load was applied. The simple flexure formula of Eqn. 2 does not take into account the added effects of transverse shear and localized stress.[25] In effect, the bending apparatus used here and the simplified flexure equations result in a conservative estimate of the true FS. A less rigorous testing method that minimized non-bending loads would likely cause an upward shift in the lines of Figures 3d and 6d.

Morphology. In general, the imperfect shape of these sections is undesirable. However, the manufacturing hardware used for this work is a first-generation experimental laboratory set-up, and better results can be expected with further improvements to the equipment. The natural tendency for the reinforcement to position itself around the perimeter of the profile is advantageous in terms of the flexure properties of the less-reinforced materials. This type of structure is routinely utilized in larger UFRP profiles to maximize specific FS and specific E_f, *i.e.*, flexure properties normalized by the material density. Accordingly, the present individual profiles could be useful as stand-alone elements in small-scale flexure applications where weight is a consideration, e.g., low-mass cables and springs for auto/aerospace or high-frequency mechanical mechanisms.

The explosive nature of the tensile failures in these materials (Fig. 4) indicates that the entire specimen was near the point of failure when rupture occurred. This is in contrast to a material that fails at a single weak location, lowering the capability of the entire specimen (as does the proverbial "weakest link in the chain"). The global nature of the failures may also be related to the propagation of shock waves that result from the release of stored elastic energy.

Comparison to other data. The E_f and FS of the materials produced here compare favorably with the properties of other glass UFRPs taken from the literature.[12,26] Table VI contrasts some of these properties for a collection of different materials at a similar level of reinforcement. The flexure properties of the miniature profiles typically match or exceed what has been reported by others. The tensile properties are clearly higher–especially the UTS, which for the glass-reinforced materials was roughly double that of the other materials.

Table VI. Comparison of this study to other glass-UFRP composites.

Matrix Material	Fiber Content (%wt)	Et (GPa)	UTS (MPa)	Ef (GPa)	FS (MPa)	Reference Study
Poly-urethane	75	—	890	—	72	26
Nylon 6	75	—	869	—	469	26
Poly-phenylene sulfide	72	—	793	—	965	26
Polyester	75	—	828	—	828	26
Phenolic	73	—	448	—	483	26
Acrylo-nitrile-butadiene-styrene	75	—	710	—	538	26
Phenolic	76-77	—	—	48.3	1210	12
Polyester	73-74	—	—	49.0-51.0	1100-1200	12
Phenolic	62-67	33.8-35.9	524-607	21.2-24.8	503-607	12
BIS-GMA/ TEGDMA quartz	75	46.6	1970	45.2	940	this study
BIS-GMA/ TEGDMA S2-glass	70	45.5	2050	47.7	1290	this study
BIS-GMA/ TEGDMA E-glass	70	37.8	1040	38.8	1140	this study

Pultruded UFRP profiles are not routinely manufactured in such small sections. In fact, these authors are only aware of one other published report of an attempt to do so, and of reports in related literature of small profiles that were produced by slitting thin pultruded tapes.[13-17] The most highly reinforced profiles produced by these methods contained less than 60%v fibers, and the stiffness and efficiency of the materials were less than what has been reported here, e.g., a 56%v (72.5%wt) S2-glass-reinforced polyethylene-terepthalate-glycol composite had an E_f that varied between 14 GPa and 45 GPa ($0.29 \leq \varphi \leq 0.92$) depending on the span to depth ratio of the flexure test. The higher value was obtained only at a ratio of about 60:1. At a ratio of 18:1 (17.5:1 was used for this study) the E_f was only about 27 Gpa, which represents a φ of 0.55.

Potential uses as "pre-composites." Small pultruded profiles could be useful as basic building blocks for the fabrication of larger composites. The successful use of a common quick-setting epoxy to form the grips of the tensile specimens illustrates the capability of the finished composite to form strong bonds with other polymers. Large curved pieces could be formed from the "lay-up" of these smaller profiles. A thin-section material can conform to a moderate curve with little or no induced stress. The handling of a pre-composite could be much simpler than handling a prepreg fabric, a plain yarn, or a roving. Woven forms of pre-composite might wet-out easier in processes like resin transfer molding, allowing larger pieces to be made with a broader range of resins.

Rectangular or square profiles would pack efficiently and easily into the die of a secondary pultrusion process, allowing the high reinforcement levels to be passed-on to the larger piece. Starting with a higher proportion of pre-polymerized matrix material could speed processing as well as reduce shrinkage in situations where high dimensional tolerance is critical. Control of variably-reinforced distributions for custom-engineered sections might be easier and more reliable with pre-composites. Starting with predetermined reinforcement levels could lead to more uniform distributions and smoother transitions between adjacent zones. Glass filaments on the interior of a pre-composite profile would be protected from damage, which could potentially speed-up processing by eliminating careful handling procedures.

Conclusions

Small profiles on the order of 0.5mm in diameter can be pultruded with fiber reinforcement levels as high as 79%v. Composites of this size can be made from a commercially available S2-glass yarn and an acrylic-based matrix, with mechanical properties that equal or exceed those of larger, more conventional profiles made from other polymers. The mechanical properties of these profiles are reproducible and predictable via strong correlations with fiber content. The finished composite can be successfully bonded with quick-setting epoxy, allowing the good properties of the material to be easily utilized. Such materials have great potential to be useful as pre-composites in the formation of larger, more-complex structural shapes.

Acknowledgment

Special thanks is extended to the Johnson & Johnson Focused Giving Program, to the North Carolina Dental Foundation, and to the University of North Carolina Department of Orthodontics for support during this period.

References

1. Bowen,R.L., J. Dent.Res., 37:1, 90 (1958).
2. Bowen,R.L., and M.S.Rodriguez, J.Am.Dent.Assoc., 64, 378-87 (1962).
3. Bowen,R.L., J.Am.Dent.Assoc., 66, 57-64 (1963).
4. Peterson,E.A., J.Am.Dent.Assoc., 73, 1324-36 (1966).
5. Macchi,R.L., and R.G.Craig, J.Am.Dent.Assoc., 78, 328-34 (1969).
6. Dennison,J.B., and R.G.Craig, J.Am.Dent.Assoc., 85, 101-8 (1972).
7. Nuckles,D.B., and C.C.Cosby,Jr., J.Dent.Res., 51:5, 1512 (1972).
8. Powers,J.M., L.J.Allen, and R.G.Craig, J.Am.Dent. Assoc., 89, 1118-22 (1974).
9. Raptis,C.N., P.L.Fan, and J.M.Powers, J.Am.Dent. Assoc., 99, 631-2 (1979).
10. Martin,J.D., Plast.Eng., 53-7 March(1979).
11. Gibson,A.G., C.Y.Lo, D.W.Lamb, and J.A.Quinn, Plast.Rub.Proc.Appl., 12:4, 191-7 (1989).
12. Dailey,Jr.,T.H., R.W.Allison, and M.W.Klett, in Proceedings 48th Annual Conf., Composites Institute, The Society of the Plastics Industry, Inc., Feb. 8-11, 1993, Session 21-B, p.1-6.
13. Jancar,J., and A.T.DiBenedetto, J.Mater.Sci.:Mater. Med., 4, 555-61 (1993).
14. Goldberg,A.J., and C.J.Burstone, Dent.Mater., 8, 197-202 (1992).
15. Patel,A.P., A.J.Goldberg, and C.J.Burstone, J.Appl. Biomat., 3, 177-82 (1992).
16. Goldberg,A.J., C.J.Burstone, I.Hadjinikolaou, and J.Jancar, J.Biomed.Mater.Res., 28, 167-73 (1994).
17. Jancar,J., A.T.Dibenedetto, Y.Hadziinikolau, A.J.Goldberg, and A.Dianselmo, J.Mater. Sci.:Mater. Med., 5, 214-8 (1994).
18. Craig,R.G., Dent.Clin.N.Am., 25:2, 219-39 (1981).
19. R.G.Craig, ed., "Restorative Dental Materials 6th ed.," p.398-9, The C.V.Mosby Co., St. Louis, Missouri (1980).
20. Kusy,R.P., and A.M.Stush, Dent.Mater., 3:4, 207-17 (1987).
21. E.P.Popov, "Introduction to Mechanics of Solids," p.183,397, Prentice-Hall Inc., Englewood Cliffs, New Jersey (1968).
22. J.F.Shackelford, "Introduction to Materials Science for Engineers 2nd ed.," p.473, Macmillan Publishing Co., New York, New York (1988).
23. Eldridge,J.I., in NASA Tech Briefs, Lew-15297 (1989).
24. Bowles,K.J., in NASA Tech Briefs, Lew-15731 (1993).
25. A.Higdon, E.H.Ohlsen, W.B.Stiles, J.A.Weese, and W.F.Riley, "Mechanics of Materials 3rd ed.," p.239-316, John Wiley and Sons, Inc., New York, New York (1976).
26. Chen,C., and C.M.Ma, J.Appl.Polym.Sci., 46, 937-47 (1992).

Improvements in the Processing and Performance
of High Temperature Laminate Components

J. Harper
Balvenie Technologies
Long Beach, California

J. Leslie
Advanced Composite Production & Technology
Huntington Beach, California

ABSTRACT

Recent changes in regulations concerning raw material usage in epoxy processing, specifically restrictions on the use of aromatic diamines such as MDA, methylene dianiline, have caused companies such as Advanced Composites Products and Technology, Inc, ACPT, to seek alternate high temperature epoxy resin systems for polymer matrix composites. One candidate system (patent pending), developed by Balvenie Technologies, is based on multi-functional epoxies, liquid and solid anhydrides, and further modified with crosslinking rubber polymers. This toughened epoxy exhibits a Tg of 266°C, features room temperature processing, and offers an amine-free work place. This preliminary study compares this new epoxy system with traditional aerospace epoxies and current bismaleimide systems.

Introduction

Resistance to impact damage while maintaining good physical properties is a common definition of toughened PMCs, polymer matrix composites. Another way to describe toughness might be to say the matrix resin is hard, yet not brittle. It is within these parameters that a new RTE (Rubber Toughened Epoxy) was formulated: a measured quantity of a very specific rubber co-cured within the hard epoxy resin phase. Because this rubber, maleinized polybutadiene or MAPBD, shown in Figure 1, crosslinks to the epoxy itself, the rubber portion or phase is chemically bonded to the epoxy and reinforces the physical strength of the cured epoxy.

Additionally, it is proposed that the vinyl functionality on the liquid rubber cyclizes, forming a highly crosslinked network or second

phase within the continuous epoxy matrix. Liquid anhydrides are used as the main hardener, by volume almost equal to the resin volume, and

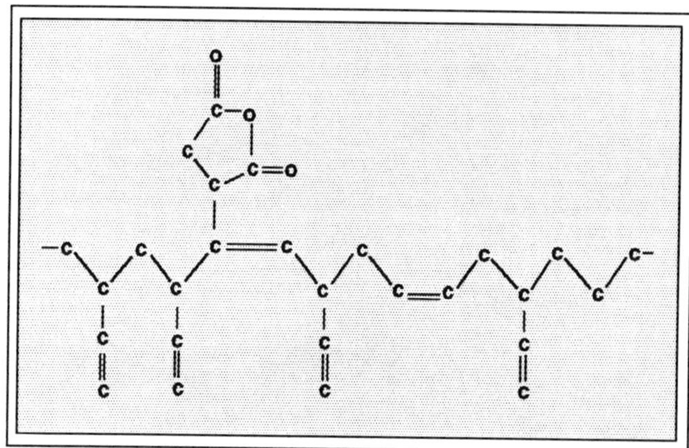

Figure 1: Maleinized 1,2 high-vinyl Polybutadiene

the complete two-component system is a low viscosity mix at room temperature. Our initial studies indicate it is necessary to crosslink the rubber with the epoxy first, at relatively low temperatures, before the epoxy-anhydride reaction is completed. This necessitates a two step cure, the last step to crosslink the vinyl moiety, not uncommon for any high temperature resin system.

Raw Materials

There are any number of raw materials that can be used to formulate the epoxy side as well as the anhydride side. One must first determine which physical and chemical properties are needed, and then formulate to obtain these properties. Standard resins are di-functional which develop long chain polymers. Multi-functional resins are used to produce higher performance epoxies through additional crosslinking of the polymer chains. Epoxy novolac resins have traditionally provided high temperature properties but they also contribute high viscosity and subsequent handling problems at room temperature. For this study, a low viscosity multi-functional resin as one of the components in the epoxy side was chosen specifically to maintain good working viscosity

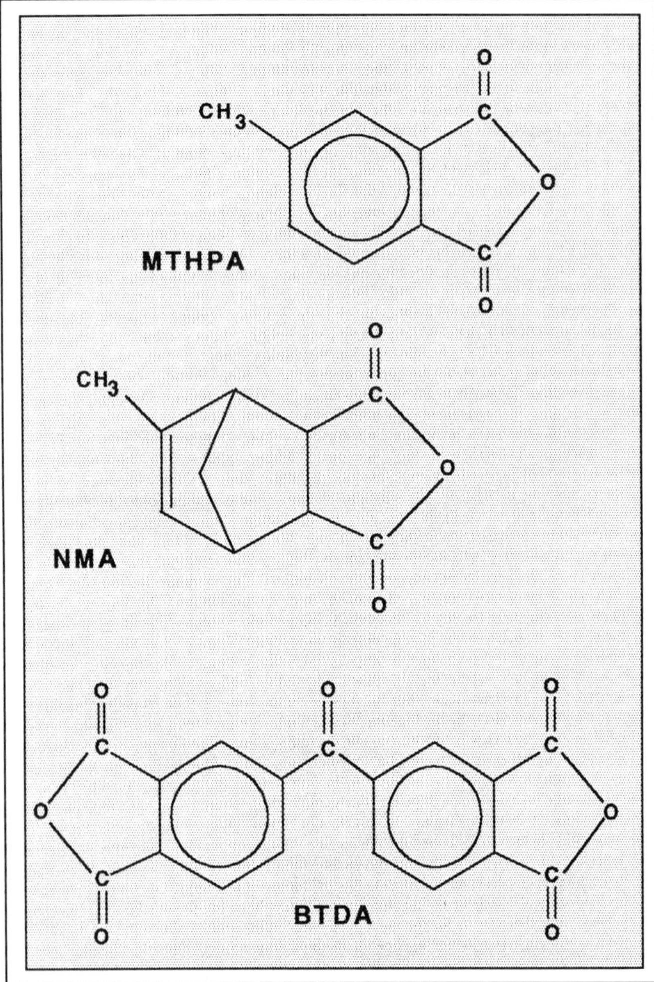

Figure 2: Anhydride Structures

at room temperature while developing a highly crosslinked epoxy network.

Anhydrides, as shown in Figure 2, are similarly described, MTHPA and NMA being mono-functional and BTDA being di-functional. The former two are very low viscosity materials, almost like reactive diluents, but actually very effective hardeners; the latter a high melting solid. Di-functional anhydrides have been used for years, producing high Tg products, primarily in a polyimide family known as PMR-15. BTDA is mostly used in epoxy powder coatings, but has proven to be a very effective crosslinker in this study.

Some basic RTE systems were introduced about two years ago(1), using MTHPA as the anhydride. Variations in the curing mechanisms created three distinct products, one for high speed RTM processing, one for long pot-life filament winding systems, and a third for intermediate time-temperature processing by either method. The Tg of these systems are all in the range of 125-135°C (255-275°F). A typical cure cycle is 30 minutes at 120°C (250°F), actually quite fast for anhydride cures.

When we sought higher temperature properties we utilized NMA for the liquid anhydride as it is well known for its improved high temperature properties, although it also cures very slowly. And by using the multi-functional resins we were able to produce Tgs in the 200-250°C (390-480°F) range. By adding BTDA, a di-functional anhydride, further advances in heat resistance were achieved, over 260°C (500°F). A system, still quite fluid, based on this combination was filament wound at ACPT in Huntington Beach, CA and evaluated for Tg and short beam shear properties.

Commercial High Temperature Systems

At this time there are a number of competing resin systems for the high temperature market, including a popular aerospace system from Hercules, 3501-6, based on MY-721/DDS, a multi-functional epoxy and high temperature crosslinking agent in powdered form, and the common bismaleimides (BMI), another solid system, that both require heat and a complicated fabricating process to use. Hercules 3501-6 is commercially available as a pre-preg, but under special arrangements, the neat resin system is available to filament winders. The Tg of this system is 210°C (410°F), and recently Hercules introduced a new system, 4502, with a higher Tg, 285°C (545°F) which is available only as a pre-preg. Recommended cure cycles are complicated, requiring 3 to 4 steps in temperature and pressure in an autoclave.

Another high temperature system is one containing a bismaleimide monomer, Figure 3, and crosslinking agent, usually MDA, but there are numerous variations available. Because of the high melting point of such systems, filament winding is exceedingly difficult without precise control over the process. It is necessary to maintain a temperature about 90°C (200°F) all the way from the resin bath through the fiber path to the heated mandrel, a complicated procedure at best. In spite of this difficult processing, it has become necessary to use a BMI to get to the higher service temperatures demanded by today's market.

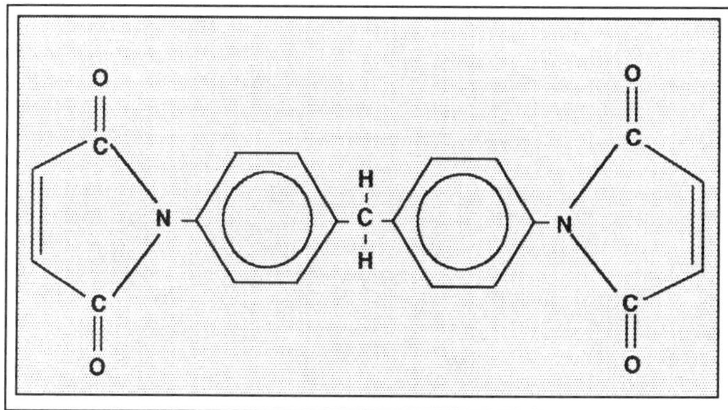

Figure 3: Bismaleimide of MDA

Using expensive pre-pregs is a viable alternative to using these solid systems that require heat to fabricate, as it will allow the fabricator to produce high temperature laminates under controlled conditions with considerably more time to work. This is not cost-effective, however, as labor and waste add to the final cost.

An additional problem with these and other established high temperature systems, both epoxy and BMI, is that certain raw materials based on MDA and now other aromatic diamines are restricted in use by certain OSHA regulations, and substitutes or alternative materials generally produce lower properties.

One such system that ACPT has used extensively in the past, originally obtained from Applied Plastics and more recently from Hexcel, Epolite 2447/2323, has been taken off the market. This particular system exhibited a HDT of 175°C (350°F) with short term use at 200°C (390°F), and was easily processed as a filament winding system at room temperature.

It was initially the purpose of this study to find a replacement system for the Hexcel product, one with properties as good as the old one, both in the handling of the wet resins and in the finished cured PMC. Developing a new replacement system with properties approaching those of the benchmark aerospace epoxy system and/or a BMI system would be considered a bonus, especially so if the system was liquid at room temperature with a long pot life, and contained no aromatic diamines. Additionally, this new system must be competitive in cost.

Costs for various systems must be balanced with the properties derived from each system. Typical aerospace systems will sell in the $5-25/lb range, BMIs approaching $20/lb, and the new cyanate esters in the $30/lb range. Of course a standard di-functional epoxy resin cured with NMA will price out about $3.00/lb, but it represents the lower range of thermal properties, Tgs around 150-175°C (300-350°F).

It is expected that these RTE systems will bridge the gap between the lower and higher cost systems, with physical properties leaning toward the higher performance side. Such systems might range in cost from about $4/lb to $12, depending upon the required physical properties.

Experimental Liquid Epoxy High-Temperature System

This paper introduces a new high temperature system incorporating multifunctional resins and NMA as well as BTDA. While the hardener side is a paste, once it is blended into the resin the mixed system becomes quite thin and manageable, about 1000 cps. It exhibits about a three hour pot life. At ACPT the filament winding of test specimens, NOL rings, proceeded quickly and efficiently at room temperature. It was generally considered at ACPT that the preparation and winding of this new high temperature system was as simple as standard epoxy systems, and nothing like the complicated BMI or aerospace epoxy are to process.

These specimens were gelled at approximately 95°C (200°F) for one hour, then immediately cured at 175°C (350°F) for 3 hours. The next day they were post-cured at 250°C (480°F) for an additional 4 hours. One sample, however, was gelled at 95°C (200°F) and removed from the oven, and a second was allowed to simply gel overnight at room temperature. Both were cured and post-cured the next day as per the above schedule. The study of these high temperature epoxies, anhydride based, cured at room or low temperature, is scheduled for future work.

The Tg was determined to be 266°C (510°F) via the TMA method at Orange County Materials Testing Labs. At Allco Chemical, the supplier of BTDA, the TGA of this system was determined to be 330°C (600°F). Their DSC could not identify a Tg, but weight loss measurements after two hours of heating in an air oven were as follows: 250°C (500°F) 1.5%, 275°C (550°F) 3%, and 300°C (600°F) 7%.

Samples were prepared for short beam testing, with the results on the following page, Table 1 and Figure 4. At room temperature, the RTE compared very well against the BMI. When compared against a 9000 Series epoxy from Shell Chemical, it was significantly better.

It should be noted here that these systems containing rubber exhibiting high temperature stability do not act as conventional epoxies in many of these analytical devices. Determining the Tg via TMA proved more satisfactory than with a DSC. Sufficiently crosslinking the rubber is necessary to obtain high Tgs, and

	Tg	Short Beam Shear, room temp
Hercules 3506-1	210°C	na
Compimide® 65FWR	280°C	8001 psi
Tufpoxy® 7105	266°C	7770 psi

Table 1: Comparing Three Resin Systems for SBS at RT and Tg

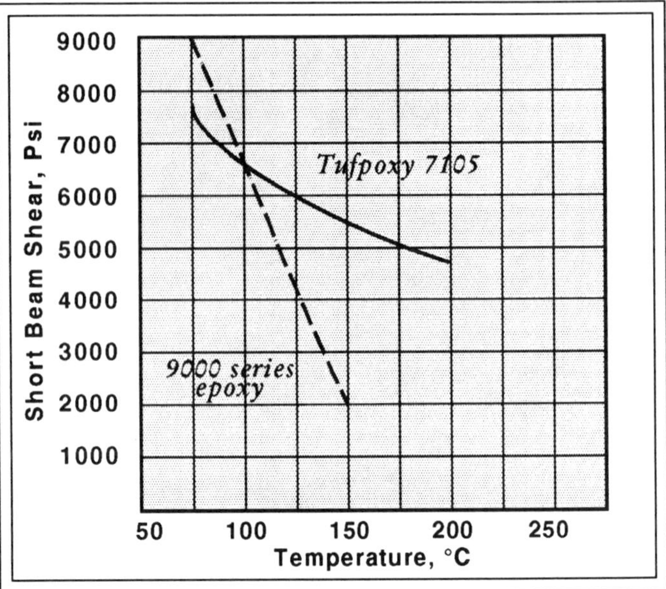

Figure 4: Comparing Two Epoxy Systems for Short Beam Shear at elevated temperatures

proper attention to the curing cycle is vital. We have found that even after 8 hours of curing/post-curing, an additional post-cure was necessary on two occasions.

Actually, a DSC study following the reactivity of the raw materials by Thermal Options of Escondido, CA shows that the entire cure can be completed, a full 100%, at 100°C (212°F) in 8 minutes. However, this does not take into consideration the necessity to develop the rubber/epoxy phase fully before the epoxy/anhydride cure is completed. The cure cycle described in this study is used as our standard cure cycle for all our high temperature work, and further laboratory work is necessary to identify the most efficient cure cycle. For example, it is vital to initiate the gel at 80°C (176°F) or lower, even at room temperature, to develop the rubber phase, before the temperature is raised to complete the epoxy cure and the vinyl crosslinking. What we have yet to learn is the precise nature of this second step and its time /temperature relationship.

For example, the specimen for this test was post-cured at Orange County Materials Testing Lab for an additional 2 hours at 230°C (450°F)

Table 2: Processing Comparison

	Hercules 3506-1	Compimide® 65FWR	Tufpoxy® 7105
Processability	as prepreg	very difficult, must keep hot	simple, at room tempeature
temperature of resin bath	na	200°F	room temperature
Ease of cure	4-5 step temp, several hours	3 step cure several hours	2-3 step temp, several hours

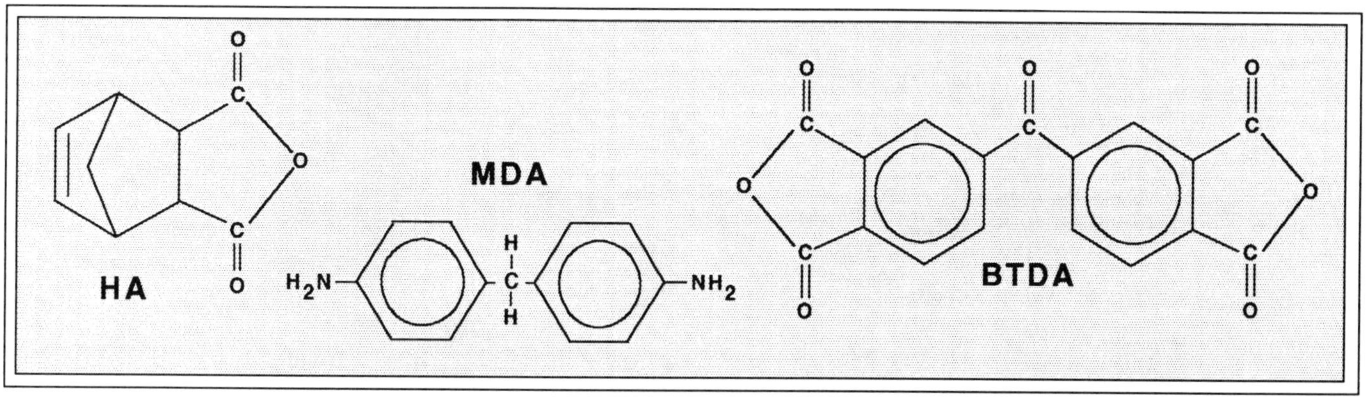

Figure 5: Basic Raw Materials for PMR-15

after 8 hours at lower temperatures in the lab. Prior to this evaluation, a circuit board manufacturer evaluated a RTE system containing no BTDA, just the NMA and MAPBD. Samples submitted were cured by our standard cure cycle, and post-cured an additional 2 hours at 220°C (424°F), yielding a Tg of 232°C (450°F). This is not to say that our cure cycle is necessarily lengthy, but rather that further work on catalyst levels, cure time and temperature, is required.

Summary

This paper introduces a new concept in high temperature polymer matrix composites, PMCs, one based on a highly crosslinked epoxy/anhydride/rubber system, in which the rubber chemically bonds to the epoxy phase, yet establishes its own crosslinked rubber phase. It features low viscosity liquid processing at room temperature, with a relatively simple cure cycle.

Raw materials used to produce PMR-15, the industry standard polyimide, are very similar to this RTE system. The mono- and di- anhydrides are the same with the exception of the methyl group on the CPDE based anhydride. MAPBD is a polymeric anhydride. PMR-15 uses an aromatic diamine to make the imide group, and this system uses a polyfunctional epoxy, in both cases to build up the polymer structure and crosslink density of the system. It is conceivable that the NMA crosslinks through the double bond in this new epoxy system as it does in the PMR-15 system: studies continue in this area.

It is also possible that room temperature gels with one elevated cure might be all that is necessary, useful for any composite application, as long as the initial impregnation provides a proper resin/fiber ratio. It also appears there is some flexibility in the curing process as a whole, which should make this new system adaptable to numerous production methods: RTM and pultrusion for fast production rates and filament winding for long open times with snap cures.

Further testing to identify a preferred cure cycle is planned, as is continuing physical testing. There exists a large gap in performance between traditional epoxies and bismaleimides. It is likely that RTE composites as described herein will find a ready market where physical properties are especially important..It is also expected that there is a significant market for commercial composites in ground transportation, tooling, and possibly in more traditional hand lay-ups products, if the room temperature cure method generates the physicals necessary, for this new cost effective resin system.

Acknowledgements

1. Monika Tomasik, ACPT, Inc. for supervisng plant trials of experimental resin.

References

1. J. Harper and H. Sessions, "Fast Curing Toughened Epoxies for RTM Processing", Proceedings of the 8th Advanced Composites Conference, (1992)

Proceedings of the 10th Annual ASM/ESD Advanced Composites Conference, Dearborn, Michigan, USA, 7-10 November 1994

A Water Based Manufacturing Method for Two Phase Matrix Composite Materials

J. Fernandes, L.T. Drzal
Michigan State University
East Lansing, Michigan

Abstract

A novel processing technique for the manufacture of continuous fibers with a two phase matrix from a well-mixed aqueous colloidal dispersion bath was developed. This process not only offers a method for preparing prepreg tape wherein each individual fiber is completely coated but also can produce coatings of multiple phases of matrix material in order to provide independent control of the microstructure and morphology of the prepreg and composite. In the results reported here, carbon fibers were spread using acoustic energy and then coated with the two particles, i.e. an epoxy resin and an elastomer, by passing through a bath of the mixture. The prepreg tapes were then consolidated into composite having different ratios of epoxy resin particles to elastomeric particles. The parts were subjected to volume fraction/void analysis to ensure no voids. SEM micrography was carried out on the fractured surfaces of the different parts. These photographs revealed a uniform dispersion of rubber particles in the epoxy matrix throughout the composite material. In this case, the two phase matrix morphology was achieved independently of the matrix thermodynamics. Mechanical testing of the consolidated composite parts also indicated presence of two phases.

SEVERAL PREPREGGING PROCESSES using powders are currently available. All of these processes involve the coating or deposition of a single type of particle from a medium (e.g. air, solvent, gel, etc.) onto the fiber surface. Significant advantages can be gained in composite properties if two phase systems can be simultaneously produced as matrices in combination with the fibers. Elastomer or thermoplastic particles are often used as the toughening phase in neat resins. When these two phase systems are used as matrices, either the particles size is too large for the inter-fiber spacing or the degree of dispersion is altered due to the presence of the fibers. Other processes available for manufacturing of two phase matrix composites, involve double pass impregnation or processing through a melt phase where the particles morphology is controlled by the dissolution and precipitation of the elastomeric phase.

It could be very cost-effective and environmentally beneficial to design a process which involves single pass impregnation, whereby the fiber tow or a fiber preform is coated by passing it through a mixture of the two colloidal dispersions and yet would result in a high degree of dispersion of particles in the continuous matrix, which is a desirable morphology. Such a process would offer complete control on the final coating of resin on the fiber surface and at the same time provide a tape of uniform dimensions and a high degree of drapability.

Process Description

In conceiving the proposed process, whereby aqueous colloidal dispersions of two materials from a mixture are used as the coating medium, all the currently available prepregging techniques like slurry processing developed by O'Connor[1] and Dyksterhouse et al.[2], aqueous foam processing developed by Chary et al.[3], and dry powder processing developed by Iyer et al.[4] were evaluated and the salient features of each process were adapted into the final process. A schematic representation of the process is given in Figure 1.

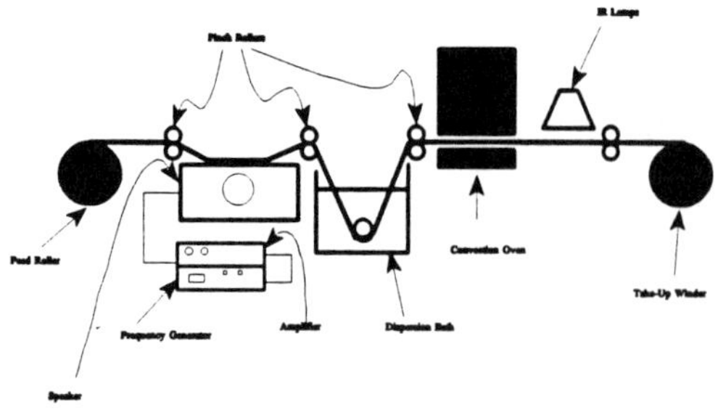

Figure 1 Schematic of Aqueous Colloidal Dispersion Prepregging Process

The continuous prepregger consists of stages for fiber motion, fiber spreading, coating and heating. Fiber motion is controlled by two motors and a take-up winder. The first motor unwinds the feed spool. The second motor draws the fibers through the spreader, bath and heating section, by means of pinch rollers, which are interconnected with chain links. In the spreader acoustic energy is used to spread the fibers. Spread fibers pass through the dispersion bath and are coated with the dispersed solid particles. Moisture is removed in the heating section and prepreg tape is finally wound on a spool mounted on a take-up winder.

Sample Preparation

Materials. Materials used in this process were Hercules Magnamite Type IM-7 12k carbon fiber tow, an aqueous based colloidal dispersion of an epoxy resin, Epi-Rez W60-3515 (manufactured by Rhône-Poulenc) and carboxylated styrene butadiene rubber latex (manufactured by Gencorp Polymer Products). The PAN-based carbon fibers are continuous, high performance, intermediate modulus, having 12,000 filament-count tows. These fibers are in an unsized state, so as to facilitate spreading. Epi-Rez W60-3515 is a non-ionic, aqueous dispersion of a bisphenol A epoxy resin with an equivalent weight of 220-260. Curing is done with dicyandiamide and 2-methyl imidazole is used as the dicy accelerator. Carboxylated styrene butadiene (CSBR) is used in controlled concentrations, as the dispersed phase in the epoxy matrix and acts as a toughening agent.

Solution Preparation. Epi-Rez W60-3515 is supplied having approximately 63% by weight of solids. The dispersion does not have any catalyst to promote cross-linking reaction during the consolidation cycle of the prepreg tape. The catalyst and the accelerator have to be added to the dispersion prior to utilization in the prepregging operation. 13 parts by weight of deionized or distilled water is heated to 71-79°C (160-175°F) and 3.5 parts by weight of solid dicyandiamide is added to it slowly. Stirring is done to dissolve the dicy. When solution is complete, add 0.1 parts by weight of 2-methyl imidazole and dissolve. Add the warm dicy solution to 100 parts by weight of epoxy dispersion (as supplied) and agitate slowly at room temperature until well mixed (30-60 minutes). If the dicy solution cools and begins to crystallize before it is added to the dispersion, it should be reheated to 71-79°C (160-175°F). An induction period of at least one hour is required prior to use.

After the induction period the dispersion containing the catalyst has to be diluted with deionized water, such that a coating level of 25-30% by weight of epoxy resin on fibers is obtained. This coating level is generally obtained by mixing 1:1 parts by volume of dispersion in deionized water.

The process was run with different levels of CSBR particles coated onto the fiber surface. These levels corresponded to the concentration of CSBR in epoxy resin dispersion. Therefore, dispersions containing 0%, 5%, 10%, 15% and 20% by weight of CSBR were made for the prepregging process.

Processing. The processing protocol is similar for each of the dispersions. A spool of unsized carbon fiber is located on the feed spool frame. The fiber tow is unwound, threaded through the first guide slot and then passed over a pinch roller. The tow is then alternately passed through the guide shafts on the spreader. From the spreader the tow passes through another pinch roller. It is then passed under the support roller in the dispersion bath, through a set of pinch rollers and through the heating zone. From the heating section the tow passes though a final guide slot and is then taken up by a spool located on the rotating arm of the automatic torque controlled winder. Once the threading of the fiber tow is done through the continuous prepregging unit, the diluted epoxy resin dispersion, with CSBR if present, is added to the dispersion bath. The convection oven is switched on and the temperature is maintained at 190°C. The infra-red lamps are switched on and the temperature at the fiber surface due to these lamps is about 120°C. The feed spool unwinding motor and the main motor are switched on. The setting of the main motor is around 1.75 on the dial, which corresponds to around 2.54 cm/sec (1 inch/sec) in terms of line speed. The speed of the feed spool motor is adjusted manually, so that there is a slight slack between the spool and the first pinch roller. The take-up winder is started and its speed is adjusted manually to match that of the main motor. The torque controller adjusts the speed of the winder so as to maintain a constant torque between the last pinch roller and the winder. Winding of the prepreg tape is done in such a way that the spool is rotating along its axis, whereas the prepreg tape is traversing the entire length of the spool. In this way the geometric configuration of the wound spool is similar to that of the feed spool. The speed of all the rollers are synchronized by inter-linking them with chains.

Once the fibers are set in motion, the spreader is switched on. The underlying principle of the spreader is that when the fibers pass over the vibrating speaker, acoustic energy assist the fibers in separating into individual filaments. Normal operating conditions of the speaker are 38Hz and 11.2-11.6V to spread the fiber tow by 7.6 cm- 10.2 cm (3"-4"). It is also important to maintain zero tension in the fiber tow through the spreader to achieve the desired effect. This is done by using two pinch rollers at either end of the speaker to control fiber tow tension.

Coating of the fibers takes place due to a positive thermodynamic driving force[5], and penetration of fluid into the inherent surface roughness of the fibers[6],[7]. Spreading of the fibers give a larger surface area for coating. Most of the moisture is removed in the convection oven and under the two infra-red lamps. Prepreg tape thus produced is very drapable and tacky and the coating is very even, with a complete wetout of the fiber surfaces. See Figure 2 for SEM of the prepreg tape.

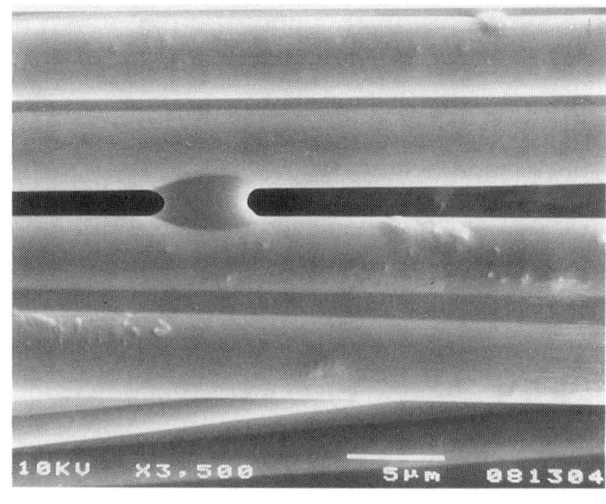

Figure 2 SEM of Prepreg Tape

Prepreg tape is then unwound as a flat multi-laminate part having dimensions of 15.2 cm - 20.3 cm (6"x 8"). A 3 cm- 20.3 cm (1.18" x 8") non-porous teflon taps is inserted at one end of the part, so that it traverses the entire width and acts as the initial crack propagator in end-notch flexure testing. The part is then placed in a vacuum bag and molded in the autoclave under a specific consolidation cycle. Samples are then cut from the finished composite part

for fiber volume and void fraction analysis, DMA analysis, 3-point flexural strength and modulus analysis, short beam shear strength analysis and end-notched flexure (ENF) analysis. Samples were prepared from composite parts containing 0%, 5%, 10%, 15% and 20% by weight CSBR in epoxy matrix.

Results and Discussion

In this study the photomicrographic technique was used to determine the fiber volume fraction and void fraction in the composite samples. Thirty images of the polished surface for each sample were scanned by the digitizer and the images were then analyzed by the Optical Numeric Volume Fraction Analyzer (ONVFA). Results of this analysis are given in Table 1. It can be seen that the void fractions of all the composite parts are well within the prescribed limit of less than ~5%.

Table 1 Volume/Void Fraction Analysis

CSBR Content (%)	Void Fraction (%)	Vol Fraction (%)
0	0.41	65.57
5	0.70	65.48
10	0.64	67.92
15	1.08	63.56
20	0.52	61.57

Three-point flexural testing was done as per ASTM D790-86. Results of the tests for flexural strength and flexural modulus are given in Figures 3 and 4 respectively.

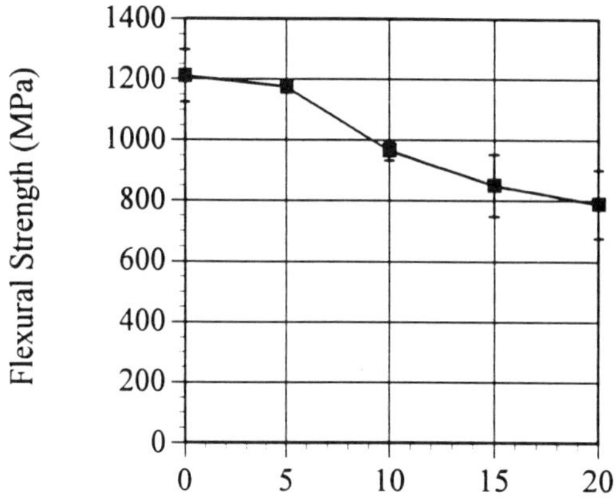

CSBR Content (by weight %)

Figure 3 Flexural Strength of Composite with varying CSBR Concentration

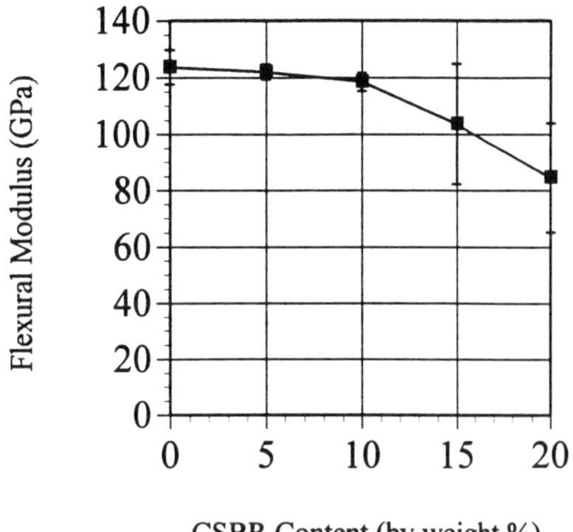

CSBR Content (by weight %)

Figure 4 Flexural **Modulus** of Composite with varying **CSBR** Concentration

Evaluation for short beam shear strength was done as per ASTM D2344-84. Results are as shown in Figure 5.

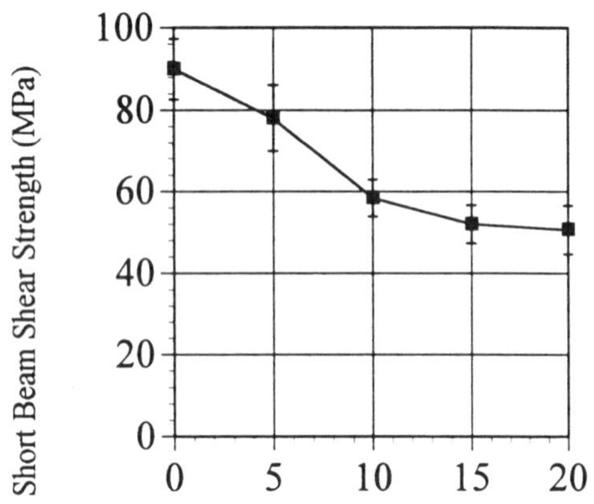

CSBR Content (by weight %)

Figure 5 Short Beam Shear Strength of Composite with varying CSBR Concentration

ENF testing was done as per experimental set-up outlined by Carlsson et al.[8]. Fracture energy, for different composite samples, as calculated by this method are shown in Figure 6.

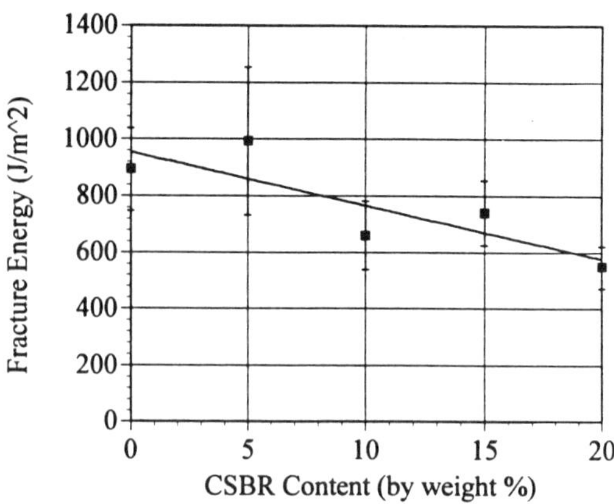

CSBR Content (by weight %)

Figure 6 Fracture Energy of Composite Parts with varying CSBR Concentration

From Figures 3, 4, 5, and 6 it can be seen that the flexural strength, modulus, the short beam shear strength and the fracture toughness all decrease with the increase in the CSBR concentration in the epoxy matrix. These results are to be expected as the increase in the volumetric amount of the "soft" rubbery inclusions proportionally lowers the mechanical properties of the matrix like flexural strength, modulus and short beam shear strength. Work done by Ting et al.,[9], [10] show similar results for a graphite epoxy system modified by a CTBN elastomer except that the fracture toughness increases with increasing elastomer concentration. As the concentration increases, the softer rubber particles present in the second phase begin to dominate the properties of the epoxy matrix by a simple rule of mixtures phenomenon. Epoxy materials in itself are very brittle and are highly susceptible to rapid crack propagation upon application of load. On the other hand rubbers are very tough and resilient materials and are capable of withstanding large loads before complete failure occurs. When these rubber particles are dispersed within the epoxy matrix, they act as crack arresters in case of a crack propagation. The softer rubber particle requires a lot more energy for the crack to propagate through it, therefore as the concentration of the rubber particles increases in the epoxy matrix the fracture energy should increase, i.e. the amount of energy required for the crack to propagate throughout the epoxy matrix. As seen in Figure 6, the fracture energy of the composite samples decrease with the increase in CSBR concentration. This behavior is quite contrary to what should have been expected from the above discussion.

An important aspect to consider for the

propagation of the crack through a rubber-toughened epoxy matrix is the interfacial adhesion between the epoxy matrix and the rubber particles. Work done by Kinloch et al.[11], [12], Mülhaupt et al.[13], Qian et al.[14] and other relevant literature place a strong emphasis on the interfacial adhesion of the rubber particle and the epoxy matrix for improvement in the toughening characteristics. In toughened resins, as the crack propagates through the matrix a dilational stress field is formed causing cavitation of rubber particles[15]. The cavitation is initiated by interfacial debonding around the particles. In case of a strong interfacial bonding, the epoxy matrix will get tougher. A possible explanation for decrease in the fracture energy of CSBR toughened epoxy/carbon composites with the increase in CSBR concentration, is the weak interfacial bonding between CSBR particles and epoxy matrix.

Dynamic mechanical analysis (DMA) was used to see if the two materials i.e. the epoxy and the rubber particles, existed as two distinct phases within the composite matrix. Results for the different CSBR samples show two distinct peaks occurring at or around the glass transition temperatures of the two materials. This clearly indicates that the CSBR particles exist as a distinct phase in the epoxy matrix.

Crack surfaces obtained after ENF testing, were studies under the SEM to identify the morphology of the rubber-toughened epoxy matrix. Figures 7, 8, and 9 show the photomicrographs of the morphology of the cracked surfaces of the ENF composite samples with 0%, 5% and 20% by weight CSBR, respectively.

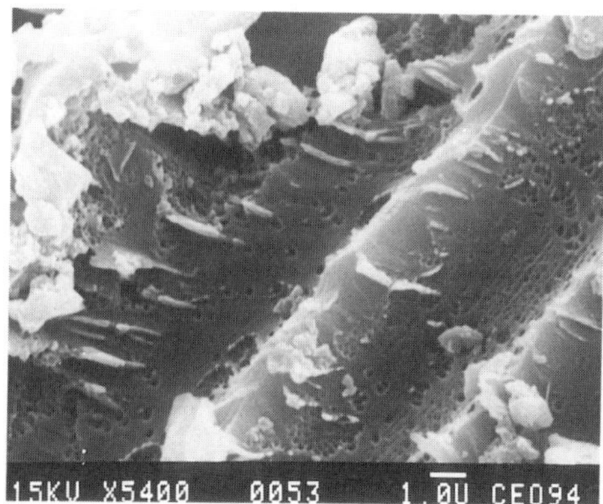

Figure 8 ENF Fracture Surface for 5% CSBR Composite Sample

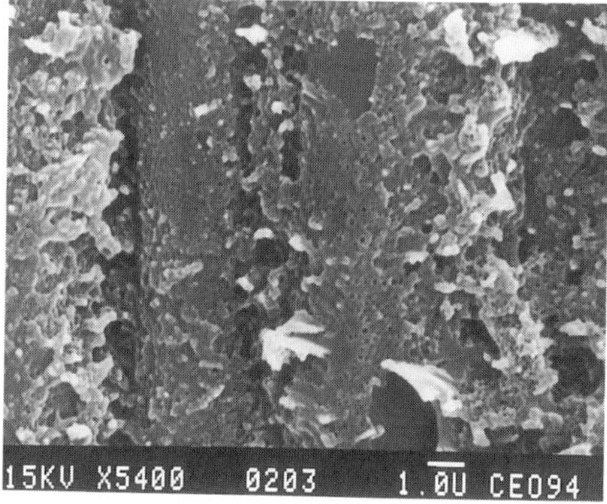

Figure 9 ENF Fracture Surface for 20% CSBR Composite Sample

Figure 7 shows the fracture surface of 0% CSBR composite part. Features of the surface are typical of a pure epoxy matrix failure. There are well defined ridges and planes. Figure 8 shows the fracture surface of 5% CSBR composite part. The features are very similar to that of the 0% CSBR part, except that there are small cavities evenly dispersed all over the surface. These cavities are an identifying feature of a rubber toughened epoxy composite. This porous structure results from the cavitation of rubber particles in response to the dilational stress field, as a crack propagates through the material[15]. Figure 9 show the fracture surface of a 20% CSBR composite part. The porous structure in this case is more pronounced and the cavities are dispersed all over the epoxy matrix indicating a larger concentration of CSBR particles in the epoxy matrix. Also the morphology has a spongy texture indicating gross plastic deformation. The morphology of all the fracture surfaces are typical of rubber toughened

Figure 7 ENF Fracture Surface for 0% CSBR Composite Sample

epoxy matrixes as cited in literature.

Conclusions

The developed water-based prepregging unit was successfully used to manufacture quality prepreg tape coated with two different polymer particles from a colloidal aqueous mixture of each. SEM study of the tape indicated an even coating spread over the entire surface of the carbon fibers. Small colloidal particles improve wetting of fibers, prevent agglomeration and coating is more even. Water, as a dispersing medium, is environmentally friendly. The prepreg tapes were consolidated to give void-free composite parts having varying concentrations of CSBR particles dispersed all through the matrix. These multiple phase matrix composites were fabricated independent of the compatibility of the matrix components. As the concentration of the CSBR particles was increased in the epoxy matrix, mechanical properties like flexural strength, flexural modulus and short beam shear strength all went down. This is because increasing the concentration of the "softer" CSBR particles causes the mechanical properties of these particles to dominate the properties of the epoxy/carbon composite. These results are consistent with other research work done in the area of rubber-toughened epoxy matrixes. A secondary objective was to study the improvement in the fracture behavior of the epoxy matrix by the increase in concentration of the elastomer. In this study it was observed that the fracture energy decreased with the increase in concentration of the rubber particles. A possible explanation could be the poor interfacial adhesion between the rubber particles and the epoxy matrix, causing the crack to propagate through the interface with relative ease. Morphological study of the fractured surface indicated distinct cavitation in the epoxy matrix due to propagation of the crack. These results are consistent with that seen in literature. The rubber particles are well dispersed through the entire volume of the matrix. Also a lot of plastic deformation was noticed in the fracture surfaces that had a higher concentration of the rubber particles.

References

1   O'Connor, Phillips, U.S. Patent 4,680,224.

2   Dyksterhouse, R., & Dyksterhouse, J.A., U.S. Patent 4,894,105.

3   Chary, R.R., & Hirt, D.E., ANTEC '92, pg 1181-1183 (1992).

4   Iyer, S.R., & Drzal, L.T., PhD Dissertation, Michigan State University (1990)

5   Zosel, A., "Adhesion and Tack of Polymers; Influence of Mechanical Properties and Surface Tensions", Colloid Polym. Sci., 263, 541 (1985).

6   Kaelble, D. H., in Treatise on Adhesion and Adhesives (R. L. Patrick, ed.), Marcel Dekker, New York, 1967, p. 169.

7   Huntsberger, J. R., Chem. Eng. News, Nov 2, 1964, p. 82.

8   Carlsson, L.A., & Pipes, B.R., "Experimental Characterization of Advanced Composite Materials", p 115-119, Prentice-Hall Inc., 1987.

9   Ting, R.Y., & Moulton, R.J., SAMPE Tech. Ser. 12:265 (1980).

10 Moulton, R.J., & Ting, R.Y., Paper presented at the Int. Conf. on Composite Structure, Paisley, Scotland, Sept. 1981.

11 Kinloch, A. J., & Young, R. J. (eds.), "Fracture Behavior of Polymers" Elsevier Applied Science, London 1983.

12 Kinloch, A. J., in "Structural Adhesives", Kinloch, A. J. (ed), Elsevier Applied Science Publ., Essex 1986, p. 127.

13 Mülhaupt, R., & Buchholz, U., "Compatibilized Liquid Rubbers as Epoxy Toughening Agents", Freiburger Materialforschungszentrum (FMF) and Institut für Makromolekulare Chemie", Stefan-Meier-Str. 31, D-79104 Freiburg e. Br., Federal Republic of Germany.

14 Qian, J. Y., Pearson, R. A., Dimonie, V. L., and El-Aasser, M. S., "Epoxy Polymers Toughened with Novel Latex Particles", J. Mater. Sci., 26 3828 (1991), 24 2571 (1989), 21 2462 (1986).

15 Bascom, W. D., Cottington, R. L., Jones, R. L., Peyser, P., J. Appl. Polymer Sci., 19:2545 (1975).

Electrostatic Powder Spray Process for In Situ Low Cost Thermoplastic Composite Manufacturing

K. Ramani, D.E. Woolard, M.S. Duvall, N. Parasnis
Purdue University
West Lafayette, Indiana

Abstract

Manufacture of thermoplastic composites is expensive due to the use of pre-impregnated materials. Powder processing shows great promise as an impregnation technique for a cost-effective manufacturing process. The overall objective of this study is to develop a continuous process for impregnating and consolidating thermoplastic composites using powder polymers. A negative corona electrostatic spray gun is used to charge and direct the powder to adhere to both sides of the spread fiber tow. The tows are then pulled through ceramic heaters to coat the fibers, and then consolidated via filament winding using hot nitrogen gas heaters.

The effect of the parameters governing the spray impregnation are investigated experimentally, as well as using a three-dimensional finite element model to predict the electric field strength. Electric field "wrap-around" as well as the distribution along the length of the fiber bundle are found to agree with experimental observations. The coating of the fibers was studied using an environmental scanning electron microscope (ESEM) and a heating stage. Distinct stages of the polymer flow were identified. The key factors governing the consolidation of in-situ coated tows in a filament winding operation are examined.

COMPOSITE MATERIALS CAN GAIN wider acceptance if the manufacturing costs are significantly reduced. Thermosetting resin-based composites require long process cycles in order to properly cure the polymer. Thermoplastic composites offer shorter cycle times, but pose difficulties in uniformly impregnating the polymer into the fibers due to high melt viscosities of the resin. Thermoplastic resins offer advantages such as reformability, recyclability, and improved chemical and impact resistance compared to thermosets. However, manufacturing of thermoplastic composite materials must be made cost-effective and reliable in order to use them in a greater number of commercial engineering applications.

This paper discusses a technique for impregnating glass fibers with thermoplastic polymer powder. This process utilizes an electrostatic powder spray gun to deposit the polymeric powder onto a continuous fiber tow (1,2). A key element of this process is that it combines mechanical entrapment of the powder with electrostatic effects. The spray gun is positioned and oriented to control the deposition and coat both sides of the tow. Although the process demonstrates the manufacture of glass fiber reinforced thermoplastics, thermosetting resins and carbon fibers may also be used.

The effect of various operating parameters on powder deposition will be discussed. A three-dimensional finite element model that predicts the electric field strength using a thermal analogy was developed. The results of the model substantiates the "wrap around" effect in coating as well as the distribution of the powder along the length of the fiber tow.

Coating studies using an ESEM and heating stage revealed coalescence of the particles, thermo-capillary, surface tension and viscosity driven flows. The various stages in coating were identified. Coated tows were compression molded and tested in three point bending. An apparatus for consolidating the coated tows in a filament winding operation was constructed to demonstrate in-situ impregnation and consolidation in filament winding. Preliminary experiments demonstrate good consolidation in filament wound cylinders.

Thermoplastic Powder Impregnation

Different techniques of impregnation have been used for manufacturing thermoplastic composites. These include melt impregnation, fiber co-mingling, film stacking, solution processing, and powder impregnation. Among these techniques, powder processing is emerging as an attractive method for achieving a low-cost manufacturing process. A review of some commercial powder processes is provided by Fisher (3). Iyer and Drzal (4) discuss some issues involved in powder processing of thermoplastic composites. Powder processes may be classified as either wet or dry. In wet powder techniques, the fibers are pulled through a liquid medium in which the polymer powder is suspended. Dry powder processes impregnate the fibers directly with the polymeric powder.

The primary methods used in wet powder processing are slurry-based techniques in which the powder is suspended in an aqueous medium. Some of the wet techniques were developed by Ramani, et al. (5), Dyksterhouse et al. (6), O'Connor (7), and Taylor (8).

Dry powder processes may be distinguished by the method used to deposit the powder on the fiber tow. Most dry powder processes rely on either mechanical entrapment or electrostatic attraction of the powder to the tow. A vast majority of the work using dry powders has used carbon fibers for reinforcement. Carbon fiber tows are easily exposed through spreading as they typically do not have a significant amount of binder that hold the individual filaments together. Muzzy and Varughese (9,10) developed and commercialized a process which uses an electrostatic fluidized bed to coat the fibers. The process produces flexible tow-preg which is then post-processed to produce a finished component. Baucom and Marchello (11) compared a similar electrostatic fluidized bed in their work with re-circulating fluidized bed coating. They observed enhanced powder deposition due to the charged powder. Throne and Sohn (12) used an electrostatic spray gun to create a charged powder cloud in an enclosure for coating carbon fibers.

Baucom and Marchello (13) used a metered screw feed mechanism to dispense powder onto the top of a spread carbon fiber tow in the powder curtain prepreg process. Previously, Baucom et al. (11,14) developed a re-circulating fluidized powder bed for depositing powder onto a spread carbon fiber tow. Similarly, Iyer et al. (15,16,17) employed acoustic energy to suspend powder particles in a vertical tube through which a spread carbon fiber tow was passed. The powder is deposited by direct interception with the fibers and electrostatic attraction due to natural tribocharging of the powder. Ogden et al. (18) developed a process which directly deposits powder onto fibers moistened by an ultrasonic humidifier. Ganga (19,20) is credited with developing a process for Ato-Chem in which fibers are impregnated with fine powder and then encased by a thin sheath of equal or lower melting point resin. A pre-preg process attributed to de Jager (21) spreads and impregnates a fiber tow by sending it through powder-laden air jets.

Process Description

A schematic of the electrostatic spray impregnation process is shown in Figure 1. The process uses a bi-directional fiber tensioner to maintain constant tension in the fiber tow. A nitrogen gas heater is used to transiently heat and soften the binder on the fibers while the gas flow separates the fiber bundle into individual filaments. A fine water mist is then applied to the fibers to make them electrically conductive. The fiber tow is then passed over a pulley which is used to electrically ground the fibers. Fiber spreading occurs by passing the fibers over a spherical surface, just prior to coating the fibers. A negative corona electrostatic spray gun is used to coat the spread glass fibers with the polymeric powder. The coated fibers are then passed through two flat ceramic fiber heaters to melt and adhere the powder to the fibers. The towpreg is then wound onto a take-up mandrel. A more detailed discussion of fiber spreading and electrostatic powder deposition follows.

Fiber Spreading. A "geodesic fiber spreader", shown in Figure 2, consisting of a tow centering pulley, grounding pulley, ceramic sphere, and a ceramic eyelet is used to spread the fibers. The fiber tow is introduced into the spreading region via a grooved centering pulley which constrains the lateral movement of the tow. The positioning pulley grounds, flattens, and stably positions the moistened fiber tow above the center of the ceramic sphere. The sphere has a hard, smooth surface to minimize fiber abrasion and resist wear. In addition, the sphere is made from a non-conductive material so the electrostatic powder deposition is not disrupted. After traversing over the spherical surface, the fiber passes through a ceramic fiber guide which constrains the final fiber position. An individual filament will follow the path of shortest length from the centering pulley to the ceramic eyelet. Since the fiber tow has a finite initial width, each filament will attempt to take the shortest path over the sphere (geodesic path), resulting in spreading the fiber tow.

The width of fiber spreading is affected by the addition of water, the geometric design of the fiber spreader, the fiber tow characteristics, and fiber tension. The water, added to increase the electrical conductivity of the fibers, acts as a lubricant when the fibers pass over the spherical surface. However, the surface tension of the water tends to hold the fiber tow together. The geometric parameters which affect the level of spreading include the radii of the grounding pulley and spherical surface, the position of the grounding pulley and ceramic eyelet with respect to the sphere, and the width of the grounding pulley.

Electrostatic Powder Spray Deposition. The electrostatic powder spray process consists of powder being pneumatically conveyed from a fluidization chamber to a spray gun. The gun nozzle contains a pointed electrode which is maintained at a high negative voltage (-30 to -100 kilovolts) to create a corona discharge at the electrode. The ions produced in the corona region charge the powder. The charged powder is then directed towards the grounded fibers under the influence of aerodynamic, electrostatic, and gravitational forces.

$$F_{net} = F_a + F_g + F_e \qquad (1)$$

A very thorough review of charging and deposition of powder coatings is presented by Wu (22). Ang and Lloyd (23) studied particle trajectories from an electrostatic spray gun. The aerodynamic force is responsible for particle transport from the gun. The charged polymeric particles, placed in the electric field between the corona and the grounded fibers, experience a force that directs them towards the fiber tow. In addition, the charged particles themselves create a space charge, that results in the particles "fanning out" when exiting from the nozzle. However the space charge is significant only in the region near the nozzle. When the particles are within a few millimeters of the fiber tow, the image field between the particles and the fiber, results in the final deposition on the tow.

The net electric field is the vector sum of the space charge field due to the charged particles and ions, the applied field generated by the corona electrode, the image field, and the field resulting from the deposited layer of charge particles on the grounded substrate.

$$\mathbf{E}_{net} = \mathbf{E}_{space} + \mathbf{E}_{electrode} + \mathbf{E}_{image} + \mathbf{E}_{deposited} \qquad (2)$$

Process Characteristics

PEKK powder supplied by Dupont with a volume average diameter of 72 microns were used in the experiments. The fibers used were Owens-Corning Type-30 rovings with an experimental sizing chemistry, average fiber diameter of 16 microns, and a yield of 450 yards/pound. The position and orientation of the gun in the experiments is shown in Figure 3. The operating parameters of the process are given in Table 1.

Static Fiber Tow Experiment. The effect of the corona voltage of the gun on powder deposition was explored by spraying onto a stationary fiber tow for 15 seconds. After deposition, the fiber tow was pulled at very low speed through an oven to melt and adhere the powder to the tow. The coated tow was cut into 13 millimeter long segments and weighed to determine the resin distribution along the fiber tow. Figure 4 shows the powder deposition along the length of the tow, with and without the corona voltage applied to the gun.

The electric field strength is strongest at the nozzle exit region, as shown by region "A" in Figure 3. In this region, the powder is strongly directed to the tow by the electrostatic forces. When the powder is not charged, it is influenced only by the air flow and gravity. Thus, the uncharged powder deposits on the fibers only through direct interception with the spread fiber tow. It can be seen from Figure 4 that half the total deposition occurs within a distance of 25 millimeters from the nozzle exit *with* corona voltage and within a distance of 65 millimeters from the nozzle exit *without* corona voltage. Clearly, the electrostatic forces cause more powder deposition on the fibers near the nozzle exit.

Powder Mass Flow versus Air Flow. Figure 5 shows the powder mass flowrate as a function of the air pressure applied to the powder pump on the fluidizing hopper. The powder pump operates on the venturi principle for entraining the powder particles into the air flow. A least squares fit for the relationship between the powder mass flow and the applied air pressure to the powder pump is

$$\dot{m}_p = 0.028 P_{flow} + 0.030 \qquad (3)$$

for applied air pressures from 6.90 to 20.7 kilopascals. The fluidizing pressure in the powder hopper did not have a major effect on the consistency of the powder flow. However, as the amount of powder in the hopper reduced, the size distribution of the powder in the hopper shifted, resulting in lower powder flow.

Powder Deposition versus Air Flow. The amount of powder deposited onto an unspread fiber tow was measured as a function of the applied air pressure to the powder pump. The air pressure was varied from 7.0 to 21 kilopascals. The amount of powder deposited onto fibers was found by measuring a twelve inch sample of coated glass fibers. The resin weight fraction was averaged for three samples at each pressure setting and is shown in Figure 6.

As the powder mass flowrate increases, the powder deposition achieves a maximum for every orientation of the gun. For the configuration shown in Figure 3, the maximum deposition occurs at 14 kilopascals, as shown in Figure 6. Further increase in the powder mass flowrate increases the exit velocities of the powder and air. High powder velocities overcome the electrostatic image force near the fiber tow while a high air velocity removes powder that is already deposited on the fibers. In addition, increase in powder flowrate results in particle shielding due to an increase in powder density in the corona region. Particle shielding reduces the charge acquired per particle. These factors contribute to a reduction in powder deposition.

Powder Deposition versus Corona Voltage. The amount of powder deposited onto an unspread fiber tow is shown as a function of the applied corona voltage in Figure 7. The corona voltage was varied from -40 to -100 kV in 15 kV increments. As shown in Figure 7, powder deposition increases until it saturates at a corona voltage of -70 kV. Initial increase of the corona voltage (-40 to -70 kV) causes a higher charge per particle as well as a stronger applied electric field. These two effects result in higher electrostatic forces, which directs the powder to the fibers, in the nozzle exit region (region "A", Figure 3). However, increasing the corona voltage results in greater ion generation from the corona region. Once the particles achieve their charge saturation limit, the increased ion generation contributes solely to the space charge density. The space charge effect counteracts the increase in electric field strength obtained by the higher corona voltage. Also, the image force, which is proportional to the square of the particle charge, reaches its maximum when the charge saturation limit is achieved. Therefore, once the particles reach their charge saturation limit, there is no significant increase in the image force or net electrostatic force to improve deposition.

Powder Deposition versus Fiber Velocity. The amount of powder deposited onto a spread fiber tow as a function of the fiber tow velocity and electrostatic gun height was monitored. The fiber velocity varied between 13 and 51 millimeters per second while the gun nozzle was positioned 13 and 38 millimeters above the fiber tow. Figure 8 shows the powder deposition as a function of fiber velocity for both the low (LO) and high (HI) gun positions, *with* (WE) and *without* (WOE) electrostatics (corona voltage) applied.

Powder deposition decreases at higher fiber velocities and increases with electrostatics. Without electrostatics, the uncharged powder falls off due to the vibration of the moving fiber tow. When the gun is closer to the fiber tow, the electric field intensity is greater resulting in strong electrostatic forces directing the particles to the fiber tow. Hence, a lower position of the gun with electrostatics is beneficial for good deposition and adherence of the powder.

Electric Field Modeling

Through use of an analogy between electrostatic and heat conduction phenomena, commercial FEA software (ANSYS) was used to model electrostatic phenomena.

Boundary Conditions and FEM Assumptions. In modeling the electrostatic spray impregnation process, a number of simplifying assumptions were made. The corona electrode is approximated as a constant negative point potential. The spread fiber bundle is modeled as a flat plate maintained at zero potential on its surface and edge effects are ignored. The permittivity of the air is taken to be equal to the permittivity of free space $(8.854 \times 10^{-14}$ F/cm). The outer boundaries of the solution domain are assumed to be located far distance from where the large gradients occur (far field approximation. For these outer boundaries of the domain, both the electric potential and electric field strength (gradient of potential) are zero.

Finite Element Model Development and Solution Domain. A three-dimensional finite element model using symmetry through the plane which passes through the corona electrode and centerline of the spread fiber bundle was developed. Figure 9 displays a schematic of the solution domain for the process geometry along with relevant boundary conditions. The solution domain was discretized into eight-noded linear brick elements. The element sizes were progressively made larger towards the outer boundaries due to smaller potential gradients. The model was run for several heights of the corona and corona voltages.

The differential equations for heat conduction and electrostatic phenomena are both governed by Poisson's equation. Table 2 displays both input and output quantity relationships for the thermal analogy.The electric field is strongest directly below the corona electrode on the top surface of fibers. The electric field decreases both along length of the fiber bundle and from along the thickness of the grounded fibers tow. As shown in Figure 10, the electric field strength decreases rapidly along the length of the fibers. The electric field strength increases with increasing voltage. Also, as shown in Figure 11, the electric field strength decreases rapidly as the point corona electrode to grounded fiber bundle distance is increased.

Fiber Coating

The applied electric field generated between the point electrode and grounded fibers is not restricted to the top surface of the fibers. This field extends to the bottom surface of the fibers as well and is shown schematically in Figure 12. Hence, in addition to coating the top side of the fiber tow, a significant amount of powder "wraps around" to coat the underside of the spread fiber tow. SEM photographs were taken of the end of a glass fiber tow coated using the spray impregnation process. Figures 13 and 14 show the polymer coating on both sides of the tow for low and high fiber spreading, respectively.

The difficulty in spreading glass fibers is due to the presence of the binder that holds the tows together, thereby increasing the amount of lateral force required to separate them. In this process a hot nitrogen gas torch softens the binder while separating the filaments. The interparticle distance, which is governed by the amount of spreading, particle size distribution, and the volume fraction of the polymer, governs the behavior of the particles when the tow is heated. Low spreading (Figure 13) causes the coated fiber tow to be 30-40 fiber diameters thick, resulting in the polymer having to flow a longer distance for impregnation during consolidation. In Figure 14, moderate spreading of the fibers results in the coated tow only being 10-15 fiber diameters thick, reducing the flow length required by the polymer during consolidation. In both the cases above the inter-particle coalescence dominates the coating process and a polymer sheath is formed around the tow. Higher spreading of the tow allows the particles to coat the fibers through thermo-capillary, surface-tension, and viscosity driven flows. A simple model that assumes that the particles are spherical with a radius equal to the volume average particle size, and are uniformly distributed on both sides of the tow can be used to relate the particle packing to the level of fiber spreading.

The polymer particles undergo morphological changes upon heating from random shaped solid phase to round liquid phase and ultimately spread on the fibers. The in-situ heating study using an environmentally scanning electron microscope (ESEM) and a heating stage showed four different stages of the polymer flow. In the first stage the particles transformed from randomly shaped particles to smooth spherical shapes. The polymer droplets coalesced with nearby droplets and formed larger droplets. With increase in temperature, some of these droplets became unstable and spread on the fibers.

Consolidation Behavior

In order to measure the performance of the electrostatic powder spray coated tows in making a finished component, the coated tows were compression molded and tested. The materials used in making the panels were glass fibers and the PEKK powder.

Experimental Procedure. The procedure used was similar to the compression molding experiments conducted by Hoyle (24). The coated tows were wound onto a metal frame and then inserted in a matched die mold for producing a 3.2 x 25 x 200 millimeter panel as shown in Figure 18. The die was then placed in a heated press for consolidation. The die was heated to the polymer processing temperature of 325°C prior to closing the die. The die was closed with an applied pressure of 5.2 MPa and held under pressure for a cycle time of 30 minutes. The die was then cooled to below the glass transition temperature of the polymer before removing the panel. The specimens were then cut to a length of 127 millimeters and tested in 3-point bending according to ASTM D790-92 Test Method I, Procedure A.

Three-Point Bending Results. Seven panels were produced with a fiber mass fraction of 69±1% in the finished panels. The void content of the consolidated panels ranged from 0 to 7%. These panels were tested in 3-point bending with the average flexural strength of 1.1 GPa with a standard deviation of 73 MPa. The average modulus of elasticity was 42 GPa with a standard deviation of 1.9 GPa. These are compared with theoretical values using the rule of mixtures of 1.0 GPa and 43 GPa for tensile strength and modulus respectively.

Microstructure Analysis. Figure 15 displays optical micrographs taken from a polished sample of a tested panel cross-section. The fractured surface of the tested panels were observed under a scanning electron microscope (SEM) to determine the effect of the sizing on the interfacial adhesion. As shown in Figure 16, PEKK resin remains adhered to the pulled out glass fibers, indicating the effectiveness of the fiber's sizing agent. Good interfacial adhesion also implies that prior processing did not remove the sizing from the fiber surface.

In-Situ Filament Winding

This process demonstrates the impregnation and consolidation of the coated tows in a single continuous process. An apparatus to perform on-line consolidation during the winding of the coated tows onto a standard commercial filament winder was designed. The apparatus uses opposing steel compression roller mounted on identical air cylinders (Figure 17). This design allows the two support rollers to counteract the consolidation force of the single roller. This design can function as an add-on component to a commercial light-duty thermoset filament winder without causing undue loading and deformation. It can also manufacture smaller diameter tubes, without the consolidation forces causing deflection in the mandrel.

The coated tows from the spray process consist of a polymer sheath encapsulating a fiber bundle. The winding process must complete the impregnation and consolidation to produce a quality, low-void part. Hot nitrogen gas heaters were used for pre-heating and final heating of the tows before consolidation. Microstructure analysis of the tows revealed that the further impregnation occurred through the thickness of the tow during the consolidation. SEM micrographs of the consolidated cylinders (Figure 18) reveal good consolidation and minimal void content. A more detailed analysis of the in-situ filament winding process will be presented in future publications.

Concluding Remarks

A new electrostatic spray impregnation process which combines the benefits of both mechanical entrapment and electrostatic adhesion of the powder has been demonstrated using glass fibers. A spherical surface locally spreads the fiber tow in the area of maximum powder deposition. The electrostatic field due to the corona voltage is used to coat both sides of the spread fiber tow. Electrostatic attraction increased the velocity component towards the fiber tow, especially below the nozzle exit. Resin to fiber weight ratios up to 50% with less than 3% variation have been achieved under normal spray and fiber spreading conditions. Process parameters which may be controlled for efficient powder deposition include position and orientation of the gun, charging of the powder, and the particle velocities and trajectories. Void-free compression molded panels have been produced from electrostatically spray coated tows and exhibit good mechanical properties. Coating of the fibers was studied using an ESEM.

Acknowledgements

The authors would like to thank Nordson Corporation for their electrostatic equipment donation and technical assistance. We would like to thank Professor R. Stroshine for the use of his Sintech testing machine. Support of National Science Foundation Grant DDM-9308498 is greatly appreciated. Donation of the PEKK powder from DuPont and glass fibers from Owens-Corning Fiberglas is also gratefully acknowledged. Finally, we would like to thank Rick Caillat and Mark Bays for the design and fabrication of the on-line consolidation apparatus.

References

1. K. Ramani and D. E. Woolard, Invention Disclosure #93036, Office of Technology Transfer, Purdue University (1993).

2. K. Ramani, D.E. Woolard, and M. S. Duvall, *An Electrostatic Powder Spray Process for Manufacturing Thermoplastic Composites*, Processing, Design, and Performance of Composite Materials, ASME Winter Annual Meeting, Chicago, Nov 13-19, 1994.

3. K. J. Fisher, *Adv. Compos.*, **8**, 30 (1993).

4. S. R. Iyer and L. T. Drzal, *J. Thermoplastic Compos. Mater.*, **3**, 325 (1990).

5. K. Ramani, M. Tryfonidis, C. Hoyle, and J. Gentry, Processing, Fabrication, and Manufacturing of Composite Materials, ASME MD-Volume **35**, 115 (1992).

6. R. Dyksterhouse and J. Dyksterhouse, U.S. Patent No. 4894105, (1990).

7. J. E. O'Connor, U.S. Patent No. 4680224, (1987).

8. G. J. Taylor, U.S. Patent No. 4292105, (1981).

9. J. D. Muzzy and B. Varughese, U.S. Patent No. 5094883, (1992).

10. J. D. Muzzy and B. Varughese, U.S. Patent No. 5171630, (1992).

11. R. M. Baucom and J. M. Marchello, NASA Technical Memorandum 102648, (1992).

12. J. L. Throne and M. Sohn, *Proc. 35th Int'l. SAMPE Symp.*, **35**, 2086 (1990).

13. R. M. Baucom and J. M. Marchello, *Proc. 38th Int'l. SAMPE Symp.*, **38**, 1902 (1993).

14. R. M. Baucom, J. J. Snoha, and J. M. Marchello, U.S. Patent No. 5057338, (1991).

15. S. R. Iyer, L. T. Drzal, and K. Jayaraman, U.S. Patent No. 5102690, (1992).

16. S. R. Iyer, L. T. Drzal, and K. Jayaraman, U.S. Patent No. 5123373, (1992).

17. S. R. Iyer and L. T. Drzal, U.S. Patent No. 5128199, (1992).

18. A. L. Ogden, M. W. Hyer, G. L. Wilkes, and A. C. Loos, *J. Thermoplastic Compos. Mater.*, **5**, 14 (1992).

19. R. A. Ganga, U.S. Patent No. 4614678, (1986).

20. R. A. Ganga, U.S. Patent No. 4713139, (1987).

21. Gui G. de Jager, U.S. Patent No. 4839199, (1989).

22. S. Wu, *Polym.-Plast. Technol. Eng.*, **7**, 119 (1976).

23. M.L. Ang and P.J. Lloyd. *Int. J. Multiphase Flow*, McGraw-Hill, London (1969).

24. C. J. Hoyle, Master's Thesis, Purdue University, West Lafayette, IN (1994).

TABLE 1. OPERATING PARAMETER CONDITIONS FOR EXPERIMENTAL INVESTIGATIONS.

Experiment	Corona Voltage (kV)	Fluidizing Pressure (kPa)	Air Flow Pressure (kPa)	Flat Heaters Temperature (°C)
Static Fiber Tow Experiments	85	41.4	13.8	500
Powder Deposition vs Powder Mass Flow	85	41.4	6.90-20.7	525
Powder Deposition vs Corona Voltage	40-100	41.4	13.8	625
Powder Deposition vs Fiber Velocity	75	41.4	13.8	575-650
PIV Experiments	100	55.2	6.90	N/A

Experiment	Fiber Tow Tension (grams)	Fiber Tow Velocity (mm/sec)	N_2 Torch Exit Temperature (°C)	N_2 Torch Flow-rate (m³/hr)
Static Fiber Tow Experiments	200	N/A	750	2.55
Powder Deposition vs Powder Mass Flow	200	25.4	750	2.55
Powder Deposition vs Corona Voltage	200	25.4	750	2.55
Powder Deposition vs Fiber Velocity	200	12.7-50.8	750	2.55

TABLE 2. RELATED QUANTITIES IN THERMAL ANALOGY

Electrostatic Phenomena		*Heat Conduction Phenomena*	
Governing Equation	$\nabla \bullet (\varepsilon \nabla \phi) = -\rho$	Governing Equation	$\nabla \bullet (k \nabla T) = -\ddot{q}$
Electric Potential	ϕ	Temperature Potential	T
Charge Density	ρ	Heat Generation Rate	$\ddot{q}$
Material Permittivity	ε	Thermal Conductivity	k
Electric Field Vector	$\mathbf{E} = -\nabla \phi$	Thermal Gradient	$-\nabla T$
Electric Current Density	$\mathbf{D} = -\varepsilon \nabla \phi$	Thermal Flux	$\mathbf{q}'' = -k \nabla T$

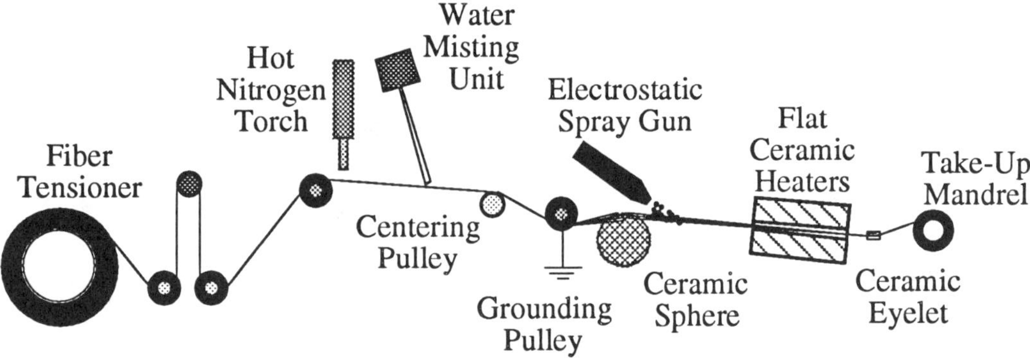

Figure 1.   Schematic of electrostatic spray impregnation process.

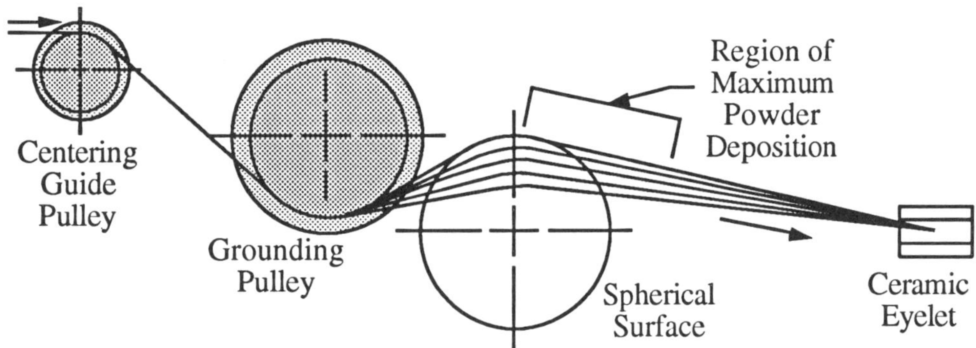

Figure 2.   Side view of geodesic fiber spreader.

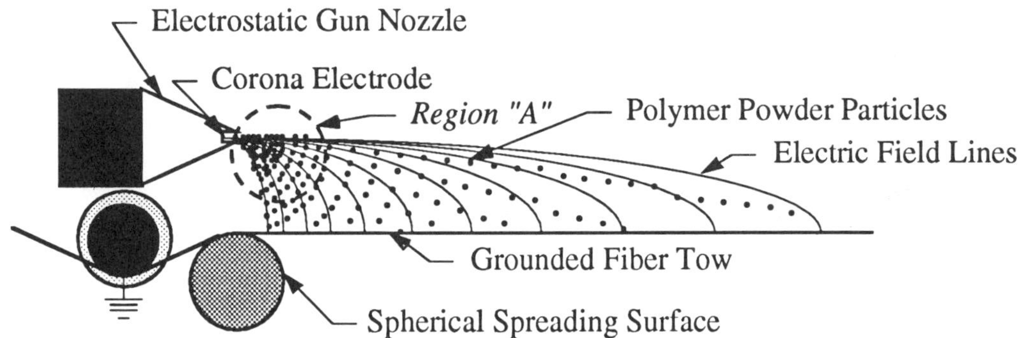

Figure 3.   Side view of spray deposition area for parallel orientation of gun to fiber tow.

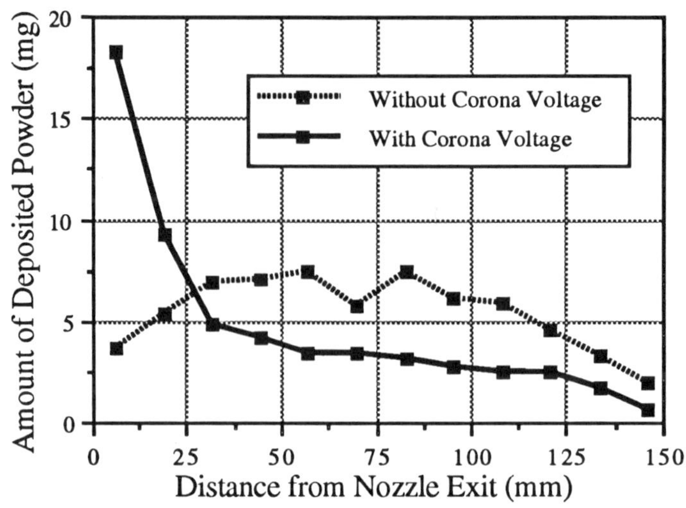

Figure 4.   Powder deposited in 15 seconds along the length of a fiber tow.

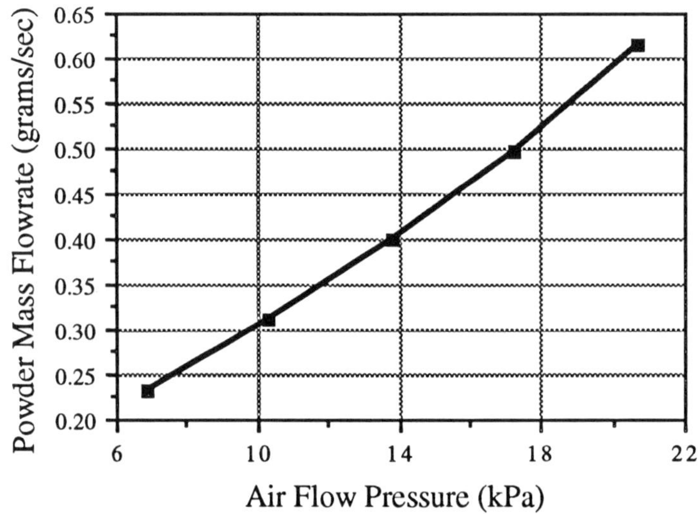

Figure 5.   Mass flowrate of 72 micron PEKK powder from the gun nozzle
as a function of air pressure to the venturi pump.

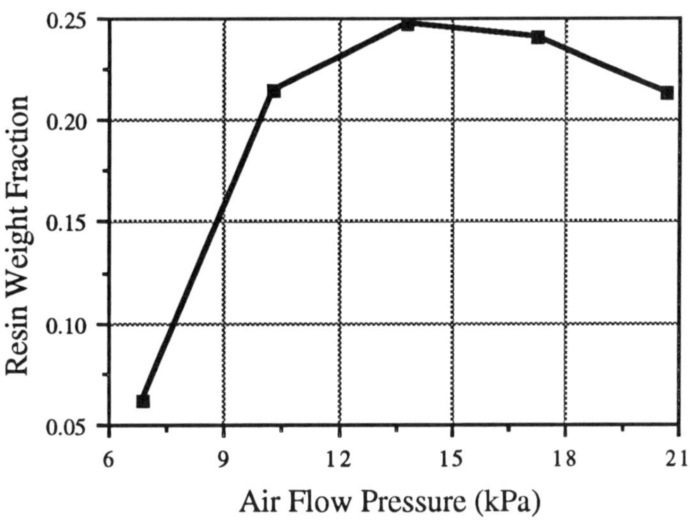

Figure 6.   Amount of powder which is electrostatically deposited onto the
fiber tow as a function of the fiber tow velocity.

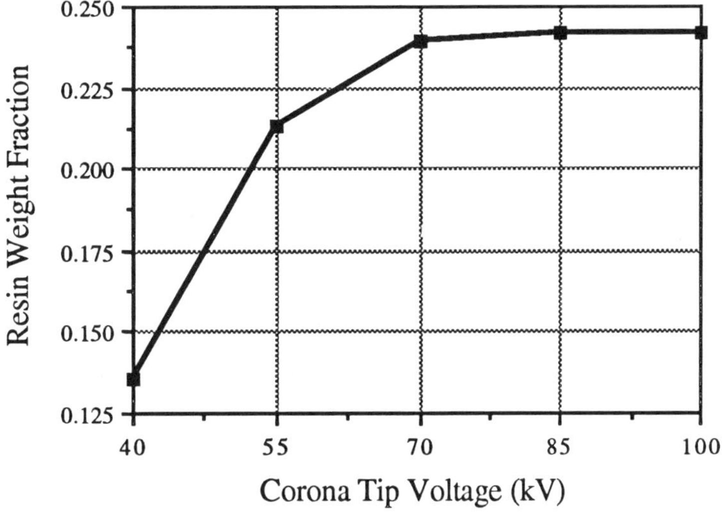

Figure 7.   Amount of powder which is electrostatically deposited onto the glass fibers as a
function of voltage which is applied to the tip electrode.

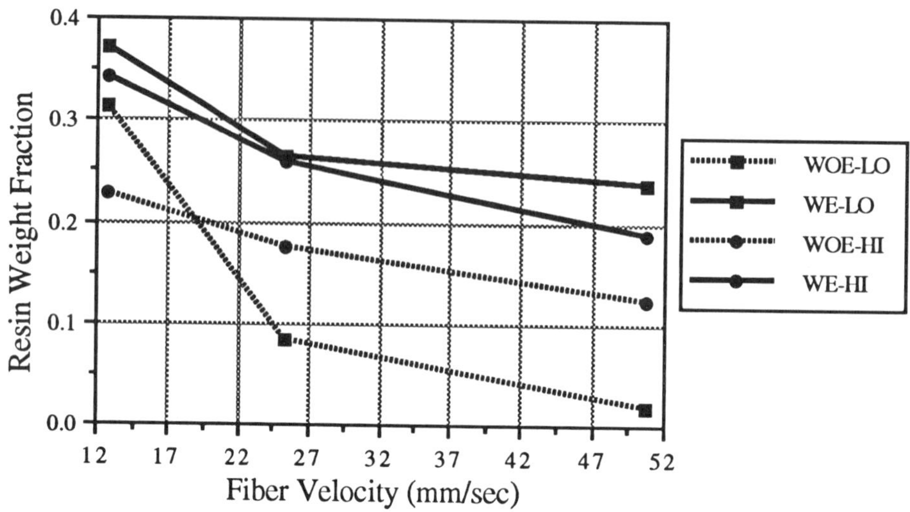

Figure 8.   Amount of powder deposited onto the fibers as a function of the fiber tow velocity.

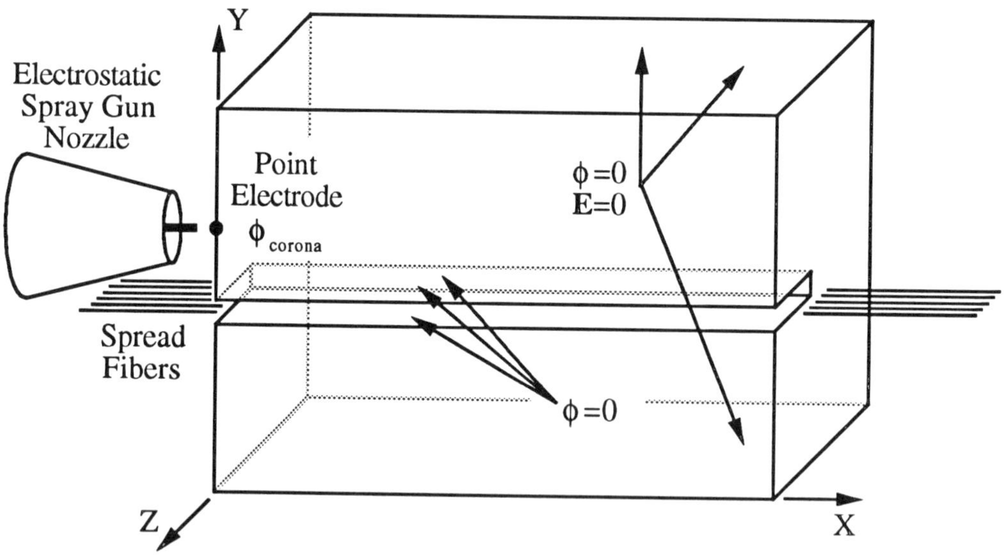

Figure 9.   Schematic of finite element model solution domain with boundary conditions identified.

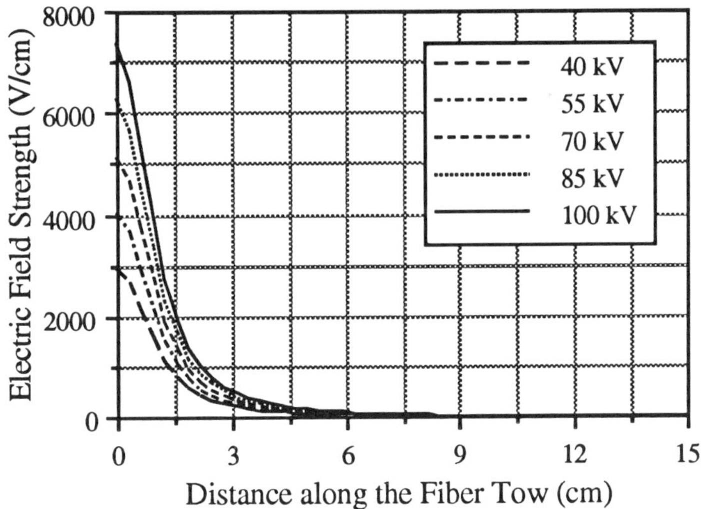

Figure 10.

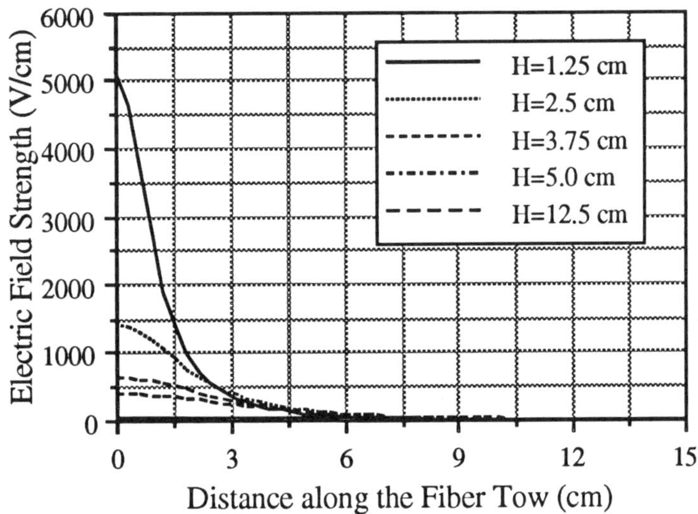

Figure 11.

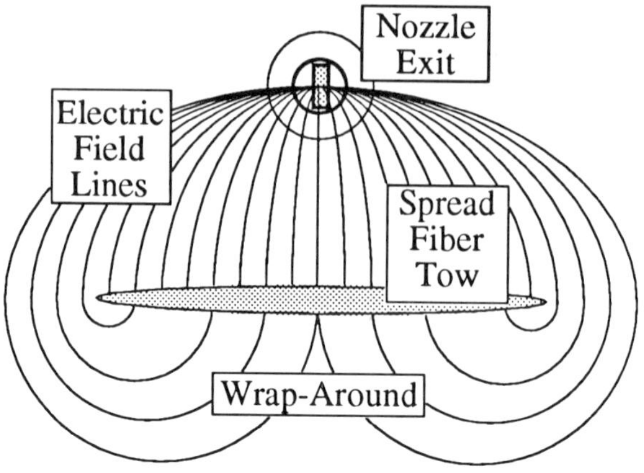

Figure 12.   Diagram depicting representative electric field lines as viewed towards the nozzle.

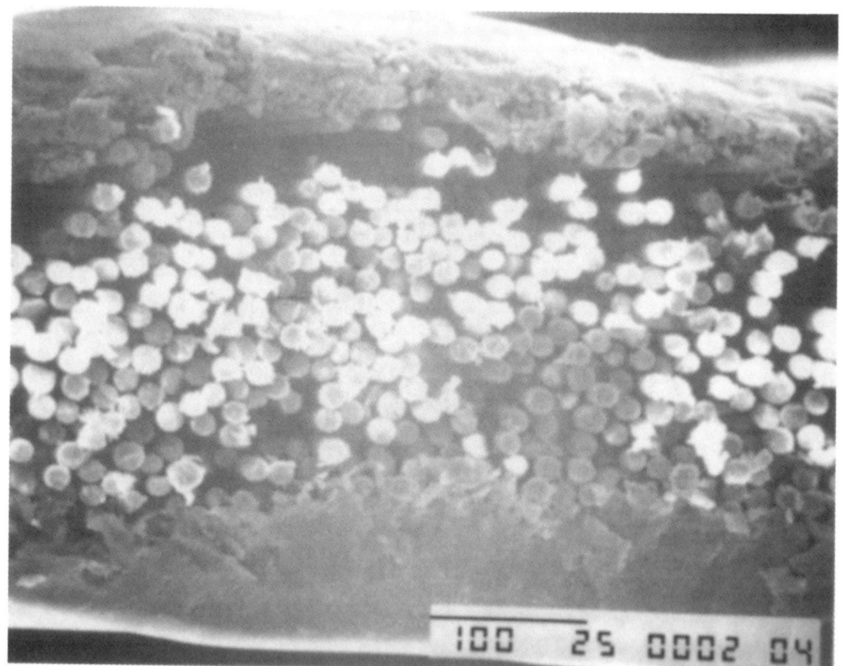

Figure 13.   SEM photograph end view of an electrostatically spray impregnated glass fiber tow with a low spreading width.

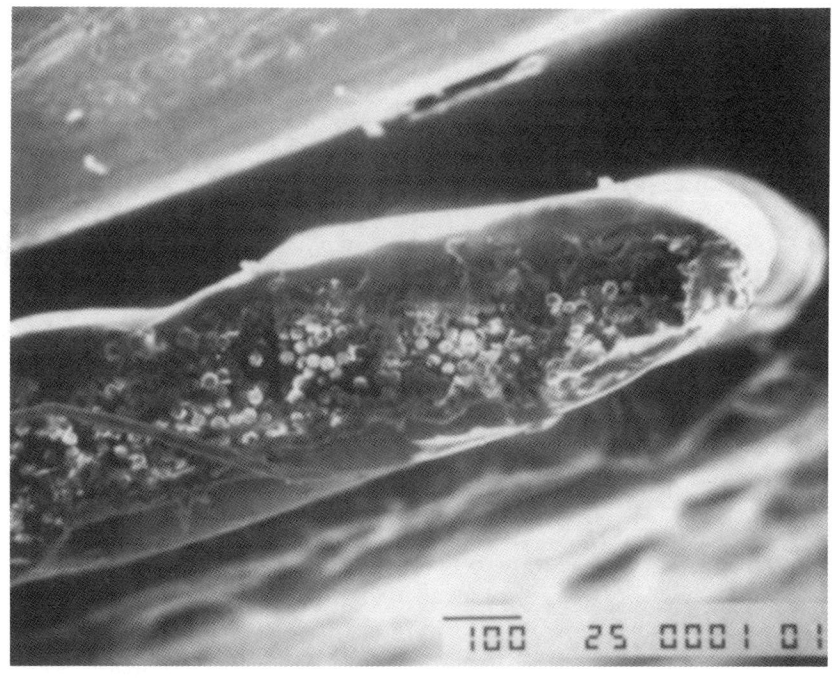

Figure 14. SEM photograph end view of an electrostatically spray impregnated glass fiber tow with a high spreading width.

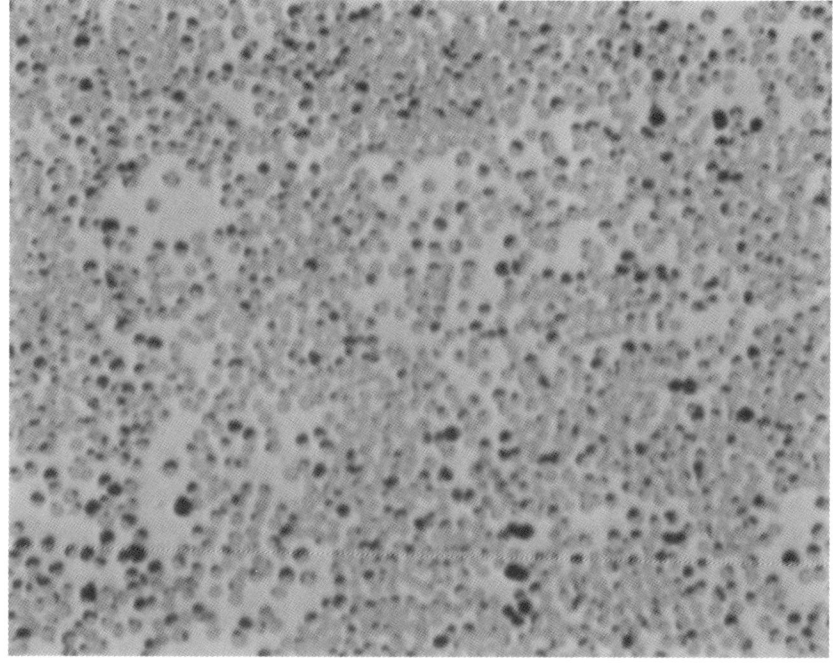

Figure 15.   Optical micrograph of compression molded panel cross-section.

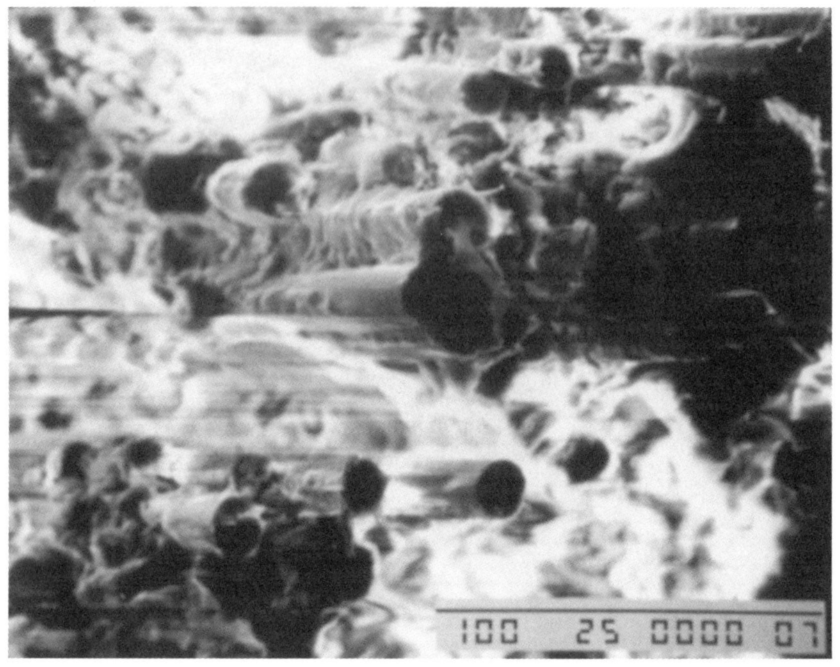

Figure 16. SEM photograph of the fractured interface after a 3-Point Bending test.

Figure 17. Filament winding and consolidation apparatus.

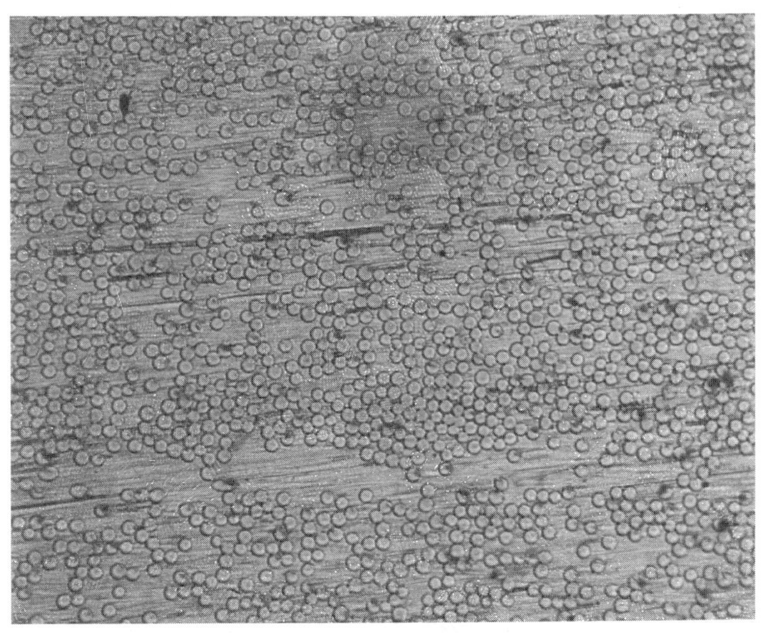

Figure 18.   Optical Micrograph of consolidated cylinder.

Proceedings of the 10th Annual ASM/ESD Advanced Composites Conference, Dearborn, Michigan, USA, 7-10 November 1994

Experiments on Compression Molding and Pultrusion of Thermoplastic Powder Impregnated Towpregs

K. Ramani, H. Borgaonkar
Purdue University
West Lafayette, Indiana

C. Hoyle
Motorola Inc.
Chicago, Illinois

Abstract

Powder impregnation techniques have been developed to increase design and manufacturing flexibility with thermoplastic composites. The effect of pressure, temperature, mold closing rate, and time on consolidation of the PEKK powder impregnated glass fiber towpregs in compression molding were studied. A design of experiments approach was used. Isothermal pultrusion experiments using nylon 11 and PEKK powder impregnated glass fiber tows were conducted. These experiments were repeated at different temperatures and pulling speeds.

The microstructural changes during coating and consolidation of the powder impregnated tows were studied. Temperature and mold closing rate were observed to be the significant parameters affecting consolidation in compression molding. Superposition of the pulling force-pulling speed data at different processing temperatures was demonstrated in pultrusion.

ECONOMICAL PROCESSES to manufacture composites must be developed to use them in a wide range of applications. Thermoplastic matrix composites offer excellent impact resistance, environmental stability, and recyclability. However, resin impregnation is one of the difficult subprocesses in manufacturing continuous fiber reinforced thermoplastic-matrix composites due to high melt viscosities exhibited by the thermoplastics. Melt impregnation has drawbacks such as the possibility of polymer degradation due to the long residence times at high temperatures. Solvent based techniques are environmentally hazardous and expensive. Powder impregnation is attractive since it ensures low residence times of the polymer during heating, localized melt flow of the polymer during consolidation resulting in better fiber wetting and less fiber damage. Slurry based powder impregnation techniques are simple, environmentally safe, and have been known for several years (1,2). In this investigation, a slurry based powder technique was used to produce towpregs for compression molding and pultrusion (3).

The powder impregnated tow is heated to fuse the powder on the tow before consolidation. The microstructure of the coated tow during heating is developed through inter-particle and particle-fiber interactions. The coated tows are then consolidated to form the composite product by application of heat and pressure in subsequent processing. The optimum degree of coating prior to processing, varies for different processes. Well-coated tows are preferred in continuous processes like filament winding and pultrusion, while partially coated flexible tows are preferred for the processes such as weaving so that handling the towpregs will be easier. The stages of the fiber coating were studied using an optical microscope with a heating stage and an environmental scanning electron microscope (ESEM) for nylon 11 impregnated glass fiber tows and PEKK impregnated glass fiber tows (4). The stages and the mechanisms of the fiber coating were identified. The polymer flow in the powder impregnated towpreg during consolidation, depends upon the extent of polymer coating and flow prior to consolidation, particle size, particle size distribution, and the flow characteristics of the polymer. Hence the effect of the processing parameters such as pressure and temperature on the consolidation of the powder impregnated towpregs is different from that of the other material forms such as melt impregnated prepregs. This was the motivation for studying the influence of the processing parameters on the consolidation behavior of the powder impregnated towpregs.

Background

Compression molding. Previous studies on thermoplastic consolidation have focused on the use of intermediate material forms such as commingled fibers (5), prepregs (6), tapes (7), and more recently on powder impregnated towpregs (8,9). Vodermayer et al. (10) calculated the optimal particle size of spherical powders in order to minimize the flow length, based on geometrical models. However the wetting behavior and the extent of coating were not considered.

In the present work, the influence of four key processing parameters on the degree of consolidation of the powder impregnated towpregs in compression molding was investigated. These parameters are consolidation time, pressure, temperature, and displacement rate. Full factorial experiments were used to determine the effect of each parameter and parameter interaction upon consolidation.

Pultrusion. Extensive experimental and analytical research has been done in the thermoset pultrusion area due to the ease of fiber impregnation with thermosets and availability of thermoset materials (11-19). Thermoplastic pultrusion has been investigated only recently. Models for thermoplastic pultrusion using melt impregnated prepregs have been presented in recent years (20-22). These models included analysis of pulling force, heat transfer in the die, and pressure due to compaction of the prepregs in the die.

Pulling force in the thermoplastic pultrusion depends on the processing parameters (temperature, pulling speed), the material parameters (type of polymer, fiber volume fraction, degree of impregnation), and the design parameters (die size and profile, die taper). Pulling force is a very important parameter. It must be studied in order to understand and design the process. Viscous resistance of the polymer layer between the fibers and the die wall, resistance due to compaction of the composite in the tapered section of the die, and frictional resistance between the die wall and the dry fibers, contribute to the total pulling resistance (21).

Pultrusion with the thermoplastic powder impregnated tows offers an attractive method to integrate impregnation and pultrusion in one process. Pultrusion with the powder impregnated tows involves coalescence of the polymer particles, followed by a film formation in the preheating stage and local (longitudinal and transverse) flow of the polymer during consolidation. In the present investigation, the effect of tow temperature and pulling speed on pulling force, in isothermal pultrusion of thermoplastic powder impregnated tows is studied.

Experimental Investigation

Material system. The thermoplastic resins used in this investigation were nylon 11, and polyetherketoneketone (PEKK). Volume average particle sizes determined using a Climet RL series Hydrocell liquid particle sensor (4) were 26.3 μm and 71.3 μm for nylon 11 and PEKK respectively. The fibers used were continuous E-glass (Owens Corning 497) with a silane based sizing. A slurry, consisting of the thermoplastic powder, a suspension agent (Hercules Klucel M), and a wetting agent (3M FC-430) was prepared for impregnating the fibers. The impregnated fiber tows were wound on the creels for pultrusion, and around a frame for compression molding as shown in Figure 1. PEKK impregnated tows (55 % fiber volume fraction) were used for the compression molding experiments. The pultrusion experiments were conducted with PEKK impregnated tows (65 % fiber volume fraction)

and nylon 11 impregnated tows (55 % fiber volume fraction).

Compression molding. The coated tows were wound to a 25 mm width on a frame to form a fiber charge as shown in Figure 1. A frame with strain gages was used to apply a known tension to the fibers before consolidation, in order to avoid fiber waviness and form a unidirectional composite. The wound fibers were placed in the consolidation apparatus, consisting of a matched die, a nitrogen chamber, and a mounting mechanism as shown in Figure 2. The apparatus was mounted on an MTS tensile testing machine to control and record the load and the displacement of the die during consolidation. A thermocouple was placed on the tow surface to monitor the composite temperature during consolidation. The wound fibers were consolidated to form a 25 mm x 203 mm x 3 mm unidirectional composite that was 70 % fiber by weight.

The consolidation apparatus was used to determine the effects of the consolidation time (time), pressure (press), displacement rate (rate), and temperature (temp) on the properties of the composite. A predetermined process cycle was used for all the experiments. The fiber charge was heated to a desired temperature in the die without applying pressure. The die was then closed using a constant displacement rate. The pressure was held constant during consolidation. The die was cooled at the end of the cycle until the polymer was sufficiently solidified before removing the composite plaque from the die.

A full 2^4 factorial study, which consisting of 16 individual experiments was conducted to determine the effect of each of the four process parameter interaction (see Table 1). High (+) and low (-) values of each of the process parameters were chosen based upon preliminary consolidation experiments (see Table 2). The values of each parameter were chosen to produce plaques with varying degrees of consolidation. The results, X_e, of the plaques were quantified through mechanical testing and determination of void content. Sixteen plaques were cut to 25 mm x 127 mm x 3 mm and tested in three-point bending to determine flexural modulus and strength. Void content of each sample was determined using Archimedes' principle. The results using both methods are listed in Table 1. The effects R_j, were evaluated using two different results criteria (X_e), void content and yield stress. The results of each of the sixteen experiments were recorded in an analysis of variance (ANOVA) table. The effect of each variable and interaction of one variable with another can be determined from this table.

The effect of a single parameter or parameter interaction, j, is determined by R_j :

$$R_j = \Sigma \, X_e|_j{}^+ - \Sigma \, X_e|_j{}^-$$

The sign of an interaction was determined by multiplying the signs of the individual parameters making up the interaction. The magnitude, $|R_j|$, determines the significance of the parameter or the parameter interaction, j.

The sign of R_j shows the influence of the parameter or parameter interaction level on the flexural strength or the void content.

Pultrusion. A pultrusion line consisting of a creel stand to mount the powder impregnated tows, a radiant tunnel preheater, a hot gas preheater, a die, and a pulling mechanism was used to conduct the experiments. A 100 mm long heated die was used to produce composite tapes with a rectangular cross section of 13 mm x 0.15 mm. The die assembly was mounted on a slide parallel to the pulling direction, and a load cell was used to measure the pulling force, as shown in Figure 3. A personal computer equipped with a data acquisition system was used to monitor and record the pulling force, radiant oven temperature, hot gas temperature, and the die temperature.

The temperatures of the radiant oven and the hot gas were adjusted to achieve the desired tow temperature at the die entrance, and the die was maintained at the same temperature as the tows. The experiments using PEKK impregnated tows were conducted at four different temperatures and seven different pulling speeds. Experiments using nylon 11 impregnated tows were conducted at three different temperatures and seven different pulling speeds (Table 3). Each experiment was conducted for five minutes and the average pulling force was recorded.

Results and Discussion

Microstructural evolution. The consolidation process consists of fiber bed compaction and polymer flow within the fiber bundle. ESEM pictures show that the polymer particles undergo morphological changes upon heating. They change from a randomly shaped solid phase to a rounded liquid phase and ultimately spreads along the length of the fiber (Figure 4). The increase in temperature causes reduction in surface tension and the viscosity. This is the reason for the phase change. Optical microscope pictures of the preheated powder impregnated tow surface show several channels available for the coalesced droplets to flow between the fibers with minimal resistance (Figure 5). Pictures of the consolidated tow surface show the channels between the fibers bridged by a smooth polymer film (Figure 5).

Scanning electron microscope (SEM) pictures of the preheated nylon tow cross section show that the polymer is distributed through the thickness of the tow (Figure 6). Smaller average particle size (20 μm) of nylon 11 results in uniform impregnation through the thickness of the tow. In addition, low melt viscosity of nylon 11 above its melting point increases viscosity driven flow. The cross section of the consolidated pultruded tape show uniform distribution of the polymer through the thickness of the tape (Figure 6).

SEM pictures of the preheated PEKK tow cross section show that a sheath of polymer is formed, while the core of the tow is not impregnated (Figure 7). Particle sizes that are large (Volume average 71.3 μm) compared to the fibers (10 μm in diameter) result in dense packing on the surface of the tow. In this case, inter-particle coalescence dominates the fiber coating causing a polymer sheath formation around the tow. Cross section of the PEKK pultruded tape shows that the polymer flow through the tow thickness occurs to consolidate the tow (Figure 7).

It is clear that local longitudinal flow occurs in the initial stages of the consolidation while transverse flow is necessary to complete the consolidation. The extent of the local longitudinal flow and the transverse flow depends on the average particle size, uniformity of the impregnation, and the viscosity-temperature relationship of the polymer.

Compression molding. Using void content and strength criteria, it was found that temperature was the most significant, displacement rate and consolidation time were equally significant, and pressure was the least significant processing parameters. The signs of all effects were positive, showing that the high (+) level of each variable positively influenced consolidation. The most significant positive interactions were pressure-displacement rate and time-displacement rate-temperature. The most significant *negative* interaction was time-pressure-displacement rate and the most *insignificant* interaction was among all four parameters. The experiment conducted with high levels of all the parameters produced the laminate with the highest degree of consolidation.

The local longitudinal flow and transverse flow should be maximized by using the optimum set of processing parameters to minimize cycle time. For a given towpreg permeability, the flow rate can be maximized by increasing the pressure gradient or decreasing the viscosity. Increasing the applied pressure causes greater fiber compaction and a decrease in permeability of the towpreg (23), explaining the relative insignificance of pressure on the consolidation.

The viscosity of the polymer can be reduced by increasing the temperature and increasing the shear rate. In compression molding, the shear rate of the polymer can be increased by increasing the displacement rate. Using the squeeze flow approximation, the average shear rate was calculated as the ratio of the displacement rate and the displacement. The viscosity of the polymer (PEKK) was measured with a capillary rheometer at high (+) and low (-) processing temperatures. Viscosity-shear rate data measured at a fixed temperature was fit to a power law expression. The viscosity of the polymer during the initial stages of compression molding was then obtained by substituting the calculated shear rate into the power law expression. The average ratio of the viscosity for high shear rate (+) and low shear rate (-) was 0.78. The average ratio of viscosity at high temperature (+) and low temperature (-) was 0.93. This reduction in the viscosity of the polymer due to increase in the shear rate and temperature explains the significance of the temperature and the displacement rate on the consolidation.

The tows occupy a greater volume before consolidation than after, due to the packing effects of the polymeric particles. Significant volume reduction of the tows occurs during consolidation. This volume reduction is due to squeeze flow, transverse flow, and coating of the fibers. Polymer flow is enhanced by the reduction in the viscosity of the polymer and coating of the fibers is driven by reduction in the surface tension. The displacement rate effect occurs over a very short distance during mold closing. The resulting squeeze flow occurs over a short time compared to the time taken for completing consolidation.

Pultrusion. Average pulling force values were plotted as function of pulling speeds at different temperatures for nylon 11 impregnated tows and for PEKK impregnated tows as shown in Figure 8 and Figure 9. Power law constants (A), power law exponents (n), and correlation coefficients (R) for each of the curve fits are listed in Table 4. The correlation coefficients were observed to be close to 1. This shows a good power law fit to the data at all the temperatures. A reference temperature 190 °C for nylon 11, and 340 °C for PEKK was selected. An arbitrary pulling force value in the observed range was then selected. The pulling speeds were calculated at the reference temperature (V_{ref}) and at the desired temperature (V_T) from the corresponding curve fit equations for the chosen pulling force. A shift factor (a_T) for a temperature (T) was then calculated as

$$a_T = V_{ref} / V_T$$

The shift factors were calculated for three different pulling forces and averaged. Hence one average shift factor for a temperature (T) was obtained. The shift factors for nylon 11 and PEKK are listed in Table 5. Log (pulling force) was plotted as a function of log ($a_T * V$) for nylon 11 and PEKK as shown in Figure 10 and Figure 11 respectively. All the points on the plots lie on the line represented by the reference temperature curves for both nylon 11 impregnated tows and PEKK impregnated tows (Figures 10 and 11). The resultant master curves shown in Figure 10 and Figure 11 clearly indicate the validity of the superposition method. Thus, by conducting a limited number of experiments for a particular material system and using the superposition method, one can predict pulling force over a wider range of temperatures and pulling speeds. Superposition of shear stress-shear rate curves for different temperatures was demonstrated in a similar manner for polymers by Mendelson (24). Validity of superposition in pultrusion indicates that the viscous resistance of the polymer film between the die wall and the outer fibers is the major component of the pulling resistance.

Some divergence of the data from the master curve was observed for nylon 11 at 190 °C and high pulling speeds (20 mm/sec) (Figure 10). This divergence from the master curve may be attributed to the shear thinning of nylon 11 and a corresponding reduction in the pulling force. Divergence of the data from the master curve was not observed at the higher temperatures for nylon 11 and at all the processing temperatures for PEKK.

Conclusions

Surface tension, capillary action, and squeeze flow contribute to the consolidation behavior of the powder impregnated tows. Localized longitudinal flow along the length of fibers dominates initial stages of consolidation of the thermoplastic powder impregnated tows, while transverse flow is necessary to complete the consolidation. The localized polymer flow during the consolidation depends on the extent of coating prior to the consolidation, fiber volume fraction, particle size, particle size distribution, and the flow characteristics of the polymer. Polymer sheath formation due to inter-particle coalescence occurs when the particle sizes are large compared to the fiber diameter.

Temperature and displacement rate were observed to be the most significant factors affecting consolidation of the thermoplastic powder impregnated towpregs based on the void content and the strength criteria. Experimentally based pulling velocity-temperature superposition of the pulling force curves for isothermal pultrusion of powder impregnated towpregs was demonstrated. Validity of the superposition process indicates that the major contribution to the pulling resistance is due to the viscous resistance of the polymer film between the die wall and the outer fibers.

References

1. Chabrier, C., Moine, G., Mavrion, R., and Szabo, R, U.S. Patent No. 4,626,306 (1986)
2. O'Connor, J. E., U.S. Patent No. 4,680,224 (1987)
3. Ramani, K., et al., *Processing, Fabrication, and Manufacturing of Composite Materials* ASME MD-Vol. 35, 115-129, (1992)
4. Ramani, K., Hoyle, C. J., and Parasnis, N.C. *Use of Plastics and Plastic Composites : Materials and Mechanics Issues* ,ASME, 46, 633-657, (1993)
5. Van West, V. P., Pipes, R. B., and Advani, S. G., *Polymer Composites* , 12, 417-427, (1991)
6. Lee, W. I., and Springer, G. S.,*Journal of Composite Materials* , 21, 1017-1055, (1987)
7. Wood, D., and Mantell, S. C., 38th *International SAMPE Symposium*, 21, 152-163, (1993)
8. Yang, Heechun, and Colton , Jonathan S. *Polymer Composites* , 15, 34-41, (1994)
9. Yang, Heechun, and Colton , Jonathan S. *Polymer Composites* , 15, 42-45, (1994)
10. Vodermayer, A. M., Kaerger, J. C., and Hinrichsen, G.*Composites Manufacturing* , 4, 123-132, (1993)
11. Chang Dae Han, and Dai Soo Lee, *Polymer Engineering and Science* , 26, 393-404, (1986)
12. Herng-Tay Wu, and Babu Joseph ,*SAMPE Journal* , 26, 59-69, (1990)
13. Price, H. L., Ph.D. Thesis, Old Dominion University, (1979)
14. Price, H. L., and Cupschalk, *Polymer Blends and Composites in Multiphase Systems*, C.D.

Han ed., American Chemical Society, (1984)

15. Batch, G. L., and Macosko, C. W., 42nd Annual Conference, Composites Institute, Society of Plastics Industry, Inc., (1987)

16. Gibson, A. G., Lo, C. Y., Lamb, D. W., and Quinn, J. A., *Processing and Applications*, 12, 191-197, (1990)

17. George Viola, et al., *35th International SAMPE Symposium*, 35, 1968-1981, (1990)

18. Lackey, E., and Vaughan, J. G., *Journal of Reinforced Plastics and Composites*, 13, 188-198, (1994)

19. Batch G. L., Ph.D. Thesis, University of Minnesota (1989)

20. Lee, W. I., and Springer, G. S., *Journal of Composite Materials*, 25, 1632-1652, (1991)

21. Astrom, T. B., and Pipes, R. B., *Polymer Composites*, 14, 173-183, (1993)

22. Astrom, T. B., and Pipes, R. B., *Polymer Composites*, 14, 184-194, (1993)

23. Gutowski, T. G., *SAMPE Quarterly*, 16, 58-64, (1985)

24. Mendelson, R. A., *Polymer Engineering and Science*, 8, 235-240, (1968)

Table 1. Experimental Matrix Used in Consolidation Experiments.

Test	Time	Press	Rate	Temp	Strength (GPa)	% Void
1	Low	Low	Low	Low	0.246	23.2
2	High	Low	Low	Low	0.610	19.7
3	Low	High	Low	Low	0.101	29.9
4	High	High	Low	Low	0.934	15.6
5	Low	Low	High	Low	0.273	28.8
6	High	Low	High	Low	0.579	26.9
7	Low	High	High	Low	0.963	14.5
8	High	High	High	Low	0.349	12.4
9	Low	Low	Low	High	1.135	17.0
10	High	Low	Low	High	0.494	15.3
11	Low	High	Low	High	0.429	20.3
12	High	High	Low	High	0.852	12.3
13	Low	Low	High	High	0.462	16.1
14	High	Low	High	High	1.704	5.2
15	Low	High	High	High	1.246	7.9
16	High	High	High	High	1.851	4.1

Table 2. Values of Low and High Parameters Used in Consolidation.

Parameter	Low	High
Time	750 sec	1500 sec
Press	2.6 MPa	5.2 MPa
Rate	0.207 mm/sec	0.415 mm/sec
Temp	295 °C	325 °C

Table 3. Experimental Matrix Used in Isothermal Pultrusion Experiments.

Pulling Speed (mm/sec)	2	5	8	11	14	18	21
Temperature PEKK (°C)	340	350	360	380			
Temperature nylon 11 (°C)	190	210	230				

Table 4. Power Law Coefficients for PEKK and nylon 11 Data.

PEKK				nylon 11			
Temperature (°C)	A	n	R	Temperature (°C)	A	n	R
340	49.96	0.198	0.944	190	89.6	0.168	0.86
350	46.84	0.186	0.97	210	45.64	0.339	0.906
360	42.24	0.219	0.977	230	31.98	0.37	0.904
380	40.67	0.214	0.992	-	-	-	-

Table 5. Shift Factors for PEKK and nylon 11.

PEKK (reference temp. - 340 °C)		Nylon 11 (reference temp. - 190 °C)	
Temperature (°C)	Shift Factor (a_T)	Temperature (K)	Shift Factor (a_T)
350	0.6085	210	0.452
360	0.570	230	0.229
380	0.447	-	-

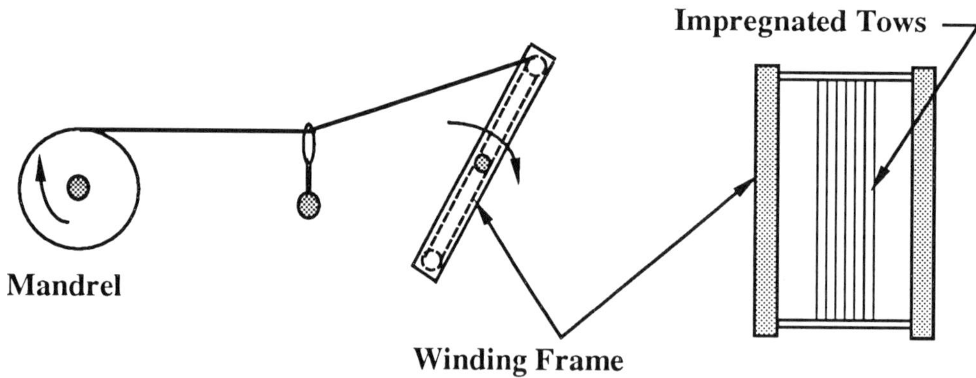

Figure 1. Winding Frame Used for Consolidation Experiments.

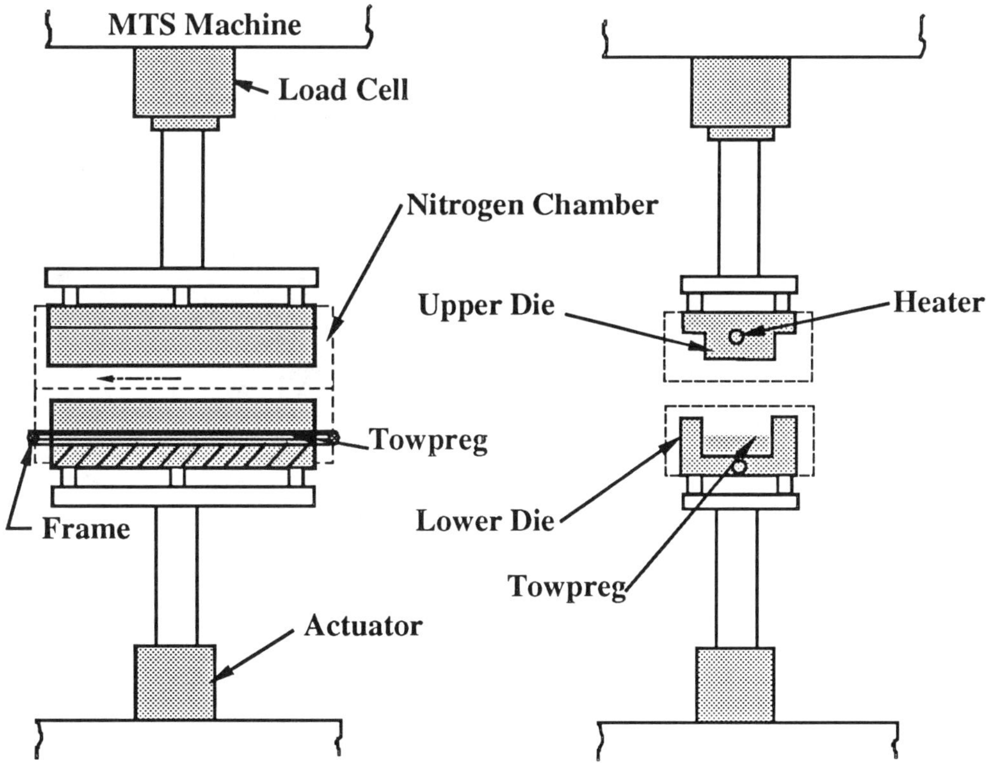

Figure 2. Consolidation Apparatus.

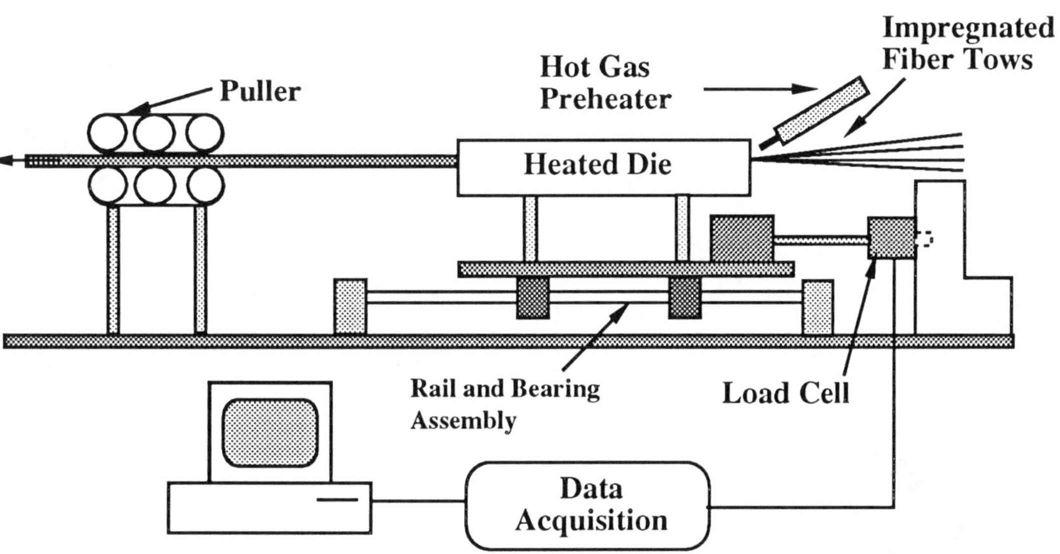

Figure 3. Schematic Diagram of the Experimental Pultrusion Facility.

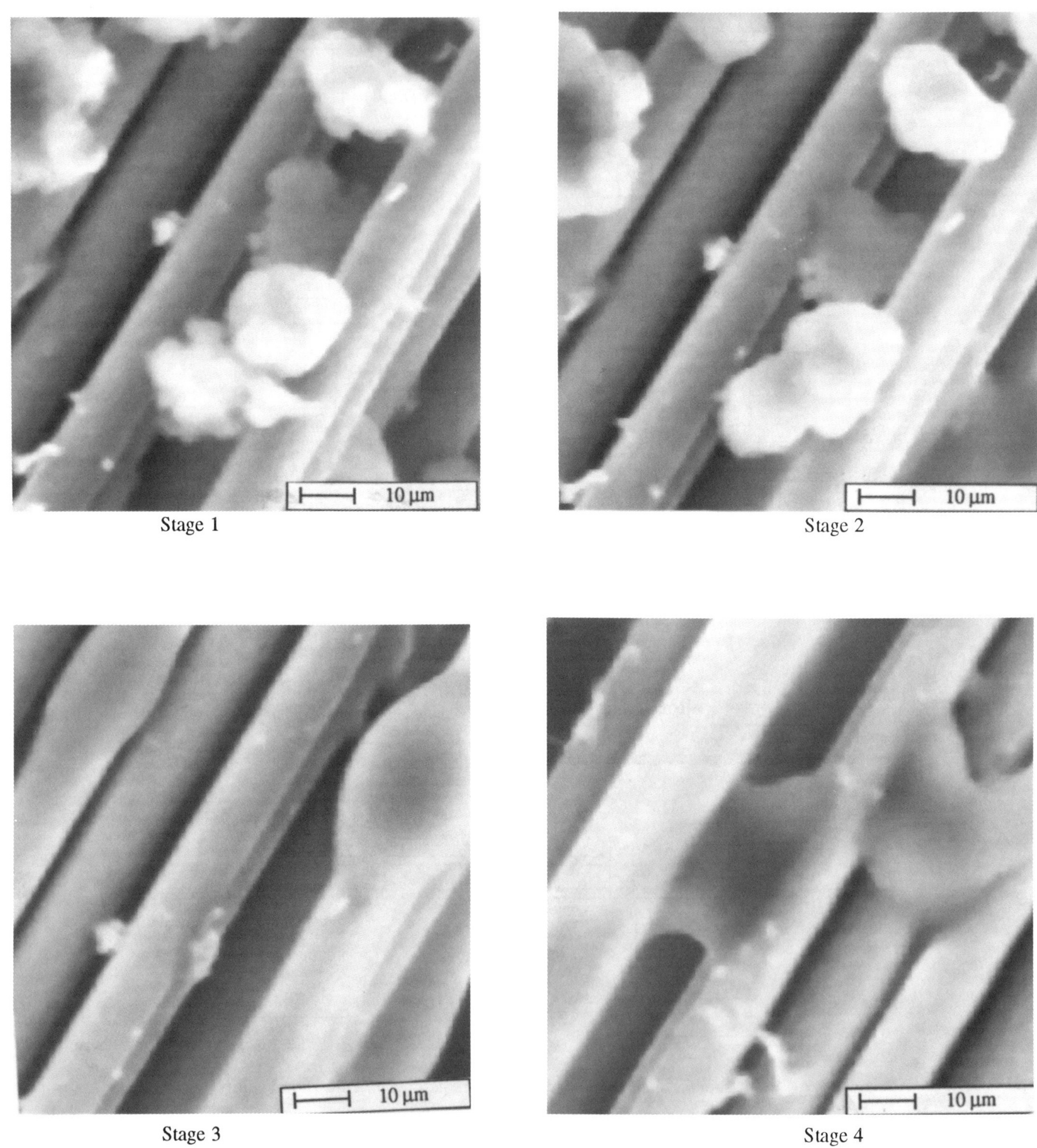

Stage 1   Stage 2

Stage 3   Stage 4

Figure 4. ESEM Pictures Showing the Stages of Polymer Flow.

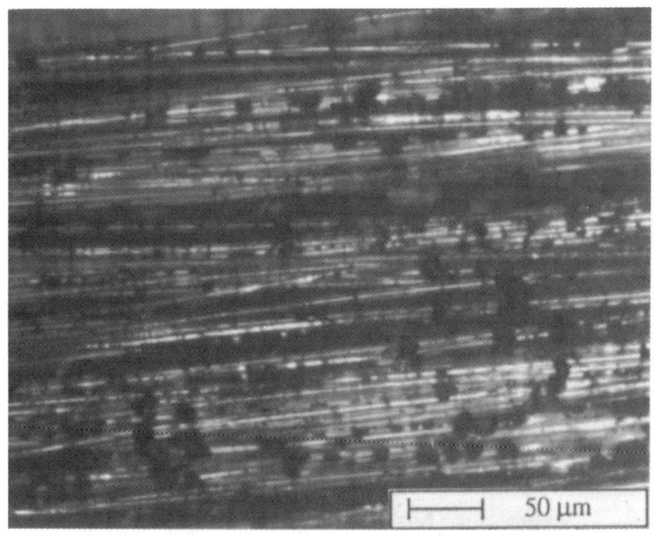

Stage 1. Dry Impregnated Tow (before Heating)

Stage 2. Preheated Tow.

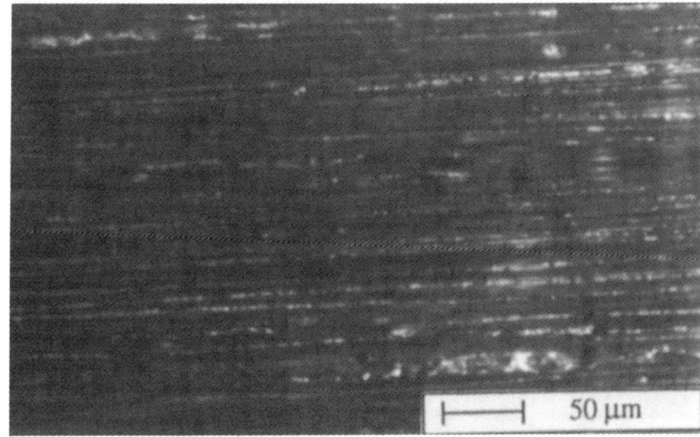

Stage 3. Consolidated Tow.

Figure 5. Changes in Towpreg Microstructure during Consolidation.

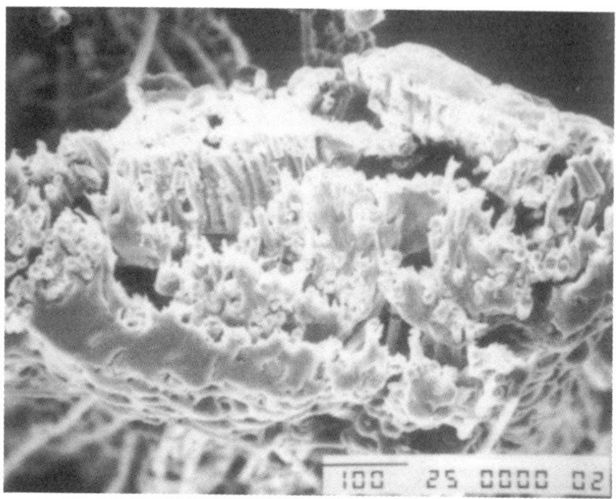

Cross Section of the Preheated Nylon 11 Tow

Cross Section of the Preheated PEKK Tow

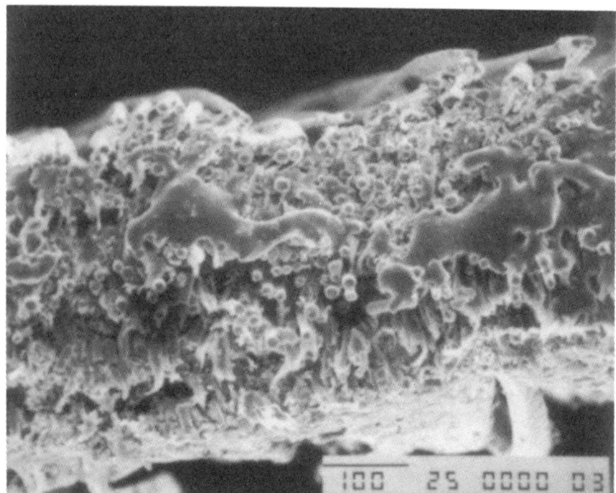

Cross Section of the Pultruded Nylon 11 Tape

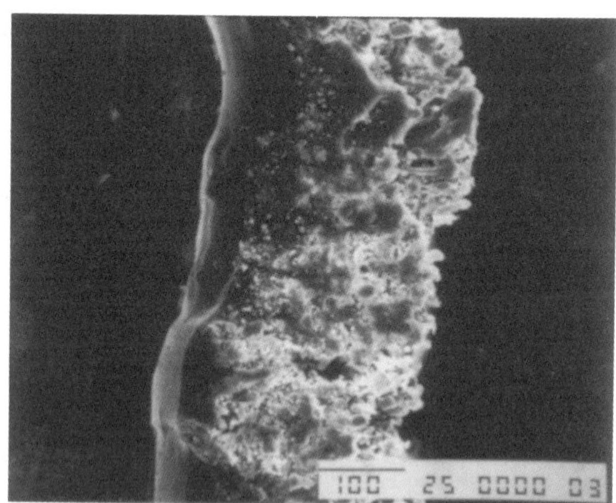

Cross Section of the Pultruded PEKK Tape

Figure 6. Cross Section of Nylon 11 Tows Before and After Consolidation.

Figure 7. Cross Section of PEKK Tows Before and After Consolidation.

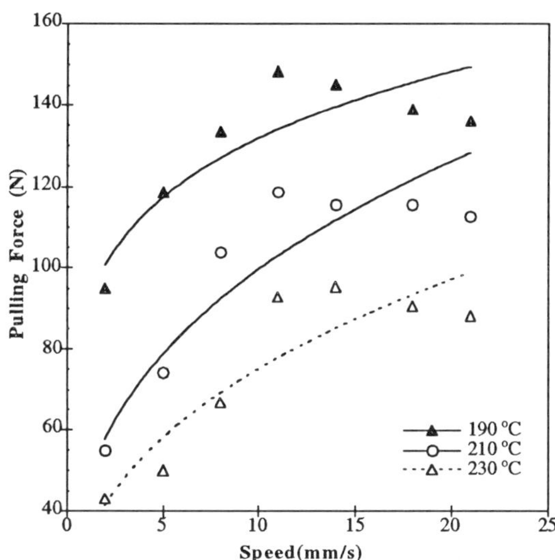

Figure 8. Pulling Force as a Function of Pulling Velocity
(Nylon 11 Impregnated Glass Fiber Tows).

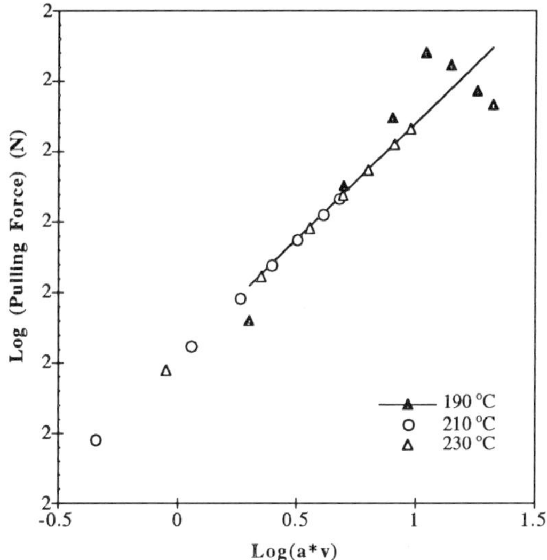

Figure 10. Master Curve for Nylon 11 (Reference
Temperature 190 °C).

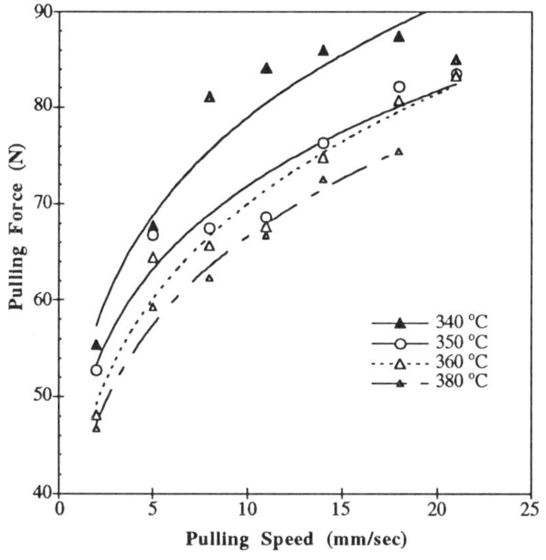

Figure 9. Pulling Force as Function of Pulling Velocity
(PEKK Impregnated Glass Fiber Tows).

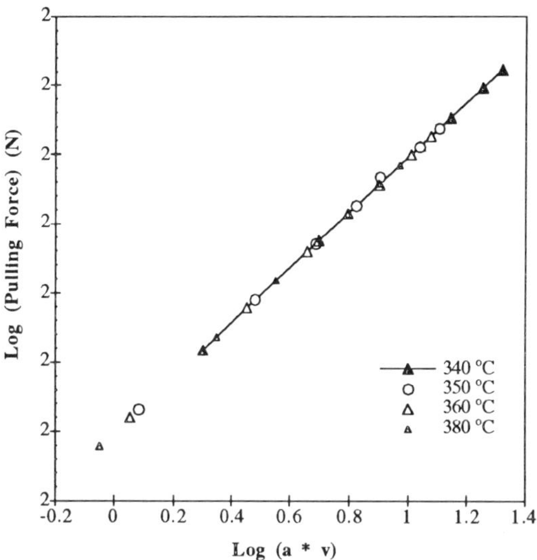

Figure 11. Master Curve for PEKK (Reference
Temperature 340 °C).

239

Proceedings of the 10th Annual ASM/ESD Advanced Composites Conference, Dearborn, Michigan, USA, 7-10 November 1994

Copper Alloy Molds Provide Quality and Economic Advantages for Composites in Transportation Applications

K.C. Apacki
Isorca, Inc.
Granville, Ohio

Abstract

Polymer composites require a mold to make parts, and recent investigations have shown that mold material selection has an significant impact on part quality and production costs. This paper summarizes and highlights research that significantly adds to our understanding of resin transfer molding (RTM) and the effects of mold materials on temperature distribution and laminate curing. Research sponsored by the International Copper Association is presented that shows how the thermal diffusivity of copper alloys enables the mold to reach steady state quickly, to maintain a more uniform temperature, to dissipate reaction exotherm, to improve temperature control, and to provide more uniform curing of the part. These advantages can be translated to economic advantages by reducing cycle time, improving production efficiencies, and lowering the initial cost of the mold. The thermal and economic benefits are available from a wide range of copper alloys that eliminate the need to compromise hardness or strength.

RESIN TRANSFER MOLDING (RTM) and structural reaction injection molding (SRIM) are composite fabrication processes that have been recognized as having high potential in several manufacturing industries. They are attractive for high performance, advanced composites for aerospace applications and are the object of considerable attention for their potential in high volume manufacture of automotive components. To this end, molders must choose mold materials to achieve production rates and quality that allow these processes to deliver the economic advantages anticipated.

Major advantages of RTM and SRIM processes include low mold filling pressure and low mold clamping force. These attributes have encouraged exploitation of mold materials which are themselves made from composites, such as aluminum filled fiberglass reinforced epoxy, that have lower material costs and shorter lead times. However, for thermally activated thermoset resins used in RTM and SRIM, mold temperature significantly influences the temperatures in the mold cavity, hence resin reaction, and thus overall cycle time. Additionally, the quality of the composite part relies on uniform cure, high ultimate resin conversion, good resin to fiber bonding, and low residual stresses.

Two studies, summarized and highlighted in this paper, have added significantly to our understanding of resin transfer molding and the influence mold materials have on productivity and laminate quality. They clearly indicate the distinct superiority of copper alloy as a mold material because of its excellent thermal properties.

Laminate Temperature Distributions During Impregnation

In the first study Lebrun, Rudd, and Gauvin (1) reported on their investigation of the effects of mold and preform temperatures, the fiber content, and mold filling time on the temperature distribution in the laminate during impregnation. Laminates were molded using a rectangular plaque aluminum mold 540 mm x 520 mm (21.25" x 20.5"). The laminate thickness was 3.5 mm (0.138"). The mold was fitted with thermocouples and pressure transducers to monitor and record temperature and pressure during the resin transfer molding. A gallery configuration for the gate was used to obtain a predominantly one dimensional flow of the resin during the mold filling phase.

A standard continuous filament random mat with a thermoplastic binder (Vetrotex Unifilo U750-450) was used as the reinforcement. The resin system was a medium reactivity orthophthalic polyester resin (Synolac 6345.001), catalyzed to produce a single part, hot setting formulation.

All of the preforms were made by stacking the mats in the same direction so that the resin flow was in the weft direction with respect to the roll of mat. Trials were made varying the fiber volume fraction, mold and preform temperatures, and resin supply pressure. Pressure at the gate was adjusted to achieve an approximately constant filling time of 30 seconds for each trial.

Results and Discussion. The results showed that the effect of fiber content on temperature distribution is not as important as the mold temperatures. A twofold increase in fiber content showed little effect on temperature distribution. On the other hand the resin heat up is heavily influenced by the mold temperatures, especially at the gate and between the gate and the first thermocouple (12.5 cm away from the gate). It was interesting to see that over this short length, the resin temperature rose to approximately 70% of the difference between the mold temperature and the gate temperature.

The results also showed that the effect of the filling time on temperature distribution was negligible for the time scale involved. The results indicate that the fiber surface area coming in contact with the fluid and the contact time are not significant compared to the temperature difference between the resin and the preform and mold temperature. This suggests that the mold *materials* have a significant effect on resin temperature distribution as shown by the following research.

Evaluation of Thermal Characteristics of Alternative Mold Materials

The second study, by Perry *et al.* (2) is an evaluation of the influence of a variety of RTM mold materials on fiber reinforcement heating, resin injection, resin curing, and subsequent mold cooling. One mold was made from each of five mold materials, aluminum filled epoxy/fiberglass composite, 1018 steel, C95400 Aluminum bronze alloy, 6061 Aluminum, and C18200 chrome copper alloy. Strip heaters were installed along the length of the top and bottom of the mold platens. Table 1 shows the thermal properties of each metal mold material.

The properties of the aluminum filled epoxy composite are difficult to estimate and only its relative qualities are shown.

The dimensions of the mold cavity were 279 mm x 127 mm x 9.53 mm (11" x 5" x 0.375"). By changing the spacer, thinner samples were molded at a thickness of 3.16 mm (0.125"). The molds were instrumented with four thermocouples arranged to allow observation of temperature gradients in the resin flow direction as well as in the source heat flux direction from the strip heaters. The cavities were instrumented with five thermocouples also arranged to measure temperature gradients in the resin flow direction and from the mold surface to the center of the laminate.

Molded specimens were made using an automotive type RTM resin (Ashland Chemical Arotran D-1479-8) and continuous strand fiber glass mat reinforcement (OCF-M8610). The mold cavity was loaded with eleven layers of fiberglass mat to yield a fiber volume fraction of 20 percent. The mats were positioned in the mold cavity 12.7 mm (0.5") from the inlet to ensure axial flow through the mat. Resin was injected into the mold at 448 kPa (65 psi). The trials were made varying the mold materials only and observing the distribution of the mold and mold cavity temperatures.

Thinner composite specimens were molded for the purpose of evaluating the influence of mold material on molded composite quality. Thinner specimens were molded with the set point temperature below the optimum recommended so that analysis of residual reactivity would give evidence of any cure gradient in the composite. Dog bone tensile coupons were cut from these thinner composite samples at the gate and outlet regions.

Mold Heating. Table 2 shows the temperature gradient experienced in the mold material as the mold was heated to steady state (60°C). Molds made of materials with higher thermal conductivity heat up faster than materials of lower thermal conductivity.

The observed behavior follows the trend in thermal diffusivity (see below). The chrome copper mold has a higher thermal conductivity than the steel mold. Thus the

Table 1. Mold Material Thermal Diffusivity

MOLD MATERIAL	Thermal Conductivity (W/m°K)	Density (gm/cm³)	Specific Heat (J/kg°K)	Thermal Diffusivity* (cm²/s)
Epoxy Composite**	much lower	much lower	similar	much lower
1018 Steel	51	7.82	486	0.134
C95400 Aluminum Bronze	59	7.45	420	0.189
6061 Aluminum	167	2.70	896	0.690
C18200 Chrome Copper	324	8.89	385	0.947

Source: Perry, *et al.* (2)
* Thermal Diffusivity = (Thermal Conductivity) / (Density * Specific Heat)
** Author's approximation of properties compared to steel

Table 2. Temperature Gradients During Mold Heating

MOLD MATERIAL	Greatest Temperature Difference (°C)	Temperature Difference at Steady State (°C)	Time to Reach Steady State (min.)
Epoxy Composite	13	5	>40
1018 Steel	3	1.75	40
C95400 Aluminum Bronze	3	1.75	40
6061 Aluminum	2	1	20
C18200 Chrome Copper	1.5	<0.5	20

Source: Perry, *et al.* (2)

mold temperatures are more uniform as heat dissipates throughout the mold more readily. The steel mold has a higher thermal mass than the chrome copper mold. This higher capacity to store heat resulted in a magnified temperature difference throughout the steel mold. The differences would have been greater for the more popular tool steels that have conductivities as low as one half that of the carbon steel used in this evaluation (see Table 5).

Thermal Diffusivity. Examination of thermal properties in Table 1 provides an explanation for many of the trends in mold temperature behavior. Thermal diffusivity, the ratio of thermal conductivity to thermal mass, is the appropriate property to consider in an analysis of thermal characteristics. Thermal conductivity is a measure of the rate at which a material can transfer heat. Thermal mass, the product of density and specific heat, is a measure of the capacity to store heat. Thermal diffusivity thus accounts for both the rate of heat transfer and the capability to store heat.

The most uniform temperatures would be observed in a material with a large thermal diffusivity; that is, having a low capacity to store heat (a small denominator) and which quickly transferred heat throughout the material and away from the heat source (a large numerator).

The chrome copper alloy has the largest thermal diffusivity and exhibited the most uniform temperatures in the mold. A larger maximum temperature difference and an observable steady state offset for the aluminum mold is explained by its lower diffusivity. Comparable thermal diffusivities of 1018 steel and C95400 aluminum bronze explain the small difference in behavior of the two molds. Even more so, the low thermal conductivity of the epoxy composite mold caused the large thermal gradients compared to the chrome copper alloy mold.

Resin Injections and Exotherm. Temperature profiles were generated by monitoring the mold and cavity thermocouples while resin was injected, cured, and cooled. Analysis of these profiles showed that the differences in the thermal properties of the mold materials had a significant effect on the laminate. Table 3 summarizes some of the observations by Perry, *et al.*

As room temperature resin is injected into the preheated mold, the temperature of the mold drops as the resin is heated and flows toward the outlet. The maximum temperature swing of the mold at the gate is seen to be the lowest for the molds with high thermal conductivity.

Table 3. Thermal Response of Mold and Laminate During Resin Injection and Exotherm

MOLD MATERIAL	Mold Temp. Swing at Inlet (°C)	Lag Time to Reaction Initiation (min.)	Peak Exotherm Temp. at Outlet (°C)	Cavity Temp. Swing from Injection to Peak Exotherm (°C)
Epoxy Composite	41	4.1	122	91
1018 Steel	10	2.9	137	107
C95400 Aluminum Bronze	10	2.6	130	100
6061 Aluminum	7	2.2	119	89
C18200 Chrome Copper	7	2.3	117	96

Source: Perry, *et al.* (2)

It was observed that the epoxy mold was significantly different from metal molds. The maximum temperature difference in the epoxy mold due to resin injection and cure exotherm was about six times higher than the difference in the high thermal conductivity aluminum and chrome copper alloy, and about four times higher than in the aluminum bronze and steel molds. Additionally, the temperatures in the epoxy mold after the exotherm was completed were much higher and took longer to cool to steady state than in the metal molds. In contrast to metal molds, the temperatures on the surface of the cavity of the epoxy mold were influenced greatly by the injected resin. The surface temperatures dropped to approximately the same temperature as the incoming resin.

The peak exotherm temperatures in the metal molds follow the trend in thermal conductivity. As the thermal conductivity decreased, the peak temperature increased. However, the epoxy mold was significantly different from the expected trend. One explanation for the lower than expected peak temperature in the epoxy mold is the extremely long lag time to reheat the resin prior to reaction onset. Table 3 shows that the lag time from injection to reaction onset in the epoxy mold was nearly double the lag time for the metal molds. Thus low mold temperatures and slow heat transfer from the epoxy mold to the resin caused a slower reaction with subsequent lower peak exotherm temperatures.

It was observed that in the epoxy mold the heat up time took the longest and temperatures remained high for some time after reaction was complete. While the metal mold cavity temperatures had nearly settled back to steady state after 10 minutes, the epoxy temperatures were still 40 degrees higher than the initial mold temperatures.

Again, the observed temperature profiles follow the trend in thermal diffusivity of each mold material. The temperature profiles of the 6061 aluminum and C18200 chrome copper alloy were observed to follow the same trends. Table 1 shows that the thermal conductivity of the C18200 chrome copper alloy is about twice that of 6061 aluminum, however the density of the former is over three times that of the latter resulting in similar thermal diffusivities.

Similarly, the temperature profiles for 1018 steel and C95400 aluminum bronze alloy were closely matched as a result of their similar thermal diffusivities. However, the slightly higher thermal diffusivity of C95400 aluminum bronze alloy compared to the 1018 steel allowed the peak exotherm temperature to be lower by about 7 degrees as shown in Table 3.

The epoxy temperature profile stood out distinctly from the metal molds because it has such a low thermal diffusivity. The temperature of the epoxy mold dropped much farther and stayed down longer than in the metal molds following resin injection. The epoxy mold just could not cool and reheat as quickly as the metal molds. In a production application in which short cycle time and temperature control are desired, this could cause a serious reduction in productivity.

Molded Part Quality. Thinner samples were molded at 3.18 mm (0.125") at an initial reaction temperature 10 degrees below that recommended by the resin supplier to emphasize the influence of the molds on molded composite quality. Small samples of cured resin were extracted from inlet and vent locations and weighed and tested in a differential scanning calorimeter for residual reaction heat.

Table 4 shows the residual heat from the outlet samples. The least amount of residual heat was measured for the samples molded in the epoxy mold. The reason for this was the temperatures were higher in the aluminum filled epoxy composite mold as the controllers had difficulty steadying the initial temperatures and the epoxy retained heat during cure exotherm.

In contrast, the composite molded in the C18200 chrome copper alloy had the greatest residual heat. The controllers readily kept the chrome copper alloy at the setpoint temperature prior to resin injection and the mold dissipated heat readily as the exotherm temperature peaked. Because of excellent temperature control and heat management, the chrome copper alloys lend themselves best to consistent molding conditions for high quality composite parts as well as increasing throughput. Intermediate residual heats were measured for the composites molded in steel and aluminum bronze.

Table 4. Influence of Mold Materials on Molded Composite Quality

MOLD MATERIAL	Residual Heat of Outlet Laminate (cal/g)	Laminate Tensile Strength Gradient (MPa)	Tensile Strength Gradient (% outlet)
Epoxy Composite	1.36	30.7	20
1018 Steel	2.49	20.7	13
C95400 Aluminum Bronze	1.67	16.1	12
C18200 Chrome Copper	3.04	15.1	10

Source: Perry, *et al.* (2)

There was a cure gradient in all composites molded in the different molds. The outlet region had less residual reactivity than the inlet area. The cure gradient is due to the thermal history of the resin. The resin was injected at room temperature into a heated fiber reinforcement. The resin gained heat as it flowed farther into the mold. Hence, resin farthest from the inlet had a higher initial reaction temperature so cure proceeded to a higher conversion. This cure gradient is reflected in the composite performance as the tensile test results in Table 4 show.

The gradient of tensile strength from inlet to outlet shows a trend that suggests more uniform properties were obtained in the composite molded in the C18200 chrome copper alloy. The tensile strength differences from inlet to outlet were largest in the composite sample molded in the aluminum filled epoxy composite. The progression of composite tensile strength difference followed the inverse order of the thermal diffusivity. As the thermal diffusivity increased from epoxy to steel to aluminum bronze to chrome copper alloy, the tensile strength gradient from inlet to outlet decreased.

Injection Molding

Similar research (3,4) was conducted for injection molding where the goals are to control temperatures and remove heat from a thermoplastic resin in the mold. The heat transfer characteristics of six copper alloys of varying thermal conductivities, strengths, and hardness were compared to P-20 steel during the injection molding cycle. With up to 70 percent of the cycle time required for cooling, heat exchange in the mold is extremely important, as are the thermal properties of the mold materials.

Stanton *et al.* (3) and Sudit *et al.* (4) showed that significantly lower cavity surface temperatures can be achieved through the use of copper alloys. Heat buildup was also considerably less in the more difficult cooling regions of the molds. Furthermore, the temperature distributions in copper alloy molds with high thermal conductivity were considerably more uniform than the distribution in the P-20 steel mold, which is an advantage since cavity temperature uniformity is the primary objective of cooling system design.

A reduction in the cooling time was also anticipated and realized for copper alloy molds. As much as 25 percent reduction in cycle time may be achieved by using molds made from highly conductive copper alloys. Reduced cycle time results in lower cost parts as hourly labor and overhead is applied over more units. In addition, the fixed capital equipment produces more revenue without the need for additional capital investment.

Copper Alloys for Molds

It is generally accepted that the cost of materials in a mold is overshadowed by the design, machining, and finishing costs of the mold (5). While copper alloys may generally cost more than conventional steel and aluminum materials, copper alloys provide advantages that may actually reduce the finished mold cost and provide economic advantages throughout the entire production lifetime.

Figure 1 shows that copper alloys can provide thermal conductivities two to three times that of steel or aluminum at any desired surface hardness. While hardness is not the only measure of wear resistance, Table 5 shows that there is no need to sacrifice mold life to get thermal performance. Copper alloys allow the thermal design to be better at a lower cost. Heating and cooling lines can be fewer and spaced farther from the mold surface because the thermal conductivity is better. For cast molds these lines can be cast into the metal, reducing machining and increasing design flexibility by cooling hard to reach locations.

Surface machining rates and finishing rates are enhanced with the use of heat treated copper alloys. Casting copper alloys to near-net-shape reduces machining time and costs to a minimum. Modifying and repairing damaged molds is facilitated with the excellent welding and brazing characteristics of copper alloys.

An investigation by Stein (6) of nickel-aluminum bronze for use in cast-to-size molds for advanced composites showed that aluminum bronze is the preferred material based on the evaluation of weldability, pattern shrinkage, castability, surface quality, thermal expansion, and corrosion resistance. These properties make copper alloys attractive for most plastic molding processes.

Molded Product Cost

In order to estimate the economic impact that reduced molding cycle time and improved molding efficiency would have on the costs of molded parts, the International Copper Association commissioned ISORCA, Inc. to develop a economic analysis of a large compression molded satellite antenna dish (5).

Using the advantages of casting aluminum bronze and designing a mold that would insert the cast aluminum bronze surface into a steel master die, the cost of the mold was compared to a machined P-20 steel mold. The results indicated that the cast copper alloy mold concept reduced cost by 19% and mold delivery time by two to four weeks.

Combining the lower initial mold cost with the thermal benefits of aluminum bronze -- shorter molding cycle time and improved molding efficiency -- the economic benefits of highly conductive molds were estimated. The estimates showed that the incremental effect of an individual benefit is relatively small. However, when all the benefits are included in an optimal case, the projected savings per part ranged from 5 percent to 15 percent. The higher volume production providing increased economic returns as the economic benefits are multiplied with each unit produced.

While this analysis points to promising economic advantages, the need to demonstrate the processing advantages is being addressed by the International Copper Association. ISORCA, Inc. is currently preparing a cast copper alloy mold for an automotive production application. Once in production, the performance will be

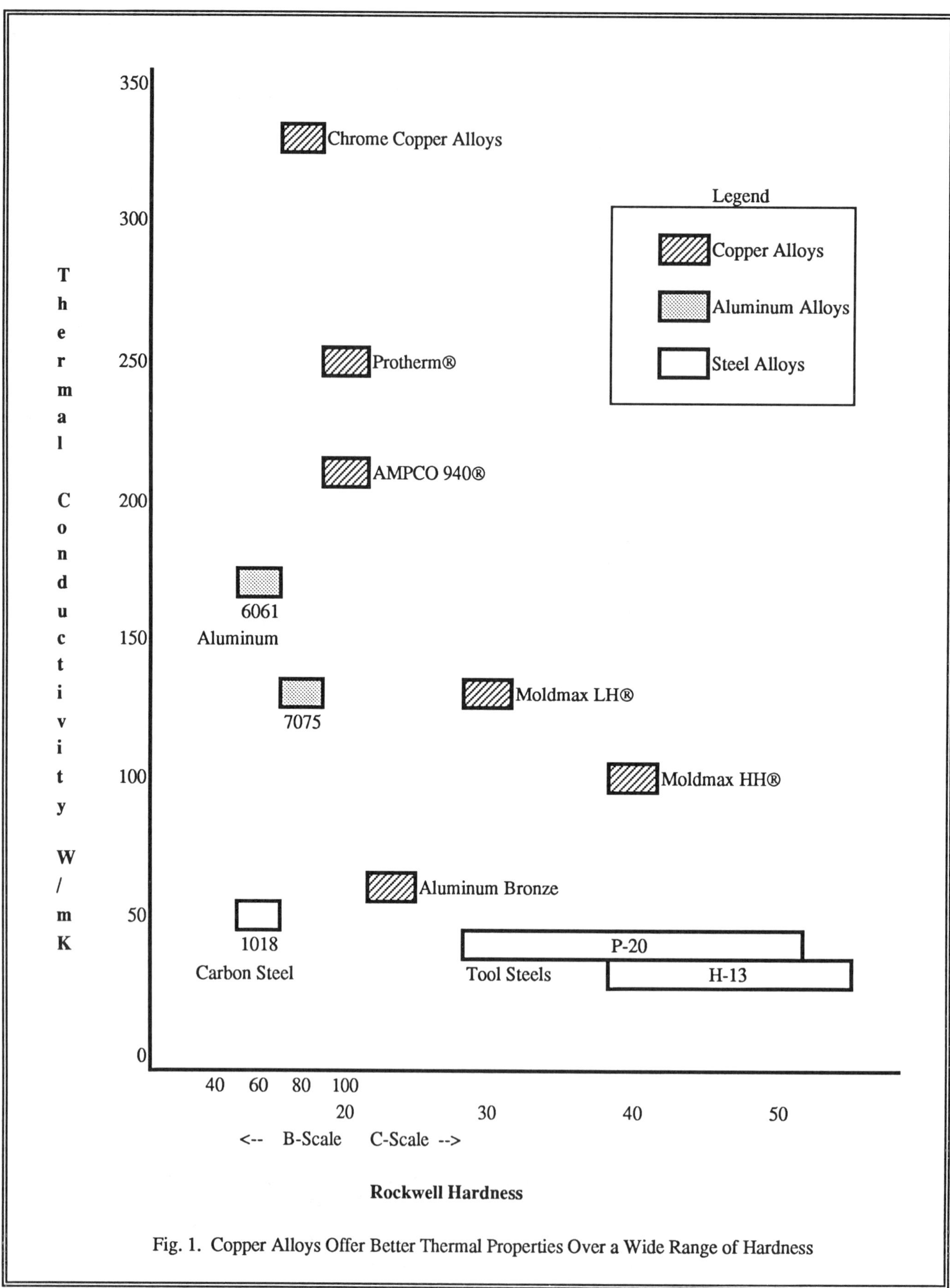

Fig. 1. Copper Alloys Offer Better Thermal Properties Over a Wide Range of Hardness

actively observed and documented and compared to conventional steel molds. The results will be presented and published upon completion of the project.

Conclusions

The work highlighted above has contributed significantly to the growing understanding of RTM and the benefits of using copper alloy molds for plastic processing. The results provide us with the following conclusions:

- The effects of mold and preform temperature are more important than fiber volume fraction or filling time on resin temperature distribution during the injection stage of resin transfer molding.

- In general, as thermal diffusivity increases, temperatures in the mold and the molded composite are more uniform, resulting in improved molded composite properties.

- Steady state temperatures are reached most quickly and uniformly in copper alloy molds. Mold length temperature gradients during dynamic mold and reinforcement heating follow the trend in thermal diffusivity. The most uniform temperatures are maintained in copper alloy and least uniform occur in aluminum filled epoxy composite.

- The large lag time to reheat an epoxy mold following resin injection significantly influences the composite cure. The large lag time causes a lower initial reaction temperature and a slower reaction which results in a lower peak exotherm temperature compared to the composites molded in metal molds. After completion of resin cure, the four metal molds dissipate the excess exotherm heat in a matter of minutes. Epoxy composite molds retain a considerable amount of heat for an extended time.

- Copper alloy had the highest degree of controllability. As thermal diffusivity of the molds decreases, the magnitude of the overshoot, oscillations and rise time increases such that control performance degrades.

Table 5. A Comparison of Mold Alloy Properties

CATEGORY	UNS Number	Common Name	Thermal Conductivity (W/m°K)	Rockwell Hardness	Tensile Strength (MPa)
Carbon Steel	G101800	1018 Carbon Steel	51	B70	400-440
Stainless Steel	S42000	Type 420	25	C27-54	1720
Tool Steel	T20813	H13	25	C38-54	1420
	T51620	P20	38	C28-50	1010
Aluminum	A96061	Type 6061 T6	167	B60	276
	A97075	Type 7075 T6	130	B88	462
Aluminum Bronze	C95400	Aluminum Bronze	59	C24	585-723
Beryllium Copper	C17200	Moldmax HH®	105	C40	1275
Alloy	C17200	Moldmax LH®	131	C30	1170
	C17510	Protherm®	253	B96	792
Chrome Copper	C18200	Chrome Copper	324	B80	234-590
Alloy					
	C18400	Chrome copper	325	B60-75	345
Chrome Silicon Copper Alloy	C18000	Ampco 940®	216	B94	620

Sources: Handbooks and Product Literature
Moldmax HH®, Moldmax LH®, and Protherm® are products of Brush Wellman, Inc.
Ampco 940® is a product of Ampco Metal

- Deliberate choice of low reaction temperature allows observation of a cure gradient along the length of the molded composite, which is reflected in the tensile strengths of the molded specimens. Tensile strengths near the outlet are higher than at the inlet. Additionally, the difference in strength is smallest for composites molded in copper alloy and the largest for aluminum filled epoxy composite.

- Using mold materials with high thermal conductivity can result in faster cycle times and reduced variability in part dimensions and properties. Reduced cycle time results in lower cost parts and increased revenues without the need for additional capital equipment.

- Molders should seize the the opportunity to exploit the productivity advantages of copper alloy with its high thermal diffusivity and without loss of mechanical properties that protect the life of the mold. It can be seen in Table 5 and Figure 1 that there is a wide selection of copper alloys available.

- The designers and buyers of molds for plastic processing should consider the use of copper alloys for molds cast to near net shape to take advantage of cast-in-place cooling and heating lines, reduced machining requirements, shorter delivery times, and lower mold costs.

Acknowledgments

The author acknowledges the International Copper Association for supporting the investigations of the thermal characteristics of mold materials and assisting with the preparation of this summary.

References

1   Lebrun, G., C.D. Rudd, R. Gauvin. *Laminate Temperature Distribution During Non-Isothermal Impregnation of Fiber Preforms*, 49th Annual Conference, Composites Institute, The Society of the Plastics Industry, Inc. February, 1994.

2   Perry, M. J., Y. Ma, J. Xu, T. J. Wang, M. J. Liou, L. J. Lee. *Evaluation of Thermal Characteristics of Alternative Tooling Materials in Resin Transfer Molding*, Engineering Research Center for Net Shape Manufacturing, The Ohio State University, June 1992.

3   Stanton, K. J., M. J. Liou. *A Study of Copper-Alloy Mold Materials for Use in Injection Molding*, Engineering Research Center for Net Shape Manufacturing, The Ohio State University, October, 1990. Report No. ERC/NSM-P-90-32.

4   Sudit, I., K. J. Stanton, G. R. Glozer, M. J. Liou. *Thermal Characteristics of Copper-Alloy Tooling in Plastic Molding*, Engineering Research Center for Net Shape Manufacturing, The Ohio State University, October 1991, Final Report ICA Project No. 397.

5   Gerson, F.T. *Fresh Approaches to Mold Steel Selection*, Society of Plastics Engineers, ANTEC '91 Conference Proceedings, May 1991, page 1769.

6   Stein, D. *Tooling Parameters of Nickel-Aluminum Bronzes*, Man-Tech Development Institute, April, 1988, Final Report INCRA Project No. 397.

7   Clark, J.O. *A Comparison of Cast Aluminum Bronze Tooling versus Machined Steel Tooling on Molded Composite Product Cost*, Isorca Associates, Inc., December 1988, Final Report ICA Project No. 410.

Proceedings of the 10th Annual ASM/ESD Advanced Composites Conference, Dearborn, Michigan, USA, 7-10 November 1994

On-Line Process Control for Liquid Molding Using Contour Maps for Identifying Process Behavior

G. Mychajluk, S. Manoochehri
Stevens Institute of Technology
Hoboken, New Jersey

R.S. Parnas
National Institute of Standards and Technology
Gaithersburg, Maryland

ABSTRACT

An on-line control algorithm that uses computer-generated contour maps to predict physical process behavior is used to control a liquid molding process. The maps are generated off-line and represent a set of feasible system operations for a wide range of injection parameters. These feasible operations are stored as set-point values in a data base for rapid on-line retrieval of information. Set-points are assigned a direction corresponding to a minimum cycle-time solution for the given input parameters. The directions are generated using a nonlinear optimization search technique that provides a minimum cycle-time solution subject to constraints imposed on the process behavior model. The specified constraints define the limits of feasible system operations by adding relationships between process set-point values and product performance specifications to the model of the liquid molding process. Optimum set-point values are output to base-level controllers that maintain these set-points during an injection cycle. These base-level controllers utilize closed-loop feedback control algorithms. On-line sensor measurements provide process inputs to a high-level controller that compares these inputs to set-points in the data base. The high-level control algorithm implements cycle-to-cycle adjustments by modifying set-point values based on an analysis of the differences between process inputs and set-points in the data base. After the control system converges to the optimum set-point values, process disturbances are categorized as either external disturbances or model inaccuracies. These disturbances are identified based on the measured discrepancies between a particular process input and a corresponding set-point value.

LIQUID MOLDING is a versatile process that has considerable potential for use in high-volume manufacturing of structural composite components. The basic process consists of injecting a prepolymer into a mold containing a pre-placed dry fiber reinforcement. During impregnation of the reinforcement, the prepolymer undergoes chemical reactions which cause the viscosity, pressure and temperature fields to change with time. After the fiber reinforcement is completely saturated with resin, the cure reactions continue past the gel-point to produce a solid material. The kinetics of the cure reactions are defined by the choice of resin system but may be manipulated during processing by controlling the temperature.[1,2] Control of the curing reactions is complicated, however, because large temperature gradients are often generated in rapidly curing systems due to the exothermic nature of the polymerization reactions. Temperature gradients across the thickness of the mold cause residual thermal stresses to develop which contribute to part warpage and excess shrinkage.[3-6] During autoclave processing, thermal stresses have been shown to initiate matrix cracks and cause delamination in thick laminated thermoset composites.[7] Cure kinetics, part geometry, heat transfer parameters, and the temperature cure cycle are variables that influence spacial temperature gradients during a process cycle. Since the temperature cure cycle is the only variable that is not necessarily defined prior to processing, the mold temperature is a key variable that can be used to manipulate the spacial temperature gradients during the cure cycle.

Patel, Rohatgi and Lee [8] studied the influence of mold temperature and pressure on tensile and interfacial bond strength for liquid molded composite parts. An important conclusion from their work is that the resin and fibers must already have bonded before the polymer gel-point is reached. They also found that void minimization and temperature control were needed to increase fiber wetting and adhesion at the interface. Mold pressure was controlled to reduce voids, and the pressure control mechanism achieved a balance between viscous and surface tension forces which controlled resin impregnation of the fibers.

The first goal of the present work is to develop a control methodology which accounts for the tight coupling among the resin flow, cure, and the final part properties of molded composite parts. The dependence of mechanical strength on process temperatures, pressures, and degree of cure will be identified and incorporated into an integrated system model. The integrated system model will consist of the feasible system operations which are defined by all combinations of processing parameters satisfying both the processing and performance constraints. Additionally, the integrated model will compute the cycle time, which affects manufacturing costs, for each choice of processing parameters. The process variables defining the feasible system operations can be used as set-point values in a base-level control system.

Current research [9] in control system design has led to the development of transfer-function-based feedback-control algorithms that are often more effective than the classical PID algorithm, and include algorithms such as Internal Model Control (IMC), Simplified Model Predictive Control (SMPC), and Conservative Model Based Control (CMBC). These algorithms assume that the feasible system operations, defined by set-point values, can be derived by numerical techniques prior to on-line implementation of the control system. The method has been successfully implemented in a closed loop control strategy through the use of a CMBC algorithm for rapid regulation of mold pressure and cavity packing for an injection molding cycle.[10] The significance of that work is that rapid process regulation is achieved based on feasible system operations obtained from an independent, off-line model of system behavior.

A limitation of any feedback control system is the inability to provide feasible system operations when arbitrary process disturbances are introduced. For example, material batch contamination is a common process disturbance. When this type of disturbance occurs it is possible for a feedback control system to maintain ideal process set-point values during an injection cycle and produce low-quality composite parts. An effective control system must have the capability to identify the cause of arbitrary process disturbances based on an on-line comparison of actual process behavior to the predicted behavior. This requires a high-level control system to implement an on-line analysis which operates independently of the feedback controller.

The second objective of this work is to develop a high-level controller to identify process disturbances, determine corrective action, and implement cycle-to-cycle adjustments for on-line process control. Process adjustments are made by modifying the set-point values that are used during the injection cycle by the base-level feedback controllers. Such adjustments can be implemented by arranging the feasible system operations generated by the integrated system model into digital process contour maps for a wide range of injection parameters. The maps are stored in a data base and are used by the high-level controller as an interactive tool for identification of process disturbances. The structure of the contour maps enables a comparison of measurements of actual process temperatures and pressures to the corresponding process set-point values. Additional comparisons are also possible between the measured degree of cure and the predicted degree of cure if a cure sensor is available. Process disturbances are identified by the location of the state of the actual process on the map. These disturbances are categorized based on measured discrepancies between a particular process input and a corresponding set-point-value. The set-points within the feasible design space are assigned a direction that corresponds to a minimum cycle-time solution for a given set of input parameters. This direction is identified using a nonlinear-optimization model which generates the digital data for the contour maps. The optimization model provides a minimum cycle-time solution in the form of process set-point values that are subject to constraints imposed on the system behavior.[11] System behavior is currently defined by mathematical models [12,13] that predict the flow and cure behavior of the resin for a specified set of material properties and injection parameters. The specified constraints are incorporated into the behavior model in the form of equations and act as the mechanism to create an integrated model of feasible system operations.

Currently, the Polymers Division of the National Institute of Standards and Technology (NIST) is conducting an experimental study for proof of concept and control algorithm development. A selected range of contour map set-points will be tested on a plaque mold filled with unidirectional glass fibers. Results of these tests will be used to determine the sensitivity of mechanical properties to processing variables. The intent is to provide a matrix of sensitivity derivatives which relate mechanical properties to processing conditions. This sensitivity matrix will be incorporated into the flow and cure models

defining process behavior in the form of constraint equations added to the existing cycle-time optimization model.

CONTROL ALGORITHM

The overall control strategy consists of both on-line and off-line components. The off-line portion is a cycle-time optimization model that is used to generate an integrated system model representing feasible system operations for a wide range of injection parameters. The optimization technique used in this work is Powell's method with a penalty function [14,15], which is a constraint-based-nonlinear-programming solution method. Constraints are imposed on mathematical models [12,13] predicting the flow and cure behavior. These constraints are specified in the form of equations and define the limits of feasible system operations based on relationships between process set-point values and product performance specifications. For example, the unconstrained process flow and cure models determine set-point values for mold inlet and holding pressures. A constraint equation evaluates void size using the calculated set-point holding pressure as an input variable. This void size is then compared to the maximum allowable void size that will satisfy product mechanical strength requirements. Mold holding pressure set-point values are included in the set of feasible system operations only if the calculated void size satisfies the mechanical strength requirements. Each constraint equation represents a particular relationship between a process parameter and product performance requirement. The integrated system model is defined as the set of feasible system operations satisfying all constraints imposed on the process behavior model.

Formulation of the liquid molding process optimization begins with transcribing the relationships between process parameters and product performance requirements into a well-defined mathematical statement. For a specified resin system, the integrated system model currently consists of relations between mold temperature and injection pressure to resin spacial temperature gradients and void volumes. Void volume is estimated based on mechanical entrapment of air located inside the fiber bundles of the reinforcement preform. This mathematical formulation is important because feasible system operations are determined from cycle-time optimization output. The processing hardware must also be identified in the optimization model. For example, a plaque mold is defined by cavity length, width, depth, and overflow dimensions. Cure kinetic parameters and material density define the resin system. The preform is defined by fiber volume fraction, permeability and geometry. These inputs are used to find processing variables that minimize cycle-time. For this work the initial resin injection temperature and pressure were chosen as design variables.

The optimization model is important to development of an on-line control strategy because, in addition to providing the initial optimal set-points, it provides a set of feasible system operations in the form of set-point values. The database of set-points is structured such that feasible system operations define a digital contour map consisting of lines of constant cycle-time. Therefore, the contour map can be used to determine the direction from any feasible system operation towards an optimal feasible system operation (i.e. minimum cycle time). This digital process contour map is the link between a base-level feedback controllers and the high-level controller. In a more general case, the digital contour map will include final part properties and cycle-time as part of the definition of the state of the system. An illustration of the control strategy is shown in figure 1. Set-point values that correspond to a minimum cycle-time solution of the optimization model are input to base level controllers. These controllers operate independently of each other in feed-back loops designed to maintain the processing parameters at the set-point values. Currently, base-level controllers are used to maintain mold temperature and inlet/outlet pressures. During an injection cycle, sensors measure the physical process parameters and these

values become available to both base-level controllers and to the high-level controller.

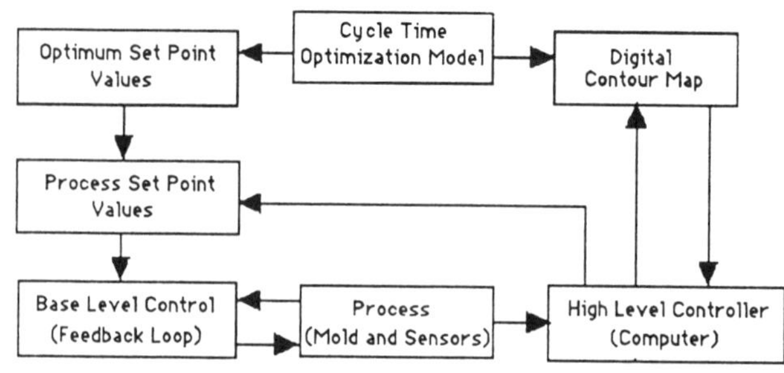

Figure 1. Control strategy schematic indicating the off-line optimization and the on-line process controllers.

Process parameters include pressure, temperature, and cure sensor measurements. Pressure will be measured at the mold inlet and outlet. Temperature data consists of mold wall, resin injection, and cavity midplane temperatures. In the high-level controller, sensor measurements are compared to set-point values stored in the contour map. The high level controller evaluates this comparison and implements cycle-to-cycle process adjustments by modifying set-point values that are input to the base-level controllers for the next injection cycle.

Figure 2 illustrates the algorithm currently under development for the high-level controller. The high level controller will utilize material, mechanical properties and process sensor measurements to make adjustments to base-level set-points. Material property inputs will include stiffness, density, permeability, etc., and are characterized prior to processing. Since the mechanical properties are expected to be a function of material properties and processing conditions, and cannot be directly measured during processing, they will be predicted by using a matrix of sensitivity derivatives relating such properties to processing behavior.

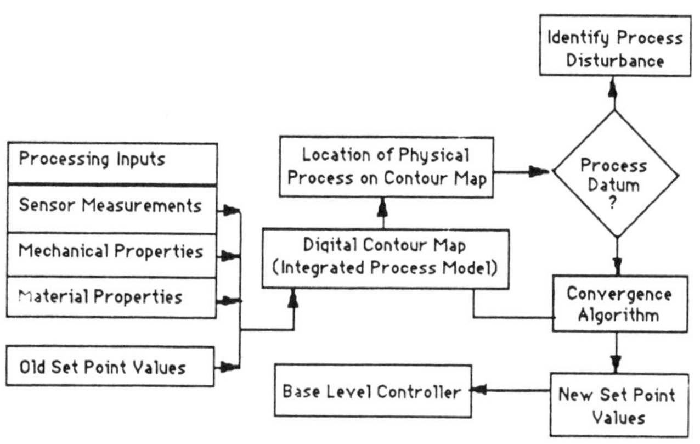

Figure 2. High-level controller design indicating the use of input data and the digital contour map for generating new setpoints.

Thus, the high level controller will use a combination of models to predict final part properties based on process measurements. Currently, the high level controller does not include a prediction of final part properties, and thus the location of the process on the contour map is directly related to processing variables only. The high level controller will be able to determine the position of the actual physical process on the contour map with respect to optimal molding parameters. This identifies a direction to adjust base-level controller set-points for the next injection cycle. A convergence algorithm is being designed to determine the magnitude of set-point adjustments. Due to time constraints required for on-line calculations, set-point adjustments will probably be performed independently of one another with a one-dimensional search algorithm.

The high-level controller also identifies process disturbances. Differentiating the two functions of set-point adjustment and process disturbance identification is critical to maintaining process stability. The task of identifying process disturbances will not be attempted until the high-level controller has converged the system on an optimal operating condition. Once converged, the control system is eliminated from consideration as a cause of process disturbances. Then, residual differences between the process optimum and the originally computed optimum are expected to be due to either external disturbances or model inaccuracies. External disturbances are categorized based on measured discrepancies between a particular process input and a corresponding set-point value. For example, during an injection cycle, it is possible for cavity-under-packing to occur while the base-level controller maintains the designated pressure set-point values. A discrepancy between the pressure set-point value and process input represented by the weight of the molded part signals a possible tooling-related disturbance. The high-level controller will determine that the mold gate dimensions should be checked based on the magnitude of this discrepancy.

HARDWARE DEVELOPMENT

The liquid molding hardware is designed for proof-of-concept and control-algorithm development. Figure 3 is an illustration of the experimental setup.

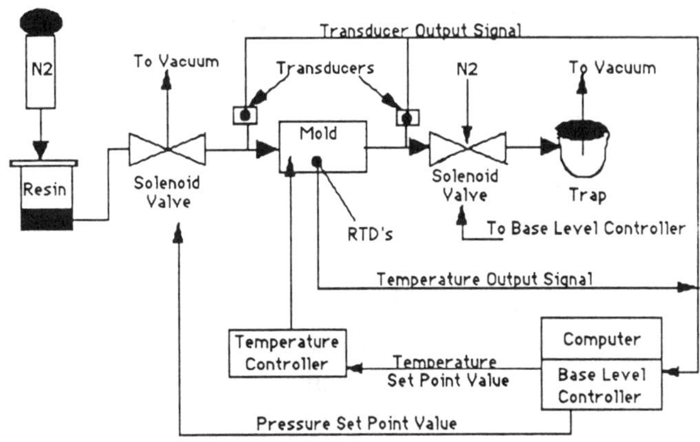

Figure 3. The liquid molding process control experiment illustrating the temperature and pressure control loops. An cure sensor will also be included and will supply a signal directly to the high-level controller in the computer.

The plaque mold is 20 cm long, 16.5 cm wide and 0.63 cm deep. Top and bottom platens of the mold are cast aluminum with cast-in electrical heating elements and cooling tubes (Watlow). The mold is also equipped with bayonet fittings to support Resistance Temperature Devices (RTD's) which measure platen temperatures. Midplane resin temperature and degree of cure are measured on-line during an injection cycle with a thin-film RTD (Omega F3105) and evanescent wave fluorescence sensor (EWFS), both manually positioned in the mold prior to an injection cycle. Mold inlet and outlet injection pressures are monitored with pressure transducers (Omega PX603) and independently controlled with base-level controllers.

The mold platen heating elements are connected to a stand-alone controller (CVL Instruments) to maintain a remote set-point value using a PID control algorithm. Air is allowed to flow through the cooling tubes during an injection and cure cycle. This allows for both mold heating and cooling as specified by the controller regulated input set-point value. The set-point value is output to the stand-alone controller from a microcomputer (50 MHz i486DX) which monitors process temperature, pressure and degree of cure via signals received from sensor inputs. These inputs are filtered through a signal conditioner (Omega OM5). The computer is programmed to output set-point values to the stand alone controller and act as a high-level control system. Mold pressure control is obtained by manipulating pressure in the resin feed container. A 3-way solenoid valve is connected to the feed container from both a high pressure source (Nitrogen Cylinder) and a low pressure source (Vacuum pump). The solenoid valves are activated by base level controllers and are manipulated between the high and low pressure sources to maintain the desired set-point value.

The computer program which implements the high-level control algorithm is written in Microsoft Visual Basic. Pressure and temperature set-point values are input to the system from the off-line cycle-time optimization model. During a process cycle these set-point values are output to base-level controllers programmed for closed-loop feedback control of these set-point values. The base-level control algorithms are programmed on a Data Acquisition Processor board (DAP 1206E CyberResearch), which has an on-board 25MHz i486SX micro-processor in addition to signal input and output capability, and is mounted in the 50 MHz i486DX host computer.

EXPERIMENTATION

This research combines theoretical development of a model system behavior with experimentation to incrementally design and test the control system. Experiments are being used to study the response of mechanical properties to time-varying processing temperature, pressure, degree of cure and viscosity. Initially, a plaque mold filled with unidirectional glass fibers is injected with an epoxy resin mixed with a curing agent. Data acquisition for degree of cure is obtained using an EWFS being developed by the NIST Polymers Division. Quantitative knowledge of relationships between processing and performance is necessary to verify the general applicability of the sensor development effort to composite processing, and to develop a knowledge base for control synthesis. As in many complex processes, sensitivity derivatives are not constants, but may vary significantly due to physical nonlinearities. In such cases, an understanding of the nonlinearity is critical to developing robust control strategies. Nonlinearities are well known in the molding of composites. Two causes are resin cure-kinetics and the formation of residual porosity. Residual porosity, or void fraction, is a critical variable in composite processing since it strongly influences a number of mechanical properties. Obtaining control of void fraction during processing is one of the major objectives of this program. Nonlinearity in the resin cure-kinetics stems from classical Arrhenius behavior of chemical reactions, coupled with heat transfer properties of the curing part and molding equipment. Thus, properties intermediate between processing variables and mechanical properties, such as void fraction and resin cure, are to be measured as part of the effort to determine the sensitivity matrix.

A selected range of contour map set-points will be tested to define a cause and effect relationship between set-point values and process inputs shown in figure 2. Data will be correlated and used to structure the contour map so that particular process disturbances are identified based on the location of the actual physical process on the map. At the time of this writing, the performance of the injection system, mold, and in-mold temperature sensors have been verified by producing high quality parts. The EWFS has been incorporated into the parts but the degree of cure was not monitored during the production cycle. However, after the molding was completed, but before the mold was opened, the optical fiber protruding from the mold was connected to the fluorescence optical system to verify that an evanescent wave signal could be obtained from the part in-situ.

SUMMARY

An on-line process control algorithm has been presented for the liquid molding process that uses contour maps to predict model behavior. The research combines theoretical development of an integrated model system behavior with experimentation to incrementally design and test the control system. Experiments are used to study effects of mechanical properties on time varying process parameters. These relationships are incorporated into the control algorithm as constraint equations and are part of a cycle-time optimization model used as an off-line tool to generate contour maps. The data base of contour maps allow comparison between time measurements of actual process temperature, pressure and degree of cure, taken during an injection cycle, to corresponding process set-point values stored in the data base. Process disturbances are identified based on the location of actual process parameters on the map and determine which set-points should be changed for the next injection cycle. One advantage of this methodology over traditional design and control techniques is a systematic approach accounting for an integrated process model is developed. On-line evaluation of a process provides the engineer with a flexible tool to predict and control multi-variable process behavior and results in faster cycle time and higher quality parts.

ACKNOWLEDGEMENT

The authors gratefully acknowledge the careful review of the manuscript and the helpful comments by Dr. Brian Dickens of the NIST Polymers Division.

REFERENCES

1. Adams,D.C."Cure Behavior of Unsaturated Polyester Resin Composites," Technical Report CCM Report 88-16, Center of Composite Materials Composite Manufacturing Science Laboratory, University of Delaware, Newark, Delaware, 1988.
2. Kamal, M.R. and S. Sourour, "Kinetics and Thermal Characterization of Thermoset Cure," Polymer Engineering and Science, January 1973, Vol. 13, No.1, pp.59-64.
3. Bogetti, Travis A., "Process-Induced Stress and Deformation in Thick-Sectioned Thermossetting Composites," Technical Report CCM Report 89-32, Center of Composite Materials Composite Manufacturing Science Laboratory, University of Delaware, Newark, Delaware, 1990.
4. Levitsky, Myron and Bernard W. Shaffer, "The Approximation of Temperature Distributions in Homogeneous Exothermic Reactions," The Mechanical Engineering Journal, vol 5, 1973, pp.235-242.
5. Shaffer, Bernard W. and Myron Levitsky, "Thermoelastic Constitutive Equations for Chemically Hardening Material," Transactions of the ASME, Sept. 1974, pp.650-652.
6. Levitsky, Myron and Bernard W. Shaffer, "Thermal Stresses in Chemically Hardening Elastic Media with Application to the Molding Process," Journal of Applied Mechanics, Sept.1974, pp. 647-650.
7. Bogetti, Travis A. and John Gillespie Jr., "Residual Stress and Deformation in Thick Laminated Composites Undergoing Chemical Hardening and Shrinkage," Technical Report CCM Report 89-34, Center of Composite Materials Composite Manufacturing Science Laboratory, University of Delaware, Newark, Delaware, 1990.

8. Patel, N, V.Rohatgi and L. James Lee, "Influence of Processing and Material Variables on Resin-Fiber Interface in Liquid Composite Molding," Polymer Composites, April 1993, Vol. 14,No. 2, pp. 161-172.

9. Nunn,Robert and C. Grolman, "Adaptive Process Control for Injection Molding", Journal of Reinforced Plastics and Composites, May 1990, Vol9, pp. 282-298.

10. Smud, Silva. D. Harper, P. Deshpande and K. Leffew, "Advanced Process Control for Injection Molding", Polymer Engineering and Science, August 1991, pp. 1081-1085.

11. Mychajluk, G. and S. Manoochehri, "An Integrated Process Model and Cycle Time Optimization for Resin Transfer Molding," Proceedings of the 1994 ASME Winter Annual Meeting, Nov. 1994.

12. Parnas, Richard S. and Frederick R. Phelan Jr., "The Effect of Heterogeneous Porous Media on Mold Filling in RTM," Sampe Quarterly, Jan. 1992, pp. 53-60.

13. Bogetti, Travis A. and John Gillespie Jr., "Two Dimensional Cure Simulation of Thick Thermosetting Composites," Technical Report CCM Report 89-23, Center of Composite Materials Composite Manufacturing Science Laboratory, University of Delaware, Newark, Delaware, 1989.

14. Fox,Richard L., Optimization Methods for Engineering Design: Addison-Wesley Publishing Company, Reading Massachusetts, 1971.

15. Arora, Jasbit., Introduction To Optimum Design; McGraw-Hill, Inc., New York,1989.

An Evanescent Wave Fluorescence Sensor for Process Control of RTM

J. Dunkers, D.L. Woerdeman, R. Parnas
National Institute of Standards and Technology
Gaithersburg, Maryland

Abstract

Fluorescence has been demonstrated to be an accurate measurement of resin cure and has been used with distal-end fiber optic sensors in liquid molding. Evanescent wave fiber optic sensors offer the additional advantage of sensing the cure very close to the fiber surface and, therefore, provide a means to determine interfacial properties in fiber reinforced composites. Evanescent wave sensors, however, require optical fibers with a refractive index larger than the refractive index of the cured resin. In the past, application of optical fiber sensor technology has been limited by the low refractive index of the optical fiber compared to the resin. An economical optical fiber sensor has been developed with a refractive index in excess of 1.61, permitting evanescent wave monitoring of most typical liquid molding resins. Experiments are being conducted with the optical fiber embedded into a high fiber volume fraction preform, and the resin cure is followed to maximum conversion. The fluorescence signal can be correlated with resin cure by comparison with infrared spectra collected at several temperatures.

RESIN TRANSFER MOLDING (RTM) is a versatile and efficient process for producing fiber-reinforced polymer composite structures ranging from small articles of simple shape to large articles with complex shape and high structural performance characteristics (1). In the basic process (Figure 1), a reactive pre-polymer is pumped into a mold with a preplaced dry fiber reinforcement. After the mold is filled, time is allowed for polymerization. When the part has gained sufficient strength, it is demolded. The chief advantage of RTM over other composite forming processes is the ability to make large parts of complex shape at a lower cost.

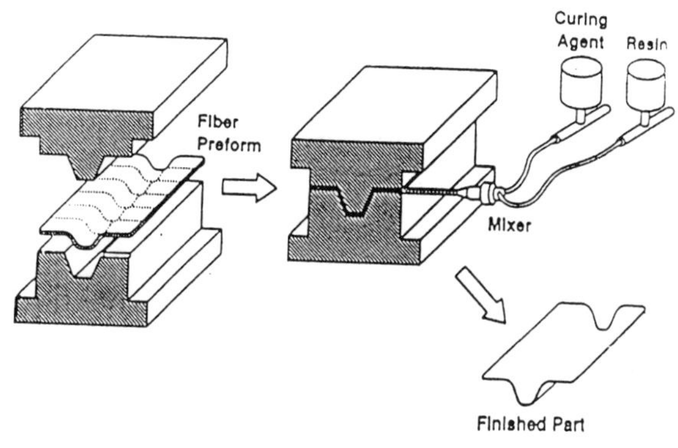

Figure 1. Schematic diagram of the RTM process.

While RTM has been in use for nearly 50 years (1) it is just now beginning to find use in high performance composite applications and mass production situations requiring fast cycle times. Process conditions are typically determined empirically with molding experiments, making process development expensive and processing parameters difficult to optimize. A part of the effort to reduce development time focuses on process control and the in-mold sensors required to measure the process variables needed by the controllers. An important process variable is the resin cure, and a number of sensors have been developed based upon ultrasonic, dielectric, microwave, and spectroscopic measurement methods (2-4). For example, fluorescence monitoring has been used with distal mode optical fiber sensors for cure monitoring of the injection molding process (5). Figure 2 displays the typical fluorescence information obtainable with a widely used viscosity sensitive probe, 1-(dimethylaminophenyl)-6-phenyl hexatriene (DMA/DPH) (6).

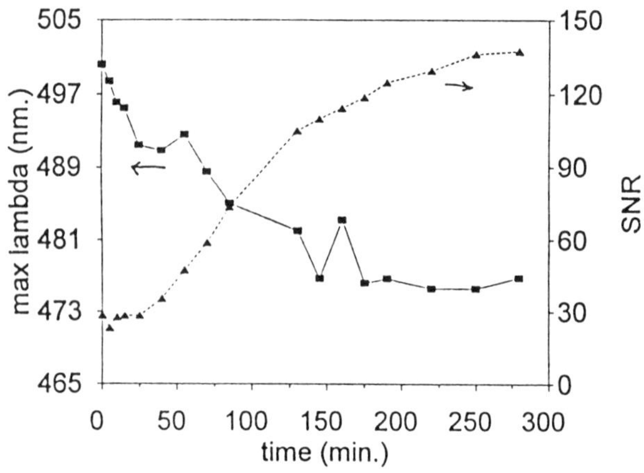

Figure 2. The response of a widely used viscosity sensitive probe, DMA/DPH, showing the change in fluorescence intensity, SNR, and peak position, max lambda, during the cure of epoxy resin.

Note that if the changes in intensity of the fluorescence signal are used to monitor resin cure, then an internal standard must be used to maintain accuracy. Identifying a dye for an internal standard that is both viscosity-insensitive and does not fluoresce in the same wavelength range as DMA-DPH has proven to be nontrivial (7). In this work,

a fluorescent probe was used that displays a large fluorescence wavelength shift as the resin cures, eliminating the need for an internal standard (8).

Some previously developed sensors suffer the drawback that they measure the degree of cure averaged over a large volume of the mold and, in many cases, averaged over the entire mold thickness. However, the resin cure kinetics can vary substantially across the thickness of the mold due to thermal effects and the cure kinetics close to the fiber surfaces may also vary substantially from the bulk resin cure kinetics due to surface induced chemical potential gradients.

Although localized sensors do exist and have been applied to cure monitoring of polymer composites, they tend to isolate the resin from the reinforcement (9). Examples include: microdielectric probes, which measure the response of the material in the capacitor gap, and the less obtrusive distal type sensor, which monitors a volume of material in the shape of a cone at the fiber terminus (9).

A surface sensitive measurement technique, evanescent wave sensing, has been applied to resin cure monitoring of graphite fiber/polyimide composites by employing infrared-transmitting optical fiber sensors (10). While there are advantages to using these fibers, a deterrent to using this infrared technique is that it requires using fibers that are sometimes large, fragile, costly, and potentially toxic.

In this work, a fluorescent evanescent wave sensor was developed and is in the process of being applied to resin cure monitoring for composites. Currently, fluorescence appears more favorable for application to a manufacturing environment since UV/visible transmitting fibers are less costly, potentially more durable and less toxic than mid-infrared transmitting fibers (9).

Evanescent wave sensors, in general, require the refractive index of the sensor to be larger than the refractive index of the surrounding medium. The refractive index of typical glass optical fibers is roughly 1.46, whereas the refractive index of epoxy resins increases from about 1.48 to 1.58 during cure (7). In this work, an economical optical fiber was used that consists of glass containing lead which increases the refractive index to approximately 1.62. It also has minimal absorption and fluorescence in the frequency range of interest, permitting the evanescent wave sensing of epoxy resins to full cure. Evanescent wave sensors consisting of small diameter glass fibers are expected to provide realistic measures of the state of cure within liquid molded parts during processing.

An appropriate measure of the resin cure during processing is required for real-time control of the liquid molding process. The degree of cure at the center of the mold may be required, or several measurements at particular locations may be required in complex parts. Optical fibers provide an opportunity to position cure sensors at any location in the mold and across the thickness of the part. Furthermore, the evanescent wave sensor has the potential to provide information concerning the fiber/matrix interface (i.e., cure, wetting, and mechanical properties).

In an evanescent wave fluorescence sensor, fluorescence probes in the resin surrounding the sensor fiber are excited by the electric field in the standing-wave situated at the fiber-resin interface, as illustrated in Figure 3 (7,11).

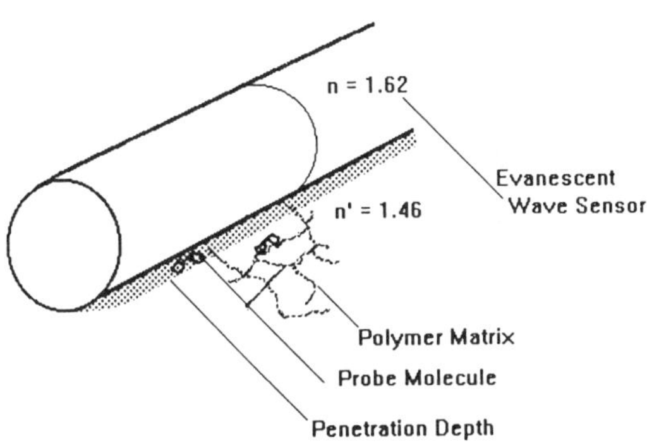

Figure 3. Schematic of the evanescent wave sensor. The shaded area indicates the measurement region near the fiber surface.

The standing, or evanescent, wave extends beyond the reflecting interface into the surrounding medium and decays approximately exponentially in

amplitude. The distance at which the electric field amplitude decreases to e^{-1} of its original value is termed the depth of penetration (7,11), beyond which minimal excitation results. The penetration depth, d_p, is given by

$$d_p = \frac{\lambda}{2\pi n_1 (\sin^2\theta - (\hat{n}_2 / n_1)^2)^{\frac{1}{2}}} \quad (1)$$

where λ is the wavelength of light, n_1 is the refractive index of the fiber, $\hat{n}_2$ is the complex refractive index of the medium, and θ is the angle of incidence of the propagating light at the interface. Values for d_p are typically of order 10^2-10^4 nm, depending upon the refractive index ratio $\hat{n}_2/n_1$, the angle of incidence, θ, and the wavelength λ of the incident radiation. The fluorescence emitted by the surrounding dye molecules is coupled back into the optical fiber and converted into an electronic signal by the photomultiplier detector (12).

Interactions between fluorophore molecules and their surroundings are known to affect the energy difference between the ground and excited states (13). The Lippert equation (Eq. 2) is often employed to estimate the general solvent effects on the fluorophore's emission spectrum, known as the Stokes' shift (13):

$$\overline{v}_a - \overline{v}_f \propto \frac{2}{hc}\left(\frac{\varepsilon - 1}{2\varepsilon + 1} - \frac{n^2 - 1}{2n^2 + 1}\right)\frac{(\mu* - \mu)^2}{a^3} \quad (2)$$

where h is Planck's constant, c is the speed of light, n is the refractive index of the solvent, ε is the dielectric constant, a is the radius of the sphere in which the fluorophore exists, $\mu*$ and μ are the dipole moments of the excited and ground states, and $\overline{v}_a$ and $\overline{v}_f$ are the wavenumbers (cm^{-1}) of the absorption and emission, respectively. Equation 2 predicts that either an increase in n or a decrease in ε will result in a decrease in the Stokes' shift, and both of these trends have been observed during epoxy cure (6,8,13,14). For example, as the dielectric constant ε is decreased the emission spectrum exhibits a blue shift and the wavelength of the emission decreases.

Experimental Details

An optical bench was constructed for the cure monitoring experiments. The optical bench (Figure 4) included an Ar^+ laser (Spectra Physics Series 2000), two mirrors, a beamsplitter (microscope slide cover slip), a beamstop, a 20X microscope objective, a focusing lens, a 1/8 m monochromator (Oriel, 77390), two cut-on wavelength filters, a photomultiplier detector (Oriel, 77348), and a lead silicate optical fiber. The optical fiber, which was developed at NIST, has the following compounds: 46.0% SiO_2, 45.3% PbO, 5.6% K_2O, 2.5% Na_2O, and 0.6% R_2O_3, where R represents the remaining substituents (15).

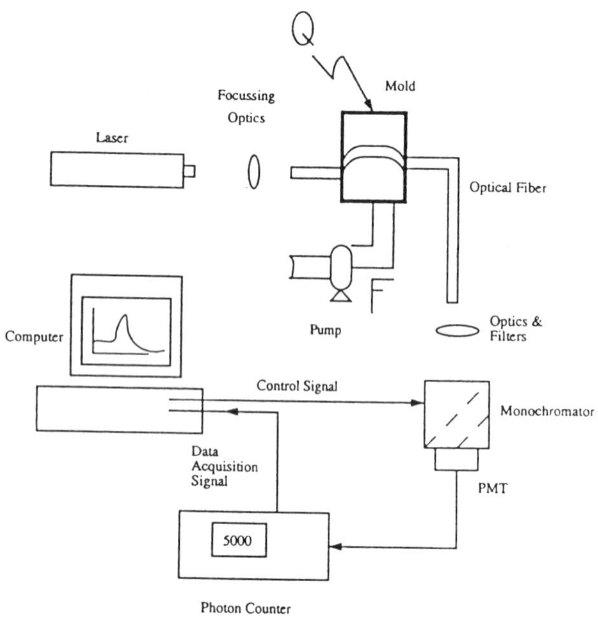

Figure 4. The optical bench constructed for evanescent wave fluorescence measurements includes an Ar^+ laser, focussing optics, a 1/8 m monochromator, a photomultiplier detector, and a lead silicate optical fiber. The diagram also includes the RTM apparatus to indicate how the entire configuration will appear after the fiber optic sensor has been incorporated into the RTM mold.

The accompanying electronics consisted of a fast preamplifier (Stanford Research Systems (SRS) SR445), a photon counter (SRS SR400), a high voltage power supply (SRS PS310), a two-axis stepper motor driver with TTL interface (Oriel, 20030), and a microcomputer.

The fluorescent dye used in this investigation (see Figure 5a) was a zwitterion, called 3-[4-(p-N, N- didecylaminostyryl 1-pyridinium) propylsulfonate], abbreviated di-10-ASPPS (Molecular Probes), and its salient feature is its strong response to the polarity of its environment.

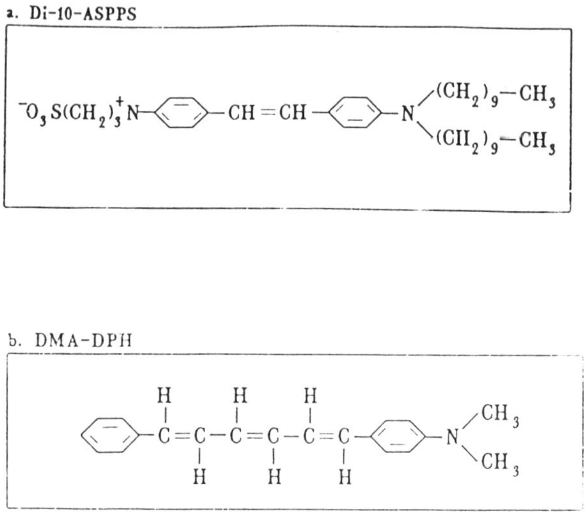

Figure 5. The fluorescent probes used in the experimental work. a. The dye that displayed a large emission wavelength shift was a zwitterion, called 3-[4-(p-N, N-didecylaminostyryl 1-pyridinium) propylsulfonate] (di-10-ASPPS). b. A viscosity-sensitive probe which has previously been employed for cure monitoring is 1-(4-dimethylaminophenyl)-6-phenyl-1,3,5 hexatriene (DMA-DPH).

A viscosity-sensitive probe that has previously been employed for cure monitoring is DMA-DPH (Figure 5b). It is also classified as a molecular rotor dye, which upon absorption, will decay to its ground state via radiation or radiationless decay. For DMA-DPH, the extent of radiationless decay is governed by rotations of the dimethylamino end group (4). As the resin viscosity increases, the dye loses its ability to undergo radiationless decay, and consequently, the intensity of the fluorescence emission increases over time.

The fluorescent dye was mixed into an epoxy resin system for the experiments. A commercial epoxy system, Tactix 123 (DOW Chemical) and Jeffamine D400 curing agent (Texaco), was used for all the experiments. The resin/dye system was characterized using a conventional spectrofluorometer (ISS, K2 multifrequency phase). Sample preparation involved dissolving the fluorescence probe, di-10-ASPPS, in the Tactix 123 diglycidyl epoxy bisphenol A (DGEBA) resin overnight at a concentration of 10^{-5} M. Just prior to an experiment, 3 g of the dyed DGEBA were mixed with D400 at a weight ratio of 1.8:1. The system was then stirred for several minutes under a nitrogen atmosphere at room temperature and was subsequently degassed. After placement in the heated cell of the fluorometer, approximately 5 minutes was required for the sample to reach the cure temperature of 60°C. The excitation wavelength for di-10-ASPPS was held constant at 488 nm to match a commonly used Ar^+ laser line. Additionally, the change in the Stokes' shift over time can mainly be attributed to the shift in the emission frequency maximum (8).

The next stage of this project will entail incorporating the fiber optic sensor into the RTM mold. To date, the evanescent wave sensor has been incorporated into a neat resin sample and the details of this study will be discussed at the conference. Additionally, it should be noted that the RTM facility is operational and liquid molded parts have been produced.

Parts were molded in a flat plaque mold indicated in Figure 4. The top and bottom platens of the mold were cast aluminum and contained cast-in electrical heating elements and cooling tubes (Watlow custom design). The platens also contained bayonet fittings to support resistance temperature devices (RTD's) which were used to measure the platen temperatures. A thin film RTD (Omega F3105) was mounted in the center of the preform to provide a direct measurement of the center temperature of the molded part during processing. One of the RTD signals was used as the process variable in a stand alone temperature controller (CVL Instruments) which varied the electrical power supplied to the platen heating elements in order to control the temperature to a setpoint. All temperature measurements were forwarded to a computer (50 MHz i486DX) via a signal conditioner (Omega OM5) to allow the use of the measured temperatures in higher level calculations. The controller permitted the setpoint to be downloaded from a computer, by that allowing for complex cure cycles that could be determined on-line by a higher level controller running in the computer.

The injection system consisted of a pressurized container to feed resin to the mold and a vacuum pump (Trivac D2A) for degassing the resin and maintaining pressure control in the mold during injection. Both mold inlet and outlet pressures were measured with pressure transducers (Omega PX603) and controlled independently with stand alone controllers. Pressure control was achieved at the mold inlet, for example, by varying the pressure in the resin feed container. A 3-way solenoid valve connected the resin feed container to both a high pressure source (N_2 cylinder) and a low pressure source (vacuum pump), and by manipulating the solenoid valve between the high and low pressure sources the pressure in the container and at the mold inlet was adjusted to the desired set point. The pressure setpoints of the controllers could be downloaded from the computer enabling complex injection strategies designed to minimize void formation.

The composite parts consisted of Knytex D155 unidirectional fabric impregnated with Tactix 123 (DOW Chemical) and Jeffamine D400 curing agent (Texaco). The parts were molded with a fiber loading of approximately 52% by volume, and a high index optical fiber will be included in the preform during the planned molding experiments. The measured permeability of the reinforcement was roughly $9.0 \cdot 10^{-8}$ cm^2 in the direction of flow during injection, and the permeability determines the relationship between the resin injection rate and the required pumping pressure. A more complex reinforcement was considered for these experiments, but the unidirectional fabric was chosen because of modeling considerations that included the microflow and process control, and the interpretation of subsequent mechanical testing data.

Infrared spectra were taken with a Nicolet Magna 550 Fourier transform infrared spectrometer equipped with a liquid nitrogen cooled MCT-A detector. The epoxy resin was placed between two potassium bromide plates. The plates were placed in a transmission hot cell at 60°C, and spectra were taken every 10 min.

Results

The fluorescence behavior of the resin/dye system was characterized and correlated with infrared cure data. Emission spectra were obtained for the resin/dye system in a conventional spectrofluorometer in the wavelength range between 495 and 700 nm, as shown in Figure 6.

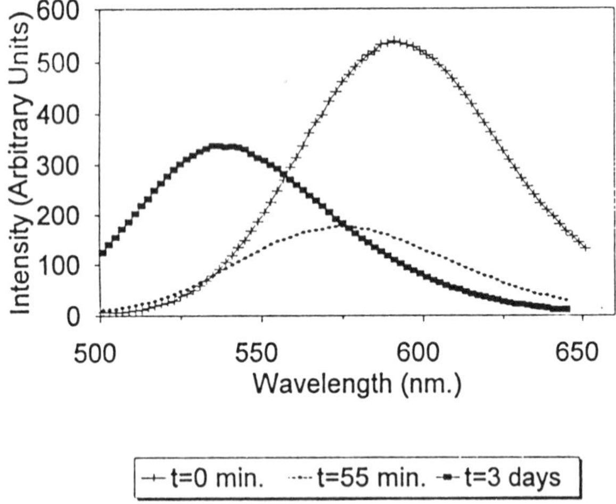

Figure 6. Fluorescence emission spectra of di-10-ASPPS at three times during the cure of Tactix 123 and D400.

The wavelengths of the maxima in the fluorescence intensity of the three spectra shown in Figure 6, as well as several additional spectra obtained during the cure, are illustrated in Figure 7. The fluorescence peak was observed at 592 nm in the uncured sample, and during an isothermal cure

at 60°C, the peak wavelength shifted nearly 45 nm to lower values. The large shift in the peak position illustrates the usefulness of di-10-ASPPS for the cure monitoring of epoxy resins.

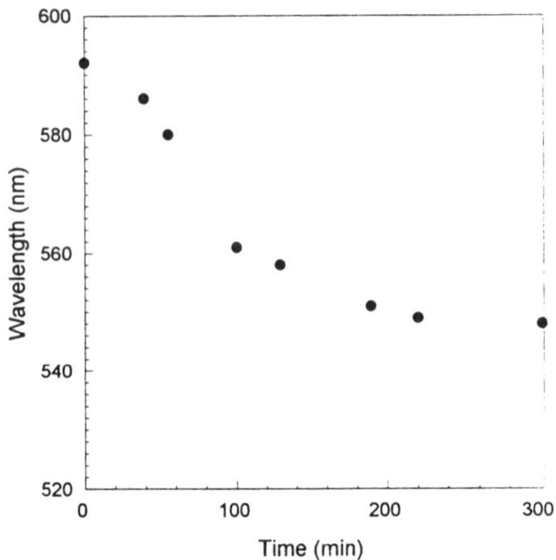

Figure 7. The shift in the fluorescence maximum of the di-10-ASPPS as the resin sample cured.

A similar experiment was conducted for DMA-DPH (Figure 2), which absorbs maximum radiation at 390 nm in the ultraviolet range. Besides a large intensity increase in the fluorescence emissions, a smaller 22 nm blue shift from 498 nm was observed. However, after the first 175 minutes of cure only small changes in the intensity could be observed, and the blue shift in the peak position is too small to reliably measure the latter stages of the cure. DMA-DPH is therefore not sensitive enough for monitoring the cure of epoxy resins in composite manufacturing.

Di-10-ASPPS appears to offer a desirable response for epoxy cure monitoring, especially at longer times into the cure cycle. Additionally, di-10-ASPPS offers the advantage that its peak excitation wavelength is in the visible wavelength range, unlike DMA-DPH, which absorbs radiation in the UV range.

Demonstrating evanescent sensing with the high index fiber and the resin/dye system discussed above is in progress as of this writing.

Preliminary results are available with the high index fiber and a calibration dye, Rhodamine B, that was used to calibrate the optical bench. Rhodamine B was dissolved in propylene glycol at a concentration of $1 \cdot 10^{-4}$ M. The fluorescence dye was excited at 488 nm and its peak fluorescence was observed at 580 nm. A series of experiments was conducted in which assorted lengths of the evanescent waveguide fiber were immersed in the dye solution (see Fig. 8), and in accord with theory (7), the fluorescence intensity was directly proportional to the exposed fiber length (Fig. 9).

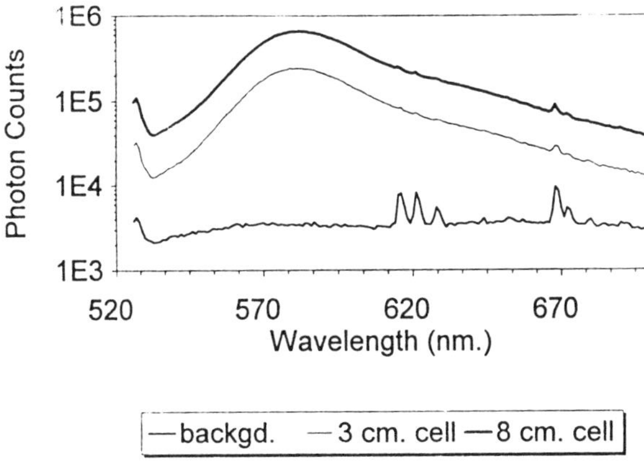

Figure 8. The optical fiber immersed in 10^{-4}M Rhodamine B in propylene glycol.

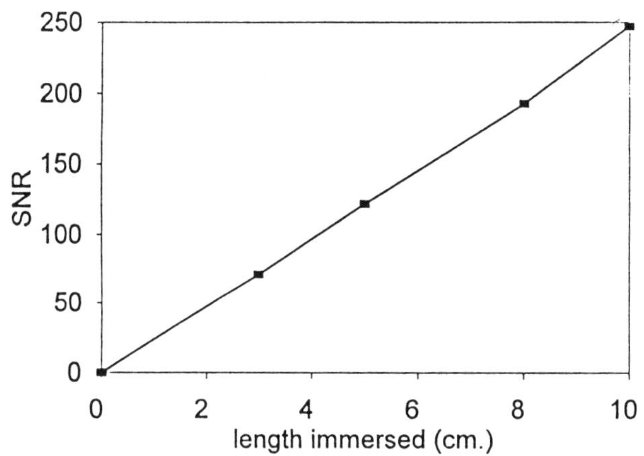

Figure 9. The fluorescence intensity versus the length of optical fiber immersed in the Rhodamine B/propylene glycol solution.

These experiments simply demonstrate the functionality of the optical bench and high index fiber for research purposes. For fluorescence to be useful a correlation must be developed between the fluorescence signal and the degree of cure. IR spectroscopy is being used to generate the correlation data (Fig. 10), and Fig. 11 depicts the conversion over time. The epoxide/amine reaction was monitored by tracking the disappearance of the oxirane or epoxide ring peak.

Once a similar set of experiments is completed to demonstrate evanescent wave sensing with the epoxy/di-10-ASPPS system, molding experiments will be conducted. Initial experiments are to be conducted in which the high index fiber will be placed between two layers of the Knytex D155 unidirectional fabric. Molding experiments with the optical fiber signal processing equipment integrated into the molding laboratory have not been completed at the time of this writing but will be presented at the conference.

A more rugged optical system is currently being designed in which the laser excitation source will be replaced by an incoherent source (xenon lamp), which should be rugged enough for an industrial environment. It has already been shown that optical fibers can be woven or braided directly into preforms. Thus, a final remaining technical hurdle to industrial application of a fluorescence cure monitor is the tooling design to allow rapid and reliable optical connections to a fiber protruding from a preform.

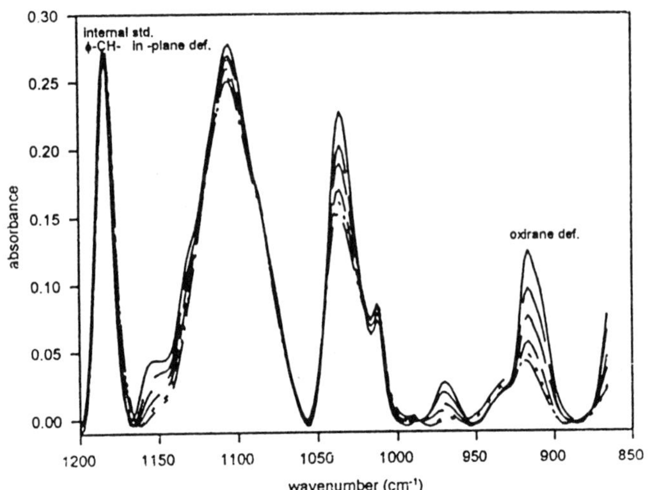

Figure 10. Infrared spectra collected during cure of DGEBA with Jeffamine D400 at 60°C.

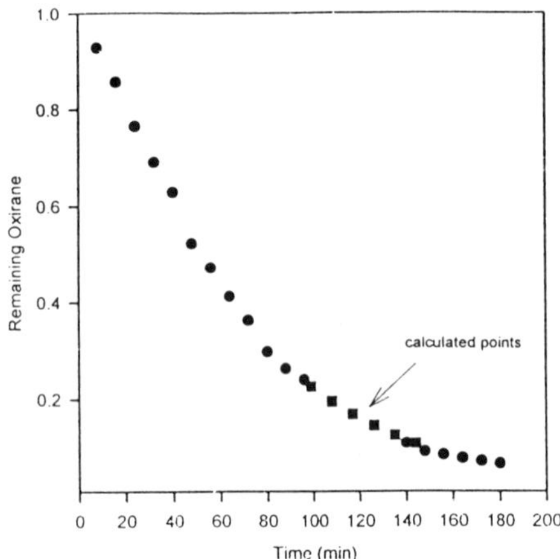

Figure 11. Degree of cure as a function of time. This plot was extracted from the data presented in Fig. 10. Conversion was measured by taking the ratio of the heights of the diminishing epoxide ring peak to the internal standard peak.

Summary

A cure monitoring sensor has been developed based upon an evanescent wave fluorescence measurement performed with an optical fiber embedded in the preform. The optical fiber is similar in cost to standard glass optical fiber, but the addition of lead raises the refractive index above that of the resins, permitting evanescent wave sensing. Evanescent wave sensing allows the measurement to be obtained at specific locations within the mold and within an approximately 1 micron radius of the fiber surface, providing a measurement of resin cure that is most appropriate for predicting final part properties. The molding system used to make the parts with the embedded optical fiber is designed to allow the use of the fluorescence signal in a hierarchical control strategy.

Acknowledgment

The authors would like to thank Mr. D. Blackburn of the Ceramics Division at NIST for drawing the high index optical fibers used in this work.

References

1. Johnson, C.F., in "Engineered Materials Handbook, Vol. 1, Composites", ASM International, Materials Park, Ohio (1987).
2. Chang, S. S., Mopsik, F. I., and Hunston, D. L., Proc. 19th Intl. SAMPE Tech. Conf., **19**, 253 (1987).
3. Kranbuehl, D., Delos, S., Yi, E., Mayer, J., and Jarvie, T., Polym. Eng. Sci. **26**, 338 (1986).
4. Bur, A. J., Wang, F. W., Thomas, C. L., and Rose, J. L., Polym. Eng. Sci. **34**, 671 (1994).
5. Levy, R. L. and Ames, D. P., Polym. Mater. Sci.& Eng., **53**, 176 (1985).
6. Lin, K. F. and Wang, F. W., Polymer, Vol. 35, 687, (1994).
7. Wang, F. W. and Fanconi, B. M., NBSIR 87-3581 (1987).
8. Wang, F. W. and Lowry, R. E., "Wavelength-Shift Fluorescent Probes for Cure Monitoring of Epoxy Resins," submitted to Polymer, (1994).
9. Hunston, D., McDonough, W., Fanconi, B., Mopsik, F., Wang, F., Phelan, F., Chiang, M., NISTIR 4514 (1991).
10. Young, P.R., Druy, M.A., Stevenson, W.A., Compton, D.A.C., SAMPE Journal, **25**, 11 (1989).

11. Harrick, N. J., "Internal Reflection Spectroscopy", Harrick Scientific Corporation, Ossining, New York (1979).

12. Carniglia, C. K., Mandel, L., and Drexhage, K. H., J. Opt. Soc. Am., 62, 479 (1972).

13. Lakowicz, J. R., "Principles of Fluorescence Spectroscopy", Plenum Press, New York (1983).

14. Mangion, M. B. M. and Johari, G. P., Journal of Polymer Science: Part B: Polymer Physics, Vol. 29, 1117-1125 (1991).

15. Blackburn, D., National Institute of Standards and Technology, Gaithersburg, MD, Private Communication, (1994).

* Identification of a commercial product is made only to facilitate experimental reproducibility and to describe adequately experimental procedure. In no case does it imply endorsement by NIST or imply that it is necessarily the best product for the experiment.

EB Manufacturing of Polymer-Fiber Composite Vehicle Structures

W.J. Chappas, N. Grossman, B. Pourdeyhimi
University of Maryland
College Park, Maryland

Abstract

The treatment of curable monomeric and polymeric systems by energetic electrons offers a high speed, low-temperature, continuous method for the large scale manufacture of vehicle structures. Based on modern EB sources, the process is proven to be extremely reliable, rugged, and easily integrated into continuous production schemes. Unlike other radiation curing techniques, the EB process uses no radioactive materials and neither the processing area, or product become radioactive. This paper describes on - going work to develop and commercialize an EB process for the manufacture of thick (e.g. 5 cm) polymer - fiber composite structures.

OVER THE PAST THIRTY YEARS there has been a continuous and significant growth in the development and application of radiation (UV, electron beam, and gamma) curable materials, primarily in coatings and adhesives applications. As this industry continues to develop innovative new products based on radiation's high efficiency and easy process control, a logical extension for this technology is in the field of composite materials. The use of radiation (typically electron beam processing) techniques to cure composite materials offers several advantages: (1) curing at ambient temperatures; (2) reduced curing times; (3) continuous operation; (4) improved resin stability; (5) less atmospheric pollution; and (6) increased design flexibility through process control.

A review of available radiation curable resins for the production of composite material was conducted and a resin-fiber combination, similar to those used by the US Army, was selected. Sample pieces were prepared and then radiation cured.

The objective of this research was to provide; (1) a practical demonstration of this technology; and (2) reference materials for the production of radiation curable composites suitable for the manufacture of support structures on the US Army Composite Armored Vehicle (CAV).

Background

In the past few years, fiber reinforced thermoset composites have steadily grown in use. Led by the Defense Department's need for light, high-performance materials, principally for aerospace applications, the manufacturers of composite materials have developed hundreds of new products. Yet the curing of high performance composites still requires the time consuming, heat-curing step. In addition, traditional autoclaving; (1) liberates low molecular weight volatiles in the uncured material, thereby producing bubbles that weaken the composite and cause out-gassing; (2) "cooks" the composite from the outside often producing less than optimal cure distributions and severe internal stresses; and (3) limits the designers' options for innovative product designs that might include optical sensors or other heat sensitive components.

The French aerospace company AEROSPATIALE has established a radiation production scheme for the manufacture of rocket booster bodies up to 12 feet in diameter and 30 feet long. The French have mastered a technology largely ignored in the US that can reduce the cure time for the casing from 100 hours to 8 hours -- and have gained an advantage in technology and international competitiveness.

Besides the applications in the manufacture of aircraft, the technique may have other applications in vehicles, the production of very low cost composite tooling, and the manufacture of composite building materials.   The advantages of radiation curing have been previously summarized by Saunders and Singh. They include:

Curing at ambient temperatures. The appropriate selection of tooling materials is required to produce a component to strict dimensions while minimizing internal stresses. During the thermal curing cycle, the tool expands and contracts, often at different rates than the molded fiber-reinforced plastic. These movements can alter the dimensions and produce internal stresses in the cure product that can decrease its strain to failure and fracture toughness. Electron beam curing at ambient temperatures eliminates thermal expansion in both the tool and composite, thus reducing dimensional changes and internal stresses in the final product. Ambient temperature processing should also allow the use of lighter tools made from less expensive materials, including composites.

Reducing curing times for individual components. A typical electron curable carbon fiber epoxy laminate can be cured with a dose of about 50 kGy. A 50 kW electron accelerator can provide this dose to about 1800 kg/h of material, assuming that 50% of the energy provided by the accelerator is absorbed by the product being irradiated (energy utilization factor). This production speed is expected to be greater than for thermal curing with a typical autoclave, even though the products are cured one at a time rather than in large batches. The overall production speed for a typical autoclave (300 cubic meters) manufacturing hand laid-up composites is about 100 kg/h, assuming a 6 hour cure cycle and an average part size a 1.2 m x 3 m x 0.1 m (density 1.69 g/cm^3; 125 parts/8 hour shift). The production speed provided by electron processing would be reduced slightly in facilities that manufacture many different shaped products, because of a lower energy utilization factor due to increased inefficiencies.

Electron beam processing is a continuous operation and components can be further processed immediately after they are produced. This continuous operation makes production scheduling and inventory control easier and reduces the number of identical molds needed to economically manufacture products. Electron beam processing is also better suited to small production runs because parts are cured one at a time.

Improved resin stability at ambient temperatures. The shelf-life of radiation curable resins can be much longer than the shelf-life of formulations designed for thermal curing because electron-curable formulations do not normally auto-cure at room temperature, making low-temperature storage unnecessary.

Reducing the amount of volatiles produced. Thermal curing of composites often produces volatile degradation products that can be hazardous and require special control procedures. Electron beam curing eliminates the production of thermal degradation products, although very small amounts of gaseous radiation products such as carbon dioxide, may be formed.

Better control of the energy absorption profile for a component, allowing greater design flexibility. In conventional processing, materials with different thermal curing cycles can not be combined in a single product. Since radiation can be controlled, materials of different cure characteristics can be combined in the same product.

Radiation processing is a powerful, energy-efficient, non-polluting method for the production of advanced materials, frequently, at a fraction of normal production costs. In many instances, there may be no conventional alternatives to radiation processing. The quest for higher performance from more sophisticated equipment places critical importance on the development of new materials in aircraft, aerospace, marine and transportation applications, vehicles, and equipment. Radiation processing offers unique advantages to all of the important applications.

Radiation processing has the potential to provide new low-cost structural materials for everything from airplanes to automobiles, boxes to building. Successful application of radiation technology could result in lighter cars manufactured at lower costs, lighter and more fuel-efficient airplanes, and cheaper and stronger building materials. All of which will help the US regain international manufacturing competitiveness, enhance employment and economic conditions, and improve our international trade deficit.

Technical Background

The ability of gamma and high energy electrons to induce chemical and physical changes in materials in general, and polymers and elastomers specifically, have been the basis for research since first reported in 1938[1]. Recently, however, has radiation been used to cure polymer-fiber composites.

Hiramoto and co-workers[2] found that in addition to the cationic mechanism, chain transfer reactions are significant (a proton transferred from the chain propagating intermediate to the monomer). However, all these studies concluded that the presence of trace amounts of water inhibit the polymerization reaction and thus most epoxies are not good candidates for industrial applications. On the other hand, the radiation-induced polymerization of the acrylated epoxies proceeds by a free radical mechanism and therefore the presence of water does not inhibit the polymerization process.[3,4]

One method for imparting good radiation-curing properties to an epoxy resin is to acrylate the terminal epoxy groups of an epoxy oligomer. Gotoda and co-workers[5,6] studied the radiation-induced polymerization of epoxy-acrylate prepolymers in the presence and absence of several vinyl monomers. They found that an acrylic acid based epoxy cured at a dose of less than 10 kGy. Electron beam curing of epoxy acrylates has been shown to give a larger amount of crosslinking than UV curing (dose rate of 0.2 kGy/s). Epoxy acrylates can be electron cured with a dose of 5 to 200 kGy, depending on the type of epoxy modification and the type of crosslinking promoters used. The resulting maximum gel fraction has been reported to be about 98%. Epoxy acrylate mixtures containing 20 to 50% aliphatic epoxy acrylates (the remainder aromatic epoxy acrylates) radiation cured with electron radiation (dose rate of 8 kGy/s) were reported to give excellent mechanical properties. The thermal stability of radiation cured acrylated epoxies increased with the number of acrylated epoxide groups present in the parent compound.

In more recent work by Saunders et.al.[7,8] focused on acrylate epoxy compounds that could produce an aircraft quality composite. The general formula that proved to meet all of the mechanical specification imposed by a major US aerospace company was; 20% dipentaerythritol monohydroxypentaacrylate; 30% polybutadiene diacrylate; and 50% epoxy diacrylate. A plain weave carbon fabric was selected for the woven roving. A solvent process using methylethylketone was used to impregnate the fabric with about 35% (by mass) of the resin. The prepared material was then irradiated at a dose-rate of 17 kGy/h to a total absorbed dose of 50 kGy.

Material Selection

Fiber: Although several fibers were considered for use in this study, most of the investigation focused on glass, carbon, and Kevlar fibers. This paper reports our preliminary results obtained using standard glass woven roving. The manufacturers include: Certainteed Corp. Fiber Glass Industries; Owens-Corning Fiberglass Corp.; and PPG Industries. The principal weavers of glass fibers are: Advanced Textiles; BGF Industries; Fiber Materials, Inc.; Hexcel Corp.; and King Fiber Glass Corp.

Weave Selection: A standard plain weave was selected and was used as received. Since there are currently no fibers available that have a radiation curable finish, a fabric with no finish was chosen. The sizing (normally applied to the yarns during weaving) was not removed.

Glass Fiber: An S2 roving fiberglass weave (Owens Corning) with a yield of 250 yds/lb was used. A color tracer yarn ran down one side of the material to indicate the warp direction. The warp direction was considered the zero direction of the weave and all the cuts from the fabric are referenced to the warp.

There are several characteristics of glass fibers that make them ideal as a reinforcing material: (1) the high tensile strength to weight ratio, which leads to superior tensile strength; (2) glass fibers are incombustible, and have a low coefficient of expansion; (3) They do not exhibit significant creep; deflection of glass fibers is directly proportional to the applied force and they return to their original size when the force is removed; (4) glass fibers do not retain moisture, therefore, they do not swell, or undergo any chemical change that could be caused by moisture; (5) glass fibers will not shrink or stretch and (6) glass fibers have a relatively low cost compared to other high performance fibrous reinforcements. A summary of the characteristics of the materials used is given below.

Table 1. Summary of Woven Roving (S2 Glass) Specifications.

	Warp	Weft	Total
Weight (oz./sq.yd.)	11.80	12.09	23.89
Fabric Count	5.0	5.12	
Strength Ratio (warp/weft)			49.5/50.5
Weight Ratio (warp/weft)			49.5/50.5
Breaking Strength (lbs/lin.in)	1423	1457	
Tensile Strength - Virgin Form (psi)	655	655	
Tensile Modulus (psi)	12.6	12.6	
Thickness (in.)			0.035

The glass fiber used in this study was treated with a commercial sizing. Sizing is commonly used to protect the fibers and hold strands together during weaving. The presence of sizing on fabric is acceptable for some applications but it is not compatible with some resins. Typically, the sizing prevents the ply from 'wetting out' and is therefore removed by burning and a finish is applied.

Resin: There are a wide variety of resins readily available for the production of conventionally cured composites. Manufacturers of conventional epoxy resins include Shell Chemical Company; Ciba-Geigy Corporation; Dow Chemical; Rhone-Poulenc; Reichhold Chemical Inc.; Anhydrides and Chemicals; Mitsubishi; Plasco Corporation; Dexter Composites; Fiber-Resin Corporation; and Hexcel.

Also, a wide range of radiation curable resins, developed for adhesives and coating applications, are commercially available and suitable for polymer fiber composite applications. Manufacturers of radiation curable epoxy resins include UCB Radcure; Sartomer; Cargill Inc.; Henkel; Polychem Corporation; Rohm and Haas; and Rahn (US distributor: Biddle Sawyer).

Chemistry: The resins used for radiation curable composites in this work were manufactured and supplied by Sartomer in Exton, Pennsylvania. The resins were:

SR 399, dipentaerythritol hydroxypenta-acrylate which is a multifunctional monomer

$$
\begin{array}{l}
\text{CH}_2=\text{CH-C(=O)-O} \\
\quad\quad | \\
\quad\quad \text{CH}_2 \\
\text{HOCH}_2-\text{C-CH}_2\text{OCH}_2\text{CCH}_2-\text{O-C(=O)-CH=CH}_2 \\
\quad\quad | \\
\quad\quad \text{CH}_2 \\
\text{CH}_2=\text{CH-C(=O)-O}
\end{array}
$$

Table 2. SR 399 Characteristics.

Chemical Description	Dipentaerythritol Pentaacrylate
Appearance	Viscous Liquid
Reactive Esters	99 %
Odor	Mild
Color (Alpha)	80
Molecular Weight	525
Specific Gravity @ 25°C	1.192
Viscosity @ 25°C	13700 cps
Inhibitor	300±100 MEHQ

SR 454, ethyoxylated trimethylolpropane triacrylate, another multifunctional hydrocarbon

$$CH_2\text{-}O\text{-}CH_2CH_2\text{-}O\text{-}\overset{\overset{O}{\|}}{C}\text{-}CH=CH_2$$
$$CH_3\text{-}CH_2\text{-}C\text{-}CH_2\text{-}O\text{-}CH_2CH_2\text{-}O\text{-}\overset{\overset{O}{\|}}{C}\text{-}CH=CH_2$$
$$CH_2\text{-}O\text{-}CH_2CH_2\text{-}O\text{-}\overset{\overset{O}{\|}}{C}\text{-}CH=CH_2$$

Table 3. SR 454 Characteristics

Chemical Diescription	Ethoxylated Trimethylolpropane Triacrylate
Appearance	Clear Liquid
Reactive Esters	99 %
Odor	Mild
Color (Alpha)	80
Molecular Weight	428
Specific Gravity @ 25°C	1.1043
Viscosity @ 25°C	62 cps
Inhibitor	300±25 HQ

and CN 120 epoxy acrylate oligomer

$$CH_2=CH\text{-}\overset{\overset{O}{\|}}{C}\text{-}O\text{-}CH_2\text{-}\overset{\overset{OH}{|}}{CH}\text{-}CH_2\text{-}O\text{-}\bigcirc\text{-}\overset{\overset{CH_3}{|}}{\underset{\underset{CH_3}{|}}{C}}\text{-}\bigcirc\text{-}O\text{-}CH_2\text{-}\overset{\overset{OH}{|}}{CH}\text{-}CH_2\text{-}O\text{-}\overset{\overset{O}{\|}}{C}\text{-}CH=CH_2$$

Table 4. Typical CN 120 Epoxy Acrylate Resin Characteristics.

Polymer Solids	100 %
Gardner Color	1 - 2
Viscosity @ 65°C	2300 cps
Specific Gravity	1.15
Functionality	2
Acid Value, mg KOH/g	1

All three chemical structures are esters of acrylic acid with at least two C=C double bonds. These unsaturated portions are reactive vinyl groups $CH_2=CH\text{-}$. Through the unsaturations, acrylates undergo radical addition polymerization reactions to form a three dimensional network. For this work, the following blend of the two acrylated epoxies and the epoxy oligomer was used:

70%	CN120
25%	SR454
5%	SR399

Kinetics:

The kinetics can be summerized as follows: Thermally cured resins are designed to polymerize by cationic reactions that are initiated at elevated pressures and temperatures. Resins designed to undergo free radical reactions can be cured by radiation at ambient conditions.

Initiation: The initiation reaction involves a complex of excited and ionized states and species. Within microseconds, especially in the presence of even a small concentration of water, the important transient species is a neutral alkyl free radical:

$$M \rightsquigarrow R\bullet$$

where M = monomer, R• = free radical

Propagation: The propagation reaction proceeds by the addition of the individual vinyl groups to the propagating free radical,

$$R\bullet + R \rightarrow RR\bullet$$

where R• represents a chain of any length with a reactive radical at least one end.

Termination: The termination reaction takes place when two free radicals react and the polymerization process ends:

$$R\bullet + R\bullet \rightarrow RR$$

An additional reaction which must be considered is the reaction of oxygen with the propagating free radical. This peroxidation reaction involves the addition of the molecular oxygen to the carbon centered free radicals to produce the corresponding peroxy radicals. The peroxy radicals can attract hydrogen atoms from adjacent molecules and undergo chain scission, or react with an alkyl free radical resulting in premature termination of a propagating chain. These forementioned reactions, can be described by three simple rate equations. The initiation rate:

$$R_i = \overset{\bullet}{D}G$$

where R_i = the rate of initiation, or in this case the production rate of the propagating free radical per unit volume; $\dot{D}$ = the dose rate; and G is the radical yield per unit energy absorbed.

The propagation rate:

$$R_p = k_p[R\bullet][M]$$

where R_p = rate of propagation
k_p = propagation rate constant;
[R] = free radical concentration; and
[M] = monomer concentration.

The termination rate:

$$R_t = k_t[R\bullet]^2$$

where R_t = rate of termination, and
k_t = termination rate constant

Material Preparation

Coupon Samples: The testing coupons were constructed in accordance with MIL-L-46197(MR), "LAMINATE: S-2 GLASS, FABRIC-REINFORCED, POLYESTER RESIN IMPREGNATED".

Layup: The composite was layed-up over a period of five to six hours. Resin was poured onto each ply and the excess was removed by a hand roller. Approximately 6 grams of resin was used per ply. As each ply was preprepared, it was stacked on top of the previous ply, then rolled to minimize voids and remove the excess resin. When the nine plies were completely assembled, the uncured composite was packed in standard aerospace quality bagging material. The composite rested on a one half inch aluminum plate that was coated with a teflon release film. The outer perimeter of the plate was outlined with tacky tape. On top of the composite, and in direct contact with the composite, was a porous release film and breather material. A quick release fixture was connected to the vacuum bag and pump hose. The material remained under vacuum ten hours before irradiation.

Irradiation: The nine ply samples were cured by a two step irradiation process; (1) a gamma ray irradiation to consolidate the resin and fiber; and (2) high energy electron beam irradiation to complete the cure

The partially gamma cured composite was taken to Irradiation Industries (Gaithersburg, MD) where it was irradiated with their 3 MeV electron beam. The first two electron beam passes deposited 10 kGy each in the sample. The subsequent passes deposited a dose of 20 kGy each in

the sample. Although each irradiation lasted only seconds, there was approximately 20 minutes between irradiations. Temperature labels were used to monitor the temperature rise during irradiation.

Dosimetry: Dosimetry was performed in accordance with a NIST traceable protocol for high dose dosimetry. This procedure involves; (1) the equilibration of NIST calibrated nylon transfer dosimeters in a 50% r.h. room at 25°C for 24 hours. The dosimeters were then sealed in an aluminized polyethylene bag to assure constant humidity during irradiation. Following irradiation, the dosimeters were opened and stored at 25°C/50% relative humidity for 24 hours before reading. The films were read on a Beckman 25 UV/VIS spectrophotometer that was calibrated using NIST standard reference materials. The dose is reported as dose in water.

Differential Scanning Calorimeter (DSC): Inititial testing for confirmation of cure and determination of the glass transition temperature was conducted with a Perkin Elmer Differential Scanning Calorimeter (DSC). An example of a typical thermogram is shown in Fig. 4.

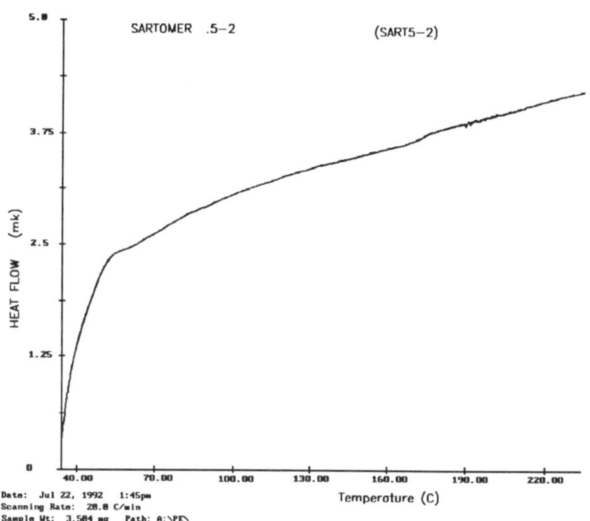

Figure. 1. DSC for radiation cured neat resin.

Testing

Mechanical Testing: Mechanical tests were performed on nine ply composite samples to evaluate mechanical properties. All the tests were performed on a Sintech 50 static tensile tester with a maximum load of 20,000 lbs. Flexural, compression, and tension tests were performed. All the samples were cut from the same 9 ply radiation cured sheet prepared in accordance with MIL-L-46197(MR), "LAMINATE: S-2 GLASS, FABRIC REINFORCED, POLYESTER RESIN IMPREGNATED". The 0° direction coincides with the direction of warp for the first ply. Each successive ply was oriented in accordance with MIL-L-46197(MR).

Flexural Properties: The ASTM D790-81 Method I Procedure A Standard Test Methods for Flexural Properties of Unreinforced and Reinforced Plastics and Electrical Insulating Materials was used as a guideline for measuring flexural properties. A three point loading system was constructed for this purpose. Two supports held a 10 inch by 1.5 inch test coupon with a support span of 7.0 inches. The original surface of the composite remained during the test procedure, no polishing or machining was performed. All tests were performed by the same operator on the same day. ASTM standard D790 recommends specific crosshead rates based on the dimensions for test specimens for support span (L) to depth ratios. The configuration for this work was the same as that of the ASTM standard. The calculated rate of crosshead motion can be determined by

$$R = \frac{ZL^2}{6d}$$

where R = rate of crosshead motion, in./min.
    L = support span, in.
    d = depth of beam, in., and
    Z = rate of straining of the outer fiber, Z shall equal 0.01

The calculated rate is equal to 0.33 in./min. In this experiment L=7.0 in., d(average)=0.25 inches, and the total length of the specimen is 10 in. The width of the samples were 1.5 inches. The parameters of these test coupons corresponded with an L/d equal to a 40:1 ratio in Table 1 of standard D790. The crosshead velocity for this experiment was 0.50 in./min.

Compressive Properties: A modified version of the ASTM D695 Standard Test Method for Compressive Properties of Rigid Plastics was used as a guideline for the compressive data. Four were tested oriented in the 0° direction and four test specimens were oriented in the 90° direction. All the test specimens were cut from the same sheet. The specimen dimensions were; length=6.5 inches, width=0.25 inches and depth (average)=0.25 inches. No compression tool was used for applying the load. Because the specimen was not in a compression tool, no support jig or micrometers were used. A compressometer determines the distance between two fixed points on the test specimen at any time during the test.

The sample was placed freestanding vertically between two flat round parallel surfaces. The top surface was the load and the bottom the support. The specimen was centered on the plates and the end surfaces were flat. The crosshead was adjusted so the top plate made contact with the sample to hold it in place but no measurable force was placed on the sample before testing. The speed of testing was 0.05 in./min. The plate pushed down on the sample until the strain on the composite reached 75%. The samples did not buckle in the plastic region. The samples buckled at the flexural yield point where they failed.

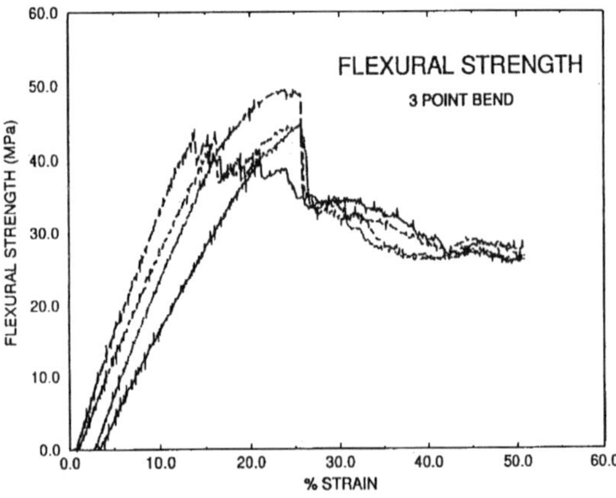

Fig. 2. Flexural Behavior.

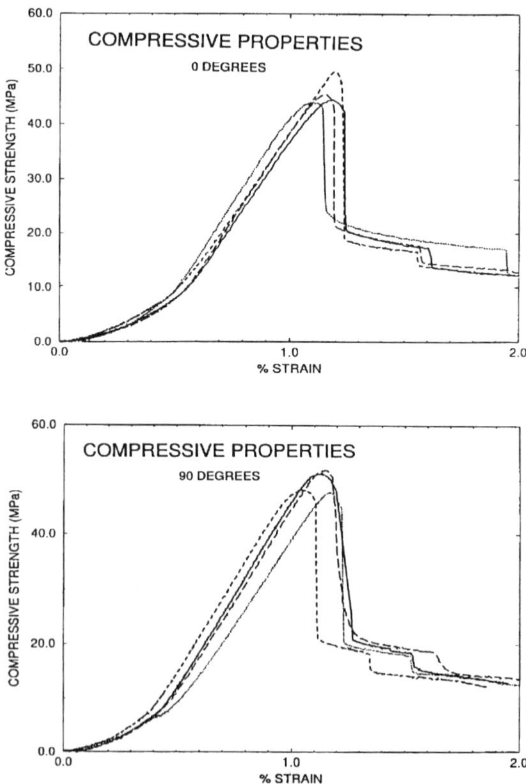

Fig. 3. Compressive Behavior.

Tensile Properties: A modified version of the ASTM D638 Specimen Type I Standard Test Method for Tensile Properties of plastics was used as a guide for measuring the tensile properties of the composite. Ten specimens were tested, all cut from the same composite sheet. The axis of the test specimens were in the 0° direction. The overall length of the specimens were 6.5 inches and the overall width was 0.75 inches. In this study the sample dimensions were:

    average width overall: 0.32 in.
    average length overall: 6.50 in.
    average thickness overall: 0.21 in.

Part of the variation from the standard was due to the thickness of the diamond blade while cutting the samples and the width of the line used to mark the test coupons. The test cross-section is equal to the width times the thickness, 0.067 in^2.

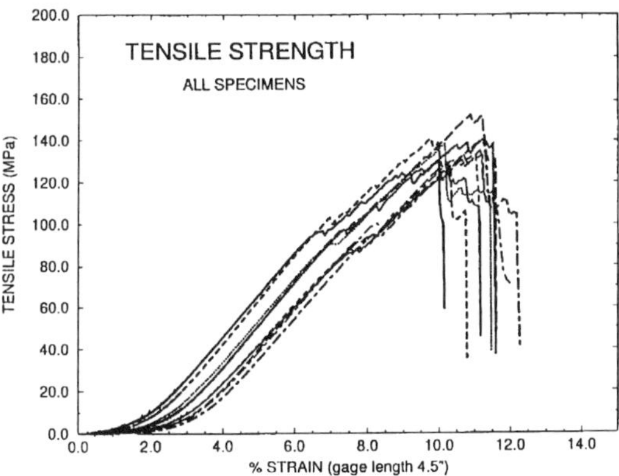

Fig. 4. Tensile Strength.

Flammability: A simple test to assess flammability of a coupon was conducted. The coupon was held in a bunsen burner flame long enough to ignite the coupon. Ignition was not rapid, but it did start after several seconds. The coupon was then removed from the flame. It continued to burn for almost one minute until the outer layer of resin had burned off. This test was repeated with a conventionally cured composite coupon containing the same fiber plys and similar in all physical dimensions. The results were identical.

Thick Section Samples

In order to extend our understanding of radiation processing, especially with regard to thick section applications, two thick section sample parts were manufactured. The materials used were identical to those used to produce the 8 ply mechanical test coupons described previously. Because of differences in geometry, and, the

availability of facilities, there was some variation in irradiation techniques.

The prototype samples were established to demonstrate how thermally induced stresses normally associated with conventional curing methods can be eliminated by radiation curing. Two "problem geometries" were addressed: (1) alternating layers of carbon and glass (having different thermal expansion coefficients) that easily delaminate and (2) thick section, right angle bends that typically experience complex and/or severe temperature gradients during cure, resulting in excess voids or delamination.

The prototype sample selected for this demonstration was a right angle bend 69 layers, alternating (10 layers each) glass and carbon fibers with an epoxy resin. The sample dimensions were selected to conveniently match readily available tooling and radiation sources. The resulting piece is as shown in Figure 5. To demonstrate that metallic components can be included in radiation cured composites, one layer of copper Faraday screen was included.

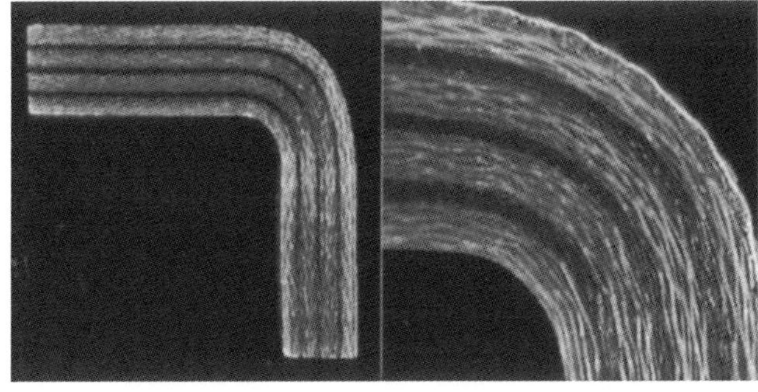

Fig. 5. 70 Ply glass/carbon prototype with Faraday screen

In addition to the right angle bend, a standard, 12" by 12", 69 ply all glass sample was produced. See Figure 6.

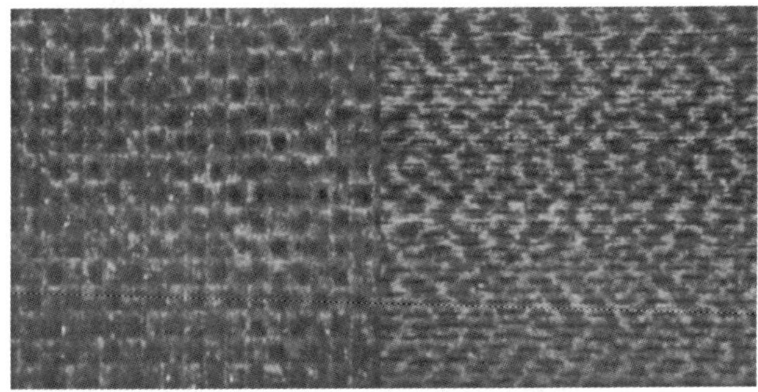

Fig 6. 69 Ply glass prototype (surface and side views)

Materials Used The selection of materials for the thick section samples was intended to replicate, as closely as possible, materials that the US Army might use for structural systems in their composite Armored Vehicle (CAV).

Glass Fiber The glass fibers selected were woven roving consistent with the requirements of MIL-L-46197(MR). The manufacturer, King Fiber Glass Corporation of Arlington, Washington, indicated the material used was consistent with the requirements of MIL-A-46165 as well. The Technical Data (Nominal) provided with the woven roving is shown in Table 1.

Carbon Fiber A plain weave Hercules AS4 graphite fiber was used. Characteristics of this fiber are listed in Table 5.

Table 5. Summary of Carbon Fiber Characteristics.

Tensile Strength	560 ksi
Tensile Modulus	30 Msi
Tensile Strain	1.6 %
Density	0.064 lb/in^3
Carbon Content	94 %
Filament Diameter	0.271 mil
Bundle Size	3k
Area Weight	5.71 oz/yd^2
Nominal Ply Thickness	10.5 mil

Resin: The acrylated epoxy blend was the same as that reported and tested in the 9 ply samples.

Irradiation: The radiation dose for the 69 ply sample parts was established during the development of the process for the 9 ply samples.

Results

The results for the 9 ply samples are summarized in Table 6.

Table 6. Summary of 9 Ply Glass-epoxy Properties

Tensile Strength	140 MPa
Flexural Modulus	9,200 MPa
Flexural Strength	90 MPa
Compressive Strength	50 MPa
Glass Transition Temperature	180° C
Resin Content	72%

These results indicate that a blend of an acrylated oligomer and multifunctional monomers can be used to produce high quality glass fiber reinforced composite materials. The results are comparable to conventional glass-epoxy systems but lower than those for polyester composites manufactured with the same woven roving. It is important to note however, that the samples were found to be free of voids. Similarly, in thick section samples, visual inspection can find no voids. There are however, resin rich and resin deficient regions present in these samples. These are most likely due to the wet-hand-lay up technique used throughout this work.

Conclusions

This work provides the first step in understanding the use of glass and glass/carbon reinforced epoxy resins for the construction of thick section support structures in future combat vehicles. The work demonstrated that existing commercial resins produce strong, void free composites with small doses of radiation. In addition, even in complex geometries with multiple fibers, the resulting composite has no observable voids or delamination as a result of internal stresses.

This work is currently being extended to other resins and geometries in an attempt to provide an engineering data base for composite engineers developing and designing new structures.

Acknowledgment

The authors greatfully acknowledge the support of Irradiation Industries (Gaithersburg, MD) and Atomic Energy of Canada (Pinawa, Canada) for the electron irradiations. We also thank Luis Hinojosa of the US Army Tank Automotive Command for his technical contributions, and Professor Joseph Silverman and Dr. Mohamad Al-Sheikhly for valuable insights regarding the fundamental radiation chemistry of these systems. The authors also acknowledge the laboratory contributors of Joseph Poindexter who performed hand lay-up activities and sample irradiations, and Professor Yusuf Ulcay for the design and construction of the mechanical test fixtures. This work was supported by the DAMILIC Corporation under contract to the US Army.

References

1. A. Chapiro, "Radiation Chemistry of Polymeric Systems", Interscience Publishers, New York (1962)
2. Hiramoto, T., M. Nishii, K. Hayachi and S. Okamura (1971). Radiation-induced polymerization of 1,2-cyclohexene oxide in the plastic crystalline state. J. Poly. Sci.: Part A-1, 9, 3647-3659.
3. Cordischi, D., A. Mele (1965). Radiation-induced polymerization of 1,2 cyclohexene oxide. J. Poly. Sci., Part A, 3, 3421-3426.
4. Cordischi, D., A. Mele and A. Somogyi (1967). Radiation induced polymerization of aliphatic and alicyclic oxides. In J. Dobo and P. Hedvig (Eds.),

Proc. 2nd Tihany Symp. Radiat. Chem., Akademiai Kiado, Budapest. pp. 483-491.

5. Gotoda, M., K. Yanagi, T. Hanyuda and K. Mori (1974) Radiation curing of epoxy-acrylate prepolymers and their mixture with vinyl monomers. II. Electron beam curing of epoxy-acrylate prepolymer/vinyl monomer mixtures. In fundamental studies on Osaka laboratory for radiation chemistry. Annual report: April 1, 1972-March, 1973., JAERI-5029, pp.114-120.

6. Gotoda, M., Y. Miyashita and K. Takeyama (1975). Radiation curing of epoxy-acrylate prepolymers and their mixtures with vinyl monomers. III. A survey on low vapour pressure vinyl monomers, suitable for use in electron beam curable coating compositions. In fundamental studies in Osaka laboratory for radiation chemistry. Annual report: April 1, 1973-March 31, 1974., JAERI-5030,pp.125-135.

7. C.B. Saunders, L.W. Dickson, A. Singh, A.A. Carmichael, and V.J. Lopata, Polymer Composites, Vol. 9, No. 6, p. 389 (1988)

8. L.W. Dikson, A. Singh, "Radiation Curing of Epoxides", Radiat. Phys. Chem., 31, 587 (1988)

9. Gillespie, J.W. and Bogetti, T.A., Cure Simulation of Thick Thermosetting Compounds, U.S. Army Ballistic Research Laboratory Report BRL-TR-3121, July, 1990.

Design Strategies for Fibre Preforms

C.D. Rudd, A.C. Long
The University of Nottingham
Nottingham, United Kingdom

Abstract

Traditional design methods for fibre preforms are based upon the fibre distribution and properties of flat plaque laminates. This neglects the effects of substantial fibre re-orientation which is known to occur during preforming operations. This paper describes how analytical methods can be used to predict fibre distributions in complex parts and how these predictions can be linked to produce distributions of preform permeabilities and elastic properties for the part. These automatically generated data sets can then be used conveniently as inputs to FE based mould flow and structural analyses to ensure that processing and structural requirements are met. Finally, a method is demonstrated for preform optimisation based upon fibre distribution.

FIBRE COMPOSITE MATERIALS have been used in the aerospace industry for a number of years, and are becoming increasingly popular for automotive applications. The design flexibility and weight reduction offered by these materials have been common motivating factors to both industries. In recent years, interest from the automotive industry has been driven by the potential reduction in tooling costs and lead times. Of the many manufacturing routes available, liquid composite moulding processes such as resin transfer moulding (RTM) and structural reaction injection moulding (SRIM) appear to offer the greatest potential for high volume automotive applications(1), and are also proving viable for many high performance aerospace components(2).

To allow access to higher production volumes it is common to preform reinforcements to the component shape prior to resin injection. Conventionally, this involves the forming of several layers of mat or fabric between matched moulds. For semi-structural applications fibres are usually randomly oriented, whilst for more demanding components it is common to use aligned fabrics. Although this technique is more appropriate for structural composites than alternatives such as chopped fibre spray-up, there are several disadvantages associated with current preform technology(3). Fibre movement induced by the forming operation can have adverse effects on the processing and performance characteristics of the preform. The formability of many reinforcement fabrics is limited by the fibre architecture, with defects such as preform wrinkling common for deep-draw components. The level of waste generated is undesirable on both economic and environmental grounds, especially in the absence of an effective recycling route. One possible technique for eliminating these problems is via direct fibre placement to produce a net-shape fibre laydown which can be formed to produce a preform with the desired characteristics(4).

In recent years, increasing interest has been shown in the use of CAE techniques for component design due to the potential reduction in development times. Structural design based on laminate theory has become routine for composites. Simulation of the moulding process, and in particular the mould filling phase, has been the focus of much attention in recent years (eg 5,6). An accurate process simulation allows optimal siting of injection and vent ports, and can also highlight possible problems such as air entrapment due to bypass flow. The predicted pressure distribution can be used as an aid to mould design, especially when using low stiffness tooling based on composite or nickel shell moulds.

The validity of any design analysis depends upon the quality of the input data. In this respect, both structural analysis and flow modelling depend upon materials data which, traditionally, are measured from flat plaque mouldings. For complex parts produced from fibre preforms, the fibre architecture will inevitably differ from that present in the flat plaque with consequent changes in the material properties. Flow modelling is usually based on D'Arcy's law, using permeability measurements carried out within a two-dimensional rectilinear or radial cavity(7). However this neglects the effects of deformation induced during preform manufacture. The same is true of structural analysis, where the properties of the component are likely to differ from simple two-dimensional laminates. If the fibre architecture in the preform can be determined, then it is likely that accurate estimations of both permeability, elastic properties and laminate strengths can be made. This is achieved most conveniently by modelling the deformation of the reinforcement during the preforming process.

This study is aimed at developing a CAE system for the design of structural preforms. This is based on a kinematic draping algorithm, capable of predicting the fibre orientations within three-dimensional preforms. By post-processing the predicted draped fibre architecture, the effect of reinforcement deformation on the processing and performance characteristics of the preform are evaluated. In this way, the design process is refined by basing analyses on improved estimates of material properties. The same models can be used to determine optimal fabric geometries for preform quality. However, the process can be taken a stage further and instead of accepting the properties which result from the fabric deformation or preforming process, the analysis could be used to work backwards and determine the starting point necessary to produce an "ideal" or optimum preform. For example, additional reinforcement can be placed in areas which are subject to thinning during forming, or fibres can be laid such that they achieve the desired orientation *after* the forming process. Such an approach implies a capability to place reinforcements as desired, not as is done conventionally by the use of tailored mat or fabric but by direct fibre placement using a high speed CNC manipulator.

Experimental Details

Experimental work, which included the measurement of properties for laminates based on sheared reinforcement fabrics, was based upon Tech Textiles E-LT 1134 Style 1047 stitch bonded glass fibre reinforcement. This contains two parallel layers of E-glass reinforcement (600 tex rovings, 15 micron filament diameter) at 0/90° which were restrained by a polyester fibre in the form of a compliant stitch (1 % by mass). Flat plaque mouldings were manufactured by RTM using Cray Valley Totale 6345.001 unsaturated polyester resin in aluminium tooling operated at 90 °C. Elastic properties were measured using an Instron 1195 universal testing machine to a method encompassing BS 2782. Reinforcement permeability measurements were carried out using a radial flow arrangement according to the method described in our previous paper(7).

Preforms for the prototype wheel-hub, used as a case study in the following sections, were produced using a screw press with epoxy tooling. These were then processed by SRIM using a Krauss Maffei RIMStar 40 machine with centre injection into a steel mould. The mould contained in-situ pressure transducers and thermocouples for process monitoring purposes. The resin system was Shell Epikote DX6008 epoxy with DX6511 hardener mixed at a ratio of 4:1. The mould was operated at 120 °C, with the resin and hardener injected at 117 °C and 86 °C respectively.

Reinforcement Deformation Modelling

A kinematic draping algorithm has been developed, based on the draping of an inextensible fibre network over an arbitrary surface defined using flat patches(8). The fibre crossovers (or stitch centres) are assumed to act as pin-joints, with inter-fibre slip assumed to be negligible. This may be the case for many practical forming operations(9), although for some fibre structures relative slippage may be the dominant mode during deep-draw. One of the aims of this study is to establish the validity of the "pin-jointed net" model for relatively complex components. The use of flat patches allows an explicit solution to the equations of intersection between the fabric and the component surface, and hence offers significant computational advantage over simulations based on curved surface elements(10).

The draping simulation has been applied to a number of complex components. Figure 1 shows the draped fibre pattern for an automotive undershield draped with a ± 45° reinforcement. The inset figure shows an enlarged view of an area of relatively deep-draw (known as the swage corner). Clearly a significant level of fibre re-orientation is anticipated, with a minimum inter-fibre angle of 26° predicted. This level of shear would be expected to result in reinforcement wrinkling during preform production, a problem which has been confirmed during manufacture.

Figure 2 compares the predicted and experimental fibre patterns for one quarter of a prototype wheel-hub draped with a 0/90° reinforcement grid. As shown in Figure 2(a), maximum deformation is predicted mid-way along the circumference of the component, with a minimum inter-fibre angle of 28°. To validate these predictions, preforms were made using five layers of Tech-Textiles E-LT 1134 0/90° fabric. Each layer was cut out to the predicted net-shape pattern shown in Figure 3, which was obtained by mapping each fibre path back to its position in the two-dimensional sheet. This proved to be an extremely accurate template, as preforms did not require trimming prior to moulding. Warp and weft tows on the outer layer were marked at 18 mm intervals to highlight the fibre paths. As shown in Figure 2(b), these were clearly visible after moulding. A qualitative comparison with the predicted pattern indicates that the model provides an accurate description for this component.

One measurable effect of reinforcement deformation is the local increase in fibre volume fraction induced by shear. This can be estimated at each fibre crossover using

$$V_f = \frac{S_0}{\rho \, t_0 \sin\beta} \tag{1}$$

The predicted variation in fibre volume fraction around the rim of the component is compared with experimental measurements in Figure 4, showing an excellent correlation. The nominal (unsheared) fibre volume fraction of 36% is increased to approximately 54% in the most highly sheared region. Any further increase is likely to be limited by the fibre architecture.

Flow Simulation

Simulations of the impregnation phase during liquid moulding are useful in component design, as they allow the optimal siting of injection and vent ports to achieve the shortest possible injection times. Problems such as air entrapment due to bypass flow can also be anticipated at the design stage, ensuring that a satisfactory moulding process is achieved. However the accuracy of such process simulations is dependent on the available permeability data, which are often obtained using a simple two-dimensional flow experiment. This neglects the effects of reinforcement deformation, which may be expected to result in a highly non-uniform permeability distribution within the preform. The permeability of a deformed aligned fibre reinforcement can be estimated by assuming the fabric to be comprised of two uni-directional (UD) layers. Experimental permeability data has been obtained for a commercially available UD reinforcement(7), Tech-Textiles E-LPb 567, giving the following empirical relationships between permeability and porosity:

$$k_1 = 50.85\,\phi^{7.18} \; , \; k_2 = 4.50\,\phi^{9.05} \qquad (2)$$

where the units of the above are m^2 x 10^{-9}. The permeability of the combined bi-directional reinforcement can then be approximated using a simple addition rule(7), giving

$$k_{avg} = k_1 + \frac{1}{2}\left(k_2 - k_1\right)\left(\sin^2\alpha_1 + \sin^2\alpha_2\right) \qquad (3)$$

Figure 5 demonstrates the effect of various degrees of shear on the permeability predicted by Equation 3, based on 5 layers of Tech-Textiles E-LT 1134 0/90° reinforcement within a 6 mm cavity. If fibre volume fraction were to remain constant (36%), the maximum permeability (corresponding to flow mid-way between the fibre directions) is expected to increase approximately linearly with shear. With fibre volume fraction defined as a function of shear using Equation 1, the maximum permeability peaks at an inter-fibre angle of 80°, and then decreases rapidly to zero as the degree of shear is increased. The minimum principal values are expected to be reduced by shear, with increasing fibre volume fraction expected to enhance this effect. These observations demonstrate that permeability is likely to be dominated by the variations in fibre content rather than local fibre orientations. This is further demonstrated by Figure 6, which shows the predicted permeability distribution for radial flow within the draped wheel hub described in the previous section. In the most highly sheared region where fibres are oriented towards the direction of flow, the permeability is expected to be practically zero (as compared to an expected value of 0.72 x 10^{-9} m^2 for this material).

To study the effect of the above predictions on the filling phase of liquid moulding, a control volume finite element flow modelling package developed by Rice was used(5). This is based on a modified version of the PAFEC-FE code, known as CFILL, which uses D'Arcy's law and in-plane mass continuity to describe the flow of resin into the preform. The boundary conditions used are constant mass flow rate (250 g/sec) at the centre node and atmospheric pressure at the flow front. The analysis in this case is isothermal since the variation in resin viscosity within the working temperature range is negligible. The output from the analyses, representing flow front positions or isochrones, are shown in Figure 7. Figure 7(a) shows the predicted filling pattern for the nominal (0/90°) reinforcement, with the circular isochrones indicating flow isotropy. Figure 7(b) corresponds to the same model using permeabilities generated using the drape analysis but with fibre volume fraction held at a constant 36 %. This results in a displacement from the isotropic case, with a clear flow leader at the centre of the rim. Comparison with Figure 7(c) shows that the effect becomes less marked when the change in volume fraction caused by reinforcement shear is included, although flow is still promoted in the same region despite the local reduction in permeability suggested by Figure 6. This effect is explained by the local decrease in porosity which implies a reduced volume of resin required for impregnation.

Short shots were carried out corresponding to the isochrones in Figure 7 to validate the above predictions, using the equipment described above. An example is given in Figure 8 which shows the position of the flow front after injection for 2.8 s. The maximum principal flow axis corresponds to the axis of maximum shear and *minimum* permeability, thus confirming the above predictions.

Structural Analysis

Structural design is conventionally based on the use of laminate analysis or finite element packages, using mechanical properties obtained from flat plaque test specimens. This approach neglects the effects of reinforcement deformation as described above, and in practice it is common to apply a conservative design strategy to account for the likely range of mechanical properties within components. To avoid over-design and to ensure efficient use of materials, it is essential to base structural design on an accurate knowledge of the deformed fibre distribution. A relatively simple approach can be taken utilising the method developed by Krenchel (11), based on the calculation of the load-bearing efficiency factor of the reinforcement. For a sheared bi-directional reinforcement, this can be written as

$$\eta = \frac{1}{2}\left(\cos^4\alpha_1 + \cos^4\alpha_2\right) \qquad (4)$$

The local modulus of the laminate can then be estimated using a modified version of the well known rule of mixtures:

$$E_c = E_m\left(1 - V_f\right) + \eta E_f V_f \qquad (5)$$

where the local fibre volume fraction is given by Equation

1. This relationship has been investigated by subjecting fabric samples to varying degrees of pre-shear prior to manufacture of flat plaque mouldings as described above. The fabric samples were clamped to one side of the laboratory bench using one side of a two bar linkage. The opposite edge of the fabric was then displaced by a predetermined amount to induce the desired shear within the fabric. Specimens were cut and tested in the directions which bisected the two fibre axes. The resulting moduli are compared with theoretical predictions in Figure 9, showing very close agreement.

This technique has been applied to the results of the drape analysis for the automotive wheel-hub as described above. Figure 10 shows the predicted variation in modulus for loading applied in the radial direction. A wide variation in properties is anticipated, with modulus increasing to 37.4 GPa in the most highly sheared region (as compared with an expected value of 14.3 GPa for laminates based on roll-stock reinforcement). The increase in fibre volume fraction associated with shear deformation serves to magnify local variations in mechanical properties. As with the flow simulation described earlier, it should be possible to simulate the effect of reinforcement shear within an FE structural analysis. This possibility is currently under investigation and will be reported in the near future.

Optimisation Strategies

The analyses described so far in this study have been applied to homogeneous reinforcement structures, ie orthogonal aligned fibre fabrics. However the likelihood of reinforcement wrinkling, coupled with the significant variations in both permeability and mechanical properties described above, would suggest that these materials are not the ideal structures required to achieve the desired properties. Using a direct fibre placement system, such as that described in our previous publication (4), allows the production of reinforcement "fibre laydowns" which can be post-formed to provide preforms with the desired characteristics. The ability to predict the reinforcement net-shape has already been demonstrated for the prototype wheel-hub. This could be used to define the fibre placement paths, ensuring that materials utilisation is maximised. An integrated design system, based on deformation modelling, flow simulation and structural analysis, is required to optimise the fibre laydown. This is likely to be based on an iterative process, involving continual assessment and re-design.

One of the major problems anticipated by the deformation model was that of reinforcement wrinkling due to excessive inter-fibre shear. The associated increase in fibre volume fraction is expected to cause a wide variation in properties within the preform. By simply re-designing the fibre laydown, it is possible to reduce the variation in fibre volume fraction and effectively eliminate the problem of preform wrinkling. This is demonstrated by Figure 11, which shows the effect of a simple local re-design of the fibre pattern within a deep-draw region of the automotive undershield described earlier. By applying a 33° shear to the fibre path on the right of the figure, the maximum predicted fibre volume fraction is reduced from in excess of 90% (certain to result in wrinkling) to 51%. Applying a similar technique to the component as a whole would allow the entire reinforcement structure to be optimised to eliminate wrinkling and to reduce the variation in processing and performance properties.

Summary

A kinematic draping model has been developed to simulate the effects of reinforcement deformation on the fibre architecture within structural preforms. The associated changes in fibre volume fraction and orientation are likely to cause a marked departure from properties anticipated using flat plaque data. Permeability is expected to be dominated by local variations in fibre content rather than re-orientation, with regions of excessive shear having virtually zero permeability. Tensile properties are expected to be dramatically influenced by deformation, as fibre volume fraction variations enhance the effects of re-orientation. An advanced preform production system, based on direct fibre placement to produce net-shape fibre laydowns with improved or optimum properties, has been developed to overcome some of the deficiencies highlighted for current preform technology. The design of reinforcement structures using this system has been demonstrated for a section of a structural component.

Acknowledgements

The mould filling simulation was performed by Dr E.V. Rice of Crescent Consultants Ltd (UK). Assistance with the experimental work was provided by Roger Smith, David Winstanley and Geoff Tomlinson (University of Nottingham). The authors are grateful to the following organisations for their continued support:

Ford Motor Company, Dowty Aerospace, DSM Resins, The Department of Trade and Industry, ICI Chemicals, PPG Glass Fibers, Science and Engineering Research Council, Shell Chemicals (all UK) and Ford Research Laboratory, Dearborn, Michigan.

References

1.   Chavka, N.G. & Johnson, C.F. "The Taming of Liquid Composite Moulding for Automotive Applications", Advanced Composite Materials: New Developments and Applications Conference Proceedings, 115-122 (Sept 1991)

2.   Robertson, F.C. "Resin Transfer Moulding of Aerospace Resins - A Review", British Polymer Journal, 20, 417-429 (1988)

3.   Owen, M.J., Middleton, V. & Rudd, C.D. "Fibre reinforcement for high volume resin transfer

moulding (RTM)", Composites Manufacturing, 1, 74-78 (June 1990)

4. Rudd, C.D., Middleton, V., Owen, M.J., Long, A.C. & McGeehin, P. "Design, Processing and Performance of Structural Preforms", 2nd Canadian International Conference on Composites (CANCOM 93), Ottawa, 139-152 (Sept 1993)

5. Owen, M.J., Rice, E.V., Rudd, C.D. & Middleton, V. "Resin Transfer Moulding for Automobile Manufacture: Reality and Simulation", CADCOMP 92, University of Delaware, 121-142 (May 1992)

6. Trochu, F., Boudreault, J.-F., Gao, D.M. & Gauvin, R. "Three-Dimensional Flow Simulations for the Resin Transfer Moulding Process", 2nd Canadian International Conference on Composites (CANCOM 93), Ottawa, 153-159 (Sept 1993)

7. Rudd, C.D., Rice, E.V., Bulmer, L.J. and Long, A.C. "Process Modelling and Design for Resin Transfer Moulding", Plastics, Rubber and Composites Processing and Applications, 20, 2, 67-76 (1993)

8. Long, A.C. "Preform Design for Liquid Moulding Processes", PhD Thesis, University of Nottingham, (May 1994)

9. Potter, K.D. "The Influence of Accurate Stretch Data for Reinforcements on the Production of Complex Structural Mouldings", Composites, 161-167 (July 1991)

10. Van West, B.P., Pipes, R.B. and Keefe, M. "A simulation of the draping of bidirectional fabrics over arbitrary surfaces", J Text Inst, 4, 448-460, (1990)

11. Krenchel, H. "Fibre Reinforcement", Akademisk Forlag, Copenhagen, (1964)

Nomenclature

E_c	Composite modulus
V_f	Fibre volume fraction
E_f	Fibre modulus
$\alpha_{1,2}$	Fibre orientation wrt flow/loading
E_m	Matrix modulus
β	Inter-fibre angle
$k_{1,2}$	Principal in-plane permeabilities
ϕ	Reinforcement porosity ($1-V_f$)
k_{avg}	Average stack permeability
η	Reinforcement efficiency factor
S_0	Stack superficial density
ρ	Fibre (glass) density
t_0	Stack thickness

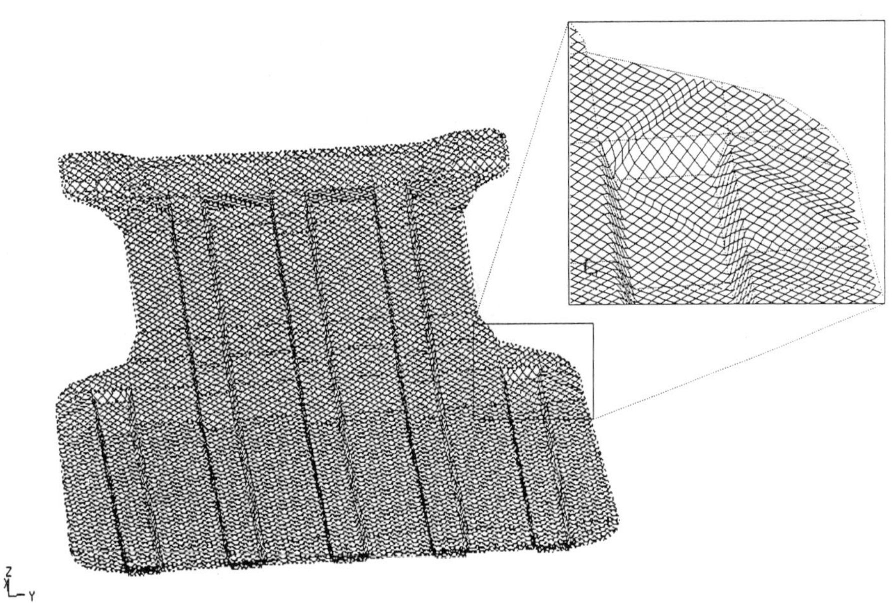

Figure 1: Ford Cosworth undershield draped with $\pm\,45°$ reinforcement

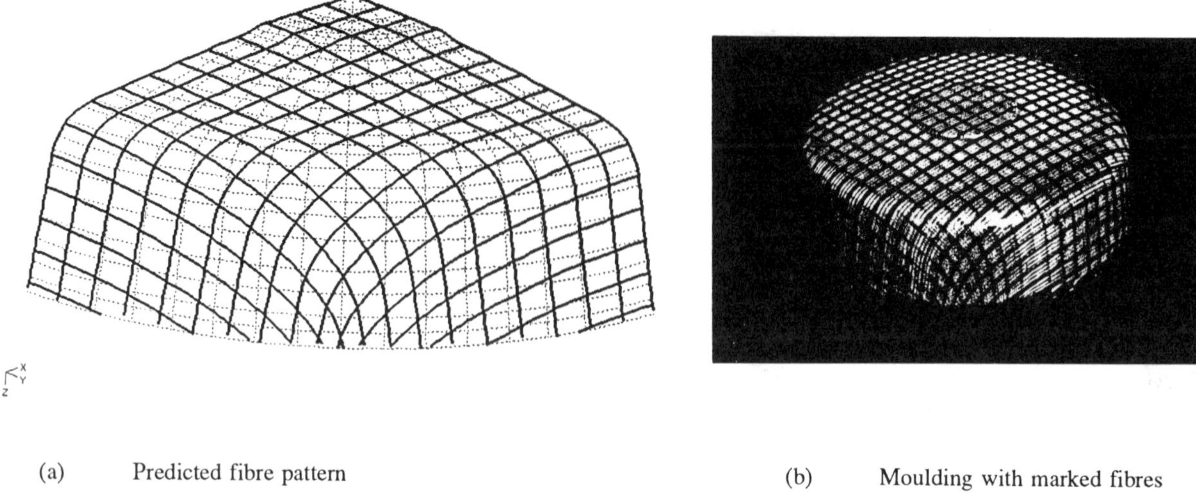

(a)   Predicted fibre pattern   (b)   Moulding with marked fibres

Figure 2: Validation of predicted fibre pattern for prototype wheel-hub

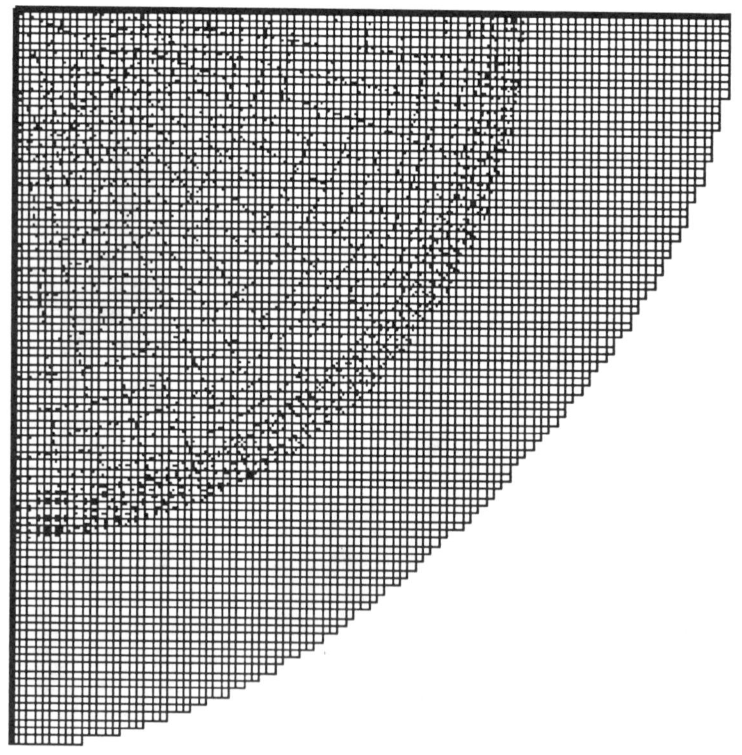

Figure 3: Predicted net-shape pattern for prototype wheel-hub

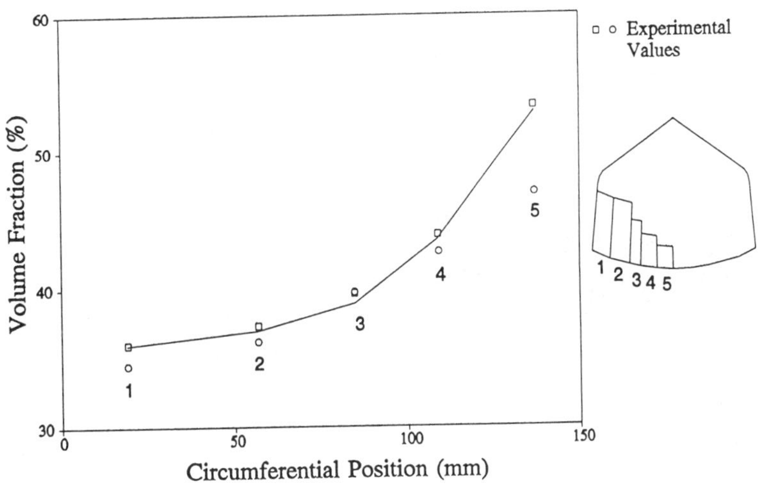

Figure 4: Circumferential fibre volume fraction variation for prototype wheel-hub

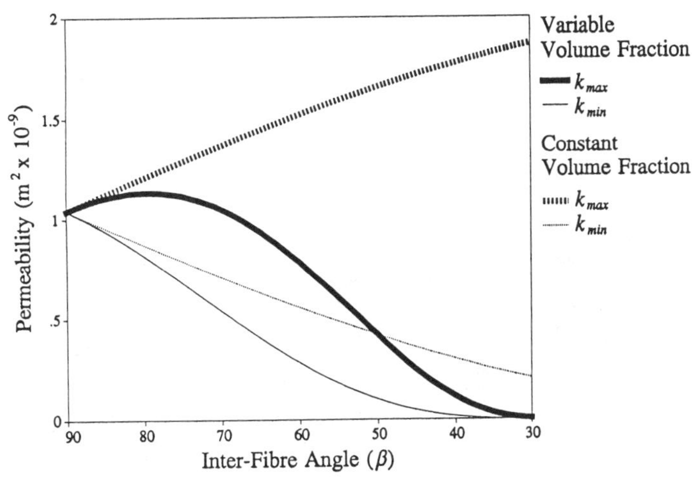

Figure 5: Effect of reinforcement shear on principal in-plane permeabilities

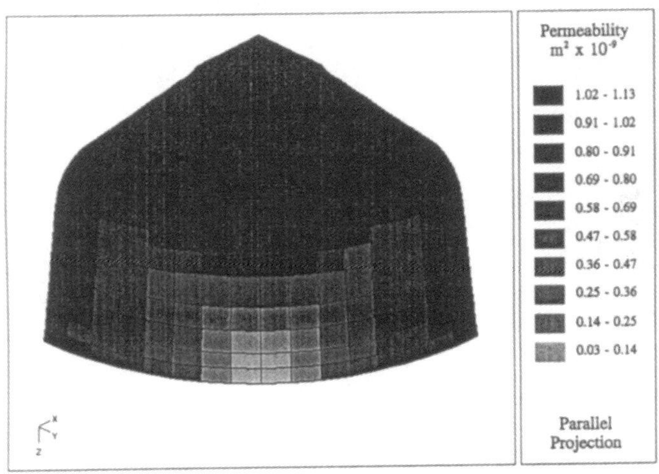

Figure 6: Predicted radial permeability distribution for prototype wheel-hub

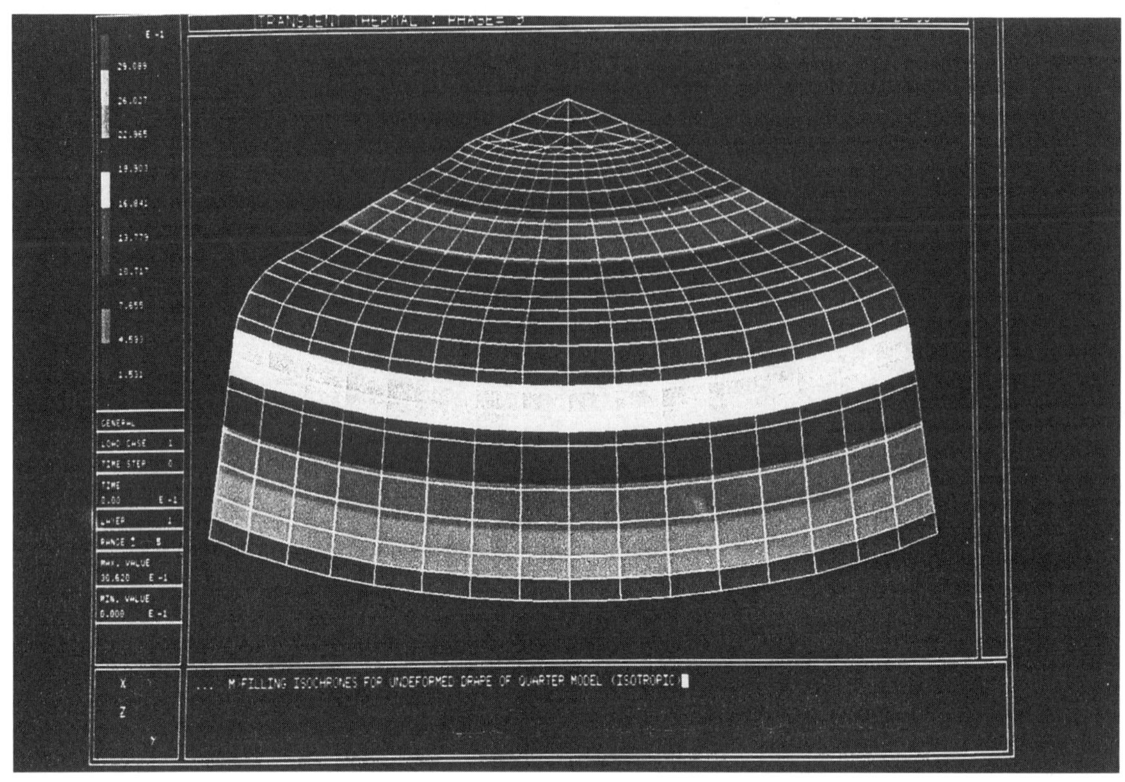

(a)   Undeformed fibre pattern

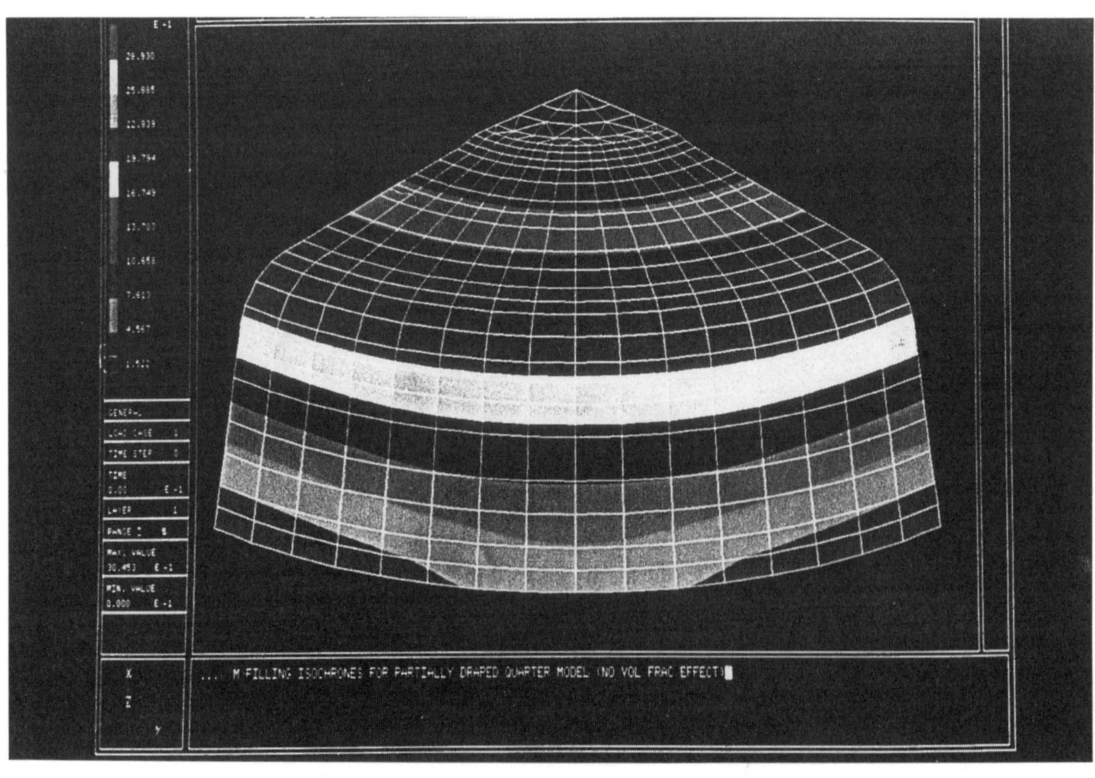

(b)   Partially deformed (constant volume fraction)

Figure 7: Predicted isochrones during impregnation for prototype wheel-hub

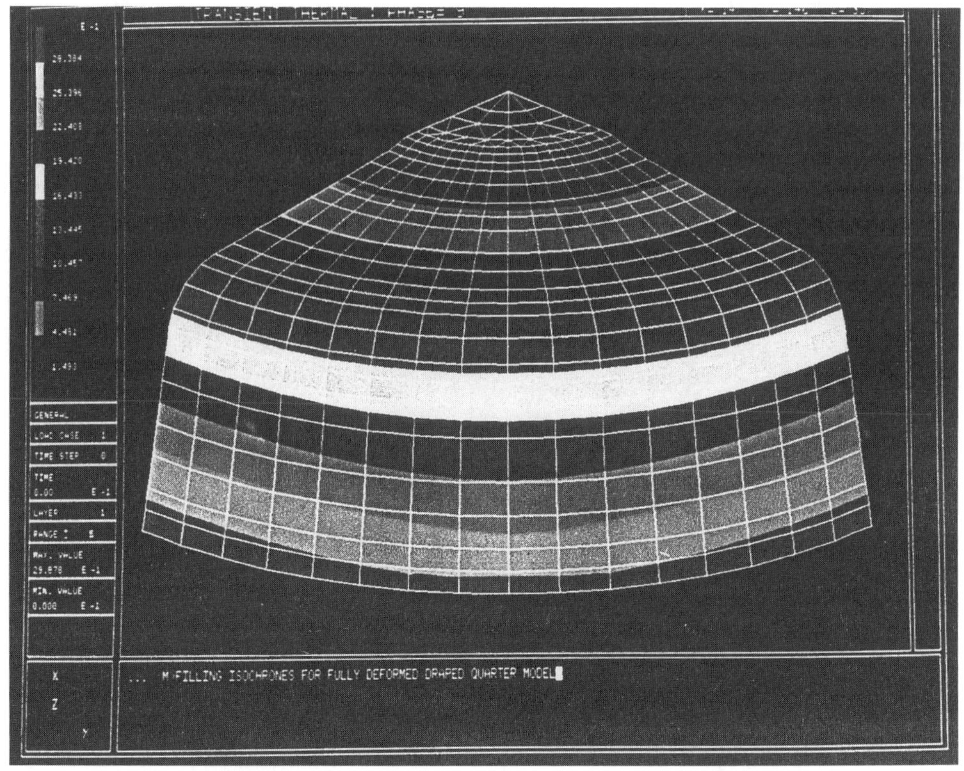

(c)   Fully deformed fibre pattern

Figure 7 (contd): Predicted isochrones during impregnation for prototype wheel-hub

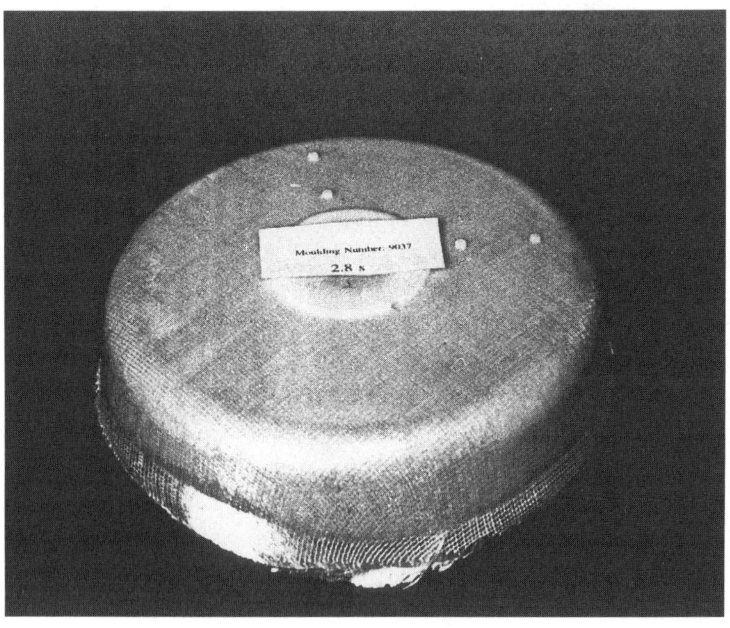

Figure 8: Short shot for prototype wheel-hub

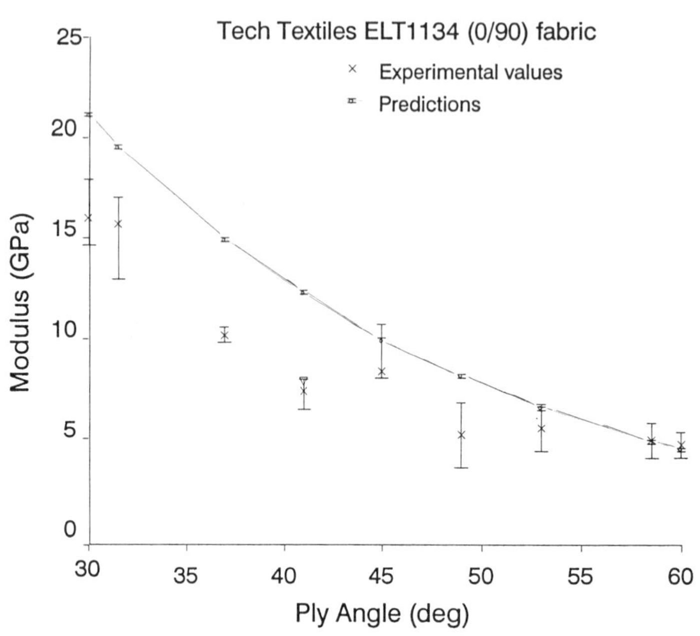

Figure 9: Effect of reinforcement shear on laminate modulus

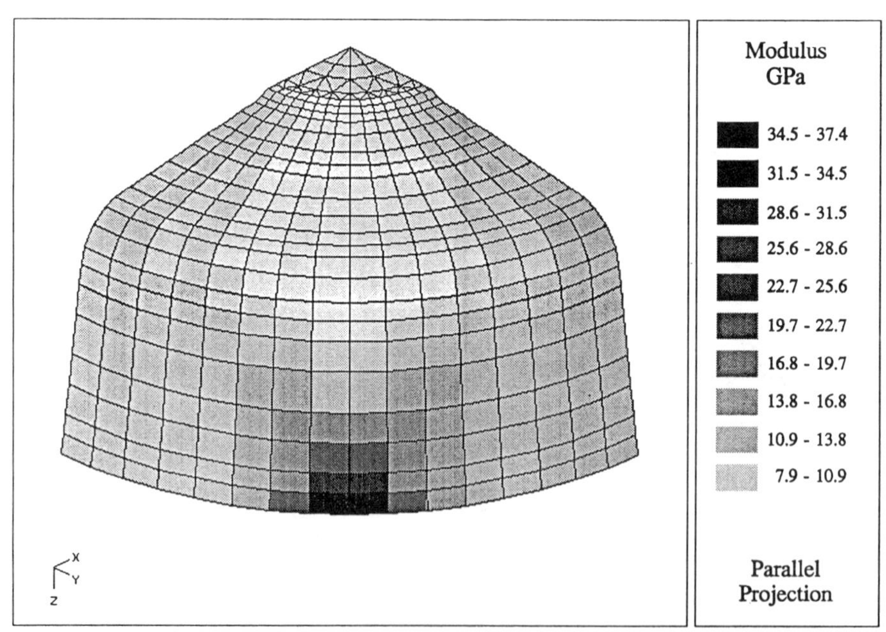

Figure 10: Predicted radial modulus distribution for prototype wheel-hub

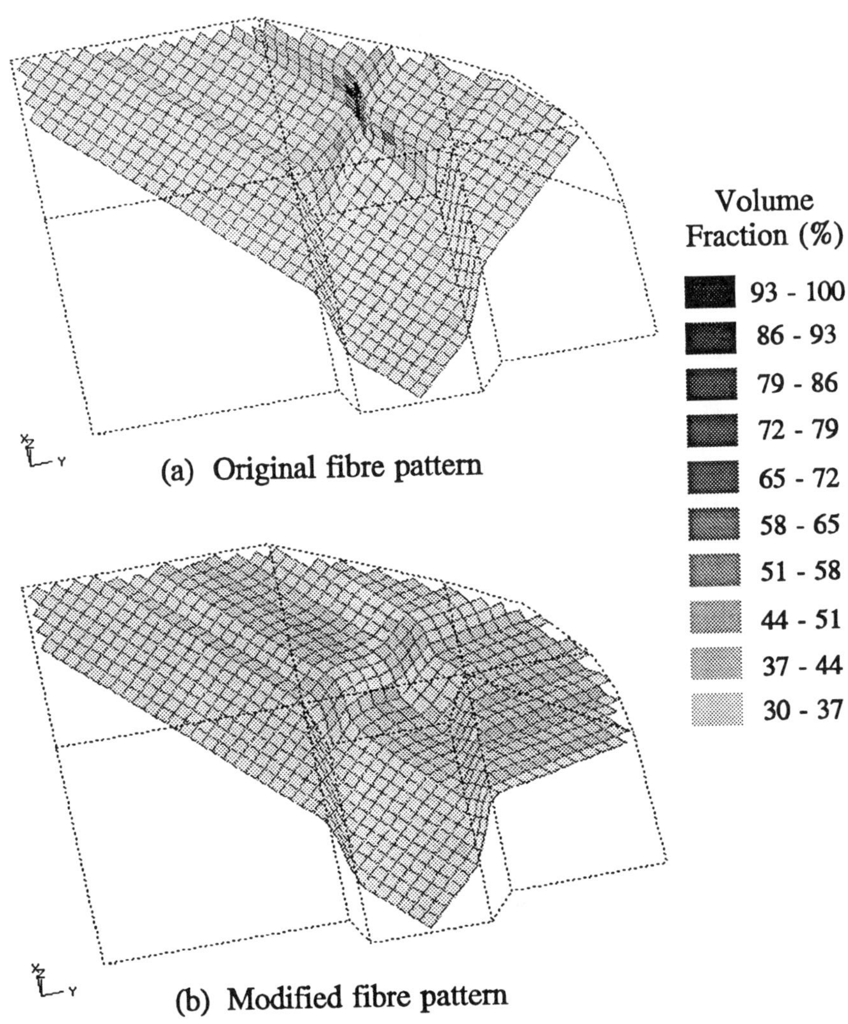

Figure 11: Optimisation of fibre structure undershield swage corner

Proceedings of the 10th Annual ASM/ESD Advanced Composites Conference, Dearborn, Michigan, USA, 7-10 November 1994

Some Practical Issues in Composite Tooling Design Modeling

H.P. Wang, E.M. Perry
GE Corporate Research & Development
Schenectady, New York

Abstract

A series of fast-acting process modeling tools has been developed that address many of the major process, design, and quality issues in the manufacture of composite parts. For example, stringent constraints on the part temperature field must be met to uniformly cure composite parts with variable thickness; a model that helps to determine the proper location, depth, and size of the heating lines in the mold has been developed. Also, in resin transfer molding, the permeability of the preform is critical to proper mold filling. Problems with the dimensional tolerance of the tooling can lead to local compression of the fiber preform, which frequently results in variations in the permeability that affect the filling behavior and gating design. In addition, bends in the part geometry, such as those that occur in tubes and ducts, can also lead to compression of the braided fiber perform structure, affecting the flow behavior. Finally, inverse analyses are a powerful tool for obtaining the material properties of composites, such as the thermal conductivity or permeability, in cases where theoretical estimates may not be adequate. Tools such as these address the some of the practical issues that the process and design engineers must face in making science-based decisions regarding process and design questions prior to cutting the mold.

PROCESS MODELING integrates three basic functions in the design-for-manufacturing concept for making composite parts: tooling design, process optimization, and material property issues. This concept is illustrated in Figure 1, showing that the process model uses as input the part and tooling geometries, the process conditions, and the material properties, in order to predict how the process will perform – whether the part can be made (its producibility), how fast it can be made (the productivity), and whether it may contain defects (its quality). The key output from models such as this is a quantitative understanding of the process.

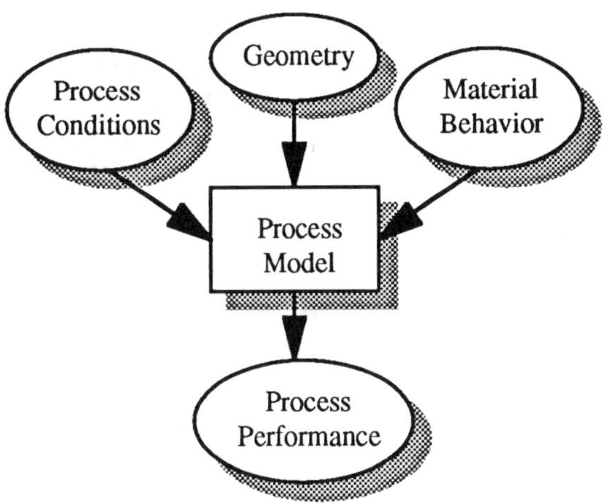

Figure 1. Schematic illustration of process modeling concept.

Parametric studies conducted on the input variables with the process model can provide guidance in process planning and lead to reductions in product development time and increases in part quality. These modeling tools can be used to answer many of the "what-if" processing questions in an interactive manner, significantly reducing the number of molding experiments required as well as shortening the costly cut-and-try procedure for mold making. Changes in the mold and tooling geometries, the mold heating cycles, the process conditions, or the materials are usually interdependent, and their effects on molding processes would be difficult to predict without such process modeling tools as those described in this paper.

In the area of tooling heat transfer, design of the heating system is always a challenge. The arrangement of the complicated heating circuitry must be based on its effectiveness in heating the part. The difference between the thermal diffusivities of the composite part and the mold is several orders of magnitude. Therefore, a decoupled analysis approach is justified for analyzing the problem: a steady-state analysis for the heating line arrangement and a transient analysis for the part thermal response. There are many 3-D finite element heat transfer analysis software packages available today, but their usage is still very limited due to two factors: the mesh generation process for the very complex tooling geometry is generally lengthy and tedious, and the associated CPU time required for the computations is excessive. A pragmatic way to analyze the tooling heat transfer from the tooling engineer's point of view is to use the shape factor concept for sizing and locating the heating lines and 1-D thermohydraulics for balancing and linking the circuitry.

In the resin transfer molding (RTM) process, the preform permeability is a critical material property that directly affects the flow behavior of the resin. However, there are many factors which can significantly change the permeability. One typical example is non-uniform over-packing due to problems with the dimensional tolerance of the tooling, and another example is the density change of the braided fiber preform structure commonly used in making tubes and ducts that results from bends in the tube geometry. Sometimes the constitutive equations for describing the relationship between the permeability and the geometry change are not readily available; empirical approaches are then required.

Another important aspect of the composite process and tooling analysis area is the lack of material property data. One very cost-effective method recommended here is to use the inverse analysis technique on the data collected from simple experiments. For example, the throughwall thermal conductivity or the permeability of multi-layered prepreg sheets is very difficult to find in the literature. There are no fundamental equations which can help to calculate these two properties accurately. Inverse analysis is one very effective way to acquire these difficult-to-get material properties.

Mold Heating Design

In order to uniformly cure composite parts with variable thickness, stringent constraints on the part temperature field must be met. The heating system design within the mold, including the location, depth, and size of each of the heating lines, must be properly laid out to reduce heating imbalances between different sections of the part, thus improving part quality by achieving more uniform cure. Also, the flow rate of the heating oil throughout the different heating circuits must be balanced in order to achieve uniform heat transfer. Once the design is completed, it must be verified that the pumping capacity of the heating units will be sufficient to meet the flow demand.

Analyses on this type of problem can be performed in two parts. In the first part, a steady state calculation on the heating oil flow throughout the mold is made. The MCAP (Mold Cooling Analysis Program) software (1,2), originally developed to optimize the cooling line configuration for the injection molding process, can be used for these calculations. The analysis results will help the mold designer to properly locate, size, and connect each of the heating lines. In the second part of the analysis, transient finite element calculations provide the time-dependent heating of both the mold and part, using the convective heat transfer coefficient of the heating oil obtained from the first analysis as a boundary condition.

Steady State Heating Line Arrangement

The first step in analyzing the mold heating design is to obtain a preliminary design for the heating line configuration. Figure 2 is a schematic representation of the cross-section of a simple mold cavity, in which the part thickness varies over its width; the part thickness has been exaggerated for better visualization. The location of each heating line, including the "depth," or distance from the part surface to the center of the line, and the "pitch," or center-to-center distance between adjacent lines, as well as the diameter of the lines, can be obtained from such a diagram. As shown in this figure, there are four heating lines located in each half of the mold.

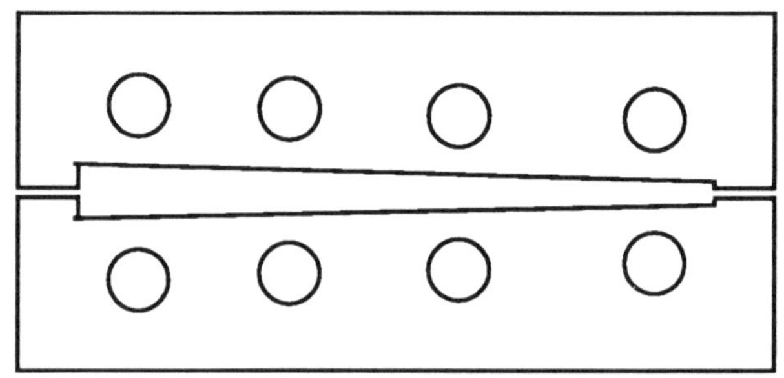

Figure 2. Schematic representation of cross-section of simple mold cavity (part thickness is exaggerated).

Figure 3 is a top view of one of the mold halves, showing one possible circuitry connection for these heating lines, in which they were connected in series. This figure also shows the location of this circuit relative to the part, which is represented by the shaded region; only a portion of each heating line runs above the part. The heating oil comes in at the left side in this figure, and runs through the four heating lines, with 90° turns at the end of each section, and exits at the right side. Both

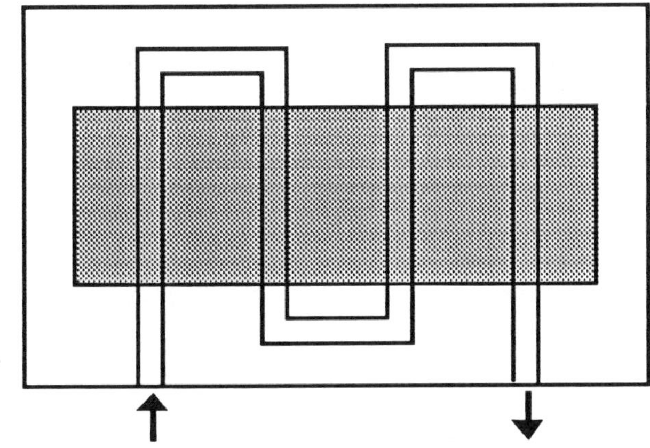

Figure 3. Schematic representation of mold half, showing heating line circuitry.

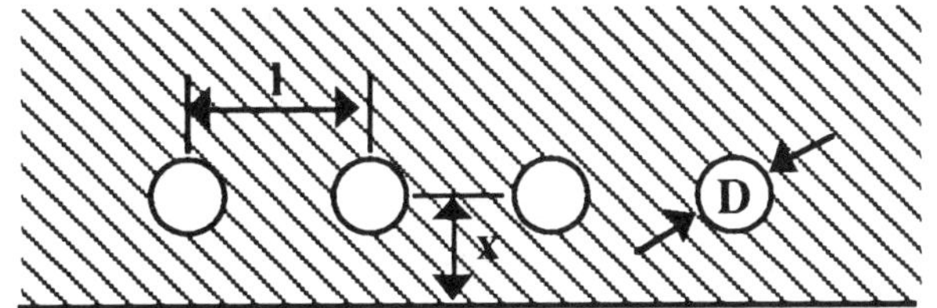

Figure 4. Geometric features needed for shape factor calculation for row of parallel lines at equal depth from part surface.

The shape factor S for this particular geometrical feature can be calculated as shown in Eq. 1:

$$S = \frac{2\pi}{\ln\left[\left(\dfrac{2l}{\pi D}\right)\sinh\left(\dfrac{2\pi x}{l}\right)\right]} \qquad (1)$$

where l is the pitch, or the center-to-center distance between adjacent heating lines, D is the line diameter, and x is the depth, or the distance from the center of the heating line to the part surface. Equations for additional geometrical features, such as a single line parallel to the part surface, a row of parallel lines between two surfaces (in the case of an insert in the mold cavity), or a line that is perpendicular to the part surface, are also included in the software library.

The convective heat transfer coefficient of the heating oil is dependent on the Reynolds number – the higher the Reynolds number, the more effective is the heat transfer. However, the heating effectiveness is limited by the size of the heating lines and the pumping capacity of the machines. The overall heat transfer rate therefore depends on both the conduction and convection terms, as well as the heating oil inlet temperature and flow rate; the geometry of the heating lines; the mold surface temperature; and the material properties of the mold, part, and heating oil.

Finally, the hydraulics of the heating oil flow account for the friction loss of uniform circular pipes running full of fluid. This value for the pressure drop in the lines is dependent on the geometry of the lines (diameter, length, angle bend, circuitry connections, and so on), as well as the flow rate, Reynolds number, and material properties of the heating oil. The heating lines can be connected together into different circuits to modify the pressures and flow rates obtained.

The MCAP software is used to evaluate the effect of the different heating line configurations and process parameters on the oil flow rates and heat transfer coefficients. The mold designer can then see whether the calculated flow rate and pressure drop is within the capacity of the pumping system. If they exceed this capacity, then the configuration of heating lines can be varied, by connecting different lines in series together

mold halves have the same circuitry connections. Many other circuitries, such as parallel lines or a mixture of parallel and series connections, should also be evaluated before any one design is chosen.

The steady-state flow analysis software uses several pieces of data to model the heating oil flow rates and the convective heat transfer within each heating line. It uses the material properties of the heating oil, the mold, and the composite, as well as the heating oil inlet temperature. For each "section" of heating line, where a section is defined as a straight line segment, the software needs the diameter, depth, pitch, total length, effective length, and final bend angle. The total length of the section is needed for the flow calculations, while the effective length, defined as that portion of the length which will actually provide some heating to the part, is required for the thermal calculations on the convective heat transfer coefficient. The circuitry connections, or the manner in which the heating lines are connected to each other, either in series or in parallel, are needed to compute the oil flow rates.

The MCAP software is composed of three parts:

- heat transfer by conduction from the heating lines to the part
- heat transfer by convection through the heating oil
- hydraulics of the heating oil flow.

Calculations of the conduction heat transfer between the heating oil lines and the part are based on the use of shape factors, which depend solely on the geometrical features of the lines and part. A library of shape factors is included in the software, for most of the geometrical configurations likely to be used. For example, Figure 4 shows a typical geometric feature, consisting of a row of parallel heating lines, all of which are at an equal depth from the part surface; this is the case illustrated in the simple mold shown in Figure 2.

or by eliminating some lines, for example, or perhaps the process parameters can be altered.

All of these analyses are conducted in an interactive manner. Many different design alternatives for real production parts were evaluated in a single day, leading to a significant number of unnecessary lines being eliminated from the mold.

Transient Heat Transfer Calculations

Once the convective heat transfer coefficient, h, of the heating oil has been calculated from the steady state analysis, a transient finite element analysis can be performed on a part section, to determine how quickly the part reaches some desired temperature. Frequently, the part geometry can be simplified based on symmetry conditions. For parts of varying thickness, the calculation should be repeated at different part sections, to determine whether there is an imbalance in the heating times throughout the part. The finite element analysis uses the geometry of the nearby heating lines, the part thickness, and the material properties of both the mold and part in its calculations. A convection boundary condition is applied, using the heat transfer coefficient from the steady state analysis, along with a constant heating oil temperature boundary condition.

Figure 5 contains a schematic representation of the geometry needed for these finite element calculations, which must include the mold, the part, and the heating line. Only one-half of the part thickness was modeled, since the other half of the part is heated by the heating line in the other part of the mold, as shown in Figure 1. Due to symmetry conditions between adjacent heating lines, only one-half of the distance between lines was modeled. The size of the mold was set to twice the depth of the heating line, in order to provide a more accurate representation of the heat sink provided by the mold, which slows the part heating.

The one-dimensional nature of the temperature response from the heating line through the mold up to the part allowed the use of a fairly coarse finite element mesh, which significantly improved the computation time. Two typical sets of finite element analysis results are shown in Figures 6 and 7. Figure 6 shows, for a "thin" section of the part, the temperature response as a function of time for locations in the mold interior, the part surface, and the part centerline. Figure 7 shows the temperature responses for a "thick" section.

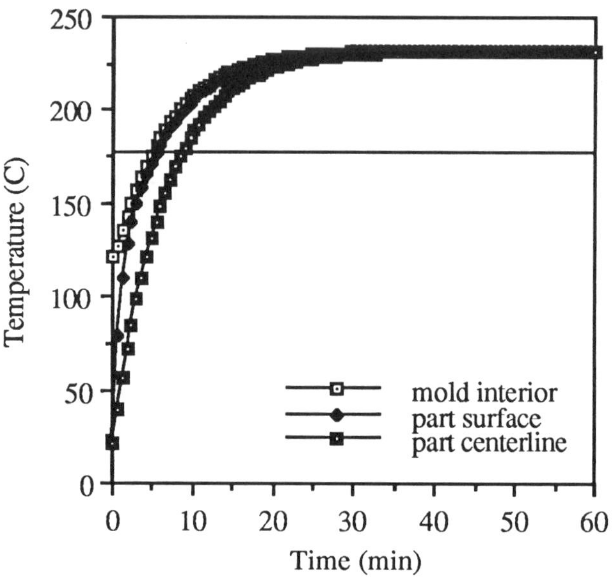

Figure 6. Temperature response of mold interior, part surface, and part centerline for a "thin" section.

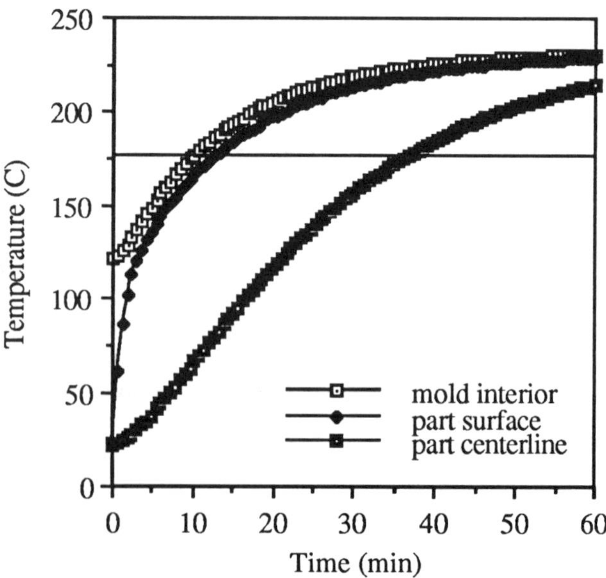

Figure 7. Temperature response of mold interior, part surface, and part centerline for a "thick" section.

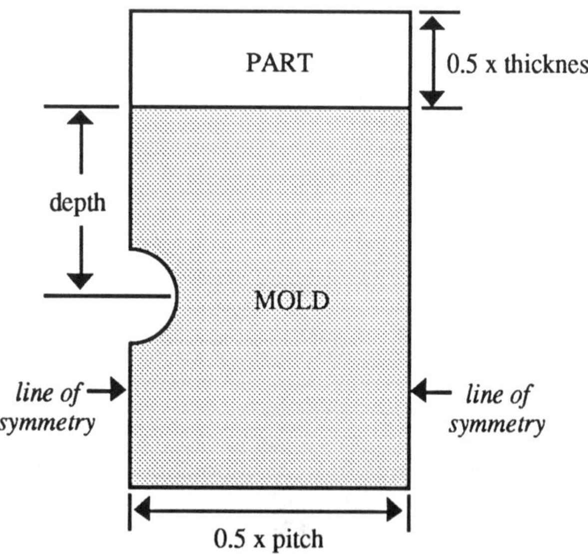

Figure 5. Schematic representation of geometry needed for transient finite element calculations.

In the thin section, the interior of the part is heated fairly rapidly by the heating oil; it requires less than 10 minutes for the part centerline to reach 177 °C (350 °F). However, in the thick section, it takes more than 35 minutes for the part centerline to reach this temperature.

As shown in these two figures, the imbalance in the time required for this part to reached 177 °C is more than 25 minutes, which could lead to non-uniform curing of the part and possible material quality problems. This type of imbalance in the time required for a variable thickness part to reach some specified temperature is one of the critical issues that can be addressed by this type of process modeling tool.

Variations in Permeability

In resin transfer molding, variations in the fiber preform permeability can significantly affect the mold filling behavior and the gating design. A finite element process model that simulates the injection, compression, and resin transfer molding processes, and predicts the mold filling behavior as a function of the part/gating design, the process parameters, and the material properties, has been developed (3-7). These three processes share common tooling, processing, and quality issues (such as gating/venting design, unfilled molds, weld lines, and mold cooling/heating design) as well as common process parameters (such as pressure, force, flow control, and temperatures). The similarities between these processes allowed consolidation of the models into one integrated software package, known as FEMAP (Finite Element Mold-filling Analysis Program).

Each of these processes deals with shell-like parts in which the part thickness is much smaller than the other physical dimensions. Therefore, complex three-dimensional geometries can be modeled as surfaces divided into small shell elements. Each element may have a different thickness associated with it to take into account any variations in the part thickness and hence the gap height of the cavity. However, because the parts are generally shell-like, the thin-walled geometry assumption is allowed, and any material flow in the thickness direction is ignored. Therefore, a thickness-averaged approach can be taken for computational efficiency.

For the resin transfer molding process, the mold filling can be modeled as flow through a stationary and incompressible porous medium within a mold, and the relationship between pressure and resin flow rate can be described by Darcy's law with anisotropic permeabilities. In general, the components of the permeability tensor may vary from element to element, since different types of fiber mat may be placed in different region of the mold. For low viscosity resin flowing in high permeability preforms, the gravity effect may be important and should be included in the governing equation (7). In the energy balance, heat conduction in the flow direction is neglected based on the assumption that the gap dimension is much smaller than the other two dimensions.

The pressure and temperature equations must be solved simultaneously to simulate the cavity filling, because of the strong temperature dependency in the viscosity and the convection and viscous heating effects in the process. A full derivation of the governing equations used in the FEMAP process model, as well as information about the various viscosity models used, can be found in the references (3-7).

Several features were incorporated in this process model to speed up the computation time, to produce a fast-acting model that could be easily used for process optimization. First, the finite element technique was chosen because it can efficiently solve the flow field in complex geometries, such as those encountered in molds with variable cavity gap thickness or with boundaries of arbitrary shape. Also, the volume-of-fluid (VOF) technique for flow front tracking (8) adds to the computational efficiency, because it uses a fixed mesh system through which the flow fronts pass, thereby eliminating the tedious remeshing requirement typical of moving boundary problems.

A multiple time step scheme was adopted for the model, where different size time steps were used for the filling and the temperature calculations, since, in general, the stable filling step is orders of magnitude larger than the stable temperature step. Finally, a further improvement in computation time was gained by using the domain decomposition technique, in which a multizoned numerical algorithm was devised. In this algorithm, different size time steps are used in different regions of the part, depending on the regional stability criteria. Typical CPU times for an isothermal calculation on a real part with several thousand elements is on the order of seconds; a non-isothermal calculation requires only minutes on a low-end workstation.

Dimensional Tolerance

Local compression of the fiber preform can result from problems with the dimensional tolerance of the tooling, which frequently leads to variations in the fiber permeability that affect the mold filling. Also, foam core RTM is a popular process, but possible dimensional changes in the core under pressure are a cause for concern. Shrinkage of the core can result in open spaces near the fiber preform through which the resin can flow rapidly, resulting in unexpected mold filling behavior.

A typical example of the dimensional tolerance problem has occurred in aerospace components. Measurements on the thickness of the mockup part, which was to be used to form the mold cavity, indicated some variations in thickness around the circumference of the part. Concerns were raised that the thickness variations would impact the resin flow to such an extent that there could be problems with filling the mold, since there might be local compression of the fiber perform in the thinner sections of the mold cavity.

To simplify the modeling effort, a rectangular 2-D part geometry of varying thickness was generated rather than a full 3-D geometry of the entire original part. A perspective view of this simplified geometry is shown in Figure 8, where the x direction dimension approximates the circumference of the part and the y direction dimension approximates the flow length for the resin. Examination of the thickness measurements on the mockup part showed that the variation appeared to be approximately ±8% of the average thickness. Therefore, the thickness in the z direction was specified to be equal to the average thickness in the center section, ±4% of this average thickness in the sections to either side of the center section, and ±8% in the outside sections; the thickness variations are exaggerated in Figure 8. In this way, the effect of local thickness variations on the fiber preform permeability and the filling patterns can be studied.

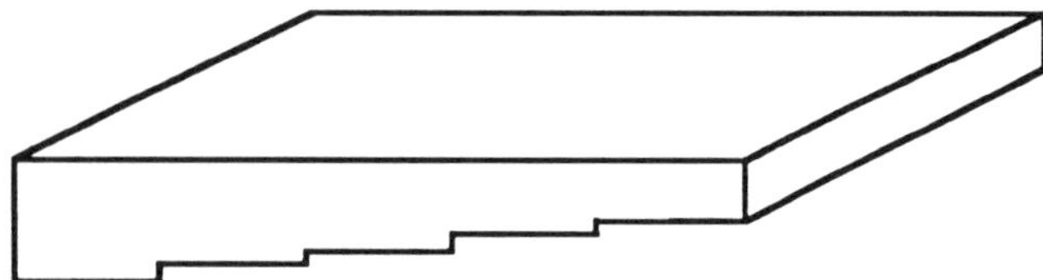

Figure 8. Perspective view of simplified geometry (part thickness variation is exaggerated).

In order to model the mold filling behavior for this part, only the fiber preform permeability is needed in addition to the part geometry. An isotropic permeability of 1.0e-6 cm^2 was used for the average-thickness center section of the geometry, and a nominal fiber volume fraction of 0.5 was used for that same section. It was assumed that as the thickness of each section varied, the fiber volume fraction would vary slightly as well, since the same fiber mat would be placed throughout the mold cavity. Therefore, the preform would be somewhat compressed, or "packed," in the thinner sections and expanded in the thicker sections. As the fiber volume fraction within the part changed, so did the permeability, according to the relationship expressed in Eq. 2, which is known as the Carmen-Kozeny equation:

$$K_{i,j} = \frac{R_f^2 \left(1 - V_f\right)^3}{4 k V_f^2} \qquad (2)$$

where $K_{i,j}$ is the permeability tensor, R_f is the radius of the yarn in the fiber preform, V_f is the fiber volume fraction, and k is the Kozeny constant. According to this equation, as the fiber volume fraction increases (in this case, as the part becomes thinner), the permeability

decreases, which means that there is more resistance to the flow of the resin. The change in permeability is amplified due to the cubic relationship in the equation. In this manner, changes in fiber volume fraction and permeability associated with the changes in part thickness are included. Table I summarizes the thicknesses, fiber volume fractions, and permeabilities for one typical section in the fan case.

Table I

thickness	thickness (cm)	fiber volume fraction	permeability (cm^2)
+ 8%	0.672	0.463	1.445e-6
+ 4%	0.647	0.481	1.211e-6
nominal	0.622	0.500	1.0e-6
- 4%	0.597	0.521	0.811e-6
- 8%	0.572	0.544	0.644e-6

For a part such as the fan case, the resin would typically enter the mold cavity from a resin delivery system that balanced the flow to the sections of different thickness. Rather than modeling the entire resin delivery system, the injection ports were set up as a "fan gate," whereby some resin entered the part at each node along the x axis in the finite element model, simulating the injection of the resin all around the circumference of the part. In order to balance the pressure at each of the injection port nodes, as would happen during actual production due to the resin delivery system, the percentage of resin entering each of the five regions of different thickness was iteratively adjusted until the pressures were balanced.

The filling patterns for the 2-D rectangular geometry with varying thickness are shown in Figure 9. In this figure, a schematic of the thickness variation was placed beneath the flow fronts to indicate the relative thickness of each of the regions; the thicknesses are exaggerated. Each line in the flow fronts represents a different time period. Therefore, more lines in some region indicates that it would take longer to fill that section.

Figure 9 indicates that a thickness variation around the circumference of the fan case would have a significant effect on the flow fronts, as the thickness varies from 108% of the nominal thickness down to 92%. The resin flows more easily through the thicker part sections, where the fiber volume fraction is lower and the permeability is larger. The resin flows more slowly through the thinner sections, due to the increased fiber volume fraction and reduced permeability.

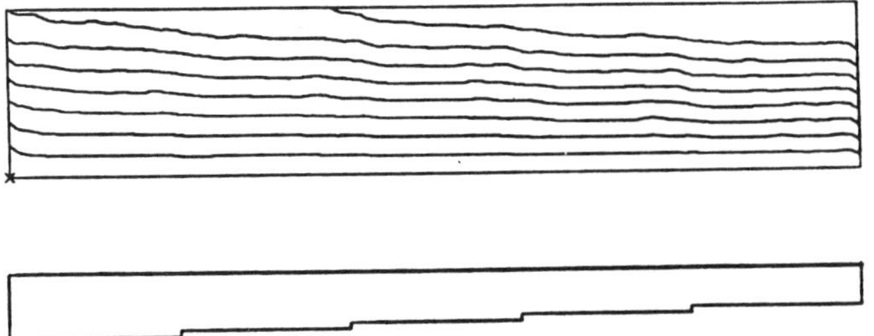

Figure 9. Filling patterns in 2-D rectangular geometry, showing effect of part thickness variations.

Because the permeability function is very sensitive to the fiber volume fraction, which is linearly proportional to the thickness change, any slight changes in the mold dimensions due to tolerance will give a significant change in flow speeds, resulting in a non-uniform flow front and potential trapped air regions.

Bends in Part Geometry

In addition to those permeability variations that result from problems with the dimensional tolerance in the tooling, certain part geometries can contain local permeability variations as well. For example, tubes and ducts used in the aerospace industry contain sharp turns and bends, which can cause local compression of the braided fiber perform structure, affecting the local preform permeability and altering the flow behavior of the resin during the RTM process. The FEMAP process model previously discussed can be used to study the effects of this local compression on the resin flow rates, pressure requirements, and flow patterns.

Figure 10 contains the finite element mesh for one portion of a typical aerospace tube geometry, which is approximately 46 cm in total length and 3.2 cm in diameter, with a constant wall thickness of approximately 1.3 mm. Only one portion of the mesh is shown because this region of the part exhibits the largest "bend" in its geometry.

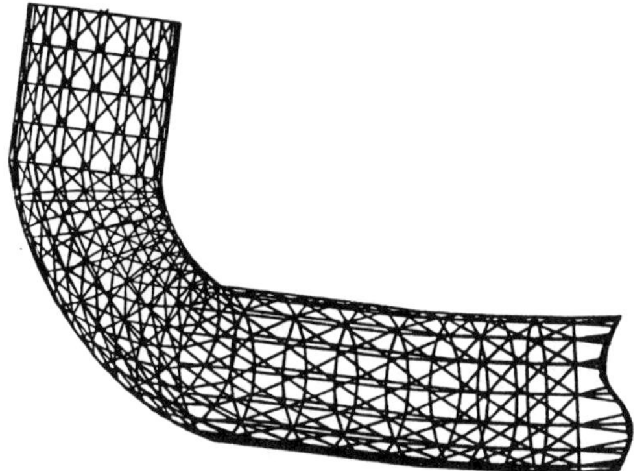

Figure 10. Finite element mesh for portion of typical aircraft engine tube geometry.

A braided fiber preform of constant isotropic permeability is placed inside the mold cavity. However, due to this bend in the geometry, the local permeability around the bend is altered. On the outside of the bend radius, the braid tow becomes more axial (that is, it runs more parallel to the resin flow direction along the length of the tube), the fiber volume fraction decreases, and the permeability increases, lowering the resistance to resin flow. However, on the inside of the bend radius, the braid tow angle becomes more oblique, the fiber volume fraction increases, and the permeability decreases, increasing the resistance to resin flow.

A fundamental theory for the change in magnitude of the permeability due to the fiber packing at the bend is not clear today. In theory, volume fraction could first be derived, through experimentation, as a function of bend angle, tube diameter, and wall thickness. Permeability could then be computed by using the Carmen-Kozeny equation previously discussed. Currently, however, the changes in magnitude for the volume fraction were made empirically. Table II shows the nominal value for the fiber volume fraction through most of the part, as well as the estimated values at the outside and inside of the bend radius; the corresponding permeabilities were calculated using the Carmen-Kozeny equation.

Table II

	Fiber Volume Fraction	Permeability (cm^2)
Nominal values	0.50	5.00e-7
Outside of bend radius	0.417	1.14e-6
Inside of bend radius	0.625	1.35e-7

Therefore, the elements in the finite element mesh located at the outside of the bend radius had their permeabilities increased by approximately 25 percent over the nominal value, while those elements at the inside of the bend radius had their permeabilities reduced to about 75 percent of the nominal value.

The data listed in Table II was used by the process model to determine the proper vent location in the part. Resin was injected at the far right end of the part, as it is viewed in Figure 10. In order to assess the effect of this permeability variation on the vent location, two cases were considered. For the first case, the permeability was assumed to be constant throughout the part; for the second case, the permeability was varied on the outside and the inside of the bend in the tube as previously described in Table II.

The flow front contours for these cases are shown in Figure 11. In Figure 11a, the constant permeability case, the part should be vented at the outside "corner" of the end of the tube, as marked in the figure. However, in Figure 11b, for the case in which the permeability is updated for the changes in fiber volume fraction near the bend, the vent is moved towards the inside "corner," depending on how much the fiber volume fractions are actually changed. In general, the flow front will move faster around the outside of the bend, due to the reduced fiber volume fraction and increased permeability there.

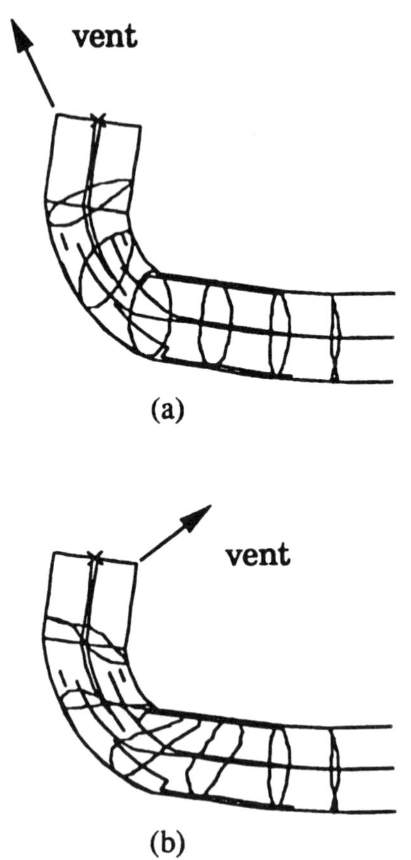

Figure 11. Finite element mesh for portion of typical aircraft engine tube geometry.

The proper location for the vent would, in fact, depend on the precise relationship between between fiber volume fraction and bend angle, tube diameter, wall thickness, and so on. This purpose of this example is to illustrate that there is indeed an effect on filling due to the geometry, although the exact magnitude of this effect cannot at present be fully quantified.

Inverse Analyses for Material Properties

Inverse analyses can effectively provide a method of obtaining the material properties of composites, such as the thermal conductivity and the permeability, for those cases where theoretical estimates may not be adequate or available, and measured data may not exist. Simple ex-periments can be conducted on simple geometries, and data can be measured for parameters such as temperature and load. The process model is then run repeatedly, and the input data for the material properties is iteratively adjusted until the predicted behavior for temperature or load matches the measured behavior. The iteration procedure can either be manual, in which the engineer adjusts the input data for a new run based on the results observed from the current run, or else it can be automat-ed, in which the model itself adjusts the input data for subsequent runs.

One material property that is typically not available for composites is the thermal conductivity, k, in the through-thickness direction. The Springer equation (9) can be used to provide a first-order approximation for this property, as a complex function of the fiber volume fraction and the fiber and matrix thermal conductivities. However, while the Springer equation provides a good approximation for k in the lateral dimensions of a thin-walled part, that is, the directions in which the fibers lay, it does not provide a good approximation in the thick-ness direction, that is, the direction perpendicular to the fiber layout. In a resin transfer molding problem, the conduction heat transfer is typically most predominate in this thickness direction. Therefore, an effective tech-nique to approximate the thermal conductivity, such as using inverse analyses, is needed.

Another material property typically not available for composites is the permeability of the fiber preform, κ. For certain fiber structures, a value for the permeability can be estimated through theoretical equations, such as the Carmen-Kozeny equation previously mentioned which relates permeability to fiber volume fraction and yarn diameter. However, for other fiber structures, such as "prepregs," which are cloth that is coated with resin and then layered in various patterns in order to build up the final part shape, the needed equations simply do not exist. Inverse analyses provide a way to approximate this value.

This technique was recently applied to determine these two material properties for a prepreg layup. Exper-iments were conducted to measure both the load and the temperature, as a function of time, for a prepreg part that was placed in a heated mold and then compressed. Finite element analyses were conducted on the part, with the FEMAP process model previously described, while values for the two material properties were iteratively adjusted. The process involved a double iteration loop, as illustrated in Figure 12. Initial guesses for the thermal conductivity, k, were used, so that the permeability, κ, could be iteratively adjusted for each value of k until the predicted load matched the measured load. Once the permeability was fixed for each thermal conductivity, k was adjusted interactively, using the corresponding value for κ, until the predicted temperature response curves matched the measured data.

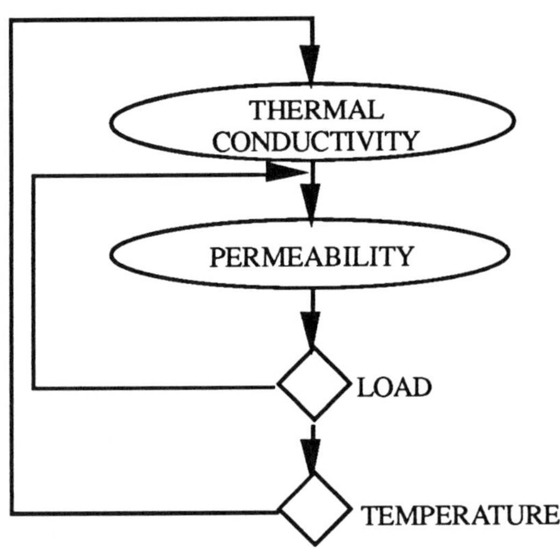

Figure 12. Double iteration loop used in inverse analyses for determining material properties.

Two values for the thermal conductivity were used as the limiting cases in these analyses: the value for the pure resin provided the lower limit, and the value computed by the Springer equation, which includes the effect of the higher conductivity fibers, provided the upper limit. First, a series of inverse analyses was conducted with the process model using these limiting values for the thermal conductivity, while iteratively adjusting the value of the permeability until there was a good match between the predicted and the measured load data. Two sets of results are shown in Table III, indicating the variation in the calibrated permeability as the composite thermal conductivity is varied. In these analyses, the thermal conductivity of the resin was 0.275 W/m-C and that for the fibers in the cloth was 3.113 W/m-C.

Table III

	case 1	case 2
Measured Load (N)	2.567e5	2.567e5
Through-thickness thermal conductivity (W/m-C)	0.275	0.709
Calibrated permeability (cm^2)	6.30e-8	5.25e-8
Predicted Load (N)	2.558e5	2.562e5

Several additional cases were also studied, for some intermediate values of the thermal conductivity, in order to determine values for the permeability as a function of conductivity.

The next step was to calibrate the through-thickness thermal conductivity of the composite by comparing the predicted temperature response curves for the part with measured thermocouple data. In these experiments, the thermocouples were mounted through the thickness of the part, for two different part thicknesses. Values for the thermal conductivity were iteratively adjusted during the finite element analyses until the predicted curves matched the measured data, using the calibrated values for permeability that were obtained during the previous set of inverse analyses.

Figure 13 shows a comparison of the measured thermocouple data as a function of time with the predicted temperature response curves, for three different sets of thermal conductivity/permeability data, for a thin test part. Figure 14 shows the comparison for a thicker part.

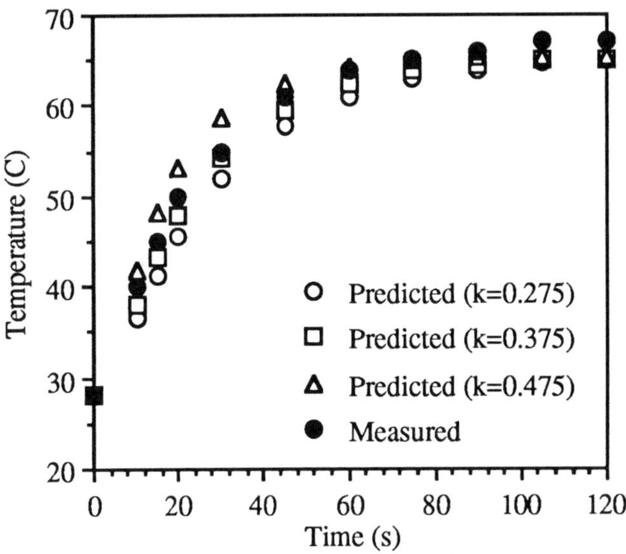

Figure 13. Comparison of thermocouple data and predicted temperature response curves for thin part.

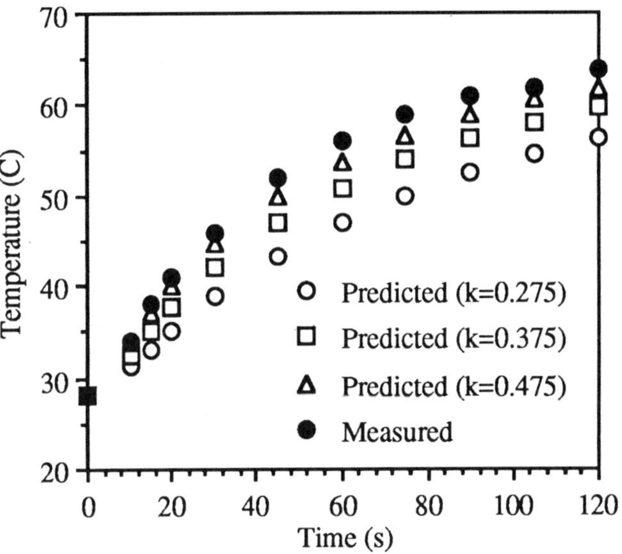

Figure 14. Comparison of thermocouple data and predicted temperature response curves for thick part.

Based on the results shown in these two figures, it can be determined that the best overall match between the predicted temperature response curves and the measured thermocouple data is obtained when the thermal conductivity is 0.475 W/m-C. For this value of k, a permeability of 5.69e-8 cm^2 provides good agreement between the predicted and measured load data.

Values for unknowns in the material property data, such as the thermal conductivity and the permeability, can be identified by conducting inverse analyses with the process models and comparing the predicted results with measured data obtained from simple experiments.

Conclusions

As illustrated by the examples described in this paper, three key elements must be included in those process models designed to aid in solving the practical problems associated with composite tooling design:

- Analysis capability must be integrated to include tooling geometry, process conditions, and material properties.
- Appropriate engineering assumptions are always needed in dealing with the complex geometries and processing conditions encountered in manufacturing.
- Fast-acting models that provide rapid solution response times are required in the process and tooling design environments.

Incorporating all three key elements in process models has always been and will continue to be a challenge for the practitioners in industry, but without all of these pieces it will be very difficult to solve practical issues in a timely manner.

Acknowledgments

The authors would like to thank K.Y. Wang and T.H. Hwang for providing the experimental data used in the inverse analysis section of this paper.

References

1   Wang, H.P., *A Computer Program for Injection Mold Cooling Analysis*, General Electric TIS Report 78CRD258 (1978).

2   Singh, K.J., *Optimization of Injection Mold Cooling Using Computer Program 'MCAP,'* General Electric TIS Report 81CRD031 (1981).

3   Wang, H.P. and H.S. Lee, *Fundamentals of Computer Modeling for Polymer Processing*, C.L. Tucker (ed.), pp. 369-401, Hanser Publications, NY (1989).

4   Wang, H.P., E.W. Liang, and E.M. Perry, *Mathematical Modelling for Materials Processing*, M. Cross, J.F.T. Pittman, and R.D. Woods (eds.), pp. 433-454, Clarendon Press, Oxford (1993).

5   Liang, E.W., H.P. Wang, and E.M. Perry, *Advances in Polymer Technology* **12** (3), pp. 243-262 (1993).

6   Wang, H.P., H.S. Lee, and E.M. Perry, "User's Manual for FEMAP (Finite Element Mold-filling Analysis Program), Version 5.5," July 1990.

7   Wang, H.P., E.W. Liang, and E.M. Perry, *Conference Proceedings for the Advanced Composites Conference and Exhibition (ACCE)*, Detroit, MI (1991).

8   Hirt, C.W. and B.D. Nichols, *J. Comp. Physics*, **39**, 201 (1981).

9   Springer, G.S. and S.W. Tsai, *Environmental Effects on Composite Materials*, G.S. Springer (ed.), pp. 7-14, Technomic Publishing Company, CT, (1981).

Advanced Fiber Placement Program for Aircraft Structures

B. Brailsford
GreatLakes Composites Consortium
Kenosha, Wisconsin

Abstract

One Of The Key Factors driving next-generation fighter and attack aircraft is affordability. With the majority of the composite structures used on today's fighter aircraft being manufactured by hand collation of individual plies of composite materials, new processes must be developed that will reduce costs and improve repeatability. A technology that meets this criteria is fiber placement. Through the U.S. Navy Center of Excellence for Composites Manufacturing Technology (CECMT), funding has been provided to establish and demonstrate the production readiness of Advanced Fiber Placement technology for Navy/DOD applications. Specifically, this includes demonstrating this technology on F/A-18E/F Engine Inlet Ducts, Horizontal Stabilator Skins and Aft Fuselage Panels and on future low observable applications. This paper will discuss the multi-disciplinary and multi-company approach, planned program milestones, and the benefits to the composites fabrication industry and to future weapon system programs.

In 1990 The GreatLakes Composites Consortium (GLCC) a member-owned not-for-profit organization was formed to manage the U.S. Navy Center of Excellence for Composites Manufacturing Technology (CECMT). This center was created to develop manufacturing processes for composite materials in both defense and commercial applications. Since its inception the consortium has grown to a membership of over eighty members.

There are three levels of membership within the consortium: Principal, Supporting and Associate. Also, there is an additional category for educational institutions to become involved as partners. Some of the larger companies that are involved include Bell Helicopter-Textron, Grumman, Lockheed, McDonnell Douglas, Northrop, and Rockwell International. Medium sized companies include Cincinnati Milacron, Hercules Composite Products and ICI Fiberite. There is a very large group of smaller companies that have also joined in this effort, while a number of educational institutions bring an academic perspective to the consortium.

With the defense budget being reduced, affordability of composite structures for future Navy aircraft has become a large concern. Most of the composite structures used on today's fighter aircraft are fabricated by hand layup of individual plies of composite materials. Studies have indicated that the largest recurring cost drivers of hand layup manufacturing processes are fabrication and assembly. In light of the reduced defense budgets, the Navy decided to investigate Advanced Fiber Placement technology as a possible new process technology that would meet their demands of more complex designs and could reduce costs associated with component fabrication. In October of 1991, the GreatLakes Composites Consortium was chartered with the task of putting together a program with the objective of establishing and demonstrating production readiness of the fiber placement process on Navy/DOD applications.

Discussion

Previous to the formation of GLCC, Northrop Aircraft Division (NAD), McDonnell Douglas Aerospace (MDA) and Hercules had entered into a technology exchange that was termed as the Tri-Party agreement in 1987. This agreement allowed NAD and MDA to apply and refine existing Hercules fiber placement technology for the purpose of aircraft components manufacturing. Although the Fiber Placement process had matured, there had been no production implementation. Certain critical technologies and implementation issues remained to be addressed.

In 1991 a Technical Direction Letter (TDL) was released by the Navy to put together a program that would address fiber placement technology and would involve members of the consortium. A program was structured and given the title of "Advanced Fiber Placement" (AFP) in which three components are addressed as follows: Task One - F/A-18E/F Horizontal Stabilator; Task Two - F/A-18E/F Engine Inlet Duct; and Task Three - Low Observable Technology for future aircraft applications. During March of 1993, another technical direction letter was released from the Navy that added Task 4 to address F/A-18E/F Aft Fuselage Panels. The tasks are represented in Figure 1 and indicate the players that are associated with each of them.

In the fiber placement process, material in the form of continuous prepreg tow is drawn from storage creels to an application head. The head is mounted on a multi-axis machine that is programmed to apply the tow material to the nip point. The application head compacts the material and allows complex contoured surfaces to be fiber placed. The head controls a band of collimated tows that can be individually cut or added. The ability to cut and add tows enables the fiber placement process to manufacture complex ply shapes very effectively using minimum weight designs. Reference Figure 2. The following describes the work effort within each of the tasks:

McDonnell Douglas Corporation is leading Task 1 as they are the contractor of this component for the F/A-18E/F program. Hercules Composite Products is teamed with them in support of this activity. This task is structured into a seven subtask building block approach for the development of the full scale article that will design, build, and test specimens at the coupon, element, and subcomponent levels. In addition, it will also prepare material and process specifications for the fiber placement process. The subtasks consist of : Integrated Product Concept development, Specification Generation, Fiber Placement Flat Panel Demonstration, Subcomponent Demonstration, Full-Scale Demonstration, Technology Transfer, and Production Transition Plan.

Northrop Corporation is leading Task 2 as they are the contractor of this component for the F/A-18E/F program. Hercules Composite Products is teamed with them in support of this effort. The scope of this activity is to demonstrate the production readiness of the fiber placement process for this component. This task is structured into a six subtask building block approach for the development of the full scale engine inlet ducts. The subtasks consist of: Duct Design for Fiber Placement, Process Generation and Validation, Quality Assurance, Cost Data, Fiber Placement Cell Readiness, and Technology Transfer.

Grumman Corporation, Rockwell International, Cincinnati Milacron and Hercules are teamed together in Task 3 to demonstrate the use of fiber placement techniques to fabricate honeycomb sandwich inlet duct designs using low observable materials. This task consists of a six step building block approach as follows: Establishment of Baseline Design Requirements, Specimen Test Matrix Development, Physical Placement Tests, Structural and Low Observable Properties Testing, Technology Transfer, and a Scale-Up Plan.

Northrop Corporation is heading up Task 4 as they are the contractor for this component. The scope of activity is aimed at demonstrating the production readiness of the fiber placement process for the F/A-18E/F aft fuselage skins. This task is structured into a eight subtask building block approach as follows: Part Selection and Design for Fiber Placement, Element/Subcomponent Fab and Test, Tool Design and Fabrication, Full-Scale Fabrication, Quality Assurance, Cost Data, Production Readiness and Technology Transfer.

In Task 1 an Integrated Product Concept Development approach will be used that will identify and evaluate a number of fiber placement concepts for additional components and identify the most promising for development and production demonstration. The evaluation will consider the parameters of weight efficiency, cost savings potential, suitability for the fabrication using the fiber placement process, and sufficient design flexibility to meet the requirements for an advanced fighter application.

Material and process specifications will be generated within the program to ensure a quality product is being produced for this program and for production implementation. This will ensure the final specifications will not set unrealistic goals as requirements. Also, this will develop a better understanding of critical process variables prior to finalizing the specifications.

A test matrix of flat panels will demonstrate the mechanical properties of laminates fabricated using the fiber placement process are equivalent to those laminates fabricated by conventional processes (hand layup of prepreg broadgoods).

Subcomponent and full-scale articles will investigate critical details of the design and manufacturing method and reduced risk for the program. Several types of test specimens will be obtained from these articles and tested.

The method of technology transfer in this program is through quarterly reviews and reports as well as a final review at the completion of each task. This is a formal way of information exchange and progress evaluation by the team and the customer. Results from low cost manufacturing methods, lessons learned, as well as cost analysis will be reported. In addition, a final report and video tape will be made of the program detailing the significant progress made and the lessons learned.

A production transition plan will be generated in each of the tasks to describe the transition of the development work into the F/A-18E/F production environment. This plan will outline items necessary for implementation such as process control, quality assurance, equipment, and specifications.

Progress

As was indicated in the introduction portion of this paper, this activity is a multi-disciplinary multi-company approach to addressing the development and implementation of fiber placement technology on Navy/DOD applications. The program has had several Quarterly Reviews at various members' locations with the members of the team to discuss the progress that has been made and to see the components that have been fabricated using fiber placement processing.

Task 1 has evaluated several composite parts on the F/A-18E/F aircraft to select high payoff candidates. The evaluation identified design and manufacturing approaches required to allow for fiber placement of the composite parts. Cost, weight, and risk impacts of the proposed changes to fiber placement were assessed. The results were correlated and reviewed and recommendations made. The conclusions indicated that not all composite parts lend themselves to be fiber placed since this technology is very geometry dependent. Also, efficient hand collated components do not show a major savings for fiber placement.

Common material and process specifications have been generated and reviewed by the team and additions and corrections made for fiber placement. These have been submitted to the customer for approval. This effort included coordination with McDonnell Douglas, Northrop, Hercules, and Fiberite to establish common fiber placement specifications for the F/A-18E/F program. Several issues were addressed and resolved such as sampling methods, identification of material, etc.

The objective of the structural equivalency test matrix was to ensure that the properties of laminates fabricated using the fiber placement process have equivalent properties to those using manual layup of broadgoods. This was a large effort requiring many test panels. The mechanical properties test data is being compared to the F/A-18E/F composite structures data base. Preliminary analysis of the mechanical properties indicates that fiber placed laminates are equivalent to manually layed up laminates.

The subcomponent article demonstrations investigated the critical design/manufacturing details reduced risk for the full-scale articles. Component designs were created along with tooling being designed and fabricated. The proof of concept article fabricated in Task 1 was successfully fabricated and demonstrated the feasibility of the basic design/manufacturing approach. Non-destructive inspection of the article indicated no significant defects detected.

The full-scale article demonstrations in Task 1 were focused on ensuring that the F/A-18E/F horizontal stabilator is producible using the fiber placement process. The primary objective of this activity was to increase the level of confidence that the design/manufacturing approach of the F/A-18E/F stabilator skin will be successful and that the projected cost savings will be achieved. Articles were fabricated by McDonnell Douglas and Hercules. Detailed fabrication data was acquired during the fiber placement process. This data is being used to review the cost effectiveness of this process and in the fabrication of the particular ply sizes and shapes. One of each of the full-scale parts fabricated at McDonnell Douglas and Hercules were destructively evaluated and the other used for demonstration. Fiber placement has been selected as the baseline approach process for F/A-18E/F horizontal stabilator skins.

Task 2 generated a producible design for a tow placed inlet duct which is representative of the F/A-18E/F inlet duct. This design takes into account the capabilities and limitations of tow placement. In addition, generation of the equivalence and structural test plans necessary to qualify the tow placement process for application on the E/F program was accomplished.

Similar to the structural equivalence work that was done in Task 1, Task 2 also generated the information necessary to demonstrate equivalence of fiber placement to the current manual layup method. This was necessary since this laminate is fabricated using a different fiber than Task 1. Based on the acceptable results, fiber placement has been selected as the baseline process for F/A-18E/F engine inlet ducts.

Two full scale inlet duct mandrels were fabricated for Task 2. They have graphite/bismaleimide skins with the backup structure attached to a steel shaft.

Full-scale inlet ducts have been fabricated by both Northrop and Hercules. The first set of ducts had panels removed from them for post buckling tests. The second set of ducts were used for Fuel/Hammershock tests. The third set of ducts are being used as demonstration articles. Each of these ducts have been subjected to dimensional and other nondestructive examinations as well as gathering data during processing.

A baseline design has been developed in Task 3 for a low observable inlet duct that consisted of a duct wall fabricated of honeycomb sandwich construction. This design is capable of attenuating electromagnetic energy. Since this program is a feasibility study and not tied to a particular aircraft or weapon platform, materials selections were based on previous experience.

Baseline hand layed up control panels and fiber placed comparison panels were fabricated and tested. The results of the mechanical and electrical testing indicates equivalency between hand layed up and fiber placed low observable laminates.

A Phase II addition to this effort is in work. This activity will identify fiber placement process limits for advanced inlet designs as well as establish a methodology to link design properties to process variation and process control requirements.

The F/A-18E/F fuselage skins were reviewed and evaluated in Task 4 to determine the best candidate for this program. The criteria used in the selection process was the skin should contain features which could cause potential problems in implementing fiber placement processing for fuselage skins. Also, the skin should have sufficient complexity to ensure that successful demonstration of fiber placement processing for this skin will result in confidence for applying this technology to any other F/A-18E/F skin. Using this criteria an F/A-18E/F Upper Side Skin was selected for use on this program.

Development test panels have been designed to incorporate those features which are considered most critical by both manufacturing and engineering personnel. Test plans are being developed to generate the necessary information to demonstrate the equivalence of the fiber placed skin to the current manual layup method.

Conclusion

In summary, this program has combined the resources of several member companies of the GreatLakes Composites Consortium to develop, fabricate, and evaluate this technology for implementation on Navy/DOD applications. Specifically this technology is being transitioned into the F/A-18E/F program where it has been baselined as the manufacturing process for fabrication of the Horizontal Stabilator Skins, Engine Inlet Ducts, and Aft Fuselage Panels on the F/A-18E/F program.

The program has generated fiber placed component designs, created a common material specification to address the impact of incoming material quality, created a common process specification, developed tooling for designs that incorporate the fiber placement process, performed structural equivalency testing to verify that fiber placed laminates are equivalent to hand layup, and fabricated and tested subcomponent and full scale hardware. These activities have produced the required data to incorporate fiber placement technology with low risk into the F/A-18E/F program. This technology lends itself as a state of the art, low cost solution, and as a substitute for existing and future composite hand layup designs.

Throughout the program technology transfer has taken place through regular (quarterly) progress reviews and through written reports. Program plans and results have been reviewed and discussed with McDonnell Douglas Corporation, Northrop Corporation, Hercules Composite Products, Grumman Corporation, Rockwell International, Cincinnati Milacron, GLCC and NAVAIR.

ADVANCED FIBER PLACEMENT PROGRAM

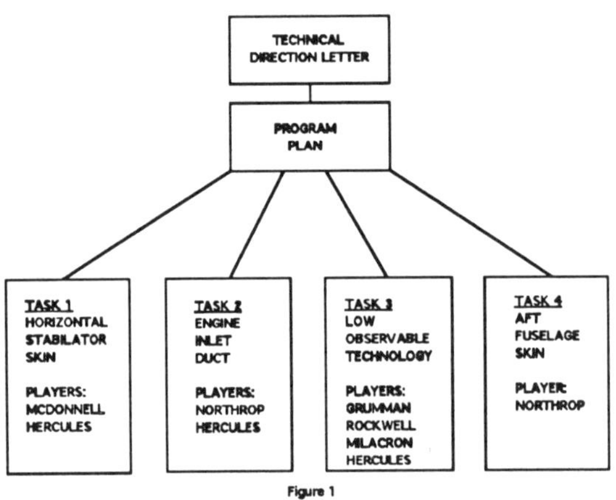

Figure 1

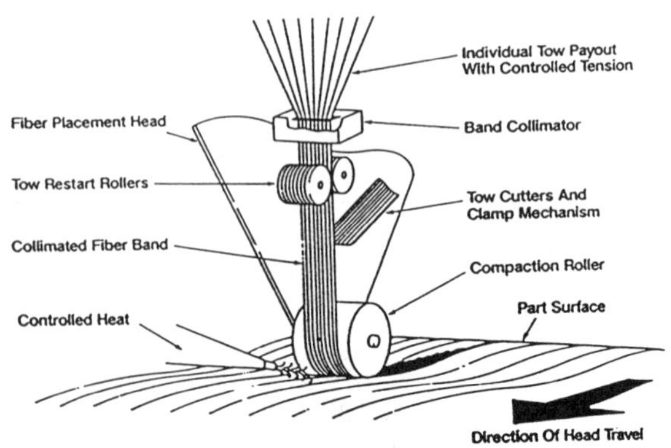

Figure 2

Slurry Process for Preform Manufacture

S.K. Soh
University of Detroit Mercy
Detroit, Michigan

Abstract

In the slurry process for manufacturing glass fiber preform for liquid molding, slurry of water and chopped fiber is filtered through a screen made in the form of the desired preform either by raising the screen through the slurry or by pumping the slurry through the screen. Orientation of fiber direction may be induced by guide vane system during the pumping-filtration stage. Random orientation is developed in the absence of such a guide vane system. The thickness distribution of preform and fiber packing density was studied. The filtered preform can be cured to green strength by a number of methods : 1) application of water-borne polyurethane binder, 2) using glass fiber mixed with thermoplastic fiber such as polyethylene terephthalate (PET) or nylon so that the filtered preform can be sintered in an oven before cooling and removal from screen, 3) application of water dispersed chemical binder based on silane type monomer.

Introduction

Advanced composite materials are gaining acceptance in the military and civilian industry as a means of achieving high performance with weight savings (5-7). The structure of advanced composite materials normally consists of reinforcing fibers (E-glass fibers, S-glass fibers, carbon fibers, polymeric fibers, etc.) and resin matrix. Therefore composite manufacturing processes may be classified into three groups based on the form of reinforcing fibers used: 1) Processes that use formable mats : thermoforming and thermostamping may be used to prepare near net preforms from woven mats (2,4). This process can achieve high fiber content and controlled fiber orientation, so that good strength properties are achieved, but the use of expensive woven mats and large amount of trimmed waste mat material puts this group of processes at a cost disadvantage. Some pultrusion processes and diaphragm forming (8) may also be included in this group. 2) Processes that use chopped fibers : directed fiber process (1,3,6) and the slurry process presented here belongs to this group. They are economical as they use less expensive chopped fiber and trimming waste is nonexistent or very small. However, the fiber packing density tends to be low and fiber orientation can not be developed easily. Thus the performance of parts may be less than other processes. The slurry process has the potential to improve part performance properties while retaining the economic advantage. and 3) Processes using unidirectional roving or tapes : Using filament winding and braiding, high fiber content and excellent fiber orientation is possible without edge trimming. However, long cycle time generally renders this process expensive and the part geometry where this process can be applied is limited to cylinders or near cylindrical shapes.

Directed fiber process and its countless modifications (1,3,6), have been used over 35 years. In it, continuous roving is fed to chopper and the resulting chopped fibers fall by gravity and suction onto screen to form a preform which may be rigidized while being held in shape by suction, by chemical curing of applied liquid binder or melting and solidification of thermoplastic fibers such as PET or nylon fibers mixed with glass in the preform. Considerable degree of fiber orientation may be developed by employing guides whose orientation can be controlled by electro-mechanical actuators. The advantages of this process is the low cost possible by 1) using the low cost roving instead of expensive woven mats (about 50 % unit cost), 2) virtual elimination of wasted fibers, with no low trimming requirement (20 % - 40% material savings), and 3) the relatively short cycle time (1 - 3 minutes) is common. A limitation of directed fiber process is the tendency of the chopped fibers to fall vertically to the screen surface, especially for thick preforms, as falling fibers gain speed by gravity and vacuum suction. In general it is preferable for fibers to lie in the direction tangent to the screen surface. Also the fiber packing efficiency may become low due to this unfavorable orientation.

The slurry process has not been treated extensively in the published literature although it has been in existence for more than 25 years. The fibers are suspended in liquid (mostly water), and the screen is raised to filter the suspended fibers. Instead of screen movement, pumping of water through screen may accomplish identical effect. Either way, the falling velocity of fibers will be much lower than those observed in the directed fiber process, so that the fibers tend to lie flat on the screen surface. Thus high fiber content may be achieved, approaching the values formerly achievable only be using woven mats or braiding.

The filtered preform must be rigidized before detaching

from screen. Chemical binders are used either as a latex which is submicron suspension in water or as a solution in a solvent. As they are heated and dried, curing reaction binds the fibers. Sometimes UV irradiation may be used instead of thermal decomposition of initiators (10,11). Preforms with mixed-in thermoplastic fibers such as PET and nylon can be rigidized by sintering, or heating near their melting point followed by rapid cooling. One advantage of thermoplastic fiber binders is that they do not coat glass fiber surface and therefore do not change the glass fiber-resin interface and the effective fiber diameter.

The Slurry Machines

Fig. 1 shows a schematic of slurry machine. Screens may be installed flush with frame as in Fig. 1(a) or be raised as in Fig. 1(b). No significant differences have been observed between the two setups. Constant air bubbling keeps the slurry well mixed. Sized glass fibers were used as received

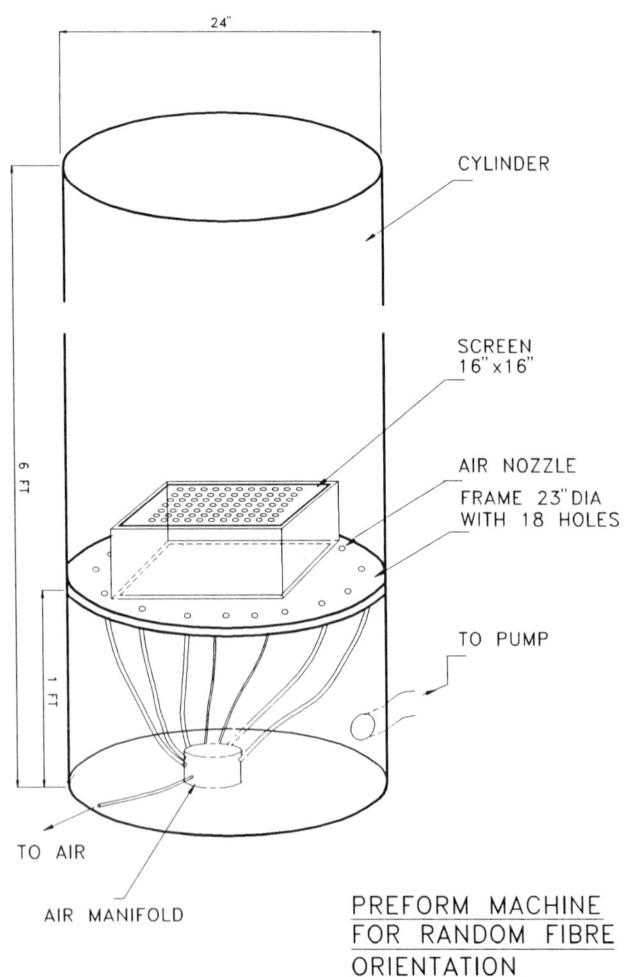

Fig. 1(a) Preform Machine with Flush Mounted Screen

Fig. 1(b) Preform Machine with Raised Screen

from manufacturers. The glass fiber bundles begin breaking up within one minute of being introduced into water and this filamentization to individual filament becomes complete in less than 10 minutes at room temperature. By controlling the contact time, it is possible to control the degree of filamentization. It is believed that more filamentization produces higher stiffness, while maintaining aggregate bundles produces higher impact strength. However, this effect of bundle size may only produce minor differences and no attempts were made to optimize the degree of filamentization.

The preform machine setup in Fig. 1 may be used to produce randomly oriented glass fibers. Fig. 2 shows a fixed guide vane setup which can develop glass fiber orientation. The space above the trap door is used to disperse glass fiber uniformly before filtration. Same effect can be achieved by pumping prepared slurry into the guide vane space.

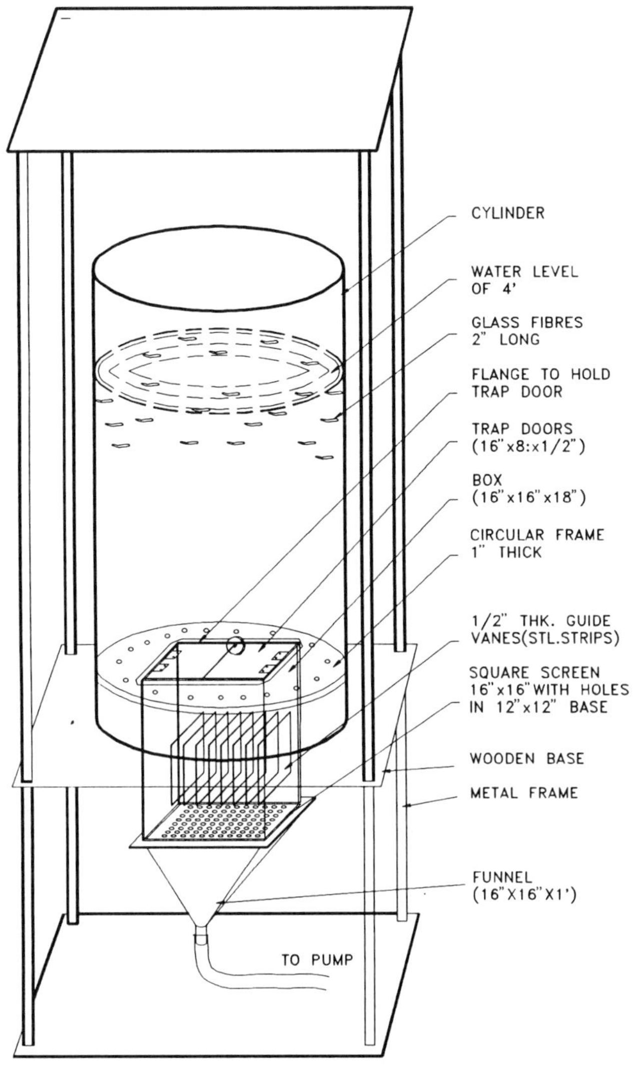

Fig. 2. Preform Machine with Fixed Guide Vanes

Results and Discussion

Fig. 3 shows glass weight distribution from a flat circular screen with 9.5 mm (3/8 in.) holes and 5469 holes/m^2 (3.5 holes/in^2) density within a 254 mm (10 in.) diameter circular area. The vertical axis is the ratio of deposition density (in g/cm^2) to the center deposition density. The two runs were made with different total fiber charge, resulting in different preform thicknesses. Reasonably uniform deposition is observed in the center area, with thinning at the edge. In some applications, this thin area may have to be trimmed. However, if the trimming is done before rigidization, the trimmed edges may be thrown back to slurry so that they can be reused without creating scrap material. Furthermore, it will be possible to adjust hole sizes and distribution to extend uniform deposition. A theoretical model of slurry process will be helpful in optimizing screen hole size and distribution to achieve more uniform and controlled deposition of glass fibers. At present the author is testing a promising model and he hopes to include some results during his presentation of this paper. The packing density measured were uniform throughout the preform and the weight distribution of Fig. 3 is directly proportional to the corresponding thickness data.

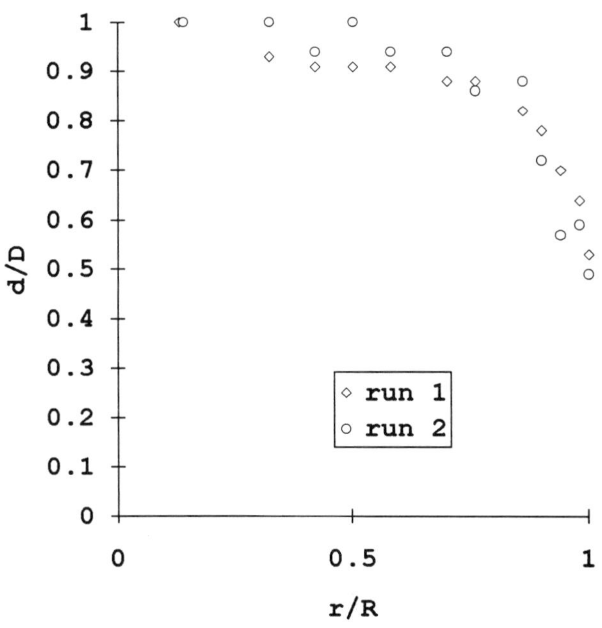

Fig. 3. Preform Weight Distribution.

The wet preforms have apparent packing efficiency of 20 - 30 %. They can be compressed in the mold to higher packing density. Fig. 4 shows compression data. Without damaging glass fibers, up to 60 % volume packing efficiency was achieved. This is quite a respectable value even when compared to labor intensive hand layout of woven mats.

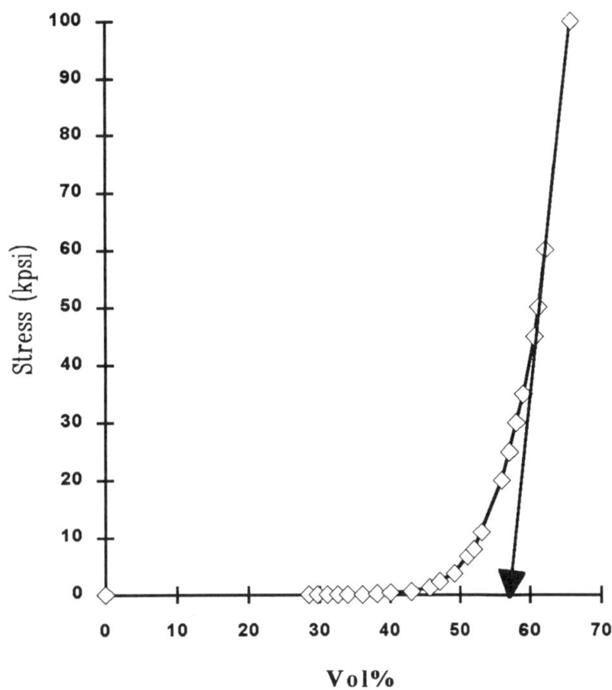

Fig. 4. Compression of Random Preforms

Rigidization

Three methods of rigidization were studied: 1) water-borne polyurethane binder, 2) PET and nylon fiber, and 3) chemical binder.

Water borne polyurethanes are generally made for surface coating applications and various hardness grades are available in the marketplace. Spersol L51 from Reichhold Chemicals worked quite well when used as 5% solution. Its binding action starts even before water is dried out. Once dried, it does not dissolve in water again. However, this binder is expensive as it need be used in high concentrations., and may reduce resin permeability.

Thermoplastic fibers are often used with directed fiber process. As these fibers do not coat glass fiber surface, the original sizing agent used for glass fiber remain available. Use of thermoplastic fiber may improve impact resistance of molded parts, at the expense of reduced stiffness. Presence of low melting thermoplastic fiber may lower maximum use temperature.

Chemical binders generally work slower than binders which does not require chemical reaction. We are trying to use binders with built-in silane group (10-13), so that the sizing and binding need may be satisfied simultaneously.

Conclusions

The slurry process under development promises to provide a preform manufacturing method capable of achieving high fiber packing efficiency and complex geometry with reduced scrap using chopped glass fiber.

Acknowledgments

The author gratefully acknowledges the support of this research by the U.S. Tank Automotive Command (TACOM) through National Automotive Center (NAC) contract DAAE07-93-C-R085.

References

1. E. P. Carley, J. F. Dockum, Jr., and P. L. Schell, "Preforming for Liquid Composite Molding", Polymer Composites for Structured Automotive Applications, SAE Congress Paper Number 900311, p115 (1990).

2. S. W. Horn and D. T. Buckley, "A New Automated Preforming Process for RTM & SRIM", ibid., Paper Number 900074, p41 (1990).

3. M. Jander, "Industrial RTM - New Developments in Molding and Preforming Technologies", Advanced Composite Materials Conference proceedings, p.29 (1991).

4. D. Weyrauch and W. Micheli, "A New Preforming Technology for RTM and SRIM", Society of Plastics Engineers Annual Technical Conference Proceedings, p761 (1992).

5. B. Miller, "Preforms Pave the Way for Lower-cost Structurals", Plastics World, no. 1, p26 (1988).

6. B. Miller, "New Methods, Material Bloom at Composites Conference", Plastics World, no. 2, p40 (1993).

7. V. Wigotsky, "Advanced Composites are Spreading their Wings", Plastics Engineering, no. 10, p25 (1988).

8. D. T. Tsahalis, S. G. Pantelakis, and V. Schulz, "Modeling of the Diaphragm Forming Technique Applied to Continuous Fiber Reinforced Thermoplastic Composites", in Sung K. Soh and A. A. Tseng eds., "Processing of Polymers and Polymeric Composites", ASME MD-Vol. 19, p91 (1990).

9. J. K. Rogers, "SRIM and RTM Sprint", Plastics Technology, no. 3, p50 (1989).

10. E. P. Plueddemann, "Silane Coupling Agents", p132, Plenum Press, New York (1982).

11. E. P. Plueddemann, "Mechanism of Adhesion through Silane Coupling Agents", in E. P. Plueddemann eds, "Interfaces in Polymer Matrix Composites", vol. 6 of "Composite Materials", p174, Academic Press, New York and London (1974).

12. S. W. Horn, "Advances in Binders for RTM and SRIM Fiber Preforming", Advanced Composite Conference Proceedings, p41 (1991).

13. B. M. Vanderbilt, Society of Plastics Engineers Annual Technical Conference, Sect. 26, paper 1 (1966).

Proceedings of the 10th Annual ASM/ESD Advanced Composites Conference, Dearborn, Michigan, USA, 7-10 November 1994

A Numerical and Experimental Study of the Permeability of Fiber Preforms

S. Ranganathan, S.G. Advani
University of Delaware
Newark, Delaware

G.M. Wise, F.R. Phelan, Jr., R.S. Parnas
National Institute of Standards and Technology
Gaithersburg, Maryland

Abstract

A finite element simulation for modeling flow in 3-D fibrous porous media has been developed for the purpose of predicting the macroscopic permeability of fiber preform structures. The simulation calculates flow in arbitrary geometries that consist of solid, open and porous regions in any combination. The Brinkman equation is used to model flow in porous intra-tow regions, and the Stokes equation to model flow in open regions between the tows. Permeability is predicted by calculating the flow rate--pressure drop relationship across a unit cell geometry that represents the building blocks of the overall preform structure. The simulation is tested by comparing theoretical predictions for the permeability of a uni-directional material in a nested stacking sequence with experimental measurements.

FLUID FLOW IN LIQUID COMPOSITE MOLDING (LCM) processes such as resin transfer molding (RTM) and structural reaction injection molding (SRIM) is generally modeled using Darcy's Law given by

$$\langle v \rangle = -\frac{K}{\mu} \cdot \nabla \langle P \rangle \qquad (1)$$

where $\langle v \rangle$ is the average (superficial) velocity in the medium, $\langle P \rangle$ is the pressure, K is a symmetric, second order tensor known as the permeability and μ is the viscosity of the fluid. In recent years, a number of numerical simulations based on Darcy's law for modeling the macroscopic mold filling process in LCM have been developed. This capability provides a time efficient and cost effective means for minimizing costly trial and error efforts in LCM process design.

In Darcy's law, all the complicated interaction that takes place between the fluid and the fiber preform structure is lumped into, K, the permeability tensor. Accurate permeability data, therefore, are a critical requirement if simulations based on Darcy's law are to be successfully used in the design and optimization of these processes. Currently, permeability values are most often obtained experimentally by measuring the directional pressure drop versus flow rate relationship as a function of volume fraction (e.g., 1). However, the number of potential preform materials and range of volume fractions is very large, and the experiments are presently difficult and time consuming. Moreover, for some materials there is often not an adequate means of getting the preform into permeability measurement molds without extensively altering the microstructure, such as for those formed via braiding or filament winding. Thus, a theoretical means of predicting permeability as a function of preform microstructure would be useful. Such a capability would help to reduce the number of experimental measurements presently required, and thus, speed up and simplify the design process. For example, preform packing leads to permeability changes (2). Modeling may be far more robust than experiment for calculating permeability as a function of deformation. A model would also lead to a better understanding of the structural features that influence the physics of the flow through such materials, and thus potentially enable microstructure to be tailored such that it has both the desired reinforcing capability and the necessary permeability to fill efficiently.

The most important requirement in developing a model for predicting permeability in fibrous porous media is to account for all the structural characteristics of real preform materials that have an effect on the microflow. The microstructure of most structural preform materials consists of a network of fiber bundles called tows. The main structural characteristics of this fibrous network can be identified as: (1) the number of fibrils per tow -- tows contain from 1000 to 12,000 individual fibrils; (2) the intra-tow fibril diameter; (3) tow cross-sectional shape -- which varies from circular to highly elliptical; (4) weave pattern --

which ranges from simple unidirectional alignment, to complex weaves and braids; (5) packing characteristics -- the relative inter-ply spacing and alignment. At a minimum, a model must account for the geometric shape, weave pattern and packing characteristics of the tows. Additionally, for fibrous materials a model must account for preform heterogeneity. The materials used in LCM preforms are heterogeneous in the sense that the network of tows which make up the global porous medium are themselves porous. Thus, in actual materials, there is flow not only around but through the tows. Modeling studies (3-4) indicate that as the tow volume fraction increases, intra-tow flow plays an increasingly significant role in determining the overall permeability. The effect of intra-tow flow on permeability has also been seen experimentally (5).

Several models have been proposed in the past to estimate the value of the permeability for various porous media. The basic approach used to develop such models is to determine the resistance of a viscous fluid to flow in an idealized model geometry, and then back-calculate the permeability from the flow rate -- pressure drop relationship. Capillary models such as the Kozeny-Carman equation are among the earliest models for predicting the permeability of a porous medium based upon an idealized medium structure. However, even though this model has been used successfully for isotropic granular media, it does not work well for either axial or transverse permeability of aligned fibrous media (3,6). Another class of models for predicting the transverse permeability in unidirectional fibrous media is based on the approximation of the medium as a regular array of cylinders with circular cross-section (6-10). Jackson and James (11) have provided a good overview of the early experimental and theoretical literature in this area. Since these models assume that the cross-section of the fibers in the porous media are circular, and do not take into account intra-tow flow, they are limited in the types of media they can model. An extension of the models above was obtained in Phelan (3). In that work, a model for the transverse permeability of rectangular arrays of porous cylinders of arbitrary cross-section was developed using the lubrication approximation to predict the flow rate -- pressure drop relationship for the unit cell. The array of porous cylinders represents the tows of the preform structure, thus, there is an accounting for the effect of intra-tow flow on the overall permeability. Phelan and Wise (12) have shown that this model gives excellent agreement with rigorous finite element calculations over a wide range of volume fractions for porous cylinders with elliptical cross section. Ranganathan et al. (13) have developed a generalized model for the transverse permeability of *regular* arrays of porous cylinders. The development is similar to the approach used in Phelan (3) and Phelan and Wise (12), but in this case no constraint on the configuration of the arrays was imposed and a much more general result was derived. A model for the permeability of wovens has also been developed (14), but it does not account for preform heterogeneity.

While analytical models are highly desirable due to their simplicity, at present, they are still greatly limited with respect to the broad variety of materials commonly used in LCM processes. Most of the relations that have been developed to date are for unidirectional materials which are geometrically simple enough to be amenable to analytical analysis. For more complex fiber architectures -- e.g., woven and braided materials -- analytical relations are much more difficult to generate without compromising the accuracy of the predictive capability. Thus, a more general approach is needed.

In this study, a finite element simulation for modeling flow in 3-D fibrous porous media has been developed. The simulation calculates flow in arbitrary geometries that consist of solid, open and porous regions in any combination. The Brinkman equation is used to model flow in porous intra-tow regions, and the Stokes equation to model flow in open regions between the tows. Permeability is predicted by calculating the flow rate--pressure drop relationship across a unit cell geometry that represents the building blocks of the overall preform structure. The numerical approach has several potential advantages. The first is that it is not limited by the geometric complexity of the preform microstructure, save the requirement that the geometry be regular such that there is a unit cell representative of the entire medium. The second is that the finite element mesh geometries used to simulate the flow may also be used to study the microstructure deformation, and thus, in a concurrent engineering approach obtain permeability as a function of deformation information in a robust manner. Such information is important for obtaining accurate local permeability values in preforms that undergo extensive, non-homogeneous deformations during placement and closure in a mold.

An important aspect of model development is validation. In this study, numerically calculated predictions for permeability are compared with experimentally measured values for flow in a unidirectional material in the nested stacking configuration shown in Figure 1. This material and configuration were chosen to try and achieve a controlled and predictable preform microstructure for the purpose of comparison between modeling and experiment. It has been found in previous studies (4,15) that it can be difficult to align the tows in unidirectional materials one on top of the other as some of the tows tend to slip and end up in the nested configuration when compressed. Such extraneous defects are hopefully minimized by placing the material in above configuration in the first place. Comparison of the numerical and experimental, axial and transverse in-plane permeabilities shows that the model consistently predicts the correct order of magnitude for the permeability values, but usually underpredicts quantitatively. Possible reasons for the quantitative discrepancies are discussed.

Numerical

Governing Equations. The modeling of microscale flow in

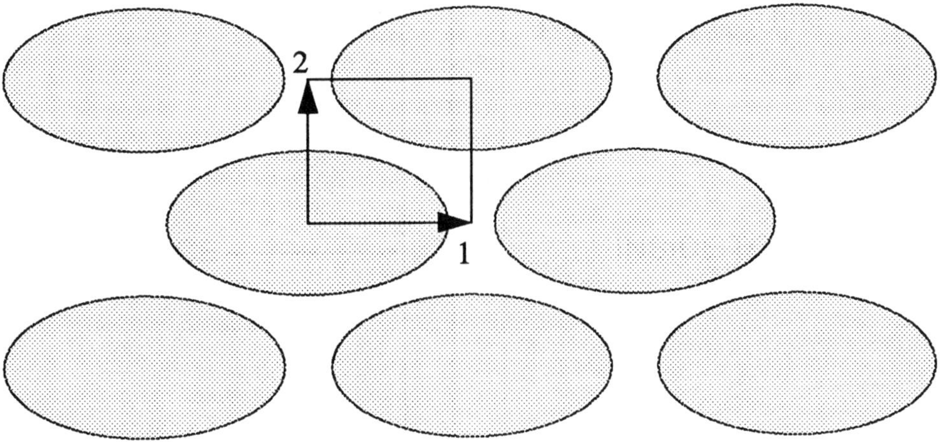

Fig. 1 - Axial view of a unidirectional material in a nested configuration. The arrow labeled 1 indicates the in-plane transverse flow direction, and the arrow labeled 2 indicates the through-thickness flow direction. The box indicates the unit cell for the arrangement. The cell is symmetric about all sides.

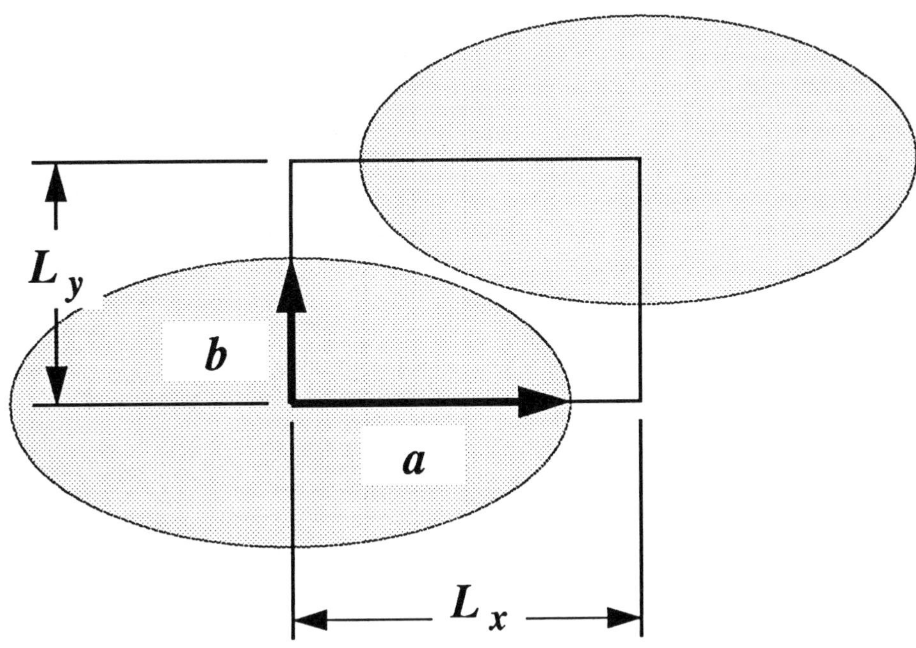

Fig. 2 - Geometric parameters for describing the unit cell of a unidirectional material in a nested configuration. The tows are approximated as ellipses. The box indicates the unit cell for the arrangement.

fibrous porous media is complicated by the existence of both an open region around the tows and a porous media inside the tows. There are a variety of approaches to such problems. In this study, the momentum balance for flow in the open media surrounding the tows is modeled using the Stokes equation given by

$$\mu \nabla^2 v = \nabla P \qquad (2)$$

The flow inside the tows is modeled using the Brinkman equation (16) which may be written as[*]

$$\mu_e \nabla^2 \langle v \rangle - \mu K^{-1} \cdot \langle v \rangle = \nabla \langle P \rangle \qquad (3)$$

where K represents the permeability of the tow, and the brackets indicate volume averaged quantities. Note that both $\langle v \rangle$ and K in this expression differ from those in Eq. (1) by a factor of ε^{-1}, where ε is the porosity of the medium. The conservation of mass equations for the respective media are given by the relations

$$\nabla \cdot v = 0, \ \nabla \cdot \langle v \rangle = 0 \qquad (4)$$

Finite Element Modeling. A 3-D Galerkin finite element scheme has been developed to solve the above set of equations. To approximate the velocity field, 27-noded Lagrangian hexahedral elements with quadratic shape functions are used, while 8-noded hexahedrons with linear shape functions are used for the pressure. The corner nodes of the element have four degrees of freedom -- the three components of the velocity and the pressure -- while all other interior nodes have just the three components of the velocity. Expressing the velocity and pressure approximations as

$$v = \sum_i v_i \phi_i, \ P = \sum_i P_i \pi_i \qquad (5)$$

the Galerkin discretization of these equations are given by

$$-\mu_e \int_\Omega \nabla \phi_i \cdot \left(\nabla v + \nabla v^t \right) dV + \int_\Omega \nabla \phi_i \, P \, dV$$

$$-\alpha \mu \int_\Omega \underline{K}^{-1} \cdot v \, \phi_i \, dV \qquad (6)$$

$$+ \int_{\partial \Omega} \hat{n} \cdot \left[\mu_e \left(\nabla v + \nabla v^t \right) - P \underline{\delta} \right] \phi_i \, dA = 0$$

$$\int_\Omega \pi_i \, \nabla \cdot v \, dV = 0 \qquad (7)$$

where the ϕ_i and π_i are quadratic and linear basis functions respectively, and α is a parameter equal to 0 in open media, and 1 in porous media. This discretization procedure yields a linear system of equation of the form $A \cdot x = f$, where A is the stiffness matrix, x the vector of unknowns, and f is the forcing vector. A and f are modified appropriately for boundary conditions. The matrix A is stored in a format appropriate for use with a banded solver which is used at the present time to solve the system of equations.

Unit Cell Geometry. Unidirectional fiber mats layed up in the nested configuration shown in Figure 1 are investigated in this study. The individual plies of the material are approximated numerically as an array of porous cylinders with elliptical cross section. In an ideal configuration, the centers of the ellipses are distributed on a regular lattice such the centers of the ellipses in one ply, are lined up with midpoints of the ellipses in the adjacent plies. Due to the repeating structure of the medium, analysis only needs to be done for the unit cell region outlined in the figure, and is extended to the entire medium by applying the appropriate periodic boundary conditions. The unit cell is described by the semi-major and semi-minor axes a and b of the ellipse, and the lengths L_x and L_y as depicted in Figure 2. The porosity of the system is given by the relation

$$\varepsilon = 1 - \frac{\pi a b}{2 L_x L_y} (1 - \varepsilon_{tow}) \qquad (8)$$

where ε denotes porosity. The intra-tow porosity, ε_{tow}, may be estimated from the relation

$$\varepsilon_{tow} = 1 - N_f \frac{\pi d_f^2}{4 A_{tow}} \qquad (9)$$

where N_f is the number of individual fibrils in the tow, d_f is the fibril diameter, and A_{tow} is the tow cross-sectional area. Once the tow porosity is known, estimates of the axial tow permeability may be obtained from results for flow parallel to arrays of solid cylinders. (e.g.,4) Estimates of the transverse flow permeability may be obtained using the relations derived by Gebart (10) for flow normal to arrays of solid cylinders. In the analysis of heterogeneous media, it is useful also to define the nominal porosity (5) which is given by the relation

$$\varepsilon_{nom} = 1 - \frac{A_{tow}}{2 L_x L_y} \qquad (10)$$

The nominal porosity is the porosity based on tow shape alone, neglecting intra-tow porosity; it is a better indicator of the closeness of tow packing in heterogeneous systems than the overall porosity, as it is not a function of intra-tow properties.

In actual experimental layups, the tows are not separated as idealized in Figure 2, but instead come into intimate contact deforming the tows. Two different geometric models are used to represent the deformed microstructure as

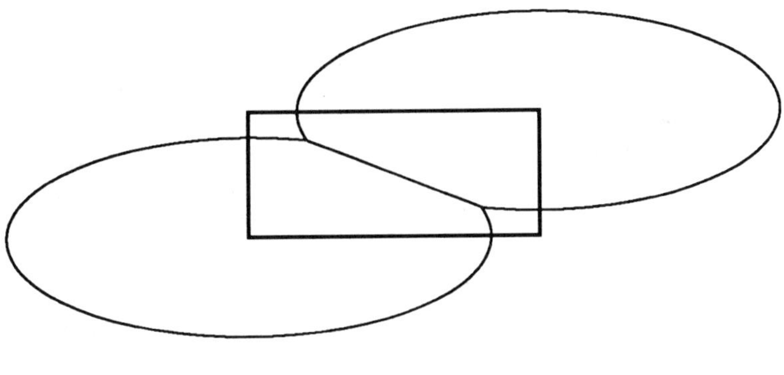

(a) Model 1

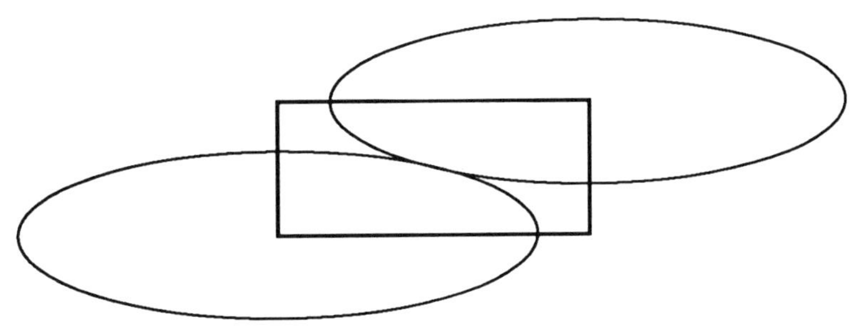

(b) Model 2

Fig. 3 - Geometric models for describing the unit cell of a unidirectional material in a nested, stacked configuration. (a) Tows touch along a line. (b) Tows touch at a point. The parameters L_x, L_y, and a are same in both cases. In the first case, the ellipse aspect ratios are kept at their nominal uncompressed values and the points of intersection of the two ellipses define the line along which the tows make contact. In the second case, the ellipse aspect ratios are reduced until they touch at a point.

shown in Figure 3. In both models it is assumed that the tow shape remains elliptical. Model 1 corresponds to a case where the tow-tow contact deforms the tows only locally in the area of contact, and the tows touch along a line defined by the intersection of the initial tow elliptical cross sections. Model 2 corresponds to a case where the tow-tow contact deforms the entire tow such that the tows remain in contact only at a point. These two models were formulated to test the sensitivity of the permeability calculation to structure. Model 2 results are reported only for the in-plane transverse flow case at the present time.

The finite element simulation described above was used to calculate the flow rate -- pressure drop relationship in this unit cell for the axial and in-plane transverse flow cases (through-thickness transverse calculations will be reported on at a later date). In the numerical procedure, a pressure difference was imposed across two opposite faces which define the flow direction, and velocity symmetry conditions were used on the other four faces. The flow rate through the unit cell was computed by integrating the velocity field over one of the faces on which the pressure is specified. From the calculated flow rate, the prescribed pressure drop, the fluid viscosity, and the cell geometry, Darcy's law was used to compute the effective permeability. The relevant tow dimensions used in the computations are given in the next section. The unit cell for an axial flow calculation, illustrating the calculated pressure contours and velocity vectors is shown in Figure 4.

Experimental

In order to compare modeling predictions with experimental results for actual fiber systems, a unidirectional material was obtained and characterized in permeability measurement experiments. The material is a Knytex D155 knit manufactured by Hexcel Corp.[**] The glass tows in the knit are held in place by polyester warp direction fibers which account for 2% of the mass of the material. In the as received state, the tows are coated and permeated with a binder material, which makes them much less permeable than when uncoated. This allowed for the planning of two sets of experiments: the first with the binder material on the tows, the second with the binder material removed by an acetone wash.

Permeability Measurement. Permeability measurements were conducted in radial flow and unidirectional in-plane flow geometries (through-thickness flow experiments, in progress at the time of this writing will be reported at a later date). A radial in-plane flow is achieved by injecting the fluid through a central 1.27 cm (0.5") diameter gate into a 30.48 cm x 30.48 cm (12"x 12") region between two parallel plates containing the reinforcement. A spacer between the parallel plates set a cavity thickness of 6.35 mm (0.25"). Each layer of reinforcement has a 1.27 cm (0.5") diameter hole, created by a steel punch, centered over the injection gate to permit the fluid to penetrate evenly through the

thickness of the sample. A 2.54 cm (1") thick acrylic upper plate permits recording the progress of the flow front. The thickness of the upper plate was chosen to permit no more than a 6.35×10^{-2} mm deflection (1% of the cavity thickness) at the center of the plate for a fluid injection pressure of 3.45×10^5 N/m² (50 psi). Since the injection pressures never exceeded 1.05×10^5 N/m² (15 psi), and clamping pressures were minimal, mold deflection in the radial flow experiments was not expected to significantly affect the results. The flatness of the acrylic plate was checked for deflection before each experiment.

The one-dimensional flow experiments were conducted in a mold with a width and height of 15.24 cm x 15.24 cm (6"x 6"), and a thickness of 1.27 cm (1/2"). The top of the mold was 2.54 cm thick acrylic to permit observation of the flow front. Two problems commonly associated with one-dimensional flow measurements, pressure gradient induced bending of the mold faces and edge effects, were carefully monitored and controlled. Measurements of the acrylic flatness confirm that the plexiglass did not deform significantly during the course of the experiments and, accordingly, a linear relationship of the flow and pressure drop was obtained in the experiments. The acrylic allowed visual observation of the progress of the flow front through the mold during the initial filling, and if edge effects were observed the mold was repacked and the experiment repeated. After the mold was filled, a 2.54 cm thick steel plate was clamped over the acrylic mold face as additional insurance against mold deflection.

Fluid was forced through the molds from a pressure-pot system. The fluid flow was measured with a calibrated differential pressure cell as it entered a mold, and it was also monitored at a mold exit with volumetric measurements. Corn syrup, diluted with distilled water, was used for the test fluid in all experiments due to its ease of handling and cleanup. The experimental fluid was a mixture of 79% corn syrup and 21% distilled water, which had a viscosity between two and three poises. Before each experiment, the fluid was filtered and its viscosity measured. From earlier work (1) the corn syrup-water mixture's viscosity was known to be constant with both time and shear rate. Although an RTM resin might seem to be more representative of injection conditions, a particular choice of resin would be no more appropriate for a given laboratory than corn syrup if that laboratory planned to use a different resin to make parts. Additionally, the most important criterion for the choice of fluid for a reliable permeability measurement, that it wet the fabric well, is well satisfied by aqueous corn syrup solutions.

Pressures at the mold inlets were measured by calibrated transducers (Omega model PX-605). The outputs from the pressure transducers and differential pressure cell were collected with an analog-to-digital board (CyberResearch ACPC 168) mounted in a personal computer (50 MHz i486) running customized data acquisition and control software. The software contained a feedback control algorithm that was used to manipulate a valve (Kammer

80057P) via a digital-to-analog board (CyberResearch ACAO 128). The controller could be user-configured to permit either constant injection pressure or constant injection flow, although only constant injection flow experiments are reported below.

In the radial flow experiment, the anisotropy of the in-plane permeability was estimated by fitting the shape of the experimental flow front to an elliptical function, and taking the ratio of the major and minor axes. The in-plane permeability components were subsequently estimated by using both the anisotropy and the time dependent mold inlet pressure as outlined in Advani et al. (18). Both the anisotropy and the measured permeabilities varied over time as the fluid penetrated the unidirectional fabric. Such variation is thought to be due to wicking, and is discussed at length in Parnas et al. (19). Due to these transients, the radial flow experiments were used only to estimate the anisotropy of the in-plane permeability, but not to provide a quantitative measurement of the permeability components.

To quantify the in-plane components of the permeability tensor, unidirectional flow experiments in which the fluid flowed in the direction of the glass tows (axial flow) were conducted. The material's permeability in the axial direction was calculated by measuring the pressure drop and flow rate as outlined in Parnas and Salem (1). These measurements provided the larger principal permeability in the plane of the material since the major axis of the flow front in the radial flow experiments was oriented in the direction of the glass tows. The transverse component of the in-plane permeability was obtained by simply combining the axial component and the anisotropy estimated from the radial flow experiments. Thus, the radial flow experiments provided the geometry of the in-plane permeability and the unidirectional flow experiments provided quantification.

In both the radial and uni-directional flow experiments, material was packed at an average value of 60 layers per 2.54 cm (1"). This corresponds to fifteen layers of fabric in the 0.635 cm thick radial flow mold, and 30 layers of fabric in the 1.27 cm thick unidirectional flow mold.

Binder Removal. To determine the permeability of the material without binder, the binder was removed with an acetone wash. In this operation, the Knytex layers were stacked onto shelves and immersed in 15-16 liters of acetone for 12-15 hours. Following this pre-soak, the acetone in the bath was circulated for several hours, and then the fabric layers were removed and dried for 12-15 hours. The layers were then weighed and used in the experiments as outlined above. It was found that the polyester crossing threads that hold the tows in place shrank in the acetone bath, reducing the width of the material by about 13%. Thus, a larger number of tows was required to fill a given size mold after the binder was removed, and the desired goal of having a material with the same overall weave structure, but with different intra-tow properties was not achieved.

To determine the mass of binder removed during the acetone bath, the soaking solution was boiled in a one-stage distillation column. When less than 20 mLs remained, the residue was poured onto a pre-weighed watch glass, further dried in a vacuum oven, and then weighed. The percent of material weight removed in the acetone bath was compared to the fractional weight loss of tows placed in a 600°C oven for two hours (an oven temperature of 600°C was found to be adequate for binder removal from the results of several thermal gravimetric analysis (TGA) experiments). Both the acetone bath and the 600°C oven consistently removed about 0.5% of the total fabric weight.

Tow Properties. Tow properties for Knytex D155 in the uncompressed state have been estimated or obtained from the manufacturer in a previous study (4). A summary is shown in Table 1 below

Table 1 - Nominal Tow Properties Knytex D155, Uncompressed			
a (cm)	b (cm)	N_f	d_f (cm)
0.0888	0.0288	2000	0.00159

In all of the simulations, the value of a is constant, but the value of b in Model 2 varies due to the deformation of the microstructure during compression. The extent of this deformation is affected by the acetone wash which shrank the polyester crossing threads that hold the material together, and changed the value of L_x by about 13%. The values of these geometric parameters for the various cases are listed in Tables (2-3)

Table 2 -- Value of b (cm) in Model 2	
As Received	Washed
0.0261	0.0246

Table 3 -- Value of L_x (cm)	
As Received	Washed
0.1041	0.0918

Using Eq. (9) above, the values in Tables (1-3) can be used to obtain the following estimates for the intra-tow porosity

Table 4 -- Estimated Tow Porosity, Knytex D155		
	Model 1	Model 2
As Received	0.487	0.454
Washed	0.463	0.424

The permeability of the tows used in the simulation was estimated based on the tow porosity, intra-tow fibril diameter, an assumed hexagonal packing configuration of the fibrils within the tow, and the overall tow shape. These values are summarized in Tables (5-6)

Table 5 -- Estimated Tow Permeability (cm^2), As Received		
	Model 1	Model 2
Axial	4.4×10^{-8}	3.1×10^{-8}
Transverse	9.1×10^{-9}	6.5×10^{-9}

Table 6 -- Estimated Tow Permeability (cm^2), Washed		
	Model 1	Model 2
Axial	3.4×10^{-8}	2.3×10^{-8}
Transverse	7.1×10^{-9}	4.8×10^{-9}

Based on the tow packing and dimensions, the nominal porosity in the FEM model for the various cases takes on the following values

Table 7 -- Nominal Porosity		
	Model 1	Model 2
As Received	0.121	0.173
Washed	0.047	0.113

The extremely low values are indicative of the high degree of tow packing in the system.

Results and Discussion

Tables (8-9) compare numerical and experimental values of the effective permeability for the case in which binder was still present on the tows. For the axial flow case, the tows were modeled as solid due to the presence of the binder. However, for the transverse flow case, intra-tow flow had to be allowed since there is no free path through the system outside the tows.

Table 8 -- Axial Permeability (cm^2), Knytex D155, Nested Configuration, With Binder		
Exp. 1	Exp. 2	Model 1
1.5×10^{-6}	2×10^{-6}	3.2×10^{-6}

Table 9 -- Transverse Permeability (cm^2), Knytex D155, Nested Configuration, With Binder			
Exp. 1	Exp. 2	Model 1	Model 2
5.0×10^{-8}	7.5×10^{-8}	2.5×10^{-8}	2.3×10^{-8}

The model predictions are of the same order of magnitude as the experimentally measured values, but good quantitative agreement is not achieved. For the axial flow case, the numerically calculated value is about 90% higher than the average experimental value. This is the only case in which the computed values were higher than the measured values. For the transverse flow case, both the Model 1 and Model 2 configurations yield similar results, and are 2-3 times lower than the experimental values.

As mentioned above, the acetone bath used to remove the binder shrank the polyester crossing threads that hold the tows together. This lowered the nominal porosity of the system, and hence, the observed permeability values were also decreased. The results are summarized in Tables (10-11)

Table 10 -- Axial Permeability (cm^2), Knytex D155, Nested Configuration, Without Binder		
Exp. 1	Exp. 2	Model 1
3.1×10^{-7}	3.3×10^{-7}	2.6×10^{-7}

Table 11 -- Transverse Permeability (cm^2), Knytex D155, Nested Configuration, Without Binder			
Exp. 1	Exp. 2	Model 1	Model 2
1.9 x 10^{-8}	2.1 x 10^{-8}	1.5 x 10^{-8}	1.4 x 10^{-8}

Note that much better agreement between the theoretical and numerical values was achieved in this case, and the scatter in the experimental data was also smaller. The numerical axial flow permeability was lower than the average experimental value by about 20%, while the numerical transverse values are lower by about 30-40%.

For a first effort, these results have to be considered quite encouraging. In all cases, the numerically predicted values of the permeability are of the same order of magnitude as the experimental values. While quantitatively the comparison is not that close for the case in which binder was not removed from the tows, the numbers are very close to being acceptable for the case in which the material was washed to remove the binder.

There are number of possible reasons for the observed discrepancies. Note that the washing of the material in the acetone bath caused the experimental scatter to decrease by an order of magnitude for both the axial and transverse flow measurements. This may be an indication that there are non-uniformities in the packed material architecture, which are homogenized to some degree (but not completely) when the tows are washed due to their becoming more soft and pliable. Note that in most cases, the experimental values are higher than the values calculated numerically. Non-uniformities -- e.g., fluctuations in tow dimensions about a mean -- may be expected to allow a greater degree of flow through a porous material as pore size diameter increases allow more flow through them than becomes retarded by equivalent pore size diameter decreases. Further evidence for this interpretation can be ascertained from the micrograph in Figure 5 (15). This micrograph is for Knytex D155, in an attempted stacked (rather than nested) packing configuration. The complexity of real materials, relative to the idealizations required in numerical computations is evident. Although the tows are somewhat elliptical in cross section, image analysis reveals a broad distribution of tow dimensions, as well as dislocations in the tow packing arrangement. Thus, to use the simulation for quantitative purposes, it may be necessary to account for statistical variations in microstructure by doing a number of simulations, and weighting the results about the mean.

Summary and Conclusions

A finite element simulation for modeling flow in 3-D fibrous porous media has been developed for the purpose of predicting the permeability of fiber preforms. Numerically calculated predictions for permeability have been compared with experimentally measured values for flow in Knytex D155 unidirectional material in a nested stacking configuration. This material and configuration were chosen to try and achieve a controlled and predictable preform microstructure for the purpose of comparison between modeling and experiment. Two sets of experiments were performed, the first with the material in the as received state, the second with the material having been washed in an acetone bath to remove the intra-tow binder.

For a first effort, the results that have been obtained are encouraging. In all cases, the numerical values are of the same order of magnitude as the experimental values. For the case in which the material was washed in an acetone bath to remove intra-tow binder, the quantitative agreement is within 20-40%. In most cases, the numerical predictions are lower than the values measured experimentally. It is conjectured that this is due to variations in the tow size dimensions. To obtain better quantitative agreement, future work may need to account for statistical variations in microstructure by doing a number of simulations, and weighting the results about the mean.

The sensitivity of the numerical predictions to microstructure was looked at in two ways. The effect of tow structure and packing was assessed by using two different geometric models to represent the material. Very little difference based upon the details of the packed tow architecture was found between Model 1 and Model 2 for the transverse flow case. However, the numerical calculations for the material with and without binder (the essential effect of which was to bring the inter-tow spacing close together) shows about a 40% difference. These results are reflective of the fact that the nominal porosity change is relatively large due to the changes in inter-tow spacing, but smaller for the changes in tow architecture.

Acknowledgements

We thank Mohamadou Diallo and Kathy Flynn for running the permeability measurement experiments.

References

1   Parnas, R. S. and A. Salem, Polymer Composites, 14(5), 383, (1993).

2   Rudd, C.D., A.C. Long, P. McGeehin, L.J. Bulmer, and J.R. Lowe, "Report on the Workshop on Manufacturing Polymer Composites by Liquid Molding", R.S. Parnas, A.J. Salem, K.N. Kendall, M.V. Bruschke, (Eds.), U.S. Department of Commerce, NISTIR-5373, p. 96, (1994).

3   Phelan Jr., F. R., "Advanced Composite Materials: New Developments and Applications", Proceedings of the 7th Annual ASM/ESD Advanced Composites Conference, ASM International, 175, (1991).

4   Phelan Jr., F. R., Y. Leung, and R.S. Parnas, J. Thermoplastic Composite Materials, to appear (July 1994).

5   Sadiq, T. A., R. S. Parnas and S. G. Advani, Submitted to Int. J of Multiphase Flow, (1994).

6   Bruschke, M. V. and S. G. Advani, J. Rheol., 37(3), 479, (1993).

7   Happel, J., AIChEJ, 5, 174, (1959).

8   Sangani, A. S. and A. Acrivos, Int. J. Multiphase Flow, 8, 193, (1982).

9   Drummond, J. E. and M. I. Tahir, Int. J. Multiphase Flow, 10(5) 515, (1984).

10   Gebart, B. R., SICOMP Technical Report 90-006, (1990).

11   Jackson, G. W., and D. F. James, Can J. Chem. Eng., 64, 364, (1986).

12   Phelan Jr., F. R. and G. M. Wise, Submitted to Composites Manufacturing, (1994).

13   Ranganathan, S., F. R. Phelan Jr., and S.G. Advani, Submitted to Polymer Composites, (June 1994).

14   Berny, C.A., "Permeability of Woven Reinforcements for Resin Transfer Molding", M.S. Thesis, Department of Mechanical Engineering, University of Illinois at Urbana-Champaign, (1993).

15   Roberts, J. and R.S. Parnas, Private Communication, Research Conducted at the National Institute of Standards and Technology, (1993).

16   Slattery, J.C., "Momentum, Energy and Mass Transfer in Continua", Robert E. Kreiger Publishing Company, Huntington, New York, 2nd Edition, Huntington, New York (1981).

17   Martys, N., D. P. Bentz and E. J. Garboczi, Phys. Fluids, to appear (1994).

18   Advani, S.G., M.V. Bruschke, and R.S. Parnas, "Flow and Rheology in Polymer Composites Manufacturing", p.465, S.G. Advani (Ed.), Elsevier, NY, (1994).

19   Parnas, R.S., J.G. Howard, T.L. Luce, S.G. Advani, Submitted to Polymer Composites, (1994).

*. The relationship between μ_e and μ has been discussed by several authors but no consensus has been reached. It has been argued that their relationship is a function of the detailed microstructure (e.g.,15). In this work, it is assumed that the ratio μ_e/μ is unity.

**. Certain commercial materials and equipment are identified in this paper in order to specify adequately the experimental procedure. In no case, does such identification imply recommendation or endorsement by the National Institute of Standards and Technology (NIST) nor does it imply that they are necessarily the best for the purpose.

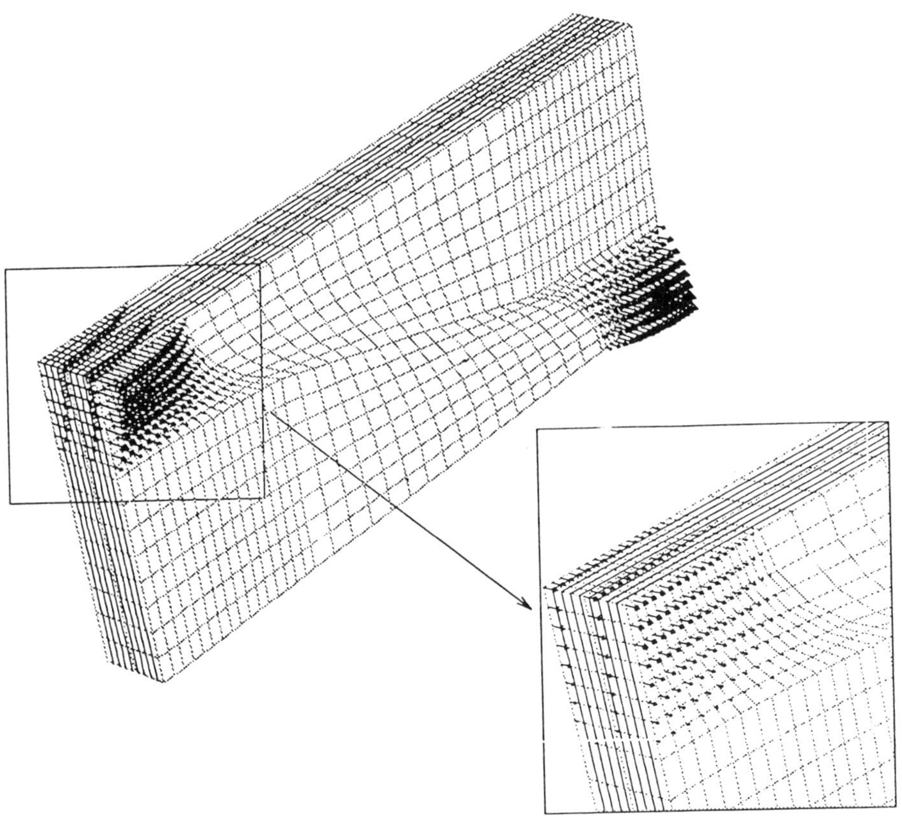

Fig. 4 - Velocity vectors and pressure contours for axial flow through the unit cell used in the finite element calculations. The pressure contours are equally spaced in the flow direction and the maximum velocity is achieved in the corner of the unit cell which corresponds to the center of the flow channel formed by the tows.

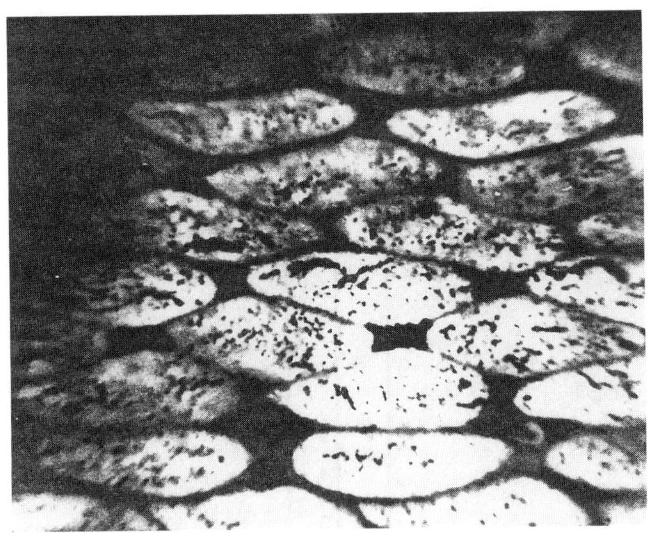

Fig. 5 - Micrograph showing the cross section of a unidirectional fiber mat.

Predicting Shrinkage and Warpage of Thin Compression Molded SMC Parts

B.A. Davis, P.J. Gramann, T.A. Osswald
University of Wisconsin
Madison, Wisconsin

R. Griffith, J. Castro
GenCorp Research
Akron, Ohio

Abstract

Fiber reinforced composites have become increasingly important in several industries because of their enhanced mechanical properties and low weight. However, the processing of these materials greatly influences the final performance of the part; both mechanically and cosmetically. Fiber orientation due to processing causes material anisotropies throughout the part. In addition, temperature imbalances through the thickness of the part result from the material curing and poor thermal layout. These processing phenomena cause the part to experience a residual stress build-up that leads to shrinkage and warpage and complicates analysis and design of the final part. Using simulation these effects can be calculated allowing an engineer to design a part and accurately take into account shrinkage and warpage - eliminating costly modifications to the final mold.

This paper presents a numerical solution to the combined stress and thermal analysis to predict the shrinkage and warpage that commonly occurs during processing. Comparison of simulation results to experiments for several cases molded under realistic processing conditions are shown. In addition, the paper shows an example of how processing conditions are modified to significantly reduce the shrinkage and warpage in a critical area of an automotive body panel.

Introduction

Technological advances in material development have made fiber reinforced composites an important material in several industries. However, when designing fiber reinforced composite parts one must account for the manufacturing process since it greatly effects the performance of the finished part. One such process is the compression molding of sheet molding compound (SMC). The major problems with this process is the prediction and control of the final shrinkage and warpage. During a compression molding process, the material properties become anisotropic which affects the performance and characteristics of a finished part. These material anisotropies make analysis and design considerably more difficult. Further problems arise as a result of curing effects during molding and also from poor thermal mold lay-out. Modeling and simulation of the compression molding process enables the prediction of shrinkage and warpage. Osswald and Tseng (1) represented the through-the-thickness variations that lead to warpage with equivalent moments, thus eliminating the thickness dimensions and significantly reducing computation costs. This paper presents a comparison of shrinkage and warpage of a automotive body panel molded under realistic processing conditions. The simulation is composed of an incremental finite element stress analysis coupled to a 1-D finite difference energy equation with curing effects.

Material Properties Of Short Fiber Composites

During the compression molding process the material properties of fiber reinforced composites become highly anisotropic. This is in great part due to the fiber orientation distribution induced by the molding process. Osswald and Tseng (1) used Halpin's laminate analogy model and the Folgar-Tucker model to calculate the stiffness of a fiber reinforced composite with an in-plane fiber orientation distribution. The laminate analogy states that the elastic properties of an oriented composite is equivalent to that of superposed continuos fiber laminates layed in every direction represented in the fiber orientation function. The elastic constants of the composite are found by a weighted addition of the elastic constants of all layers

$$\overline{E}_{opqm} = \sum_{n} E_{ijkl}^{n} \cdot h^{n} \tag{1}$$

where $\overline{E}_{opqm}$ represents the elastic constants of the composite and h^{n} the fraction of fibers that are in the n'th layer.

To calculate the thermomechanical properties of SMC plates with arbitrary fiber orientation distributions and SMC plates with unidirectional continuous fibers, the laminate analogy model with Schneider's (2) approximation was implemented. Using the thermomechanical properties of the fiber, f, and the matrix, m, Schneider approximated the thermal expansion coefficients of unidirectional fiber reinforced composites as

$$\alpha_1 = \alpha_f + \frac{\alpha_m - \alpha_f}{(\phi E_f / (1-\phi)E_m)+1} \tag{2}$$

$$\alpha_1 = \alpha_3 = \alpha_m - (\alpha_f - \alpha_m)$$

$$\left(\frac{2(1+\upsilon_m)(\upsilon_m^2-1)C}{(1+1.1\phi)/(1.1\phi-1)-\upsilon_m+2\upsilon_m^2 C} - \frac{\upsilon_m E_f/E_m}{1/C+E_f/E_m} \right) \quad (3)$$

where ϕ represents the fraction of glass fiber content and $C=1.1\phi/(1-1.1\phi)$. The subscript 1 denotes the direction parallel to the fibers and 2 and 3 directions perpendicular to them. As with the elastic constants, the thermal expansion coefficients are transformed from the fiber direction of the n'th layer to the main coordinate system of the structure (3).

Theoretical Background Of Thermomechanical Stress Analysis

The compression molding process of SMC begins when a charge at room temperature is placed inside a heated mold. The mold is then closed, forcing the material to flow over the surface of the mold. An order-of-magnitude analysis shows that for thin parts the heat conduction across the thickness is much larger than the conduction along the larger dimensions of the part (4-5). Therefore, the energy equation reduces to a one dimension and can be solved by the finite difference method. However, when dealing with thermosetting polymers, the energy equation has an extra term, Q, that deals with the exothermic curing reaction. This heat generation due to cure can be expressed as

$$\dot{Q} = Q_t \frac{dc}{dt} \quad (4)$$

where c is the degree of cure, Q_t is the total amount of heat released during the reaction and dc/dt can be represented with a model such as the one developed by Barone and Caulk (4)

$$\frac{dc}{dt} = (d_1 + d_2 \cdot c^m)(1-c)^n \quad (5)$$

The terms d_1 and d_2 contain the temperature dependence of the curing reaction rate

$$d_1 = a_1 \cdot e^{-b_1/RT} \quad (6a)$$
$$d_2 = a_2 \cdot e^{-b_2/RT} \quad (6b)$$

where R is the gas constant and b1, b2, a1 and a2 are constants which can be obtained by fitting Eqs.(6.a-b) to data measured in a differential scanning calorimeter (3).

Once the part begins to solidify, residual stresses start to build up. The governing equations for the stress analysis during solidification and part removal are derived using the principle of virtual work. The principle states that for a body which is in static equilibrium, the virtual internal work created by the stresses and the virtual strains, equals the virtual external work done by the external forces and the virtual displacements. Equating the internal and external virtual work results in

$$\int_V \{\delta\varepsilon\}^T \{\sigma\} dV = \{\delta u\}^T \{f\} \quad (7)$$

where $\{\delta\varepsilon\}$ are the virtual strains, $\{\sigma\}$ the stresses, $\{\delta u\}$ the virtual displacements and $\{f\}$ the externally applied forces. The

stresses in the model are represented as a function of local strain and the residual stress $\{\sigma_0\}$

$$\{\sigma\} = [E]\{\varepsilon\} - [E]\{\varepsilon_o^{tot}\} + \{\sigma_0\} \quad (8)$$

In Eq. (8) the material tensor $[E]$ is anisotropic and dependent on the local degree of cure and $\{\varepsilon_o^{tot}\}$ is the total internal strains that occur due to curing, cooling or heating during a time step. Two kinds of internal strains should be included in simulating the thermomechanical behavior of thermoset composites. One is thermal strain caused by a temperature change and the other is curing strain resulting from cross-linking polymerization of thermoset resins. The thermal strains can be represented in terms of temperature change and thermal expansion coefficients.

$$\{\varepsilon_o\}^{Thermal} = \Delta T \{\alpha_{xx} \ \alpha_{yy} \ \alpha_{zz} \ \alpha_{xy} \ 0 \ 0\} \quad (9)$$

Tseng and Osswald (1) neglected the effect of the shrinkage due to curing, since warpage in thin sections is mainly caused by the difference between thermal expansion coefficients of the fiber and matrix. However, for thicker sections this may not be overlooked since the shrinkage due to curing is the main cause of internal cracking.

Since the thickness of compression molded SMC parts is much smaller than the dimensions along the surface, the three-dimensional shell elements that are used to calculate the mold filling can also be used to analyze the stress field of this structure. This allows the usage of the same finite element mesh that was used for both the mold filling simulation and the fiber orientation calculation - eliminating considerable computation and user interaction costs. These elements that are used to calculate the stress field are formed by combining the constant strain triangular element (CST) (5) and the discrete Kirchhoff triangular element (DKT) (6). To approximated the variations through the thickness, Tseng and Osswald (1) used thermal forces and moments.

The material properties of thermoset composites vary as the part cures. The material properties are governed by competing mechanisms between chemical hardening and viscoelastic relaxation (7-8). However, Hahn and Kim (9) have reported a linear-like correlation between the material properties and degree of cure. At the beginning of gelation (c=5%) the composite is regarded as a fluid, and toward the end of cure (c=80%) the material is regarded as having its final strength (5). Once the whole part reaches a degree of cure of at least 80%, the part is removed from the mold and allowed to cool down to room temperature.

Due to the large fiber-length/part-thickness ratio, most fibers in SMC parts are oriented in the planar direction. This leads to higher thermal expansion coefficients in the thickness direction as compared to those in the surface direction. It is well known that angular distortion can occur when forming a curved-sectioned part, a consequence of the anisotropy of the composites, as shown in Fig. 1. This anisotropy induced curvature change is generally called the "spring-forward" effect. O'neill and Rogers (10) reported the analytical solution for a cylindrical shell with three-dimensional anisotropies that exist in composite laminates. Their approach indicated that through-thickness thermal strains, caused by different thermal expansion coefficients, could lead to an angle distortion of a cylindrical shell experiencing a temperature change. As demonstrated in Fig. 1, when a curved part experiences a temperature change of ΔT, the curved angle, θ, will change by $\Delta\theta$ (3). Sometimes only a slight angle change can cause significant warpage in a larger part. Therefore, it is of great importance to take this effect into account.

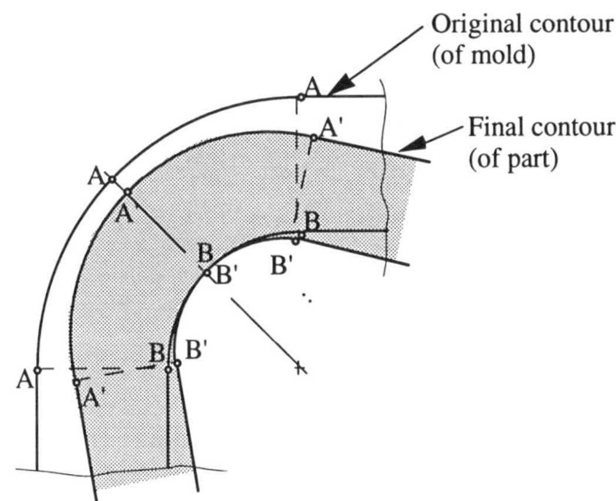

Fig. 1 - "Spring-forward" phenomena caused by difference of thermal expansion coefficient in the surface and thickness directions.

Unfortunately, conventional thin shell element formulation neglects the stress through the thickness of the part since it is much smaller than the in-plane stresses. To account for the through-thickness warping effect the through-thickness stress distribution is approximated by a linear function. This allows the stress variations across the thickness of the part to be replaced by an equivalent moment, obtained by integrating the stress distribution through the thickness.

The simulation begins with the input of the part's geometry, initial conditions, process parameters and boundary conditions. Next, the steps mentioned above are executed. The calculations in the filling and curing stage are done in several time steps and since relaxation is neglected, the cooling stage is a single time step calculation.

Results And Discussions

Comparison Between Numerical and Experimental Results of Compression Molded Parts. In order to verify the numerical model presented here, several experiments for different thin compression molded parts have been conducted using several different kinds of SMC. These experiments were done to better understand the mechanisms of shrinkage and warpage during compression molding of SMC parts and to also verify the simulation model. Since the amounts of shrinkage and warpage were quite small, cautious and accurate measurements were required. Fortunately, there are several straight lines along the edges of the molds, which aided in measuring the shrinkage and warpage in the final parts.

To measure the shrinkage and warpage of the final parts, a coordinate measurement machine was used. The accuracy of the machine was 20 μm; accurate and convenient for most of the measuring cases. However, the warpage in some cases was relatively small and hard to measure, therefore, manual measurements were required. Here, a micrometer and a straight edge were used to measure warpage along straight lines. All the experimental results were compared with the simulation results and will be discussed in the next section.

Compression Molded Automotive Body Panel. The compression molding of the automotive fender, shown in Fig. 2. was simulated and compared to experimental results. The length and width of the part was 2150 mm and 800 mm, respectively, with a thickness of 3 mm, except at the edge around the tire which was 4 mm. The glass fiber content by volume of the part was 21% (ϕ=0.21). The finite element mesh used for the simulation had 748 three-noded shell elements. The initial charge configuration and the resulting mold filling pattern are shown in Fig. 2. The fiber orientation distribution for the body panel was computed using the Folgar and Tucker model and is shown in Fig. 3.

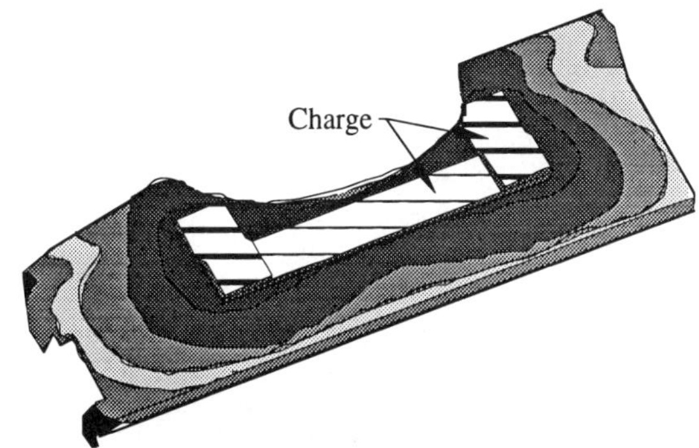

Fig. 2 - Initial charge location and simulated mold filling pattern of an automotive body panel.

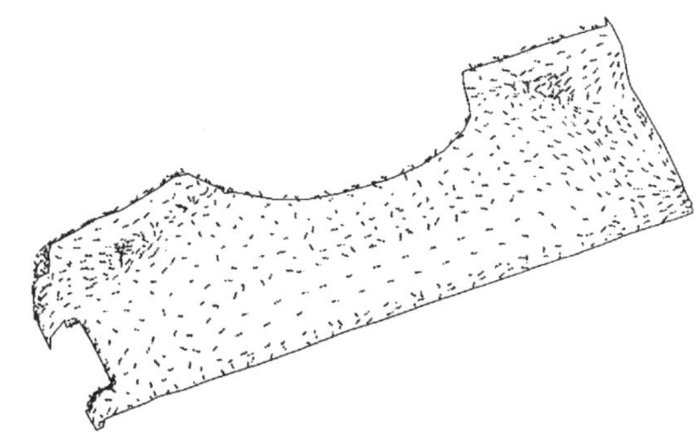

Fig. 3 - Simulated fiber orientation distribution of an automotive body panel.

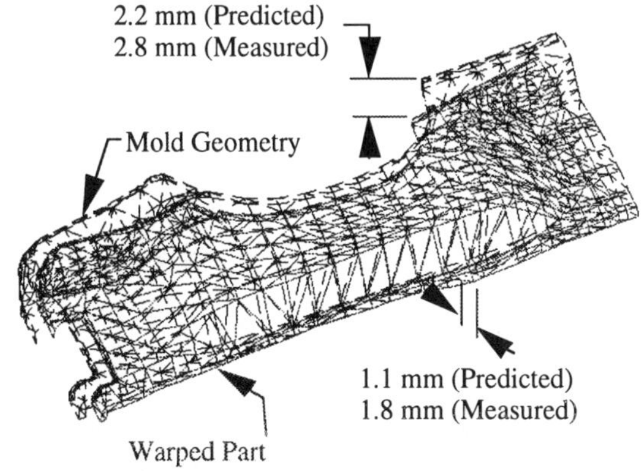

Fig. 4 - Comparison between measured and predicted shrinkage and warpage of an automotive body panel.

Since the geometry of the body panel is quite complicated, the comparisons will be limited to several critical dimensions and displacements. The temperatures of the two mold halves were set at 150°C. The resulting shrinkage and warpage with a 20x enlargement factor is shown in Fig. 4. As shown, the mode of deformation was well predicted. Also from the figure, a twist in the part was measured experimentally to be 2.8mm and a bump on a critical straight edge to be 1.8mm. The simulation calculated the displacement in these areas to be 2.2mm and 1.1mm, respectively. Using the simulation, this part was then optimized to reduce this twist and bump that occurred (3).

Conclusion

From the results displayed, it is clear that the model and simulation developed within this research project agree quite well with the actual process. Although the results are encouraging, there are still some factors that need to be addressed further. Among them are the uneven density distribution throughout the part. The issue of fiber-matrix separation is especially crucial when molding parts with large curvature and ribs. Also, the shrinkage and residual stress build-up in ribbed or thick sections needs to be further addressed. These are areas that the Polymer Processing Research Group at the University of Wisconsin-Madison are now investigating in their research. However, compression molding simulation has reached the state were the part and process can actually be designed and optimized before ever cutting a piece of metal.

Acknowledgments

The authors would like to thank GenCorp Research, Fiat USA and the National Science Foundation under grant numbers DDM 9009158 and DDM 9215287 for their financial support.

References

1. Osswald, T. A. and S. C. Tseng. ,1993, "Orientation and Warpage Prediction in Polymer Processing," a chapter in *Innovation in Polymer Processing.*

2. Schneider,W.,1971, "Wärmeausdehnungskoeffizienten von Glasfaser-Kunststoff-Verbunden, "Kunststoffe, Vol. 61, p.273.

3. Tseng, S. C., 1993., "Simulation of the Thermomechanical Behavior of Fiber Reinforced Thermoset Composites," Ph.D. *Thesis,* Dept. of Mech. Eng., University of Wisconsin-Madison.

4. Barone, M. R. and D. A. Caulk., 1979, "The Effect of Deformation and Thermoset Cure on Heat Conduction in a Chopped-Fiber Reinforced Polyester during Compression Molding,"*Int.J.HeatMass Transfer,* Vol. 22, p.1021.

5. Osswald, T. A.,1991,"A Finite Element Analysis of the Thermomechanical Behavior of Fiber Reinforced Composites," *J. Thermoplast. Comp. Mater.,* Vol. 4, p.173.

6. Jeyachadrabose, C. and J. Kirkhope, 1987, "An Alternative Explicit Formulation for the DKT Plate-Bending Elements," *Int. J. Num. Eng.,* Vol. 21, p.1289.

7. Bogetti, T. A. and J. W. Gillespie, 1989., "Process Induced Stress and Deformation of Thermoset Composites,"*21st Int. SAMPE Tech. Conf.*

8. Bogetti, T. A. and J. W. Gillespie, 1990, "Residual Stress and Deformation in Thick Laminated Composites Undergoing Chemical Hardening and Shrinkage," *45th SPI Conf. Proc.,* Vol. 19A, p.1.

9. Kim, K. S. and H. T. Hahn, 1989, "Residual Stress Development during Processing of Graphite/Epoxy Composites," *Comp. Sci. Tech.,* Vol. 36, p.121.

10. O'Neill, J. M., T. G. Rogers and A. J. M. Spencer, 1988, "Thermally Induced Distortions in the Molding of Laminated Channel Sections, *Math. Eng. Ind.,* Vol. 2, p.65.

Effect of Capillary Pressure on Mold Filling and Fiber Wetting in Liquid Composite Molding

N. Patel, L.J. Lee
The Ohio State University
Columbus, Ohio

Abstract

Mold filling in liquid composite molding (LCM) processes has been traditionally modeled based on Darcy's law, which is valid for low Reynolds number laminar flow of a Newtonian fluid through porous media fully saturated with that fluid. This assumption neglects the effects of capillarity. However, for mold filling at low flow rates and/or for high fiber loading, the capillary pressure may have to be included in the flow analysis. In addition, the micro scale flow pattern in fiber preform and trapping of air are strongly affected by capillarity. This paper addresses the role of capillary forces during mold filling in LCM with an aim to understand and simulate air entrapment in the form of voids. Capillary pressure of fiber reinforcements is measured and correlated to wettability. A phenomenological model has been developed to simulate micro scale flow pattern and air entrapment during mold filling. A criterion for removal of trapped air is also developed.

THE need in recent years for light weight, high strength parts with complicated geometry has made liquid composite molding (LCM) an attractive alternative to manufacture polymer composite parts. LCM processes such as resin transfer molding (RTM) and structural reaction injection molding (SRIM) are matched mold processes involving injection of liquid thermosetting resin into a mold cavity containing dry fiber preform. Resin displaces air from the mold and impregnates the fiber tows. After the mold filling is complete, the resin is allowed to react and solidify. This is followed by demolding and trimming to obtain the final part.

Although a substantial amount of R&D effort is being spent in industry, automotive and aerospace industries in particular, to develop liquid composite molding (LCM) technology, current applications of these processes are limited due to many processing related difficulties. There is a great need to improve these processes through a better technical understanding and through process innovation. Some work has been carried out in recent years to develop LCM process technology, but most of this is aimed at macro scale analysis for predicting flow front progression and inlet pressure during mold filling, and temperature distribution during curing. Very little has been done for micro scale analysis. In LCM, the resin must quickly fill the mold cavity and wet all the individual fibers before much reaction occurs. If dry spots or large voids remain after mold filling, the part is ruined. Dry spot is a large area of resin starved fiber preform. In this paper air pockets trapped between fiber tows, or crossing many filaments are referred to as macro voids, whereas interstitial voids trapped within fiber tows are referred to as micro voids. The presence of voids and dry spots degrade performance and quality of composite part. Thus, molding induced defects are a major issue in these processes. As a result, there is a growing interest in this area. Several researchers have carried out theoretical [1-4] and experimental [4-9] analysis of void formation in LCM. However, there has been no systematic study which quantifies the air entrapment both experimentally and theoretically. In our recent papers [10-15] we reported the mechanisms of void formation and quantified the relationships between the void fraction, liquid velocity and properties for different types of fiber reinforcements used in LCM industry. The purpose of this paper is to present a theoretical investigation and modeling of micro scale flow behavior and air entrapment in LCM during mold filling. Most models reported in literature assume an ideal, regular arrangement of fibers or fiber tows. Such modeling effort is useful for fundamental understanding of flow behavior, but is unlikely to result in a useful analytical tool which can handle various real fiber preforms which are heterogeneous and non-uniform in nature. Our approach is to develop a hybrid model which depends on the micro scale flow behavior and several constitutive relationships, which are determined experimentally. The effect of various fiber architectures is represented by the parameters of the constitutive relationships. Such an approach is widely used in soil science, petroleum engineering and flow through packed beds [16].

Micro Scale Flow Behavior and Void Formation

Theory

Mold filling in LCM processes has been traditionally described as flow through porous media. The pores act like many tiny capillary tubes. Thus the fluid moves like a plug into the mold and there is no fountain flow. Several models are available in literature to describe this type of behavior. The most common and simple one is based on Darcy's law, which is valid for low Reynolds number (Re) laminar flow of a Newtonian fluid through porous media fully saturated with that fluid [16]:

$$\overline{u_s} = -\frac{1}{\eta}\overline{\overline{k}}\,\overline{\nabla}P \qquad \dots\dots\dots(1)$$

where $\overline{u_s}$ is the average (superficial) velocity in the porous media, $\overline{\overline{k}}$ is the permeability of the porous media, η is the viscosity of fluid, and $\overline{\nabla}P$ is the imposed pressure gradient. Deviations from this law at Re greater than 10 are attributed to inertial forces. Its application to transient mold filling in LCM is based on the assumption that the residual saturation of air (non wetting phase) in the region infiltrated by the resin (wetting phase) is zero. Saturation of fluid phase is defined as fraction of accessible pore space occupied by that fluid. This assumption permits the effect of capillarity to be neglected in the modeling of mold filling process. However, for mold filling at low flow rates and/or for high fiber loading the capillary pressure may have to be included in the flow analysis. In addition, the micro scale flow pattern in fiber preform and trapping of air are strongly affected by capillarity.

For porous media with a pore size distribution shown in Fig. 1, displacement of the non wetting phase by the wetting phase (transient two phase flow) has been described by writing Darcy's law for both phases [16]. A fiber preform, however, consists of two distinguishable pore structures: micro pores consisting of gaps between fiber filaments in a fiber tow and macro pores consisting of gaps between the tows (Fig. 2). These gaps are often non-uniform in an actual fiber preform. There are also obstructions that interfere with resin flow: binder, stitches, crimps, and irregular fiber tow orientations. Most of these obstructions are necessary for holding fiber tows together in a fiber preform. Because of these alternate paths available for resin flow, there is fingering or lead-lag in the primary flow front [10-15].

The fingering in the primary flow front depends on the balance of capillary and hydrodynamic pressures (Fig. 3). Because of very small interstitial spaces within fiber tows, capillary pressure is much larger in the tows as compared to in the gaps between the tows. At low flow rates, due to significant contribution by capillary pressure, flow leads in the tows or stitches which are in the direction of global flow. In addition, there is wicking of liquid ahead of the primary flow front due to capillary action. When the leading flow front reaches an obstruction such as a stitch or a transverse tow, there is a cross flow by capillary action. This often leads to the macro void formation. For mold filling at high flow rates, on the other hand, flow leads in the gaps between the axial tows or in the gaps around the crimps. This is because the permeability of these gaps are high and the contribution of the capillary pressure to the driving force for flow is relatively little as compared to the hydrodynamic pressure. Since flow leads in the gaps between the tows, liquid wicks into the tows after the primary flow front has passed. This often leads to trapping of air in the fiber tows in the form of micro voids. No macro voids are formed at high flow rates. In our previous work [10-14] we experimentally correlated void formation with modified capillary number (Ca*).

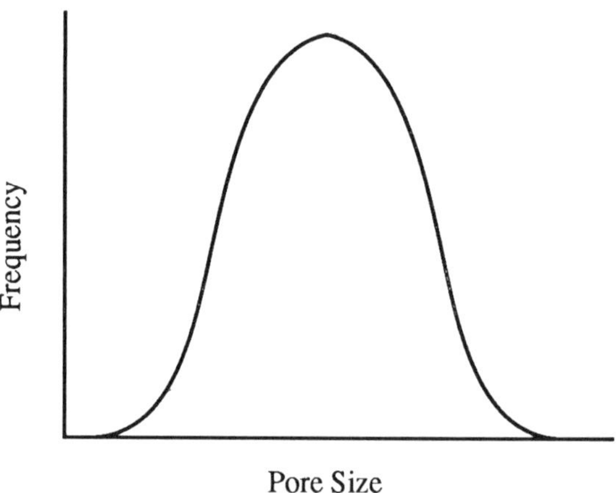

Fig. 1   Pore size distribution of a typical porous medium

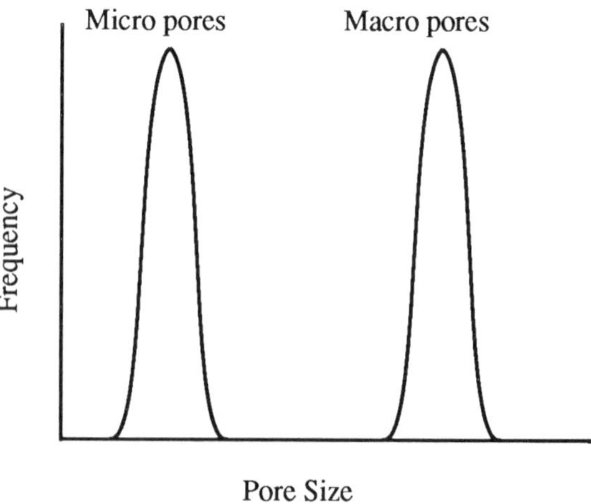

Fig. 2   Pore size distribution of fiber reinforcements used in LCM

$$Ca^* = \frac{\eta u_s}{\gamma_{LV} \cos\theta} \qquad \ldots\ldots\ldots(2)$$

where γ_{LV} is the surface tension of liquid and θ is the fiber-liquid contact angle.

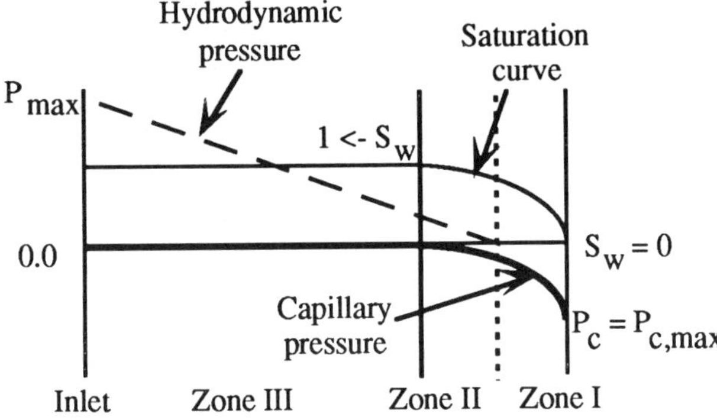

Fig. 3 Mechanism of micro scale flow in a fiber reinforcement

The capillary pressure in porous media is a function of saturation. In the dry area, where the saturation of wetting phase (resin) is zero, the capillary pressure is at its maximum (Fig. 3). Moving towards the primary flow front, saturation increases and hence, the capillary pressure decreases. At the primary flow front, saturation is still not 1.0, that is, the fibers are not completely wet. Hence there is a finite capillary pressure. As the saturation approaches 1.0, the capillary pressure becomes negligibly small and the only driving force is the hydrodynamic pressure. Based on these, the flow through porous media can be divided into three regions. In Zone I (the wick front zone), only the capillary pressure is operative. In Zone II (the primary front zone), both the capillary and the hydrodynamic pressures are important. In Zone III (the fully developed flow zone), only the hydrodynamic pressure is the driving force.

Model Development

As mentioned earlier, the key features of void formation in LCM are the flow front lead-lag in the two pore structures and the transverse flow. If there is no flow lead-lag, no void would form. For an actual fiber preform, the micro scale flow pattern is very complicated. We, however, may simplify the flow based on a conceptual model shown in Fig. 4. Here, macro and micro pores are displayed as two channels parallel to each other with flow transfer occurring between the two channels. These two flow channels have different porosity and different permeability. In other words, the fiber preform is treated as

consisting of two interacting continuous porous media: macro pores and micro pores. The two porous media communicate by means of transfer terms. Depending on the permeabilities and capillary pressures, liquid leads in one channel as compared to the other at a particular flow rate. At a distance L, there is a "bridge", where the leading flow from one channel can quickly invade the other channel and round up voids in the trailing channel. This characteristic distance L depends on fiber architecture, e.g. stitches, binder, crimps and transverse tows, and is determined experimentally. Since the two channels are two porous media, there is a resin saturation distribution within each channel.

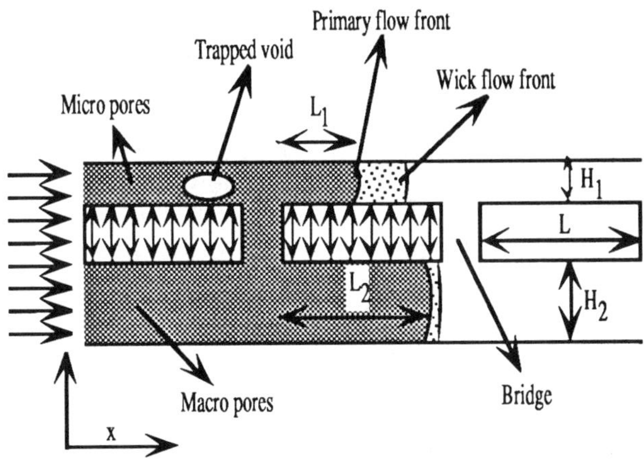

Fig. 4 Proposed conceptual model for void formation during mold filling in LCM

The flow behavior in each channel is computed using the equations based on the two phase Darcy's law and continuity written for each phase in each channel. For one dimensional horizontal flow the equations for the micro pores (tows) are [16-19]:

$$\phi_t \frac{\partial Sw_t}{\partial t} = \frac{\partial}{\partial x}\left(\lambda w_t \frac{\partial Pw_t}{\partial x}\right) - \lambda w_{trans}\, \xi\left(Pw_t - Pw_g\right) \qquad \ldots\ldots(3)$$

$$\phi_t \frac{\partial\left(Snw_t\, \rho nw_t\right)}{\partial t} = \frac{\partial}{\partial x}\left(\rho nw_t\, \lambda nw_t \frac{\partial Pnw_t}{\partial x}\right)$$
$$- \rho nw\, \lambda nw_{trans}\, \xi\left(Pnw_t - Pnw_g\right) \qquad \ldots\ldots(4)$$

where, S_{wt} and S_{nwt} are the wetting and non wetting phase saturations and are related by: $S_{wt} + S_{nwt} = 1$. P_{wt} and P_{nwt} are wetting and non wetting phase pressures, which are related by: $P_{nwt} - P_{wt} = P_{ct}$, P_{ct} being the capillary pressure. ρ_{wt} is the density of air calculated by ideal gas law. ϕ_t is the porosity of micro pores (tows) based on total volume. ξ is the shape factor for the transfer term, which depends on the specific surface area of tows. $\lambda_{wt} = K_{it} \, k_{rwt} \, / \, \mu_w$ and $\lambda_{nwt} = K_{it} \, k_{rnwt} \, / \, \mu_{nw}$, where K_{it} is the intrinsic permeability of tows. Intrinsic permeability is the permeability of completely saturated porous medium. k_{rwt} and k_{rnwt} are relative permeabilities, and μ_w and μ_{nw} are viscosities of wetting and non wetting phases respectively. Other terms have similar meaning. In eqs. 3 and 4, the left hand side is the rate of change of saturation with respect to time, the first term on the right hand side is the flow due to pressure drop in micro pores, and the second term on right hand side represents the transfer of fluids between two types of porous media. Similar equations can be written for macro pores (gaps) also. Thus four partial differential equations have to be solved simultaneously to get saturation distribution in the two porous media.

Model Implementation

Simulations were carried out for injection of DOP oil into a unidirectional stitched fiberglass mat (CoFab A0108). This fiber mat has fiber tows oriented in one direction only, which are held together by continuous stitches. The global flow was in the direction of fiber tows. The properties of DOP oil and unidirectional mat are reported in Table 1.

Table 1 Parameters used in simulation runs

Parameter	Value
Viscosity of DOP oil	45×10^{-3} kg/m sec
Surface tension of DOP oil	25.4×10^{-3} N/m
Viscosity of air	18×10^{-6} kg/m sec
Axial permeability of micro pores, K_{it}	8.0×10^{-12} m^2/sec
Permeability of macro pores, K_{ig}	80.6×10^{-12} m^2/sec
Permeability of fiber mat, K_i	88.6×10^{-12} m^2/sec
Transverse permeability of micro pores, $K_{it, \, trans}$	1.0×10^{-12} m^2/sec
Porosity of micro pores, ϕ_t	0.31
Porosity of macro pores, ϕ_g	0.26
Porosity of fiber mat, ϕ	0.57
Shape factor for transfer term, ξ	630915 m^{-2}

Capillary pressure versus saturation relationship for the case under study was measured by a centrifuge device developed in our laboratory [20]. In this method, a fiber mat sample completely saturated with test liquid is subjected to centrifugal force at various speeds of rotation. Centrifugal force tries to force liquid out of the sample, whereas capillary pressure tries to contain it in the pores. At each rotational speed equilibrium is attained at different liquid saturations. From the weight of liquid remaining in the porous sample at equilibrium and the weight of liquid

contained in completely wet sample (i.e. at the beginning of the experiment), liquid saturation can be calculated. The capillary pressure is calculated from the speed of rotation. Detailed design of the centrifuge device and experimental procedure are given in reference [20].

Fig. 5 shows the relationship between capillary pressure and liquid saturation for DOP oil / unidirectional stitched fiberglass mat. At low saturations, the capillary pressure is high. Liquid is primarily in the smaller pores [16], i.e. in the fiber tows (Region I). With increase in saturation, the capillary pressure drops rapidly, since liquid enters relatively larger pores. The lower portion of the curve (Region II) corresponds to the gaps between the tows. The fractional pore space in the tows and the gaps was determined by an optical microscope. Measurement of pore dimensions by statistical scans of microphotographs of porous medium cross sections is a direct technique for obtaining an average pore body size [21]. Thus, P_c versus S_w curves for the two types of pores can be determined.

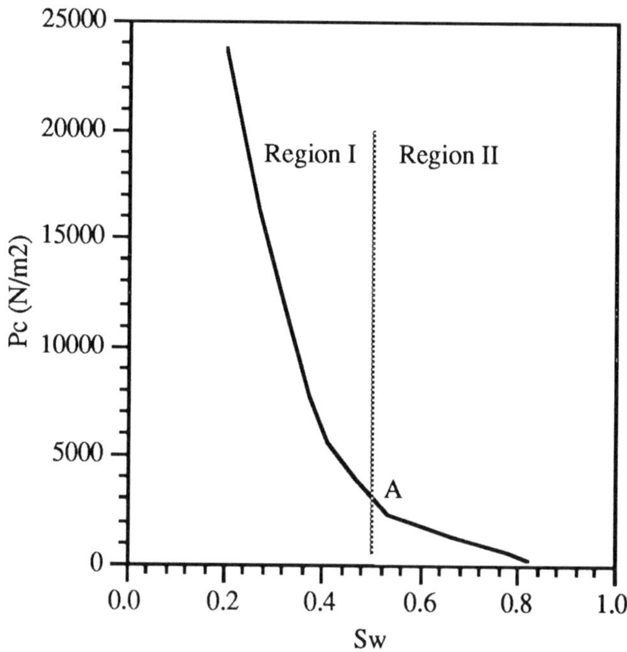

Fig. 5 P_c-S_w relationship for DOP oil / unidirectional stitched fiberglass mat

Such P_c versus S_w curves have also been generated for other types of fiber mats and other liquids in our laboratory. Fig. 6 shows results obtained for various liquids in unidirectional stitched fiberglass mats. These P_c versus S_w curves can be correlated with pore structure and liquid properties using a single dimensionless function known as Leverett J function [16, 21]:

$$J\left(S_w\right) = \frac{P_c\left(S_w\right)}{\gamma_{LV} \, T(\theta)} \sqrt{\frac{K}{\phi}} \qquad \qquad \dots\dots\dots(5)$$

where, K and ϕ are permeability and porosity of the porous sample. Leverett J function depends on pore entry radii that appears implicitly in the capillary pressure term of eq. 5. Because the relationship between pore size and permeability is not known, an average permeability of the entire porous media is used in this equation [21]. Thus, J is a dimensionless capillary pressure rather than a constant. $T(\theta)$ is the geometric factor for the contact angle of the liquid-air interface passing through the pore, where θ is the contact angle. For a straight capillary tube, $T(\theta)$ reduces to $\cos \theta$. In a fiber mat, however, it may differ from $\cos \theta$. Since the detailed pore structure of a fiber mat is not known, it is often approximated by $\cos \theta$. We propose a procedure wherein P_c versus S_w curves are normalized with respect to that for DOP oil. That is, $T(\theta)$ for DOP oil is taken as 1 and $T(\theta)$ for all other liquids are computed with respect to DOP oil. Figs. 7 and 8 show typical normalized curves obtained by the above mentioned procedure.

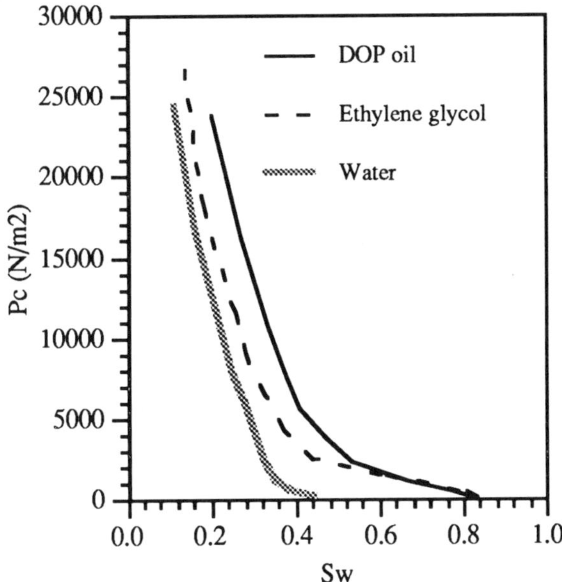

Fig. 6 P_c-S_w relationships for various liquids in unidirectional stitched fiberglass mat

The relative permeability of the porous media is also a function of saturation [16]. Because of the relatively narrow pore size distribution of macro and micro pores, the relative permeability of any phase is equal to its saturation (i.e. relative permeability functions are linear).

Fig. 9 shows the simulated saturation profiles in (a) the gaps and (b) the tows at various time instances for the injection of DOP oil at 1.0 cm/sec in the absence of cross flow. The wetting phase saturation (S_w) of 0 indicates completely dry fibers. S_w --> 1, on the other hand, represents wet fibers. The saturation gradient in the gaps is quite sharp (Fig. 9 a) indicating a plug-like shape of the wetting front. The wetting phase saturation in the tows (Fig. 9 b) changes gradually from 0 to the final steady state value (--> 1), i.e. the wetting front in the tows is

spread out. The wetting front in the fiber tows trails the front in the gaps. The wetting of the fiber tows continues even after the primary flow front has passed, primarily because of the transfer of liquid from the gaps to the tows.

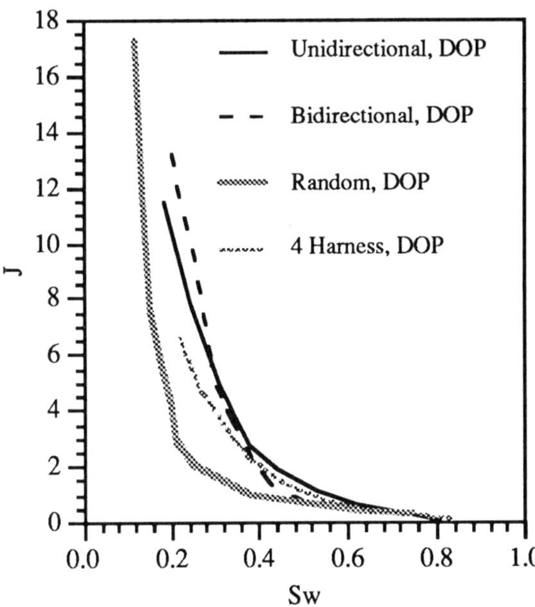

Fig. 7   J-S_w relationships for DOP oil in 4 different types of fiberglass mats

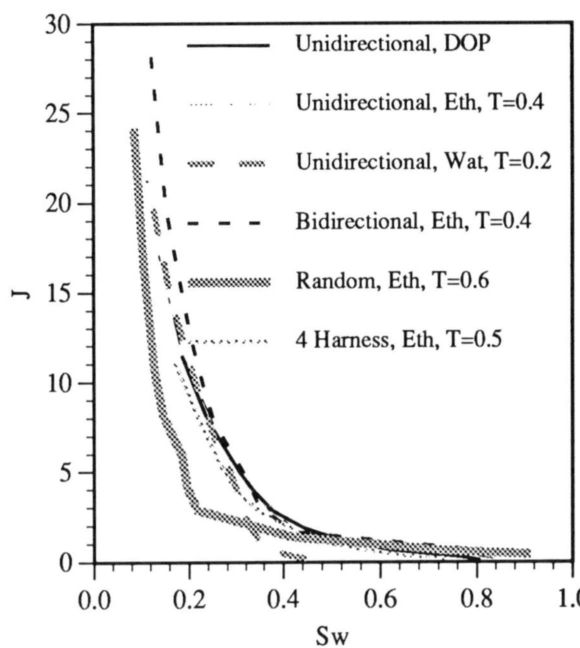

Fig. 8   J-S_w relationships for various liquids in 4 different types of fiberglass mats

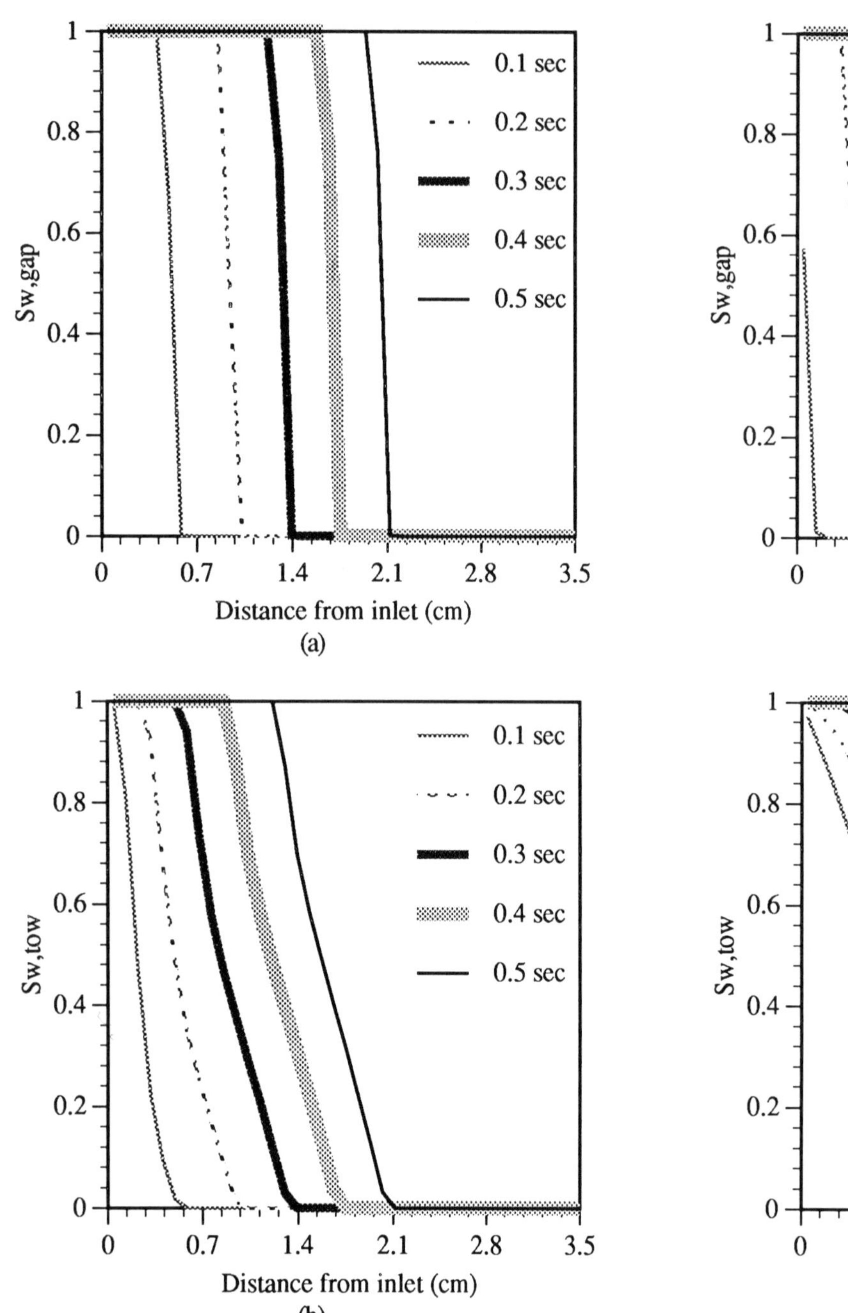

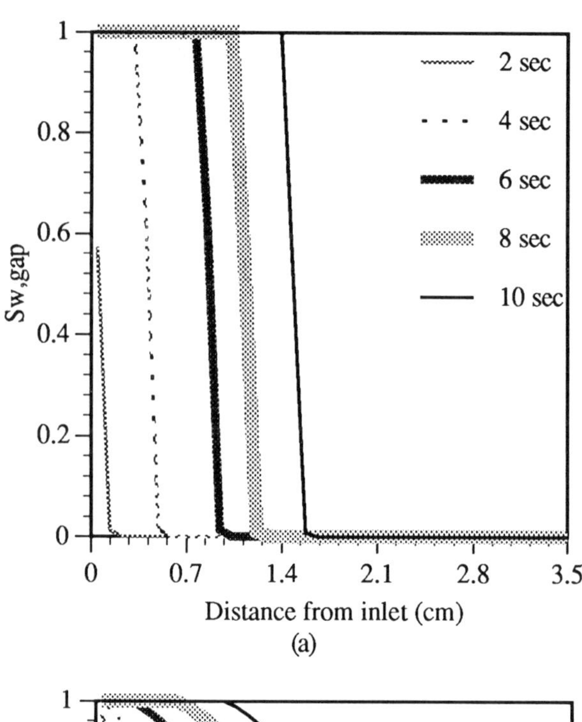

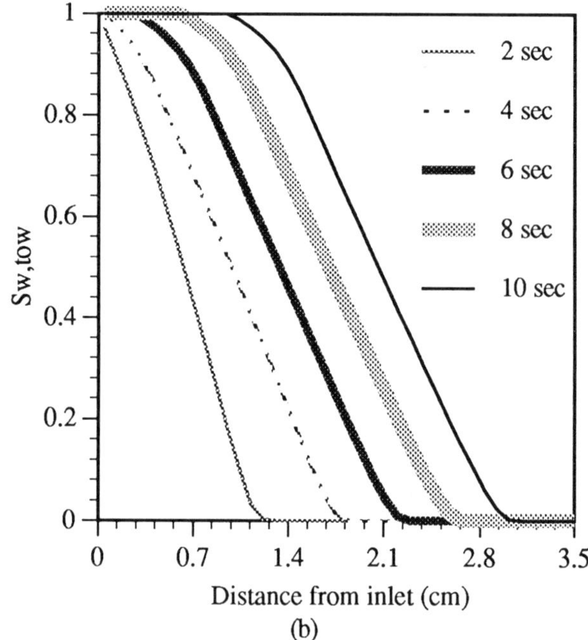

Fig. 9   Wetting phase saturation profiles for injection of DOP oil at 1 cm/sec

Fig. 10   Wetting phase saturation profiles for injection of DOP oil at 0.05 c:n/sec

Fig. 10 shows the simulated saturation profiles in (a) the gaps and (b) the tows at various time instances for the injection of DOP oil at 0.05 cm/sec in the absence of cross flow. Again the saturation profiles in the gaps exhibit a relatively sharp gradient (Fig. 10 a) and the wetting front in the tows is spread out (Fig. 10 b). In this case, the saturation front in the fiber tows leads the front in the gaps. Thus, the simulation results support the proposed mechanism of micro scale flow pattern.

The simulation results for both high and low flow rates indicate that, in the absence of cross flow, the saturation of wetting phase will become equal to 1 everywhere, if the liquid injection is continued for sufficient time. That is, no voids would form if there is no mechanism for cross flow. The presence of "bridges" (obstructions), however, provides an alternate path for liquid flow (cross flow), thereby interrupting the development of saturation profiles. If the cross flow from

330

the leading front is completed before the lagging front can drive the air out, voids would form. The actual procedure for calculating the void fraction is as follows. If saturation in the micro pores (tows) at the bridge, i.e. at distance L, reaches critical value ($S_{w,cri}$) before the flow front in the macro pores (gaps) reaches the bridge then macro voids are formed and the macro void fraction is given by:

$$V_{macro\ void} = \frac{1 - \int_0^L \phi_g\, S_{w_g}\, dx}{L} \qquad \ldots\ldots\ldots(6)$$

On the other hand, if the saturation in the macro pores at the bridge, i.e. at distance L, reaches the critical value before the micro pores at the bridge are completely filled with liquid, then micro voids are formed and the micro void fraction is given by:

$$V_{micro\ void} = \frac{1 - \int_0^L \phi_t\, S_{w_t}\, dx}{L} \qquad \ldots\ldots\ldots(7)$$

To obtain values of model parameters L and $S_{w,cri}$, experimentally determined void fractions at two different velocities were used. Simulations were carried out for these velocities and values of L and $S_{w,cri}$ that gave experimentally determined void fractions were obtained. For axial flow of DOP oil into a unidirectional stitched fiberglass mat, L ~ 2.1 cm and $S_{w,cri}$ ~ 0.61. As mentioned earlier, the parameter L represents the effects of fiber mat architecture, specifically the cross flow characteristics of the fiber mat. For axial flow in unidirectional stitched fiberglass mat, the stitches provide the path for cross flow. For comparison, the distance between two consecutive stitches is ~ 0.5 cm. Using the calculated values of these parameters, simulations were carried out at various velocities and void fractions were computed at these velocities.

Fig. 11 shows percentage voidage versus log (superficial velocity) obtained by simulation and flow visualization experiments for the case under study. For details of flow visualization experiments reader is referred to our previous work [10-14]. At low velocities, macro voids are formed. As velocity increases, the macro void fraction decreases and above a certain velocity no macro voids are formed. At high velocities, however, micro voids are formed. The micro void fraction increases with velocity. Both macro and micro voids exhibit logarithmic relationship with velocity.

The agreement between simulated and experimental results is good for macro voids. However, the model predicts higher micro void fractions than experimental results. This is because of the following reasons. In the developed model, it is assumed that when the leading flow front reaches the bridge, it quickly invades the trailing porous medium. In other words, the dynamics of the cross flow was neglected. This is not the case in reality as was observed in flow visualization experiments.

At high flow rates, wetting of fiber tows continued even after the flow front had passed [10-14]. The micro voids formed could coagulate to form meso voids. These meso voids were then removed. In the developed model once micro voids are trapped, no further flow from gaps to tows is allowed. The removal of voids is thus, not included in the simulation. The void removal increases with increase in velocity. In our flow visualization experiments, the removal of macro voids during mold filling was not observed for axial flow in unidirectional stitched fiberglass mats. Thus the developed void formation model is adequate to describe the relationship between the macro void fraction and velocity. However, it over-predicts the micro void fractions. Adding the effect of void removal to the model may make the predictions better.

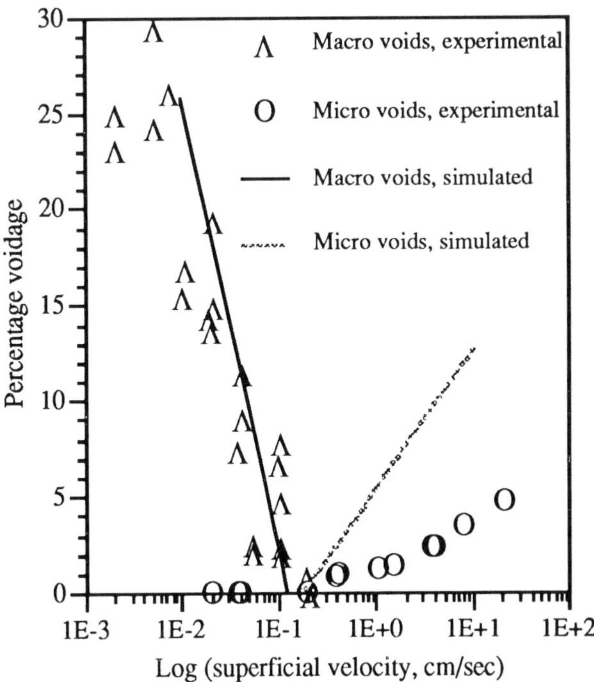

Fig. 11 Percentage voidage versus log (superficial velocity) for axial flow of DOP oil into unidirectional stitched fiberglass mat

Void Removal

Voids formed during mold filling are stabilized by capillary forces. In our previous study [10-12], we reported the mechanism of movement of macro and micro voids for different types of fiber mats. In this paper, criteria for the mobilization of voids are developed. The mobilization of trapped discontinuous non wetting phase from porous media has been studied extensively by many researchers [16, 21-26]. A trapped void will move when the capillary pressure difference between the ends of the void is less than the viscous forces exerted by the flowing fluid around the void. Chatzis & Morrow [24] derived the expression for the pressure required for the onset of mobilization of a void in terms of the pore throat diameter (D_e) and the pore body diameter (D):

$$\Delta P_m = 4\gamma_{LV}\left(\frac{\cos\theta_r}{D_e} - \frac{\cos\theta_a}{D}\right) \qquad ....(8)$$

where θ_r and θ_a are the receding and advancing contact angles.

If l_{void} is the distance between two ends of the void (length of the void) and α is the angle the line between them makes with the average flow direction (Fig. 12), then the pressure difference between them resulting from viscous flow is [22]:

$$\Delta P_{visc} = \frac{\eta\, u_s}{K_i\, k_{rw}} l_{void} \cos\alpha \qquad ....(9)$$

based on Darcy's law.

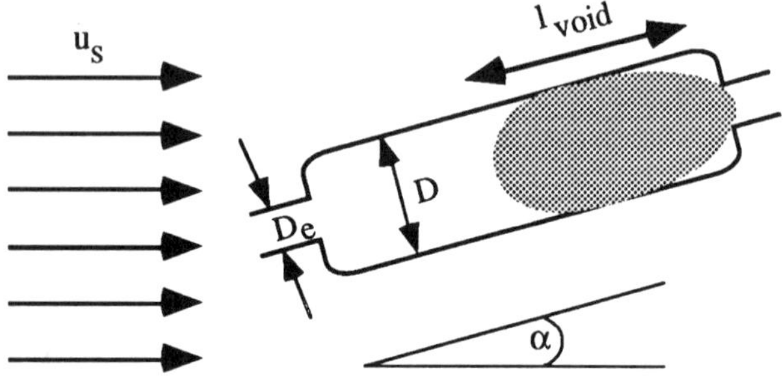

Fig. 12 Schematic diagram showing a void trapped in a pore of neck diameter D_e and body diameter D

The critical capillary number for mobilization (Ca_{cri}^{mob}) is obtained by equating eqs. (8) and (9) and rearranging:

$$Ca_{cri}^{mob} = \left(\frac{\eta\, u_s}{\gamma_{LV}}\right)_{cri}^{mob} = \frac{4\, K_i\, k_{rw}}{l_{void}\cos\alpha}\left(\frac{\cos\theta_r}{D_e} - \frac{\cos\theta_a}{D}\right) \qquad ....(10)$$

At capillary number slightly larger than the critical capillary number for mobilization, the voids line up with and move parallel to the macroscopic flow direction. This involves stretching of voids and may necessitate snake-like or zig-zag motion of the voids. The voids may fission into smaller voids due to the randomness of the flow path. This has been classified as quasi-static displacement [25, 26]. At capillary numbers substantially higher than the critical capillary number for mobilization, the voids are more prone to break up into smaller voids. This type of void movement has been referred to as dynamic displacement [25, 26]. Dynamic displacement may be disastrous if the small voids obtained by void break-up become trapped. However, if the capillary number is larger than the critical value for all void sizes then this type of void movement is advantageous since the void velocity would be much higher under dynamics displacement conditions.

In the theory discussed above the rate of void movement was not considered. The detailed simulation of velocity and path of void movement is a formidable task due to the random nature of the flow channels in fibrous porous media, distribution of voids, and pore constriction and body sizes, and dynamic break-up of voids during movement. The developed theory also ignores internal circulation inside the void. However, eq. 10 can serve as a guideline regarding capillary numbers required to move voids. For example, consider a macro void of diameter 0.067 cm and length 0.067 cm trapped in a unidirectional stitched fiberglass mat. Voids of similar sizes are formed during mold filling using DOP oil for a capillary number of $\sim 3.39 \times 10^{-3}$. The smallest pore size, i.e. pore throat available for the movement of this macro void corresponds to the point A on Fig. 5. From the capillary pressure at point A, the pore throat diameter (D_e) can be calculated using Laplace equation:

$$P_c = \frac{\gamma_{LV}\cos\theta}{D_e} \qquad ....(11)$$

Drainage capillary pressure versus saturation relationship obtained by injection of air into liquid filled porous sample (as was done in the present study), or injection of mercury into evacuated porous sample (mercury intrusion porosity) is a common tool for determining pore entry diameters of porous media [21].

For DOP oil, both advancing and receding contact angles are $0°$. For $\alpha = 0°$ and $k_{rw} = 1$, the critical capillary number for mobilization is $\sim 1.09 \times 10^{-2}$, which is about an order of magnitude higher than that for formation. Similar analysis for several other types of fiber mats gives the results shown in Table 2. Thus, except in 4 harness woven mats, macro voids formed during mold filling at low flow rates remain trapped in the fiber mat. The elimination of these macro voids require injection of liquid at high flow rates (bleeding). For 4 harness woven mats, the critical capillary number for the formation and movement are similar, indicating that void movement is possible even during mold filling at low flow rates. These phenomena were observed experimentally and have been reported in our previous publications [10-12]. Let's now consider a micro void of length 100 μm and diameter 4 μm trapped inside a fiber tow. From eq. 10, the critical capillary number for the mobilization of this micro void is $\sim 7.1 \times 10^{-1}$, indicating that elimination of micro voids from the fiber tows is very difficult. Similar analysis for micro voids has also been reported by Chen et al. [4]. However, in our previous study, we observed that several micro voids can coagulate, come out of the fiber tows and form relatively larger meso voids, which are then subsequently removed [11-14]. Using eq. 10, the critical capillary number at the onset of mobilization of meso voids of typical size, i.e. of length ~ 0.02 cm and diameter ~ 0.0067 cm, is $\sim 1.3 \times 10^{-2}$. This is similar to the capillary numbers at high flow rates for which micro voids are formed. Thus elimination of meso voids is relatively easy. In addition to injection of liquid at high flow rates, high capillary numbers can also be generated by lowering capillary forces (by using surfactants). The developed theory for void mobilization, thus explains the experimentally observed results, at least qualitatively.

Table 2 Criterion for macro void formation and movement

Type of Fiber Mat	$Ca_{cri}^{mobilization}$ (from eq. 10)	$Ca_{cri}^{formation}$ (by experiments)
Unidirectional Stitched	1.09×10^{-2}	3.39×10^{-3}
Bidirectional Stitched	1.13×10^{-2}	1.25×10^{-3}
Continuous Random	2.02×10^{-2}	2.44×10^{-3}
4 Harness Woven	7.2×10^{-4}	1.03×10^{-3}

Conclusions

The void formation in LCM requires two types of micro flows: lead-lag or fingering in the flow front and cross flow. A phenomenological model was developed in this work to simulate the micro scale flow pattern and void formation during LCM mold filling. The flow front lead-lag was predicted by the dual porosity, dual permeability model based on two phase Darcy's law. The developed model illustrates the effects of capillarity on mold filling and fiber wetting. Results of flow visualization study were used to characterize the parameters of the model. These parameters were then used to predict void formation as a function of inlet velocities. The model predicted macro void formation very well, but over-predicted micro voids fractions, primarily because the micro void removal was not included in the simulation study. A criterion for the movement of trapped air bubbles was also developed based on force balance. This void removal model can be used to obtain the liquid velocity required to eliminate trapped voids from the fiber preform.

References

1   A. W. Chan and R. J. Morgan, *Polym. Compos.*, **14(4)**, 335 (1993).
2   R. S. Parnas and F. R. Phelan Jr., *SAMPE Q.*, **22(2)**, 53 (1991).
3   A. W. Chan and R. J. Morgan, *SAMPE Q.*, **23(3)**, 48 (1992).
4   Y. T. Chen, H. T. Davis, and C. W. Macosko, *AIChE J.*, in press (1994).
5   W. R. Stabler, G. B. Tatterson, R. L. Sadler, and A. H. M. El-Shiekh, *SAMPE Q.*, **23(2)**, 38 (1992).
6   A. D. Mahale, R. K. Prud'homme, and L. Rebenfeld, *Polym. Eng. Sci.*, **32(5)**, 319 (1992).
7   R. C. Peterson and R. E. Robertson, *Proc. of 8th ASM/ESD Adv. Composites Conf.*, 63 (1992).
8   R. C. Peterson and R. E. Robertson, *Proc. of 7th ASM/ESD Adv. Composites Conf.*, 203 (1991).
9   T. J. Wang, M. J. Perry, and L. J. Lee, *SPE ANTEC Tech. Papers*, **50**, 756 (1992).
10   N. Patel, V. Rohatgi, and L. J. Lee, *Polym. Eng. Sci.*, in press (1994).
11   N. Patel and L. J. Lee, *Polym. Eng. Sci.*, in press (1994).
12   N. Patel, V. Rohatgi, and L. J. Lee, *Proc. of 49th Ann. SPI Compos. Inst. Conf.*, Session 10 - D (1994).
13   V. Rohatgi, N. Patel, and L. J. Lee, *Proc. of 9th ASM/ESD Adv. Composites Conf.*, 81 (1993).
14   V. Rohatgi, N. Patel, and L. J. Lee, paper submitted to *Polym. Compos.* (1994).
15   N. Patel, V. Rohatgi, and L. J. Lee, *Polym. Compos.*, **14(2)**, 161 (1993).
16   F. A. L. Dullien, in *Porous Media Fluid Transport and Pore Structure*, Academic Press, Inc., San Diego, CA (1992).
17   T. D. Van Golf-Racht, in *Fundamentals of Fractured Reservoir Engineering*, Elsevier Scientific Publishing Co., New York (1982).
18   K. Aziz and A. Settari, in *Petroleum Reservoir Simulation*, Applied Science Publishers Ltd., London (1979).
19   G. P. Willhite, in *Waterflooding*, Society of Petroleum Engineers, Richardson, TX (1986).
20   K. Han, C. Wu, and L. J. Lee, *Proc. of 9th ASM/ESD Adv. Composites Conf.*, 19 (1993).
21   G. L. Stegemeir, in *Improved Oil Recovery by Surfactant and Polymer Flooding*, edited by D. O. Shah and R. S. Schechter, Academic Press Inc., New York (1976).
22   K. M. Ng, H. T. Davis, and L. E. Scriven, *Chemical Engineering Science*, **33(8)**, 1009 (1978).
23   J. C. Melrose and C. F. Brandner, *Journal of Canadian Petroleum Technology*, **13(4)**, 54 (1974).
24   I. Chatzis and N. R. Morrow, *Society of Petroleum Engineers Journal*, 555, October (1984).
25   R. E. Hinkley, M. M. Dias, and A. C. Payatakes, *PhysicoChemical Hydrodynamics*, **8(2)**, 185 (1987).
26   A. C. Payatakes, *Annual Review of Fluid Mechanics*, **14**, 365 (1982).

Development of a Process and Consolidation Model for Powder Prepreg Composites

S. Padaki, L.T. Drzal
Michigan State University
East Lansing, Michigan

Abstract

The MSU Powder Prepreg Process is capable of making prepreg tapes from varying sizes of particles and utilizing either thermoplastic or thermoset powders. A number of factors (e.g. time, pressure, particle size, volume fraction and viscosity) affect the consolidation of these tapes. The efficiency of the consolidation for any given tape would also depend on the material properties of the polymer, particle size, temperature, consolidation pressure, and volume fraction of the tape. The consolidation process in composites made out of powder impregnated tapes also differs from other composite forms because of the size and distribution of polymer powder on the fiber in the unconsolidated states. This paper reports the development of a mathematical process model required for the consolidation of a given powder prepreg tape. For the purposes of modeling, the consolidation process is split up into sub- models. The sub-model for heat transfer uses a heat transfer through a slab approach to predict the heating and cooling times. The sub-model for flow of polymer uses a Hele-Shaw type creeping flow model. These times are combined with the times for autohesion and crystallization to give an overall consolidation time.

POLYMERIC COMPOSITES ARE being increasingly used in today's aircraft, automobile, sporting goods and other related industries. A number of processing routes have been developed to manufacture these composites. These include fiber co-mingling, film stacking, melt impregnation, slurry processing, solution processing and dry powder processing [1,2,3,4,5,6,7]. Of these, dry powder impregnation has been identified as the technique with the most potential for success as a viable alternative for the manufacture of *thermoplastic polymer* prepreg tape [8]. A dry powder composite prepregging process using continuous fiber and polymer powder has been developed at Michigan State University. This process is capable of making prepreg tapes from varying sizes of particles and utilizing either thermoplastic or thermoset powders.

The final product of almost all thermoplastic composite processes is a prepreg. This is a material containing the reinforcing fibers and polymer in an unconsolidated form and can be in the form of tape, laminates or fabric. They have been transformed into thermoplastic composite parts by a variety of methods categorized under the broad heading of consolidation. These methods have been empirically successful and mathematical models for these processes have been developed by a number of authors [9,10,11,12,13]. However, the process of consolidation of *powder impregnated tapes* is still not well understood. A consolidation model for these prepregs will increase the acceptance of these materials in industry. The present study aims to develop a consolidation model based on a fundamental understanding of material and process parameters.

Model Development

The modeling of the consolidation step is complex due to the number of simultaneous steps that occur during the process. A typical consolidation process would involve heating of the layup to processing temperature, application of a consolidation pressure for a predetermined time, holding at a predetermined crystallization temperature (if required) and finally cooling down to room temperature under pressure.

The most widely used approach in modeling the process is to write sub-models for each individual step and then combine them together in an overall model. Some key sub-models that have to be developed are for heat transfer, fiber deformation, flow, and crystallization. The objective

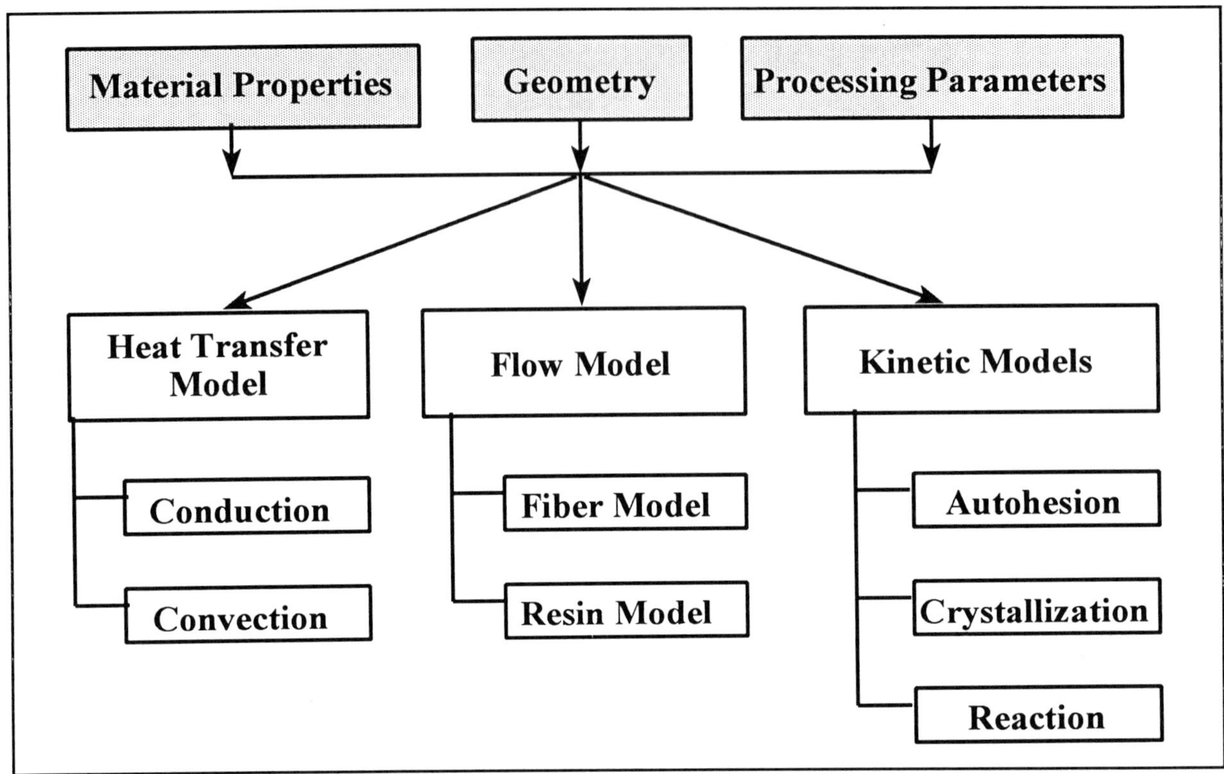

Figure 1 Layout of Consolidation Model

in this study was to develop these sub-models independently for powder impregnated materials and conduct controlled experiments to test each of them. These could then be combined in an overall model that would predict a material specific consolidation envelope. Figure (1) shows the layout of the overall model. The sub models for the heat and flow aspects of the consolidation are discussed below. Additional sub-models for autohesion, crystallization and reaction are in the developmental stage. The input to the overall model are the material parameters (density, thermal diffusivity, viscosity, etc.), geometry (particle size, fiber diameter, part size, etc.) and process parameters (pressure, temperature and time). Depending on the input process parameters, the model would generate the remaining parameters.

Heat transfer sub-model. The composite layup under consolidation conditions can be approximated as unsteady state heat conduction through an infinite flat slab as shown in Figure (2). Heat transfer to the slab occurs via convection to a conductive cowl plate that helps in the application of pressure. The cowl plate could also be replaced by a mold surface without changing the boundary conditions on the model. Heat transfer then occurs by conduction that causes the slab to heat up and eventually reach an uniform steady temperature, T_∞. The differential equation that governs the heat transfer slab is given by [14]

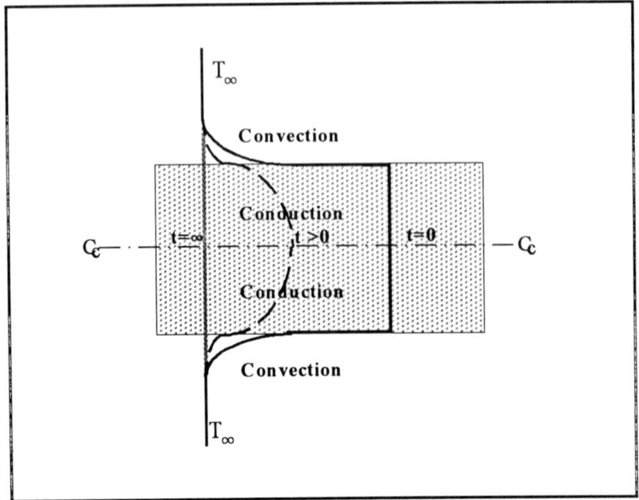

Figure 2 Heat transfer through a composite slab

$$\frac{\partial T}{\partial t} = \alpha \frac{\partial^2 T}{\partial x^2} \qquad (1)$$

where T is the temperature, t is the time, α is the thermal diffusivity of the slab and x is the position within the slab of thickness $2L$. The applicable boundary condition are

$$at \quad x = \pm L, \quad -k \frac{dT}{dx}\bigg|_{x=\pm L} = h(T-T_\infty), \quad t>0$$
$$at \quad x = 0, \quad \frac{dT}{dx}\bigg|_{x=0} = 0, \quad t>0 \tag{2}$$

where h is the heat transfer coefficient that is dependent on the heating equipment. The initial condition necessary to solve the above equation analytically is

$$at \quad t=0, \quad T = T_i, \quad -L \leq x \leq +L \tag{3}$$

The solution of equation (1) is

$$\frac{T_\infty - T}{T_\infty - T_i} = \sum_{n=1}^{\infty} B_n \exp\left(-\frac{\lambda_n^2 \alpha t}{h^2}\right) \cos\left(\frac{\lambda_n x}{L}\right) \tag{4}$$

where

B_n are the eigen coefficients and have the form $2\sin \lambda_n / (\lambda_n + \sin 2\lambda_n/2)$

λ_n are the eigen values which are the solutions of $\lambda_n \tan \lambda_n$ = Biot number

The only unknown parameter in equation (4) is the heat transfer coefficient which can be experimentally determined as discussed later.

The thermal properties required for the model can be approximated using a rule of mixtures type relation as shown in equations (5) and (6)

$$K_c = \frac{1}{\dfrac{v_f}{K_f} + \dfrac{(1-v_f)}{K_m}} \tag{5}$$

$$C_{p_c} = w_f C_{p_f} + (1-w_f) C_{p_m} \tag{6}$$

where the subscripts c, f, and m refer to the properties of the composite, fiber and matrix respectively, K is the thermal conductivity and C_p is the specific heat capacity.

Flow sub-model. The distribution of polymer within a powder prepreg composite is unique. The particles are essentially uniformly coated on every fiber in the layup. Therefore the approach to modeling needs to be slightly different from those that have been used for traditional materials. Most composite processing models use a modified form of the Darcy equation for flow through porous media. The parameter of most concern in these models is the permeability of the layup. Since the flow

direction for polymer in most traditional prepregs is normal to the fiber, the permeability is estimated in this direction. This transverse permeability is extremely small when compared with the axial permeability. Gutowski, et al. [15] report the axial permeability on the order of 10^{-12} m^2 for aligned graphite fibers and 10^{-15} m^2 for the transverse direction. Moreover, the permeability is a function of the

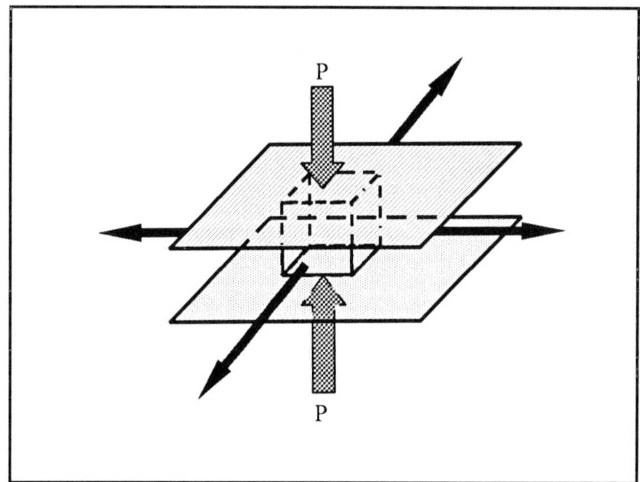

Figure 3 Schematic of the Hele-Shaw flow model

applied pressure and decreases rapidly with applied pressure.

Powder prepregs have the distinct advantage that flow occurs predominantly parallel to the fiber axis. This results in shorter flow times and consequently small consolidation times. In order to model the flow behavior, a Hele-Shaw type flow model [16] for the compressive flow of a single particle between two parallel plates was considered as shown in Figure (3). The spherical particle was approximated as an equivalent cube using a mathematical transformation. A rectangular coordinate geometry was used for the purposes of modeling and a creeping flow assumption combined with a fluid that was incompressible and of constant viscosity was used to simplify the problem. The equations of continuity and motion become

$$2\frac{\partial v_x}{\partial x} + \frac{\partial v_z}{\partial z} = 0 \tag{7}$$

$$\frac{\partial P}{\partial x} = \mu \left[\frac{\partial^2 v_x}{\partial x^2} + \frac{\partial^2 v_x}{\partial z^2}\right] \tag{8}$$

$$\frac{\partial P}{\partial y} = \frac{\partial P}{\partial x} \qquad (9)$$

$$\frac{\partial P}{\partial z} = \mu \left[\frac{\partial^2 v_z}{\partial z^2}\right] \qquad (10)$$

Equation (9) implies that the in-plane flow is assumed equal in both x and y directions.

The boundary conditions applicable to this process are

$$
\begin{array}{llll}
at & x = 0, & v_x = 0; & \dfrac{\partial v_y}{\partial x} = 0; \quad \dfrac{\partial v_z}{\partial x} = 0 \\[2ex]
at & z = 0, & v_x = 0; & v_y = 0; \quad v_z = 0 \\[2ex]
at & y = 0, & v_y = 0; & \dfrac{\partial v_x}{\partial y} = 0; \quad \dfrac{\partial v_z}{\partial y} = 0 \\[2ex]
at & z = h, & v_x = 0; & v_y = 0; \quad v_z = -v_w
\end{array}
\qquad (11)
$$

where v_w is the velocity of one of the surfaces as it moves towards the other stationary surface. These equations were simultaneously solved to yield

$$P - P_0 = \frac{3\mu v_w}{h}\left[\frac{(a^2-x^2)}{h^2} + \frac{(b^2-y^2)}{h^2} - \frac{2z}{h}\left(1-\frac{z}{h}\right)\right] \qquad (12)$$

This equation can be integrated over the area of the cube to give the total force, F. The velocity of the plate is a difficult parameter to measure and it can be eliminated by

equating it to $-\dfrac{dh}{dt}$ and integrating the resulting equation

over the time domain of the consolidation. This results in

$$F = \frac{8\mu a^4}{t}\left[\frac{1}{h^2} - \frac{1}{h_0^2}\right] \qquad (13)$$

where a is the side of the equivalent cube, μ is the viscosity of the polymer at the processing temperature, h_0 is the initial spacing between the plates, and h is the spacing at time t.

In order to be able to use the above model the initial and final spacings need to be determined. The initial spacing is assumed to be the particle size since the fibers are spaced apart by powder particles. The criterion for complete consolidation determines the final spacing. The criterion assumes a hexadecimal packed structure at the final

volume fraction of the composite and can be mathematically written as

$$h = \sqrt{\frac{\pi d_f^2}{4 V_f \sin(60)} - d_f} \qquad (14)$$

The theoretical limit for the volume fraction under these conditions is 90.69% fiber when h=0.

During the consolidation of the tape, the applied load distributes itself between the fiber and the matrix. This distribution is predicted by Gutowski, et al. [15] and an equation for the fiber side pressure as a function of current volume fraction has been developed by the authors and is given by

$$\sigma = \frac{3\pi E}{\beta^4}\frac{\left(\sqrt{\dfrac{V_f}{V_0}}-1\right)}{\left(\sqrt{\dfrac{V_a}{V_f}}-1\right)^4} \qquad (15)$$

where σ is the fraction of the total pressure applied to the system that is carried by the fibers, V_0 is the original fiber volume fraction, V_a is the available fiber volume fraction, E is the bending stiffness of the fiber and β is the typical "(span length)/(span height)" for the fiber beam network. At $V_f = V_0$, $\sigma=0$ (load completely carried by resin) and at $V_f \rightarrow V_a$, $\sigma \rightarrow \infty$ (load completely carried by fiber network). The matrix pressure, P_m calculated using equation (11) is added to this fiber pressure to give an overall effective consolidation pressure as shown below

$$P_t = P_m + \sigma \qquad (16)$$

Equation (16) relates the time and applied pressure. This time can be added to the heat transfer and autohesion times to give a total overall time.

Materials

The materials used in this investigation were unsized Hercules AS4 carbon fiber and Orgasol polyamide-12 supplied by Atochem. The carbon fiber were in 3K, 6K and 12K tows but this difference was not thought to influence the modeling of the consolidation since the properties and size remained the same for each tow. The polyamide powder was in different mean sizes of 5, 10, and 20 microns. These differences were incorporated in the model since fiber spacing is governed by the size of the particles.

The heat transfer coefficient determination required the use of a high conductivity metal for reasons explained later. A cast aluminum slab of known geometry was used for this purpose.

Experimental Procedure

The equations developed above were incorporated into a FORTRAN program. In its current state, the sub-models exist independently. The overall model would however, use these sub-models as subroutines to a main program.

Heat Transfer Sub-model. The heat transfer coefficient was the most important variable of interest. This was determined by conducting an experiment with an aluminum slab of known geometry and with boundary conditions similar to those discussed in the model. The slab was placed in a preheated Carver Press and the temperature at the center of the slab was monitored. The data from this experiment was fitted to equation (4) and the value of h was determined.

The next step in the experimentation was to predict heat transfer profiles through a slab of composite. Slabs of the composite were cut into known dimensions. A fast response thermocouple was sandwiched between two identical slabs. The entire setup was then placed in the preheated Carver Press and the temperature at the center of the slab was monitored. The experiment was repeated for two different volume fractions and thicknesses of composite. The results were compared with the predictions of the model.

Flow Sub-model. Prepreg tapes were made from AS4-3K carbon fiber and different particle size polyamide powders. The tapes from a single prepreg run were cut and laid up in a two-piece, matched-die, chrome steel mold and then consolidated in a hydraulic press with heating and cooling capabilities. A typical consolidation cycle used for these experiments consisted of a 30 minute heating cycle to above the melt temperature of the polymer and a cooling cycle of 15 minutes. The pressure was maintained at a predetermined level throughout the entire consolidation cycle. The composite parts thus produced were cut, mounted in acrylic holders, polished to a fine grit and then characterized for voids in an optical imaging system (ONVFA) [17]. An operating line was drawn on a plot of volume fraction and consolidation pressure that separated void free parts from parts with voids. This was repeated for different particle sizes. The experimentally predicted operating line was compared with the model predictions.

Results and Conclusions

Heat Transfer Sub-model. Figure (4) shows the heat transfer profile for the slab of aluminum heated from both sides. The value of h for the Carver Press as determined by this experiment was 200 W/m^2 K. This is an accurate estimate of the heat transfer coefficient for

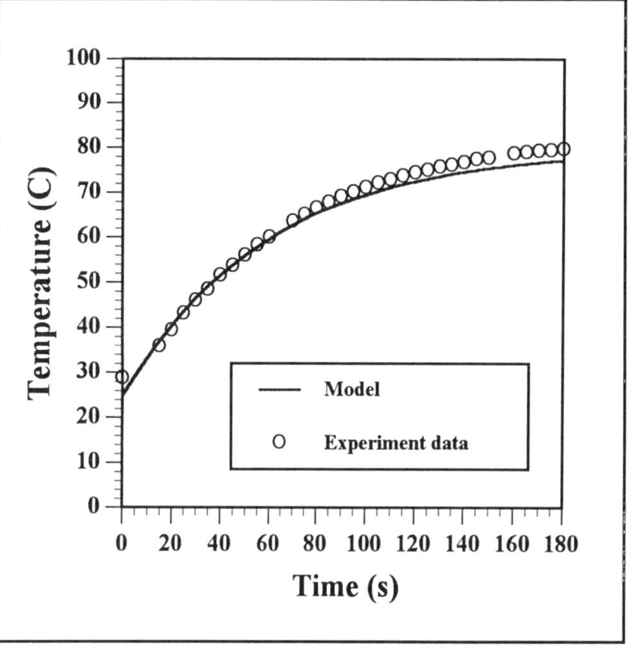

Figure 4 Determination of heat transfer coeffecient

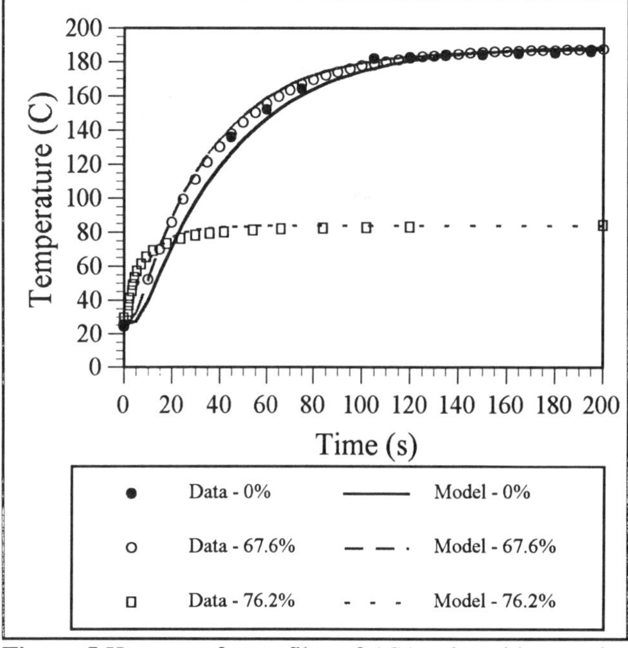

Figure 5 Heat transfer profiles of AS4-polyamide samples

materials that have thermal properties poorer than aluminum. Therefore, for all practical cases this value can be assumed to be a constant. A similar experiment can be performed for any other heating source such as an autoclave in order to determine the heat transfer coefficient. Also, a heat transfer coefficient for cooling can be similarly determined. Experiments are underway to determine these constants.

Figure (5) shows the data and model fit for two different composite samples. The data for pure polyamide is also shown which corresponds to 0% fiber volume fraction. The case for 100% fiber was not performed. In all cases, the heat transfer coefficient used in the calculations was 200 W/m^2 K. This shows that the heat transfer model performs well especially in predicting the time for heat transfer. Some differences in the profile still occur and this could be due to uneven sample heating in the press which results in a two dimensional heat transfer situation.

Flow Sub-model. Figure (6) shows the experimental operating lines and the model predictions for three particle sizes. The experimental results show no signi-

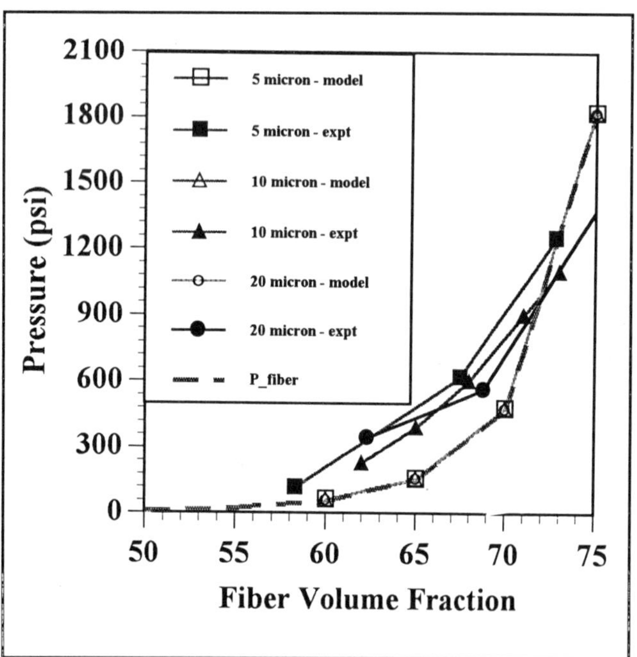

Figure 6 Results of the flow model

ficant difference for the three particle sizes. The model predictions correlate closely with the experimental data. In addition, the experimental data follows the trend for the fiber side pressure. This suggests that the pressure required for polymer flow is much smaller than that for fiber deformation. In terms of consolidation time, this implies a short consolidation time since deformation of fiber is almost instantaneous. Additional experimentation is underway to develop better correlations for the model.

Acknowledgements

The authors would like to acknowledge the financial support of the NSF Center for Low-Cost, High-Speed Polymer Composite Processing, Michigan State University during the course of this study.

References

[1]   Clemans, S. R., Western, E. D., and Handermann, A. C., Materials Engineering, Vol. 105, 27-30 (1988).

[2]   Carlsson, L. A., "Thermoplastic Composite Materials," Elsevier Science Publishers (1991).

[3]   Lind, D. J., and Coffey, V. J., U.K.Patent 1,485,586.

[4]   Muzzy, J. D., Varughese, B., Thammongkol, V., and Tincher, W., SAMPE Journal, 25 (5), 15-21 (1989).

[5]   Price, R. V., U.S.Patent 3,742,106.

[6]   O'Connor, J. E., U.S.Patent 4,680,224.

[7]   Turton, N., and McAinsh, J.,U.S.Patent 3,785,916.

[8]   Iyer, S. R., L. T. Drzal, and Jayaraman, K., U.S. Patent 5,102,690.

[9]   Ye, L., Klinkmüller, V. and Friedrich, K., Journal of Thermoplastic Composite Materials, Vol. 5, 33-48 (1992).

[10] Lee, W. I., and Springer, G. S., Journal of Composite Materials, Vol. 21, 1017-1055 (1987).

[11] Seo, J. W., and Lee, W. I., Journal of Composite Materials, Vol. 25, 1127-1142 (1987).

[12] Loos, A. C., Howes, J. C., and Dara, P. H., Adhes. Sci. Rev., Proc. 5th Annual Program Rev./Workshop, 263-281 (1987).

[13] Dave, R., Kardos, J. L., and Duduković, M. P., Polymer Composites, Vol. 8, No. 1, 29-38 (1987).

[14] Kreith, F., "Principles of Heat Transfer", 4th edition, Harper & Row (1986).

[15] Gutowski, T. G., Cai, Z., Bauer, S., Boucher, D., Kingery, J., and Wineman, S., Journal of Composite Materials, Vol. 21, 651- 669, (1987).

[16] Churchill, S. W., "Viscous Flows: The Practical Use of Theory", Butterworth Publishers, (1988).

[17] Waterbury, M. C., Drzal, L. T., J. Reinforced Plastics and Composites, 8, 627-636, (1989).

Computer Modeling of Liquid Composite Molding
for 3-Dimensional Complex Shaped Structures

A.W. Chan, R.J. Morgan
Michigan State University
Midland, Michigan

Abstract

Polymer composite components are increasingly being used in durable goods and automotive applications. Design tools, such as computer models and simulators, can help speed up preliminary design and reduce the time frame from design to final production. This work focuses on the development of a model and solution algorithm for liquid composite molding simulation. The model was based on sound engineering principles and the solution algorithm was formulated for computational stability and efficiency. A computer code was developed in this work which permits liquid composite molding simulations to be conveniently performed on workstations and high-ended microcomputers. The code is applicable to the total molding cycle (mold fill and in-situ mold cure) in liquid molding of thin-shelled composite parts and structures with complex shapes and features.

LIQUID MOLDING is an important manufacturing technique for polymer composite parts and structures. The liquid molding process essentially consists of injecting a thermoset resin into a mold cavity with the fiber preform or reinforcement in place. The mold is usually heated and the filling process is accompanied by resin reaction and heat transfer among the components. The resin is cured in-situ until the part reaches the desired stiffness. After demolding, the part can be heated in a conventional oven for further cure.

To facilitate analysis of the liquid molding process, a number of theoretical models and numerical algorithms have been developed [1-8]. The use of control volume finite element approach in modeling isothermal mold filling has led to a simpler scheme for tracking resin flow front movement [1-3]. Algorithms for non-isothermal mold filling were also reported [4-7]. The goal of these efforts in modeling is to develop capabilities for predictive studies of the behavior of selected materials under different molding conditions, for a variety of mold geometries. To attain this goal, the model has to mathematically describe the physical events as closely as possible, while the algorithm used to simulate the model has to be efficient and stable. The scheme proposed in this work was developed to satisfy both requirements in simulating non-isothermal fill and in-situ cure in molds with complex geometries. An example is included to illustrate the application of the proposed scheme in simulating liquid composite molding.

Theory

Flow front tracking in this work is based on the control volume finite element approach [1-4]. This approach simplifies the task of flow front tracking in complex shaped molds. The formulation in this work will be limited to composite parts that can be described as thin shelled structures with 3-dimensional spatial features. The part can possess geometric features such as cutouts and bends (or curved surfaces), and the shell thickness can vary. A coordinate transformation by rotation is used to transform the 3-dimensional surface into an "equivalent 2-dimensional surface". The global normal in this transformation is rotated with respect to the global axes to have it aligned with the normal to the finite element [4]. The transport equations are written for the local coordinates (x,y,z) obtained after the rotation transformation. These equations are written in terms of volume averaged variables [8,9].

Flow in the mold cavity is modeled using Darcy's

law:

$$\langle v \rangle = -\frac{1}{\mu}[S]\cdot\nabla P \qquad (1)$$

where v is the superficial velocity vector, μ is the resin viscosity, $[S]$ is the in-plane preform permeability tensor [10] and ∇P is the intrinsic fluid pressure gradient. Quantities enclosed between the bracket $< >$ are volume averaged values. For incompressible flow, the continuity equation is written as:

$$\nabla\cdot\left\{\frac{1}{\mu}[S]\cdot\nabla P\right\} = 0 \qquad (2)$$

The mass balance equation for resin conversion is written as:

$$\epsilon\frac{\partial C}{\partial t} + \langle v \rangle\cdot\nabla C = \epsilon S_c \qquad (3)$$

where C is the intrinsic resin conversion, ϵ is the preform porosity, and S_c is a source term due to resin reaction. The form for S_c depends on the reaction kinetics of the resin.

The energy equation, based on the assumption of local thermal equilibrium, is written as:

$$\langle \rho C_p \rangle\frac{\partial\langle T\rangle}{\partial t} + (\rho C_p)_f\langle v\rangle\cdot\nabla\langle T\rangle =$$
$$\nabla\cdot[k]\cdot\nabla\langle T\rangle + \mu\langle v\rangle\cdot[S^{-1}]\cdot\langle v\rangle \qquad (4)$$
$$+ \epsilon\rho_f(-\Delta H)S_c$$

where ρ is density, C_p is heat capacity, $[k]$ is the lumped thermal diffusivity tensor (includes the effective molecular thermal conduction and thermal dispersion), $[S^{-1}]$ is the inverse permeability tensor, and ΔH is the resin heat of reaction. Subscript f refers to fluid (or resin) properties. The terms in eq. 4 represent heat effects due to accumulation, convection, "effective" conduction, viscous dissipation and contribution from resin reaction, respectively.

Upon completion of mold filling, further resin cure takes place in-situ in the mold. The appropriate transport equations for describing in-situ mold cure are eq. 3 with the convection term dropped, and eq. 4 with both convection and viscous dissipation terms dropped.

Appropriate boundary conditions for the transport equations, details on the control volume finite element approach, and material parameters appropriate for the model are summarized in previous papers [1-8,10,11].

Subsequent discussions will focus on the numerical procedure used to solve the model.

Solution Procedure

The mold cavity gap is discretized into a number of layers of finite element meshes. The number of elements and nodal connectivities in each mesh layer are the same. For heated molds, only half of the mold thickness is used in the simulation. In this case, the topmost mesh layer corresponds to the surface in contact with the mold, while the bottom mesh layer corresponds to the mid-plane halfway between both mold surfaces. At each time step, the pressure and velocity fields are estimated using previous time values of temperature and conversion. This velocity field is used to compute new time values of temperature and conversion. The updated temperature and conversion is used to compute the current time pressure and velocity fields. Flow front movement corresponding to the current time interval, as well as the value of the subsequent time interval, are computed from the updated velocity field. The time integration is continued until mold filling is complete.

Hele-Shaw flow is assumed in the solution of the pressure and velocity fields. The pressure field, from eq. 2, is solved using the Galerkin finite element method. Gap averaged viscosities are used at each finite element. The velocity field is computed from eq. 1 and the pressure gradient field.

The energy equation is solved using the operator splitting technique [12] on eq. 4. The equation is split into two parts: one corresponding to the 1-dimensional gap or mold thickness direction, and another to the 2-dimensional in-plane surface corresponding to the finite element mesh at each layer. The finite difference method is used to solve the transient conduction over the gap dimension. This is followed by solution for the in-plane mesh at each layer. The least squares finite element method is used to solve the convection-conduction equation at each layer. Details on least squares finite element solution of convection dominated equations are given in references [5-6]. Solution of resin conversion from eq. 3 is done using the least squares finite element method. The solution is obtained over each finite element layer in succession.

Simulation Example

An example case study will demonstrate the use of the model and algorithm in simulating liquid composite molding. An idealized part shape is used in this study. The finite element mesh at the mid-plane of the part is shown in fig. 1. Several layers of the finite element mesh are placed along the thickness dimension. Materials and conditions simulated are typical for structural reaction injection molding (SRIM) with glassfiber preforms. Material parameters are given below:

	Resin	Fiber
ρ (kg/m^3)	1120.	2540.
Cp (J/kg·K)	1400.	840.
k (W/m·K)	0.13	0.1/0.01
-ΔH (kJ/kg)	230.	
ε		0.7

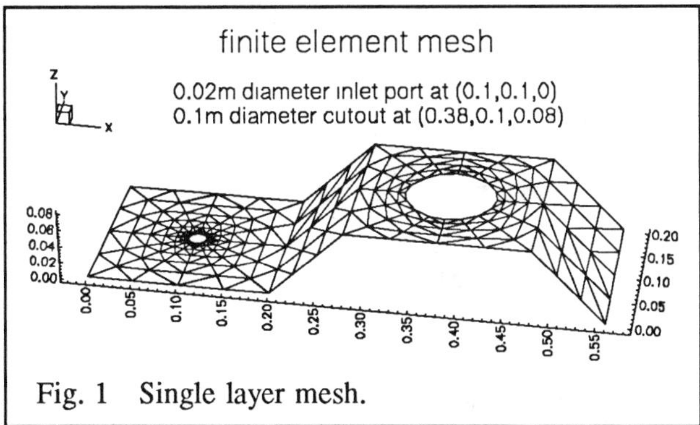

Fig. 1   Single layer mesh.

The two numbers for fiber thermal conductivity refer to the longitudinal and transverse thermal conductivities, respectively. Resin kinetics and viscosity are taken from the work of Reboredo and Rojas [11]:

$$S_C \ (s^{-1}) = 300 \exp\left[-\frac{3277}{\langle T \rangle}\right](1-C)^2 \qquad (5)$$

$$\mu \ (Pa \cdot s) = \mu_\infty \exp\left[\frac{E_\mu}{\langle T \rangle}\right] \exp\left[\frac{a}{C_g - C} - b\right] \qquad (6)$$

where $\mu_\infty = 2.56 \times 10^{-21}$, $E_\mu = 1.45 \times 10^4$, a = 0.695, C_g = 0.487 and b = 1.502. SI units are used in the simulations. The preform permeability tensor is given by:

$$[S] \ (m^2) = c \begin{bmatrix} 1 & 0 \\ 0 & 1 \end{bmatrix} \qquad (7)$$

where c = 1×10^{-8}. Molding Parameters are:

Injection rate (constant), q = 6.5×10^{-5} m^3/s
Resin inlet temperature, T_{in} = 310 K
Initial preform temperature, T_o = 350 K
Mold surface temperature (constant, mold heated on both sides), T_w = 370 K
Mold thickness, h = 3 mm

Simulation results are summarized in figs. 2 to 7. Flow front movement during mold filling is shown in fig. 2, which is a contour plot of the time at which each control volume is filled. Formation of a weld line after flow across the cutout can be seen from the plot. Fig. 3 shows the pressure field in the mold at end of fill. The contour plot reveals a high pressure gradient near the inlet region. This is typical in radial flow geometries. The presence of the cutout leads to some distortion in the pressure distribution across the flow direction.

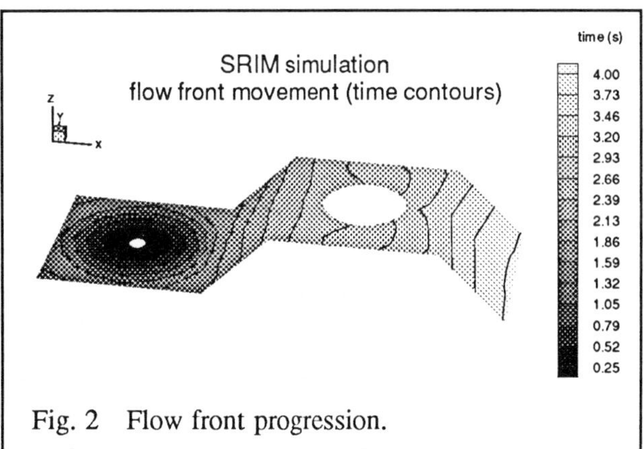

Fig. 2   Flow front progression.

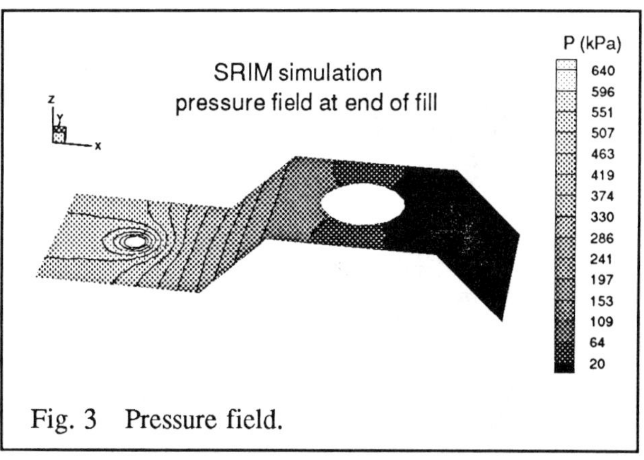

Fig. 3   Pressure field.

The temperature field at end of fill is shown in fig. 4 as contours at three mesh layers. The contour plot shows approximately the 3-dimensional temperature field. As expected, the layer near the mold surface exhibits a higher temperature field as a result of mold heating. The effect of convection (from resin entering the mold at a lower temperature) is evident in the contour plots. At the flow front region, the temperature field is strongly influenced by the initial preform temperature.

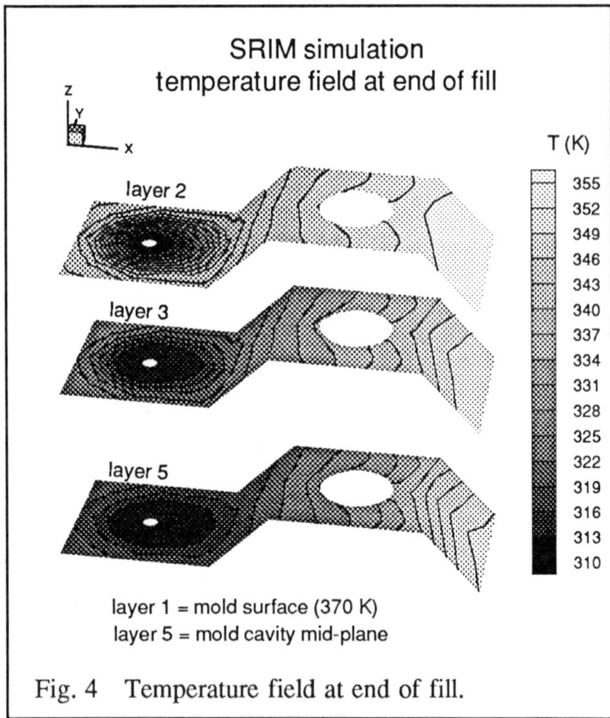

Fig. 4   Temperature field at end of fill.

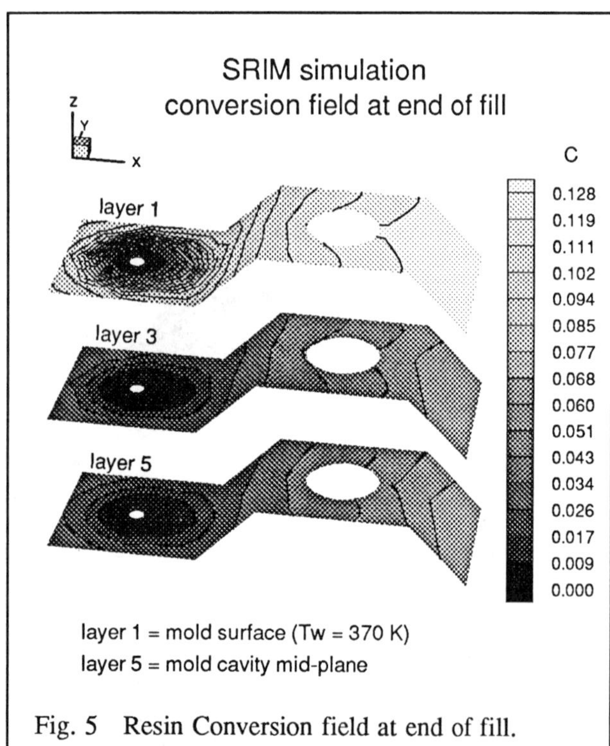

Fig. 5   Resin Conversion field at end of fill.

The resin conversion field, again at end of fill, is shown in fig. 5. The conversion is highest at the mold surface, where the temperature is also highest. Convective effects resulted to a lower conversion at the mold interior. At completion of fill, the conversion field is substantially below the gel point ($C_g = 0.487$).

The contours for both temperature and conversion fields in figs. 4 and 5 are relatively smooth and free from oscillations. This is due to the numerical stability of the solution algorithm. For the convection dominated problem, the use of an optimal weighing scheme in the least squares finite element formulation leads to excellent numerical stability.

Compared to the total molding time, the time to fill the mold is relatively short. In the example, the mold is filled in 4 s (fig. 2). At completion of fill, resin stops flowing and undergoes in-situ mold cure. The 3-dimensional temperature and conversion fields are obtained by solving eqs. 3 and 4 with all convective effects neglected.

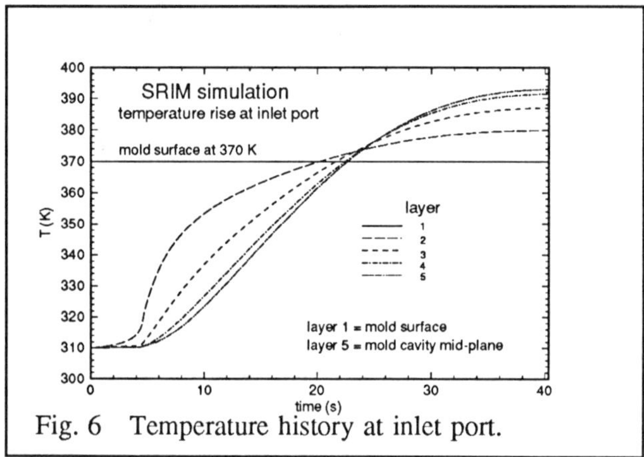

Fig. 6   Temperature history at inlet port.

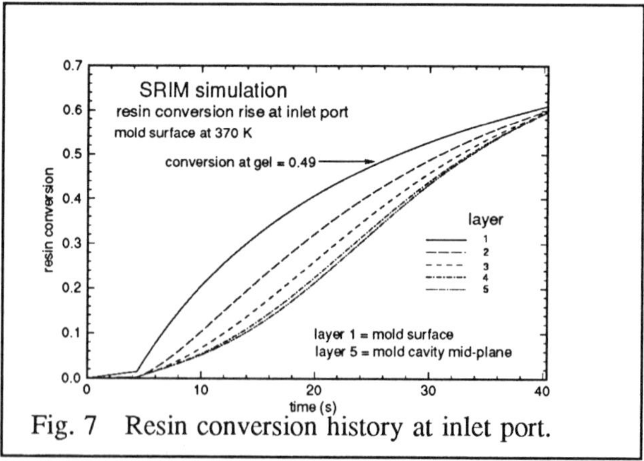

Fig. 7   Resin conversion history at inlet port.

In the example, resin enters the mold at a temperature lower than the mold and initial preform temperatures. It is of interest to simulate the rise in temperature and conversion at the inlet port during in-situ mold cure.

These are shown in figs. 6 and 7. Fig. 6 shows the temperature history at different locations along the thickness direction. Due to the reaction exotherm, temperature at the mold mid-plane eventually exceeds the mold surface temperature. Control of this overshoot is important in preventing resin degradation and non-uniform cure. The corresponding conversion history at the inlet port is shown in fig. 7. Gel occurs at the inlet port in about 30 s. At 40 s, conversion at the inlet port is about uniform. The temperature and conversion histories at locations other than the inlet port can be obtained from the simulated results in the same manner. In general, a detailed picture of the full molding cycle can be conveniently obtained from the simulation results.

The coarse mesh was used in this example to demonstrate the computational stability of the algorithm. A finer mesh should be used in performing liquid composite molding simulations. Both spatial and time discretization errors are made smaller with a fine mesh. A finer mesh, however, requires more computational time. An objective of this work is to bring the computational requirements for liquid molding simulations to the level of workstations and high ended microcomputers. The proposed algorithm meets this objective. This example, with 5 finite element layers and 368 elements per layer, consists of 177 time integration steps during mold filling and another 100 time steps for in-situ mold cure computations. The CPU time for this example was 7 minutes on a PC with a 486-33MHz microprocessor (or 40 s on a SGI workstation).

Conclusion

A model and algorithm have been developed for simulating liquid molding of polymer composite parts and structures. The model takes into account the coupled phenomena of flow, resin reaction, viscosity buildup, and heat generation and transfer among fiber, resin and mold components. The algorithm used to solve the model was formulated for both computational stability and efficiency. The proposed scheme can be used to simulate liquid molding of thin shelled composite components with relatively complex shapes and geometries.

Acknowledgement

The support of this work by the National Science Foundation, as part of the NSF High Speed Composites Processing Center at Michigan State University, is acknowledged. The Authors would also like to thank Dr. Selim Yalvac of Dow Chemical Company for helpful insights and suggestions.

References

1. M.V. Bruschke and S.G. Advani, **Polym. Composites**, 11, 398 (1990).
2. W.B. Young, K. Han, L.H. Fong, L.J. Lee and M.J. Liou, **Polym. Composites**, 12, 391 (1991).
3. F.R. Phelan, **Proc. 7th Tech. Conf. of American Soc. Composites**, 90 (1992).
4. W.B. Young, **Ph.D. Dissertation**, The Ohio State University (1991).
5. A.W. Chan and S.T. Hwang, **Polym. Eng. Sci.**, 32, 310 (1992).
6. A.W. Chan and S.T. Hwang, **J. Mater. Proc. Manuf. Sci.**, 1, 105 (1992).
7. M.V. Bruschke and S.G. Advani, **SPE ANTEC Paper**, 834 (1993).
8. C.L. Tucker III and R.B. Dessenberger, "**Flow and Rheology in Polymer Composites Manufacturing (S. Advani, Ed.)**", Elsevier Science Publishers (1993).
9. J.C. Slattery, **AIChE J.**, 15, 866 (1969).
10. A.W. Chan, D.E. Larive and R.J. Morgan, **J. Composite Mater.**, 27, 996 (1993).
11. M.M. Reboredo and A.J. Rojas, **Polym. Eng. Sci.**, 28, 485 (1988).
12. W.H. Press, B.P. Flannery, S.A. Teukolsky and W.T. Vetterling, **Numerical Recipes**, Cambridge University Press, U.K., Chap. 17 (1986).

Tool Heat Transfer Analysis in Liquid Composite Molding

T.J. Wang, J. Guan, L.J. Lee
The Ohio State University
Columbus, Ohio

INTRODUCTION

Resin transfer molding (RTM) is a process sensitive to heat transfer. During preform loading, resin injection and the early part of curing, the mold halves provide heat to initiate the resin reaction. On the other hand, during and after the reaction exotherm the mold halves have to remove heat generated from the reaction. In order to properly control the mold filling and curing processes, an appropriate mold design and thermal management (such as the arrangement of heating/cooling lines) is necessary.

Low molding pressure is a major advantage of the RTM process since it allows for the use of low cost, non-metal materials as molds. Non-metal mold materials, however, have low thermal diffusivities, which is difficult for temperature control. Even when metal molds are used, heat transfer may still be a problem because many RTM parts have a foam core in the design. One side of the fiber preform (part) may contact the metal mold with external heating/cooling capability, while the other side of the preform (part) may contact the foam core which is essential a thermal insulator. The largely different thermal properties between the two sides of the preform (part) may affect the heat transfer inside the mold cavity and, consequently, the product quality. A comprehensive simulation of the mold filling and curing processes in RTM should include modeling of heat transfer in the mold halves and foam core.

In this study, the heat transfer in the mold halves and that in the mold cavity are treated separately with an interface condition to link them together. At the interface the temperature and the heat flux are assumed the same for both domains. Four methods to treat the heat transfer in mold halves are discussed.

The simplest approach is to neglect the tool temperature distribution by assigning a known heat transfer condition at the interface between the part and the mold halves. This can be a specified temperature, a specified heat flux, or a specified heat transfer coefficient depending on the thermal properties and geometry of the mold halves. For molds made of highly conductive materials with a large thermal mass, heat can be transferred through the mold quickly, consequently, the mold temperature may remain largely unchanged. Assigning a constant temperature at the mold interface would be appropriate.

The second method considers the temperature change inside the mold halves as an initial spatial temperature distribution (or a cycle averaged temperature) superimposed with a local time variation. The variation is only a function of time in the thickness direction. The boundary element method is used to solve the initial steady state spatial temperature distribution and the finite difference method is used to solve the one dimensional transient temperature change. This simplifies a three dimensional unsteady heat transfer problem to a one dimensional case. For processes with short cycle time, such as injection molding, the time dependent temperature change is often limited to the area next to the interface. Such an approach can result in satisfactory prediction [1].

The third method uses a dual reciprocity boundary element method to simulate the time dependent temperature and heat flux at the surface of mold halves. As a result, a three dimensional (volume) unsteady heat transfer problem becomes a two dimensional (surface) one. In principle, this method can handle processes with long cycle time and large temperature variations in the mold halves. However, when the temperature change inside the mold halves varies significantly from location to location, the number of internal nodes needed to well represent this temperature change becomes very large, which limits the usage of this method.

The last method is to use a control volume finite element method to solve the transient three dimensional heat transfer in the mold halves. It has least limitation, however, needs large computing time and computer space.

FORMULATION OF HEAT TRANSFER IN MOLD HALVES

Heat transfer in the mold halves can be considered as a transient heat conduction process without heat generation. The interfaces between the part and the mold halves, and the part and the cooling/heating lines are treated as boundary.

The governing equation for heat transfer with constant thermal properties can be written as

$$\frac{\partial T_m}{\partial t} = \frac{k_m}{\rho_m c_{pm}} \nabla^2 T_m \qquad (1)$$

where subscript m stands for the mold halves, T is temperature and k, ρ and c_p are thermal conductivity, density and thermal capacity, respectively.

During a production cycle, the temperature of the mold halves keep changing. Depending on the thermal properties of the mold material, the geometry of the part and mold halves, and the processing conditions, the heat transfer in the mold halves can be quite different. Two dimensionless parameters, Biot number and Fourier number, can be used to characterize tool heat transfer.

Biot number is defined as

$$Bi = \frac{hL}{k}$$

where h is the heat transfer coefficient between the mold halves and the cavity, L is the characteristic length of the mold halves and k is the thermal conductivity of the mold material. Biot number can be considered as the ratio of the heat convection strength at the interface to the heat conduction strength in the mold halves. When Biot number is much smaller than 0.1, either there is only a small amount of heat transferred from the cavity to the mold halves (small h) or the heat conduction in the mold halves is very efficient (large k). Any heat transferred from the cavity can be distributed to the entire mold half easily and the temperature in the mold halves can be assumed uniform. The thermal conductivity of RTM resins is about 0.25 W/mK. The heat transfer coefficient between the cavity and the mold is around 25 to 250 W/m^{2}K. The thermal conductivities for aluminum, steel and epoxy are about 200, 50 and 0.25 W/mK, respectively. In this range, the characteristic mold thickness that the mold temperature can be considered uniform (i.e. Bi $\leq$ 0.1) is 0.08 ~ 0.8, 0.02 ~ 0.2 and 1.0×10^{-4} ~ 1.0×10^{-3} m for aluminum, steel and epoxy, respectively. This indicates that for metal molds the temperature distribution in the mold halves is quite uniform, while when epoxy mold is used the spatial temperature variation can be substantial.

The other dimensionless parameter, Fourier number, is defined as

$$Fo = \frac{\alpha t}{L^2}$$

where α is the thermal diffusivity of the mold material and t is time. Fourier number can be explained as the ratio between the rate of heat conduction and the rate of heat absorption in the mold halves. A large Fourier number means more heat is transferred by conduction through the mold halves instead of stored inside the mold halves, consequently, the temperature variation originated from the boundary can propagate much farther from the boundary. For a chosen mold geometry and mold material, the cycle time effect can be reflected by this number. For thermoplastic injection molding, the cycle time is in seconds (i.e. small Fo), while for RTM, the cycle time is in minutes or hours (i.e. large Fo). Therefore, for the injection molding process, the temperature variation at the boundary will not propagate to the entire mold half and we can analyze the temperature change near the interface without creating too much error, i.e. the second method is appropriate. This is generally not true for the RTM process.

Method 1:

When Biot number is small, such as metals with high thermal conductivity being used as mold materials, any temperature variation at the interface can be dissipated quickly and to the entire mold halves. Consequently, the temperature change of the entire mold halves (including the interface) would be very small. We can simply assign the interface temperature as the initial mold temperature. This is especially true for processes in which the incoming resin temperature is not too much different from the initial mold temperature and the reaction exotherm is small. As a result, the tool heat transfer effect can be treated as a boundary condition at the cavity interface, i.e.

$$T_m = T_{interface} = constant \qquad (2)$$

For mold halves with simple geometry, a shape factor between the interface and each heating/cooling line can be calculated and an equivalent heat transfer coefficient can be obtained and used as a boundary condition at the interface [2]. However, for mold halves with complicated geometry or with complicated arrangement of heating/cooling lines, the calculation of shape factor becomes very tedious, if not impossible.

Method 2:

When Fourier number is small, the mold temperature can be considered as the summation of a cycle averaged temperature distribution and a time variation in the thickness direction near the mold interface [1]. The heat transfer formulation can be written as

$$T_m(t, x, y, z) = T_{m,t}(t, z) + T_{m,ss}(x, y, z) \qquad (3)$$

where

$$\nabla^2 T_{m,ss} = 0 \qquad (4)$$

and

$$\frac{\partial T_{m,t}}{\partial t} = \frac{k_m}{\rho_m c_{pm}} \frac{\partial^2 T_{m,t}}{\partial z^2} \qquad (5)$$

The subscripts t and ss indicate transient and steady state respectively. Here, z is chosen as the thickness direction coordinate.

As shown in Figure 1, the boundary element method is used to solve the cycle averaged temperature distribution of the mold halves and the one dimensional finite difference method is used to solve the time variation of temperature. The basic principles of boundary element method (BEM) to solve heat transfer problems can be found elsewhere [3-10]. BEM is one of the weighted residual methods which minimize the residual of the integral for a given differential equation with a particular weight function. The most commonly used weight function in the boundary element formulation is the fundamental Green's function which is the temperature distribution of a steady state heat conduction with a point source at location S*. For three dimensional problems, the fundamental Green's function and its derivative are

$$G^* = \frac{-1}{4\pi k_m r}, \qquad \frac{\partial G^*}{\partial n} = \nabla G^* \cdot \bar{n} = \frac{\bar{r} \cdot \bar{n}}{4\pi k_m r^3} \qquad (6)$$

where r is the distance between the source point and any location (field point) in this domain as shown in Figure 2. To form a boundary integral equation, the governing differential equation is first integrated with a weight function with respect to the entire volume domain. For steady state heat conduction, the resultant integral residual equation can be reduced to a surface (boundary) integral equation by applying the integrating by part method and Green's theory, and it can be written as

$$\omega T_m(S^*) - \int_{\Gamma_m} k_m \frac{\partial G^*}{\partial n} T_m(S)d\Gamma(S)$$

$$+ \int_{\Gamma_m} k_m \frac{\partial T_m(S)}{\partial n} G^* d\Gamma(S) = 0 \qquad (7)$$

Here S^* is the source point and S is the field point on the mold half boundary Γ_m or inside the mold half and ω is a fractional factor [4]. It can be considered as there is a very small sphere at location S^*. If the sphere is inside the mold half, $\omega = 1$. If S^* is on a smooth mold half boundary Γ_m, only half of the sphere will be occupied by the mold half and $\omega = 0.5$. If the mold half is a rectangular parallelepiped and S^* is at one of the corners, $\omega = 1/8$ since only 1/8 of the sphere is occupied by the mold half. We can assign all the source points on the surface such that all the unknowns are only located on the boundary. Equation 7 is then discretized

numerically. In this study, a linear (three node) triangular element is used to generate the mold surface mesh in order to match the element used in the mold filling and curing calculation in the mold cavity.

The procedure for the one dimensional finite difference method is trivial. Of the calculation domain, one end is at the interface and the other end is either a heating/cooling line or an edge of the mold half.

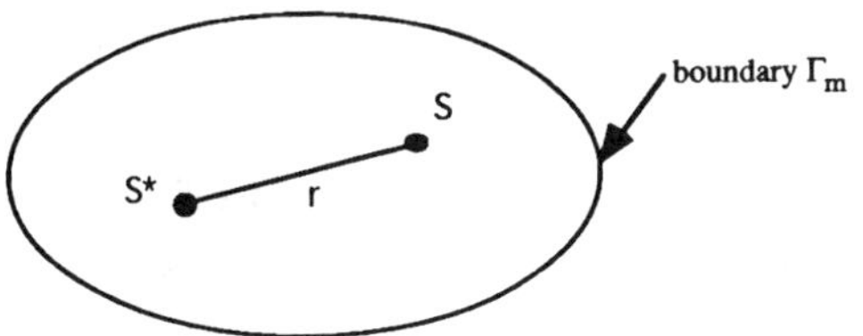

Figure 2. Point source at S* (source point) and any location S (field point).

Method 3:

For RTM, the tool heat transfer is generally not limited to the mold interface because of long cycle time and reaction exotherm. A dual reciprocity boundary element method which has been demonstrated in a two dimensional unsteady heat conduction problem [9] is introduced here to solve the heat transfer in the mold halves. Similar approach is followed and a three dimensional algorithm is developed.

First step is to approximate the time derivative term which is not in a divergence form by a set of linear independent functions $f^j(S)$ and their time dependent coefficients $\Lambda^j(t)$, i.e.

$$\frac{\partial T_m(S)}{\partial t} = \sum_{j=1}^{M} \Lambda^j(t) f^j(S) \qquad (8)$$

And $f^j(S)$ can be chosen in such a way that

$$f^j(S) = \nabla^2 \varphi^j \qquad (9)$$

This procedure allows the Green's theorem being also performed on the time derivative term. The multiplication of Green's function and Equation 1 can be integrated over the entire mold half. After the Green's theorem is carried out, the final equation can be written as

$$\omega T_b(S^*) - \int_{\Gamma_m} k_m \frac{\partial G^*}{\partial n} T_b(S)d\Gamma(S)$$

$$+ \int_{\Gamma_m} k_m \frac{\partial T_b(S)}{\partial n} G^* d\Gamma(S) = \sum_{j=1}^{M_b} \frac{\rho_m c_{pm}}{k_m} \Lambda^j(t)$$

$$\left\{ \omega \varphi^j(S^*) - \int_{\Gamma_m} k_m \frac{\partial G^*}{\partial n} \varphi^j(S)d\Gamma(S) \right.$$

$$\left. + \int_{\Gamma_m} k_m \frac{\partial \varphi^j(S)}{\partial n} G^* d\Gamma(S) \right\} \qquad (10)$$

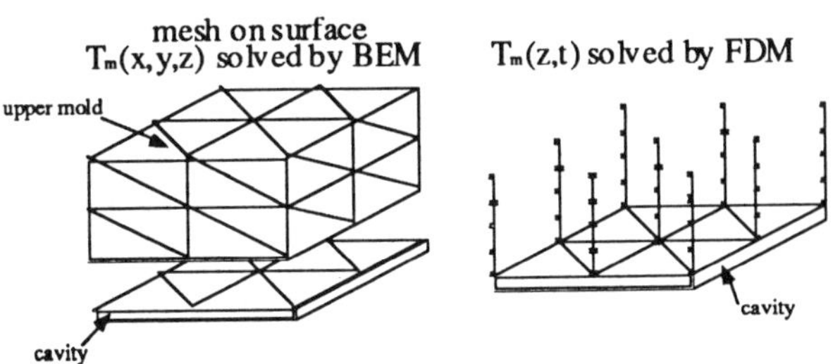

Figure 1. Schematic of boundary element method and finite difference method used on the upper mold half.

in which only surface integrals remain.

The functions $f_b^j(S)$ can be chosen as any linear independent functions [9]. In this study, the functions are chosen as

$$f_b^j = r_j, \tag{11}$$

and

$$\varphi^j = \frac{r_j^3}{12} \tag{12}$$

where $r_j = |S^j - S|$, S is any point on the boundary Γ_m and S^j is where the jth node is located. Having constructed the coordinate functions, Equation 10 is ready for numerical discretization.

The advantage of this method to other numerical techniques such as finite element and finite difference methods is that the mesh only needs to be generated on the mold surface. Consequently, the mesh preparation time is short. However, when the temperature change inside the mold halves varies significantly from location to location, a large number of internal nodes need to be used in order to assure the validity of the solution. This will make the formulation and computation much more complicated, and, consequently, make this method much less useful than other methods.

Method 4:

A three dimensional control volume finite element method is also used to solve the transient heat conduction of the mold halves. Wedge type elements are used as shown in Figure 3. For each control volume, $\Omega_{m,i}$, a volume integration is carried out. The integration of the diffusion term on the RHS of Equation 1 can be further reduced to a surface integration by applying the Green's theorem. As a result, only flux terms across the control volume surface $\Gamma_{m,i}$ need to be calculated. The final equation can be written as

$$\int_{\Omega_{m,i}} \frac{\partial T_m}{\partial t}(S)d\Omega(S) = \int_{\Gamma_{m,i}} \frac{k_m}{\rho_m c_{pm}} \frac{\partial T_m(S)}{\partial n} d\Gamma(S) \tag{13}$$

For all these four methods, the temperature distribution inside the mold cavity is calculated first. Then the heat flux at the cavity interface can be obtained from the multiplication of the temperature gradient at the interface and the conductivity of the part. This is used as a boundary condition

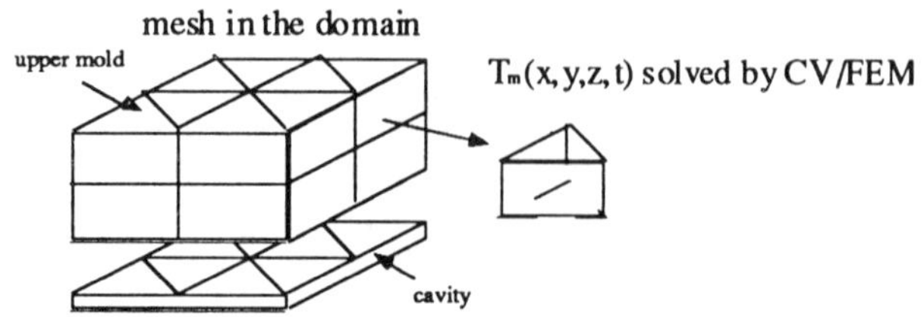

Figure 3. Schematic of control volume finite element method used on the upper mold half.

for the temperature calculation in the mold halves. After the temperature distribution is calculated inside the mold halves, the temperature at the interface is then used as the boundary condition for the calculation of cavity temperature at the next time step.

THE COMPARISON OF DIFFERENT METHODS

A simple case representing part of a mold half is used to illustrate the usage and limitation of different numerical methods. A parallelepiped (1m x 1m x 0.5m) is shown in Figure 4. Thermal conductivity is assumed 1 W/mK and thermal diffusivity is assumed 1 m^2/s. The bottom, front and back surfaces are assumed insulated (i.e. q=0). The top surface representing the interface between the cavity and the mold half is specified with a constant flux (i.e. q=-1). Both right and left surfaces have convective boundary conditions with a heat transfer coefficient h=1. Initially, the right side temperature $(T2_0)$ is at 60°C (e.g. initial temperature of a heating channel) and the left side temperature $(T1_0)$ is at 30°C (e.g. ambient temperature at an edge of the mold). When the computation starts, the right side temperature $(T2_0)$ is changed to 100°C (e.g. increasing the heating channel temperature for post-cure).

Since the initial temperature distribution has an x coordinate dependency, method 1 which specifies a constant temperature at the interface is not appropriate. Method 2 only considers the temperature change in the thickness direction, which also can not handle this case. The results from methods 3 and 4 are shown in Figure 5. T1 and T2 represent the temperatures on the left and right surfaces respectively. Method 3 tends to over-predict the temperature slightly.

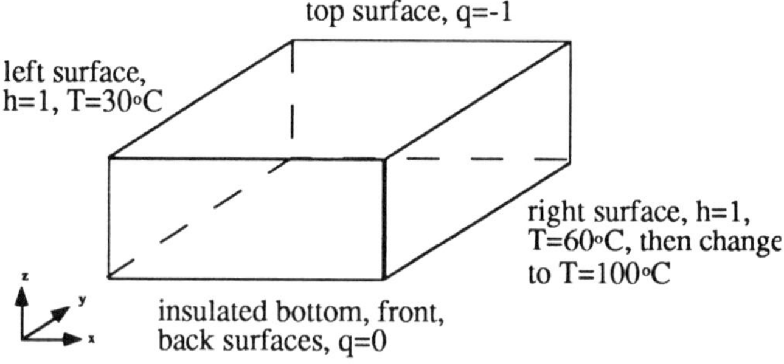

Figure 4. Schematic of a parallelepiped (1m x 1m x 0.5m) and its boundary condition.

FORMULATION OF MOLD FILLING AND HEAT TRANSFER IN MOLD CAVITY

For thin RTM parts whose thickness is much smaller than the planar dimensions, the velocity in the thickness direction is neglected and the planar velocity is averaged in the thickness direction during the mold filling process. Inside the cavity, resin is considered to be a incompressible fluid and a continuity equation can be written as

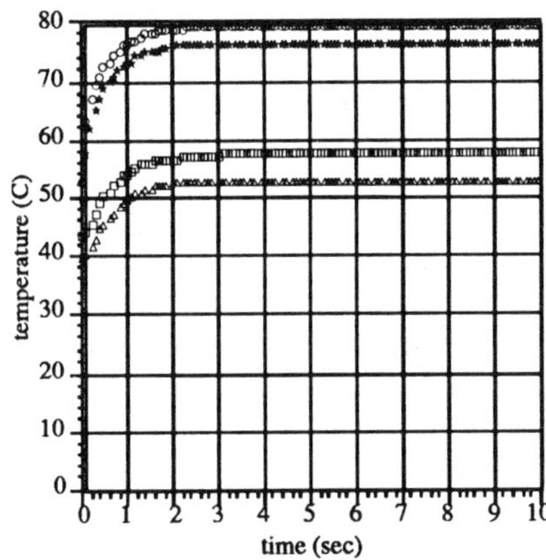

Figure 5. Temperature profiles at two side walls using methods 3 and 4.

$$\frac{\partial \overline{u}}{\partial x} + \frac{\partial \overline{v}}{\partial y} = 0 \tag{14}$$

The reinforcing fiber preform is considered to be rigid porous media and if the flow is considered to be slow in the mold filling process, Darcy's law

$$\begin{bmatrix} \overline{u} \\ \overline{v} \end{bmatrix} = -\begin{bmatrix} S_{xx} & S_{xy} \\ S_{yx} & S_{yy} \end{bmatrix} \begin{bmatrix} \dfrac{\partial P}{\partial x} \\ \dfrac{\partial P}{\partial y} \end{bmatrix} \tag{15}$$

is used to represent the flow without considering the inertia effect, in which

$$\begin{bmatrix} S_{xx} & S_{xy} \\ S_{yx} & S_{yy} \end{bmatrix} = \frac{1}{h_z} \int_{-\frac{h_z}{2}}^{\frac{h_z}{2}} \frac{1}{\mu(x,y,z)} \begin{bmatrix} K_{xx} & K_{xy} \\ K_{yx} & K_{yy} \end{bmatrix} dz \tag{16}$$

and $\overline{u}$ and $\overline{v}$ are the components of gapwise average velocity in the planar directions, h_z is the thickness of the thin cavity, μ is viscosity, [K] is the permeability matrix of the fiber mat defined in the local element coordinate system.

During the molding process, resin flow, heat transfer and resin conversion are coupled together because of reaction exotherm and viscosity changes with respect to temperature and conversion. Under the thin part assumption, the resin flows along a two dimensional surface. On the other hand, the heat transfer in the thickness direction (the interaction between the mold halves and the part) plays an important role, a three dimensional heat transfer problem needs to be considered. The heat transfer equation for the mold cavity can be written as

$$\rho c_p \frac{\partial T}{\partial t} + \rho_r c_{pr} (\overline{u} \frac{\partial T}{\partial x} + \overline{v} \frac{\partial T}{\partial y})$$
$$= k(\frac{\partial^2 T}{\partial x^2} + \frac{\partial^2 T}{\partial y^2} + \frac{\partial^2 T}{\partial z^2}) + \phi \Delta H \dot{m} \tag{17}$$

A dispersion effect due to the contribution of joint variations in temperature and velocity may be added to Equation 17 [11]. The conversion (the mass balance of chemical species) equation can be written as

$$\phi \frac{\partial \alpha}{\partial t} + (\overline{u} \frac{\partial \alpha}{\partial x} + \overline{v} \frac{\partial \alpha}{\partial y}) = \phi \dot{m} \tag{18}$$

where T is temperature, α is resin conversion, $\dot{m}$ is the mass generation rate, ΔH is the heat of reaction. And

$$\begin{aligned} c_p &= c_{pr} w_r + c_{pf} w_f \\ \rho &= (\rho_r \rho_f) / (\rho_f w_r + \rho_r w_f) \\ k &= (k_r k_f) / (k_f w_r + k_r w_f) \\ w_r &= \frac{\phi}{\rho_f} / (\frac{\phi}{\rho_f} + \frac{1-\phi}{\rho_r}) \\ w_f &= 1 - w_r \end{aligned} \tag{19}$$

where ρ is density, c_p is specific heat coefficient , k is thermal conductivity, and ϕ is the fiber mat porosity. The thermal parameters used in Equations 17 and 19 are represented by using subscript "f" for the fiber mat and "r" for the resin fluid. The assumption made here is that the resin density does not change despite chemical reaction. In the mold filling stage of RTM, the viscous dissipation is usually insignificant compared to other contributions such as heat transfer between the resin and the fiber reinforcement or the reaction exotherm.

For our interest, the mass generation will be assumed only via the polymerization process. The polymerization process can be either a step growth polymerization (usually for saturated monomers), a chain growth polymerization (typically for unsaturated monomers), or both. The mass diffusion is also negligible during the polymerization process because the chemical reaction rate is often much higher than the mass diffusion rate. In this study, an unsaturated polyester resin is used as the matrix resin. The kinetic model chosen was proposed by Stevenson [12]. It includes Initiation:

$$\frac{d[I]}{dt} = -k_d[I] \tag{20}$$

Radical Concentration:

$$\frac{d[R\cdot]}{dt} = 2fk_d[I] \tag{21}$$

Propagation:

$$\frac{d\alpha}{dt} = k_p(1-\alpha)[R\cdot] \tag{22}$$

where

$$k_d = A_d \exp\left\{\frac{-E_d}{RT}\right\} \tag{23}$$

$$k_p = k_{p0}(1 - \frac{\alpha}{\alpha_f})^m \tag{24}$$

$$k_{p0} = A_{p0} \exp\left\{\frac{-E_{p0}}{RT}\right\} \tag{25}$$

In above equations, [I] and [R·] are initiator and radical concentrations; f is the initiator efficiency; k_d and k_p are initiator and propagation rate constants; E_{p0} and E_d are activation energies; A_{p0}, A_d are pre-exponential factors; R is the gas constant, and T is the absolute temperature. The inhibition step was neglected.

The resin viscosity is assumed to be a function of temperature and conversion only. Initially, the viscosity decreases with the increase of resin temperature. When the reaction starts, the resin viscosity increases rapidly due to the formation of polymer network. The resin viscosity change for a thermosetting polymer can be expressed by a widely used model [13]:

$$\mu = A_\mu e^{E_\mu / RT} (\frac{\alpha_g}{\alpha_g - \alpha})^f \tag{26}$$

$$f = a_1 + a_2\alpha' + a_3\alpha'^2 \tag{27}$$

$$\alpha' = \alpha - \frac{c_1(T - T_{ref})}{c_2 + (T - T_{ref})} \tag{28}$$

where α_g is gel conversion, E_μ is the activation energy, and A_μ, a_1, a_2, a_3, c_1 and c_2 are constants. The momentum and energy equations are coupled through the viscosity since the viscosity depends on the resin temperature and conversion. The energy equation and balance of species are also coupled through the rate of reaction. Inside the mold cavity, a control volume finite element method is used. The pressure and velocity fields are obtained using the procedures described by Young et. al. [14]. Temperature and conversion equations are solved following the method used by Lin et. al. [15] in which the convection terms are treated implicitly, and the conduction and reaction terms are treated explicitly. The heat conduction in the gapwise direction is discretized by using the collocation method [15].

GEOMETRY AND MATERIAL PROPERTIES

In this study, one metal mold (aluminum) and one non-metal mold (epoxy) were used. The mold halves were made of two rectangular thin plates. The dimension of the aluminum mold is 30.5cm x 15.2cm x 1.91cm and the dimension of the epoxy mold is 30.5cm x 15.2cm x 1.41cm. Mold cavity was formed by placing a spacer with 0.95 cm thickness between the two mold plates. Inlet and outlet are at two ends as shown in Figure 6. Runners were used near the inlet and outlet such that one dimensional flow could be obtained. The dimension of the mold cavity is 25.4cm x

12.7cm x 0.95cm. A random fiber mat (OCF M8610) with a porosity of 0.8 was used. The thermal properties of the fiber, resin and mold halves and the parameters used in the kinetic model are shown in Table 1. The parameters used in viscosity models are shown in Table 2. The fiber mat and mold halves were preheated to 57°C when the aluminum mold was used and to 53°C when the epoxy mold was used. The inlet resin temperature was set at 29°C. The inlet pressure was set at 4.48×10^5 N/m^2 (65 psi) during the mold filling process. More experimental details can be found elsewhere [16].

Table 1. Thermal and kinetic parameters used in RTM simulation

$$\rho_r = 1100 \ (kg/m^3)$$
$$\rho_f = 2560 \ (kg/m^3)$$
$$c_{pr} = 1680 \ (J/kg \cdot K)$$
$$c_{pf} = 670 \ (J/kg \cdot K)$$
$$k_r = 0.168 \ (W/m \cdot K)$$
$$k_f = 0.0335 \ (W/m \cdot K)$$
$$f = 0.1$$
$$m = 4$$
$$\alpha_f = 0.68$$
$$A_d = 5.55 \times 10^{16}$$
$$A_{p0} = 3.84 \times 10^{10}$$
$$E_d = 1.41 \times 10^5 \ (J/mole)$$
$$E_{p0} = 4.27 \times 10^4 \ (J/mole)$$
$$[I](t=0) = 0.005208$$
$$[R\cdot](t=0) = 0$$
$$\Delta H = 2.687 \times 10^8 \ (J/m^3)$$

Table 2. Rheological parameters used in RTM simulation

$$A_\mu = 2.16 \times 10^{-4} \ (N \cdot sec/m^2)$$
$$E_\mu = 26214 \ (J/mole)$$
$$a_1 = -0.9621$$
$$a_2 = 6.482$$
$$a_3 = -3.205$$
$$c_1 = 2.131$$
$$c_2 = 374.6$$
$$\alpha_g = 1.0$$
$$T_{ref} = 328 \ °K$$

EXPERIMENTAL RESULTS

Five thermocouples were placed inside the mold cavity as shown in Figure 6. The temperature history at these five locations are shown in Figures 7 and 8. At each location, before the resin arrived, the temperature showed the initial preform temperature. Once the resin reached this location, the temperature began to decrease. This also served as a flow front location indicator. Once the resin passed, the temperature at this location kept decreasing. When the outlet was closed, the curing stage began and the temperature increased because of the heat transferred from the mold halves by conduction. At the later stage, when the reaction exotherm occurred, the temperature inside the cavity became higher than that of the mold halves. Heat inside the cavity needed to be transferred out through mold halves.

From Figure 7, it can be seen that the temperatures at locations near the interface (T1 and T3) only dropped slightly, when aluminum mold halves were used, since the aluminum mold can transfer the energy to the cavity quite efficiently. When epoxy mold halves were used, the temperature at these two locations dropped more substantially as shown in Figure 8, because of low conductivity of the mold halves. The temperature near the outlet (T2) was higher than the initial resin temperature because the resin absorbed energy from the fiber mat and mold halves as it traveled inside the cavity.

After mold filling, the cavity kept receiving energy from the mold halves by conduction. This period lasted about 100 seconds when the aluminum mold is used and it lasted 200 seconds when the epoxy mold was used.

After the conduction stage, the reaction exotherm occurred. This stage lasted about 100 seconds. The temperatures near the interface were higher for the case using epoxy mold halves. This is because the reaction exotherm could not be transferred out of the cavity fast enough when mold halves had a low thermal conductivity.

After the temperature reached the peak, it took 500 seconds for the aluminum mold to cool down to the initial mold temperature. In the same time period, the temperature inside the epoxy mold dropped much less.

The temperature histories inside the mold half are shown in Figures 9 and 10. For the aluminum mold, the temperature variation is less than 6°C, while for the epoxy mold the temperature variation reached 40°C.

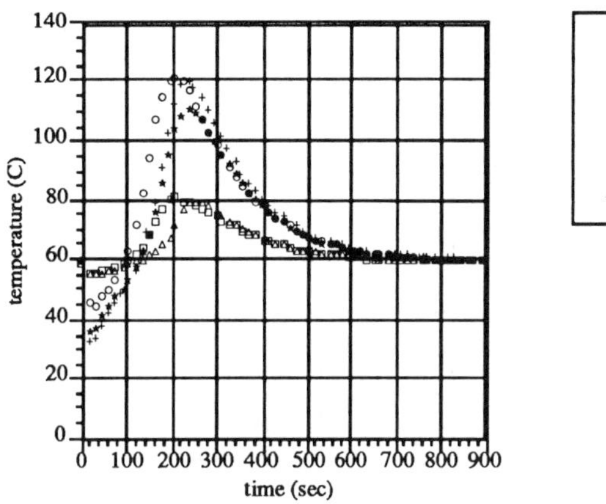

Figure 7. Temperature profiles in composite molded in aluminum tool.

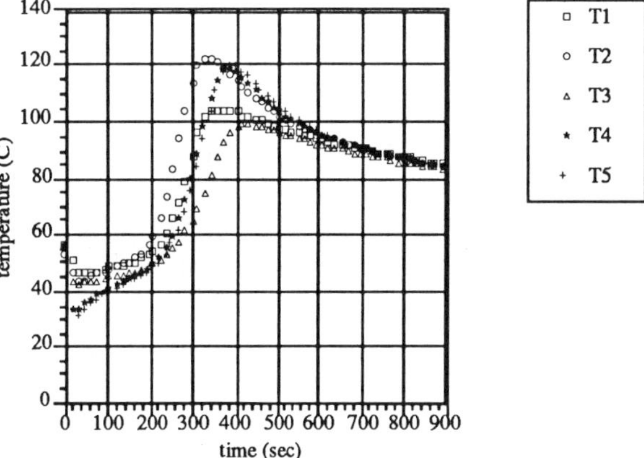

Figure 8. Temperature profiles in composite molded in epoxy tool.

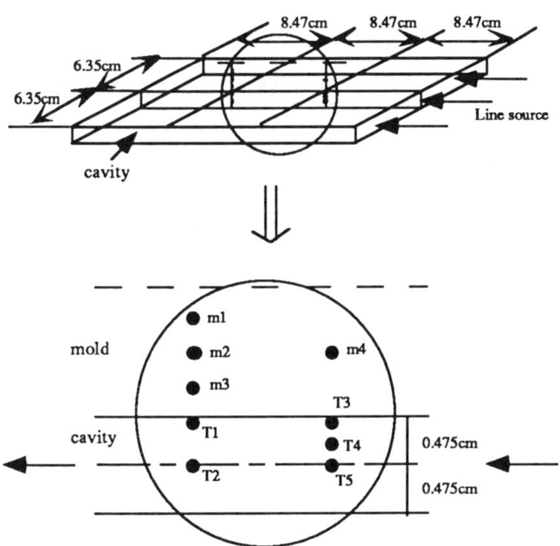

Figure 6. The schematic of mold cavity and the locations of five thermocouples inside the cavity and four thermocouples inside the mold half.

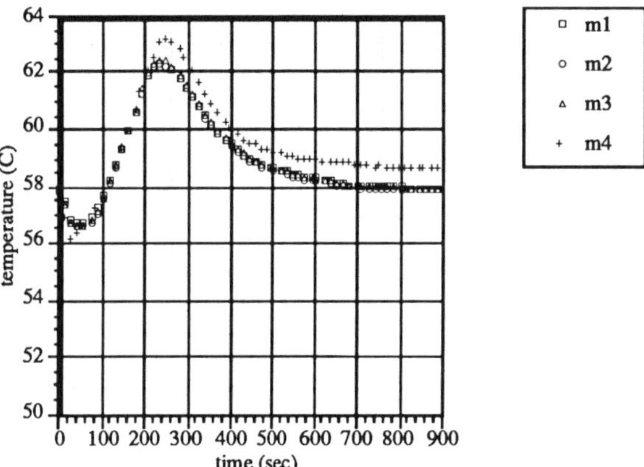

Figure 9. Temperature profiles in aluminum tool.

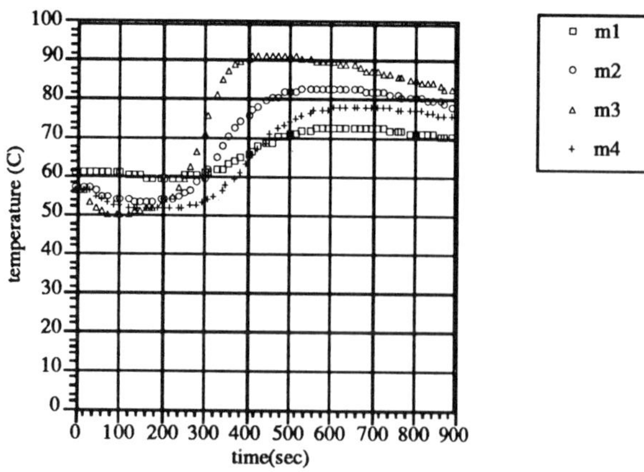

Figure 10. Temperature profiles in epoxy tool.

SIMULATION RESULTS

For each mold half, the temperature on the opposite side of the cavity interface was assumed constant. A free convective boundary condition was assigned to the four side walls of the mold halves.

Aluminum Mold

The temperature histories at the five corresponding locations of a case specifying constant temperature at the boundary are shown in Figure 11. Compared with the experimental results, a specified temperature at the interface seems to be able to give reasonably good prediction. From the molding experiments, it was found that the temperature of the mold halves varied little when the aluminum mold was used. To the mold cavity, this is equivalent to a constant temperature at the interface.

The numerical results of method 2 are shown in Figure 12. The temperature histories are almost the same as those using method 1. It is because the temperature of the aluminum mold halves changed very little.

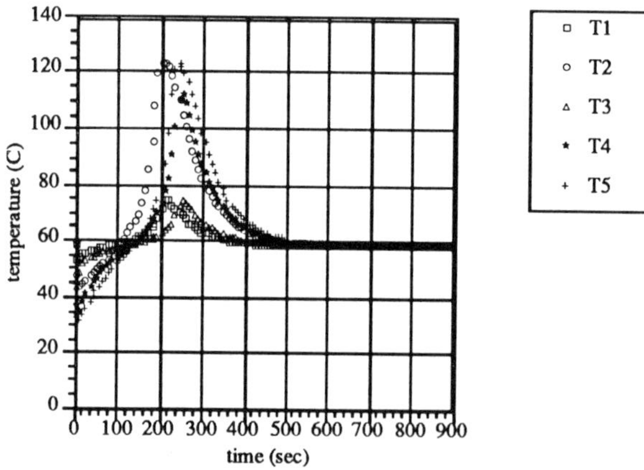

Figure 11. Simulated temperature profiles in composite molded in aluminum tool using method 1.

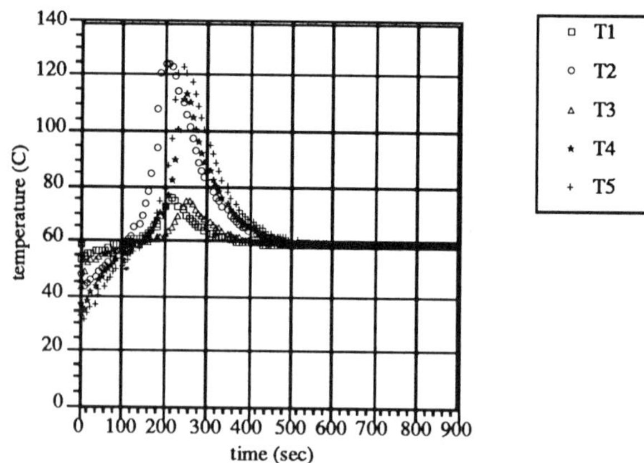

Figure 12. Simulated temperature profiles in composite molded in aluminum tool using method 2.

Epoxy Mold

The numerical results of method 2 are shown in Figure 13. Although they agree with the experimental data in trend, there is a large difference regarding peak temperature and cooling rate. For the epoxy mold, the temperature difference inside the mold cavity is quite significant and the heat transfer in the planar direction may also be important since the cycle time is long. Method 2, however, does not include these effects.

When the temperature change varies greatly from location to location inside the mold half, as in the epoxy mold, method 3 is difficult to use because boundary nodes and linear expansion functions can not show such complicated temperature change. Consequently, the integration on RHS of equation 10 creates a large error and the numerical results become meaningless. Higher order expansion functions and more internal nodes are needed, however, this requires complicated formulation and longer computation time.

Simulation results of method 4 will be discussed during presentation.

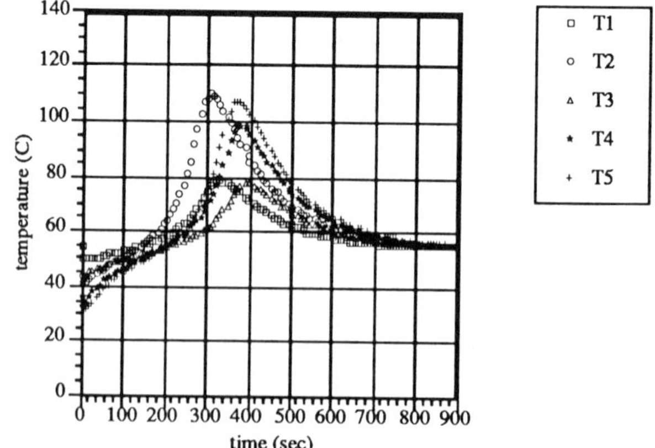

Figure 13. Simulated temperature profiles in composite molded in epoxy tool using method 2.

REFERENCES

1. Himasekhar K., J. Lottey and K. K. Wang, J. Eng. Ind., <u>114</u>, 213 (1992).
2. Incropera F. P. and D. P. DeWitt, Introduction to Heat Transfer (1985).
3. Um M. K. and W. I. Lee , 35th Int. SAMPE Symposium, 1905, April (1990).
4. Brebbia C. A., The Boundary Element for Engineers, Pentech, London (1984).
5. Azevedo J. P. S. and L. C. Wrobel , Int. J. Numer. Meth. Eng., <u>26</u>, 19 (1988).
6. Pasquetti R. and A. Caruso, Numer. Heat Transfer, <u>17</u>, 83 (1990).
7. Osswald T. A. and C. L. Tucker, Polym. Eng. Sci., <u>28</u>, 413 (1988).
8. Rencis J. J. and K. Y. Jong, Computer Methods in Applied Mechanics and Engineering, <u>73</u>, 295 (1989).
9. Wrobel L.C and C.A. Brebbia , Finite Elements in Water Resources VI, Springer, Berlin, 147 (1986).
10. Lera S. G., E. Paris and E. Alarcon, Appl. Math. Modeling, <u>6</u> (1982).
11. Tucker, C. L. and R. B. Dessenberg, Flow Phenomena in Polymeric Composites (1993).
12. Stevenson J.F., Poly. Eng. Sci., <u>26</u>(11), 746 (1986).
13. Castro J.M. and C.W. Macosko, AIChE J. <u>28</u>(2), 250 (1982).
14. Young W. B., K. Han, L. H. Fang, L. J. Lee, and M. J. Liou, Polym. Composites, <u>12</u>(1), 30 (1991).
15. Lin R. J., L. J. Lee and M. J. Liou, Intl. Polym. Proc., <u>6</u>(4), 356 (1991).
16. Perry M. J., J. Xu, Y. Ma, L. J. Lee and M. J. Liou, SAMPE Quarterly, <u>23</u>(4), 20, (1992).

Compaction and Creep Behavior of Glass Reinforcement for Liquid Composites Molding

R. Gauvin, P. Clerk, Y. Lemenn, F. Trochu
Ecole Polytechnique
Montréal, Québec, Canada

Abstract

Continuous fiber reinforced polymeric composites can be manufactured by injecting a reactive liquid into a mold with preplaced reinforcement. The two best known processes are Resin Transfer Molding (RTM) and Structural Reaction Injection Molding (SRIM). In these processes, the fiber materials are first compressed in the cavity and then the resin is injected. The knowledge of the compaction and creep behavior are particularly important to evaluate the mold closing forces and also to evaluate the compresses thickness of each layer for permeability assessment of a multilayered preform. When a single type of fiber reinforcement is used throughout the mold, thickness, the flow front is readily predicted assuming uniform permeability in the thickness direction. However, when several types of reinforcements are laid on top of one another, an average permeability across the mold thickness is needed and analysis of filling is more complex.

In this work, the compaction and creep behavior of five continuous strand mats and six fabrics are reported. The effect of the number of layers and compaction speed are examined. Results show that the relaxation under pressure or creep occurs in the first 10 to 20 sec. and that for all the reinforcements tested the creep for a given number of layers is more important for a low initial pressure and a slow compaction speed.

CONTINUOUS FIBER REINFORCEMENT THERMOSET POLYMERIC COMPOSITES can be produced by injecting reactive liquid resins into a mold with preplaced reinforcement. The two best known liquid composite molding (LCM) processes are Resin Transfer Molding (RTM) and Structural Reaction Injection Molding (SRIM). In these processes the reactive liquid is pumped into the mold where fiber reinforcement has been preplaced.

Typically, in RTM, resin reactivity is lower than for SRIM and viscosity higher. In these processes, injection pressure could be as low as 100 KPa and as high as 3000 KPa. In many cases, low cost fiberglass-polyester or epoxy tools are used. In these processes, the reinforcement could take several forms, for instance, it could be a chop strand, or a stitched preform, a continuous strand mat or any type of fabrics such as a woven roving, a no-crimp stitched (NCS) fabric, a weaved fabric, or a combination of two or more of these reinforcements.

The necessary pressure to compress the reinforcements into the mold and the resin pressure being developed during mold filling are two important parameters in mold design and equipment selection. Mold filling softwares currently developed such as the RTMFLOT software available from our laboratory can predict the pressure distribution in the cavity during filling. However the accuracy of the prediction is a direct function of the accuracy of the permeability values describing the reinforcement. To eliminate the necessity to measure the permeability on each reinforcement combination, for multi-layered composites, permeability models for stacks of non similar reinforcements must be developed. Such a model was previously proposed [1] and was later modified by others [2] to take into consideration the interface between adjacent layers. In these models the compaction pressure on each layer is assumed to be the same. If one knows the compaction pressure (P) versus fiber volume fraction (V_f) for each type of reinforcement in the stack, the V_f for each layer can be obtained with data for permeability versus volume fraction for each reinforcement. Than the global permeability of the multilayer can be calculated with a simple rule of mixture or any other model that may better fit the reality of a specific case.

In this paper, compaction behavior at room temperature of dry reinforcements is analyzed. Data for five continuous strand mats and six fabrics from various suppliers are given. The effect of the number of layers and the closing speed are analyzed. The relaxation behavior of the fabrics is also examined.

Background

In the RTM and SRIM processes, the mold must be rigid enough to prevent undesirable thickness variation due to forces acting on the mold face during the filling and curing phases. As mention previously, there are two forces acting on the mold face during filling. One comes from the fluid pressure required to force the liquid resin through the fibrous reinforcement. It varies during the filling phase and can be determined with Darcy's law which relates the flow, the pressure gradient, the resin viscosity and the permeability. For unidirectional flow in a constant cross section this law is expressed by the following equation:

$$v = \frac{K}{\mu} \frac{\Delta P}{\Delta L} \tag{1}$$

where v is the flow front speed, μ the fluid dynamic viscosity, $(\Delta P/\Delta L)$ the pressure gradient and K the permeability of the porous medium. In this equation, K is generally expressed in m^2.

The second force acting on the mold face comes from the compaction pressure needed to compress the reinforcement to the cavity thickness. In literature, compaction behavior of reinforcements was studied. In 1987, T.G. Gutowski et al. [3] have studied the consolidation of laminates and proposed the following semi-empirical model for the stiffness behavior of a bundle of confined fibers, assuming the fibers to act as bending beams between multiple contact points:

$$P = A_s \frac{\sqrt{\dfrac{V_f}{V_0}} - 1}{\left[\sqrt{\dfrac{V_m}{V_f}} - 1\right]^4} \tag{2}$$

where V_f is the actual fiber volume fraction, V_0 the original fiber volume fraction and V_m the maximum attainable volume fraction for the particular arrangement.

A_s is a spring constant which depends on the stacking geometry and the mechanical properties of the fibers:

$$A_s = \frac{3\pi E}{\beta^4} \tag{3}$$

where E is the fibers flexural modulus and ß is defined has the span length over the span height of the fibers which quantify their straightness.

In 1988, R. Gauvin and M. Chibani [1] proposed the following empirical model to link the compaction pressure (P) and the compressed thickness of the reinforcement (h):

$$\frac{h}{h_0} = A_0 P + A_1 \ln P + \frac{A_2}{P} + A_3 \tag{4}$$

where h_0 is the original reinforcement thickness and A_0, A_1, A_2 and A_3 are curve fitting parameters. At the time, for this model, 20 layers of reinforcement were used to collect compaction data. As can be seen on Figure 1 for the NICO758, this model overestimates the necessary pressure to reach a given fiber volume fraction for a small number of layers.

In 1990, G.L. Batch and S. Cuminskey [2] have developed a semi-empirical approach based on the compressibility behavior of one layer of reinforcement. A Hookean and non-Hookean regime where suggested and their model can be expressed by the following equation:

$$P = B (V_f - V_0)$$

where:

$$B = Bo \quad \text{if} \quad V_f \leq V_{f_{cont}} \tag{5}$$

$$B = B_0 \frac{(1 - \eta)}{\eta} \frac{\dfrac{1}{V_0} - \dfrac{1}{V_f}}{\dfrac{1}{V_f} - \dfrac{1}{V_\infty}} \quad \text{if} \quad V_f > V_{f_{cont}}$$

In this model, B is the fiber compaction spring constant, B_0 the Hookean fiber compaction spring constant, η the fiber packing efficiency, V_0 the fiber volume fraction at a near zero pressure, V_∞ the ultimate fiber fraction at maximum pressure and $V_{f\,cont}$ the volume fraction at the transition between the Hookean and non Hookean regime. At the same conference in 1990, L. Trevino et al. [4] proposed the following empirical model which was based on experiments with one random fiber mat and two stitched mats:

$$\log V_f = A_0 \log P - A_1 \tag{6}$$

where A_0 and A_1 are curve fitting parameters which are different for each type of reinforcement.

A few months later, Y.R. Kim [5] reported that a model previously proposed by Taylor [6] for soil could be used for fiber reinforcements:

$$V_f = V_1 + C_c \log P \qquad (7)$$

where V_1 is the volume fraction at 1 Kgr./cm^2 and C_c a curve fitting parameter. Both, V_1 and C_c vary for each reinforcement.

None of these models take into account the number of layers and the compaction speed. Results from this study indicate something different.

In 1992, L.J. Piechowski and K.N. Kendall [7], in a study on preform manufacturing, used designed experiment analysis techniques to determine the effects of tool closing speed, temperature, number of layers and dwell time at maximum displacement on compressibility and relaxation of preforms made with random continuous strand E-glass mat. They also studied the interaction of these parameters. Their results showed that preform tool closing speed has the largest effect on material compressibility and temperature has the largest effect on preform relaxation after it has been removed from the preforming tool.

Experimental

In this study, compaction measurements were done between two steel plates fixed on a servohydraulic testing machine. The test specimens were cut at 12.5 cm x 12.5 cm and stacked between the plates. Force and displacement are recorded and converted into pressure (P) and fiber volume fraction (V_f) to report the results. The compaction speed for the random mat was fixed at 5 mm/min. For the fabrics, compaction speed of 2 mm/min, 240 to 250 mm/min and 865 to 1030 mm/min were used. All the tests were run at room temperature on reinforcement as received from the supplier. The reinforcement reported here are:

A) Continuous strand glass fiber mats

	From	Surface density
OCF 8610	Owens Corning	300 g/m^2
NICO758	Nico Fibers	450 g/m^2
U-101	Vetrotex CertainTeed	450 g/m^2
U-812	Vetrotex CertainTeed	450 g/m^2
U-814	Vetrotex CertainTeed	450 g/m^2

B) Glass fabrics

	From	Surface density
0-90° no-crimp stitched		
NCS 82675A	J.B. Martin	315 g/m^2
NCS 81053A	J.B. Martin	618 g/m^2
Woven roving		
WR 24 oz	Armkem	814 g/m^2
0-90° no-crimp stitched		
BTI C 24 oz	Brunswick Tech. Inc.	814 g/m^2
Injectex 0-90° special stitched thermoformable		
EB 315 E02 120	Brochier	315 g/m^2
± 45° stitched		
EBX 936 HD	Tech Textile	1000 g/m^2

Results and Discussion

For the continuous strand mats, Figure 1 shows the compaction pressure (P) as a function of fiber volume fraction (V_f) for one to six layers of each mat. For each one, except the NICO758, the curves for various number of layers superpose reasonably well up to a V_f of 0.3. We also notice that in general, the necessary pressure to reach a specific fiber volume fraction is higher for a greater number of layers even though the U-814 does not follow that rule which is more applicable to the NICO758 and the OCF 8610.

To ease the use of these data, they are easily model by the following exponential equation:

$$V_f = a P^b \qquad (8)$$

where V_f is the fiber volume fraction, P the pressure in Pascals and (a) and (b) are curve fitting parameters. This model is valid for $P > 0$.

Table 1 shows the coefficients for the model of Equation 8 for one to six layers for each continuous strand mat tested.

Their is a limited amount of results of that nature in literature. Nevertheless, we were able to find some for the OCF 8610, the two models proposed by Batch and Trevino are compared to our measurements on Figure 1. We observe that their models are good up to a V_f of 0.3 which covers most of the practical range. For larger volume fractions however they overestimate significantly the necessary compaction pressure. For the NICO758, the experimental results are compared with the Trevino's model and our old empirical model [1] which was obtained for 20 layers. In this case both models have a bad fit. It can be notice however that for an unknown reason the NICO758 has a much more dispersed behavior for the various number of layers than the other four mats.

In an attempt to evaluate the presence of some hysteresis effect in the compaction behavior of mats, a second compaction was performed immediately after the first one. Figure 2 shows the results for the first and second compactions of three and six layers of the OCF 8610, NICO758 and U-101 mats. We observe that an hysteresis effect is present which could be the results of fiber breakage and fiber rearrangement. Fiber breakage is particularly significant for the NICO758 which has much finer strands than the other two. These finer strands are used for a better surface finish.

Results on fabrics are presented in Figure 3 to Figure 6. Figure 3 shows the results of the first compaction done at a rate of 2 mm/min for three, six or 12 layers for the six fabrics tested. Because of the minimum thickness requirements, the NCS 82675A could not be tested for three layers and six and 12 layers were chosen instead.

In all cases, the number of layers does not affect the compaction pressure for the lower values of the fiber volume fraction. The high fiber content however is more difficult to obtain with a larger number of layers. This could partly be explain by the fact that in a three layers stack, two of the layers are in contact with the steel plate which could help the rearrangement of fibers. Nevertheless, it is not fully understand why. This trend was also observed on some of the continuous strand mat shown in Figure 1.

Since all fabrics were tested with six layers, we used that number to study the effect of the compaction speed on the compaction pressure versus fiber volume fraction. Results are presented in Figure 4. Here, their is no uniform behavior. The architecture of the fabric, the size of the fiber strand as well as the binder, thermoformable or not, have an effect on the capacity of a specific fabrics to rearrange itself under pressure. The ones that do rearrange the most, like the WR 24 oz and the NCS 82675A, are doing it more easily at high speed. The Brochier EB 315 E02 120, is a thermoformable fabric with a low melt temperature epoxy for binder. This might explain the reverse trend observed at high speed for this fabric.

The creep behavior, which is the relaxation of the reinforcement under pressure in the mold, is an other important phenomenon that was also observed by Y.R. Kim [5]. Figure 5 shows creep data for two fabrics, the WR 24 oz and the Brunswick C 24 oz. The ordinates of all the graphs are normalized with a nominal unit pressure P_0. The top two graphs show the influence of the initial pressure on the creep behavior for a compaction speed of 2 mm/min. In both cases, the creep is much more important at low pressure than it is at high pressure. That phenomenon was also observed by Y.R. Kim [5]. At a lower pressure, the fiber bundles have more freedom to slip and to rearrange themselves. It is also observed that in all cases, 60% to 70% of the creep occurs in the first 10 to 20 seconds.

To evaluate the effect of the compaction speed used to reach the maximum pressure before the machine was stopped and the reinforcement was allowed to relax, tests were run at three compaction speeds: 2, 250 and 1000 mm/min. The initial pressure was chosen at about 850 KPa which corresponds to a V_f in the range of 0.6 for both fabrics when six layers are compressed. In both cases the creep is much more important at high speed but the speed effect seems to be more sudden for the WR 24 oz while it is more progressive in the case of the BTI C 24 oz. Finally, the bottom two graphs illustrates the influence of the number of layers on the creep behavior. In both cases the initial pressure, in the range of 850 KPa, was reach at a compaction speed of 2 mm/min. We observe that the creep is more important for a large number of layers. The results for the other four fabrics tested are similar to those presented in Figure 5.

To sum up these results, one can say that most of the creep occurs in about 10 to 20 sec and that the maximum creep will occur for a low compaction pressure, a slow compaction speed and a small number of layers in the stack.

Finally, Figure 6 compares the experimental results with models form literature for two fabrics for which we were able to compute the parameters needed in each models. These parameters are given in Table 3. As one would expect the five parameters model previously published by Gauvin and Chibani [1] fit perfectly the experimental results. The other ones do not fit as well for the high fiber volume fraction but they have the advantage of using fewer parameters. The simple empirical model proposed in this paper, Equation 8 and Table 2, does provide a very good fit throughout the spectrum of experimental values with only two parameters and seems to be the most interesting one.

Conclusion

From this study, we can draw the following conclusions. Generally speaking continuous strand E-glass mats and E-glass fabrics behave similarly in a dry compaction test at room temperature. Their behavior can be model by a simple exponential equation of the following form:

$$V_f = a\,P^{\,b}$$

where V_f is the volume fraction and P the compaction pressure which must be larger than zero to avoid indetermination. (a) and (b) are curve fitting parameters but could also be seen as a combination of a material parameters and stack characteristics. For instance, (a) would be the volume fraction of the stack for a nominal unit pressure and (b) the compaction stiffness index of the material for a given number of layers.

The creep of the reinforcement while being compressed in the mold is also an important phenomenon. It occurs rapidly in 10 to 20 sec. and its amount is inversely proportional to the applied initial pressure. The compaction speed and the number of layers also influence significantly the creep values.

The temperature which was not part of this study is most likely to have an effect on the compaction behavior since usually the reinforcements have binders that will soften with an increase in temperature and will change the slipperage characteristics of the fiber bundles.

Finally we can say that the compaction and creep behavior of reinforcements is an issue that should not be overlook because it plays a major role in permeability assessment and the pressure being developed should be taken into account in tool design and equipment sizing.

Acknowledgements

This work was funded by grants received by the National Research Council of Canada, from the Québec Government (Fonds FCAR) and from The Institut de recherche en santé et sécurité du travail (IRSST). We also thank J.B. Martin, Armkem, Vetrotex CertainTeed and Brochier for providing free materials.

References

[1] Gauvin, R. and M. Chibani, 43th Ann. Conf. Comp. Inst., Soc. of Plast. Ind., Session 22C(1988).

[2] Batch, G.L. and S. Cumiskey, 45th Ann. Conf. Comp. Inst., Soc. of Plast. Ind., Session 9A(1990).

[3] Gutowski, T.G. and T. Morigaki and Z. Cai, J. of Comp. Mat., 21,172-188(1987).

[4] Trevino, L., J.L. Lee, K. Rupe and M.J. Liou, 45th Ann. Conf., Comp. Inst., Soc. of Plast. Ind., Session 9E(1990).

[5] Kim, Y.R., S.P. McCarthy, and J.P. Fanucci, ANTEC,1252-1256(1990).

[6] D.W. Taylor, "Fundamentals of Soil Mechanics", John Wiley and Sons, New York(1948).

[7] Piechowski, L.J. and K.N. Kendall, 8th Ann. ASM-ESD Adv. Comp. Conf. ,(1992).

Table 1 Coefficients for the Eq. 8 model

Number of layers	Coeff.	OCF 8610	NICO758	U-101	U-812	U-814
1	a	0.0021	0.0007	0.03	0.0014	0.0021
1	b	0.4	0.48	0.21	0.43	0.39
2	a	0.0024	0.0021	0.0066	0.0063	0.004
2	b	0.38	0.38	0.26	0.30	0.35
3	a	0.0051	0.001	0.018	0.006	0.004
3	b	0.33	0.42	0.23	0.31	0.34
4	a	0.004	0.0054	0.02	0.0055	0.0031
4	b	0.34	0.28	0.22	0.31	0.37
5	a	0.0052	0.0021	0.02	0.0057	0.0036
5	b	0.32	0.37	0.22	0.31	0.36
6	a	0.006	0.0052	0.02	0.005	0.004
6	b	0.31	0.28	0.21	0.31	0.34

Table 2 Coefficients for the Eq. 8 model

Fabrics	Number of layers	a	b
EBX 936 HD	3	0.13	0.125
EBX 936 HD	6	0.17	0.10
Brunswick C 24 oz	3	0.09	0.14
Brunswick C 24 oz	6	0.136	0.105
WR 24 oz	3	0.08	0.15
WR 24 oz	6	0.13	0.11
NCS 81053A	3	0.07	0.17
NCS 81053A	6	0.11	0.126
NCS 82675 A	6	0.067	0.16
NCS 82675 A	12	0.09	0.14
Brochier EB 315 E02 120	3	0.016	0.26
Brochier EB 315 E02 120	6	0.044	0.18

Table 3 Coefficients for the models from literature and Eq. 8 for six layers

Model	Constant	NCS 82675	C 24 oz
Gauvin	V_0	0.33	0.4
Gauvin	A_0 (10^{-2})	-1.54	-0.26
Gauvin	A_1 (10^{-2})	-9.1	-7.7
Gauvin	A_2 (10^{-2})	0.17	0.1
Gauvin	A_3 (10^{-2})	75	87
Taylor	C_c	0.175	0.13
Taylor	V_1	0.435	0.5
Taylor	P_1 (KPa)	105	250
Gutowski	A_s	3.6	5.6
Gutowski	V_m	0.91	0.85
Gutowski	V_0	0.3	0.4
Batch	K_0	91	290
Batch	$V_{f\,cont}$	0.41	0.47
Batch	η	0.4	0.29
Batch	V_m	0.91	0.85
Eq. 8	a	0.06	0.13
Eq. 8	b	0.175	0.11

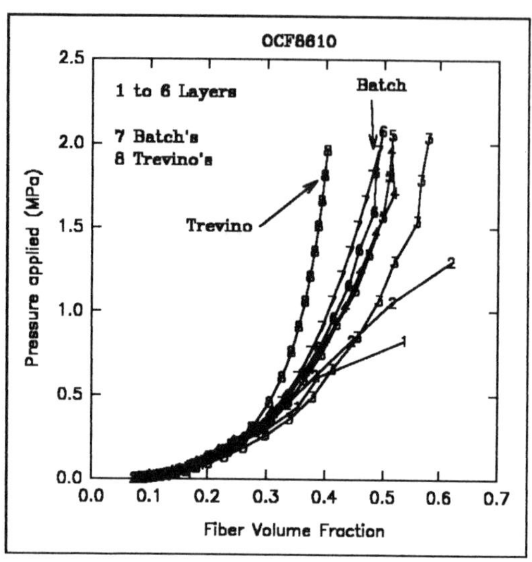

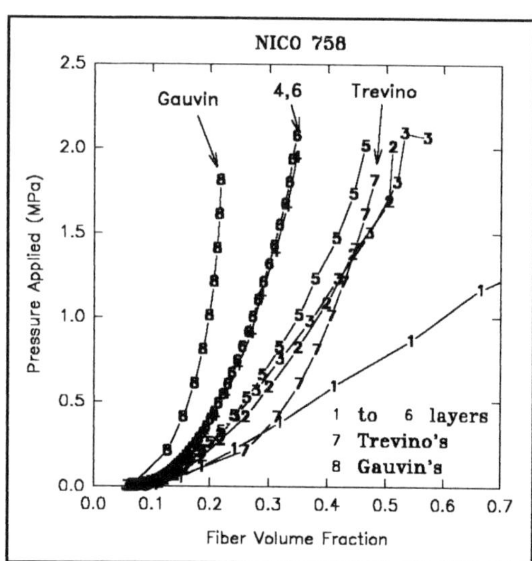

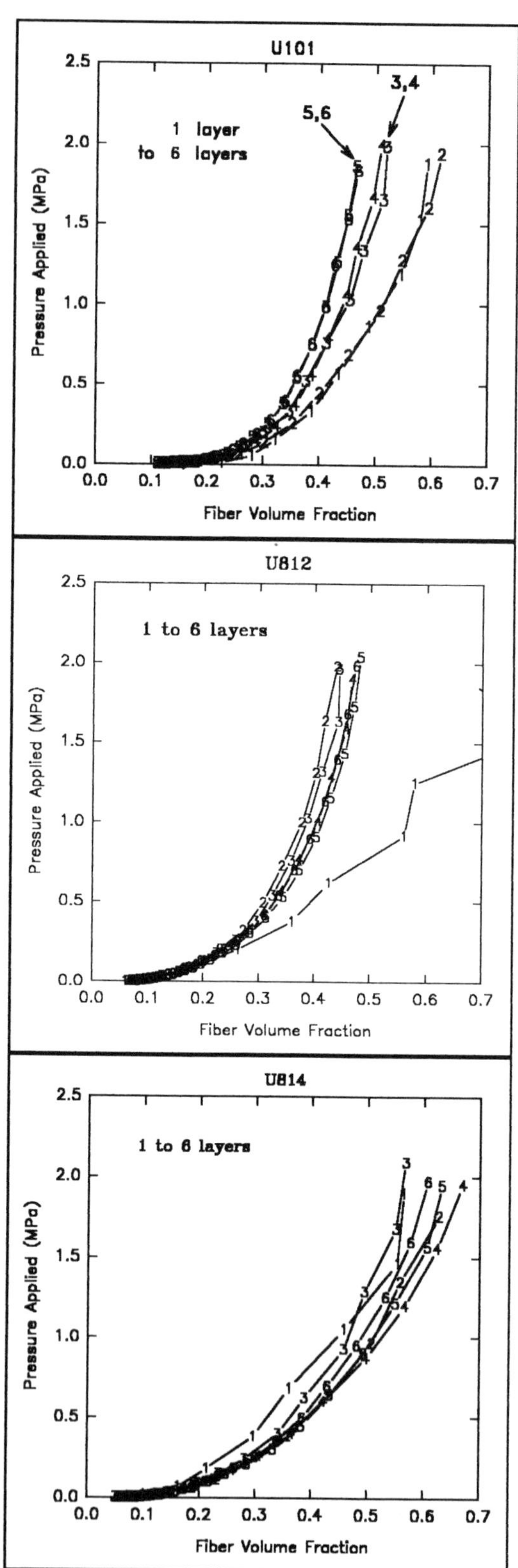

Fig. 1 - Compaction pressure vs Fiber volume fraction for 1 to 6 layers of mat and comparison with models (OCF8610, NICO758)

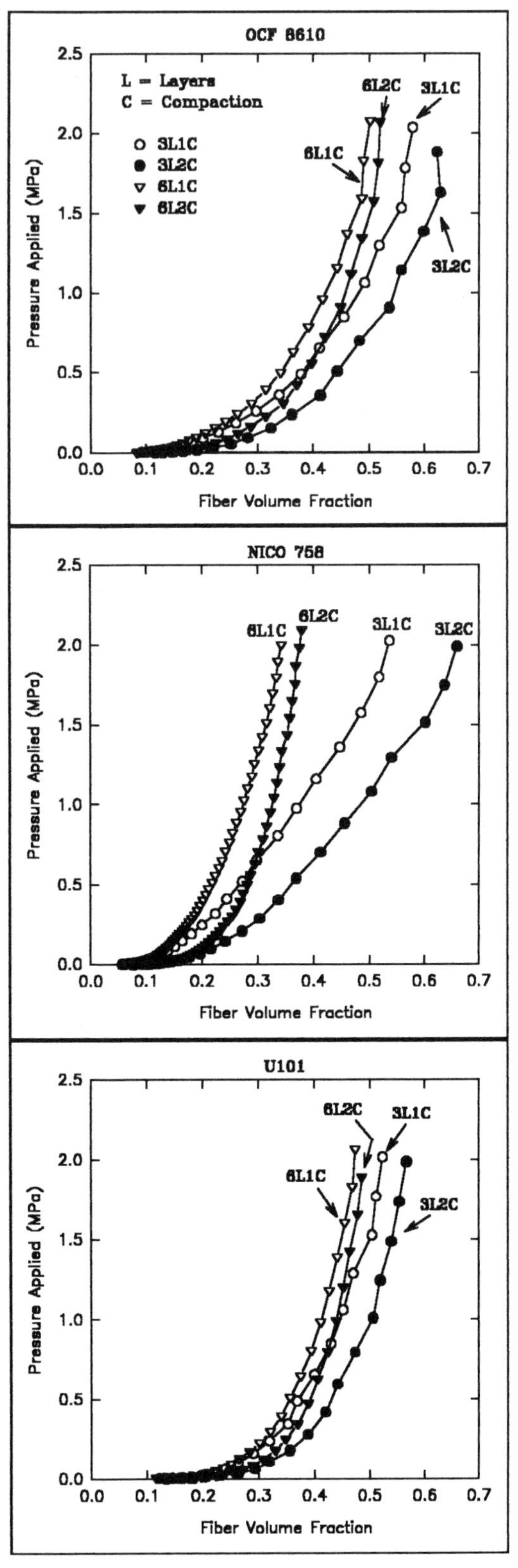

Fig. 2- Effect of successive compactions on mats

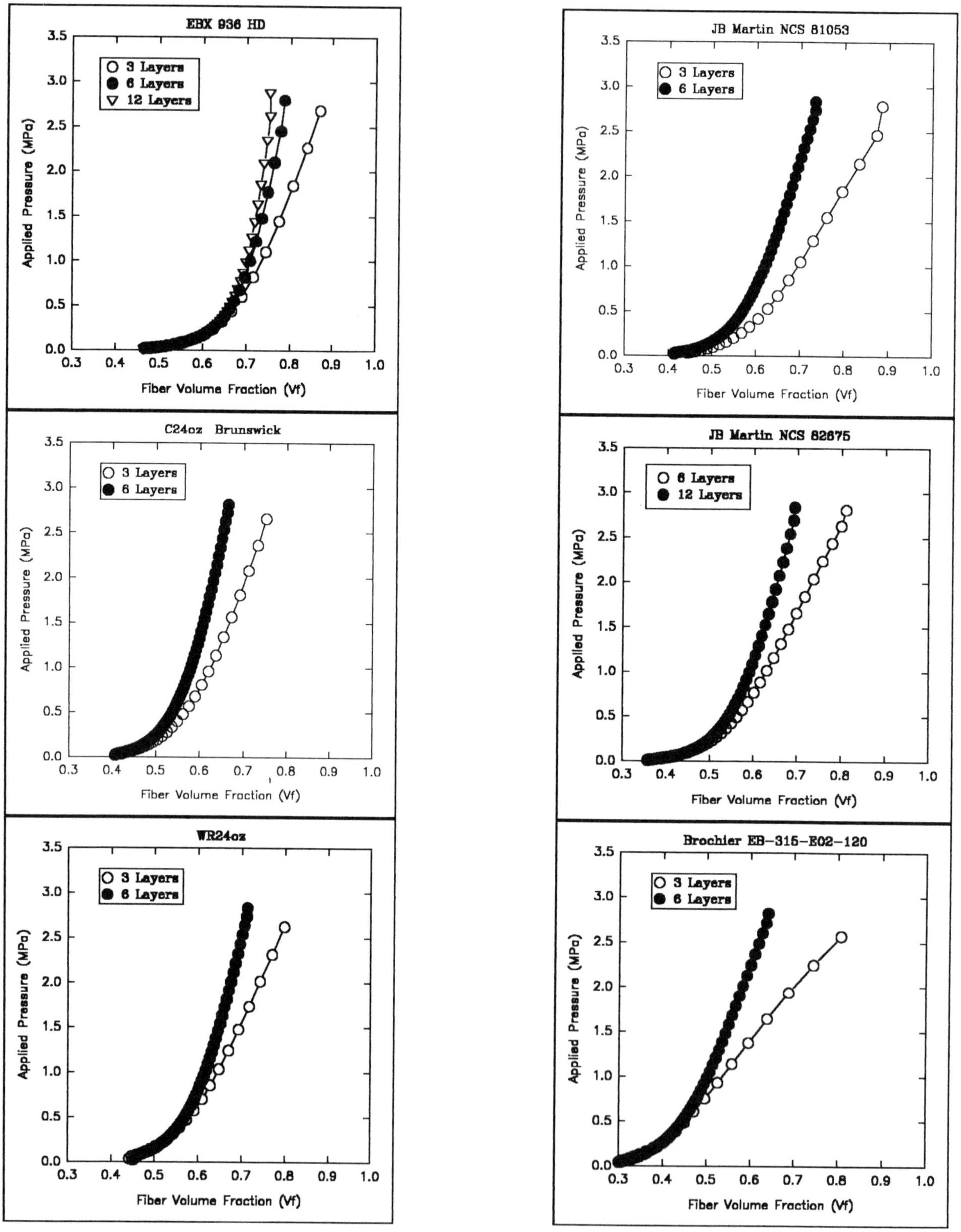

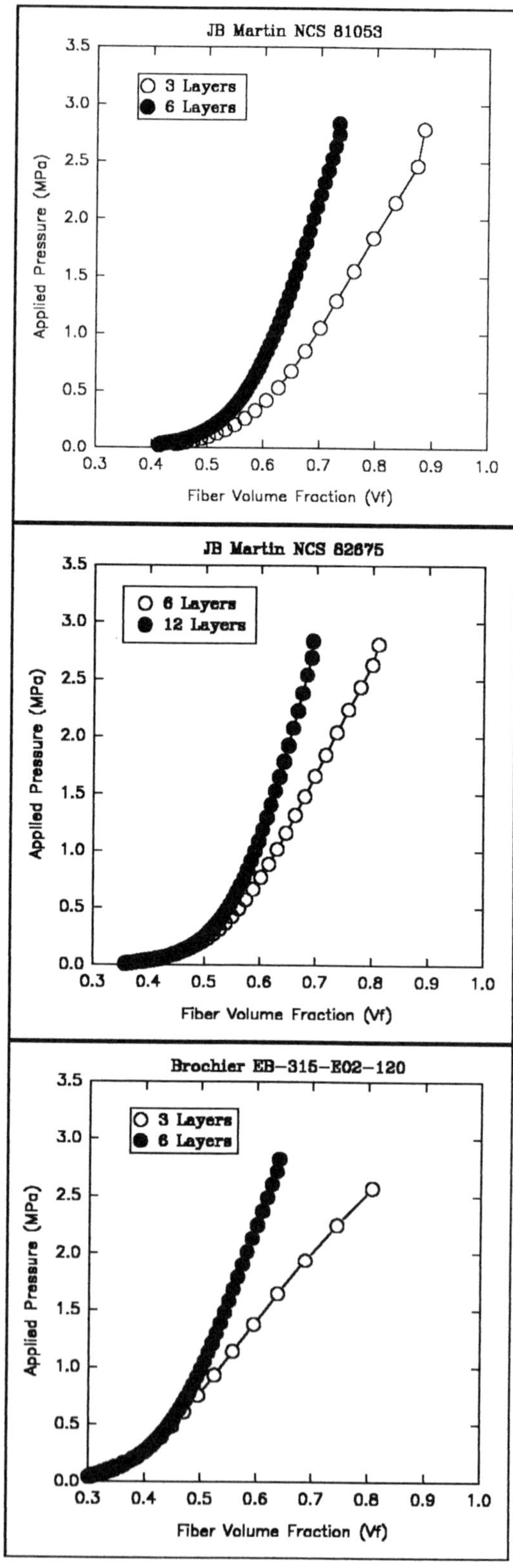

Fig. 3- Effect of the number of layers on the compaction behavior of fabrics (compaction speed = 2 mm/min)

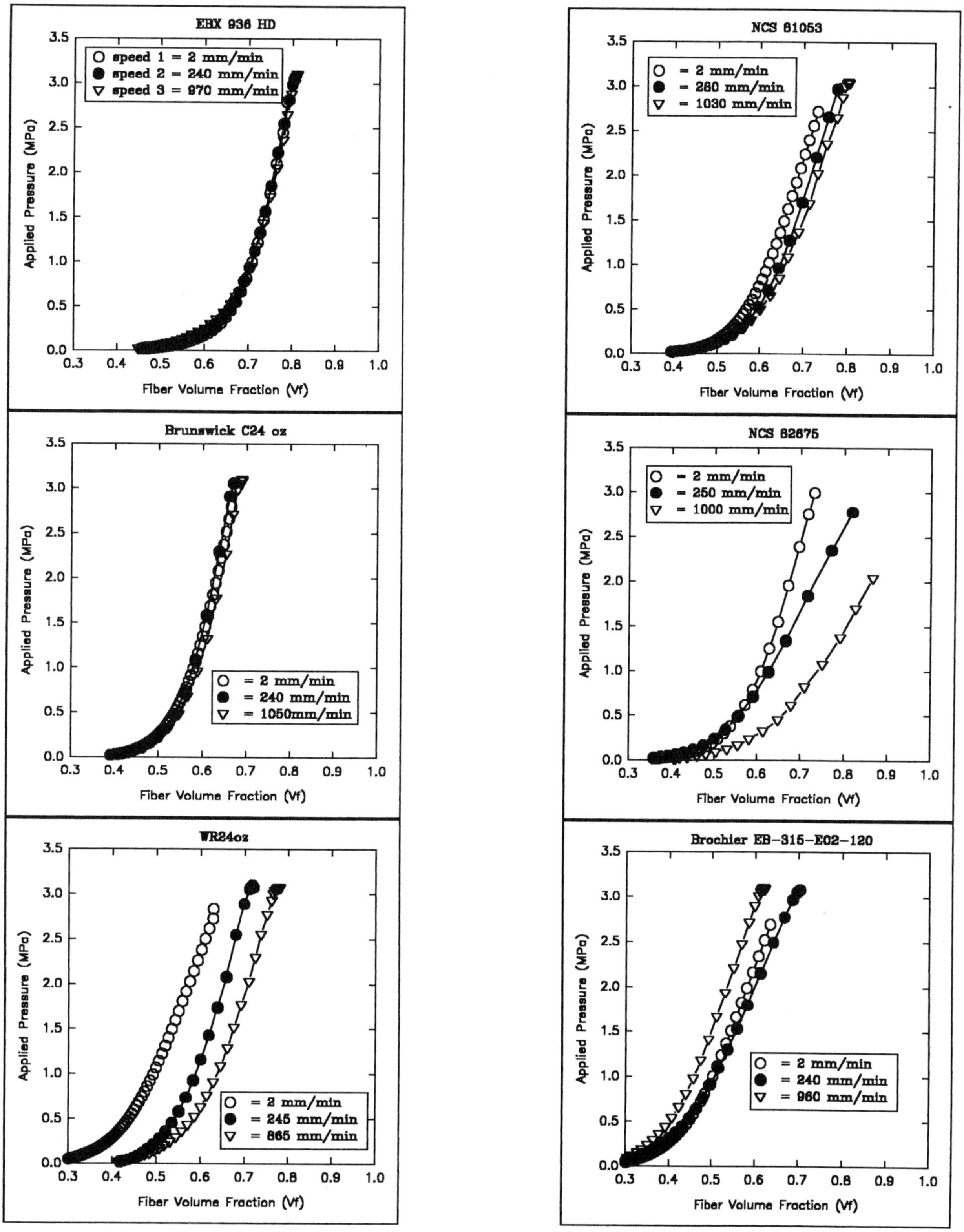

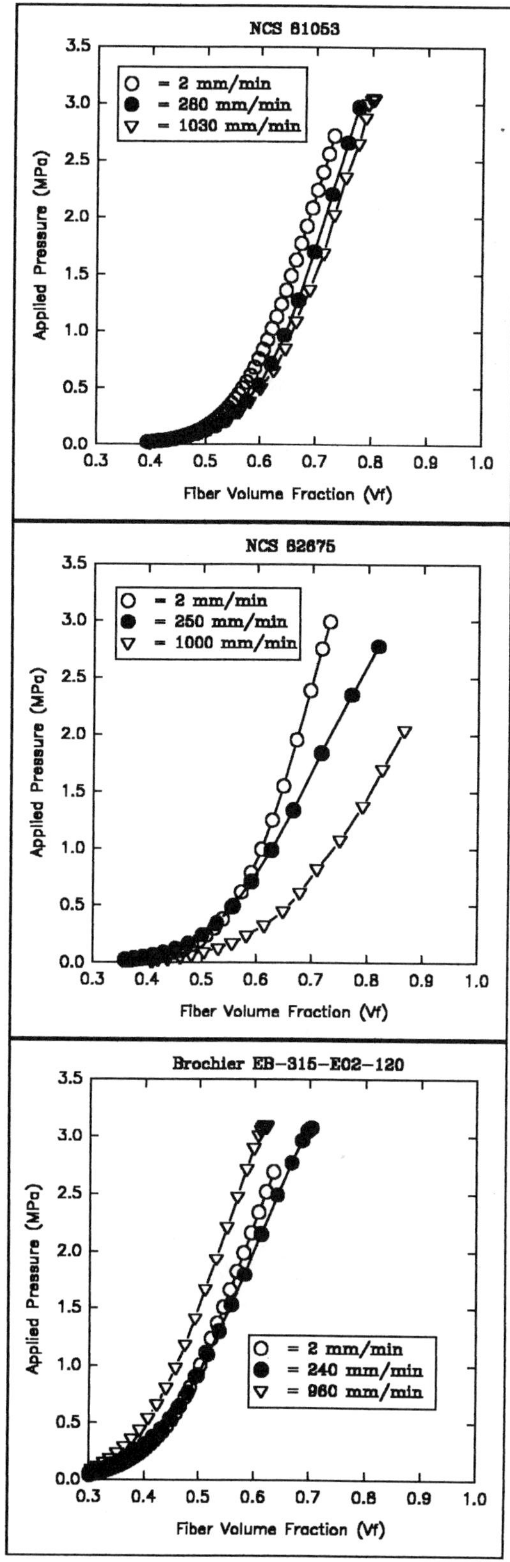

Fig. 4 - Effect of compaction speed on the compaction behavior of 6 layers

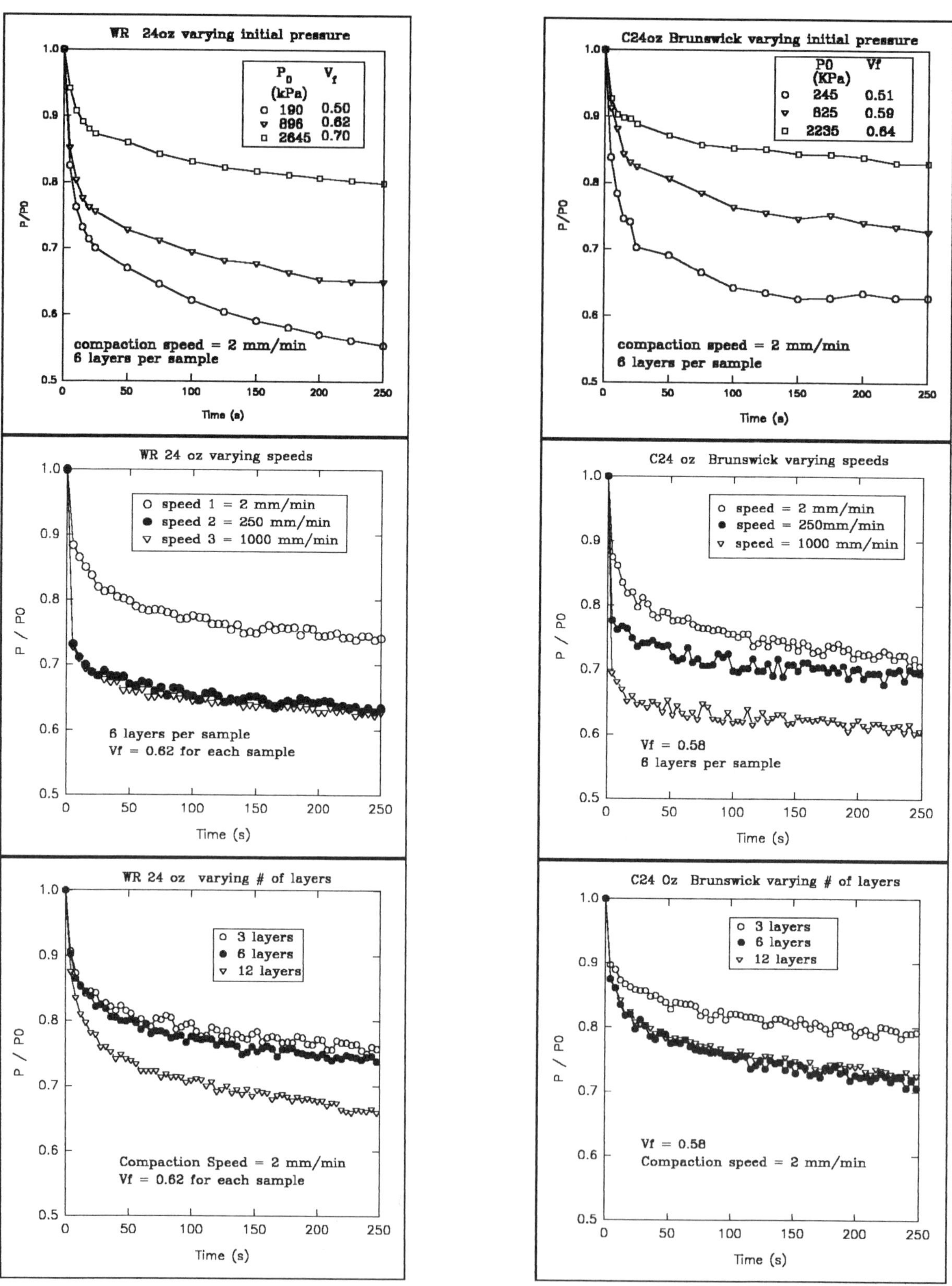

Fig. 5- Effects of Initial Pressure, Compaction Speed and # of Layers on the creep behavior of two fabrics

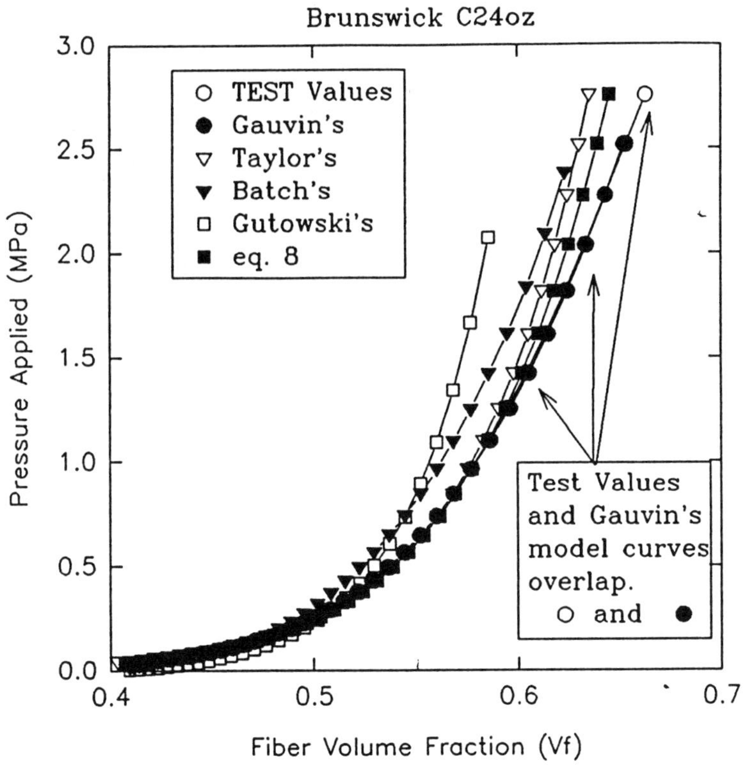

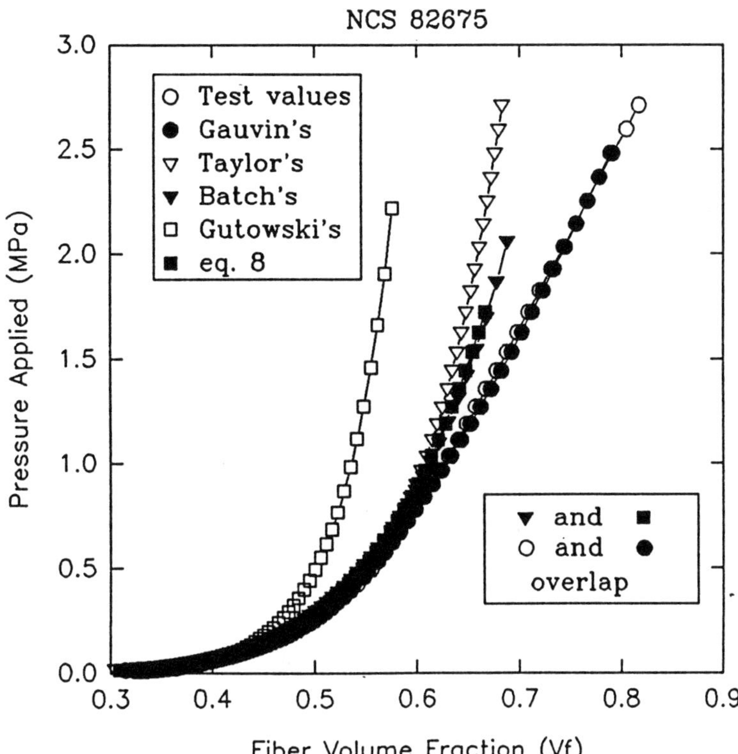

Fig. 6- comparison of compaction test values on 6 layers of C24 oz and NCS82675 with models from literature

In Situ Characterization of Liquid Crystalline Polymer Flow Using Neutron Scattering

M. Dadmun
University of Tennessee
Knoxville, Tennessee

Abstract

The alignment during processing of a liquid crystalline polymer (LCP) results in the unusual ultimate properties and unique physics of LCPs. We are interested in understanding the coupling of shear flow to the orientation of a liquid crystalline polymer in solution. We will describe a method for determining the change in orientation of an LCP by shear flow with neutron scattering. We will discuss results of the application of this technique to solutions of poly (benzyl L-glutamate) (PBLG) in deuterated benzyl alcohol and hydroxypropylcellulose (HPC) in deuterated water. It will also be shown that the neutron scattering results correlate well to the simultaneously measured shear viscosity, which allows the use of this technique to a greater audience as the neutron and viscosity measurements can be completed once and then the viscosity can serve as a secondary standard in the future.

POLYMERS that exhibit liquid crystalline behavior (LCP) have attracted enormous commercial interest in applications such as precision injection molded parts, high strength fibers, optical devices, and flow enhancing agents. The unique properties that LCPs exhibit, which have sparked this commercial interest, is primarily due to their ability to form mesoscopically ordered domains. These ordered domains result in increased tensile strengths, minimal shrinkage, high temperature stability, and anisotropic optical properties, which in turn allow their utilization in many advanced materials applications, such as electrical connectors, sporting equipment, very thin wall parts, and optical fiber connectors. The alignment of the LCP in commercial products usually occurs during processing, *i.e.* shear induced alignment is responsible for the molecular orientation. Unfortunately, the rheology of LCPs is complex and not fully understood. Anomalies such as director tumbling, a negative first normal stress difference, and extremely long transients add to the uncertainty in understanding the coupling of the flow of LCP solutions to the desired molecular alignment.

Experiments to describe the orientation of two model liquid crystalline polymers have been completed using small angle neutron scattering. Using this technique, the change in the orientation with shear rate of poly (benzyl L-glutamate) (PBLG) in deuterated benzyl alcohol (DBA) and hydroxypropylcellulose (HPC) in deuterated water (D2O) have been determined. PBLG is a model liquid crystalline polymer whose backbone is very stiff. Therefore this system and its response should be similar to LCPs that have a rod-like structure, such as Kevlar, which is used in bulletproof jackets and sporting equipment. The structure of HPC is slightly more flexible and therefore models many other high strength materials that have some flexibility incorporated in the molecular structure to increase their solubility and, therefore, their blending ability as well as moderate the required processing conditions.

Experimental

The samples used are a 21 wt.% solution of PBLG in DBA and a 50 wt. % solution of HPC in D2O. The molecular weight of the PBLG is 236,000 and that of HPC is 100,000. The scattering experiments were completed on the NG5 8 m small angle scattering instrument in the Cold Neutron Research Facility at the National Institute of Standards and Technology (NIST) in Gaithersburg, MD. The shear cell that was used is an in house design which allows the *in-situ* scattering of a fluid while under shear flow in a couette geometry as well as the simultaneous measurement of the viscosity. Its particulars are described in detail elsewhere.[1] In this geometry the incident radiation enters parallel to the shear gradient and the collected scattering pattern is in the flow and neutral directions. The PBLG experiment was completed at 65 °C and the shear rate was varied from 0.07 s^{-1} to 1081 s^{-1}. The HPC experiment was completed at room temperature and the shear rate was varied from 0.02 s^{-1} to 10 s^{-1}. All scattering patterns were taken under steady state conditions. In other words, data were only

collected after at least the first 100 strain units,

$\dot{\gamma}$t > 100, so that the transient did not contribute to the scattering pattern.

The collected scattering data were corrected for empty cell scattering, detector non-linearity, electronic noise and stray scattering, sample transmission, and incoherent scattering. This corrected data were radially averaged over defined azimuthal sectors and then reduced to absolute units using a silica gel as a secondary standard to obtain scattering curves of I(q) (cm^{-1}) vs. q (Å^{-1}). q is the scattering wavevector (q=4π/λ sin(θ/2)), where λ is the wavelength of the incident radiation and θ is the scattering angle.

In order to analyze the scattering pattern along separate axes, it is possible to complete the radial averaging over an azimuthal sector. For the PBLG experiment, the radial averaging has been completed over 10° sectors that are centered parallel and perpendicular to the shear flow direction, while for the HPC results the sectors for averaging were 20° in breadth.

Results

Figure 1 is a plot of viscosity of the PBLG/DBA solution vs. shear rate as determined

simultaneously with the scattering experiments. This plot clearly shows that the shear rates studied include the plateau and the shear thinning regions of the viscosity curve (Regime II and III of the Onogi and Asada scheme).[2] Unfortunately, the torque transducer in the shear cell used to ascertain the viscosity was not sensitive enough to determine

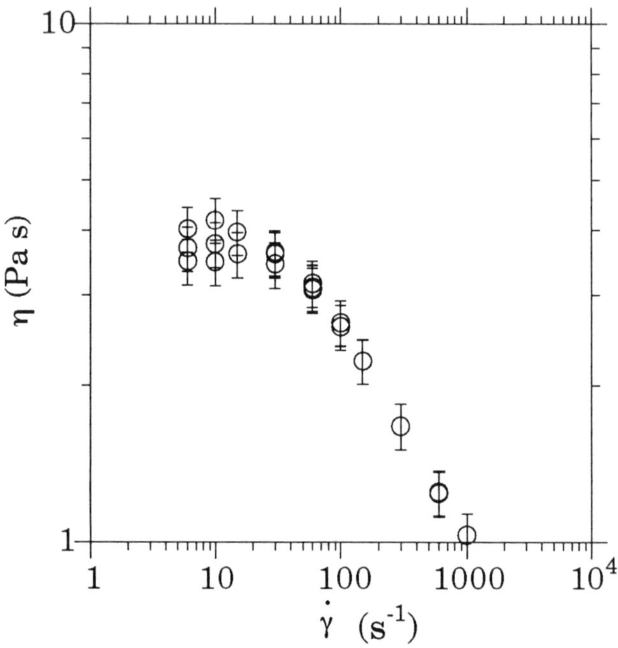

Fig. 1 - Viscosity of PBLG in DBA for the conditions of the scattering experiments.

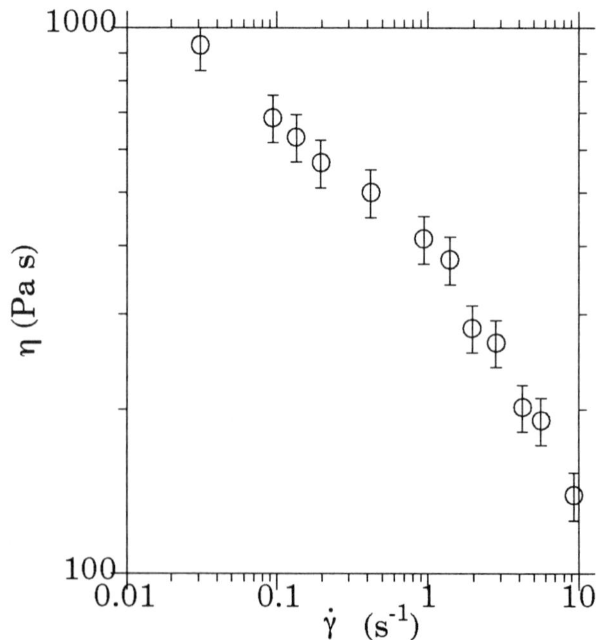

Fig. 2 - Viscosity of HPC in D2O.

the viscosity at lower shear rates. Figure 2 is a similar plot of viscosity vs. shear rate for the HPC solution. These results agree with previous studies of the rheology of HPC solutions in terms of the change of viscosity with shear rate.

Figure 3a and 3b show, respectively, the two dimensional scattering patterns for the PBLG solution and the HPC solution under shear flow at *ca.* 5 s^{-1}.

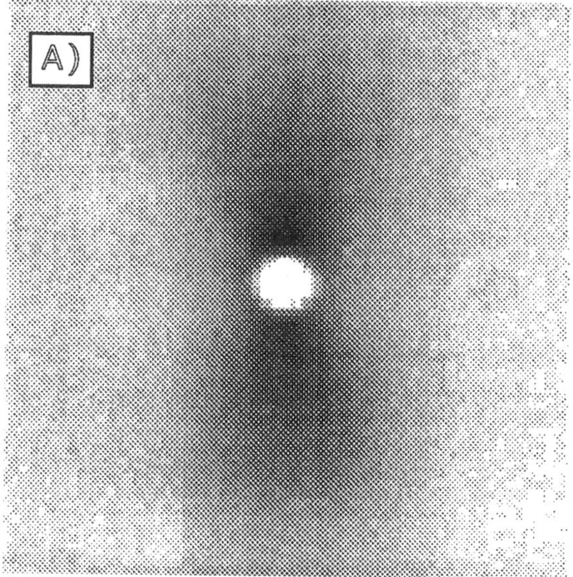

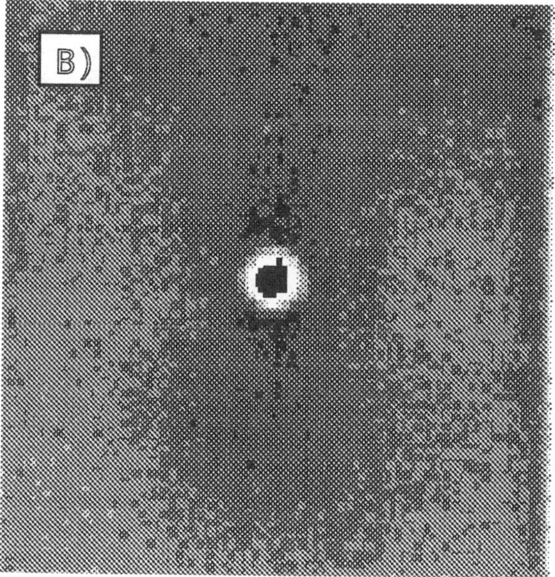

Fig. 3 - Representative two dimensional scattering patterns for a) the PBLG solution at 6.11 s^{-1} and b) the HPC solution at 5.65 s^{-1}.

The scattering patterns are clearly anisotropic demonstrating the alignment of the LCPs by the shear field. On a qualitative level, the scattering pattern of the PBLG solution shows a clearly defined lobe whose width and shape becomes more well defined with increasing shear rate. The scattering pattern of the HPC solution is also anisotropic, however, here the lobe is poorly formed with diffuse boundaries. This contrast is a result of the molecular structural differences, the rodlike PBLG molecule is more readily aligned than the less rigid HPC molecule.

However, we would like to quantify the alignment of the LCP molecules by the flow and demonstrate their change in alignment with shear rate. To do this, the fact that neutron scattering gives information of the solution on a molecular length scale is utilized. A quiescent LCP solution will have an isotropic 2-dimensional scattering pattern because the fluctuations that scatter the radiation are isotropic due to the polydomain nature of the solution. However, as the flow field interacts with the LCP, the fluctuations become anisotropic. This can be seen with the help of figure 4. Figure 4a shows the polydomain structure of an LCP solution at rest while figure 4b shows the same solution that has been partially oriented by a flow field. As the LCP molecule is aligned along the flow direction there is an increase in the concentration fluctuations along the vorticity direction and a decrease along the flow direction. This results in a decreased correlation length, *i.e.* average fluctuation size, normal to the flow direction and an increase in the correlation length parallel to the flow. It follows that the 2-dimensional scattering pattern of an LCP solution under shear that is aligned by the flow field will have an increased intensity perpendicular to flow and a suppression parallel to flow. Furthermore, the extent of the alignment of the LCP by the flow field can be characterized by the scattering intensity at low angles, which is proportional to the number of fluctuations present, in the flow and vorticity directions and their change with shear rate as well as a measure of the correlation length along the two orthogonal directions.

By completing an Ornstein-Zernike (OZ) style analysis, *i.e.* fitting the scattering profile to

$$I(q) = \frac{I(0)}{1 + q^2 \xi^2}$$

in the vorticity and the flow direction, a correlation length, ξ, and the intensity at q = 0, I(0) in both directions can be determined from the scattering patterns.

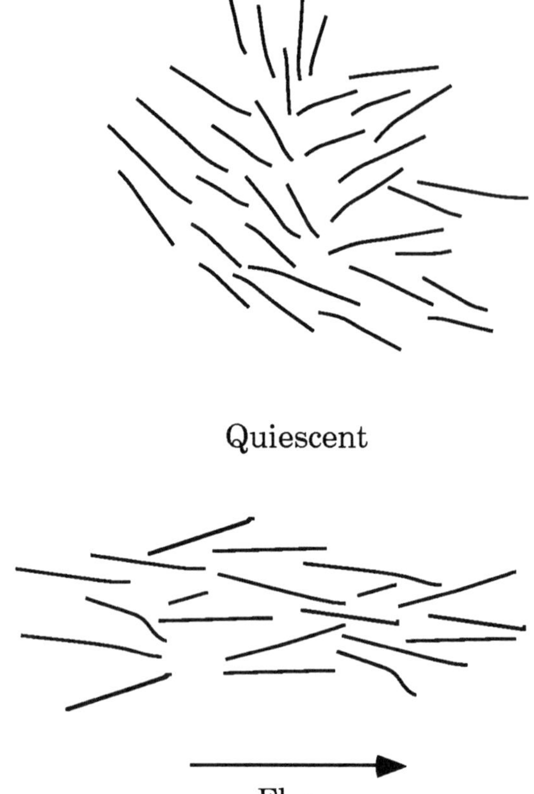

Fig. 4 - Schematic diagram of an LCP solution in the quiescent state (top) and partially aligned by shear (bottom).

This equation can be rewritten as $1/I(q) = 1/I(0)(1+ q^2\xi^2)$. It is this form of the equation that is fit to the data.

From the previous paragraph it can be seen that due to the anisotropic nature of the aligned LCP solution, the intensity of the scattering at $q=0$ in the vorticity direction, $I(0)_v$ will be greater than in the flow direction, $I(0)_f$. The magnitude of this difference, $\Delta I = I(0)_v - I(0)_f$, is a quantifiable measure of the alignment. Figures 5 and 6 show the change in ΔI with shear rate for the PBLG and the HPC solutions, respectively.

Figure 5 demonstrates the change with shear rate in the orientation of the rod-like PBLG molecule in solution. At low shear rates there is an increase in ΔI with shear rate, signifying an increase in the molecular alignment with flow velocity until a first critical shear rate is reached, $2 \text{ s}^{-1} < \dot{\gamma}_1 < 5 \text{ s}^{-1}$. As the shear rate passes this critical shear rate, the alignment of the PBLG

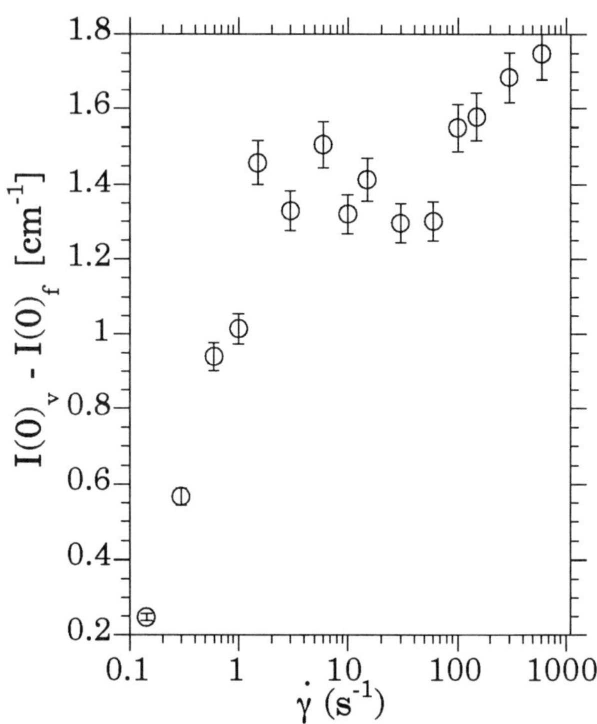

Fig. 5 - The change in $I(0)_v - I(0)_f$ with shear rate for the PBLG solution.

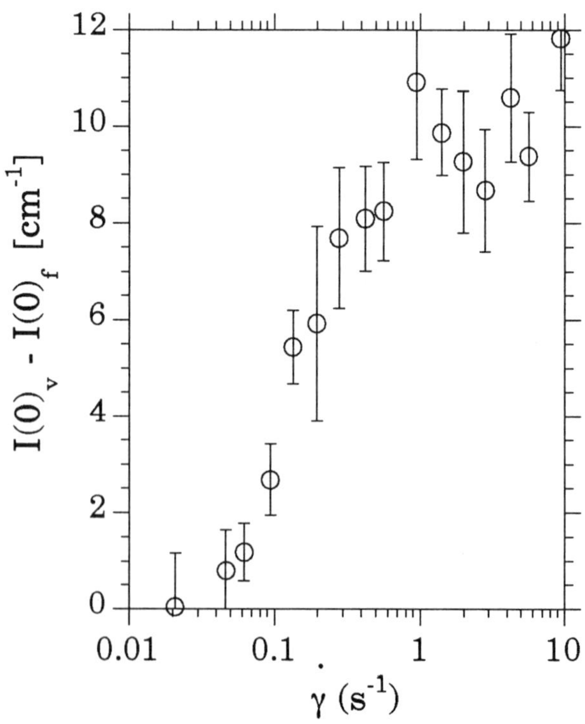

Fig 6 - The dependence of $I(0)_v - I(0)_f$ on shear rate for the HPC solution.

molecule does not change much with shear rate until a second critical shear rate is reached,

$20 \text{ s}^{-1} < \dot{\gamma}_2 < 50 \text{ s}^{-1}$. Above this second critical shear rate, the orientation of the LCP molecule continues to increase with shear rate. Figure 6, on the other hand demonstrates the change in the orientation of the less rigid HPC molecule in solution with shear rate. At low shear rates, there is again an increase in ΔI, and therefore the molecular alignment, with shear rate until a first

critical shear rate is reached, $\dot{\gamma}_1 \approx 1 \text{ s}^{-1}$. Above this first critical shear rate, the alignment of the HPC solution does not change much with shear rate. Unfortunately, it is not possible to complete the experiment for higher shear rates as the flow becomes unstable above $ca.$ 10 s^{-1} and there is sample loss from the cup.

To ensure reproducibility and reliability, it is desirable to have a second, independent measure of the molecular alignment from the obtained scattering pattern. The anisotropic lobe of the two dimensional scattering pattern is a consequence of the anisotropy of the solution, *i.e.* the LCP alignment. Therefore a quantification of the lobe anisotropy will also quantify the orientation in the solution. To do this, the width and height of the lobe at a given q were determined. This was completed by taking an azimuthal cross section of the two dimensional scattering pattern at a given q and fitting this cross section to a Lorentzian to garner the height and width of the lobe. In this analysis, the peak height is proportional to the number average of molecular segments that are aligned along the flow direction and the peak width is a measure of the distribution of the LCP alignment about the average value. Figure 7 shows the change in the peak height of the lobe with shear rate for both the PBLG and the HPC solutions on a double Y plot.

The changes in the orientation of the two LCP solutions with shear rate mirror those found in the previous ΔI analysis for both solutions, namely an increase in alignment with shear rate at low shear until a first critical shear rate is reached, a leveling off of the change of alignment with shear rate in the intermediate regime, and finally, for the PBLG solution, a further increase in the alignment with shear rate after a second critical shear rate is passed.

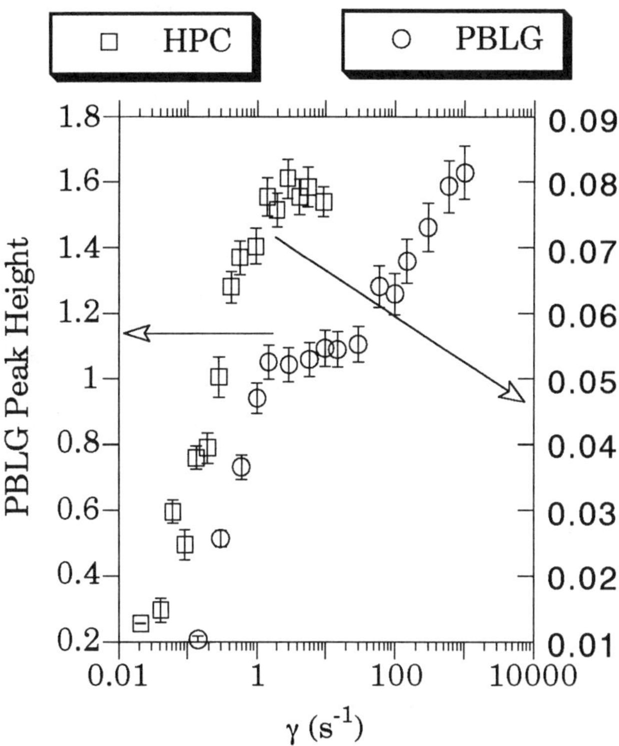

Fig. 7 - The change in the peak height of the anisotropic lobe with shear rate for the PBLG and HPC solutions

Discussion

There is a striking similarity of the alignment of the rod-like PBLG molecule by shear flow to the alignment of the less rigid HPC molecule. In both cases, there is an increase in the alignment of the LCP molecule with shear rate at low velocities. As a critical shear rate is surpassed the alignment does not change much with shear rate with further increased rate. Moreover, the magnitude of these

first critical shear rates are similar, $\dot{\gamma}_1$ for this HPC

solution is $ca.$ 1 s-^1 while $\dot{\gamma}_1$ for this PBLG

solution is $ca.$ $2\text{-}5 \text{ s}^{-1}$. It is tempting to draw conclusions regarding the proximity of these two first critical shear rates. However, there is evidence that the physical origin of the first critical shear rate for the rodlike PBLG molecule is the suppression of the longest molecular relaxation mechanism, the end over end rotation of the rod, by the shear flow.[3] As the shear rate exceeds the inverse of the rotational relaxation time, the flow field overcomes and suppresses the rotation. This suppression results in a leveling off of the alignment magnitude as the rod proceeds from a tumbling mechanism to wagging. For this physical mechanism to also hold true for the HPC solution, the rotational relaxation time of the semiflexible HPC molecule would have to be on the order of magnitude of the PBLG molecule. As the HPC molecule is shorter (smaller molecular weight) and more flexible it seems improbable that this is true and a physical origin for this response of the semiflexible molecule is being sought. Therefore, it seems merely fortuitous that the first critical shear rates for the two solutions are similar.

On a more practical point, the use of neutron scattering as a method for determining the orientation of an LCP by shear flow is not trivial. Therefore, we have also attempted to correlate the orientation of the LCP as measured by neutron scattering to a more common experimental method, viscosity measurement. Figures 8 and 9 show the relation of the orientation of the PBLG and HPC solutions, respectively, as measured by the peak height to the simultaneously measured viscosity. In both cases, the measured changes in the orientation of the LCP molecule correlate to observed changes in the viscosity. As seen in figure 8, the change in the viscosity from the Newtonian regime to shear thinning of the PBLG solution is accompanied by a transition from no change to an increase in the alignment of the LCP with shear rate. As was described previously, the torque transducer is not sensitive enough to accurately determine the viscosity of this solution at lower shear rates. This is unfortunate as the lower limit of the transducer is near the first critical shear rate, and therefore the viscosity and alignment can not be correlated in this regime. Further experiments are planned to overcome this shortcoming.

Figure 9 demonstrates how the change in alignment of HPC with shear rate correlates with the measured viscosity. Again, it is seen that the change in the alignment of the HPC molecule from an increase to leveling off with shear rate correlates to the change in the shear thinning behavior of the LCP solution. The slope of the shear thinning behavior increases as the alignment transition is surpassed.

Conclusion

A method to ascertain the orientation of a liquid crystalline polymer using neutron scattering has been described. This method has been used to determine the change in the alignment with shear rate of poly (benzyl glutamate) in deuterated benzyl alcohol, a model rodlike polymer, and

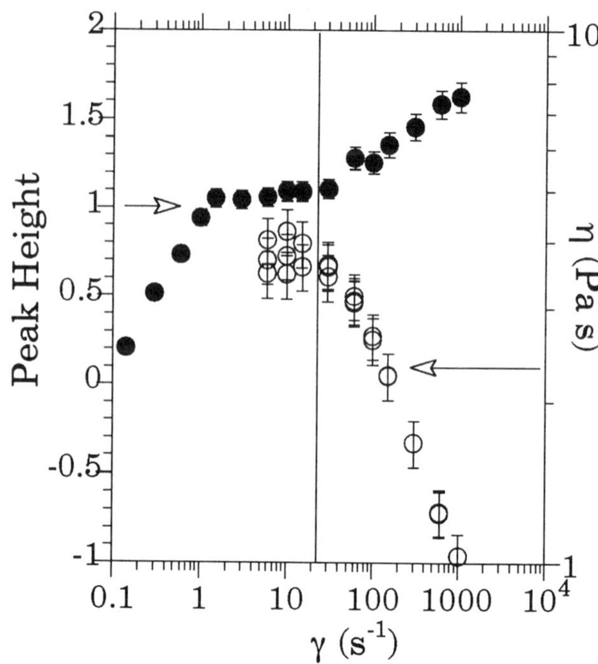

Fig. 8 - The correlation of the molecular alignment of PBLG in solution as determined by the azimuthal peak height to the viscosity.

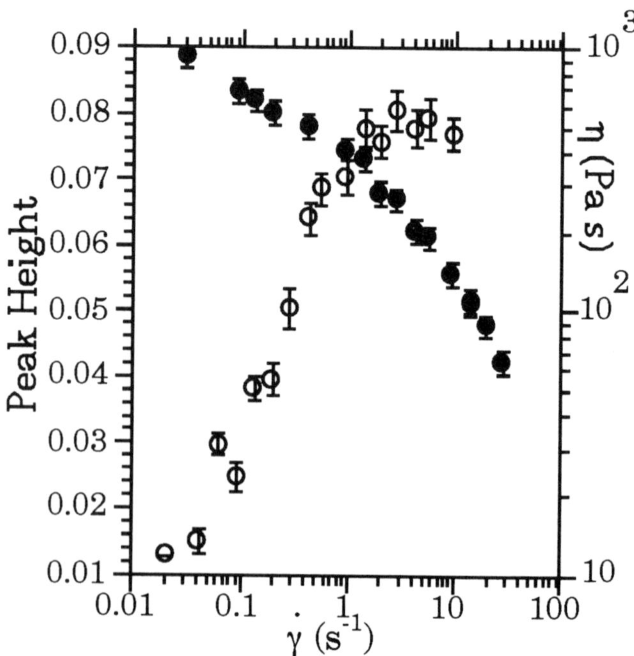

Fig 9. - The correlation of the molecular alignment of HPC in solution as determined by the azimuthal peak height to the viscosity

hydroxypropylcellulose in deuterated water, a model semiflexible liquid crystalline polymer. Though the rigidity of these two liquid crystalline polymers is very different, this technique has proven to be effective in determining the coupling of the shear field to the orientation of either LCP in solution. The results demonstrate different regimes in shear rate of the response to the shear flow for both LCPs studied here. An understanding of the relation of processing conditions to these regimes is required to optimize the desired properties of a commercial LCP product. Evidence of correlation of the alignment of the LCP as measured by neutron scattering to the shear viscosity is also presented. This allows the access of this technique to a greater community, as the neutron scattering and viscosity experiment can be completed once and then the viscosity can be utilized as a secondary standard to determine the molecular alignment of an LCP in solution in the future.

Acknowledgments

I would like to thank Dr. Charles Han and Dr. Alan Nakatani for help in completion of the described experiments and useful discussion during the data analysis. I would also like and the National Research Council for funding as a postdoctoral associate during which some of these experiments were completed.

References

1   A.I. Nakatani, H. Kim, and C.C. Han,
    J. Res. Natl. Inst. Stand. Technol., 95, 7
    (1990)
2   S. Onogi and T. Asada In "Rheology",
    G. Astarita, G. Marruci, and L. Nicolais Eds.,
    Plenum Press, New York (1980)
3   M.D. Dadmun and C.C. Han, Submitted to
    Macromolecules

Proceedings of the 10th Annual ASM/ESD Advanced Composites Conference, Dearborn, Michigan, USA, 7-10 November 1994

Characterization of Thermally Crosslinkable PPTA-co-XTA Copolymers

D.C. Martin, M-C. Jones, T. Jiang, J. Rigney
University of Michigan
Ann Arbor, Michigan

D. Mielewski
University of Michigan and Ford Motor Company
Dearborn, Michigan

Abstract

The mechanical properties of PPTA-co-XTA fibers were characterized as a function of XTA content and post-spinning heat treatment. Copolymer and PPXTA homopolymer fibers were dry-jet wet spun and exposed to a range of heat treatments to induce structural reorganization and crosslinking. Crosslinked copolymer fibers showed an improvement in tensile properties over as-spun fibers over the range of compositions. Strength and toughness tended to decrease in the PPXTA homopolymer with increasing length and temperature of the heat treatment. This is thought to be due in part to a free-radical chain scission process following crosslinking. Compressive properties of these fibers were also invesitgated. PPXTA and PPTA-co-5XTA fibers exhibited greater creep resistance in the crosslinked state than in the as-spun state. Crosslinked fibers exhibited increased resistance to lateral deformation.

PREVIOUS PAPERS HAVE DISCUSSED the synthesis[1-3], processing[3], and structural characterization[4] of crosslinkable PPTA-co-XTA fibers. It has been shown that the addition of even large amounts of XTA does not disrupt the ability of the copolymer to form anisotropic solutions or to be spun into highly oriented fiber, and that this order is not lost upon crosslinking. In this paper, we investigate the mechanical properties of these fibers as a function of XTA content and heat treatment. The creep behavior of these fibers is also examined. Other attempts to crosslink highly oriented fibers have included work on crosslinking PPTA and PBZT copolymers using aryl halide comonomers[5]. While improvements in the compressive properties of the materials were observed, the reaction chemistry involved the elimination of halogens which resulted in a significant loss of mass and may lead to voids and other structural defects. Rickert et al. described a crosslinking scheme for PPTA in which a disulfone substituted diamine is incorporated into the backbone during synthesis and activated during a post-spinning heat treatment[6]. This process also requires the elimination of small molecules (SO_2). They find that the 5% copolymer has compressive strength still comparable to that of PPTA, while the addition of more of the substituted constituent leads to a decrease in compressive strength as well as tensile properties[7].

Materials and Experimental Procedure

The PPTA-co-XTA copolymers used in this study were synthesized in the Chemistry department at the University of Michigan. The chemical structures of PPTA, and PPXTA are given in Figure 1. Details of the synthesis and spinning are given elsewhere[3]. We have found it possible to control the orientation, crystallinity, crosslinking and degradation of these fibers. Some of the fibers were annealed in a tube furnace at 260°C under slight tension. Both the spinning and tensioned heat treatment steps were designed to maximize the orientation of the molecules within the fiber. Residence times in the furnace were on the order of 10 seconds. Other as-spun fibers were heat treated at 400°C for

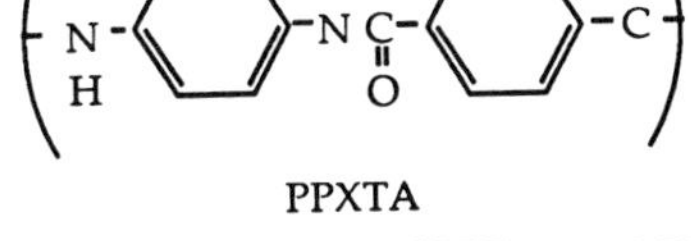

PPTA

PPXTA

Figure 1: The chemical structures of PPTA and PPXTA.

10 seconds to induce crosslinking. These heat treatments took place in batches rather than in a continuous manner in order to increase the possible residence times, and to allow small amounts of fiber to be heat treated. The batch-processed fibers were wound around a ~2 inch wire frame, and slid into a Lindberg tube furnace. The oven was calibrated with a Cole-Parmer Digisense thermocouple.

In addition to fibers spun at the University of Michigan, PPTA-co-5XTA (5 mole% XTA copolymer) and PPXTA homopolymer were sent to DuPont for spinning. PPTA-co-5XTA as-spun fibers and fibers heat treated at 450°C for approximately 6 seconds were returned to us. In the case of PPXTA, only as-spun fiber was sent back to us. Because the PPXTA fiber was not spooled, heat treatment took place in batches as described above. The PPXTA fibers were heat treated at 330°C, 410°C, and 530°C for 10, 30, and 120 seconds at each temperature.

Testing. Single fibers were swollen in sulfuric acid (96%, Aldrich) under a nitrogen atmosphere to determine whether they had been crosslinked during heat treatment. A small piece of fiber (200-400 μm in length) was placed on a glass slide and a few drops of sulfuric acid were placed near the fiber. The slide was placed in a glass chamber that was flushed with dry nitrogen and sealed to prevent the absorption of water by the acid. The chamber was placed on the stage of a Leitz optical microscope and tilted so that the acid came in to contact with the fiber. The behavior of the fiber in the acid was videotaped using a Sony CCD camera attached to the microscope.

Tensile tests were carried out on an Instron 4204 testing frame equipped with an Omega 1000-g capacity load cell and fiber grips. A crosshead speed of 1 mm/min and a sample gage length of 1 inch were used in all tests; tests were run to fiber failure. The load-displacement curves were recorded on an Omega strip chart recorder. The stress on the fiber was calculated from the load and from measurements of the fiber diameter. Diameter measurements were carried out using a Leitz optical microscope hooked up to a Macintosh Quadra 700. The image analysis program Image 1.53 was then used to measure the widths of the fibers. The machine compliance was determined according to ASTM standard D3379.

Toughness measurements were done by scanning the load-displacement curves into Adobe Photoshop with a Xerox Datacopy 730 GS scanner. The digitized curves were "filled in" and then opened in Image, where the number of pixels under the curve was measured. This value was divided by the number of pixels in one square of the chart paper in order to get the area under the curve, which could then be translated to known units of energy (J/m^3).

Recoil tests, as described by Allen[8], were carried out using the same equipment as the tensile tests. The fibers were loaded below their breaking stress and carefully cut. The fiber halves were then inspected for kinks. This procedure was repeated for a number of different stress levels. The compressive strength of the fiber was taken as the stress at which the fiber started to kink after cutting.

Elastica tests were carried out to get an estimate of the fibers' compressive strength. The test was developed in 1950 by Sinclair to test the compressive properties of glass fibers[9]. A loop is made in a fiber and drawn down. As the loop shrinks, the ratio of the major and minor axes should remain constant at about 1.34 until the fiber is damaged, either by kinking (in the interior of the loop), or tensile failure (on the exterior). The strain at either surface of the fiber may be calculated from the major axis of the loop and the radius of the fiber according to the formula:

$$\varepsilon = \pm \frac{R}{0.350 M} \tag{1}$$

where R and M are the radius and major axis, respectively. Thus an estimate of the critical strain (the strain at the point when the ratio of the axes deviates from its original value) can be made.

For our elastica tests, fiber diameters were first measured as previously described. Then a single fiber was taped at one end to a glass microscope slide, and a small piece of tape attached to the other end to make it easier to grasp. The fiber was given a half-twist to form a loop. A drop of microscope immersion oil was placed at the center of the loop and a glass coverslip placed on top, covering the loop but leaving the free end of the fiber uncovered. The slide was placed under a Nikon SMZ-27 stereoscope attached to the Sony CCD camera. The videotape was started, the free end of the fiber grasped with tweezers and slowly pulled to draw down the loop. The test was continued until the fiber broke or the loop was completely drawn down. The videotape was later analyzed by grabbing images of the loop with a RasterOps color card into Image, and measuring the major and minor axes of the loop.

Creep tests were performed on fibers as well. Cardboard tabs were epoxied to the ends of long (~15 cm) fibers and held in place with clothespins until the epoxy was dry. Reference marks were made in ink on each fiber approximately 10 cm apart. The fibers were hung from a stand and preloaded with a paperclip (~0.4 g) before measurements of their initial lengths were made with a PTI traveling microscope. The load on each fiber was chosen according to fiber diameter; generally, the DuPont-spun fibers were loaded with 5-10 g and the University of Michigan-spun fibers were loaded with 50 g. These loads correspond to stresses of approximately one-third of the tensile strength. Once the fiber was loaded, the distance between the marks was measured as a function of time. Fiber diameters were measured as described above to determine the stress on each fiber.

Experiments were carried out to study the transverse compressive behavior of fibers as a function of crosslink density. Sample fibers were mounted on a flat surface (glass microscope slide) and compressed laterally by a pipette with a series of dead loads. The deformed samples were examined in the optical microscope and SEM to determine the change in diameter and morphology upon compression.

Results and Discussion

UM-spun copolymers. *Swelling:* Swelling experiments confirmed that as-spun fibers and those that had been heat treated at 260°C were not crosslinked, as they dissolved in the acid as neat PPTA does. Those that had been heat treated at 400°C swelled in the acid but did not dissolve, indicating the presence of crosslinks.

Tensile testing: Information on a number of properties of the fibers was gathered by analyzing the load-displacement curves generated by tensile tests. Among these were modulus, tensile strength, and toughness. Results of these analyses are shown in Figures 2, 3, and 4, respectively. These plots demonstrate that XTA may be substituted into the neat PPTA structure without degrading fiber properties, and may even improve them slightly in the as-spun and annealed states. The tensile properties increase almost linearly with XTA content. In contrast, Glomm et al. found that the best properties of DSDA-substituted PPTA are found at low DSDA concentrations and decrease sharply with additional substitution[7]. Upon crosslinking, PPTA-co-XTA properties tend to follow the same trends as they do from the as-spun state to the annealed. Modulus and strength increase even more, while elongation decreases. The toughness of

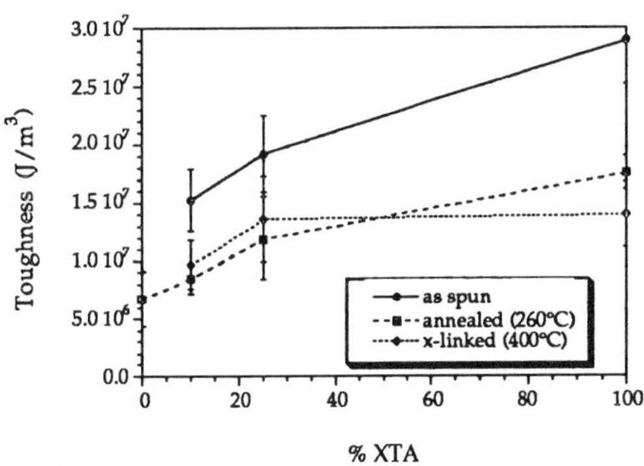

Figure 4: Toughness increases with XTA content.

crosslinked copolymer fiber shows a slight improvement over the annealed fiber at low XTA concentrations, but decreases for the PPXTA homopolymer.

Elastica: Figure 5 shows the critical bending strains of as-spun, annealed, and crosslinked copolymers. In these fibers, the critical strain is first reached on the side of the fiber in compression. The fact that there is little change in the critical values upon crosslinking suggests that the compressive strength does not change. However, more information about the compressive modulus of the fibers and the effect of fiber anisotropy in this test is needed for clearer interpretation of the results.

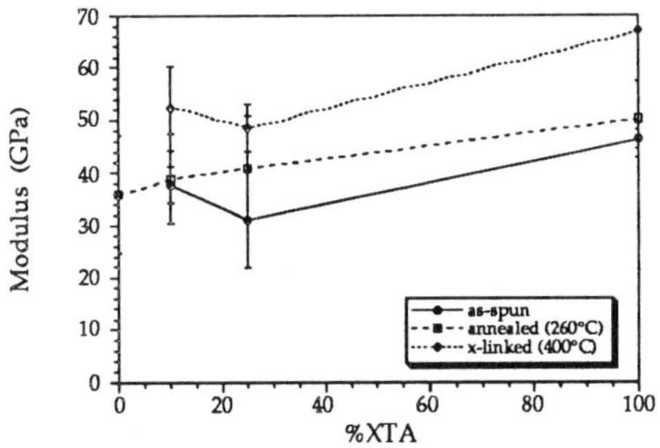

Figure 2: Modulus increases slightly with crosslinking and XTA content.

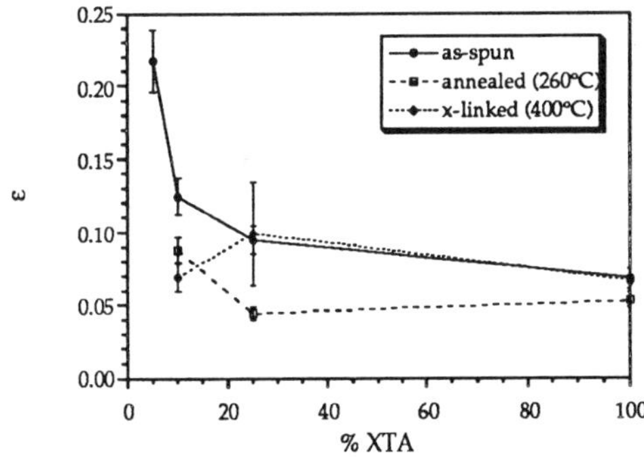

Figure 5: Critical bending strain does not change greatly upon crosslinking.

DuPont-spun PPTA-co-5XTA. The fibers which had been heat treated at 450°C were found to be crosslinked upon swelling in sulfuric acid, while the as-spun fibers dissolved. Fiber properties of the as-spun and crosslinked copolymers are given in Table 1. Values in italics are those reported by DuPont. Torsional modulus was obtained through a procedure described by DeTeresa et al.[10]

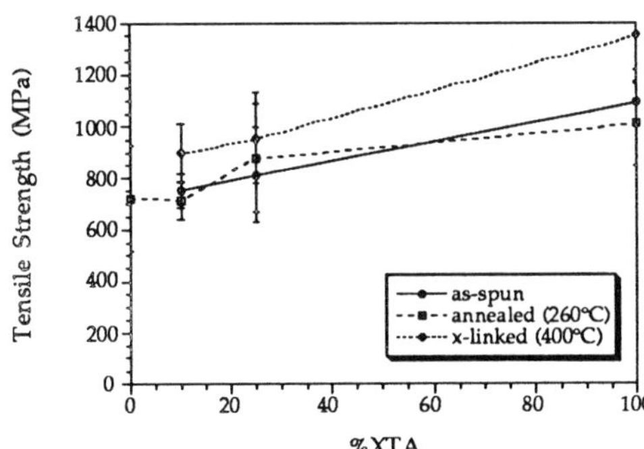

Figure 3: Strength increases with crosslinking and XTA content.

	As-Spun	Crosslinked
Tensile Modulus (GPa)	49.2	66.57
	40.18	*64.08*
Tensile Strength (MPa)	2.36	1.64
	2.03	*1.65*
Elongation (%)	4.34	2.40
	5.2	*2.36*
Toughness(J/m^3)	4.13 10^7	1.52 10^7
Critical Bending Strain	0.0513	0.0227
Torsional Modulus (GPa)	2.21±0.50	2.04±0.13
Compressive Strength (MPa)	253	288
Creep Rate Parameter (ε/decade)	0.00103	0.00039

DuPont-spun PPXTA. *Swelling*: Swelling experiments confirmed that those fibers that had been heat treated at 330, 410 and 530 °C were crosslinked, as they did not dissolve in the sulfuric acid, while the as-spun fibers did dissolve. A quantitative analysis of PPXTA fiber swelling will be presented in another paper.

Tensile testing: It was evident merely from handling the PPXTA fibers during sample preparation that the fibers became more and more brittle as heat treatment time and

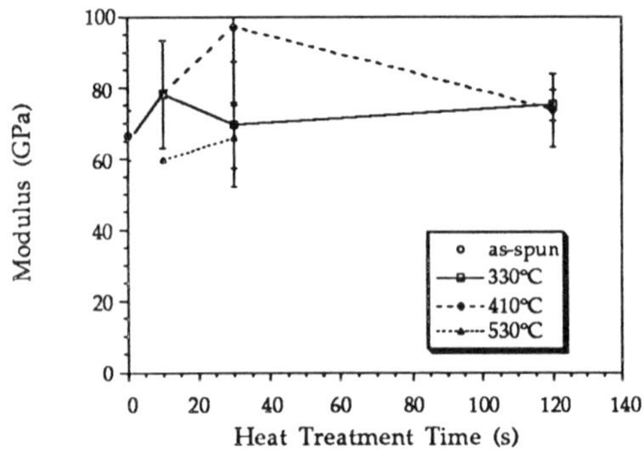

Figure 6: Tensile modulus of PPXTA fibers increases slightly upon crosslinking.

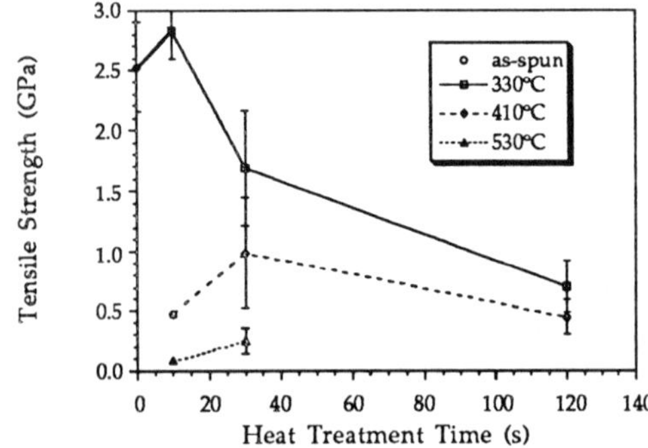

Figure 7: Strength of PPXTA fibers decreases upon all but the least rigorous of crosslinking heat treatments.

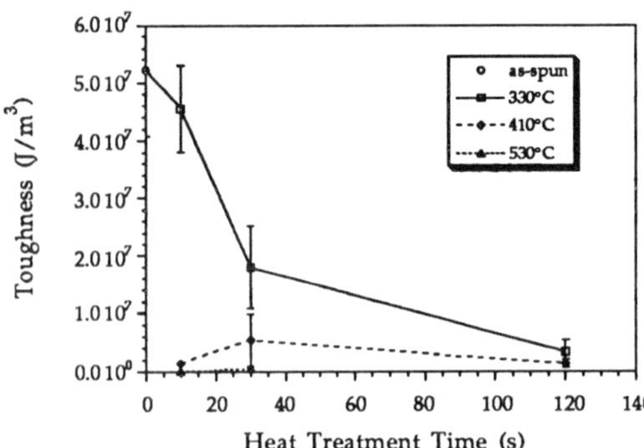

Figure 8: Toughness of PPXTA fibers decrease upon crosslinking.

temperature were increased. Results of the tensile tests, presented in Figures 6 through 8, show that while the tensile modulus tends to increase slightly after heat treatment, the strength and toughness of the crosslinked fibers are dramatically lower than those of the as-spun fiber. A possible explanation for this loss of strength is the production of free radicals after the crosslinking reaction, which could lead to chain scission. FTIR results show a loss of the hydrogen-bonded amide functionality[11], while electron spin resonance (ESR) spectroscopy indicates the formation of free radicals in crosslinked samples[12]. Sweeny has suggested an isocyanate degradation product[5] consistent with our FTIR data[11].

Observation of the fracture surfaces in the SEM showed a transition from a fibrillar mode of failure in the as-spun and lower heat treatment conditions to a brittle fracture mode at higher heat treatments. This transition from a slower, energy-absorbing fibrillar failure mode to a fast brittle failure could also contribute to the fibers' loss of toughness. A similar transition was seen by Feldman et al. in a study of the tensile properties of poly(*p*-phenylene benzobisthiazole) (PBZT) films[13]. In it, they found that the morphology of tensile samples after failure exhibited a fibrillar-brittle transition around 500°C, accompanied by a decrease in tensile strength. They attributed this behavior to thermo-oxidative degradation. Sweeny also mentions a change in "tensile failure appearance" of crosslinked PBZT fibers from fibrillar to brittle after the loss of 30% of the original halogen concentration, corresponding to 0.15 crosslinks per unit, and a decrease in tensile strength[5]. Taken together, these observations support the hypothesis that the morphological transition and decrease in strength are due to degradative chain scission.

The effect of aging on the tensile properties of PPXTA fibers was a concern, as Glomm et al. reported a degradation of tensile properties of ther fibers over the course of a year[7]. One group of PPXTA fibers that had been heat treated at 430°C for 30 seconds was tensile-tested three days after heat treatment, and another group 38 days afterwards. The differences in the values of tensile strength, modulus, elongation and toughness of the two groups of fibers were found to be equal within experimental error over the course of a month. More work on this phenomenon is in progress.

Creep: The heat treated PPXTA fibers exhibited increased resistance to creep when compared with the as spun fibers. It was found that for samples subjected to similar stress levels; the amount of creep and the creep rate are higher in the as-spun material. When comparing the creep rate of fibers subjected to different heat treatment temperatures, the creep rate of the crosslinked material (heat treated at 330°C) was less than that of the as-spun and annealed materials (heat treated at 260°C). However, the mechanisms of this enhanced creep resistance need further study in order to separate the effects of crosslinking, orientation and recrystallization.

Elastica: Definite transition points could not be seen in the plots of M/m vs. m for the PPXTA samples. However, it was possible to see a transition of fiber failure mode from kinking to brittle fracture, roughly corresponding to the degree of heat treatment the samples had received. That is, samples that had been heat treated at low temperatures and times tended to kink, while those heat treated at high temperatures and long times tended to break. The transition in failure mode is indicative of the competition between increasing compressive strength and decreasing tensile strength.

Lateral Deformation: Figure 9 shows the change of diameters of laterally deformed PPXTA fibers heat treated at 330°C for various lengths of time. Fibers which had been heat treated longer deformed less. It is our understanding that covalent crosslinks would restrain the relative movement between molecules and crystallites which could take place during lateral deformation. Longer heat treatment leads to

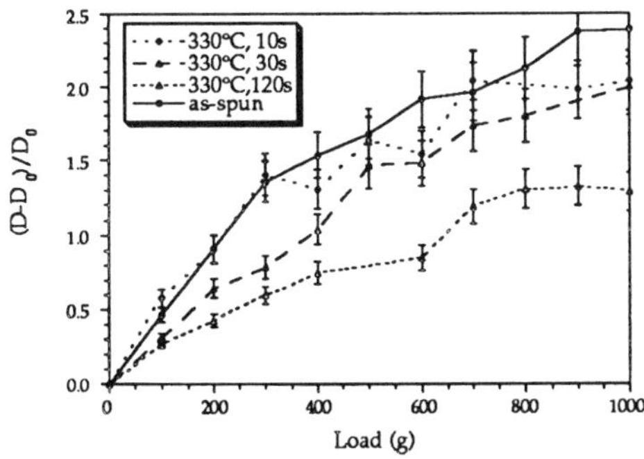

Figure 9: Change in fiber diameter after lateral compression increases with applied load and decreases as a function of heat treatment time.

higher crosslink density and, therefore, higher resistance to lateral compression.

When the laterally deformed fibers were examined in the optical microscope and the SEM, we observed strikingly different morphologies between the crosslinked and uncrosslinked fibers. For the uncrosslinked samples the deformed areas were flat and smooth while for the crosslinked fibers we saw crack-like textures in compressed areas. This is believed to be the consequence of the microfibrillar structure within the fibers. The microfibrillar boundaries in uncrosslinked fibers played no role under external compression and the fibers deformed uniformly. When the fibers were crosslinked, however, the boundaries confined the crosslinking within the individual fibrils and created a structural heterogeneity. These microfibrils tended to fall apart under lateral compression and resulted in a different deformation morphology.

Conclusions

The substitution of XTA in PPTA has been shown to occur without a degradation of tensile strength, modulus, and creep resistance, both in the as-spun and annealed states. Crosslinking improves the modulus, strength, and toughness of the UM-spun fibers relative to the annealed state. In DuPont-spun fibers, modulus is retained upon crosslinking, but a decrease in toughness and strength is also seen. This is believed to be due at least in part to an oxidative chain scission reaction which follows crosslinking. A decrease in strength and transition in tensile failure mode similar to what we observe has been found in other systems as they degrade. Also, the presence of free radicals and products of an attack on the amide bond in the backbone have been observed. Recoil and elastica tests have been inconclusive in determining the compressive strengths of these materials. Lateral compression of PPXTA fibers, however, indicates that crosslinking reduces the ability of the chains to move past each other. The creep and creep rate parameter of crosslinked PPXTA fibers are smaller than those of as-spun and annealed fibers, but the contributions of various mechanisms to this decrease are not yet fully understood.

Acknowledgments

This research was funded by the U.S. Army through grant no. DAAK60-92-K-0005. D.C.M. acknowledges support from the National Science Foundation in the form of a Young Investigator award. We thank Jospeh Manista and Robert Irwin of DuPont for spinning fibers, and Prof. J. C. Bilello for use of the Instron. Additional support came from DuPont, Hoechst-Celanese, and Ford Motor Company.

References

1. Walker, K.A.; Markoski, L.J. and Moore, J.S. *Synthesis*, 1992, 165-168.
2. Markoski, L.J.; Walker, K.A.; Deeter, G.A.; Spilman, G.E.; Martin, D.C. and Moore, J.S. *Chemistry of Materials*, 1993, **5**, 248-250.
3. Jiang, T.; Markoski, L.J. and Martin, D.C. Submitted to *Macromolecules*
4. Jones. M.-C. G.; Jiang, T. and Martin, D.C. Submitted to *Macromolecules*.
5. Sweeny, W. *Journal of Polymer Science, A: Polymer Chemistry*, 1992, **30**(6), 1111-1122.
6. Rickert, C.; Neuenschwander, P. and Suter, U.W. *Macromol. Chem.Phys.*, 1994, **195**, 511-524.
7. Glomm, B.; Rickert, C.; Neuenschwander, P. and Suter, U.W. *Macromol. Chem. Phys.*, 1994, **195**, 525-537.

8. Allen, S.R. *Journal of Materials Science*, 1987, **22**, 853-859.

9. Sinclair, D. *Journal of Applied Physics*, 1950, **21**, 380-386.

10. DeTeresa, S.J.; Allen, S.R.; Farris, R.J. and Porter, R.S. *Journal of Materials Science*, 1984, **19**, 57-72.

11. Jiang, T. The University of Michigan, unpublished results.

12. Mielewski, D.F.; Martin, D.C. and Bauer, D.R. submitted to *Polymer Preprints*, Fall 1994.

13. Feldman, L.; Zihlif, A.M.; Farris, R.J. and Thomas, E.L. *Journal of Materials Science*, 1987, **22**, 1199-1205.

Aging Characteristics of PBV in Interlayer Glass Composites

B.J. Love, R.J. May
Virginia Polytechnic Institute and State University
Blacksburg, Virginia

Abstract

PVB is used as the polymeric glue layer between glass panes in safety glass. In this laminated glass composite structure, by analyzing individual components and by characterizing the interface between them it is reasonable to model composite behavior under various load conditions. From this model, improvements can be made to increase various aspects of composite performance. PVB used in laminated composites actually consists of a copolymer structure and includes additional additives such as plasticizers. These affect the structure/properties of the innerlayer. There appear to be a number of problems associated with its long term use under photoaging conditions as an optical film. We will discuss photoaging and its effect on the impact and pressure loading capabilities of laminated glass based on load transfer adhesion, and durability.

LAMINATED SAFETY GLASS is made from two panes of glass held together by a transparent adhesive material. There has been renewed interest in characterizing and modeling the behavior of this material to determine its merit relative to other types of safety glass such as tempered or wire reinforced glass[1]. The performance of laminated composites is heavily dependant on the properties of this interlayer material[2,3]. Mechanical, optical, electrical and even acoustical properties of the adhesive all determine how the composite structure fulfills its functions. To fully characterize laminated glass and model its behavior in various loading conditions, it is first necessary to characterize the interlayer which holds it together.

The interlayer is a complex formulation that can include photo-oxidants, plasticizers/anti-plasticizers, colorants, as well as the polymer adhesive. The most prevalent polymer currently in use is Poly(Vinyl Butyral-co-Vinyl Alcohol) (PVB), a copolymer made from the reaction of Poly(Vinyl Alcohol) with Butyraldehyde. The primary additive to the formulation is a low molecular weight solvent, referred to as a plasticizer, which softens the polymer. Plasticization of PVB results in a pliable adhesive material that absorbs large amounts of energy on impact.

The primary role of safety glass used in transportation is to resist impact. To perform this function properly, the glass composite must absorb large amounts of energy during loading without allowing penetration of the impacting object which would occur if the composite failed. This resistance is directly related to the mechanical properties of the adhesive as well as the strength of the adhesive bond between the polymer and the glass panes.

Vallabhan *et al*[1] have done some work on characterizing PVB with the goal of modeling the composite behavior for use in architectural applications. Others[3,4] have studied the adhesive characteristics of this polymer for similar reasons. Earlier work[5,6] has shown that PVB changes chemically under the influence of UV radiation present in sunlight. This work will attempt to correlate the aging effects of solar radiation on the viscoelastic and adhesive characteristics of PVB with the goal of modeling laminated glass composite response to impact loadings.

VISCOELASTIC MECHANICAL PROPERTIES

All polymers undergo molecular rearrangement in the amorphous region during loading. The result is a lowering of the load carrying ability of the material while also absorbing energy. This relaxation is a time dependant phenomenon and a function of the thermal energy present in the material. There exists a temperature at which the relaxation process is greatly enhanced, ie the molecules are free to rearrange in the same manner as a liquid. This transition from a glassy solid to an equilibrium liquid occurs at the glass transition temperature (Tg). Above Tg the modulus and yield strength decrease rapidly. Most importantly, a peak in the energy dissipation with loading occurs around this temperature.

As the rate of loading increases, the thermal energy required to allow this relaxation is increased and thus the glass transition is raised relative to a slow loading process. Even below Tg, some relaxation mechanisms exist such that the mechanical properties are still rate dependant but to a lesser extent. This phenomenon is referred to as visco-elasticity and is the reason why understanding the effect of the loading rate is so critical to modeling polymer mechanical behavior.

Although still not fully understood, impact is known to be a high frequency process. The time of loading is on the order of microseconds to milliseconds and even faster. A polymer subjected to high frequency loading will appear to have the behavior of a material at a much lower temperature or correspondingly higher Tg.

Despite this effect, indications of the polymer response to impact can be determined using stress relaxation and variable rate tensile tests. The stress relaxation data can be used to determine inherent relaxation times which govern molecular rearrangement. Similar information can be generated by observing the effect of loading rate on the stress/strain characteristics of the polymer.

A further reason for mechanically testing PVB is that the polymer can absorb energy in two different ways. Molecular rearrangement and the loss peak at Tg are explained above. The second mode of absorption is gross yielding of the polymer. A typical stress/strain curve for PVB is shown in Figure 1 below. The area under the curve is the amount of work required to cause failure. Any strain above the yield point contributes to energy which is irrecoverable upon unloading. Thus the more ductile the polymer and the higher the strain to failure the more energy absorbed during loading.

The load carrying ability of a polymer is a function of plasticizer content and the existence and degree of cross-linking among other things. Plasticizers, when added to a polymer, dramatically affect the load carrying ability by decreasing polymer-polymer interactions. This can increase the free volume present in the polymer and lead to an increase in molecular motion as evidenced by an lowering of Tg. Secondly, the stress/strain characteristics can be drastically altered by the presence of plasticizer since the molecules are more free to rearrange. These effects can be studied using stress relaxation and variable rate tensile testing.

Cross-linking binds groups of chains together and reduces the ability of the polymer to flow during loading. A high degree of cross-linking can drastically reduce the energy absorption during impact whereas small amounts can be beneficial by allowing the polymer to undergo large strains without failing.

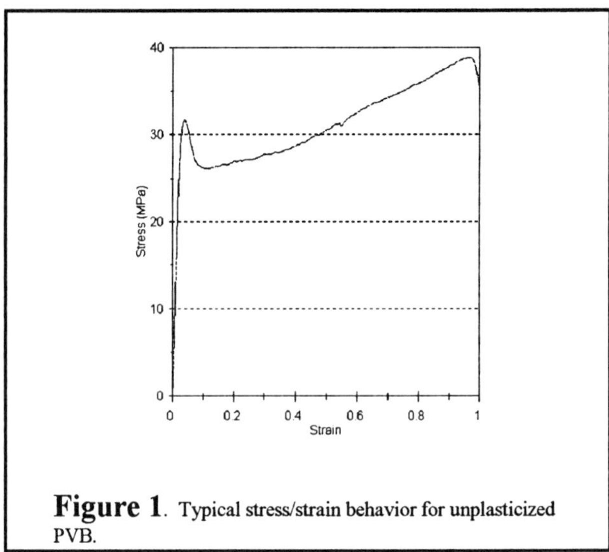

Figure 1. Typical stress/strain behavior for unplasticized PVB.

ADHESION CHARACTERISTICS

The load transfer mechanism of the adhesive layer is not only dependant on the mechanical properties of the PVB layer but on the interfacial adhesion between the glass and the polymer. This adhesion is greatly dependant on a number of factors, most prominent of which are the temperature of the composite and the surface characteristics of the adherends.

There are a number of theories to explain PVB/Glass adhesion. Chehimi and Watts[7] have speculated that the adhesion is due to Acid-Base interactions present. Chugunov et al[4] investigated the effect of surface roughness on PVB/Glass adhesion and concluded that mechanical interlocking is not a significant factor. Huntsberger[3] studied the effect of water at the interface and concluded that water greatly lowers the adhesion of PVB to glass. In this work we will be investigating role of aging in adhesion.

Peel and lap shear testing have been a common form of mechanically testing adhesion. Both yield a force required to separate the adherands but neither can distinguish the force required to cause yielding in the adherends and the strength of the bond. This problem can be circumvented by the use of a properly run Blister Test as shown schematically in figure 2.

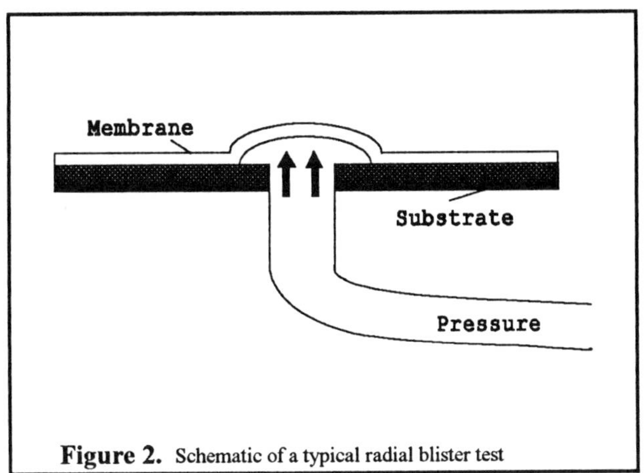

Figure 2. Schematic of a typical radial blister test

The radial blister test involves bonding a film of adhesive onto a substrate and applying pressure using air or other gas through a hole in the substrate. This pressure causes a blister to form in the film which transfers the stress to the bond interface. If the test is designed properly, debond occurs prior to yielding in the film. From the pressure and the resultant volume changes, a critical strain energy release rate, G_c, can be calculated. G_c is a measure of the energy required to cause rapid, unstable debonding of PVB from the glass substrate. It is given by the following equation

$$G_c = \frac{3(1-\nu^2)}{32E*H^2} * p^2a^4 \qquad (1)$$

where ν is the poisson's ratio, E the elastic tensile modulus, p the pressure, H the height at the center of the debond and a the radius of the initial debond. H can be calculated by the volume of gas required to cause the debond. Lai and Dillard[7] have proposed a method of designing the blister test such that yielding does not occur. By adjusting the initial debond radius and thickness of the film, the stress in the film membrane can be lowered below the yield stress of the polymer. Thus accurate measurement of the strength of an adhesive interface between the film and the substrate is possible without introducing errors caused by yielding of the adherends.

AGING EFFECTS

The above consideration of time and temperature dependent mechanical properties as well as the adhesion characteristics of the glass/PVB system are generally considered when first designing the composite and assumed to hold true throughout the life of the part. Unfortunately PVB is not stable and suffers from aging effects. Work done by Reinohl et al[5] identified spectroscopic changes in PVB when it was irradiated by a light source corresponding to changes in the structure of the polymer. They have also reported some limited effects on mechanical properties.

Mikhailik et al[6] have reported the formation of free radicals using ESR in PVB radiated with 240-270nm wavelength light. They showed that at extremely low temperatures, -100 C, these free radicals will not recombine due to insufficient mobility but that at room temperature, it is possible for these radicals to react with other chains to form crosslinks.

Radicals form when the polymer absorbs significant amounts of radiative energy. UV light absorption causes chain scission and the formation of reactive endgroups. It is possible, if there is sufficient mobility, for these endgroups to react with adjacent chains and form links. These links are chemical bonds which are much stronger than the secondary bonds which normally link polymer chains. The formation of these bonds will change the resulting mechanical properties and energy absorption capability. Such cross-linking may also affect adhesion. Mechanical interlocking will be promoted but any chemical interactions will be significantly altered due to the change in chemical structure of the polymer chain. If cross-linking does not occur, the presence of free radicals will still alter the chemical characteristics of the chains and may significantly lower the molecular weight. If these changes appear at the interface between the polymer and the glass they will alter the polymer/glass adhesion which would lead to changes in impact performance.

EXPERIMENTAL

PVB was obtained in granular form from PolySciences and had a published Mw of 180,000 to 270,000 g/mol. Butyl Benzyl Phthalate was used in various concentrations ranging from 0 to 40 % as the plastizicer. A 5% by weight solution of polymer and plasticizer was prepared in a 20/80 mixture of Toluene/Ethyl Alcohol by weight. This solution was then shaken on a Gyratory@ Shaker-Model G2 until the polymer completely dissolved, normally 1.5 - 3 days. This solution was then cast onto a clean glass plate and the solvent allowed to evaporate. The films were peeled off the glass substrate after 12 hours and any remaining solvent allowed to evaporate over several days. The neat PVB, 5% and 10% plasticized samples were removed from the plate without difficulty at room temperature. The more heavily plasticized films required freezing before removal.

The surface quality of the solution cast films and the possibility of residual stresses prompted the use of a hot press to obtain some of the required films. The films were pressed between either Kapton or Teflon release film and shims of the desired thickness. The films were pressed at a temperature between 140°C and 200°C and 30 psi load.

Glass transition measurements were done using Dynamic Mechanical Spectroscopy (DMS) on a Seiko Dynamic Mechanical Analyzer. A temperature sweep at a frequency of 0.1hz was run from -150°C to 100°C. The temperature at which a peak in the storage modulus corresponding to a dropoff in elastic modulus was taken to be Tg. The stress relaxation experiments were performed on a Polymer

Laboratories Miniature Materials tester with a 200 N load cell. The required dogbone samples were cut using a die conforming to ASTM D1708-84. These samples were initially strained 2 to 4% at a rate of 4 mm/mm/sec depending on the plasticizer content and allowed to relax for up to 30 hours. The more heavily plasticized samples were initially strained up to 70% to generate higher initial stress values however this caused the formation of cross-links in the samples. Tensile modulus measurements were also taken using the Miniature Materials tester at rates of 0.5 to 5 mm/mm/sec.

RESULTS AND DISCUSSION

A typical stress relaxation curve for the PVB samples is shown in Fig 3. This behavior can be modeled using a Maxwell-Weichert Model which is a combination of a number of Maxwell elements connected in parallel. Maxwell elements are mathematical equivalents of mechanical springs and dashpots. By using a set of perfectly elastic (springs) and perfectly viscous (dashpots) elements it is possible to model viscoelastic response. From these elements a time dependant elastic modulus is generated which is a summation of the contributions from each. This model has been shown to accurately model polymer behavior if enough elements are used[9].

The model is defined by the following equation.

$$E(t)=\sum E_i * \exp(-t/\tau_i) \qquad (2)$$

Where E_i is the elastic spring constant of the ith element. τ_i is the relaxation time for the ith elements.

For this work, a 3 element, 6 parameter model was first fitted to the data from a sample with 10% plasticizer. This model fitted the very short and the very long time behavior well but inadequately described the relaxation behavior between 5 and 50 minutes. For this reason a 4 element model was chosen which described all three regions well. Similar analyses are being performed on samples with different plasticizer content to determine the effect of plasticizer on relaxation mechanisms. The parameters used to generate the curve in Figure 3 are given in Table 1.

Table 1. Model Parameters

	1	2	3	4
E (MPa)	613	100	100	37
τ (min)	0.2	10	75	2000

The neat resin shows similar behavior but with significantly longer relaxation times.

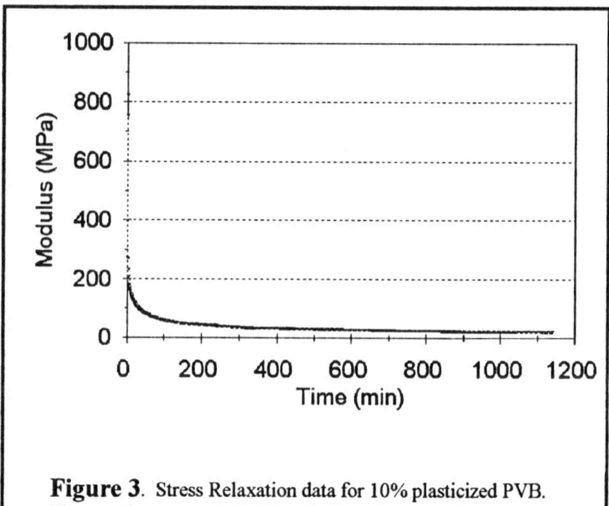

Figure 3. Stress Relaxation data for 10% plasticized PVB. The sample was strained 4% and allowed to relax for 19 hours. The solid line represents the fit of the Maxwell-Weichert model.

The results for the stress/strain tests are given in Table 2. As expected the modulus values decrease with increasing plasticizer content. This corresponds to the decreasing Tg. As Tg approaches the test temperature, the load carrying ability decreases accordingly. These experiments show the modulus increasing with increased rate, typical of a visco-elastic polymer tested near Tg. Thus, it can be seen that viscoelasticity contributes to the material response of this polymer.

We will present results from aging experiments run on the neat polymer and a 30% plasticized polymer. These experiments will include initial modulus and stress relaxation data and similar data after exposing the samples to simulated sunlight.

CONCLUSIONS

From the stress relaxation and DMS data it can be clearly seen that plasticizer affects the time dependant characteristics of the polymer. The polymer chains are more mobile and capable of more molecular rearrangements at lower temperatures with increased plasticizer content. This is clearly beneficial for impact conditions since the chains can respond more quickly to the high speed loading.

At present the aging studies are incomplete but it is believed that exposure to solar radiation will instigate molecular changes in PVB which may lead to cross-linking and/or Molecular Weight degradation. PVB has already been shown to generate free radicals when exposed to UV radiation and this may lead to cross-linking if the chains are mobile enough to allow reaction and link formation between adjacent chains. If this cross-linking continues unabated, PVB may stiffen under aging conditions and change its impact energy absorption characteristics with time.

Table 2. Initial Modulus (MPA) at the four loading rates and Tg Data ($^{\circ}$C)

% Plast	0.1 mm/min	1 mm/min	10 mm/min	100 mm/min	Tg
Neat	--	1700	--	--	76
5	--	700	200	--	60.1
10	500	675	890	--	55.1
15	--	--	520	--	48.2
20	--	--	115	70	37
30	--	--	--	9	22.4
40	--	--	--	2	36.6

ACKNOWLEDGEMENT

We would like to thank Dr. G. Wilkes and Mr. S. Srinivas for their help in obtaining the DMS data.

REFERENCES

1) C. V. G. Vallabhan, Y. C. Das, and M. Ramasamudra., "Properties of PVB Interlayer Used in Laminated Glass," J. Mat. Civil. Eng., Vol 4., No 1., p 71-76, February (1992).

2) B. J. Love, "Discussion of 'Properties of PVB Interlayer Used in Laminated Glass,'" ASCE J. Mat. Civil. Eng., Vol 5, No 4, p 546, (1993)

3) J. R. Huntsberger, "Adhesion of Plasticized Poly(vinyl Butyral) to Glass," J. Adhesion, Vol 13, No 2, p 107-129 (1981).

4) A. M. Chugonov, A. I. Kaprov, and V. N. Gusel'nikova., "Adhesive Bond in Triplex Glass," Steklo i Keramika, Vol 3, p 11-12, March (1985).

5) V. Reinohl, J. Sedlar and M. Navratil, "Photo-oxidation of Poly(Vinyl Butyral)," Polymer Photochemistry, Vol 1, No 3, p 165-175, July (1981).

6) O. M. Mikhailik, Y. N. Seropegi, M. Y. Melnikov, and N. V. Fock, "The kinetics and mechanism of Photo-ageing of Poly(Vinyl Butyral)," Eur. Polym. J., Vol 17, No 9, p 1011-1019, (1981).

7) Chehimi, M. M. and Watts, J. F. "X-ray Photoelectron Spectroscopy Investigations of Acid-Base Interactions in Adhesion. Part 2. The Determination of a Scale of Polymer Basicity by a Solid State Acid-Base Titration Method," J. Adhesion, Vol 14, No 1-4, p 81-91, (1993).

8) Y. Lai and D. A. Dillard., "Fracture Efficiency of Tests for Adhesive Bonds," Submitted to Intl. J. Solids and Structures. (1994).

9) J. J. Aklonis and W. J. MacKnight, Introduction to Polymer Viscoelasticity, Ch. 8, Wiley Interscience, New York, NY (1983).

3-D Stress Analysis in Textile Composite Plates Using a Combination of ANSYS and Sub-Element/Deficient Approximation Function Analysis

C.M. Pastore, A.E. Bogdanovich
North Carolina State University
Raleigh, North Carolina

V. Kumar, M. German
General Electric Corporate R&D
Schenectady, New Jersey

Abstract

One of the opportunities to improve the 3-D stress calculations for textile composite structures is to incorporate a novel 3-D Sub-Element/Deficient Approximation Function (SEDAF) analysis in commercial finite element codes. The SEDAF analysis allows one to provide a uniquely accurate and computationally efficient stress prediction, but this is currently limited to a rather simple geometries: rectangular plates, shallow shells, and circular cylinders. However, its combination with commercial finite element codes (using SEDAF analysis as a post-processor) seems to be a promising option for improving the local stress predictions in textile composite structural parts.

The methodology proposed is demonstrated using SOLID 46 element of ANSYS code. Nodal displacements calculated with SOLID 46 at the exterior surface of a selected local volume inside the structure, are used as input data for the SEDAF analysis. The nodal displacements are first interpolated along the coordinates on the exterior surfaces in order to generate the necessary input data. The interpolated displacement values are then used as boundary conditions in the SEDAF analysis.

The numerical example addresses simply supported rectangular plate made from a stitched laminated plane weave composite exposed to a transverse bending. The results obtained from the combination of ANSYS and SEDAF analysis are compared to the results from the direct calculations with ANSYS, and the results from the direct application of SEDAF analysis. It is concluded that the ANSYS SOLID 46/SEDAF combination improves radically the stress predictions provided by ANSYS SOLID 46 alone.

THREE-DIMENSIONAL ANALYSIS OF composite plates can be provided using different analytical and numerical approaches. A comprehensive discussion on a 3-D analysis of laminated plates can be found in [1, 2]. It was emphasized in these works that any displacement approximation that possesses a continuous (with respect to the through-thickness coordinate) first derivative at the interfaces between distinct layers provides necessarily discontinuous transverse stresses (in other words, the equilibrium of an interfacial element is violated). The continuity of transverse stresses can be satisfied only if transverse strains are discontinuous at the interfaces, as follows from Hooke's law. This means, in turn, that the first derivatives of displacements must be discontinuous with respect to the through-thickness coordinate at each interface. Although such a discontinuity is the necessary requirement for the correct and accurate solution, this is not the sufficient one [1, 2].

Textile reinforced structural parts which

1. are characterized with the elastic property variation in all three coordinate directions and

2. usually belong to a more general class of anisotropy than orthotropy,

provide a new challenge to a structural analyst. Some special analytical and computational tools need to be developed to address this novel complex problem. Commonly used commercial finite element codes like ANSYS, ABAQUS, ADINA, MSC NASTRAN, PATRAN-II, DYNA-3D all have 3-D orthotropic solid element solvers. However, the solid elements are often used for the analysis of inhomogeneous (laminated, specifically) structural parts. The accuracy of results obtained in such a way is very quastionable. From the conceptual point of view, this seems to be inconsistent because the continuity of transverse normal and shear stresses at the interfaces is not required. The question must be asked: "What is the quantitative effect of this inconsistency?" Until now this problem has not been investigated. From the general point of view, it is expected that the quantitative effect might have not be substantial in the "global-type" problems (when only displacements are of interest), but it would certainly result in significant error when stresses (particularly interlaminar stresses) are calculated for the purposes of failure analysis. An attempt to address this very practical question is taken in the present paper.

The basic methodology for a 3-D analysis of textile composite structural parts was presented in [1, 3, 4, 5]. One possible approach is based on the meso-volume concept. This allows one to reduce the problem to the analysis of a mosaic brick-type structure, considering a textile reinforced composite part as an assemblage of 3-D volumetric elements, each of them possessing individual, but uniformly distributed anisotropic elastic properties. In this case, one faces a step-wise variation of elastic properties throughout the part. The mathematical approach and computational procedure for this problem were presented in [5]. Another approach is to introduce a continuous variation of elastic properties along all three coordinates [3, 4]. The variation can be represented using the continuously interpolated reinforcement geometry throughout the volume of a part. The structural analysis problem is then reduced to the minimization of a total energy function containing a coordinate-dependent stiffness coefficients. The third approach incorporates some special interphase volumetric elements which are aimed to smoothen local step-wise variations of elastic properties in a reinforced structure [1]. The solution procedure in this case would be a combination of the first two. Several applications of the first approach were demonstrated by some of the authors. A 3-D woven rectangular plate was

considered in [3, 4]. Triaxially braided rectangular plate and cylindrical panel were treated in [5, 6]. A plain weave unit cell modelling was performed in [7]. Triaxially braided T-section plate was analyzed in [6].

The SEDAF analysis developed in [5] for a mosaic brick-type structure will be utilized in the present paper as a post-processor for the ANSYS SOLID 46 code. The example of a simply supported rectangular plate made from a stiched laminated plain weave composite will illustrate applicability of the combined ANSYS/SEDAF analysis.

Material Modelling

A 3-D SEDAF analysis presented in [5] is applicable to the composite structures which can be characterized at the meso-level as an assemblage of orthotropic rectangular parallelepipeds. This means that the whole structure can be discretized into a number of orthotropic meso-volumes each having uniformly distributed elastic properties. The set of 9 stiffnesses characterizing each meso-volume may be unique or may be identical for some of the meso-volumes. Thus, the following problems have to be addressed in the first two steps of the analysis:

1. an appropriate discretization of the structure into meso-volumes has to be established;

2. each of the meso-volumes should be characterized with 9 stiffness constants. (*i.e.* each meso-volume is orthotropic)

Problem (1) requires a special model to be developed for each specific type of a textile composite. For example, in the case of a traditional cross-ply laminate, its ply stiffnesses are independent of the in-plane coordinates, and the only stiffness variation is a step-wise function of the through-thickness coordinate. In this case, there is a simple choice: to consider each individual layer as a meso-volume.

In the more complex case of 3-D woven composite, for example, there is a significant local stiffness variation in all three directions. Therefore modelling this material as a brick-type mosaic with each individual brick being an orthotropic solid (condition 2), requires a careful selection of discretization. Clearly, there are infinite options for the division into bricks, and the question is: which of them provide meso-scopically orthotropic (or at least, quasi-orthotropic) bricks?

This basic problem recurs in most detailed analysis of complex reinforcement schemes. Currently the solutions to these problems are based upon an intimate knowledge of the reinforcement system and an understanding of the appropriate subdivision from this knowledge.

Assuming that the problem (1) is solved, the elastic properties of each individual meso-volume have next to be calculated. This can be done using some micro-mechanics analysis, averaging procedures utilizing the known yarn and resin properties and local spatial variation of the reinforcement architecture. Some work in this area has been presented previously by some of the authors and their colleagues. For example [8] illustrates a wide range of models available in the literature, and [9] shows an efficient methodology for predicting the local meso-volume elastic properties.

3-D Boundary Problem Analysis

Using the elastic properties defined previously for each meso-volume in a mosaic brick-type structure, the global type analysis of a structurally inhomogeneous composite part can be addressed. Consider a rectangular plate composed of $L \times M \times N$ orthotropic bricks (Figure 1). Any brick, while related to the "global" coordinate system x, y, z, has three planes of elastic symmetry perpendicular to the x, y, and z-axes. The stress-strain equations for a composite plate as a whole are written in the form:

$$\sigma_i(x, y, z) = Q_{ij}(x, y, z)\varepsilon_j(x, y, z) \qquad (1)$$

where $\sigma_i = \{\sigma_x, \sigma_y, \sigma_z, \tau_{yz}, \tau_{xz}, \tau_{xy}\}$ and $\varepsilon_j = \{\varepsilon_x, \varepsilon_y, \varepsilon_z, \gamma_{yz}, \gamma_{xz}, \gamma_{xy}\}$. The 9 stiffnesses $Q_{ij}(z)$ are considered step-wise functions of the x, y and z-coordinates. The strain-displacement equations are taken in the linear form

$$
\begin{aligned}
\varepsilon_x &= \frac{\partial u_x}{\partial x} \\
\varepsilon_y &= \frac{\partial u_y}{\partial y} \\
\varepsilon_z &= \frac{\partial u_z}{\partial z} \qquad (2) \\
\gamma_{xy} &= \frac{\partial u_x}{\partial y} + \frac{\partial u_y}{\partial x} \\
\gamma_{xz} &= \frac{\partial u_x}{\partial z} + \frac{\partial u_z}{\partial x}
\end{aligned}
$$

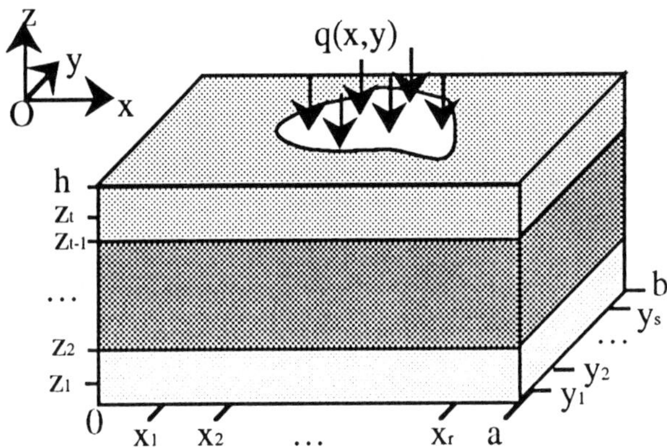

Figure 1: Illustration of Coordinate system used for analysis

$$\gamma_{yz} = \frac{\partial u_y}{\partial z} + \frac{\partial u_z}{\partial y}$$

The strain energy of a plate is given as

$$P = \frac{1}{2} \int \int_V \int Q_{lm}(z)\varepsilon_l\varepsilon_m \, dxdydz \qquad (3)$$

Assuming that distributed normal forces $q_z^+(x, y)$ and $q_z^-(x, y)$ are applied on some parts S^+ and S^- of the top $(z = h)$ and bottom $(z = 0)$ surfaces of the plate, the work of external forces is expressed as:

$$
\begin{aligned}
W &= \int int_{S+} q_z^+(x, y)u_z(x, y, h)dxdy \\
&+ \int\int_{S-} q_z^-(x, y)u_z(x, y, 0)dxdy \qquad (4)
\end{aligned}
$$

The required solution has to satisfy (1), (2), and the minimum total energy principle

$$\delta(P - W) = 0 \qquad (5)$$

under a number of specific constraints. The constraints are obtained from the external and internal boundary conditions imposed at the face surfaces, interfaces, and the side edges of a plate. Before discussing each of these constraints, a general view on the proposed computational algorithm has to be taken.

Let us introduce the division of a plate into brick sub-elements using three sets of mutually parallel planes: $x_\alpha = x_1, x_2, \ldots, x_{r-1}$; $y_\beta = y_1, y_2, \ldots, y_{s-1}$; and $z_\gamma = z_1, z_2, \ldots, z_{t-1}$ (as shown in Figure1). This division can be arbitrarily non-uniform with respect to all of the three coordinates. It is important to note that each of the interfaces in a brick-type mosaic structure must coincide with some plane from the sets above.

After the sub-element mesh has been established, the appropriate system of basis functions has to be derived. Let us represent displacement field in a plate in terms of the following triple series

$$u_x(x, y, z) = \sum_{i=1}^{l_x}\sum_{j=1}^{l_y}\sum_{k=1}^{l_z} U_{ijk} X_i^u(x) Y_j^u(y) Z_k^u(z)$$

$$u_y(x, y, z) = \sum_{i=1}^{l_x}\sum_{j=1}^{l_y}\sum_{k=1}^{l_z} V_{ijk} X_i^v(x) Y_j^v(y) Z_k^v(z) \quad (6)$$

$$u_z(x, y, z) = \sum_{i=1}^{l_x}\sum_{j=1}^{l_y}\sum_{k=1}^{l_z} W_{ijk} X_i^w(x) Y_j^w(y) Z_k^w(z)$$

where $X_i^u(x), X_i^v(x), X_i^w(x)$; $Y_i^u(y), Y_i^v(y), Y_i^w(y)$; and $Z_i^u(z), Z_i^v(z), Z_i^w(z)$ are systems of polynomial basis functions, and $U_{ijk}, V_{ijk}, W_{ijk}$ are unknown coefficients.

Under the specific loading conditions on the top and bottom surfaces of a plate introduced in (4), the set of boundary conditions on the face surfaces take the form:

$$\begin{aligned}
\sigma_z(x, y, h) &= q_z^+(x, y) \\
\sigma_z(x, y, 0) &= q_z^-(x, y) \quad (7)\\
\tau_{xz}(x, y, h) &= \tau_{yz}(x, y, h) = \\
&= \tau_{xz}(x, y, 0) = \tau_{yz}(x, y, 0) = 0
\end{aligned}$$

In the present approach, using only the approximation (6) for the displacements, the boundary conditions (7) are not addressed explicitly. The presence of the non-zero integrals in (4) indicates that the corresponding stress component is a non-zero function of x and y coordinates along the S^+ and S^- areas respectively. It was shown in our previous works on laminated plates that when using increasingly finer meshes of sub-layers, the calculated $\sigma_z(x, y, h)$ function converges to the prescribed functions on the outer surfaces of a plate. The calculated $\tau_{xz}(x, y, h)$, $\tau_{yz}(x, y, h)$, $\tau_{xz}(x, y, 0)$, and $\tau_{yz}(x, y, 0)$ stresses converge to zero at any point of the corresponding face surface. Thus, the boundary conditions (7) can be satisfied in a limiting sense.

The following two sets of internal continuity conditions have to be satisfied on any interface $t = t_i$ between the ith and $(i + 1)^{st}$ bricks (here t may be the x, y or z coordinate):

- the continuity of displacements:

$$\begin{aligned}
u_x^i(t_i) &= u_x^{i+1}(t_i) \\
u_y^i(t_i) &= u_y^{i+1}(t_i) \quad (8)\\
u_z^i(t_i) &= u_z^{i+1}(t_i)
\end{aligned}$$

- the continuity of transverse stresses

$$\begin{aligned}
\sigma_z^i(t_i) &= \sigma_z^{i+1} \\
\tau_{xz}^i(t_i) &= \tau_{xz}^{i+1}(t_i) \quad (9)\\
\tau_{yz}^i(t_i) &= \tau_{yz}^{i+1}(t_i)
\end{aligned}$$

The conditions (8) are satisfied automatically when using any continuous approximation functions in (6). More sophisticated considerations are needed when trying to satisfy the conditions (9). As discussed in [1, 2, 3, 4, 5],, only those displacement approximations are appropriate for the problem under consideration, which possess selective discontinuous first derivatives with respect to the x, y, or z-coordinate, depending on the particular interface. When applying "deficient" splines (in the terminology of [10]), for example, as the approximation functions and using increasingly finer meshes of the brick sub-elements, two distinct transverse stress magnitudes calculated from both sides of the interface tend to the same value. This means that the set of numerical solutions obtained when increasing number of sub-elements, converges to the unique solution which satisfies all the necessary equations and boundary conditions, including the continuity of transverse stresses at the interfaces. It should be pointed out that the situation is totally different if continuously differentiable polynomials are used for the displacement approximation in (6). In this case, two distinct limits for each transverse stress component are obtained from the two sides of an interface when using increasingly finer sub-element mesh. Accordingly, in this case a continuous distribution of transverse strains is obtained in the limit. This emphasizes the necessity of incorporating selective "deficiencies" in the approximation functions for a brick-type mosaic structure.

Appropriate systems of basis functions having selective discontinuous first derivatives can be derived in different ways. In particular, the recursive procedure was developed and used for generating deficient splines of an arbitrary degree in [11, 12, 2]. These basis functions are used in the present analysis. Analogous functions can be obtained by using deficient B-splines [10]. If the spline functions having degrees m_x, m_y, and m_z are used in the x, y, and z-coordinate approximations, the upper summation indices in (6) are: $l_x = r + 1 + L(m_x - 1)$, $l_y = s + 1 + M(m_y - 1)$, and $l_z = t + 1 + N(m_z - 1)$. Here L, M, and N are the numbers of meso-volumes and r, s, and t - the numbers of sub-elements in the plate in the x, y, and z-directions accordingly.

A general set of "deficient" basis functions to be applied in (6) has to be specified for the particular boundary conditions at the side edges of the plate. This aspect of the solution methodology was comprehensively described in [2, 5]. The primary systems of basis functions (derived without any constraints coming from the side edge boundary conditions) are first represented in a vector form: $\{X_i^s\} = \mathbf{X}^s$ and $\{Y_i^s\} = \mathbf{Y}^s$, with $s = \{u, v, w\}$. Then the "specified" basis vectors $\mathbf{X}_*^u = \mathbf{U}^x \mathbf{X}^u$, $\mathbf{X}_*^v = \mathbf{V}^x \mathbf{X}^v$, $\mathbf{X}_*^w = \mathbf{W}^x \mathbf{X}^w$, $\mathbf{Y}_*^u = \mathbf{U}^y \mathbf{Y}^u$, $\mathbf{Y}_*^v = \mathbf{V}^y \mathbf{Y}^v$, $\mathbf{Y}_*^w = \mathbf{W}^y \mathbf{Y}^w$ are obtained. Here, $\mathbf{U}^x$, $\mathbf{V}^x$, $\mathbf{W}^x$, $\mathbf{U}^y$, $\mathbf{V}^y$ and $\mathbf{W}^y$ transformation matrices which have to be specified for any particular set of boundary conditions.

In this paper, the traditional boundary conditions of a "simply supported edge" will be used:

$$
\begin{aligned}
u_z(0, y, z) &= u_z(a, y, z) = 0 \\
u_z(x, 0, z) &= u_z(x, b, z) = 0 \\
u_y(0, y, z) &= u_y(a, y, z) = 0 \\
u_x(x, 0, z) &= u_x(x, b, z) = 0 \\
\sigma_x(0, y, z) &= \sigma_x(a, y, z) = 0 \\
\sigma_y(x, 0, z) &= \sigma_y(x, b, z) = 0
\end{aligned}
\tag{10}
$$

After the specified basis functions corresponding to some set of boundary conditions at the side surfaces of a plate have been derived, the solution procedure is rather traditional and can be briefly described in the general terms. The triple series (6) which includes the specified basis functions, are substituted in (2), then the displacements and strains are substituted in (3) and (4). And finally, these energy expressions are substituted in the variational equation (5). This results in a system of linear algebraic equations (its order can be equal to or less than $3 \times l_x \times l_y \times l_z$). The system has to be solved with respect to the unknown coefficients U_{ijk}, V_{ijk}, and W_{ijk} using some common numerical technique. After that, the summation of (6) provides displacements. The strains are calculated from (2) using term-by-term differentiation of the series (6). The stresses are then calculated from (1). As it is seen, the presented approach does not use any assumptions rather than the displacement representation in the form of (6). The procedure does not incorporate any post-processors for strain/stress calculations. The 3-D stress/strain state in any specific point x^*, y^*, z^* can be directly obtained using the three coordinates input.

3-D Analysis of the Local Volumetric Element

The solution algorithm described for a 3-D global boundary problem in the previous section can be used, after some modifications, for the case of a mosaic brick-type volumetric element representing some part of a composite structure. This is a necessary step in using SEDAF analysis as a post-processor for a commercial finite element code. The mathematical approach and computational algorithm developed for laminated composite structures can be found in [13]. Actually, this does not need any modification when applied to a mosaic brick-type structure. The only difference is that not only the vectors $\mathbf{Z}^u$, $\mathbf{Z}^v$, and $\mathbf{Z}^w$, but also the vectors $\mathbf{X}^u$, $\mathbf{X}^v$, $\mathbf{X}^w$, $\mathbf{Y}^u$, $\mathbf{Y}^v$, and $\mathbf{Y}^w$ in the primary set of basis functions have to possess selective discontinuities in their first derivatives when a mosaic structure is under consideration. This peculiarity does not influence the specification of basis functions for any external boundary conditions. Thus we can refer to [13] for the details and restrict ourselves here with some main features of the algorithm.

There are three sets of the boundary condition matrices $\mathbf{U}^x$, $\mathbf{V}^x$, $\mathbf{W}^x$, $\mathbf{U}^y$, $\mathbf{V}^y$ and $\mathbf{W}^y$ which have to be defined for a local composite element. Consider the same element shown in Figure 1 assuming that now this represents some internal part of the whole plate. The values of u_x, u_y and u_z displacements are prescribed at the set of nodal points which belong to the surfaces $x = 0$, $x = a$, $y = 0$, $y = b$, $z = 0$, and $z = h$. The nodal points are defined by the discterization meshes introduced in the x, y, and z-directions. Let us consider u_z displacement and represent its discrete variation along the surfaces $x = 0$ and $x = a$ in the form

$$
\begin{aligned}
u_z(0, y_m, z_n) &= \alpha_{w0}^{mn} \\
u_z(a, y_m, z_n) &= \alpha_{wa}^{mn}
\end{aligned}
\tag{11}
$$

where $m = 0, 1, \ldots, s$ and $n = 0, 1, \ldots, t$. Here, s and t are characteristics of the discretization meshes in the y and z- directions (see Figure 1). This allows one to treat an arbitrary non-uniform distribution of displacements prescribed at the nodal points $\{y_m, z_n\}$ belonging to these surfaces. For example, elements of the boundary condition matrix $\mathbf{W}^x$ can be written, for any $\{m, n\}$ pair, in the following form:

$$
\begin{aligned}
w_{pq}^x(m, n) &= \delta_{pq} + (\alpha_{w0}^{mn} - 1)\,\delta_{p1}\delta_{q1} + \\
&\quad + (\alpha_{wa}^{mn} - 1)\,\delta_{pl_x}\delta_{ql_x}
\end{aligned}
\tag{12}
$$

The components of a specified basis vector $\mathbf{X}_*^w(x)$, corresponding to the line $\{y_m, z_n\}$, now can be represented as follows

$$
\begin{aligned}
X_{*1}^w(x, y_m, z_n) &= \alpha_{w0}^{mn} X_1^w(x) \\
X_{*2}^w(x, y_m, z_n) &= X_2^w(x) \\
&\cdots \\
X_{*l_x-1}^w(x, y_m, z_n) &= X_{l_x-1}^w(x) \\
X_{*l_x}^w(x, y_m, z_n) &= \alpha_{wa}^{mn} X_{l_x}^w(x)
\end{aligned}
\tag{13}
$$

where $m = 0, 1, \ldots; s$ and $n = 0, 1, \ldots, t$.

Note that the components $X_{*2}^w, \ldots, X_{*l_x-1}^w$ are the same as in the primary basis vector $\mathbf{X}^w$. Hence only $2 \times (s + 1) \times (t + 1)$ components:

$$
X_{*1}^w(x, y_0, z_0) = X_{*1}^{w00}(x),
$$

$$
\begin{aligned}
X_{*1}^w(x, y_0, z_1) &= X_{*1}^{w01}(x), \\
&\cdots \\
X_{*1}^w(x, y_0, z_t) &= X_{*1}^{w0t}(x), \\
X_{*1}^w(x, y_1, z_0) &= X_{*1}^{w10}(x), \\
&\cdots \\
X_{*1}^w(x, y_s, z_t) &= X_{*1}^{wst}(x),
\end{aligned}
$$

and

$$
\begin{aligned}
X_{*l_x}^w(x, y_0, z_0) &= X_{*1}^{w00}(x), \\
&\cdots \\
X_{*1}^w(x, y_s, z_t) &= X_{*1}^{wst}(x),
\end{aligned}
$$

depend on the boundary conditions and have to be specified.

Each of these two groups of terms can be represented in a matrix form: $\mathbf{X}_{*1}^w(x) = \{X_{*1}^{wmn}(x)\}$ and $\mathbf{X}_{*l_x}^w(x) = \{X_{*l_x}^{wmn}(x)\}$. The functional matrices $\mathbf{X}_{*1}^w(x)$ and $\mathbf{X}_{*l_x}^w(x)$ are related to the primary basis functions $X_1^w(x)$ and $X_{l_x}^w$ through the first and the last sets of equations in (13) which can be written in the following matrix form:

$$
\begin{aligned}
\mathbf{X}_{*1}^w(x) &= X_1^w(x) \circ A_0^w \\
\mathbf{X}_{*l_x}^w(x) &= X_1^w(x) \circ A_a^w
\end{aligned}
\tag{14}
$$

Here, A_0^w and A_a^w are $(s + 1) \times (t + 1)$ matrices with the elements $\{\alpha_{w0}^{mn}\}$ and $\{\alpha_{wa}^{mn}\}$ accordingly.

The above considerations for the u_z displacement demonstrates a basic idea for the derivation of specified basis vectors. The general derivation for all three displacement components and all three coordinate directions can be found in [13].

If the volumetric element is divided into r, s, and t intervals in x, y, and z- directions correspondingly, the total number of the input displacement components is

$$
\begin{aligned}
K &= 6\,((s + 1)(t + 1) + \\
&\quad + (r + 1)(t + 1) + (r + 1)(s + 1))
\end{aligned}
\tag{15}
$$

When applying this formula for the case $r = s = t = 1$, one obtains $K = 72$. Actually, in this case the volumetric element contains 8 nodal points only, and 3 displacement components are prescribed at each of them. Hence the total number of the necessary input displacements in this case is 24. Another 48 are excessive. This example demonstrates that when developing a computational algorithm for the problem under consideration, one has to be especially careful in imposing displacement values at the corners and on the intersections of the exterior surfaces of the element. Another important requirement on the algorithm is that the element must be in the equilibrium under the prescribed nodal displacements. Also, any rigid body translations or rotations should be eliminated.

After the complete system of basis functions has been derived, the solution procedure follows the description presented previously. It has to be pointed out that for the problem considered in this section, the work of external force defined by (4) is identically zero.

Interpolation of Displacements Calculated With ANSYS

The ANSYS code provides a set of displacement vectors (or force vectors) associated with a set of discrete points on or within the volume of the part. For the purpose of this work, the displacement vectors were only used. As described above, it is necessary to establish boundary conditions (in terms of displacements) to carry out the SEDAF analysis. This can be done using various approaches. A displacement field can be directly calculated using the shape functions of the finite elements, and the boundary conditions needed for the SEDAF analysis can be then determined by evaluating the displacements and displacement gradients. This requires a substantial amount of information from ANSYS, and restricts the input data for SEDAF analysis according to the initial data obtained directly from ANSYS.

The method adopted for this work is to use the displacement values at a set of points which coincide with "pseudo-nodes" of the SEDAF analysis. The displacement values are only taken at the surfaces of the volume to allow the SEDAF analysis more freedom in calculating internal stresses, which is the ultimate objective of the approach. The initial set of displacement values were those at the nodes provided by ANSYS. Since the SEDAF analysis can require more boundary conditions, it is necessary to develop

some method for interpolating the values to complete the necessary set of input data. It is important to note that we know in advance the location of all pseudo-nodes $\{x_i, y_j, z_k\}$ in the SEDAF analysis which need the displacement boundary conditions.

The interpolation technique used here involves 3 steps, schematically illustrated in Figure 2. First, the nodal displacement values obtained from ANSYS are linearly interpolated along the exterior edges of the internal volumetric element. This interpolation follows the direction of the edge:

$$
\begin{aligned}
\mathbf{u}(x_i, y_j, z_k) \;=\; & \; t \times \mathbf{u}(x_{i-i'}, y_j, z_k) + \\
& + (1 - t) \times \mathbf{u}(x_{i+i''}, y_j, z_k)
\end{aligned} \qquad (16)
$$

where the interpolation is carried out on the line $\{y = y_j, z = z_k\}$, with $x_{i-i'}$ and $x_{i+i''}$ representing the "left" and "right" nearest neighbors to x_i on this line which possess displacement values obtained from ANSYS; and $t = \frac{x_i - x_{i-i'}}{x_{i+i''} - x_{i-i'}}$. An analogous method is incorporated for the y and z direction linear interpolations.

The set of displacement values obtained in step 1 is then used for the interpolation in step 2. The values are linearly interpolated along the material boundaries associated with the volume. The algorithm follows equation (16) with $x_{i-i'}$ and $x_{i+i''}$ representing again the "left" and "right" nearest neighbors to x_i on this line which have displacement values available after applying step 1.

In the step 3, the set of displacement values obtained in step 2 is used. Bilinear interpolations are carried out along the face surface to which the point belongs, following the general rule of:

$$
\begin{aligned}
\mathbf{u}(x_i, y_j, z_k) \;=\; & \; t \times \mathbf{u}(x_{i-i'}, y_j, z_k) + \\
& + (1 - t) \times \mathbf{u}(x_{i+i''}, y_j, z_k) + \\
& + s \times \mathbf{u}(x_i, y_{j-j'}, z_k) + \\
& + (1 - s)\mathbf{u}(x_i, y_{j+j''}, z_k)
\end{aligned} \qquad (17)
$$

where $x_{i-i'}$, $x_{i+i''}$ and $y_{j-j'}$, $y_{j+j''}$ are the "left", "right", and "top", "bottom" nearest neighbors to the point (x_i, y_j, z_k), t is as before (equation 16) and $s = \frac{y_j - y_{j-j'}}{y_{j+j''} - y_{j-j'}}$. After completing this three step process, all of the displacement

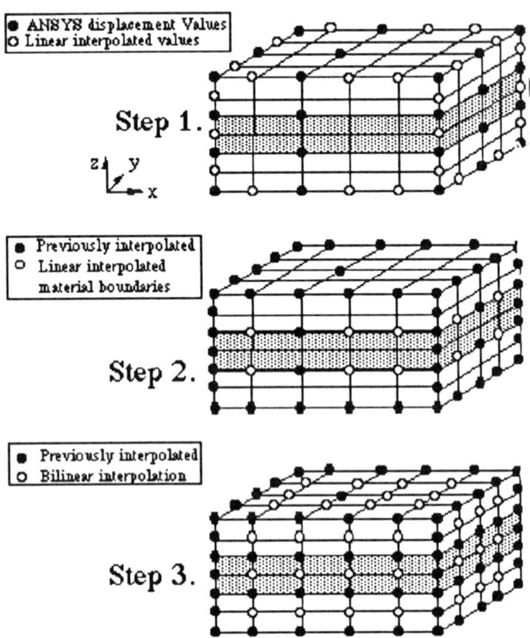

Figure 2: Illustration of interpolation scheme for transferring ANSYS output into SEDAF input.

boundary values are defined, and these values are converted into input data for the SEDAF analysis following the methodology presented previously.

Numerical Example

Consider a rectangular simply supported plate made from a stitched laminated plain weave composite. The material is modelled as an orthotropic solid containing stitches throughout. Figure 3 illustrates the material layout. The darkened regions represent areas of stitching with the remaining material consisting of plain weave.

With such a model, there are only two distinct material properties to be attributed to the structure. These will be denoted A for the background plain weave and B for the stitched plain weave component.

The material properties were derived through

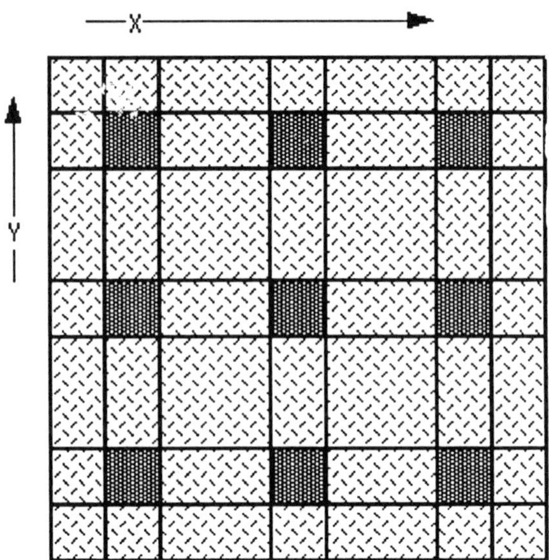

Figure 3: Schematic of stitched plain weave material modelled for numerical example results.

application of the model presented in [9] assuming a 12K AS-4 yarn woven into a 5 harness satin weave at a texture of 12 ends per inch and 12 picks per inch. The stitching was assumed to be carried out with a 24K AS-4 graphite yarn at a density of 3 stitches per inch in both directions. The effects of local fiber volume fraction increases were included in the model. The principal elastic constants determined and used for modelling are presented in Table 1.

Table 1: Principal Elastic Constants of Meso-Volumes Used in Stitched Plain Weave Analysis.

Material	E_{11} (GPa)	E_{22} (GPa)	E_{33} (GPa)
A	65	65	5
B	63	63	61

A square plate with width to thicknes of 9 was considered. 3 stitches were placed in both x and y directions symmetrically to represent a repeating volume. The plate was subject to a normal load applied on the top surface:

$$q_z^+(x,y) = q_0 sin\left(\pi\frac{x}{a}\right) sin\left(\pi\frac{y}{b}\right) \qquad (18)$$

The edge boundary conditions require no normal stresses on the x and y edges, and no transverse displacements on these same edges.

The stress analysis was provided using three different methods:

3-D SEDAF analysis using symmetry and modelling one quarter of plate, using cubic splines in the plane and quadratic through the thickness. A total of 8 sublayers in each of x and y were used, and 2 sublayers in the thickness direction.

ANSYS with 364 *SOLID 46* elements in the plate region. (the stresses were directly calculated with the finite element code).

ANSYS-SEDAF ANSYS SOLID 46 used for the displacement calculation and SEDAF analysis used for the stress calculation.

STRESS RESULTS   All of the previous work showing SEDAF results have demonstrated the ability to match within arbitrary precision the analytical results of laminated plates subject to transverse loading. Extending these results, the SEDAF solution for this stitched plate is considered as the baseline. As described previously, the kinematics of the problem are correct. Thus a comparison is made amongst the three solution methods, and it is assumed that the SEDAF result is closest to correct.

Since the difficulty in stress solutions is associated with jumps in material properties, it is insightful to consider stress distributions which pass through different material boundaries. In the present case, the stresses passing through both "background" weave and stitches should be of interest. The plane $x = 1.5h$ passes directly through two stitches, so is chosen as a region of comparison. Additionally, the "smooth" path, $x = 2.5h$, which does not pass through any stitches was examined.

To check the general validity of applied boundary conditions, the normal stress σ_z was examined as a function of x along the line $\{y = 2.5h, z = h\}$. As shown in Figure 4, all three of the solution methods give reasonable results for these values. However, the ANSYS solution does present a negative value of σ_z at the edge of the plate, $x = 0$. This value is no shown in the graph, but reaches a value of -0.2 at this point. The strange "dip" at the end of the ANSYS/SEDAF graph is a numerical fluctuation associated with imposition of symmetry. This can be removed with additional sub-layers added to the model.

When considering the same stress distribution but along a line which passes through a stitch

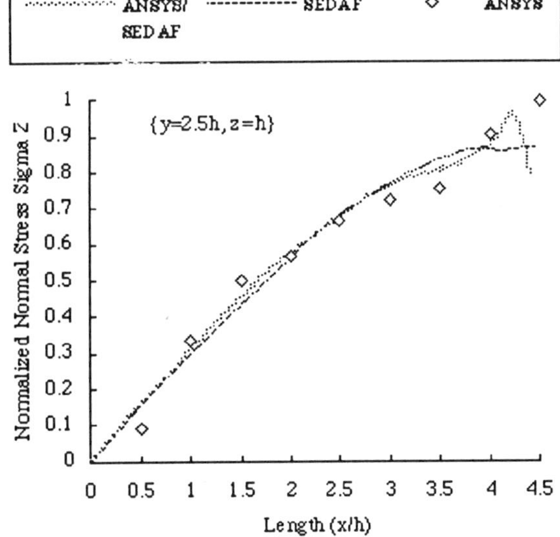

Figure 4: Distribution of Normal Stress σ_z along the line $\{y = 2.5h, z = h\}$ for SEDAF, ANSYS, and ANSYS/SEDAF analyses.

($\{y = 1.5h, z = h\}$, (Figure 5), it can be seen that the ANSYS results are rather poor, SEDAF does a good job of matching the prescribed values, and ANSYS/SEDAF improves the original ANSYS results considerably. It should be noted that there are obvious stress jumps in the ANSYS/SEDAF post-processed results, which could be eliminated with the use of additional sub-layers.

A more complex value to match is the transverse stress, τ_{xz} along the same two lines used above. Even when passing through a constant material ($\{y = 2.5h, z = h\}$), as illustrated in Figure 6, ANSYS does a rather poor job in predicting these stresses. Of course this is also related to the fact that only a single element was used through the thickness of the part. When post-processing the ANSYS results through SEDAF, the ANSYS/SEDAF results give much better values compared with SEDAF.

On the line $\{y = 1.5h, z = h\}$, as shown in Figure 7, the discrepencies are much more distinct. Even the post-processed ANSYS/SEDAF results show significant stress jumps at the stitch interfaces. These jump values could be reduced by increasing the number of sub-layers used in the analysis, however for the purposes of this paper, the same number of sub-layers was used for SEDAF and ANSYS/SEDAF to give computationally similar results.

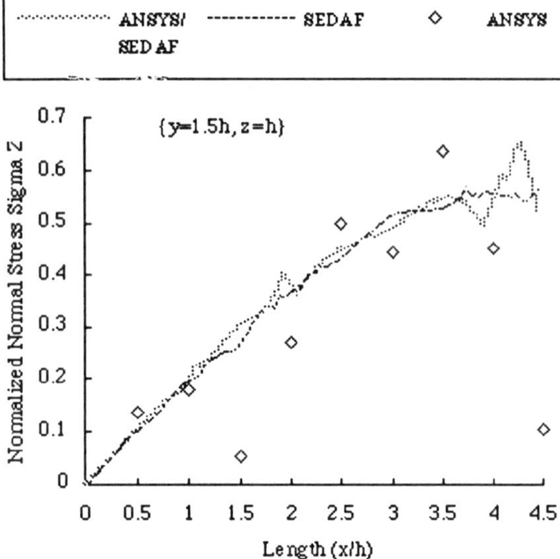

Figure 5: Distribution of Normal Stress σ_z along the line $\{y = 1.5h, z = h\}$ for SEDAF, ANSYS, and ANSYS/SEDAF analyses.

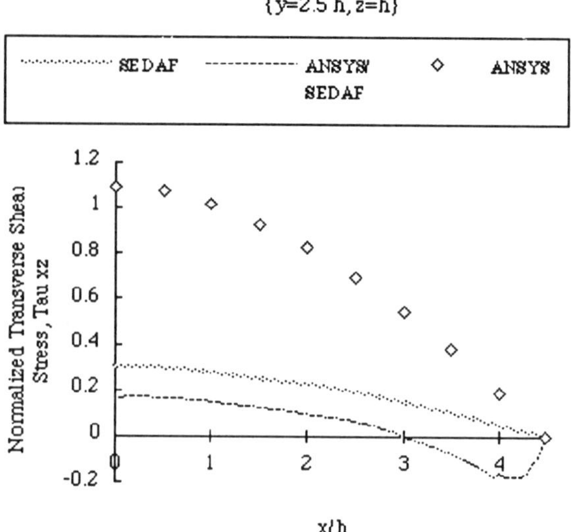

Figure 6: Distribution of Transverse Stress τ_{xz} along the line $\{y = 2.5h, z = h\}$ for SEDAF, ANSYS, and ANSYS/SEDAF analyses.

Conclusions

The developed ANSYS/SEDAF algorithm incorporates the primary displacement calculation using the commercial finite element code ANSYS and the successive stress calculation using 3-D SEDAF analysis. This combination allows one to obtain substantially higher accuracy when calculating both the in-plane and transverse stresses as demonstrated on the example of a stitched laminated plain weave plate. After some generalization, the presented ANSYS/SEDAF algorithm will be able to provide an accurate 3-D stress calculations for any structural analysis problem which allows for the application of ANSYS code.

Acknowledgments

The authors would like to thank Mr. B. P. Deepak and Mr. A. B. Birger for their assistance in providing the computational work.

References

[1] Alexander E. Bogdanovich. Spline function aided analysis of inhomogeneous materials and structures. In J. N. Reddy and K. L. Reifsnider, editors, *Local Mechanics Concepts for Composite Material Systems*, pages 355–382, Berlin, October 1991. International Union of Theoretical and Applied Mechanics, Springer-Verlag.

[2] Alexander E. Bogdanovich. Three dimensional analysis of laminated composite plates. In *Proceedings of Advanced Composites '93*, pages 79–91, Wollongong, Australia, February 1993. TMS, TMS Minerals, Metals, Materials.

[3] Alexander E. Bogdanovich and Christopher M. Pastore. On the structural analysis of textile composites. In *Topics in Composite Materials and Structures, AE-Vol 26, AMD-Vol. 133*, pages 13–91. ASME, ASME Publications, 1992.

[4] Christopher M. Pastore, Alexander E. Bogdanovich, and Yasser A. Gowayed. Applications of a meso-volume based analysis for textile composite structures. *Composites Engineering*, 3(2):181–194, 1993.

[5] Alexander E. Bogdanovich. Three dimensional analysis of aniostropic spatially reinforced structures. *Composites Manufacturing*, 4(4):173–186, 1993.

[6] A. Bogdanovich, C. Pastore, and A. Birger. *Analysis of Composite Shallow Shell Structures Reinforced with Textiles*, volume 2, pages 35–44. Editions Pluralis, Paris, 1992.

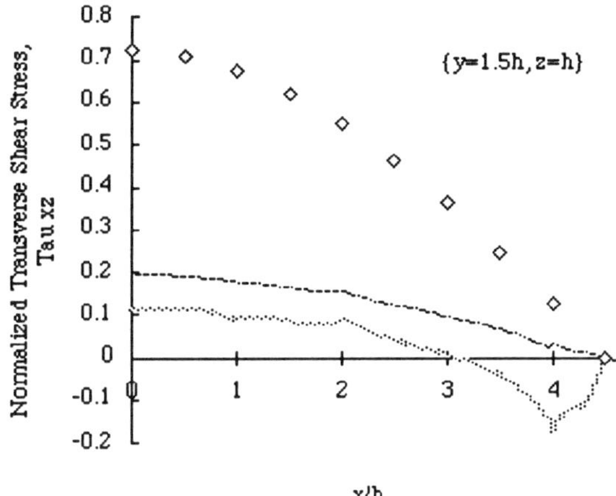

Figure 7: Distribution of Transverse Stress τ_{xz} along the line $\{y = 1.5h, z = h\}$ for SEDAF, ANSYS, and ANSYS/SEDAF analyses.

[7] Alexander E. Bogdanovich, Christopher M. Pastore, and Alexander B. Birger. Three dimensional deformation and failure anlaysis of textile reinofrced composite structures. In *Proceedings of the Ninth International Conference on Composite Materials*, pages 495–501, Madrid, July 1993. ICCM, Woodhead Publishing Limited.

[8] Yasser A. Gowayed and Christopher M. Pastore. Analytical techniques for the prediction of elastic properties of textile reinforced composites. *Mechanics of Composite Materials*, 5:579–596, Dec 1993.

[9] Christopher M. Pastore and Yasser A. Gowayed. A self-consistent fabric geometry model: Modification and application of a fabric geometry model to predict the elastic properties of textile composites. *Journal of Composites Technology and Research*, 16:32–36, January 1994.

[10] Carl DeBoor. *A Practical Guide to Splines*. Springer-Verlag, New York, Heidelberg, Berlin, 1978.

[11] A. E. Bogdanovich and E. V. Yarve. Numerical analysis of impact deformation and failure in laminated composite plates. *Journal of Composite Materials*, 26(4):520–545, 1992.

[12] Alexander E. Bogdanovich and Alexander B. Birger. Application of spline functions to the 3-d analysis of laminated and composite plates subject to static bending. In *Proceedings of the Second International Symposium on Composite Materials and Structures*, pages 719–725, Beijing, P.R.C, August 1992. Peking University Press.

[13] Alexander E. Bogdanovich, Viren Kumar, Christopher M. Pastore, and Marge German. 3-d stress analysis in laminated plates using a combination of ansys and sub-element deficient approximation function analysis. In *Proceedings of the ASC 9th Technical Conference on Composite Materials*, page . Technomic Publishers, September 1994.

Proceedings of the 10th Annual ASM/ESD Advanced Composites Conference, Dearborn, Michigan, USA, 7-10 November 1994

Thermal Expansion Coefficients of Textile Composites with Arbitrary Preform Structures

Y. Gowayed
Auburn University
Auburn, Alabama

Abstract

In this research activity, an analytical technique is developed to predict the thermal expansion coefficients of textile composite materials with arbitrary preform structures. The model starts with a comprehensive description of the preform geometry utilizing a processing science model and a graphical rendering tool. Volume intersection techniques are utilized to identify the composite total and relative volume fractions and yarn spatial orientations. The composite is then descritized into repeat "unit cells". Finite element method and the method of virtual work are integrated to calculate the properties of the unit cell. Hybrid hexahedra brick elements with fibers and matrix around each integration point are employed in the Finite Element formulation via micro-level homogenization. A computer algorithm was developed for the analytical model. Model predictions matched well with experimental results.

Introduction

Ceramic Matrix Composite materials are usually utilized for high temperature applications. At such high temperatures, a mismatch between the Coefficients of Thermal Expansion (CTE) of fibers and matrix could cause excessive thermal stresses. Such phenomena is not only evident in ceramic matrix composites but also in Polymer Matrix Composites. Thermal stresses caused by resin curing could alter the static and dynamic characteristics of the composite. From an engineering standpoint, the determination of the CTE's of composite materials is crucial for a safe and accurate design.

The problem of predicting the coefficients of thermal expansion for composites and multi-phase heterogeneous materials has gained researchers attention since late 60's. Extensive research efforts were invested in prediction of CTE for unidirectional composites [1, 2, 3, 4, 5, 6, 7] and laminated composites [8, 9, 10]. This effort succeeded in quantifying, with an acceptable degree of accuracy, the CTE for different unidirectional and laminated composites.

Prediction methods of CTE for textile composite materials has also been documented in literature. Ishikawa and Chou [11, 12] studied the CTE for plain weaves and satin weaves by modifying the classical laminate theory. Three of their models were employed; the Mosaic Model, the Fiber Undulation Model and the Bridging Model. They concluded that only predictions using the Bridging Model gave results that matched experimental data for a 5 harness satin weave. Tarnopol'skii, Polyakov and Zhigun [13, 14] developed a method to predict the CTE for Composites reinforced along cube diagonals (a special case of 3-D braids) employing a stiffness averaging technique. They reported disagreement between experimental and analytical results. They concluded that the disagreement is due to either a poor geometrical representation of the fabric (i.e. neglecting the intersection of the cube diagonal elements at the center point of the cube) or an experimental error. Kregers [15] investigated possible bounds for the solution (i.e. a fork). Kregers modeled the yarns as composite rods with a proportional distribution of matrix. He proved that the upper bound was represented by a stiffness averaging approach while the lower bound could be derived from a compliance averaging. He reported that the upper bound conformed well with experimental results for hybrid carbon fiber-organic fiber/epoxy composite.

The following observations could be drawn from the quick overview of the prediction techniques shown above : i) massive homogenization approaches were usually adopted; ii) poor geometrical characterization of the fabric preform, if any, was included in these models. Although [13, 14] as well as [15] adopted similar approaches, results in [15] conformed better with experimental data due to a better geometrical representation for the composite under consideration; and iii) stiffness averaging techniques gave good results for simple fabric structures [11, 15].

Furthermore, Foye [16] introduced a hybrid finite element model to predict the CTE's for textile composite materials with arbitrary preform structures (e.g. weaves, knits and braids). This model is based on a unit cell representation of the preform structure. Each unit cell is surrounded by 6 similar unit cells. The properties of the unit cell was assumed to represent the properties of the composite. This model was tested for a plain weave, Oxford weave and satin weaves T300/epoxy composites. Predicted results had an average difference of 20% for the 5 test data points reported. Foye [16] concluded that the disagreement is due to the low degree of accuracy of the testing method. He also stressed the need for a comprehensive geometrical representation of the preform structure. Another factor could be attributed to the low degree of agreement between experiment and prediction. It is the numerical instabilities in the Finite Element solution due to the considerable difference in the properties of the fiber and resin that originated form the hybrid nature of the analysis.

Based on the observations outlined in the previous paragraphs, it could be seen that the first step towards a successful model to predict the CTE's of textile composite has to include a comprehensive geometric model able to represent the "waviness" nature of textile preforms. In this paper, such geometric model is employed and complimented with a numerical solution to predict the Coefficients of Thermal Expansion for textile composite materials. The numerical model is a enhancement of the original hybrid model presented in [16]. This is done by incorporating a stiffness averaging micro-level homogenization technique to overcome the problem of numerical instabilities.

The ensuing sections provide details of this model.

Geometrical Modeling

In order to understand and predict the properties of the composite, a comprehensive determination of the location of all preform yarns is essential. Two methods could be adopted to determine the preform structure : i) experimental method using imaging techniques [17] and ii) analytical techniques for geometrical modeling [18].

The experimental method [17] is fruitful for limited number of specimens of arbitrary preform structure. It is able to measure the exact location and orientation of each yarn in the composite space. The drawbacks to the experimental approach are the time and expense constraints. On the other hand, geometrical modeling could provide a good approximation of the fabric preform. The method adopted in this research activity [18] was verified using the experimental method [19].

A good geometrical model would start from a complete understanding of the textile machinery operation. Processing science modeling could be used to model the machine motion [20]. It could relate machine forming parameters to the final product spatial geometry. A set of fixed points in space "knots" that identify the yarn path would form the output of such model.

From the set of knots identified through process science modeling, the yarn's path could be identified throughout the preform. This is done by incorporating a B-spline function to construct the yarn center line's path in space through the previously calculated knots. The use of the B-spline functions, beside providing a smooth curve, minimizes the curvature of the yarns while maintaining its C^2 continuity. Figure 1 illustrates the use of B-splines to produce a smooth yarn centerline path.

B-splines are usually used to approximate open poly-

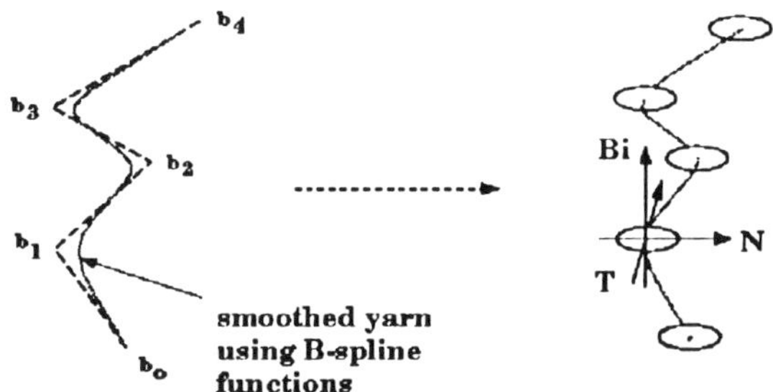

Figure 1: Solid modeling construction

gons into smooth curves. For $n + 1$ vectors b_i ($i = 0, 1, 2, .., n$), the parametric curve segment of degree n for a Bézier curve could be represented as follows [21, 22] :

$$P(t) = \sum_{i=0}^{n} b_i B_{i,n}(t) \qquad 0 \le t \le 1 \tag{1}$$

where : $\quad B_{i,n}(t) = C_i^n t^i (1 - t)^{n-i}$

and : $\quad C_i^n = \frac{n!}{i!(n-1)!}$

$B_{i,n}$ constitutes a density function of a binomial distribution. If the b_i vectors are joined, a Bézier open polygon is obtained.

It is now possible to define some of the yarn characteristics which control its mechanical response. The vector tangent T_i to the yarn at point b_i is determined as follows (see Figure 1) :

$$T_i = v_{10} + v_{12} \tag{2}$$

where : $\quad v_{10} = b_i - b_{i-1}$

and : $\quad v_{12} = b_{i+1} - b_i$

b_{i-1}, b_i and b_{i+1} are used here in vector notation, v_{10} is the vector joining points b_{i-1} and b_i, v_{12} is the vector joining points b_{i+1} and b_i.

The bend angle of the yarn at point b_i could be calculated as the dot product between vectors T_i and T_{i+1}. Also, the twist angle associated with the yarn path is determined as the angle between successive normal vectors N_i (where N_i is the cross product of the vector T_i

and v_{12}).

The next step is to identify the yarn cross section and sweep it through the proposed center line in a way to conform with the expected bends and twists. The yarn cross section can vary from a circle to a very flat elliptical shape. Since a circular section is a special case of an ellipse with $r_1 = r_2$ (where r_1 and r_2 are major and minor ellipse radii), a basic elliptical shape is proposed for the yarn cross section. From the computer graphics point of view the ellipse is actually represented as n-gon shape. These points are calculated on intervals π/n. Where n is an arbitrary integer. The higher the value of n, the closer the cross sectional shape would look like an ellipse. The space points pp_i that represents the external surface of the yarn in space around a center point b_i could be calculated as follows in vector notation for each value of π/n :

$$pp_i = b_i + r_1 . \cos(\frac{\pi}{n}) . N_i + r_2 . \sin(\frac{\pi}{n}) . B_i \tag{3}$$

where, ($i = 0, 1, 2, 3, \ldots, N$), and B_i is the bi-normal vector at point b_i which is equal to the cross product of vectors T_i and N_i.

This approach provides a numerical representation of yarns spatial orientations coupled with a graphical rendering technique to characterize and display composite preform structures. Figure 2 illustrates the micrographical image and geometrical model for 3-D angle interlock woven fabric preform.

403

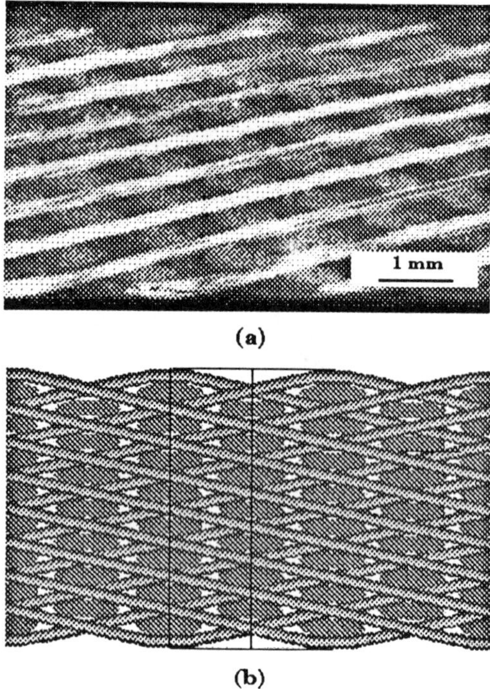

(a)

(b)

Figure 2: a) Micrographic image and b) Geometric model of 3-D angle interlock woven composites

Thermal Analysis

Mathematical Formulation

Thermal expansion coefficients are the average strains of a composite material resulting from a unit increase in the temperature of this material. From a continuum mechanics point of view, the values of the Coefficients of Thermal Expansion (CTE) are not only dependent on the increase in temperature but also on the stiffness properties of the composite [8, 14, 15, 16]. This adds another level of complexity to the problem. Consequently, calculation of stiffness properties of the composite constitutes the first step in the solution. The thermo-elastic model of composite materials could be represented as follows :

$$\epsilon_{ij} = C_{ijkl}\, \sigma_{kl} + \alpha_{ij}\, \Delta T \qquad (4)$$

$$\sigma_{ij} = S_{ijkl}\, \epsilon_{kl} - \Gamma_{ij}\, \Delta T \qquad (5)$$

where, $\Gamma_{ij} = C_{ijkl}\, \alpha_{kl}$; $i, j, k, l = x, y, z$; ϵ and σ are

strain and stress tensors; S and C are compliance and stiffness tensors; α is the CTE tensor; Γ is the tensor of thermal stress coefficients and ΔT is the temperature difference.

There are six different CTE's for the general anisotropic case; three extensional and three shear components which constitutes a 3-D strain state. Figure 3 illustrates these strain cases.

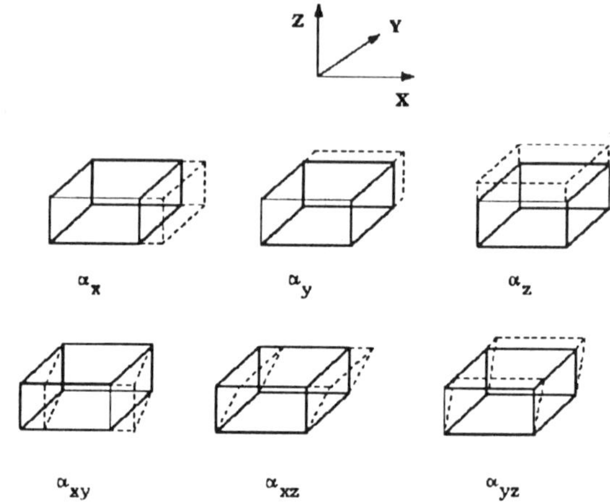

Figure 3: Thermal Expansion Coefficients Strain Cases

Finite Element Analysis

The numerical solution for this problem was carried out based on the Modified Unit Cell Continuum Model [23]. This approach is based on structural mechanics principles and independent of size and scale. In this approach, a unit cell is assumed to represent the composite. Each unit cell is surrounded by six identical unit cells. The unit cell is divided into hybrid hexahedra brick elements called "subcells". Micro-level homogenization tool [24] is used to evaluate the properties around integration points of each subcell. This technique prevents solution instability and asserts directionality of preform elements [25]. Figure 4 gives a schematic representation of the micro-level homogenization scheme.

A virtual work technique is utilized to quantify the six Coefficients of Thermal Expansion. A single thermal load case is applied in each of the six directions

404

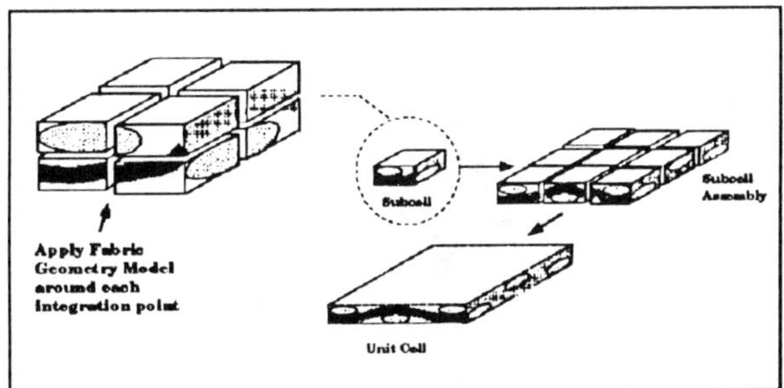

Figure 4: Micro-level homogenization for 8 noded hex-ahedra brick element

illustrated in Figure 3.

Boundary Conditions

Since all repeating unit cells are assumed to be similar, their response will also be similar. This can be seen through considering the boundary conditions on the surfaces of such unit cells. Consider the upper and lower surface of a unit cell. For every point on one face there is an "image" point on the opposite face as shown in Figure 5. This image point is actually a similar point on the neighboring unit cell. Since the two unit cells are identical, they will deform similarly :

$$\begin{aligned} \delta_x^i &= \delta_x^j \\ \delta_y^i &= \delta_y^j \\ \delta_z^i &= \delta_z^j \end{aligned} \qquad (6)$$

where δ_x, δ_y and δ_z are displacements in the X, Y and Z directions respectively; i and j refer to the image points. The upper equations provide three equations in six unknowns. The other three equations can be determined from the stress continuity conditions at these points assuming that surface stresses are similar for identical unit cells. Therefor, stresses at image points are equal in magnitude and opposite in direction :

$$\begin{aligned} \sigma_{zz}^i &= \sigma_{zz}^j \\ \tau_{xz}^i &= \tau_{xz}^j \\ \tau_{yz}^i &= \tau_{yz}^j \end{aligned} \qquad (7)$$

Unit cell edges need a different argument. Any edge point is shared by four similar unit cells. Consider a plane perpendicular to an edge as in Figure 5. Since all four unit cells sharing the corner point are similar, displacement of the unit cells at these points can be related as follows :

$$\begin{aligned} \delta_x^i &= \delta_x^j = \delta_x^k = \delta_x^l \\ \delta_y^i &= \delta_y^j = \delta_y^k = \delta_y^l \\ \delta_z^i &= \delta_z^j = \delta_z^k = \delta_z^l \end{aligned} \qquad (8)$$

Three more equations are needed to completely specify the boundary conditions for the corner nodes. These three equations are calculated from the stress equilibrium conditions at these points as follows :

$$\begin{aligned} \sum_{i,j,k,l} (\tau_{xy} + \tau_{xz})ds &= 0 \\ \sum_{i,j,k,l} (\tau_{yz} + \sigma_{yy})ds &= 0 \\ \sum_{i,j,k,l} (\tau_{yz} + \sigma_{zz})ds &= 0 \end{aligned} \qquad (9)$$

where ds is an incremental side length over each of the four sides of the four unit cells which meet at that edge.

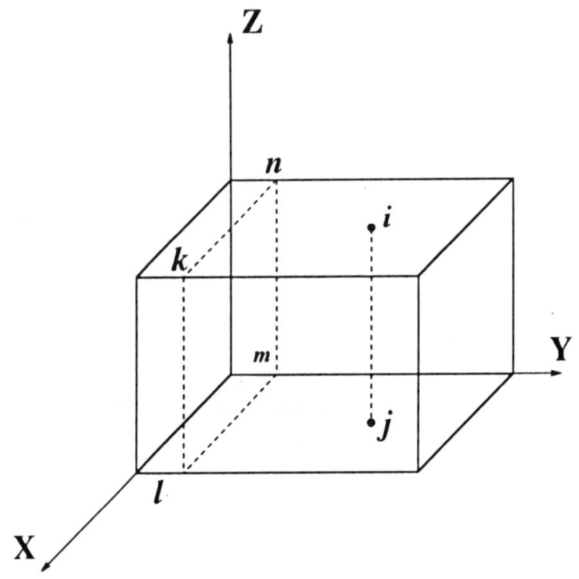

Figure 5: Surface and corner points for boundary conditions

405

Experimental Observations

A computer algorithm for predicting the thermal expansion coefficients for textile composite materials was developed. It is entitled "Graphical Integrated Numerical Analysis (GINA)". The following sections give details for experimental data and its comparison with analytical predictions using GINA. This is carried out for ceramic and polymer matrix textile composites.

Plain Weave Carbon/Epoxy Composites

The balanced plain weave fabric preform was made from Morganite Type II carbon yarns. DLS351/BF$_3$400 epoxy resin was used as the matrix. Table 1 lists the elastic and thermal properties of the constituent materials. Details of the test procedures is found in [10, 26].

Table 1: Constituent material properties for plain weave carbon/epoxy composite

Material	Carbon fiber	Epoxy matrix
E (GPa)	262	4.9
G (GPa)	109.2	1.81
ν	0.20	0.35
α	$-9x10^{-7}$ long. dir. $12x10^{-6}$ transv. dir.	$4.8x10^{-5}$

The geometrical model was constructed using a 2-D weave processing science model for plain weaves. The number of subcells used in the Finite Element analysis was 16 as shown in Figure 6. Table 2 lists the experimental and analytical predictions for the composite.

Table 2: Thermal expansion coefficients for plain weave carbon/epoxy composite

Property	Experimental	Predicted
$\alpha_x, \alpha_y (10^{-6}/^{\circ}C)$	3.26	3.55
$\alpha_z (10^{-6}/^{\circ}C)$	3.46	3.40

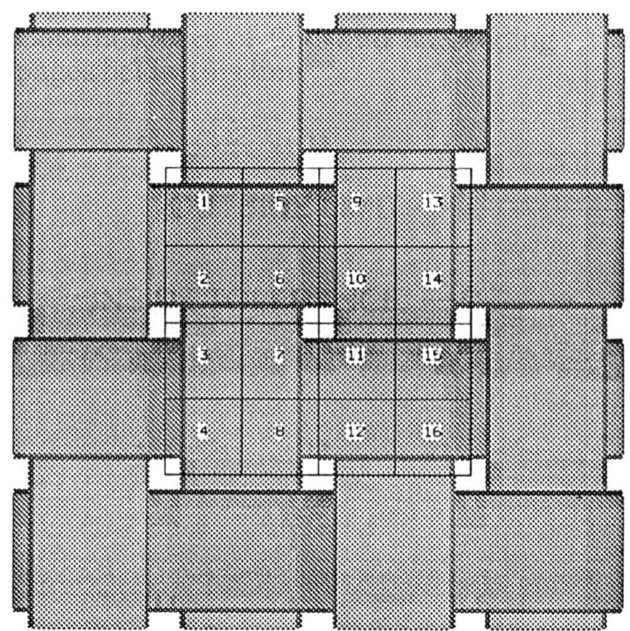

Figure 6: Geometric model of plain weave carbon/epoxy composite

Satin Weave SiC/SiC Ceramic Composites

This SiC/SiC composite was manufactured by BP Chemicals (Hitco). This composite was made by CVI of Nicalon® balanced 8 harness satin fabric. The matrix is a Refractory CVI Silicon Carbide. Fiber and matrix properties are listed in Table 3.

Table 3: Constituent material properties for satin weave SiC/SiC composite

Material	SiC fiber	SiC matrix*
E (GPa)	193.1	241.3
G (GPa)	76.5	89.6
ν	0.20	0.15
α_x	$4x10^{-6}$	$0.5x10^{-6}$

* estimated

The composite specific density was 2.53. The fiber volume fraction was 38% and the void volume fraction was 12%. The geometrical model was constructed using the satin weave processing science model for 8 harness satin weave. The number of subcells used in the Finite Element analysis was 256. Table 4 lists the experimental and analytical predictions for the composite.

Table 4: Thermal expansion coefficients for satin weave SiC/SiC composite

Property	Experimental	Predicted
$\alpha_x, \alpha_y (10^{-6}/°C)$	1.62	1.78

Conclusions

In this paper, an integrated approach for geometrical and thermal modeling of textile composites is presented. This model takes into consideration a comprehensive understanding of fabric geometrical characteristics into the thermal analysis. The geometrical model is based on processing science modeling of textile machinery operations combined with graphical rendering techniques. Geometrical data obtained from the geometrical model is used as input for the thermal analysis.

The thermal analysis constitutes a Finite Element approach (Unit Cell Continuum Model) to predict the properties of a composite repeat unit cell. The basic concept presented in the Unit Cell Continuum Model [16] provides a framework for the prediction of thermal properties of textile reinforced composites. However, a micro-level homogenization technique was incorporated to overcome prediction inconsistency originated from the use of hybrid elements, especially when the difference in fiber and matrix properties is substantial.

The integrated model was experimentally verified for thermal expansion coefficients of ceramic and polymer composite using a Graphical Integrated Numerical Analysis Model (GINA). GINA combines processing science modeling, geometrical modeling and graphical rendering along with the modified UCCM. Predictions of thermal expansion coefficients matched well with experimental results for ceramic and polymer matrix composites.

Acknowledgment

The author would like to acknowledge Pratt & Whitney and the National Textile Center for providing funds to this research project.

References

[1] V. Levin, "Thermal expansion coefficients of heterogenous materials," *Meckhanika Tverdogo Tela*, vol. 2, pp. 88–94, 1967.

[2] R. Schapery, "Thermal conductivity of unidirectional materials," *Journal of Composite Materials*, vol. 2, pp. 380–404, 1968.

[3] W. Rosen and Z. Hashin, "Effective thermal expansion coefficients and specific heats of composite materials," *Journal of Engineering Science*, vol. 8, pp. 157–173, 1970.

[4] T. Ishikawa, K. Koyama, and S. Kobayashi, "Thermal expansion coefficients of unidirectional composites," *Journal of Composite Materials*, vol. 12, pp. 153–168, 1978.

[5] Z. Hashin, "Analysis of properties of fiber composites with anisotropic constituents," *Journal of Applied Mechanics*, vol. 46, pp. 543–550, 1979.

[6] G. Dvorak and T. Chen, "Thermal expansion of three-phase composite materials," *Journal of Applied Mechanics*, vol. 56, pp. 418–422, 1989.

[7] R. Valisetty and J. Teply, "Overall instantaneous viscoplastic properties of composites," *Journal of Composite Materials*, vol. 26, 1992.

[8] C. Chamis, "Simplified composite micromechanics equations for hygral thermal and mechanical properties," *38th Annual Conference, Reinforced Plastics/Composite Institute, The Society of the Plastic Industry, Inc.*, 1983.

[9] K. Rogers, L. Phillips, D. Kingston-Lee, B. Yates, M. Chandra, and S. Parker, "The thermal expansion of carbon fiber-reinforced plastics : Part 6, the influence of fiber weave in fabric reinforcement,"

Journal of Material Science, vol. 16, pp. 2803–2818, 1981.

[10] B. Yates, M. Overy, B. Sargent, B. McCalla, D. Kingston-Lee, L. Phillips, and K. Rogers, "The thermal expansion of carbon fiber-reinforced plastics : Part 2, the influence of fiber volume fraction," *Journal of Material Science*, vol. 13, pp. 433–440, 1978.

[11] T. Ishikawa and T. Chou, "In-plane thermal expansion and thermal bending coefficients of fabric composites," *Journal of Composite Materials*, vol. 17, pp. 92–104, 1983.

[12] T. Ishikawa and T. Chou, "Thermoelastic analysis of hybrid fabric composites," *Journal of Material Science*, vol. 18, pp. 2260–2268, 1983.

[13] I. Zhigun and V. Polyakov, "Composites reinforced along cube diagonals: Part 1, shear and compression resistance," *Mechanics of Composite Materials*, vol. 3, pp. 303–309, 1988.

[14] Y. Tarnopol'skii, V. Polyakov, and I. Zhigun, "Composites reinforced along cube diagonals: Part 2, coefficients of linear thermal expansion," *Mechanics of Composite Materials*, vol. 6, pp. 758–765, 1989.

[15] A. F. Kregers, "Mathematical modeling of the thermal expansion of spatially reinforced composites," *Polymer Mechanics*, vol. 3, pp. 316–325, January 1988.

[16] R. L. Foye, "Thermal expansion of fabric reinforced composites," in *Fiber Tex '90* (J. Buckley, ed.), (Hampton, VA), pp. 45–53, NASA Langley Research Center, 1990.

[17] Y. Gowayed and J. C. Russ, "Geometric characterization of textile composite preforms using image analysis techniques," *Journal of Computer Assisted Microscopy*, vol. 3, pp. 189–200, 1991.

[18] C. Pastore, Y. Gowayed, and Y. J. Cai, *Applications of Computer Aided Geometric Modelling for Textile Structural Composites*, pp. 45–53. Southampton, UK: Computational Mechanics Publications, 1990.

[19] Y. Gowayed, "Experimental verification of textile composites geometric modeling using image analysis techniques," in *Fiber Tex '91* (J. Buckley, ed.), (Hampton, VA), NASA Langley Research Center, 1991.

[20] Y. Gowayed and C. Pastore, "Processing science model of textile machinery : Theory and application to 3-dimensional braiding," 39th *International SAMPE Symposium and Exhibition*, pp. 857–867, April 1994.

[21] P. Bézier, *Numerical Control Mathematics and Applications*. USA: John Wiley and Sons, 1972.

[22] P. Bézier, "Mathematical and practical possibilities of unisurf," *Computer Aided Geometric Design*, pp. 127–152, 1974.

[23] Y. Gowayed and C. Pastore, "An integrated approach to the geometrical and mechanical modeling of textile reinforced composites," 39th *International SAMPE Symposium and Exhibition*, pp. 2841–2854, April 1994.

[24] C. Pastore and Y. Gowayed, "A self consistent fabric geometry model : Modification and application of a fabric geometry model to predict the elastic properties of textile composites," *Journal of Composites Technology and Research*, vol. 16, pp. 32–36, January 1993.

[25] Y. A. Gowayed, "Thermo-mechanical modeling of textile reinforced ceramic matrix composites," *Project Final Report, Pratt & Whitney*, Dec. 1993.

[26] K. Rogers, D. Kingston-Lee, L. Phillips, B. Yates, M. Chandra, and S. F. Parker, "The thermal expansion of carbon fiber-reinforced plastics : Part 6, the influence of fiber weave in fabric reinforcement," *Journal of Material Science*, vol. 16, pp. 2803–2818, 1981.

Role of the Molecular Weight of the Thermoplastic Matrix on Fiber-Matrix Adhesion in BPA-Polycarbonate/Carbone Fiber Composites

V. Raghavendran, L.T. Drzal
Michigan State University
East Lansing, Michigan

Abstract

The interfacial adhesion in thermoplastic polymer composites depends on a number of complex factors such as fiber and matrix morphology, surface topography and chemistry, presences of reactive functionalities, modulii of the fiber and matrix, additives and impurities and the processing conditions. Many of the matrix properties like viscosity, modulus and chain entanglement are dependent upon the molecular weight of the resin. In the present paper, the influence of the molecular weight of the amorphous thermoplastic matrix on the interfacial adhesion in BPA-Polycarbonate/carbon fiber composite was investigated. The effects of the consolidation temperature was also examined. The micro indentation technique was employed to measure the interfacial shear strength (ISS). Polycarbonate resin in four different molecular weight ranges were used. Molecular weight of the matrix was found to have significant influence on the ISS. The adhesion level showed increases of more than 100 % across a critical molecular weight range. Above this critical level there is a monotonous increase in adhesion with increase in the molecular weight, but the magnitude of the increase is not on the same scale. Increases of 20 % to 30 % in ISS were seen for different molecular weight ranges with increasing consolidation temperature. Increases of 8 % to 15 % in ISS were seen with increasing molecular weight for different consolidation temperatures. It was also observed that the interfacial failure mode changed from adhesive to cohesive across the critical molecular weight range.

THE INTERFACIAL ADHESION in the thermoplastic matrix composites is dependent on a number of complex factors both intrinsic to the fiber and the matrix such as the morphology and mechanical properties, crystallinity or crosslinking of the matrix and extrinsic such as the processing conditions, presence of additives and impurities at the interface[1,2].

A number of matrix properties of a thermoplastic polymer is related to its molecular weight. The molecular weight is known to affect the strength, viscosity, crystallinity and the entanglement density of the matrix. In linear chain thermoplastic polymers such as polycarbonates, the mechanical properties start showing an increase only after the chain starts entangling. Below the critical molecular weight M_C, which is approximately twice the entanglement molecular weight M_E, the matrix is brittle and would exhibit low strength and elongation to break. It is well known that the chain ends act as imperfections in the chain adversely affecting the strength properties. The number of chain ends are higher in concentration in low molecular weight chains and consequently lower the strength. Above the minimum molecular needed to get satisfactory mechanical behavior, the strength and elongation increases towards a limiting value due to the increase in the chain entanglements. This increase in the entanglement density with increase in molecular weight also contributes to the amorphousness of the long chain polymers[3].

Though the effects of molecular weight of a thermoplastic polymer on its other properties is well documented in the literature very little is known about its effects on the interfacial adhesion in polymer composites. The present investigation addresses this problem. A quantitative investigation of the effects of molecular weight and the processing conditions on the interfacial strength in polycarbonate carbon fiber composites was conducted to gain a better understanding of interfacial phenomena in thermoplastic composites.

The interphase region between the fiber and matrix is known to be of great importance in the development of the mechanical properties of a composite material[4-6]. Though due to the complex physiochemical interactions occurring at the interface, the interphase region might have properties very different from the bulk properties of the fiber and the matrix, the trend seems to indicate a correlation between the interphase properties and the constituent properties. Theoretical relations based on elastic stress transfer mechanisms between interfacial properties and bulk properties have been proposed by many authors. Cox[7] and later Cooke[8] have developed models in which an elastic fiber of length l, embedded in an elastic matrix under general strain ε. The model assumes perfect bonding between the fiber and the matrix. Using the assumption of load transfer through the ends of fiber, leads to Eq.1

$$\tau = E_f \, \epsilon_m \left(\frac{G_m}{2E_f \, \ln\left(\frac{R}{r}\right)} \right)^{0.5} \frac{\sinh \beta \, (0.5L - x)}{\cosh \beta \frac{L}{2}} \tag{1}$$

Where

E_f	=	tensile modulus of the fiber,
ε_m	=	strain in matrix,
G_m	=	shear modulus of matrix,
R	=	interfiber spacing,
r	=	radius of the fiber,
β	=	scaling factor,
L	=	length of the embedded fiber,
x	=	radial distance outward,
τ	=	interfacial shear strength at a fixed point

It is clearly evident that, for a fixed specimen geometry, fiber and polymer type, the theory predicts a direct dependence of the interfacial shear strength on the matrix strain. Since the matrix yield strain increases with increase in molecular weight due to higher entanglement density, therefore the maximum interfacial strength at yield would be proportional to the molecular weight.

$$\epsilon_m \propto MW \; ; \quad \therefore \; \tau \propto MW \tag{2}$$

A second approach on the atomistic level dealing with the chain conformation and adsorption at the fiber matrix interface can also be used to understand the interfacial interactions. The measure of adhesion between the polymer chain and the fiber can be thought of as being dependent on the number of contact points between the active sites on the surface and the chain. Due to the constraints imposed on the number of conformations a polymer chain can have at the interface, most of the chains tend to flatten out to occupy a lower energy state[9,10]. This chain segmental orientation and conformations are seen to extend up to 3.5 times the radius of gyration R_g of the longest polymer chain. Even though, the number of polymer chains per unit

area of the surface will be lower in case of the higher molecular weight polymer chains due to their longer chains lengths, the number of contact points would be more. The next layer adsorbed on top of these polymer chains would also be able to entangle more with the loops and tails and create a strong adhesive and cohesive bonding between the surface and the bulk matrix medium. The lower molecular weight polymer chains on the other hand do not lose as many number of conformations as a higher molecular weight chain[11]. Consequently the lower molecular weight chains tend to still retain a major part of their isotropic bulk conformation. This would reduce the number of contact points between the polymer chain and the surface leading to lowering of the interfacial adhesion.

Materials

Hercules Magnamite[*], PAN based type II, high strength AS4 6K carbon fiber tow with standard commercial oxidative treatment was used as the reinforcing fiber medium. Bisphenol-A polycarbonate which is an amorphous thermoplastic was used as the polymer medium. The polycarbonate of different molecular weight ranges were obtained from the GE Plastics Inc. Three of the four grades were received in pure form directly from their reactors in Mt. Vernon, Ind. plant, without any additives or stabilizers being added to them. The fourth grade, Lexan 8040 is FDA approved commercial grade with no UV stabilizer. Pure grades were used to preclude any effect these agents might have on the interfacial adhesion.

Table 1 gives the details of the different grades of polycarbonate, their nomenclatures and the molecular weight ranges determined by gel permeation chromatography (GPC) based on light scattering absolute weight average molecular weight standards for polystyrene done on a Perkin Elmer LC30 Chromatograph.

Table 1　Nomenclature and molecular weights of the Bisphenol-A Polycarbonate

Resin Grade	Author's nomenc--lature	Weight Average MW M_W	Number Average MW M_N	M_W/M_N
ML5721-111	PC100	31,135	14,382	2.165
Lexan 8040	PCFF	24,894	11,865	2.098
ML5221-111	PCHF	22,963	10,310	2.227
ML5832-111	PCOQ	16,152	7,623	2.119

Experimental

Thermoplastic matrix microindentation test specimens were fabricated by hot pressing fibers between pairs of matrix sheets. Prior to the fabrication of the final test specimens, preform sheets one half the thickness of the final specimens were first produced by hot pressing the dried polycarbonate powder between aluminum foil sheets with the thickness controlled by a surrounding dam of stainless steel. Times and temperatures for this preliminary step were kept as low as possible to minimize chances of thermal degradation, the assembly was kept sealed inside the foil throughout the process to minimize oxidation. Aluminum foil pressing sheets were employed to avoid the use of release agents which might cause surface contamination.

The low volume fraction composite laminates for microindentation test were fabricated in the following manner. A 1.6 mm thick preform sheet of dimensions 15 cm x 2.5 cm was draped with three spread tows of AS4 carbon fiber and held in place using a high temperature tape. The assembly was surrounded by a 3 mm thick stainless steel dam. The upper preform sheet was then place over the assembly followed by an aluminum sheet and sealed in an aluminum foil, A schematic sketch of the assembly is shown in figure 1. Hot pressing then proceeded by applying a light load equal to the weight of the upper platen of the carver press on the assembly and ramping the temperature to 125°C. The assembly was allowed to soak at this temperature for 1 hour to remove moisture.

Following the 125°C soak, the temperature was raised at approximately 5°C/minute to the desired processing temperature. Upon reaching the processing temperature, a 7 MPa pressure was applied and held throughout the pressing cycle. Cooling was accomplished by a standard platen water cooling system at approximately 1°C/second. Sample specimens having the dimensions of 2cm x 1cm were cut from the fabricated composite specimens for microindentation test and mounted and polished using standard metallographic techniques.

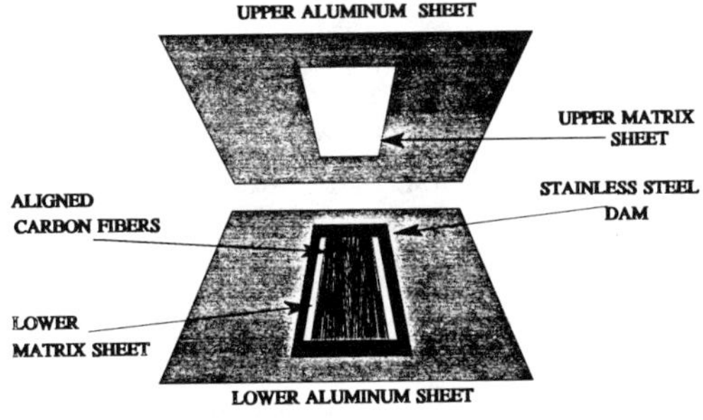

Figure 1   Hot Press Specimen Fabrication Assembly

The microindentation tests were done on a commercially available Interfacial Testing System (ITS) developed by Dow Chemical Co. Inc.[12]. This fully automated instrument is designed to test real composite specimens. A diamond tipped indenter mounted on the objective lens of a high power optical microscope is used to indent single fibers in the composite specimen as shown in the figure 2. A television camera monitor is used to observe the debonding. The force required to debond is input to a computer program which utilizes a finite element analysis algorithm derived from Mandell, *et al* [13] to calculate the interfacial shear strength.

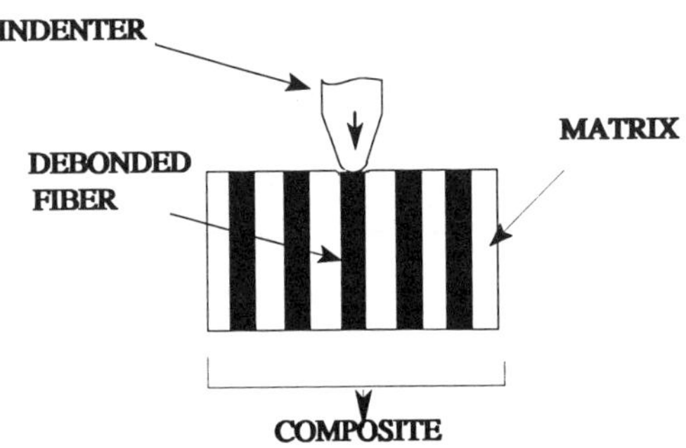

Figure 2 Schematic sketch of the microindentation test

Results and Discussions

As can be seen from table 2 and figure 3, with increased molecular weight, there is an increase in ISS at each of the processing temperature, with a major increase between the low molecular weight grade which is below the critical molecular weight range and the next higher molecular weight grade. Above the critical limit the increase was monotonic but to a lesser degree. It was also seen that the increase in the consolidation temperature also produced increased level of adhesion except in the case of PCOQ grade, which had very poor adhesion and did not show any change in ISS with any of the variables. For the other grades, the increase was significant between specimens fabricated at 250°C and 275°C, but was not much between 275°C and 300°C. It can therefore be inferred that the optimal processing temperature for polycarbonate carbon fiber composites lies between 275°C and 300°C, requiring lower times at higher temperatures for the attainment of similar level of interfacial adhesion.

The increase in the ISS with the increasing molecular weight can be explained using both the elastic stress transfer mechanism and the physiochemical interactions occurring at the interface. Eq. 1 shows that the interfacial adhesion is dependent upon the matrix strain. The matrix yield strain increases with increase in molecular weight due to higher entanglement density, therefore the maximum interfacial strength at yield would be proportional to the

Table 2   ISS of the Different Molecular Weight Grades of Polycarbonate Consolidated at Different Time and Temperatures.

Specimen Designation	Weight Average MW	Processing Temperature °C	Processing Time (min.)	ISS MPa (Std. Dev.)
PC16K1	16,152	250 °C	45 minutes	20.22 (2.51)
PC16K2	16,152	275 °C	15 minutes	21.23 (2.72)
PC16K3	16,152	300 °C	8 minutes	21.13 (2.76)
PC23K1	22,963	250 °C	45 minutes	34.07 (5.08)
PC23K2	22,963	275 °C	15 minutes	37.54 (4.61)
PC23K3	22,963	300 °C	8 minutes	42.01 (5.09)
PC25K1	24,894	250 °C	45 minutes	34.23 (3.75)
PC25K2	24,894	275 °C	15 minutes	41.30 (3.16)
PC25K3	24,894	300 °C	8 minutes	43.14 (3.71)
PC31K1	31,135	250 °C	45 minutes	36.43 (4.31)
PC31K2	31,135	275 °C	15 minutes	47.80 (5.68)
PC31K3	31,135	300 °C	8 minutes	48.32 (6.83)

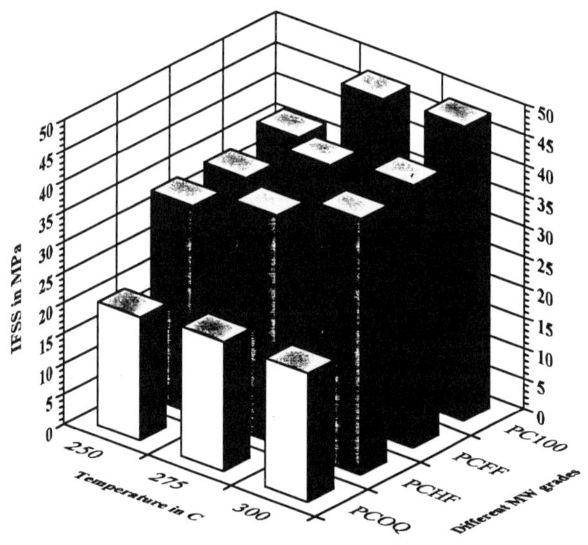

Figure 3 A Three Dimensional Plot of the ISS vs Molecular Weight at Different Consolidation Temperatures

molecular weight.It was also noted during the tensile testing of pure matrix specimens that the low molecular weight specimen PCOQ, failed before yielding at around 1.3 % elongation, while the elongation to yield increased from 5.5 to 5.9 % for the grades PCHF and PC100 respectively.

Thermoplastic polymer matrices are generally inert and do not react chemically with the fiber surface, so the driving force for the adhesion to take place comes from enthalpic interactions like dispersive interaction between the polymer molecules and the molecules on the fiber surface. The other type of interactions which are possible are polar interactions and hydrogen bond formations. The BPA polycarbonate contains polar carbonyl groups surrounded by bulky phenyl and methyl groups, so tends to behave like a non polar polymer in bulk conditions. As mentioned earlier, the polymer chains start losing conformations as they near the interface, tending to flatten out to occupy a lower energy state. This is more so in case of the higher molecular weight chains compared to the lower molecular weight chains, which tend to retain a major part of their isotropic bulk conformations. The uncoiling and flattening of the longer chains of higher molecular weight at the interface would allow more polar components to interact with the fiber surface functionalities present on the fiber surface and increase the work of adhesion, thereby increasing the interfacial shear strength for higher molecular weight specimens.

The increase in ISS with increasing consolidation temperature can be due to the lowering of the matrix viscosity at elevated temperature and the consequent improvement in the fiber wetting by the matrix. The large increase in the ISS between 250°C and 275°C might be due to the fact that the polycarbonate melts at 265°C and specimens consolidated at higher temperatures would be able to completely uncoil and diffuse through the fiber bundles and adsorb more strongly on to the fiber surface. Work done by Brady et al.[19] and Stone et al.[20] leads to a similar conclusions. Brady et al. did annealing experiments on the unidirectional polycarbonate/carbon fiber composites and found that the interfacial toughness increased with annealing time and temperature. Stone et al. observed similar results for the compressive strength developent.They attributed the increase in strength and toughness above the melting point to adsorption of the matrix onto the fiber.

Conclusions

The polycarbonate/carbon fiber composite fabricated from four different molecular weight grades and at different processing temperatures were investigated. It was observed that, the interfacial shear strength increases with increase in both molecular weight and processing temperature. The increase is seen to be significant between the low molecular weight PCOQ grade and the next higher molecular weight grade. This increase in the ISS ranged from 70-100%. Above this critical molecular weight range, the increase was seen to be monotonic but to a lesser extent, ranging from 8-15% for different consolidation temperatures and 20-30% for different molecular weight ranges with increasing consolidation temperatures. A rationale for the increase in the level of adhesion was proposed based on the polymer wettability, reduction in viscosity, the chain size, its conformations in the interphase region and also the interactions between fiber surface and the polymer chains.

References

1   Rao, V., and Drzal, L.T., *Polym. Compos.*, 12, 48-56 (1991)

2   Monette, L., Anderson, M.P., and Grest, G.S., Polym. Compos., 14, 101-115 (1993)

3   Nielsen, L.E., "Mechanical Properties of Polymer Composites," Marcel Dekker Inc., New York (1974)

4   Theocaris, P.S., "Mesophase Concept in Composites," Springer-Verlag Press, New York (1987)

5   Drzal, L.T., J. of Mater. Sci. and Engg., A126, 289-293 (1990)

6   Bascom, W.D., and Jensen, R.M., J. Adhesion, 19, 219-239 (1986)

7   Cox, H.L., Br. J. Appl. Phys., 3(1), 122 (1952)

8   Cooke, T.F., J. of Polym. Proces., 7(3), 197-254 (1987)

9   Mansfield, K.F., and Theodorou, D.N., Macromolecules, 22, 3143 (1989)

10 Chakrabarti, A., Toral, R., Macromolecules, 23, 2016 (1989)

11 Pangelinan, A.B., "Surface Induced Molecular Weight Segregation in Thermoplastic Composites,"Ph.D dissertation, University of Delaware , 1991

12 Tse, M.K., U.S. Patent 4,662,228

13 Mandell, J.F., Grande D.H., Tsiang, T.H., and McGarry, F.J., "A Modified Microdebonding Test for Direct In-Situ Fiber/Matrix Bond Strength Determination in Fiber Composites," Research Report R84-3, Massachussetts Institute of Technology, December 984

14 Brady, R.L., and Porter, R.S., J. Thermo. Compos., 2, 164 (1989)

15 Stone,P.R., and Nairn, J.A., Polym. Compos., 15, 197-205 (1994)

High Strength Glass Fibers

D.R. Hartman, M.E. Greenwood, D.M. Miller
Owens-Corning Science and Technology Center
Granville, Ohio

Abstract

Continuous glass fibers, first conceived and manufactured during 1935 in Newark, Ohio, started a revolution in reinforced composite materials which has grown to a global annual consumption of about 3 billion kilograms. During 1942 glass fiber reinforced composites were first used in structural aerospace parts. In the early 1960's high-strength glass fibers, S GLASS, were first used in joint work between Owens-Corning and the United States Air Force. Later in 1968 S-2 Glass® products began evolving into a variety of commercial applications. High-strength glass fibers combine high temperature durability, stability, transparency, and resilience at a very reasonable cost-weight-performance. The utility of high-strength glass fiber compositions are compared by physical, mechanical, electrical, thermal, acoustical, optical, and radiation properties. Recent revolutionary glass fiber developments such as HOLLEX™ (S-2 GLASS hollow fiber) and evolutionary glass compositions are developing niches for new incremental global opportunities.

COARSE FIBERS DRAWN FROM HEAT SOFTENED GLASS were used by ancient Egyptians to make containers. The French scientist Reaumur considered the potential of forming fine glass fibers for woven glass articles as early as the 18th century. Continuous glass fibers were first manufactured in substantial quantities by Owens-Corning in the 1930's for high temperature electrical applications. Revolutionary and evolutionary technology continues to improve manufacturing processes for continuous glass fiber production illustrated in Figure 1. Raw materials such as silicates, soda, clay, limestone, boric acid, fluorspar or various metallic oxides are blended to form a glass batch which is melted in a furnace and refined during lateral flow to the forehearth. The molten glass flows to platinum/rhodium alloy bushings and then through individual bushing tips with orifices ranging from 0.76 to 2.03 mm (0.030 to 0.080 in) and is rapidly quenched and attenuated in air (to prevent crystallization) into fine fibers ranging from 3 to 24 μm (120 to 950 μin). Mechanical winders pull the fibers at lineal velocities up to 61 m/s (200 ft/sec) over an applicator which coats the fibers with an appropriate chemical sizing to aid further processing and performance of the end product.

High strength glass fibers like S-2 GLASS are compositions of aluminosilicates attenuated at higher temperatures into fine fibers ranging from 6 to 14 μm (230 to 550 μin). Several other types of silicate glass fibers are manufactured for the textile and composite industry. Various glass chemical compositions described below from ASTM C 162 were developed to provide combinations of fiber properties directed at specific end use applications.

A-GLASS - Soda lime silicate glasses used where the strength, durability, and good electrical resistivity of E-glass are not required.
C-GLASS - Calcium borosilicate glasses used for their chemical stability in corrosive acid environments.
D-GLASS - Borosilicate glasses with a low dielectric constant for electrical applications.
E-GLASS - Alumina-calcium-borosilicate glasses with a maximum alkali content of 2 wt.% used as general purpose fibers where strength and high electrical resistivity are required.
ECRGLAS - Calcium aluminosilicate glasses with a maximum alkali content of 2 wt.% used where strength, electrical resistivity, and acid corrosion resistance are desired.
AR-GLASS - Alkali resistant glasses composed of alkali zirconium silicates used in cement substrates and concrete.
R-GLASS - Calcium aluminosilicate glasses used for **reinforcement where added strength and acid corrosion** resistance are required.
S-2 GLASS - Magnesium aluminosilicate glasses used for

textile substrates or reinforcement in composite structural applications which require high strength, modulus, and stability under extreme temperature and corrosive environments.

Glass Fiber Chemical Compositions

Chemical composition variation within a glass type is from differences in the available glass batch raw materials, or in the melting and forming processes, or from different environmental constraints at the manufacturing site. These compositional fluctuations do not significantly alter the physical or chemical properties of the glass type. Very tight control is maintained within a given production facility to achieve consistency in the glass composition for production capability and efficiency. Table 1 provides the oxide components and their weight ranges for eight types of commercial glass fibers (1-6).

Glass Fiber Properties

Glass fiber properties, such as tensile strength, Young's modulus, and chemical durability, are measured on the fibers directly. Other properties, such as dielectric constant, dissipation factor, dielectric strength, volume/surface resistivities, and thermal expansion, are measured on glass that has been formed into a bulk sample and annealed (heat treated) to relieve forming stresses. Properties such as density and refractive index are measured on both fibers and bulk samples, in annealed or unannealed form. The properties presented in Tables 2 and 3 are representative of the compositional ranges in Table 1 and correspond to the following overview of glass fiber properties.

Physical Properties. Density of glass fibers is measured and reported either as formed or as bulk annealed samples. ASTM C 693 is one of the test methods used for density determinations (7). The fiber density (in Table 3) is less than the bulk annealed value by approximately 0.04 g/cc at room temperature. The glass fiber densities used in composites range from approximately 2.11 g/cc for D-glass to 2.72 g/cc for ECRGLAS. The S-2 GLASS density of 2.46 g/cc is reduced to an apparent fiber density of 1.8 g/cc with the revolutionary continuous hollow fiber called HOLLEX.

Tensile strength of glass fibers is usually reported as the pristine single-filament or the multifilament strand measured in air at room temperature. The respective strand strengths are normally 20 to 30% lower than the values reported in Table 2 due to surface defects introduced during the strand-forming process. Moisture has a detrimental effect on the pristine strength of glass. This is best illustrated by measuring the pristine single-filament strength at liquid nitrogen temperatures where the influence of moisture is minimized. The result is an increase of 50 to 100% in strength over a measurement at room temperature in 50% relative humidity air. The maximum measured strength of S-2 GLASS at liquid nitrogen temperatures is 11.6 GPa (1.68×10^6 psi) for a 12.7 mm gauge length, $10\mu m$ diameter fiber. The loss in strength of glass exposed to moisture while under an external load is known as static fatigue (4,8). The pristine strength of glass fibers decrease as the fibers are exposed to increasing temperature. E-glass and S-2 GLASS fibers have been found to retain approximately 50% of their pristine room-temperature strength at 538°C (1000°F) and are compared to organic reinforcement fibers in Figure 2.

The Young's modulus of elasticity of unannealed silicate glass fibers ranges from about 52 GPa (7.5×10^6 psi) to 87 GPa (12.6×10^6 psi). As the fiber is heated, the modulus gradually increases. E-glass fibers that have been annealed to compact their atomic structure will increase in Young's modulus from 72 GPa (10.5×10^6 psi) to 84.7 GPa (12.3×10^6 psi)(4). For most silicate glasses, Poisson's ratio falls between 0.15 and 0.26 (9). The Poisson's ratio for E-glasses is 0.22 ± 0.02 and is reported not to change with temperature when measured up to 510°C (950°F) (10).

High strength S-2 GLASS annealed properties measured at 20°C are as follows:

Young's Modulus	93.8 GPa (13.60 x 10^6 psi)
Shear Modulus	38.1 GPa (5.53 x 10^6 psi)
Poisson's Ratio	0.230
Bulk Density	2.488 g/cc

Chemical Resistance. The chemical resistance of glass fibers to the corrosive and leaching actions of acids, bases, and water is expressed as a percent weight loss. The lower this value, the more resistant the glass is to the corrosive solution. The test procedure involves subjecting a given weight of 10 micron diameter glass fibers, without binders or sizes, to a known volume of corrosive solution held at 96°C (205°F). The fibers are held in the solution for the time desired and then are removed, washed, dried, and weighed to determine the weight loss. The results reported are for 24-hr (1 day) and 168-hr (1 week) exposures. As Table 3 shows, the chemical resistance of glass fibers depends on the composition of the fiber, the corrosive solution, and the exposure time.

It should be noted that glass corrosion in acidic environments is a complex process beginning with an initial fast corrosion rate. (Note the similarity in weight loss between the 1-day and 1-week samples treated with acid in Table 3.) With further time, an effective barrier of leached glass is established on the surface of the fiber and the corrosion rate of the remaining unleached fiber slows, being controlled by the diffusion of compounds through the leached layer. Later, the corrosion rate slows to nearly zero as the non-silica compounds of the fiber are depleted. For a given glass composition, the corrosion rate may be influenced by the acid concentration (Figure 3), temperature, fiber diameter, and the solution volume to glass mass ratio.

In alkaline environments weight loss measurements are more subjective as the alkali effects the network and reprecipitates the metal oxides. Tensile strength after exposure is a better indicator of the residual glass fiber properties as shown in Figure 4 for 24-hour exposure at 96°C (205°F).

Electrical Properties. The electrical properties in Table 3 were measured on annealed bulk glass samples according to the testing procedures cited (11-13). The dielectric constant or relative permittivity is the ratio of the capacitance of a system with the specimen as the dielectric to the capacitance of the system with a vacuum as the dielectric. Capacitance is the ability of the material to store an electrical charge. Permittivity values are affected by test frequency, temperature, voltage, relative humidity, water immersion, and weathering.

The dissipation factor of a dielectric is the ratio of the parallel reactance to the parallel resistance, or the tangent of the loss angle, which is usually called the loss tangent. It is also the reciprocal of the quality factor, and when the values are small, tangent of the loss angle is essentially equal to the power factor, or sine of the loss angle. The power factor is the ratio of power in watts dissipated in the dielectric to the effective volt-amperes. The dissipation factor is dimensionless. In almost every electrical application, a low value for the dissipation factor is desired. This reduces the internal heating of the material and keeps signal distortion low. The dissipation factor is generally measured simultaneously with permittivity measurements, and is greatly influenced by frequency, humidity, temperature, and water immersion.

The loss factor, or loss index, as it is some times called, is occasionally confused with the dissipation factor, or loss tangent. The loss factor is simply the product of the dissipation factor and permittivity and is proportional to the energy loss in the dielectric.

The dielectric breakdown voltage, is the voltage at which electrical failure occurs under prescribed test conditions in an electrical insulating material that is placed between two electrodes. When the thickness of the insulating material between the electrodes can be accurately measured, the ratio of the dielectric breakdown voltage to the specimen thickness can be expressed as the dielectric strength in kv/cm. Breakdown voltages are influenced by electrode geometry, specimen thickness (because dielectric strength varies approximately as the reciprocal of the square root of the thickness), temperature, voltage application time, voltage

wave form, frequency, surrounding medium, relative humidity, water immersion, and directionality in laminated and inhomogeneous plastics.

Thermal Properties. The viscosity of a glass decreases as the temperature increases. Figure 5 shows the viscosity-temperature plots for E-glass and S-2 GLASS. Note that the S-2 GLASS temperature at viscosity is 150 - 260°C (300-500°F) higher than that of E-glass, which is why S-2 GLASS has higher use temperatures than E-glass.

Several reference viscosity points are defined by the glass industry as used in Table 2. The softening point is the temperature at which a glass fiber of uniform diameter elongates at a specific rate under its own weight when measured by ASTM C 338 (14). The softening point is defined as the temperature at which glass will deform under its own weight; it occurs at a viscosity of approximately $10^{6.6}$ Pa.s ($10^{7.6}$ P). The annealing point is the temperature corresponding to either a specific rate of elongation of a glass fiber when measured by ASTM C 336 (15), or a specific rate of midpoint deflection of a glass beam when measured by ASTM C 598 (16). At the annealing point of glass, internal stresses are substantially relieved in a matter of minutes. The viscosity at the annealing point is approximately 10^{12} Pa.s (10^{13} P). The strain point is measured following ASTM C 336 or C 598 as described above for annealing point. At the strain point of glass, internal stresses are substantially relieved in a matter of hours. The viscosity at the strain point is approximately $10^{13.5}$ Pa.s ($10^{14.5}$ P).

The mean coefficient of thermal expansion over the temperature range from -30 to 250°C (-22 to 482°F) is provided in Table 3. The expansion measurements were made on annealed bars using ASTM D 696 (17). A lower coefficient of thermal expansion in the high strength glasses allows higher dimensional stability at temperature extremes.

The specific heat data in Table 3 was determined using high-temperature differential scanning calorimetry techniques. In general, the average specific heat values can be represented as follows: 0.94 kJ/kg·K (0.224 Btu/lb·°F) at 200°C (390 °F), 1.12 kJ/kg·K (0.267 Btu/lb·°F) just below the transition point, and 1.40 kJ/kg·°K (0.333 Btu/lb·°F) in the liquid state above the transition. These values are accurate to about 5%. Above the transition temperature, no further increase in specific heat was observed. The transition temperature is nearly identical to the annealing temperature of bulk glass.

Thermal conductivity characteristics in glasses differ considerably from those found in crystalline materials. For glasses, the conductivity is lower than that of the corresponding crystalline materials. Also, the conductivity of glasses drops steadily with temperature and reaches very low values, near absolute zero. For crystals, the conductivity continues to rise with decreasing temperature until very low temperatures are reached (18). Thermal conductivity data for glass varies among investigators for materials which are normally identical (19). In general fused silica glass and the alkali and alkaline earth silicate glasses have relatively similar conductivities at room temperature, whereas conductivities of borosilicate and glass that contain lead and barium are somewhat lower. Near room temperature, the thermal conductivity for glasses ranges from 0.55 W/m·°K (3.8 Btu·in/hr·ft²·°F) for lead silicate (80% lead oxide, 20% silicon dioxide) to 1.4 W/m·°K (9.6 Btu·in/hr·ft²·°F) for fused silica glass (20). E. H. Ratcliffe developed property coefficients for predicting thermal conductivity from the percentage weight compositions of component oxides making up the glass (20). Using this calculation, it is found that the approximate thermal conductivity of C-glass is 1.1 W/m·°K (7.9 Btu·in/hr·ft²·°F), E-glass is 1.3 W/m·°K (8.9 Btu·in/hr·ft²·°F), and S-2 GLASS is 1.45 W/m·°K (10.1 Btu·in/hr·ft²·°F) near room temperature.

Optical Properties. Refractive index is measured on either unannealed or annealed glass fibers. The standard oil immersion techniques are used with monochromatic sodium D light at 25°C (77°F). In general, the corresponding annealed glass will exhibit an index that will range from approximately 0.0030 to 0.0060 higher than the as-formed glass fibers given in Table 2.

Radiation Properties. S-2 GLASS and E glass fibers have excellent resistance to all types of nuclear radiation. Alpha and beta radiation have almost no effect, while gamma radiation and neutron bombardment produce a 5 to 10% decrease in tensile strength, a less than 1% decrease in density, and a slight discoloration of fibers. This data was true of 10^{20} NVT neutrons or gamma radiation up to 10^5 J/g. Glass fibers resist radiation because the glass is amorphous, and the radiation does not distort the atomic ordering. Glass can also absorb a few percent of foreign material and maintain the same properties to a reasonable degree. Also, because the individual fibers have a small diameter, the heat of atomic distortion is easily transferred to a surface for dispersion.

E-glass, ECRGLAS, and C-glass are not recommended for use inside atomic reactors because of their high boron content. S-2 GLASS is suitable for use inside atomic reactors. Because quite a wide variety of organic products are used in diverse radiation environments, it is usually necessary to try out most products in simulated conditions to determine whether the organics will be satisfactory.

Glass Fiber Size Treatments

The surface treatment chemistry of glass fiber follows the necessary product function. Textile size chemistries based on starch or polyvinyl alcohol film formers are capable in weaving, braiding, or knitting processes. Typically the weaver then scours or heat cleans the glass fabric and applies a finish compatible with the end product. Nonwoven size chemistries often include dispersants compatible with white water chemistry for wet formed mats or additives compatible with dry or wet binder chemistry for dry formed mats. Reinforcement size chemistries must be compatible with a multitude of processes and with the composite material end use performance criteria. Processes such as injection molding require chopped fibers with compatibility for thermoplastic compounds. Filament winding and pultrusion require continuous fibers with utility in thermoset and thermoplastic compounds. Typically three basic components are used with high strength glass size chemistries: a film former, lubricant, and coupling agent. Table 4 outlines evolutionary research for glass fiber size chemistry by each components role.

Glass Fiber Composite Utility

Composite Properties. Application of glass fiber composite materials depend on proper utilization of glass composition, size chemistry, fiber orientation, and fiber volume in the appropriate matrix for desired mechanical, electrical, thermal and other properties. Table 5 gives typical mechanical properties for high strength S-2 GLASS in epoxy with unidirectional fiber orientation. The elastic constants and strain allowables are used for design input. The effect of glass composition and fiber volume in epoxy are shown in Figure 6 for the coefficient of thermal expansion and in Figure 7 for dielectric constant. The S-2 GLASS has lower dielectric constant and therefore the potential for better radar transparency. HOLLEX is also compared with an apparent dielectric constant of 3.8 at 10 GHz which further improves radar transparency (21).

Environmental Durability. The durability of glass fiber composite materials is one of the features that attract users to them. Composites do not corrode like metals and are low maintenance materials. However, the durability and reliability in specific applications is often requested for design input. In load bearing structures, the long-term behavior of the material is needed to complete the design of the structure. How much load will the structure hold and survive for a given period of time? Or, how thick must the part be to handle a load for a given period of time?

These questions are usually addressed empirically, with accelerated testing. ASTM D 2992 and ASTM D 3681 for example are methods often incorporated in the practice of evaluating and designing composite pipe. In these methods, actual prototype products are manufactured and loaded in simulated environments for which the product is intended. To accelerate testing, the loads are significantly higher than operating conditions to induce failure in a relatively short period of time. The extrapolation of short-term data to the expected life of the product allows engineers to predict safe operating loads (stresses) for the purpose of design. This practice has served the composites industry well for over 30 years.

To demonstrate this process, the results of several test series are reported. Tests were conducted using pultruded rods made of glass fibers and thermoset resins. These tests are similar to those referenced in ASTM except the loading condition is pure tension and under a constant load. Figure 8 summarizes stress rupture testing using S-2 GLASS in epoxy with and without an adverse environment, a calcium hydroxide solution with a pH of 13. For reference, the initial tensile strength of the composite rod was 2070 MPa (300 x 10^3 psi). The stress rupture behavior of an S-2 GLASS/epoxy rod in this test indicates a long-term stress capability of 65% of the initial ultimate tensile stress. As expected, the stress rupture behavior of the composite material is effected by the presence of the environment. The long-term stress capability of this material in the high pH environment is roughly 50% of the initial ultimate tensile strength.

A second test series compared the stress rupture performance of E glass and S-2 GLASS reinforcements in epoxy resin with the high pH environmental exposure. Figure 9 summarizes this data. The S-2 GLASS reinforced composite rod had a higher initial tensile stress than the E glass reinforced rod. The influence of the combined effects of stress and environment were similar for the two materials.

Acknowledgments

The authors thank Lynn McIlyar for assistance with the manuscript.

References

1. D. M. Miller, Glass Fibers, Engineered Materials Handbook, Vol. 1 - Composites, ASM International, 1987, pp. 45-48.

2. G. J. Mohr, and W. P. Rowe, "Fiber Glass," Van Nostrand Reinhold Co., (1978), pp. 207.

3. K. L. Loewenstein, "The Manufacturing Technology of Continuous Glass Fibers," Elsevier, (1973), pp. 28-30.

4. R. D. Lowrie, Glass Fibers for High-Strength Composites, in Modern Composite Materials, Addison-Wesley Publishing Co., (1967), pp. 270-323.

5. W. W. Wolf, The Glass Fiber Industry -- The Reason for the Use of Certain Chemical Compositions, Seminar at University of Illinois, Urbania, IL, (Oct. 1982).

6. J. C. Watson, and N. Raghupathi, Glass Fibers, Engineered Materials Handbook, Vol. 1 - Composite, ASM International, (1987), pp. 107-111.

7. Standard Test Method for Density of Glass by Buoyancy, C693, "Annual Book of ASTM Standards," American Society for Testing Materials.

8. P. K. Gupta, Examination of the Textile Strength of E-Glass Fiber in the Context of Slow Crack Growth, Fracture Mechanics of Ceramics, Vol. 5, (1983), pp. 291.

9. J. R. Hutchins, III and R. W. Harrington, Glass, in "Encyclopedia of Chemical Technology," Vol. 10, 2nd Ed., pp. 533-604.

10. R. T. Brannan, Am. Ceram. Soc., Vol. 36, (1953), pp. 230-231.

11. Standard Test Methods for A-C Loss Characteristics and Permittivity (Dielectric Constant) of Solid Electrical Insulating Materials, D. 150, "Annual Book of ASTM Standards," American Society for Testing and Materials.

12. Standard Test Methods for D-C Resistance or Conductance of Insulation Materials, D. 257, "Annual Book of ASTM Standards," American Society for Testing and Materials.

13. Standard Test Methods for Dielectric Breakdown Voltage and Dielectric Strength of Solid Electrical Insulating Materials at Commercial Power Frequencies, D. 149, "Annual Book of ASTM Standards," American Society for Testing and Materials.

14. Standard Test Methods for Softening Point of Glass, C. 338, "Annual Book of ASTM Standards," American Society for Testing and Materials.

15. Standard Test Methods for Annealing Point and Strain Point of Glass by Fiber Elongation, C. 336,

"Annual Book of ASTM Standards," American Society for Testing and Materials.

16.   Standard Test Methods for Annealing Point and Strain Point of Glass by Beam Bending, C. 598, "Annual Book of ASTM Standards," American Society for Testing and Materials.

17.   Standard Test Methods for Coefficient of Linear Thermal Expansion of Plastics, D. 696, "Annual Book of ASTM Standards," American Society for Testing and Materials.

18.   C. Kittel, Phys. Rev., Vol. 75, (1949), pp. 972.

19.   C. L. Babcock, Symposium on Heat Transfer Phenomena in Glass, J. Am. Ceram, Soc., Vol. 44 (No. 7), (July 1961).

20.   E. H. Ratcliffe, Thermal Conductivities of Glass Between -150°C and 100°C, Glass Technology, Vol. 4 (No. 4), (August 1963).

21.   D. R. Hartman, and S. Schuster, Dielectric and Structural Properties of Hollex™ Fiber Composites, SAMPE Proceedings, Vol. 38, (May 1993), pp. 1529.

TABLE 1. COMPOSITION RANGES FOR GLASS FIBERS								
	A-GLASS	C-GLASS	D-GLASS	E-GLASS	E-CR GLASS	AR-GLASS	R-GLASS	S-2 GLASS®
OXIDE	%	%	%	%	%	%	%	%
SiO_2	63-72	64-68	72-75	52-56	54-62	55-75	55-65	64-66
Al_2O_3	0-6	3-5	0-1	12-16	9-15	0-5	15-30	24-25
B_2O_3	0-6	4-6	21-24	5-10		0-8		
CaO	6-10	11-15	0-1	16-25	17-25	1-10	9-25	0-0.1
MgO	0-4	2-4		0-5	0-4		3-8	9.5-10
ZnO					2-5			
BaO		0-1						
Li_2O						0-1.5		
Na_2O+ K_2O	14-16	7-10	0-4	0-2	0-2	11-21	0-1	0 - 0.2
TiO_2	0-0.6			0-1.5	0-4	0-12		
ZrO_2						1-18		
Fe_2O_3	0-0.5	0-0.8	0-0.3	0-0.8	0-0.8	0-5		0 - 0.1
F_2	0-0.4			0-1		0-5	0-0.3	

TABLE 2. PROPERTIES OF GLASS FIBERS

PHYSICAL PROPERTIES

	A-GLASS	C-GLASS	D-GLASS	E-GLASS	E-CR GLASS	AR-GLASS	R-GLASS	S-2 GLASS®
DENSITY, gm/cc (lbs/ci)	2.44 (153)	2.52 (158)	2.11 (132)	2.58 (161)	2.72 (170)	2.70 (169)	2.54 (159)	2.46 (154)
REFRACTIVE INDEX	1.538	1.533	1.465	1.558	1.579	1.562	1.546	1.521
SOFTENING POINT, C (F)	705 (1300)	750 (1382)	771 (1420)	846 (1555)	882 (1619)	773 (1424)	952 (1745)	1056 (1932)
ANNEALING POINT, C (F)		588 (1090)	521 (970)	657 (1215)				816 (1500)
STRAIN POINT, C (F)		522 (1025)	477 (890)	615 (1140)				766 (1410)
TENSILE STRENGTH,MPa (Ksi)								
at 196C		5380 (780)		5310 (770)	5310 (770)			8275 (1200)
at 23C	3310 (480)	3310 (480)	2415 (350)	3445 (500)	3445 (500)	3241 (470)	4135 (600)	4890 (700)
at 371C				2620 (380)	2165 (314)		2930 (425)	4445 (645)
at 538C				1725 (250)	1725 (250)		2140 (310)	2415 (350)
YOUNG'S MODULUS, GPa (Msi)								
at 23C	68.9 (10.0)	68.9 (10.0)	51.7 (7.5)	72.3 (10.5)	72.3 (10.5)	73.1 (10.6)	85.5 (12.4)	86.9 (12.6)
at 538C				81.3 (11.8)	81.3 (11.8)			88.9 (12.9)
ELONGATION %	4.8	4.8	4.6	4.8	4.8	4.4	4.8	5.7

TABLE 3. PROPERTIES OF GLASS FIBERS

	A-GLASS	C-GLASS	D-GLASS	E-GLASS	E-CR GLASS	AR-GLASS	R-GLASS	S-2 GLASS®
DURABILITY (% weight loss)	CHEMICAL PROPERTIES							
H_2O: 24 hr	1.8	1.1	0.7	0.7	0.6	0.7	0.4	0.5
168 hr	4.7	2.9	5.7	0.9	0.7	1.4	0.6	0.7
10% HCl: 24 hr	1.4	4.1	21.6	42	5.4	2.5	9.5	3.8
168 hr		7.5	21.8	43	7.7	3.0	10.2	5.1
10% H_2SO_4: 24 hr	0.4	2.2	18.6	39	6.2	1.3	9.9	4.1
168 hr	2.3	4.9	19.5	42	10.4	5.4	10.9	5.7
10% Na_2CO_3 24 hr		24	13.6	2.1		1.3	3.0	2.0
168 hr		31	36.3	2.1	1.8	1.5	1.5	2.1
ELECTRIC PROPERTIES								
DIELECTRIC CONSTANT 1MHz	6.2	6.9	3.8	6.6	6.9	8.1	6.4	5.3
10 GHz			4.0	6.1	7.0			5.2
DISSIPATION FACTOR 1MHz		0.0085	0.0005	0.0025	0.0028		0.0034	0.0020
10 GHz			0.0026	0.0038	0.0031		0.0051	0.0068
VOLUME RESISTIVITY (ohm-cm)	1.0E+10			4.02E+14	3.84E+14		2.03E+14	9.05E+12
SURFACE RESISTIVITY (ohms)				4.20E+15	1.16E+16		6.74E+13	8.86E+12
DIELECTRIC STRENGTH (volts/mil)				262	250		274	330
THERMAL PROPERTIES								
SPECIFIC HEAT J/G C (BTU/LB F) 23C	0.796 (0.190)	0.787 (0.188)	0.733 (0.176)	0.810 (0.193)				0.737 (0.176)
200C		0.900 (0.215)		1.03 (0.247)	0.97 (0.232)			
THERMAL EXPANSION COEFFICIENT (x 10-7)	C (F)	C (F)	C (F)	C (F)	C (F)	C (F)	C (F)	C (F)
-30C to 250C	73 (40.6)	63 (35)	25 (13.9)	54 (30)	59 (33)	65 (36.1)	33 (18.3)	16 (8.9)

TABLE 4. GLASS FIBER SIZE CHEMISTRY SUMMARY

NOMENCLATURE	TYPICAL CHEMISTRY	ROLE	EVOLUTIONARY TECHNOLOGY
Film Formers	Epoxies, polyesters, PVAc, EVAc, polyolefins, and polyurethanes, etc.	Fiber protection, strand integrity, wetting and solubility.	Improved thermal stability, low surfactant, controlled solubility, etc.
Lubricants	Imidazolines, tetraethylene amide, mineral oil/amide ester, acid amide, polyethylene glycols, etc.	Strand integrity, surface friction, improved fiber forming.	"Non-migrating" lubes, functionalized lubes.
Emulsifiers	Polyoxyethylene nonylphenyl ether, EO/PO condensate, polyoxyethylene octylphenyl glycol ether, etc.	Render film former and lubes water compatible.	Fugitive emulsifiers, functionalized emulsifiers.
Coupling Agents	Silanes, titanates, zirconates.	Resin/glass bonding.	Improved thermal stability, bonding to new matrices.
Other Additives			
• Antistats	Metal halides, quat. ammonium, etc.	Increase conductivity.	Less-hygroscopic species.
• pH Control	Organic acids/bases.	Control pH.	Improved coupling stability.
• Nucleating Agents	See patent literature.	Surface nucleation.	Encourage transcrysallinty.

PROPERTY	ASTM STANDARD	75°F(22°C)	
Elastic Constants		<u>Msi</u>	<u>GPa</u>
Longitudinal Modulus, E_L	D3039	7.7-8.5	53 - 59
Transverse Modulus, E_T	D3039	2.3-2.9	16 - 20
Axial Shear Modulus, G_{LT}	D3518	0.9-1.3	6 - 9
Poisson's Ratio, v_{LT}	D3039	0.26 - 0.28	1.8 - 1.9
Strength Properties		<u>Ksi</u>	<u>MPa</u>
Longitudinal Tension, F^{tu}_L	D3039	230 - 290	1590 - 2000
Longitudinal Compression, F^{cu}_L	D3410	100 - 180	690 - 1240
Transverse Tension, F^{tu}_T	D3039	6 - 12	41 - 82
Transverse Compression, F^{cu}_T	D3410	16 - 29	110 - 200
In-Plane Shear, F^{su}_{LT}	D3518	9 - 24	62 - 165
Interlaminar Shear, F^{isu}	D2344	8 - 15	55 - 103
Longitudinal Flexural	D790	180 - 250	1240 - 1720
Longitudinal Bearing	D953	68 - 80	469 - 552
Ultimate Strains			
Longitudinal Tension, ϵ^{tu}_L	D3039	2.7 - 3.5%	
Longitudinal Compression, ϵ^{cu}_L	D3410	1.1 - 1.8%	
Transverse Tension, ϵ^{tu}_T	D3039	0.25 - 0.50%	
Transverse Compression, ϵ^{cu}_T	D3410	1.1 - 2.0%	
In-Plane Shear, γ^{su}_{LT}	D3518	1.6 - 2.5%	
Physical Properties			
Fiber Volume (%)	D2734	57 - 63%	
Density	D792	<u>lb/in³</u>	<u>g/cm³</u>
		0.071 - 0.073	1.96 - 2.02

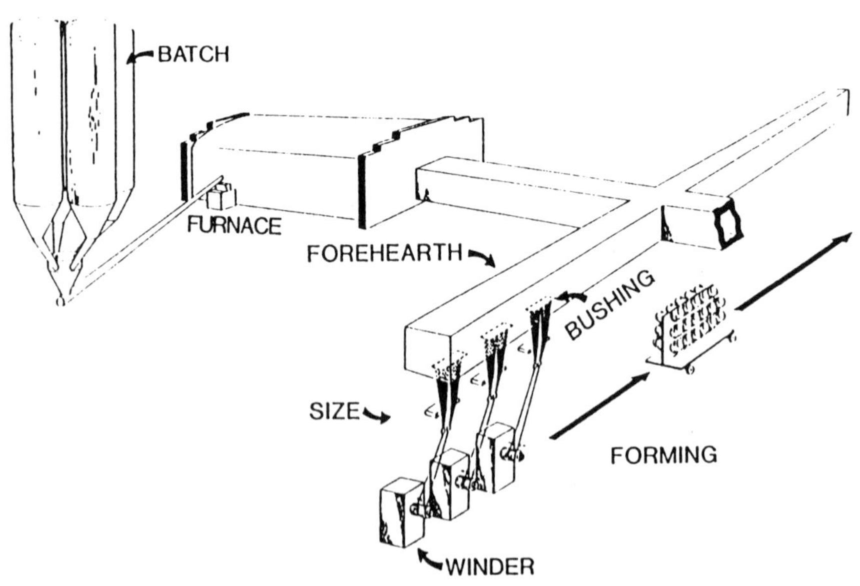

FIGURE 1. CONTINUOUS GLASS FIBER MANUFACTURING PROCESS

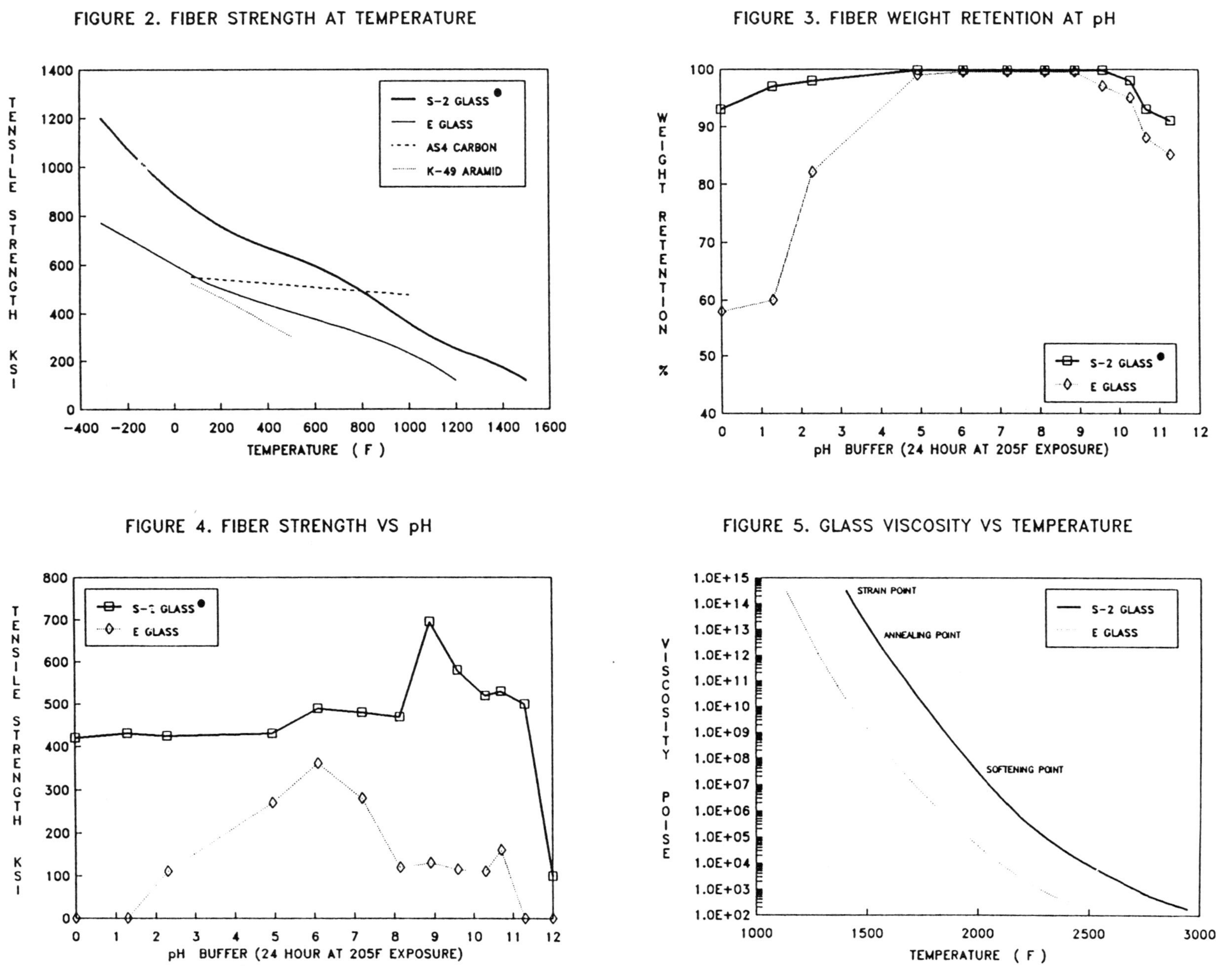

FIGURE 2. FIBER STRENGTH AT TEMPERATURE
S-2 GLASS
E GLASS
AS4 CARBON
K-49 ARAMID
TENSILE STRENGTH KSI
1400
1200
1000
800
600
400
200
0
TEMPERATURE (F)
-400
-200
0
200
400
600
800
1000
1200
1400
1600

FIGURE 3. FIBER WEIGHT RETENTION AT pH
WEIGHT RETENTION %
100
90
80
70
60
50
40
S-2 GLASS
E GLASS
pH BUFFER (24 HOUR AT 205F EXPOSURE)
0
1
2
3
4
5
6
7
8
9
10
11
12

FIGURE 4. FIBER STRENGTH VS pH
S-2 GLASS
E GLASS
TENSILE STRENGTH KSI
800
700
600
500
400
300
200
100
0
pH BUFFER (24 HOUR AT 205F EXPOSURE)
0
1
2
3
4
5
6
7
8
9
10
11
12

FIGURE 5. GLASS VISCOSITY VS TEMPERATURE
VISCOSITY POISE
1.0E+15
1.0E+14
1.0E+13
1.0E+12
1.0E+11
1.0E+10
1.0E+09
1.0E+08
1.0E+07
1.0E+06
1.0E+05
1.0E+04
1.0E+03
1.0E+02
STRAIN POINT
ANNEALING POINT
SOFTENING POINT
S-2 GLASS
E GLASS
TEMPERATURE (F)
1000
1500
2000
2500
3000

FIGURE 6. THERMAL EXPANSION VS VOLUME

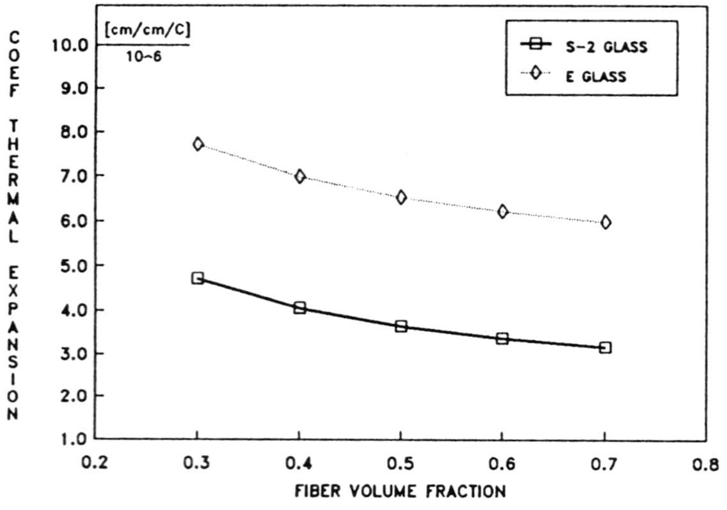

FIGURE 7. DIELECTRIC VS FIBER VOLUME

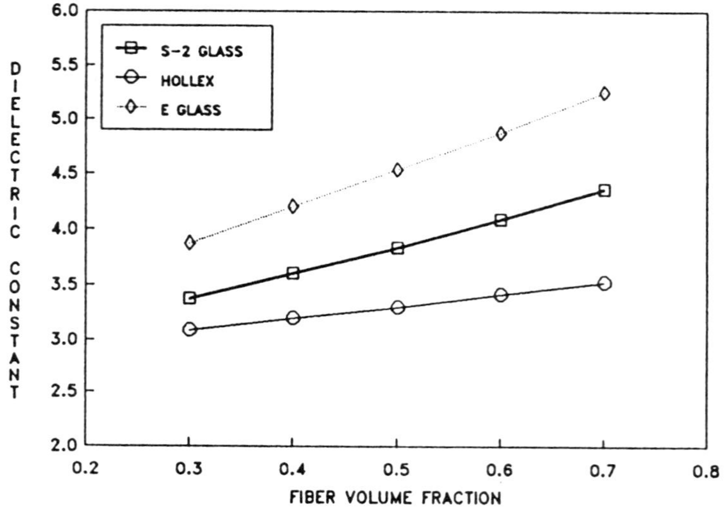

FIGURE 8. ENVIRONMENTAL STRESS RUPTURE

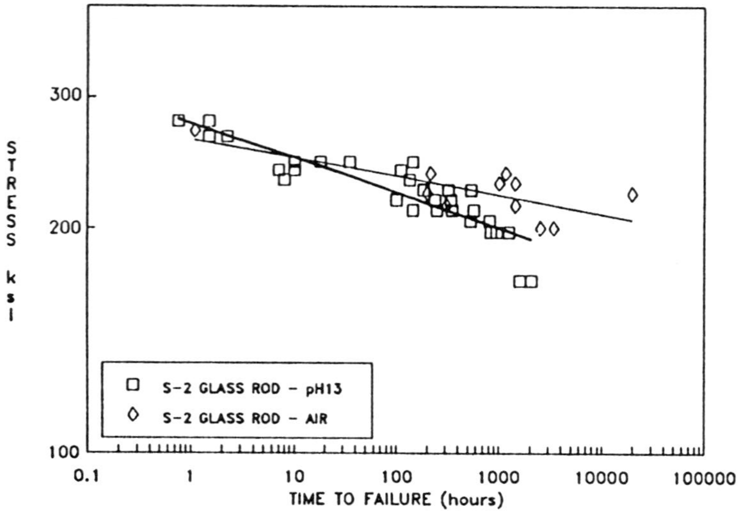

FIGURE 9. ENVIRONMENTAL STRESS RUPTURE

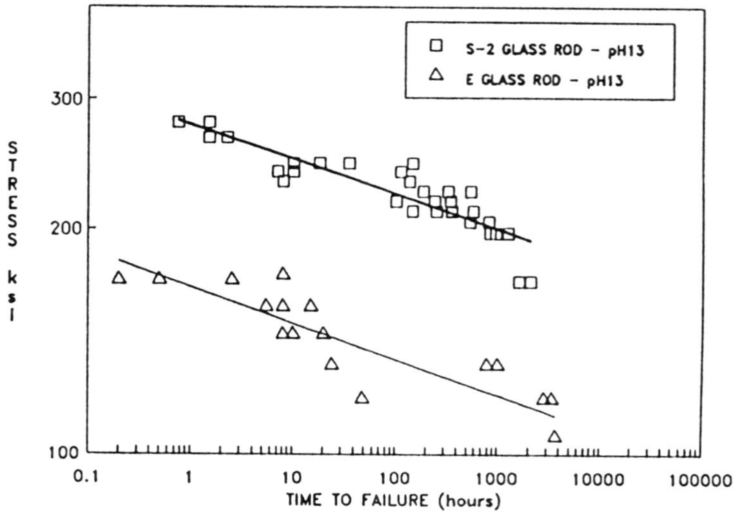

Novel Isocyanate-Based Matrix Resins for
High Temperature Composite Applications

V. Sendijarevic, A. Sendijarevic, K.C. Frisch
University of Detroit Mercy
Detroit, Michigan

P. Reulen
University of Technology in Eindhoven
Eindhoven, The Netherlands

Polyoxazolidone composites were prepared from polymeric isocyanate (PAPI 901) and epoxides (Epon 828 and DEN 431) in the presence of an oxazolidone forming catalyst, triphenylantimony iodide. The effects of isocyanate index, type of epoxide and amount of fiberglass reinforcement on the composite properties were studied as well as the effects of post-curing temperature and time. Increasing the fiberglass content of the polyoxazolidone composites resulted in an improvement of the thermal and mechanical strength properties. The heat deflection temperature of all polyoxazolidones was >250°C. The retention of the tensile strength at 150°C was excellent, ca. 90% or higher. Polyoxazolidone composites based on DEN 431 at 120 isocyanate index with 70% of fiberglass and post-cured at 150°C for 48 hours exhibited the best properties. According to the results of DMA, TMA and DSC, the maximum operating temperature for polyoxazolidone composites is around 200°C. The TGA data showed that the decomposition temperature was around 330°C.

INTRODUCTION

The best known and commercially most important isocyanate-based heat resistant structure is heterocyclic isocyanurate. This structure is thermally stable up to 400°C. Although polyisocyanurates have excellent thermal and hydrolytic stability, 100% polyisocyanurates are not of commercial significance because of their brittleness as a result of the high crosslink density. In typical commercial applications (in rigid foams), polyisocyanurates are modified with polyols leading to the formation of a urethane-isocyanurate polymer network [1-6]. The thermal stability of such polymer structure is limited by the thermal stability of the urethane groups, which are less thermally stable. A number of studies on the thermal degradation of isocyanate-based linkages with model monomers and polymers confirmed the following order of thermal stability: isocyanurate > oxazolidone > urea > urethane [6-12].

Polyoxazolidones are formed from isocyanates and epoxies in the presence of oxazolidone forming catalysts [13,14]. The crosslink density of polyoxazolidones can be controlled by selection of polyfunctional components (both isocyanates and epoxies) and by varying the isocyanate index (isocyanate to epoxy equivalent ratio multiplied by 100). An excess of isocyanate leads to formation of an oxazolidone-isocyanurate polymer network.

However, oxazolidone formation is a complex reaction which depends on a number of factors, such as a catalyst, temperature and type of reactants. Just recently the selective bulk polymerization of polyoxazolidones was reported using triphenylantimony iodide as a catalyst [15,16]. This catalyst was first described in a Dow Chemical patent in 1987 [17]. Polyoxazolidones of different crosslink densities were prepared in our laboratories by reacting NCO-terminated oxazolidone quasi-prepolymers with diglycidyl ether of bisphenol A. The NCO-oxazolidone quasi-prepolymers were prepared utilizing different types of isocyanates: 4,4'-methylene bis(phenyl isocyanate) (MDI), polymeric MDI (PMDI) and 4,4'-methylene bis(cyclohexyl isocyanate) (H$_{12}$MDI). The heat resistance of those materials was excellent which was confirmed by thermal analysis and retention of mechanical properties at elevated temperatures. According to dynamic mechanical analysis, the storage shear modulus, G', of the compression molded polyoxazolidones with an isocyanate index of 200 did not change up to 200°C and the first transition temperature was 224°C. The mechanical properties of compression molded polyoxazolidones were very good. The tensile strength was high, 86 MPa for polyoxazolidones based on cycloaliphatic isocyanate (H$_{12}$MDI) and in the range of 100 MPa for polyoxazolidones based on aromatic isocyanates (MDI and PMDI). All polyoxazolidones, regardless of the type of isocyanate and isocyanate index, retained over 50% of the room temperature tensile strength at 120° and 150°C. All polyoxazolidones exhibited excellent retention of properties after aging for seven days at 150°C [15,16]. The good thermal properties of polyoxazolidones were also confirmed in adhesive applications. The lap shear strength of the polyoxazolidone adhesive (steel plates) based on PMDI did not change significantly with temperatures up to 150°C and after immersion in boiling water for three hours [16].

In this paper, the preparation and properties of polyoxazolidone composites are reported. Polyoxazolidone composites were prepared utilizing PMDI (PAPI 901) at an isocyanate index of 105, 120 and 200. The effect of the fiberglass load on the composite properties was studied as well as the effects of the type of epoxide and post-curing temperature and time.

EXPERIMENTAL

Materials. PAPI 901 (polymeric MDI; Dow Chemical Corporation), triphenylantimony (Aldrich) and iodine (Aldrich) were used as received from the supplier. EPON 828 (diglycidyl ether of bisphenol A; Shell Chemical Company) and DEN 431 (epoxide novolac resin; Dow Chemical Corporation) were dried over night under vacuum (1 mm Hg) at 80°C prior to use.

Preparation of NCO-terminated Oxazolidone Quasi-prepolymers. NCO-terminated oxazolidone quasi-prepolymers were prepared at a 5/1 isocyanate to epoxy ratio by reacting PAPI 901 and an epoxide (Epon 828 or DEN 431). As an example, a mixture of EPON 828 (103. 7 g) , triphenylantimony (4.67 g) and iodine (3.33 g), preheated to 80°C, was added to 380 g of PAPI 901, preheated to 80°C. The reaction was started by heating the mixture to 120°C under continuous stirring and a steady flow of dry nitrogen in order to prevent the isocyanate to react with moisture from the air. After a reaction time of approximately 2 hours, the percent NCO was determined by the di-n-butylamine titration method (ASTM-D1638-74) and the quasi-prepolymer was immediately used in preparing the compression molded polyoxazolidones and/or polyoxazolidone composites. The remaining amount of oxazolidone quasi-prepolymer was placed in a glass bottle and sealed under nitrogen. The chemical structure of the polyoxazolidone quasi-prepolymers was characterized by Fourier transform infra-red spectroscopy (FTIR, Nicolet Impact 400).

Preparation and Properties of Compression Molded Polyoxazolidones. Due to the high viscosity of the quasi-prepolymers at room temperature, they were preheated to 80°C before a certain amount was put in a reaction kettle. The amount of epoxide that had to be added to obtain the desired isocyanate index was calculated from the percent NCO of the quasi-prepolymer and equivalent weight of the epoxide. The blend of isocyanate and epoxide, preheated to 80°C, was mixed for five minutes and then trapped air was removed by degassing the resin for 20 minutes under vacuum at 80°C. The resin was poured in an aluminum mold covered with Teflon sheets, preheated at 150°C and put in a press. The press was closed after reaching the gel time. The gel time was determined as the string formation time (approximately 45 minutes). The polyoxazolidone resins were cured for 24 hours at 150°C. Immediately after demolding, the polyoxazolidones were post-cured in an oven at 150°C for an additional time period (24, 48 and 168 hours). Compression molded polyoxazolidones were prepared at three different isocyanate indices: 105, 120 and 200. The chemical structure of the polyoxazolidone composites was characterized by means of FTIR.

The thermal properties of the compression molded polyoxazolidones were measured by means of a TA Instrument 900/9900: a Thermomechanical Analyzer (TMA) Model 2940, Dynamic Mechanical Analyzer (DMA) Model 983, Differential Scanning

Calorimeter (DSC) Model 910 and a Thermogravimetric Analyzer (TGA) Model 951.

The following mechanical properties of the compression molded polyoxazolidones were measured: tensile strength and modulus (ASTM D 412-83) at room temperature, 120°C and 150°C, impact resistance (ASTM D 256-84, method A, Cantilever beam - Izod type), flexural strength and modulus (ASTM D 790-84), and Shore hardness (ASTM D 2240-75). The heat deflection temperature was measured according to ASTM D 648-82.

Preparation and Properties of Polyoxazolidone Composites. The preparation of polyoxazolidone composites was carried out in the same way as the preparation of the compression molded polyoxazolidones, as describe above. The only difference was that these materials were reinforced with Owens Corning continuous strand mat M8610 (445 g/m^2) which is a continuous fiberglass, non-woven mat. Polyoxazolidone composites were prepared at different isocyanate indices (105, 120 and 200) with different fiberglass percentages and different post-curing conditions (temperature and time).

The tensile strength and modulus, flexural strength and modulus, impact strength and heat deflection temperature of polyoxazolidone composites were determined following the same test procedures used in molded polyoxazolidones.

RESULTS AND DISCUSSION

Compression molded polyoxazolidones based on PAPI 901 and Epon 828 were prepared as references at different isocyanate indices (105, 120 and 200) and post-curing times (24 and 48 hours). Increasing the isocyanate index, the crosslik density of the polyoxazolidones increased, and as expected, the glass transition temperatures increased as well (Table 1). It appeared that the glass transition temperatures of the polyoxazolidones, prepared at low isocyanate indices (105 and 120), did not change with an increase of the post-curing time from 24 to 48 hours. However, the glass transition temperature of the polyoxazolidone with a 200 isocyanate index increased 10°C with increase in the post-curing time, indicating that the

Table 1. Effect of isocyanate index and post-curing time on the glass transition and decomposition temperatures of compression molded polyoxazolidones

Designation	1	2	3
Isocyanate index	105	120	200
Post-curing: 24 hours at 150°C			
Glass transition temp. (°C)			
TMA	197	199	215
DMA	205	205	218
DSC	207	209	–
Decomposition temp. (°C)			
% weight loss			
2%	304	304	300
10%	365	365	364
50%	418	421	430
Post-curing: 48 hours at 150°C			
Glass transition temp. (°C)			
TMA	193	205	227
DMA	202	209	228
DSC	195	207	–
Decomposition temp. (°C)			
% weight loss			
2	338	341	334
10	386	387	382
50	438	439	441

post-curing for 24 hours at 150°C was not sufficient. This was also confirmed by an increase in decomposition temperature, as measured by TGA, which increased over 30°C from approximately 300° to 330°C with increase in the post-curing time.

Even though the polyoxazolidones were not completely cured after post-curing for 24 hours, the mechanical properties were good and the retention of the room temperature tensile strength was more than 60% at 150°C (Table 2). The heat deflection temperatures which were measured according to ASTM D 648-82 (Table 2), were very close to the glass transition temperatures, as measured by TMA, DSC and DMA (Table 1).

As expected, polyoxazolidone composites with 40% fiberglass reinforcement, prepared under the same conditions as compression molded polyoxazolidones, exhibited significant improvement in both, mechanical and thermal properties (Table 3). All properties improved including the heat deflection temperature which was higher than 250°C for all

Table 2. Effect of isocyanate index on properties of compression molded polyoxazolidones, post-cured at 150°C for 24 hours

Designation	1	2	3
Isocyanate index	105	120	200
Tensile strength (MPa)			
RT	57	69	66
120°C	56	57	55
150°C	44	49	41
Elongation (%)			
RT	9	13	12
120°C	15	18	17
150°C	15	17	17
Modulus of elasticity (MPa)			
RT	1092	962	984
120°C	759	641	623
150°C	661	628	626
Flexural strength (MPa)	87	124	171
Flexural modulus (MPa)	2798	2513	3003
Impact strength (J/m)	104	144	152
Heat deflection temp. (°C)	194	195	198
Hardness (Shore D)	85	85	86

Table 3. Effect of isocyanate index on properties of polyoxazolidone composites, containing 40% of fiberglass, post-cured at 150°C for 24 hours

Designation	Isocyanate index		
	105	120	200
Tensile strength (MPa)			
RT	127	113	126
120°C	130	141	155
150°C	119	124	143
Elongation (%)			
RT	13	12	13
120°C	14	14	16
150°C	15	15	15
Modulus of elasticity (MPa)			
RT	2177	1986	809
120°C	633	761	707
150°C	579	659	1054
Flexural strength (MPa)	233	253	263
Flexural modulus (MPa)	7722	6747	8515
Impact strength (J/m)	657	653	783
Heat deflection temp. (°C)	>250	>250	>250

polyoxazolidone composites (250°C was the limit of the testing equipment).

With an increase in the post-curing time from 24 to 48 and 168 hours all properties of the polyoxazolidone composites with 40% fiberglass reinforcement improved significantly (Tables 3-5). In general, the polyoxazolidone composites post-cured at 150°C for seven days exhibited better tensile strength, modulus of elasticity, flexural strength, flexural modulus and impact strength than polyoxazolidone composites post-cured for 24 and 48 hours at 150°C. The retention of mechanical properties at elevated temperatures was excellent for all polyoxazolidone composites, regardless of the isocyanate index and the post-curing time. In the case of polyoxazolidone composites post-cured for seven days, the room temperature tensile strength did not change significantly at 150°C, and the retention of modulus of elasticity was higher than 60%.

It appeared that the isocyanate index did not have a significant effect on the composite properties with 40% fiberglass reinforcement, post-cured for seven days. The room temperature tensile strength of these composites was approximately 150 MPa, the modulus of elasticity higher than 1100 MPa, the flexural strength around 300 MPa, the flexural modulus around 10 GPa and the impact strength between 850 and 950 J/m.

Previously it was reported that the thermal properties of compression molded polyoxazolidones improved when subjected to a longer period of time at elevated temperatures [15,16]. An improvement of overall properties of the compression molded polyoxazolidones and polyoxazolidone composites with heat aging can be ascribed not only to the completion of cure, but also to a microstructural reordering (annealing) of the polyoxazolidones.

In order to determine the effect of the fiberglass content on the properties of the polyoxazolidone composites, the

Table 4. Effect of isocyanate index on properties of compression molded polyoxazolidones, containing 40% of fiberglass, post-cured at 150°C for 48 hours

Designation	Isocyanate index		
	105	120	200
Flexural strength (MPa)			
RT	143	145	176
120°C	153	153	193
150°C	138	142	168
Elongation (%)			
RT	13	13	13
120°C	14	14	15
150°C	15	14	15
Modulus of elasticity (MPa)			
RT	823	1027	933
120°C	781	795	896
150°C	702	705	761
Flexural strength (MPa)	256	228	301
Flexural modulus (MPa)	8191	7474	9488
Impact strength (J/m)	643	665	950
Heat deflection temp. (°C)	>250	>250	>250

Table 5. Effect of isocyanate index on properties of polyoxazolidone composites, containing 40% of fiberglass, post-cured at 150°C for 7 days

Designation	Isocyanate index		
	105	120	200
Tensile strength (MPa)			
RT	148	158	158
120°C	168	150	174
150°C	153	166	172
Elongation (%)			
RT	12	13	13
120°C	13	13	14
150°C	14	15	14
Modulus of elasticity (MPa)			
RT	1278	1110	1256
120°C	969	848	948
150°C	796	832	843
Flexural strength (MPa)	328	299	310
Flexural modulus (GPa)	10.6	10.1	9.7
Impact strength (J/m)	884	857	955
Heat deflection temp. (°C)	>250	>250	>250

polyoxazolidone composites with an isocyanate index of 120 and post-curing time of 48 hours were prepared with different fiberglass contents (40, 50, 56, 64 and 70%). With an increase of the fiberglass content from 40 to 70%, all properties improved significantly; the tensile strength increased from 145 MPa to 261 MPa, the flexural strength from 228 to 467 MPa, the flexural modulus from 7.5 to 14.5 GPa and the impact strength from 665 to 2669 J/m (Table 6). In all cases, the heat deflection temperature was >250°C and the room temperature tensile strength and modulus of elasticity changed only slightly with temperature.

The properties of the polyoxazolidone composites prepared with 70% fiberglass content, and post-cured at 150°C for 48 hours, did not change significantly with change of the isocyanate index. The only significant change was observed in the impact strength which increased from 2358 J/m to 2669 J/m and 2839 J/m by increasing the isocyanate index from 105 to 120 and 200 (Table 7). A similar phenomenon was observed in polyoxazolidone composites with 40%

fiberglass (Tables 3-5). With an increase in isocyanate index from 105 to 120 and 200, the overall increase in isocyanurate concentration and crosslink density is too small to have a significant effect on the composite properties, even though the isocyanate index had an effect on the properties of the compression molded polyoxazolidones (Tables 1 and 2).

In addition to the post-curing time (Tables 3-5), the post-curing temperature had an effect on the properties of the polyoxazolidone composites (Table 8). Polyoxazolidone composites, post-cured at 180°C, exhibited somewhat better properties than polyoxazolidone composites post-cured at 150°. Both polyoxazolidone composites were prepared with 40% of fiberglass reinforcement at 120 isocyanate index and were post-cured for the same time period (48 hours).

The type of epoxide had also an effect on properties of polyoxazolidone composites. Polyoxazolidone composites prepared from DEN 431 at 120 isocyanate index with 70% of fiberglass and post-cured at 150°C for 48 hours exhibited better properties than the polyoxazolidone composites prepared

Table 6. Effect of fiberglass content on properties
of polyoxazolidone composites, prepared at 120 isocyanate
index, and post-cured at 150°C for 48 hours

Designation	Fiberglass content (%)				
	40	50	56	64	70
Tensile strength (MPa)					
RT	145	189	198	252	261
120°C	153	178	185	225	254
150°C	142	199	170	225	237
Elongation (%)					
RT	13	14	15	16	16
120°C	14	14	15	16	16
150°C	14	15	16	16	16
Modulus of elasticity (MPa)					
RT	1027	1150	1011	1129	1131
120°C	795	889	870	967	1108
150°C	705	893	752	921	1035
Flexural strength (MPa)	228	350	395	457	467
Flexural modulus (GPa)	7.5	10.3	12.0	13.7	14.5
Impact strength (J/m)	665	1351	1718	2562	2669
Heat deflection temp. (°C)	>250	>250	>250	>250	>250

Table 7. Effect of isocyanate index
on properties of polyoxazolidone
composites, containing 70% of fiberglass
and post-cured at 150°C for 48 hours

Designation	Isocyanate index		
	105	120	200
Tensile strength (MPa)			
RT	275	261	275
120°C	258	254	260
150°C	230	237	261
Elongation (%)			
RT	16	16	17
120°C	17	16	16
150°C	17	16	17
Modulus of elasticity (MPa)			
RT	1188	1131	1167
120°C	972	1108	1059
150°C	920	1035	1020
Flexural strength (MPa)	487	467	459
Flexural modulus (GPa)	14.2	14.5	14.4
Impact strength (J/m)	2358	2669	2839
Heat deflection temp. (°C)	>250	>250	>250

from EPON 828 (Table 9). Better properties of the polyoxazolidone composites based on DEN 431 can be ascribed to the higher crosslink density of the polyoxazolidone matrix compared to that based on EPON 828. DEN 431 is an epoxide with epoxy functionality higher than two while EPON 828 has a functionality of two. So far, the polyoxazolidone composite based on DEN 431 exhibited the best properties; the tensile strength was 292 MPa, the modulus of elasticity 1550 MPa, the elongation 19%, the flexural strength 498 MPa, the flexural modulus 15.1 GPa and the impact strength 2802 J/m. The glass transition temperature of the polyoxazolidone composites based on DEN 431, as measured by TMA, DMA and DSC, was also higher than that of polyoxazolidone composites based on EPON 828, as a result of higher crosslink density.

An exact comparison of the results of the study on the polyoxazolidone composite properties with the properties of commercially available composites is difficult, since the fiberglass used in reinforcing polyoxazolidones can be very different from the type of reinforcement used in these commercial composites. Not only type of reinforcement, the amount of reinforcement could also have been

Table 8. Effect of post-curing temperature on properties of polyoxazolidone composites, containing 40% of fiberglass, prepared at 120 isocyanate index and post-cured for 48 hours

Designation	Post-curing temperature (°C)	
	150	180
Tensile strength (MPa)		
RT	145	158
120°C	153	158
150°C	142	140
Elongation (%)		
RT	13	14
120°C	14	15
150°C	14	15
Modulus of elasticity (MPa)		
RT	1027	1316
120°C	795	789
150°C	705	657
Flexural strength (MPa)	228	277
Flexural modulus (MPa)	7474	8750
Impact strength (J/m)	665	861
Heat deflection temp. (°C)	>250	>250

Table 9. Effect of type of epoxide on properties of polyoxazolidone composites, containing 70% of fiberglass, prepared at 120 isocyanate index and post-cured at 150°C for 48 hours

Designation	Type of epoxide	
	Epon 828	DEN 431
Tensile strength (MPa)	261	292
Elongation (%)	16	19
Modulus of elasticity (MPa)	1131	1550
Flexural strength (MPa)	467	498
Flexural modulus (GPa)	14.5	15.1
Impact strength (J/m)	2669	2802
Heat deflection temp. (°C)	>250	>250
Glass transition temp. of corresponding polyoxazolidones (°C)		
TMA	205	219
DMA	209	227
DSC	207	212
Decomposition temp. of corresponding polyoxazolidones (°C) % weight loss		
2	341	338
10	387	388
50	439	444

different. However, the properties of the polyoxazolidone composites can be qualitatively compared with typical properties of commercially available composites, published in technical literature. In Table 10, typical properties of fiberglass reinforced polyesters, polyepoxides and polyimides (the amount of fiberglass was not reported) published in technical literature [18], are compared with properties of the polyoxazolidone composites with 70% fiberglass reinforcement prepared from polymeric MDI PAPI 901 and epoxide DEN 431 (epoxy functionality higher than two) at 120 isocyanate index (Table 9).

The tensile strength of the polyoxazolidone composites (292 MPa) is comparable to the tensile strength of the polyester composites (303 MPa).

The flexural strength of the polyoxazolidone composites (498 MPa) is comparable to the flexural strengths of polyepoxide (517-654 MPa) and polyimide (517-695 MPa) composites, and

significantly better than the ultimate flexural strength of the polyester composites (214 MPa).

The impact strength of the polyoxazolidone composites (2800 J/m) is excellent, almost twice as high as that of polyepoxide (1600 J/m) and more than twice higher than that of polyester composites (750-960 J/m). Polyimide-based materials are known for their poor impact resistance.

The heat deflection temperature of the polyoxazolidone composites as well as the retention of the tensile strength at elevated temperatures are much better than the heat deflection temperatures and retention of tensile strength of polyester and polyepoxide composites reported in reference 18. According to DMA, DSC and TMA results, the maximum

Table 10. Ultimate properties of polyester, polyepoxide, polyoxazolidone and polyimide composites

Designation	Type of composite			
	Polyester	Polyepoxide	Polyoxa-zolidone	Polyimide
Tensile strength (MPa)	303	379–517	292	345–565
Flexural strength (MPa)	214	517–654	498	517–695
Impact strength (J/m)	750–960	1600	2800	–
Heat deflection temp. (°C)	177	204	>250	–
Max. operating temp. (°C)	121–149	149	ca. 200	288–371
Tensile strength at elevated temp. (MPa)	103 (149°C)	262 (149°C)	261 (150°)	447 (288°C)

operating temperature for polyoxazolidone composites is around 200°. TGA data showed that the decomposition temperature was around 330°C. Taking all these data into consideration, the polyoxazolidone composites appear to be as good or better than polyester and polyepoxide composites. Although, the thermal properties of polyimide composites are superior to those of the polyoxazolidone composites, the greatest advantage of the polyoxazolidone composites over the polyimides are a superior impact strength, relatively inexpensive raw materials and ease of processing.

CONCLUSIONS

Polyoxazolidone composites can be readily prepared from polymeric MDI and epoxides at variable isocyanate to epoxide ratios in the presence of the oxazolidone forming catalyst, triphenylantimony iodide. Post-curing temperature and time had an effect on the properties of the polyoxazolidone composites as well as the type of epoxide. Increasing the fiberglass content of the polyoxazolidone composites resulted in an improvement of the mechanical strength properties.

The mechanical properties of polyoxazolidone composites with 70% fiberglass are comparable to the ultimate properties of commercially available polyester and polyepoxide composites. However, the impact strength of the polyoxazolidone composites with 70% fiberglass, as well as thermal properties are better than those of commercially available polyepoxide and polyester composites. The heat deflection temperature was higher than 250°C for all polyoxazolidone composites. The retention of mechanical properties for all polyoxazolidone composites was excellent at 150°C, and all of them exhibited improvement in properties with prolonged exposure at 150°C.

Even though polyoxazolidone composites are less thermally stable than polyimide composites, they are good alternatives to very expensive, commercially available polyimide composites in applications with maximum operating temperature in the range of 200°C. The impact strength of the polyoxazolidone composites is superior to commercially available polyimide composites. An additional advantage of the polyoxazolidone over polyimide composites, is the relatively lower price of the polyoxazolidone raw materials compared to that of polyimides. The ease of processing could be considered an additional advantage of the polyoxazolidone composites.

ACKNOWLEDGEMENT

The authors would like to gratefully acknowledge the financial support from the Northrop Corporation.

REFERENCIES

1. Hayash, E.F., Reymore H.E. and Sayigh A.A.R., U.S. Pat. 3,673,128 (to the Upjohn Co) **1972**
2. Ashida K. and Frisch K.C., J. Cell. Plastics, 8, **1972**, 160
3. Ashida K. and Frisch K.C., J. Cell. Plastics, 8, **1972**, 194
4. Kresta, J.E. and Frisch, K.C., J. Cell. Plastics, 11, **1975**, 68
5. Nawata T., Kresta J.E. and Frisch, K.C., J. Cell. Plastics, 11, **1975**, 267
6. Reymore H.E. Jr., Carleton P.S., Kolakowski R.A. and Sayigh A.A.R., J. Cell. Plastics, 11, **1975**, 328.
7. Ashida K. and Yagi T., British Pat. 1 1,155,768 (to Nisshin Spinning Co.) **1969**
8. Farrissey W.J., Rose J.S. and Carleton P.S., J. Appl. Polymer Sci., 14, **1970**, 1093
9. Sayigh, A.A.R., "Advances in Urethane Science and Technology", Vol. 3, K.C. Frisch and S.L. Reegen, eds., Westport, CN: Technomic Publishing Co., Inc., **1974**, p.141
10. Kresta J.E. and Hsieh K. H., Macromol. Chem., 179, **1978**, 2997
11. Kordomenos P.I., Frisch K.C. and Kresta J.E., J. Coat. Technol., 55 (700), **1983**, 59
12. Kordomenos P.I., Kresta J.E. and Frisch K.C., Macromolecules, 20 (9), **1987**, 2077
13. Speranza P. and Peppel W.J., J. Org. Chem., 23, **1958**, 1922
14. Pankratov V.A., Frenkel Ts.M. and Fainleib A.M., Russion Chemical Reviews, 52 (6), **1983**, 576
15. Frisch K.C., Sendijarevic A. and Sendijarevic V., Progress in Rubber and Plastics Technol., 9 (2), **1993**, 89
16. Sendijarevic V., Sendijarevic A. Lekovic Ha., Lekovic Hu. and Frisch K.C., The Polyurethane Industry's International Conference UTECH 94, The Hague, The Netherlands, March 22-24, **1994**
17. Marks M.J. and Plepys R.A., U.S. Pat. 4,658,007 (to Dow Chemical) **1987**
18. Lubin G., "Handbook of Composites", Van Nostrand Reinhold Company, New York, **1982**, Appendix A

Solid State Processing of Thermoplastic Polymers

R.G. Kander, L.W. Vick
Virginia Polytechnic Institute and State University
Blacksburg, Virginia

Abstract

The difficulties encountered in melt processing and the environmental problems associated with solution processing of high-performance thermoplastic polymers have made solid-state processing methods more attractive. This paper describes three areas of solid-state processing: the production of thermoplastic polymer powders, the formation of polymer blends and polymer-based composites from powders, and the consolidation of powders. Two research topics being pursued by our group, the use of mechanical alloying to form polymer blends and composites and consolidation of powders through compaction and sintering, will be discussed in greater detail.

Introduction

Thermoplastic polymers with superior mechanical properties, thermal stability, and solvent resistance have been developed for high-performance engineering applications. However, high viscosity, thermal degradation near the melting temperature, and processing temperatures exceeding the limits of conventional processing equipment can make these polymers very difficult and expensive to process.

Solution processing, which involves dissolving a polymer in a compatible solvent, is an alternate processing method which avoids the difficulties involved with melt processing. With recent efforts to make manufacturing more environmentally-friendly, some of the most effective solvents for solution processing can no longer be used. The chlorofluorocarbon solvents, which are nonflammable, nontoxic, and highly solvating at low temperatures for many polymers, have also been determined to be the most harmful to the environment and are being banned from use.

With the difficulties encountered in melt processing and the environmental problems associated with solution processing, solid-state polymer processing is becoming more attractive. The production of fine, high-performance thermoplastic polymer powders has facilitated new processing techniques for polymers which are difficult to process using conventional methods. For example, dry electrostatic prepregging [1] and aqueous suspension prepregging [2,3] of fiber tows was recently demonstrated as a means of uniformly coating graphite with high-temperature polymers to produce fiber-reinforced composite materials. Selective laser sintering of thermoplastic polymer powders is currently being performed with the aid of computer-aided design systems to produce rapid prototypes.[4-7] Other techniques, including slip casting of near net shapes and powder infiltration of braided and woven fiber preforms can be envisioned to take advantage of fine powders.

In addition to allowing new processing methods to be developed, polymer powders can lower the required processing pressures (as in powder prepregging) [1-3] and temperatures (as in compaction and sintering of metals and ceramics) [8], making processing more cost-effective. Brief descriptions of powder processing methods, the formation of polymer blends and composites from powders, and consilidation of powders will be presented, along with more detailed discussions on two aspects of solid-state polymer processing being investigated by our research group: the use of mechanical alloying to form blends and composites and the consolidation of powder through compaction and sintering.

Thermoplastic Powder Formation

There are three general methods of producing thermoplastic polymer powders from bulk material: grinding, air impingement, and precipitation from solution. The first two involve mechanically decreasing the size of existing polymer powders, pellets, or bulk material, while the latter involves forming the particles after the polymer has been dissolved in a suitable solvent. All methods yield particles in a range of sizes, with the breadth of the particle size distribution depending on the technique and processing conditions used. Mechanical sorting, or sieving through screens of known mesh sizes, can be performed to decrease the range of particle sizes present in a powder.

In grinding, conventional mechanical techniques are used to reduce the size of polymer particles to as small as 30 microns. Air impingement utilizes high pressure streams of gas to generate high energy impacts between the particles. Particle sizes as small as 4 microns can be produced using this method. Both grinding and air impingement are most successful in processing brittle, high modulus polymers that fragment readily. Because these methods rely on mechanical deformation and fracture to reduce particle size, some degradation of molecular weight is possible.

The third method, precipitation from solution, can be used to form fine particles, provided a suitable solvent can be found to dissolve the polymer. The solvent can be a liquid, gas, or supercritical fluid, and is preferably nonflammable, nontoxic, and highly solvating at low temperatures. When a polymer is soluble in a solvent, a limited amount of the polymer can be dissolved into solution. To form a powder, the solubility of the polymer in solution is then reduced by adding another component to the solution or by dropping the temperature or pressure of the solution. This causes the polymer to precipitate from the solution. Note that polymerization carried out in solution directly results in powder formation, as the high polymer precipitates from solution at a specific molecular weight.

The particle size and morphology of the precipitated powder are highly dependent on the processing conditions. Precipitation methods can be used to form powders with particles as small as 0.2 microns.[9,10] It is possible to make powders with more narrow particle size distributions and without the molecular weight degradation associated with mechanical methods. However, special means of preventing the particles from agglomerating, or sticking to one another, may be required.

Polymer Blends and Composites through Mechanical Alloying

Polymer blending, the physical mixing of two or more polymers or copolymers, has two primary motivations: to enhance the properties of the polymer matrix or to reduce the cost of a high-performance resin with minimal reduction in performance. The most common methods of producing polymer blends and alloys are solution and melt blending. Problems associated with melt processing polymer blends include degradation of the blend due to excessive heating or chemical reaction during processing [11-13], formation of non-uniform skin-core morphologies as in systems that include liquid crystalline polymers [13-16], and the inability to combine polymers with disparate melting regimes. The mechanical alloying process is being investigated as a means of overcoming some of these problems and the environmental concerns associated with polymer solution blending.[17,18]

Mechanical alloying (MA) is a solid-state process developed for the production of oxide dispersion strengthened superalloys.[19] It is used extensively for microstructural control of metals. Ball milling, a form of mechanical alloying, involves placing powders of two or more materials in a vial with several hardened metal or ceramic balls. As the vial is agitated, powder particles trapped between the colliding balls in the mill are repeatedly fused, deformed, and fractured as a result of the high energy impacts. This solid-state process is typically carried out near or below room temperature.

Mechanical alloying of polymers overcomes some of the difficulties associated with melt blending, as the mixing is performed in the solid state no more than 10 to 20 degrees above ambient conditions, and does not require the use of solvents as in solution blending. Mechanically alloyed powders need only be heated to the lowest processing temperature of the constituent components during consolidation, allowing a material with a high processing temperature to be combined with a material with a much lower processing temperature. For example, a Vectra™ B950 liquid crystalline polymer (LCP)/polypropylene (PP) blend was produced in our lab through mechanical alloying and hot pressing at 180°C [17,18], much lower than the 295°C required to process the blend through injection molding.[20]

In addition to lowering the required processing for some systems, mechanical alloying can be used to alter the microstructure of a polymer blend. Mechanically alloyed powders consist of intimately mixed components having a structure much finer than the dimensions of the particle. Bulk material can be made from these powders having more homogeneous and more isotropic structures than can be produced through conventional melt processing.

In the case of the LCP/PP material [17,18], a submicron structure, as shown in Figure 1, was achieved by hot pressing the mechanically alloyed powders. Mechanical testing and scanning microscopic examination revealed a more isotropic and homogeneous structure than obtained through injection molding [20]. Differential scanning calorimetry (DSC) also suggested changes in the structure brought about by the very small domain size of the milled powder. Milling appeared to cause some type of interaction between the LCP and PP which disrupted crystallization in the PP-rich phase. Further results of this study will be published in a masters thesis [21].

In addition to producing polymer blends, the mechanical alloying process can be used to form polymer-based composites. Two types of composites have been produced by our group: recycled wood/recycled polyethylene (PE) [20] and copper particle-reinforced poly(ether ether ketone) (PEEK) [22]. As in the case of the LCP/PP polymer blend [17,18], the resulting microstructure of the wood/PE composite was finer, more homogeneous, and more isotropic than the structure resulting from melt-extrusion, as observed through scanning electron microscopy and mechanical property testing.[17,23] As expected, the isotropic strength and modulus of the mechanically alloyed wood/PE composite were between the values reported for the extrusion and transverse directions of Mobil's Timbrex™, which is made of the same starting materials, but in unmilled form.[23]

The copper/PEEK composite was also very isotropic and homogeneous. The scanning electron micrograph in Figure 2 shows a sample of copper/PEEK powder milled together in a ball mill for 8 hours. The size and shape of the copper particles were not significantly altered by the milling operation, however, mechanical testing did detect an increase in strength and flexural modulus and a decrease in ultimate elongation with ball milling time.[22]

Consolidation of Thermoplastic Powders

The concept of using fine particles to decrease processing temperatures was first used in powder metallurgy. This technique has been highly developed for metals and has been modified for processing ceramics below their melting temperatures. The success in powder metallurgy rests on the fact that increasing the surface area of a material raises its free energy, creating a large driving force for the particles to chemically bond to each other at elevated temperatures. As the particle size is decreased, the driving force is increased, and processing below the melt temperature is possible.[8]

As the particle size of a polymeric powder decreases, a higher percentage of the polymer chain segments are located on the surface of the particle, where they experience lower constraints than those at the center of the particle. The average mobility of the polymer segments in the powder is higher, lowering the observed glass transition temperature below that of the bulk polymer. As a result of the increased mobility of the polymer chains and the increased free energy of the powder, it is believed that the processing temperature of thermoplastic powders can possibly be lower than the temperatures required to melt-process the bulk material if the particle size was sufficiently small.

A number of studies on polymer compaction (applying pressure to form contacts) and sintering (applying heat to promote bonding) have been performed, and a good review of this research is presented by Jog [24]. A fair amount of the literature is concerned with compaction and not sintering, although it has been known since the earliest studies that heat treatment greatly improves the mechanical properties.[25]

Experiments have shown that processing variables (time, temperature, pressure) and powder characteristics (size, size distribution, and morphology) affect the processing of polymer powders as they do in metals and ceramics. Out of all the literature published on polymer compaction and sintering, it has not been demonstrated that polymers can be processed below their melt temperature without a sacrifice in mechanical properties, as can be done with metals and ceramics. In fact, a study using a range of controlled particle sizes of PEEK showed that reducing particle size below a given range may actually decrease the ease of processing.[22]

The majority of polymer sintering literature is experimental in nature and does not consider sintering below the melt temperature. The basic approach used to theoretically model polymer sintering is based on early work by Frenkel [26]. Kuczynski [27] modified Frenkel's expression which predicts the neck radius between two spheres during the early stages of viscous sintering. Kuczynski's model predicts the neck radius between two spherical particles as a function of time for a given particle radius, surface tension, viscosity, and an empirically determined rheology constant. Other studies by Rosenzweig and Narkis [28] have experimentally and theoretically addressed the sintering of two spherical polymer particles. These types of models which look at free sintering of spherical particles are not always applicable, as they do not take into account:

- differences in geometry resulting from either non-spherical particles or distortion of spherical particles due to compaction, a step which helps raise the strength and density of the sintered part by inducing plastic deformation of the particles,
- sintering of particles of different sizes, which is important, as a distribution of particle sizes is needed to improve packing of the material prior to sintering,
- sintering below the melt temperature, a possibility being explored in the current research,
- sintering with an applied external pressure which may help densification, and
- development of mechanical properties during sintering, which cannot be directly correlated to neck radius.

The original motivation for studying polymer compaction and sintering was to avoid the processing difficulties associated with high viscosity polymers, such as ultra-high molecular weight polyethylene [29,30], or high melting temperature polymers, such as polyimide [31,32] and PEEK [33,34]. Metal particle-reinforced composites formed by compaction and sintering that have desirable electrical [35] or magnetic [36] properties has also been investigated.

A new application for the sintering of polymer powders is selective laser sintering (SLS), a process used to produce rapid prototypes.[4-7] In this process, a bed of particles is heated to within a few degrees of the glass transition temperature. Computer aided design equipment, controls a laser which is used to selectively heat and sinter a thin slice of a three-dimensional object. New layers of powder are added and selectively sintered to complete the object. Commercial SLS equipment is currently being sold along with powders specifically designed for selective laser sintering.

Current Research:
A Fundamental Study of Polymer Sintering

Although there is abundant literature on polymer compaction and sintering, the molecular mechanisms that control the behavior of polymers during compaction and sintering are not clearly understood. However, some of the approaches used to predict polymer healing [37] and physical aging [38,39] may be of use in understanding polymer sintering. It is the goal of the current research to investigate the fundamental mechanisms of polymer sintering. Powders donated by DTM which were designed for use with selective sintering equipment, including wax [40], polycarbonate [41], and nylon-11 [42,43] will be used for sintering experiments, along with poly(phthalamide) powder donated by Amoco.

Basic sintering experiments will be performed to determine the variables affecting void elimination and chain diffusion, to determine what factors limit the development of mechanical properties. Experiments will include microscopy, laser dilatometry, and mechanical testing. Results which were not available at the time of publication will be presented at the conference.

Summary

Solid-state processing of thermoplastic polymers is becoming more attractive as higher temperature polymers are being developed that are difficult to melt process and as environmental regulations make solution processing more difficult. Three aspects of solid-state processing were discussed: the production of thermoplastic powders, the formation of polymer blends and composites through mechanical alloying, and the consolidation of thermoplastic polymer powders. The current research focusing on the fundamental mechanisms of polymer sintering was introduced, although results were not available at the time of this publication.

Acknowledgments

The authors would like to acknowledge the financial support of the Virginia Institute for Material Systems.

References

1. J. Muzzy, Varughese, B., Thannongkol, V., and Tincher, W., *SAMPE Journal*, **25**(5), 15-21 (1989).
2. A. Texier, Davis, R.M., Lyon, K.R., Gungor, A., McGrath, J.E., Marand, H., and Riffle, J.S., *Polymer*, **34**(4), 896-906 (1993).
3. A. Gonzalez-Ibarra and Davis, R.M., **Proceedings**, *Virginia Tech CASS and CCMS Program Review*, held Oct. 3-5, 1993 at Blacksburg, VA, II.D.1-II.D.12 (1993).
4. L.L. Kimble, **Solid Freeform Fabrication Proceedings**, proceedings of the Solid Freeform Fabrication Symposium held Aug. 3-5., 1992, at Austin, Texas, 212-219 (1992).
5. X. Deng, Zong, G., and Beaman, J.J., **Solid Freeform Fabrication Proceedings**, proceedings of the Solid Freeform Fabrication Symposium held Aug. 3-5., 1992, at Austin, Texas, 102-109 (1992).
6. J.C. Nelson, Xue, S., Barlow, J.W., Beaman, J.J., Marcus, H.L., and Bourell, D.L., *Ind. Eng. Chem. Res.*, 32, 2305-2317 (1993).
7. U.S. Patent 5304329, issued April, 1994.
8. J.S. Hirschhorn, **Introduction to Powder Metallurgy**, American Powder Metallurgy Institute, New York, 1969.
9. Lin, T., K.W. Stickney, Rogers, M., Riffle, J.S., McGrath, J.E., Marand, H., Yu, T.H., and Davis, R.M., *Polymer*, **34**(4), 772-777 (1993).
10. A.E. Brink, Gutzeit, S., Lin, T., Lyon, K., Hua, T., Davis, R., and Riffle, J.S., *Polymer*, **34**(4), 825-829 (1993).
11. S.H. Jung and Kim, S.C., *Polymer Journal* **20**(1), 73-81 (1988).
12. P.L. Magagnini, Paci, M., LaMantia, F.P., and Valenza, A., *Polymer International*, **28**, 271-275 (1992).
13. M.R. Nobile, Amendola, E., Nicolais, L., Acierno, D., and Carfagna, C., *Polym. Engr. and Sci.*, **29**(4), 244-257 (1989).
14. G. Kiss, *Polym. Engr. and Sci.*, **27**(6), 410-423, (1987).
15. J. Seppala, Heino, M., and Kapanen, C., *J.Appl. Polym. Sci.*, **44**, 1051-1060 (1992).
16. A. Siegmann, Dagan, A., and Kengi, S., *Polymer*, **26**, 1325-1330 (1985).
17. R.G. Kander, Aning, A.O., Farrell, M.P., and Vick, L.W., **Proceedings**, *Virginia Tech CASS and CCMS Program Review*, held Oct. 3-5, 1993 at Blacksburg, VA, II.E.1-II.E.7 (1993).

18. M.P. Farrell, Kander, R.G., and Aning, A.O., submitted for publication, July, 1994.

19. J.S. Benjamin, *Metall. Trans.*, **1**, 2943 (1970).

20. H.J. O'Donnel, Chen, H.H., and Baird, D.G., **Proceedings**, *Virginia Tech CASS and CCMS Program Review,* held Oct. 3-5, 1993 at Blacksburg, VA, II.B.1-II.B13 (1993).

21. M.P. Farrell, *Masters Thesis,* Virginia Polytechnic Institute and State University, in preparation.

22. S.L. Butsch, "**Mechanical and Physical Properties of Particulate Reinforced Composites,**" *Masters Thesis,* Virginia Polytechnic Institute and State University, Dec., 1993.

23. Technical data sheet on Timbrex™ from Mobil Chemical Company, Composite Products Division (1993).

24. J.P. Jog, *Adv. Polym.Technol.*, **12**(3), 281-289 (1993).

25. T. Maeda and Matsuoka, S., *J. Fac. Eng., Univ. of Tokyo,* **XXXII**(3), 191-226 (1975).

26. J. Frenkel, *J. Phys. (USSR)*, **9**, 385 (1945).

27. G. Kuczynski, Neuville, B., and Toner, H.P., *J. Appl. Polym. Sci..*, **14**, 2069-2077 (1970).

28. N. Rosenzweig and Narkis, M., *J. Appl. Polym. Sci.*, **26**, 2787-2789 (1980). *Polym. Engr. and Sci.*, **21**(10), 582-585 (1981); *Polymer* **21**, 988-989 (communication), (1980); *Polym. Engr. and Sci.*, **21**(17), 1167-1170 (1981)

29. G.W. Halldin and I.L. Kamel, *Polym. Eng. and Sci.*, **17**(1), 21-26 (1977).

30. K.S. Han, Wallace, J.F., Truss, R.W., and Geil, P.H., *J. Macromol. Sci. - Phys.*, **B19**(3), 313-349 (1981).

31. K.W. Rausch, Jr., and Farrissey, W.J., *Soc. of Plastics Engineers Annual Tech. Conf., 34th,* 644-646 (1976).

32. J.L. Throne, *Advances in Polym. Tech.*, **9**(4), 281-291 (1989).

33. J.J. Reilly and Kamel, I.L., *Polym. Eng. Sci.*, **29**(20), 1456-1465 (1989).

34. AE Brink, K.J. Jordens, and J.S. Riffle, submitted for publication.

35. J.J. Reilly and Kamel, I.L., *Polym. Eng. Sci.*, **29**(20), 1446-1455 (1989).

36. V Krishnamurthy and Kamel, I.L., *Polym. Eng. Sci.*, **29**(8), 564-572 (1989).

37. R.P. Wool, Yuan, B. L., and McGarel, O.J., *Polym. Eng. and Sci.*, **29**(19), 1340-1367 (1989).

38. M.R. Tant and Wilkes, G.L., *Polym. Eng. and Sci.*, **21**(14), 1981.

39. J. Mijovic, *Polym. Eng. and Sci.*, **34**(5), 1994.

40. Technical data sheet for Laserlite™ LWX2010 Wax Compound, DTM Corporation, Austin, TX (1994).

41. Technical data sheet for Laserlite™ LPC Polycarbonate Compound, DTM Corporation, Austin, TX (1994).

42. Technical data sheet for Laserlite™ LN4010 Nylon Compound, DTM Corporation, Austin, TX (1994).

43. Technical data sheet for Laserlite™ LNF 5000 Nylon Compound Fine Grade, DTM Corporation, Austin, TX (1994).

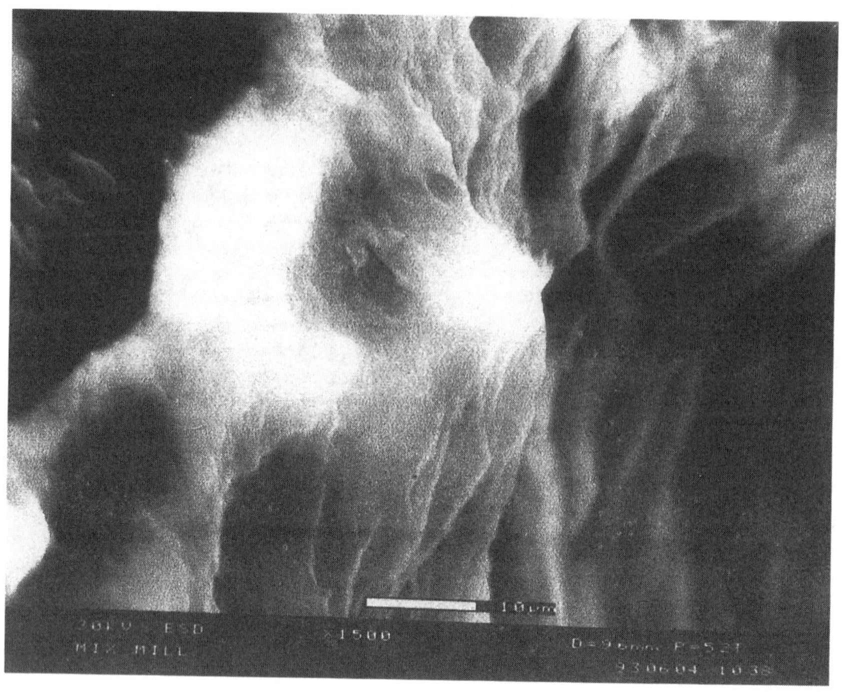

Figure 1. Scanning electron micrograph showing fracture surface of a PP/LCP blend with submicron structure hot-pressed from mechanically alloyed powder.[18]

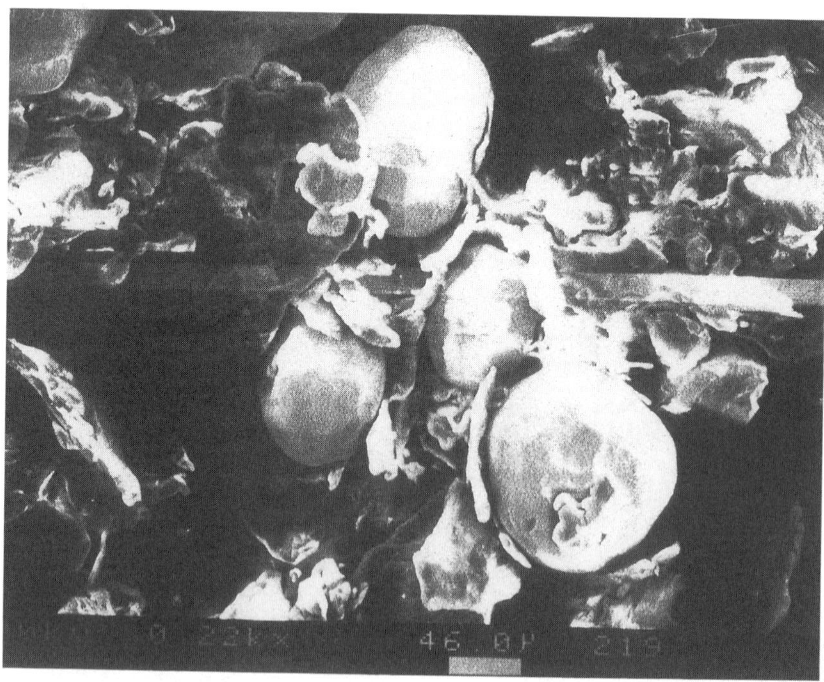

Figure 2. Photomicrograph of mechanically alloyed copper/PEEK powder.[22]

Proceedings of the 10th Annual ASM/ESD Advanced Composites Conference, Dearborn, Michigan, USA, 7-10 November 1994

Low Cost Molding with Low Pressure Molding Compound (LPMC)

J.A. Neate
National Composites Inc.
South Bend, Indiana

J.J. Young, S.P. Hardebeck
Owens-Corning
Valparaiso, Indiana

Abstract

Low Pressure Molding Compound (LPMC) offers potential advantages over conventional Sheet Molding Compound (SMC). Foremost, it has the ability to mold under low pressures, as low as 1.0 MPa (150 psi). Also, even under these low pressures, LPMC displays excellent flow characteristics. Additionally, LPMC displays consistent maturation viscosities at extended molding windows; molding windows measured in months instead of weeks. The key to LPMC's unique properties and characteristics is LPMC's thickening process. Based on crystalline polyester technology, it, unlike normal SMC, contains no earth oxide thickener, and therefore does not chemically thicken. LPMC is compounded at elevated temperatures where the crystalline polymer melts and compounding takes place much like conventional SMC. Handleable compound is achieved by recrystallization of the polyester at lower temperatures. The lack of chemical thickening leads to LPMC's ability to display superior molding characteristics at low pressures. Formulations ranging from Class A, to zero-shrink, to structural, to pigmentable have been successfully compounded and molded in LPMC. This versatility, as well as the attributes mentioned above, lends itself to a very bright future for LPMC.

THE ADVANTAGES OF THE NEW LPMC TECHNOLOGY will vary from molder to molder, depending on their present process capability. Traditional SMC molders will be able to mold at lower molding pressures; this translates to lower tooling costs, lower maintenance costs, and better competitiveness at lower part volumes. Other advantages include better flow, better rib fill, greater compound shelf life, the ability to use more elaborate inserts, and the capability of molding with strategically placed dry glass preforms.

RTM and cold press molders will benefit in some additional important areas such as eliminating secondary trimming with fixtures or robots by molding net-sized parts. Other advantages include the complete elimination of the dry glass preform operation (with the associated waste factors), dramatically reduced cycle times, and a user friendly process (easy operator training). However, the most important attribute to this type of molder is that this technology, because of its competitive economics and cycle times, has the potential to compete for higher volume jobs with existing presses.

Open molders recognize that this new technology enables them to make a quantum leap in process technology without excessive capital investments. LPMC has already enabled two major open molding companies to take advantage of the benefits associated with the closed mold process. Along with the previously mentioned advantages of LPMC, open molders will recognize the important additional benefits of dramatically reduced styrene emissions and good housekeeping capabilities.

Finally, the versatility of LPMC is virtually limitless as demonstrated by the wide array of existing and potential applications.

- Components for cars, trucks, school buses, motor homes, motorcycles and fork lifts.
- Military shipping containers.
- Hospital bed bases.
- Composite door skins.
- Personal water craft.
- Tank seat (Army) structural.
- Manhole cover (with dry reinforcements).

This paper covers the following topics concerning LPMC: the chemistry/theory of LPMC, a property comparison of similar LPMC and SMC formulations, a brief overview of the LPMC compounding process, and finally a discussion of press and mold technology with LPMC.

Chemistry/Theory of LPMC

The key ingredient in LPMC is the patented crystalline unsaturated polyester produced by Scott Bader called Crystic Impreg. This polyester is a solid at room temperature with a melting point of approximately 65°C. LPMC is compounded much like traditional SMC with formulations similar to SMC; however, LPMC is compounded at elevated temperatures, above 65°C. At these elevated temperatures the crystalline polyester is a liquid and disperses homogeneously throughout the LPMC. Following impregnation of the reinforcement, the LPMC is cooled to room temperature, where the crystalline polyester returns to the solid phase. This return to the solid phase "thickens" the LPMC, resulting in handling characteristics similar to matured SMC. Since LPMC contains no earth oxide thickener, "maturation" is complete when the material reaches room temperature, and will remain at that viscosity for weeks, months, and even years.

LPMC can be designed for any traditional SMC application using typical raw materials. However, compared to a traditional SMC formulation, an LPMC formulation actually displays two main differences. The first difference is the aforementioned absence of earth oxide thickener, and the second is the addition of the crystalline polyester. The crystalline polyester is not a true addition to the formulation, but a replacement for part of the traditional unsaturated polyester. The amount of traditional polyester replaced is dependent on the intended application and desired level of "thickening"; however, a minimum amount of crystalline polyester is needed to achieve the desired sheet handling and integrity.

The main benefit, however, of LPMC is its ability to mold under low pressures. The ability to mold under low pressure is again due to the crystalline polyester. When placed on a heated mold, the crystalline polyester melts -- greatly reducing the viscosity of the LPMC. This phenomenon allows for superior material flow in the mold under low pressures. Traditional SMC with earth oxide thickening does not exhibit the same magnitude of viscosity drop as LPMC under similar temperature conditions. SMC flow is inhibited by the increase in apparent molecular weight, resulting from the polyester/thickener interaction, that remains viable under typical molding conditions.

LPMC Versus SMC

Experimental Description. In this study a general purpose SMC formulation was compared to a similar general purpose LPMC formulation. An Owens-Corning VIBRIN® polyester and low profile additive were used in both systems; the formulations are described in Table 1.

There are two main differences between these formulations. In the LPMC formulation, part of the VIBRIN® polyester is replaced by Crystic 772 (crystalline unsaturated polyester) and styrene. Also, the LPMC formulation contains no earth oxide thickener.

Table 1. LPMC and SMC Formulations in PHR

Material	LPMC	SMC
OC VIBRIN® Polyester	23.6	55.0
OC VIBRIN® LPA	40.0	40.0
Crystic 772	22.0	0.0
Styrene	14.4	5.0
29B75 Catalyst	1.2	1.2
BYK W-995	1.0	1.0
Zinc Stearate	3.2	3.2
PDI 1805 Black Pigment	0.1	0.1
Gamaco II CaCO$_3$ Filler	175.0	175.0
Marinco H Mg(OH)$_2$	0.0	4.0

These formulations were compounded in the laboratory and evaluated in the following manner: maturation viscosity, spiral flow distance, Linear Voltage Displacement Tranducer (LVDT) cure (i.e. platen dip cure), shrinkage, surface analysis, mechanical properties, coating adhesion and adhesive bonding. The experimental parameters are briefly described below.

<u>Molding Parameters.</u> 30.5 x 30.5 x 0.254 cm (12 x 12 x 0.001 inch) and 30.5 x 45.7 x 0.254 cm (12 x 18 x 0.001 inch) flat plaques were molded of each system. The mold temperature was 152°/146° C (305°/295°F). SMC charges were molded at 6.89 MPa (1000 psi), while LPMC panels were molded at 2.76 MPa (400 psi).

<u>Maturation Viscosity.</u> Viscosity of the SMC paste was monitored at room temperature for 21 days using a Brookfield HBT viscometer with a TF spindle at 1 rpm.

<u>Spiral Flow.</u> Spiral flow data was generated for both systems at a mold temperature of 149°C (300°F). However, the SMC systems were evaluated at a ram pressure of 5.38 MPa (780 psi), while LPMC systems were evaluated at a ram pressure of 2.07 MPa (300 psi). Both systems were molded at 3, 7, 14 and 21 days. Also, spiral flow data was generated for 3-day-old LPMC at different ram pressures.

<u>LVDT Cure.</u> Monitoring platen position is a way to measure the cure of SMC (or LPMC) in the press on an actual part. In this case the cure measurements were performed for both systems on 30.5 x 30.5 cm flat plaques.

<u>Shrinkage.</u> Shrinkage measurements were performed on the 30.5 x 30.5 cm flat plaques. Shrinkage values were determined by cold part versus cold tool calculations. Negative shrinkage value denotes expansion.

444

Surface Analysis. A LORIA® Surface Analyzer was used to evaluate the surface of the 30.5 x 45.7 cm flat plaques.

Mechanical Properties. Standard ASTM test methods were used to determine moisture absorption (ASTM D570) specific gravity (ASTM D792) flexural properties (ASTM D790) at 22°C (72°F) and 150°C (302°F) tensile strength (ASTM D638) at room temperature and unnotched Izod impact strengths (ASTM D265).

Coating Adhesion. In-Mold Coating (IMC) and primer adhesion were checked on both systems. Sherwin Williams' BC 1902 IMC and Seibert Oxidermo's SO 9471 primer were tested. Five different adhesion tests were performed using the Ford crosshatch method -- initial IMC adhesion, IMC plus primer adhesion, IMC plus primer adhesion after ELPO bake, primer adhesion on bare substrate and primer adhesion on bare substrate after ELPO bake.

Bond Adhesion. Bond Adhesion was checked on both systems by way of Lap Shear testing. Two different adhesive systems were used -- Fusor® 320/322 epoxy system and Pliogrip® 7000/7020 polyurethane system. Lap Shears were tested under 5 different conditions -- 25°C (77°F), 82°C (180°F), -30°C (-22°F), 54°C (130°F) water soak 7 days (immediate), and 54°C (130°F) water soak 7 days (recovery). Lap Shear strengths and failure modes were recorded.

Experimental Results. The following describes the experimental results.

Maturation Viscosity. The maturation viscosity of the LPMC remained at a constant 10MM cps from three to 21 days at 27°C (80°F). The SMC system displayed a typical SMC thickening profile, increasing to over 80 MM cps in 21 days (See Table 2).

Table 2. Maturation Viscosity Data @ 27°C (80°F)

Viscosity	LPMC	SMC
3 day	10 MM cps	55 MM cps
7 day	10 MM cps	58 MM cps
14 day	10 MM cps	65 MM cps
21 day	10 MM cps	+ 80 MM cps

Spiral Flow. The LPMC system displayed greater spiral flow values (i.e. longer flow) at lower pressures than the SMC system at higher pressures. Also, the SMC system displayed a more dramatic decrease in spiral flow values from three to 21 days, 70.0 cm to 53.0 cm (27.5 inches to 21.0 inches), than the LPMC system, 74.0 cm to 69.0 cm (29.0 inches to 27.0 inches). See Table 3. For the sake of comparison, the LPMC system displayed a dramatic increase in spiral flow values at higher pressures, 114 cm @ 5.38 MPa (45 inches at 780 psi). See Figure 1.

Table 3. Spiral Flow Data

Spiral Flow	LPMC @ 2.07 MPa (300 psi)	SMC @ 5.38 MPa (780 psi)
3 day	74.0 cm	70.0 cm
7 day	69.0 cm	58.0 cm
14 day	69.0 cm	53.0 cm
21 day	69.0 cm	53.0 cm

LVDT Cure. The LPMC system displayed a lower LVDT cure value than the SMC system (23 seconds vs. 38 seconds). This indicates that the LPMC system is capable of achieving fast cycle times. See Table 4.

Shrinkage. The LPMC system displayed less shrinkage (i.e. more expansion) than the SMC system (- 0.042% vs. 0.027%). See Table 4.

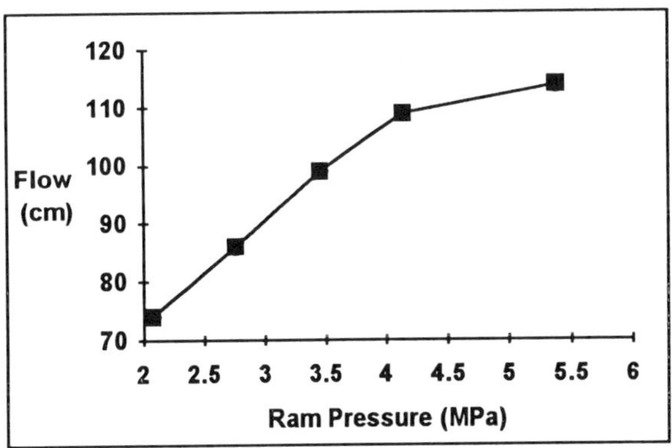

Figure 1. LPMC Flow at Different Ram Pressures.

Surface Analysis. The LPMC system displayed dramatically lower LORIA® Index values than the SMC system (73 vs. 212). Combined with the above mentioned shrinkage data, the LORIA® data emphasizes that this LPMC system is capable of good surface smoothness. See Table 4.

Table 4. LVDT Cure, Shrinkage and Surface Analysis

Test	LPMC	SMC
LVDT Cure (seconds)	23	38
Shrinkage (%)	-0.042	0.027
Surface Analysis -- LORIA® Index	73	212

Mechanical Properties. Both the LPMC and SMC systems displayed satisfactory mechanical properties. See Table 5.

Coating Adhesion. Both the LPMC and SMC systems passed the primer adhesion tests on bare substrate tests using SO 9471 before and after ELPO simulation. Also, both systems passed the three adhesion tests that were performed in this study with Sherwin Williams' BC 1902 IMC. See Table 6.

<u>Bond Adhesion</u>. Both the LPMC and SMC systems displayed similar bond adhesion performance in the five different Lap Shear tests performed in this study. See Tables 7 and 8.

Table 5. Mechanical Properties

Test	LPMC	SMC
Glass Content (%)	27	27
Moisture Absorption (%)	0.30	0.24
Specific Gravity	1.844	1.824
Flexural Modulus @ 22°C (72°F)		
0.5 mm deflection (MPa)	11840	12600
2.5 mm deflection (MPa)	7971	8302
Flexural Modulus @ 150°C (302°F)		
0.5 mm deflection (MPa)	5151	5641
2.5 mm deflection (MPa)	3905	4561
Flexural Strength (MPa)	169	161
Tensile Strength (MPa)	82.0	76.9
Unnotched Izod Impact (J/m)	1084	915

Table 6. Coating Adhesion (Ford Crosshatch)

Test	LPMC	SMC
Initial IMC		
Sherwin Williams BC 1902	Pass	Pass
IMC + Primer (SO 9471)		
Sherwin Williams BC 1902	Pass	Pass
IMC + Primer + ELPO Bake		
Sherwin Williams BC 1902	Pass	Pass
Primer on Bare Substrate		
Initial SO 9471	Pass	Pass
Primer + ELPO Bake	Pass	Pass

Table 7. LPMC Bond Adhesion

Test	Strength (kPa)	Failure Mode (%)
Lap Shear @ 25°C		
Fusor® 320/322	3916	100 FT
Pliogrip® 7000/7020	4156	100 FT
Lap Shear @ 82°C		
Fusor® 320/322	1375	100 CF
Pliogrip® 7000/7020	3328	100 FT
Lap Shear @ -30°C		
Fusor® 320/322	3726	100 FT
Pliogrip® 7000/7020	3627	100 FT
Lap Shear 54°C Water Soak 7 days (immediate)		
Fusor® 320/322	2415	100 CF
Pliogrip® 7000/7020	3670	100 FT
Lap Shear 54°C Water Soak 7 days (recovery)		
Fusor® 320/322	3617	100 FT
Pliogrip® 7000/7020	3701	100 FT

Table 8. SMC Bond Adhesion

Test	Strength (kPa)	Failure Mode (%)
Lap Shear @ 25°C		
Fusor® 320/322	4099	100 FT
Pliogrip® 7000/7020	3656	100 FT
Lap Shear @ 82°C		
Fusor® 320/322	1537	100 CF
Pliogrip® 7000/7020	3724	100 FT
Lap Shear @ -30°C		
Fusor® 320/322	3849	100 FT
Pliogrip® 7000/7020	3593	100 FT
Lap Shear 54°C Water Soak 7 days (immediate)		
Fusor® 320/322	3023	100 CF
Pliogrip® 7000/7020	4018	100 FT
Lap Shear 54°C Water Soak 7 days (recovery)		
Fusor® 320/322	3884	100 FT
Pliogrip® 7000/7020	3675	100 FT

Note, FT is Fiber Tear and CF is Cohesive Failure

LPMC Manufacturing and Processing

Description. The basis of the physical thickening process in LPMC is the unsaturated crystalline polyester. It is supplied in a crystal/granular form that melts when heated and which, when later cooled, reverts back to a solid -- "thickening" the LPMC. Conventional in-line SMC mixing equipment is unable to handle the high temperatures, approximately 90°C (~200°F), required to compound LPMC. Therefore, a continuous mixer resembling a thermoplastic extruder was designed by National Composites, Inc. (NCI) for LPMC paste preparation. This method is cost effective, versatile and optimizes the Crystalline polyester technology. In this process, all of the ingredients (except fiber glass) of the LPMC formulation are accurately metered and continuously fed into individual ports located along the barrel of the continuous mixers. The paste is heated (or cooled) in the mixer, as required, thoroughly mixed and delivered into the doctor boxes of an SMC machine. This SMC machine is modified to control compound temperature from beginning to end. The fiber glass is added (similar to traditional SMC) and the subsequent rolls of LPMC are produced.

Equipment. A Smoot pneumatic conveying system was selected to handle the delivery of the fillers and the crystalline polyester into the metering system. Acrison loss-in-weight metering and feeding equipment was selected to handle all the formulation raw materials. Two PTI special extruders were designed to receive the metered materials, melt/mix as required, and deliver the blended "premix" into the doctor boxes. A Schmidt &

Heinzmann modified SMC machine was selected to add the chopped fiber glass (and other reinforcement materials as required) and produce the LPMC in 38 inch (96 cm) wide rolls. Since this process was designed to run continuously and to make formulation changes "on the fly", a computer control system was installed to rapidly interface between each of the individual pieces of equipment. Because everything is controlled by the computer system there is a 100% tracking of the process, providing substantial in-process Quality Assurance and Quality Control data.

LPMC Mold/Tooling Presses and Support Equipment

One of the tremendous assets and advantages of the LPMC Technology is the low cost investment factor. Typical molding pressures for LPMC are in the 1.38-3.45 MPa (200-500 psi) range versus 5.51-8.27 MPa (800-1200 psi) for conventional SMC. This leads to dramatic decreases in the tonnage requirements for presses and most importantly to lower cost molds and tooling. NCI has worked closely with two press manufacturers; Hepburn and Jesse Engineering/Wallace Coast. Specially designed, lower cost, presses are now available to produce molded parts to meet Class A Automotive and large industrial molding standards.

The real breakthrough has come in the area of low cost molds/tooling. Again NCI has selected to work with three major tooling companies in the industry to optimize the advantages of LPMC -- Ace Mold, Autodie International and R&R Tool & Mold. The results of this development, even at this early stage, are showing reduction in tooling costs of over 50%--with added improvements and specifications developing all the time.

In addition to these lower cost steel tooling developments, great progress has been made in other lower volume tooling areas such as Vapor Formed Nickel, Cast Steel and Aluminum and even Epoxy type "soft tooling" for prototypes and pilot production requirements. Lower molding temperature, 93°/85°C (200°/185°F) LPMC has been developed to allow moldings to be made satisfactorily in "soft tooling" for limited volume runs.

Conclusions

The viability of LPMC has been demonstrated in a wide array of markets. This technology offers potential benefits to hand lay-up/spray-up, RTM, cold molding, and SMC/BMC molders. The benefits include reduced emissions of volatile organic compounds (voc), reduced tooling costs and lower capital investments for presses. Other benefits include improved material consistency, greater compound shelf life, reduced cycle times, and equal or better end-use performance. LPMC formulations have been developed for a broad spectrum of applications ranging from structural military components to Class A automotive body panels. The combination of LPMC's versatility and competitive economics will undoubtedly result in new opportunities for the FRP industry well into the next decade.

Acknowledgment

®LORIA is a registered trademark of Ashland.
®Fusor is a registered trademark of the Lord Corporation.
®Pliogrip is a registered trademark of Ashland Chem. Inc.

Effects of Sizings on Microscopic Flow Encountered in RTM Processing of E-Glass Fibers

G.R. Palmese, J.M. Deputy, V.M. Karbhari
University of Delaware
Newark, Delaware

Abstract

Sizings significantly affect the processing and final performance characteristics of resin transfer molded (RTM) parts. Manufacturers of reinforcements often use sizings to enhance mold filling. In this work the effect of commercially available S-2 glass sized fiber systems on the microscopic flow behavior of vinyl ester and epoxy systems was investigated. This was done by using a previously developed technique to monitor the effects of fiber-sizing-surface interactions on the flow of resin around fibers aligned in the direction of bulk flow. Surface energy effects were found to play an important role in determining microscopic flow behavior. It was found that surface free energies of the investigated fibers are significantly varied, and that the resulting differences in resin-fiber compatibility influence flow behavior. In particular a significant retardation in the flow of resin into fiber bundles was noticed for fiber-resin systems expected to possess lower capillary driving forces. The results contained herein indicate that fiber-sizing interactions will affect the rate of tow impregnation and that the selection of a sizing-resin pair for RTM must take into account such considerations.

Introduction

When selecting materials for manufacturing fiber reinforced polymeric composites, choices include not only resin and fiber selection but also the type of reinforcement surface treatment and/or coating employed. Most glass and carbon fibers used commercially for the production of polymeric composite are coated with sizings which may serve several purposes. These include facilitation in handling, protection of the fibers, aid in processing, and enhancement of composite behavioral characteristics. Such sizings affect the degree and type of interactions between the reinforcement and matrix phase of a composite. In fact, manufacturers often claim to design the thickness and composition of sizings to optimize performance.

Examples of sizing materials include epoxies, silanes, and chrome complexes designed to serve as coupling agents. Generally, commercial fiber sizings range in thickness up to .1 μm. Even for moderately filled systems, sizings may comprise a significant portion of the matrix phase (1). Indeed under appropriate conditions sizings have been shown to cause the formation of interphase regions which possess gradients in material properties and affect composite behavior (1,2).

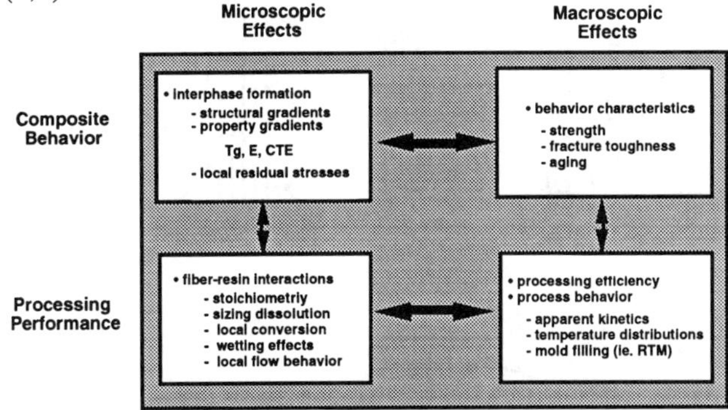

Figure 1: Effects of sizings on processing and performance of polymeric composite materials

As represented by Figure 1, sizings will affect two aspects of composite performance: composite behavioral characteristics and processing performance. In both cases microscopic phenomena are responsible for observed macroscopic effects. On a microscopic level, the behavior of a formed part is affected by the formation of interphase zones that can lead to variations in local properties such as modulus and the glass transition temperature. These variations are linked to microcracking and interfacial strength which can affect macroscopic behavioral characteristics such as strength and fracture toughness (3). Similarly, on a microscopic level processing performance may be affected by fiber-sizing-resin interactions through the development of stoichiometric imbalances and compatibility effects which influence wet-out and local flow behavior. Such microscopic effects may in turn influence process performance by affecting apparent cure kinetics and macroscopic flow behavior.

As shown schematically in Figure 1, it is important to note that the effects of sizings on the formed composite behavioral characteristics and on the processing performance are not independent. In fact, these two aspects of composite behavior are coupled at the microscopic level. For example, the molecular interactions between the constituents of a composite which lead to the formation of interphase zones are affected by processing conditions such as temperature. On the other hand, heat evolution and resin rheology may be affected by the stoichiometric imbalances resulting from interphase formation during processing.

In RTM a highly tailorable reinforcement system known as a preform is placed in a mold and is subsequently infiltrated with resin by injection. A distinct advantage of RTM is that unlike most traditional composites manufacturing techniques (i.e. prepeg lay-up) resin is added to the preform after the reinforcement architecture has been designed and formed. This allows the designer to achieve a "materials-by design" approach to the development of a component. However, it comports that the preform design (including materials selection) not be decoupled from the infusion and curing stages of RTM since material selection, among other factors, will influence key aspects of the infusion and curing process (4). Such aspects include the microscopic and macroscopic flow characteristics during RTM infusion which are the subject of this work.

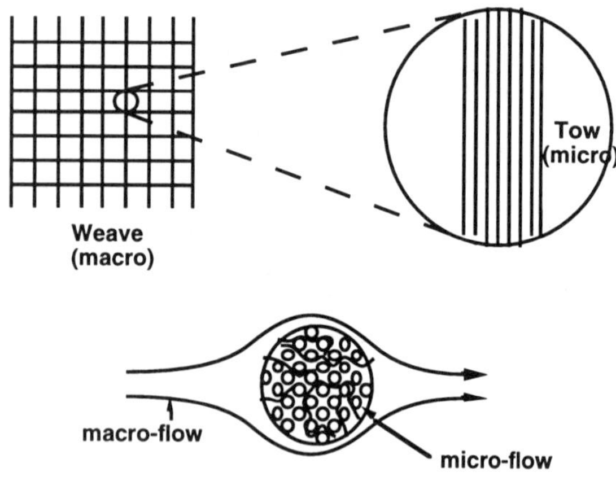

Figure 2: Schematic representation of macro- and micro-flow in RTM preform structures.

During the infusion stage of RTM flow on the macroscopic as well as microscopic level may be identified as illustrated by Figure 2. The flow of resin around the fiber tows (i.e. macro-flow) is responsible for distributing the resin throughout the mold cavity, while the flow of resin between the fibers in a tow (i.e. micro-flow) is responsible for attaining proper wet-out of the fiber bundles. It is often assumed that provided resin and fiber surface are compatible, that the microscopic infusion of the fiber bundles occurs on a time scale similar to that of macro-infusion. In a previous

communication a technique was developed to evaluate the flow of resin around and within carbon fiber bundles aligned in the direction of bulk flow (5). Results suggest that in fact the characteristic time for micro-infusion can be longer than that for mold filling. This results in poor fiber wet-out and inadequate composite behavior. It was also found that the presence of sizings significantly affected flow behavior. In this work the flow behavior of epoxy and vinyl ester resins in contact with sized S-2 glass systems is investigated paying particular attention to issues concerning resin-sizing compatibility.

Experimental

Materials. Dow Derakane 411-C50 vinyl ester and Shell Epon 862 epoxy were used for this investigation. The room temperature viscosities of these materials are 100 cps and 3500 cps respectively. Owens Corning sized S-2 glass rovings designated by the manufacturer as 365, 933, 449, and 463 were utilized. All rovings were 750 (yards/pound) in weight. However, 365 rovings are comprised of 13 μm diameter filaments (~2000 filaments per roving) while 933, 449, and 463 are comprised of 9 μm diameter filaments (~4200 filaments per roving). These fiber systems have been sized with coatings claimed to have been designed for use with specific resin systems. The 933 fiber system is to be used with high temperature materials such as BMI and PEEK; 365 with vinyl ester and polyester; and 449 and 463 with epoxy.

An attempt was made to gain some insight into the composition of the sizings by stripping the fibers with acetone and then analyzing the dissolved material using Fourier transform infrared spectroscopy (FTIR). Figure 3 shows FTIR spectra of the material stripped from the four fiber types.

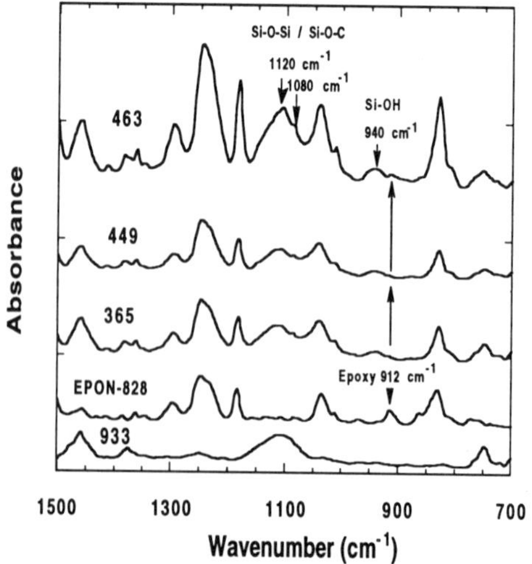

Figure 3: FTIR spectra of sizing material stripped from Owens Corning S-2 glass fibers using acetone.

The spectra for 463, 449 and 365 fibers show peaks at 1120 cm^{-1}, 1080 cm^{-1}, 940 cm^{-1}, and 3400 cm^{-1}. The peaks at 1120 and 1080 cm^{-1} can be associated with the Si-O-Si asymmetric stretch (6,7,8) and Si-O-C asymmetric stretch (6). The absorption peaks at 940 cm^{-1} and 3400 cm^{-1} are due to Si-OH stretch (6,7,8). The spectrum for 933, however, only displays one (1120 cm^{-1}) of the adsorption peaks associated with silicon containing species described above. Therefore, although all fibers show the presence of silicon, it is felt that 463, 449, and 356 fibers contain some partially hydrolyzed material while 933 appears to contain more fully condensed silane. Moreover, it is clear from comparison of the spectra for 463, 449, 365 and 933 with that of Shell Epon 828 (also in Figure 3) that 463, 449, and 365 contain a substantial amount of "loose" epoxy which does not appear in the spectrum for 933. These observations are consistent with the high temperature specifications of the 933 S-2 glass fiber. Note that stripping the fibers with acetone does not necessarily remove all sizing material. It is in fact expected that a substantial part of the condensed silane may remain bound to the fiber surfaces.

Flow Characterization.. The effects of sizing on flow were investigated by measuring the cumulative flow of resin through a stainless steel tube containing axially oriented fiber rovings as described elsewhere (5). Such experiments measure the outflow of resin as a function of time for a given injection pressure. A typical flow trace for such experiments is shown in Figure 4. Such a trace shows two distinct flow regimes. First, a transient region characterized by an initial rapid flow rate, and second, a steady state regime possessing a significantly lower flow rate. As discussed in a previous communication, the duration of the transient period is an indication of the efficacy of resin-sizing interactions in promoting micro-flow infiltration of the fiber bundles. Stronger interactions lead to higher capillary pressures which allow for more rapid intra-bundle infiltration.

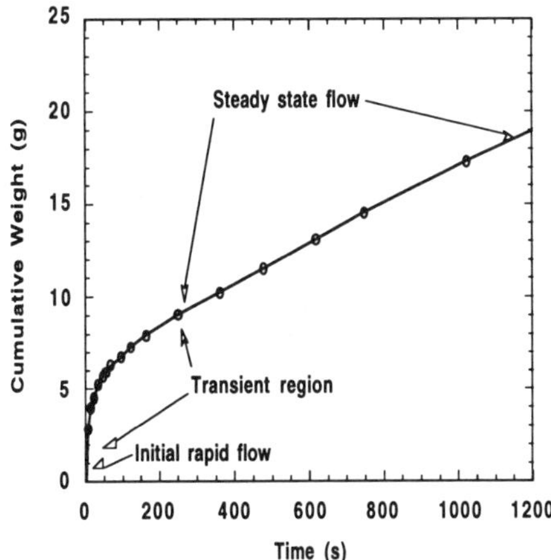

Figure 4:Typical cumulative flow plot showing transient flow regime with initial rapid flow rate and a steady state flow regime characterized by a slower rate.

The flow of Shell Epon 862 epoxy resin and Dow Derakane 411-C50 was investigated in conjunction with the four fiber types described above. All experiments were carried out at room temperature in the absence of catalysts and curing agents to avoid cure effects on viscosity. A 15 cm long tube having .38 cm internal diameter was employed. The fiber loading and injection pressures utilized are given in Table I. In using this technique it is important that conditions are selected such that the transient region is apparent. Thus fiber loading and injection pressures were selected according to resin viscosity.

Table I: Fiber loading and injection pressures for resins employed in this study.

Resin	# Tows	Volume Fraction (%)	Pressure (psig)
Derakane 411-C50	11	26	15
Epon-862	6	14	20

Surface Energy Characterization. The polar and dispersive components of surface energy for the Owens Corning fibers were obtained by measuring the contact angles of these fibers in a series of test fluids of known polar and dispersive surface tensions and analyzing the results using a method proposed by Kaeble (9). The contact angles were obtained using the micro-Wilhelmy technique with a DCA-322 microbalance from Cahn Instruments, Inc. The fibers were cut and glued to a wire using a cyanoacrylate adhesive, and the wire was suspended from the measuring arm of the microbalance. Advancing and receding contact angles were recorded for immersion rates of 20 μm/s. Only the advancing contact angle values were used for the analysis. Values of the fiber radii (which are essential for determining contact angles) were obtained using the DCA and octane as the wetting fluid. For each fiber ten fiber specimens were analyzed.

Results

Figures 4 and 5 show the effects of Owens-Corning fiber type on the flow behavior of Shell Epon 862 epoxy resin and Dow Derakane 411-C50 vinyl ester respectively. The cumulative flow traces represent the total mass of resin which has flown through the tube as a function of time. The characteristic shapes of the curves are a result of two flow regimes: i] a transient period characterized by an initial rapid flow rate; and, ii] a steady state regime possessing a significantly lower flow rate.

Observation of changes in the transient flow behavior as a function of fiber type allows for the evaluation of the effects of sizings on micro-flow. During the transient period the flow rate is attenuated until it reaches a significantly lower steady state value. It is suspected that radial flow into the fiber tows induced by capillary forces is primarily responsible for this effect (5). The infiltration of resin into the fiber bundles spreads the fibers apart and decreases macroscopic channeling volumes. Although the volume utilized for flow increases during this period, the ratio of surface area to volume increases and the flow rates through the tube decrease.

It is apparent from Figures 5 and 6 that fiber type significantly influences flow behavior. For the epoxy system

Figure 5 shows that the order of transient period duration times is as follows: 933 > 463 ~ 449 > 365. However, the order of the value of the steady state flow rates is: 365 > 933 ~ 463 ~ 449. Evaluation of the duration of the transient periods for the vinyl ester/S-2 glass combinations is more difficult. Nevertheless, Figure 6 shows that the transient period for 933 is significantly longer than that of the other fiber systems. On the other hand, the order of the values of steady state flow rates for the vinyl ester system is as follows: 365 >> 463 > 933 ~ 449.

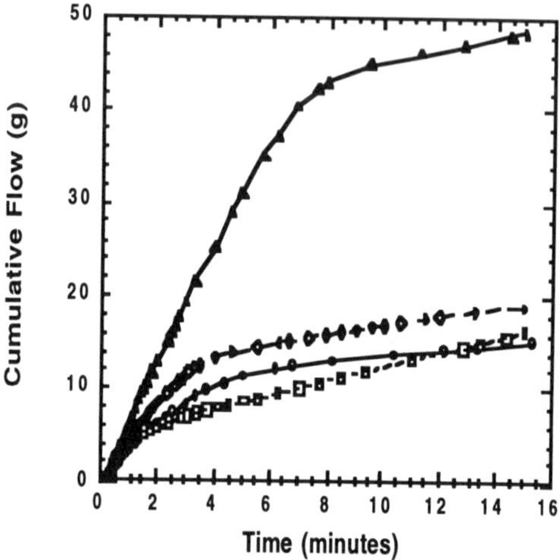

Figure 5: Cumulative flow traces showing the effects of Owens Corning 933 (triangles), 449 (circles), 463 (diamonds), and 365 (squares) on the flow of Shell Epon 862 epoxy resin.

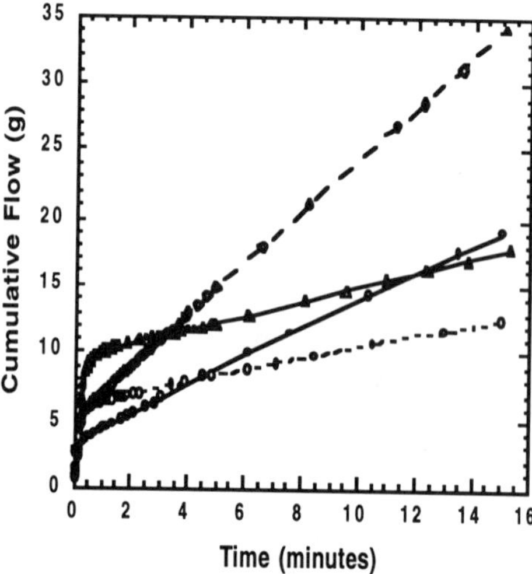

Figure 6: Cumulative flow traces showing the effects of Owens Corning 933 (triangles), 449 (circles, dashed line), 463 (circles, solid line), and 365 (diamonds) on the flow of Dow Derakane 411-C50 vinyl ester.

Discussion

Steady State Regime: The steady state flow rates for the 365 fiber systems are significantly higher for the epoxy and vinyl ester systems. This is perhaps due to the difference in fiber dimensions. Although, the volume fractions for experiments using 933, 463, 449 and 365 are the same for a given resin system, the ratio of fiber surface area to resin volume is significantly lower for 365 as a result of larger fiber diameters. Calculated surface area to volume ratios are tabulated in Table II. Such differences could account for the observed steady state flow rate differences.

Table II: Ratio of fiber surface to resin volume for investigated systems.

Resin # Tows	Volume Fraction (%)	Surface/Volume (1/m)
Derakane 411-C50 365	26	1.059×10^5
Derakane 411-C50 933, 449, 463	14	1.540×10^5
Epon-862 365	26	4.988×10^4
Epon-862 933, 449, 463	14	7.263×10^4

However, the more rapid steady state flow rate of the 463/vinyl ester pair in Figure 6 cannot be explained as easily. It is possible that sizing dissolution effects may be responsible in this case.

Transient Regime: It has been proposed that capillary forces may be the governing factor in determining the duration of the transient period (5). Such capillary forces are directly affected by the distance between fibers and by physical and chemical interactions between the fiber surface and the impregnating resin.

Equation 1 is an expression for the capillary pressure (ΔP_c) associated with a fluid in contact with a fiber bundle (10). We assume that the transverse spaces between fibers may be treated as a set of cylindrical tubes.

$$\Delta P_c = \frac{2\gamma_l}{R} \cos \theta \qquad (1)$$

In this expression R represents an effective radius, γ_l is the resin surface tension and θ is the fiber-resin contact angle.

The duration of the transient region should be shorter for higher values of capillary pressure. In light of this discussion sizings are expected to influence the capillary pressure by changing resin-substrate interactions. Equation (2), the Young-duPree equation, relates the contact angle (θ) to the resin surface tension (γ_l), the fiber surface tension (γ_s), and the interfacial tension (γ_{sl}).

$$\cos \theta = \frac{\gamma_s - \gamma_{sl}}{\gamma_l} \qquad (2)$$

Equations 1 and 2 are combined in equation 3 to yield an expression for pressure as a function of surface and interfacial tensions.

$$\Delta P_c = \frac{2}{R} (\gamma_s - \gamma_{sl}) \qquad (3)$$

The interfacial tension (gsl) can be related to the solid and liquid surface tensions by equation 4 which takes into account the polar and dispersive components of the interactions (11).

$$\gamma_{sl} \approx \gamma_s + \gamma_l - 2\sqrt{^P\gamma_s{}^P\gamma_l} - 2\sqrt{^d\gamma_s{}^d\gamma_l} \qquad (4)$$

By combining equations 4 and 5 an expression for capillary pressure as a function of surface tensions is obtained, viz.;

$$\Delta P_c \approx \frac{2}{R}\left(2\sqrt{^P\gamma_s{}^P\gamma_l} + 2\sqrt{^d\gamma_s{}^d\gamma_l} - \gamma_l\right) \qquad (5)$$

Provided values of the polar and dispersive components of the fiber and resin surface tensions are available, equation 5 can be used to assess the effect of sizings on capillary pressure. Table III reports measured values of the polar and dispersive components of the surface free energy of the Owens Corning S-2 fibers investigated herein. Table III also contains surface tension measurements for the Dow Derakane 411-C50 resin and calculated values of the surface tension for the Epon 862 epoxy. The surface tension of the epoxy was not measured due to its high viscosity. The technique employed for calculating surface tension using group additivity methods is described elsewhere (12).

Table III. Surface free energies for Owens Corning S-2 glass fiber systems.

Fiber	γ^p (dyne/cm)	γ^d (dyne/cm)	γ^t (dyne/cm)
449	10.88±2.51	44.44±4.52	55.33±5.17
463	6.94±1.40	45.07±2.8	52.00±3.13
933	12.60±3.02	32.50±3.95	45.10±4.97
365	8.60±1.18	33.09±1.92	41.69±2.25
Dow 411-C50	1.2±1.0	34.2±1.0	35.4±0.7
Shell 862	12.2	34.6	47.2

Using equation 3 and the surface energy values reported in Table III, capillary pressures for the investigated resin-fiber combinations have been computed and are given in Table IV. The calculations were performed for a 1μm capillary radius and are intended solely for comparing the magnitudes of fiber-resin interactions.

Table IV: Computed values of capillary pressure for the investigated fiber-resin combinations (1 μm effective radius)

Fiber-Resin Pair	ΔP_c (dyne/cm^2)
S2-365/Derakane 411-C50	82.7 x 10^4
S2-933/Derakane 411-C50	90.1 x 10^4
S2-463/Derakane 411-C50	101.2 x 10^4
S2-449/Derakane 411-C50	110.3 x 10^4
S2-365/Epon 862	76.6 x 10^4
S2-933/Epon 862	78.1 x 10^4
S2-463/Epon 862	97.8 x 10^4
S2-449/Epon 862	99.6 x 10^4

The capillary pressures computed for the epoxy resin are significantly higher than those computed for the vinyl ester fiber-resin combinations. Furthermore, the same relative ordering of capillary pressure values is found for both resin systems: 449 > 463 > 933 > 365. Based on these values it would be predicted that the duration of the transient period should be shorter for the 463 and 449 systems than for the 365 and 933 fibers. With the exception of the 365 system such predictions are corroborated, albeit qualitatively, by the experimental results. It is possible that the larger size of the 365 glass fiber is responsible for the observed differences.

As was previously discussed, the duration of the transient period is a measure of the efficacy of fiber wet-out by the resin. The above results indicate that indeed fiber-sizing interactions will affect the rate of tow impregnation and that the selection of a sizing-resin pair for RTM must take into account such considerations.

Concluding Remarks

In this work the effect of commercially available S-2 glass sized fiber systems on the microscopic flow behavior of vinyl ester and epoxy systems was investigated by using a previously developed technique to monitor the effects of fiber-sizing-surface interactions on the flow of resin around fibers aligned in the direction of bulk flow.

It was found that surface free energies of the investigated fibers are significantly varied, and that the resulting differences in resin-fiber compatibility influence flow behavior. In particular a significant retardation in the flow of resin into fiber bundles was noticed for fiber-resin systems expected to possess lower capillary driving forces.

It is known that sizing-resin compatibility plays a significant role in ensuring complete wet-out of a composite. This work further demonstrates that the rate at which fiber wet-out occurs is strongly dependent on sizing-resin interactions. Hence, care must be taken to ensure that for a given sizing-resin combination adequate time is allowed for tow impregnation during RTM processing.

References

1. G.R. Palmese and R.L McCullough, J. Adhesion, in press (1994).
2. T.P. Skourlis and R.L. McCullough, Compos. Sci. Tech.,49, 363 (1993).
3. G.R. Palmese et al., J. Adhesion, in press (1994).
4. D.A. Steenkamer et al., Process. Advanced Materials, 3, 89 (1993).
5. G.R. Palmese and V.M. Karbhari, Polym. Compos., in press (1994).
6. D.L. Leyden and J.B. Atwater,"Silanes and Other Coupling Agents," p 143, K.L. Mittal Ed., VSP Utrecht, The Netherlands (1992).
7. J.M. Park and R.V. Subramanian,"Silanes and Other Coupling Agents," p 473, K.L. Mittal Ed., VSP Utrecht, The Netherlands (1992).
8. F.J. Boerio et al., J. Adhesion, 13, 159 (1981).
9. D.H. Kaeble, J. Adhesion, 2, 66 (1970).

10. J.N. Israelachvili, "Intermolecular and Surface Forces,"
    Academic Press, New York, pp. 213-220, (1985).
11. D.H. Kaelble, J. Adhesion, 2, 50 (1970).
12. G.R. Palmese, CCM Report 92-05 (1992)

Proceedings of the 10th Annual ASM/ESD Advanced Composites Conference, Dearborn, Michigan, USA, 7-10 November 1994

Effect of Fiber Sizing on Cure Kinetics of Vinyl Ester RTM Composites

G.R. Palmese, O. Andersen, V.M. Karbhari
University of Delaware
Newark, Delaware

Abstract

Fiber surface treatments and sizings are known to affect the cure characteristics of resin systems such as epoxies and polyesters which are commonly used for the manufacture of composites. The effects of sizings on resin cure behavior may be of particular importance in resin transfer molding (RTM) where such variations may affect mold filling and part performance. The effect of commercially available sized S-2 glass on the cure of vinyl ester systems was studied. Results show a significant increase in the cure rate of the glass modified systems. Furthermore, a relationship between the surface energy characteristics of the fibers and the degree of cure acceleration was established. It is apparent, therefore, that sizing selection may significantly affect cure processes for vinyl ester systems and the processing behavior of RTM parts.

Introduction

When selecting materials for manufacturing fiber reinforced polymeric composites choices include not only resin and fiber selection but also the type of reinforcement surface treatment and/or coating employed. Most glass and carbon fibers used commercially for the production of polymeric composites are coated with sizings which may serve several purposes. These include facilitation in handling, protection of the fibers, aid in processing, and enhancement of composite behavioral characteristics. Such sizings affect the degree and type of interactions between the reinforcement and matrix phase of a composite. In fact, manufacturers often claim to design the thickness and composition of sizings to optimize performance.

Examples of sizing materials include epoxies, silanes, and chrome complexes designed to serve as coupling agents. Generally, commercial fiber sizings range in thickness up to .1 μm. As shown in Figure 1, even for moderately filled systems, sizings may comprise a significant portion of the matrix phase. Indeed, under appropriate conditions sizings have been shown to cause the formation of interphase regions which possess gradients in material properties and affect composite behavior (1,2).

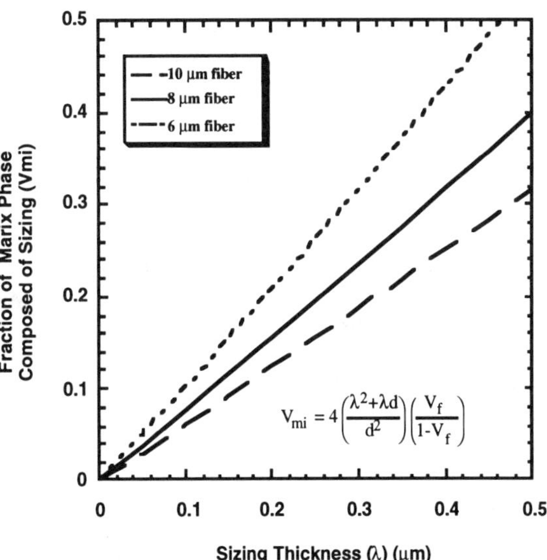

$$V_{mi} = 4\left(\frac{\lambda^2 + \lambda d}{d^2}\right)\left(\frac{V_f}{1-V_f}\right)$$

Figure 1: Fraction of matrix phase composed of sizing for hexagonally arranged fibers of varying diameters (d) having a 60% fiber volume fraction (Vf) (after Palmese (1))

Sizings will affect two aspects of composite performance: composite behavioral characteristics and processing performance. In this work we investigate the degree to which sizing selection may affect the resin transfer molding (RTM) process through changes in apparent cure kinetics.

In RTM, a highly tailorable reinforcement system known as a preform is placed in a mold and is subsequently infiltrated with resin by injection. A distinct advantage of RTM is that unlike most traditional composites manufacturing techniques (i.e. prepeg lay-up) resin is added to the preform after the reinforcement architecture has been designed and formed (preform). This allows the designer to achieve a "materials-by design" approach to the development of a part. However, it comports that the preform design (including materials selection) not be decoupled from the infusion and curing stages of RTM since material selection, among other factors, will influence key aspects of the infusion and curing processes. These aspects include the mico- and macro-flow characteristics during infusion (3), and the reaction kinetics during cure which are the subject of this work.

It is well known that fiber surface treatments affect the cure kinetics of anhydride and amine cured epoxy systems (1,4,5,6). Surface and sizing treatments are also known to affect polyester cure kinetics (7,8,9). In general, one observes changes in the overall rate of cure and final conversion. The mechanisms by which such changes are effected are not clear. Processing characteristics such as viscosity, heat evolution, time to gelation, and time to vitrification are all strongly influenced by changes in the kinetics of cure, and each of these are factors which may influence RTM processing. In this investigation we will evaluate the effects of commercially available S-2 glass sizings on the cure of vinyl ester systems.

Experimental

Materials. Dow Derakane 411-C50 vinyl ester in conjunction with Witco USP-245 bisphenol peroxide (1.75 wt%) were used in this investigation. The Derakane 411-C50 vinyl ester is a low viscosity material containing 50 wt.% styrene.

Owens Corning sized S-2 glass rovings designated by the manufacturer as 365, 933, 449, and 463 were utilized. All rovings were 750 (yards/pound) in weight. However, 365 rovings are comprised of 13 μm diameter filaments (~2000 filaments per roving) while 933, 449, and 463 are comprised of 9 μm diameter filaments (~4200 filaments per roving). These fiber systems have been sized with coatings claimed to have been designed for use with specific resin systems. The 933 fiber system is to be used with high temperature materials such as BMI and PEEK; 365 with vinyl ester and polyester; and 449 and 463 with epoxy.

An attempt was made to gain some insight into the composition of the sizings by stripping the fibers with acetone and then analyzing the dissolved material using Fourier transform infrared spectroscopy (FTIR). Figure 2 shows FTIR spectra of the material stripped from the four fiber types.

The spectra for 463, 449 and 365 fibers show peaks at 1120 cm^{-1}, 1080 cm^{-1}, 940 cm^{-1}, and 3400 cm^{-1}. The peaks at 1120 and 1080 cm^{-1} can be associated with the Si-O-Si asymmetric stretch (10,11,12) and Si-O-C asymmetric stretch (10). The absorption peaks at 940 cm^{-1} and 3400 cm^{-1} are due to Si-OH stretch (10,11,12). The spectrum for 933, however, only displays one (1120 cm^{-1}) of the adsorption peaks associated with silicon containing species described above. Therefore, although all fibers show the presence of silicon, it is felt that 463, 449, and 356 fibers contain some partially hydrolyzed material while 933 appears to contain more fully condensed silane. Moreover, it is clear from comparison of the spectra for 463, 449, 365 and 933 with that of Shell Epon 828 (also in Figure 2) that 463, 449, and

365 contain a substantial amount of "loose" epoxy which does not appear in the spectrum for 933. These observations are consistent with the high temperature specifications of the 933 S-2 glass fiber. Note that stripping the fibers with acetone does not necessarily remove all sizing material. It is in fact expected that a substantial part of the condensed silane may remain bound to the fiber surfaces.

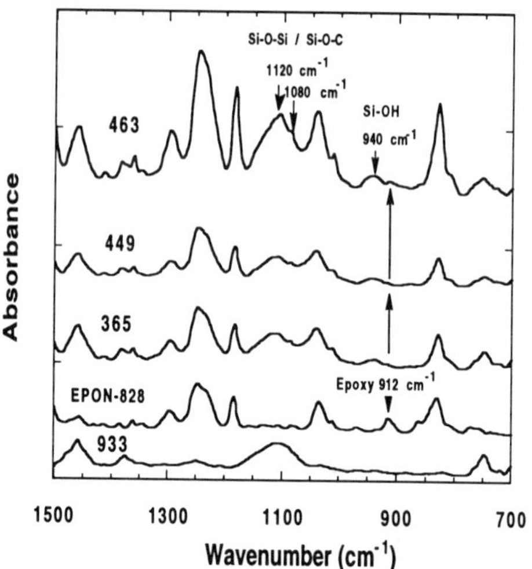

Figure 2: FTIR spectra of sizing material stripped from Owens Corning S-2 glass fibers using acetone.

Procedures. The following procedures were carried out to evaluate fiber affected cure kinetics and fiber surface free energies.

Cure Monitoring. The cure kinetics of the neat and fiber modified Dow Derakane 411-C50 vinyl ester were investigated using a DuPont Model 9900 differential scanning calorimeter (DSC). Isothermal experiments were conducted at 90°C, 100°C, 110°C, and 120°C for the neat and fiber modified systems to measure heat evolution as a function of time. At least three runs were performed per temperature and material pair. A nominal fiber loading of 20 volume % was used for the experiments using 933, 365, 463, and 449 glass fibers. Such samples were prepared by adding the proper amount of resin to a DSC sample pan containing a pre-weighed amount of chopped fiber (~ 4 mg). Care was taken to ensure the sample pans were hermetically sealed in order to avoid styrene evolution during cure. Following isothermal cure, temperature scans from ambient temperature to 200°C at 10°C per minute were conducted to measure residual heats of reaction.

Data Analysis. The DSC provides the rate of heat evolution as a function of time, q(t), for a sample of mass, m, which is assumed to be proportional to the rate of polymerization. Thus the rate of change of dimensionless conversion, α(t), is related to q(t) by equation 1 where ΔH_r is the specific heat of reaction.

$$\frac{da(t)}{dt} = \frac{q(t)}{\Delta Hr\ m} \tag{1}$$

The total heat of reaction for a sample of mass m (Q_{tot} = ΔH_r m) can be obtained from DSC measurements.and if not known. It is obtained by summing the heat evolved during isothermal experiments (Q_i) with the residual heat (Q_r) obtained from subsequent temperature ramps, viz:

$$Q_{tot} = Q_i + Q_r \qquad (2)$$

The heat of reaction for the Dow Derakane 411-C50 was taken as the average value of multiple runs under reaction conditions which yielded the highest total heat. The heat of reaction for Dow Derakane 411-C50 was found to be 425±15 J/g. This value was was used throughout the analysis.

Simplified second order autocatalytic rate expressions have been used previously to model the cure of styrene based thermoset resins (7). Although such expressions are empirical and much work is ongoing to develop more sophisticated mechanistically based rate expressions, the autocatalytic rate expression given by equation 3 where m+n =2 and k is the reaction rate constant has been used successfully by many (7,13,14). This rate expression was employed in this work for the purpose of comparing fiber sizing effects on cure.

$$\frac{d\alpha(t)}{dt} = k\ \alpha(t)^m\ (1-\alpha(t))^n \qquad (3)$$

Data obtained from DSC can be used directly to obtain best-fit values for k, m, and n in equation 3. Such fitting was done for all DSC experiments and typically yielding good agreement with the data.

Surface Energy Characterization. The polar and dispersive components of surface energy for the Owens Corning fibers were obtained by measuring the contact angles of these fibers in a series of test fluids of known polar and dispersive surface tensions and analyzing the results using a method proposed by Kaeble (15). The contact angles were obtained using the mico-Wilhelmy technique with a DCA-322 microbalance from Cahn Instruments, Inc. The fibers were cut and glued to a wire using a cyanoacrylate adhesive, and the wire was suspended from the measuring arm of the microbalance. Advancing and receding contact angles were recorded for immersion rates of 20 μm/s. Only the advancing contact angle values were used for the analysis. Values of the fiber radii (which are essential for determining contact angles) were obtained using the DCA and octane as the wetting fluid. For each fiber ten fiber specimens were analyzed.

Results

Multiple isothermal experiments were carried out for the temperature conditions and resin/fiber pairs discussed in the previous section. Figure 3 shows thermograms obtained for neat vinyl ester at 90°C, 100°C, 110°C, and 120°C. Three sets of data are shown per temperature to demonstrate the repeatability of the DSC experiments. Similarly Figure 4 shows thermograms obtained for 365 S-2 glass filled vinyl ester at 90°C, 100°C, 110°C, and 120°C. These data are also representative of experiments carried out with other fiber types. Note that the repeatability suffers in the presence of fibers and particularly at the lower temperatures.

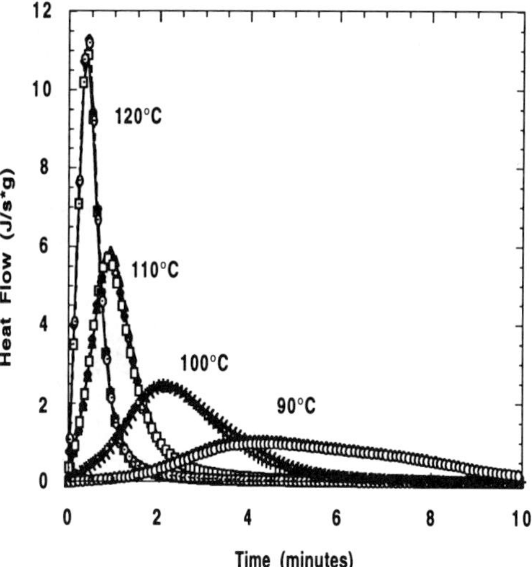

Figure 3: Dow Derakane 411-C50 vinyl ester isothermal cure thermograms showing multiple runs at 90°C, 100°C, 110°C and 120°C.

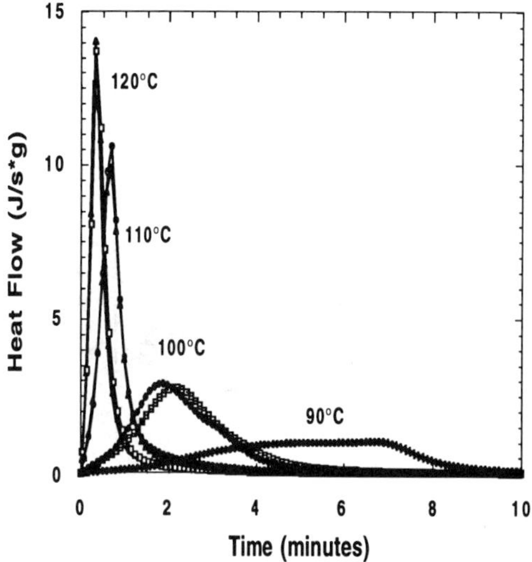

Figure 4: Owens Corning 365 S-2 glass modified Dow Derakane 411-C50 vinyl ester isothermal cure thermograms showing multiple runs at 90°C, 100°C, 110°C and 120°C.

Table I shows average values of k and m obtained by fitting DSC data to equation 3 for neat and 365 filled vinyl ester. As demonstrated by the values of the correlation coefficients reported above and with the exception of the filled system at 90°C, the data are generally amenable to analysis using the empirical rate expression given by equation 3 discussed in the previous section. Furthermore, as shown by the errors reported in Table I, rate constant variations are greater for the filled system. However, it is clear that the rate constants for the 365 filled system are higher than those of the neat resin.

Table I: Average rate constants for neat and Owens Corning 365 S-2 glass filled Derakane 411-C50 vinyl ester

Temperature °C	k (1/min)	m	Correlation Coefficient
Neat Resin			
90	0.711±.05	0.772±0.05	99.18±0.2
100	1.316±0.1	0.717±0.2	99.85±0.2
110	2.63±0.1	0.617±0.1	99.67±0.1
120	5.05±0.05	0.528±0.1	99.71±0.1
365 S-2			
90	---	---	---
100	1.494±0.7	0.721±0.2	99.69±0.1
110	3.77±0.1	0.54±0.1	97.67±0.1
120	5.25±0.2	.414±0.3	99.04±0.2

Arrhenius plots may be used to obtain values of the activation energy and pre-exponential factors so as to more easily compare reactivities over wider ranges of temperature. Figure 5 shows the Arrhenius plots for the neat and 365 filled systems. While the linear relationship between the reciprocal of temperature and the logarithm of the rate constant is clear for the neat resin, the 365 filled material exhibits uncharacteristic behavior. The apparent change in slope may indicate competing cure processes.

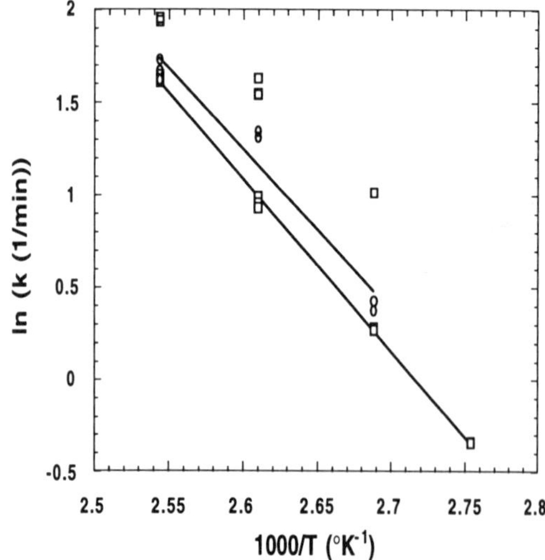

Figure 5: Arrhenius plots for neat (squares) and Owens Corning 365 S-2 glass filled (circles) Dow Derakane vinyl ester resin.

Figure 6 shows DSC thermograms comparing the reaction rate of neat vinyl ester to that of filled systems containing 449, 463, 933, and 365 S-2 glass at 110°C. It is apparent that in all cases the filled systems cure significantly

more quickly than their neat counterpart. Although more pronounced at 110°C, the same effect is noticed at 100° and 120°C. However, Figure 7 indicates that at 90°C the fibers tend to affect the cure behavior by retarding polymerization. The data in these thermograms, with the exception for neat vinyl ester, are clearly not amenable to analysis by equation 3.

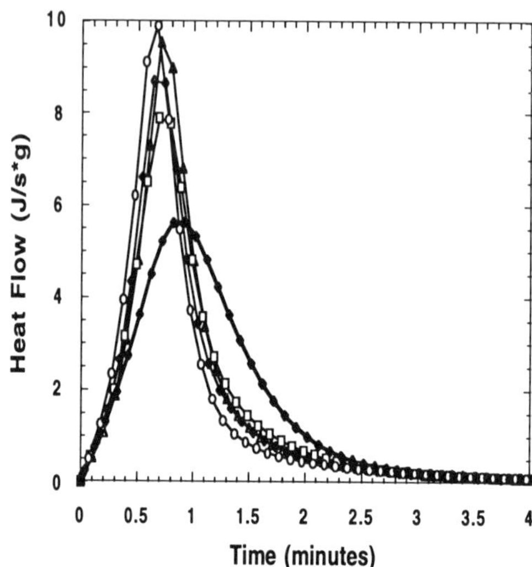

Figure 6: Isothermal cure thermograms for neat (filled diamonds) Dow Derakane 411-C50 vinyl ester and Owens Corning 449 (triangles), 463 (open diamonds), 933 (squares), 365 (circles) S-2 glass filled Dow Derakane 411-C50 vinyl ester at 110°C.

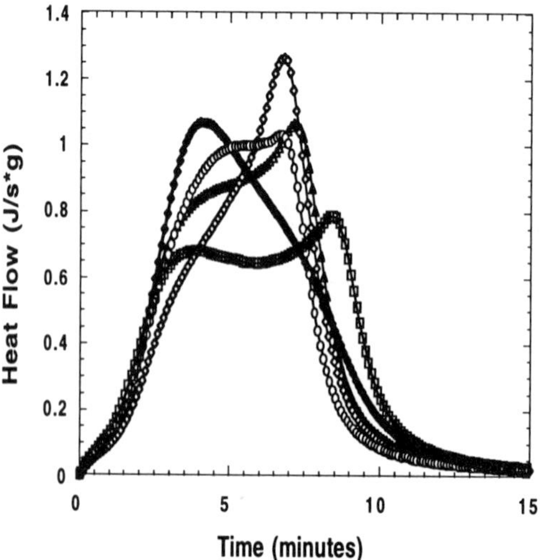

Figure 7: Isothermal cure thermograms for neat (filled diamonds) Dow Derakane 411-C50 vinyl ester and Owens Corning 449 (triangles), 463 (open diamonds), 933 (squares), 365 (circles) S-2 glass filled Dow Derakane 411-C50 vinyl ester at 90°C.

Table II reports values of activation energy (Ea) and pre-exponential factors (A) obtained by Arrhenius analysis of data between 100° and 120°C for the filled systems and 90° and 120°C for neat vinyl ester. The activation energies are generally substantially lower for the glass fiber modified systems. This analysis suggests that the fibers decrease the apparent activation energy significantly and hence accelerate cure.

Table II: Arrhenius parameters for neat and Owens Corning S-2 glass filled Dow Derakane 411-C50 vinyl ester

Material	Ea (J/mol)	ln A (1/min)
Neat 411-C50	77407±1000	25.29±0.32
449	69058±5252	22.87±1.65
463	63447±4409	21.14±1.39
933	79828±4981	26.19±1.57
365	72482±6957	23.91±2.18
365 stripped	54146±2320	25.31±0.65

Included in Table II are Arrhenius parameters for stripped 365 fibers. In this case 365 fibers were washed in acetone in an attempt to remove the sizing. The 365 fibers treated in this manner significantly reduced the apparent activation energy for cure. The increased cure rates are shown by the thermograms in Figure 8.

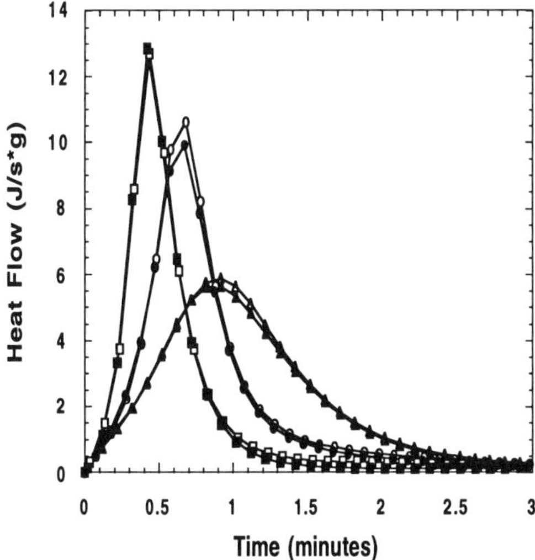

Figure 8: Isothermal cure thermograms for neat (triangles) Dow Derakane 411-C50 vinyl ester and Owens Corning 365 (circles) and stripped 365 (squares) S-2 glass filled Dow Derakane 411-C50 vinyl ester at 110°C

Discussion

Three major points can be drawn from the results presented above. First, at typical cure temperatures (100-120°C) the S-2 glass fibers increase the cure rate of the Dow Derekane 411-C50 vinyl ester substantially. Second, the type of fiber sizing or surface treatment affects the degree of acceleration. Third, at lower temperatures (90°C) the enhancing effect is diminished and in some instances replaced by a retardation effect (Figure 7).

Several investigators have noted increased and decreased reaction rates for filled polyester systems. Lee and Han (8) investigated the effects of silane treated and untreated glass beads on the cure of polyester systems. They found that bare glass beads reduced the reactivity of the polyester system while a silane coating produced a lesser decrease. Ishida and Koenig (9) found that bare E-glass retards the cure of polyester systems. They attributed this retardation to a decrease in free radical concentration resulting from the formation of charge transfer complexes between the free radicals and the inorganic oxides. However, such surface termination cannot account for the accelerated cure behavior observed in this work with vinyl ester/sized S2-glass systems.

Lem and Han (7) investigated the effects of inorganic fillers on the cure behavior of polyester systems and found that particularly calcium carbonate accelerates polyester cure. They attributed this behavior to an apparent enhancement of the peroxide decomposition rate.

Radical Chain Polymerization. The vinyl ester cure process is a free radical chain polymerization. For such systems the course of polymerization can be broken into three stages: a] the inhibition stage; b] the propagation stage; and, c] the diffusion limited stage. During the inhibition stage, radicals, which are generated by the decomposition of initiator, are consumed by inhibitors. Once the inhibitor is consumed the polymerization rate increases throughout the propagation stage. Finally, the diffusion limited regime is reached when increases in molecular weight limit the mobility of the reacting species and hinder propagation.

The kinetics of free radical polymerization (particularly linear polymerizations) have been studied extensively (16). Several phenomena which influence the rates of chain polymerization may provide clues to the mechanisms by which sized S-2 glass fibers affect the cure of vinyl ester systems. The following is a brief discussion of such phenomena and their possible relation to vinyl ester fiber affected cure.

Inhibitors. Radical chain polymerization may be suppressed by the presence of inhibitors. These substances act by reacting with initiating and propagating free radicals and converting them to species which will not sustain propagation. Inhibitors are often used by resin manufacturers to prolong resin shelf life. Such inhibitors could adsorb onto the surfaces of the fibers thus increasing initiation and propagation rates. Note that oxygen is a common inhibitor that will be present in resins unless measures are taken to eliminate it.

Diffusion Effects. Typically, free radical polymerizations exhibit an acceleration in the rate of polymerization such as the one exhibited by the vinyl ester system being investigated. This behavior is a result of hindered translational mobility of the reacting species. As a result of such diffusion limitations bimolecular termination is hindered and the resulting increase in free radical concentration accelerates the reaction. This effect is known as the "Trommsdorff" or "gel" effect (17). The presence of fibers may physically reduce molecular mobility and thus enhance such an effect.

On the other hand, using a similar argument, reduced mobility resulting from the presence of fiber surfaces can aid radical trapping (unimolecular termination) (18) and decrease initiator efficiency (18) which might suppress polymerization rates.

"Matrix" Polymerization. The addition of stereoregular polymers to monomers has been found to accelerate polymerization rates (19). In general the growing radical associates with the added polymer and propagation occurs with the radical bound to the "matrix". As a result, segmental mobility of the growing chain is reduced and the termination rate is decreased. It is possible that vinyl ester radicals associate with the fiber surfaces reducing the segmental mobility necessary for termination and increasing the observed polymerization rate.

Sizing Dissolution. The material which comprises the sizing may be prone to dissolve in the reacting resin. Depending on the nature of interactions between the dissolving material and reacting resin the reaction rate could be affected.

Fiber Surface Energy. Enhanced surface termination, adsorption of inhibitors, surface induced facilitation of peroxide decomposition, "matrix' polymerization, and sizing dissolution are all mechanisms which are in one way or another affected by fiber surface (i.e. sizing) characteristics. Fiber surface free energy depends on the composition of the fiber surface. Polar and dispersive components of the surface free energy can be measured. Such measurements provide further insight into the nature of the surface in question.

Table III shows values for the dispersive (γ^d) and polar (γ^p) components of the surface free energy for the fibers investigated herein. In general, for these fibers the polar component is significantly lower than the dispersive component of the surface energy. Particularly interesting is the very low polar contribution to the surface free energy of the stripped 365 fibers. This is probably due to the removal of the relatively more polar epoxy from the sizing as was discussed earlier.

Table III. Surface free energies for Owens Corning S-2 glass fiber systems.

Fiber	γ^p (dyne/cm)	γ^d (dyne/cm)	γ^t (dyne/cm)
449	10.88±2.51	44.44±4.52	55.33±5.17
463	6.94±1.40	45.07±2.8	52.00±3.13
933	12.60±3.02	32.50±3.95	45.10±4.97
365	8.60±1.18	33.09±1.92	41.69±2.25
st365	3.39±1.43	35.48±3.85	38.87±4.11
Epoxy	12.2	34.6	47.2

The previous discussion of possible mechanisms for the acceleration of vinyl ester cure suggests that surface free energy should play a significant role in determining the degree of cure rate enhancement for glass filled vinyl ester systems. Particularly, this should be the case for mechanisms which rely on the adsorption of resin components onto fiber surfaces such as the adsorption of inhibitors, "matrix" polymerization, and surface termination.

In fact, based on the systems studied, it appears that a there exists a relationship between polar surface free energy and the degree of vinyl ester cure activation.

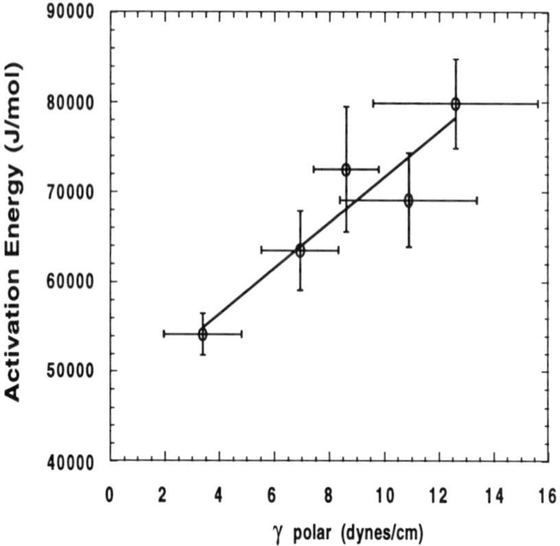

Figure 9: Activation energies for the S-2 glass fiber modified Dow Derakane 411-C50 vinyl ester cure found in Table I ploted as a function of the polar component of the fiber surface free energies reported in Table III.

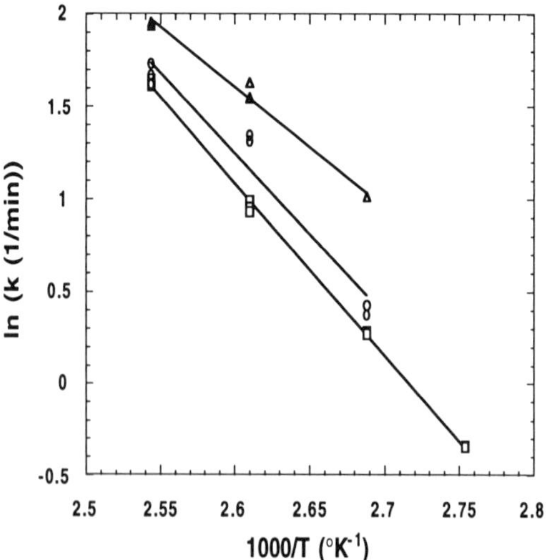

Figure 10: Arrhenius plots for neat Dow Derakane vinyl ester resin.(squares), Owens Corning 365 S-2 glass filled (circles), and stripped (triangles) 365 S-2 glass filled Dow Derakane vinyl ester resin.

Figure 9 shows activation energies for the fiber modified vinyl ester cure found in Table I plotted as a function of the polar component of the fiber surface free energies reported in

Table III. A strong relationship exists between the polar surface free energy and the activation energy. In general, a lower polar surface energy comports a decrease in activation energy and a more pronounced acceleration of the vinyl ester cure.

While it is not possible from this relationship to determine which mechanism discussed above is responsible for the affected cure behavior, it becomes clearer that the mechanism must be closely linked to interactions between resin components and the surface chemistry of the fibers. Furthermore, the possibility of several mechanisms acting in concert should not be discounted. For example, sizing dissolution effects may also play a role in affecting cure as is shown by the comparison of Arrhenius plots for neat vinyl ester resin, 365 modified, and stripped 365 modified systems. While the 365 modified material exhibits uncharacteristic behavior, stripped 365 follows the Arrhenius relationship more closely.

Concluding Remarks

In this work the effect of commercially available S-2 glass sized fiber systems on the cure of vinyl esters was investigated. It was found that these fibers significantly accelerate the cure of vinyl ester. Similar behavior has been reported for polyester systems. However, the mechanisms by which the cure behavior is affected is not well understood. A relationship between the polar component of the surface free energy of the fibers and the cure enhancement has been established for the systems investigated herein. This suggests, perhaps, that an adsorptive surface effect should be considered as a possible mechanism for cure acceleration in glass modified vinyl ester systems.

Nevertheless, it is clear from this work that glass sizing composition has a significant effect on the cure of vinyl ester systems and that the sizing-resin interactions need to be taken into account when selecting processing conditions for RTM applications.

References

1. G.R. Palmese and R.L McCullough, J. Adhesion, in press (1994).
2. T.P. Skourlis and R.L. McCullough, Compos. Sci. Tech.,49, 363 (1993).
3. G.R. Palmese and V.M. Karbhari, Polym. Compos., in press (1994).
4. A. Garton and W.T.K Stevenson, J. Polym. Sci. Polym. Chem. Ed., 26, 541 (1988).
5. J. Mijovic and H.T. Wang, J. App. Polym. Sci., 37, 2661 (1989).
6. A. Garton et al., British Polym. J., 19, 459 (1987).
7. C.D. Han and K-W Lem, J. App. Polym. Sci., 28, 3185 (1983)
8. D-S Lee and C.D. Han, J. App. Polym. Sci., 33, 419 (1987).
9. H. Ishida and J.L. Koenig, J. Polym. Sci. Polym Phys. Ed., 17, 615 (1979).
10. D.L. Leyden and J.B. Atwater,"Silanes and Other Coupling Agents," p 143, K.L. Mittal Ed., VSP Utrecht, The Netherlands (1992).
11. J.M. Park and R.V. Subramanian,"Silanes and Other Coupling Agents," p 473, K.L. Mittal Ed., VSP Utrecht, The Netherlands (1992).
12. F.J. Boerio et al., J. Adhesion, 13, 159 (1981).
13. M.R. Kamal and S.Sourour, Polym. Eng. Sci, 13, 59 (1973).
14. P.W.K. Lam, Polym. Compos., 8, 427 (1987).
15. D.H. Kaeble, J. Adhesion, 2, 66 (1970).
16. G. Odian, "Principles of Polymerization," John Wiley & Sons, p 179, New York (1981).
17. T.J. Tulig and M. T. Tirrell, Macromolecules, 14, 1501 (1981).
18. G.L. Batch and C.W. Macosko, J. App. Polym. Sci., 44, 1711 (1992).
19. J. Gons et al., J. Polym. Sci. Polym. Chem. Ed. 15, 771 (1977).

Proceedings of the 10th Annual ASM/ESD Advanced Composites Conference, Dearborn, Michigan, USA, 7-10 November 1994

Characterizing the Fracture Behavior of Structural Adhesive Bond

D. Hunston
National Institute of Standards and Technology
Gaithersburg, Maryland

Abstract

The automotive industry has considerable interest in using adhesives to join structural components. Moreover, this desire has been accelerated by the possibility of composite structural parts since bonding is an ideal joining methods for composites. With structural adhesives, however, the need for testing and optimized design become critical since the consequences of failure are so severe. Over the past 30 years, there has been much research into this area, and a major focus has been on the fracture mechanics approach. Although fracture testing does not provide absolute answers to either materials selection or joint design, the results are useful for ranking of materials by toughness and important for developing design guidelines to minimize the chances of failure in a bonded joint. The purpose of this paper is to give a general overview for some of the most significant results found by the previous studies in this area. Topics covered include the effects of loading type (cleavage vs shear), adhesive toughness, and bond geometry. Moreover, the role of adhesive microstructure and morphology will be mentioned, and the effects of temperature and loading rate on failure behavior will be discussed.

JOINING IS AN IMPORTANT ISSUE FOR THE AUTOMOTIVE INDUSTRY. Adhesives are widely used because they offer many advantages. For example, the adhesive bonding process can be automated and bonded joints can distribute the loads over a wide area. Adhesives also provide gap filling and sealing capabilities. The vast majority of bonded joints in automotive applications, however, are not structural; i.e. they do not carry the major loads in the vehicle so their failure does not dramatically reduce or eliminate the usefulness of the structure. Now, there is significant interest in moving adhesives into structural applications. This desire is accentuated by the possibility that structural composite parts will be introduced since bonding is an ideal way to join composites. As a result, the areas of testing adhesives, design of bonded joints, and prediction of joint performance and lifetimes are receiving considerable attention. One approach that has been used is fracture mechanics. Over the last 30 years, there has been extensive research on the fracture of adhesive joints. This paper will summarize some of the most significant conclusions from this work.

Early Studies

The idea to apply fracture mechanics to the study of adhesive joints probably originated in the late 50's or early 60's when the general area of fracture mechanics was very active. One of the first major efforts on adhesive fracture was sponsored by the Office of Naval Research, and this program resulted in a series of publications throughout the 60's and early 70's by Ripling, Mostovoy, Corten, and coworkers (1-3). These papers examined the idea of applying fracture mechanics to adhesive joints and developed a number of test methods for this purpose. The general conclusion was that fracture mechanics provided an excellent tool to study bonded joints.

Much of the initial work was with "brittle" adhesives, but in the late 60's and early 70's, studies started to address the toughened adhesive systems that began to appear on the commercial markets in the mid 60's. A number of groups participated in this effort, but particularly noteworthy were the studies of McGarry et. al (4-6) at MIT who looked at the fracture behavior of the adhesives materials as bulk resin samples, a team from BFGoodrich (7-9) who developed one of the most common toughening agents, and Bascom et. al. (10-13) at the Naval Research Laboratory who initiated the studies on adhesive bonds. The results showed that toughened systems have much more complex behavior than brittle resins, and much of the work in the 70's and 80's was aimed at understanding such complications. Through these studies, the research on adhesive bonding has continued to grow. In addition to the Navy-sponsored research, there were major programs initiated by the Air Force, NASA, and NSF. These programs plus the work of numerous companies scientists and individuals investigators have created a solid base for the application of fracture mechanics to adhesive joints.

The Fracture Mechanics Approach

The objective in a fracture mechanics approach is to characterize the resistance to crack growth from a preexisting flaw or stress concentration point by parameters that are material properties; i.e. their values are independent of specimen geometry (14-16). This is important because much adhesive testing produces results that depend not only on the particular test specimen used but even on the size of that specimen. Consequently, the data provide little insight for what will happen with other geometries.

Fracture Modes: The fracture mechanics approach focuses on three basic types of failure: Mode I, Mode II, and Mode III (see Figure 1). Many test specimens have been developed to impose either one of the basic loading modes onto a sample or some specific combination of these modes. Many of the most common test specimens are discussed in refs. 17 & 18. A precrack is produced in the specimen which is then loaded until the crack begins to propagate. The load required to cause this is measured, and by using a stress analysis or an energy analysis, this load is related to a parameter that characterize the materials resistance to crack growth. The stress analysis produces a parameter called the critical stress intensity factor, K_C, while the energy analysis gives the critical strain energy release rate, G_C.

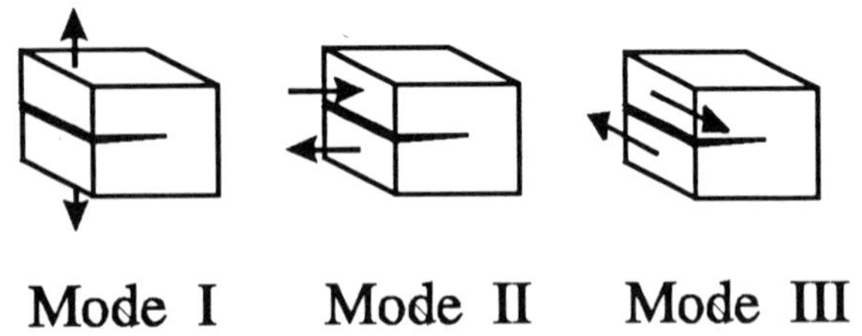

Figure 1: The three loading modes.

A number of assumptions are involved in these analyses. For example, the stress approach is based on the spatial distribution of the stresses and strains at the crack tip being invariant and depending only on the presence of the crack (crack-tip autonomy). For a linear elastic material, the stress intensity factor, K, describes the strength of the crack-tip stress field in terms of the conditions at the external boundary. K_C is the critical value of the stress intensity factor that produces crack growth and is also called the fracture toughness. A tougher material will require higher stresses to produce crack growth and therefore have a higher value for K_C.

In a similar way, G_C describes the energy per unit area of new crack that is needed to initiate crack propagation so larger values mean the material is more resistant to crack growth initiation. The critical strain energy release rate is also called the fracture energy. For a linear elastic material, there is a simple relationship between G_C and K_C. Although these analyses were originally based on a linear elastic materials, they have since been extended to include a small amount of localized plastic deformation at the crack-tip.

Failure Characterization: To characterize a material, the fracture parameter for each of the three basic modes must be determined (K_{IC}, K_{IIC}, and K_{IIIC} or G_{IC}, G_{IIC}, G_{IIIC}) and an appropriate mixing scheme must be developed to predict the value of the fracture parameter for various combinations of the three modes. The key factor is that all specimens that produce a given mix of loading should give the same answer for the fracture parameter. _As a result, these parameters provide a unique failure criterion that can be used in design and failure prediction as long as the underlying assumptions remain valid._

A further complication is that the fracture parameters depend on the width of the specimen. For example, the Mode I fracture energy is high at low widths but decreases as the width is increased. Above a minimum width, however, the fracture energy becomes constant for further incases in width. Bulk samples of a relatively tough adhesive will exhibit a constant fracture energy down to widths somewhat below a centimeter while the fracture energy for bulk samples of a brittle adhesive will be invariant to widths less than a millimeter. With bonded joints, the width is usually well above the minimum value for constant fracture energy so it is only the fracture parameters associated with thick specimens, termed the plane-strain value, that is usually of interest.

As shown in Figure 1, the loading mode is defined in terms of the relationship between the loads and the precrack, but with adhesives, there is also the direction of the bond-line to consider. The accepted practice is to insert the precrack down the bond-line so that the planes of the precrack and bond-line are the same.

A distinction should be made between the classic definition of the three fracture modes and the general usage now. Originally, Modes I, II, and III were defined as failure modes, and the assumption was that co-planer crack growth occurred; i.e. the crack advanced in the same plane as the precrack (14-16,19). In practice, the researcher can control what loads are applied to the sample but not what direction the crack grows. The actual experiments show that co-planer crack growth is seen only in Mode I fracture. (Even here, close examination may show deviations from the plane due to factors such as material inhomogeneities.) As a result, it is now customary to talk about loading modes rather than failure modes.

Bulk vs Adhesive Fracture: This change in definition is important because in brings up an major distinction between homogeneous materials and adhesive joints for other than Mode I loading. In the former, the crack is free to propagate in any direction and will generally advance along the plane where the tensile load perpendicular to that plane is greatest. In an adhesive joint, however, the crack will generally propagate down the bond-line because the adherends are often much tougher than the adhesive.

The situation is even more complicated than this, however (19-22). Many fracture-test geometries measure only the initiation of crack growth because once the crack begins to grow it propagates rapidly so the experiment is over. In this case the important question is what direction the crack growth starts. If the precrack is along the adhesive-adherend interface, it may be forced to grow down the bond-line in a co-planar manner, but a precrack that ends in the center of the bond may start to grow somewhat across the bond-line. If the growth is unstable, this becomes the direction of crack initiation. In some cases, however, the crack may advance over to the adhesive-adherend interface and stop. The load must then be increased further to cause the crack to propagate down the interface. Here the direction of the crack growth is parallel to the bond line but the crack has undergone a step change in direction.

What makes these situations important is that they represent deviations from the simple picture of fracture testing and a barrier to comparing data on bulk adhesive samples and bonded joints for anything other than Mode I. A sophisticated fracture analysis may be able to handle these complications, but in most cases the details of this crack growth initiating process and its effect on the fracture parameters are unknown. This represents a difficult but important area of research. At present, most studies ignore these questions and simply report the data. As will be discussed later, there are some results which suggest that these questions may not be as big a concern with tough resins.

Brittle and Tough Adhesives

Structural adhesives are usually based on highly cross-linked materials like epoxies, polyesters, and polyimide because they have many desirable properties like low creep, high modulus, and good behavior at high temperatures. At the same time these base materials are brittle. This makes the application of linear elastic fracture mechanics more straight forward, but also gives the problem illustrated in Figure 2 which shows the Mode I fracture energics for common materials (19). The brittle resins are only slightly tougher than inorganic glass. The adhesive companies realized this in the mid 60's and began developing toughened systems, such as the elastomer-modified epoxies. These materials have fracture energies that are one to two

orders of magnitude better than the unmodified materials. This is much more acceptable for structural applications.

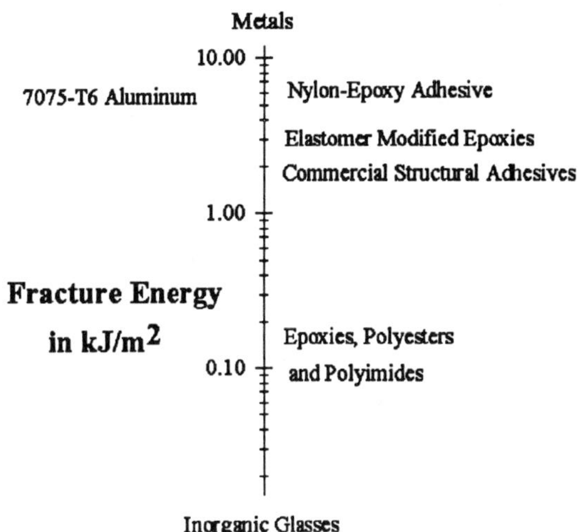

Figure 2: Mode I fracture energies for common materials and bulk samples of structural adhesives.

These toughened materials are almost always two-phase systems; i.e. the modifier is present as a second phase dispersed in and bonded to the main component or matrix as small, micron-sized particles. Since the bulk properties like modulus and glass transition temperature depend largely on the matrix, the toughened system retains most of its desirable bulk properties. The presence of the second phase, however, has a dramatic effect on the fracture behavior which depends on the response of the material in a localized region near the tip of the precrack. In tough materials, a large zone can develop at the crack tip. In this region, the material deforms by a variety of mechanisms like yielding, plastic flow, cavitation, and dilatation (4,5,11,23-25). This blunts the crack tip and decreases the local stress concentration so fracture is suppressed (23,25).

One consequence of this is that the fracture behavior depends not only on the nature of the materials involved but also on the morphology and microstructure of the adhesive. Figure 3 which is an electron micrograph of a rubber toughened epoxy, illustrates the two phase nature of such systems. The rubber particles are stained to make them more visible.

By changing the formulation and chemistry, it is possible to generate materials with a wide variety of morphologies. It is common to see samples with particles in the 1 to 2 micron range or in the range below 0.2 microns, or both. Each morphology has different fracture performance so structure-property relationships are a prime topic for study. Over the last 20 years, this area has received a great deal of attention, and the results are summarized in several recent books (26-32). Consequently, this subject will not be discussed here.

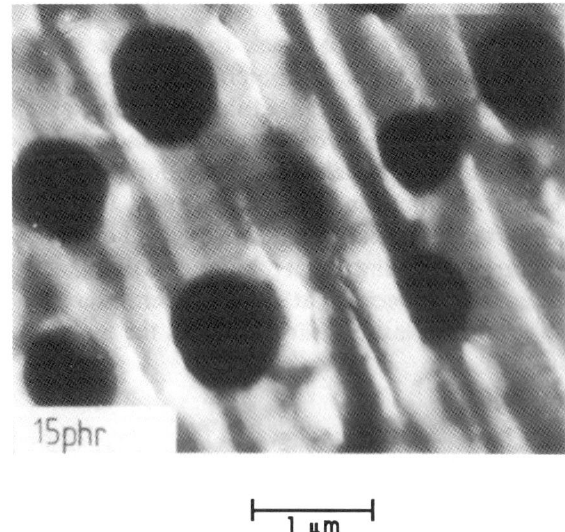

Figure 3: Electron micrograph showing the morphology of a two phase system.

General Results

Figure 2 illustrates only the Mode I aspects of failure. A general picture suggesting how an adhesive bond might behave under a variety of loads can be proposed based on the limited data in the literature (1-3,10-13,17-26,33-35). Most fracture testing has focused on Mode I loading or a combination of Modes I and II. As a result, this type of failure is will documented. In addition, data for composites delamination under Mode I, II, and combinations of the two can provide some insight because the tests are in many ways similar to the adhesive bond experiments (36-39). Efforts (33-35) have been made to test pure Mode II loading and some studies have examined pure Mode III. These experiments are difficult, however, because alignment becomes critical since the specimens generally tend to twist. Few if any studies have investigated mixed mode loading involving Mode III. What Mode III data does exist seems to suggest that the resistance to fracture for Mode II and Mode III loadings are similar (34,35). If this is true, and if it is assumed that the same trend also holds for Mixed-Mode experiments involving Mode III, it is possible to schematically indicate how the fracture might behave as a function of loading. This is done in Figure 4.

The Figure shows estimated behavior for both a brittle adhesive like a simple epoxy and a tough adhesive such as a rubber-toughened epoxy. Although only schematic, the Figure does suggest a number of important points. First, the worst case is Mode I loading; i.e. this generally gives the lowest or close to the lowest resistance

to fracture. As the component of Mode II or Mode III loading increases, the fracture resistance generally goes up. This trend is much greater, however, for the brittle systems than for tough materials. Thus toughening does most to improve the Mode I fracture behavior. This result is a major reason why much of the research has focused on Mode I. It also suggests that such tests provide the most sensitive experiments for resin selection and development. If a new formulation does not show an increase in Mode I behavior, it's very unlikely to have better properties for any other type of loading. Mode I values can also be used for design analysis since they seem to represent a worst case.

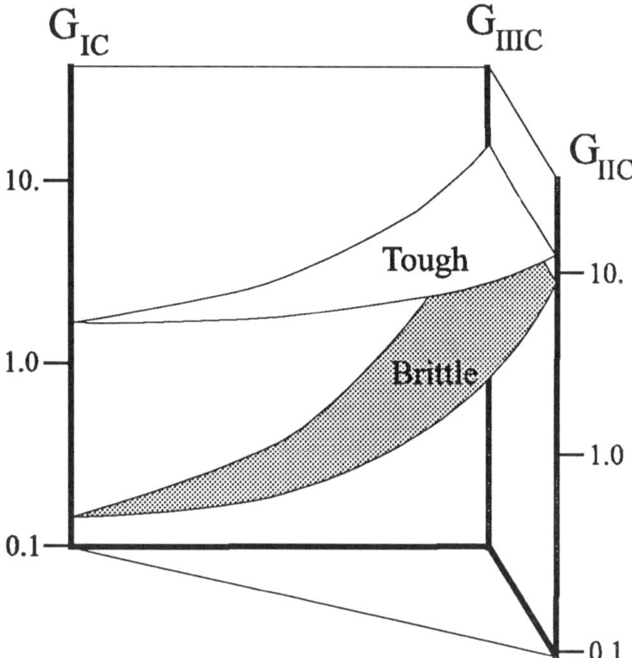

Figure 4: Schematic for the estimated behavior for bond fracture as a function of loading type for a brittle and a tough adhesive.

Figure 4 also illustrates why the designs for practical adhesive joints generally try to minimize the Mode I component of the loading. The more the design loads move to the right in the Figure, the better the resistance to crack growth. This also means that a design analysis based on Mode I measurements may be very conservative. As a result, considerable work in the last ten years has focused on specimens designed to produce a mixture of Modes I and II loads. This more accurately reflects the loads generally present in practical joint designs.

Another interesting observation is that the nature of the loading seems to make less difference for tough adhesives that for the more brittle resins. There is no clear explanation for this effect, but the following speculation can be offered (19). The tough materials exhibit a large, crack-tip deformation zone. Under these conditions, one might speculate that failure is related to processes that occur through out the zone so the direction of the precrack related to the loading may be less significant. If this is true, it may say something related to the questions raised earlier about uncertainties in the direction of crack-growth initiation for mixed-mode tests. If directionality is less important, ignoring such details may be less of a problem than would be expected.

Model System for a Structural Adhesive

To this point, the discussion has been essentially qualitative. The remainder of the paper focuses on Mode I results and quantitative examples will be used. Some of the mode I questions that will be discussed here have also been examined for other types of loading, but the results are much more limited and a great deal still remains to be done. The large number of literature studies that have been conducted in this area have used both commercial and model adhesives. To illustrate the results from these studies literature data for a commonly used model, elastomer-modified epoxy will be employed. Details on the preparation of this system can be found in the literature (10-13) and will be summarized here only briefly.

The epoxy, a diglycidyl ether of bisphenol A (DGEBA), was cured for 16 hours at 120°C with 5phr (parts by weight per hundred parts by weight of epoxy) of piperidine. When toughened, various concentrations of an elastomer, carboxyl-terminated polybutadiene acrylonitrile (CTBN), were included. After degassing the epoxy or epoxy/CTBN mixture, the piperidine was added. The liquid was then poured into molds to make plates for tests on bulk samples or between adherends for fabrication of bonded joints. Fracture tests on bulk resin samples were conduced using compact tension specimens machined from the plates while adhesive fracture experiments utilized the tapered double cantilever beam geometry. Literature studies have been conduced with a wide range of elastomer concentrations, but the results discussed here will be illustrated with data on the unmodified epoxy and two toughened systems: one with 15phr of CTBN and the other with 18.5phr of CTBN.

Adhesive Fracture Results for Brittle Materials. The literature results for brittle materials, like unmodified epoxy, show relatively straight forward behavior in Mode I failure (11,19,26). The fracture energy for an adhesive bond has been shown to be quite similar to the value obtained from experiments on bulk resins samples of the same material so long as the failure was cohesive, i.e. within the adhesive layer. In many cases where differences have been seen, one or more complicating factors were present. For example, if the adhesive fails to wet the adherend during bond formation, a weak interface can be formed. This can produce a bond that fails prematurely along the interface (adhesive failure). If the cure shrinkage is high, residual stresses can be present, and these are usually not adequately accounted for in the stress analysis used to calculate the fracture energy. Thus the true stress in the bonded region can be different than what is expected. Finally, the temperature history seen by the resin during cure may be quite different for the adhesive in a bonded joint than for the same material in a bulk sample because the adherends are generally metal. This could mean the adhesive itself is different and may fail differently.

Experiments with brittle resins have been conducted over a range of geometry variables such as bond thicknesses. The results have shown that the adhesive fracture energy is independent of these variables. This is illustrated in Figure 5 with results for the model epoxy (unmodified). Moreover, when such tests have been performed at a variety of different temperatures and loading rates, the resulting fracture energy changes only slightly so long as the conditions remain well below the glass transition temperature. The changes that are observed are consistent with a viscoelastic process; i.e., higher test temperatures or slower loading rates produce slightly higher fracture energies. Here again, the behavior of the resin in bulk specimens and adhesive joints is similar. Consequently, it seems that in general simple thermosets in an adhesive bond fail in much the same way as they do in bulk samples (11,19,26).

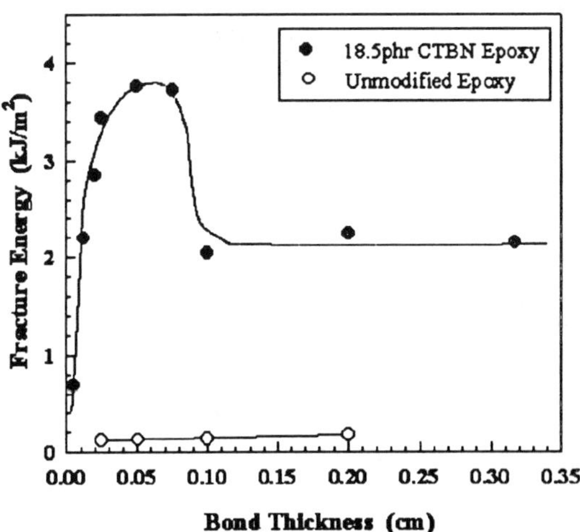

Figure 5: The adhesive fracture energy is shown as a function of bond thickness for the unmodified epoxy and a rubber-modified epoxy (from ref. 11).

Adhesive Fracture Behavior for Tough Adhesives. Toughened adhesives often exhibits very different behavior. First, the fracture energy can show a marked dependence on geometry variables such as the thickness and width of the adhesive layer. This is illustrated in Figure 5 which shows data reported for fracture energy as a function of bond thickness for the model epoxy with

18.5phr of CTBN (11). At very high thicknesses, the adhesive fracture energy is approximately the same as that for bulk samples. This is reasonable since at very large thicknesses, the adhesive bond is essentially a bulk sample. As the thickness decreases, the fracture energy goes up and reaches a value that is almost twice that of the bulk material; i.e. the bonded structure is actually tougher than the adhesive material itself. As the thickness continues to decrease, the toughness goes down to levels well below that of the bulk material.

There are two ways to interpret this result. First, it can be asserted that the fracture energy for an bonded joint made with a tough adhesive is not a material property since it depends on the geometry of the structure and not just the materials. Alternatively, it can be argued that the parameter determined is not a valid fracture energy since the crack-tip stress field is not consistent with the assumptions in linear-elastic fracture mechanics. In either case what is important here is that a single value for the fracture parameter does not provide a unique criterion for failure of an adhesive bond. If the fracture toughness, K_{IC}, is used instead, the results are similar to those shown in Figure 7. A more recent study (19,26) has been conducted using the J-integral approach which is designed to deal with materials that display plasticity. It was hoped that this would provide a unique failure criterion, but here again J_C was found to depend on bond thickness. Consequently, _none of the fracture parameters provides a unique failure criterion that can be used to characterize the failure of a bonded joint or to design joints with a classic fracture mechanics approach_.

If the usual design approach does not work, the next best hope would be to develop an understanding of the geometry effects so that they could be predicted. A number of studies have focused on this problem and developed a variety of possible explanation. The most popular is based on the size (or more specifically the volume) of the crack-tip deformation zone. As mentioned earlier, the mechanisms for toughening are generally associated with the ability of these materials to form a large deformation zone at the crack-tip. A very simple approximation for this situation is provided by the elastic-plastic model (11) which relates the fracture energy to the radius of cylindrical, crack-tip deformation zone, r_c, and the adhesive's yield stress, σ_y, and yield strain, ε_y.

$$G_{IC} = 2\pi\sigma_y\epsilon_y r_c \qquad (1)$$

This is the solution for plane-stress which has been shown to give slightly better agreement with experimental results than the plane strain equation although the differences are small. The deformation processes in the crack-tip region have been studied extensively and involve much more complex deformations than simple plasticity. Moreover, the zone shape generally has an elliptical cross section rather than a circular shape. Nevertheless, eq. (1) has been extremely useful in describing the behavior of these systems

(11,19,26).

Bond Thickness Hypothesis. How does eq. (1) help explain the bond thickness dependence? The hypothesis offered in the literature suggests that there are two effects operating on the deformation zone (11,40-44). First, in very thin bonds, the bond thickness becomes less than $2r_c$ so that the adherends physically constrain the size of the zone to smaller values than the material is capable of developing. This reduces r_c and hence the fracture energy.

The second effect is associated with the unique stress field that is present in an adhesive bond. Stress analyses (42-44) have shown that as the bond thickness decreases, the crack tip stress field changes in a way that the high stress levels extend further down the bond line ahead of the crack tip than they do in a bulk sample or a thick adhesive bond. It was postulated that this changes the shape of the deformation zone by increasing its length (down the bond line) while not changing its height (perpendicular to the bond line). The result is an increase in size (or volume) of the deformation zone and hence the fracture energy.

The bond thickness hypothesis combines these two effects as illustrated in Figure 6. In very thick bonds the deformation zone is similar to that in bulk samples and so the fracture energies are the same. As the bond thickness decreases, the zone first becomes larger by virtue of the stress field effect. As a result, the fracture energy becomes larger than that for the bulk adhesive. Eventually, the bond thickness reaches the height of the deformation zone so that further decreases constrict the zone. Even though the stress effect may continue to extend the zone further down the bond line, the predominant effect is the constraint on zone height so the fracture energy decreases.

A number of experiments have been performed in an effort to test this hypothesis (11,42-46). These include direct measurements of the zone size and shape with high speed movies of the crack-tip region taken during loading. The results seem to support the hypothesis. Another interesting test is to determine the optimum bond thickness h_{max}, i.e. the thickness where the fracture energy is a maximum, G_{ICmax}. This value can then be compared with the size of the zone in a bulk sample where the zone is unconstrained. G_{ICmax} and h_{max} are determined from data such as shown in Figure 5. While r_c is calculated from eq. (1) using values of the parameters measured on bulk specimens for the adhesive. When this is done, there is excellent agreement between h_{max} and $2r_c$. As a result, the hypothesis described above, offered a possible explanation for the observed behavior. _To date, however, there is no procedure to quantitatively predict the curve shown in Figure 5 so a fracture mechanics based design is not possible._

It is interesting to note that the bond thickness effect appears to be a consequence of high toughness in the resin. The unmodified epoxy does not show a thickness dependence presumably because the deformation zone is so small that any effect would occur at extremely low thicknesses. Although the number of different adhesives formulations that have been examined in detail is small, the results generally support this picture. With a high temperature system, for example, the bond thickness dependence may not appear at all because the materials are normally very brittle and may only appear at high temperatures where the material exhibits significant toughness.

Thick Bond

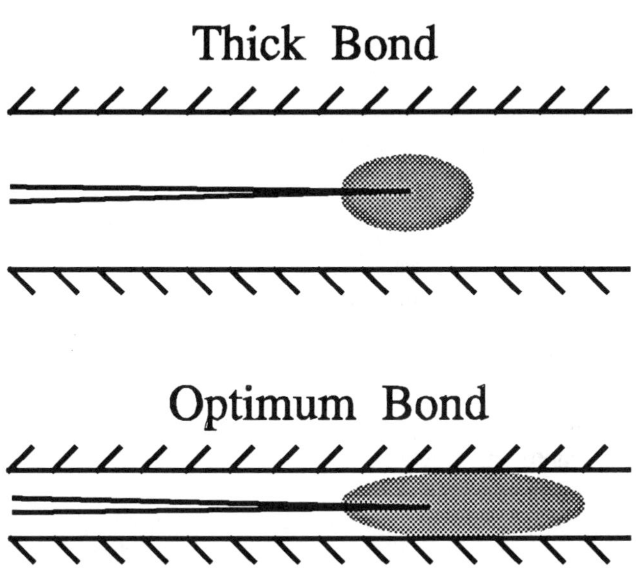

Optimum Bond

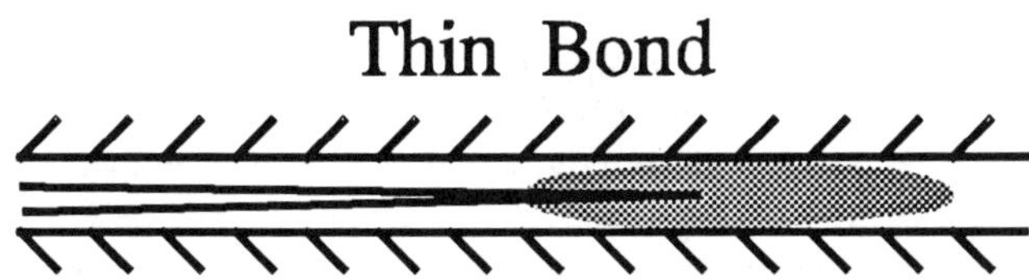

Thin Bond

Figure 6: Schematic of proposed variation in crack-tip deformation zone as a function of bond thickness.

Loading Rate and Temperature Effects. The bond thickness effect is further complicated by the fact that it changes with temperature and loading rate (42-46). This can be illustrated with literature results for the model system (42). The loading rate, S, was defined as the cross head speed used to load a double cantilever beam of a given size with precracks of approximately the same length. The results showed that the primary effect of loading rate and temperature is to change the optimum bond thickness, h_{max}. Increasing the temperature or decreasing the loading rate causes h_{max} to increase. This can be seen in Table I which lists h_{max} and G_{ICmax} for a variety of conditions (42). The data in the top half of the Table (constant cross-head speed at various temperatures) were from tests on samples containing 18.5phr of CTBN while the data in the bottom half (one temperature and various cross head speeds) involve samples with 15phr CTBN.

The results for h_{max} in Table I can be compared

with predictions from the model described above. For each set of test conditions in the Table, values of $2r_c$ were calculated from measurements on bulk samples of the epoxy and eq. (1). The results are listed in the last column of the Table and the agreement is very good.

Table I: Adhesive Fracture (ref. 42)				
Temp. °C	Log (S) m/s	G_{ICmax} kJ/m^2	h_{max} mm	$2r_c$ mm
50	-4.66	3.0	1.1	1.6
37	-4.66	2.9	0.9	1.2
25	-4.66	3.8	0.6	0.6
0	-4.66	3.0	0.5	0.4
20	-6.08	3.8	1.0	0.8
20	-4.78	3.6	0.8	0.7
20	-3.78	3.6	0.6	0.5
20	-3.08	3.2	0.4	0.4

The results in Table I illustrate two important observations about adhesive fracture behavior. It is customary in describing adhesive behavior to give the fracture energy as a function of temperature for a recommended bond thickness. This generally leads to a curve that goes through a maximum, and the region around the peak is recommended for optimum performance. Although this behavior is well known, it is difficult to explain because the peak position doesn't usually relate to any other property of the system. The problem is that data focus on a single bond thickness without realizing that thickness is an important parameter. The reason for the behavior can be understood by considering the material in Figure 5. If the manufacturer happened to pick 0.6mm as the recommended bond thickness, the optimum behavior would occur around room temperature. If 1.0mm were chosen as the recommended thickness, the fracture behavior would be better at higher temperatures, while recommending a thickness of 0.2mm would give the best performance at low temperatures.

The second observation concerns the effect of loading rate on the fracture behavior. If one consults the literature about loading rate effects, the results are directly contradictory (40). For example, when one asks what is the effect of increasing the loading rate, some studies have said that the adhesive fracture energy increases, others assert that it decreases, while still others show that there is no effect at all. In fact all of these observations are true. If the material in Figure 5 is tested at a bond thickness of 0.2mm and the loading rate is increased, the fracture energy will go up because h_{max} shifts to lower values. On the other hand, tests on a bond with thickness of 0.4mm may show relatively little change in fracture energy as long as the increase in loading rate is so large that it shifts the peak to thicknesses well below 0.4mm. Experiments on bonds with thicknesses of 0.8mm will show a sizable decrease in toughness with increasing loading rate.

In essence, the effect is the same as that discussed above for temperature. If the rate were changed over a wide enough range, the data would go through a peak just as they do for temperature. As with any viscoelastic effect, however, very large changes in rate are needed to get the same result as a modest change in temperature. Consequently, changing the rate by 2 or 3 decades produces only a modest shift in h_{max}.

Bulk Adhesive Behavior: In light of the highly viscoelastic behavior of the bonded joint, an obvious question is how the fracture energy of the bulk adhesive behaves as a function of rate and temperature. Experiments on bulk samples also show a dramatic effect (28,46). For example, Figure 7 gives the fracture energy for a rubber-toughened sample (18.5phr CTBN) over a range of temperatures at three cross-head speeds. The variation in G_{IC} is well over an order of magnitude. As with the unmodified epoxies, increasing the temperature changes the behavior in the same direction as decreasing the loading rate.

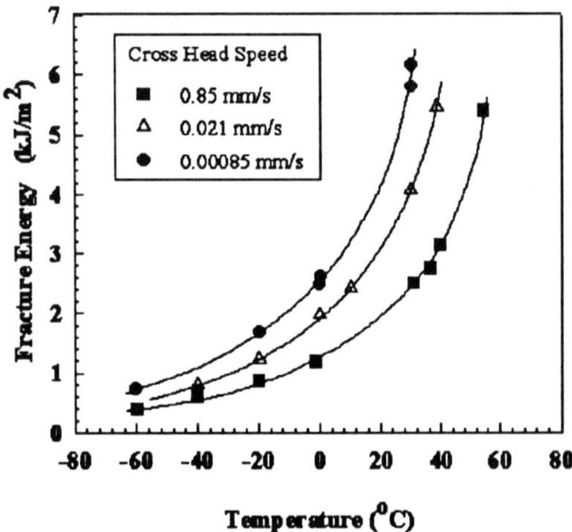

Figure 7: The fracture energy of a rubber-toughened epoxy shown as a function of temperature and loading rate (ref. 28).

The properties of bulk resin samples have been examined for a variety of formulations. All of the toughened systems show this dramatic increase in G_{IC} as the temperature approaches the glass transition temperature, T_g. For example, the material shown in Figure 7 has a T_g of approximately 100°C. This suggests that it is useful to compare toughened materials at temperatures that differ by the same amount from each material's T_g.

One important consequence of behavior such as that shown in Figure 7 is that no single parameter can characterize the toughness of the system. Moreover, when two different formulations are characterized in this way, the curves can cross so that one resin is tougher when measured at a particular set of test conditions while the other resin is tougher in tests at a different set of conditions. Consequently, simple comparisons can be very misleading or even dangerous.

Failure Models for Bulk Adhesive Samples: Several approaches have been used to model the fracture behavior of bulk resins (23,29,40,46). Since the results suggest a viscoelastic process, one procedure that has been used successfully is time-temperature superposition. One problem in this approach was how to characterize the loading rate. The procedure chosen was to use time to failure, i.e. the time period from the initial application of the load until the fracture point. The fracture energy was plotted against time to failure, t_f, at a series of temperatures, and the curves shifted to see if superposition was possible. When this procedure was used, the data appeared to coalesce to a single curve although the large uncertainty that is characteristic of fracture data makes it impossible to prove that a true master curve was obtained. Nevertheless, the results do suggest a simple empirical model for describing the data. The fracture energy vs time to failure relationship was modeled with a power law while the temperature dependence was expressed as shift factors, a_T, and described with an exponential equation. The result was the following expression for fracture energy:

$$G_{IC} = A\left(\frac{t_f}{a_T}\right)^m + G_{ICB} \qquad (2)$$

where

$$a_T = e^{\frac{\Delta E}{R}\left(\frac{1}{T} - \frac{1}{T_o}\right)} \qquad (3)$$

A, m, ΔE, and G_{ICB} are empirically determined parameters, R is the gas constant, and T_o is a reference. To normalize the data with the glass transition temperature, T_o was taken to be T_g - 80°C. The 80°C was selected for the model systems tested in ref. (23,29,40,46) since this choice places T_o in the temperature range where the fracture measurements were made. For other materials, a different choice might be more appropriate. To compare different materials, however, the same value must be used.

Although the parameters in eqs. (2) and (3) are strictly empirical, they do provide a useful way to compare the bulk fracture behavior of different adhesive formulations. The parameter G_{ICB} provides a measure of the limiting toughness at low temperatures and/or high loading rates. The parameter m indicates how sensitive the material is to loading rate effects. The value of ΔE indicates how quickly the system becomes more brittle as the temperature is lowered. The constant A provides a measure of toughening in the sense that a larger value means the fracture energy increases more rapidly with decreases in loading rate and/or increases in temperature. These relationships also provide a useful prediction tool as long as the conditions remain significantly below the glass transition point.

Conclusions

With brittle adhesive, a fracture mechanics approach to the characterization and analysis of bonded joints may be straight forward, but when the joint carries the primary loads in the structure, a tough adhesive is usually required. With such resins, the fracture behavior of an adhesive bond depends on structural variables such as the bond thickness. Although there is a qualitative understanding of the factors which control the bond thickness effect, it is not currently possible to predict this dependence quantitatively. Consequently, none of the usual fracture parameters, K_C, G_C, or J_C, can provide a unique fracture criterion for design or failure prediction.

Nevertheless, the fracture approach can be very useful in developing guidelines and insights for improved adhesives and better joint designs. For example, the effects of loading rate and temperature on the performance of bulk samples of adhesive resins are at least partially understood, and this has provided new tools and procedures for characterizing and comparing different adhesive formulations and relating their toughness to the microstructure and morphology. For bonded joints, the current knowledge indicates that better designs are obtained when the percentage of the more desirable mode II and mode III loads are maximized. Finally, significant progress has been made in correlating the toughness of the adhesive itself with the behavior of the bonded joint so the possibility of formulating adhesives for optimum performance is closer to reality. Although much remains to be done, there is little doubt that the fracture approach is a very useful tool for the field of adhesives.

References

1. Mostovoy, S. and Ripling, E.J. *J. Appl. Polym. Sci.* **10**, 1351 (1966) and **13**, 1083 (1969).
2. Mostovoy, S., Ripling, E.J., Corten, H.T. and Bersch, C.F. *J. Adhes.* **3**, 107, 125, & 145 (1971).
3. Mostovoy, S. and Ripling, E.J. *J. Appl. Polym. Sci.* **15**, 641 & 661 (1971).
4. Sultan, J.N., Laible, R.C. and McGarry, F.J. *J. Appl. Polym. Sci.* **6**, 127 (1971).
5. Sultan, J.N. and McGarry, F.J. *J. Polym Eng. Sci.* **13**, 29-34 (1973).
6. Laible, R.C. and McGarry, F.J. *Polym.-Plast. Technol. Eng.* **7**(1), 27-47 (1976).
7. Rowe, E.H., Siebert, A.R. and Drake, R.S. *Mod. Plast.* **47**, 110-117 (1970).
8. Rowe, E.H. and Riew, C.K. *Plast. Eng.* pp. 45-47, March (1975).
9. Riew, C.K., Rowe, E.H., and Siebert, A.R. p. 326 in **Toughness and Brittleness of Plastics**, Advances in Chemistry Series 154, Deanin, R.D. and Crugnola, A.M., Eds., American Chemical Society, Washington, DC, 1974.
10. Bascom, W.D., Timmons, C.O. and Jones, R.L. *J. Mater. Sci.* **10**, 1037 (1975).
11. Bascom, W.D., Cottington, R.L., Jones, R.L. and Peyser, P. *J. Appl. Polym. Sci.* **19**, 2545-2562 (1975).
12. Bascom, W.D. and Cottington, R.L. *J. Adhes.* **7**, 333 (1976).
13. Bascom, W.D., Cottington, R.L. and Timmons, C.O. *J. Appl. Polym. Sci., Appl. Polym. Symp.* **32**, 165 (1977).
14. Liebowitz, H. **Fracture, An Advanced Treatise**, Vols. I-VII, Academic Press, New York, 1968.
15. Tetelman, A.S. and McEvily, A.J. **Fracture of Structural Materials**, Wiley, New York 1067.
16. Corten, H.T. *J. Adhes.* **3**, 103 (1971).
17. Kinloch, A.J. and Shaw, S.J. "A Fracture Mechanics Approach to the Failure of Structural Joints," Chap. 3 in **Developments in Adhesives-2**, Kinloch. A.J., Ed., applied Science Pub., London, 1981.
18. Kinloch, A.J. and Young, R.J. Fracture Behavior of Polymers, Applied Sci. Publ., London, 1983.
19. Bascom, W.D. and Hunston, D.L. "Fracture of Elastomer-Modified Epoxy Polymers, A Review,", Chap. 6 in **Rubber-Toughened Plastics**, Advances in Chemistry Series 222, Riew, C.K., Ed., American Chemical Society, Washington, DC (1989).
20. Trantina, G.G. *J. Compos. Mater.* **6**, 191 & 371 (1972).
21. Trantina, G.G. "Combined Mode Crack Extension in Adhesive Joints," T&AM Report No. 352, University of Illinois, Urbana, 1971.
22. Bascom, W.D., Jones, R.L. and Timmons, C.O. p. 501 in Vol. 9B of **Adhesion Science and Technology**, Lee, L.H., Ed., Plenum, new York, 1975.
23. Kinloch, A.J., Shaw, S.J., Tod, D.A. and Hunston, D.L. *Polymer* **24**, 1341 & 1355 (1983).
24. Hunston, D.L. and Bascom, W.D. "Failure Behavior of Rubber-Toughened Epoxies in Bulk, Adhesive, and Composite Geometries," Chap. 7 in **Rubber-Modified Thermoset Resins**, Advances in Chemistry Series 208, Riew, C.K. and Gillham, J.K., Eds., American Chemical Society, Washington, DC (1984).
25. Kinloch, A.J. and Williams, J.G. *Journal of Mat. Sci.* **15**, 987-996 (1980).
26. Bascom, W.D. and Hunston, D.L. "The Fracture of

Epoxy and Elastomer-Modified Epoxy Polymers," Chap. 4 in **Treatise on Adhesion and Adhesives, Vol. 6**, Patrick, R.L. Ed., Marcel Dekker, New York (1989).

27. Riew, C.K. and Kinloch, A.J. **Toughened Plastics I: Science and Engineering**, Advances in Chemistry Series 233, American Chemical Society, Washington, DC, 1993.

28. Manzione, L.T., Gillham, J.K. and McPherson, C.A. *J. Appl. Polym. Sci.* **26**, 889-905 & 907-919 (1981).

29. Hunston, D.L. and Bullman, G.W. *Int. J. Adhesion and Adhesives* **5**(2), 69-74 (1985).

30. Kinloch, A.J. and Hunston, D.L. *J. Mater. Sci. Lett.* **6**, 131-139, (1987).

31. Wu, W.L., Hunston, D.L., Yang, H. and Stein, R.S. *Macromolecules* **21**, 756-764 (1988).

32. Pearson, R.A. and Yee, A.F. *J. Mater. Sci.* **26**, 3828-3844 (1991).

33. Ripling, E.J., Mostovoy, S. and Patrick, R.L. ASTM Spec. Tech. Publ. **5**, 360 (1963).

34. Mulville, D.R., Hunston, D.L. and Mast, P.W. *J. Eng. Materials and Tech.* **100**, 25-31 (1978).

35. Chai, H. *Int. Journal of Fracture* **37**, 137-159 (1988).

36. Johnson, W.S. and Mangalgiri, P.D., "Influence of the Resin on Interlaminar Mixed-Mode Fracture," pp. 295-315 in **Toughened Composites**, ASTM Special Technical Publication 937, Johnston, N.J., Ed., ASTM, Philadelphia (1987).

37. O'Brien, T. K., Johnston, N. J., Raju, I. S., Morris, D. H. and Simonds, R. A. "Comparisons of Various Configurations of the Edge Delamination Test for Interlaminar Fracture Toughness," pp. 199-221, in **Toughened Composites**, ASTM Special Technical Publication 937, Johnston, N.J., Ed., ASTM, Philadelphia (1987).

38. Crews, J.H. and Reeder, J.R. "A Mixed-Mode Apparatus for Delamination Testing," NASA Technical Memorandum 100662, Langley Research Center, Hampton, 1988.

39. Hashemi, S. Kinloch, A.J. and Williams, J.G. *J. of Composite Mat.* **24** 918-956, (1990).

40. Bitner, J.L., Rushford, J.L., Rose, W.S., Hunston, D.L. and Riew, C.K. *J. Adhesion* **13**, 3-28 (1981).

41. Kinloch, A.J. and Shaw, S.J. *J. Adhesion* **12**, 59-77 (1981).

42. Hunston, D.L., Kinloch, A.J., Shaw, S.J. and Wang, S.S. "Characterization of the Fracture Behavior of Adhesive Joints," pp. 789-807 in **Adhesive Joints**, Mittal, K.L., Ed., Plenum, New York, 1984.

43. Hunston, D.L., Kinloch, A.J. and Wang, S.S. *J. Adhesion* **28**, 103-114 (1989).

44. Hunston, D.L., Rushford, J.L., Wang, S.S. and Kinloch, A.J. **Proc. 37th Conf. Reinforced Plastics/Composites Institute**, The Society of the Plastics Industry, New York, Jan. 1982.

45. Hunston, D.L., Mizumachi, H. and McDonough, W. **Proc. Int. Adhesion Conf.**, PRI, London, Sept. 1990.

46. Hunston, D.L., Bitner, J.L., Rushford, J.L., Oroshnik, J. and Rose, W.S. *J. Elastomers and Plastics* **12**, 133-149 (1980).

Fracture Testing and Analysis of Adhesively Bonded Joints for Automotive Applications

R.G. Boeman, C.D. Warren
Oak Ridge National Laboratory
Oak Ridge, Tennessee

Abstract

In 1992, the Oak Ridge National Laboratory (ORNL) began a cooperative effort with the Automotive Composites Consortium (ACC) to conduct research and development that would overcome technological hurdles to the adhesive bonding of current and future automotive materials. This effort is part of a larger Department of Energy (DOE) program to promote the use of lighter weight materials in automotive structures for the purpose of increasing fuel efficiency and reducing environmental pollutant emissions. In accomplishing this mission, the bonding of similar and dissimilar materials was identified as being of primary importance to the automotive industry since this enabling technology would give designers the freedom to choose from an expanded menu of low mass materials for component weight reduction.

Early in the project's conception, five key areas were identified as being of importance to the automotive industry. (1) The development of appropriate methods for determining the properties of the adherends and adhesives independent of one another. (2) The determination of accurate, highly standardized fracture test methods for quantifying, not just qualifying, an adhesive/adherend system's resistance to crack growth. (3) Modeling of joints so that designers would be able to examine the effects of minor design changes without entering into an expanded testing program. (4) Non-destructive inspection of production bonds either during the bond formation, after adhesive curing or after component completion. (5) Mechanisms for increasing the manufacturability and reducing the production costs of bonded composites.

This program is in its second year. The tasks under this program are being performed by industry, university and government researchers and are being managed in a joint effort between the ACC Joining Group and ORNL staff members. Plans for expansion of this research project to meet future research needs are also being considered.

This paper concentrates on the details of developing accurate fracture test methods for adhesively bonded joints in the automotive industry. The test methods being developed are highly standardized and automated so that industry suppliers will be able to pass on reliable data to automotive designers in a timely manner. Mode I fracture tests have been developed that are user friendly and automated for easy data acquisition, data analysis, test control and test repeatability. The development of this test is discussed. In addition, materials and manufacturing issues are addressed which are of particular importance when designing adhesive and composite material systems.

IN THE FUTURE, automobiles will be forced to travel further between refuelings while discharging lower levels of pollutants [1]. Currently automobiles account for just under two-thirds of the nation's gasoline usage and about one-third of the total United States energy consumption. By improving automotive fuel efficiency, the United States can lessen the impact that foreign oil prices have on our economy and lives. In addition, decreased emissions from reduced fuel consumption will provide a cleaner environment for future generations. At current usage rates, a 25% weight reduction in current United States vehicles would save 750,000 barrels of oil each day, reduce the yearly domestic fuel consumption by 13% and prevent 101 million tons of CO_2 from being emitted into our atmosphere each year. [2]

A significant reduction in fuel consumption can only be achieved by one of three means: (1) improving engine and drivetrain efficiency; (2) reducing automotive component mass and thus vehicle weight; or (3) reducing the size and thus weight of an automobile. Engine efficiency improvements are being studied by a wide variety of industry and government organizations, and great strides are being made in this area. Vehicle down-sizing has been undertaken since the early '70s and is still occurring, however, consumers are reluctant to purchase smaller and smaller vehicles because their transportation requirements dictate the necessity for a family-size car. Reducing component weight and thus vehicle weight, while not sacrificing vehicle size, reducing safety or increasing vehicle cost, can only be accomplished by the use of alternate, lighter weight materials. The goal of this project is to provide one enabling technology, adhesive bonding, which will allow for the use of alternate materials, particularly reinforced polymer composites.

The commercial application of composites has an extensive history in the marine, aerospace and construction industries, but has evolved relatively slowly in the automotive industry during the past 20 years [3,4]. Composite use has traditionally been limited to secondary structures like appearance panels and dash boards, but as the evolution of the automobile continues, fiber-reinforced polymers are being considered for weight reduction in future automotive structures and load-bearing components [5]. A critical aspect of using these materials is the manner in which they are joined. Adhesive bonding is potentially an economical and structurally sound means of joining reinforced polymers and other alternative automotive materials and may overcome a major obstacle to the incorporation of lighter weight materials into automobiles.

As with composites, the major problems limiting the utility of aluminum in automotive structures have been related to joining technologies [6]. Reliable joining methods for aluminum alloys are needed to make the lightweight metal more attractive for structural applications [7].

While much work has been conducted in adhesive bonding for the aerospace, construction, and some consumer goods industries, the automotive industry does not currently have a complete set of processes and methods for evaluating candidate adhesives for use in bonding structural automotive components. The charter of this project is to develop those processes and methods. Emphasis is placed on deriving designer usable test data and models from industry-ready standardized test methods. Since this work is concerned with developing processes and not simply evaluating specific materials for specific applications, only a few materials have been selected and will be subjected to the entire method and process development. After completion of this step, the processes, methods and standards developed will then be verified using other materials. The materials used for the initial phase of this program are: one urethane based adhesive; one epoxy based adhesive; one structural reaction injection molded (SRIM), glass-fiber reinforced urethane composite; a standard E-coated steel; and a standard aluminum alloy. The adhesives are experimental and are being developed and refined by two industry suppliers. The SRIM composite is made with an experimental resin developed by a supplier and the steel and aluminum are standard industry stock.

Experimental Needs

Polymer based composites have historically found their greatest usage in the aerospace and military markets. These industries have expended tremendous resources in developing test methods and test standards for material evaluation and selection. Due to the high performance environments that structural composites were subjected to in these applications and the low factors of safety that were allowable, the test methods were highly involved (and thus expensive) and highly specific to the end use application of the material under evaluation. As a result, when one surveys current aerospace and military industry standards, it becomes apparent that there are so many individual standards for arriving at a specific material property that is fair to say that there are few standardized standards.

An example of this that the one of the authors recently noted was experimental data being derived by more than 10 members of a consortium involved in basic composite materials research. Each company had it's own set of standards for measuring certain material properties. The member of the consortium who was responsible for consolidating the data had a nearly impossible task in drawing conclusions due to the differences in the test methods that were employed.

In the early part of this century the metals industries were forced in adopt a single set of standards. This was due primarily to the limited number of steel producers and the size of their industry. When those producers decided to use a set of standards (ASTM standards) for reporting data, the rest of the industry and other related industries had to follow suit. The structural composites industry has not developed with the same limited number of mega-producers.

High production rate consumer goods industries, such as the automotive industry, cannot bear testing costs in the same manner as the aerospace and military equipment industries. They cannot afford the time required for full-scale, multi-year prototype testing of each material before making material selection and moving into production. High production rate consumer goods also have a greater variability in material properties from one batch of material to another, or from one location in a component to another due to the increased rate of productivity and the need to use less expensive composites. Additionally, consumer goods tend to be made from random chopped or swirled fibers where the aerospace industry relies more heavily on laminates and uni-directional fiber placement. All of these factors point to the need for testing standards that cater to the needs of these industries and can be performed at a cost and schedule that is within acceptable limits.

After consultation with members of the automotive industry, it was determined that standardized and automated test methods need to be developed for the evaluation of composites joined by adhesive bonding. The single-lap shear strength values that are currently employed yield a qualitative comparison between adhesively bonded joints, but do not produce specific material property values that an engineer can use in designing structural components of an automobile.

The first issue to be tackled in this effort is the development of standards and methods for accurately and precisely predicting the fracture behavior of adhesively bonded joints. The standards must yield designer-usable fracture toughness numbers. Three fracture modes are being considered: opening mode, in-plane shear and mixed opening and shear. The opening or Mode I test method is nearly finalized, and the other two test methods have received preliminary consideration. Before development of the test procedures, certain material use and specimen fabrication issues had to be resolved.

Materials Issues

When the first composite samples were bonded using the epoxy based adhesive, the composite blistered and the adhesive "blew" out of the joint during the adhesive curing cycle, yielding warped samples and joints with little adhesive on the interior. After some quick analysis, the source of the problem became readily apparent.

The composite resin is a polyisocyanurate which has a large affinity for absorbing atmospheric moisture. Upon heating at 150°C (the adhesive cure temperature), the absorbed moisture was constrained from escaping into the atmosphere due to microscopic, localized, thermal constriction of the voids and capillaries in the composite. This allowed sufficient pressure to build inside the composite to produce blistering. Similarly, the thixotropic adhesive was constraining the surface and subsurface moisture from escaping due to its high viscosity. As heating progressed, the adhesive's viscosity decreased, and the steam pressure increased until the adhesive was literally blown out of the joint by the escaping gas. This resulted in a large percentage of disbonds in the joint.

The obvious solution to these problems was to eliminate the water before bonding. After an extensive series of tests, it was determined that a 48 hour, 101°C pre-drying treatment would remove more than 95% of the absorbed moisture. Twelve inch square material plaques were then bonded using this composite pre-drying treatment prior to application of the adhesive. After the 45 minute, 150°C adhesive curing cycle, no material problems were noted. In other efforts to reduce the drying time by boosting the drying temperature, it was also determined that 125°C was the highest temperature that the composite could be subjected to for extended periods of time (> 4 hours) without suffering degradation.

The evaluation of the effect of drying time, drying temperature and moisture content on the mechanical strength of adhesive joints was the final step in this evaluation. Single lap shear samples were used to obtain an idea of the relative quality of adhesive joints prepared by pre-drying at two different temperatures for different lengths of times. Samples were prepared by pre-drying one batch of samples at 101°C and

a second batch of samples at 125°C. Drying times for each batch of material were 1, 2, 3, 4, 8, 16, 24, 36 and 48 hours. After drying, single lap shear plates were bonded using the epoxy adhesive (30 mil bondline thickness) and cured for 45 minutes at 150°C. For comparison, a third set of samples were prepared that had undergone no pre-drying treatment. Next, the plates were sectioned into one inch wide lap shear samples which were tested in a conventional Instron using a crosshead speed of 1.27 mm/min. (0.05 in./min.) All samples failed by composite fiber pullout and fiber tear.

Figures 1, 2 and 3 show typical load displacement curves for samples dried for 3, 16 and 48 hours, respectively. From these curves it is apparent that the slope of the "elastic" (polymers are not truly elastic) curve is approximately the same regardless of the drying treatment. Drying the composite at 101°C tends to produce a slight decrease in the apparent "yield strength" of the joints when compared to samples not dried. Increasing the temperature further to 125°C produces and even greater decrease in the apparent "yield strength".

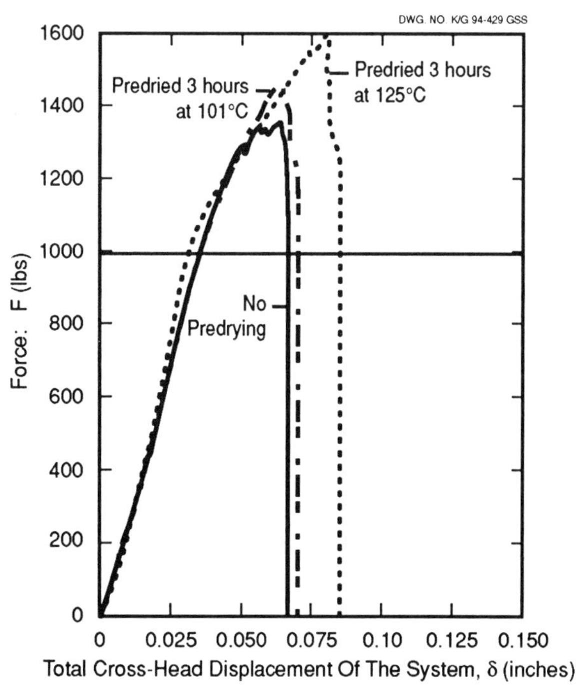

Figure 1. Load vs. Displacement for Composite/Adhesive Single Lap Shear Samples Pre-Dried for 3 Hours.

While decreases in apparent "yield strength" are noted with increasing drying temperatures, the opposite effect is seen on the "ultimate strength" of the samples. Drying the samples at 101°C produces an increase in "ultimate tensile strength" of the joint, and boosting the drying temperature to

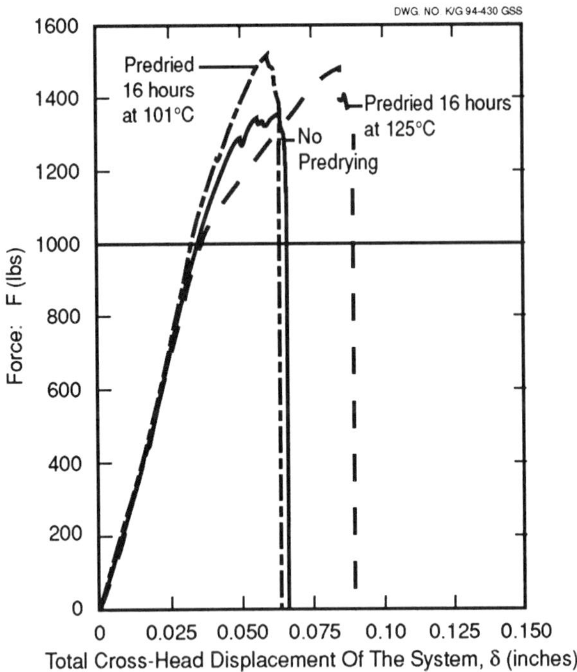

Figure 2. Load vs. Displacement for Composite/Adhesive Single Lap Shear Samples Pre-Dried for 16 Hours.

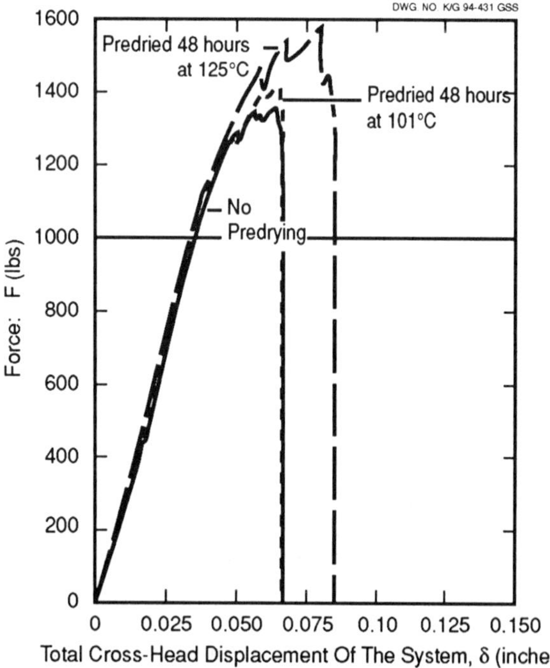

Figure 3. Load vs. Displacement for Composite/Adhesive Single Lap Shear Samples Pre-Dried for 48 Hours.

125°C further increases this system property. The total crosshead displacement, and thus system deformation of the joint, was approximately the same between samples dried at 101°C and those not dried. Samples dried at 125°C had a significantly increased plastic zone which indicates that the composite may have been annealed by the drying treatment.

In conclusion, subjecting the composite to a 101°C drying treatment reduces the apparent "yield strength" but increases the "ultimate strength". Increasing the temperature to 125°C further exaggerates these changes. Using the higher temperature drying treatment also increases the energy absorbing ability of the joint, but at the expense of lowering the apparent "yield strength". Regardless of whether or not a drying treatment is used, satisfactory bonds can be formed with this material combination. By satisfactory it is meant that the strengths of the adhesive and the adhesive/substrate interface exceed the strength of the composite in the near interface region.

Mode I Test Development

Mode I fracture toughness is a mechanical property that defines a material's resistance to crack propagation for a crack acted upon by tensile forces directed normal to the crack surface. The typical test specimen for adhesively bonded joints, the uniform double cantilever beam (UDCB or DCB), is the subject of ASTM Standard Practice D3433.[8] The standard was developed for testing adhesive joints with metallic adherends, but has gained broader acceptance including the determination of the fracture toughness for laminated composites. It has been demonstrated to work quite well for aerospace-grade composites.

Of interest here however, are bonded joints in which the adherends are an automotive-grade composite. Specifically, the composite is made SRIM panel made with a polyisocyanurate resin and randomly oriented continuous glass strand mats. It is a low-cost and rapid process that results in a composite having a higher void content and a lower fiber volume fraction than the typical aerospace-grade composites (Figure 4). Furthermore, due to the randomness of the fiber placement, the uniformity is significantly less than "high-tech" composites resulting in random zones of high-fiber content and resin-rich pockets.

The applicability of DCB testing practices, as typically found in the literature, was investigated with specimens made by bonding two 3 mm (0.125 in.) thick SRIM panels with an epoxy to form a 0.75 mm (30 mil) bondline. Specimens 25.4 mm by 241 mm (1 in. by 9.5 in.) were machined from the bonded panels after the adhesive was cured for 1 hour at 150° C. Hinges for load introduction were bonded to the sample with Hysol®XEA 9359.3 structural adhesive. The specimens were loaded in a 5 kN (1000 lbs.) electromechanical testing machine with a cross-head speed of 5 mm/min.(0.2 in./min.).

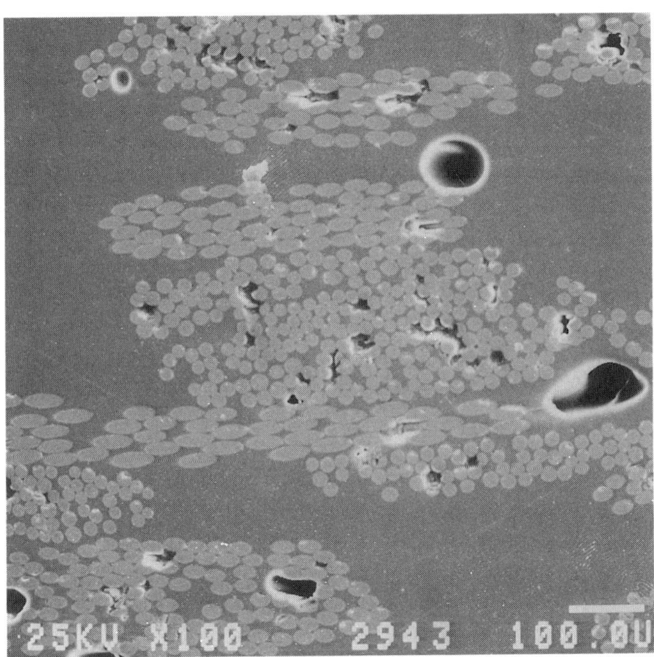

Figure 4. Micrograph illustrating the variability of the SRIM composite. Fibers are concentrated in bundles which are randomly dispersed with resin-rich pockets. These panels exhibit a high degree of porosity in the matrix as well as in the fiber bundles. (magnification=100x)

Figure 5. SRIM composite adherends failed prematurely due to excessive bending during traditional DCB test.

Crack extension in the adhesive was preempted by damage accumulation in the composite adherends resulting in one of the specimen's arms failing prematurely due to bending as shown in Figure 5. As a result, fracture toughness measurements were not possible, and it was determined that the "standard" DCB geometry was not appropriate for these materials.

In order to conduct a successful fracture toughness test for bonded joints with these SRIM composite adherends, a modified specimen is required. Since the adherends fail prematurely due to excessive bending, it is concluded that stiffening the adherends by bonding on "backing-beams" would be beneficial. Although an earlier paper by the authors [9] espoused the backing-beam concept for reasons discussed below, it appears to be necessary to prevent damage for these materials. Others have used this approach also. Byun, Gillespie, and Chou [10] reported the use of aluminum backing-beams for three-dimensional fabric composite DCB specimens. River and colleagues [11,12] used aluminum and wood backing-beams to test wood DCB specimens. Whitney and Short [13] used steel backing-beams for similar reasons to test the interlaminar shear strength of graphite/epoxy composites using a modified Short Beam Shear (SBS) specimen.

Backing-Beam Concept. The motivation for this work is to develop test procedures that would help resolve theoretical and experimental issues dealing with specimen design and data reduction and be valid for a wide range of both adherend and adhesive properties using standardized geometries, sizes, fixtures and procedures. To that end, a contoured shape as developed by Mostovoy and colleagues [14-16] was employed for the backing-beams as shown in Figure 6. The Mostovoy specimen, the height-tapered double cantilever beam (HTDCB) is also the subject of ASTM D3433. Employing backing-beams with the Mostovoy contour has advantages for the following reasons:

Small Displacements. In many applications of the DCB, large displacements of the cantilever ends are encountered. This introduces two primary error sources that must be accounted for in the analysis of the results. Firstly, large deflections cause an effective shortening of the cantilever. Secondly, if end blocks (rather than hinges) are used to introduce the load and if deflection is measured at the load-line then end block rotation reduces the deflection. Correction factors can be applied to account for these effects. As a practical testing matter, the correction factors are troublesome, but correction factors can be circumvented by incorporating the backing-beam concept. With this concept the deflections are governed by the stiffer backing beam thereby limiting the deflection to acceptably-small values. In addition, since the backing beams provide the majority of the overall stiffness, the deflections from tests with a wide range of adherend stiffnesses will exhibit a much narrower range avoiding the need to change the test setup for the variety of different adherends of interest to the automotive industry.

Crack Length Measurement. The HTDCB test is designed such that the determination of the strain energy

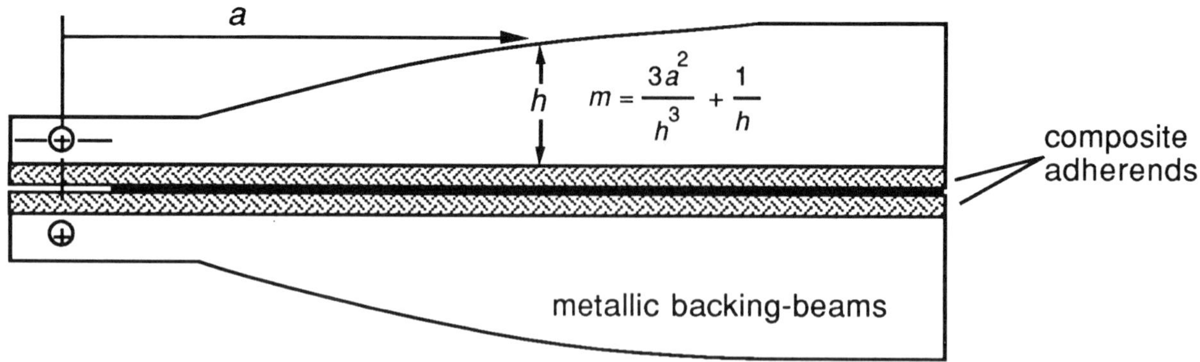

Figure 6. Backing Beam Concept using the Mostovoy Contour.

release rate is independent of the crack length — it is only necessary to measure the load required to drive the crack. In the present case the stiffness of the backing beam is modified by the adherend, and thus crack length independence is lost. If however, the stiffness of the backing beam dominates the overall stiffness, then the toughness should become only weakly dependent on the crack length. Thus the sensitivity of the experimental results to errors in the measurement of the crack length has been minimized. This is a particularly desirable feature when the crack length varies through the width of the specimen.

Anticlastic Curvatures. It has been reported [17] that thin (perpendicular to the crack surface) adherends develop anticlastic curvatures. As a result, strong width-variations of the strain energy release rate develop. By bonding the backing beams to the specimen, it is expected that the curvature and the subsequent variation in the strain energy can be significantly diminished. It is further believed that this would result in crack lengths that are more uniform through the thickness.

Experiments. Backing-beams, 12.7 mm (0.5 in.) wide by 254 mm (10 in.) long with a contour parameter m = 3.543 1/mm (90 1/in.), were machined from 17-4PH stainless steel. SRIM panels (approximately 3 mm thick) were bonded with an experimental epoxy to form a 0.75 mm (30 mil) bondline with an inserted Teflon film to serve as a crack initiator. Composite-epoxy-composite specimens, 12.7 mm (0.5 in.) by 241 mm (9.5 in.), were machined from the panel after the adhesive was cured at 150°C for approximately 1 hour. The backing beams were then bonded to the joint with 3M AF-163-2 film adhesive. When cured, the specimens were loaded in an electromechanical testing machine under displacement control with a cross-head speed of 2.54 mm/min (0.1 in./min). Data acquisition equipment was used to collect and process the data in real-time.

Under these conditions this adhesive exhibited the "run-arrest" response indicative of rate-sensitive adhesives as shown in Figure 7. Neglecting the stiffness of the composite, the initiation and arrest fracture toughnesses, G_{1c} and G_{1a}, are

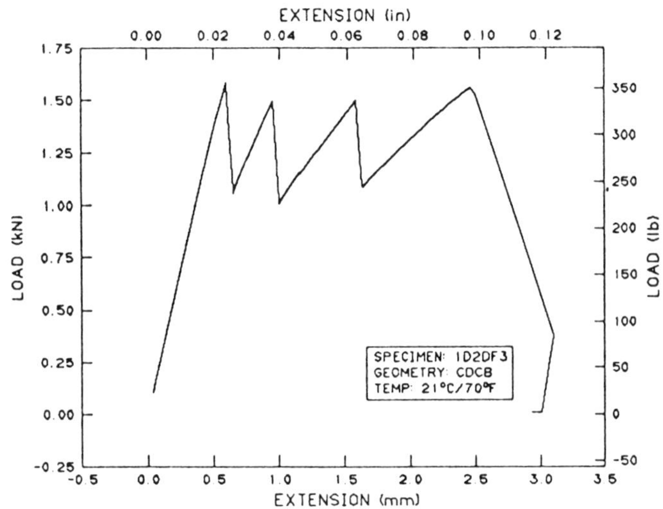

Figure 7. Load-deflection curve for composite/adhesive/composite specimen tested using backing-beams. The sawtooth nature of the curve indicates a run-arrest response.

calculated by

$$G_{1c} = \frac{4L^2(\text{max})\,m}{EB^2}$$

$$G_{1a} = \frac{4L^2(\text{min})\,m}{EB^2}$$

where L(max) is the load at crack initiation, L(min) is the load at crack arrest, E is the modulus of the backing-beam, and B is the width of the specimen and backing-beam. For the data plotted in Figure 8, the average initiation toughness was 993

478

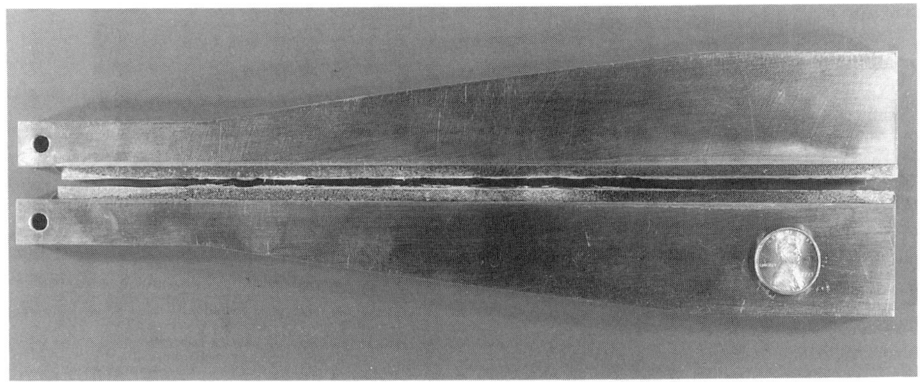

Figure 8. The failure is constrained to the near interface region when employing the backing-beams.

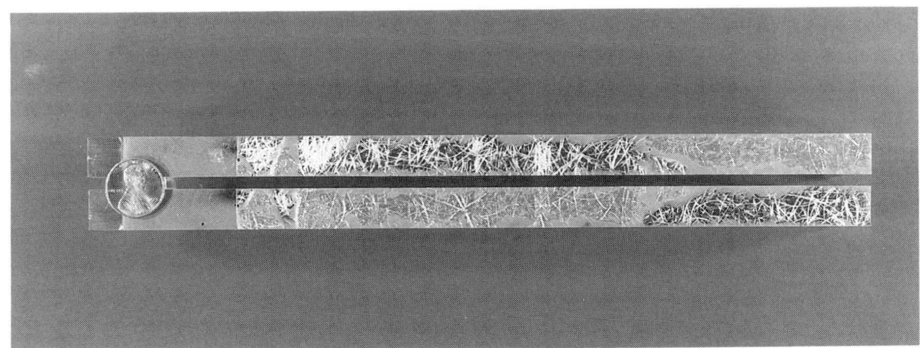

Figure 9. The failure exposes bare fibers indicating that the fiber/matrix interface may be the dominate factor in the failure process.

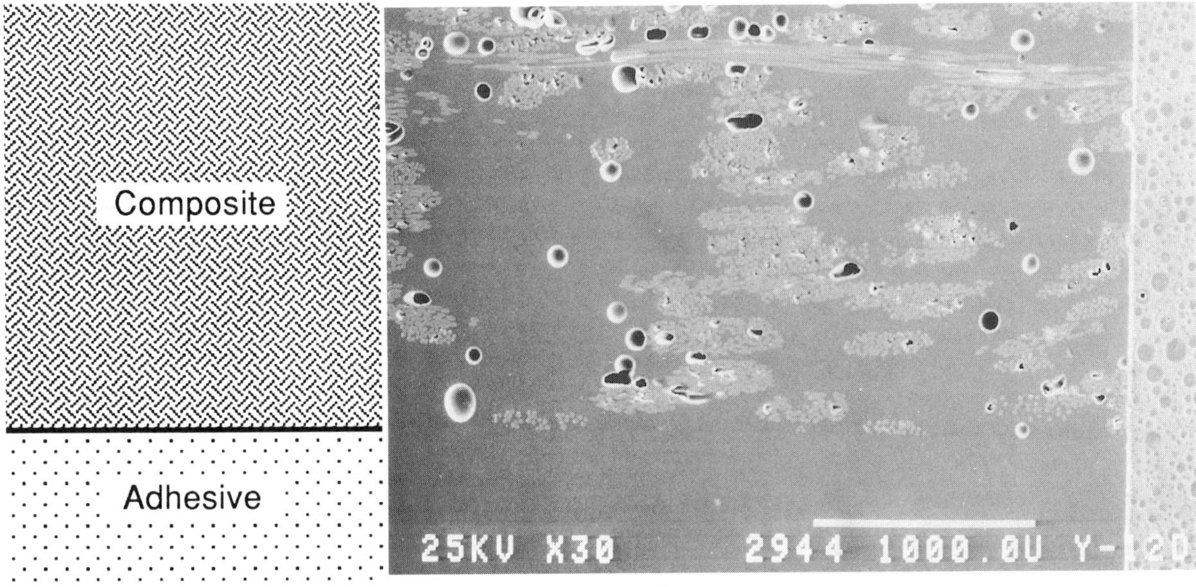

Figure 10. Micrograph indicating a very good composite/adhesive interface. Note the porosity, fiber-bundle concentrations and the resin-rich pockets. (magnification=30x)

J/m^2 (5.67 in-lb/in^2), whereas the average arrest toughness was 471 J/m^2 (2.69 in-lb/in^2). A side view of the failed specimen is shown in Figure 8, and the failure surface is shown in Figure 9. Note that the adhesion to the adherend is excellent (Figure 10), and that the failure generally takes place in the composite. This often exposes the glass fiber in areas of high fiber content near the adhesive/composite interface indicating that the fiber/matrix interface may dominate the response. It has been observed that in some specimens the "crack" location changes from the near interface in one adherend to the near interface in the other. It is hypothesized that this is because the crack follows a path that takes it to the "weakest" interface, that is the interface with the highest local concentration of fibers near the interface. Figure 5 certainly illustrates the variability of the fiber content near the interface and thus leads credence to the hypothesis. It is also quite probable that the distributions of voids, particularly near the high fiber content zones, affects the path of the "crack."

Future Work. In the experiments described in the previous section, the contribution of the composite adherends to the specimen compliance was neglected requiring only the load to be known to calculate the toughnesses G_{1c} and G_{1a}. This is an approximation. In future work the crack length and compliance as a function of the crack length will be measured and used to determine G_{1c} and G_{1a}. Tests will also be conducted on specimens where the adherends are an E-coated steel. The entire test method will be automated including the crack length measurement. Then the complete process will be repeated for Mode II (shear mode) and Mixed-Mode (opening and shear) test development. Throughout the process, analytical and numerical studies will be conducted to assess the advantages of the backing-beam concept and to define optimal configurations.

Conclusions

Test methods for determining the Mode I fracture toughness of adhesive joints containing automotive-grade SRIM composite adherends were developed. Standard double cantilever beam techniques were found to be inadequate because the adherends failed prematurely due to excessive bending. A backing-beam concept was successfully employed to prevent the adherend failures. Very good adhesion between the epoxy adhesive and the polyurethane, resin-based composite adherends was achieved. The failure surface was observed to expose the glass fibers in the composite near the adhesive/adherend interface indicating a weak fiber matrix interface as a leading factor in the failure. The failure was observed to randomly jump from one interface region to the other, and it was hypothesized that the "crack" followed the path toward the highest local concentration of fibers near the interface.

Acknowledgements

The authors wish to thank the following individuals for their diligent effort in assisting with this project: Felix Paulauskas, Fahmy Haggag, and Ronny Lomax of Oak Ridge National Laboratory; Thomas Meek, University of Tennessee; Dallas Smith, Tennessee Technological University; Carl Weber, BF Goodrich Adhesives and the entire crew of the ACC's Joining Group.

This project is sponsored by the U. S. Department of Energy, Office of Transportation Materials, Lightweight Materials Project. Oak Ridge National Laboratory is operated by Martin Marietta Energy Systems Inc. under contract DE-AC05-84OR21400.

References

1. Baxter, D. F., "Plastics Beat the Heat in Underhood Components", *Advanced Materials & Processes*, May (1990) pp. 36-41.

2. Gjostein, N. A. "Automotive," *Advanced Materials & Processes,* January (1990) pp. 73-74.

3. Reindl, J. C. "Commercial and Automotive Applications," pp. 832-835.

4. Beardmore, P., "Automotive Components: Fabrication," pp. 24-31.

5. McConnell, V. P., "In the Fast Track: Composites in Race Cars," *Advanced Composites*, March/April (1991) pp. 23-35.

6. Hunt, M., "Automotive Aluminum," *Materials Engineering*, May (1991), pp. 26-29.

7. Baker, H., "The Road Ahead for Metals in Autos," *Advanced Materials & Processes*, May (1990) pp. 27-34.

8. "Standard Practice for Fracture Strength in Cleavage of Adhesives in Bonded Joints," D 3433-75 (Reapproved 1985), *Annual Book of ASTM Standards,* American Society for Testing and Materials, Philadelphia.

9. Warren, C. D., R. G. Boeman, and F. L. Paulauskas, "Adhesive Bonding of Polymeric Materials for Automotive Applications," *Proceedings of the Annual Automotive Technology Development Contractors' Coordination Meeting 1993*, Dearborn, MI, October 1993, pp. 1-11.

10. Byun, J.-H., J. W. Gillespie, and T.-W. Chou, " Mode I Delamination of a Three-Dimensional Fabric Composite," *Journal of Composite Materials*, May 1990, pp. 497-518.

11. Scott, C. T., B. H. River, and J. A. Koutsky, "Fracture Testing Wood Adhesives with Composite Cantilever Beams," *Journal of Testing and Evaluation*, Vol. 20, No. 4, July 1992, pp. 259-264.

12. River, B. H. and E. A. Okkonen, "Contoured Wood

Double Cantilever Beam Specimen for Adhesive Joint Fracture Tests," *Journal of Testing and Evaluation*, Vol. 21, No. 1, January 1993, pp. 21-28.

13. Whitney, J. M., and S. R. Short, "A Modified Short Beam Shear Test," *Proceedings of the American Society for Composites, Second Technical Conference*, 1987, pp. 3-8.

14. Mostovoy, S., E. J. Ripling, and C. F. Bersch, "Fracture Toughness of Adhesive Joints," *Journal of Adhesion*, Vol. 3, 1971, pp. 125-144.

15. Ripling, E. J., S. Mostovoy, and H. T. Corten, "Fracture Mechanics: A tool for Evaluating Structural Adhesives," *Journal of Adhesion*, Vol. 3, 1971, pp. 107-123.

16. Mostovoy, S., P. B. Crosley, and E. J. Ripling, "Use of Crack-Line-Loaded Specimens for Measuring Plain-Strain Fracture Toughness," *Journal of Materials*, Vol. 2, No. 3, September 1967, pp. 661-681.

17. Crews, J. H., Jr., K. N. Shivakumar, and I. S. Raju, "Strain Energy Release Rate Distributions for Double Cantilever Beam Specimens," *AIAA Journal*, October 1991, pp. 1686-1691.

Selection of Adhesive System for Radio-Frequency Heating of Structural SMC Composites

G.W. Malaczynski
General Motors North American Operations
Warren, Michigan

G.J. Cinpinski
General Motors Corp.
Warren, Michigan

Abstract

A series of laboratory tests and computer simulations of adhesive curing enhanced by radio-frequency heating indicated that not all adhesive systems can be effectively used for a RF-based curing process due to the danger of thermal run-away. Our investigation of the process mechanism also resulted in several other findings which could lead to further improvement of the processing cycle. Both experimental and theoretical results indicated a very high sensitivity of the adhesive exothermal reaction kinetics to the thickness of the bond line, which varies with changes of the thickness of the processed sheet molding compound (SMC) components. Thus, regardless of the type of adhesive system employed, tighter dimensional tolerance would insure better quality of the bond. Finally we proved that the initial surface temperature of the SMC members substantially affects the process effectiveness and joint quality, and should be strictly monitored.

IT IS GENERALLY KNOWN that the manufacturing cycle for SMC assemblies should not exceed one minute to be cost effective. To meet the time constraint, the thermoset adhesive curing station may augment the conventional heat diffusion from the preheated tool base with dielectric heating (RF irradiation). The dielectric heating, which is an almost adiabatic process [1], is applied directly into the bond line.

Conventional curing of a thermoset adhesive applied to the joint of a plastic structure is a slow process. It relies on heat conduction from the hot press platens to the joint through the material of the components being bonded together (Fig. 1). The thicker the material being cemented, the slower the process becomes. In fact, ignoring the additional heat evolved by the thermosetting process, the time required to cure the joint can be estimated as the characteristic diffusion time, τ_d, derived from the heat diffusion equation:

$$\tau_d = d^2 \rho C_p / \kappa_T \tag{1}$$

where d is the distance from the heat source (platen) to the center of the joint line, and ρ, C_p, and κ_T are the material density, specific heat, and thermal conductivity, respectively.

High frequency electromagnetic energy, with its vastly superior penetration through an insulating material, offers a unique opportunity for enhancing the rate of temperature buildup in the adhesive, thereby accelerating the bonding process (see eg.[2]). Unlike conventional heating, the electromagnetic energy is absorbed throughout the volume of the adhesive layer and, thus, gives rise to a volumetric heat source in the thermoset material [1,3]. However, heat diffusion cannot be totally ignored, since the temperature buildup inside the adhesive will be reduced by the escape of heat into the cold structures being bonded together. To some extent this heat loss can be prevented if the structure is preheated or if it happens that the structure also heats up together with the adhesive due to electromagnetic energy absorption [2].

The most important factor reflecting heat loss at the joint, which may degrade the kinetics of the thermoset reaction, is the ratio of the adhesive layer thickness to the area of the joint, since electrical energy dissipation is a volumetric process while heat conduction depends on the temperature gradient and contact area [2]. Finally, the amount of electric power dissipated may depend on the degree of curing. In other words, the heating process will terminate if the solid bond structure created no longer absorbs any electric energy. However, if the cured adhesive still responds well to the electromagnetic field a self-destructive thermal run away may occur. The conversion of electromagnetic energy into heat, heat conduction, and the kinetics of the chemical reaction of a specific adhesive system, must all be included in a valid model of the curing process.

Mathematical Model

To analyze the time varying temperature distribution produced by internal heating caused by electric-to-heat energy conversion in electrically lossy material (i.e. dielectric heating) enhanced by the exothermal chemical reaction of the adhesive system, the initial and boundary conditions for heat diffusion must be formulated together with a mathematical description of the heat sources involved. The complexity of the problem, in our case, is somewhat reduced since the geometry of the joint allows for a two-dimensional approximation. This is well justified since the third dimension of the joint in most practical assemblies is much bigger than the dimensions of the cross section in the plane perpendicular to the bond line (see Fig. 1). Therefore, the time varying temperature distribution in the joint cross section will be described by a two-dimensional diffusion equation with spatially distributed heat sources, $P_E(x,y,t)$, representing the power density induced by electromagnetic irradiation, and $\partial Q/\partial t$, the heat density rate, representing the exothermal polymerization reaction within the adhesive.

$$\rho C_p \partial T/\partial t = \kappa_1(x,y,t)\partial^2 T/\partial x^2 + \kappa_T(x,y,t)\partial^2 T/\partial y^2 + \partial Q/\partial t + P_E(x,y,t) \qquad (2)$$

P_E, the time average electric power dissipated per unit volume can be expressed in terms of the dielectric properties of involved materials, frequency f, and strength, E, of applied electric field (root mean square value) [1]:

$$P_E = 2\pi f \varepsilon_0 \kappa' E^2 \tan\delta \qquad (3)$$

where both, relative dielectric constant κ', and electric power dissipation factor, $\tan\delta$, are time dependant since the dielectric properties of all plastic components involved vary with temperature. In addition, the adhesive dielectric properties usually vary with the degree of cure (see eqs. 4 and 5). The temperature dependance of SMC component of the system was ignored for the sake of simplicity of calculations, while the adhesive response appeared to be an important factor in the system behavior (see Fig. 3) and was considered in our calculations. The dependance of dielectric properties on frequency of applied electric field, E, well known from

numerous experiments and widely described in the literature (see e.g.[4,5]), is not explored in our work since, the Federal Regulations limits selection of frequencies to 13.56MHz or 27.12MHz which are allocated for industrial applications.

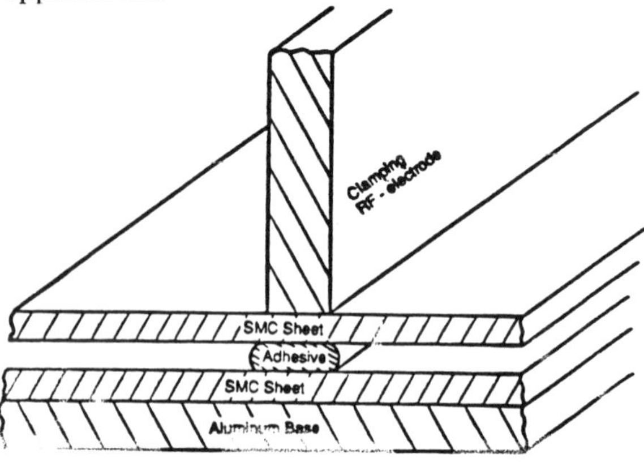

Fig. 1 Geometry of the joint in the clamping fixture.

The evolving heat of the thermoset reaction is described by an equation of state since the degree of curing, α, determines the instantaneous amount of heat being released. Since $\partial Q/\partial t$ depends explicitly on the degree of cure, it is a path-dependent function of temperature. Therefore, the degree of cure can be defined as the ratio of the heat of cure generated, Q, to the total heat density of the exothermal reaction, Q_T, that will eventually be generated

$$\alpha = Q/Q_T \qquad (4)$$

The empirical input data obtained from differential scanning calorimetry (DSC) experiments defines the form of the cure rate equation [6,7,8]. To demonstrate differences in the behaviors of various adhesive systems, the DSC results for 320/322 epoxy adhesive and Tyrite 7525 A/E urethane adhesive were selected for further considerations. The DSC routine was supported by the conversion of experimental results into specification of the set of reaction kinetic coefficients, provided by the vendor of the adhesives [9], following the Borchardt and Daniels representation [9] of the exothermal reaction:

$$d\alpha/dt = k(T)[1-\alpha]^n \qquad (5)$$

where n is the reaction order and k(T) is the specific rate constant at temperature T.

The Borchardt and Daniels approach assumes the Arrhenius behavior:

$$k(T) = Z\, e^{-E/RT} \qquad (6)$$

where E - activation energy [J/mole]

    Z - pre-exponential factor or Arrhenius frequency factor [1/min]

    R - gas constant = 8.314 J/mole-K

The specific rate constant and reaction order can be calculated given the standard DSC experimental results (routine data reduction) which provide the pre-exponential factor, activation energy, and total heat of reaction, Q_T.

Due to the nonlinearity of the equation set (2),(5) the temperature and degree of cure development in time cannot be obtained in an analytical way. With specified boundary and initial temperature conditions (see Fig. 6), however, the solution can be found with the assistance of the TWODEPEP finite element computer program (IMSL software library).

Experimental Results

It can be easy demonstrated that the temperature buildup in the bond line depends on the thickness of the adhesive exposed to electromagnetic irradiation [2]. The inherent heat losses from the thin adhesive layer dramatically delays the curing process unless they are compensated by a higher dissipated electric power density or by preheating the structural members of the joint. The varying thickness of the bond line resulting from the large allowed tolerance of the thickness of the bonded SMC may result in degraded cure conditions due to a reduced temperature rise within the adhesive. To the contrary, if the RF energy is increased to compensate for the heat escape, the bond line may experience overheating. Typically, but not always, a partially cured adhesive is less responsive to electromagnetic irradiation than an uncured one, i.e., the adhesive response, if designed for RF treatment, is self-limiting. However, if a polymerized adhesive still responds well to an electromagnetic field, the overexposure may lead to a weakened joint due to overbaking or even total self-destruction of the joint. Consequently, the first step in our assessment of the existing routine was to establish how the SMC thickness tolerance affects the peak temperature experienced by the adhesive exposed to a given dose of electromagnetic irradiation.

With constant tool clamping distance, the specified tolerance in the SMC thickness converts directly to a variation in bond line thickness. To assess a potential sensitivity of the temperature buildup in the bond line due to this adhesive volume variation, one may run a simplified computer simulation which does not fully account for the timing sequence of assembling SMC panels on a real production line. For that purpose it was assumed that one SMC sheet is kept in contact with the preheated tool's base temperature which is 65 °C, the temperature of the adhesive is 32 °C, and the temperature of the other SMC sheet is 37 °C. A more detailed accounting of the actual time and temperature sequence which may be encountered on the production line will be covered later in this paper. Finally, it was assumed that a one-dimensional model provides fairly accurate representation of the processes, at least for the initial evaluation of the system response and would eliminate unnecessary investigation of procedures having only minor impact on the final product performance.

Computer simulation results shown in Fig. 2a and 2b indicate the following:

1) - The allowed (arbitrarily) tolerance for the SMC sheets leads to substantial variation of the temperature-time profile within the adhesive which, in our opinion, does not allow for a consistent processing of the joint in a statistical sense.

2) - In the extreme case when both SMC sheets are 0.75 mm thick and the bond line thickness is at its maximum, the temperature inside the adhesive may reach the upper limit, and, in the event of increased RF exposure time, the system may be overbaked and loose its expected bonding ability. In the other extreme, when the volume of the adhesive is at its minimum, it may happen that some of the bonded components are undercured due to

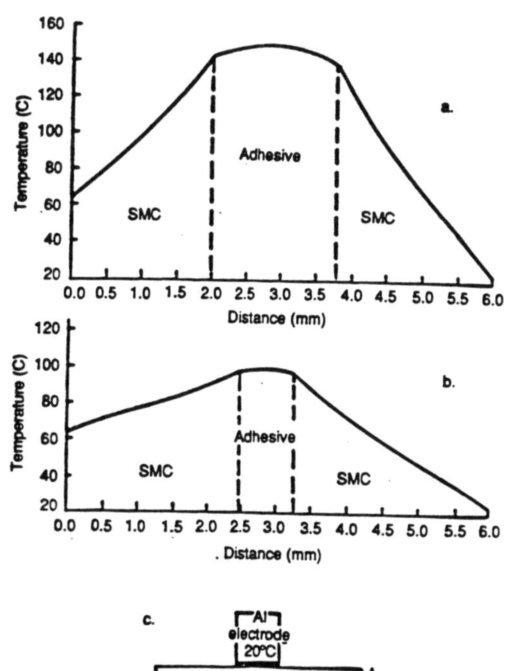

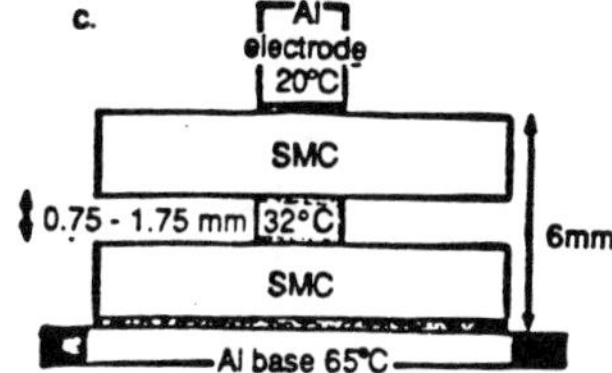

Fig. 2 Temperature distribution through the bondline cross-section (one-dimensional model) for epoxy 320/322 adhesive system

a - adhesive line thickness 1.75 mm

b - adhesive line thickness 0.75 mm

c - system geometry with marked initial temperature conditions insufficient heat generation. Attempting to compensate for this with an RF exposure would result in overbaking areas where the adhesive is over-applied because of the SMC thickness variation.

The overheating caused by RF over-exposure occurs only if the increase in degree of the resin polymerization is not followed by a gradual decrease in efficiency of electric-to-thermal energy exchange. Since, the decrease in response to dielectric heating with the degree of polymerization represents rather typical behavior of a wide range of polymers, this rather unusual response of the epoxy 320/322 system indicates that this adhesive should not be considered for processes involving dielectric heating.

Adhesive selection for bonding structural components is no doubt a very complicated, time consuming process. It involves bond line strength tests under various climatic conditions, requires guaranteed performance regardless of adhesive aging, vibrations, temperature and humidity variation, response to impacts, moisture, etc., etc. The available performance data of polyurethane adhesive Tyrite 7525A/7525E provided by Lord Corporation [9] indicated clearly that this polyurethane adhesive meets requirements expected from cemented structural composites. In addition, the lower viscosity of the Tyrite 7525 system when compared with the epoxy 320 system makes it very competitive for a mass-production application due to the less demanding construction of the adhesive feeding system.

Experimental verification of the more favorable response of the Tyrite 7525A/7525E to the RF electric field vs. the epoxy 320/322 adhesive is given in Fig. 3. The temperature at the center of each sample's volume was monitored while both adhesive samples were simultaneously exposed to an electromagnetic field of the same strength (see Fig. 4). The epoxy 320/322 adhesive clearly has a tendency to overheat when exposed to the electric field for a prolonged period of time while the

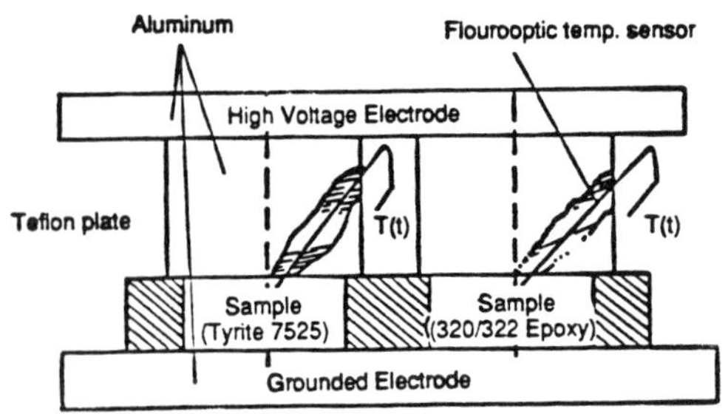

Fig. 4 Experimental fixture for simultaneous monitoring of the temperature buildup inside two different materials exposed to identical RF dose.

polyurethane adhesive response is self-controlled. Once the polymerization process takes off, the polyurethane coupling to the electric field diminishes. The thermal runaway behavior of the epoxy 320/322 adhesive disqualifies it for application when processing involves dielectric heating unless the exposure time is strictly enforced and the adhesive volume is precisely controlled by enforcing strict dimensional tolerances. To enhance the evidence provided by the side-by side exposure test described in Fig. 4, samples of the SMC material were cemented with the epoxy 320/322 adhesive. Two extreme cases of thick and thin bond line were processed in an RF field of the strength required for proper heating of the sample containing the low adhesive volume. A subsequent shear stress test (result shown in Fig. 5) clearly indicates degraded performance of the sample containing excess adhesive as occurs at the thinnest allowable SMC dimension. This is explained in the previously mentioned thermal run-away effect specific to the epoxy 320/322 system.

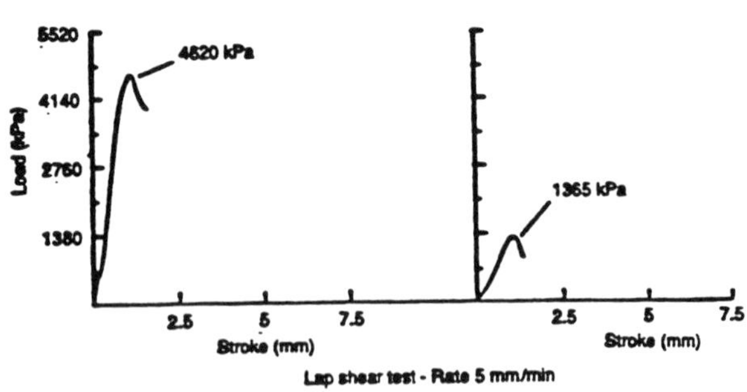

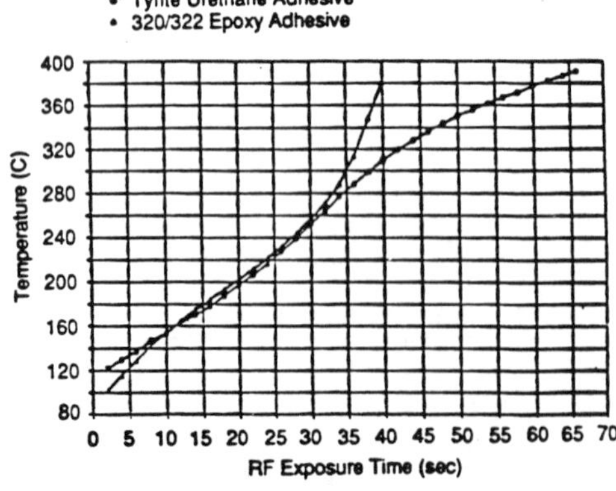

Fig. 3 Response of Tyrite 7525 and epoxy 320 adhesive systems to identical RF dose. Clearly, the response of the epoxy 320 system indicates the danger of thermal run-away.

Fig. 5 Results of the lap shear test for the sample of the material bonded with the epoxy 320 adhesive system exposed to a proper RF radiation dose (at left) and sample overheated due to excessive amount of adhesive (at right).

Computer Simulation

A full picture of workpiece sensitivity to initial temperature conditions, electromagnetic irradiation dose, and bonding operation timing can be obtained with the help of two-dimensional computer based analysis of heat generation and diffusion. It was assumed, that the thickness of the applied adhesive (Tyrite 7525A/7525E) could vary between 0.8 and 1.8 mm. The possible manufacturing routine of reinforced SMC structure was approximated by the following sequence of operations:

step 1: The outer SMC sheet which is to be bonded to its reinforcing structure is subjected to a 10 second preheating period. It is assumed that the SMC sheet is laid on a preheated aluminum base whose surface is preheated to 65.5 °C, that the applied initial adhesive temperature is equal to 32 °C, and that the reinforcing member and SMC sheet are at room temperature, 20 °C.

step 2: This workpiece, clamped between the preheated base and cold (room temperature) clamping electrode, is exposed to a radio-frequency (RF) irradiation of constant strength for a duration of 20 seconds.

step 3: A 10 second post-RF clamping period precedes the next step in which the inner to outer SMC panels are actually bonded. The initial temperature conditions for this step are those found at the end of the previous operation.

step 4: Adhesive preheated to a temperature of 32 °C is applied to the cold (room temperature) inner panel and 10 seconds later the whole assembly is clamped and exposed to the RF field in the manner described in the step no. 3.

Schematically all steps are depicted in figures 6a through 6d. At each step, new boundary conditions are entered. They are defined by aluminum base and aluminum HF-electrode temperatures which are considered to be perfect heat sinks. For the portion of the system which interfaces air, the heat convection was assumed, i.e., boundary condition can be formulated by balancing the heat flux normal to the interface, $(\mathrm{grad}T)_\perp$, with the convective heat flux which is proportional to the difference between the surface and air temperatures (Newton's law of cooling):

$$- (\mathrm{grad}T)_\perp = h(T_s - T_\infty) \tag{7}$$

It was found that the temperature buildup within the adhesive layer is practically independent of the assumed free convection coefficient, h, within the reasonable range 5-25 W/m^2 degK [10].

Calculated temperature maps at the end of each RF exposure cycle (i.e. at the end of step #2 and step #4) are shown in figures 7a and 7b for adhesive thickness at its lower limit (i.e. 0.8 mm) and in figures 8a and 8b for adhesive thickness at its upper limit (i.e. 1.8 mm).

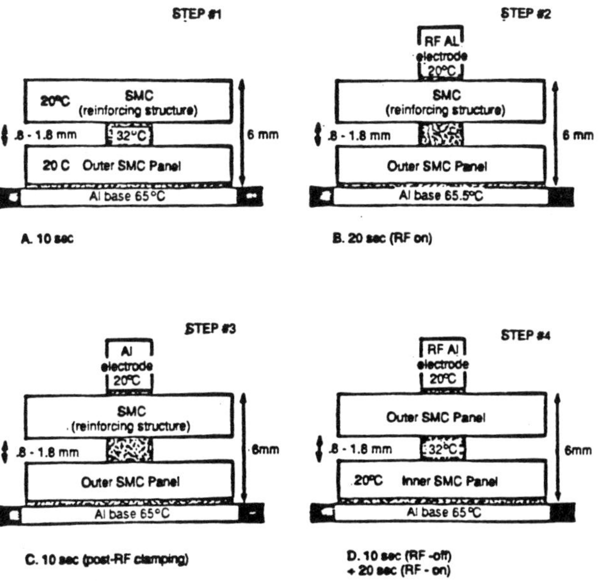

Fig. 6 System geometry and temperature initial conditions at the beginning of steps 1 through 4 described in the text.

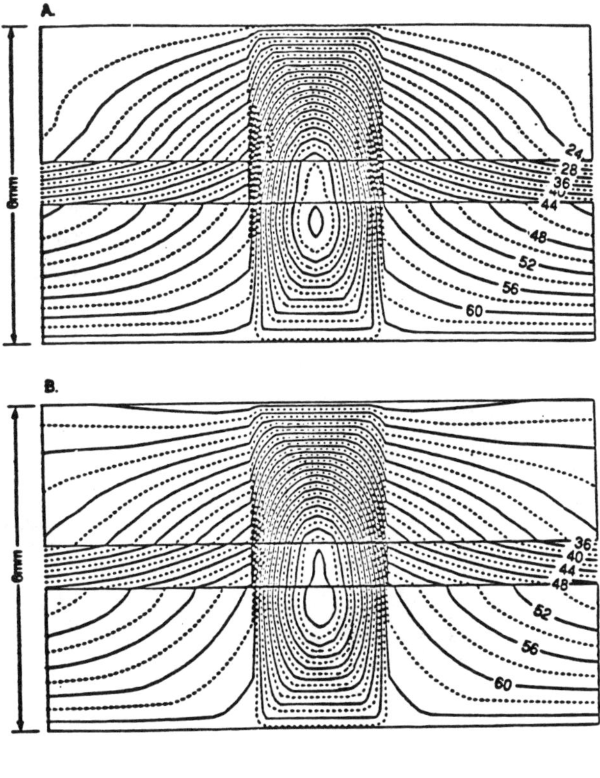

Fig. 7 Two-dimensional computer simulation of the temperature distribution in the SMC sheets, adhesive, and air gap at the end of the step #2 and step #4 for adhesive thickness in its lower limit defined by the SMC sheet tolerance.

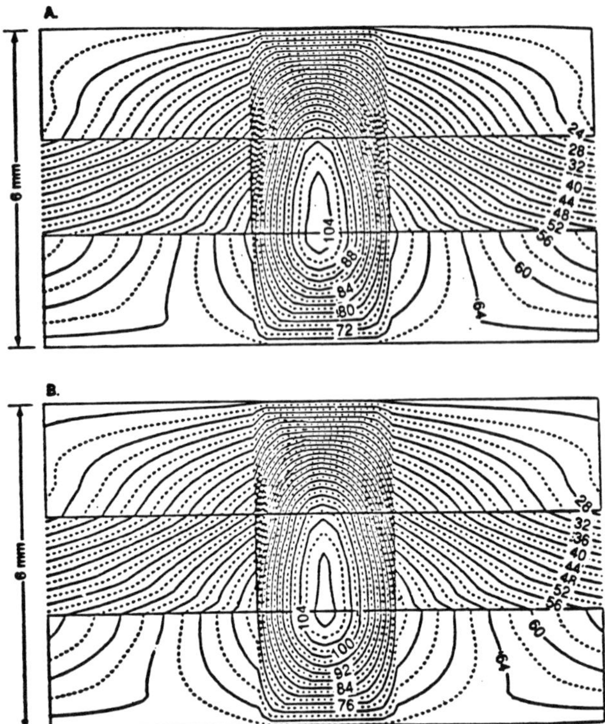

Fig. 8 Two-dimensional computer simulation of the temperature distribution in the SMC sheets, adhesive, and air gap at the end of the step #2 and step #4 for adhesive thickness in its highest limit defined by the SMC sheet tolerance.

Figures 9 and 10 represent the degree of cure at the end of the corresponding processing cycles (i.e. at the end of the step #3 and 10 seconds after the end of step #4) for thin and thick adhesive line respectively, and figures 11 and 12 represent the corresponding temperature distributions in the cross-section of the bond line. Evidently, variation of the bond line thickness together with the additional heat released by the exothermal chemical reaction of the urethane system polymerization leads to substantial differences in peak temperature and speed of cure. Therefore, even when applying the proposed improved adhesive material, the processing time may be critical. It is worthwhile to note that the process timing would depend on initial temperature of the involved components since the reaction kinetics are augmented by the elevated SMC surface temperature and the lower temperature gradients which prevent heat escape from the bond line. The significance of this effect may be appreciated by comparing the results of the outer-to-reinforcement and the inner-to-outer operations featuring different initial temperature conditions.

As results presented in Figs. 7 and 8 indicate, the outer SMC surface temperature is practically not affected by the temperature buildup inside the adhesive. The minimal temperature rise at this point is attributed to heat diffusion from the adhesive and the direct heating of

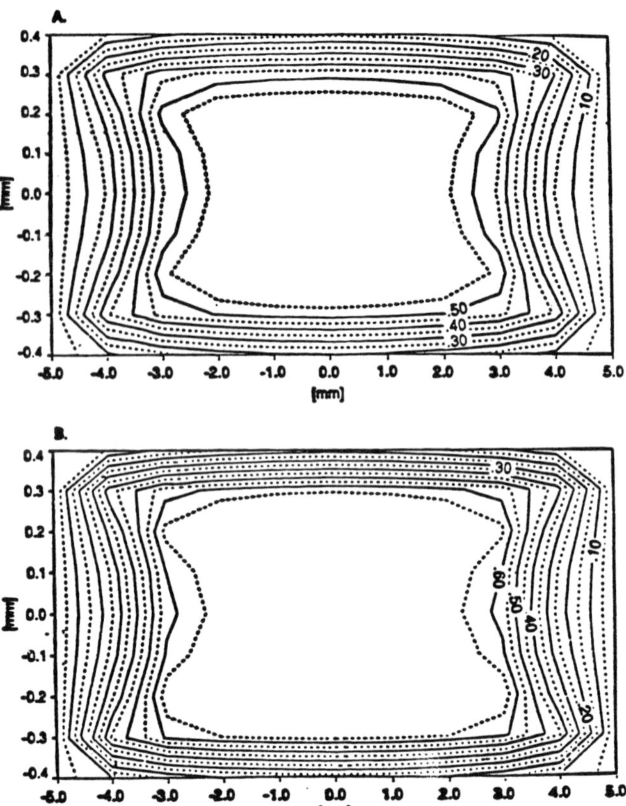

Fig. 9 The degree of cure distribution inside the adhesive for the bond line thickness at its lower limit.
a - at the end of the step #3
b - 10 seconds after the end of the step #4

the SMC in response to the RF irradiation. This response is much weaker than that of the adhesive but nevertheless not equal to zero. When considering the heat capacity of the clamping structure which tries to enforce a constant temperature at the SMC surface and the total heat which may potentially escape from the adhesive, it is anticipated that no accountable temperature rise, indicative of the heating status of the adhesive, will occur.

The situation would change however, if the interface between the SMC and the metal clamp were separated by a thermal barrier which was transparent to RF irradiation. This, in practice, may be created by introducing a teflon shim or by coating the metal electrode. Then, the temperature distribution at the end of the curing cycle would dramatically differ (see Figs. 13a and 13b and compare with Figs. 7 and 8). Now the temperature at the interface is well related to the thermal conditions inside the adhesive. Nevertheless, monitoring temperature at the clamping electrode-SMC interface is not a straightforward method for development of cure control. It requires an expensive fiberoptic temperature sensor since a regular thermocouple would not serve reliably in a strong electromagnetic field. Moreover, sensing the temperature at one point does not guarantee

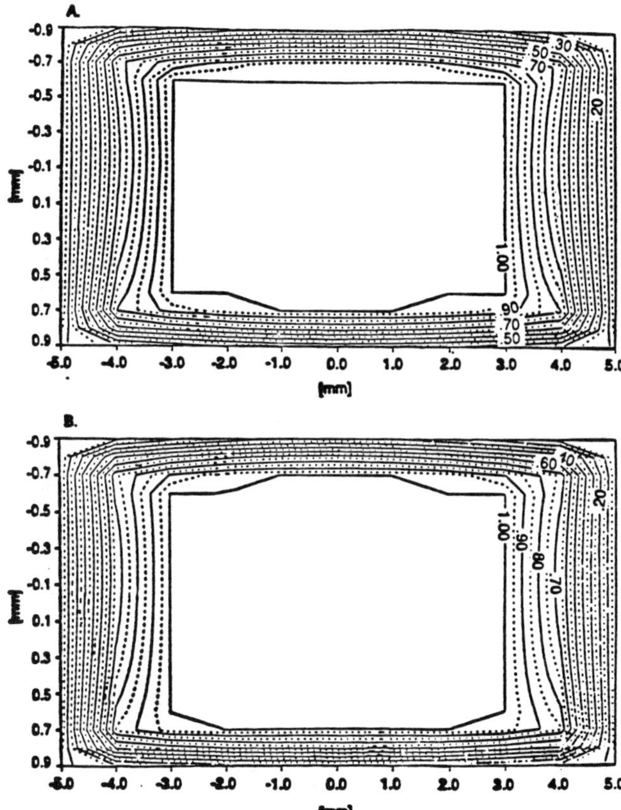

Fig. 10 The degree of cure distribution inside the adhesive for the bond line thickness at its highest limit.
  a - at the end of the step #3
  b - 10 seconds after the end of the step #4

a proper control of the process development along the bond line since the adhesive thickness may vary from point to point. Consequently, a clamping electrode temperature sensor with an effective thermal barrier (Fig. 14) may serve as a very effective, one-point monitoring device to warn of radical change in the preset processing parameters (partial loss of RF-field strength, radical difference in initial components or adhesive temperature, etc.). However, this generic feature of radical temperature redistribution which resulted from the thermal barrier (compare Fig. 7 or 8 and 13) can be utilized as a potentially effective, quick control of the cure process completeness.

One should remember that the minute the workpiece is released from the clamping fixture, the heat escape into the clamping tool ceases to exist and some time later the initial temperature distribution represented in Figs. 7 and 8 starts to reshape in the direction depicted in Fig. 13. This happens because the heat accumulation in the adhesive will escape from the bond line and reach a room temperature equilibrium in just a few minutes. This effect is clearly illustrated by following the temperature distribution development in the post-RF stage under the assumption that the heat conductive clamp was removed upon the RF-exposure termination (see Fig. 15). Now, since the whole workpiece surface is available for the

inspection, an infrared photograph at one glance provides the information on the cure uniformity along the bond line. This inspection method may be further enhanced by introducing an effective thermal barrier in between the metal clamping electrode and the workpiece (a thin teflon coating would do the job), since no doubt, a careful control of the heat escape flow from the very beginning would affect the final outcome (see the computer simulation result presented in Figs. 16 and 17).

One must remember, that the proposed infrared based temperature inspection (either instant photograph or scanning) is not an effective, reliable technique of bond line processing status control unless it is used in conjunction with other methods. The absolute temperature reading will depend on room temperature, air flow which affects the heat escape from the surface due to the heat convection, the time elapsed since the RF-exposure termination, etc. Therefore, a one-point fiber-optic sensor, which would provide a temperature reading during the RF heating and a signal for the heating process termination, in conjunction with the proposed post-heating bond line infrared temperature inspection appears to be a perfect combination. Thereupon, not only is the RF radiation dose well controlled but in one shot the degree of cure uniformity,

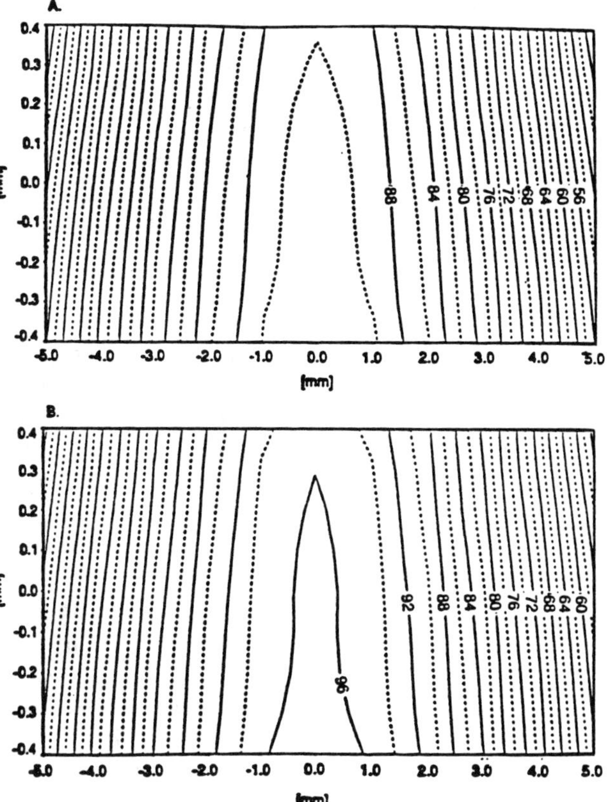

Fig. 11 The temperature map inside the adhesive for the bond line thickness at its lower limit.
  a - at the end of the step #3
  b - 10 seconds after the end of the step #4

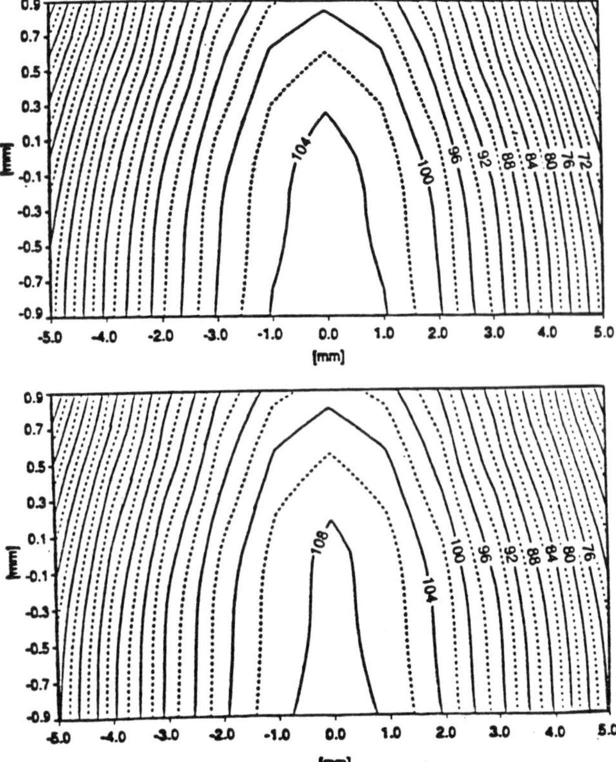

Fig. 12 The temperature map inside the adhesive for the bond line thickness at its highest limit.

    a - at the end of the step #3

    b - 10 seconds after the end of the step #4

by whatever reason it is affected, may be inspected. Now, based on the detection of a departure from an experimentally defined standard in the relative temperature non-uniformity along the bond line, infrared workpiece examination becomes an effective method for allowing for the early detection of a manufacturing failure.

As was demonstrated, both experimental and theoretical evidence indicates for strong dependance of the heat generation caused by RF exposure on the bond line thickness (see also [2]). Consequently, variation in thickness of the bond line results in noticeable differences in adhesive degree of cure and thus, bond line performance. In addition, it was demonstrated that the bond quality and process automation requires identical initial surface temperature of both cemented components at the beginning of RF cycle. Accounting for possible fluctuation in the performance of the source of electromagnetic energy due to the aging process, tool's wear, etc., it is essential that the one point temperature sensor gathering information about the temperature rise during RF exposure be employed to provide the temperature reading taken as a reference base for post cure infrared inspection along the whole bond line. Such a system would enable the detection of any non-uniformity along the bond line and provide a reference

for on-line correction of the electromagnetic energy dose required by the fluctuations of preset processing system parameters. Finally, it is suggested that the workpiece clamping period should be extended beyond the time required by an RF dose to account for the time lag of the thermosetting process behind the temperature buildup induced by electromagnetic irradiation.

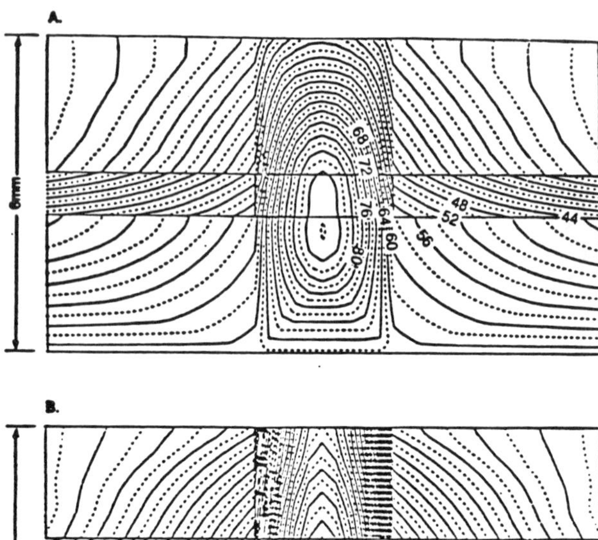

Fig. 13 Temperature map at the end of the step #3 in the case of the RF clamping electrode supplied with the insulating cladding preventing heat escape from the processed workpiece into well conducting aluminum electrode (compare with Figs 7 and 8).

    a - adhesive thickness at its lowest limit

    b - adhesive thickness at its highest limit

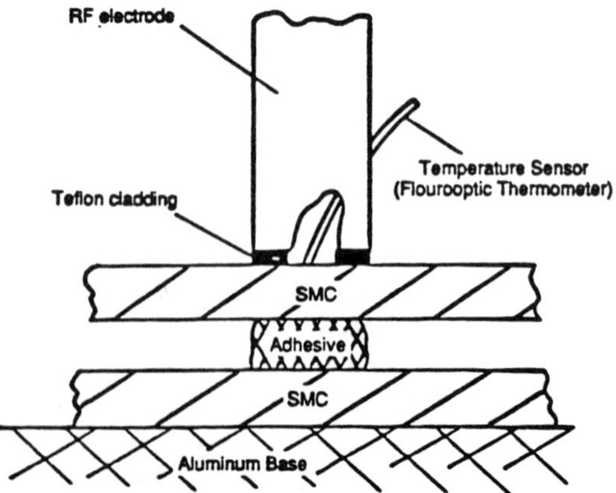

Fig. 14 Proposed fixture enabling temperature buildup at the outer surface of the bonded element. Monitored temperature buildup correlates well with the heating status of the adhesive.

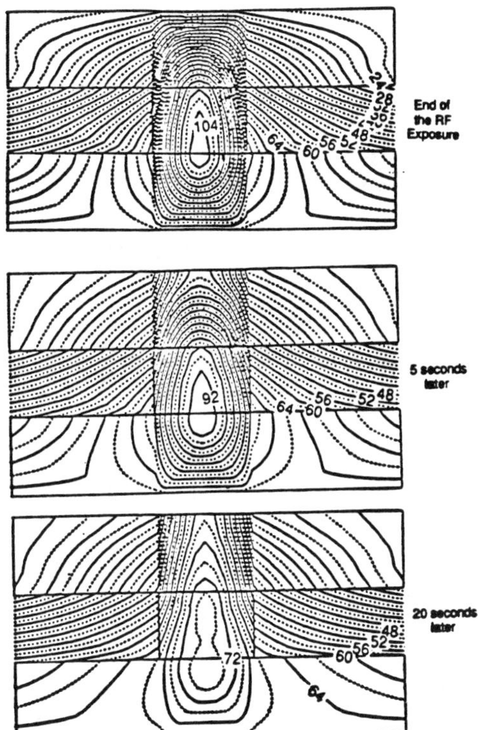

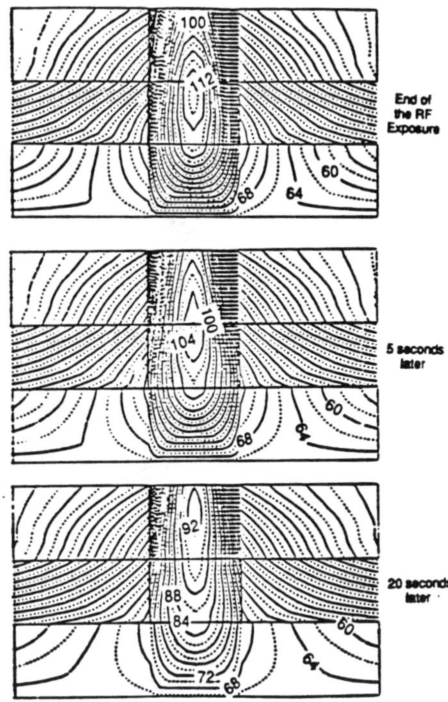

Fig. 15 Temperature buildup at the outer surface of the bonded element in the post-RF stage if the heat conductive clamp (RF electrode) is removed upon the RF-exposure termination. Due to the limited heat conduction through air, the heat escape is terminated and the isotherms are rearranged indicating almost solely lateral heat escape from the hot area (constant temperature lines are perpendicular to heat flow direction).

Fig. 16 Further enhancement of the outer surface temperature buildup may be achieved by an insulating cladding of the clamping RF electrode ((see Fig. 14). Now the surface temperature even better correlates with the degree of cure evolvement at the bond line (compare with the result calculated for different bond line thickness, same RF exposure and same initial temperature conditions shown in Fig. 17).

Conclusions

In general, the temperature buildup in time is governed by three factors:
- thermal diffusion from a preheated molding tool at the initial stage and heat escape into the tool once the adhesive temperature becomes higher than the clamping tool and SMC sheets,
- heat released by the exothermal chemical reaction,
- the electric energy absorption and, hence, conversion to heat which takes place uniformly in the whole bulk of the adhesive, and to a certain extent, depending on SMC electric power dissipation factor in the volume of cemented plastic sheets.

Consequently the following steps must be followed in any practical application of the discussed process:

1. Adhesive selection
Adhesive processing enhanced by radio-frequency heating requires an adhesive system which responds well

to electromagnetic irradiation, and thus efficiently converts the electromagnetic energy into heat. At the same time, once cured, the energy absorption should be limited to avoid the danger of a thermal run-away effect. This conclusion is well exemplified when comparing some behaviors of Tyrite 7525 urethane and epoxy 320 adhesive systems.

2. SMC sheet thickness tolerance
Variation in the SMC sheet thickness combined with a constant clamping distance affect the bond line dimension. Temperature rise inside the adhesive caused by the electric-to-heat energy conversion strongly depends on the bond line thickness. In case of thick bond lines, the temperature rises faster and reaches higher levels which, in case of thickness nonuniformity, leads to a variation in the degree of cure along the bond line. Also, with constant exposure to the RF field, some workpieces may be undercured (minimum adhesive thickness) while others may be overbaked (maximum adhesive thickness).

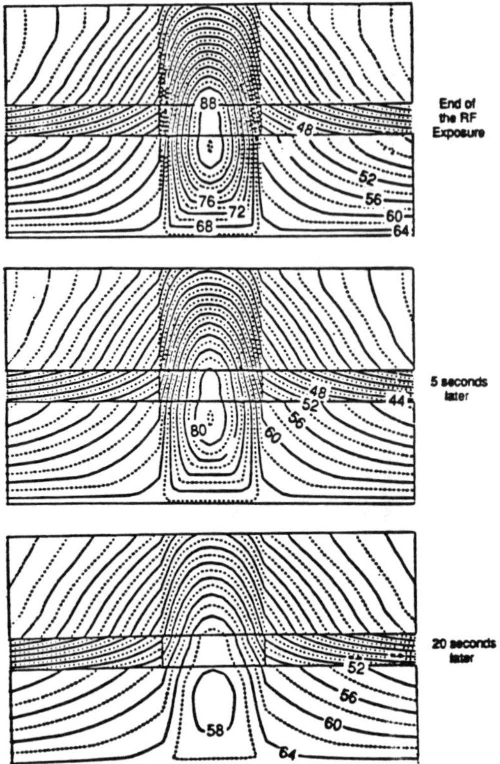

Fig. 17 Clearly, different adhesive thickness and, thus, different rate of temperature rise inside the adhesive can be observed at the outer surface of the processed workpiece (compare Figs. 16 and 17).

Consequently, a tight tolerance of the workpiece's related dimensions is recommended.

3. Initial temperature of the bonded components

The initial surface temperature of the SMC members contributes substantially to process effectiveness and joint quality, and **should be strictly monitored.** Further improvement of the bond quality may be obtained by insuring **identical surface temperature of both cemented components** at the beginning of RF cycle.

4. Process monitoring

The most effective process monitoring seems to be the combination of **a one point fiberoptic temperature sensor** gathering information about the temperature development during the curing cycle which in turn would provide the temperature reading taken as a reference base for a subsequent, **post cure infrared inspection along the whole bond line.** Such a combination would probably require **thermal insulation of the clamping electrode** to prevent heat escape from the area being monitored. Lack of the thermal barrier would make one-point temperature sensing impractical and compromise the infrared post-cure inspection.

5. Workpiece clamping time

It is known that **the kinetics of the thermosetting process lags behind the temperature buildup induced by intensive electromagnetic irradiation;** the rapid temperature rise may be considered as a fast, almost adiabatic heating process. Therefore, **the workpiece clamping period should be extended beyond the time required by an electromagnetic irradiation dose** so that the adhesive reaches a solid state phase before the clamping pressure is released. Premature part discharge may disrupt the final shape of the bond line due to a not fully developed adhesion.

References

1. G. M. Brown, C. N. Hoyles, and R. A. Bierwirth, _Theory and Application of Radio-Frequency Heating,_ D. Van Nostrand Co., Inc., (1947).
2. G. W. Malaczynski, _Polym. Eng. Sci.,_ **28**, 1270 (1988).
3. A. C. Metaxas and R. J. Meredith, _Industrial Microwave Heating,_ IEE Power Engineering Series (1983).
4. A. R. von Hippel, _Dielectric Materials and Applications,_ J. Wiley, New York 1954.
5. A. K. Jonscher, _Dielectric Relaxation in Solids,_ Chelsea Dielectric Press, London (1983).
6. M. R. Kamal and S. Sourour, _Polym. Eng. Sci.,_ **13**, 59 (1973).
7. S. Sourour and M. R. Kamal, _Thermochimica/Acta,_ **14**, 41 (1976).
8. M. R. Kamal, S. Sourour, and M. Ryan, _SPE ANTEC Tech. Papers,_ **19**, 187 (1973).
9. Lord Corporation - Adhesive Studies: Tyrite 7525 system.
10. F. P. Incropera, D. P. DeWitt, _Fundamentals of Heat and Mass Transfer,_ J. Wiley, New York (1985).
11. G. W. Malaczynski, C. A. DiNatale, G. J. Cinpinski, _Polym. Eng. Sci.,_ **33**, 240 (1993).

Collapse Strength Characteristics of Foam Filled Laminated Composite Components

A. Bhimaraddi
DCED
Clawson, Michigan

H.F. Mahmood
Ford Motor Company
Dearborn, Michigan

Abstract

The prediction of crush strength characteristics are presented for foam filled laminated composite components subjected to axial and bending crush loads. Foam filling has a definite influence on the mode of failure and strength of a component. A thin-walled component (unfilled) that exhibits predominant local buckling under axial load, changes its buckling mode to a higher mode. This change results in higher load carrying capacity and stable crush of the filled component. On the other hand, a thick-walled (unfilled) component which does not exhibit local buckling shows marginal increase in the load carrying capacity of the filled section. The amount of load increase in the filled component, as compared to the unfilled component, strongly depends on the density and strength of the foam being used. Strength characteristics of the filled component are predicted by examining the relationship between the buckling load and the maximum load of an unfilled component and the foam density. The predicted results show good correlation with the static test data for axial crush and bending crush.

THE WEIGHT REDUCTION plays an important role when fuel efficient, and hence economically and environmentally feasible, vehicles are to be made. Occupant protection is an important factor to be considered in the design of vehicles for their crashworthiness. Mostly, structural members such as, rails and pillars absorb the crash energy in a crash event. These components are of thin-walled cross-section made of either metal (steel or aluminum) or composite material (such as E-Glass/Epoxy laminates). Recent advances of composite technologies in both manufacturing and design areas have substantially increased the acceptance of these materials in automotive industries.

Composite materials were previously used only in the secondary load carrying members. More recently, they are being considered for the primary load carrying and energy absorbing structural members in automotives (such as rails and pillars). A major advantage of composite materials is their ability to be tailor-made to suit the load environment, in addition to their high-strength and low-density characteristics. The impact energy is absorbed in metal structural components by folding mechanism; and whereas, it is absorbed by matrix cracking, delamination, and fibre rupture in the case of composite structural components.

Theoretical knowledge in the area of crashworthiness analysis of composite structures is very limited and is not yet implemented, successfully, in the general purpose finite element codes such as, DYNA3D, PAM-CRASH, and RADIOSS. Due to this lack of understanding of crashworthiness of composite structures, the design procedure relies, to a large extent, on the experimental data. Some discussion on various failure mechanisms and how to accommodate them in a finite element procedure can be found in the papers by de Rouvray and Haug [1] and Mahmood and Bhimaraddi [2].

The type and nature of the failure mode is a very complex function of the material constituents, fiber and resin content, lamina orientation, crushing speed, triggering mechanism, and the geometry of the component as may be seen from the experimental studies of Hull [3], Thornton [4], and Farley [5]. When using any analytical technique to predict crush characteristics, it is important that the failure modes are identified accurately. Then, some kind of failure criteria are normally employed as may be seen in references [1] and [2]. Implementation of these criteria in a finite element procedure leads to large amount of computer time and hence such an analysis is not suitable during the preliminary stages of design.

During the early design stages several parameters such as, shape and size of the components; material and manufacturing alternatives have to be tried before arriving at the final layout. This initial stage involves several iterations. Sometimes the cost of preparing and running a finite element code may become comparable to that of conducting a full scale test. In such situations it is desirable to have simplified analyses tools that can predict quickly, with a reasonable accuracy, the behavior of the structure and guide the early design. This would help in utilizing the detailed finite element codes only in the final stages of the design, to verify the performance of the structure prior to testing. In this direction, Mahmood and his colleagues (see references [6-10]) have made significant contribution towards predicting the crush strength characteristics in axial and bending crush of composite component.

Mostly, automotive components (made of metal or composite materials) have thin-walled closed cross-section. Local buckling is a dominant mode of failure in axial crush of such components. Foam filling of these components prevents early local buckling and hence improves the load carrying capacity. As noted in references [11-14], foam filling will only find general usage if it aids in the solution of a complex set of problems, viz., (1) improved energy absorption, (2) damageability-dent resistance of panels, (3) reduced vibration through higher stiffness, (4) weight reduction and (5) rust prevention. For foam filled structures to be effective energy absorbers they must offer an increased mean crush load that is not attainable by other means. One obvious way of increasing the mean crush load is to increase the wall thickness. Detailed discussion on the influence of foam filling, from the crashworthiness point of view, on the crush characteristics of components is dealt with in the following sections.

It is to be noted here that the above mentioned studies on the crashworthiness of composite components deal with unfilled components. There is neither theoretical nor experimental information available, in the open literature, on the crashworthiness of foam filled composite components. However, there is limited amount of information available on the crashworthiness of foam filled metal (steel) components [11-14]. It is the purpose of this paper to shed some light on the crashworthiness prediction of foam filled composite components. In this work analytical prediction of crush characteristics of foam filled composite components is given. This study is an extension of the authors' earlier efforts on the subject of crashworthiness prediction of composite structural components [6-8]. Good correlation between the theoretical predictions and the experimental results is observed. Before discussing the procedure to account for the presence of foam in the component, some relevant details on the failure mechanisms and collapse mode of filled and unfilled components is given in what follows next.

Influence of Foam on the Crush Characteristics

The objective of the crashworthiness design of a component is to absorb, as much as possible, the crash energy with minimum component. Crush distance and the mean crush load are the two major parameters that determine the amount of crush energy that can be absorbed. For the same total energy to be absorbed, one can design the component to have higher mean crush load and a lower crush distance, or vice versa. For example, a thin-walled steel component that buckles locally and folds regularly under axial crush results in a larger crush distance but a lower mean crush load. Whereas, a thick-walled steel component that does not show local buckling may experience higher mean crush load but has smaller crush distance due to global bending.

Major differences, apart from the mean crush load and crush distance, in the crush characteristics of thin and thick-walled components are: (1) thin-walled component is more weight effective, (2) thin-walled component is softer and hence may cause intrusion, (3) thick-walled component is stiffer and hence may cause high deceleration rates, and (4) very thin-walled components fold irregularly and hence may result in unstable collapse mode, absorbing little energy. Lampinen and Jeryan [14] have identified four types of failure modes in foam filled steel columns. One of the interesting findings of this work is the ability of a foam filled spot-welded section to absorb higher energy due to the increase in both the mean crush load and crush distance. This was attributed to the fact that the failure of spot-welds resulted in the separation of foam which in turn suppressed the global buckling and hence causing a regular folding mechanism of the column.

Foam filling a very thin-walled component delays the local buckling and stabilizes the collapse mode resulting in a more regular folding mechanism. This ultimately results in a higher mean crush load and a larger crush distance and hence large amount of energy absorption. In this case foam filling changes the structure from a non-absorber to an absorber of energy. Foam filling a thick-walled component suppresses any local buckling and hence results in an Euler-type of global buckling and in little energy absorption. In this case foam filling becomes weight ineffective and changes the structure from a possible absorber to a non-absorber of energy. Thus, for the foam filled components to be effective energy absorbers they must offer an increased mean crush load and crush distance that is not possible by means other than increasing the wall thickness.

Furthermore, experiments on steel columns [14] have shown that high density (20 pcf) foams increase the load carrying capacity to the Euler buckling load which reduces the axial crush distance. Thus, higher density foams are weight ineffective as energy absorbers. Low density (< 4 pcf) foams do not help in increasing the mean crush load and hence are again not effective as energy absorbers. However, medium density (6 pcf - 13 pcf) foams are found to help energy absorption and are weight effective.

In contrast to the above observations on foam filled metal components, the failure mechanisms in composite components are different. Composite structures absorb energy by breaking into pieces and corners being zipped open as against the folding mechanism of metal structures. This kind of failure allows the foam to separate and move out of the confined space within the component. As a result of corners being zipped open, and also due to the presence of crush initiators, the global buckling is eliminated in thick-walled composite components.

When thick-walled components are filled with foam, they may not show Euler-type global buckling because of the fact that the corners zip open before the onset of global buckling and hence a stable crush mode can still be obtained. However, there may not be any appreciable increase in the mean crush load since foam does not participate in the suppression of local buckling which does not exist. Hence, in this situation foam filling may not result in the weight effective energy absorbing structure, as is also the case in metal structures. It must be noted here that the failure mechanism is in total contrast to that occurring in a thick-walled metal structures.

The thin-walled composite components have a higher tendency toward local buckling. Thus, when thin-walled components are loaded axially they undergo local buckling during the early stages of loading and a noticeable buckling wave pattern can be seen along the length. With a further increase in the load, these components break along the wave pattern accompanied by the tearing of corners. When these components are filled with foam, the local buckling will be delayed and thus allowing the component to buckle at a higher load and with a shorter wave pattern. With a further increase in the load, the column breaks at the wave pattern. This change in the failure pattern results in the increase of higher mean load and more stable crush mode.

It is clear from the above discussion that it is necessary to know the crush strength characteristics of unfilled composite components in order to predict those of the foam filled components. The methodology to predict crush characteristics (critical buckling load, maximum crush load, mean crush load, minimum crush load, and mode of collapse) of unfilled composite sections has been dealt with extensively in References [6-8]. In the following section the methodology to predict crush characteristics of foam filled sections, knowing the characteristics of the unfilled section and the foam properties, will be presented.

Prediction of Crush Strength Characteristics

As noted earlier most components encountered in automotive structures have closed cross-sections. Usually, these cross-sections are made-up of flat plate segments. There is a wealth of information available in the open literature on the behavior of composite flat plates as regard to buckling, strength, and deformation behavior. This information has been altered and has been used effectively to suit the current design problem.

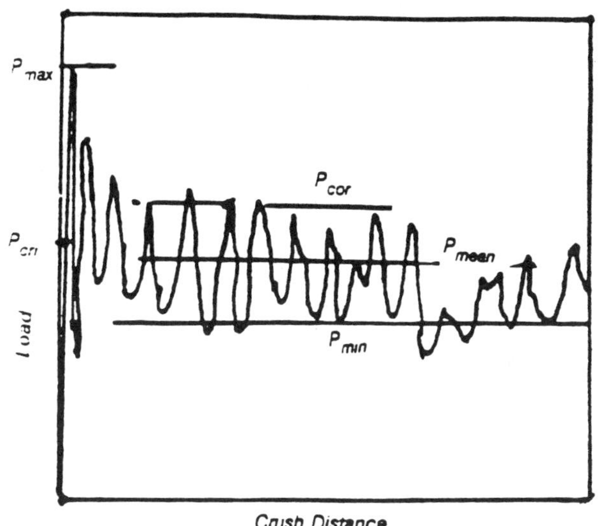

Fig. 1 **Typical Load-Deflection Curve in Axial Progressive Crush**

As noted earlier, the collapse strength predictions of filled sections are based on those of the unfilled sections. It is appropriate, here for the sake of continuity and completeness, here to describe briefly the procedure used for unfilled sections. Equations (1) through (7) below give brief description of strength characteristics of unfilled sections. The reader may obtain details on the development of these equations from References [6-8]. Load or stress terminology used in this paper is described in Figure 1. In general, the equation for critical buckling stress of an anisotropic long plate of thickness (t) and width (b) can be written as follows:

$$\sigma_{cri} = \phi \pi^2 E^* \left(\frac{t}{b}\right)^2 \qquad (1)$$

In this equation, ϕ is the coefficient which depends on the material properties, lay-up configuration, and the degree of restraint at the unloaded long edges; and E^* is an equivalent elastic modulus computed as follows:

495

$$E^* = \frac{\sqrt{D_{11}D_{22}}}{t^3} \qquad (2)$$

Here D_{ij} are the laminate flexural rigidities. Ultimate strength (σ_{ult}) of any laminate, under uniaxial compression, can be computed using the tensor polynomial failure criteria which are well documented in literature. The general form of these failure criteria can be written as,

$$f(\sigma) = F_i\sigma_i + F_{ij}\sigma_i\sigma_j = 1 \qquad i,j = 1,2,...6 \qquad (3)$$

Here F_i, F_{ij} are the second and fourth order lamina strength tensors; and σ_i are the stress components referring to the lamina coordinate system. It may be noted here that the laminate strength was computed based on the notion of first-ply failure in conjunction with a chosen failure criterion.

The maximum strength (σ_{max}) of a composite laminated plate has been computed using the following formula:

$$\sigma_{max} = \sigma_{ult}\left[\frac{1.22\pi^2 K_p\theta E^*(t/b)^2}{\sigma_{ult}}\right]^{0.43} \qquad (4)$$

In equation (4) K_p and θ depend on the material properties, thickness, width, and the restraint conditions of the plate segment. Corner (σ_{cor}), mean (σ_{mean}) and minimum (σ_{min}) crush strengths are computed by using the following relations,

$$\sigma_{cor} = \beta_1\,\sigma_{max} \qquad (5)$$

$$\sigma_{min} = \beta_2\,\sigma_{max} \qquad (6)$$

$$\sigma_{mean} = \frac{\sigma_{cor} + \sigma_{min}}{2} \qquad (7)$$

In the above equations, β_1 and β_2 are coefficients which depend on the ply angle and the restraint conditions of the plate segment (see Reference [8]). Having predicted the collapse strength characteristics for unfilled section we move on to the predictions for foam filled sections. Before doing so it is essential to discuss foam characteristics such as stress-strain relations, density and strength relationships. Figure 2 shows (schematic only) the stress-strain relation in

uniaxial compression for urethane foam with higher and lower density. From this figure one may note that stress increases with strain in portion A, stress is almost constant in portion B, and whereas stress increases tremendously with little increase in strain in portion C. The foam strength referred to in this paper corresponds to the constant stress level of portion B. Further, it may be seen from this figure that higher density foams have higher strength. Figure 3 shows the relationship between the foam density (ρ_f) and the foam strength (σ_f). It is very useful to have some kind of analytical equation describing the relation between the density and strength. One such relation, for predicting the foam strength when the foam density is known, is given by Thornton [13] and Lampinen and Jeryan [14] as follows:

$$\sigma_f = 9.07\,\rho_f^{1.6} \qquad (8)$$

In the above equation σ_f is expressed in psi and ρ_f is expressed in pcf.

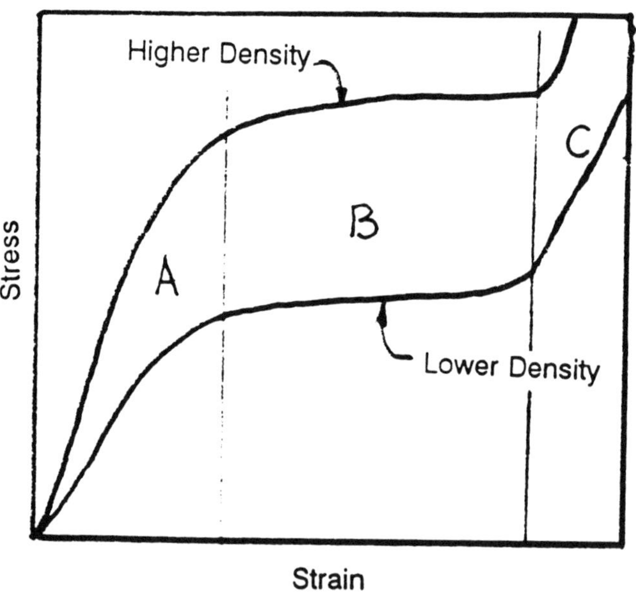

Fig. 2   **Typical Stress-Strain Curve for Urethane Foam in Uniaxial Compression**

Thus far all the information on the crash behavior, individually, of an unfilled section and the foam are known to us. During the initial stages of axial loading of the component to its maximum capacity, load is resisted by the composite shell, enclosing the foam, and the foam itself. Also it is to be said here that the amount of load resisted by the composite shell is greater than it would have resisted were it not filled with foam. This increase in the load resistance of the composite shell is attributed to the load sharing interaction that takes place between the shell and the foam. Increased levels in the mean crush load can be expected due to the increase in the maximum load for a foam filled component. Initially, in this work the buckling and maximum stresses for each segment, in an unfilled component, were predicted. Then the ratio of buckling to

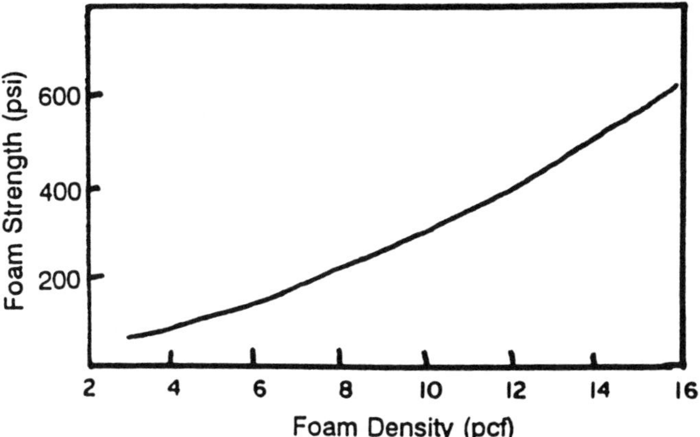

Fig. 3   **Relationship Between Density and Strength for Urethane Foam**

maximum stress ($R = \sigma_{cri}/\sigma_{max}$) for each segment was studied. The segments, for which this ratio was less than one, were observed to be the buckling segments and those for which this ratio was greater than one were observed to be the restraining segments. The segments that have a higher tendency to buckle do so at a higher stress level when the component is filled with foam; their maximum stress capacity also increases.

Following this line of thought and also based on the experimental observations on foam filled components, an empirical relation was derived for the factor to increase the maximum stress of a segment in the component as follows:

$$f = 1.5 + 0.1102\ R - 1.4752\ R^2 + 1.2411\ R^3 - 0.3037\ R^4 \tag{9}$$

where, $R = (\sigma_{cri}/\sigma_{max})_{unfilled}$. Equation (9) is valid for $\sigma_f = 200$ psi.

Figure 4 shows graphical representation of equation (9) for various foam strengths. Maximum and mean strengths of any segment in a foam filled section is computed

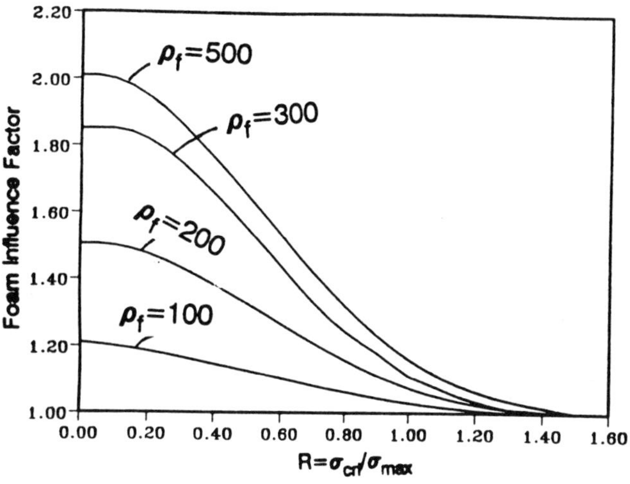

Fig. 4   **Plot of Foam Influence Factor for Various Foam Strengths**

as follows:

$$\left(\sigma_{max}\right)_{filled} = f\left(\sigma_{max}\right)_{unfilled} \tag{10}$$

$$\left(\sigma_{mean}\right)_{filled} = f\left(\sigma_{mean}\right)_{unfilled}$$

Having computed the stresses for each segment in the section, maximum and mean loads for the section are computed as follows:

$$\left(P_{max}\right)_{filled} = \sum_{i=1}^{n} t_i \times \left(\sigma_{max}\right)_{filled} + \sigma_f A_{foam} \tag{11}$$

$$\left(P_{mean}\right)_{filled} = \sum_{i=1}^{n} t_i \times \left(\sigma_{mean}\right)_{filled} + \sigma_f A_{foam}$$

In equation (11) t_i is the thickness of the ith segment, A_{foam} is the area of the foam, and n is the total number of plate segments the section is made-up of.

The above prediction methodology is for pure axial crush characteristics. The computation of maximum bending capacity (M_{max}), of an unfilled section, has been given in Reference [9]. When the component is loaded in pure bending crush, part of the section, experiencing maximum moment, will be under tension and the rest will be under compression. With a gradual increase in load the plate segment in compression buckles and the further increase in load causes this plate segment to crush. Thus, the maximum crushing strength (σ_{max}) of the compressive plate can be computed using the above relations. Knowing the strength of the tension plate laminate and the σ_{max} of the compression plate, the stress distribution across the section can be predicted. Maximum bending capacity of the section, in pure bending crush, is then computed using the simple principles of mechanics, namely, balancing the compressive and tensile forces across the section and taking the moment of these forces about the neutral axis.

Discussion of Numerical Results

The aforementioned methodology has been used in the crush strength prediction of foam filled components tested in axial and bending crush. The lamina type, their elastic moduli and strength properties are given in Table 1 and 2, respectively. In Table 1, E_L and E_T are the elastic moduli in longitudinal and transverse directions; G_{LT} and μ_{LT} are the shear modulus and the Poisson's ratio. In Table 2, X_c and X_t are the compressive and tensile strengths in the longitudinal direction; Y_c and Y_t are the compressive and tensile strengths in the transverse direction; S is the in-plane shear strength.

Table 1. Lamina Moduli (Msi)

Lamina	E_L	E_T	G_{LT}	μ_{LT}
CSM	2.0	2.0	0.76	0.31
Braided	2.67	0.94	0.72	0.5

Table 2. Lamina Strengths (ksi)

Lamina	X_c	X_t	Y_c	Y_t	S
CSM	36.9	26.4	36.9	26.4	21.0
Braided	41.4	45.7	21.0	5.4	16.7

Figure 5a and 5b show the geometric details of the sections considered in this study. Static axial and bending crush tests were conducted on the components with these sections. Continuous Strand Mat (CSM) material was used for Section A (Fig. 5a). Three wall thicknesses (3mm, 5mm and 7mm) were considered for this component. For Section B (Fig. 5b) Braided configuration with a braid angle of ± 30 was used and the wall thickness was 3mm. Foam properties were 6 pcf (density) and 228 psi (strength).

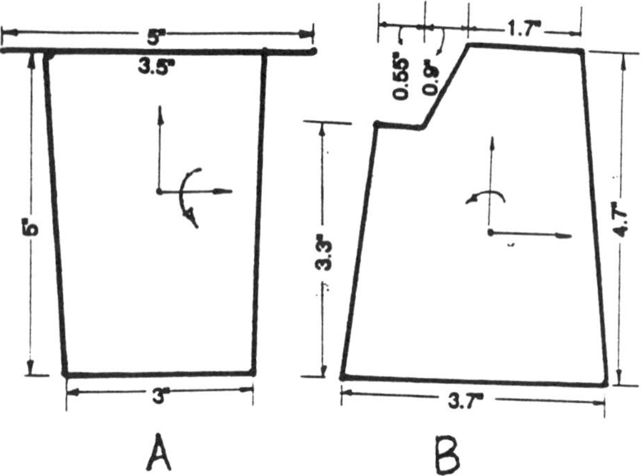

Figure 5. **Geometric Details of Composite Sections Tested in Axial and Bending Crush**

Table 3. Axial Crush Results for Section A and B of Fig. 5

	P_{max} (pounds)		P_{mean} (pounds)			
	Unfilled		Unfilled		Filled	
	(Predi.)	(Predi.)	(Test)	(Predi.)	(Test)	(Predi.)
Sec. A (3mm)	16150	23230	—	5466	12187	11717
Sec. A (5mm)	43085	60756	13273	11764	22750	23615
Sec. A (7mm)	95822	99565	—	19203	28965	32553
Sec. B (3mm)	16385	24461	1517	5478	12600	13156

Table 3 shows the axial crush results from test as well as from current predictions. Test results reported in this paper are taken from the experimental studies conducted by Loosle et al. [15] on the crush performance of composite components. Tests were conducted on the triggered components and hence test results for P_{max} values are not included in Table 3. It may be said here that unfilled Section-A and Section-B with wall thickness 3mm showed pronounced local buckling in axial crush. Whereas, unfilled

Section-A with wall thicknesses 5mm and 7mm did not show the local buckling phenomenon. Comparing the results between the filled and the unfilled sections for P_{max}, it may be said here that the maximum load carrying capacity of 3mm section increased by 45% and whereas only 4% increase may be seen for 7mm thick section.

The effect of foam filling on P_{mean} load level is the most important factor from the energy absorption point of view. Comparing the predicted P_{mean} values for filled and unfilled sections it is evident from Table 3 that there is 115% - 240% increase in the mean crush load for 3mm wall thickness and whereas, there is 70% - 100% increase for 5mm and 7mm wall thicknesses. Furthermore, good correlation between the test and predicted results may be observed from this table. In the case of unfilled Section B, the test did not go as expected due to the sensitivity of the thin-walled sections in buckling and in load eccentricities. Test Sample did show some bending and broke into large pieces after the peak load was reached and hence resulted in a smaller mean crush load. Due to this a large difference may be noticed between the test and predicted results for unfilled component (Section B) in Table 3. These results also indicate that sections with thinner wall thickness, when filled with foam, become very efficient energy absorbing structures due to the marked increase in their maximum and mean crush load levels. It is possible, as there is lesser increase in the load levels, that the thicker wall thicknesses (like 7mm) may not be efficient energy absorbers because of the weight increase with foam filling.

Table 4. Maximum Bending Capacities (M_{max} lbs-in)

	Unfilled		Filled	
	(Test)	(Predi.)	(Test)	(Predi.)
Section A (3mm)	—	20694	47256	50191
Section B (3mm)	27744	30490	55500	52094

Loosle et al. [15] conducted the experimental study on the bending crush performance of the composite components shown in Figure 5. Static bending crush test results reported in Table 4 are taken from this work. Bending axes are shown in these figures. A four point bending test was conducted for Section-A whereas, a three point bending test was conducted for Section-B. Test and predicted results are given in Table 4. It may be seen from this table that the bending capacity increases two-fold for filled sections when compared with the unfilled section. It may further be noted here that there is good agreement between the predicted and test results.

Conclusions

The prediction methodology for computing the collapse strength of foam filled composite thin-walled components in static axial and bending crush has been described in this work. It is well-known that, when the section is filled with foam, its maximum and mean crush load carrying capacities increase. This increase is attributed to the change in local buckling mode. Thus, the proposed method relies on the prediction of local buckling and strength characteristics of an unfilled component. Depending on the amount of local buckling in an unfilled component, as compared to its maximum strength, the foam influence factor is determined. This factor is used to augment the crush strength characteristics of the corresponding unfilled component. Predicted results are compared with the test results for two typical foam filled automotive components. It is observed from these results that the components with thinner wall thickness, when filled with foam, showed marked improvement in the collapse strengths whereas, those with thicker wall thickness showed marginal improvement. Further, good agreement between the predicted and the test results is observed.

Acknowledgements

Authors would like to thank Mr. Richard A. Jeryan for his valuable comments, criticisms, and encouragement during the course of this work.

References

1. de Rouvray, A. and E. Haug, "Failure of brittle and composite materials by numerical methods," Chapter 4 in the book 'Structural Failure', Proceedings of the International Symposium on Structural Failure, held at MIT, June 6-8, (1988)

2. Mahmood, H. F. and A. Bhimaraddi, "Failure of composite structures in crush environment," Proceedings of the 8th Advanced Composites Conference, Chicago, Illinois, USA, 2-5 November, (1992)

3. Hull, D., "Energy absorption of composite materials under crush conditions," Progress in Science and Technology of Composites, ICCM-IV, Tokyo, Japan, (1982)

4. Thornton, P. H., "The crush behavior of glass fiber reinforced plastic sections," Composite Science and Technology, 27, 199-223, (1986)

5. Farley, G. L., "Energy absorption of composite materials and structures," 43rd Annual forum and technology display, Proceedings of the American Helicopter Society, St. Louis, Missouri, USA, 18-20 May, (1987)

6. Mahmood, H. F., J. H. Zhou and M. S. Lee, "Axial strength and mode of collapse of composite components," Proceedings of the 7th International Conference on Vehicle Structural Mechanics, Warrendale, Pennsylvania, USA, SAE Paper No. 880891, April (1988)

7. Mahmood, H. F. and J. H. Zhou, "Local buckling and maximum strength of plate type composite components," Proceedings of the 4th Annual Conference on Advanced Composites, Dearborn, Michigan, USA, 13-15 September, (1988)

8. Mahmood, H. F. and Bhimaraddi, A., "Prediction of deep collapse strength characteristics of components made-up of E-glass/epoxy laminates," Proceedings of the 9th Annual ASM/ESD Advanced Composites Conference, Dearborn, Michigan, November 8-11, (1993)

9. Mahmood, H. F. and J. H. Zhou, "Bending crush analysis of thin-walled composite components," Forensic Engineering, 2(3), 403-414, (1990)

10. Mahmood, H. F., P. H. Thornton and J. H. Zhou, "Material strengths of glass fiber-reinforced epoxy laminates," Proceedings of the 6th Annual ASM/ESD Advanced Composites Conference, Detroit, Michigan, USA, 8-11 October, (1990)

11. Braess, H., "Examples of the development of ESV subsystems - Porsche Kg.," Fourth International Technical Conference on Experimental Safety Vehicles, March, 1973.

12. Magee, C. L. and Thornton, P. H., "Design considerations in energy absorption by structural collapse," SAE Paper # 780434, Detroit, Michigan, February, 1978.

13. Thornton, P. H., "Energy absorption by foam filled structures," SAE Paper # 800081, Detroit, Michigan, February, 1980.

14. Lampinen, B. E. and Jeryan, R. A., "Effectiveness of polyurethane foam in energy absorbing structures," SAE Paper # 820494, Detroit, Michigan, February, 1982.

15. Loosle, D. G., Bonnett, R. E. and Carpenter, R. A., "Crush performance of composite components," Proceedings of the 10th Annual ASM/ESD Advanced Composites Conference, Dearborn, Michigan, November 7-10, (1994).

Comparison of Static and Impact Energy Absorption
of Glass Cloth/Epoxy Composite Tubes

H. Hamada, S. Ramakrishna, M. Nakamura, M. Maekawa
Kyoto Institute of Technology
Kyoto, Japan

Abstract

Glass cloth/epoxy composite tubes with two different fiber surface treatment have been investigated. One set of tubes contained glass cloth treated with aminosilane coupling agent and another set of tubes contained glass cloth treated with acrylsilane coupling agent. Composite tubes were axial compression tested under quasi-static and impact conditions. The aminosilane-treated tubes displayed higher specific energy absorption capability (E_S) than the acrylsilane-treated tubes. This was attributed to the superior interfacial bonding of aminosilane-treated material than the acrylsilane-treated material. In acrylsilane-treated tubes the E_S increased by 15% with the change in crushing speed from quasi-static to impact. The E_S of aminosilane-treated tubes increased only marginally with the change in crushing condition. This difference was attributed to the crush zone morphology. The aminosilane-treated tubes crushed progressively by splaying mode independent of testing condition. The acrylsilane-treated tubes crushed by fragmentation mode under quasi-static conditions and by splaying mode under impact conditions. The changes in the crush zone morphology were related to the changes in the frictional mechanisms and mechanical properties.

THE ABILITY TO ABSORB IMPACT ENERGY in the event of accident is an issue to be considered in the design of structural members of transportation vehicles. Current research suggests that when properly designed, polymer composite materials are suitable for crashworthy structural applications [1-4]. Much of the experimental work on the energy absorption performance of composite materials has been carried out using axisymmetric composite tubes

with a chamfer trigger at one end. Tubes were axially compressed between flat platens at a constant speed under quasi-static conditions. Failure starts at the chamfer tip and the damage zone propagates down the tube without catastrophic fracture, and the energy is absorbed in a controlled manner. Some researchers have carried out compression tests at impact speeds to simulate the actual crash conditions. Farley [5] investigated carbon/epoxy, Kevlar/epoxy, and glass/epoxy composite tubes with $[0/\pm\theta]_4$ fiber architecture at speeds of quasi-static and 7.6 m/sec impact. These results suggested that the E_S of composite materials was not a function of crushing speed. Farley [6] repeated experiments on carbon/epoxy and Kevlar/epoxy tubes with $[\pm\theta]_3$ fiber architecture and found a 35% increase in E_S with the change in crushing condition from quasi-static to impact. Bannerman and Kindervater [7] also reported an increase in E_S of carbon/epoxy and Kevlar/epoxy tubular and beam specimens with the crushing speed. In contrast, Schmueser and Wickliffe [8] reported a decrease of up to 30% in E_S of carbon/epoxy, glass/epoxy and Kevlar/epoxy tubes with $[0_2/\pm45]_S$ fiber architecture, as compared to static test results. Thronton [9] studied the energy absorption behavior of pultruded glass/polyester and glass/vinyl ester tubes in the crushing speed range from 2.1 X 10^{-4} m/s to 15 m/s. He reported a 20% increase of E_S with testing speed in the case of glass/polyester tubes and a 10% decrease in the case of glass/vinyl ester tubes. This study indicates that the E_S could be significantly changed with specimens composed of different resin matrices. Hamada *et al* [10] reported that the E_S of carbon/PEEK tubes decrease by 50% with the change in testing condition from quasi-static to impact.

All the above mentioned investigators did not specify the type of fiber surface treatment given. Recently, we found that the energy absorption performance of glass/epoxy tubes is influenced by the type of fiber/matrix bond strength [11]. In this paper we compared the energy absorption characteristics of glass/epoxy tubes with different fiber surface treatments tested at quasi-static and impact speeds.

Materials and Experimental Procedure

Two types of woven glass cloth/epoxy prepreg sheets were obtained from Arisawa Mfg Co Ltd, Japan. One type contained glass cloth treated with gamma-aminopropyltriethoxysilane (aminosilane) coupling agent and the other type contained glass cloth treated with gamma-methacryloxypropyltrimethoxysilane (acrylsilane) coupling agent. Tubes were fabricated by mandrel wrapping of prepreg sheets with the warp and weft directions parallel to the hoop and axial directions of the tubes respectively. The fiber content was estimated to be 43% by volume. All the tubes had an internal diameter of 50 mm and a wall thickness of 2.5 mm. The fabricated tubes were cut into specimens of 55 mm long. To initiate progressive crushing a 45° chamfer was ground onto one end of each tube specimen.

Quasi-static Crush Test. Quasi-static tests were performed using a 250 kN Mand servo-hydraulic testing machine at a constant cross-head speed of 1.67×10^{-5} m/s. Composite tubes were axially crushed between parallel steel flat platens, one static and one moving. The fixed platen was fitted with a load cell from which the load signal was taken directly to the x-y plotter. In each test the crush load was plotted on the y-axis and the cross-head displacement on the x-axis.

Impact Crush Test. Impact tests were performed using drop test facility designed and built for this purpose by Japan Automobile Research Institute, Inc., Japan. Schematic diagram of the test facility is shown in Figure 1. Test specimen was mounted on the impact plate such that the axis of the tube was parallel to the direction of the travel of the dead weight. The drop platform weight 60 kg can be raised to a maximum height of 7 m. For the present test program, the drop platform was raised to a predetermined height of 3.54 m which gave an impact velocity of 8.5 m/s. The drop platform was released with a grip latch pin mechanism that was extracted with two manually activated electronic solenoids. The signals from the load cells located beneath the impact plate were fed to the analogue-digital converter. The digitised data representing the load-time and displacement-time histories of the impact test was processed and load-displacement (P-d) response was recorded using an x-y plotter.

Specific Energy Absorption. The total energy absorbed during progressive crushing of the composite tube is defined as the area under the P-d trace (Figure 2). For composite tubes of uniform cross-section, the mean crush load ($\overline{P}$) is independent of the crush distance. Hence the specific energy absorption (E_s) is given by

$$E_s = \frac{\overline{P}}{A\,\rho} \qquad \text{Eq. 1}$$

where A and ρ are the cross-sectional area and density of the tube respectively. This equation can be rewritten as

$$E_s = \frac{\overline{P}}{m/l} \qquad \text{Eq. 2}$$

where m/l is defined as mass per unit length of the tube. Therefore, E_s is calculated as the ratio of the mean crush load from the P-d curves and the mass per unit length of the uncrushed tube.

Microscopy. The crushed tubes were cast in polyester resin to preserve the crush zone morphology. The microfracture processes in the crush zone were determined by cutting sections through the embedded material. The sections were then polished and photographed on an optical microscope. Scanning electron microscopy (SEM) was used to study the fracture surfaces.

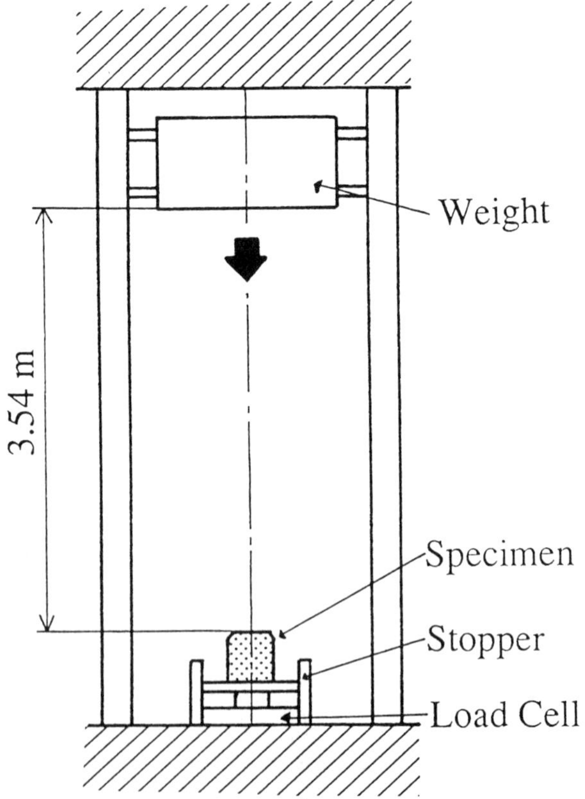

Fig. 1. Schematic diagram of impact testing machine.

Results

Quasi-static Crush Test Results. Typical load-displacement traces obtained from quasi-static testing of aminosilane-treated and acrylsilane-treated tubes are shown in Figure 2. Intially the load increased rapidly to a maximum value. This rapid load increase was associated with fracture of chamfered portion of the tube wall. After the initial rapid increase the load fluctuated around a mean value known as mean crush load ($\bar{P}$). The P-d trace of acrylsilane-treated tubes was more serrated than the P-d trace of aminosilane-treated tubes. In the case of aminosilane-treated tubes progressive crushing occurred at a mean crush load of 45.7 kN. This is higher than the 37.5 kN mean crush load observed in the case of acrylsilane-treated tubes. The energy absorption performances of both the types of tubes are given in Table 1. The aminosilane-treated tubes absorbed 25% higher energy than the acrylsilane-treated tubes.

The basic features of crushing of both types of tubes are illustrated in Figures 3 and 4. Detailed photographs of sections cut through the crush zones parallel to the axis of the tubes are shown in Figures 5 and 6. For the aminosilane-treated material the stable crush zone is typified by crushed material which has produced fronds inside and outside of the tube wall. This is called the 'splaying' mode of crushing [2]. The material splayed to the inside of the tube is termed as internal frond and the material splayed to the outside of the tube is termed as external frond. Fronds have been bent through sharp radii of curvatures at the crushing platen. A well defined debris wedge splitting the tube wall into internal and external fronds was observed. Below the debris a longitudinal crack of 7 mm long was also observed. A completely different form of crush zone developed with acrylsilane-treated tubes in which the formation of thin rings of material can be seen. This is called the 'fragmentation' mode of crushing. Shear cracking of tube wall can be seen clearly in Figure 6. This successive shear cracking of tube wall produced fragments of the tube wall which separated from the inside and outside of the tube wall (Figure 4). The contact regions of crush zone with the crushing platen were rougher in the case of acrylsilane-treated tubes than in the aminosilane-treated composite tubes.

Table 1. Energy absorption performances of quasi-static tested glass cloth/epoxy tubes.

Tube Type	Mean Crush Load $\bar{P}$ (kN)	Mean Crush Stress (MPa)	Specific Energy (kJ/kg)
Aminosilane	45.7	108.4	66.6
Acrylsilane	37.5	90.2	53.0

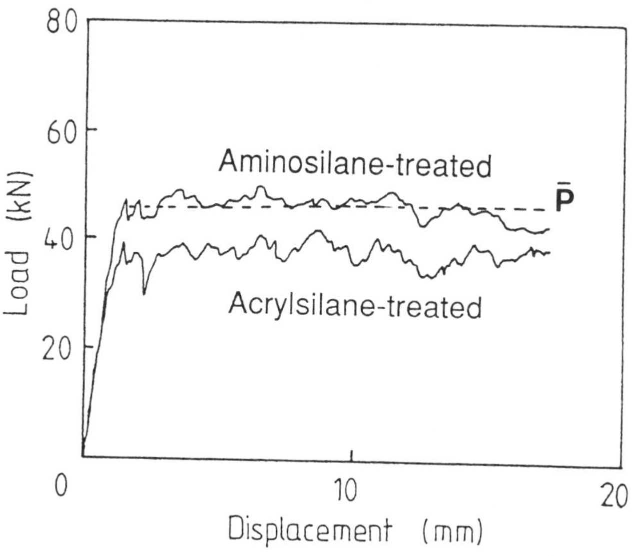

Fig. 2. Typical load-displacement curves for quasi-static tested glass cloth/epoxy tubes.

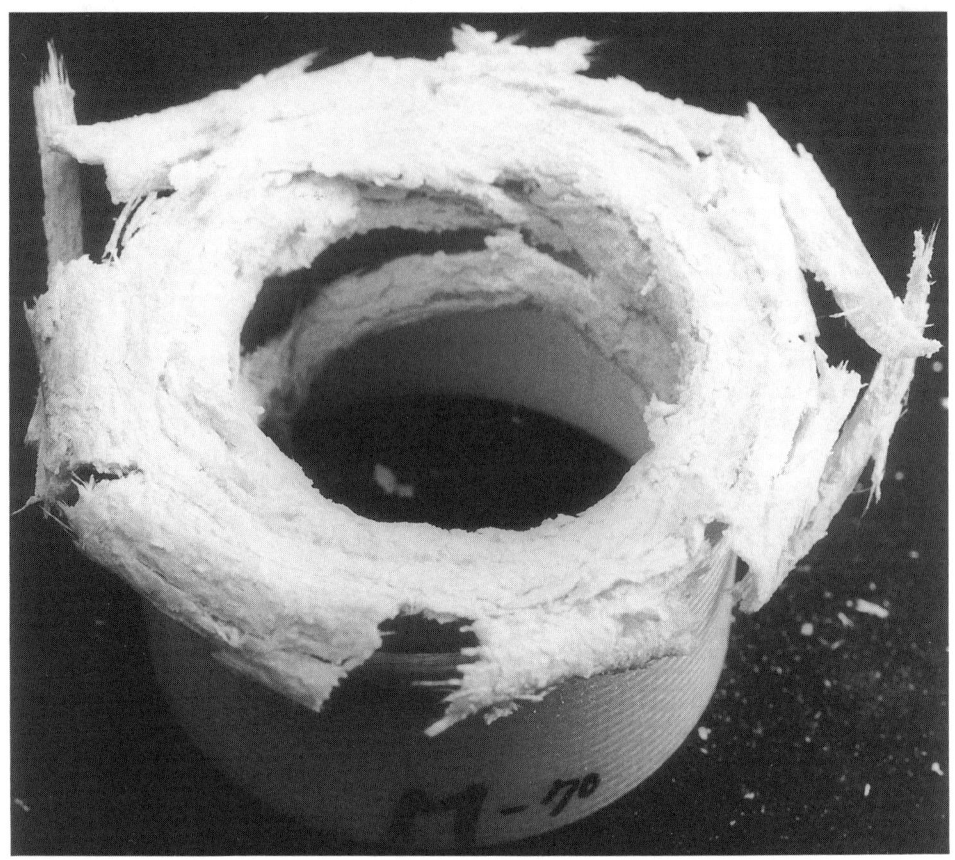

Fig. 3. Optical photograph of quasi-static tested aminosilane-treated glass cloth/epoxy tube.

Impact Crush Test Results. Typical P-d traces obtained from impact testing of aminosilane-treated and acrylsilane-treated tubes are shown in Figure 7. Compared to the quasi-static P-d traces the impact P-d traces were less serrated. The mean crush loads obtained from the P-d traces and the specific energies of impact tested tubes are summarised in Table 2. The aminosilane-treated tubes displayed 15% higher E_s than the acrylsilane-treated tubes. In the case of aminosilane-treated tubes the E_s increased only marginally by changing the testing speed from quasi-static to impact. However, the E_s of acrylsilane-treated tubes increased by 15% with the change in testing condition from quasi-static speed to impact speed.

Optical photographs of crush zones of impact tested aminosilane-treated and acrylsilane-treated tubes are shown in Figures 8 and 9 respectively. Typical photographs of sections cut through the crush zones of both the types of tubes are shown in Figures 10 and 11. Unlike the quastic-static tested tubes, the appearance of impact crush zones of both the types of tubes was similar. The tubes have been split into a number fronds along the circumference of the tube wall. Both the aminosilane-treated and acrylsilane-treated tubes crushed progressively by splaying mode. Tube wall has been splayed into internal and external fronds. Internal fronds bent through shaper radius of curvature than the external fronds. A number of hackles were observed on the concave side of fronds. A debris wedge and a longitudinal crack below the debris wedge were also observed.

Table 2. Energy absorption performances of impact tested glass cloth/epoxy tubes.

Tube Type	Mean Crush Load $\overline{P}$ (kN)	Mean Crush Stress (MPa)	Specific Energy (kJ/kg)
Aminosilane	51.5	116.1	69.5
Acrylsilane	46.0	106.1	61.4

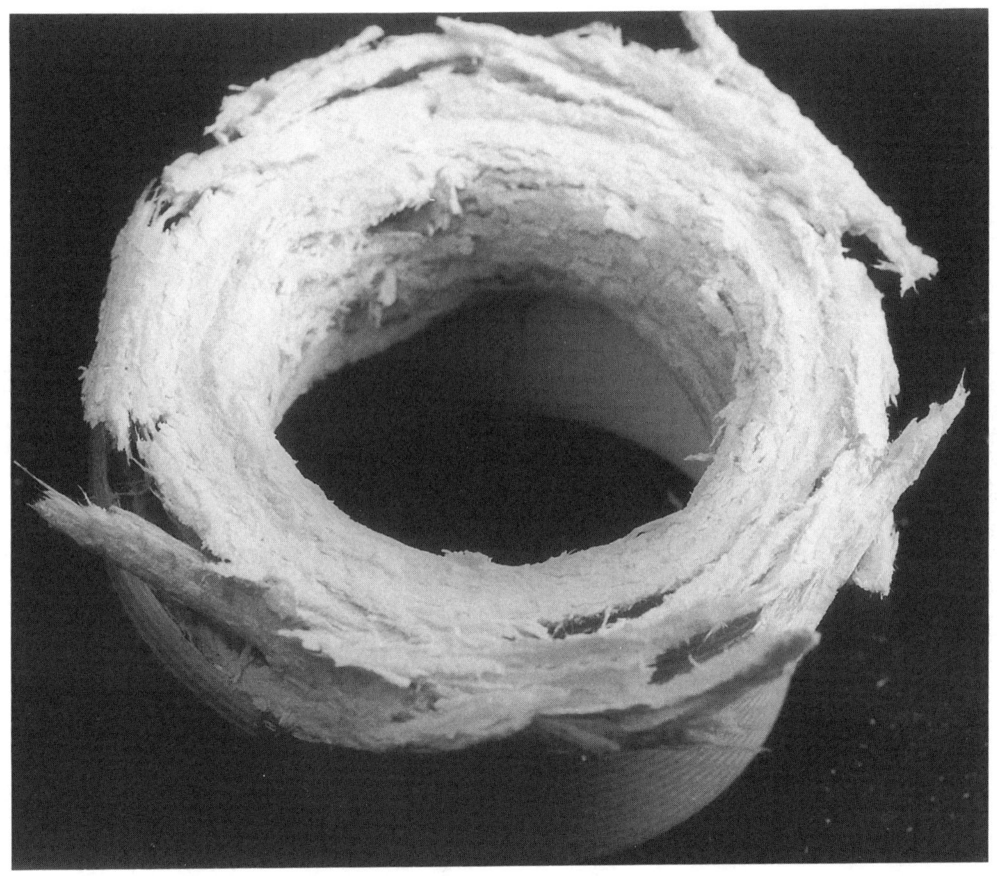

Fig. 4. Optical photograph of quasi-static tested acrylsilane-treated glass cloth/epoxy tube.

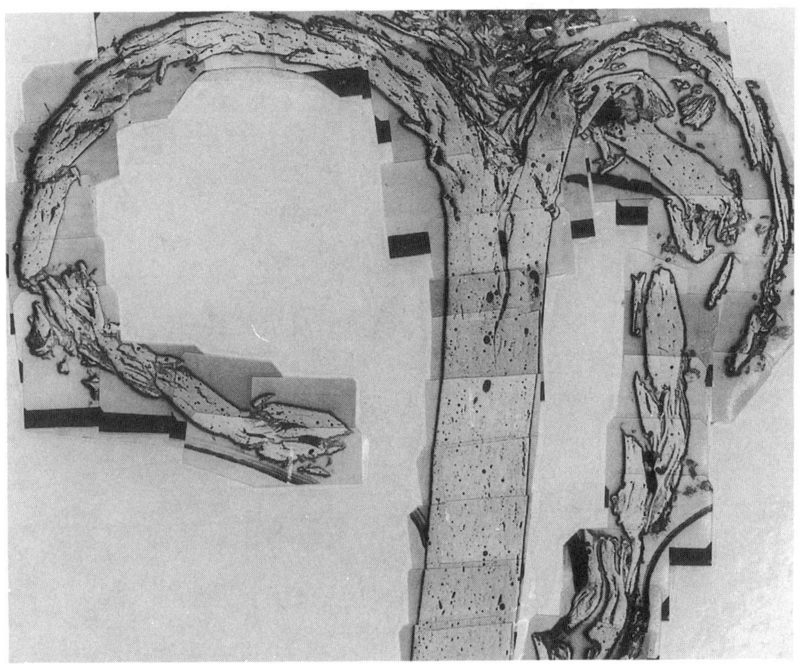

Fig. 5. Photomicrograph of a cross-section through the crush zone of quasi-static tested aminosilane-treated tube.

505

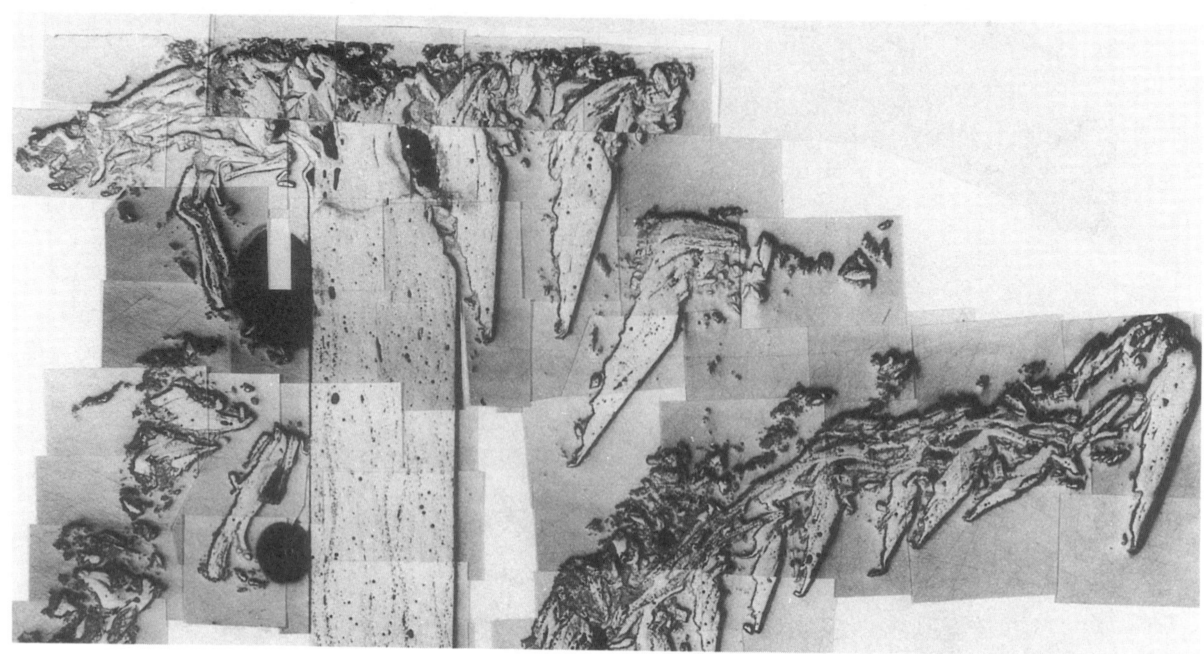

Fig. 6. Photomicrograph of a cross-section through the crush zone of quasi-static tested acrylsilane-treated tube.

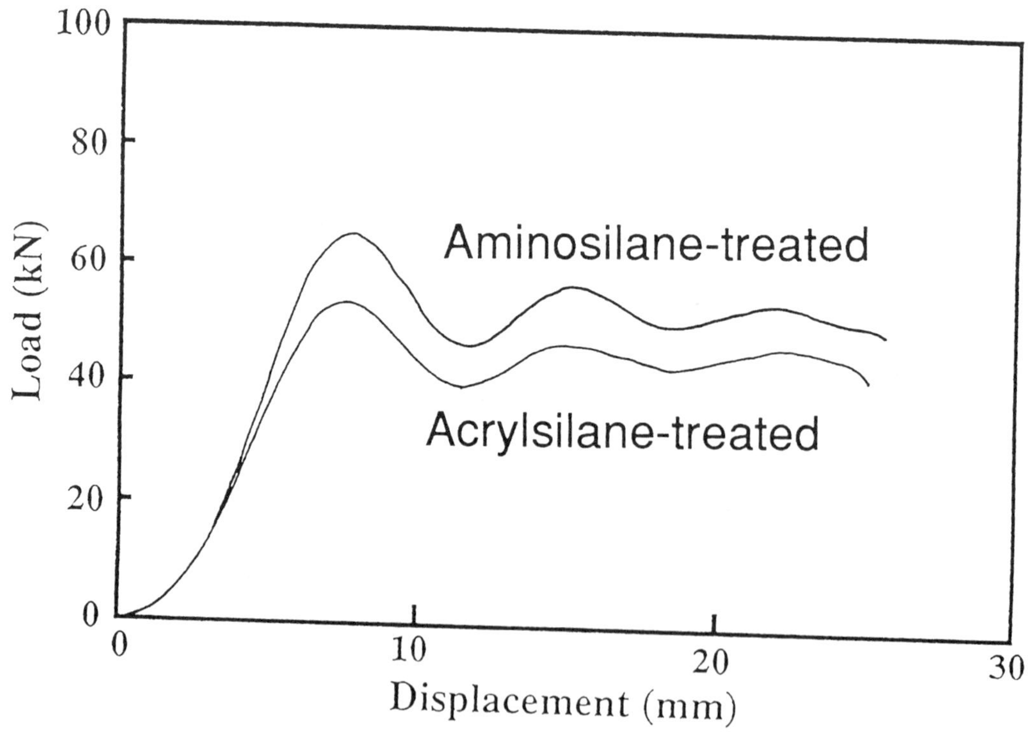

Fig. 7. Typical load-displacement curves for impact tested tubes.

Fig. 8. Optical photograph of impact tested aminosilane-treated glass cloth/epoxy tube.

Fig. 9. Optical photograph of impact tested acrylsilane-treated glass cloth/epoxy tube.

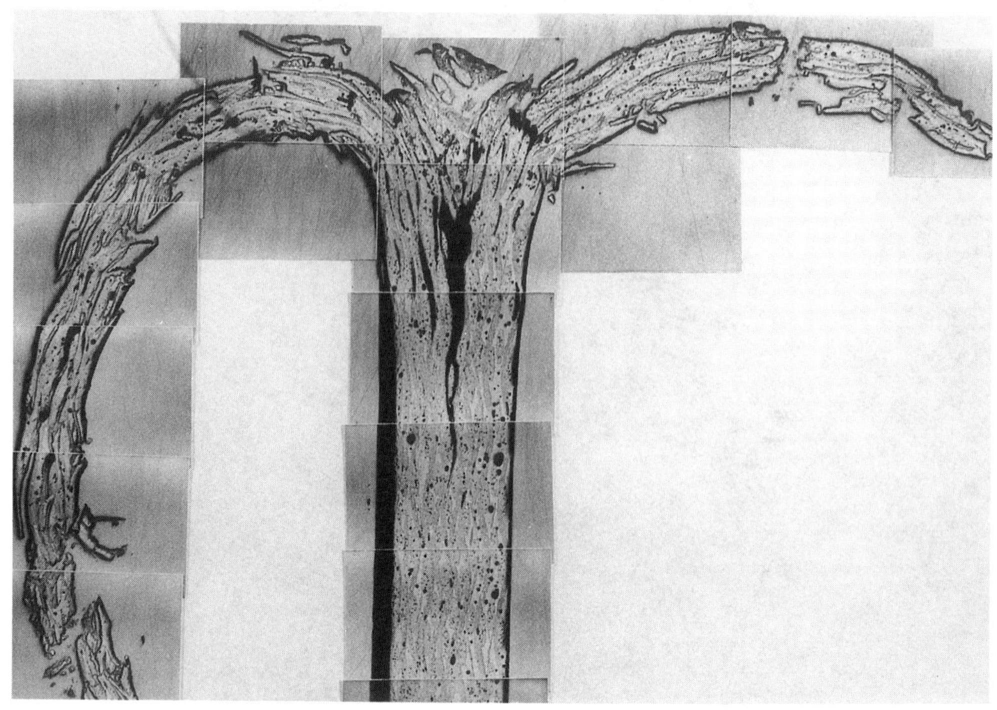

Fig. 10. Photomicrograph of a cross-section through the crush zone of impact tested aminosilane-treated tube.

Fig. 11. Photomicrograph of a cross-section through the crush zone of impact tested acrylsilane-treated tube.

Discussion

In general, the aminosilane-treated tubes displayed higher energy absorption performance than the acrylsilane-treated tubes (Tables 1 and 2). This may be due to the different fiber/matrix bonding. For further understanding we measured the mechanical properties of both the types of materials (Table 3). Aminosilane-treated materials displayed higher mechanical properties than the acrylsilane-treated materials, indicating the good interfacial bonding between the aminosilane-treated glass fibers and the epoxy resin matrix.

Table 3. Mechanical properties of glass cloth/epoxy composite materials.

Property	Aminosilane Treated	Acrylsilane Treated
Tensile Modulus (GPa)	21.2	20.5
Tensile Strength (MPa)	403.1	315.6
Bending Modulus (GPa)	19.1	16.6
Bending Strength (MPa)	511.9	360.4
Compressive Strength (MPa)	272.8	211.3
Fracture Toughness G_{IC} (J/m^2)	1201.7	1008.3

The energy absorption capability of both the types of composite tubes increased with increasing testing speed from quasi-static to impact. However, the increase was significant in the case of acrylsilane-treated tubes compared to the aminosilane-treated composite tubes. This may be due to the change in progressive crushing mode in the case of acrylsilane-treated tubes. The aminosilane-treated tubes crushed by splaying mode under quasi-static as well as impact testing conditions. In the case of acrylsilane-treated tubes the crushing mode changed from fragmentation to splaying mode with the change in testing condition from quasi-static to impact.

The energy absorption capability of a composite tube is determined by various fracture processes in the crush zone. Fairfull and Hull [12] showed experimentally that the friction between the fronds and crushing platen, and the friction between the fronds and debris wedge are major sources of energy absorption. These frictional mechanisms contribute 60% of the total energy absorption. Compared to the fragmentation crush zone, the fronds of splaying mode crush zones were large and continuous. Larger contact surfaces lead to higher frictional forces. The presence of higher frictional forces might have contributed for the superior energy absorption capability of the tubes crushed by the splaying mode. The general increase in energy absorption capability with increasing testing speed may be attributed to the strain rate sensitivity of the composite material. It is known that the tensile and compressive properties of composite material increase with increasing strain rate [13-14]. This may lead to higher energy absorption capability. However, the fracture toughness and the resistance to the crack growth decrease with increasing strain rate which may lead to reduction in energy absorption capability. Detailed speculation is not possible since systematic experimental data relating the specific energy with the mechanical properties is not available.

Conclusions

The energy absorption capability of glass/epoxy composite tubes increases with increasing testing speed from quasi-static to impact. In general, the crushing performance of aminosilane-treated tubes is higher than that of acrylsilane-treated tubes. The splaying mode crushing of aminosilane-treated tubes is independent of testing speed. In the case of acrylsilane-treated tubes the crush zone morphology changes from fragmentation mode to splaying mode with increasing testing speed.

References

1   Hull, D. Science and Technology Review, University of Wales, UK, 3, 22-30 (1988).

2   Hull, D. Composites Science and Technology, 40, 377-421 (1991).

3   Thronton, P.H. and P.J. Edwards, J. Composite Materials, 16, 521-545 (1982).

4   Ramakrishna, S and D. Hull, Composites Science and Technology, 49, 349-356 (1993).

5   Farley, G.L. J. Composite Materials, 17, 267-279 (1983).

6   Farley, G.L. J. Composite Materials, 25, 1314-1329 (1991).

7   Bannerman, D.C. and C.M. Kindervater, in Proceedings of the Fourth International SAMPE European Chapter, Bordeaux, France, 1984, p. 155-167.

8   Schmuesser, D.W. and L.E. Wickliffe, J. Engineering Materials and Technology, 109, 72-77 (1987).

9   Thornton, P.H. J. Composite Materials, 24, 594-615 (1990).

10   Hamada, H., S. Ramakrishna, M.Nakamura, Z.Maekawa, H.Sato and D.Hull, in Proceedings of 12th Symposium on Composite Materials: Testing and Design, ASTM, Montreal, Canada, May 1994.

11   Hamada, H., S. Ramakrishna, M. Nakamura, Z. Maekawa and D. Hull, Composite Interfaces, 2, 2, 127-142 (1994).

12   Fairfull, A.H. and D. Hull, "Energy absorption of polymer matrix composite structures: frictional effects," p 255, Structural Failure, T.Weirzbicki and N.Jones, Eds, John Wiley & Sons (1989).

13   Pink, E. and J.D.Campbell, J.Materials Science, 9, 658-664 (1974).

14   Harding, J. and L.M.Welsh, in Proceedings of ICCM-IV, Tokyo, 1982, p. 845-852.

Proceedings of the 10th Annual ASM/ESD Advanced Composites Conference, Dearborn, Michigan, USA, 7-10 November 1994

Energy Absorption Behavior of Hybrid Composite Tubes

H. Hamada, S. Ramakrishna, M. Nakamura, Z. Maekawa
Kyoto Institute of Technology
Kyoto, Japan

Abstract

Majority of the energy absorption research work has been carried out using glass fiber and carbon fiber reinforced composite tubes. It has been shown experimentally that when properly designed these composite materials absorb impact energy in a controlled manner. Poor post crushing integrity of these composite materials is an issue to be addressed before they can be considered for crashworthy structural applications. This problem can be overcome by using ductile reinforcing fibers such as high performance polyethylene fibers (Dyneema SK60). However, these fibers show poor properties in compression. Hence, to obtain post-crushing integrity with good energy absorption capability, hybrid composite tubes reinforced with both carbon and Dyneema fibers have been proposed. This paper describes the effects of fiber orientation, stacking sequence and proportions of fibers on the energy absorption characteristics of hybrid composite tubes.

THERE IS NOW a considerable amount of published data on the energy absorption characteristics of polymer composite materials. It is well established that when properly designed, polymer composite materials display higher specific energy absorption than the conventional metals [1-2]. This unique potential of polymer composite materials has generated considerable interest in recent years for crashworthy structural applications in aerospace and automotive vehicles. It has been shown that the crushing characteristics of composite tubes are influenced by tube geometry [3-9], fiber architecture [10-11], fiber type [12], matrix resin type [13] and fiber/matrix bonding strength [14]. Majority of energy absorption research has been carried out using glass fiber and carbon fiber reinforced thermoset composite tubes. Under axial compression these composite tubes crush progressively from one end by either splaying mode or fragmentation mode. Often a mixture of both the crushing modes are also observed in the crush zones. These progressive crushing processes are associated with a large amount of microfracturing of tube wall. The fine debris produced during crushing may cause environmental problems. Also the loss of structural integrity after crushing is a major drawback with respect to their use in primary structures, as the structure must remain intact to provide protection after a crash. One of the ways this problem can be over come is by reinforcing with ductile fibers such as aramid and high performance polyethylene fibers instead of brittle glass and carbon fibers. Farley [12] reported that Kevlar/epoxy tubes crushed by progressive folding and with good post crush integrity. However, the Kevlar/epoxy tubes displayed lower energy absorption capability than the carbon/epoxy or glass/epoxy tubes. To obtain both post crush integrity and reasonable energy absorption Peijis and Vanklinken [15] and Thuis and Metz [16] suggested hybrid composite tubes in which both high performance polyethylene fibers and carbon fibers are used as reinforcement. These studies were confined to particular stacking sequences. This paper describes the effects of fiber orientation, stacking sequence and proportions of fibers on the energy absorption characteristics of hybrid composite tubes.

Materials

Composite tubes used in this study were supplied by Toyobo Co, Japan. High performance polyethylene fibers (Dyneema SK60), carbon fibers (T300B) and epoxy resin were used in hybrid composite tubes. Composite tubes were fabricated using two different techniques. One set of tubes were manufactured by filament winding technique using unidirectional prepregs. The other set of tubes were fabricated by sheet winding technique using woven fabric prepregs. Tubes fabricated using filament winding technique are summarised in Table 1. The value given in the brackets gives the fiber orientation with respect to the tube axis. For example, in D(10) and D(30) tubes only Dyneema fibers were used and they were oriented at $\pm10^0$ and $\pm30^0$ with respect to the tube axis respectively. $[C\overline{D}C]_S(10)$ and $[C\overline{D}C]_S(30)$ tubes were made by sandwitching of layers of Dyneema fibers between two layers of carbon fibers. In otherwords all the carbon fibers were on the inside and outside layers of the tube wall and the Dyneema fibers were in the core layer. $[CD]_S(10)$ and $[3C1D]_S(10)$ tubes were fabricated by alternate winding of Dyneema and carbon fibers at $\pm10^0$ to the tube axis. In $[CD]_S(10)$ tubes the ratio of carbon fibers to Dyneema fibers was 1:1. In the case of $[3C1D]_S(10)$ tubes the ratio of carbon fibers to the Dyneema fibers was 3:1.

Table 1. Details of filament wound tubes.

Tube Nomenclature	Fiber Orientation	Density (g/cc)
D(10)	$\pm10^0$	1.09
D(30)	$\pm30^0$	1.04
$[C\overline{D}C]_S(10)$	$\pm10^0$	1.39
$[C\overline{D}C]_S(30)$	$\pm30^0$	1.41
$[CD]_S(10)$	$\pm10^0$	1.25
$[3C1D]_S(10)$	$\pm10^0$	1.37

C and D mean carbon fibers and Dyneema fibers respectively

Three types of woven fabrics were used (Figure 1): first type is the plain woven fabrics containing only Dyneema fibers in warp and fill directions (Figure 1a); second type is hybrid fabrics containing carbon fibers and Dyneema fibers in the warp and fill directions respectively (Figure 1b); and third type is also hybrid fabrics made by alternate weaving of carbon fibers and Dyneema fibers (Figure 1c). Different composite tubes were fabricated by varying the orientation of woven fabric with respect to the tube axis (Table 2). (0) and (90) indicate axial and hoop directions of the tube. The tube nomenclature D(0)/D(90) means that Dyneema fibers of woven fabric were aligned in the axial and hoop directions of the composite tube. Similarly the nomenclature D(45)/D(-45) means that Dyneema fibers were aligned at $\pm45^0$ to the tube axis. In the case of C(0)/D(90) tubes, the woven fabric was arranged such that the carbon fibers were parallel to

tube axis and the Dyneema fibers were in the hoop direction. Similarly in C(90)/D(0) tubes the woven fabric is arranged such that the Dyneema and carbon fibers were in the axial and hoopdirections of the tube respectively. The CD(0/90) and CD(±45) tubes were fabricated using third type woven fabrics (Figure 1c). In CD(0/90) tubes the warp and fill directions of the woven fabric were parallel to the axial and hoop directions of the tube respectively. In CD(±45) tubes the warp direction of the woven fabric is aligned at 45^0 to the tube axis.

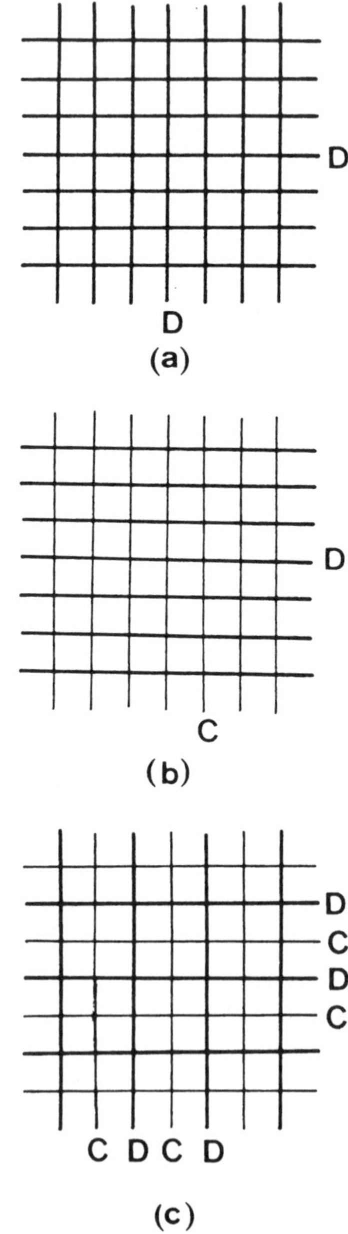

Fig. 1. Arrangement of carbon fibers (C) and Dyneema fibers (D) in woven fabrics.

Table 2. Details of woven fabric tubes.

Tube Nomenclature	WovenFabric Type	Density (g/cc)
D(0)/D(90)	Plain Fabric	1.06
D(45)/D(-45)	Plain Fabric	1.06
C(0)/D(90)	Hybrid Fabric	1.22
C(90)/D(0)	Hybrid Fabric	1.21
CD(0/90)	Hybrid Fabric	1.25
CD(±45)	Hybrid Fabric	1.25

C and D mean carbon fibers and Dyneema fibers respectively

Experimental Procedure

All the tubes had an internal diameter of 50 mm and a wall thickness of 2.5 mm. The fabricated tubes were cut into specimens of 55 mm long. To initiate progressive crushing a 45^0 chamfer was ground onto one end of each tube specimen. Crush tests were carried out using a 100 kN Shimadzu Autograph servo-hydraulic testing machine at a constant cross-head speed of 1mm/min. Composite tubes were axially crushed between parallel steel flat platens, one static and one moving. The fixed platen was fitted with a load cell from which the load signal was taken directly to the x-y plotter. In each test the crush load was plotted on the y-axis and the cross-head displacement on the x-axis. Specific energy absorption capability (E_s) is calculated as the ratio of the mean crush load from the load-displacement (P-d) curves and the mass per unit length of the uncrushed tube.

The crushed tubes were cast in polyester resin to preserve the crush zone morphology. The microfracture processes in the crush zone were determined by cutting sections through the embedded material. The sections were then polished and photographed on an optical microscope.

Results and Discussion

Filament Wound Composite Tubes. Typical P-d curves for filament wound composite tubes are shown in Figure 2. The corresponding failure modes of the crushed tubes are shown in Figure 3. The P-d curves of D(10) and D(30) tubes were smooth compared to the P-d curves of other tubes. Similarly the appearance of crush zones of D(10) and D(30) tubes was different from the other crush zones. The D(10) and D(30) composite tubes were reinforced with Dyneema fibers only. Whereas the other tubes were reinforced with carbon fibers as well as Dyneema fibers. The D(10) and D(30) tubes crushed by local buckling of tube wall from the chanfered end. This local buckling of tube wall was similar to the pregressive folding exhibited in metallic, thermoplastic and aramid/epoxy tubes. In the case of hybrid tubes, the P-d curves showed an initial rapid load increase followed by a large load drop. After the load drop, the load saturated at an average value. This average value is called as mean crush load. The mean crush load of $[C\overline{D}C]_S(10)$ tube was higher than that of the other hybrid composite tubes. All the hybrid composite tubes crushed progressively from the chamfered end in a similar manner. Tube wall has been splayed to the inside and outside of the tube wall. This mode of crushing is known as 'splaying mode' crushing [2]. Figure 4 shows optical photograph of typical section through the crush zone of $[C\overline{D}C]_S(10)$ hybrid composite tube. The outer carbon fiber layers splayed to the inside and outside of the tube wall. Whereas the core Dyneema fibers buckled and formed as a debris wedge. Breakage of carbon fibers in the outer layers can also be seen. The features of crush zone of $[C\overline{D}C]_S(30)$ tube were similar to those observed in the crush zone of $[C\overline{D}C]_S(10)$ tube. Optical photograph of a typical section through the crush zone of $[CD]_S(10)$ tube is shown in Figure 5. The crush zone consisted typical features of splaying mode crush zone such as a debris wedge splitting the tube wall into internal and external fronds, and longitudinal crack below the debris wedge. The fronds were large and continuous. The fracture of carbon fibers in the fronds was less prominent compared to that observed in the crush zone shown in Figure 4. Debris wedge consisted both buckled Dyneema fibers and fractured carbon fibers. The crush zone of $[3C1D]_S(10)$ tube showed a large number of fractured carbon fibers compared to the crush zone of $[CD]_S(10)$ tube. In other words, eventhough the over all mode of crushing of hybrid composite tubes was similar there exist differences regarding microfracture processes in the crush zone.

The energy absorption performances of filament wound tubes are summarised in Table 3. The hybrid composite tubes displayed higher specific energy absorption than the pure Dyneema composite tubes D(10) and D(30). This may be attributed to the absence of certain fracture mechanisms in pure Dyneema composite tubes. No significant fiber fracture was observed in the crush zones of pure Dyneema composite tubes. The progressive folding process of tube wall is associated with fracture processes such as matrix cracking, fiber/matrix debonding and interlaminar delamination. Only these fracture processes determine the energy absorption capability of pure Dyneema composite tubes. In the case of hybrid composite tubes, fracture of carbon fibers was observed. Also the carbon fibers are stiffer than the the Dyneema fibers. The presence of carbon fibers enhances the resistance to the advancement of crushing platen. These factors might have resulted in superior energy absorption capability of hybrid composite tubes. The $[C\overline{DC}]_S(10)$ tubes displayed 67.1 kJ/kg specific energy higher than the 31.0 kJ/kg of $[C\overline{DC}]_S(30)$ tubes. This is a direct result of the effect of fiber orientation with respect to the tube axis. In $[C\overline{DC}]_S(10)$ tubes the fibers were aligned closer to the axis of the tube and hence higher specific energy. Among all the hybrid composite tubes, the $[3C1D]_S(10)$ tubes absorbed highest specific energy of 76.4 kJ/kg. This may be due to the higher proportion of carbon fibers used in these particular composite tubes. It may be concluded that energy absorption capability of hybrid composite tubes can be optimised by aligining more number of carbon fibers in the axial direction of the tube.

Woven Fabric Reinforced Composite Tubes. Typical P-d curves for woven fabric reinforced composite tubes are shown in Figure 6. The corresponding failure modes of the crushed tubes are shown in Figure 7. Similar to the filament wound tubes, the pure Dyneema woven fabric reinforced tubes displayed lower mean crush loads than the hybrid woven fabric composite tubes. In D(0)/D(90) tubes, progressive folding of tube wall occurred from the chamfered end (Figure 7a). In the case of D(45)/D(-45) tubes, buckling of tube wall occurred away from the chamfered end (Figure 7b). Intially the chamfered portion of the tube wall was compressed. With further advancement of the crushing platen the tube wall close to unchamfered end buckled. This led to gradual drop of load in the load-displacement curve (Figure 6). This implies that no trigger is necessary for initiating progressive

folding in composite tubes reinforced with Dyneema fibers only. But the mean crush load of these composite tubes will be lower that of the hybrid composite tubes. The P-d curves of hybrid woven fabric reinforced tubes showed no initial peak load. After the initial rapid load increase, the load fluctuated around a mean value. Among the hybrid woven fabric tubes the C(0)/D(90) tubes displayed highest mean crush load of 23.9 kN. Optical photographs of sections through the crush zones of hybrid woven fabric composite tubes are shown in Figures 8 to 11. The hybrid C(0)/D(90) tubes crushed progressively from the chamfered end by splaying mode (Figure 8). Splitting of tube wall into internal and external fronds can be seen. The debris wedge consisted buckled and fractured Dyneema fibers and carbon fibers. Fronds were continuous and did not separate from the tube wall. In the case of hybrid C(90)/D(0) tubes tube wall has been folded suceessively from the chamfered end (Figure 9). This progressive folding was similar to that observed in certain pure Dyneema composite tubes. The difference between the crushing modes of C(0)/D(90) and C(90)/D(0) tubes may be attributed to the carbon fiber orientation with respect to the tube axis. In C(0)/D(90) tubes the carbon fibers were aligned parallel to the tube axis whereas they were aligned in hoop direction in C(90)/D(0) tubes. Composite tubes with axial carbon fibers posses higher stiffness and hence they crushed by splaying mode. The C(90)/D(0) tubes were less stiffer in the axial direction and hence they crushed by progressive folding mechanism. Hybrid composite tubes CD(0/90) crushed progressively from the chamfered end by splaying mode (Figure 10). Both the internal and external fronds have been bent through sharp radii of curvatures. In the case of internal fronds compressive buckling of plies on the concave side was observed. However no fiber fracture was noticed. Fronds were large and continuous. The crush zone of hybrid CD($\pm$45) tube contained features of unstable splaying mode crushing (Figure 11). In the crush zone, tube wall has been split and splayed to the inside and outside of the tube. Away from the crush zone buckling of tube wall was observed. In other words, initially a splaying mode crush zone was formed by crushing of chamfered portion of the tube wall. With further advancement of crushing platen the crush load may be sufficently high to produce buckling of tube wall. This was associated with gradual load drop in the P-d curve (Figure 6).

The energy absorption performances of woven fabric reinforced composite tubes are given in Table 4. Similar to the filament wound tubes, the pure Dyneema woven fabric reinforced tubes displayed lower energy absorption capability than the hybrid woven fabric composite tubes. Among the hybrid woven fabric composite tubes, the C(0)/D(90) and CD(0/90) tubes displayed higher energy absorption capabilities than C(90)/D(0) and CD($\pm$45) tubes. This may be attributed to the difference in crushing mode.

Table 3. Energy absorption performances of filament wound tubes.

Tube Type	Mean Crush Load (kN)	Mean Crush Stress (MPa)	Specific Energy (kJ/kg)
D(10)	11.5	31.6	29.0
D(30)	6.0	16.5	15.1
$[C\overline{DC}]_S(10)$	42.0	94.8	67.1
$[C\overline{DC}]_S(30)$	18.9	43.1	31.0
$[CD]_S(10)$	27.3	59.7	48.0
$[3C1D]_S(10)$	43.6	104.6	76.4

C and D mean carbon fibers and Dyneema fibers respectively

The former tubes crushed by well defined splaying mode whereas the later tubes crushed by progressive folding mechanism. In other words the tubes crushed by splaying mode display higher energy absorption performance than the tubes crushed by progressive folding. In both the crushing modes the post crush integrity of the tubes was good. This may be due to the presence of ductile Dyneema fibers. The highest energy absorption recorded for filament wound hybrid composite tube $[3C1D]_s(10)$ was 76.4 kJ/kg. Whereas the highest specific energy was 49.5 kJ/kg in the case of hybrid woven fabric composite tube C(0)/D(90). The superior energy absorption performance of filament wound tubes may be due to the higher proportion of carbon fibers compared to those in the woven fabric composite tubes. Also the fibers in the woven fabrics were wavy in nature at the cross-over points of the warp and fill yarns. In the case of filament wound tubes the fibers were straight without any wavyness. The reinforcement efficiency of straight fibers will be higher than that of the wavy fibers. Hence, the filament wound tubes displayed higher energy absorption capabilities than the woven fabric composite tubes.

Table 4. Energy absorption performances of woven fabric reinforced tubes.

Tube Type	Mean Crush Load (kN)	Mean Crush Stress (MPa)	Specific Energy (kJ/kg)
D(0)/D(90)	13.4	33.8	32.1
D(45)/D(-45)	6.5	16.4	16.0
C(0)/D(90)	23.9	60.3	49.5
C(90)/D(0)	16.4	42.8	35.5
CD(0/90)	23.0	58.9	47.3
CD(±45)	19.4	49.5	40.0

C and D mean carbon fibers and Dyneema fibers respectively

Summary

Tubes reinforced with all Dyneema fibers displayed a maximum specific energy absorption of 30 kJ/kg and crushed by progressive folding of tube wall. The hybrid composite tubes displayed higher energy absorption capability than the tubes reinforced with Dyneema fibers only. In general, the hybrid composite tubes crushed progressively by splaying mode. Filament wound hybrid tubes with carbon and Dyneema fibers in 1:1 proportion displayed a specific energy of 50 kJ/kg. The specific energy was further increased to 80 kJ/kg by changing the proportion of carbon: Dyneema fibers to 3:1. For the same fiber proportions, the stacking arrangement of Dyneema and carbon fibers affected the energy absorption capability. Tubes with a layer of Dyneema Fibers sandwitched between two layers of carbon fibers displayed 65 kJ/kg compared to the 50 kJ/kg displayed by the tubes made by alternate winding of carbon and Dyneema fibers. The energy absorption capability decreased with increasing fiber orientation with respect to the longitudinal axis of the tube. The filament wound tubes displayed higher energy absorption capability than the woven fabric tubes. It has been shown that the energy absorption and crushing mode can be tailored by controlling the fiber architecture.

References

1   Hull, D., Science and Technology Review, University of Wales, UK, 3, 22-30 (1988).

2   Hull, D., Composites Science and Technology, 40, 377-421 (1991).

3   Thronton, P.H. and P.J. Edwards, J. Composite Materials, 16, 521-545 (1982).

4   Fairfull, A.H. and D. Hull, in Proceedings ICCM-VI & ECCM 2, F.L. Matthews, N.C.R.Buskell, J.M.Hodgkinson and J.Morton, Eds, Elsevier Applied Science, London, 1987, p 3.36-3.45.

5   Farley, G.L. and R.M. Jones, J. Composite Materials, 26, 78-89 (1992).

6   Thornton, P.H., Composites Science and Technology, 27, 199-224 (1986).

7   Price, J.N. and D. Hull, Composites Science and Technology, 28, 211-230 (1987).

8   Farley, G.L., J. Composite Materials, 20, 390-400 (1986).

9   Farley, G.L. and R.M. Jones, J. Composite Materials, 26, 1741-1751 (1992).

10   Ramakrishna, S and D. Hull, Composites Science and Technology, 49, 349-356 (1993).

11   Ramakrishna, S. and D. Hull, in Proceedings of International Conference on Advances in Structural Testing, Analysis and Design, B.Dattaguru, Ed., Tata McGraw-Hill Publishing Company Ltd, New-Delhi, 1990, p 69-74.

12   Farley,G.L., in Proceedings of the 43rd American Helicopter Society Annual Forum, St.Louis, 1987, p.613-627.

13   Thornton, P.H. J. Composite Materials, 24, 594-615 (1990).

14   Hamada, H., S. Ramakrishna, M. Nakamura, Z. Maekawa and D. Hull, Composite Interfaces, 2, 2, 127-142 (1994).

15   Peijis, A.A.J.M. and E.J. Vanklinken, J.Materials Science Letters, 11, 520-522 (1992).

16   Thuis, H.G.S.J. and V.H.Metz, Composite Structures, 25, 37-43 (1993).

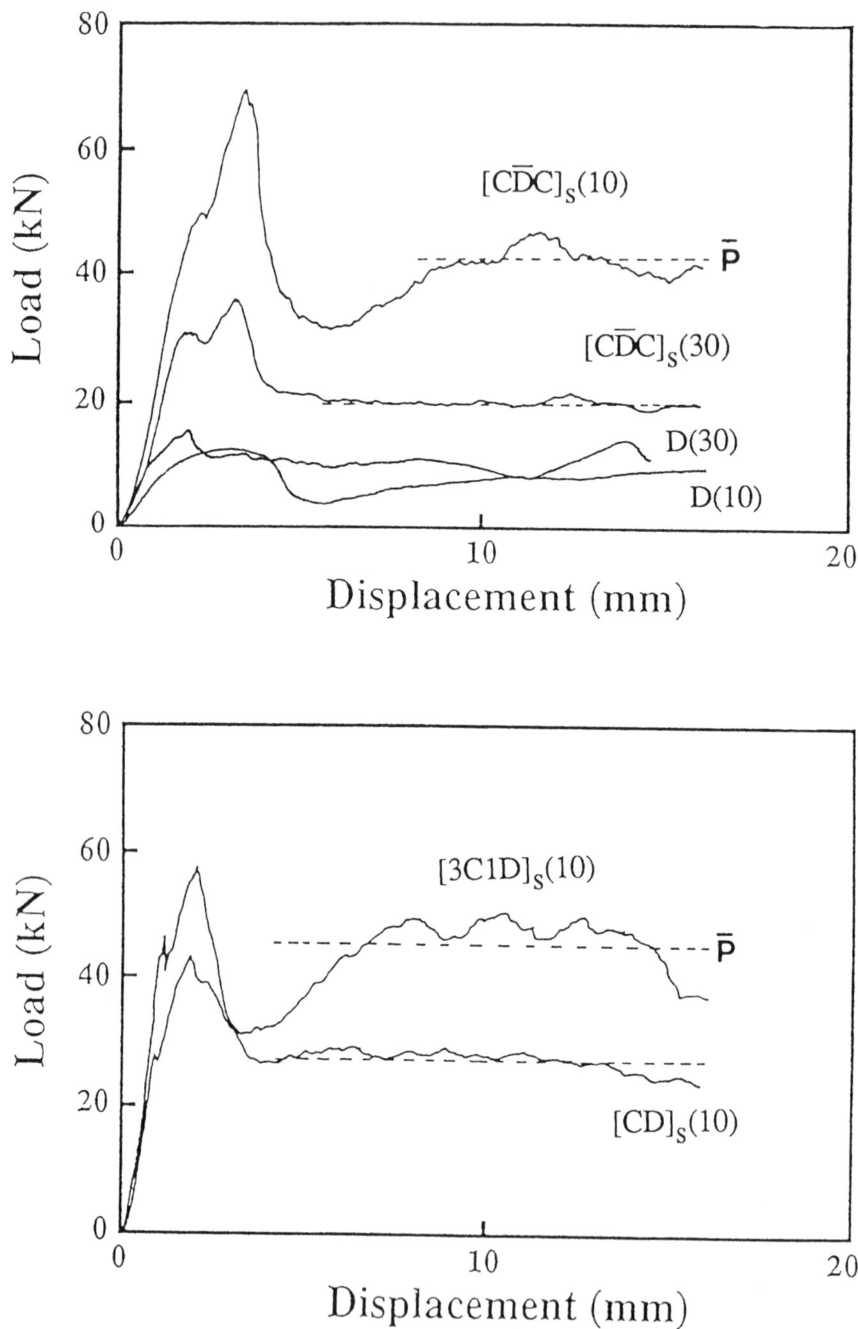

Fig. 2. Typical load-displacement curves for filament wound tubes.

Fig. 3. Optical photographs of crushed tubes a) D(10), b) D(30), c) $[C\overline{D}C]_S(10)$, d) $[C\overline{D}C]_S(30)$, e) $[CD]_S(10)$ and f) $[3C1D]_S(10)$.

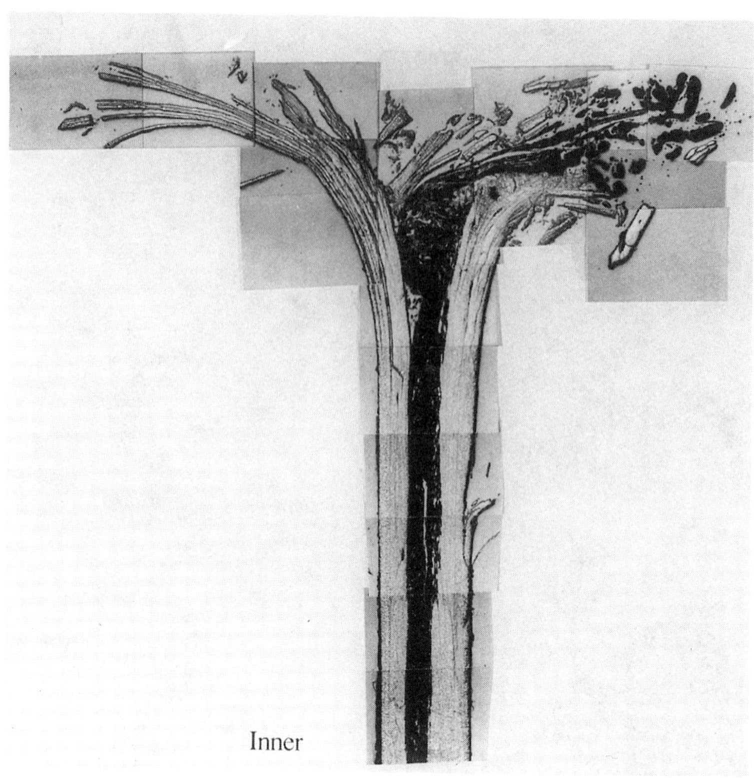

Fig. 4. Photomicrograph of section through the crush zone of $[C\overline{D}C]_S(10)$ hybrid tube.

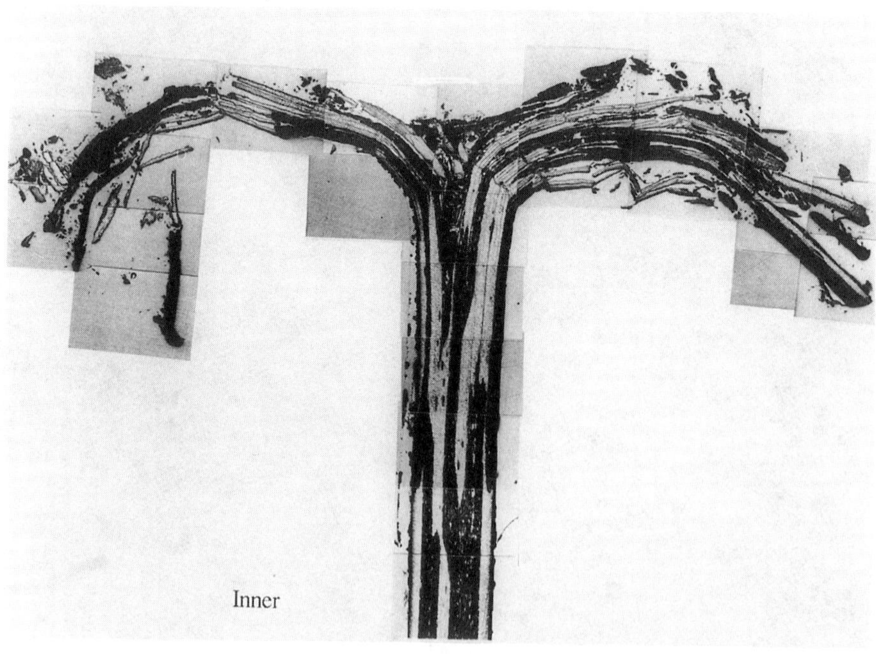

Fig. 5. Photomicrograph of section through the crush zone of $[CD]_S(10)$ hybrid tube.

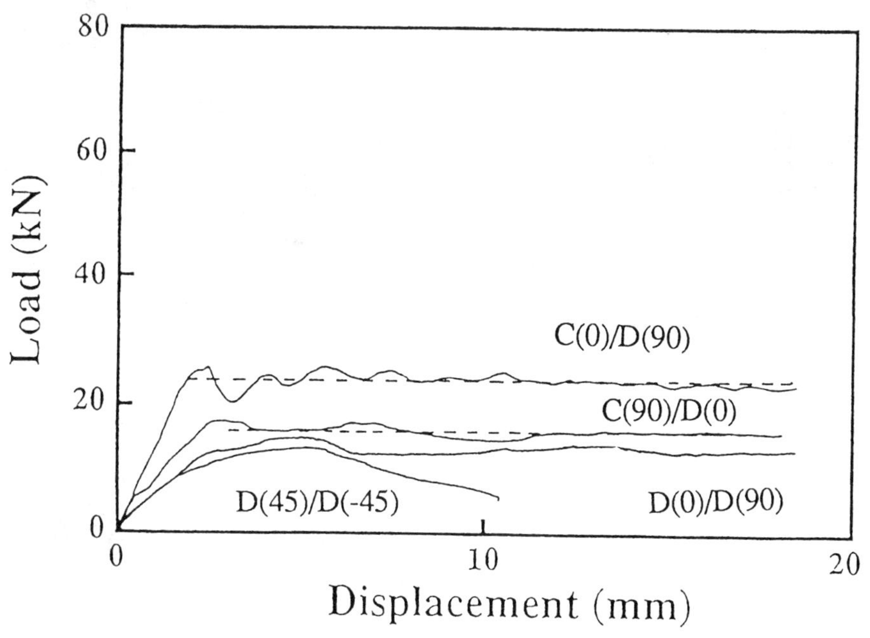

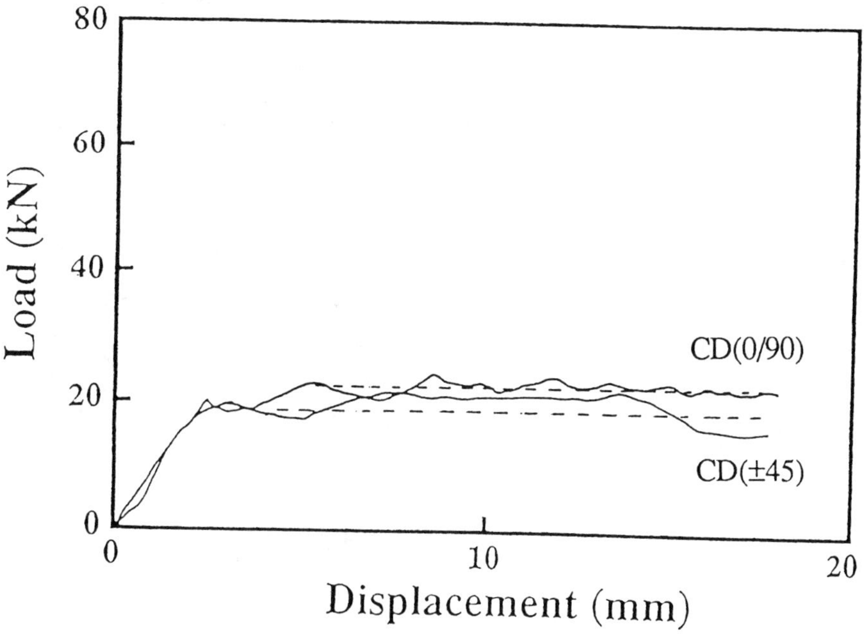

Fig. 6. Typical load-displacement curves for woven fabric reinforced tubes.

Fig. 7. Optical photographs of crushed tubes a) D(0)/D(90), b) D(45)/D(-45), c) C(0)/D(90), d) C(90)/D(0), e) CD(0/90) and f) CD(±45).

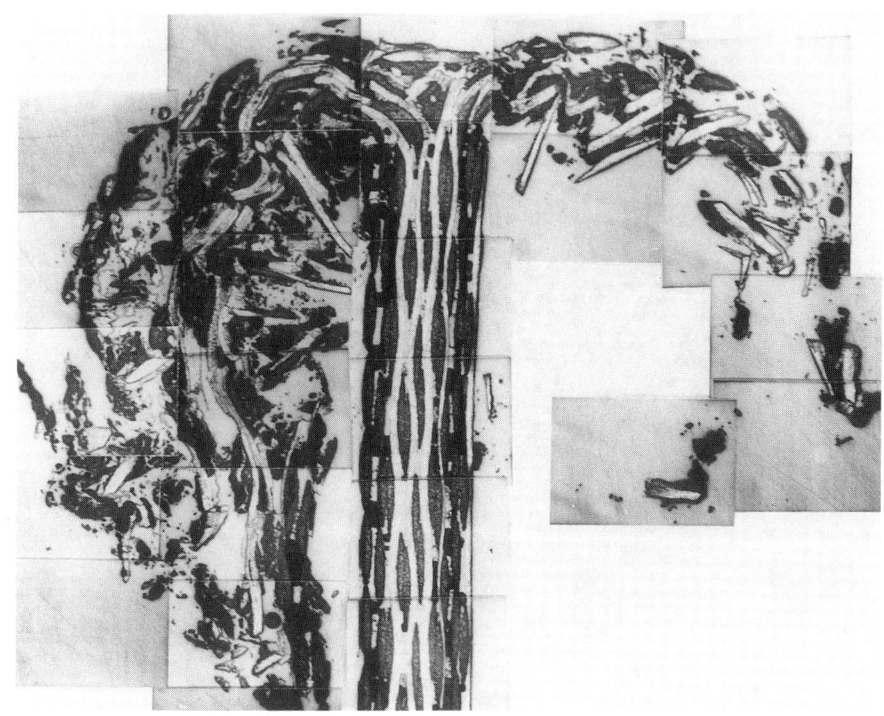

Fig. 8. Photomicrograph of section through the crush zone of C(0)/D(90) hybrid tube.

Fig. 9. Photomicrograph of section through the crush zone of C(90)/D(0) hybrid tube.

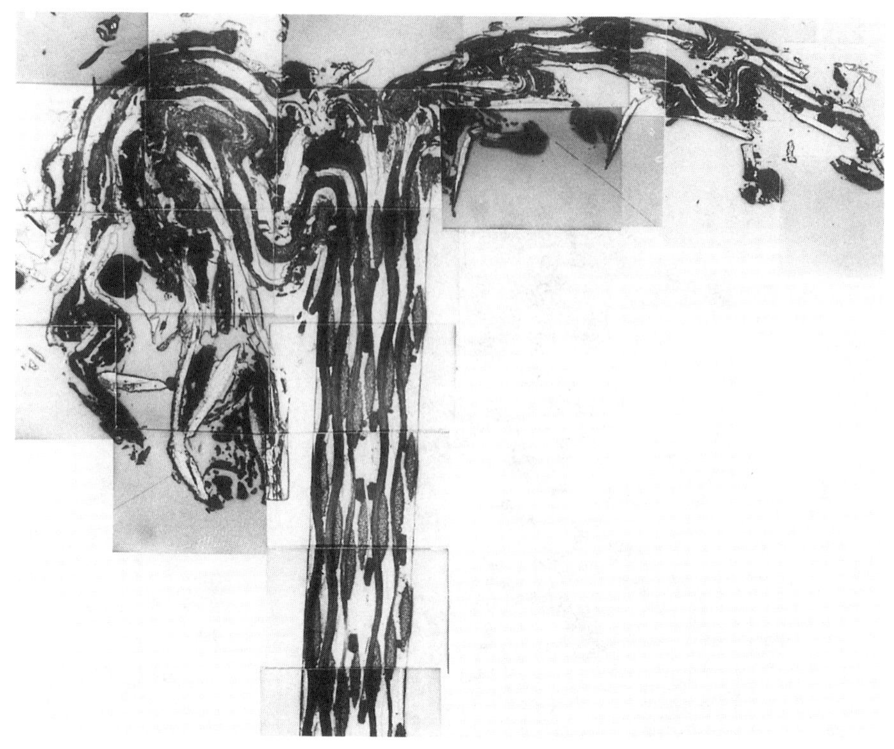

Fig. 10. Photomicrograph of section through the crush zone of CD(0/90) hybrid tube.

Fig. 11. Photomicrograph of section through the crush zone of CD(±45) hybrid tube.

Proceedings of the 10th Annual ASM/ESD Advanced Composites Conference, Dearborn, Michigan, USA, 7-10 November 1994

Energy Absorption Characteristics of Composite Tubes with Different Cross-Sectional Shapes

H. Hamada, S. Ramakrishna, Z. Maekawa, M. Nakamura
Kyoto Institute of Technology
Kyoto, Japan

T. Nishiwaki
ASICS Corporation
Kobe, Japan

Abstract

Much of the energy absorption research work has been performed on axisymmetric cylindrical composite tubes. However, in actual application the geometry of the energy absorbing structure may not be axisymmetric nature. This paper describes the energy absorption characteristics of composite tubes with different cross-sectional shapes. Glass cloth/epoxy tubes and carbon fiber/PEEK tubes with cross-sectional shapes full circle, three-quarter circle, half circle and quarter circle were investigated. Quasi-static tests were performed by axial compression between two flat platens. All the specimens crushed progressively from the chamfered end. The aminosilane-treated glass cloth/epoxy and carbon fiber/PEEK specimens crushed by splaying mode whereas the acrylsilane-treated glass cloth/epoxy specimens crushed by fragmentation mode. The mode of progressive crushing was independent of the cross-sectional shape. However, the energy absorption capability was influenced by the cross-sectional shape of the tube.

It has been shown experimentally that composite structural elements, when propoerly designed, display energy absorption potentials as good as, and in many cases, better than those of metals. Many investigations have been carried out to evaluate the energy absorption performances of structural elements such as composite tubes, stringer stiffened, sandwich, and corrugated (sine wave) web sections [1-2]. Hanagud *et al* [3] reported that the sine web composite material exhibit good energy absorption characteristics when crushed in the web direction. They demonstrated that the energy absorption efficiency of web specimens with 180° included angle is equal to corresponding tube specimens. The crushing mode changed from stable local crushing to unstable global buckling by reducing the included angle of the sine web to 90°-60°. Farley and Jones [4] repeated experiments and reported that the energy absorption capability increases by 10 to 30% with decrease of included angle from 180° to 90°. Small and Ansell [5] investigated the energy absorption behavior of sandwich panels with cellular composite structure as the core. The energy absorption performance of celluar structure is slightly higher than that of the single composite cylinder with comparable aspect ratios. The superior performance of cellular structure is attributed to the stability imparted to each tube element by its neighbor. Pafitis and Hull [6] and Fleming and Vizzini [7] investigated the energy absorption behavior of truncated composite cones. It has been shown that progressive crushing can be produced in composite cones from the wide end by suitable variation in the wall thickness along the length of the cone. The crush load depends on the cross-sectional area. Thornton and Edwards [8] investigated the effect of section shape on the specific energy absorption of axisymmetric composite tubes. They concluded that for a given fiber lay-up and tube geometry, the specific energy absorption increased in the order, rectangle < square < circular. Considerable published data is available on the effects of variables such as tube geometry [8-13], fiber architecture [14-15], fiber material [16], resin material [17], testing conditions [18-20], etc on the energy absorption characteristics of composite tubes. All these data is generated for axisymmetric cylindrical composite tubes. In actual application the geometry of the energy absorbing structure may not be axisymmetric nature. Also the energy absorption performance of a slotted composite tube is not known. In this paper the energy absorption characteristics of composite tubes with cross-sectional shapes full circle, three-quarter circle, half circle and quater circle are presented. Two types of composite materials were investigated. One type is a thermoset composite material of epoxy resin reinforced with woven glass cloth. The other type is a thermoplastic composite material of polyetheretherketone (PEEK) matrix reinforced with carbon fibers.

Materials

Glass Cloth/epoxy Composite Tubes. Two types of glass cloth/epoxy composite tubes were fabricated. One type of tubes contained glass cloth treated with an aminosilane coupling agent and the other type of tubes used glass cloth treated with acrylsilane coupling agent. Aminosilane is a suitable coupling agent for epoxy resin whereas acrylsilane is not. Composite tubes were made by mandrel wrapping of prepreg sheets with the warp and weft directions parallel to the hoop and axial directions of the tubes. Composite tubes contained 43% volume fraction of fibers. All the tubes had an internal diameter of 50 mm and a wall thickness of 2.5 mm. To initiate progressive crushing a 45° chamfer was lathe machined onto one end of each specimen.

Carbon Fiber/PEEK Composite Tubes. Carbon fiber/PEEK tubes were fabricated from unidirectional fiber prepreg sheets of AS4/APC-2 supplied by ICI-Fiberite Co. The prepreg has 61% volume fraction of carbon fibers. Tubes were made by wrapping the prepreg onto a variable diameter PTFE mandrel and then inserting into an outer mould. Expansion of the inner mandrel provided compression during fabrication and ensured high dimensional accuracy and reproducibility with low void content. The tubes were consolidated at 380^OC for two hours in conventional oven and then water cooled. The carbon fibers were aligned parallel to the tube axis. All the tubes had an outer diameter of 55 mm and a wall thickness of about 2.65 mm. The fabricated tubes were cut into specimens of 55 mm long. To initiate progressive crushing a 45° chamfer was lathe machined onto one end of each specimen.

Experimental Details

Composite cylindrical specimens with three-quarter circle, half circle and quarter circle cross-sections were made by cutting parallel to the tube axis (Figure 1). Specimens were crush tested in axial compression between parallel flat platens of a servo-hydraulic testing machine. All the specimens were tested at a constant cross-head speed of 1 mm/min. During each crush test the load-displacement (P-d) response was recorded using an x-y plotter. Specific energy absorption capability of progressively crushed specimens was calculated as the ratio of mean crush load from the P-d curve and the mass per unit length of uncrushed specimen. The crushed tubes were cast in polyester resin to preserve the crush zone morphology. The microfracture processes in the crush zone were studied by cutting sections through the embedded material. The sections were then polished and photographed on an optical microscope.

Results

Glass Cloth/Epoxy Tubes. All the glass cloth/epoxy specimens crushed progressively from the chamfered end. Typical load-displacement curves for aminosilane-treated and acrylsilane-treated composite tubes with different cross-sectional shapes are given in Figures 2 and 3 respectively. The mean crush loads obtained from the load-displacement curves and the specific energy absorption capabilities are summarised in Tables 1 and 2. The appearance of load-displacement curves was similar. Irrespective of the tube cross-sectional shape, all the load-displacement curves showed initially a rapid increase in load. After reaching a maximum value the load fluctuated around a mean crush load value. There was no significant change in the number of serrations and the amplitude of serrations with the change in the cross-sectional shape of the tube. However, the rate of increase of initial load was dependent on the tube cross-section. Composite tubes with full circle cross-section displayed higher rate of load increase than the tubes with quater circle cross-section. Compared to the tubes with full circle cross-section the tubes with other cross-sectional shapes showed lower mean crush loads. The tubes with full circle cross-section displayed highest energy absorption capability and the tubes with quater circle cross-section displayed lowest energy absorption capability (Tables 1 and 2). The tubes with three-quarter and half circle cross-sections displayed similar energy absorption capabilities. In the case of aminosilane-treated tubes, the specific energy decreased by 20% with the change in tube cross-section from full circle to quarter circle. In the case of acrylsilane-treated tubes, the tubes with quarter circle cross-section displayed 23% lower specific energy than the tubes with full circle cross-section. For a given tube cross-section, the aminosilane-treated specimens displayed higher specific energy absorption than the acrylsilane-treated specimens.

Optical photographs of crushed aminosilane-treated and acrylsilane-treated specimens are shown in Figures 4 and 5 respectively. All the specimens crushed progressively from the chamfered end. Optical photograph of the section through the crush zone of aminosilane-treated tube with half circle cross-section is shown in Figure 6. This figure shows splaying of tube wall to the inside and outside of the tube. The material splayed to the inside of the tube wall is called as internal frond and the material splayed to the outside of the tube wall is known as external frond. Both the fronds have been bent through sharp radius of curvature. Crush zones also contained a debris wedge and a longitudinal crack below the debris wedge. The crush zone with these features is known as 'splaying mode' of crushing. All the aminosilane-treated specimens independent of the tube cross-sectional shape crushed by splaying mode. Optical photograph of the section through the crush zone of acrylsilane-treated tube with half circle cross-section is shown in Figure 7. Shear cracking of tube wall and sliding of broken fragments to the inside and outside of the tube can be seen. Shear cracking along the circumference of the specimen resulted in the formation of internal and external rings. This type of progressive crushing is known as 'fragmentation mode' of crushing. A similar kind of features were observed in the crush zones of specimens with other cross-sectional shapes. In other words, the mode of crushing of acrylsilane treated specimens is always fragmentation irrespective of the cross-sectional shape of the tube.

Carbon Fiber/PEEK Tubes. Typical load-displacement curves for carbon fiber/PEEK specimens are shown in Figure 8. Corresponding optical photographs of the crush zones are shown in Figure 9. The load-displacement curves showed four different stages. In the first stage, the load increased rapidly to a peak value. After the initial peak, the load dropped and then increased gradually. During the fourth stage the load saturated and fluctuated marginally. This fourth stage was associated with progressive crushing of tube wall. Irrespective of the tube cross-sectional shape all the specimens crushed by splaying mode. Splaying of tube wall into well defined internal and external fronds can be seen in the Figure 9. An optical photograph of the section through the crush zone of carbon fiber/PEEK tube with half circle cross-section is shown in Figure 10. A well defined debris wedge splitting the tube wall can be seen. Unlike the aminosilane-treated glass cloth/epoxy tubes, the longitudinal cracks below the debris wedge were short. The fronds have been bent through sharp radius of curvature at the crushing platen. During bending the fronds have been further split into thin beams. Fractured carbon fibers were also observed in the fronds. The energy absorption performances of various carbon fiber/PEEK specimens are summarised in Table 3. The specific energy absorption capability of carbon fiber/PEEK tubes is much higher than that of the glass cloth/epoxy tubes. Among the carbon fiber/PEEK specimens investigated the tubes with full circle cross-section displayed highest specific energy of 194.1 kJ/kg. Whereas the tubes with quarter circle cross-section displayed a specific energy of 187.1 kJ/kg. In other words the specific energy of carbon fiber/PEEK tubes decreased marginally by 5% with the change in cross-section from full circle to quarter circle.

Analysis and Discussion

One of the imporant results of this study was that the mode of progressive crushing is independent of the tube cross-sectional shape. It is mainly determined by the constituant materials of the composite. Good interfacial bonding of glass fibers to the epoxy resin matrix resulted in the splaying mode crushing of aminosilane-treated composite tube specimens. The presence of higher frictional forces in the crush zone resulted in the fragmentation mode of progressive crushing of acrylsillane-treated composite tube specimens. More details on the effect of fiber surface treatment on the progressive crushing behavior of glass cloth/epoxy tubes can be found in Reference 18. The carbon fiber/PEEK tube specimens crushed progressively by splaying mode. This is mainly due to the orientation of carbon fibers parallel to the tube axis. Under axial compressive loads, the fibers buckle and fracture. The debris wedge produced during initial stages of crushing splits the tube wall into internal and external fronds, resulting in splaying mode crush zone. Another impartant result of this study was that the specific energy absorption capability is dependent on the cross-sectional shape of the tube. In the case of glass cloth/epoxy tubes the specific energy decreased by 20% with the change in tube cross-section from full circle to quarter circle. For the same change in cross-sectional shape the specific energy of carbon fiber/PEEK tube specimens decreased by 5% only. Since the crush zone morphology and the fracture processes remained unchanged with the change in cross-sectional shape, the variation of specific energy may be attributed to the changes in the axial and hoop strains of the specimens.

We tried to measure the axial and hoop strains by mounting biaxial strain gauges on the tube specimens. These experiments were unsuccessful due to the failure of strain gauges. The strains in the viscinity of the crush zone were much higher than those in the portions of the tube wall away from the crush zone. During crushing as the crush zone approaches the strain gauges, they failed. Hence the analysis of strain distributions in the tube specimens has been carried out by means of finite element code ADINA. This finite element package is suitable for the linear and non-linear analysis of three-dimensional solids and structures. Using solid elements and rigid supports as boundary conditions the hoop and axial strains were computed. The strains at the center of tube wall have been compared. The ratios of hoop to axial strains on the inside and outside of the tube wall are summarised in Table 4 for tubes with different cross-sectional shapes. The ratio of hoop to axial strains was different for tubes with different cross-sectional shapes. The tubes with a cross-section of full circle displayed lowest hoop to axial strain ratio. Whereas the tubes with a cross-section of quarter circle displayed highest hoop to axial strain ratio. In other words, the cross-sectional shape of the tube influences the strain distributions. It is speculated that these variations in strain distributions might have changed the energy absorption capability with the change in cross-sectional shape of the tube specimens. It is not clear why the change of specific energy with cross-sectional shape was more pronounced in the case of glass cloth/epoxy tubes than the carbon fiber/PEEK tubes. Further studies are necessary to analyse the effects of strain variations on the mean crush load and the energy absorption capability.

Conclusions

The mode of progressive crushing was independent of the cross-sectional shape of the composite tubes. The aminosilane-treated glass cloth/epoxy specimens crushed by splaying mode whereas the acrylsilane-treated glass cloth/epoxy specimens crushed by fragmentation mode. The carbon fiber/PEEK specimens crushed progressively by splaying mode. The specific energy absorprtion capability of composite tubes was influenced by the cross-sectional shape of the tube. In the case of glass cloth/epoxy specimens, the specific energy decreased by 20% with the change in tube cross-section from full circle to quarter circle. This decrease in specific energy was small in the case of carbon fiber/PEEK specimens.

References

1 Kindervater, C.M, in Proceedings of the American Helicopter Society National Specialist Meeting on Composite Structures, Philadelphia, USA, March 1983.

2 Kindervater, C.M, in Proceedings of 30th National SAMPE Symposium, Anaheim, USA, 1985, p. 1191-1201.

3 Hanagud, S., J.I.Craig, P.Sriram and W.Zhou, J.Composite Materials, 23, 448-459 (1989).

4 Farley,G.L. and R.M. Jones, J. Composite Materials, 26, 12, 1741-1751 (1990).

5 Small, G.D. and M.P. Ansell, J. Materials Science, 22, 2717-2722 (1987).

6 Pafitis, D.G. and D. Hull, SAMPE Journal, 27, 3, 29-35 (1991).

7 Fleming, D.C. and A.J.Vizzini, J. Composite Materials, 26, 4, 486-499 (1992).

8 Thornton, P.H and P.J.Edwards, J. Composite Materials, 16, 521-545 (1982).

9 Fairfull, A.H. and D.Hull, in Proceedings of ICCM VI & ECCM2, F.L.Matthews, N.C.R.Buskell, J.M.Hodgkinson and J.Morton, Eds, Elsevier Applied Science, London, 1987, p. 3.36-3.45.

10 Farley, G.L and R.M. Jones, J.Composite Materials, 26, 78-89 (1992).

11 Thornton, P.H., Composites Science and Technology, 27, 199-224 (1986).

12 Price, J.N. and D.Hull, Composites Science and Technology, 28, 211-230 (1987).

13 Farley, G.L., J.Composite Materials, 20, 390-400 (1986).

14 Ramakrishna, S. and D.Hull, Composites Science and Technology, 49, 349-356 (1993).

15 Ramakrishna, S., H. Hamada and D.Hull, in Proceedings of European Symposium on Impact and Dynamic Fracture of Polymers and Composites, Porto Cervo, Italy, September 1993.

16 Farley, G.L., in Proceedings of the 43rd American Helicopter Society Annual Forum, St.Louis, USA, 1987, p. 613-627.

17 Thornton, P.H. J.Composite Materials, 24, 594-615 (1990).

18 Hamada, H., S.Ramakrishna, M.Nakamura, Z.Maekawa and D.Hull, Composite Interfaces, 2, 2, 127-142 (1994).

19 Bannerman, D.C and C.M.Kindervater, in Proceedings of the Fourth International SAMPE European Chapter, Bordeaux, France, 1984, p. 155-167.

20 Schmuesser, D.W. and L.E.Wickliffe, J. Engineering Materials and Technology, 109, 72-77 (1987).

Table 1. Energy absorption performances of aminosilane-treated glass cloth/epoxy tubes.

Sectional Shape	Mean Crush Load (kN)	Mean Crush Stress (MPa)	Specific Energy (kJ/kg)
Full circle	45.7	108.4	66.6
3/4 circle	33.0	101.8	60.6
1/2 circle	21.7	101.8	60.1
1/4 circle	9.5	89.9	53.5

Table 2. Energy absorption performances of acrylsilane-treated glass cloth/epoxy tubes.

Sectional Shape	Mean Crush Load (kN)	Mean Crush Stress (MPa)	Specific Energy (kJ/kg)
Full circle	37.5	90.2	53.0
3/4 circle	22.5	70.9	41.3
1/2 circle	15.1	71.9	41.8
1/4 circle	7.4	70.1	40.7

Table 3. Energy absorption performances of carbon fiber/PEEK tubes.

Sectional Shape	Mean Crush Load (kN)	Mean Crush Stress (MPa)	Specific Energy (kJ/kg)
Full circle	131.6	303.4	194.1
3/4 circle	98.1	301.2	192.9
1/2 circle	64.4	298.0	190.6
1/4 circle	30.5	292.7	187.1

Table 4. The ratios of hoop to axial strains of different specimens.

Sectional Shape	Hoop Strain/Axial Strain Ratio	
	Inside	Outside
Full circle	-0.649	0.715
3/4 circle	-0.663	-0.730
1/2 circle	-0.663	-0.727
1/4 circle	-0.694	-0.757

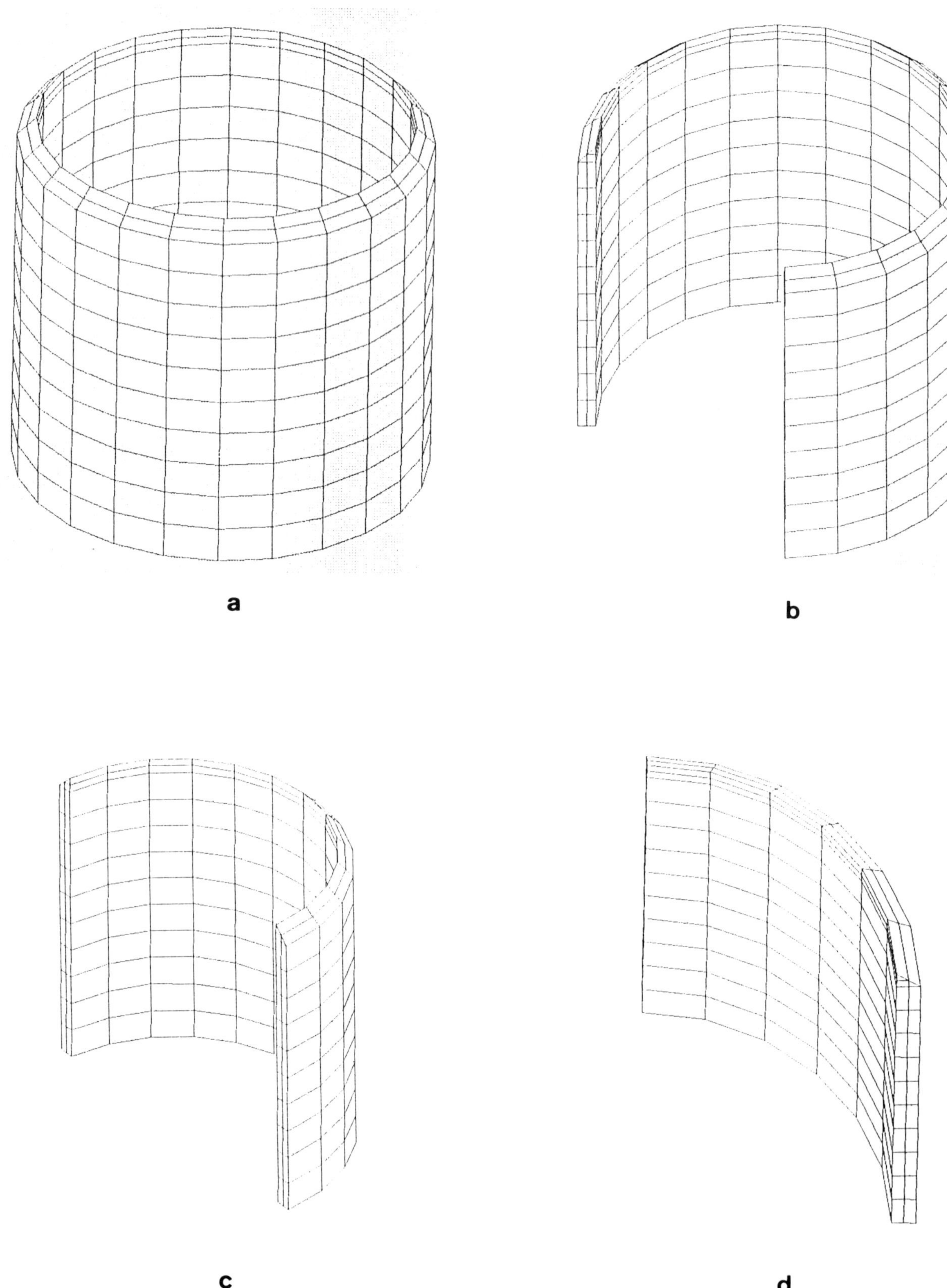

Fig. 1. Composite tube specimens with a) full circle, b) three-quarter circle, c) half circle and d) quarter circle cross-sections.

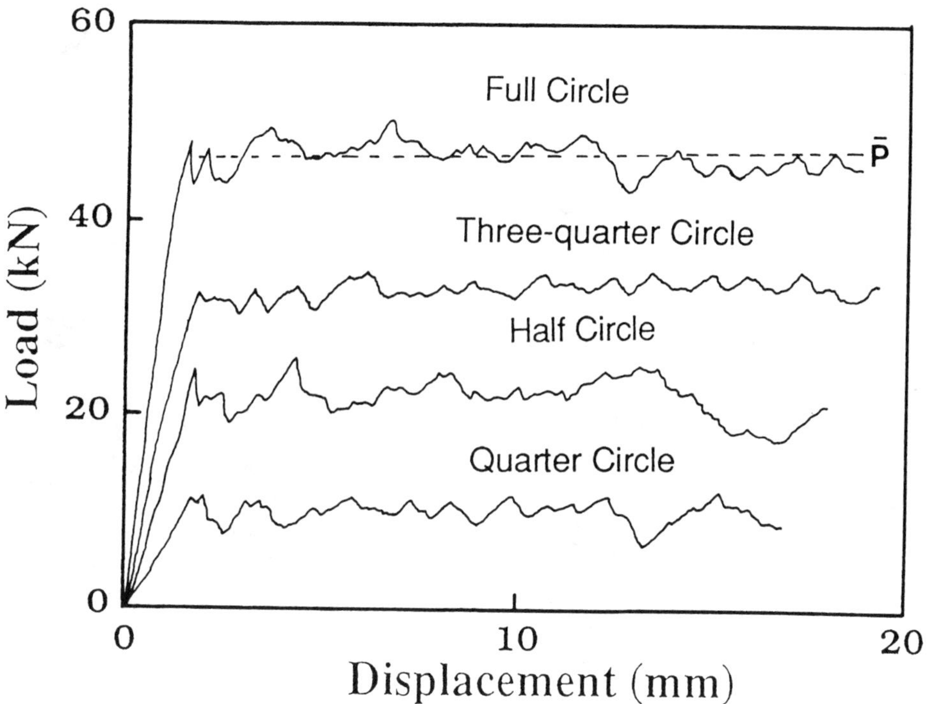

Fig. 2. Typical load-displacement curves for aminosilane-treated glass cloth/epoxy specimens.

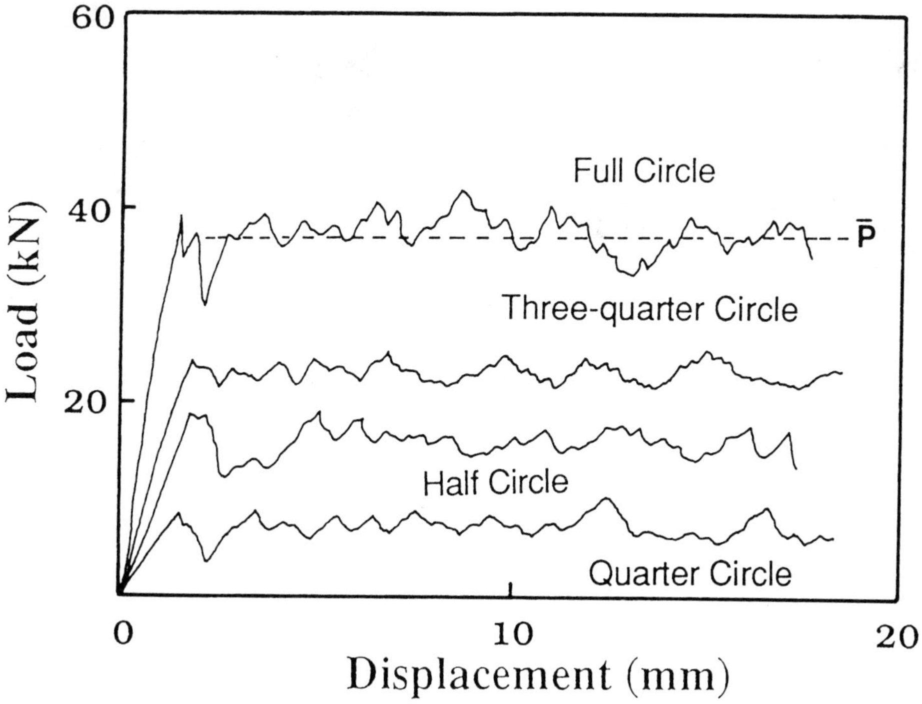

Fig. 3. Typical load-displacement curves for acrylsilane-treated glass cloth/epoxy specimens.

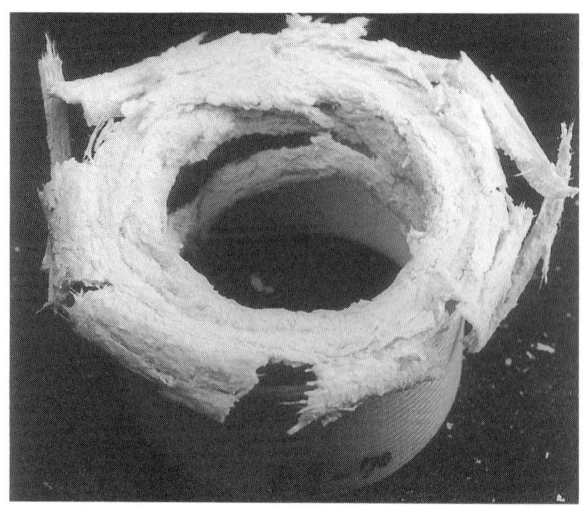
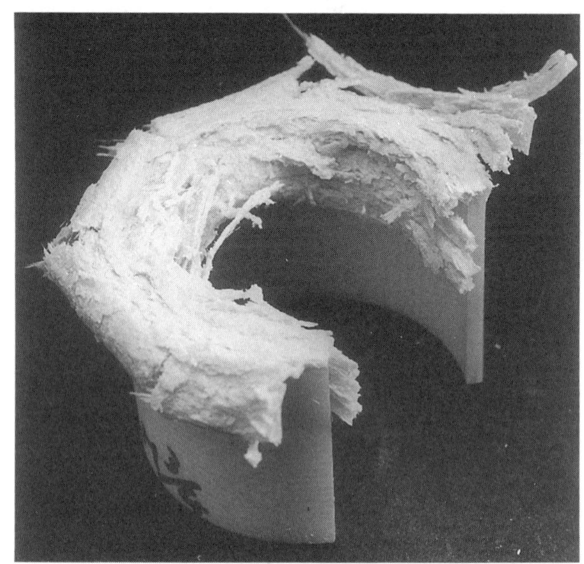
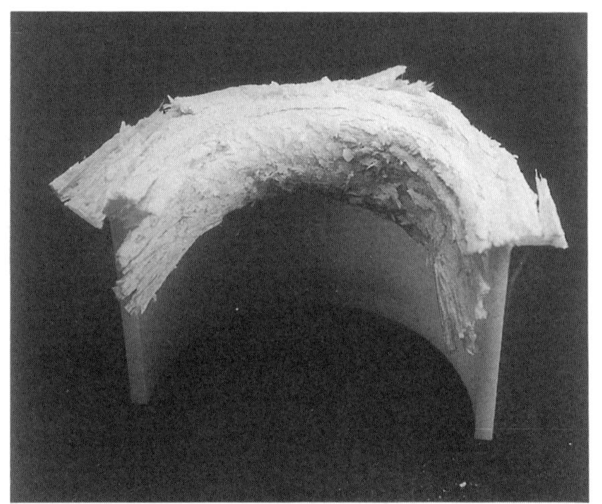

Fig. 4. Optical photographs of crush tested aminosilane-treated glass cloth/epoxy specimens.

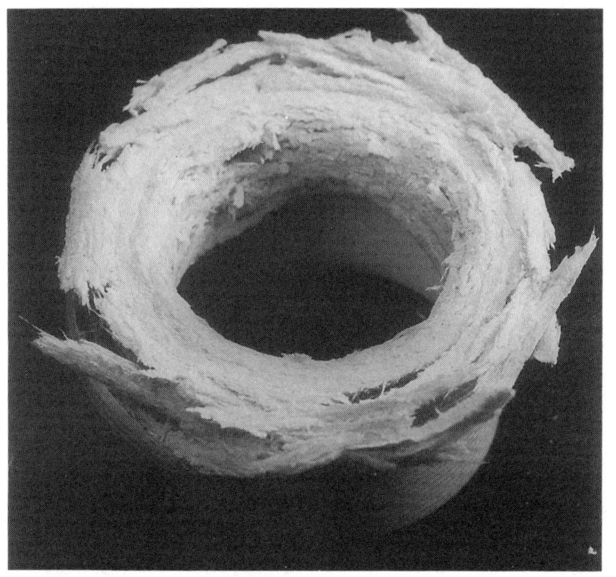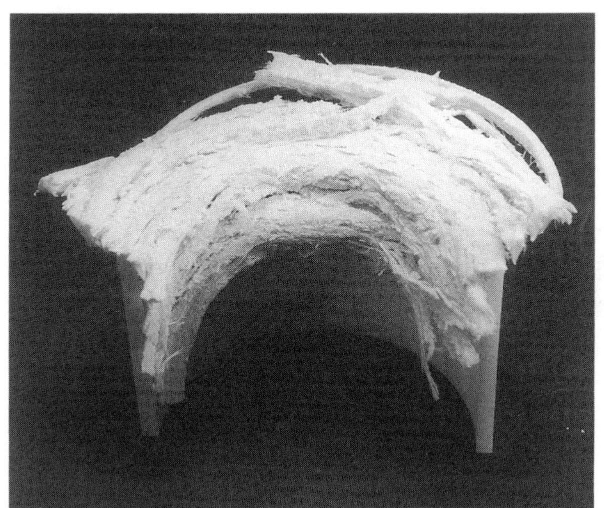

Fig. 5. Optical photographs of crush tested acrylsilane-treated glass cloth/epoxy specimens.

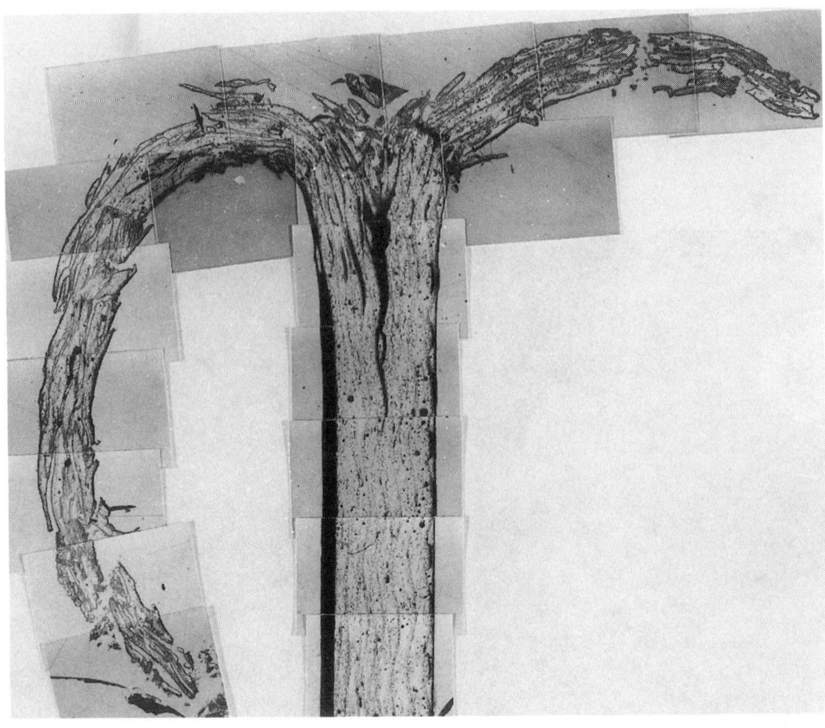

Fig. 6. Photomicrograph of a section through the crush zone of aminosilane-treated glass cloth/epoxy tube with half circle cross-section.

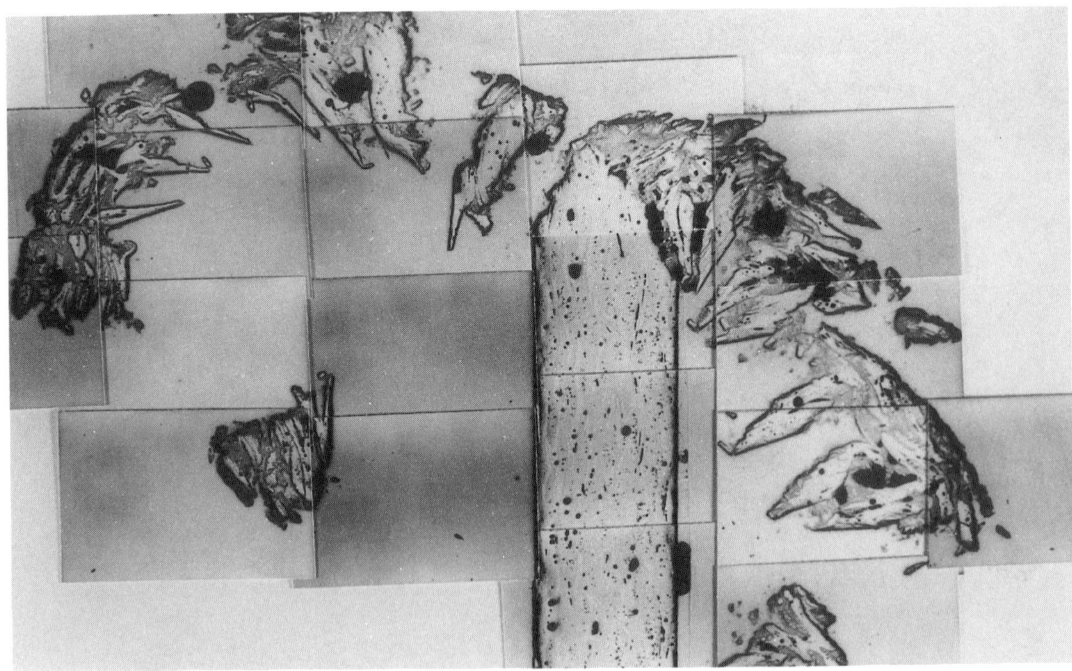

Fig. 7. Photomicrograph of a section through the crush zone of acrylsilane-treated glass cloth/epoxy tube with half circle cross-section.

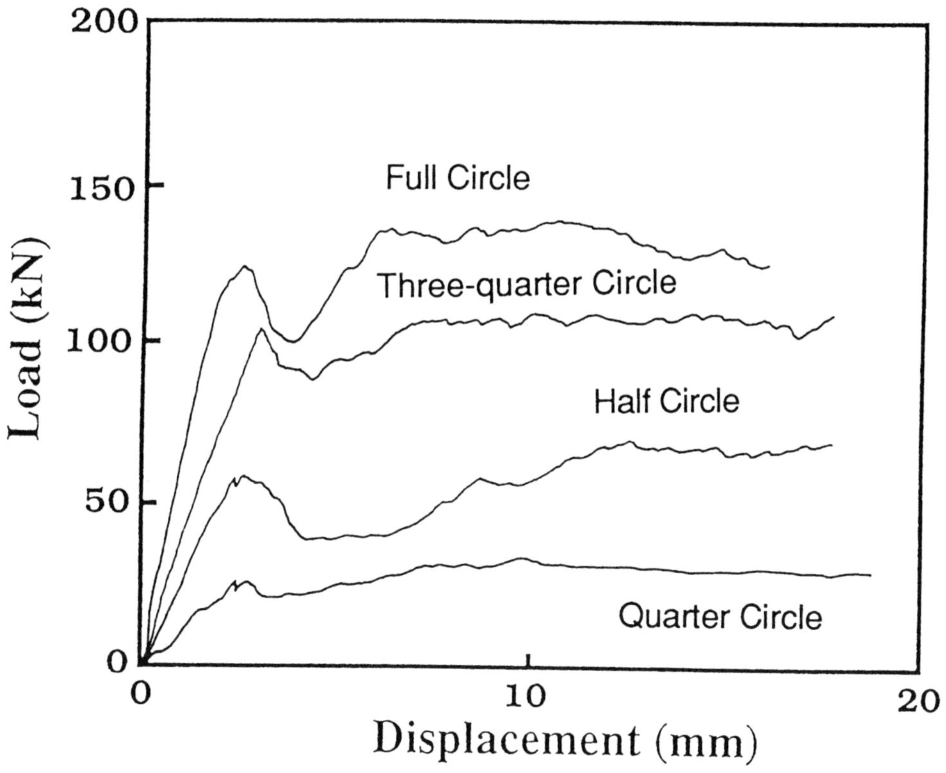

Fig. 8. Typical load-displacement curves for carbon fiber/PEEK specimens.

Fig. 9. Optical photographs of crush tested carbon fiber/PEEK specimens.

Fig. 9. Optical photographs of crush tested carbon fiber/PEEK specimens.

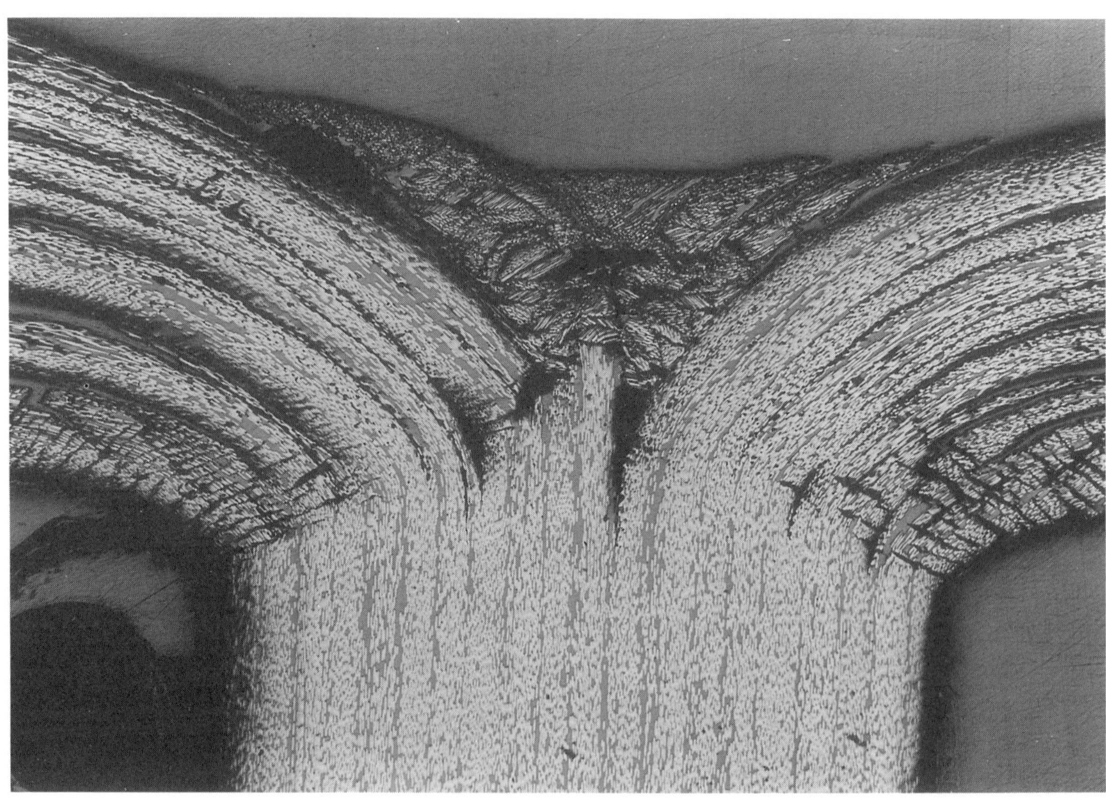

Fig. 10. Photomicrograph of a section through the crush zone of carbon fiber/PEEK tube with half circle cross-section.

Processing and Testing Issues for a Model Polyisocyanurate/Glass Single Fiber Composite

W.G. McDonough, C.L. Schutte, C.K. Moon, C.R. Schultheisz, J.P. Dunkers
National Institute of Standards & Technology
Gaithersburg, Maryland

Abstract

The National Institute of Standards and Technology (NIST) has an ongoing program to investigate durability issues in polymeric composites. An important component of this program is a collaboration with the Automotive Composites Consortium (ACC) to study the effect of moisture on the fiber-matrix interface. Besides experiments on composites, NIST is using the single fiber fragmentation test to study the interface itself and gain insight in understanding the behavior of the full composite. The initial work is underway using an epoxy/glass system as a model, but in response to the interests of the ACC, a model isocyanurate/glass system is now being developed. With normal sample preparation methods, it is impossible to make acceptable specimens for the fragmentation test with polyisocyanurates. The reaction is too rapid for processing samples, and the samples will foam. By changing the processing conditions, however, polyisocyanurate samples were fabricated successfully. Studies were then conducted to determine if these specimens were chemically and mechanically equivalent to materials made by structural reaction injection molding (SRIM). In addition, tests

have also been performed to determine if the specimens meet the strain-to-failure requirement for the single fiber fragmentation test.

FOR THE PAST FEW YEARS, the National Institute of Standards and Technology (NIST) has initiated and is conducting a program to study durability issues in composite materials. One part of this durability program is a collaboration with the Automotive Composites Consortium (ACC) to study the effect of moisture on the fiber-matrix interface in polymeric composites. The material used in this program is an epoxy/glass system, but, in response to the interests of the ACC, a model isocyanurate/glass system is being developed. The goal of this paper is to describe the processing and testing issues that occurred when developing this polyisocyanurate/glass system. One issue was to decide if void-free single fiber fragmentation samples could be made, and if so, did they have the necessary extensibility to allow the fiber breaks to reach "saturation," that is, the point where no breaks are observed upon further straining of the sample in a single fiber test. Furthermore, if the process used to make single fiber fragmentation samples is different from the structural reaction injection molding (SRIM) process, then are the polyisocyanurates from the different processes chemically and mechanically equivalent? The results presented in this paper will show that the isocyanurate samples can be strained to allow the glass fiber specimens to be taken to saturation in the single fiber fragmentation test. In this paper, we present some background on the durability program at NIST and describe our findings. Subsequently, we discuss a brief background on polyisocyanurates with our focus on processing issues. This information will be followed by an experimental section in which we describe how we made and tested our polyisocyanurate/glass samples. We finish with a discussion and some concluding comments.

In developing the program on durability, we decided to focus on the response of the fiber-matrix interface to long-term exposure to moisture (1). More specifically, this program uses the single-fiber fragmentation test as a way to investigate interfacial phenomena, since a bulk composite would have so many interactions as to render an unambiguous interpretation of the results difficult. The single fiber test is one of several techniques that try to reveal more clearly events at the fiber-matrix interface (2,3). As a fiber is loaded in tension during this test, the axial stresses in the fiber rise until the fracture strength is reached, then the fiber breaks at a point that depends upon the fiber defect distribution. As the load increases, the fiber continues to break similarly until the fragment lengths of the fiber become too short for sufficient stress to be transferred into the fiber to cause further breakage. This length is considered the critical length (3). Kelly and Tyson (4) used this method on tungsten and molybdenum wires in copper matrices, and they developed an equation for specimens that had plastic deformation of the resin and

elastic deformation of the fiber. The equation is as follows:

$$\frac{L_c}{D} = \frac{1}{2} \times \frac{\sigma_f}{\tau} \qquad \textbf{(1)}$$

where L_c is the critical length of the fiber, D is the fiber diameter, σ_f is the tensile strength of the fiber at a length equal to L_c, and τ is the interfacial shear strength. Typically, the fiber strength is estimated by conducting many tensile tests on individual fibers of various lengths and then doing a statistical analysis on the data. Since σ_f depends on the fiber length, an extrapolation is used to determine the value at L_c. As a further development, Waterbury and Drzal (6) and Wagner et al. (7) have used the single fiber test itself to estimate the tensile strength of fibers at their critical lengths. This novel use of the test method is significant when one considers the problem of exposure to moisture. Exposure to moisture should weaken the interface, if the interface is susceptible to moisture, and result in lower interfacial shear strengths as expressed by longer aspect ratios, L_c/D. However, if the glass fiber is also weakening through mechanisms such as leaching ions out of the fiber, then it may break into shorter fragments and decrease the aspect ratio, thus making the interface appear stronger unless the correct σ_f can be determined and used. Shioya et al. (8) have also considered this problem and has used the single fiber test method to estimate both the strength of the fiber in the resin and the strength of the interface as a function of exposure to moisture. This is critical because it would be difficult, if not impossible, to expose the fiber to a moisture environment like it sees when it is encased in resin and exposed to moisture.

Schutte and coworkers studied the interfacial interactions between E-glass and an epoxy resin cured with a stoichiometric amount of m-phenylene diamine (mPDA) (1,9). They found that not only the strength of the interface decreased upon exposure to moisture, but also the strength of the fiber degraded with this exposure. Some results of this work are shown in Figures 1 and 2. Figure 1 shows how the interfacial region degraded when the samples underwent long term exposure to distilled water at 75 °C , yet did not degrade when only exposed to the high temperature condition.

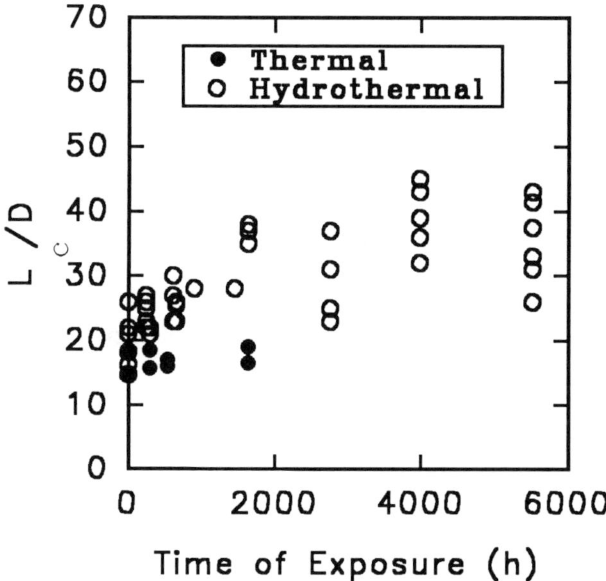

Fig. 1 - Plot of the critical length/diameter (L$_c$/D) as a function of time of exposure to thermal (75 °C, black circles) or hydrothermal (75 °C distilled water, open circles) conditions.

The results shown in Figure 2 were derived by using an equation by Wagner et al. (10) in which Weibull statistics were applied to the weakest link model to calculate the strength of the fiber from the single fiber fragmentation test. The probability of failure of the fiber, (F(σ)), has been related to the breaking stress of the fiber, σ, the fiber length-dependent scale parameter, α_L, and the fiber length-independent shape parameter, β.

$$F(\sigma) = 1 - \exp\left\{-\left(\frac{\sigma}{\alpha_L}\right)^\beta\right\} \qquad (2)$$

When the ln[-ln(1-F(σ))] is plotted against lnσ, the slope is β and the y intercept is βlnα_L. From this one can calculate the strength of the fiber from the fragmentation profile. Schutte and co-workers then estimated the swelling strains caused by the exposure to moisture, and, with the apparent strength of the glass fiber, they applied Eq. 1 and calculated the apparent interfacial shear strength. One of their conclusions was that the interfacial shear strength degraded faster than the strength of the fiber.

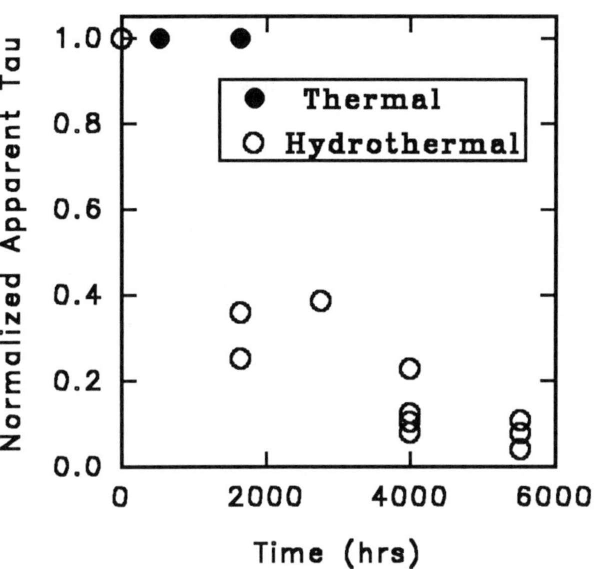

Fig. 2 - Plot of the normalized apparent interfacial shear strength as a function of time of exposure to 75 °C heat or time of immersion in 75 °C distilled water.

This initial work used an epoxy/glass system as a model as stated above, but in response to the interests of the ACC, we are developing a model polyisocyanurate/glass system. The ACC makes polyisocyanurate/glass composites via structural reaction injection molding (SRIM). This processing method involves combining the ingredients in a mixing head, with high speed impinging of the components responsible for the mixing (11). The resin system is then injected into the cavity of the mold that was already loaded with reinforcement with the processing cycle being in the range of seconds or minutes. Unfortunately, the high speeds and pressures used in this process are incompatible with the usual techniques employed to make single fiber fragmentation samples. For example, the epoxy/mPDA samples mentioned earlier are made by first laying down a single fiber in a cavity of a mold, then casting resin into the cavity and curing the resin in an oven.

The problem with polyisocyanurates, as well as polyurethanes, is that they foam very rapidly at room temperature and atmospheric pressure. Further, the catalyst plays an important role in setting the speed of the foaming process, and if one were to reduce the level of the catalyst, one should be able to extend the gel time, thus resulting in longer times to process the samples. However, even solving the problem of processing time would still leave the problem of the bubbles caused by the reaction with water. Consequently, it will be shown that, with the addition of pressure using an inert gas, N_2, it is possible to suppress the formation of bubbles in making a fragmentation sample.

One final note on polyisocyanurates is that, for the system and catalyst used in this paper, the isocyanurate trimer linkages are formed in addition to the urethane linkages. The isocyanurate linkages provide excellent thermal insulation, very good hydrolytic stability, and good resistance to burning (13). The polyisocyanurates are formed by mixing an isocyanate, based on methylene diphenyl diisocyanate, with a polyol, for this work based on a polyether polyol blend. But before these two components are mixed, an amine based catalyst is mixed into the polyol. Different catalysts favor certain linkages, i.e., urethane formation vs. urea formation (11), and the catalyst used in this work promoted both urethane and isocyanurate groups. For this work, when comparing the infrared spectra of samples made by different methods, the isocyanurate (1410 cm^{-1}) was identified (12). Assuming proper mixing, the isocyanate peak (2250 cm^{-1}) should diminish as the reaction progresses. Finally, water changes the nature of the linkages and produces carbon dioxide (11). Water will react with an isocyanate to form unstable carbamic acid, which immediately breaks down to form an amine end group and carbon dioxide. Subsequently, the amine end group can react with isocyanate end groups to form substituted ureas. Water is ubiquitous in the SRIM process and is the primary cause for bubbles in finished parts.

Materials and Experimental Procedure

As stated earlier, the SRIM process involves impingement mixing of the polyol, isocyanate, and catalyst, and injecting the mixture into a heated and filled mold cavity. The cycle time is usually a few minutes. The problem with trying to make single fiber composites from the isocyanurate system used in the SRIM process described above is that SRIM entails very fast processing at high injection pressures. This process is impractical for single fiber composites because the fibers would not survive the injection pressures. In addition, making single fiber coupons from thermoset resins is usually done by casting under atmospheric conditions - conditions under which polyisocyanurate resins will react with water and foam. Therefore, to make void-free single fiber composites required us to consider an alternate processing route.

The first step in making single fiber fragmentation samples was to make an eight-cavity mold out of RTV-664 from General Electric (14) following the procedure described by Drzal et al. (5). Single filaments (Vetrotex U-750 glass from Certainteed) were separated from a fiber bundle with care being taken to touch only the ends of the fiber. Each cavity in the mold has sprue slots in the center of each end to aid in aligning the fiber in the center of the cavity. Once the fibers were in the cavities, we fixed them in place by putting a small drop of five-minute epoxy (Hardman Adhesives) at the far end of each sprue slot. Once the epoxy had hardened, we were ready to prepare the resin.

Although we tried many ways to make single fiber test samples, the simplest procedure was to keep the ratio of isocyanate (Spectrim (R) MM 364-A from the Dow Chemical Company) to polyol (Spectrim (R) MM 364-B from the Dow Chemical Company) at the same level as in the SRIM process: 2.4/1. However, the catalyst (Dabco 33-LV from Air Products and Chemicals, Inc.) level was reduced from 1.5% by weight polyol to between 0.3-0.5%. A catalyst level of 0.5% was the highest catalyst level that still permitted enough time to process the samples. Both the polyol and isocyanate were kept in a desiccator over phosphorus pentoxide for at least four weeks before processing to minimize moisture that may affect the processing. First, we added the catalyst to the polyol and stirred the liquid for 30 seconds, then we added the isocyanate to the mixture and stirred for 30 seconds. Subsequently, we poured the resin

into the mold cavities and transferred the mold to an autoclave (Minibond from United McGill Corporation).

The cure cycle was to pressurize the chamber to 0.34 MPa above atmospheric pressure, while holding the temperature to 24 °C. Once the pressure inside the autoclave reached 0.34 MPa, we raised the temperature and pressure simultaneously. The temperature was raised at 2.5 °C/minute to 93 °C as the pressure was increased to 0.55 MPa above atmospheric pressure. The mold was held at 93 °C and 0.55 MPa for 30 minutes, then the pressure was reduced quickly to atmospheric pressure and the temperature was quickly cooled to 66 °C, and then the mold was cooled slowly overnight to room temperature. Subsequently, the samples were postcured in an oven (Blue M model number MP-256C-1) for one hour at 150 °C and then slow cooled to room temperature. Finally, we removed the samples from the mold and examined them. Valid samples were regarded as those that had no air bubbles, and whose fibers remained straight.

The single fiber fragmentation tests were carried out on a small, hand operated testing machine such as that described by Drzal (5). This machine was attached to a Nikon Optiphot-Pol polarizing microscope. During the test, the microscope, to which a video camera (Optronics LX-450A RGB Remote-Head microscope camera) and an optical micrometer (VIA-100 from Boeckeler) were attached, allowed us to measure the fiber diameters and the fragment lengths. Fragment lengths were measured with the aid of a transducer (Trans-Tek, Inc. model 1002-0012). Once we placed the sample in the jig, we induced a tensile force on the test sample along the axis of the fiber and loaded the sample at a rate of approximately 0.2% strain every five minutes. Subsequently, as the fragmentation proceeded, we recorded the data as the number of breaks at a measured strain, with the strain being estimated from markers that we put on the sample. We continued this process of loading the sample and recording the number of breaks until the fragmentation process ceased. At this point, the fiber breaks reached saturation, and we stopped the test.

Besides using the fragmentation test, we used differential scanning calorimetry (DSC-7 Differential Scanning Calorimeter from Perkin Elmer) and dynamic mechanical analysis (DMA) (Dynastat Mark II from Imass, Inc.) to estimate glass transition temperatures (T_g). We conducted DMA studies on neat polyisocyanurates, made by autoclaving and SRIM, and on an SRIM composite. The material was measured from 100 °C to 230 °C at a heating rate of 0.5 °C/minute and at frequencies ranging from 0.5 HZ to 50 HZ. All samples were tested in displacement control. In addition, we used a Fourier Transform Infrared Spectrometer (FTIR) (Nicolet Magna-IR 550 Spectrometer) to examine the chemical structure of samples made by SRIM and autoclaving. We used transmission and internal reflectance techniques for this examination. For the transmission samples, we either cast thin films between two potassium bromide (KBr) crystals or ground cured resin in with KBr and formed a pellet. For the internal reflectance technique, we used a KRS-5 crystal from the Harrick Scientific Corporation.

Results and Discussion

The goal of this work was to make void-free fragmentation samples whose resin allowed the fiber to be fragmented to saturation and had properties similar to resin made by SRIM. As stated earlier, the process by which the polyisocyanurates are made in the automotive industry, SRIM, is too violent for the manufacture of single fiber samples. Consequently, we needed to develop a different processing route. Early attempts to make samples with a catalyst level of 1.5%, the level used in the SRIM process, revealed that this level was too high: the resin foamed immediately upon mixing. By reducing the amount of catalyst, we were able to increase the pot life so that we could pour the resin into the mold cavities that contained the single fibers. However, upon placing them into an oven, we found that the resin foamed and then gelled.

Clearly, we needed an alternate approach, and so we decided to use an autoclave to provide pressure to suppress the formation of bubbles and to provide heat to drive the reaction. We should note that we tried applying pressure at room temperature to see if the resin would cure without foaming. Interestingly enough, the pressure had the effect of delaying the reaction: when we removed the pressure, the samples started to foam. Subsequently, we believed that if we applied pressure to the samples, and then heated them, it was possible to gel and cure the samples and suppress foaming. This belief proved to be correct. We settled on the cure cycle described in the experimental section because we found that if we started with a lower pressure, we would not eliminate all of the voids before gelation occurred. Many of our test samples were made with 0.3% (weight % of polyol) catalyst, but we were able to determine that 0.5% catalyst can be taken as an upper limit. We noted some bubbles on the top surface of our fragmentation samples that were made with the higher catalyst level, but these bubbles could be removed easily by sanding the samples. One final note on processing the polyisocyanurates is that we needed to keep the silicone molds at room temperature (24 °C). We found that this lower temperature had the effect of inhibiting the reaction: the thin resin castings in the silicone mold remained liquid long after the bulk resin in the mixing cup gelled. In our previous work with epoxies, we had to preheat the molds to 75 °C before adding the resin to reduce the chance for air to be entrapped along the silicone/resin interface.

Some results from the fragmentation test are presented in Figure 3. It shows that the samples could be taken to saturation. Results on the interface tests will be presented in a future paper. We also tried to compare the T_g's of the polyisocyanurates molded via SRIM and via autoclaving. We estimated values of 172 °C ± 5 °C for the neat polyisocyanurate made via SRIM, 182 °C ± 5 °C for the neat autoclaved resin, and 190 °C ± 5 °C for the SRIM composite. These results are presented in Figure 4. In all three cases, the peak in the Tan δ curve was very broad which may be characteristic of this material. The data shown for the NIST resin sample was taken at 1 Hz, while the data shown for the ACC resin sample and the ACC composite sample were taken at 0.79Hz. The discontinuity in the data for the NIST resin sample reflect adjustments of the loading conditions of the dynamic mechanical analyzer.

We tried using the DSC to measure the T_g's, however, we did not detect any deviations from the baseline which seems consistent with the broad tan δ peaks. Additionally, we used the thermo-mechanical analyzer (TMA) to try to get comparative T_g's. Unfortunately, the signal from the pure resin sample was too broad and flat to get an accurate number, and the fibers in the composite sample made via SRIM dominated the signal for that material, thus, we were unable to measure accurate T_g's via this technique.

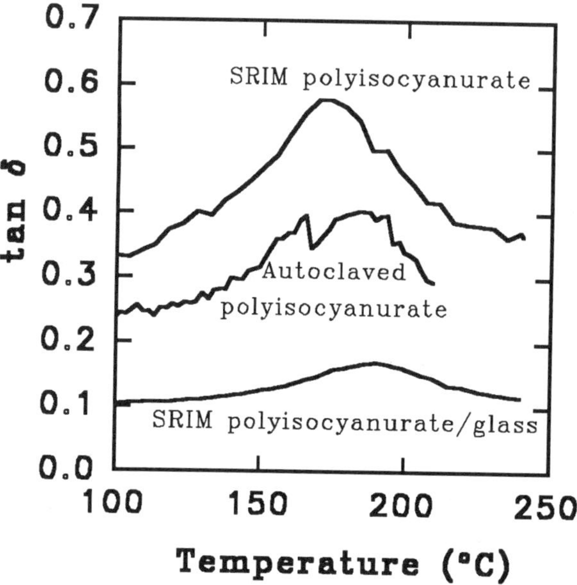

Fig. 4 - Plot of dynamic mechanical analysis data for polyisocyanurate resins samples made via autoclaving and SRIM. The SRIM resin samples were made following the process for the composite. NIST data was taken at 1 Hz, whereas the ACC data was taken at 0.79 Hz.

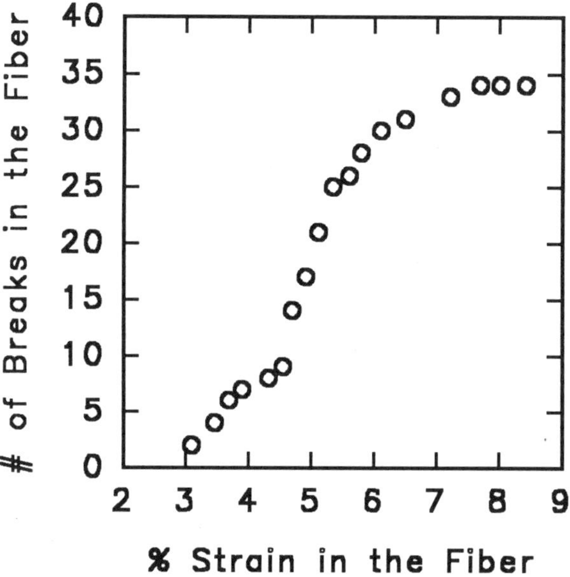

Fig. 3 - Plot of the number of breaks in a single fiber test sample as a function of % strain in the fiber for a polyisocyanurate/glass sample.

Preliminary results from our FTIR analysis shown in Figures 5 and 6. The spectrum in Figure 5 is of the resin made via SRIM, whereas the spectrum in Figure 6 is of the resin made in the autoclave. We can see that isocyanurate linkages were formed in the resins

(1410 cm^{-1}). In addition, the isocyanate peak at 2266 cm^{-1} in both of the cured resin samples appears to have disappeared. However, both spectra for the cured resin samples were taken by internal reflectance. Although these spectra indicate the strong similarity between the resins made in an autoclave and via the SRIM process, we should try to get good transmission spectra, since internal reflectance is done by taking a solid piece of resin and pressing it up against a crystal. The spectrum, then, is of the surface of the material and depends strongly on the contact between the sample and the crystal. The residual isocyanate near the surface may have reacted with water. In addition, if the surface of the sample was exposed to mold release agents or was ground down to fit in the internal reflectance cell, then these surface effects may show up in the spectra. Nevertheless, the preparation of samples is much easier for internal reflectance than it is for transmission. Transmission samples, at least for samples made via SRIM, needed to be microtomed and formed into potassium bromide (KBr) pellets. Although a clear spectrum could be obtained for the SRIM resin, we were unable to obtain a clear spectrum for the resin made in the autoclave with transmission.

Concluding Comments

We have shown that we can make void-free single fiber fragmentation samples and bring the fibers in these samples to saturation. In addition, preliminary results on the T_g's of the resin systems show that they are similar. Furthermore, the results from the FTIR analyses show that the spectrum for polyisocyanurates made via autoclaving is similar to the spectrum for polyisocyanurates made via SRIM. By being able to make void-free fragmentation samples that can be taken

to saturation and that can be considered comparable to SRIM polyisocyanurates, we have established this product as a viable material to add to our durability data base. Future work will include optimizing the autoclave processing cycle, obtaining transmission spectra for resins made via SRIM and autoclaving, and comparing the mechanical properties on the resins made by the two processing routes. Concurrently, we will be exposing single fiber fragmentation samples made via autoclaving to elevated temperature and humidity conditions.

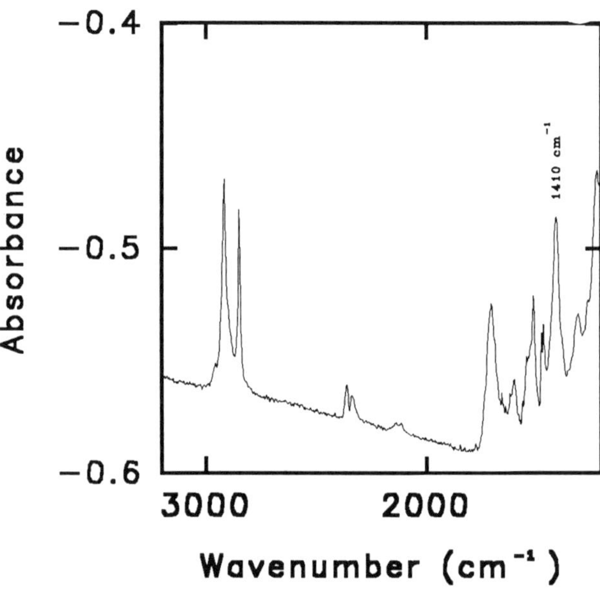

Fig. 5 - FTIR spectrum for a polyisocyanurate sample made via SRIM.

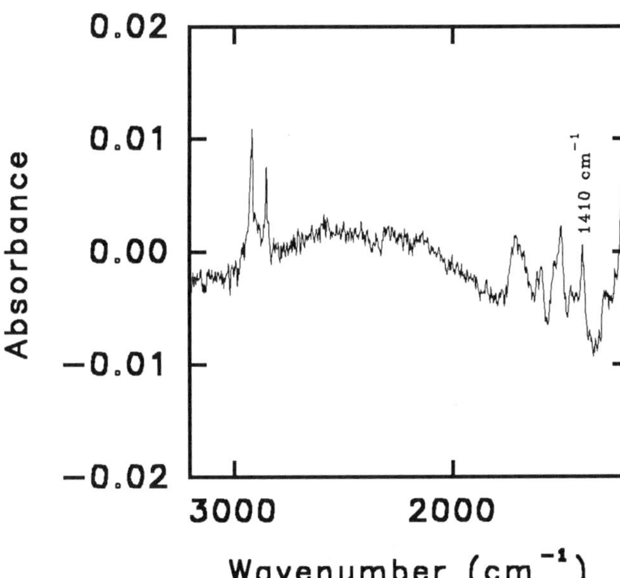

Fig. 6 - FTIR spectrum for a polyisocyanurate sample made via autoclaving.

Acknowledgments

We would like to thank Mr. Eric Byrd for his help in the FTIR work and Mr. Bill Roberts for his work on the TMA. In addition, we would like to thank the Dow Chemical Company for supplying the Spectrim (R) resin and Mr. Dan Houston of Ford for supplying the Vetrotex chopped strand mat and the neat resin and composite samples made via SRIM.

References

1   Schutte, C., W. McDonough, M. Shioya, M. McAuliffe, and M. Greenwood, Composites (special issue dedicated to the papers presented at the International Phenomena in Composite Materials '93), in press (1994).

2   Herrera-Franco, P. J. and L. T. Drzal, Composites, 23 (1), 2-27, (1992).

3   McDonough, W. G., P. J. Herrera-Franco, W. L. Wu, L. T. Drzal and D. L. Hunston, in Proceedings of the 23rd Intl. SAMPE Tech. Conf. Oct 21-24, 247-258, (1991).

4   Kelly, A. and W. R. Tyson, J. Mech. Phys. Solids, 13, 329 (1965).

5   Drzal, L. T., M. J. Rich, J. D. Camping and W. J. Park, in Proceedings of the 35th Annual Technical Conference, Reinforced Plastics/Composites Institute, 20C-1, 1-5, (1980).

6   Waterbury, M. C. and L. T. Drzal, J. Composites Tech. Res., 13 (22), (1991).

7   Yavin, B., H. E. Gallis, J. Sherf, A. Eitan and H. D. Wagner, Polymer Composites, 12 (6), 436-446, (1991).

8   Shioya, M., W. G. McDonough, C. L. Schutte and D. L.Hunston, in Proceedings of The Adhesion Society, 248-251, (1994).

9   Schutte, C. L. and M. McAuliffe, Proceedings of American Society for Composites, to be presented.

10   Wagner, H. D., J. R. Wood and G. Maron, Advanced Composites Letters, 2 (5), 173-176

11   F. Melvin Sweeney, "Reaction Injection Molding Machinery and Processes," Marcel Dekker, Inc., New York, New York, (1987).

12   de Haseth, J. A., J. E. Andrews, J. V. McClusky, R. D. Priester Jr., Ma A. Harthcock and B. L. Davis, Applied Spectroscopy, 47(2), pp 173-179, (1993).

13   Brydson, J. A., Plastics Materials, Butterworth & Co. Ltd, London, England, (1975).

14   Certain commercial materials and equipment are identified in this paper in order to specify

adequately the experimental procedure. In no case does such information imply recommendation or endorsement by the National Institute of Standards and Technology, nor does it imply necessarily that the items are the best available for the purpose.

Performance Evaluation of Glass Fiber-Epoxy Composites for Durable Goods Application

E. Shin, C. Dunn, R. Morgan
Michigan State University (AMEES)
Midland, Michigan

Abstract

In order to optimize design decisions of composite panels for use in durable goods applications, especially automotive application mechanical performance of a woven glass fabric-epoxy composite has been evaluated as a function of geometry, temperature, stress and time. Flexure of bars and plates is taken as a reflection of service reality with data related to anisotropy, temperature dependence, failure mode, damage and long term strength. In addition, creep rupture behavior using double cantilever beam (DCB) specimen is examined for two types of woven fabric composites. The failure data help with design decisions for practical service performance.

IN RECENT YEARS there has been a mushrooming demand of polymer matrix composites for use in automotive applications from structural or semi-structural application to under-the-hood application. However, main problems that the auto makers and the materials suppliers are facing immediately are lack of reliable and meaningful design data representing both short-term and long-term performance and especially, lack of effective test methodology to produce simple but meaningful design data. The same applies to screening tests for the many available PMC materials.

Obtaining design data for strength using conventional theoretical models requires a large amount of expensive data whilst leaving still unresolved the underlying question of actual relevance of the test specimens to the service article, the engineering structure. It is the authors contention that there is an alternative approach. A methodology to provide information on practical performance through specimen types which are regarded as simplified reflections of the range of service reality rather than theoretically ideal.

Important ingredients are:

a) Use of simple ratios as a 'Derating' to reduce complex continua of data to a few targeted numerical values to be handled in a data base for design.

b) Test protocols that use specimens which are simplified versions of critical parts of service structures and directly relate to the needs for the design of articles with regard to both processing and performance.

c) Establishment of real time long term tests on simple specimens and the linking of these data to reality via b.

It is understood that precision tests on ideal specimens are vital in setting up "core data," the data needed as a main reference framework. It is proposed that data from simple procedures and more practical composite structures be used to economically link the framework with service. Simplification and economy are certainly major concerns in the assessment of long term durability; only with low cost equipment and frugal test protocols can the necessary investment be kept realistic.

Some actions of the testing community are encouraging. A recent paper [1] assessed design data in the context of 'Derating'. Test systems that have a

* An earlier shorter version of this paper was presented at the 1994 SEM Spring Conference on Experimental Mechanics and Exhibit, 6-8 June, 1994, Baltimore, Maryland.

coordinated set of standard methods are operating [2] and proposals have been made [3] to extend them to involve a broad range of tests and specimen types. Also, the use of stylized parts of service structures is already an integral part of normal testing practice; the value of using flat plates is recognized in the Post-Impact Compression test and Instrumented Falling Weight Impact is Standard [4]. The latter embodies many of the features considered important in this paper. It is a quick and easy test of a sample sub-component that provides several types of design related data. A flat plate is fairly simple yet included can be biaxiality and damage that develops in accordance with natural weakness [5]. Also, a plate can incorporate features of the processing of a more complex structure, engineering geometry including holes and joints and changes imposed by a service environment.

One goal of this paper is to demonstrate some simple but effective methods on one particular composite that can be related to both aero and auto industries. Another is to generate meaningful performance data on composites for use in the durable goods industries that allow optimization of component design decisions. This paper is concerned with the physical, thermal and mechanical factors that control the strength failure modes of bars and plates in flexure.

Materials and Experimental Procedure

The main composite used in this study was 60v/v E-glass fiber woven fabric in an anhydride-cured bisphenol-A epoxy matrix (Lytex®); 20 layers of pre-impregnated plain weave glass fiber were compression molded into 4mm thick flat plates and fully cured. Another composite made of a coarse glass mat preform impregnated with the epoxy by resin transfer molding (RTM) was also used but only in studying some creep rupture behavior. The T_g of Lytex® composite was 173°C from Dynamic Mechanical Analysis shown in Figure 1. Composite test temperatures were also determined from this analysis; one at 23°C and the other at 120°C, closer to the softening point.

Various mechanical test specimens were cut from the composite plates using a purpose built circular-saw with diamond blade (D.R.Bennett LTD.). For 3-point flexure a 10mm wide bar cut in 0/90° fiber orientation was used with various span-to-depth ratios. In cases of the double cantilever beam (DCB) creep rupture and the end notch pre-damaged flexure specimen, PTFE film was inserted at mid-plane as a crack starter during lamination process. The crack initiator was pulled to propagate crack to the starting position before the specimen was transferred to constant load. Measurements were made at 23°C

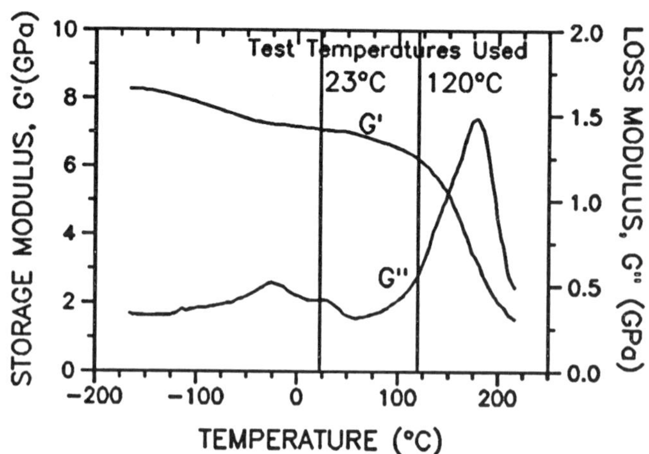

Figure 1. Dynamic Mechanical Analysis of Lytex® Composites.

or/and 120°C and all material were stored dry prior to testing. At 23°C the specimens were exposed to 50% RH on loading; at 120°C specimens were conditioned at the test temperature for 1 hour prior to loading. For creep rupture simple purpose build apparatus was used that had dead weight loading; lifetime only was measured, that is the time for complete fracture. Other tests were at short times with stiffness and strength obtained from an MTS Universal Testing Machine.

Sections for optical microscopic investigation of the DCB creep rupture specimen which had been loaded for more than 3 years were prepared by a slow speed diamond saw (Isomat®, Buehler). In order to observe in-situ deformation, the crack tip opening was frozen by embedding in a room temperature curing epoxy under load and before sectioning.

Results and Discussion

In-plane Stiffness Anisotropy. It was reported [7] that a simple method could quickly show the pattern of effective anisotropy. A specimen is taken in the form of a circular plate, parallel circular rods apply flexure in the same basic configuration as standard 3-point bending of specimen bars; low strain stiffness is given by the ratio of force applied to the resulting small deflection. Because of the low strain, the disc can be repeatedly tested without damage and a small rotation of the disc between each measurement provides a polar diagram of the anisotropy of disc stiffness uncomplicated by inter-specimen variability. Figure 2 shows some results where the rotation was 360° in 15° increments and the plate stiffness is expressed as Kilograms per mm deflection. The plate is a laminate of 0/90° woven fibers so it is natural to plot the change 0 to 90° with 4 sets of data superposed. There is indeed excellent symmetry in this

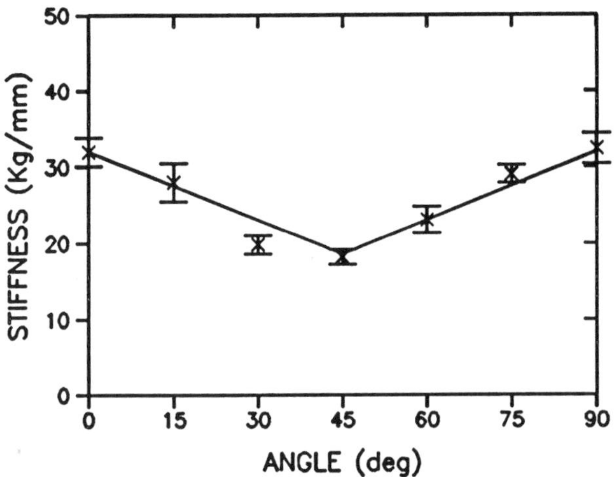

Figure 2. Anisotropy - rapid measurement by disc rotation from one specimen. Flexure by 3 line loading in MTS at 120°C.

case so the data are further reduced by combining all 8 parts 0 to 45°.

To generalize from 'plate stiffness' via plate theory requires many assumptions. These may be warranted but in accordance with a main theme of this paper the generalization here is through relative stiffness change. Figure 3 shows the effect of temperature on anisotropy in terms of a relative stiffness calculated from the reference stiffness at the 0/90° angle. For each temperature the change with angle was expressed as a change relative to the reference stiffness at that temperature. The resulting ratios can be variously described as 'Relative Stiffness,' 'Derating stiffness for angle' or simply 'Anisotropy Factor'.

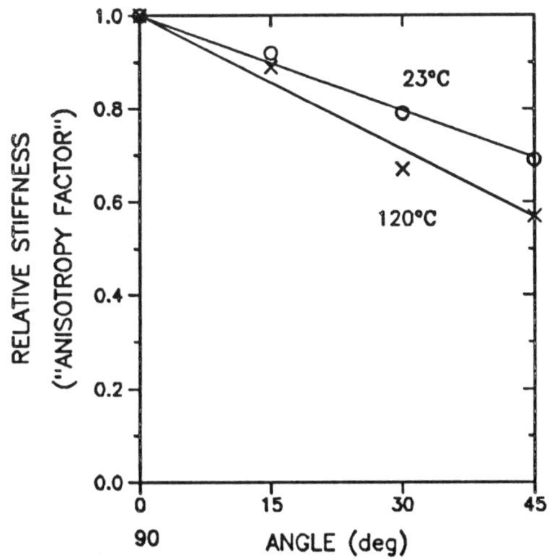

Figure 3. Anisotropy of a plate: effect of temperature on relative disc flexural stiffness.

In a realization of the composite as a plate, a circular plate with flexure in particular, there is

pronounced anisotropy. As expected that is greater at the higher temperature because of the increasing influence of the fibers as the modulus of the matrix decreases. The data shows that the 0/90 simplicity of fiber arrangement is undisturbed by the processing. In many other situations such perfection will not occur and such plate tests are useful low cost tools in general assessments, especially representing the most realistic cases.

Effects of Geometry and Damage in Flexure. Flexural stiffness and strength was measured on rectangular bars in the 0/90 direction with the fibers running along and across the bars. The test was conventional in that constant rates of deformation were imposed on bars 10 mm wide and 4 mm thick [6] and that they broke in about a minute. The geometry variation came from using a set of span values; there was a constant ratio of overhang. The thickness and width varied little between specimens so stiffness was taken as the initial slope of the force/deflection curve (kg/mm) and 'strength' as the maximum force (kg); five specimens were used at each condition. The combined effects of bar geometry and damage on strength, especially, delamination, was carried out with specimens similar to End-Notched Flexure. A planar crack on the mid-plane was manually opened up so that the naturally formed crack tip lay at a span/4 position, i.e. its starting point was always halfway between an outer support and the central load whatever the actual span used.

Figure 4 shows typical load-deflection curves for bar flexure of Lytex® with and without the end notch pre-damage and at two temperatures, 23°C and 120°C. All curves in the figure are mean lines through set of 5 curves at each condition and the span/thickness ratio was 10. The most dramatic effect

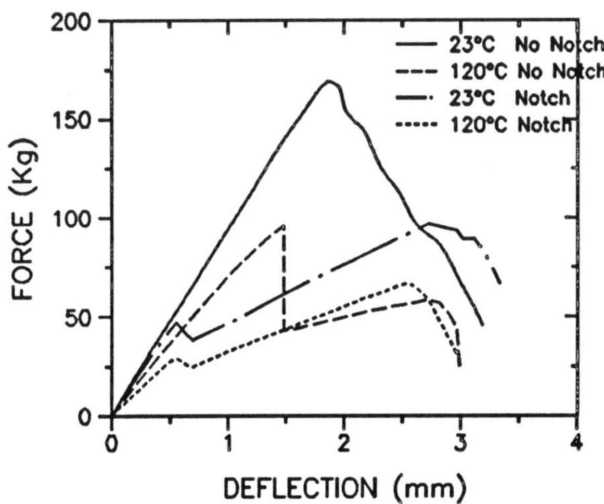

Figure 4. Typical load-deflection curves of flexure and end notch pre-damaged flexure of bars at two temperatures.

547

in the load-deflection responses can be seen from temperature change of the undamaged flexure. Increasing temperature to 120°C causes early load drop in the undamaged flexure as in the pre-damaged one; this is indicative of interlaminar delamination in both. It is obvious that the interlaminar shear strength is governed by the matrix and thus, strongly affected by temperature. However, in all cases, at the maximum load the failure of a flexed bar always involves combinations of complicated failure processes with tension, compression, and shear.

Figure 5 shows that how the change of stiffness between 120°C and 23°C varies with the bar geometry. The stiffness of bar at 120°C relative to that at 23°C is plotted against log 1/span and the relative stiffness of a disc at span/thickness ratio of 32 is compared. From this figure it is clear that as the span/thickness ratio increase the shear contribution decreases and so does temperature dependency. Therefore, in an ideal case, there will be little or no change in stiffness going to 120°C if the span/thickness ratio is large enough to minimize matrix shear contribution. Once the ratio decreases below 8 no further decrease in relative stiffness is noticed; the shear contribution reaches a maximum. Perhaps the most striking effect of temperature is the considerably lower values of disc plate than that of bar. Though at a high span/ thickness ratio the disc behaves as if it has a shear component equivalent to a 9:1 bar. However, the behavior of a plate is complex and more data are needed for significant conclusions.

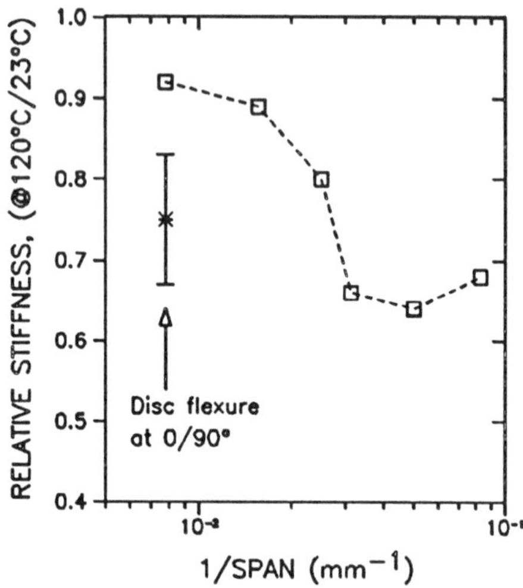

Figure 5. Relative stiffness in terms of test temperature at various span/thickness ratios of bars and disc.

The tests that gave the stiffnesses were continued to final fracture but as noted earlier, in some cases the complex damage evolutions started to occur at an early stage. Figure 6 shows the results as log ('strength') varying with log (1/span); a slope of unity shows a

connection with the simplest formula for tensile/compression stress and a horizontal line relates to a shear stress. These two slopes have been drawn on the graph to correspond to the experimental data. The transition from tensile to shear dominance is clear. It is interesting to note the shift on going to the higher temperature that indicates the increase in the influence of shear. For conventional strength calculation a failure mode would have to be assumed so that the necessary model could be used. It is important to state that here the complex phenomena associated with failure have been ignored and the practical 'strength' is taken as the maximum force supportable; constant specimen dimensions allow geometrical trends to be observed without reference to a failure mode even though this can be some combination of fiber failure and longitudinal splits in plane.

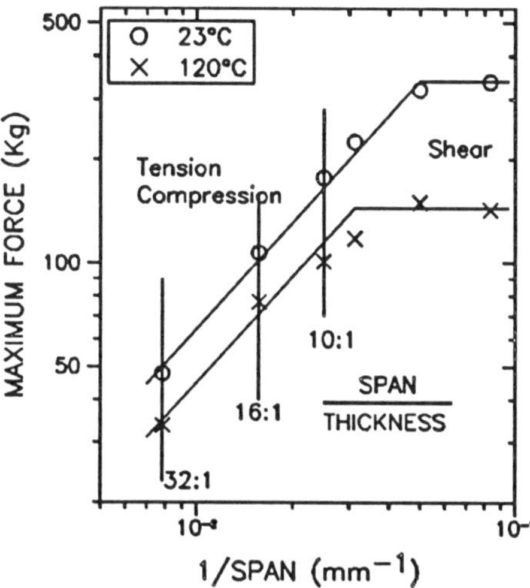

Figure 6. Effect of stress state in bar flexure on effective short term strength at two temperatures. Each point mean of 5 test results. Shown also is bar geometry selection for creep rupture test.

Combined effects on the flexural strength of bar geometry and pre-damage, as well as some temperature dependence, are shown in Figure 7. From the force-deflection curve two characteristic peaks were determined, the initial low peak when the crack propagated on the mid-plane and the maximum force supportable by the bar thus damaged. The parallelism of the curves in the figure despite the variation of failure mode suggests a unity of effective behavior that should allow normal design rules to accommodate damage effects. This is in agreement with other work [8,9] which suggested that some impact damage could be modeled by a simple hole so that it should be easy to incorporate within basic design and this despite or possibly because of the inevitable complex crack generation around woven fabric [10]. In addition to the room temperature behavior some data points at

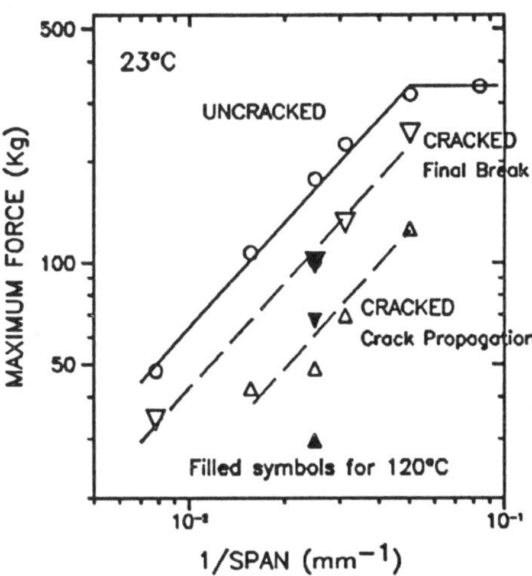

Figure 7. Effect of pre-damage in bar flexure on short term strength. Each point mean of 5 results.

120 °C show proportional shifts similar to those at room temperature. Another way to analyze the parallelism between damaged and undamaged flexure is in terms of damage sensitivity. The sensitivity can be defined as percent ratio of change in maximum force due to damage to maximum force if undamaged and such results are summarized in Table I. It is of interest to note that the sensitivities of both peaks are reasonably constant regardless of the bar geometry or temperature.

Table I. Sensitivity of strength to damage in flexure of Lytex® composite.

Span/Thickness Ratio	Sensitivity, %	
	First Peak	Maximum Peak
5:1	61	24
8:1	69	42
10:1	73	44
16:1	60	–
32:1	–	32
<u>120 °C</u>		
10:1	71	33

Creep Rupture in Flexure. The final fracture of a flexed bar of the Lytex® composite was the accumulation of various failure processes; these develop in heterogeneous zones with a complex of tension, compression and shear. A chosen test geometry can bias this. The previous section presented a picture of simplicity in the pattern of actual ultimate load supportable despite such complications. That was

for processes occurring at times of the order of 1 minute (10^{-2} hours). Measurement of the creep rupture behavior will indicate if this still holds at long times under load, if there is some transition to the detriment of service performance; if any changes in the winner from the various processes produces an actual decrease in load bearing capacity.

Creep rupture in flexure of the Lytex® composite was measured at two temperatures and at three span/thickness ratios. At a span to thickness of 32:1 the behavior should be tension/compression dominated; at the other extreme, at 10:1 there may be significant shear effects (see Figure 6). Figures 8 and 9 present the data; each point is from the final fracture of one specimen. The force applied to the individual specimens has been normalized for small differences in thickness and width by using Force/Area of cross-section. This is not the place for a detailed justification of that; though important to the philosophy of testing, in practice it makes little difference here to the overall picture that emerges. The wide range of lifetimes at one condition results from not only the applied load level but also specimen variability combined with that of the fundamental failure process; i.e., strength and lifetime of materials are the statistical quantities due to the physical or chemical structural variability within the material. The data suggest that there may be some trend to shorter lifetimes with temperature increase but many more data are required to be certain. Even with such a trend it seems likely that the effect is small considering the apparent general variability.

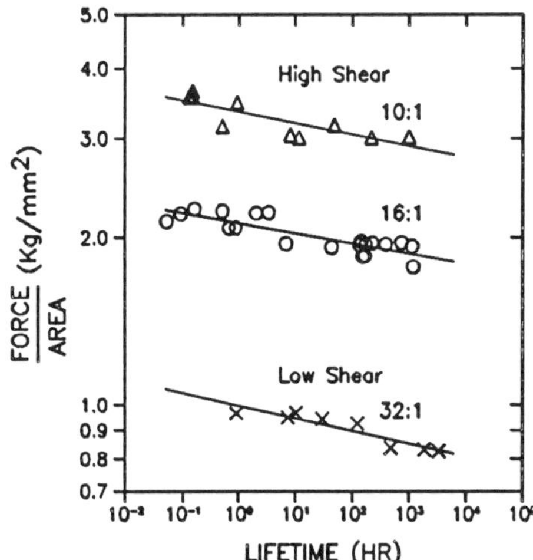

Figure 8. Effect of geometry on creep rupture in bar flexure of Lytex® composite at 23 °C.

Just as previously anisotropy was clarified by derating stiffness for direction and the effect of an increase in temperature on stiffness was shown as a proportional change, so creep rupture can be expressed as a 'Derating of Strength for Time

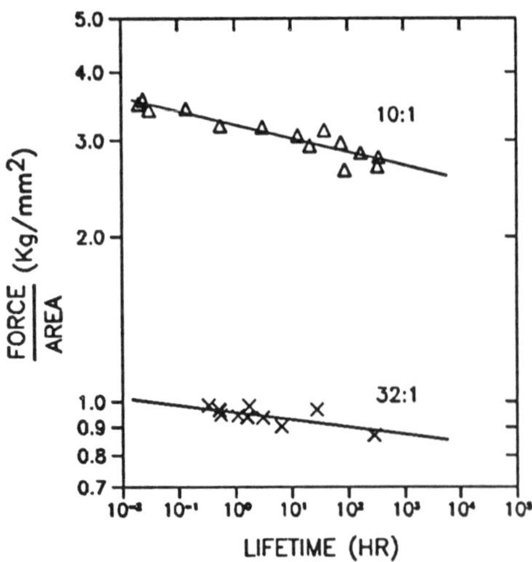

Figure 9. Effect of geometry on creep rupture in bar flexure of Lytex® composite at 120°C.

Under load'. In engineering design terms the 'Derating' shows how the measured short time strength must be reduced if loads must be supported for long times. A reference strength is chosen which is the best estimate of the value for a lifetime of 10^{-2} hours. The actual Force/Area values applied to the individual specimens are expressed as a proportion of that. Re-plotting the original creep rupture data shows 'Derating Factor' as a function of Lifetime. All five creep rupture sets can be thus treated, each referred to it own 'short term strength.' Figure 10 results.

Figure 10 shows 60 individual failures from widely differing geometrical arrangements and temperatures but for clarity only the two temperatures are distinguished; a common band is a reasonable representation. Though there may be the small trend with temperature mentioned earlier, there is no indication of any catastrophic shift of the ultimate load supportable, it also can be implied that design data based on low cost short term tests are likely to need only minor shifts to accommodate longer times under load. The variations measured suggest that one way forward is to look for trends in the statistical distribution of lifetimes through careful choice of the appropriate number and type of test result. The short time tail is of obvious prime importance.

Creep Rupture in DCB. A key element in the structural performance of a laminated composite is the resistance to various modes of interlaminar crack growth. A conventional double cantilever beam specimen was used with constant load to provide a measurement of the creep rupture behavior in the opening mode. Two types of composite, so-called Lytex® and an RTM panel, described in materials section were tested. The DCB specimen dimension was 10mm wide, 90mm long and 4mm thick for Lytex® and 12.7mm wide, 152.4mm long and 4mm thick for RTM specimen. A natural crack was formed by manually extending the defect formed by the molded in "teflon". In an MTS machine a load was applied perpendicular to the crack through a hinge type tab attached to the end of the beam. During that process the force required to propagate the original crack was recorded for individual specimen and used to calculate the each load applied for the creep rupture test. Figure 11 shows the applied load vs. lifetime of both RTM and Lytex® composite tested at room temperature and 50%RH. The applied load was normalized by the initial crack propagation load. The arrows mark data points where failure has not occurred. In case of Lytex® the load up to 92% of the critical load has been applied for more than 3 years without catastrophic failure while the RTM composite shows a quite broad distribution of lifetime at the load above 80%. It

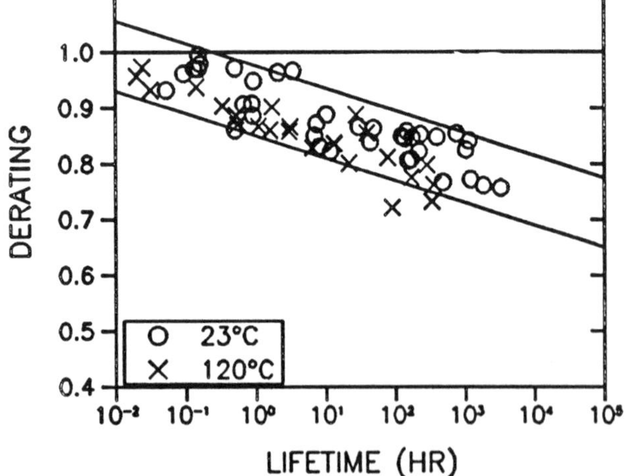

Figure 10. Overall creep rupture behavior of Lytex® composite in flexure including all data from figure 8 and 9 in terms of derating of short term strength.

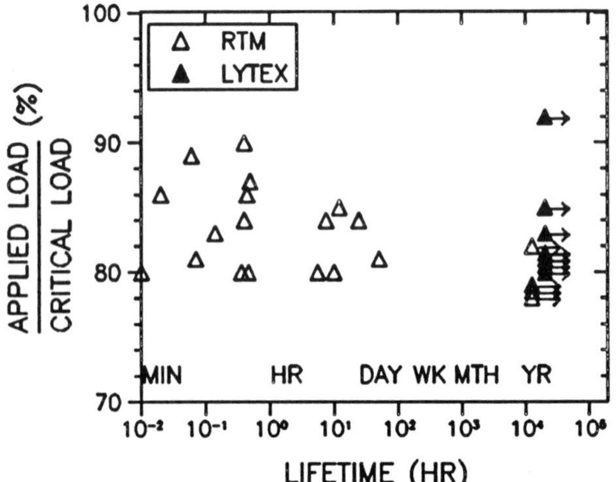

Figure 11. Creep rupture in DCB of both RTM and Lytex® composites.

should be noted that the comparison made in this study is not to compare the two materials including processing condition, but to validate the test methodology for the performance evaluation.

The result from Lytex® composite is very striking because the 92% of the critical loading is very high to bear more than 3 years without major crack propagation. This minimal change over such a long time was confirmed by subsequently increasing the load. Some specimens that for 3 years had sustained a load at 80% of that for the initial cracking, were loaded over a period of a few minutes to measure the level needed for fracture. Table II shows the results. The average difference in force to propagate the initial crack and the crack after 3 year was only 2.4%.

Table II. Critical force to propagate cracks before and after long-time creep rupture test in Kg unit.

Force to Propagate Initial Crack	Force Applied for Creep Rupture	Force to Propagate Crack after 3yrs Creep Rupture
2.83	1.98	2.96
4.52	3.16	4.28
2.83	1.98	2.78
4.07	2.84	3.74
4.29	3.00	4.24

In order to identify any possible mechanisms of this remarkable load carrying capability of the composite, firstly, the fracture surface generated from a specimen used for Table II was examined by a scanning electron microscope. The fracture surface was generally stepwise showing ups and downs through the warp and fill strands and crossover regions between them which is typical of woven fabric composite [10]. One interesting finding from the fracture surface analysis is that there were two type of failure in the crossover region. Figure 12 shows a clean delamination by fiber debonding in resin poor region, a, and gross resin plastic deformation and considerable fiber breakage in resin rich region, b. This indicates that the crossover region containing enough matrix resin makes crack difficult to propagate.

Secondly, the in-situ crack tip deformation and crack propagation mode was investigated from thin sections made from one of on-going the creep rupture specimens. The DCB specimen used for sectioning had been under 80% of critical load for more than 3 years and the crack tip deformation was frozen before sectioning as described in Materials and Experimental Procedure section. The sections were cut from one

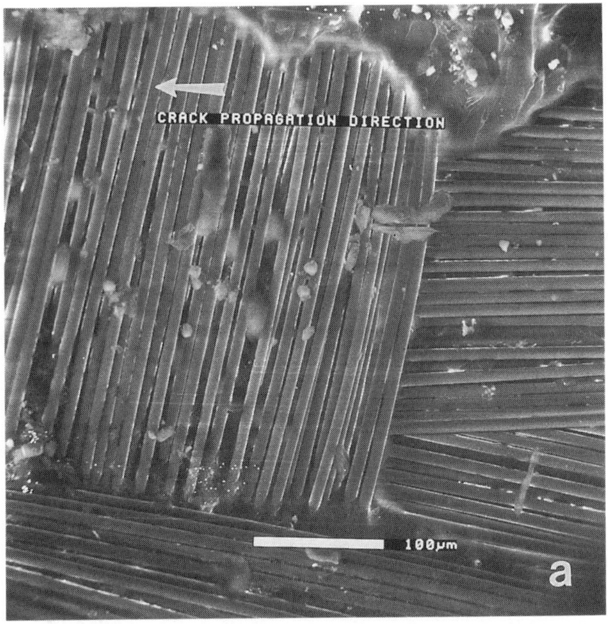

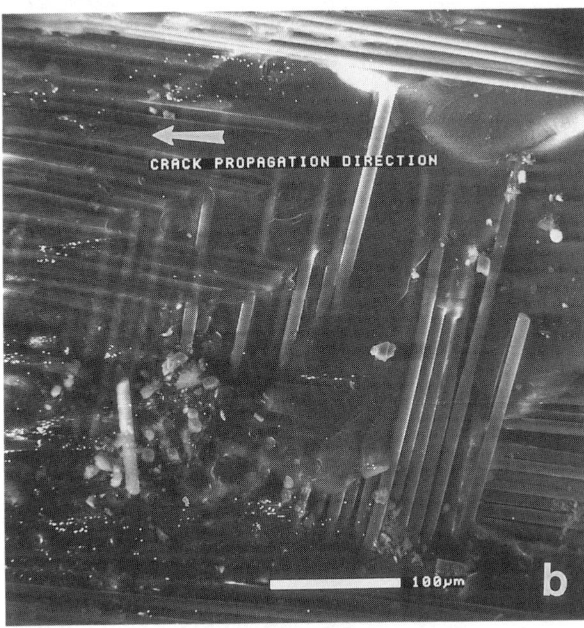

Figure 12. SEM micrographs showing crossover regions between the warp and fill strands, a: resin poor region, b: resin rich region.

side parallel to the length direction, first thick section of about 1.5mm, and then thin section of about 0.3mm and continued alternatively and named by TK1, TN1, TK2, TN2, ..., respectively. Figure 13 shows optical micrographs taken from side view of the cross-sections at various positions across the width. The vertical dark lines on the micrographs, a, b and c indicate the original crack tip position before creep rupture loading. It is clear that the crack had propagated only a few millimeters from the beginning of the creep rupture test and it had grown longer at the inside than at the edges; this can be explained by the internal dilation triaxial stress build-up or plane strain condition. Higher magnification micrographs are also shown in Figure 14. Thick sections were examined under reflective light while thin sections were viewed with transmitted light. Several observations were made from this study; In some part of the crack, secondary cracks were originated near the original crack tip as shown in Figure 13a & b and Figure 14a; the crack propagates not only interlamina but intralamina; when the crack encountered the crossover intersection, it tended to bifurcate, (see Figure 13d or 14c). All of these can be considered as strain energy absorbing mechanisms or crack arresting mechanisms. On the other hand, surprisingly enough, there was no indication of any fiber bridging which was often observed in many woven fabric composites [11]. No microcracking was found at near the crack tip or elsewhere.

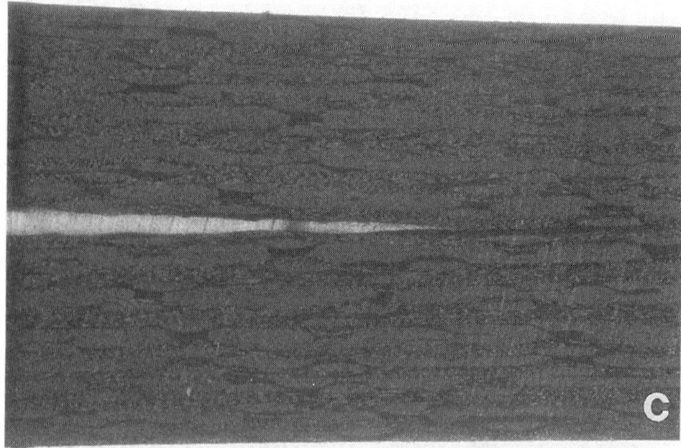

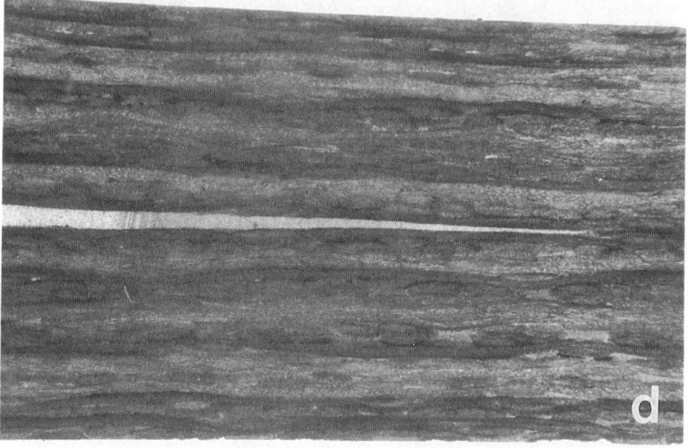

Figure 13. OM micrographs showing side-view of cross-sections around crack tips, a: TK1, b: TN1, c: TK3, and d: TN3

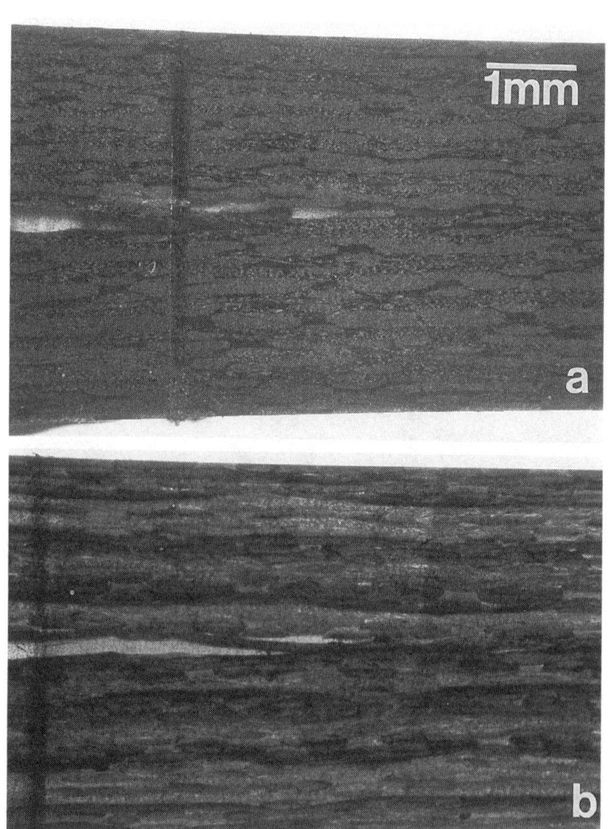

Conclusion

The data presented demonstrate the value of using simple low cost methods and their place in general materials evaluation and design. Practical estimates of stiffness and strength behavior have been obtained using specimens that constitute simplified reflections of service reality. The phenomena involved for the woven glass fabric composite were anisotropy, temperature dependence, failure mode, damage and long term strength from simple flexure of bars or plates and DCB configuration. Several specific conclusions were drawn from the experimental results.

They are:

- In-plane stiffness anisotropy of the woven glass fabric composite (Lytex®) was revealed by simple flexure of a circular plate. The anisotropy was greater at the higher temperature because of the increasing influence of the fibers as the modulus of the matrix decreased.

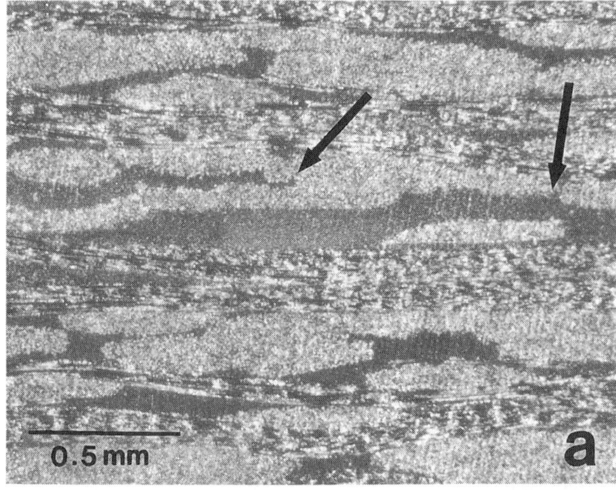

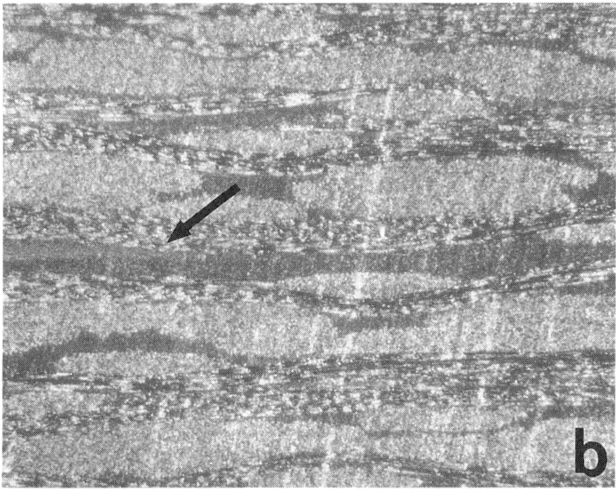

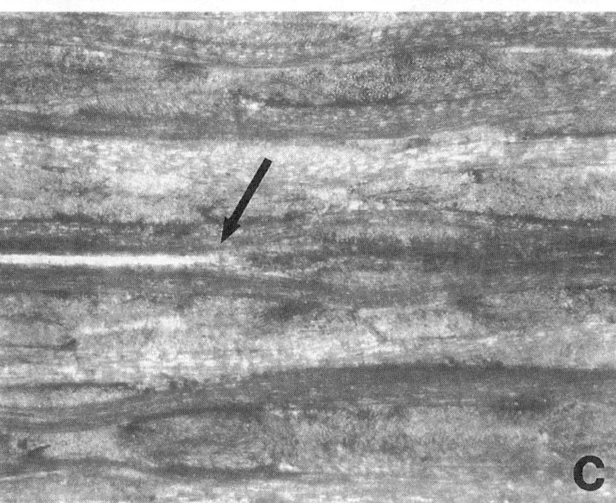

Figure 14. Higher magnification OM micrograph showing crack tip area, a: TK1, b: TK3, and c: TN3. Arrow marks indicate crack tips.

- Unnotched flexure of the Lytex® composite at 120°C, behaved like the end notch pre-damaged material both with interlaminar delamination.

- The analysis of relative stiffness, the stiffness of at 120°C relative to that at 23°C, showed that the effect of temperature increased with decreasing span/thickness ratio. At the same span/thickness ratio a disc plate showed a greater temperature susceptibility than bars.

- Decreasing the span/thickness ratio produced a change in failure mode, a transition from a tensile/compression mode which was manifested by a slope of unity in strength vs. 1/span plot, to a shear dominating mode which was horizontal line in the plot. At the higher temperature, the transition shifted to longer span.

- The characteristic force peaks from the end notch pre-damaged specimens showed a similar dependence on span to those from unnotched specimen. The damage sensitivity were reasonably constant regardless of the bar geometry or temperature.

- Long term strength of Lytex® was evaluated using creep rupture of flexed bars; the equivalent using a DCB specimen involved both Lytex® and an RTM material. The analysis showed the effects of temperature, span/thickness ratio and failure mode on the lifetime distribution.

- Possible mechanisms involved in the long term load carrying capability of Lytex® had been revealed from SEM fractographic study and an accurate sectioning analysis.

- The complex performance of a complete article will be reflected in the behavior of simpler engineering structures if these contain major features of the article itself and have related processing. The inference is that these are a set of simplified structures, 'Critical Basic Shape' or 'Sub Components,' [12] that can be used as test specimens to efficiently cover the engineering range and be utilized generally as 'common practice'.

Acknowledgment

The authors thank Eric Fouch for producing much of the experimental data in this paper.

References

1. Moore, D.R., J.M.Smith, and S.Turner, Plastics, Rubber and Comp. Proc. and Applications, 21, 19-31 (1994).

2. Automotive Composites Consortium Materials Work Group, "Test Procedures for Automotive Structural Composite Materials", 1990.

3. Dunn, C.M.R., K.Battjes, E.English, E. Fouch, P.J.Hogg, R.J.Morgan, and S. Turner, "Durability of Polymer Based Composite Systems for Structural Applications", A.H.Carlan, and Z. Verchenz, Eds., Elservier, p 129 (1991).

4. International Standard ISO 6603, "Plastics-Determination of Multiaxial Impact Behavior of Rigid Plastics", (1989).

5 Dunn, C.M.R. and M.J.Williams, Plastics and Rubber: Materials and Applications, 90-6, May (1980).

6 International Standard ISO 178, Plastics-Determination of Flexural Properties of Rigid Plastics, (1975).

7 Stephensen, R.C., S.Turner, and M.Whale, Composites, July (1979).

8 Babic, L., C.Dunn and P.J.Hogg, Plast. and Rubber Proc. and Appl., 12, 199-207 (1989).

9 Chamis, C.C. and C.A.Ginty, "Fiber Composite Structural Durability and Damage Tolerance: Simplified Predictive Models," Composite Materials: Fatigue and Fracture, Second Volume, ASTM STP, 1012, p 338-55 (1989).

10. Bascom, W.D., J.L.Bitner, R.J.Moulton, and A.R.Siebert, Composites, 9-18, January (1980).

11. Saghizadeh, H and C.K.H.Dharan, J. Eng. mater. & Tech., 108, 290-5 (1986).

12. Stephenson, R.C., S.Turner, and M.Whale, Polym. Eng. & Sci., 19(3), 173-80 (1979).

Development of Standardized Tests for an
Automotive Structural Composites Database

T.J. Dearlove, D.L. Denton, E.M. Hagerman, D. Houston, S.N. Kakarala
Automotive Composites Consortium
Troy, Michigan

ABSTRACT

The principle goal of the Automotive Composites Consortium (ACC), an R&D partnership between Chrysler, Ford and General Motors, is to accelerate the development of structural polymer composites technology for the American automotive industry. One of the major objectives necessary to achieve this goal is to assure the availability of reliable data for the design of structural composite parts. The ACC published a manual in 1990, "Test Procedures for Automotive Structural Composite Materials," that has been successfully used by industry in many characterization studies[1,2]. A second edition of that manual has now been published and has been expanded to include structural SMC and structural thermoplastic materials in addition to the liquid molded composites covered in the first edition. Test methods for determining various types of durability properties are also under development and include cyclic fatigue, compressive bolt load retention, mechanical damage tolerance, and property changes as a result of exposure to automotive fluids. In developing these test methods some initial durability characteristics of a number of liquid molded composite materials were determined.

INTRODUCTION

AS THE INTEREST IN IMPROVING the fuel economy of motor vehicles increases, the necessity of using lower mass materials also increases. However the integration of lighter weight structural polymer composites into vehicles has progressed slowly. The most notable application is the fiberglass leaf spring which first appeared on the Chevrolet Corvette in 1981 and resulted in a mass reduction of over 75% (18.6 to 3.6 kg). Another significant structural composite component on high production vehicles is the bumper beam. In spite of the slow implementation, there continues to be a strong interest in utilizing composites in structural automotive applications [3-7]. There are a number of reasons why new uses for structural composites have been slow in evolving. In part it is because of the unanswered questions related to the high volume, low cost manufacture of components with consistent part-to-part properties and known durability. Answering these questions is a key goal of the ACC.

The ACC selected liquid molding as its initial processing approach for manufacturing structural composites. Liquid molding encompasses both resin transfer molding (RTM) and structural reaction injection molding (SRIM) and has the potential to produce large complicated assemblies that achieve a high degree of parts consolidation. Some examples of these types of components are floor pans, body side frames, crossmembers and front structural members. To produce components of this complexity, with minimum mass and which meet the required performance criteria, will require very efficient design based on reliable material properties. To facilitate the acquisition of accurate, consistent design data for the liquid molding materials, the ACC published and distributed a test procedures manual in 1990 [1]. A second edition has now been published which expands its applicability to a wider range of structural composites.

A more complex and challenging set of test methods are currently being developed by the ACC to determine various aspects of durability. This includes

* General Motors, Warren, MI
** Chrysler Corporation, Auburn Hills, MI
*** Ford Motor Company, Dearborn, MI

fatigue properties, impact damage tolerance, compression under bolt load conditions, and property retention after exposure to automotive fluids such as gasoline, oil, windshield fluid, etc. The purpose of this paper is to report on the updated test manual, to review the status of test methods for durability and to present some of the results of durability testing on liquid molded composites.

TEST PROCEDURES MANUAL

In order for designers and structural analysts to have confidence in their structural composite designs, they must have access to accurate material properties. Also, they need comparable information to determine which material is best for a particular application. With this in mind, the ACC published a manual to serve as a guide for material and part suppliers for the measurement of key composite properties. Tests are specified for "static" tensile, compressive and shear properties at -40°, 23° and 120° C. Procedures are also provided to determine dynamic mechanical properties and linear coefficients of thermal expansion as functions of temperature.

Over the past several years the ACC has recieved a number of data sets from suppliers which were generated using this manual. Along with the data we received many positive comments on its intent and its content. We also received suggestions for improvement, including its expansion to cover other types of structural composites. There are many potential applications for structural composites in vehicles and liquid molded parts will not be the most effective for some of them, especially smaller, simpler ones. Design data for structural composites manufactured by processes such as compression molding are just as critical as is the data for liquid molded composites.

The second edition of the ACC Test Procedures Manual has now been published [8] and, in response to user suggestions, is significantly different than the first edition. It has been simplified to allow easier location of critical information, dimensions have been converted to "hard" metric, and the scope has been expanded to cover both structural SMC and thermoplastic composites. Finally as a reference point, we have defined structural composite materials as having a tensile modulus greater than 7 GPa and a tensile strength greater than 100 MPa.

TEST METHODS - DURABILITY

In addition to the basic design information discussed previously, it is important that the susceptability of a composite to undergo a loss in properties as a result of in-service use or being exposed to potentially harmful automotive fluids be considered in the design process. The ACC is in the process of developing standard test methods to assess the durability of structural polymer composites under cyclic fatigue, bolt load retention (stress relaxation), impact damage tolerance and resistance to automotive service fluids. The general approach of each of the durability test method development programs are presented in this section.

FATIGUE -The ACC is preparing tensile-tensile fatigue test specifications with the input and technical support of supplier companies. The goal of the specifications is to standardize the procedures for generating design data and to provide a common basis for the comparison of fatigue performance of different composite materials. Key characteristics to be quantified are modulus retention under cyclic loading and fatigue life as a function of stress level (S-N relationship).

Separate specifications are proposed for composites containing randomly oriented fibers (chopped or continuous) and aligned continuous (unidirectional or crossplied) fibers. Differences in fiber architecture and fiber volume fraction in the test material affects factors such as specimen width, use of tabs and failure criteria. The specimen geometry is consistent with that specified by the ACC for "static" tensile tests (25 mm wide, straight-sided coupons).

Monotonic properties are measured at two different testing rates, the rate specified by the ACC for "static" tests (5 mm/min), and the testing rate used in the fatigue tests (approximately 1,000 mm/min). The difference in strength at these two rates provides a indication of the rate sensitivity of the material. Polymer composites generally exhibit a higher tensile strength when the testing rate is increased.

A relatively constant rate of fatigue stressing is prescribed, which is achieved by varying the test frequency according to the following equation:

$$f = (k\,S_{ult})/(S_{max}-S_{min}) \qquad \text{Eq 1}$$

where:

f = frequency used for a particular stress level
k = "constant" for a given S-N curve
    (3 Hz is recommended)
S_{ult} = mean tensile strength determined at k Hz
S_{max} = maximum stress level for a particular fatigue test
S_{min} = minimum stress level for a particular fatigue test
    ($R = S_{min}/S_{max}= 0.1$ is specified)

Fatigue specimens are tested at four stress levels that yield fatigue lives evenly spaced over the range of 10^2 to 10^6 cycles. Any specimens surviving 2×10^6 cycles are tested monotonically for residual strength. A fifth stress level, which is half of the lowest stress level expected to cause fatigue failure, is also specified. The latter tests are intended to measure modulus retention only and fatigue testing of these specimens may be terminated at 2×10^5

cycles. A typical S-N plot that results from this test protocol is shown in Figure 1.

An example graph showing tensile modulus, retention versus log N is presented in Figure 2. It is anticipated that most relationships can be adequately represented by a linear regression. Once the data is normalized to the initial modulus the data from different specimens can be averaged. Average modulus retention data for different stress levels can be combined on one plot.

The use of constant stress rate provides a number of advantages. Previous work has shown constant stress rate tests for polymer composites usually yield linear S-N relationships over the range from the monotonic strength (determined at the fatigue rate) up to 10^6 cycles [9-11]. In some cases, a change in slope of the S-N relationship is observed at low stress levels [10-12]. Constant stress rate testing uses higher test frequencies at low stress levels (Table I), so the time needed to conduct high cycle fatigue tests is reduced. Lower test frequencies are used for high stress fatigue cycling, which reduces the temperature rise from hysteretic heating.

Table I

Test Frequencies Calculated for
Various Fatigue Stress Levels [+]

Maximum Stress Level (percent of UTS)	Frequency (Hz)
70	4.8
60	5.6
50	6.7
40	8.3
30	11.1

[+] Calculated from Eq 1 where k=3

An experiment was coordinated by the ACC to determine the effect of test frequency on fatigue life of specimens tested at a low stress level. Test specimens were cut from three RTM flat panels containing a vinyl ester resin reinforced with approximately 25 volume percent glass fiber continuous stand mat (CSM). A total of 53 specimens were fatigue tested at 48 MPa (38 % of the ultimate tensile strength) at three different supplier laboratories (Ashland Chemical, Dow Chemical and Owens-Corning). Approximately half of the specimens (25) were run at 15 Hz and the balance were tested at 3 Hz.

Analysis of the data showed that the mean fatigue life (log N) of the specimens tested at 15 Hz was 5.19 (155,000 cycles) while the specimens tested at 3 Hz failed at a log N of 4.80 (63,100 cycles). For these tests the pooled standard deviation was 0.38. No significant temperature rise was measured for either group of specimens. The

longer fatigue life at higher frequency has been previously observed and has been attributed to the higher tensile strength exhibited at the higher rate of stressing [9]. While the effect of time at mean stress could potentially give the same results, workers have found this not to be a factor for thermoset polymer composites [9]. Since rate of stressing can effect the fatigue life, the approach to keep this variable constant is supported.

The ACC is in the process of organizing round-robin testing to validate the specification for random fiber composites. Once its applicability is verified a specification will be issued. The ACC is also planning experiments to resolve technical issues identified in the testing of unidirectional fiber composites.

BOLT LOAD RETENTION - A test method is currently under development for determining the through thickness stress relaxation of mechanically fastened structural composites. The test, referred to as bolt load retention, has as its goal the measurement of the stress relaxation properties of composites under various loads and at various temperatures.

The experimental details describing the apparatus, data acquisition and software have been reported elsewhere [13]. In general a 25 x 25 x 3.2 mm composite test specimen is cut and its corners rounded to fit in a cylindrical steel pipe. A 6.3 mm hole is then diamond drilled directly in the center of the specimen in order to accept the 6.3 mm diameter strain bolt with 120 ohm strain gages mounted in the bolt interior (Figure 3). Two 19 x 3.2 mm stainless steel washers are used to distribute the load on the surfaces of the specimen. Once assembled the specimen is loaded under compression to a predetermined level (either 5, 10 or 15 MPa) and then immersed in an oil bath at either ambient, 80 or 120°C. The stress is continuously monitored, usually for 10 days, before a final stress reading is taken at ambient temperature.

In Figure 4 the change in compressive stress for a SRIM isocyanurate resin (Tg >150°C), and designated as resin A, reinforced with 25 volume percent CSM at various temperatures is shown. Note that the aluminum control showed very little loss is stress, as would be expected, thus indicating the accuracy of the equipment. With increasing temperature the retention of stress is 86, 25 and 14% respectively. In spite of the reported high Tg for this system, there is a significant effect of temperature on the stress relaxation. Some indication for this behavior can be found when viewing the DMA curve for this composite where an initial drop in storage modulus is observed at temperatures of 70-80°C. The effect of different loads was measured at 80°C (Figure 5) and as the load increased, the retention of stress went from 25 to 28 to 19%. The intermediate 10 MPa load is a common value used in automotive bolt load applications.

In Figure 6 there is shown a comparison of two SRIM isocyanurate resin systems (A and B) and a polyurethane SRIM bumper beam material (C), all reinforced with 25 volume percent CSM. They were all loaded to 15 MPa followed by heating to 120°C. The data for isocyanurate A was shown previously in Figure 4 and had 14% retention of stress. Isocyanurate B was found to have completely unloaded (0% retention) while the "low Tg" polyurethane bumper beam system (C) had 27% retention. In Figure 7 are the results for isocyanurate B tested under initial loading of 15 MPa at 80°C for 5 days at which time it had approximately 15% retention of stress. After cooling to ambient it was retorqued to 15 MPa and reheated to 80°C. This time it gave a retention value of about 75% after 5 days.

The equipment for this test is relatively simple and the experimental procedure has been shown to provide valid and accurate results. The reduction in compressive stress at relatively low temperatures is a concern and indicates that fastener designers need to account for the phenomenon. There is a need to expand this development to cover other structural composites and most importantly is to use this apparatus with various washer designs to determine what system best leads to an acceptable retention in bolt load stress. It is expected that this work will lead to a written test specification that suppliers and outside laboratories can use to assemble the required apparatus and run the desired tests.

IMPACT DAMAGE TOLERANCE - Composite materials in general are significantly more "brittle" than most metals and when loaded to failure, fracture at low strains without yielding. Because of this, there is a necessity to understand the impact behavior of these materials and to develop tests to identify the ability of these materials to be designed to function after receiving impact damage. Existing tests such as the aerospace Boeing compression-after-impact procedure quantifies the residual strength of materials containing impact damage. The compression after impact procedure has been faulted as being expensive to run, subject to high variability and unable to distinguish among various mateials.

Automotive composites tend to be predominately randomly oriented glass mat reinforced materials with relative low modulus. And, automotive structures tend to be designed to stiffness requirements. These two factors, low material modulus and design to stiffness requirements, result in automotive parts which are over designed in strength in order to fulfill the stiffness requirement. This generates the requirement that a test for automotive impact damage tolerance should identify the stiffness decay of parts containing minor impact damage. And implies that during normal usage of the automobile after the impact event, damage could propagate and part mechanical properties could be further degraded. Minor damage, in the sense of this report, refers to damage caused by rocks thrown up from the road and/or damage from inadvertent impacts from automotive tools, not from car-to-car or car-to-object impacts.

To address the above factors, the ACC has experimented with several procedures as a means to develop an understanding of the impact tolerance of automotive composites. In these experiments the impact specimens were 150 x 200 x 3.2 mm plaques of SRIM materials containing 25 volume percent CSM. The impact event was applied using a Rheometrics drop weight tester with a 12.5 mm diameter impactor. The energy of the impact event was controlled by varying the drop height and was always less than the energy necessary to puncture the specimen. The plaques were ultrasonically C-scanned to determine the dimensions of the impact damage. (Figure 8). From the plaques, 25 x 200 mm specimens were machined such that one of the specimens contained the impact damaged area. These specimens were subjected to tensile-tensile fatigue at 30 % of ultimate stress and 11 Hz frequency. At intervals during the fatiguing subsequent C-scans were produced.

In addition to the tensile-tensile fatigue described above, a similar procedure was used to produce 25 x 175 mm flexure specimens. These were exposed to four-point flexural fatigue at 30 percent of ultimate stress at 5 Hz.

Samples of isocyanurate A material used in the tensile-tensile fatigue testing had recieved impacts ranging from 18.4 to 3.8 Joules. This produced C-scan damage areas ranging from 89 to 28 % of the cross-section area of the fatigue test specimens. Severe damage, as in the 89 % case, resulted in virtually instant failure, at 33 cycles. Moderate damage, 28 to 40 % of the cross section, resulted in minor initial modulus reduction, 0-10 %, and fatigue life in the 36,000 to 200,000 cycles. Similar results were observed in four-point-flexural fatigue, with the flexural modulus decay rate of a specimen containing 400 mm^2 (60 % of the cross section area) of impact damage equaling that of an undamaged control. (Figure 9) The flexural fatigue did produce an extension of the damage area but the mechanical effect was minimal.

The conclusions from this set of experiments are somewhat ambiguous. First, these SRIM materials are very easily damaged in impact. Visible damage is produced with as little as 3 J of impact energy. However, the important mechanical property, stiffness decay, is relatively insensitive to the presence of the damage. These conclusions appear to be true for both simple tensile-tensile fatigue as well as the more complex flexural loading. If these conclusions hold, after continuing experimentation, they indicate that structures which may contain barely visible damage will not grow into catastrophic failures. Further if a damaged part is destined to fail it will do so quickly with some warning.

All of the materials tested so far, including "brittle" vinyl esters to "tough" bumper beam polyurethanes, behave in a similar manner in the stiffness decay damage tolerance procedure. This raises questions as to the overall utility of a test that cannot distinguish among somewhat disparate material behaviors.

ENVIRONMENTAL - Durability evaluations of composites in the laboratory require accelerated tests which accurately describe long term behavior under real life service conditions. Test methods are needed for environmental testing to measure the effect of long term exposure of structural composites to automotive fluids such as water, windshield washer solvent, road salt, fuels, oils, battery acid, anti-freeze, brake fluid and transmission fluid.

Simple tensile and flexural test fixtures with stainless steel springs were built to maintain constant load on samples during fluid exposure. Because of design differences, the tensile fixture had a capacity of loading a 25 volume percent CSM reinforced sample to about 15 % of its ultimate breaking stress while the flexural fixture could achieve 50 % of ultimate. Both stressed and unstressed samples were soaked in fluids for 100 hrs to determine the amount of fluid absorption and its effect on properties. Stress levels used in these tests ranged from 10 to 80 MPa which represents 5 to 40 % of the ultimate failure strength of the SRIM composites undergoing test.

Isocyanurate A material was exposed to two fluids, water and windshield washer solvent. in both the stressed and unstressed states in the flexural fixture. Weight gain with water is about 0.5 %, in comparison. this same material. absorbed about 3 percent by weight windshield washer solvent. The alcohol, amine, surfactant and other chemicals in the washer solvent appear to have a drastic effect on the SRIM material tested. Fluid absorption changes resulting from stressing the samples are fairly small.

Retention of tensile properties after 100 hrs fluid soak is shown in Figure 10. Results are normalized with "as molded" composite set at 100. Water absorption reduces properties, however, the property changes are not highly significant. On the other hand, washer solvent has a drastic effect on properties with a greater than 30 % reduction in tensile modulus. The effect of stress, up to 10 % of ultimate. during the absorption period is small but measurable in both solvent cases.

Flexural deformation provides a combination of tension, compression and shear loading on the sample and may be more discriminating. Both stressed and unstressed samples were soaked in automotive fluids as shown in Figure 11. Flexural modulus and strength were measured on "as molded" samples and on samples after a 100 hr soak in both the stressed and unstressed states. As with Figure 10, the flex results are normalized with the "as molded" sample set at 100. Flexural properties are more significantly attacked than are the tensile properties. Samples stressed to 40 % of ultimate during the soak period

lost more than 40 % of their flexural modulus after 100 hrs of soaking in washer solvent.

In order to achieve the goal of providing a meaningful environmental test procedure, we are evaluating the effect of different loading modes on fluid absorption and the ultimate mechanical properties are being evaluated. Also under evaluation are additional automotive fluids in combination with other composite compositions.

SUMMARY

The second edition of the ACC manual, "Test Procedures for Automotive Structural Materials", has been issued. The latest offering includes new sections on SMC and thermoplastic based structural composites as well as a simplified and streamlined format. Substantial material testing has been completed in the development and evaluation of test procedures in the durability areas of fatigue, bolt load creep, impact damage tolerance and the effect of automotive fluids. In addition to establishing the boundaries of the ACC durability test specifications, the results of this testing has furnished much insight into the mechanical behavior of materials based on polyurethane/isocyanurate SRIM and vinyl ester RTM composites. Test procedure evaluations are near completion in the areas of fatigue, bolt load creep and environmental exposure and draft versions of test specifications in these areas are being written. Impact damage tolerance procedures evaluations are continuing.

ACKNOWLEDGEMENT

The authors acknowledge the contributions of ACC members Carl Johnson at Ford and Brian Vinson at Chrysler in the revision of the test procedures manual. Also acknowledged for their timely review of the second edition of the test procedure manual prior to its release are Don Forster (Delsen Testing Laboratories), Charles Griffen (Univirsity of Dayton Research Institute), Scott Coguill (University of Wyoming), Diane Barron (Dow Chemical), Tom Lo and Tom Grentzer (Ashland Chemical) along with Ross Lichtenstein, Chris Gill and Alan Isham (Owens-Corning).

We also express appreciation to the following individuals and companies who have supported the development of the ACC fatigue specifications: Diane Barron (Dow Chemical), Surendra Joneja (Hercules), Al Krause (Ford), Tom Lo (Ashland Chemical) and John Sutton (Owens-Corning). Greg Bretz (Ford) is also acknowledged for his assistance in the bolt load retention activity.

REFERENCES

1. Automotive Composites Consortium, Materials Work Group, Test Procedures for Automotive Structural Composite Materials, ACCM-T-001, January 1990.

2. T. J. Dearlove, E. M. Hagerman, S. N. Kakarala, D. L. Denton, D. G. Peterson, J. C. Lynn, C. F. Johnson and D. J. Melotik, "Standardization of Test Methods for an Automotive Database," *Sixth Annual ASM/ESD Advanced Composites Conference and Exibition,* Conference Proceedings, p. 101, 1990.

3. C. F. Johnson , N. G. Chavka, R. A. Jeryan, C. J. Morris and D. A. Babbington, "Design and Fabrication of a HSRTM Crossmember Module,"*Third Annual ASM/ESD Advanced Composites Conference and Exposition,* Conference Proceedings p. 97, 1987

4. N. G. Chavka and C. F. Johnson, " The Taming of Liquid Composite Molding for Automotive Applications," *Seventh Annual ASM/ESD Conference on Advanced Composites Conference and Exibition,* Conference Proceedings, p. 115, 1991.

5. S. M. Abouzahr, N. R. Rao, D. E. Jay, C. H. Mao, L. M. Lalik and D. G. Peterson, "Composite Pickup Box: Design, Fabrication and Analysis," *ibid*, p.415.

6. R. E. Bonnett, R. A. Carpenter and S. W. Gallagher, "Two Piece Composite Truck Cab," *Truck Engineering* p.11, April, 1990.

7. R. L. Frutiger and S. Baskar, "Structural Composite Floorpans: Design Synthsis, Prototype, Build and Test", SAE Paper 921096, *8th International Conference on Vehicle Structural Mechanics and CAE,* Traverse City, Michigan, June, 1992.

8. Automotive Composites Consortium, Materials Work Group, Test Procedures for Automotive Structural Composite Materials: Second Edition, ACCM-T-02, July 1994.

9. G. D. Sims and D. G. Gladman, "Effect of Test Conditions on the Fatigue Strength of a Glass-Fabric Laminate: Part A-Frequency", *Plastics and Rubber: Materials and Applications,* p. 41, May 1978.

10. J. F. Mandell, D. D. Huang and F. J. McGarry, "Tensile Fatigue Performance of Glass Fiber Dominated Composites," *36th Ann. Conf. Reinf. Plastics/Composite Inst.,* Paper 10-A, 1981.

11. D. L. Denton, "Tensile Fatigue of Continuous and Random Fiber SMC Composites," *Composites: Materials and Engineering Symposium,* Center for Composite Materials, University of Delaware, 1984.

12. D. L. Denton, "Mechanical Properties Characterization of an SMC-R50 Composite," *SAE Passenger Car Meeting,* Dearborn, MI, Paper No. 790671, June, 1979.

13. G. L. Sullivan, G. T. Bretz, C. F. Johnson, E. J. Blais, and D. Verma, "Load Retention Predictions for Bolts Fastened to Composites", *SixthAnnual ASM/ESD Conference on Advanced Composites Conference and Exibition,* Conference Proceedings p.31, 1990

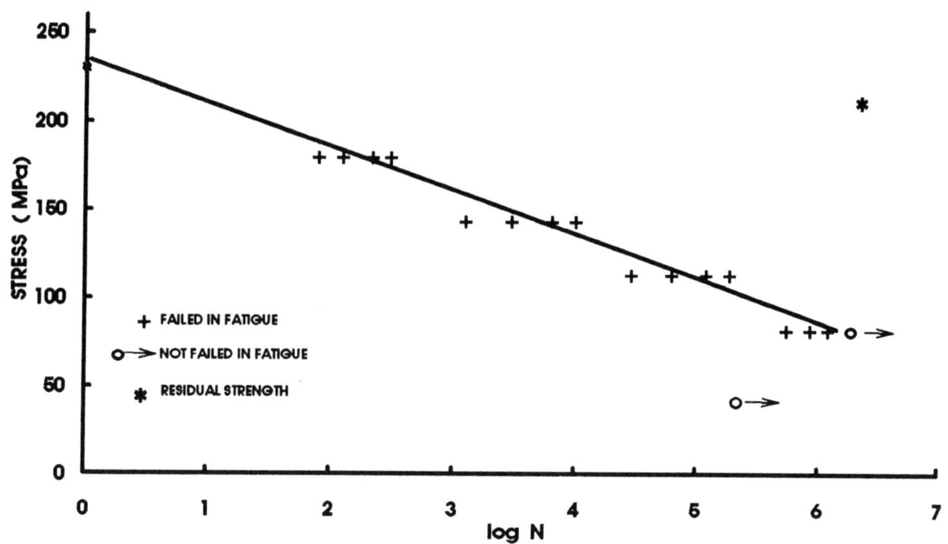

Figure 1. Typical S-N Plot Resulting From Fatigue Test Specification

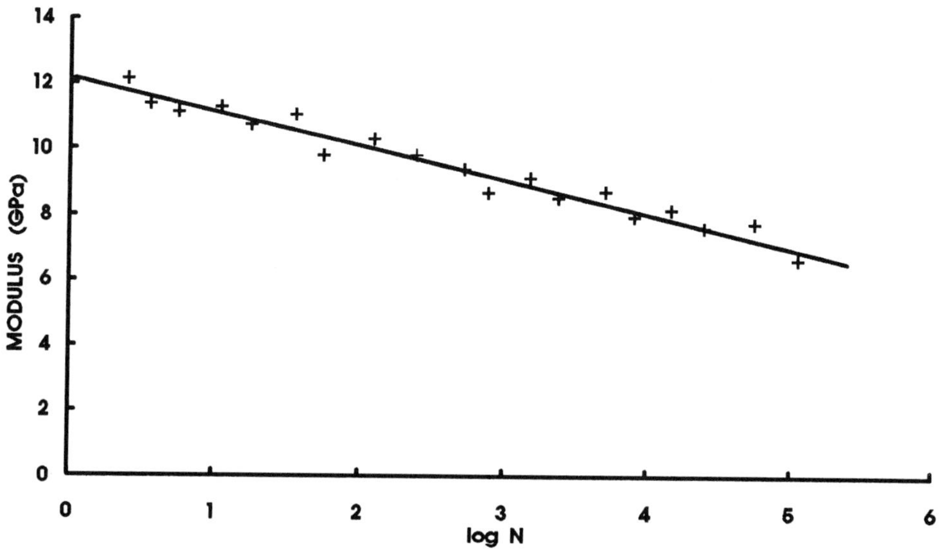

Figure 2. Modulus Retention as Function of Number of Fatigue Cycles

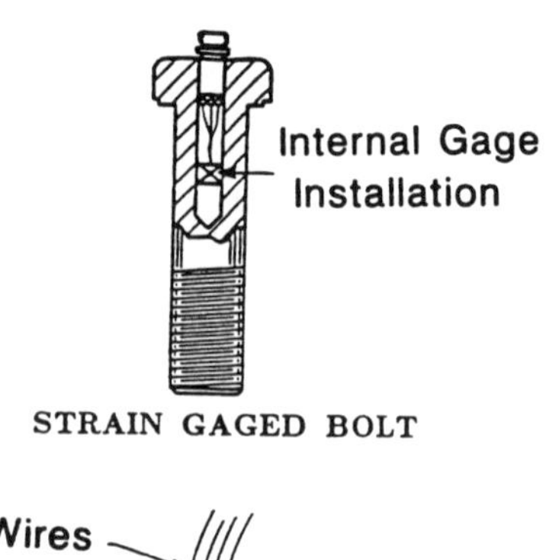

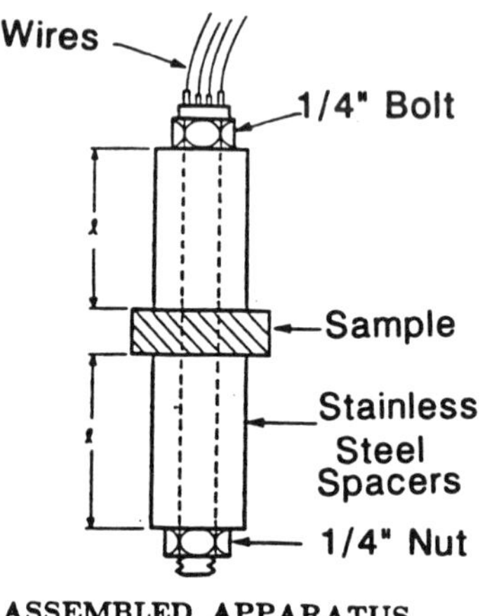

Figure 3. Bolt Load Retention Apparatus

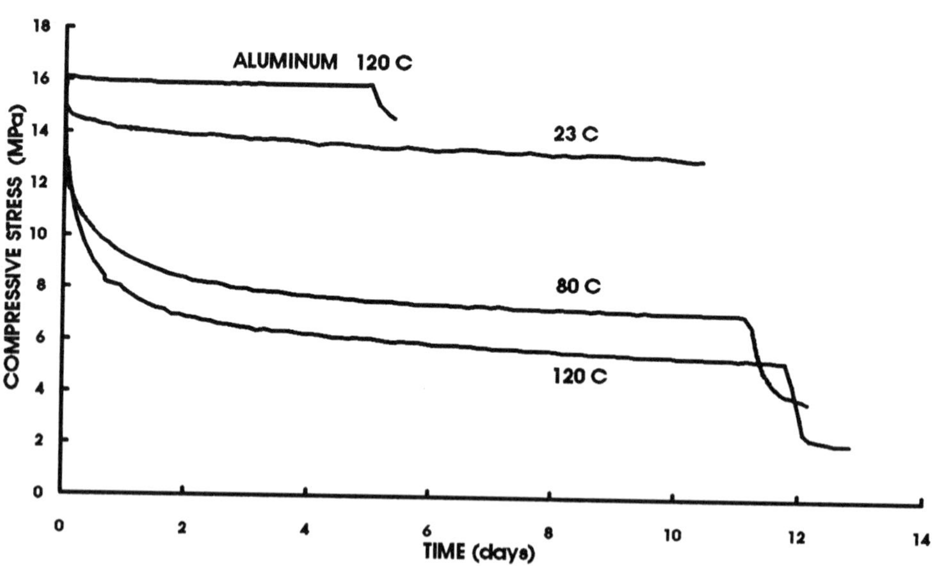

Figure 4. Compressive Stress Relaxation as Function of Temperature for Isocyanurate A

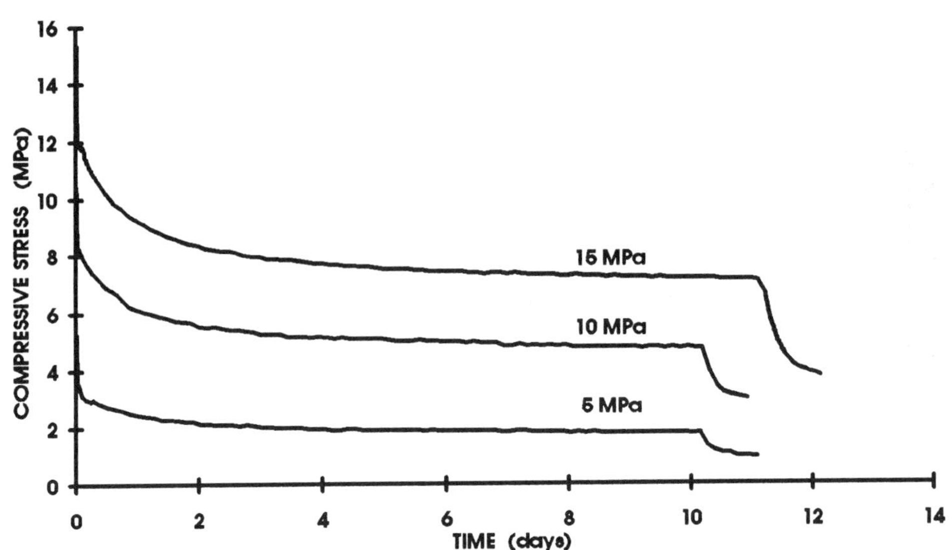

Figure 5. Compressive Stress Relaxation for Isocyanurate A as Function of Initial Load

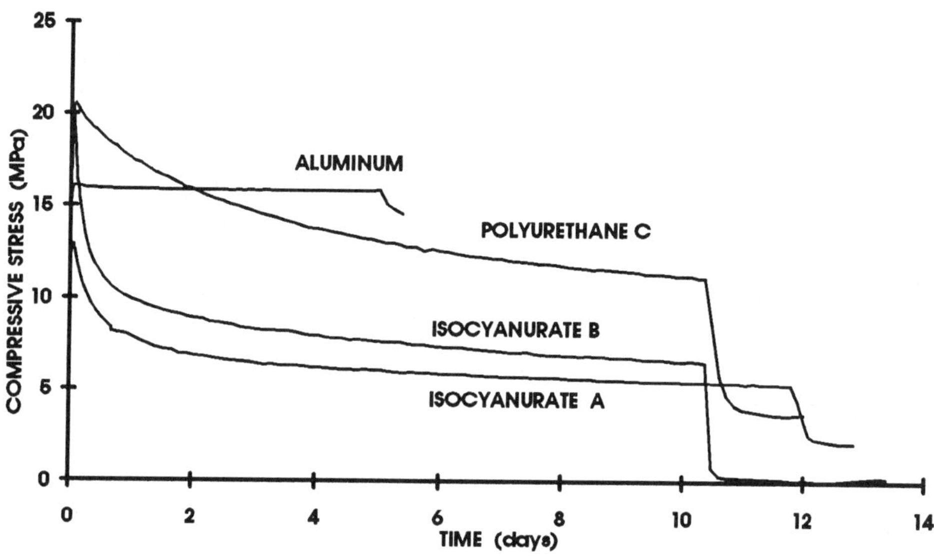

Figure 6. Comparison of Stress Relaxation Behavior of Various SRIM Materials

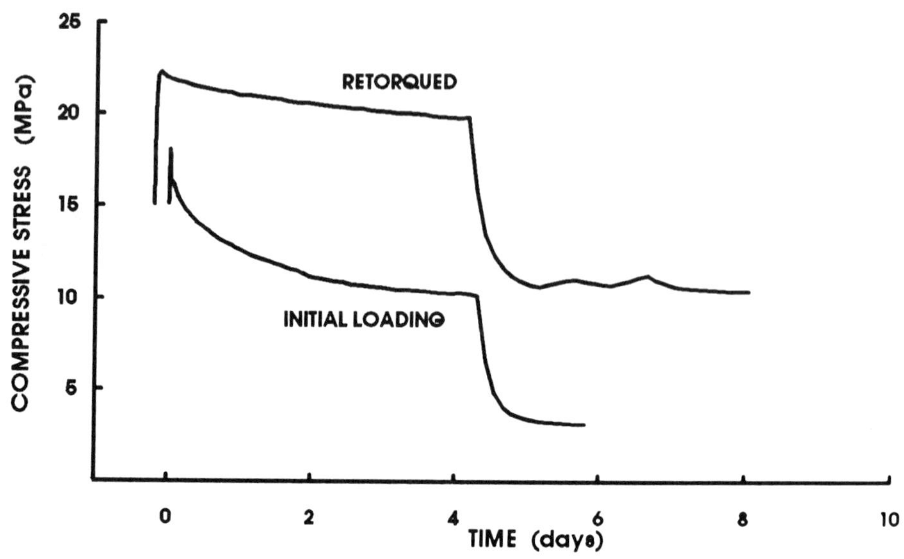

Figure 7. Effect of Retorquing Isocyanurate B

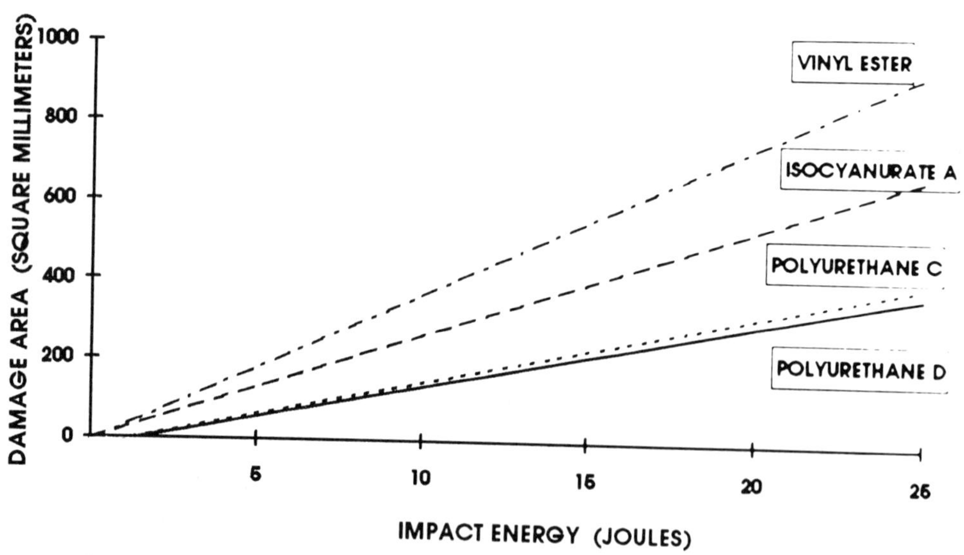

Figure 8. Damage Area as Function of Impact Energy for Various Composites

564

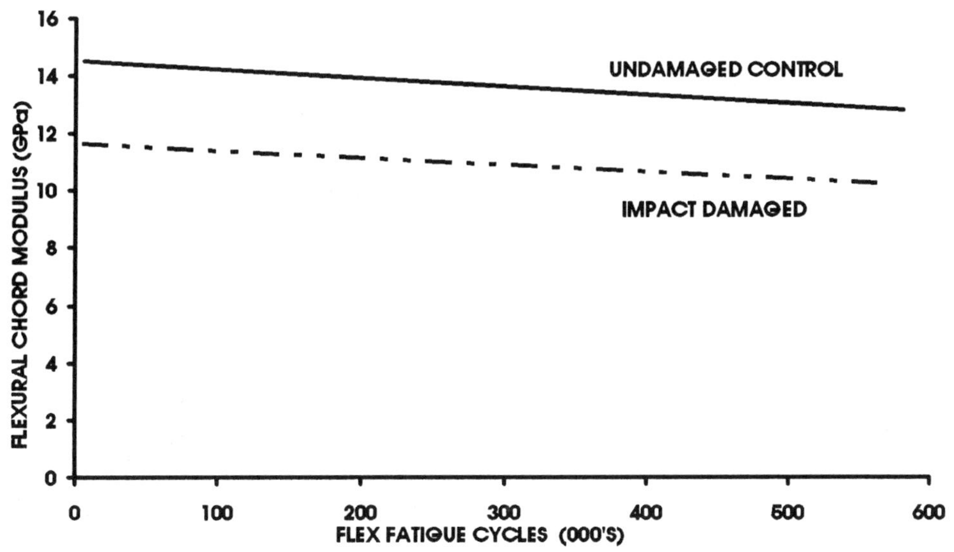

Figure 9. Flexural Modulus Decay - Undamaged vs Damaged Isocyanurate A

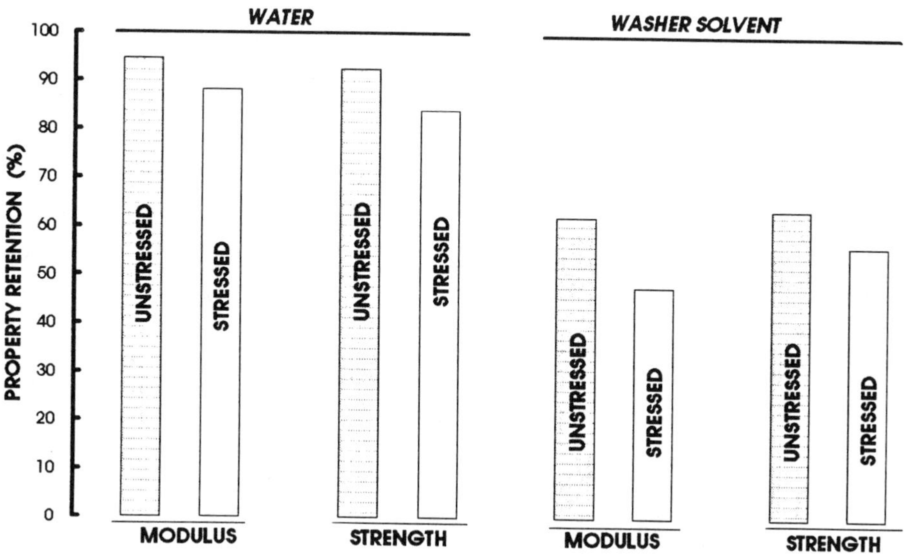

Figure 10. Retention of Tensile Properties After Environmental Exposure

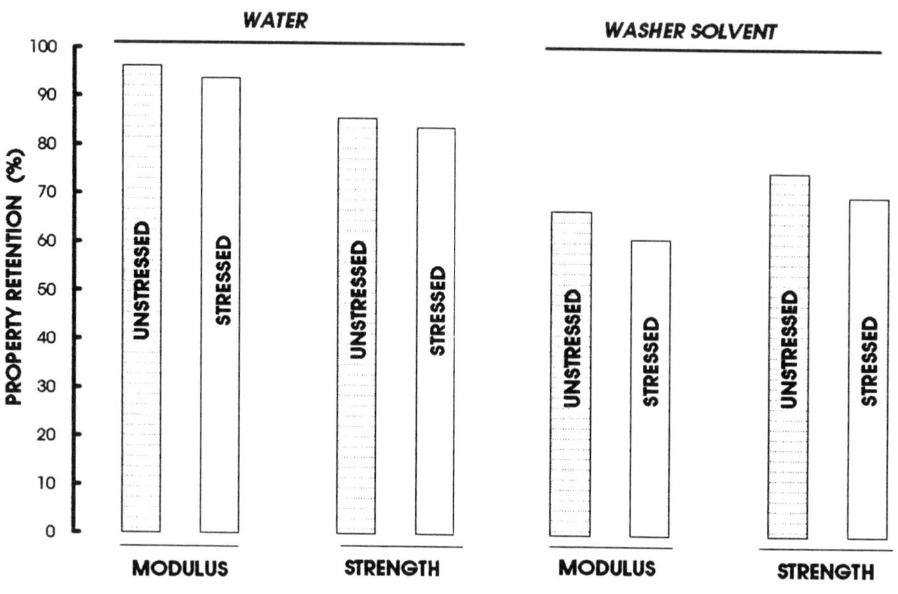

Figure 11. Retention of Flexural Properties After Environmental Exposure

Advanced Composite Design and Repair Data Models

E.L. Stanton, T.E. Mack
PDA Engineering
Costa Mesa, California

Abstract

Performance gains from advanced composites in aircraft structures have come at the expense of high initial costs and higher repair costs. The work described in this paper supports an effort to transfer composite data and models to design-analysis software tools for simulating product producibility and repairability. A key element of the program is the capture of aircraft CAD and CAE information in design and repair databases to reduce the time and effort now required to locate geometry, material, and loads data. Emerging STEP/Express information models for composite data are used to model a laminated composite C141b aircraft door for an interactive design-analysis solution to process induced spring-back deformations. This application helps to make the abstract concepts in the Express schemas concrete and identifies strengths and weaknesses in the STEP standards as they exist today for materials information.

RECENT PROGRESS IN STEP/EXPRESS tools helped in developing CAD and CAE product data exchange models of composite structures for the applications reported here. These are advanced composite design and repair models for laminated aircraft structures. The work is based in part on a STEP Application Protocol, AP 209, and experience gained from composite projects for the Air Force Logistics Command. Our goal is to make reliable CAE simulation tools available to the CAD designer in an interactive environment. The benefit is real time simulation for producibility, repairability as well as composite performance.

An aircraft composite life raft door developed for the C141b, Figure 1, is used to illustrate the composite data exchange requirements between CAD and CAE systems. This project used CAE simulations to solve process induced spring-back problems during the design phase. The association of CAD composite design data with CAE composite material models follows the schema now being standardized in STEP AP 209, [1]. We use Part_Laminate_Table Entities from the current AP to desribe

the C141b door design-analysis data exchange although the actual project work preceded the current draft of AP 209. On the project, CAE simulations of cure cycle deformations allowed the designer to see the spring-in/out associated with a particular laminate design and make laminate changes to reduce the deformation to within dimensional tolerances. We were also able to bias the residual deformations in a direction that ensured snug door closure.

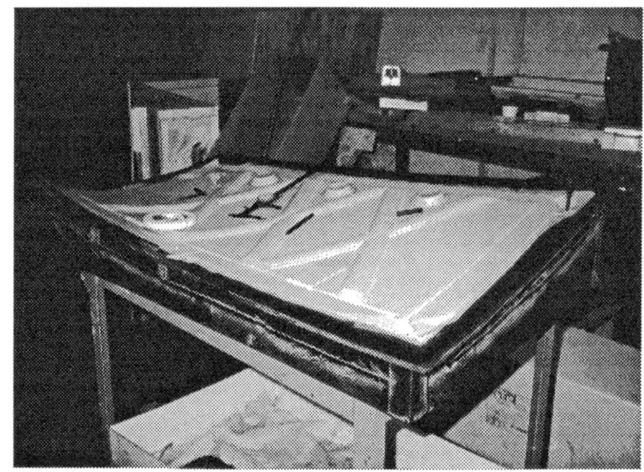

Fig. 1 - C141b Prototype Composite Door

A second Air Force project to develop an advanced composite repair analytical tool (ACRAT) illustrates composite repair data models and repair methods for a composite wing structure. This is work in progress that could provide requirements useful to AP 209's AP3 development phase. In order to repair or extend the service life of a structural design the maintenance engineer needs to recapture the original data and models used to produce the aircraft. Today this is largely a paper expedition to find drawings and reports that may no longer exist. Not surprisingly a major task on the ACRAT project is developing engineering databases to capture original equipment design data and repair technical orders.

Composite Design-Analysis Data Model – AP1

Design requirements associated with managing composite preform geometry and selecting constituent materials that will process well together requires attention to a lot of detail. Individual organizations have developed a variety of Product Data Management tools for meeting this requirement. However, exchanging this data with suppliers and customers remains difficult [2] and CAD systems offer few tools today for material data exchange. The International Standards Organization (ISO) and industry have been working for several years on a standard to exchange composite product data between designers and analysts who use CAD and CAE software, Figure 2. What we describe here is an instantiation of the C141b composite door design and FE analysis data using a modified form of the Express schema for AP 209. A few design details are missing in the current standard and there are instances where the standard is not easy to implement but all things considered the standard is a powerful tool for exchanging composite product data.

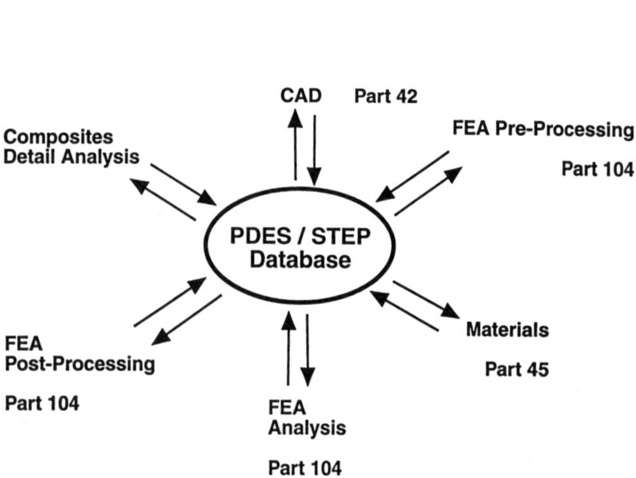

Fig. 2 - AP 209 Composite and Metallic Structural Analysis and Related Design

AP 209 supports bi-directional CAD-CAE product data exchange which is formally indicated in the standard using Assertions, cf Figure 3, that associate the design representation with the analysis representation for the same composite part. The standard does not provide procedures for making transformations between these two representations. The standard defines ENTITY's (objects) in the Express modeling language [3] that capture the attributes of both the design model and the analysis model. We developed the following Express material model from AP 209 and STEP Part 104.

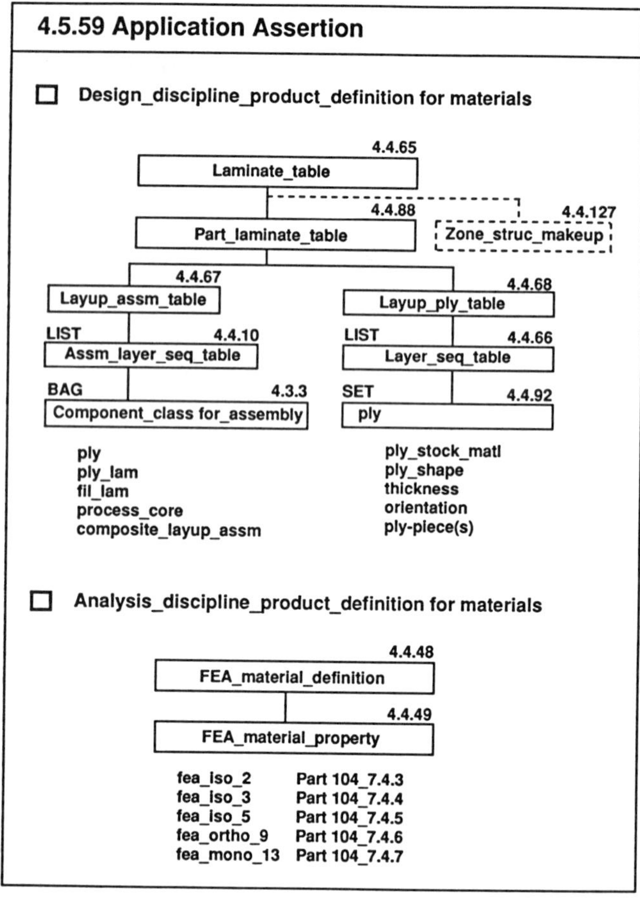

Fig. 3 - AP 209 CAD-CAE Material Application Assertion

```
(*
* AP 209 CAD-CAE Material Model.
*
* Purpose and Description:
*
*   This model is an Express Schema derived from AP 209
*   Assertion 4.5.59 and reconciled with STEP Part 104
*   FEA material models.  Modifications to the AP 209
*   FEA Entities were required.
*
*
* Revision Log:
*   Initial Version 94_06_26 ELS
*          Revised 94_07_05 ELS
*)

SCHEMA AP209_composite_material;

REFERENCE FROM measure_schema
    (length_measure,
    measure_value,
    ratio_value,
    angle_measure
    );
```

REFERENCE FROM Part104_schema
    (iso_sym_2t_3d,
        ortho_sym_2t_3d,
        aniso_sym_2t_3d,
        aniso_sym_4t_3d,
        iso_sym_4t_3d,
        iso_o_sym_4t_3d,
        iso_t_sym_4t_3d,
        ortho_sym_4t_3d
    );

REFERENCE FROM AP209_schema
    (constituent_parts,
    environment,
    definition_element,
    laminate_table,
    ply_stock_material,
    ply_shape,
    ply_piece,
    layup_assembly_table
    );

TYPE math_form = SELECT (
(*
* Coefficients of Thermal Expansion
* Part 104 Sections 7.3.8-10
*)
        iso_sym_2t_3d,
    ortho_sym_2t_3d,
    aniso_sym_2t_3d,
(*
* Elastic Constants for Stress Strain Equations
* Part 104 Sections 7.4.2-7
*)
    aniso_sym_4t_3d,
        iso_sym__3d,
    iso_o_sym_4t_3d,
    iso_t_sym_4t_3d,
     ortho_sym_4t_3d
    );
END_TYPE;

(*
* The CAD design data for a composite laminate_table
*)

ENTITY part_laminate_table
    SUBTYPE OF (laminate_table)
    SUPERTYPE OF (layup_ply_table);
END_ENTITY;

ENTITY layup_ply_table
    SUBTYPE OF (part_laminate_table);
    sequence : LIST [2:?] OF layer_sequence_table;
END_ENTITY;

ENTITY layer_sequence_table;
    plies_in_sequence : SET [1:?] OF ply;
END_ENTITY;

ENTITY ply
    SUBTYPE OF (constituent_parts);

material_type   : ply_stock_material;
shape_definition : ply_shape;
ply_thickness   : OPTIONAL length_measure;
material_orientation : OPTIONAL angle_measure;
constituents   : OPTIONAL LIST [1:?] OF
                        ply_piece;
END_ENTITY;

(* The FEA analysis property data associated with a
* laminate_table through the AP 209 Assertion 4.5.59
*)

ENTITY FEA_material_definition;
    material_identification : part_laminate_table;
(*
* The through thickness 3d analysis material model
* contains one FEA material property set for each
* part_laminate defined through the thickness of the
* composite_layup_assembly.
*)
    description : STRING;
    element   : SET [1:?] OF definition_element;
    ambient   : environment;
END_ENTITY;

ENTITY FEA_material_property;
    property_use   : FEA_material_definition;
    property_data : math_form;
END_ENTITY;

END_SCHEMA;

Composite Aircraft Door Application

The C141b composite door application is one in which PDA participated in all aspects of the design-to-manufacturing process including the production plan and delivery of the first ship set to the Air Force. We revisit that work to describe the simulation based design changes using the STEP Entities where possible. A parametrized CAD model of the composite structure was constructed from the outer skin mold surface and the trim lines for the stiffener ply patterns, Figure 4. Here we focused on the through thickness 3D models needed to simulate cure cycle spring-in/out behavior [4]. A simpler 2D model can be used to simulate structural behavior once the part is successfully manufactured.

The initial laminate design was divided into four (4) ply_laminates through the thickness for which four 3D fea_orthotropic material models were developed from the part_laminate tables. The FE model was created in PATRAN and the analysis of cure cycle deformations performed using P3/COMPOSITE. The baseline design warped 180 mils and deformed away from the aircraft at the door seals. Using analysis simulations to guide the redesign only laminate changes were required to reduce the warping to 30 mils and bias the deformations into the aircraft at the door seals. The data flow in this case was not automated and the computational experiments necessary to find a solution to the spring-in/out problem required a senior engineer but even with these schedule costs the entire laminate redesign was done in a few days with no tooling changes.

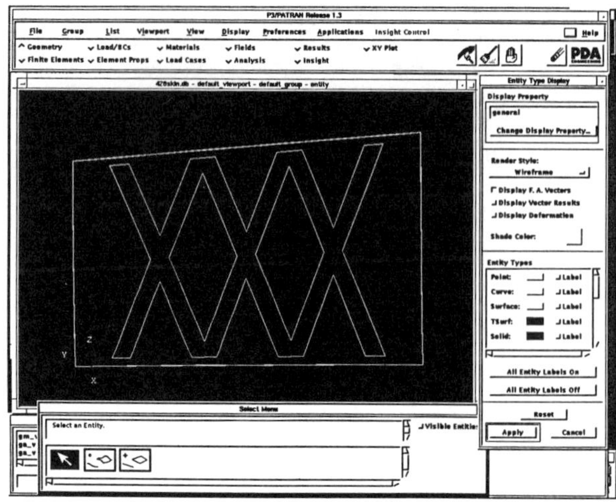

Fig. 4 - C141b CAD Geometry Model Import

The only product data from this application that have been placed in AP 209 data structures today are the material property data, cf. Appendix. In AP 209 material orientation data are mapped using control points, lines and shared coordinate frames between the design and analysis models for each ply; this is very similar to the way some CAD systems define laminate design data for a tape laying machine. This is more comprehensive than the way most FEA programs model material orientation using a collection of point laminate tables. A few analysis modeling tools are beginning to appear, Figure 5, that drape ply patterns onto part surfaces and so define material orientation for the group of finite elements that pave the ply surface. In the present door analysis model we need a 3D material geometry model which was derived from the part_laminate_table and oriented on the layup_surface by the analyst, Figure 6. Material properties were derived for each part_laminate and placed in the project database, see Appendix, using the AP 209 Entities described in this paper.

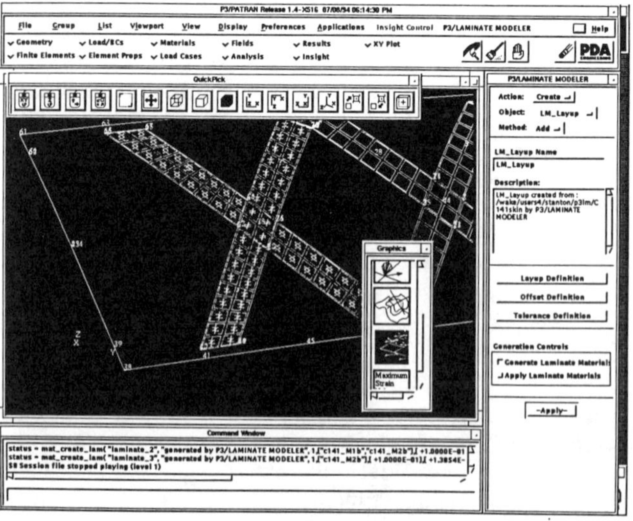

Fig. 5 - Ply Pattern Mapping Defines Finite Element
Material Orientation

Composite_door schematic

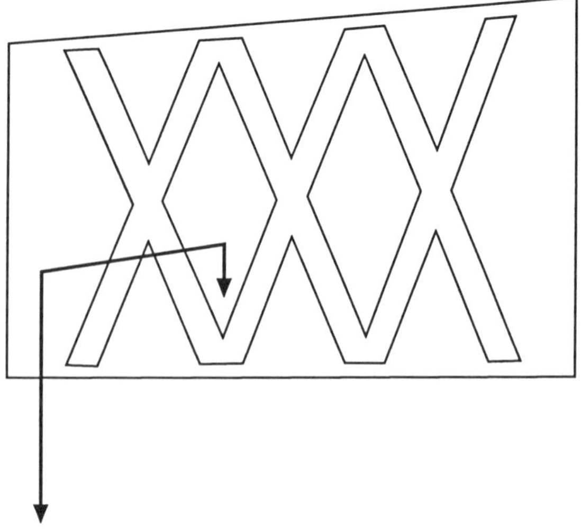

Laminate_table schematic

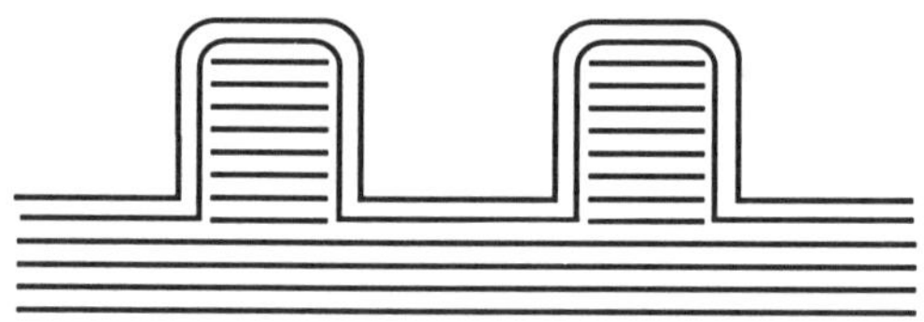

FEA_material_property schematic

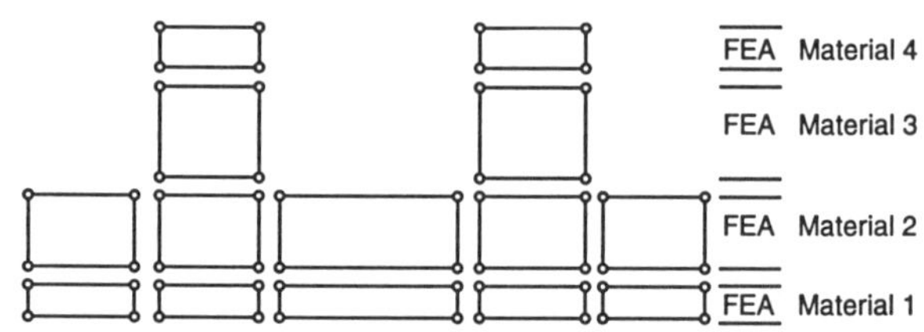

Fig. 6 - C141b 3D CAE Material Model

Composite Repair Design Data Model

Composite data models for repair can require all of the design-analysis information used in the C141b door application plus information about damage and repair details. It is a very complex design problem that is subdivided into repairs that can be made in the field and repairs that must be made at a depot or in the factory, [5]. The analytical repair tool project at PDA is scoped to include depot level repairs and includes twelve levels of information, Figure 7, in the repair database schema. Composite damage must be described in enough detail to support damage assessment and to recommend a repair type, bonded or bolted, and to supply repair details.

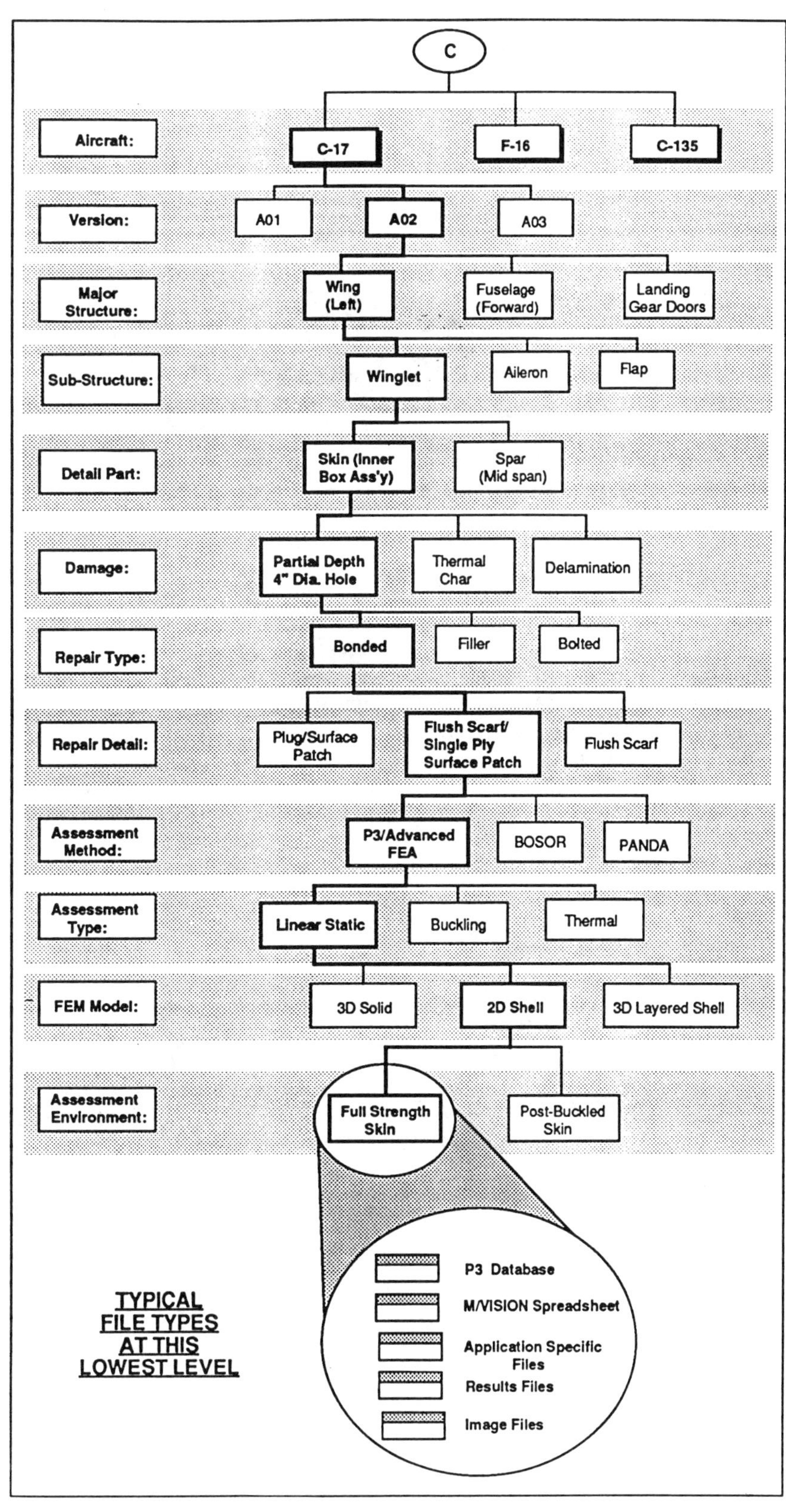

Fig. 7 - Aircraft Composite Repair Data Structure

The data levels associated with the structure, sub-structure and detail part in the database contain geometry data imported from CAD part files and laminate_tables that define part material geometry data. Unfortunately no comprehensive standard for part_laminate_tables was widely available before AP 209. Today no commercial product contains all the part material geometry information available in this emerging STEP standard. It requires a very substantial effort to transfer paper data to composite databases with each ply located in the part by coordinate frame and ply_shape representation keyed to a design surface, Figure 8. We are constructing model databases for the aircraft component selected by the Air Force for ACRAT acceptance testing but the scope of the present effort is limited to just a few composite components. The system is being designed to capture images of paper documents in files that are keyed to the database in a way similar to on-line documentation. The construction of analysis models for A4EI, PGLUE, CREPAIR and other repair programs in ACRAT can proceed without leaving the system. More comprehensive 3D finite element analysis programs are also accessible from ACRAT.

AP 209 Entities for material geometry such as ply_shape do not have local surface attributes to account for fabric distortion during layup or changing preform geometry in braided and other textile preforms. The ability to represent spatially varying material attributes in ply_stock_material also can be important for modeling damage. In future versions of AP 209 attributes for this class of composite applications are recommended.

data representations that determine material orientation from ply_shape and layup_ply_table data to concentrate on the laminate material design and analysis model data. Bi-directional material design-analysis data exchange was sufficient for the C141b producibility application we described and AP 209 data structures worked well with minor adjustments. The use of multiple 3D material models through the thickness was not anticipated by Assertion 4.5.59 and we recommend a simple change in the text to allow "one or many" FEA_materials per laminate_table. This is a minor point; overall we were impressed with the level of detail in AP 209 and its bi-directional design-analysis capability.

We did have difficulty translating our organizations way of doing things to the way AP 209 associates laminate design data with FEA material property data. The database we describe in the Appendix follows the Part 104 schema for FEA_material_property math forms and we believe that it is consistent with AP 209. The way we associate the FEA properties to the part_laminate_table data and FEA volume_element groups seems consistent but we had to make several assumptions that need review. The omission of data source and data quality indicators was easy to correct and we understand that the long form AP 209 schema will include these entities. The ply entity and other constituent material entities are assumed to have material properties that are constant over the layup surface. This was true in our application but many textile preforms, such as braids, will vary continuously over the composite part surface. Textile preforms are likely to require significant enhancements that can be deferred to future versions of AP 209. There is much work to be done just making the current AP 209 available to CAD and CAE users.

PLY_SHAPE

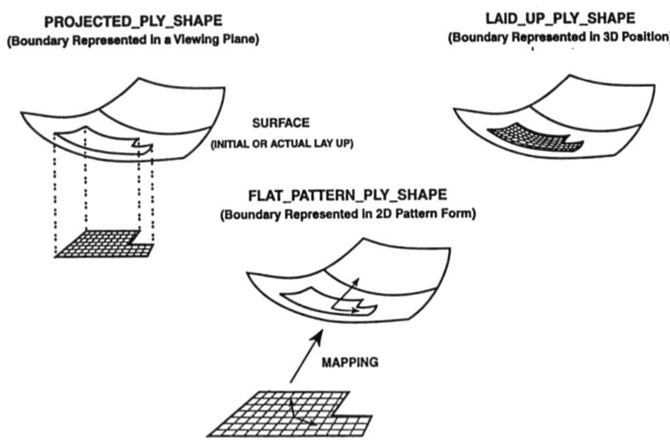

Fig. 8 - AP 209 Ply_shape Subtype Representations

Conclusions

The present effort to map composite aircraft design and analysis materials information to AP 209 data exchange structures must be considered heuristic and only a first attempt to instantiate a small part of a very comprehensive Express schema. We omitted instantiating the design shape

Appendix

```
STEP;
HEADER;
FILE_NAME( 'AP209.data', '1994-07-05T17:46:00', $, $,
      $, 'mv_to_exp 1.2', $, $ );
FILE_DESCRIPTION( ('This file was produced by
      mv_to_exp 1.2 from M/VISION databank AP209 )
      ( 'Exp_data_to_mv 1.2 is required to translate this file
      into a M/VISION database' ) );
FILE_SCHEMA( ( 'AP209' ) );
ENDSEC;
DATA;
/* DATA */

#1  = !PART_ID( 'C141b SK10 layup_ply_table', $ );
#2  = !MATL_CLASS( 'PMC composite', $ );
#3  = !CNAME( 'Fiberglass_Epoxy', $ );
#4  = !STAGE( 'Cured', $ );
#5  = !FEA_ID( 'MAT_1B', $ );
#6  = !MODEL_TYPE( 'ortho_sym_4t_3d', $ );
#7  = !FEA_MODEL( 'C141 MLG Aft Door', $ );
#8  = !ELEM_LIST( 'vol_3D_elem 1:1161', $ );
#9  = !MODEL_TEMP( 7.000000e+01, $, $ )'
#10 = !FEA_MATERIAL_DEF_RECORD( $, #1, #2, $,
      #3, $, #4, #5, #6, #7, #8, $, #9);
```

#11 = !REPORT('PDA TR 92-1641-0000', $);
#12 = !AUTHOR('T.E. Mack' $);
#13 = !TABLE('One (Plies 1:2)', $);
#14 = !QAL_IND('Nominal (Computed)', $);
#15 = !USE_IND('CET Design Tasks', $);
#16 = !DATA_TYPE('Product Data Sheets', $);
#17 = !MV_REFERENCE('UCRL 15517', $);
#18 = !FEA_DATA_SOURCE_RECORD($, #11, #12,
        #13, $, #14, #15, #16, #17);

#19 = !MVISION_METADATA('DERIVED');
#20 = !E11(3.340000e+00, #19, $);
#21 = !E22(3.340000e+00, #19, $);
#22 = !E33(7.000000e-01, #19, $);
#23 = !NU12(5.400000e-02, #19, $);
#24 = !NU31(6.300000e-02, #19, $);
#25 = !NU13(3.000000e-01, #19, $);
#26 = !NU23(3.000000e-01, #19, $;
#27 = !G12(2.800000e-01, #19, $);
#28 = !G31(2.700000e-01, #19, $);
#29 = !G23(2.700000e-01, #19, $);
#30 = !CTE11(7.300000e+00, #19, $);
#31 = !CTE22(7.300000e+00, #19, $);
#32 = !CTE33(4.140000e+01, #19, $);
#33 = !FEA_MATL_PROP_RECORD($, $, $, $, $, $,
        $, $, $, $, $, $, $, $, #20, #21, #22, #23, #24, #25,
        #26, #27, #28, #29, #30, #31, #32);
#34 = !MVISION_record(#10, #33, #18) ;

/* The database continues for all the laminate and stock
   material property data records used on the C141b project.
   The FEA material properties were computed from the
   Part laminate tables specified in the PART_ID
   specifications. This Express physical file contains AP
   209 data BUT not in a form ready for conformance
   testing. That requires a translator to filter out the data
   source records and change the title of data like the
   ELEM_LIST to definition_element. Also, the Version
   information needs to be added to the database.   */
ENDSEC;
ENDSTEP;

Acknowledgements

The authors wish to thank the organizations and staff at the Battelle Memorial Labs and the Warner Robins ALC Center for permission to use the C141b data in this STEP AP 209 test application. Additionally, we want to thank the Advanced Composites Program Office at the Sacramento ALC Center for their support in designing databases for use in the analysis of composite aircraft repairs.

References

1   ISO TC184/SC4/WG3/T9 N233A Committee Draft for Comment for AP 209: Design through Analysis of Metallic and Composite Structures, K. A. Hunten, Editor.

2   G. E. Hansen and J. J. Banisaukas, "SACMA Recommended Carbon Fiber Test methods and Lot Acceptance Procedures", Proceedings of the Sixth ASM/ESD Advanced Composites Conference, p. 113-124, October 1990.

3   D. A. Schenck and P. R. Wilson, "Information Modeling: The EXPRESS Way", Oxford University Press, 1994.

4   S. C. Nolet and P. M. Sandusky, Impact Resistant Hybrid Composite for Aircraft Leading Edges, 31st SAMPE Symposium, p. 832-843, April 1986.

5   I. H. Marshall and E. Demuts, Eds., "Supportability of Composite Aircraft", Elsevier Applied Science, 1988.

Proceedings of the 10th Annual ASM/ESD Advanced Composites Conference, Dearborn, Michigan, USA, 7-10 November 1994

A Model for the Transmission of Ultrasound During the Fusion Bonding of Composites

K.D. Tackitt, R.C. Don, J.W. Gillespie, Jr.
University of Delaware
Newark, Delaware

Abstract

This paper presents a basic model for predicting the through transmission ultrasonic (TTU) amplitude response of multilayered composite media undergoing a fusion bonding process as a function of temperature and degree of intimate contact at the interface of interest. The model response to frequency, layer thickness, and degree of intimate contact was investigated. The model predications are compared to the on-line TTU results obtained from a resistance welded PEEK / APC 2 weld.

FUSION BONDING IS an attractive technology for joining thermoplastics and thermoplastic matrix composites. There are many advantages in using this technique such as good environmental resistance, high strength (ideally that of the parent materials), and the lack of any mechanical fasteners.

Resistance welding is a fusion bonding technique that employs an embedded heating element to heat, via Joule heating, the interface between adherends. A typical weld stack is shown in Figure 1. Prior work has shown that the processing history has a significant effect on the performance of the joint produced [1]. In order to achieve optimum bond quality and performance, it is necessary to precisely control the process parameters. Currently, thermocouples implanted at the interface are used to monitor and control the interfacial temperature. This is undesirable due to the defect that this introduces.

Nondestructive sensing techniques are being investigated for their suitability to be used on-line as a replacement for implanted sensors. The possibility of using these sensing techniques for on-line inspection and control would improve the quality control of manufactured parts and joints. Developing on-line control and inspection techniques are necessary for full automation of the process.

Previous studies have also shown that through transmission ultrasonics (TTU) is sensitive to the process history of a resistance weld [2]. A typical TTU trace is shown in Figure 2. Recent work has been focused on relating the TTU signals to the various process models that have been developed to describe the heat transfer and evolution of intimate contact at the weld interface. To gain a basic understanding of the sensor-process interaction, current efforts include modeling how the signal changes with the material properties as function of temperature and the way in which complete intimate contact is achieved.

Review of Theory

To understand how the ultrasonic waves are affected by the welding process, it is necessary to look at how sound waves propagate in solids. Since the mathematics describing wave propagation in solids can be quite complex, any simplifications that can be made will be beneficial. The first simplification comes from the geometry of the problem. Figure 3, which illustrates the experimental setup used for resistance welding, shows the transducer configuration for through transmission. The transducers used in our studies are longitudinal wave transducers, so the only waves of interest are longitudinal waves at normal incidence to the series of interfaces in the weld stack. This will reduce the number of equations needed, and mode conversion can be ignored.

The second simplification that will be used is to treat the composite adherends as transversely isotropic and to treat the weld stack as semi infinite in the xy plane. This will remove complexity by providing isotropy through the thickness and allowing us to treat the wave as a plane longitudinal wave. This will mathematically uncouple the wave equation for longitudinal waves from the general three dimensional wave equation where there is coupling between the longitudinal wave and the vertically and horizontally polarized transverse waves.

Transmission Across Interfaces. The theory describing the transmission of longitudinal waves across an interface has been well developed [3-7]. Eqs. 1, 2, and 3 are harmonic solutions to the one dimensional wave equation describing the incident, reflected, and transmitted longitudinal waves at normal incidence to an interface between continuous media.

The one dimensional plane wave equation is:

$$c^2 \frac{\partial^2 u}{\partial z^2} = \frac{\partial^2 u}{\partial t^2} \qquad (1)$$

where: c is the velocity of sound, u is displacement, and t is time

Incident wave:

$$p_{1i}(z,t) = A_1 \, exp[i(\omega t - k_1 z)] \qquad (2)$$

Reflected wave:

$$p_{1r}(z,t) = B_1 \, exp[i(\omega t + k_1 z)] \qquad (3)$$

Transmitted wave:

$$p_{2t}(z,t) = A_2 \, exp[i(\omega t - k_2 z)] \qquad (4)$$

where: p = acoustic pressure, k_n = the wave numbers, given by :

$$k_n = 2\pi f / c_n \qquad (5)$$

where c_n is the velocity of sound in layer n, f is frequency in Hertz, and A_1, B_1, A_2 are the real pressure amplitudes.

The boundary conditions that must be satisfied for all times and at all points along the boundary are [3]:
1. Acoustic pressure or normal stress must be continuous at the boundary
2. Particle velocity or displacements must be continuous at the boundary

For the case of welded contact, there is continuity of shear and normal stresses, but for the case of intimate contact without bonding, only normal stresses can be transmitted. Longitudinal waves at normal incidence can only supply normal stresses so no shear stresses need to be considered here.

The most significant parameter that affects the transmission of sound across a boundary is the specific acoustic impedance, z, of the material on either side of the interface. Acoustic impedance is the acoustic analog of the optical index of refraction. The higher the impedance mismatch between two media, the stronger the reflection from a boundary separating the two will be, and the transmitted wave will be correspondingly weaker. Given the velocity of sound in the material and its density, the acoustic impedance can be found in the following way [5]:

$$z = \rho c \qquad (6)$$

where: z is acoustic impedance, ρ is density, and c is the velocity of sound.

The specific acoustic impedance is related to the particle velocity by the following relation [8]:

$$z = \frac{p}{u} = \rho c \qquad (7)$$

where: z is the acoustic impedance, ρ is the density, c is the velocity of sound, and u is the particle velocity.

Maximum transmission is approached when the media are acoustically matched, i.e. their impedance difference becomes small.

Modeling Transmission Through a Weld Stack

The approach given by Kinsler was used to model the transmission of longitudinal waves through multiple, finite layers bounded by semi infinite layers [8]. Figure 4 shows the seven layers of material that a sound wave must pass through for a typical resistance weld. There are the composite panels, the two neat interlayers, and the heating element. The final two layers represent the two 'semi infinite' delay lines used to protect the transducer from high process temperatures. Starting with the wave in the delay line, whose pressure amplitude will be called A_1, the first interface encountered is the delay line to APC 2 boundary (the couplant layer is very thin and can be neglected). The depth coordinate of this interface will be defined as z = 0. At the boundary, the wave will be partially transmitted and partially reflected. The pressure amplitude of the wave transmitted into layer 2 will be called A_2, and the amplitude of the wave reflected back into layer 1 will be called B_1. Reflected and transmitted waves at subsequent interfaces will be labeled in a similar manner.

Due to the reflection of B_1 type waves from the first interface, additional incident A_1 type waves will suffer interference. This will be encountered in each layer up to the final interface. To illustrate this, the wave interaction at the first two interfaces will be discussed below.

Wave A_1 travels through the delay until it reaches the composite layer. Here it is reflected and transmitted. The transmitted wave travels in the composite layer until it reaches the interface between the panel and the neat layer, located at z = M. Here again the wave will split into transmitted and reflected parts, A_3 and B_2. The reflected part, B_2, will travel back to the first interface at z = 0 where it will in turn be split into transmitted and reflected waves. These waves will continue this process until damped out of existence by the attenuating characteristics of the layer.

The B_2 waves reflected from point M will interfere with incoming A_2 waves transmitted from the interface at z = 0. Eventually, a steady state of acoustic energy transfer will be achieved after enough reflections in the layer. This steady state wave thus produced will be phase shifted with respect to the initially reflected and transmitted waves.

This process is repeated in each succeeding layer until the wave travels through all interfaces. The interference will be either constructive or destructive depending on the layer thickness and the frequency of the sound.

To determine the amplitude response in the seventh layer (the receiving delay line) due to the intermediate interfaces, the boundary conditions must be satisfied at each interface in succession .

Following Kinslers approach, we start with a incident wave of the form of Eq. 2. The reflected wave has the form of Eq. 3. In the second layer, the equations will be of similar form. The transmitted wave will be in the form of Eq. 4 and the reflected wave will have the form of Eq. 3 but with the amplitude replaced by B_2 and the wave number by k_2.

Satisfying the continuity of pressure at z = 0 in Figure 4 gives:

$$A_1 + B_1 = A_2 + B_2 \qquad (8)$$

Satisfying the continuity of particle velocity, from Eq. 7:

$$u_j = \frac{p_j}{\rho_j \, c_j} \qquad (9)$$

Continuity of velocities gives:

$$\rho_2 c_2 \left(A_1 - B_1 \right) = \rho_1 c_1 \left(A_2 - B_2 \right) \qquad (10)$$

In the same fashion, at z = M in, the following are obtained:

$$A_2 e^{-i(k_2 M)} + B_2 e^{+i(k_2 M)} = A_3 + B_3 \qquad (11)$$

$$\rho_3 c_3 \left(A_2 e^{-i(k_2 M)} - B_2 e^{+i(k_2 M)} \right) = \rho_2 c_2 \left(A_3 - B_3 \right) \tag{12}$$

Now, solving for A_1, and B_1 at $z = 0$ in terms of the other quantities gives:

$$B_1 = A_2 - A_1 + B_2 \tag{12}$$

$$A_1 = \frac{\left(\rho_2 c_2 + \rho_1 c_1 \right) A_2 + \left(\rho_2 c_2 + \rho_1 c_1 \right) B_2}{2 \rho_2 c_2} \tag{14}$$

Similarly, at $z = M$:

$$B_2 = \frac{1}{2\rho_3 c_3} \left(\rho_3 c_3 - \rho_2 c_2 \right) A_3 e^{-i(k_2 M)} + \frac{1}{2\rho_3 c_3} \left(\rho_3 c_3 + \rho_2 c_2 \right) B_3 e^{+i(k_2 M)} \tag{15}$$

$$A_2 = \frac{1}{2\rho_3 c_3} \left(\rho_3 c_3 + \rho_2 c_2 \right) A_3 e^{-i(k_2 M)} + \frac{1}{2\rho_3 c_3} \left(\rho_3 c_3 - \rho_2 c_2 \right) B_3 e^{+i(k_2 M)} \tag{16}$$

As can be seen from the derivations above, the amplitude in each layer is dependent on the amplitude in the next layer. Solving this set of equations analytically for anything more than three layers is quite tedious, even if an analytical computer program is used. To simplify the computation and allow the results to obtained numerically, the following recursive relationship was deduced. This formula allows one to calculate, given the number of layers, the amplitude of the transmitted and reflected waves at an arbitrary boundary:

$$A_2 = \frac{1}{2\rho_{j+1} c_{j+1}} \left(\rho_{j+1} c_{j+1} + \rho_j c_j \right) A_{j+1} e^{-i(k_j t_j)} + \frac{1}{2\rho_{j+1} c_{j+1}} \left(\rho_{j+1} c_{j+1} - \rho_j c_j \right) B_{j+1} e^{+i(k_j t_j)} \tag{17}$$

$$B_2 = \frac{1}{2\rho_{j+1} c_{j+1}} \left(\rho_{j+1} c_{j+1} - \rho_j c_j \right) A_{j+1} e^{-i(k_j t_j)} + \frac{1}{2\rho_{j+1} c_{j+1}} \left(\rho_{j+1} c_{j+1} + \rho_j c_j \right) B_{j+1} e^{+i(k_j t_j)} \tag{18}$$

where t_j is thickness of the jth layer, k_j is the wave number of the jth layer, A_j is the real pressure amplitude of the transmitted wave in the jth layer, B_j is the real pressure amplitude of the reflected wave in the jth layer, r_j is the density of the jth layer, and c_j is the velocity of sound in the jth layer.

The last layer will be a semi infinite layer like the first one, and therefore B_j for that layer is zero.

When the back substitutions are made, it then becomes possible to determine the ratio of the pressure amplitudes of the wave in the first and last layers. This number is complex, and we are only interested in the real portion, found by taking the square root of the modulus of the number.

A computer code was developed for this algorithm that uses the thermal histories of resistance welds to calculate the amplitude ratio of the pressure in the transmitting to the receiving delay lines. This was accomplished by programming the above equations into the computer and using look up tables to match a sonic velocity and density to each temperature. The sonic velocity data for PEEK and APC2 were taken from a sonic analysis performed by Woo and Seferis [9] on PEEK and APC2/AS4 composites . The variation of density with temperature for PEEK was calculated by the topological method for property predictions from Bicerano [10]. The density variation of APC2 with temperature was obtained from the literature [11]. Figures 5 and 6 show how the properties were assumed to change with temperature for this model.

The velocity curve used for the APC 2 was that for the velocity 90° to the fiber direction, corresponding to the transverse direction of composite plate. The acoustic impedance of the delay lines was determined from manufactures data for modulus and density.

The data available from the literature only covered the temperature range of 300 to 514 K eliminated the ability to predict the signal at higher temperatures. This deficiency is a major concern and much work is currently being done to expand this range.

It is important to note here that the sound speeds were assumed to not be dependent on frequency. A quick check in the laboratory found that this was a fairly good assumption for PEEK at room temperature, but the presence of a composite material and temperature effects on wave speed (through density and modulus) complicate this issue. This will be addressed in further work.

The Role of Intimate Contact. Losses associated with the transmission of a signal across a perfectly coupled interface are related in part to changes in the properties of the media which may change their acoustic impedances. Other factors that contribute to signal losses are scattering from inclusions or microstructure (if they are of comparable size to the wavelength), and damping (attenuation) which is a characteristic of the material tested. In this model there are no terms accounting for attenuation or scattering.

In the welding application, there is an additional contribution to losses in signal strength. The signal amplitude across the weld area is related to the type and amount of contact across the interface of the heating element and the polymer interlayers. Since only longitudinal waves have been used so far in testing, it will be impossible to detect any type of contact other than intimate, or completely flush, contact.

The heating element and part surface roughness will determine the fraction of overall area that is in "intimate contact" (no gap) at the beginning of the welding process. As the area fraction increases due to softening and flow of the interfacial components, the amount of area that can transmit the sound waves increases, and theoretically, so should the transmitted signal strength.

Lee, et. al. [12] have proposed a method for predicting the "degree of intimate contact" or D_{ic} (area fraction in intimate contact) for the consolidation of APC2 PEEK/CF prepreg tape. The formulation for D_{ic} is [13] :

$$D_{ic} = \left[(a) \int_0^t \frac{P_{app}}{\mu(T)} dt \right]^{1/5} \qquad (19)$$

where P_{app} is the applied pressure, m(T) is the temperature dependent viscosity of PEEK and a is an initial parameter based on a idealized surface roughness. For the PEEK APC2 system, Mantell and Springer [13] report this 'lumped' parameter as 0.29.

This model is being used as a straightforward area fraction. Using the thermal history, it is possible to calculate D_{ic} at each temperature, which is then multiplied with the signal predicted for 100% intimate contact to predict the signal that should be observed as intimate contact develops during the welding process.

There is debate whether or not this model as used accurately describes what occurs during a resistance weld. This debate is centered on the initial condition According to the model, at time t = 0, D_{ic} = 0.29, and therefore some transmitted amplitude should be present. Work is being conducted to characterize the response of the sensors to the initial condition and how D_{ic} develops during welding. This is a topic that will be revisited later in this paper in the discussion of results.

Model Sensitivity to Input

A thermal history that has a similar heating rate as an actual weld is shown in Figure 7. This was used as a simulated input to test the model response to frequency, heating element thickness, PEEK interlayer thickness, and initial degree of intimate contact. Different heating rates were also examined, and are shown in Figure 8 Since the look up tables used for material properties correspond to set values for given temperatures, the signal is similar in shape for all heating rates. A high heating rate has the effect of compressing the predicted signal vs. time, while a lower heating rate broadens the response.

A range of frequencies from 1.5 to 3.0 MHz was used to calculate the effect of frequency on the predicted signal. Figure 9 shows the general trend over this frequency range. The reference chosen for comparison was 2.25 MHz was because this is the center frequency of our laboratory transducers. It was found that when the frequency is reduced, the signal shifts towards higher temperatures, and the amplitude decreases. Peak A_0 is located at about 470 K in the reference signal. As the frequency decreases to 1.5 MHz, A_0 moves to position A_1 at about 480 K. Peak B_0, located at 493 K in the reference signal moves to the right, and disappears off the upper end of our current range of validity.

As the results for decreasing values of frequency were displayed in succession, the amplitudes at each temperature went through a regular change, rising and then falling off as though the signal were shifting up along the temperature axis. The basic signal shape was similar except at the upper end of the temperature range, where the shape changed periodically. This happened as the series of peaks from 490 K to 514 K shifted to the right and out of range and were formed again by new A_0 peaks moving to the right.

These signal fluctuations can be understood as a result of the frequency changes since the wavelength is a function of frequency for fixed phase velocity. The wavelength governs the interference conditions in a given thickness, and if the frequency is changed, the interference conditions in that thickness will then change.

As the frequency increases to 3.0 MHZ, the signal went through a similar sort of shift as noted before, and the amplitude also decreased. However in this case, the signal was shifted to the left, as would be expected based on the results of discussed above. This is illustrated by the A_0 peak in the 2.25 MHz signal moving to position A_2 at about 463 K in the 3.0 MHz signal.

The 3.0 MHz signal was a good example of the shift of the signal peak from the lower temperatures to the upper. Peak B_0 in the reference signal was seen to follow from the shift of peak B_2 in the 3.0 MHz signal to higher temperatures.

The effect of thickness of the heating element layer on the model predictions was also studied, and some results are presented in Figure 10. The frequency used for the model input was 2.25 MHZ for each case. The thickness range covered from 1.27×10^{-4} m to 3.81×10^{-4} m.

The changing thickness affected the signal in a similar fashion as the frequency changes. This is also not surprising since a change in thickness for fixed frequency and wavelength will also affect the interference pattern.

The trend was for the signal to shift to higher temperatures as the heating element thickness decreased. At the upper end of the temperature range, the signal shape again changed as the peaks moved out of range.

Figure 11 shows the result of changing the PEEK interlayer thickness. Comparisons were made to a reference thickness of 1.27×10^{-4} m. As the thickness of the layer is increased from the reference, the amplitude falls off and the signal is shifted to lower temperatures. As the thickness decreased, the signal was shifted to higher temperatures and the amplitude fell off, but to a lesser extent. As the interlayer thickness decreased, the shift along the temperature axis was evident, but the amplitude changes were less severe than when frequency or heating element thickness is changed.

The final parameter investigated was the initial degree of intimate contact, D_{ic}. As Figure 12 shows, the initial D_{ic} acts as scaling factor, shifting the signal upwards or downwards in amplitude.

These results clearly show that the model is highly sensitive to the parameters: frequency, layer thickness, and initial degree of intimate contact.

Experimental Results

The thermal history of a resistance weld monitored by 2.25 MHz transducers was input into the model. The amplitude of the predicted signal was much larger than the experimental one. This is most likely due to the lack of any attenuation mechanisms in the current model. However, when the signals are normalized relative to their respective maximum amplitudes, the shapes of the curves can be compared. Frequencies used in the model were taken from an FFT of the TTU signal through the weld which was captured on a digital oscilloscope. Figure 13 shows that there is excellent agreement between the predicted and normalized response.

Conclusions and Future Work

This paper presents a model for the transmission of sound through a resistance weld as a function of temperature. The model has been shown to be sensitive to the inputs: frequency, layer thickness, and initial degree of intimate contact. This suggests that tools such as spectrum analysis and on-line measurement of thickness changes may be useful in further understanding the TTU response of a weld. Future work will be focused on obtaining the material properties at higher temperatures, and the inclusion of attenuating terms are necessary to fully complete this model. In addition, the effects of thermal gradients in each layer need to be addressed to determine whether or not heating rate is a factor in reality. The surface roughness of the heating element needs to be characterized, and the intimate contact model must be verified for this application. Additional experiments are necessary for verification of the model. The results of the simulated inputs, and the excellent agreement with actual data, raise expectations that it will be possible to use the model to select an optimum frequency for a given thickness and thermal history to provide the greatest sensitivity to thermal history and intimate contact. Results show that through transmission ultrasonics shows promise as a non-intrusive sensor for fusion bonding and other similar processes.

References

1. Holmes, S., Gillespie, Jr. J.W. *Thermal Analysis and Experimental Investigation of Large Scale Resistance Welded Composite Joints.* in *25th International SAMPE Technical Conference.* 1993. Philadelphia, Pa.:
2. Tackitt, K.D., Don, Roderic C., Holmes, Scott T., and Gillespie Jr., John W. *Through-Transmission Ultrasonic Sensing for Process Control of the Fusion Bonding of Thermoplastic Composites.* in *SPE ANTECH 1993.* 1993. New Orleans:
3. Cracknell, *Scientific Applications of Ultrasound,* in *Ultrasonics,* P.S.N. Mott and G.R. Noakes, Editors. 1980, Crane, Russak, & Company: New York. p. 160-171.
4. Krasil'nikov, V.A., *Sound and Ultrasound Waves in Solids.* Third Edition ed. Sound and Ultrasound Waves in Air, Water, and Solid Bodies, ed. R. Hardin. 1963, Jerusalem: Israel Program For Scientific Translations. 271-325.
5. Krautkramer, J., Krautkramer H., *Ultrasonic Testing of Materials.* 3rd ed. 1983, Berlin Heidelberg, New York.
6. Kuttruff, H., *Ultrasonics Fundamentals and Applications.* 1991, London: Elsevier Science Publishers Limited.
7. Harker, A.H., *Elastic Waves in Solids with Applications to Nondestructive Testing of Pipelines.* 1988, Portsmouth: Grosvenor Press.
8. Kinsler, L.E., Frey, Austin R., *Fundamentals of Acoustics.* 2nd ed. 1962, New York: John Wiley & Sons, Inc. 136-139.
9. Woo, E.M., Seferis, J. C., *Thermal and Sonic Analysis of Polymer Matrices and Composites.* Journal of Composite Materials, 1987. **21**: p. 263-279.
10. Bicerano, J., *Prediction of Polymer Properties.* 1993, New York: Marcel Dekker, Inc.
11. Blundell, D.J., Willmouth, F.M., *Crystalline Morphology of the Matrix of PEEK- Carbon Fiber Aromatic Polymer Composites.* SAMPE Quarterly, 1985. **17(2)**: p. 50-57.
12. Lee, W., Springer, George S., *A Model of the Manufacturing Prcess of Thermoplastic Matrix Composites.* Journal of Composite Materials, 1987. **Vol. 21**(November).
13. Mantell, S.C., Springer, George S., *Manufacturing Process Models for Thermoplastic Composites.* Journal of Composite Materials, 1992. **26**(16): p. 2348-2401.

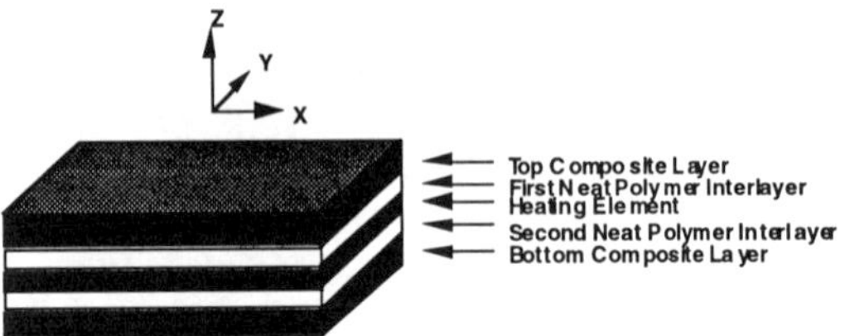

Fig. 1 PEEK/APC 2 weld stack.

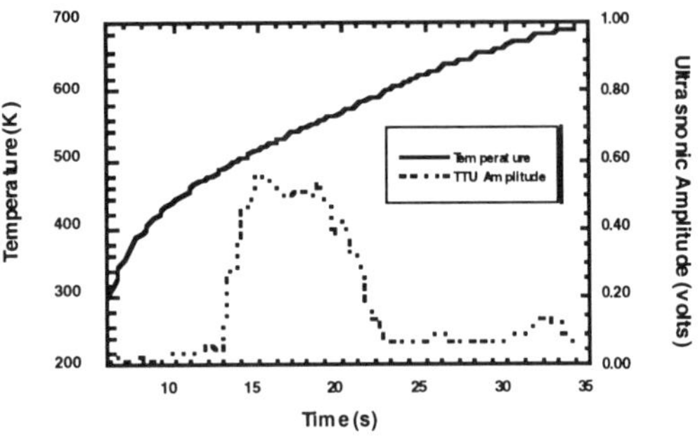

Fig. 2 Through Transmission ultrasonic Amplitude versus time and temperature for a PEEK/ APC 2 resistance weld.

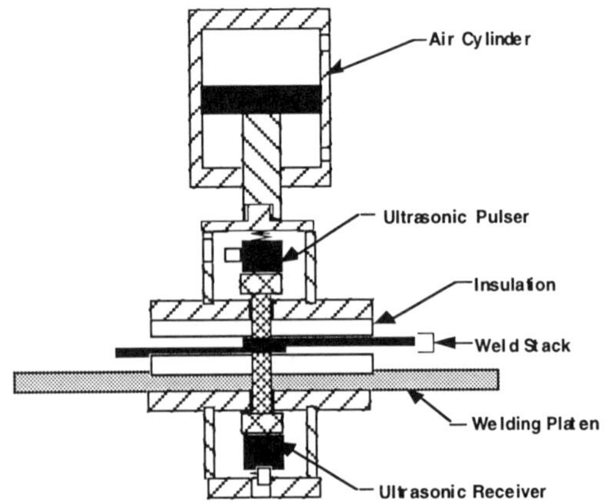

Fig. 3 Experimental tooling for resistance welding.

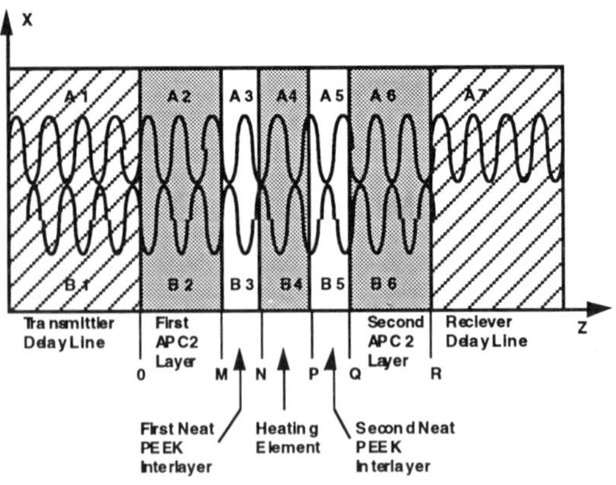

Fig. 4 Wave propagation through seven layers of a resistance weld and the ultrasonic delay lines.

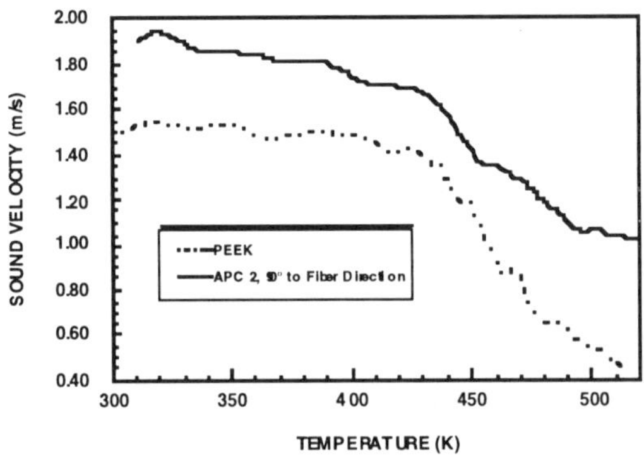

Fig. 5 Sonic velocity vesus temperature for : 90° to fiber in APC 2 and PEEK polymer.

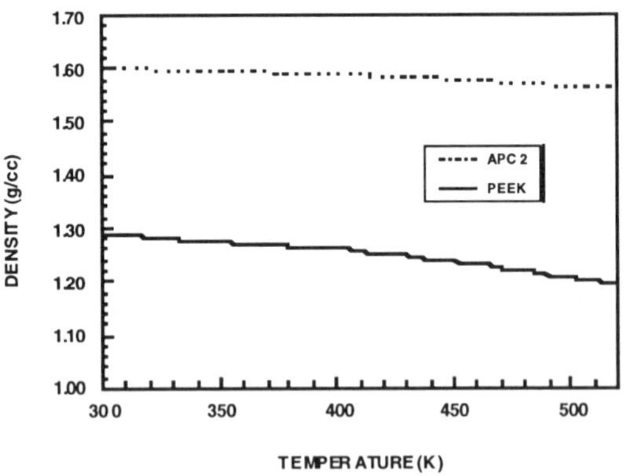

Fig.6 Density versus temperature for APC 2 and PEEK polymer

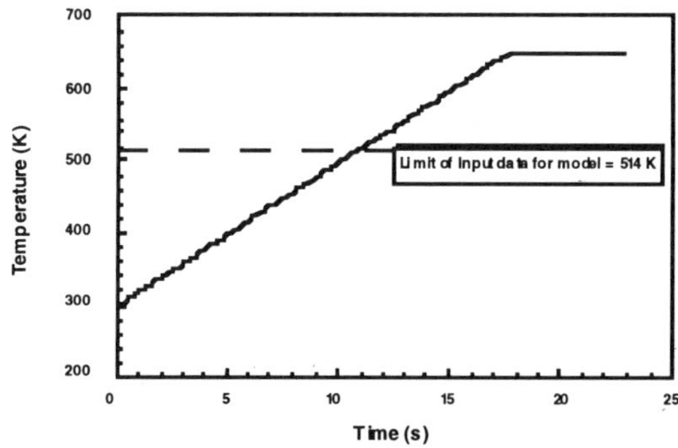

Fig. 7 Simulated thermal history used as model input.

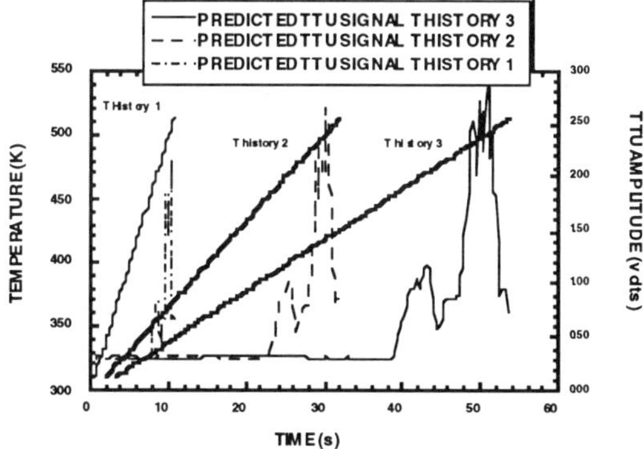

Fig. 8 The effect of increased heating rate on the predicted TTU signal at 2.25 MHz.

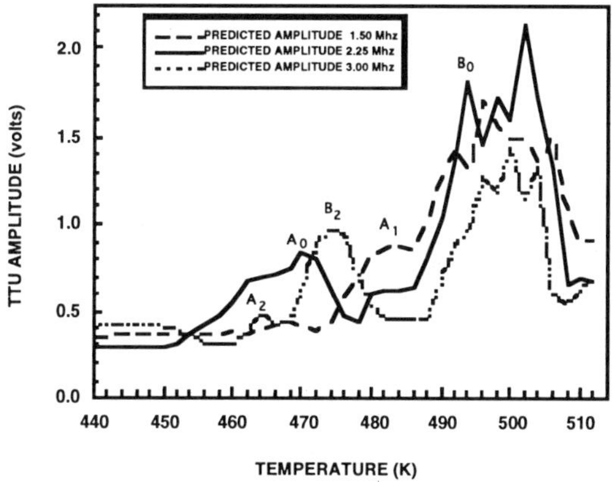

Fig. 9 Predicted TTU signals for a resistance weld as a function of temperature and frequency . The other inputs were a heating element thickness of 3.048×10^{-4} m and 1.27×10^{-4} m thick PEEK interlayers.

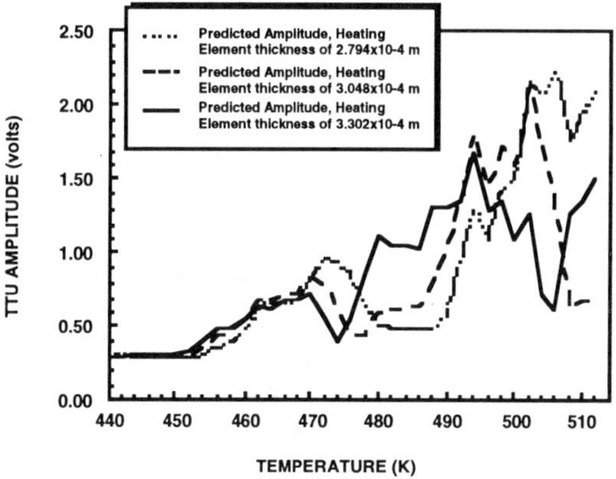

Fig. 10 Predicted TTU signal for a resistance weld as a function of temperature and heating element thickness. The other inputs were a frequency of 2.25 MHz and 1.27×10^{-4} m thick PEEK interlayers.

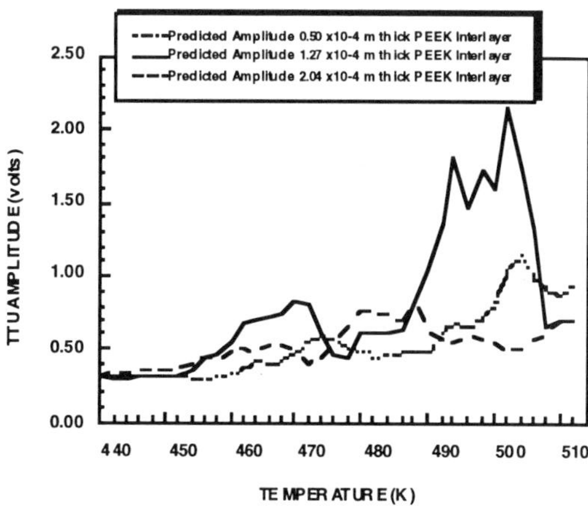

Fig. 11 Predicted TTU signal for a resistance weld as a function of temperature and   PEEK interlayer thickness. The other inputs were a frequency of 2.25 MHz and a 3.048×10^{-4} m thick heating element.

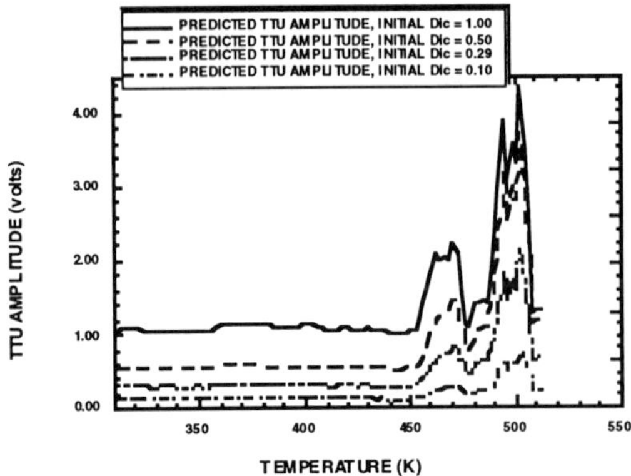

Fig. 12 Predicted TTU signal for a resistance
weld with various intial degrees of intimate
contact. The other inputs were a frequency
of 2.25 MHz, a heating element thickness of
3.048×10^{-4} m and 1.27×10^{-4} m thick PEEK
interlayers.

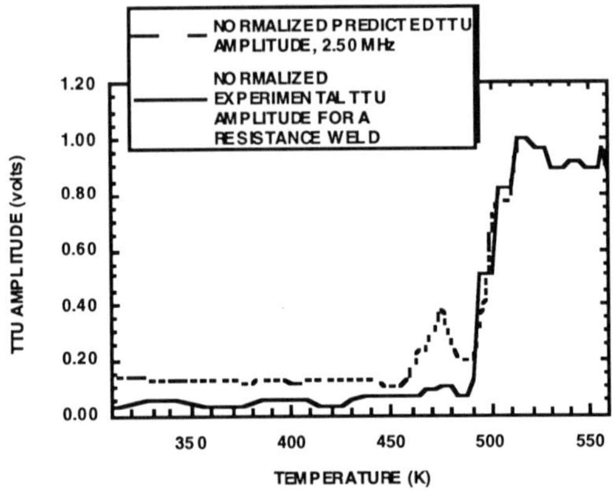

Fig. 13 Comparison of model prediction at
2.05 MHz versus experimental TTU amplitude
versus temperature for a PEEK/APC 2
resistance weld.

Proceedings of the 10th Annual ASM/ESD Advanced Composites Conference, Dearborn, Michigan, USA, 7-10 November 1994

The Role of Surface Preparation on the Performance of Metal to Polymer Adhesive Joints

S.H. McKnight, J.W. Gillespie, Jr., P.E. Bourban
University of Delaware
Newark, Delaware

ABSTRACT

The effect of some metal adherend surface treatments on the performance of metal-to-polymer adhesive bonds is presented. The strength of bonded steel and aluminum adherends were evaluated with a number of different adhesives, including thermosetting epoxies and various thermoplastic polymers. Additionally, the hot-wet durability of these bonds was studied using the wedge-crack extension test. In this work, an emphasis was placed on evaluating non-chromate based surface treatments. Sulfuric-Boric Acid Anodizing (SBAA) and the use of silane coupling agents were evaluated as alternative surface treatments for adhesive bonding of aluminum alloys, where Phosphoric Acid Anodizing (PAA) was used as a benchmark surface treatment. Both SBAA and silane treatments on 2008 aluminum were seen to provide strength and durability comparable to PAA in aluminum-epoxy joints. Silane coupling agents were shown to be excellent adhesion promoters for steel adherends in some instances. Furthermore, the effectiveness of fusion bonded thermoplastics as adhesives for metal adherends was demonstrated. In some cases the performance was superior to that obtained using toughened epoxy adhesives.

INTRODUCTION

Recently, the use of composite materials has been increasing in several applications. These materials are no longer limited to the aerospace industry, and may continue to diversify into new areas of opportunity. The use of composites, however, will most likely occur through integration with traditional structural materials, such as metals. Near-term and future structural applications will most likely require the combination of aluminum, steel, and polymeric based materials. Understanding and developing joining techniques for multi-material assemblies or structures involves many technology areas. A list of these areas would include: surface preparations that can be used for dissimilar adherend materials; joining methods suitable for combining materials of differing physical, thermal and chemical properties and compositions; durability/reliability issues; and specific component requirements such as a painting operation which could exacerbate some of the differences between the materials. For that reason, it is critical that issues surrounding the joining and assembly of hybrid components be addressed.

Currently, there are several areas where the importance of the composite-to-metal joint can be seen. For example, the use of composites to rehabilitate steel structures has be proposed as an attractive alternative to complete replacement of the structure. The bond between the existing steel and the composite will determine the overall effectiveness of the repair solution. Also, aluminum is playing an increasing role in automotive design, since the inherent corrosion resistance, high strength-to-weight ratio, and good formability of aluminum make it an attractive material for many automotive components. Forecasts consistently predict rising aluminum usage as weight reduction and fuel efficiency programs continue. For example, the Aluminum Association has forecast an increase in the use of aluminum from the current per-car average of 85 kg to 160 kg by the turn of the century [1]. Adhesive bonding of composite parts to aluminum will be necessary in certain applications, and the ultimate performance of these hybrid structures relies on the integrity of the composite-to-aluminum bond.

Traditional methods to join composites to metals include mechanical fastening and thermosetting adhesives. However several problems arise when thermosetting adhesives are used to join some thermoplastic composites. Elaborate surface preparation of the thermoplastic adherends is often necessary to achieve any bonding to thermosetting adhesives [2]. Alternatively, thermoplastic components have the ability to be fusion bonded (welded) to metal adherends.

Recent work has used thermoplastic welding technology to join dissimilar materials such as polypropylene (PP) and polypropylene/glass composites to surface treated aluminum [3-5]. Previous work [3] has shown that polypropylene can be an effective 'hot-melt' adhesive for aluminum adherends. When PP was used to join aluminum adherends, lap-shear strengths of 20 MPa were achieved. As a comparison, the lap-shear strengths of aluminum joints bonded with epoxy adhesives range from 14-21 MPa. Also, the use of thermoplastic fusion bonding was shown to be a very effective technique for joining dissimilar materials. Glass-reinforced polypropylene (GRPP) and aluminum adherends were joined using both hot-press and resistance-heated fusion bonding techniques. The bonds formed in these sandwich single-lap joints were strong enough to cause cohesive failure of the composite, well removed from the interface [4], thereby demonstrating the potential performance achieved with thermoplastic fusion bonding.

Steel and aluminum joints bonded with thermoplastics have also been investigated elsewhere [6-8]. Much work has been focused at improving strength of the metal-thermoplastic bond [6,7]. Although advances have been made in improving the initial strengths of these bonds, the durability of the joints has not been extensively researched. The long term environmental resistance of any adhesive joint is a critical issue which must be addressed. For example, limited research has demonstrated that the Al/PP bond is very durable in hot-wet environments, when the aluminum is treated using phosphoric acid Anodizing (PAA) [5]. Unlike studies on thermoplastic-metal adhesive bonds, the environmental stability of metal-epoxy joints has been reported at length elsewhere [2,9], and several methodologies to determine durability have been previously established. Surface preparation of the adherends has been shown to be the primary factor controlling joint durability in metal to epoxy adhesive joints.

The durability of aluminum adhesive bonds is especially sensitive to adherend surface preparation. In chemical and electrochemical surface treatments, the oxide morphology and chemistry have been shown to have a profound effect on the long term durability of these joints [9,10]. Specifically, phosphoric acid anodizing has been shown to provide superior joint durability when compared to other surface treatment techniques, such as grit-blasting or the Forest Products Laboratory (FPL) Etch. The structure of the PAA oxide layer aids bonding for two reasons: (1) bonding surface area is increased; (2) the porous structure promotes mechanical interlocking [9]. The excellent environmental stability of the PAA oxide has been attributed to the presence of a thin aluminum phosphate layer on the oxide surface, which inhibits the hydration of the surface oxide and degradation of the adhesive joint under the presence of water [9,10].

Surface preparation of steel adherends is also necessary to obtain maximum performance. However, less elaborate procedures are usually used with steel adherends. Unlike aluminum, the oxide structure on steel can not be controlled to promote adhesion. In fact, excess oxide must be removed for successful adhesive bonding to occur. Generally, abrasion followed by solvent cleaning are adequate to prepare the steel surface prior to adhesive bonding. However, improvements can undoubtedly be made. Previous research has shown that the durability of steel adhesive bonds can be significantly improved by changing the surface preparation technique [2].

Many currently used surface preparation techniques may not be suitable in the future. As concerns develop over environmental considerations in manufacturing processes, surface preparation techniques which employ chromate containing solutions have become the subject of new regulations. Hexavalent chromium (Cr^{+6}), which is present in the FPL etch, poses a health risk to personnel as well as to the environment. To comply with the new regulatory measures, the development of new surface treatments for aluminum has begun, and potential replacements for the FPL etch and PAA have been identified. Two notable processes are sulfuric boric acid anodizing (SBAA) and abrasive-silane treatments [11-13]. These treatments have been shown to provide strong durable bonds with epoxy based structural adhesives [11-13]. Environmental concerns are also important in steel rehabilitation applications. The use of toxic and hazardous chemicals will not be acceptable in infrastructure rehabilitation programs. The need for environmentally compliant methods for metal adherend surface preparation is therefore very pressing.

Previous investigations have revealed that the metal-adhesive interphase in the critical region in metal-composite joints [3-5]. Hence, this work examines the effect of surface preparation techniques on the adhesion of polymers to metal adherends. Both steel and aluminum are used as metal adherends. Thermosetting epoxy adhesives and thermoplastic polymer films are used to bond the adherends. Performance is assessed through tests of joint strength and hot-wet durability.

EXPERIMENTAL TECHNIQUES

Materials

Aluminum: Much attention has been placed on investigating the adhesive bonding of the aerospace grades of aluminum alloys. However, due to increasing automotive interest in non-aerospace aluminum alloys, 2008-T4 aluminum was selected for this effort because of its attractiveness in automotive applications.

Steel: High-strength low-alloy (HSLA) steels are used in several structural applications where composites may be used for rehabilitation. Therefore, the alloy A242 in a cold rolled condition was chosen for adhesive evaluations in this study.

Adhesives: Thermosetting epoxies and various thermoplastic films were used in this study. Tables 1 and 2 list the adhesives evaluated with each metal. PP was chosen as the sole thermoplastic resin used with aluminum due to previous success with this polymer. Additionally, maleic anhydride modified PP was investigated as an adhesion promoting agent for aluminum bonding. More different types of thermoplastics were investigated with the steel adherends, due to the lack of available data on thermoplastic fusion bonding of steel. The polymers listed in Table 2 were all investigated as potential adhesives for steel. Durability evaluations were performed on the most successful materials.

Table1: Adhesives used with Aluminum

Adhesive	Designation
Thermosets	
Lord Fusor 320/322 Epoxy	Epoxy A
Thermoplastics	
Polypropylene	PP
Maleic Anhydride modified Polypropylene	MA-PP

Table 2: Adhesives used with Steel

Adhesive	Designation
Thermosets	
Lord Fusor 320/322 Epoxy	Epoxy A
Ciba Geigy Araldite AV138/HV998	Epoxy B
Tonen Epoxy FR-E3P	Epoxy C
Thermoplastics	
Polypropylene	PP
Polyetherimide	PEI
Polyetheretherketone	PEEK
Polyarylene Sulfide	PAS
Polyphenylene Sulfide	PPS
Polyether sulfone	PES
Polysulfone	PSU

Surface Preparation Techniques   A wide variety of surface treatments are available for both steel and aluminum adherends. Tables 1 and 2 display the techniques employed in this study. They are by no means inclusive, but do represent a wide array of different physical, chemical, and electrochemical methods for surface preparation of metals prior to adhesive bonding. A brief discussion of the methods and their implementation is discussed below.

Numerous surface treatment techniques have been developed for adhesive bonding of aluminum and steel. However, a surface preparation technique has not been identified for the use of thermoplastic polymers as a hot-melt adhesives. Due to their non-reactive nature, adhesion improvements in metal to thermoplastic bonds have been previously realized by increasing the surface roughness to promote mechanical interlocking. For example, earlier studies on bonding aluminum with PP have had success with acid etches and anodizing [3-5,15]. The penetration of PP into the microporous cavities in the oxide has been observed [15], and it was hoped that a similar microstructure would be achieved here. Beyond interlocking, some polymers may react with the metal surface and this local interfacial reaction may increase the bond strength [8] Unfortunately, this is not often the case. Most thermoplastic resins have low adhesion to clean but untreated metal surfaces.

Table 3: Surface Treatments Used on Steel

Surface Treatments	Designation
Grit/Bead Blasting and Degreasing	Blasting
Gas Flame Treatment	Flame
Blasting then application of glycidoxymethoxy silane (Z-6040[1])	GB-GPS
Mixed silane primer (Q1-6106[1])	6106

[1]**Dow Corning** ®

Table 4: Surface Treatments Used on Aluminum

Surface Treatments	Designation
Blasting and Degreasing	Blasting
Forest Products Laboratory Etch	FPL
Phosphoric Acid Anodizing	PAA
Sulfuric-Boric Acid Anodizing	SBAA
Blasting and application of glycidoxymethoxy silane (Z-6040[1])	GB-GPS
Wet Abrasion Application of GPS	WA-GPS
Cationic Styrylamine Silane (Z-6032[1])	CSS

[1]**Dow Corning** ®

Altering the metal surface chemically is another interesting method for surface preparation. Abrasion-silane treatments have been proposed as viable surface preparation techniques for metal adherends [12,13]. These methods make use of organo-functional silane coupling agents (commonly used in glass fiber sizings and with inorganic polymer fillers) for adhesion enhancement. Previous research has shown silanes can be effective at promoting adhesion between metals and thermosetting adhesives, such as epoxies. Limited research has been reported on adhesion enhancement of non-reactive thermoplastic resins to metals [16,17]; and much of this work has concentrated on qualitative assessments of bond strength [17]. In this study, silanes were investigated as effective surface treatments for the epoxy and thermoplastic adhesives. Based on earlier findings [2,17] the silanes chosen for this investigation were Z-6040, (3-glycidoxypropyltrimethoxy (GPS) coupling agent), Z-6032 (Cationic-Styrylamine (CSS) coupling agent), and an experimental adhesion promoter Q1-6106 (GPS and melamine resin) all graciously donated by Dow-Corning.

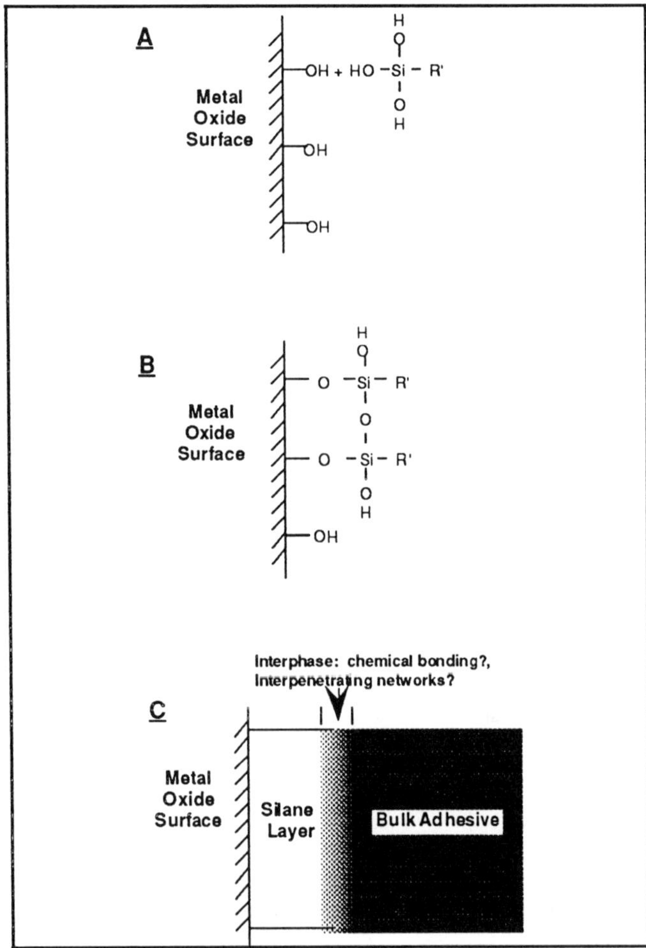

Figure 1: Schematic Representation of silane coupling to a metal oxide surface. (A) The silane solution is applied, and the silanol groups orient towards the hydrated metal oxide. (B) Upon drying, the silanol bonds condense to form a polysiloxane network that is chemically bonded to the oxide. (C) The adhesive is applied to the deposited siloxane layer.

As the use of silanes for adhesion promotion continues to develop, several silane application methods have been proposed. All techniques have some features in common. An alkoxy silane compound is mixed with slightly acidic water to form a dilute (1-2 wt. %) solution. The solution is stirred for some time to allow the silane to hydrolyze forming a silanol. (This hydrolyzed solution may next be added to an alcohol solution.) The aqueous (or alcohol) solution is then applied to a metal surface and the layer is dried. The silanol groups chemically react with the hydrated oxide layer on the metal, forming a M-O-Si linkage where M is a metal atom, as the -OH bonds condense. The organo-functional groups in the polysiloxane layer that is formed on the surface are free to react with the primer or adhesive when it is applied. This is schematically shown in Figure 1. The functional group is selected based on solution compatibility and reactivity with the chosen adhesive system. For example, epoxy-based silane compounds work very well with epoxy based primers and adhesives. The successful use of 3-glycidoxypropyltrimethoxy silane (GPS) has been reported for adhesive bonding of aluminum alloys using epoxy adhesives [17].

Specific Surface Treatments: Aluminum: The PAA was performed according to BAC-5555 [16], and the SBAA was performed following guidelines established in reference 11. The abrasion-GPS silane treatments were accomplished via two different techniques, each using 1 wt. % solution of GPS in water, which had been hydrolyzed under continuous stirring for 1 hour. In both methods, the as-received panels were first thoroughly degreased using acetone. The degreased surfaces were then abraded in two different ways. One method used grit-blasting to produce a macroscopically rough surface. Following the grit-blast, the surfaces were cleaned and the GPS solution was applied using a soft brush. The other roughening technique utilized abrasive pads and silane solution simultaneously. The surface of the aluminum was flooded with solution, and the pad was used the abrade the surface. When the surface finish appeared uniform, moistened tissues were used to remove the debris. This process continued until no trace of debris was observed on the tissues. The grit-blasted (GB) and wet-abraded (WA) samples were then dried at 93 °C for 1 hour, which has been reported as an effective drying cycle for these silane coupling agents [13]. . The CSS solution consisted of 5% hydrolyzed silane in methanol. In this case, the amount of CSS deposited was controlled to give a final silane layer thickness of about 1.5 micrometers. This has been reported as the optimum thickness for promoting adhesion to PP [17].

Steel: Grit/Bead-blasting followed by degreasing in acetone was the baseline treatment used with steel. Additionally, it was found that after this treatment oils and greases were left on the surface. Therefore, a flame treatment was used to further clean these surfaces. In this method, a propane torch was passed rapidly over the steel surface. Silanes were applied to blasted and cleaned surfaces. 1% aqueous GPS solutions and 5% 6106 in methanol solutions were applied in the same manner as the aluminum specimens.

Joint Fabrication   Joints were assembled according to lap-shear (ASTM D-1002) and wedge-crack extension (ASTM D-3762) test specimen geometries. The epoxy adhesives were mixed and applied according to manufacture's specifications. In some cases different cure cycles were investigated. (These cycles will be identified with the results.) A hot press was used to bond the thermoplastic polymers to the metal substrates. Processing conditions for each type of polymer were chosen on the basis of prior experience and the processing characteristics of each polymer. Processing pressure was fixed at 0.35 MPa for the aluminum adherends and 2.0 MPa for the steel specimens. Wire spacers (placed in the trim areas) were used to maintain a constant bondline thickness.

RESULTS AND DISCUSSION

Static Performance (Lap-Shear Testing and Initial Crack Lengths)

Epoxy - Aluminum Specimens   The lap-shear strengths of aluminum bonded with Epoxy-A have been reported elsewhere by the authors [19]. The initial crack lengths of the epoxy-A bonded specimens are displayed in Figure 2. The length of the initial crack is an indication of the toughness and interfacial strength of the adhesive-aluminum bond. At a constant crack opening displacement (imparted by the wedge), shorter crack lengths are indicative of tougher bonds. Figure 2 indicates that for the epoxy adhesive, the GPS-silane treatments yielded the highest initial toughness and interfacial strength. SBAA-treated, PAA-treated, and blasted samples displayed slightly larger initial crack lengths.

as the adhesive. Figure 3 shows the initial crack lengths for the wedge crack specimens bonded with PP as a function of surface treatment. PAA and SBAA treated specimens displayed very small initial crack lengths, but these values were greater than any of the initial crack lengths observed with the epoxy bonded specimens. Additionally, only anodizing treatments consistently enhanced strong bonding of PP to aluminum. The wet-abrasion GPS treatment did improve bonding compared to blasting, which failed completely upon wedge insertion. However, the average initial crack length was 121 mm which was nearly the entire joint length. The PP is non-reactive with the epoxy based silane, and the incompatibility of the GPS and PP does not allow formation of an interpenetrating network, which serves as the bonding mechanism for non-reacting polymers and silane coupling agents [16,17]. Better results were obtained when CSS was used. CSS is recommended for use with polypropylene [16] and is compatible as shown by its similarity in solubility parameter and surface energy to that of PP. (See Table 5) However, the adhesion found in PP-CSS treated aluminum was not consistent. This can be seen by the very large variation in initial crack lengths. Furthermore, upon examination of the failure surfaces it was seen with each sample that there were areas which were not adhered and others which were bonded very strongly. This demonstrates the potential for CSS to promote adhesion with PP, but the consistency of the bonding must be improved to use the CSS treatment successfully. The mechanisms for CSS-PP adhesion promotion and are not well understood. A fundamental understanding of these mechanisms may help to resolve these problems

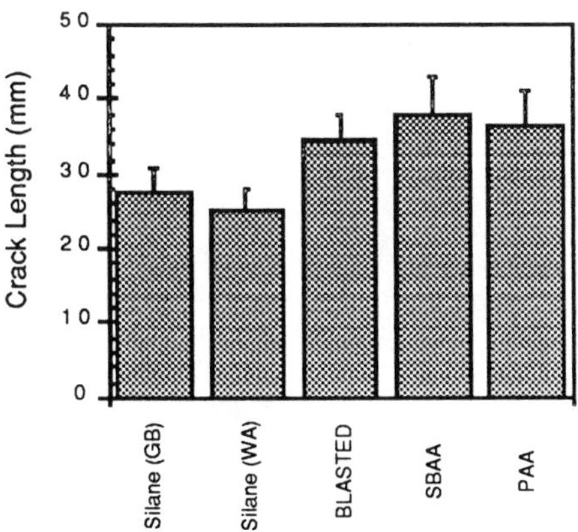

Figure 2: Initial crack lengths as a function of surface treatment for Epoxy-A bonded aluminum specimen.

PP- Aluminum Specimens   More variability in initial joint toughness and interfacial strength was seen between the different surface treatments when PP was used

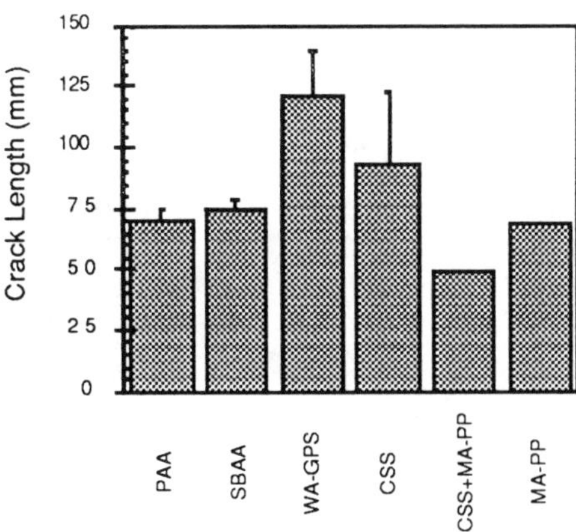

Figure 3: Initial crack lengths as a function of surface treatments for PP-bonded aluminum specimen.

The effect of the modified PP can also be seen. in Figure 3 With both CSS treated and Blasted surfaces the effect of the modified PP was to increase the degree of interaction and the bond strength. The initial crack lengths for these specimen were very small, and improved compared to the anodized specimens. The modified PP is a very interesting approach, and these results indicate that the use of this polymer may minimize the degree of surface treatment needed for bonding PP to aluminum. Further study on the use of the modified PP is warranted and currently ongoing.

Table 5: Surface and Solubility Properties of Silanes

Silane	γ (mJ/m^2)	δ (J/cm^3)$^{1/2}$
GPS	43.0	19.4
CSS	36.9	17.2
Polypropylene	32-35	16-17

Steel-Epoxy Specimens: The baseline (blasted only) lap-shear strengths for steel bonded with epoxy are shown in Figure 4. Epoxy A displayed the best lap shear strength among the three epoxy adhesives that were evaluated. This value was obtained when the adhesive was cured at elevated temperature (100° C for 1200 seconds). Generally it was found that increasing the curing temperature increased joint performance. This trend can be seen in Figure 5.

Furthermore it was found that for the A242 steel, the flame treatment significantly increased the performance of the epoxy-steel joints. This can be seen in Figure 6 for both epoxy-A and epoxy-B. Apparently, the flame is very effective in removing contaminants left after the blasting and degreasing steps. As can also be seen in Figure 5, the GPS silane did not seem to have any appreciable effect on the lap-shear strength of the epoxy-steel joints.

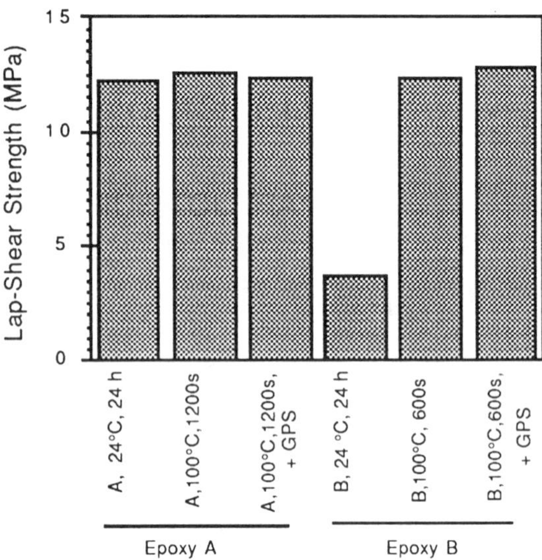

Figure 5: Effect of cure temperature and silane treatment on lap-shear strength of steel/epoxy joints.

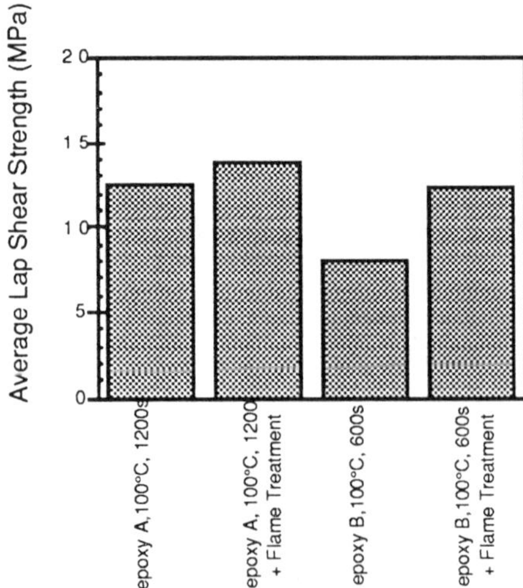

Figure 6: The effect of flame treatment on the lap-shear strength of epoxy-bonded steel specimens.

The initial crack lengths for the steel wedge crack extension specimens are shown in Figure 7. The silane treatment seems to have a more pronounced effect on these results. In specimens bonded with either epoxy-A or epoxy-B, the crack lengths decreased. This indicates that the silanes act to increase the initial joint toughness through adhesion enhancement.

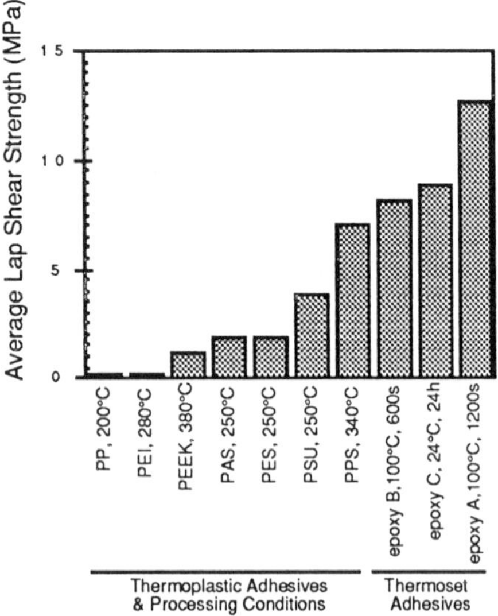

Figure 4: Lap-Shear Strength for bonded steel specimen that had been only blasted and degreased.

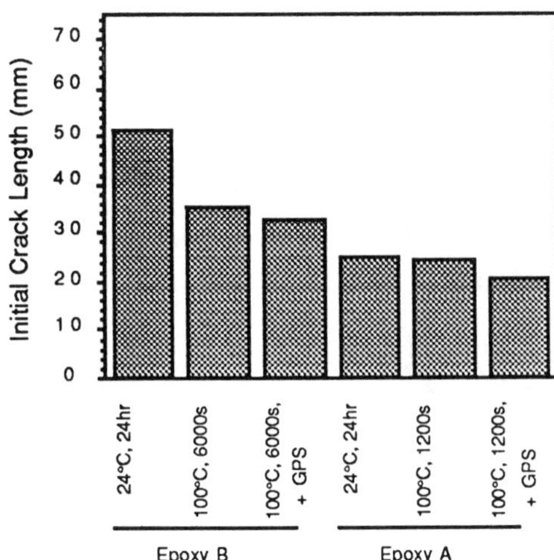

Figure 7: Initial crack lengths for epoxy-steel wedge crack specimens.

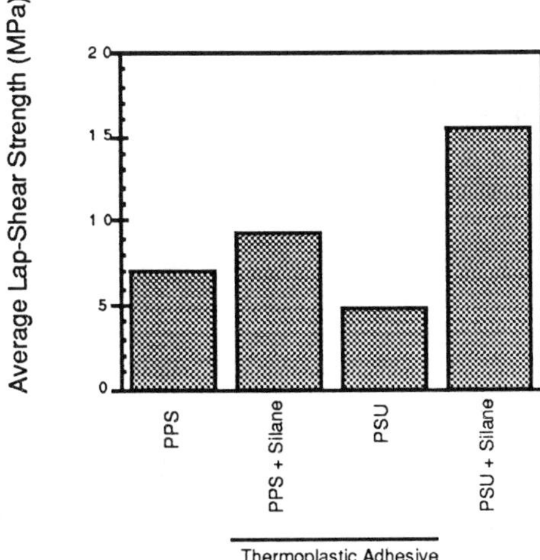

Figure 8: The effect of silane treatment (Q1-6106) on steel specimens fusion bonded using thermoplastic polymers.

Steel-Thermoplastic Specimens   Figure 4 also displays the lap-shear strengths of the fusion bonded thermoplastic joints where the steel was only blasted and degreased. As can be seen, most of the thermoplastics do not bond as well to steel as do the thermosetting epoxies. These results indicate PPS and PSU produced the highest lap-shear strengths. The lap shear strength of the PPS is comparable to the values obtained with the thermoset adhesives. Furthermore, the 6106 primer had a dramatic effect on the adhesion of both PPS and PSU to steel. (See Figure 8) In both cases, an increase in strength was seen in the 6106

treated specimens. Both silane treated samples produced lap-shear strengths comparable to the epoxy values, and the PSU bonded specimen had a larger lap-shear strength than any of the epoxy bonded samples. This is a very significant finding, for it demonstrates the potential effectiveness of thermoplastic polymers as adhesives for joining steel adherends.

Hot-Wet Durability (Wedge Crack Propagation in 65° C Water)

Epoxy - Aluminum Specimens   Crack extension versus time is displayed graphically in Figure 9, for the epoxy-bonded 2008 aluminum specimens.   The results for the epoxy bonded specimens were very interesting. Other than the grit-blasted samples (which failed after 15 minutes of exposure), the non-chromate based surface treatments demonstrated excellent durability. The SBAA and WA-GPS treated specimens performed identically to the PAA benchmark.   The GB-GPS treatment did not achieve equivalent durability, which may have been due to the method of application of the silane solution. These results indicated that the ability to promote strong durable bonds is possible in the new non-chromate surface treatments for bonding with epoxy adhesives.

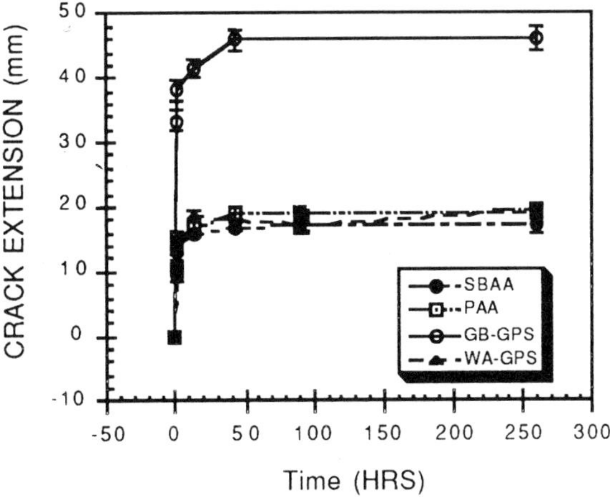

Figure 9: Crack growth ($\Delta\alpha$) vs. exposure time (65°C H_2O) for aluminum samples bonded with Epoxy-A.

PP - Aluminum Specimens The results of the wedge crack extension tests for the PP bonded aluminum samples were also very encouraging. The SBAA treatment demonstrated better durability when compared to PAA (see Figure 10), with final crack extension of 17.1 mm.   However, the silane treatments did not impart any consistent durability enhancement for the PP-bonded aluminum. The WA-GPS treated samples completely debonded after 15 minutes of exposure. The durability of such joints should improve if a compatible silane coupling agent is used in place of GPS. This was seen in some of the CSS specimens, but the results were not consistent. - Long periods of no crack growth were followed by rapid and complete failure of the joint in all CSS treated samples. However, qualitatively the performance was superior to the GPS treated joints, which demonstrates the potential for the CSS treatments to enhance PP-aluminum bonding. The modified PP bonded samples did show improvements over the silane treated samples, but the performance was not comparable to the anodized specimen.

Table 6 Summary of Crack Propagation in
Aluminum Wedge-Crack Specimens.

Sample (Surface Treatment)	Initial Crack a_0 (mm)	Final Extension Δa_f (mm)	Total Length a_f (mm)
Adhesive: Epoxy A			
Grit Blast	34.54 ± 3.53	Full	Full
PAA	36.58 ± 4.35	19.7 ± 1.50	56.28 ± 5.85
SBAA	37.79 ± 5.08	17.8 ± 1.40	55.59 ± 6.48
WA-GPS	25.40 ± 2.54	19.57 ± 0.81	44.97 ± 3.35
GB-GPS	27.68 ± 3.18	45.94 ± 2.32	73.64 ± 5.50
Adhesive: PP Film			
Grit Blast	Full	-	Full
PAA	71.48 ± 4.25	19.3 ± 0.78	90.78 ± 5.03
SBAA	73.86 ± 4.69	17.1 ± 1.20	90.96 ± 5.89
WA-GPS	121.3 ± 19.5	Full	Full
GB-GPS	Full	-	Full
CSS	92.9 ± 32.5	Full	Full
CSS/MA-PP	49.5	Full	Full
MA-PP	67.32	59	126.32

*"Full" indicates total debonding along the entire crack length, meaning part separation had occurred. Initial crack length and crack extensions were measured as described in the test method.

To compare the overall durability of epoxy-aluminum with PP-aluminum bonds, the total crack length must be examined. The total crack length observed at constant crack opening displacement (COD) is an indication of overall joint toughness. Table 6 summarizes the total crack lengths for all of the wedge crack extension specimens. The epoxy bonded specimens exhibited smaller crack lengths, and

therefore impart greater overall joint toughness and interfacial strength.

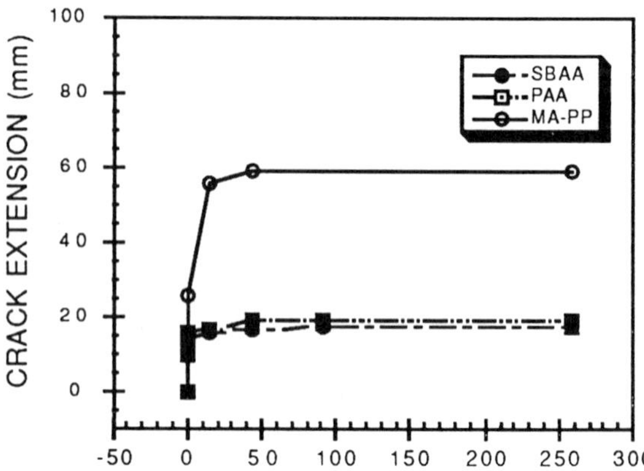

Figure 10: Crack extension (Δa) vs. exposure time (65°C H_2O) for aluminum samples bonded PP. (Specimens whose cracks grew to full extension not displayed.)

Steel Specimens The durability of the epoxy-steel joints was very good. Figure 11 shows the results for the durability measurements made using the wedge crack test for the aluminum specimens. It can be seen that the durability of joints bonded with epoxy-B is better than those joined with epoxy-A. These durability tests showed that epoxy-C to impart extremely poor durability. Full crack extension occurred in several specimens bonded with epoxy-C.

Additionally, the effect of silane treatment on the hot-wet durability of the steel-epoxy joints can be readily observed. The GPS treatments are extremely effective at increasing joint durability in hot water. This effect was observed in both epoxy-A and epoxy-B bonded specimen. In fact, the GPS treated specimens bonded with epoxy-B experienced no crack growth over the period of exposure. The absence of crack growth indicates the unusually high resistance of the bond to deterioration due to exposure to hot-water.

Unfortunately, at press time, the durability results for the thermoplastic-steel adhesive joints were incomplete and therefore only limited results can now be presented. These results are shown in Figure 12. The results were very encouraging. The durability of the PPS bonded samples with no silane treatment are equivalent to the durability demonstrated by some of the epoxies. This result further reveals the potential for bonding steel adherends effectively using thermoplastic polymers. The 6106 treated samples which were bonded with PSU also showed excellent durability. Crack growth was not extensive, and much less than the extension observed in the PPS bonded samples. The PSU+6106-bonded samples actually exhibited less crack growth during hot-wet exposure than some of the epoxy-bonded specimens. (See Figure 11.) A summary of crack lengths can be found in Table 7.

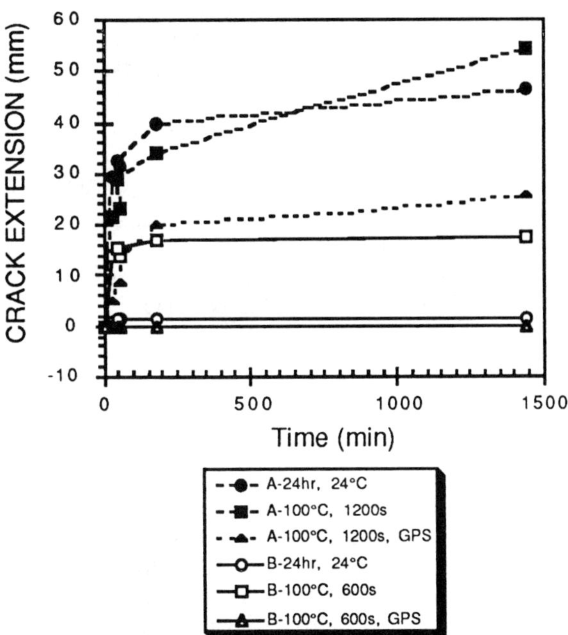

Figure 11: Crack growth (Δa) vs. exposure time (65°C H_2O) for steel samples bonded with epoxy.

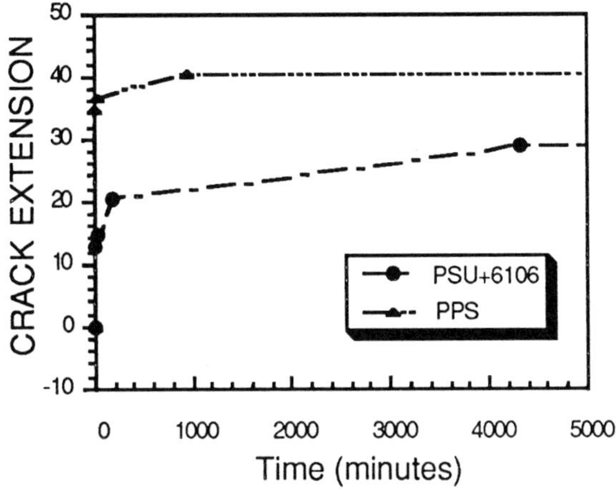

Figure 12: Crack growth (Δa) vs. exposure time (65°C H_2O) for steel samples bonded with thermoplastic polymers.

Locus of Failure The region of failure in the adhesive joints was also investigated for the steel and aluminum adhesive joints. The loci of failure for the various joints can be used to identify crucial regions in the bond. This work is still ongoing but several important findings have already been made. For example the importance of the thickness of the deposited silane layer was revealed. In the epoxy-A bonded aluminum wedge-crack specimens prepared by the GB-GPS method, the locus of failure visually appeared to be interfacial. However, upon further examination using reflectance infrared techniques, the presence of silane molecules on both sides of the failure surface was detected. This indicates a failure through the silane layer, and not at the interface. Likewise, similar behavior was seen in the CSS treated specimens bonded with PP. In this case, traces of CSS were found on the aluminum and PP sides of the failure surfaces. This may explain the spotty adhesion which was noted in the mechanical testing. In the PAA and SBAA treated samples, cohesive failure in the adhesive was observed in the epoxy and PP bonded specimens. A cohesive failure was also observed in WA-GPS treated specimens. (Further information regarding the loci of failure of these specimens will be subsequently presented in future publications.)

The effect of surface treatment on bonding of a structural steel was also presented. The steel specimens also exhibited changes in loci of failure. This was particularly seen in the difference in loci of failure for epoxy-B bonded specimens. Blasted samples failed in an adhesive manner, but flame treatment moved the locus of failure away from the interface into the adhesive. The majority of the thermoplastic-steel bonds failed adhesively. PPS and PSU lap-shear specimens failed cohesively. The loci of failure for the wedge specimen is currently being evaluated, and will be presented in the future.

Table 7: Summary of crack propagation in bonded steel wedge specimens.

Sample	Initial Crack a_0 (mm)	Final Extension Δa_f (mm)	Total Length a_t (mm)
Adhesive:	Epoxy A		
24h,24°C	24.2	46.4	70.6
1200s, 100°C	22.23	54.51	76.74
1200s,100°C+GPS	20.88	28.83	49.71
Adhesive:	Epoxy B		
24h, 24°C	51.35	0	51.35
600s, 100°C	35.88	17.3	53.18
600s,100°C +GPS	33.0	0	33.0
Adhesive:	PPS		
300s, 340°C	66.5	40.50	107
Adhesive:	PSU		
600s,250C+6106	29.58	28.96	58.54

CONCLUSIONS

The performance of metal-polymer joints was investigated for a series of different surface treatments. Both SBAA and wet-abrasion 3-glycidoxypropyltrimethoxy (GPS) silane treatments on 2008 aluminum were shown to provide equal durability to PAA in aluminum-epoxy joints. Furthermore, SBAA and PAA was shown to be an effective non-chromate based surface treatment for bonding aluminum with PP. Durability of joints using SBAA treated aluminum was slightly better than that exhibited by PAA-treated adherends. GPS did not enhance the bond between the aluminum and PP, due to the lack of reactivity and compatibility between the PP and epoxy based silane, as evidenced by poor durability and the interfacial failure loci. The use of organo-functional silanes with improved compatibility (CSS) was shown to improve the bonding to PP. However, the improvement was not consistent and very scattered results were observed.

Flame treatment of abraded and degreased steel samples was shown to increase the joint strength of steel-epoxy specimens. Furthermore, the use of silanes was seen to have little effect on the static properties of steel-epoxy bonds; but marked improvement in hot-wet durability was observed in silane treated specimens. The performance of various thermoplastic polymers for bonding steel was also evaluated. PPS and PSU were found to give the highest lap-shear strengths. The use of an silane based adhesion promoter was seen to be very dramatic, enabling the strengths of PSU-steel joints to be superior to those observed with epoxy-steel joints.

ACKNOWLEDGMENTS

The authors would like to thank the University/Industry Consortium "Application of Composite Materials to Industrial Products" and the Army Research Office (University Research Initiative Program) at the University of Delaware Center for Composite Materials, as well as the Swiss National Science Foundation for their support of this project. Additionally, the authors would like to express gratitude to ALCOA for donation of 2008-T4 aluminum; Dow-Corning for the silanes; and Turco Products Inc. for donation of the alkaline cleaners used in the aluminum surface treatment process. Furthermore the assistance of S. Wood and A. Monab with the joint fabrication and testing was most appreciated.

REFERENCES

[1]   ASM International, Advanced Materials and Processes, (2) 14, p. 143 (1993)

[2]   A.J. Kinloch, "Adhesion and Adhesives Science and Technology", Chapman Hall Ltd., London, (1987)

[3]   S.M. McKnight, P. Franco , and J.W. Gillespie, Jr., Proceedings: Society of Experimental Mechanics 1993 Annual Meeting (50) 243 (1993)

[4]   S.H. McKnight, M.G. McBride, and J.W. Gillespie, Jr., Proceedings: American Society of Composites, Cleveland, OH, (1993)

[5]   S.H. McKnight, M.G. McBride, and J.W. Gillespie, Jr., Proceedings: 25th SAMPE Technical Conference, Philadelphia, PA, 25 (1993) p 443

[6]   J. Schultz, et.al., J. Mat. Sci., 24, 4363, (1989)

[7]   W. Chinsiriksul, et.al. Proceedings: ANTEC '93, 2450, (1993)

[8]   Sugama, T., Carciello, N.R. International Journal of Adhesion and Adhesives, (11), 97, 1991

[9]   J.D. Venables, J. Mat. Sci., 19, 2431 (1984)

[10]   Kinloch, A.J., "Durability of Structural Adhesives", Chapman and Hall Ltd., London, (1982)

[11]   J. Mnich and C Schoneman, Proceedings: 1993 International Sampe Symposium, (38) 819 (1993)

[12]   R.J. Kuhbander and J.J. Mazza, Proceedings: 1993 International Sampe Symposium, (38) 1225 (1993)

[13]   F.L. Keohan, et. al., Proceedings: 1993 International Sampe Symposium, (38) 1181 (1993)

[14]   "Aluminum: Properties and Physical Metallurgy," J.E. Hatch editor, American Society for Metals, (1984)

[15]   G. Ramarathnam., et.al., Proceedings of the 1990 ANTEC Meeting, 1778 (1993)

[16]   Sung, N.H., Kaul, A., Chin, I., Sung C.S.P., Polymer Eng Sci, 22, 637 (1982)

[17]   E.P. Plueddeman, "Silane Coupling Agents", Plenum Press, New York (1991)

[18]   Boeing Specification BAC 5555, Boeing Aircraft Company, Seattle, Washington

[19]   McKnight, S.H., J.W. Gillespie, Jr., and C.L.T Lambing, Proceedings Sampe Symposium, (37) 1994

Proceedings of the 10th Annual ASM/ESD Advanced Composites Conference, Dearborn, Michigan, USA, 7-10 November 1994

Coupled Effects of Healing and Intimate Contact on the Strength of Fusion-Bonded Thermoplastics

C.A. Butler, R. Pitchumani, J.W. Gillespie, Jr.
University of Delaware
Newark, Delaware

A.R. Wedgewood
DuPont
Wilmington, Delaware

Abstract

Healing and intimate contact are two important mechanisms that contribute to the strength development at the polymer-polymer interface during fusion bonding of thermoplastic materials. Understanding these mechanisms is important in many areas of polymer processing such as laminate consolidation including the tow placement process and various polymer welding techniques such as resistance and induction welding. Theoretical and semi-empirical models for intimate contact and healing have been independently developed in the literature. To more accurately assess how the bond strength develops with time, the mechanically driven process of intimate contact and the thermodynamically driven process of interfacial healing must be considered to occur concurrently instead of sequentially. This paper presents a coupled model which accounts for the concurrent nature of these two mechanisms. The coupled model presented in dimensionless form provides a deeper understanding of the fundamentals of strength development at the interface, and is shown to yield better strength predictions in practical situations. Since the governing mechanisms are not process specific, this model can be used to assess the quality of a bond made through various fusion bonding techniques.

FUSION BONDING during thermoplastic processing is attained by inducing intimate contact between the surfaces to be joined, eliminating the interface by interdiffusion of the molecular chains (or healing), and minimizing the void content within the matrix in order to produce a monolithic structure with the joint strength approaching that of the parent material. Fusion bonding incorporates heat, mass, and momentum transfer occurring simultaneously throughout the process. Aside from intimate contact, molecular interdiffusion, and void consolidation, other phenomena such as resin flow across

and in between the plies, adhesion, fiber deformation, and diffusion of air and volatiles in the molten matrix occur simultaneously during the bonding process (1). One must also consider the thermal effects on polymer degradation and the effects of heating and cooling rates on crystallization for semi-crystalline materials. The interactions among these phenomena control the final quality of the consolidated part.

Lin et al. (1) investigate the interactions between mass, momentum, and energy transfer by classifying them on the macro, micro, and molecular levels. On the macrolevel, a relation is established between the processing parameters (temperature, pressure, duration of process) and the degree and quality of the bond. The microlevel concentrates on void formation, resin flow, and fiber deformation on the scale of the fiber diameter. Voids can form at the resin/fiber interface because of insufficient wetting of the fiber by the resin or within the resin because of the release the moisture and volatiles. The molecular level focuses on intimate contact and wetting between adjacent surfaces to be bonded and the consequent interdiffusion of molecular chains across the interface. All three levels are interrelated on how they effect the bond quality and are discussed in further detail in the literature (1, 2).

In order to assess the quality of a joint, the fusion bonding process must be fundamentally understood and the mechanisms appropriately modeled. Lin et al. (1) discuss the models that have been developed in the literature for the various phenomena. They divide the models into three major groups. The first group deals mainly with resin flow and fiber deformation behavior. They predict composite thickness, resin pressure, and fiber pressure as a function of time. The second group of models relates the process variables to thermal, chemical, and mechanical behavior of material and enables one to predict the degree of intimate contact and healing (or autohesion) as a function of time. The third group deals with thermal analysis and the effects of heat transfer. The focus of this work will be on molecular level activities

595

which include intimate contact and healing as they directly contribute to the strength development at a bond interface.

Strength development at an interface during a fusion bonding process has been studied extensively in the literature. As previously mentioned, two key phenomena occurring during bonding are intimate contact and healing. The mechanism of intimate contact is dependent upon the relative surface roughness, the temperature profile, and the pressure field at the interface of the polymer surfaces. Healing is a temperature dependent phenomenon that is governed by the migration of polymer chains across the interface in intimate contact. Full healing across the interface is essential to obtain a bond with maximum performance. Each of these mechanisms has been described in the literature as being the limiting step governing strength development. Many investigators attribute the development of mechanical properties entirely to intimate contact (3, 4, 5, 6, 7) while others claim healing is the sole mechanism (8, 9, 10). In this paper, a model is developed that couples these two mechanisms together in order to account for the variability in initiation times of healing due to the growth of area in intimate contact.

The next section of this paper presents an overview of intimate contact and healing models from the literature. The theoretical background is followed by the development of a new model that couples the effects of healing and intimate contact. A nondimensional analysis of the coupled model is presented identifying key dimensionless groups. Experimental procedures to obtain parameters for the model are given and simulated results showing parametric effects on healing and intimate contact are then illustrated. A preliminary verification of the model using strength data from the tow placement process is discussed in the section on presentation of results

Theoretical Background

Theoretical and semi-empirical models for intimate contact and polymer healing have been developed in the literature (6, 7, 9, 10, 11, 12, 13, 14, 15, 16, 17). Intimate contact models are based on a fluid flow analysis and geometrical considerations (6, 7, 18). The healing theory uses the result of reptation theories proposed by DeGennes (19, 20, 21). These two mechanisms are discussed in further detail.

Intimate Contact. Intimate contact characterizes the amount of physical contact that has been established between two surfaces after they are brought together. In reality, materials never have perfectly smooth surfaces; they contain many surface asperities. Because of these asperities, perfect contact is not established by merely bringing the two surfaces together. Upon the application of pressure and/or temperature, the asperities can be deformed increasing the relative amount of area in contact. The degree of intimate contact, D_{ic}, is a measure

of the relative amount of surface area in contact at any time. It is a function of the applied temperature and pressure, the duration of the application, and the relative surface roughness which characterizes the asperities.

Dara and Loos (5) have developed a model to describe intimate contact by representing the irregular surface by a wave of rectangular elements of varying size. Lee and Springer (18) have followed the development of Dara and Loos, with the surfaces idealized using identically sized elements as shown in Figure 1.

The Lee-Springer model describes the effects of the processing variables (i.e. temperature, pressure) on the degree of intimate contact. The application of pressure causes the elements to deform and spread along the interface as illustrated in Figure 1. In the case of thermoplastics which have a high viscosity, a large pressure is needed to cause the asperities to flow. Because of the temperature dependence of the viscosity, the application of temperature facilitates the flow allowing for greater contact area.

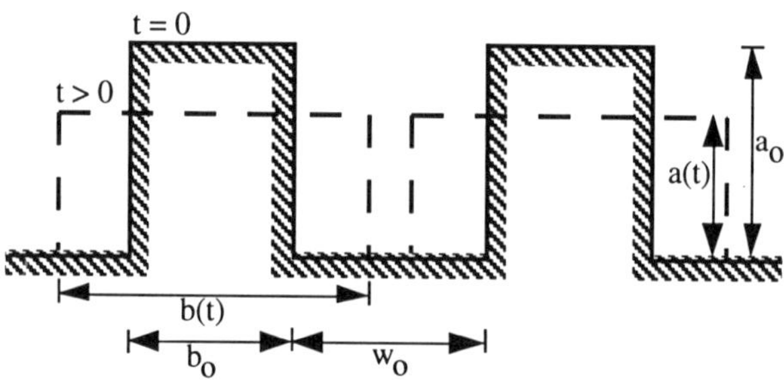

Figure (1): Schematic of idealized surface elements being deformed to the dotted line under the application of pressure

The Lee-Springer model shows the dependence of intimate contact on the pressure and the temperature-dependent viscosity but the model assumes constant temperature and pressure with time. Mantell and Springer (6, 7) have extended the same model development to account for time varying properties and conditions. Norpoth et al. (22) use a simpler model of intimate contact based on the composite height but do not give explicit dependencies of D_{ic} on pressure, temperature, and time. In this work, the Mantell-Springer model will be used to characterize the intimate contact at the bonding surface.

According to the Mantell-Springer development, the applied force causes the rectangular elements to spread to the dashed line in Figure 1 thereby increasing D_{ic}. Based on this idealization, D_{ic} can be expressed as:

$$D_{ic} = \frac{b(t)}{w_0 + b_0} \quad (1)$$

where w_o and b_o are given in Figure 1 and $b(t)$ is the width after pressure has been applied for a time t.

By assuming the volume of each element to remain a constant and applying the law of conservation of mass to a control volume of a surface element, the following expression for the degree of intimate contact can be obtained (6):

$$D_{ic} = D_{ico}\left[1 + C_1 \int_0^{t_p} \frac{P}{\mu_{mf}}\, dt\right]^{1/5} \qquad (2)$$

with :

$$D_{ico} = \frac{1}{1 + \dfrac{w_o}{b_o}} \qquad (3)$$

$$C_1 = 5\left(1 + \frac{w_o}{b_o}\right)\left(\frac{a_o}{b_o}\right)^2 \qquad (4)$$

where D_{ico} represents the amount of area that is initially in contact before any force is applied and C_1 is a function of the size of the rectangular elements. μ_{mf} is the fiber-matrix viscosity, t_p is the duration of the pressure cycle, and a_o, b_o, and w_o are given in Figure 1. If the scale of the deformation is greater than the distance of a fiber from the surface, the following expression for μ_{mf} can be used (23):

$$\mu_{mf} = \frac{1}{1 - \sqrt{\dfrac{f}{F}}}\,\mu \qquad (5)$$

where f is the fiber volume fraction, F is the packing factor, and μ is the resin viscosity. In the present study, it is assumed that a resin rich matrix layer is close to the surface as a result of which $\mu_{mf} \cong \mu$ (i.e., locally, at the interface, $f \cong 0$).

Frequently, the second term in the brackets in Eq. (2) is large compared to unity and D_{ic} can be simplified as:

$$D_{ic} = R_c\left[\int_0^{t_p} \frac{P}{\mu_{mf}}\, dt\right]^{1/5} \qquad (6)$$

where R_c, referred to as the roughness parameter, is given by:

$$R_c = D_{ico}C_1^{1/5} \qquad (7)$$

In the case of short processing times, the initial contact area cannot be neglected and the simplification in Eq. (6) is not valid and Eq. (2) in its entirety must be used.

Healing. When two amorphous thermoplastic parts are brought into contact above the glass transition temperature of the resin (or above the melt temperature for the semi-crystalline materials), interdiffusion of the polymer chains across the interface takes place contributing to the strength development at the bond area. This is referred to as healing. This section deals with a review of the previous work on describing the diffusion process on a molecular level and relating the interpenetration of the molecules to the strength at the interface. The first part of this section covers a review of the reptation theory describing chain mobility. The remainder discusses the various stages of healing with a concentration on the wetting and molecular diffusion stages.

Reptation Theory: For polymers in a melt above a critical molecular weight, such as most of the commercially available thermoplastics, the motion of a chain in an amorphous material has been modeled by the reptation theory (19, 24). According to the reptation theory, the chains in the melt can move by local Brownian motion and change shape but cannot intersect each other. The entanglement with neighboring chains imposes a topological constraint upon the chain restricting lateral motion; the chain is confined to move backward and forward along the curvilinear length of the tube and only the chain ends can exit through the ends of the tube as illustrated in Figure 2. This can be visualized in the motion of a snake where the head and the tail lead the rest of the body. Because the chain ends are free to move in a random direction away from the tube, the memory of the initial tube position in space is gradually lost. However, since the chain is in a dense system it is always in a new tube but only parts of the chain, usually towards the center, retain the memory of the initial tube (13, 15). The chain completely loses its original configuration after a time equal to the reptation time.

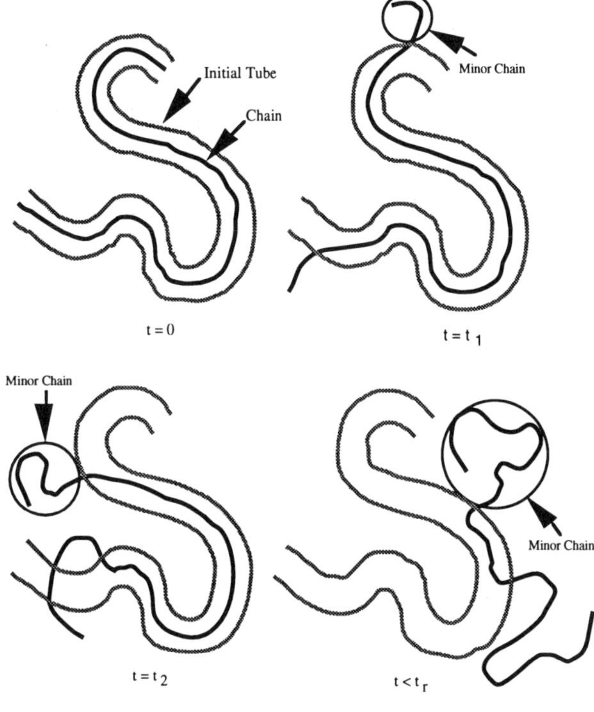

Figure (2): The reptation model for a random coiled chain in an entangled melt *via* the chain relaxing from its original configuration. The shaded tube represents the topological constraints and the minor chain is the portion that has relaxed from its original configuration (25).

Wool and co-workers have summarized the motion of a linear polymer chain in an entangled melt exhibiting reptation dynamics (13, 15).

$$\left\langle l^2 \right\rangle = 2D_r t \tag{8}$$

$$D_r \propto M^{-1} \tag{9}$$

$$\left\langle \chi^2 \right\rangle \propto \left\langle l^2 \right\rangle^{1/2} \tag{10}$$

$$\left\langle \chi^2_{cm} \right\rangle = 2Dt \tag{11}$$

$$D \propto M^{-2} \tag{12}$$

$$t_\infty \propto M^3 \tag{13}$$

where $<l^2>$ is the mean square curvilinear path diffused by the chain and also represents the length of the minor chain; D_r is the curvilinear one-dimensional diffusion coefficient; $<\chi^2>$ is the mean square monomer displacement of the minor chains; $<\chi^2_{cm}>$ is the mean square for the diffusion distance; D is the center of mass diffusion coefficient; and t_∞ is the relaxation time and it is equal to t_R, the time to achieve full healing at the interface.

Stages of Healing. When two surfaces are brought into contact under pressure at a temperature above the glass transition temperature, the polymer chains diffuse across the interface. The interface gradually disappears and the strength develops at the bondline.

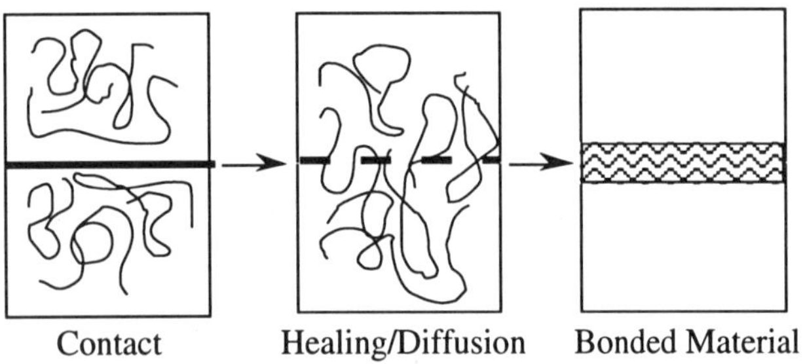

Figure (3): Illustration of the formation of a bond.

The theory of healing at an interface has been extensively studied by Wool and O'Connor. They consider healing to occur in five stages: (a) surface rearrangement, (b) surface approach, (c) wetting, (d) diffusion, and (e) randomization. These steps are illustrated in Figure 4 (13).

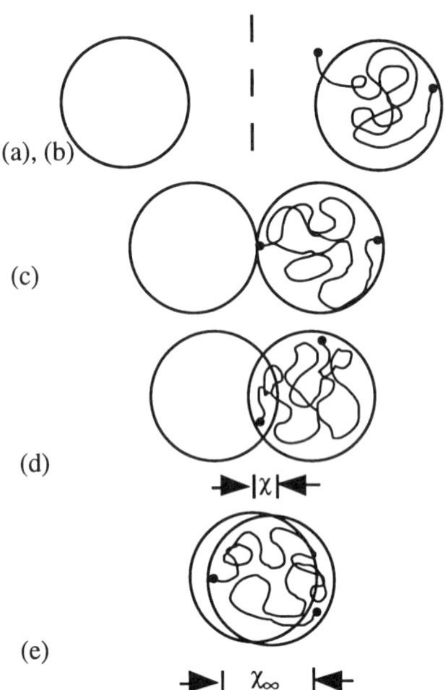

Figure (4): Stages of healing:(a) rearrangement, (b) surface approach, (c) wetting, (d) diffusion to distance χ, and (e) diffusion to distance χ_∞ and randomization (13).

Rearrangement and surface approach account for surface changes, chemical reactions, and molecular orientation changes at the surface, and the time-dependent contact of the different parts of the surface. Once these two stages are complete, the interdiffusion of the polymer chains across the interface can occur as illustrated by the wetting and diffusion stages. The randomization stage refers to the conformational relaxations contributing to equilibrium at the healed interface. This paper focuses on the influence of interdiffusion on mechanical properties. For further discussion of the total mechanism, see reference (13).

Wool has presented an intrinsic healing function where the mechanical properties that have developed are a sum of the wetting and diffusion processes initiated at different times (13). The wetting and diffusion stages are briefly reviewed below. The interested reader is referred to (13) for a more detailed discussion.

Wetting. The wetting phenomenon occurs when the surfaces are close enough for the establishment of intermolecular forces. In other words, the surface asperities are on a small enough scale for the intermolecular forces to have enough effect on the interface as to spread the asperities and establish full contact. In such a case, Wool (13, 14, 16) has described wetting as a nucleation and growth process where the fraction of wetted area is given by:

$$\Phi(t) = 1 - e^{(-kt^m)} \tag{14}$$

where k and m are constants depending on the nucleation function and spreading constants. The decaying exponential shows how wetting is initially retarded because it must overcome viscous and attractive forces.

The fundamental difference between intimate contact phenomenon previously described and wetting just mentioned is the size scale of the surface asperities. When the asperities are on a scale significantly larger than the range where the intermolecular forces have an effect, an outside force must be applied to deform the asperities in order establish interfacial contact. Intimate contact is prevalent when the asperities are large enough to warrant an external force to cause flow or deformation of the asperities. In such a case, the effects of wetting are overshadowed by the external force and the wetting phenomenon can be assumed to be instantaneous. This is generally the case in practical fusion bonding applications and will be assumed in this work.

Molecular Diffusion. Reptation theory has shown that the average interpenetration distance of the chain ends across the interface, χ, increases with time and can be seen from Eqs. (8) - (10) to follow the relation:

$$\chi \propto \left(\frac{t}{M}\right)^{1/4} \qquad (15)$$

Kim and Wool (25) show how the strength development is proportional to the average interpenetration distances of the polymers across the interface. From this one can get the relation for strength development as:

$$\sigma \propto \chi \qquad (16)$$

$$\sigma \propto \left(\frac{t}{M}\right)^{1/4} \qquad (17)$$

It has also been shown that when the time approaches the reptation time, the diffusion and randomization stages are complete and the full strength of the material, σ_∞, has been reached. The ultimate strength of the material after full healing can be represented by:

$$\sigma_\infty \propto \left(\frac{t_R}{M}\right)^{1/4} \qquad (18)$$

where t_R is the reptation time at a given temperature, which is also the characteristic time for full healing to occur. If the degree of healing is a measure of the amount of the maximum obtainable strength after a certain time, it can be represented by the ratio:

$$D_h = \frac{\sigma}{\sigma_\infty} \qquad (19)$$

where Eqs. (17) and (18) give

$$D_h = \frac{\sigma}{\sigma_\infty} \propto \left(\frac{t}{t_R}\right)^{1/4} \qquad (20)$$

Since reptation time is proportional to the cube of the molecular weight, the degree of healing can also be expressed as:

$$D_h = \frac{\sigma}{\sigma_\infty} \propto \frac{t^{1/4}}{M^{3/4}} \qquad (21)$$

The dependence of the strength buildup as time to the one quarter power has been proven in the literature by several different investigators (9, 10, 26, 27) and is well accepted.

Proposed Bonding Model

Strength development at an interface during a fusion bonding process has been studied extensively in the literature. As previously mentioned, two key phenomena occurring during bonding are intimate contact and healing. Each of these mechanisms have been modeled in the literature as the limiting step governing strength development. Many investigators attribute the development of mechanical properties entirely to intimate contact (3, 4, 5, 6, 7), while others claim healing is the sole mechanism (8, 9, 10).

Under certain material/processing conditions (for example, high pressures or low viscosities) where the time for intimate contact is negligible compared to the time for healing, healing may be regarded as the rate controlling step. On the other hand, different processing conditions could make the healing time negligible compared to the intimate contact time in which case intimate contact is the controlling mechanism. However, for some combination of material/process parameters between the two extremes, diffusion time and intimate contact time are of the same order of magnitude, for which both mechanisms are important, and a coupled analysis becomes necessary.

Therefore a model that incorporates both mechanisms and reduces to the proper controlling mechanism of either healing or intimate contact under special processing conditions would provide a more realistic description of the fusion bonding process. Such an analysis will also provide the criterion for determining operating conditions that result in either a purely healing-controlled or intimate-contact-controlled process. The rest of this work deals with the development of such a coupled model.

Model Development. The model for healing discussed in the previous section assumes that the surfaces are smooth and are initially in complete contact ($D_{ic} = 1$). In reality, the amount of area in contact increases with time, and only the surface that is in contact at any time (which is in general a fraction of the total area) can heal. An effective degree of healing may be defined that couples the two processes, and provides a more accurate measure of the resulting bond strength. As seen in Figure 1, the application of pressure causes the rectangular surface elements to deform, thus introducing a new elemental area into contact at a time t. This new area now begins to heal, continuing through the duration of application of temperature above T_g. The variability of the initiation times for healing across the bond area due to the development of intimate contact can be expressed by a convolution integral of the form:

$$D_h^* = \int_0^{t_p} D_h(t-t')\left(\frac{dD_{ic}}{dt}\right)_{t'} dt' \qquad (22)$$

where t' is an integration variable and represents the time when healing was initiated for the elemental area that came into intimate contact during the application of pressure. $D_h^*(t)$ is the effective degree of healing at any time t during the process, the derivative of D_{ic} accounts for the new area that has come into contact after a time t', and $D_h(t-t')$ is the amount of healing that has occurred across the new area that has been in contact for a time t-t', given by Eq. (20) with t replaced by t-t'. The effective degree of healing is essentially an area average of the healing over the surfaces in contact.

Non-Dimensional Analysis. Eq. (22), together with Eqs. (20) and (6) suggests that the effective healing is a function of the material and process parameters namely, pressure, temperature, time, relaxation time, viscosity, the glass transition temperature, and the roughness parameter. The influence of these parameters on the effective healing may be concisely represented by means of a non-dimensional analysis of the coupled effects.

In the present analysis, the following dimensionless groups are employed: Temperature is non-dimensionalized with respect to the glass transition temperature below which no molecular mobility occurs.

$$\theta = \frac{T - T_g}{T_g} \qquad (23)$$

Time is scaled with respect to the temperature dependent relaxation time, which is the characteristic time for full healing. It can be represented by a relaxation time at a reference temperature (t_r^*) and a shift factor $(a(\theta))$:

$$\tau = \frac{t}{t_r^* a(\theta)} \qquad (24)$$

The equations for healing, intimate contact, and the effective degree of healing can be expressed in terms of the above non-dimensional groups. In this analysis, an isothermal, constant pressure process is assumed in the interest of obtaining a closed-form expression. The application of the model to cases with nonisothermal and varying pressure fields will be discussed in the section on presentation of results.

$$D_h = \int_0^{\tau_e} d(\tau^{1/4}) = \tau_e^{1/4} \qquad (25)$$

$$D_{ic} = \left[\int_0^{\tau_p} \Omega^5 d\tau\right]^{1/5} = \Omega \tau_p^{1/5} \qquad (26)$$

Note that Eq. (26) introduces a dimensionless group, Ω, which is defined as

$$\Omega = \left[\frac{P t_r^* a(\theta)}{\mu_{mf}(\theta)}\right]^{1/5} R_c \qquad (27)$$

Physically, Ω is a function of the ratio of the characteristic time for healing to the characteristic time for intimate contact. Ω is an important non-dimensional group that characterizes the combined parametric effects.

Substituting Eqs. (25) and (26) into Eq. (22) yields the following equation for the effective degree of healing in terms of non-dimensional parameters for the special case of an isothermal process.

$$D_h^* = \frac{1}{5}\int_0^{\tau_p} (\tau_e - \tau)^{1/4}\, \Omega \tau^{-4/5} d\tau \qquad (28)$$

Notice that from Eq. (25), the non-dimensional time to achieve full healing $(D_h=1)$ is 1, and from Eq.(26), the non-dimensional time for full contact $(D_{ic}=1)$ is Ω^{-5}.

Experimental Parameters

The parameters needed as input to the coupled model are the processing temperature and pressure, Tg of the resin (or the melt temperature for semicrystalline materials), temperature dependent resin viscosity, reptation time and shift factor, and the roughness parameter. Two parameters need to be determined experimentally, the reptation time and roughness parameter. The rest of the parameters are material properties.

Roughness Parameter, R_c. The intimate contact model presented idealizes the surface as being made up of rectangular elements of equal size, which is a simplification of the actual random asperities. Nonetheless, the model has been widely accepted in describing the pressure and temperature effects on the degree of intimate contact (1). The main ambiguity in the utilization of the model is in determining the value of the roughness parameter, R_c. Mantell and Springer report that R_c ranges from zero for the roughest surface to unity for an idealistically smooth surface. The roughness parameter is essentially a characterization of an ideal surface and needs to be determined experimentally. Independent experiments such as profilometry or examination of micrographs of the actual surface would not be beneficial because they would give an indication of the actual shape of the surface and not the rectangular shape from which the model was derived. R_c has to be determined from experimental conditions because independent measurements are still needed. As an aside, the Norpoth model that was briefly mentioned also contains an empirical parameter characterizing the surface roughness.

The effect of the roughness parameter on the time necessary to achieve full intimate contact can be seen in Figure 5. The time varies by five orders of magnitude over the range of R_c. This plot was obtained using the material parameters for AVIMID®K 3 A (28) (a thermoplastic commercially available from DuPont Company) under isothermal processing conditions of 50 psi and 230°C.

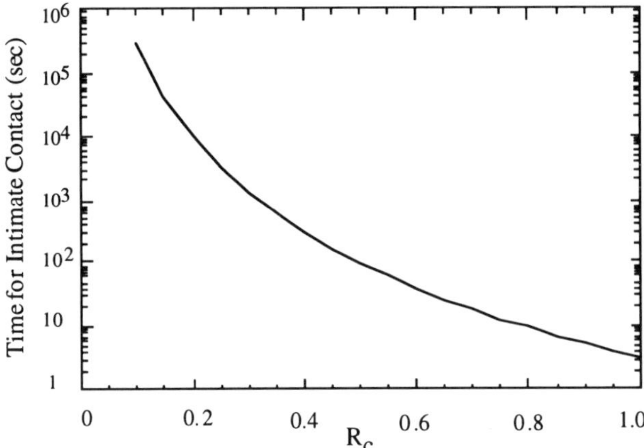

Figure (5): Sensitivity analysis for Rc showing the effects of Rc on the time for intimate contact.

Reptation Time and Shift Factor. According to the reptation theory, the time for the chain to completely lose its original configuration is the reptation time. This reptation time has been shown to be the relaxation time of the material which is the characteristic time to achieve full healing.

Because a polymer melt behaves viscoelastically, it has associated with it a characteristic frequency which separates the viscous domain from the elastic domain. Using Dynamic Mechanical Analysis (DMA), one can obtain a plot of the loss modulus of the material versus the frequency of the sinusoidal loading. The elastic to viscous transition is evidenced by the peak of the DMA curve shown schematically in Figure (6). The characteristic frequency is the inverse of the relaxation time of the polymer at a certain temperature.

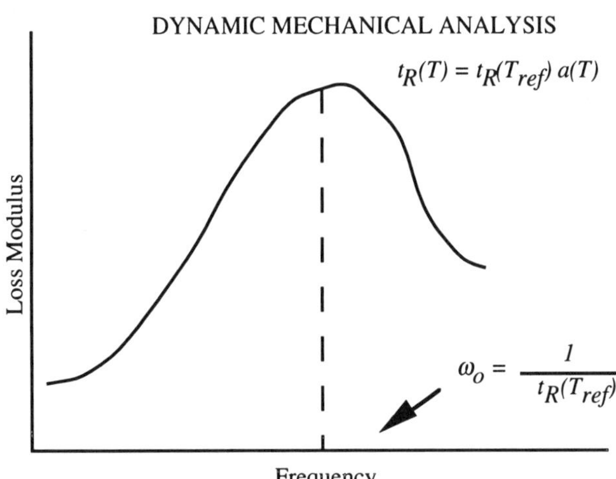

Figure (6): A schematic plot of loss modulus versus frequency obtained using DMA.

By performing the stress relaxation experiments over a range of frequencies and temperatures, and applying a time-temperature superposition to a reference temperature, one can obtain a shift factor (29). The relaxation time at any temperature can be obtained by multiplying the relaxation time at the reference temperature by the temperature dependent shift factor. The shift factor obeys a WLF relationship between T_g and $T_g+50°C$ and obeys an Arrhenius relation above that.

WLF:
$$ln(a(T)) = -\frac{C_1(T - T_{ref})}{(C_2 + T - T_{ref})} \quad (29)$$

Arrhenius
$$ln(a(T)) = \frac{E_a}{R}\left(\frac{1}{T} - \frac{1}{T_{ref}}\right) \quad (30)$$

The constants in the WLF equation or the activation energy in the Arrhenius equation are obtained from the amount of shift of each relaxation curve to obtain one overlapping curve.

Results and Discussion

Parametric effects on Intimate Contact and Healing. The degree of healing is a function of the applied temperature (T_o), the duration over which it is applied (t_e), and the reptation time of the material (t_R). The development of intimate contact is a function of the applied pressure (P_o), the duration over which it is applied (t_p), the temperature dependent viscosity (μ), and the surface roughness parameter (R_c). Parametric studies that show the effects of the various parameters on D_h and D_{ic} have been carried out using Eqs. (25) and (26) for the case of an isothermal process. Material parameters for PEKK (commercially available from ICI America) were used in Eqs. (23), (24), and (27). Trends of the effects are illustrated in Figures (7) and (8). Figure (7) shows how healing develops over a contact area with increasing temperature, time (t_e), and decreasing reptation time, all of which result in increased degree of healing. In Figure (7), the pressure time (t_p), and R_c have been held constant. Figure (8) shows that the degree of intimate contact increases with increasing pressure, time (t_p), R_c, and decreasing viscosity for constant temperature, time (t_e), and reptation time. It is evident that the area under each curve is the effective degree of healing (Eq. (28)) that has occurred.

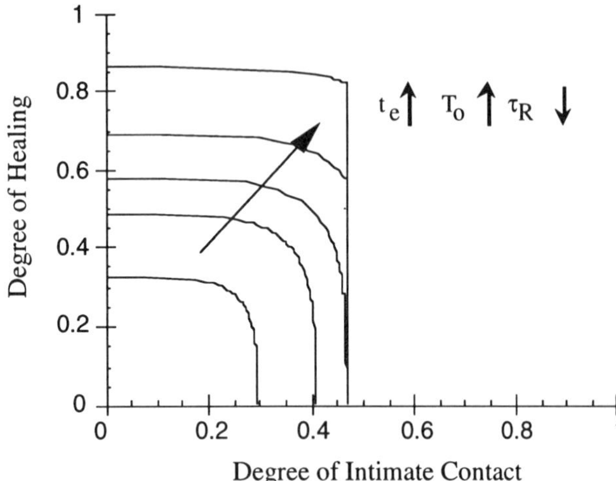

Figure (7): Parametric study showing the effects of various parameters on the extent of healing. The large arrow shows the effect on D_h as the parameters are varied according to the small arrows.

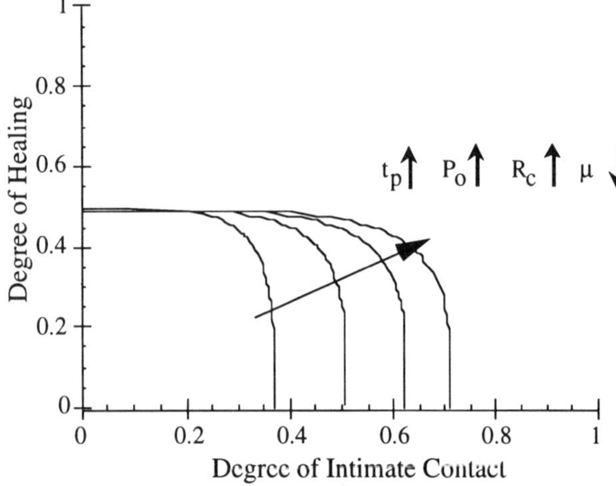

Figure (8): Parametric study showing the effects of various parameters on the extent of intimate contact. The large arrow shows the effect on D_{ic} as the parameters are varied according to the small arrows.

The parametric effects illustrated in Figures (7) and (8) have been combined together through the non-dimensional groups in Eqs. (23), (24), and (27). Eq. (28) was solved numerically using the Simpson's quadrature formula for the case of constant temperature and pressure using the material properties of PEKK. The effects of the non-dimensional parameters on the effective degree of healing (Eq 28)) are illustrated in Figure (9). The trend shows that the strength continuously develops until the curve reaches a maximum value corresponding to the maximum strength obtainable under those processing conditions. Also shown on the plot is the line corresponding to the predicted bond strength based on the healing model only (Eq. (25)) as used in the literature.

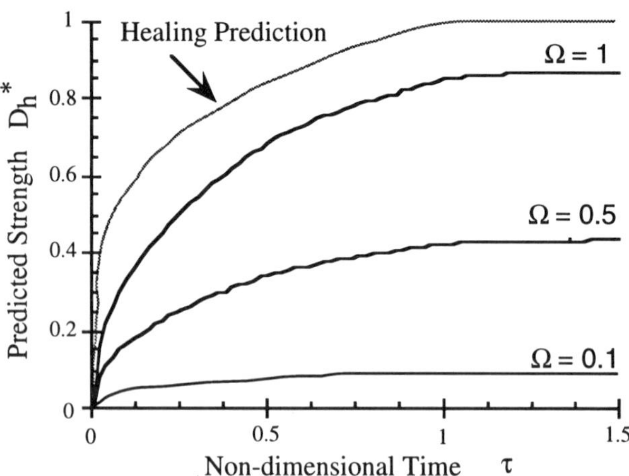

Figure (9): Parametric study showing the effects of the nondimensional parameters on the bond strength predicted using Eq. (28).

Recall that Ω is essentially a function of a ratio of the characteristic time for healing to the characteristic time for intimate contact. A large Ω could correspond to a short intimate contact time due to a high pressure, low viscosity, or high R_c. Therefore, an increase in Ω corresponds to intimate contact happening more rapidly, and the curve approaches the predictions based on healing with instantaneous contact. A large Ω could also correspond to a fast characteristic time for healing due to a high temperature or small relaxation time. This is evident in Figure 7 where at any τ, the strength at low Ω is closer to its maximum value than at high Ω.

Experimental Validation: The equations and parametric studies presented were for isothermal processes. However, real processes such as tow placement are highly nonisothermal. For nonisothermal processes, the methodology of coupling the healing and intimate contact mechanisms remains the same with the exception that the isothermal healing and intimate contact models must be replaced by nonisothermal models in Eq. (22). The nonisothermal healing model used was developed by Agarwal (9) and the Mantell-Springer model (Eq.(26)) was used to model nonisothermal intimate contact. The effective healing was calculated using a computer program as described in reference (30).

Presented here is a preliminary validation of this approach for the tow placement process. The process is described in (30). Statistically designed experiments (31) were carried out on this process with AS4/PEKK tows. The experiments covered a broad range of material and processing conditions including line speeds of 5 mm/s to 100 mm/s, temperatures between 700 and 900°C, and consolidation pressures from 40 to 350N. The strength was determined from short beam shear tests. The details are given in the reference (31) and the results are summarized here.

Figure (10) is a plot of the normalized measured bond strength (D_h^*) (normalized with respect to the strength of an autoclave processed sample, which is a representation of the maximum realizable strength) versus the bond strength predictions from the healing theory alone. Each data point represents different processing conditions. The solid line corresponds to exact agreement between the predicted and measured bond strength. All the data points lie below the solid line (the majority between 40% and 60% error) indicating that the healing model grossly overestimates the bond strengths for most of the processing conditions.

The data was also compared with predictions from the coupled model, the results of which are shown in Figure (11). As expected, the coupled model which accounts for the simultaneous development of intimate contact and healing does indeed give better strength predictions as revealed by the points lying much closer (within 20% error) to the exact agreement line.

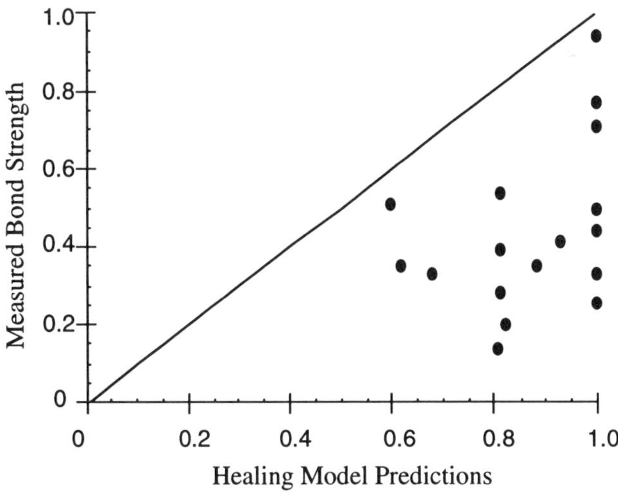

Figure 10: Plot comparing measured bond strength to predictions from the healing model. The solid line represents exact agreement.

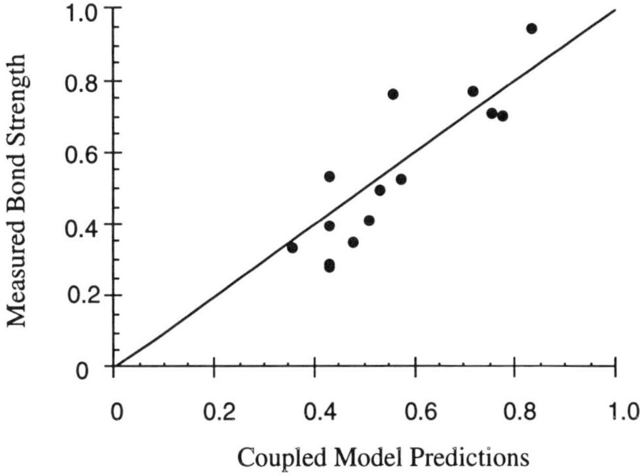

Figure 11: Plot comparing measured bond strength to predictions from the coupled model. The solid line represents exact agreement.

The data presented was for the tow placement process. Since the governing mechanisms for fusion bonding are not process specific, the models presented can also be used to assess the quality of a bond made through other joining techniques. Investigations are currently underway towards further validation of the approach for thermoplastic welding processes. The results of the study will be reported in future work.

Summary and Conclusions

This paper presented a background on the mechanisms and models previously developed in the literature that characterize the development of material properties across a bond interface. The models presented deal with the phenomena of intimate contact and healing separately, although in reality they happen concurrently. A new model that couples the phenomena together giving a more realistic representation of the physical description of fusion bonding was developed in a dimensionless form. Parametric studies revealed that Ω is an important group capturing the effects of all other relevant physical parameters.

Strength data obtained from PEKK samples in the tow placement process were used to illustrate the discrepancy between the measured strength values and the values predicted using the healing model alone. Most of the data points fell between 40% and 60% error indicating the model overpredicted the strength. The comparison of the data with the coupled model predictions shows much better agreement (within 20% error). These results indicate that neglecting intimate contact results in an overestimation of the effective degree of healing. The coupled model that accounts for the concurrent nature of the two mechanisms yields better strength predictions at the bondline because it gives a better physical description of what is actually occurring. Experimental validation of the coupled model for other processes such as resistance welding is in progress. The model is currently being extended to incorporate nonisothermal processes, and better representations of the surface irregularities in the intimate contact model.

Acknowledgments

The authors would like to thank the DuPont Company for providing financial support for this work, and Dr. R. McCullough..

References

1.   Lin, H. H., Ranganathan, S., Advani, S. G., "Flow and Rheology in Polymer Composites Manufacturing", S. G. Advani ed., Elsevier Science BV., (1994) pp. 325.

2.   Lin, H. H., Advani, S. G., CCM-92-53, Univ. DE, (1992).

3.   Muzzy, J., Norpoth, L., Babu, V., SAMPE J., **25**, 23 (1989).

4.   Howes, J. C., Loos, A. C., J. Thermoplastic Comp. Mat., **1**, (1988).

5.   Dara, P. H., Loos, A. C., CCMS-85-10, Virginia Polytech. Inst.,(1985).

6.   Mantell, S. C., Springer, G. S., J. Comp. Mat., **26**, 2348 (1992).

7.   Mantell, S. C., Wang, Q., Springer, G. S., J. Comp. Mat., **26**, 2378 (1992).

8.   Loos, A. C., Li, M. -C., SAMPE , 557-570 (1990).

9.   Agarwal, V. , Thesis Ph.D. Mat, Sci., CCM-91-39, Univ. DE (1991).

10.   Bastien, L. J., M.S. Mat. Sci., CCM-90-23, Univ. DE (1990).

11.   Kausch, H. H., Makromol. Chem.,Macromol.Symp., **41**, 1-8 (1991).

12.   Prager, S., Tirrell, M., J. Chem. Phys., **75**, 5194-5198 (1981).

13.   Wool, R. P., O'Connor, K. M., J. App. Phys., **52**, 5953 (1981).

14.   Wool, R. P., O'Connor, K. M., J. Polymer Sci., **20**, 7-16 (1982).

15.   Wool, R. P., Rubber Chem. Tech., **57**, 307 (1983).

16.   Wool, R. P., Yuan, B. L., McGrarel, O. J., Polymer Eng. Sci., **29**, 1340-1366 (1989).

17.   Wool, R. P., Macromolecules, **26**, 1564 (1993).

18.   Lee, W. I., Springer, G. S., J. Comp. Mat., **21**, 1017 (1987).

19.   de Gennes, P., J. Chem. Phys., **55**, 572 (1971).

20.   de Gennes, P., Physics Today (1983), pp. 33.

21.   de Gennes, P., "Physics of Polymer Surfaces and Interfaces", I. C. Sanchez ed. (1992) pp. 55.

22.   Norpoth, L. R., et al., SAMPE Int. Symp. (1988), **33**, pp. 1331.

23.   Pipes, R. B., Hearle, J. W. S., Beaussart, A. J., Sastry, A. M., Okine, R. K., J. Comp. Mat., **25**, 1204 (1991).

24.   Doi, M., Edwards, S. F., Faraday Trans., **74**, 1789 (1978).

25.   Kim, Y. H., Wool, R. P., Macromolecules, **16**, 1115-1125 (1983).

26.   Tirrell, M., Rubber Chem. Tech., **57**, 523 (1984).

27.   Kausch, H. H., Ann. Rev. Mat. Sci., **19**, 341 (1989).

28.   Wedgewood, A. R., DuPont Composites, Melt Processible Polyimides for High Performance Composite Application (1992).

29.   Ferry, J. D., "Viscoelastic Properties of Polymers", Wiley, New York, (1980).

30.   Pitchumani, R., Don, R. C., Gillespie, J. W., Jr., Ranganathan, S., Heat and Mass Transfer, In Press, (1994).

31.   Bauer, B. M., Steiner, K. V., CCM-94-24, Univ. DE, (1994).

Proceedings of the 10th Annual ASM/ESD Advanced Composites Conference, Dearborn, Michigan, USA, 7-10 November 1994

A Study of Improved Bonding Techniques for High Performance Thermoplastic Composites

R.C. Don, J.W. Gillespie, Jr., S.H. McKnight
University of Delaware
Newark, Delaware

Abstract

Joining of subassemblies fabricated from high performance thermoplastic matrix composites remains a critical issue in the application of those materials. For adhesive bonding, extensive surface pretreatment is normally required to develop an active surface on the thermoplastic to promote adhesion. In this study, a novel method of enhancing thermoset adhesive bonding of thermoplastic composites without surface pretreatment is described and evaluated. For comparison, a nonadhesive joining technique, resistance heated fusion bonding, was used. Both methods described here could be applied with little modification to bonding of thermoset matrix composites either to like materials or to dissimilar materials such as metals or thermoplastics.

DUE TO THEIR INHERENTLY LOW REACTIVITY, surface energies, and polarities, most thermoplastic (TP) polymers are not easily bonded using thermosetting (TS) structural adhesives such as epoxies. These difficulties can be resolved in a number of ways. The use of surface treatments to increase the wettability and polar nature of the thermoplastics will increase the potential for adhesive bonding to these materials. Techniques such as corona discharge, plasma treatments, acid etches, and oxidizing flame treatments [1] have all been proposed as suitable surface preparation methods for thermoplastics which are not easily joined with thermosetting adhesives. While these approaches can be very effective, they have constraints. Most notably, the treated surfaces do not maintain their characteristics for an infinite period. If the adhesive is not placed on the treated surface within a certain time period, the active surface's ability to promote strong bonding may be diminished or potentially eliminated due to contamination. Additionally, the equipment used in some of these procedures is quite expensive, and the maximum part size which can be treated may be restricted. Thus, much care must normally be taken when bonding thermoplastics with thermoset adhesives. A novel technique with virtually unlimited shelf life is under development to greatly ease the bonding of these materials.

Diffusion-Enhanced Adhesive (DEA) Bonding

Another approach that has recently been developed is the use of compatible thermoplastic films to enhance bonding of thermosets to otherwise difficult to bond thermoplastic adherends. In this approach, another thermoplastic polymer is used to create an interphase between the thermoplastic adherend and the thermoset adhesive. The compatible interphase concept is shown schematically in Figure 1. However, this method for creating a readily bondable thermoplastic layer (TP1) on a thermoplastic substrate (TP2) comes with the constraint of choosing compatible materials. A methodology for selection of compatible TP1-TP2 and TP2-TS polymers is currently under development, and is described below. There are two interfaces which must be considered when identifying potential compatible polymers: the TP2-TS and the TP1-TP2 interface. Though essentially the same conditions must be met with both interfaces, there are some subtle differences which should be addressed.

Thermoplastic-Thermoset Interphase.
The first consideration is the development of the bond at the TP2-TS interface. A starting point for this concept is work done on "interleaving" of TP polymer films in thermosetting composite laminates during cure [2]. Interleaving has been proposed as a method of improving the fracture toughness of TS laminates. The interleaved films are typically plasma treated to improve the film/thermoset interfacial bond. This type of bond is primarily an adhesive one. More desirable would be true interdiffusion of the TP polymer and TS pre-polymers, and curing of the TS while in the TP polymer. A diffuse interphase would permit the formation of a semi-interpenetrating network (semi-IPN), and the TP-TS bond strength could be enhanced by the entanglements of the TP molecular chains and the TS-network structure found in this region. This requires miscibility and hence true chemical compatibility of the polymers, versus simple wetting and bonding due to secondary van der Waals interactions as is the case for, say, bonding to the plasma-etched material. Here, two important phenomenon can be identified as critical to the development of a strong, diffuse thermoplastic-thermoset interphase: solution compatibility, and interdiffusion across the original interface.

Solution Compatibility. For mixing to take place at the TS-TP interface there must exist a thermodynamic driving force for the process to occur. The solution behavior of polymers has been widely investigated and several theories of varying complexity have been developed [3, 4, 5]. An excellent starting point for this phenomenon is the lattice theory for polymer solution compatibility developed by Flory and Huggins. For the case of uncured thermosetting pre-polymers with molecular weights much less than those of the thermoplastics, it is convenient to consider the uncured thermoset as a solvent which will swell the thermoplastic polymer as

mixing occurs. The thermodynamic condition for molecular mixing is that the Gibb's free energy of mixing, ΔG_m, is less than or equal to zero:

$$\Delta G_m = \Delta H_m - T\Delta S_m \leq 0 \qquad (1)$$

One can use the Flory-Huggins lattice model to express the free energy of mixing as follows [4]:

$$\Delta G_m = [\frac{1}{2}zN(\Phi_a\Phi_b(2w_{ab} - w_{aa} - w_{bb}))]$$
$$+[kT(N_a ln\,\Phi_a + N_b ln\,\Phi_b)] \qquad (2)$$

where z is the number of nearest neighbors in the lattice (the lattice coordination number), N is the total number of sites, Φ is the volume fraction of a and b respectively, and w represents the pairwise interaction energies of two adjacent sites. In many instances, the change in interaction energies can be easily related to the differences in the solubility parameters of the molecules using the Hildebrand-Scatchard equation [4] to give the enthalpy of mixing term in equation (2):

$$\Delta H_m = V\Phi_a\Phi_b(\delta_a - \delta_b)^2 \qquad (3)$$

where δ_a and δ_b are the Hildebrand solubility parameters of species a and b, and V is the total volume. The maximum solubility of polymer-a in polymer-b can be determined by evaluating the minimum of the free energy of mixing. One-phase mixtures (mixing at the interface) are possible when the second derivative of ΔG_m with respect to volume fraction is greater than or equal to zero. This occurs when the usually positive enthalpy of mixing term is small compared to the negative entropic term; i.e., when the solubility parameters of the two polymers are close in value. Infinite solubility occurs when the solubility parameters are identically matched. When the thermoset is in an uncured state, the entropy term will be relatively large, However, as the crosslinking reaction starts, the entropic term will diminish due to conformational effects. Although this is not accounted for in this treatment, adjustments could be made to the basic theory to relate phase behavior to the increasing thermoset molecular size [4].

However, the outlined approach is a first approximation into predicting which thermosetting pre-polymers have thermodynamic compatibility with any given thermoplastic polymer.

Kinetic Considerations-Interdiffusion. The interfacial mixing which may be predicted by thermodynamic considerations is a necessary but insufficient condition for providing a compatible TS-TP interface. Additionally, the ability to reach or sufficiently approach the equilibrium mixed state must exist. The mobility of the TS monomer must be high enough in the TP to reach the equilibrium penetration distance before the cure concludes. Ideally, this would occur when the TP polymer is in the amorphous or melt state. The mobility of penetrants in semi-crystalline polymers is extremely small, and this prevents the formation of an interpenetrating network to provide adhesive strength. Also, if the TP is reactive with the TS, this provides another mechanism for adhesive bond formation at the TS-TP interface. All of these conditions must be evaluated for each pair of materials to determine the best possible compatibility.

Presently, there is very little data available for the interdiffusion of thermoplastic and thermoset polymers. Work in our laboratories has begun on evaluating the diffusivity of epoxy and amine pre-polymers in different polymers [6]. The diffusivity of an epoxy pre-polymer into polystyrene at room temperature (T = 23°C) has been measured using a novel Fourier infra-red spectroscopy reflection technique (FTIR-ATR). The value of the diffusivity of the epoxy in this case was measured to be 1.05×10^{-12}. This value is extremely small, indicating that at room temperature, diffusion of the thermoset pre-polymers can be very slow and a limited amount of interdiffusion should be expected. Recently, diffusion at elevated temperatures has been studied. Qualitatively, diffusion was seen to be much more rapid at modest temperature increases above room temperature.

Further studies continue on diffusivity evaluation of thermosetting pre-polymers in thermoplastic resins in our laboratory. However, these measurements are currently being evaluated on the pre-polymer and curing agents independently. Implementation of these diffusivity results to predict interphase size becomes complicated

when one considers the effect of reaction in an actual adhesive system on the diffusion of the pre-polymers. The effect of reaction on the mobility of the thermoset is an important issue and will be addressed in future investigations. It can be easily envisioned that as temperature increases, the characteristic time for diffusion decreases, indicating further penetration for a given time period. At the same time the characteristic time for network gelling is also decreased, meaning that the ultimate penetration may not be reached due to mobility restrictions. Most likely, there will be a temperature at which the final penetration depth is maximized. These conditions should be evaluated to produce the optimal degree of interfacial entanglement.

Our work has found that the polysulfone family of TP's are compatible with epoxy-based resins, and epoxy/polysulfone (PS) and epoxy/polyethersulfone (PES) pairs have been successfully bonded. Table 1 displays ranges of the solubility parameters (δ's) for PES, PS, and typical epoxy and curing agent pre-polymers, demonstrating their compatibility via the criteria just outlined. Additionally, it was determined experimentally that both PS and polyetherimide (PEI) appear to be compatible with bismaleimide (BMI), although the solubility parameters those polymers are not readily available and have yet to be determined.

Thermoplastic-Thermoplastic Interphase (TP1/TP2). Compatibility at thermoplastic TP1 to TP2 interface is governed by the same thermodynamic considerations which yield compatibility relations for thermoplastic-thermoset pre-polymers. The major difference that arises when considering thermoplastics, is the extremely large molecular weights. The long chain nature of thermoplastic polymers creates conformational effects which reduce the entropic driving force to mix at the interface, and complete mixing of different high polymers is not frequently observed. However, segmental mixing at the interface can occur, and the degree of segmental motion across the interface can be related to strength. The length of any chain which protrudes from the TP1 to the TP2 side of the interface (or vice versa) can be related to the probability of creating a molecular entanglement. The degree of entanglement is directly related to interface and thus bond strength and toughness. De Gennes has proposed a simple relation to predict fracture toughness for TP1/TP2 polymer-polymer interfaces [7]:

$$G_{1c} \propto exp(-2N_e(\delta_a - \delta_b)^2) \qquad (4)$$

where N_e is the chemical distance between polymers (not necessarily constant for all pairs), and the solubility parameters are as defined earlier. From Eq. 4, it can be seen that there is a very sharp drop off in fracture toughness as the solubility parameters begin to differ. For that reason, it is very important that the solubility parameters of TP1 and TP2 be closely matched to achieve optimum joint performance.

Dual Polymer Thermoplastic (DPT) Fusion Bonding

The primary advantage of the dual polymer technique for fusion bonding is that the parent material of the parts to be joined does not need to be fused directly, hence (in the case of thermoplastic matrix parts) the risk of extensive melting of the parts and ensuing fiber motion and possible part distortion is eliminated. A thermoplastic layer can also be added to the surface of a compatible thermoset matrix composite, either by direct cocuring or by a hybrid interlayer technique, to allow fusion bonding. The methodology for selection of compatible polymers for thermoset cocuring was described above. Regardless of the parent polymer, the preparation of the thermoplastic bonding layer is reduced to a quick wiping with a solvent such as alcohol to remove mold release agents and other possible contaminants. Additional plies of neat resin could be added for gap filling if needed. A disadvantage of the DPT technique using amorphous polymers such as PS is the reduced solvent resistance as compared to thermosetting or semicrystalline thermoplastic polymers. This drawback is not significant, as simple steps can be taken to provide physical barriers to chemical attack of the PS, such as a fillet of silicone caulking compound applied to the exposed regions of the bond (the solvent resistance of the semi-IPN bond region as examined here has not been characterized, but may provide superior properties as compared to the amorphous polymer alone).

Fusion Bonding by Resistance Heating.
For fusion bonding of a thermoplastic interface, the bond area must be heated to a temperature above the glass transition (T_g) of the amorphous polymer surface layer, and below the melt temperature (T_m) of the "parent" matrix, in the case of thermoplastic adherends. Given time at temperature above T_g, the interface comes into intimate contact and heals, developing interfacial strength [8]. The difference between the T_g of the bond layer and the T_m of the adherends provides a window for the process temperature, provided the proper choice of interlayer polymer is made. For the case of PS mated with a PPS parent matrix, the window is as follows: T_g(PS) = 192°C; T_m(PPS) = 280°C. An upper limit on the bonding temperature of about 260°C would then be appropriate.

Resistance Welding. A resistive implant technique ("resistance welding") for direct heating of the interface was used for this evaluation. In resistance welding of composites, a heating element is placed between the parts to be joined, and remains permanently embedded in the joint. This can actually be advantageous, because the joint can be "unwelded" by reheating if needed for repair or other reasons, either by resistance or, with the use of a metallic element, induction heating. The heating element is typically comprised of either a single ply of unidirectional carbon fiber prepreg or a sheet of expanded stainless steel foil mesh. Current is passed through the heating element, causing Joule heating. The heating power needed to bond very large regions can be reduced by breaking the area up into smaller, more manageable regions that are heated separately. Fusion bonding by resistance welding has been shown in several studies to provide joint strengths approaching the strength of the parent material itself [8, 9, 10, 11].

For all experiments in this study, a surface layer (TP2) of polysulfone (PS) was used with a parent matrix (TP1) of polyphenylene sulfide (PPS).

Sample   Preparation   and   Testing. Samples were prepared for both adhesive and fusion bonding from $[0, \pm45, 90]_{2s}$ (16 ply quasi-isotropic) laminates of Quadrax S-2 glass/PPS prepreg tape, and had a single 127 μm film of neat PS, made from Amoco's Udel P1700-NT11 resin, coconsolidated to one face. The laminates were consolidated using the autoclave process cycle recommended by Quadrax for their PPS prepregs. Bonding coupons were cut from the autoclaved laminates with a diamond saw. Surface preparation was simply a thorough wiping with a rag dampened with isopropyl alcohol, followed by air drying before bonding. Examination of micrographs of bonded specimens (see discussion of Figs. 5 and 6 below) shows a very rough surface on the top (0° ply) at the interface between the S-2/PPS and PS layers.

For the DEA bonds and the hot press fusion bonded baseline samples, a double notched shear specimen was used following ASTM D3165-73 [12]. Those coupons were 19.1 cm long by approximately 14 cm wide, and were cut after bonding into 5 2.54 cm wide specimens for testing. For the resistance welded DPT experiments, single lap shear tests (ASTM D1002 [12]) were conducted, with 10.2 cm long by 8.9 cm wide coupons which were cut into 2.54 cm wide specimens after welding. All testing was performed on an Instron 1125 test machine.

Experimental   Procedure.
DEA Bonds. The adhesive chosen for the DEA bonding study was Cytec (American Cyanamid) FM300, an amine cured modified epoxy film adhesive. The 200 μm thick open knit carrier form, FM300K, was used as Cytec claims it offers the highest overall performance with that particular adhesive. Other forms of FM300 are available and are claimed to provide better bondline thickness control and gap filling ability with slightly reduced properties.

All adhesive bonding with the exception of one vacuum bagged trial took place in a Wabash heated platen hydraulic press. Following Cytec's recommended cure cycle,

the adhesive samples were subjected to a slow ramp to 177°C, then a dwell for 60 minutes, followed by cooling to room temperature, all under 276 kPa pressure. In an attempt to enhance diffusion of the epoxy into the PS, some double ramp cures were performed using intermediate temperatures of 121°C and 143°C, with dwell times of 60 and 120 minutes at the first soak temperature, before a final ramp to 177°C for the one hour final cure as before. A single vacuum bagged trial was done in an oven, using only atmospheric pressure against the vacuum bag for applying consolidation pressure, with a cure at 177°C for one hour, then cooling to room temperature with vacuum applied.

Fusion Bonding Experiments. Stainless steel mesh heating elements were preimpregnated with PS prior to use for resistance welding. To enhance bonding between the PS and the steel, the mesh was pretreated by cleaning with acetone, then dipping in a silane solution (Dow Corning Q1-6106, 5% in methanol). The pieces were then dried in an oven at 50°C for 20 minutes. To preimpregnate the mesh, it was sandwiched between 127 μm films of neat polysulfone (Udel P1700-NT11, as used in the laminate preparation), and was processed in a hot press at 260°C under 689 kPa pressure for 1 hour. Sample size was 8.9 cm wide by 10.2 cm long, with an actual bond (overlap) 2.54 cm across. The samples were welded in the Automated Resistance Welder at the University of Delaware's Center for Composite Materials (UD-CCM). Heating intensity was set at 60 kW/m^2. Consolidation pressure was varied from 172 to 690 kPa to investigate the influence of pressure on bond strength. The soak temperature was 260°C for all of the resistance welds.

A fusion bonded baseline strength was established by hot pressing the PS face of two 8.9 cm wide by 17.8 cm long laminates directly together (i.e., without a heating element) at 240°C for 1 hour under 690 kPa pressure, then cooling to room temperature under pressure. This sample was cut and tested identically to the DEA bonds.

Experimental   Results

Adhesive   Bonds.
Excellent shear strengths were attained with the DEA bonds, with a maximum of 28 MPa

(from the one hour intermediate soak at 121°C), and fairly low scatter overall. For comparison, the hot press fusion bonded baseline gave a strength of 16.8 ± 1.7 MPa. A trial of bonding the PPS side of identical laminates using FM300K with no surface preparation other than a solvent wipe gave essentially zero strength: the samples failed while being mounted in the test machine, showing very poor adhesion, with clean separation of the FM300 from the PPS surface.

Figure 3 shows shear strength plotted versus the final adhesive thickness as measured in micrographs of the bondline. There appears to be an optimum adhesive thickness, without taking into account other factors such as void content, which was not measured for this study. Figure 4 shows shear strength plotted versus soak time at intermediate temperature. There is evidently some enhancement of strength due to the longer time available for diffusion of the epoxy into the PS layer, as anticipated. The drop-off in strength at the longest soak time (120 minutes) is due in part to the reduced thickness of adhesive from flow out of the bondline. Thus the development of methodology to select an optimized intermediate soak must take into account bondline thickness effects due to the reduced viscosity of the adhesive at slightly elevated (below cure) temperatures. The effect of applied pressure on flow and void growth must be taken into account as well, particularly if only a reduced pressure for bonding was available due to geometric or load constraints. Alternatively, perhaps a reduced pressure, just sufficient to ensure good contact, during the initial "diffusion-enhancing" soak would reduce flow and excessive bondline thickness reduction.

The vacuum bagged adhesive cure had the least reduction of adhesive thickness from processing, although there was a large amount of flow from the bondline. The result was a very high void content in the adhesive, plainly visible to the naked eye in the failed specimens, from entrapped air during layup and outgassing during cure, which tend to expand in the partial vacuum. Still, a 21.6 MPa mean shear strength was attained, but with the largest amount of scatter. The open knit of the FM300K creates a dimpled surface texture which tends to trap air during layup. The other forms of FM300 film which are readily available, a tighter knit and a random mat material, are claimed to reduce the amount of entrapped air and allow closer control of final bondline thickness. These other forms are being acquired for further investigation. Cytec [13] claims that another form could be produced given demand, a "one side tacky" (OST) form that consists of the random mat which is impregnated with the FM300 adhesive on only one side, aiding layup and further ensuring release of outgases and eliminating entrapped air as the epoxy flows through the mat and wets the "dry" side during the cure.

Figure 5 is a micrograph of a typical adhesive bondline at 100x magnification, showing clearly the very uneven PS to composite interface (the 0° fiber direction is out of the page in the micrograph).

Resistance Welded Bonds.
The strengths of the resistance welded fusion bonds were less than thought possible with that technique, although they were similar to the directly fusion bonded baseline performance, with a maximum of 14.7 ± 0.1 MPa. Table 2 shows the results of each of the welds with the corresponding weld time and applied pressure. Figure 6 is a micrograph of Trial 4, showing the essentially void-free bondline, as typical for all of the fusion bonds. Although in all cases the failure surfaces were partially driven into the composite, the strengths achieved were much lower than those attained with adhesive bonding of the same adherends. The first cause that comes to mind is degradation of the PPS parent matrix or delamination of the composite at the bond temperature used (260°C). The fusion bond baseline was exposed to that temperature for 60 minutes, whereas the welds were held at temperature for only 120 to 180 seconds. Indeed, some improvement in strength and a large reduction in scatter was noted at the longest dwell time of 180 seconds. Perhaps there is some effect from the stainless steel heating element used for the welds. Other heating element materials, such as AS-4 (carbon)/PPS prepreg tape with comolded PS, have yet to be tried with this material system. The single lap shear test used may be the cause of some difference when compared to the double notch tests of the adhesive.

Discussion and Future Work

The diffusion-enhanced adhesive bonding has shown itself to be a very promising way to virtually eliminate the extensive surface preparation normally required for adhesive bonding of thermoplastic composites. Much work remains to be done in the modeling of the diffusion kinetics of thermoset adhesives into amorphous thermoplastic interlayer materials. This will enable both selection of candidate materials and prediction of optimized cure cycles to provide enhanced bond strength for this joining method.

For resistance welding with this material system, other heating element materials as well as further exploration of the bounds of the process window will be tried. This will help to develop a process map of time/temperature/pressure as a starting point for optimizing that fusion bonding technique. Bonding temperatures closer to the T_g of PS will prevent potential degradation of the PPS parent matrix.

Acknowledgements

The authors would like to thank Mr. Mark Scott for his tireless help with the experimental investigations reported here.

References

[1] Kinloch, A.J. "Adhesion and Adhesives Science and Technology" Chapman Hall, London, 1987

[2] Armstrong-Carroll, E., "The Influence of Interleaf Deformation Behavior and Film-Resin Adhesion on the Fracture Toughness of Interleaved Composites," Composite Materials: Fatigue and Fracture, Fourth Volume, ASTM STP 1156, ASTM, Philadelphia, 1993, pp. 299-317

[3] Fredrickson, G.H. in "Physics of Polymer Surfaces and Interfaces: I.C. Sanchez ed., p. 1, Butterworth-Heinemann, Boston, (1992)

[4] Palmese, G.R., Ph.D. Dissertation, Dept. of Chemical Engineering, University of Delaware, 1992

[5] Wu, S., Polymer Interface and Adhesion, M. Decker, New York, NY, 1982, pp. 380-387

[6] McKnight, S.H., T.P. Skourlis, R.L. McCullough, and J.W. Gillespie, Jr. , To Appear in Proceedings of the Annual ASC Conference, Newark, DE, September 1994

[7] de Gennes P-G, in "Physics of Polymer Surfaces and Interfaces: I.C. Sanchez ed., p. 55, Butterworth-Heinemann, Boston, (1992)

[8] R.C. Don, L. Bastien, T. Jakobsen, and J. W. Gillespie, Jr., "Fusion Bonding of Thermoplastic Composites by Resistance Heating," Proc 21st International SAMPE Tech Conf, Atlantic City, New Jersey, September 27, 1989

[9] Don, R.C., C.L.T. Lambing, and J.W. Gillespie, Jr., "Experimental Characterization of Processing-Performance Relationships of Resistance Welded Graphite/Polyetheretherketone Composite Joints," Polymer Engineering and Science, **32**, 9, pp. 620-631

[10] Holmes, S., and J.W. Gillespie, Jr., "Thermal Analysis and Experimental Investigation of Large-Scale Resistance Welded Thermoplastic Composite Joints," Proc 25th Intl SAMPE Tech Conf, Philadelphia, PA, October 26-28, 1993

[11] Howie, I., and J.W. Gillespie, Jr., "Resistance Welding of Graphite-Polyarylsulfone/Polysulfone Dual Polymer Composites," Journal of Thermoplastic Composites, **6**, July 1993

[12] ASTM Standard Test Methods D3165-73 and D1002, "Standard Test Method for Strength Properties of Adhesives in Shear by Tension Loading of Laminated Assemblies," Annual Book of ASTM Standards, Vol 15.06, ASTM, Philadelphia, 1989

[13] Personal conversation with John Paxton, Cytec Engineered Materials, Havre de Grace, Maryland, July 1994

[14] Gordon, J.L., *Cohesive Energy Density*, in *Encyclopedia of Polymer Science and Technology*, J.W. Wiley and Sons, NY, 1964, p. 833

<u>Table 1</u>: Solubility parameters, glass transition and processing temperatures for example thermoplastics and typical epoxies with their amine curing agents

Material	δ $(J/cm^2)^{1/2}$	T_g (°C)	T_p (°C)	Physical State
Polystyrene	17-19	100	-	Amorphous Polymer
Polysulfone	19.5-22*	192	-	Amorphous Polymer
Polyethersulfone	20 - 23*	216	-	Amorphous Polymer
Epoxy Monomers	17 - 21†	n/a	170 - 200**	Reacting Pre-polymer
Amine Curing Agents	17 - 22†	n/a	170 - 200**	Reacting Pre-polymer

* Calculated from values of surface tension using method proposed by Gordon in [14]
** Ranges of epoxy cure temperatures for typical systems (various sources)
† From [4]

<u>Table 2</u>: Results of resistance welds

Trial #	Consolidation Pressure (kPa)	Weld Time (s)	Shear Strength (MPa)	Standard Deviation (MPa)
1	690	120	14.4	1.7
2	345	120	12.9	2.9
3	172	120	12.5	1.9
4	690	180	14.7	0.09
5	690	180	13.6	0.02

Figures

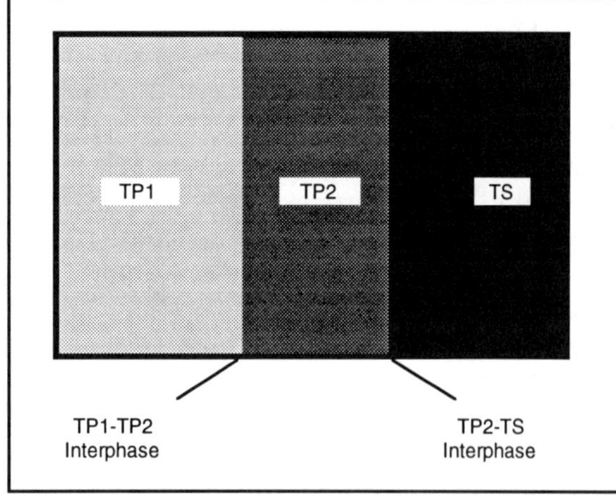

Fig. 1: Schematic representation of the compatible polymer interphase concept for enhancing bonding between thermoplastic and thermosetting polymers.

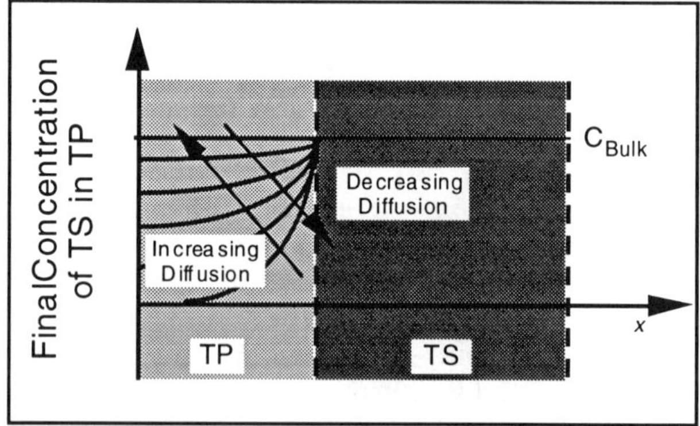

Fig 2: Conceptual figure displaying the effects of diffusion on final concentration of TS pre-polymers in thermoplastic (TP) film. Competing mechanisms govern ultimate penetration. Higher temperatures increase diffusivity but also tend to decrease the reaction time. Optimum penetration can be determined by coupling the diffusion and reaction kinetics.

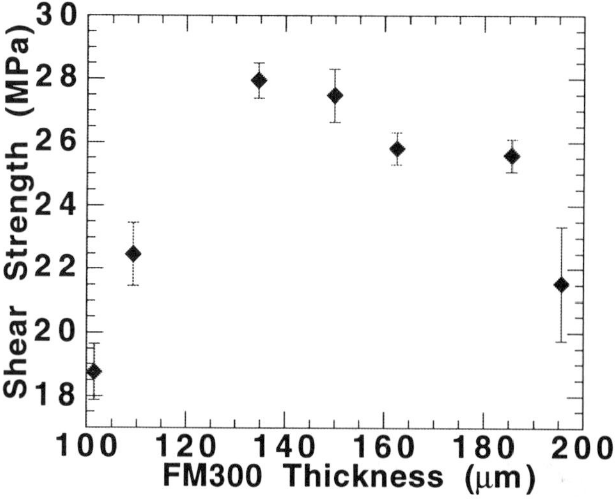

Fig. 3: Shear strength of FM300K adhesive bonds vs adhesive bondline thickness. Error bars indicate ± 1 standard deviation.

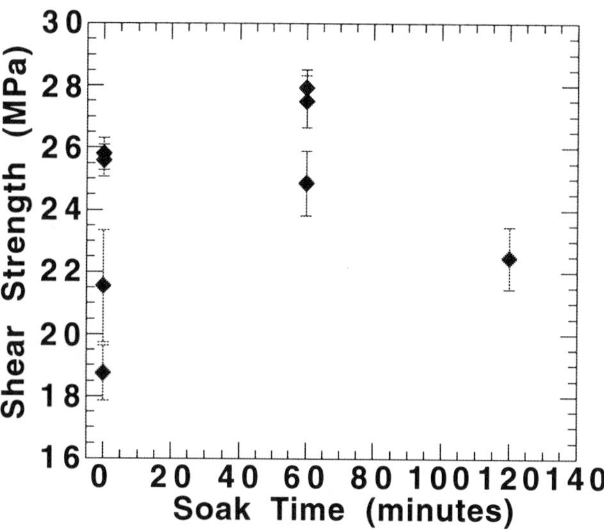

Fig. 4: Shear strength of FM300K adhesive bonds vs soak time at intermediate temperature. Error bars indicate ± 1 standard deviation.

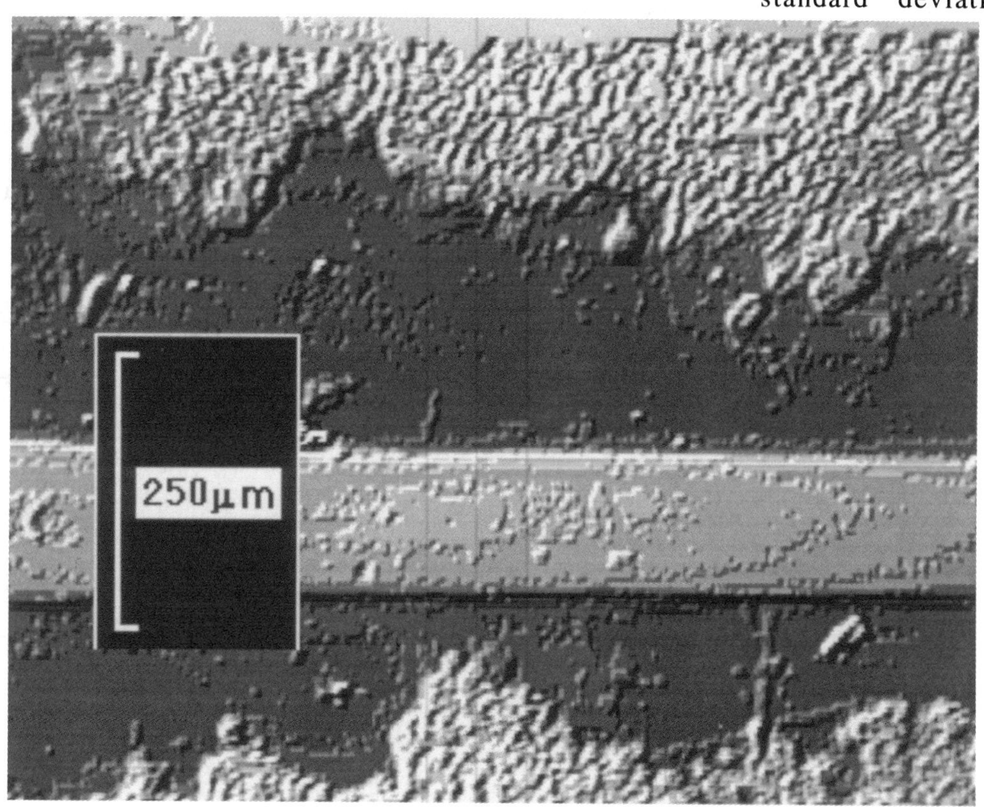

Fig. 5: Micrograph of typical adhesive bonded specimen at 100x magnification. Central band is FM300 adhesive; adjoining dark region is PS polymer cocured to S-2/PPS composite

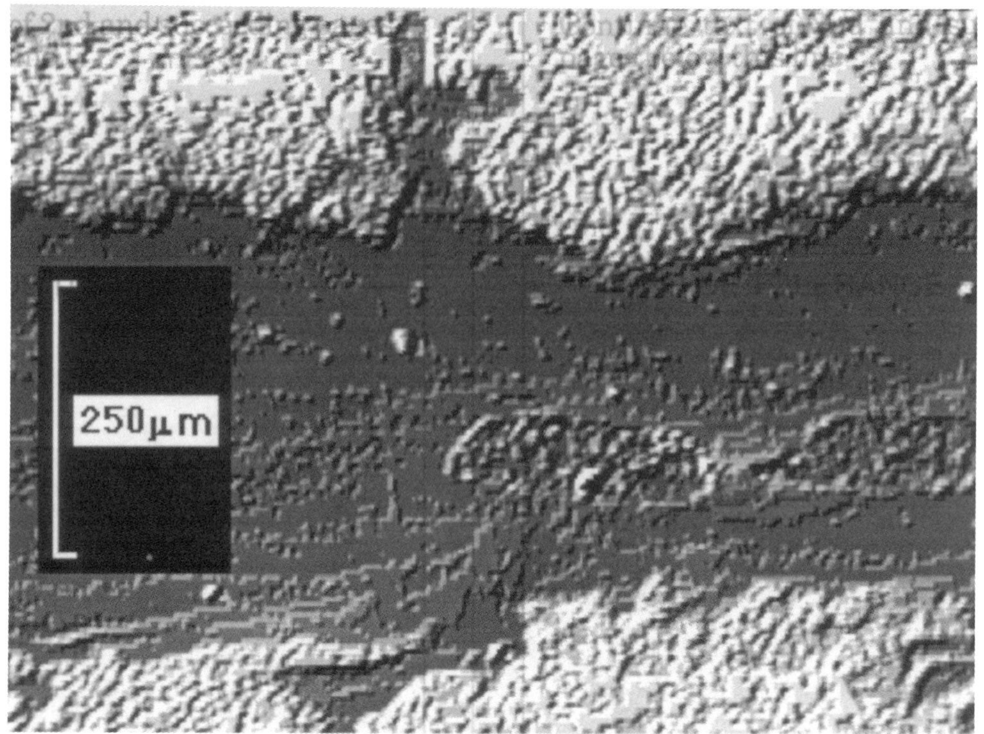

Fig. 6: Micrograph of typical resistance welded specimen (Trial 4) at 100x magnification. Central dark region is PS polymer cocured to S-2/PPS composite; two portions of the heating element mesh can be seen in the right half of the micrograph, just below center